ARTIFICIAL INTELLIGENCE, BLOCKCHAIN, COMPUTING AND SECURITY, VOLUME 1

This book contains the conference proceedings of ICABCS 2023, a non-profit conference with the objective to provide a platform that allows academicians, researchers, scholars and students from various institutions, universities and industries in India and abroad to exchange their research and innovative ideas in the field of Artificial Intelligence, Blockchain, Computing and Security.

It explores the recent advancements in the field of Artificial Intelligence, Blockchain, Communication and Security in this digital era for novice to profound knowledge about cutting edges in Artificial Intelligence, financial, secure transaction, monitoring, real time assistance and security for advanced stage learners/ researchers/ academicians. The key features of this book are:

- Broad knowledge and research trends in AI and Blockchain with security and their role in smart living assistance
- Depiction of system model and architecture for clear picture of AI in real life
- Discussion on the role of AI and Blockchain in various real-life problems across sectors including banking, healthcare, navigation, communication, security
- Explanation of the challenges and opportunities in AI and Blockchain based healthcare, education, banking, and related industries

This book will be of great interest to researchers, academicians, undergraduate students, postgraduate students, research scholars, industry professionals, technologists and entrepreneurs.

PROCEEDINGS OF THE INTERNATIONAL CONFERENCE ON ARTIFICIAL INTELLIGENCE, BLOCKCHAIN, COMPUTING AND SECURITY (ICABCS 2023), GR. NOIDA, UP, INDIA, 24–25 FEBRUARY 2023

Artificial Intelligence, Blockchain, Computing and Security
Volume 1

Edited by

Arvind Dagur
School of Computing Science and Engineering, Galgotias University, Gr. Noida

Karan Singh
School of Computer & Systems Sciences, JNU New Delhi

Pawan Singh Mehra
Department of Computer Science and Engineering, Delhi Technological University, New Delhi

Dhirendra Kumar Shukla
School of Computing Science and Engineering, Galgotias University, Gr. Noida

CRC Press
Taylor & Francis Group
Boca Raton   London   New York   Leiden

CRC Press is an imprint of the
Taylor & Francis Group, an **informa** business

A BALKEMA BOOK

First published 2023
by CRC Press/Balkema
4 Park Square, Milton Park, Abingdon, Oxon, OX14 4RN
and by CRC Press/Balkema
2385 NW Executive Center Drive, Suite 320, Boca Raton FL 33431

CRC Press/Balkema is an imprint of the Taylor & Francis Group, an informa business

© 2024 selection and editorial matter, Arvind Dagur, Karan Singh, Pawan Singh Mehra &
Dhirendra Kumar Shukla; individual chapters, the contributors

British Library Cataloguing-in-Publication Data
A catalogue record for this book is available from the British Library

Library of Congress Cataloging-in-Publication Data
A catalog record has been requested for this book

SET
ISBN: 978-1-032-66966-3 (hbk)
ISBN: 978-1-032-68590-8 (pbk)

Volume 1
ISBN: 978-1-032-49393-0 (hbk)
ISBN: 978-1-032-49397-8 (pbk)
ISBN: 978-1-003-39358-0 (ebk)
DOI: 10.1201/9781003393580

Volume 2
ISBN: 978-1-032-67841-2 (hbk)
ISBN: 978-1-032-68498-7 (pbk)
ISBN: 978-1-032-68499-4 (ebk)
DOI: 10.1201/9781032684994

Typeset in Times New Roman
by MPS Limited, Chennai, India

Table of Contents

Communications

General track

Recent advances and future technologies in IoT, blockchain
and 5G

Recent advancements and challenges in Artificial Intelligence, machine learning, cyber security and blockchain technologies

Electronics and scientific computing to solve real-world problems

Security and privacy in the cloud computing

Preface

On the behalf of organising committee, I would like to extend my heartiest welcome to the first international conference on Artificial Intelligence, Blockchain, Computing and Security (ICABCS 2023).

ICABCS 2023 is a non-profit conference and the objective is to provide a platform for academicians, researchers, scholars and students from various institutions, universities and industries in India and abroad, to exchange their research and innovative ideas in the field of Artificial Intelligence, Blockchain, Computing and Security. We invited all students, research scholars, academicians, engineers, scientists and industrialists working in the field of Artificial Intelligence, Blockchain, Computing and Security from all over the world. We warmly welcomed all the authors to submit their research in conference ICABCS 2023 to share their knowledge and experience among each other.

This two-day international conference (ICABCS 2023) was organized at Galgotias University on 24th and 25th February 2023. The inauguration was done on 24th February 2023 at Swami Vivekananda Auditorium of Galgotias University. In the inauguration ceremony, Professor Shri Niwas Singh, Director, Atal Bihari Bajpai Indian Institute of Information Technology and Management, Gwalior attended as Chief Guest. Professor Rajeev Tripathi, former Director Motilal Nehru National Institute of Technology Allahabad and Professor D.K. Lobiyal, Jawaharlal Nehru University attended as Guests of Honour. In the inauguration ceremony of the program, the Vice-Chancellor of the University, Professor K. Mallikarjuna Babu, Advisor to Chancellor, Professor Renu Luthra, Dean SCSE, Professor Munish Sabharwal welcome the guests with welcome address. The Registrar, COE and Deans of all the Schools were present. Conference Chair Professor Arvind Dagur told that in this conference more than 1000 research papers were received from more than ten countries, on the basis of blind review of two reviewers, more than 272 research papers were accepted and invited for presentation in the conference. The Chief Guest, Honorable Guests and Experts delivered lectures on Artificial Intelligence, Block Chain and Computing Security and motivated the participants for quality research. The Pro Vice-Chancellor, Professor Avadhesh Kumar, delivered the vote of thanks to conclude the inauguration ceremony. During the two-day conference, more than 272 research papers were presented in 22 technical sessions. The closing ceremony was presided over by Prof. Awadhesh Kumar, Pro-VC of the University and Conference Chair Professor Arvind Dagur, on behalf of the Organizing Committee. Conference Chair, Professor Arvind Dagur thanked Chancellor Mr. Sunil Galgotia, CEO Mr. Dhruv Galgotia, Director Operation Ms. Aradhana Galgotia, Vice Chancellor, Pro Vice-Chancellor, Registrar, Dean SCSE, Dean Engineering and university family for their co-operation and support.

Finally, once again I would like to thank to all participants for their contribution to the conference and all the organising committee members for their valuable support to organise the conference successfully. I highly believed that this conference was a captivating and fascinating platform for every participant.

On the behalf of editors
Dr. Arvind Dagur

Acknowledgements

It gives me immense pleasure to note that Galgotias University, Greater Noida, India is organizing the International Conference on Artificial Intelligence, Blockchain, Computing and Security (ICABCS 2023) on 24th and 25th February 2023. On behalf of the organizing committee, I would like to convey my sincere thanks to our Chief Patron, Honorable Shri Sunil Galgotia, Chancellor, GU and Hon'ble Shri Dhruv Galgotia CEO, GU for providing all the necessary support and facilities required to make ICABCS-2023 a successful conference. I convey my thanks to Prof. (Dr) K. Mallikharjuna Babu, Vice Chancellor and Prof. (Dr) Renu Luthra advisor to the chancellor for their continuous support and encouragement, without which it was not possible to achieve. I want to convey my sincere thanks to them for providing technical sponsorship and for showing their confidence in Galgotias University to provide us the opportunity to organize ICABCS-2023 and personally thank to all the participants of ICABCS 2023. I heartily welcome all the distinguished keynote speakers, guest, session chairs and all the authors presenting papers. In the end, I would convey my thanks to all the reviewers, organizing committee members, faculty and student volunteers for putting their effort into making the conference ICABCS 2023 a grand success.

Thank you,
Prof.(Dr.) Arvind Dagur
Organizing Chair ICABCS 2023,
Galgotias University

Committee Members

Scientific committee

Prof. Valentina Emilia Balas
Aurel Vlaicu University of Arad, Romania

Prof. Toshio Fukuda
Nagoya University, Japan

Dr. Vincenzo Piuri
University of Milan, Italy

Dr. Ahmad Elngar
Beni-Suef University, Egypt

Dr. Malik Alazzam
Lone Star College – Victory Center. Houston, TX, United States

Dr. Osamah Ibrahim Khalaf
Profesor, Al-Nahrain University, College of Information Engineering, Baghdad, Iraq

Dr. TheyaznHassnHadi
King Faisal University, Saudi Arabia

Md Atiqur Rahman Ahad
Osaka University, Japan, University of Dhaka, Bangladesh

Prof. (Dr) Sanjay Nadkarni
Director of Innovation and Research, The Emirates Academy of Hospitality Management, Dubai, UAE

Dr. Ghaida Muttashar Abdulsahib
Department of Computer Engineering, University of Technology, Baghdad, Iraq

Dr. R. John Martin
Assistant Professor, School of Computer Science and Information Technology, Jazan University

Dr. Mohit Vij
Associate Professor, Liwa College of Technology, Abu Dhabi, United Arab Emirates

Dr. Syed MD Faisal Ali khan
Lecture & Head – DSU, CBA, Jazan University

Dr. Dilbag Singh
Research Professor, School of Electrical Engineering and Computer Science, Gwangju Institute of Science and Technology, South Korea

Dr. S B Goyal
Dean & Director, Faculty of Information Technology, City University, Malaysia

Dr. Shakhzod Suvanov
Faculty of Digital Technologies, Department of Mathematical Modeling, Samarkand State University, Samarkand Uzbekistan

Dr. Upasana G Singh
University of KwaZulu-Natal, South Africa

Dr. Ouissem Ben Fredj
ISSAT, University of Kairouan, Tunisia

Dr. Ahmad Elngar
Beni-Suef University, Egypt

Dr. Omar Cheikhrouhou
CES Lab, ENIS, University of Sfax, Tunisia

Dr. Gordon Hunter
Associate Professor, Mekelle University, Kingston University, UK

Dr. Lalit Garg
Computer Information Systems, Faculty of Information & Communication Technology, University of Malta, Malta

Dr. Sanjeevi Kumar Padmanaban
Aarhus University, Denmark

Prof (Dr.) Alex Khang
Professor of Information Technology, AI Expert and Data Scientist, GRITEx VUST SEFIX EDXOPS, Vietnam and USA

Dr. Jiangtao Xi
1st degree connection 1st, Professor, Head of School of Electrical, Computer and Telecommunications Engineering at University of Wollongong, Greater Sydney

Dr. Rabiul Islam
Senior Lecturer at University of Wollongong, Australia

Prof. Lambros Lambrinos
Cyprus University of Technology, Cyprus

Dr. Xiao-Zhi Gao
University of Eastern Finland, Finland

Dr. Sandeep Singh Sanger
University of Copenhagen

Dr. Mohamed Elhoseny
University of Sharjah, United Arab Emirates

Dr. Vincenzo Piuri
University of Milan, Italy

Artificial Intelligence, Blockchain, Computing and Security – Dagur et al. (Eds)
© 2024 The Editor(s), ISBN: 978-1-032-49393-0

National Advisory Committee

Dr. S. N. Singh, IIT Kanpur
Dr. Rajeev Tripathi, MNNIT, Allahabad
Dr. R. S. Yadav, MNNIT Allahabad
Dr. Satish Chand, JNU, New Delhi
Dr. M. N. Doja, IIIT Sonepat
Dr. Bashir Alam, JMI, New Delhi
Dr. Shailesh Tiwari, KEC, Ghaziabad
Dr. Mansaf Alam, JMI, New Delhi
Dr. Ompal, DST, New Delhi
Dr. Rajeev Kumar, DTU, New Delhi
Dr. Parma Nand, Sharda University, India
Dr. Pavan Kumar Mishra, NIT Raipur
Dr. Nagendra Pratap Singh, NIT Hamirpur
Dr. Santarpal Singh, Thapar University
Dr. Samayveer Singh, NIT Jalandhar
Dr. Ankur Chaudhary, Sharda University
Dr. Ranvijay, NIT Allahabad
Dr. Manu Vardhan, NIT Raipur
Dr. Pramod Yadav, NIT Srinagar
Dr. Vinit Kumar, GCET Gr. Noida
Dr. Anoop Kumar Patel, NIT Kurukshetra
Dr. Suyash Kumar, DU Delhi
Dr. Hitendra Garg, GLA University Mathura
Dr. Chanchal Kumar, JMI New Delhi
Dr. Vivek Sharma, GLBITM, Gr. Noida
Dr. Anand Prakesh Shukla, DTE, UP
Dr. Biru Rajak, MNNIT Allahabad
Dr. Gopal Singh Kushwaha, Bhopal
Dr. Rajeev Pandey, SRMS Brailly
Dr. D. Pandey, KIET Ghaziabad
Dr. D.S. Kushwaha, MNNIT Allahabad
Dr. Sarsij Tripathi, MNNIT Allahabad
Dr. Shivendra Shivani, Thapar University, Punjab
Dr. Divakar Yadav, NIT Hamirpur
Dr. Pradeep Kumar, NIT Kurukshetra
Dr. Anand Sharma, AIT Aligarh
Dr. Udai Pratap Rao, SVNIT Surat
Dr. Vikram Bali, JSSATE Noida
Dr. Gaurav Dubey, Amity University, Noida

Artificial Intelligence, Blockchain, Computing and Security – Dagur et al. (Eds)
© 2024 The Editor(s), ISBN: 978-1-032-49393-0

Organizing committee

Chief Patron
Shri Suneel Galgotia,
Chancellor, Galgotias University, Greater Noida, India

Patrons
Shri Dhruv Galgotia,
CEO, Galgotias University, Greater Noida, India
Prof.(Dr.) Mallikharjuna Babu Kayala,
Vice-Chancellor, Galgotias University, Greater Noida, India
Ms. Aradhna Galgotia,
Director Operations, Galgotias University, Greater Noida, India

General Chairs
Prof. (Dr.) Avadhesh Kumar,
Pro-VC, Galgotias University, Greater Noida, India
Prof. (Dr.) Munish Sabharwal,
Dean, SCSE, Galgotias University, Greater Noida, India

Conference Chairs
Prof. (Dr.) Arvind Dagur,
Professor, Galgotias University, Greater Noida, India
Dr. Karan Singh,
Professor, JNU New Delhi, India
Dr. Pawan Singh Mehra, DTU, New Delhi

Conference Co-Chairs
Prof. (Dr.) Dr. Amit Kumar Goel,
HOD (CSE) and Professor, Galgotias University, Greater Noida, India
Prof. (Dr.) Krishan Kant Agarwal,
Professor, Galgotias University, Greater Noida, India
Dr. Dhirendra Kumar Shukla,
Associate Professor, Galgotias University, Greater Noida, India

Organizing Chairs
Dr. Abdul Aleem,
Associate Professor, Galgotias University, Greater Noida, India
Dr. Vikash Kumar Mishra,
Assistant Professor, Galgotias University, Greater Noida, India

Technical Program Chairs
Dr. Shiv Kumar Verma, Professor, SCSE, Galgotias University
Dr. SPS Chauhan, Professor, SCSE, Galgotias University
Dr. Ganga Sharma, Professor, SCSE, Galgotias, University
Dr. Anshu Kumar Dwivedi, Professor, BIT, Gorakhpur

Finance Chair
Dr. Aanjey Mani Tripathi, Associate Professor, Galgotias University
Dr. Dhirendra Kumar Shukla, Associate Professor, Galgotias University

Artificial Intelligence

Artificial Intelligence, Blockchain, Computing and Security – Dagur et al. (Eds)
© 2024 The Author(s), ISBN: 978-1-032-49393-0

An ensemble learning approach for large scale birds species classification

Harsh Vardhan, Aryan Verma & Nagendra Pratap Singh
Department of Computer Science and Engineering, National Institute of Technology, Hamirpur

ABSTRACT: Birds are vertebrate animals that are adapted for flight due to the presence of hollow bone structures. The entire population of birds contributes 0.08 % to the total animal biomass. In the past two decades, there has been a continuous loss and degradation of natural habitats resulting in a threat to bird population survival. The United Nations calculates that 49% of the bird population is declining, and some 1500 species have already gone extinct in the last 100 years. Researchers are studying the behavior and morphological characteristics of different bird species to understand them so that necessary steps can be taken for their protection. It is evinced that manually classifying bird species is a very inefficient and time-consuming task. Through the use of Automatic Bird Species Classification, this time can be reduced from hours to minutes. This paper features an automatic bird species classification system utilizing an ensemble of deep neural networks. Our proposed method trains individual state-of-the-art architectures like VGG 19, DenseNet 201, and ViT to classify 400 bird species. Further, the performance of these models is evaluated using metrics like F1 scores, precision, and recall. Our developed ensemble is better generalized and adapted to the problem with excellent accuracy of 99.40%. Results have stated that our approach is notably much better than existing works on bird species classification.

Keywords: Ensemble Learning, Birds Species Classification, Image Classification, Deep Learning, Pretrained Model: DenseNet-201, VGG-19, ViT

1 INTRODUCTION

Birds are members of classes *aves*; their feathers distinguish them from other classes. According to evolution theory, birds evolved from dinosaurs (Brusatte *et al.* 2015) They are a crucial member of the ecosystem due to their vital role-playing in functioning as natural pollinators, maintaining ecological balance, and keeping the pest population under control (Sekercioglu *et al.* 2016) Moreover, birds act as essential indicators for studying the state of the environment due to their susceptibility towards habitat change and the fact that they are accessible for census. These features make them an ecologist's favorite tool.

According to a report (Lehikoinen *et al.* 2019) One in every seven birds is under threat of extinction. A recent study (Pimm *et al.* 2018) highlights that there are more than 10400 living species of birds at present on this entire planet. In the past decade, bird populations had been severely affected due to many factors, such as global warming, deforestation, and the spreading of the communication network. Taking into concern, much research on wildlife bird monitoring has taken up the pace, and lots of government and semi-government programs have been initiated to protect the bird population. For this task, advanced technologies such as AI and IoT are aiding researchers in protecting the bird population. Authors in (Huang & Basanta 2019) have recognized endemic bird species through the deep learning algorithm CNN with skip connections.

Authors of (Tóth & Czeba 2016) have used a convolutional neural network-based approach to classify birds' songs in a noisy environment. Researchers in (Gavali & Banu

2020) have combined DCNN and GoogleNet to classify bird species. Advanced Technologies are helping in a task such as classifying birds, monitoring the migratory birds' status, establishing the pattern, conserving endangered species etc.

| Kingfisher | Cuban Tody | American Goldfinch | Wall Creeper |

Figure 1.   Sample Images from data set.

1.1   *Main contributions*

- The ensemble model is developed that can classify 400 bird species by training the pre-trained networks like DenseNet-201, VGG-19, and VIT.
- The dataset is cleaned and precisely pre-processed to remove the excessive noisy images that cause huge loss in feature extraction.
- Extensive performance evaluation on a wide range of parameters like accuracy, F1 score, precision, and recall is performed for drawing comparison between models.

The rest of our paper draws out the following structure. Related work is highlighted in section 2, the proposed methodology is defined in section 3, section 4 explains materials and methods, section 5 covers results and discussion, and finally section 6 includes comparative analysis section, the conclusion and future work part is covered under section 7, conflict of interest statement is in section 8 along with funding status in 9 and finally references are covered in last section.

2   RELATED WORK

Traditionally bird classification is performed by hand-picking the features after physically examining the bird image. Specifying bird species through physical examination requires excellent experience, which only expert ornithologists possess. Further, this task is very time-consuming and prone to numerous errors. Today, the biggest challenge for researchers in studying birds is that multiple species appear similar in initial appearance, which causes a delay in further research examination. Even expert ornithologists have limited study and exposure to rare bird species throughout their career. Much research has been done to solve this problem, and many papers have been published employing different tools and technologies. A technique for automating bird classification with the help of CNN has been proposed in (Gavali & Banu 2020) by employing DCNN(Deep Convolutional Neural Network) on Google Net framework. This system works by converting bird images into a grayscale format through autograph technology. A transfer learning approach has been presented in (Kumar & Das 2018) in which training is done using a multistage process and an ensemble model was formed, which consists of Inception Nets and Inception Res-Nets from localization. An existing VGG 16 architecture was implemented in (Islam *et al.* 2019) to extract the features for initiating bird species classification. In this paper comparison was made between different classification approaches such as Random Forest, K-nearest neighbor, and SVM. (Huang & Basanta 2021) Developed a new Inception ResNet v2-based transfer learning method to detect and classify endemic bird species. Their technique involves swapping missed classified data between training and test sets and then implementing it to validate the model performance. A comparison of existing recurrent convolutional networks for large-scale bird classification on acoustics has been drawn in (Gupta *et al.* 2021) it examines hybrid modeling approach that includes CNN and RCNN. An approach using a practical classification of bird

4

species by transfer learning was implemented in (Alswaitti *et al.* 2022) this paper assesses the performance of traditional machine learning and deep learning by forming comparison between different groups of classifiers.

3 PROPOSED METHODOLOGY

Our methodology highlights the use of a pre-trained deep neural network for automatic feature extraction. Let us discuss their structure to understand more about them.

3.1 *Deep neural network*

Deep neural networks (Samek *et al.* 2021) are developed by stacking up more than two neural networks. This neural network automatically extracts relevant features for the classification task. Our proposed ensemble learning model consists of VGG-19, DenseNet 201, and ViT. These three models are entirely different in their architecture and working.

3.1.1 *VGG-19*
VGG-19 is a variation of VGG-16 that consists of 19 layers instead of the standard 16 layers. Out of which (16 layers are convolution, three are fully connected, 5 are MaxPooling layers, and the remaining single layer is the Softmax function layer. Derived from Alex Net (Alom *et al.* 2018) this model improvised the traditional convolution neural network by a relatively large extent. It takes an image of size 224* 224 as input along with 3 * 3 kernel size. We trained it on our dataset by taking pre-trained weights of ImageNet itself. (Deng *et al.* 2009)

3.1.2 *DenseNet-201*
DenseNet-201 (Huang *et al.* 2017) works by connecting every layer with the other one in the network. It is often characterized as a Densely Connected Convolutional Network (Zhu & Newsam 2017) It uses transition layers between the DenseNet blocks. The transition layers consist of a batch-norm layer, then a 1x1 convolution layer, followed by a 2x2 average pooling layer. DenseNet architecture aims to make the connection between input and output layers deeper but shorter because it governs more accessible training and better feature extraction. Our method utilizes DenseNet-201 because it is much more efficient and easier to train than its other versions.

3.1.3 *VIT transformer*
Vision Transformer applies a transformer (Cho *et al.* 2014) based architecture over the image patches. It splits the image into fixed-sized patches, then connects it with a transformer-encoder stacked with multi-layer perceptron (MLP), layer norm (LN), and multi-headed self-attention layer. The resultant self-attention layer is implemented to spread out the information globally. To perform the image classification, it uses a standard approach consisting of an extra learnable "classification token" with the initial sequence.

3.2 *Ensemble learning*

The ensemble learning approach combines the performance of several other models to generate one optimal model. It explores a predictive model that is supposed to perform better than any constituting predictive model alone. Technically there are numerous ways to generate an ensemble model, but three main famous classes of ensemble learning are bagging, boosting, and stacking. Through the combination of models, many benefits are acquired, such as improved predictive accuracy, precision, recall, and better statistics. In our work, we have sketched a complex voting-based ensemble learning model that combines the prediction of each class label with the predicted class label having the most votes. The hard voting is outlined in the mathematical equation (1). 1

$$x_i \, \epsilon \, \Omega = \{x_1, x_2 \ldots \ldots \ldots x_n\} \tag{1}$$

(a) Lets specify no. of iterations by integer T
(b) Now randomly drawing F percent of V by taking itself replica of T
(c) Represent weak-learn with V_t and receive the hypothesis(classifier) h_t Now simply evaluate the ensemble E

$$E = \{h_1, h_2 \ldots \ldots \ldots h_T\} on \ x \tag{2}$$

$$Let \ a_t, j = \{h_1, h_2 \ldots \ldots \ldots h_T\} on \ x \tag{3}$$

4 MATERIAL AND METHODS

4.1 *Dataset*

This paper utilized the dataset taken from Kaggle *400 Species Image Classification* for testing and training the effectiveness of the developed model. The dataset consists of more than 50000 different images of birds separated into 400 species. This dataset comes from a public source that is constantly being updated. The moment we used this dataset for our research, it consisted of 400 classes (this may or may not be the same afterwards). Each image is of 224 x 224 size format. The dataset is already pre-divided into the test, train, and validation sets. During the data cleaning step, the images with excessive noise are removed due to their influence on extracting relevant features.

4.2 *Tools used*

We have employed NVIDIA GPU (Tesla P100 16GB HBM2), Python 3.9, and PyTorch 1.12.1 for training our model. However, the actual implementation of neural networks was done using PyTorch 1.12.1. The complete training and testing procedures are done entirely on the Google Colab platform. This platform features free GPU access for max 12 straight hours. The pre-trained weights of the model are imported from the torchvision models package.

5 RESULT AND DISCUSSION

This section illustrates a detailed analysis and discussion of the obtained results of our experiments performed on Kaggle Dataset(Wah *et al.* 2011) using the optimal ensemble learning model of three different neural networks VGG 19, DenseNet201, and ViT. Based on Table 1, it is proffered that the ViT outperformed other models on both the training and test sets. It achieved an accuracy of 99.75 and 99.40 on the training and test sets, respectively.

To examine the overall characteristics of the proposed method result, a detailed exploration of metrics like precision, recall, and accuracy are also expressed in the Table 2. All models are

Table 1. Summary of performance of different models used for birds classification in this work.

Model	Train Accuracy	Train Loss	Test Accuracy	Test Loss
VGG 19	75.27	0.912	90.50	0.350
DenseNet 201	93.52	0.303	97.20	0.124
ViT	99.61	0.008	99.50	0.016
Ensemble Model	99.75	0.005	99.40	0.012

Table 2. Performance of the models on some evaluation metrics.

Technique	Precision	Recall	F1 Score
VGG 19	0.95	0.85	0.90
DenseNet 201	0.94	0.90	0.97
ViT	0.99	0.98	0.99
Ensemble Model	0.99	0.99	0.99

trained for ten epochs. Hyper-parameters like Adam Optimizer, Cross Entropy Loss function, and Learning rate 10^{-3} are kept constant during the entire training process. The average time taken for one epoch completion is 550s for VGG 19. Similarly, it is 480 for DenseNet 201, and for ViT, it is 615s. In the end, the model is saved in h5 format for further research examination.

6 COMPARATIVE ANALYSIS

In this section, we have presented a comparison between the classification result of our approach with other approaches used for classifying bird species. Table 3 contrasts bird classification results comparison between our approach and existing one. In (Gavali & Banu 2020) authors have applied an approach to convert the bird image into an autograph from the grayscale format, then examined each autograph to calculate the score of a particular bird species. Authors of (Y.-P Huang & Basanta 2019) have implemented a skip connection-based CNN network to improve feature extraction accuracy. A novel method is proposed in the paper (Marini *et al.* 2013) that extracts colored features from unconstrained images. Table 3, simply corroborates that our method is significantly much more accurate than other developed classification models.

Table 3. Comparison of accuracy achieved using different approaches for birds classification.

Technique	Classification Method	No. of Classes	Accuracy
(Gavali & Banu 2020)	Deep CNN (Google Net)	200	88.33
(Huang & Basanta 2019)	Skip Connections CNN	27	99.00
(Marini *et al.* 2013)	Colour + Segmentation	200	90.00
Proposed Method	Ensemble Approach	400	99.40

7 CONCLUSION AND FUTURE WORK

This paper advances the potential of deep learning and ensemble learning for automatic bird species classification. This paper illustrated an optimal ensemble (Sagi & Rokach 2018) model from individual trained networks, i.e., VGG-19-bn, Dense Net 201, and ViT. The experimental results derive the f1 score from being (90.50,97.20,99.50,99.40) for VGG 19, Dense Net 201, ViT Transformer, and Hard Voting Ensemble, respectively. This ensemble learning technique is a promising approach for automatic bird species classification. We plan to implement data augmentation to increase the training dataset size in the future. Also, some aspects of the deep neural network and its underlying filters are expected to be modified to achieve much better performance, reducing the time and cost outlays.

CONFLICT OF INTEREST STATEMENT

The authors declare no conflict of interest.

FUNDING STATUS

The author states that no funding was received for this work.

REFERENCES

Alom, Md Zahangir, Tarek M Taha, Christopher Yakopcic, Stefan Westberg, Paheding Sidike, Mst Shamima Nasrin, Brian C Van Esesn, Abdul A S Awwal, and Vijayan K Asari. 2018. "The History Began From Alexnet: A Comprehensive Survey on Deep Learning Approaches." *arXiv* preprint arXiv:*1803.01164*.

Alswaitti, Mohammed, Liao Zihao, Waleed Alomoush, Ayat Alrosan, and Khalid Alissa. 2022. "Effective Classification of Birds' Species Based on Transfer Learning." *International Journal of Electrical & Computer Engineering (2088-8708)* 12 (4)

Brusatte, Stephen L Jingmai K O'Connor, and Erich D Jarvis. 2015. "The Origin and Diversification of Birds." *Current Biology* 25 (19) R888–R898.

Cho, Kyunghyun, Bart Van Merrienboer, Dzmitry Bahdanau, and Yoshua Bengio. 2014. "On The Properties of Neural Machine Translation: Encoder-decoder Approaches." *arXiv preprint arXiv:1409.1259*.

Deng, Jia, Wei Dong, Richard Socher, Li-Jia Li, Kai Li, and Li Fei-Fei. 2009. "Imagenet: A Large-scale Hierarchical Image Database." In *2009 IEEE Conference on Computer Vision and Pattern Recognition*, 248–255. IEEE.

Gavali, Pralhad, and J Saira Banu. 2020. "Bird Species Identification Using Deep Learning on GPU Platform." In *2020 International Conference on Emerging Trends in Information Technology and Engineering (ic-ETITE)* 1–6. IEEE

Gupta, Gaurav, Meghana Kshirsagar, Ming Zhong, Shahrzad Gholami, and Juan Lavista Ferres. 2021. "Comparing Recurrent Convolutional Neural Networks for Large Scale Bird Species Classification." *Scientific Reports* 11 (1) 1–12.

Huang, Gao, Zhuang Liu, Laurens Van Der Maaten, and Kilian Q Weinberger. 2017. "Densely Connected Convolutional Networks." In *Proceedings of the IEEE Conference on Computer Vision and Pattern Recognition*, 4700–4708.

Huang, Yo Ping, and Haobijam Basanta. 2021. "Recognition of Endemic Bird Species Using Deep Learning Models." *IEEE Access* 9:102975–102984.

Huang, Yo-Ping, and Haobijam Basanta. 2019. "Bird Image Retrieval and Recognition Using a Deep Learning Platform." *IEEE Access* 7:66980–66989.

Islam, Shazzadul, Sabit Ibn Ali Khan, Md Minhazul Abedin, Khan Mohammad Habibullah, and Amit Kumar Das. 2019. "Bird Species Classification from An Image using VGG-16 network." In *Proceedings of the 2019 7th International Conference on Computer and Communications Management*, 38–42.

Kumar, A., & Alam, B. (2016). Real-Time Fault Tolerance Task Scheduling Algorithm with Minimum Energy Consumption. In *Proceedings of the Second International Conference on Computer and Communication Technologies: IC3T 2015*, Volume 2 (pp. 441–448). Springer India.

Kumar, Akash, and Sourya Dipta Das. 2018. "Bird Species Classification Using Transfer Learning with Multistage Training." In *Workshop on Computer Vision Applications*, 28–38. Springer.

Lehikoinen, Aleksi, Lluıs Brotons, John Calladine, Tommaso Campedelli, Virginia Escandell, Jiri Flousek, Christoph Grueneberg, Fredrik Haas, Sarah Harris, Sergi Herrando, *et al.* 2019. "Declining Population Trends of European Mountain Birds." *Global Change Biology* 25 (2) 577–588.

Marini, Andréia, Jacques Facon, and Alessandro L Koerich. 2013. "Bird Species Classification Based on Color Features." In *2013 IEEE International Conference on Systems, Man, and Cybernetics*, 4336–4341. IEEE

Mehra, P. S., Mehra, Y. B., Dagur, A., Dwivedi, A. K., Doja, M. N., & Jamshed, A. (2021). *COVID-19* Suspected Person Detection and Identification Using Thermal Imaging-based Closed Circuit Television Camera and Tracking Using Drone in Internet of Things. *International Journal of Computer Applications in Technology*, 66(3–4), 340–349.

Pimm, Stuart L Clinton N Jenkins, and Binbin V Li. 2018. "How to Protect Half of Earth to Ensure it Protects Sufficient Biodiversity." *Science Advances* 4 (8) eaat2616.

Sagi, Omer, and Lior Rokach. 2018. "Ensemble Learning: A Survey." *Wiley Interdisciplinary Reviews: Data Mining and Knowledge Discovery* 8 (4) e1249.

Tóth, Bálint Pál, and Bálint Czeba. 2016. "Convolutional Neural Networks for Large-Scale Bird Song Classification in Noisy Environment." In *CLEF (Working Notes)*, 560–568.

Wah, Catherine, Steve Branson, Peter Welinder, Pietro Perona, and Serge Belongie. 2011. "*The Caltech-ucsd Birds-200-2011 Dataset.*"

Zhu, Yi, and Shawn Newsam. 2017. "Densenet for Dense Flow." In *2017 IEEE International Conference on Image Processing (ICIP)* 790–794. IEEE

Artificial Intelligence, Blockchain, Computing and Security – Dagur et al. (Eds)
© 2024 The Author(s), ISBN: 978-1-032-49393-0

Comprehensive analysis of human action recognition and object detection in aerial environments

Mrugendrasinh Rahevar
Chandubhai S. Patel Institute of Technology, Charotar University of Science and Technology, Changa, Anand, Gujarat, India

Amit Ganatra
Parul Univerity, Vadodara, India

Hiren Mewada
Prince Mohammad Bin Fahd University, Al Khobar, Kingdom of Saudi Arabia

Krunal Maheriya
Chandubhai S. Patel Institute of Technology, Charotar University of Science and Technology, Changa, Anand, Gujarat, India

ABSTRACT: Drone-based aerial view analysis is the newly emerging technique helpful in topological, regional analysis and interpretation of objects and features. Due to its relevance to environment monitoring, Human action recognition (HAR) and Object detection (OD) from aerial view to search and rescue is the technical challenges. They are difficulties owing to diverse views, the tiny size of persons and objects, and involved constraints in processing. The deep learning models are proven accurate in image and video processing applications. However, the impact of Drone ego-motion on object identification, human activity detection, and crowded backgrounds may weaken the deep learning applicability in aerial view analysis. This assessment establishes current trends and progress in HAR and OD. Initially, the study on various datasets, including UCF-ARG, Okutama, BirdsEye View, and DOTA, is presented. Then, a summary and comparative analysis of various areal perspective algorithms are discussed. Finally, challenges and new directions in aerial view-based HAR and OD are discussed in depth.

1 INTRODUCTION

HAR and OD are computer vision applications where HAR identifies human behavior, for example, running, walking, fighting, sky-diving, etc. And OD is a technique to locate an object's instance and provide its labeling like trucks, cars, animals, humans, etc. Identifying human actions and objects from the ground camera is easy as we can see objects and humans correctly. Still, it will be difficult in aerial view because of large variations in human body pose, differences in the appearance of interacted objects, occlusions, motions of cameras, and minimal size of the objects make it difficult to identify objects and recognizes human behavior. Also, there are fewer datasets in aerial view for both action recognition and OD. The spatiotemporal and motion aspects are the most significant to recognition actions since they influence the learning of spatial-temporal representations to comprehend the category of an action class. The spatiotemporal records the link between spatial information at distinct timestamps, whereas the motion captures features between surrounding frames. CNNs have demonstrated powerful effectiveness in collecting high level representational features in pictures unique to a given task. As a result of its versatility and high modeling capabilities, it has been widely used for picture classification (Chollet 2017; Jmour *et al.* 2018) tasks, allowing to learn spatial representations (Zuo *et al.* 2015) from visual data for tackling the challenge of human action detection.

DOI: 10.1201/9781003393580-2

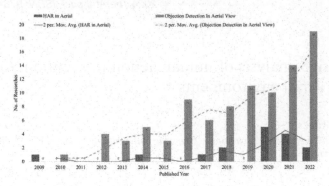

Figure 1.   Number of researches done in both for OD and HAR aerial view perspective according to year.

In order to simulate human vision and cognition, OD focuses on techniques for recognizing multiple sorts of objects within a shared framework. Image classification has advanced significantly since launch ImageNet and offered AlexNet. By multiplying the layers in the network, demonstrated VGGNet, introduced GoogLeNet, ResNet Using a residual network in the design of its image categorization, which outperforms an average person by 3.57%. HAR has wider applications, such as driving assistance, sports analysis, and video surveillance. Many problems in jobs using computer vision, such as captioning images, and object tracking, rely on OD. Nowadays, drones are used widely for an assortment of intention, which may include search and rescue, sports analysis, agriculture, and surveillance, because of their ability to capture wide areas and reach difficult arias. For example, we can use drones at the country's border to identify any suspicious activity of terrorists and detect weapons. Such surveillance cannot be done through any human or ground camera, so using drones to identify objects and human behavior at the country border is easy.

HAR and OD are important steps of search and rescue operations, their in-depth study and investigation supplement its development and better progress. This survey paper focuses on recent research to identify human actions and detected objects from an aerial view as well as which dataset is available for both studies, OD and recognition of human activities. In this paper, we focus on

- Available Dataset: Discussion on an available dataset for both studies action recognition and OD in aerial view.
- Recent research: Discussion of recent research of the past 4 to 5 years on HAR and OD in aerial view.
- Application and challenges: What applications may be made for aerial view object identification and HAR, and what problems will we have in identifying items and human activities from an aerial perspective.

Figure 2.   Methods for recognizing human actions and detecting objects in aerial views (best of our knowledge).

10

we first examine the datasets that are available for OD and HAR, and then we talk about recent OD and HAR research. Finally, challenges and applications are discussed.

2.1 Dataset for HAR in aerial view

Some research on aerial view human action has been conducted, most of which utilizes the UCFARG dataset. This section will provide an overview of aerial view HAR datasets.

Table 1. Most used dataset for aerial view HAR.

Dataset	No. of action classes	Published year
UCF-ARG	10	2008
Okutoma Action (Barekatain et al. 2017)	12	2017
Game Action Dataset	7	–
YouTube Aerial	8	–
Drone-Action (Perera et al. 2019)	13	2019

UCG-ARG Dataset collection, which is a Multiview dataset of human activity, is made up of aerial, rooftop, and ground cameras from the University of Central Florida. The set if 10 actions, such as walking, throwing, digging, boxing, carrying, and clapping, were executed by 12 different performers forming the UCG-ARG team. The acts were recorded using three different cameras: one places on the ground, another at a height of 100 feet on a rooftop, and the third on the payload platform of a Kingfisher Aerostat helium balloon, which is 13 feet high. Each actor performed each action four times, except for the opening and closing of trunks, which were performed three times on three different parked cars. The footage was captured in high quality at a resolution of 1920 x 1080, with a frame rate of 60 frames per second.4

Okutoma Dataset (Barekatain et al. 2017) A video dataset is used to find concurrent human actions in aerial perspectives. There are 12 action classes in its 43-minute-long, fully annotated sequences, including human-to-human, human-to-object, and non-interaction. Several problems unique to OkutomaAction include dynamic action transitions, rapid camera movements, dramatic size and aspect ratio shifts, and characters with multiple labels. This dataset is more complicated than others as a result, and it will help advance the discipline and enable practical applications.

Game Action Dataset to gather game activity datasets, the games GTA-5 and FIFA are used (Sultani & Shah 2021). Record the same action from several angles. Seven human behaviors—cycling, fighting, soccer kicking, running, walking, shooting, and skydiving—were recorded. The game's kicking action is included, while GTA-5 is used for the other motions. For a total of 14000 footage over seven activities, the dataset consists of 200 movies (100 ground and 100 aerial) for each activity.

The YouTube Aerial Dataset was compiled using drone footage clips sourced from YouTube. The dataset focuses on eight distinct activities, such as golf swinging, skateboarding, horseback riding, kayaking, and surfing, among others. The aerial videos in the dataset were filmed at different altitudes and feature rapid and extensive camera movements. There are a total of 50 videos for each activity. The dataset is partitioned into three subsets: 60% of the videos are reserved for training, 10% for validation, and 30% for testing.

Drone-Action dataset action recognition data for drones was collected using the Drone-Action dataset (Perera et al. 2019). The 13 actions include clapping, hitting with a bottle, hitting with a stick, jogging f/b, jogging f/b, kicking, punching, running side stabbing, walking side, and waving hands. 24 high definition video clips totaling 66,919 frames are part of the collection. To capture as many human position details in a relatively good quality, the entire movie was taken at a low height and slowly.

2.2 *Dataset of OD in aerial view*

This section provides current datasets that may be accessed and used for OD tasks in aerial views. Drones, Google Earth, and satellites are used to collect some datasets.

Table 2.   Most recent dataset of OD in aerial view.

Dataset	No of images/ Video clips	Published year
BirdsEye View (Qi *et al.* 2019)	5000	2019
DOTA (Xia *et al.* 2018)	1,793,658	2021
UAVDTBenchmark (Du *et al.* 2018)	80,000	2018
Visdrone (Zhu *et al.* 2018)	2.5 million images	2018

BirdsEye View Dataset7 for Object Classification this dataset comprises 5000 photos, each thoroughly annotated according to the PASCAL VOC criteria. The dataset contains diverse situations for which they used different datasets such as UCF-ARG and PNNL Parking dataset and selected an appropriate (i.e., can be used for OD). Parking Lot, Action Test, Routine Life, Outdoor Living, Harbour, and Social Party are the scenes. Captures frames from over 70 films as well as photos from various scenes.

DOTA More than 1,793,658 annotated object instances are included in this dataset, which is divided into 18 different categories8. These categories include airplanes, ships, tanks, baseball diamonds, tennis courts, basketball courts, ground track fields, harbors, bridges, large vehicles, helicopters, roundabout soccer fields, swimming pools, container cranes, airports, and helipads. Google Earth, the JL-1 and GF-2 satellites of the China Centre for Resources Satellite Data and Application, and other sources provided the images for this collection. Due to the vast amount of instances of an item, random orientations, many categories, a variety of aerial scenes, and a density distribution, DOTA is challenging. Nevertheless, DOTA's features make it worthwhile for real-world applications.

The UAVDT Benchmark dataset (Du *et al.* 2018) primarily concerns difficult challenging scenarios. It includes over 80,000 sample frames from 10 hours of raw video thoroughly labeled with bounding boxes. Introduce 14 characteristics for the core assignments in computer vision, namely identifying objects, tracking a single object, and tracking multiple objects, involve a range of factors such as weather conditions, altitude of flight, camera angle, classification of vehicles, and obstructions.

Visdrone Dataset (Zhu *et al.* 2018) comprises of 10,209 pictures and 263 videos with annotated frames such as bounding boxes, item occlusion, truncation ratios, and classifications, etc. 2.5 million annotations were found in 179,264 image/video frames. This dataset spans 14 distinct countries in China, from north to south. The dataset may be utilized for the following four tasks: single-object tracking, multi-object tracking, video detection, and image detection.

2.3 *HAR in aerial view*

HAR from an aerial perspective is a tough issue; however, due to the increased usage of deep learning methods, numerous types of studies have been conducted in recent years. Mmekreki's *et al.* (2021) research employed the pre-train YOLOv3 model, with re-searchers adjusting its configuration file to make the model compatible with identifying human actions from an aerial perspective. Different video frames are sent into the yolov3 model as input and a label text file. They achieve a high validation accuracy with the aid of this approach. Ketan Kotecha *et al.*

(2021) provides a solution that will deal with this issue. This approach first takes video as input, which is of this complexity level, and then takes an image frame as input. Faster motion feature modeling was utilized to identify persons. After identifying humans, accurate action recognition was used. SoftMax function was used for the final layer to classify human actions. Because this aerial view study is not widely explored, the dataset is insufficient. So, how will the model be trained in the absence of data? Sultani & Shah (2021) proposes one model for this problem that uses GAN generated dataset from ground camera features. They begin by extracting the features of various datasets, including Game Aerial Videos, Aerial Videos dataset, and Ground Videos. After extracting features from ground videos, GAN Network will be utilized to build aerial features, and all the features will be fed into the Feed-Forward Network. The authors obtained an average validation accuracy using this approach.

Authors (Mliki *et al.* 2020) separated the process of recognizing humans and human actions into two phases: offline and inference. In the offline phase model development is done for identifying humans/nonhumans and recognizing human actions from an aerial perspective. The datasets obtained using potential motion recognition are utilized to construct human/ non-human models. The inference phase is used to classify human actions into two categories: instant classification and entire classification. Instant classification classifies human activity frame by frame, whereas the whole classification produces an average of instant classification. With this approach, they achieve very promising accuracy detection and good accuracy for both instant classifications and entire classifications with this approach. EfficientDetD7 was used to identify humans with high accuracy, EfficientNetB7 to extract features, and LSTM to classify human actions. Get an average accuracy in recognizing activities under diverse conditions such as blurring, noise addition, lighting, and darkness.

A fully autonomous UAV-based activity detection system based on aerial photography has been developed by Peng & Razi (2020). It overcomes issues with aerial imaging technologies like camera vibration and motion, small human size, and poor resolution. With this method, they were able to identify every video level with excellent accuracy. A Lightweight Action Recognition Method for Unmanned-Aerial-Vehicle Video (LARMUV) (Ding *et al.* 2020) was presented. The approach was built on a teacher-student network (TSN) and employed MobileNetv3's backbone. Self-Attention was used to gather temporal information across many frames.

2.4 *OD from aerial view*

We will have difficulty identifying anything from an aerial perspective due to the items' diminutive size. Because of the restricted dataset, it will not be easy to recognize things from an aerial perspective, particularly the top view, and road view, an aerial view. As a result, we discovered issues challenging cases. To overcome this issue, Hong *et al.* (2019) presented a hard chip mining approach in patch-level augmentation for object recognition in an aerial view study. The first multiscale chip was designed to impart object-detecting knowledge. To create an object pool, they extract patches from the dataset in the second stage. To address the issue of class imbalance, these modifications will be included in the dataset. The final model is then trained after hard chips are formed from misclassified locations. However, after a calamity, like a flood or a tsunami, identifying items from an aerial perspective will be challenging. The main problem is recognizing and mapping things of interest in real time. (Pi *et al.* 2020) presented CNN detecting objects from an aerial view. These models were widely used for image classification tasks and could detect roofs, automobiles, and flooded areas.

OD in aerial images is also difficult because pixel occupancy varies across varied object sizes, the non-uniform distribution of items in aerial photographs, variations in an object's appearance due to different view angles and lighting conditions, and variations in the number of objects, and even when they are of the same type, across images (Chalavadi *et al.* 2022). Therefore, Chalavadi *et al.* proposed a hierarchical dilated convolutions operation

13

and developed a mSODANet network for multiscale object recognition in aerial images. They used parallel dilated convolutions to learn the context information of various sorts of objects at diverse sizes and fields of vision. As a result, it helped to get visual information more efficiently and improved the model's accuracy.

We discovered that identifying a car from an aerial view image was more challenging than a ground view image due to the tiny vehicle size and complicated background. Michael Ying Yang employed a Focal Loss convolutional neural network (DFL-CNN) in vehicle recognition in aerial images (Yang *et al.* 2019) to recognize a vehicle from an aerial perspective. There are skip connections used in CNN structures to improve feature learning. In addition, the focal loss function is used in the region proposal network and the final classifier to replace the usual-cross-entropy loss function (Yang *et al.* 2019).

Traffic, urban planning, defense, and agriculture all depend heavily on object identification, and convolutional neural network-based research is excellently detecting pictures. Still, high density, tiny object size, and complicated backdrop fundamental models are not performing well (Long *et al.* 2019). To appropriately identify things, Hao Long presented a method called feature Fusion Deep Network in his research on object recognition in high-altitude images using feature fusion deep learning (Long *et al.* 2019). The problem of positive and negative anchor boxes is solved by the horizontal key point-based object detector in the paper, oriented OD using boundary box-aware vectors in aerial images (Yi *et al.* 2021). To capture the oriented bounding boxes, they first determine the object's center, and then they employ the determined center BBAVectors (Yi *et al.* 2021). Fuyan Lin's study (Lin *et al.* 2020) improved the YOLOv3 model. To enhance the identification of tiny objects, the YOLOv3 model was updated by changing the anchor values and building the 4x down sampling prediction layer (Lin *et al.* 2020).

3 CHALLENGES AND APPLICATIONS

3.1 *Challenges in HAR and OD in aerial view*

The first challenge is the human's small size. The size of a human appears to be the smallest in the image from an aerial view. Because of their small size, humans cannot be seen properly by UAVs (Unmanned Aerial Vehicles), and also, we cannot see humans properly with our own eyes. Because we can't see humans properly, human parts like legs and hands aren't correctly identified, which makes it difficult to find human actions from an aerial view. Another problem is UAV camera motion; movies obtained by a UAV cannot be stabilized, and video stabilization is required to detect human movement reliably. From an aerial view, Human activities like walking and running appear practically identical. It will be impossible to distinguish that activity. Aerial view HAR research has fewer datasets available. Most analyses are based on the UCF-ARG dataset. Because the dataset is nearly 14 years old, the background environment may impair model accuracy.

Furthermore, models of deep learning needs hundreds of films for training in human air action, and gathering a huge amount of action data is challenging. One of the challenges in distinguishing human actions from an overhead perspective is the style changes, variation in view, human changes, and changes in clothes, tracking complexity of various objects and identifying anomalies and aberrant crowd behavior. It is difficult to identify many persons' actions from a single image. Objects seem different as humans from various perspectives such as top view, road view, and aerial view. Helicopters and unmanned aerial vehicles (UAVs) were employed to detect disaster damages. It is more challenging due to the tiny size of items from an aerial view. Sometimes a photograph obtained from a high height (i.e., an aerial view) has reduced pixel density or is blurry, making item identification harder. There is also a limited quantity of datasets available for aerial view object identification. Background clutter, diverse types of things, such as more than one object in one image, make detecting more than one object from one image difficult, especially from an aerial view perspective.

3.2 Applications for human actions and OD in aerial view

One of the most promising applications for OD and action recognition is a surveillance system. We can cover more ground area from an overhead view since we can see things and people from a higher height. We can, for example, conduct surveillance in a retail mall or a fair to detect suspicious behavior. While it is hard to trace terrorists' movements from the ground, aerial views allow for simple detection of suspicious activities near borders utilizing drones (UAVs). We may use aerial views to identify damage caused by natural disasters such as earthquakes and tsunamis. Strange occurrences like a lone individual loitering, many people interacting (like fights and personal assaults), people interacting with vehicles (like vehicle injury), and people interacting with facilities or locations (e.g., Object left behind and trespassing).

Use of HAR is to use surveillance systems to detect people walking. Using a double helical signature approach, Identify human walking activity in surveillance footage. Using DHS characteristics including human size, viewing different angles, camera motion, and extreme occlusion, crowded scenes may simultaneously separate people in frequent motion and identify body parts. Occlusion makes it difficult to see and count people in thick crowds. Yilmaz et al. (2006) conducted a thorough analysis of tracking techniques and divided them into groups based on the object and motion representations they employed. In addition, gender may be categorized using security cameras. Using patch characteristics to represent various body parts, Cao et al. (Cao et al. 2008) developed a part-based gender reverberation system that could accurately identify from a single frontal or rear shot, the gender picture. One of the main uses of HAR and OD is the identification of pedestrians and the prevention of falls in the elderly.

4 CONCLUSION

This article discusses current research in human action identification and OD, some of the most extensively used and promising datasets in OD and HAR, and some of the issues we face when identifying objects and human activity from an aerial perspective. We are only focused on a current study from the last 5 to 8 years, thus our article contains re-cent research most promising for an aerial perspective in recent years. There are some challenges such as changes in human appearance, different objects, and changes in camera view which need to be resolved and are addressed in this survey. We also found that a broad range of applications exists, such as in the field of surveillance, where utilizing people for monitoring costs more in terms of time and money than using a UAV. Therefore, we can create a model that can accomplish the same thing, and the studies are ideal for that. Most studies use pre-trained models, RNNs, and LSTM-based models, but we now have a new, SOTA model called a transformers model that can produce results that are more effective because it incorporates a self-attention model. Through this research, we have also found that research on attention and this self-attention mechanism is not as prevalent as it could be.

REFERENCES

Barekatain, M., Martí, M., Shih, H.-F., Murray, S., Nakayama, K., Matsuo, Y., & Prendinger, H. (2017). Okutama-action: An Aerial View Video Dataset for Concurrent Human Action Detection. *Proceedings of the IEEE Conference on Computer Vision and Pattern Recognition Workshops*, 28–35.

Cao, L., Dikmen, M., Fu, Y., & Huang, T. S. (2008). Gender Recognition From Body. *Proceedings of the 16th ACM International Conference on Multimedia*, 725–728.

Chalavadi, V., Jeripothula, P., Datla, R., Ch, S. B., & C, K. M. (2022). mSODANet: A Network for Multi-scale Object Detection in Aerial Images using Hierarchical Dilated Convolutions. *Pattern Recognition*, *126*, 108548. https://doi.org/10.1016/j.patcog.2022.108548

Chollet, F. (2017). Xception: Deep Learning with Depthwise Separable Convolutions. *Proceedings of the IEEE Conference on Computer Vision and Pattern Recognition*, 1251–1258.

Ding, M., Li, N., Song, Z., Zhang, R., Zhang, X., & Zhou, H. (2020). A Lightweight Action Recognition Method for Unmanned-Aerial-Vehicle Video. *2020 IEEE 3rd International Conference on Electronics and Communication Engineering (ICECE)*, 181–185.

Du, D., Qi, Y., Yu, H., Yang, Y., Duan, K., Li, G., Zhang, W., Huang, Q., & Tian, Q. (2018). The Unmanned Aerial Vehicle Benchmark: Object Detection and Tracking. *Proceedings of the European Conference on Computer Vision (ECCV)*, 370–386.

Hong, S., Kang, S., & Cho, D. (2019). Patch-level Augmentation for Object Detection in Aerial images. *Proceedings of the IEEE/CVF International Conference on Computer Vision Workshops*, 0–0.

Jmour, N., Zayen, S., & Abdelkrim, A. (2018). Convolutional Neural Networks for Image Classification. *2018 International Conference on Advanced Systems and Electric Technologies (IC_ASET)*, 397–402.

Kotecha, K., Garg, D., Mishra, B., Narang, P., & Mishra, V. K. (2021). Background Invariant Faster Motion Modeling for Drone Action Recognition. *Drones, 5*(3), 87.

Lin, F., Zheng, X., & Wu, Q. (2020). Small Object Detection in Aerial View Based on Improved YoloV3 Neural Network. *2020 IEEE International Conference on Advances in Electrical Engineering and Computer Applications (AEECA)*, 522–525.

Long, H., Chung, Y., Liu, Z., & Bu, S. (2019). Object Detection in Aerial Images Using Feature Fusion Deep Networks. *IEEE Access, 7*, 30980–30990.

Mliki, H., Bouhlel, F., & Hammami, M. (2020). Human Activity Recognition from UAV-captured Video Sequences. *Pattern Recognition, 100*, 107140.

Mmereki, W., Jamisola, R. S., Mpoeleng, D., & Petso, T. (2021). YOLOv3-based Human Activity Recognition as Viewed from a Moving High-altitude Aerial Camera. *2021 7th International Conference on Automation, Robotics and Applications (ICARA)*, 241–246.

Peng, H., & Razi, A. (2020). Fully Autonomous UAV-based Action Recognition System using Aerial Imagery. *International Symposium on Visual Computing*, 276–290.

Perera, A. G., Law, Y. W., & Chahl, J. (2019). Drone-action: An Outdoor Recorded Drone Video Dataset for Action Recognition. *Drones, 3*(4), 82.

Pi, Y., Nath, N. D., & Behzadan, A. H. (2020). Convolutional Neural Networks for Object Detection in Aerial Imagery for Disaster Response and Recovery. *Advanced Engineering Informatics, 43*, 101009.

Qi, Y., Wang, D., Xie, J., Lu, K., Wan, Y., & Fu, S. (2019). Birdseyeview: Aerial View Dataset for Object Classification and Detection. *2019 IEEE Globecom Workshops (GC Wkshps)*, 1–6.

Sultani, W., & Shah, M. (2021). Human Action Recognition in Drone Videos Using a Few Aerial Training Examples. *Computer Vision and Image Understanding, 206*, 103186.

Xia, G.-S., Bai, X., Ding, J., Zhu, Z., Belongie, S., Luo, J., Datcu, M., Pelillo, M., & Zhang, L. (2018). DOTA: A Large-scale Dataset for Object Detection in Aerial Images. *Proceedings of the IEEE Conference on Computer Vision and Pattern Recognition*, 3974–3983.

Kumar, A., Yadav, R. S., & Ranvijay, A. J. (2011). Fault Tolerance in Real Time Distributed system. *International Journal on Computer Science and Engineering, 3*(2), 933–939.

Tripathi, A. M., Singh, A. K., & Kumar, A. (2012). Information and Communication Technology for rural Development. *International Journal on Computer Science and Engineering, 4*(5), 824.

Kumar, A., Mehra, P. S., Gupta, G., & Jamshed, A. (2012). Modified Block Playfair Cipher using Random Shift Key Generation. *International Journal of Computer Applications, 58*(5).

Zhu, P., Wen, L., Du, D., Bian, X., Ling, H., Hu, Q., Nie, Q., Cheng, H., Liu, C., & Liu, X. (2018). Visdrone-det2018: The Vision Meets Drone Object Detection in Image Challenge Results. *Proceedings of the European Conference on Computer Vision (ECCV) Workshops*, 0–0.

Zuo, Z., Shuai, B., Wang, G., Liu, X., Wang, X., Wang, B., & Chen, Y. (2015). Convolutional Recurrent Neural Networks: Learning Spatial Dependencies for Image Representation. *Proceedings of the IEEE Conference on Computer Vision and Pattern Recognition Workshops*, 18–26.

Artificial Intelligence, Blockchain, Computing and Security – Dagur et al. (Eds)
© 2024 The Author(s), ISBN: 978-1-032-49393-0

Brain tumour detection using deep neural network via MRI images

Shadmaan, Rajat Panwar, Prajwal Kanaujia, Kushal Gautam, Sur Singh Rawat &
Vimal Gupta
Department of Computer Science and Engineering, JSS Academy of Technical Education, Noida

ABSTRACT: The brain's own aberrant and unregulated cell division is what causes brain tumors. The patient cannot heal if the growth increases by more than 50%. the identification of Brain tumor diagnosis must be swift and precise. The capture of a brain MRI scan is the initial step, after which digital imaging techniques are used to determine the precise position and size of the tumor. Gray and white matter make up MRI pictures, and the tumor-containing area is more intense. In order to enhance the given brain MRI scan, noise filters are initially utilized to remove background noise. This study aims to give a comprehensive review on detection of brain tumors.

1 INTRODUCTION

Cell proliferation that is not under control leads to tumours. Because they don't penetrate the tissues around them, these tumours could only grow in smaller areas. However, if these tumours grow close to a vital location, they may pose a risk. Malignant cancers, on the other hand, can evolve and spread in such a way that they ultimately create a fatal kind of cancer. MRI is recommended over other medical imaging modalities because it yields the most contrast images of brain tumours. In instance, it has been demonstrated in numerous studies that the transfer learning technique improves classification performance on the target dataset by applying the knowledge acquired from one task to another that is similar [1,2]. A deep convolutional neural network (DCNN) model must often be trained using a huge dataset, which includes a high level of computational complexity.

2 LITERATURE REVIEW

Using MR tests, numerous researchers created a number of procedures algorithms, and tactics to identify brain tumours, strokes, and other forms of variations in the human brain. Brain Tumor Identification and Segmentation [4–6] outline methods for the detection of brain tumours including segmentation, histograms, thresholding and morphology.

Fuzzy C means (FCM) does a good job of precisely segmenting tumour tissue. Svm [11,12] was used to recognise segmentation. Utilising learning algorithm, fundamental component analysis, as well as the wavelet transform, a hybrid technique is shown in [13,14] algorithms, wherein the accuracy of brain tumour detection is attained 98.6%.

Three multi-resolution images from [17,18] includes the various methods. Here, the suggested study asserted thatit had achieved an accuracy of 96.05%.The following approaches are presented in the Table 1 for detecting structural abnormalities, including tumours.

DOI: 10.1201/9781003393580-3

Table 1. Descriptions of some brain tumor detection algorithms.

Dataset	Method for Extracting Features	Classification Method	Accuracy
70 MR Images	Hybrid Technique Discrete Wavelet Transformation	Artificial Neural Network with Feedforward Backpropagation Algorithm	97%
T2 weighted 255 MRI images	Transform(DWPT), Shanon Entropy(SE) and Tsallis Entropy(TE)	Generalized Eigenvalue Proximate Support Vector Machine (GEPSVM)	99.61%
1800 MRI Images	CNN	CNN	98.60%
239 MRI Images	(SGLD)	ANN	99%
T2 weighted brain MR images. Dataset: 66 and Dataset: 255	Wavelet Transform Curvelet Transform and Shearlet Transform.	Support Vector Machine and Particle Swarm Optimization	97.38%
253 MRI Images	Hyper Colum Attention Module and residual block.	CNN	96.05%
500 MR Images	Fully Automatic Heterogeneous Segmentation(FAHS).	SVM	98.51%
253 MR Images	Improved ResNet50	CNN	97.01%
253 MR Images	Deep CNN	Deep CNN	98%
250 MR Images	DenseNet-169	Multiple Classifier Ensemble Multiple Classifier Ensemble Multiple Classifier Ensemble	92.37%
3000 MR Images	ResNext-101	Ensemble of multiple Classifier	93.13%

3 WORKING OF THE MODEL

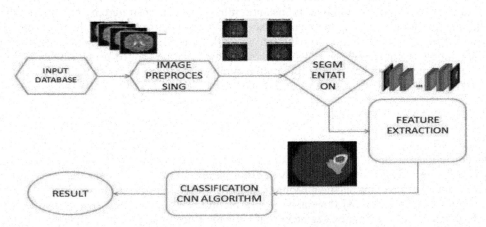

Figure 1. Analysis of brain tumor segmentation using CNN with MRI image [19].

An executive control when processing medical images is the convolutional neural network. A model of CNN as shown in the above figure is a type of machine learning used for image analysis that focuses on learning technique component knowledge. The image processing stage of this study comprises a variety of operations.

18

To begin the pre-processing process, the original gray level MR pictures are retrieved in a variety of sizes. Step 2 determines the region of interest utilising the active contours-based segmentation technique by establishing one of the largest contour. A contour is made up of an accumulation of points that have been interpolated utilising different residual methods to approximate the curve in a picture, such as linear, polynomials, or splines.

The third step in a thresholding method is to select the extreme spots. Thresholding is an essential non contextual segmentation process that generates a binary area map with a single threshold by trying to convert a greyscale or coloured image to an image pixels [16]. Models are used to extract features using the MR brain. image library. Data from the major ImageNet dataset is used to train the pre-trained CNN models [20]. InceptionResNetV2 [19], 'VGG-16, and VGG-19, Xception, ResNet50, and InceptionV3, as well as DenseNet201, are a few of CNN models that have been trained and used in the project.

4 CONCLUSION

Healthcare image processing and classification methods have gotten a lot of interest recently. A dramatically higher accuracy can be accomplished by getting a superior dataset with high-resolution images that were clearly taken from the MRI scanner. To further improve quality, classifier boosting technologies can also be used, trying to make this tool a meet its objective for any medical facility treat brain tumors.

To further reduce the noise, classifier boosting technologies could be used, making this tool a meet its objective for any medical facility treated brain tumors. MRI is the imaging technique that is most advantageous for detecting brain tumors. This study aims to give a comprehensive review on detection of brain tumors.

REFERENCES

[1]   Deepak C. Dhanwani, Mahip M. Bartere, "Survey on Various Techniques of Brain Tumour Detection from MRI Images", *IJCER*, Vol.04, issue.1, Issn 2250-3005, January 2014, pg. 24–26.
[2]   Megha A joshi, Shah D. H., "Survey of Brain Tumor Detection Techniques Through MRI Images", *AIJRFANS*, ISSN:2328-3785, March–May 2015, pp.09
[3]   Gupta, V. and Bibhu, V., 2022. Deep Residual Network Based Brain Tumor Segmentation and Detection with MRI Using Improved Invasive Bat Algorithm. *Multimedia Tools and Applications*, pp.1–23
[4]   Manoj K Kowear and Sourabh Yadev, "Brain Tumor Detection and Segmentation Using Histogram Thresholding", *International Journal of engineering and Advanced Technology*, April 2012.
[5]   Rajesh C. patil, Bhalchandra A.S., "Brain Tumor Extraction from MRI Images Using MAT Lab", *IJECSCSE*, ISSN: 2277- 9477, Volume 2, issue 1.
[6]   Vinay Parmeshwarappa, Nandish S, "A Segmented Morphological Approach to Detect Tumor in Brain Images", *IJARCSSE*, ISSN: 2277 128X, volume 4, issue 1, January 2014
[7]   Preetha R., Suresh G. R., "Performance Analysis of Fuzzy C Means Algorithm in Automated Detection ofBrain Tumor", *IEEE CPS*, WCCCT, 2014.
[8]   Amer Al-Badarnech, Hassan Najadat, Ali M. Alraziqi, "A Classifier to Detect Tumor Disease in MRI Brain Images", IEEE Computer Society, ASONAM. 2012, 142
[9]   Palvika, Shatakshi, Sharma, Y., Dagur, A., & Chaturvedi, R. (2019). Automated Bug Reporting System with Keyword-driven Framework. In *Soft Computing and Signal Processing: Proceedings of ICSCSP 2018*, Volume 2(pp. 271–277). Springer Singapore.
[10] Kumar, A., & Alam, B. (2019). Energy Harvesting Earliest Deadline First Scheduling Algorithm for Increasing Lifetime of Real Time Systems. *International Journal of Electrical and Computer Engineering*, 9(1), 539.
[11] Gudigar A., Raghavendra U., San T. R., Ciaccio E. J.,and Acharya U. R., "Application of Multiresolution Analysis for Automated Detection of Brain Abnormality Using MR Images: A Comparative Study," *Future Gener. Comput. Syst.*, vol. 90, pp. 359–367, Jan. 2019.

[12]   Toğaçar M., Ergen B., and Cömert Z., "BrainMRNet: Brain Tumor Detection Using Magnetic Resonance Images with a Novel Convolutional Neural Network Model," *Med.*

[13]   Jia Z. and Chen D., "Brain Tumor Identification and Classification of MRI Images Using Deep Learning Techniques," *IEEE Access*, early access, Aug. 13, 2020, doi: 10.1109/ACCESS.2020.3016319.

[14]   Zhuang F., Qi Z., K. Duan, Xi D., Zhu Y., Zhu H., Xiong H., and He Q., "A Comprehensive Survey on Transfer Learning," *Proc. IEEE*, vol. 109, no. 1, pp. 43–76, Jul. 2020

[15]   Szegedy C., Ioffe S., Vanhoucke V., and Alemi A. A., "Inception-v4, Inception-ResNet and the Impact of Residual Connections on Learning," in *Proc. 31st AAAI Conf. Artif. Intell.*, San Francisco, CA, USA, 2017, pp. 4278–4284.

[16]   He K., Zhang X., Ren S., and Sun J., "Deep Residual Learning for Image Recognition," in *Proc. IEEE Conf. Comput. Vis. Pattern Recognit. (CVPR)*, Las Vegas, NV, USA, Jun. 2016, pp. 770–778.

[17]   Chollet F., "Xception: Deep Learning with Depthwise Separable Convolutions," in *Proc. IEEE Conf. Comput. Vis. Pattern Recognit. (CVPR)*, Honolulu, HI, USA, Jul. 2017, pp. 1800–1807.

[18]   Szegedy C., Vanhoucke V., Ioffe S., Shlens J., and Wojna Z., "Rethinking the Inception Architecture for Computer Vision," in *Proc. IEEE Conf.nComput. Vis. Pattern Recognit. (CVPR)*, Las Vegas, NV, USA, Jun. 2016, pp. 2818–2826.

[19]   Huang G., Liu Z., van der Maaten L., and Weinberger K. Q., "Densely connected convolutional networks," in Proc. IEEE Conf. Comput. Vis. Pattern Recognit. (CVPR), Honolulu, HI, USA, Jul. 2017, pp. 2261–2269.

Artificial Intelligence, Blockchain, Computing and Security – Dagur et al. (Eds)
© *2024 The Author(s), ISBN: 978-1-032-49393-0*

A review on wildlife identification and classification

Kartikeyea Singh, Manvi Singhal, Nirbhay Singh, Sur Singh Rawat & Vimal Gupta
Department of Computer Science and Engineering, JSS Academy of Technical Education, Noida, India

ABSTRACT: Decisions on conservation and management must be supported by effective and trustworthy observation of wild creatures in their native environment. Automatic covert cameras, sometimes known as "camera traps," are becoming a more and more common method for animal surveillance due to their efficiency and dependability in quietly, regularly, and in great quantities gathering data on wildlife. However, manually processing such a massive number of photos and movies taken with camera traps is very expensive, tedious, and time-consuming. This is a significant barrier for ecologists and scientists trying to observe wildlife in a natural setting. This study suggests that present developments in deep learning techniques can be used in computer vision. This work presents a comprehensive review using current breakthroughs in deep learning methods

Keywords: Image classification, CNN, SVM.

1 INTRODUCTION

Ecology's primary goal is to learn about wild creatures in their natural habitats. By overusing natural resources, the rapid increase in population of people and the never-ending need to pursue economic growth are prompting quick, innovative, and significant updates to the systems of life on earth. Human activity has altered the population, habitat, and behaviour of wildlife on a growing area of land surface. Overusing natural resources to cause rapid, innovative, and significant changes to the Earth's ecosystems. In response to these alterations, contemporary techniques for observing wild animals, such as tracking by satellite and GPS, wireless sensor network, radio, and movement cam trap, surveillance have been developed. Because of their unique qualities, greater commercial accessibility, and simplicity of setup and performance, an increasing number of people are using remote motion-activated cameras, also known as "camera traps," to monitor wildlife.

A standard model of a hidden camera, for example, is capable of taking pictures that are not just in high definition but also collecting image data such as the moon phase, the time, and the temperature, as well as information about the day and the night a shown in the Figure 1. The enormous image collections, however, and the limitations of low-quality photographs, have a significant impact on the speed and, at times, accuracy of human classification. Images taken in a field setting present a difficult classification problem because they appear in a field setting.

Figure 1. Example of wildlife.

Variable lighting and weather conditions, a crowded background, a different pose, human photographic flaws, different perspectives, and occlusions are all factors. Due to all of these challenges, an effective algorithm for classification with the highest level of accuracy is required. Convolutional neural networks are a kind of novel simulated neural network and deep learning algorithm designed for efficient image processing. In recent years, multilayer neural networks have been successfully applied in decision-making, learning, pattern recognition, and classification.

2 LITERATURE SURVEY

This section contains papers on all-inclusive object recognition and focus mechanisms in image manipulation, and identification and categorization of animal species. Several image classification techniques were covered in this study. The most common methods for classifying images are classifiers that are object-oriented, spectral, contextual, spectral-contextual, per-pixel, per-field, as well as robust and weak categories. The most often utilised techniques are covered in this section. This survey provides theoretical information about categorization techniques as well as recommendations for the best ones. The author proposed a model called "Machine Learning, Neural Networks, and Convolutional Neural Networks" in [1]. The study of digital models that are planned to improve effortlessly via training on example data is known as machine learning. In [2] the author proposed a model called neural systems which was constructed by the means of artificial neurons. In [3] the author proposed a model titled "Challenges of Camera-trap Images for Convolutional Neural Networks." Like all machine learning models, CNN animal classification for camera-trap photos has to be taken into account and also generalization issues. They demonstrated that when classifying images from untrained camera sites, unfamiliarity with cutting-edge neural network classifiers accuracy drops.

In [4] the author proposed a model titled "Deep Convolutional Neural Networks for Automated Wildlife Monitoring: Animal Recognition and Identification." In this instance, they applied a convolutional neural network. It automatically detects important features without the need for human intervention. [5] Will focus on observing animal behavior in the wild employing facial detection and tracing. It will show a formula for detecting and tracing fauna faces in biota videos. Vincent Miele describes how efforts are still being made to develop species identification criteria that are based not only measures of the craniofacial region as well as external morphology, particularly on the cranium and maxilla, in [6]. In this case study, three different mouse species—the house mouse Mus musculus, the European wood mouse Apodemus sylvaticus, and the Cairo spiny mouse Acomys cahirinus—were examined. In [7], researchers used transfer learning to classify and forecast images in a google collaborative for the ImageNet collected information. In this study, MobileNet, MobileNetV2, VGG16, VGG19, and ResNet50 are employed as transfer learning models The Google Colab notepad served as aid for picture categorization and prediction. The main goal in [8] was to see if they could limit a sample of manually annotated camera-trap photos, educate a deep learning machine to recognize animals, including a particular species. In [9] features five elements of the detection were compared to wild, a fresh dataset of actual animal sightings focusing on difficult detection scenarios. In [10], to develop a sensor that examines video images captured by camera traps in real time. The classes to identify are rhinos, people and a collection of six typical big animals on the savannah of Africa. The scope of this project also includes the extraction of significant events. Peiyi Zeng uses Python and a simple 2D CNN to classify similar animal images in [11]. Its main focus is on the dichotomous division between snub-nosed monkeys and other monkey species. A Python crawler is used to build the database, each class having 800 and 200 photos serving as test and training data respectively. The training accuracy is 96.67% when no anomalies exist. The algorithm developed using the Missouri Hidden Cameras database [12] in Nawaz Sheikh's MobileNet architecture and the model appears to have performed well having an F1 score of 0.68 as opposed to InceptionV3's and VGG-16's respective scores of 0.70 and 0.62. According to [13], the overarching goal of this research was to provide managers with guidance on the most of the project objectives, accurate models for camera-trap image analysis are required. In [14] the goal of this study, however, is to find the creature even before hunt, not while it is going on. We propose a new method employing machine learning techniques to separate predators from non-predators by extracting animal traits. [15]. In [16], the proposed model was written in Python and tested in Visual Studio on a dataset containing 12,984 images from six different animal species belonging to six different animal classes/kingdoms. For six animal classes, the model had an accuracy of 87.22 percent.

3   MATERIAL AND METHODOLOGY

3.1   Convolutonal Neural Network (CNN)

We used Convolutional Neural Network (CNN) approach as shown in the Figure 2. This is a widely used technique in applications for computer vision. It is a type of deep neural network used to analyse visual imagery. This architectural style predominates when identifying objects in a photograph or video. Applications such as neural language processing, image or video recognition, and so on.

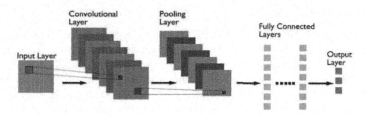

Figure 2.   CNN Architecture layers.

Five layers make up a convolution neural network: input, convolution, pooling, fully connected, and output layers. Warning capabilities, massive data capacity and picture perfect. Slower operation and long training period.

3.2   Deep learning

Categorization of camera-traps automatically in Nilgai by utilizing wildlife conservation in-depth learning A branch of machine learning called deep learning, attempts to take out knowledge from large information sets by understanding the underlying levels of more significant depiction or characteristics (Chollet 2018). A neural network is a deep learning model composed of several layers which are taught on labelled Information and pattern extraction in a hierarchical fashion. In order to generate predictions that will be used to inform subsequent layers. Using predicted and actual values, the neural network computes a rating of failures, which is then transmitted returning via the system to modify weighing value. Knowledge is acquired in a iterative manner on new unlabelled data by adjusting weights such that it maximises the capacity to lower it's error causing modest score, weights from previous layering are kept and used.

3.3   K-Nearest Neighbor (KNN) classifier

Fuzzy K-Nearest Neighbour Classifier is used to classify cattle. No training period, simple implementation, and new data may be added at any moment without altering the model since there is no preparation stage. Does not work well with large datasets, high dimensionality, or noisy or missing data.

3.4   Transfer learning

In colab notebook, image classification and prediction are accomplished by utilising transfer learning. Adaptive learning allows developers to prevent the requirement for an abundance of new data. The One of the most important restrictions on transfer learning is the problem of negative transfer.

3.5 *Random forest*

Classification of animal fibre imagery employing a mix of deep learning and random forest techniques. In addition to handling missing information automatically, Random Forest [19] is capable of solving classification and regression issues. Complexity and a longer time span.

4 DATASET

The dataset contains about 15K medium quality animal images belonging to 5 classes: elephant, horse, lion, cat and dog. This dataset is present on Kaggle and has a size of 936MB.
  Train folder
  ELEPHANT folder = 2740 files,HORSE folder = 2709 files,LION folder = 2684 files,Cat folder = 2737 files,Dog folder = 2627 files,Val folder,ELEPHANT folder = 300 files,HORSE folder = 300 files,LION folder = 300 files,Cat folder = 300 files,Dog folder = 300 files.

5 PROPOSED ALGORITHM

We used Convolutional Neural Network (CNN) approach. The five layers make up a convolution neural network: input, convolution, pooling, fully connected, and output layers.The proposed approach is motivated by the methods appeared in [17–20] for image processing and the detection of object of interest.The fully connected layer is among the finest fundamental kind of layer used in convolutional neural network. Like the title implies, every neuron in a layer that is completely interconnected is coupled to just about every neuron inside the layer underneath it. Near the conclusion of a CNN, fully linked layers are often employed to employ the predictions made by the preceding layers' learned features.

6 WORKFLOW

First a CNN network is given a source picture.
  After that, a feature map is produced using a multitude of kernels.
  Then, to make our photos more non-linear, we apply and the activate the relu. function.
  The pooling layer is then applied to every feature map in order to minimise its dimension. The pooled pictures are then flattened into a solitary, lengthy vector.
  A completely linked layer is now attached to the vector.
  Softmax is then used to determine the probabilities on every class. Choose the likelihood with the largest value after that.
  We now do forward and back – propagation algorithm training throughout a number of epochs. Continue doing this until the networks are able to classify the picture correctly.

7 CONCLUSION

As a result, here Convolutional Neural Network or (CNN) algorithm is utilised in this project to detect wild animals. This algorithm effectively and accurately categorises animal images. Convolution neural networks is used to create an image classifier that is more effective than existing approaches for classifying images and is used to quickly classify a huge number of images. The future scope of the paper are:Images can be categorized and stored in different folders. This job can be furthered extended by notifying the nearby forest office via messaging as soon as the animal is spotted.

ACKNOWLEDGEMENT

We want to show our sincere appreciation to Dr. Sur Singh Rawat, our mentor and guide, for guiding us with correct approaches and continuous feedback. We are also grateful to Mr. Vimal Gupta (coordinator) and Mr. Mahesh G for conducting regular project reviews.

REFERENCES

[1] Shalika, A.U. and Seneviratne, L., 2016. Animal Classification System Based on Image Processing & Support Vector machine. *Journal of Computer and Communications*, 4(1), pp.12–21.

[2] Miele, V., Dussert, G., Cucchi, T. and Renaud, S., 2020. Deep Learning for Species Identification of Modern and Fossil Rodent Molars. *bioRxiv*.

[3] Singh, S., Rawat, S.S., Gupta, M., Tripathi, B.K., Alanzi, F., Majumdar, A., Khuwuthyakorn, P. and Thinnukool, O., *Hybrid Models for Breast Cancer Detection via Transfer Learning Technique*.

[4] Kutugata, M., Baumgardt, J., Goolsby, J.A. and Racelis, A.E., 2021. Automatic Camera-Trap Classification Using Wildlife-Specific Deep Learning in Nilgai Management. *Journal of Fish and Wildlife Management*, 12(2), pp.412–421.

[5] Parham, J., Stewart, C., Crall, J., Rubenstein, D., Holmberg, J. and Berger-Wolf, T., 2018, March. An Animal Detection Pipeline for Identification. In *2018 IEEE Winter Conference on Applications of Computer Vision (WACV)* (pp. 1075–1083). IEEE.

[6] Kumar, A. & Alam, B. (2018). Task Scheduling in Real Time Systems with Energy Harvesting and Energy Minimization. *Journal of Computer Science*, 14(8), 1126–1133.

[7] Kumar, A., & Alam, B. (2014, February). Real Time Scheduling Algorithm for Fault tolerant and Energy Minimization. In *2014 International Conference on Issues and Challenges in Intelligent Computing Techniques (ICICT)* (pp. 356–360). IEEE.

[8] Chaturvedi, R., Kumar, S., Kumar, U., Sharma, T., Chaudhary, Z., & Dagur, A. (2021). Low-cost IoT-enabled Smart Parking System in Crowded Cities. In *Data Intelligence and Cognitive Informatics: Proceedings of ICDICI 2020* (pp. 333–339). Springer Singapore.

[9] Larson, J., 2021. *Assessing Convolutional Neural Network Animal Classification Models for Practical Applications in Wildlife Conservation* (Doctoral dissertation, San Jose State University).

[10] Singh, S., Rawat, S.S., Gupta, M., Tripathi, B.K., Alanzi, F., Majumdar, A., Khuwuthyakorn, P. and Thinnukool, O., *Deep Attention Network for Pneumonia Detection Using Chest X-Ray Images*

[11] Mahmoud, H.A., El Hadad, H.M., Mousa, F.A. and Hassanien, A.E., 2015, June. Cattle Classifications System Using Fuzzy k-nearest Neighbor Classifier. In *2015 International Conference on Informatics, Electronics & Vision (ICIEV)* (pp. 1–5). IEEE.

[12] Taheri, S. and Toygar, Ö. 2018. Animal Classification Using Facial Images with Score-level Fusion. *IET Computer Vision*, 12(5), pp. 679–685.

[13] Zhu, Y., Duan, J. and Wu, T., 2021. Animal Fiber Imagery Classification Using a Combination of Random Forest and Deep Learning Methods. *Journal of Engineered Fibers and Fabrics*, 16, p.15589250211009333.

[14] Zualkernan, I.A., Dhou, S., Judas, J., Sajun, A.R., Gomez, B.R., Hussain, L.A. and Sakhnini, D., 2020, December. Towards an IoT-based Deep Learning Architecture for Camera Trap Image Classification. In *2020 IEEE Global Conference on Artificial Intelligence and Internet of Things (GCAIoT)* (pp. 1–6). IEEE.

[15] Mahmoud, H.A., El Hadad, H.M., Mousa, F.A. and Hassanien, A.E., 2015, June. Cattle Classifications System Using Fuzzy k-nearest Neighbor Classifier. In *2015 International Conference on Informatics, Electronics & Vision (ICIEV)* (pp. 1–5). IEEE.

[16] Gupta, V. and Bibhu, V., 2022. Deep Residual Network Based Brain Tumor Segmentation and Detection with MRI Using Improved Invasive Bat Algorithm. *Multimedia Tools and Applications*, pp. 1–23.

[17] Rawat, S.S., Verma, S.K. and Kumar, Y., 2021. Infrared Small Target Detection Based on Non-convex Triple Tensor Factorisation. *IET Image Processing*, 15(2), pp.556–570.

[18] Rawat, S.S., Verma, S.K. and Kumar, Y., 2020. Reweighted Infrared Patch Image Model for Small Target Detection Based on Non-convex $\mathcal{L}$p-norm Minimisation and TV Regularisation. *IET Image Processing*, 14(9), pp.1937–1947.

[19] Mehra, P. S., Mehra, Y. B., Dagur, A., Dwivedi, A. K., Doja, M. N., & Jamshed, A. (2021). COVID-19 Suspected Person Detection and Identification Using Thermal Imaging-based Closed Circuit Television Camera and Tracking Using Drone in Internet of Things. *International Journal of Computer Applications in Technology*, 66(3–4), 340–349.

[20] Kumar, A., & Alam, B. (2016). Real-Time Fault Tolerance Task Scheduling Algorithm with Minimum Energy Consumption. In *Proceedings of the Second International Conference on Computer and Communication Technologies: IC3T 2015, Volume 2* (pp. 441–448). Springer India.

25

Artificial Intelligence, Blockchain, Computing and Security – Dagur et al. (Eds)
© 2024 The Author(s), ISBN: 978-1-032-49393-0

Image caption for object identification using deep convolution neural network

Sarthak Katyal
Department of Information Technology, Dr. Akhilesh Das Gupta Institute of Technology & Management, Delhi, India

Dhyanendra Jain
Department of CSE-AIML, ABES Engineering College, Ghaziabad

Prashant Singh
Department of Computer Science Engineering, Sunder Deep Engineering College, Ghaziabad

ABSTRACT: Our brains can readily identify what a picture is about when we view it, but doing it for a machine was thought to be impossible. However, employing computer vision's visual recognition characteristics is now achievable. The main purpose is to produce a short description using Deep Learning about the depiction of the image. To train our model, we leverage complex datasets like Fickle8k. To begin, Convolutional Neural Network is used to extricate information from the dataset's photos. The Recurrent Neural Network takes these characteristics as input and creates logical and useful descriptions for images.

Keywords: Visual Recognition, Convolutional Neural Network, Train, Computer Vision, features, Deep Learning, Dataset, Recurrent Neural Network

1 INTRODUCTION

For us to survive, we need to be able to recognize images. While a computer is incapable of doing the same feat. Models that evaluate and create captions for a picture are more efficient due to the evolution of Deep Learning techniques, the accessibility of feature-rich datasets, and processing capacity. Image caption generation is a combination of picture processing and natural language processing principles that detect an image's characteristics and define them in a user-friendly language e.g. English [1,2].

The following issues of inaccurate outcomes and limited internal memory plague this picture caption generator, which employs a traditional neural network.

The major goal of this study is to ply CNN and LSTM to create a workable model for the Image Caption generator. The current approach employs a traditional Recurrent Neural Network with internal memory (short memory), which is neither popular nor strong and may provide erroneous results [3]. Image caption generator utilizes OpenCV and alike libraries and natural language processing methodology to comprehend the structure of a picture and describe it in a neural language like French, English, or Chinese [4].

Captioning of images is used in a variety of applications, including editing software suggestions, virtual assistants, social networking, image indexing, as well as a slew of additional natural language processing products.

2 PROBLEM STATEMENT

The major problem we're working on is creating a deep learning model that allows a computer to assess an image and predict a syntactically and grammatically accurate caption for it.

DOI: 10.1201/9781003393580-5

The problem should be executed in four stages:

(a) detection and identification of objects
(b) feature modeling
(c) scene categorization
(d) creation of a description

3 METHODOLOGY

3.1 *Data set*

The Flickr 8K dataset was utilized. For each photograph in the collection, there are five captions, which totals 8000 photos. The caption of a single image assists in comprehending all of the varied circumstances. There are 6,000+ images in the training dataset Flickr 8k.trainImages.txt, 999+ images in the test dataset Flickr 8k.testImages.txt, and 999+ images in the development dataset Flickr 8k.devImages.txt. Jason Brownlee has made these datasets (size 1GB) available for direct download.

3.2 *Preprocessing*

To train the model, it takes an image and its descriptions as input. These must be thoroughly preprocessed before being fed into the model. The beginning step is to import libraries mainly NumPy, pandas, string, pickle, and others. We use keras.preprocessing to import load img, img to array, Tokenization, pad sequences, and other functions for image and text preprocessing. Image feature extractors, such as the Xception model, must also be imported. We must also import RNN model layers such as dense, LSTM, embedding, and others.

First, we import the Flickr 8K data set as text and image separately. Working on the text, we open and import all the text using inbuilt file management libraries with file.read(). Then we separate images with their text and store them separately for our comfort. The next step is to remove punctuations, and words with numbers and convert the text to lowercase. Python loops like for and while with Python conditions like if and if-else is used to clean the text. After cleaning the text, a new vocabulary is built with all the unique words.

We employ an embedded convolutional neural network model called Xception for picture preprocessing. We resize the photographs to a model-acceptable format and provide them as input to the chosen model. The characteristics of these photos are output by this model, which we keep as features.p.

3.3 *Methods*

Deep learning uses an artificial neural network with numerous levels organized in a hierarchy to carry out the machine learning process [5]. The model is constructed on deep networks, in which information flows from the initial stage, where the model has trained something basic and then dispatches the result to stage two, where the input is integrated into something extra complicated and passed on to the third stage. As each stage of the network amplifies the complicatedness of the input it receives from the ascending stage, the process continues [6].

3.3.1 *Convolutional Neural Networks (CNN)*

CNN is a technique under Deep Learning which takes an image as an input, designates significance to various visual elements, and sets them apart from each other. Pre-processing time needed for a ConvNet is significantly less than it is for other classification approaches [7].

Convolutional-NN is a type of network working technique that can handle a 2D matrix kind of data. Images are transformed into a 2D matrix, and CNN is a highly efficient tool for working along with them.

3.3.2 *Recurrent neural networks*

Recurrent Neural Networks (RNN) are a Deep Learning technique for modeling sequential data. RNN uses the same weights for every sequence element, reducing the number of parameters and enabling the model to generalize to varied sequence lengths. RNNs falls under neural network that models data in a series [8]. Feedforward network-based RNNs exhibit behavior akin to that of human brains. Simply said, RNNs are better at anticipating sequential input than other algorithms.

3.3.3 LSTM networks

These are a kind of RNN that is particularly near perfect for forecast challenges. The upcoming word can be predicted based on the previous write-up. It has proved to be more successful than regularly occurring RNNs by circumventing the short-term memory restrictions of RNNs. Long-term dependencies can be learned via LSTM. To overcome the long-term reliance issues LSTMs are developed specifically [9].

3.4 Resnet VS xception VS VGG

3.4.1 Resnet

Residual Network (ResNet) is a renowned deep learning model that was introduced by Jian Sun, Shaoqing Ren, Xiangyu Zhang, and Kaiming He. Residual blocks were introduced to tackle the problems associated with the training of very deep networks [10]. These Residual blocks are the main element of ResNet models. In the architecture, influenced by VGG-19, there is a 34 leveled simple network with shortcuts and skip links. The architecture is then converted into the residual network by these skip connections or residual blocks.

3.4.2 Xception

Xception is deep-CNN architecture with Depthwise Separable Convolutions. It outperforms Inception-v3 with a modified depthwise separable convolution. The Xception architecture consists of 36 convolutional levels or layers that serve as the network's feature-determining strength [11]. These 36 layers are organized into 14 sub-modules, all of which contain linear residual interconnections surrounding them. The Xception architecture is essentially a sequential pile of depthwise dissociable layers with residual connections. At only 91MB, Xception has the least weight serialization [12].

3.4.3 VGG16 & VGG19

VGG (Visual Geometry Group) is a Convolutional Neural Network (CNN) architecture that is multilayered in nature. The 16 and 19 tell the number of convolution layers in the network. The VGG network concept was defined by Simonyan and Zisserman in their research, 'Very Deep Convolutional Networks for Large Scale Image Recognition' [13]. They constructed a 16-layer network that had fully linked layers and convolutional layers. To keep it basic, only 33 layers were placed on top of each other [14].

3.5 Model

Defining the caption model is a major task. We need to combine the outputs of CNN and the modified text into an RNN model. The features from the CNN model squeezed from 2048 to 256 nodes using dropout and dense functions. We then define the LSTM model for text. We then merge the two using the Dense function with action as relu followed by the Dense function with activation SoftMax. The model is compiled using categorical crossentropy as loss and Adam as optimizer. The model is trained for 15 epochs for around 2.5 hours.

4 RESULT

The model ran and predicted the result for the images provided as input.

Here, the model predicts *"man in yellow kayak paddles through the water"* for the image on the left. This prediction is quite similar to what is shown in the picture.

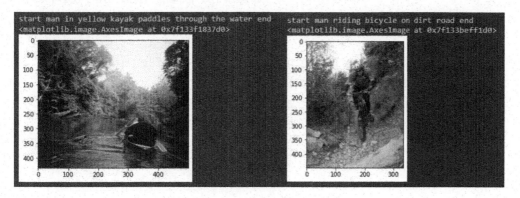

Again, the model predicts *"man riding bicycle on dirt road"* for the image on the right. This prediction is quite similar to what is shown in the photo.

5 CONCLUSION

Examining the outcoming, we can conclude that the deep learning technology used gave positive outcomes. The combination of LSTM and CNN was able to determine the connection between various objects in pictures by working in synchronization.

This presentation introduced us to a variety of breakthroughs in machine learning and artificial intelligence and the fields associated with them. While this article attempts to cover the elements required to develop an image caption generator, several of the topics covered are open to further inquiry and refinement.

REFERENCES

[1] *How to Develop a Deep Learning Photo Caption Generator from Scratch* https://machinelearningmastery.com
[2] Image Caption Generator – Gunjan Sethi-https://www.academia.edu/73881784/Image_Caption_Generator
[3] *Image Caption Generator* – Vivek Kamble, SayamKoul, Abhishek Chaudhari, Rajashri Sonawale https://www.academia.edu/52426598/Image_Caption_Generator
[4] *Image Caption Generator Using Deep Learning* – A. V. N. Kameswari, B. Prajna https://www.academia.edu/67593261/Image_Caption_Generator_Using_Deep_Learning
[5] Gerber R. and Nagel H. Knowledge representation for the generation of quantified natural language descriptions of vehicle traffic in image sequences. In *ICIP*. IEEE, 1996.
[6] Chen J., Dong W. and Li M., *"Image Caption Generator using Deep Neural Networks"*, March 2018.
[7] Andrej K, Li F-F *Deep Visual-semantic Alignment for Generating Image Descriptions*. https://cs.stanford.edu/people/karpathy/cvpr2015.pdf
[8] Xu, Kelvin, *et al.* "Show, *Attend and Tell: Neural Image Caption Generation with Visual Attention.*" *Computer Science.* (2015)
[9] Johnson J., Karpathy A., and Fei-Fei L. Densecap: Fully Convolutional Localization Networks for Dense Captioning. In *Proceedings of the IEEE Conference on Computer Vision and Pattern Recognition*, 2016.
[10] Iashin, V., & Rahtu, E. (2020). Multi-modal Dense Video Captioning. In: Proceedings of the IEEE/CVF Conference on Computer Vision and Pattern Recognition Workshops, pp.
[11] Flick, Carlos. "ROUGE: A Package for Automatic Evaluation of Summaries." *The Workshop on Text Summarization Branches Out2004*:10. (2014)
[12] Vinyals O., Toshev A., Bengio S., and Erhan D. Show and Tell: A Neural Image Caption Generator. In *Proceedings of the IEEE Conference on Computer Vision and Pattern Recognition*, 2015.
[13] Vedantam, R., Lawrence Zitnick, C., & Parikh, D. (2015). *Cider: Consensus-based Image.*
[14] Das S., Jain L. and Das A., "Deep Learning for Military Image Captioning," *2018 21st International Conference on Information Fusion (FUSION)*, 2018, pp. 2165–2171, doi: 10.23919/ICIF.2018.8455321.

Artificial Intelligence, Blockchain, Computing and Security – Dagur et al. (Eds)
© 2024 The Author(s), ISBN: 978-1-032-49393-0

Brain tumor detection using texture based LBP feature on MRI images using feature selection technique

Vishal Guleria, Aryan Verma, Rishabh Dhenkawat, Uttkarsh Chaurasia & Nagendra Pratap Singh
Department of Computer Science and Engineering, National Institute of Technology, Hamirpur, H.P. India

ABSTRACT: This study analyzes the utility of texture-based local binary pattern features to spot brain cancers. Local Binary Pattern is a highly effective texture feature that uses a three-by-three sliding window to label the pixels of an image by thresholding the adjacent pixel. To pre-possess images, they are first converted from RGB to Lab to narrow the color space, and then this image is operated with Principal component analysis. This processed image is used to extract the LBP features. Finally, Support vector machines are trained on the extracted features from the dataset and compared against the deep neural network results. We found that the deep neural network achieves 98.6% accuracy while the SVM approach achieves an accuracy of 98.4%. Finally, a comparison has been drawn between these approaches using various comparison metrics.

Keywords: MRI images, LBP feature, Thresholding, Feature selection, Classifiers

1 INTRODUCTION

A feature can be thought of in a more general sense as any piece of data that is helpful in completing the computational task that is connected with a certain application. The practice of working with datasets that contain hundreds of attributes is becoming more common. Overfitting happens when the number of features in the dataset is greater than the total number of observations that are included in the dataset. There is no doubt that regularization can help lower the risk of overfitting, but feature extraction techniques can also bring other benefits, such as increased accuracy and a decreased risk of overfitting. Accelerating the training process and making the data more visually appealing are the two steps that may be taken to boost the explain ability of our model.

A feature can be generated using either a feature detection approach, which is then applied to an image, or a general neighbor procedure. A texture descriptor known as" local binary pattern" can be utilized to characterize the local texture pattern of an image. LBP has become a prominent method in a variety of applications both as a result of its ability to discriminate and the ease with which it may be computed. In this study, a total of 34 features are retrieved from MRI scans of the brain, and then those features are used to classify patients according to whether they have a brain tumor and what kind of tumor they have (meningioma or glioma). The tumor known as meningioma originates from the meninges (It is the membrane that surrounds both our brain and spinal cord). Although it is not a brain tumor in the strictest sense, because it can squeeze or compress the neighboring region of the brain as well as the veins and nerves, it is included in this category. Meningiomas are the most frequent type of tumor that can develop in the head. Meningiomas are the most common type of tumor that can develop in the head (Siegel *et al.* 2022) Gliomas are types of brain tumors that originate from cells that, under normal circumstances, would have progressed to become healthy glial cells in the brain.

DOI: 10.1201/9781003393580-6

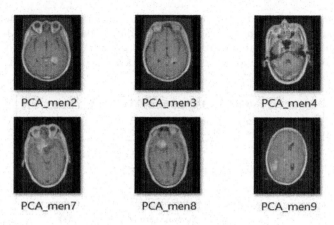

PCA_men2   PCA_men3   PCA_men4

PCA_men7   PCA_men8   PCA_men9

Figure 1.   A sample of the dataset showing images with mening tumor.

2   LITERATURE SURVEY

In the past, several techniques, such as edge detection, classification using decision trees, various segmentation methods, and genetic algorithms were utilized for the purpose of detecting brain tumors.

(Naik & Patel 2014) has done work on Tumor detection and classification using decision trees. They used image mining concepts with implicit knowledge extraction, image data relationships, and other patterns. Two classification algorithms were used: Naive Bayesian and Decision Tree. The decision tree classifier gives better performance than the Naive Bayesian Classifier. (Singh *et al.* 2015) detected tumor using a combination of Fuzzy C-Means and SVM. A hybrid methodology combining support vector machines and fuzzy means yields accurate results for brain tumor detection through clustering for classification. (Kaur & Sharma 2017) proposed work on Brain tumor detection using self-adaptive K-means clustering. Self-adaptive k-means clustering mitigates the user from a selection of a number of clusters. Clusters are computed based on the maturity of image in terms of its histogram.

(Abdalla & Esmail 2018) used an Artificial Neural Network. It consists of three steps: MRI image pre-processing, image post-processing such as segmentation, morphological computation, feature extraction, and image feature implementation for pattern recognition for tumor detection. In this texture analysis, the Haralik feature method was used on the SGLD matrix (decomposing the input image into texture features). (Na *et al.* 2019) used multimodal MRI data based on the Tamura texture feature and an ensemble SVM classifier. From each voxel, it takes 124 features, including grayscale elements and Tamura texture, and then ranks them using SVM. To build the ensemble SVM classifier based they used batch random sampling on a weighted voting mechanism. There are several approaches similar to these machine-learning techniques (Chandra & Rao 2016; Kadam & Dhole 2017; Kumar *et al.* 2017; Ramaswamy Reddy *et al.* 2013; Sharma *et al.* 2012; Shanthakumar & Ganeshkumar 2015) which have been employed for the detection of brain tumors from MRI scans.

3   FEATURES EXTRACTION

The texture is a physical or visual feature of a surface. Texture analysis is to provide a unique method for encapsulating the fundamental qualities of textures in a simplified yet distinct form so that they can be used to categorize and segment objects reliably and accurately.

Despite the significance of texture in image analysis and pattern recognition, few designs provide onboard textural feature extraction. Local binary pattern, GLCM, Tamura, and Wavelets are among the available feature extractors; however, we have only used the LBP (Local Binary Feature) feature to detect brain cancers in this study.

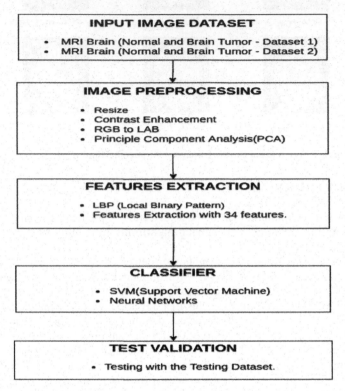

Figure 2. Summary of method used in this research work.

4 LOCAL BINARY PATTERN (LBP)

LBP is renowned for its extraction of texture-based features. As a result of its differentiating capability and computational simplicity, LBP has grown used in a variety of applications. It is computed by taking the difference in intensities between the central pixel and its local neighbors. If the difference is higher than zero, the binary bit is assigned the value 0. If smaller than zero, 1 is supplied as the binary bit.

Let C_0 describe the intensity of the center pixel and let C_{rn} be its neighbors, where r is defined as the neighborhood's radius and n is defined as the location of the neighbor, $0 \leq n \leq 7$, with n = 0 representing the center pixel's immediate right neighborhood and n is increasing in the counterclockwise direction. Considering r = 1 neighborhood for any pixel, LBP is calculated as $LBP_{rn} = 1$, if $C_{rn} \geq C_0$, 0, if C_{rn} ¡ C_0. After computation of the pattern, We get the LBP value by processing this pattern as an 8-bit binary number. where n = 8 represents the most significant bit and this 8-bit binary is then converted to its decimal equivalent in the range from [0-255].

$$LBP = \sum_{n=0}^{7}(LBP_{rn})2^{n} \qquad (1)$$

32

This decimal value is generated for each pixel in the image, and then an LBP feature is constructed using a histogram of these values based on their distribution. The LBP is calculated as:

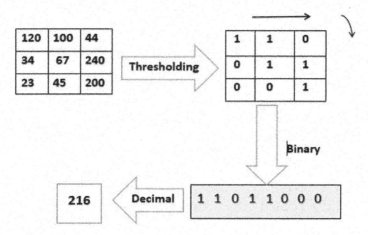

Figure 3.   Thresholding in LBP.

Algorithm 1 given below is the algorithm used to compute LBP discussed above.

5   FEATURES SELECTION

Every image is made up of different components. Therefore, taking into account all of the characteristics in picture classification may result in the overfitting problem. In addition, the computational cost can be decreased by lowering the total number of features while simultaneously raising the level of the model's performance. In this study, we selected 34 features to apply to the frequency and grey values obtained from the histogram after thresholding. These features were chosen based on their ability to differentiate between similar images. Following the application of thresholding, each feature is collected.

- min(f):- The first feature is taken as the minimum frequency after thresholding.

$$min_f = min(f_i) \text{ where } i \geq 1 \text{ and } i \leq n \tag{2}$$

- max(f):- The next feature is the maximum frequency value after thresholding.

$$max_f = max(f_i) \text{ where } i \geq 1 \text{ and } i \leq n \tag{3}$$

- means (f):- The next feature is taken as the mean value of frequency.

$$mean_f = \frac{1}{n}\sum_{i=1}^{n} f_i \tag{4}$$

- mod(f):- frequency value with higher occurrences after thresholding.
- median(f):- It is used to find the median of the frequency that is obtained from the histogram after thresholding after ascending order.

$$median_f = \left[\frac{1}{2}(n+1)\right]^{th} term, \text{ else } \frac{1}{2}\left[\left(\frac{n}{2}\right)^{th} term + \left(\frac{n}{2}+1\right)^{th} term\right] \tag{5}$$

33

Algorithm 1. LBP (Local Binary Pattern).

```
input : MRI images of size m × n
output: Features of the image
1 Function main()is
2     foreach i: image do
3     |   calculate LBP(i)
4     end
5 end

6 Function calculate LBP(image) is
7     B = im 2gray (image)
8     [m,n ]=  Size (B )
9     X =[ − 1, 0, 1, 1, 1, 0, − 1, − 1]
10    Y =[ − 1, − 1, − 1, 0, 1, 1, 1, 0]
11    foreach l in m do
12        foreach j in n do
13            foreach K in 8 do
14                x = i + X (k)
15                y = j + Y (k)
16                if B (i,j ) > B (x,y ) then
17                |   pi(k)=0
18                end
19                else
20                |   pi(k)=1
21                end
22            end
23            foreach x =1:8    do
24            |   value = value +(2 (8− x) · pi(x))
25            end
26            net(i,j )=  value
27        end
28    end
29 end
```

- std(f):- std(f) will return the standard deviation of the list of frequency values that are obtained from the histogram after thresholding.

$$std_f = \sqrt{\frac{1}{N}\sum_{i=1}^{N}(f_i - \bar{f})2}$$

(6)

- cov(f):- cov(f) will return the covariance of the vector of values of frequency, where f is a matrix whose columns represent the random variable while rows represent the frequency histogram values after thresholding. In the given equation A=vector of random variables, F=vector of frequency histogram values after thresholding, x=mean of the A, y=mean of the B values.

$$cov_{A,F} = \frac{\sum_{i=1}^{N}(A_i - \bar{x})(F_i - \bar{y})}{N - 1}$$

(7)

- median(xcov(f)):- median(xcov(f)) returns the median of the autocovariance sequence of f, where f represents the frequency histogram values after thresholding.

- median(xcorr(f)):- median(xcorr(f)) returns the median of the autocorrelation sequence of the f where f represents the frequency histogram values after thresholding.
- var(f):- var(f) returns the variance of the sequence of f where f represents the frequency histogram values after thresholding.

$$var_f = \frac{1}{N}\sum_{i=1}^{N}(f_i - \bar{f})^2 \tag{8}$$

- median(zscore(f)):- median(zscore(f)) returns the median of z-score for every element of the sequence of frequency.
- median(cummax(f)):- median(cummax(f)) returns the median of a cumulative maximum of the sequence f.
- median(cummin(f)):- median(cummax(f)) returns the median of a cumulative minimum of the sequence f.
- kurtosis(f):- kurtosis(f) returns the kurtosis of f where f represents the frequency histogram values. If f is a vector, then kurtosis(f) returns a scalar value that is kurtosis of the elements in f, where, f=frequency values sequence, $\bar{f}$ = mean of the sequence, σ is the standard deviation, E is the expectation.

$$k_f = \frac{E(f - \bar{f})^4}{\sigma^4} \tag{9}$$

- mod(xcorr(f)):- mod(xcorr(f)) returns the mod of frequency with the highest count in the autocorrelation sequence of f where f is the frequency.

$$R_{xy} = E[f_{m+n}y_n^*] = E[f_n y_{n+m}^*] \tag{10}$$

- mod(xcov(f)):- mod(xcov(f)) returns the mod of frequency with the highest count in the autocovariance sequence of g, where g represents the gray value. All the values are calculated and then the mode of these values is taken.

6   IMAGE PREPROCESSING

We are having datasets with images of varying sizes, whereas for classification and machine learning purposes, we require all the images to be the same size. 2 separate datasets that can use this one for training and testing purposes, respectively.

In order to reduce the severity of the class imbalance issues, we increased the number of images by adding noise in the form of salt, paper, and Gaussian distributions.

Following this step, the images are there in RGB color space are changed into LAB format, and then the PCA(Principal Component Analysis) analysis is then performed on them. After that, the application of adaptive histogram equalization to improve the contrast of the image.

7   EXPERIMENTAL RESULTS

Features are calculated using Local Binary Pattern from each MRI image in our dataset. A total of 34 features are selected that are discussed earlier. These 34 features of each image are stored in Excel sheets and then passed to SVM and Neural Network classifiers and the results are calculated.

Table 1.   Results using neural network.

Class	Precision	Recall	F1-score	Accuracy
Glioma	0.992	0.993	0.993	
Meningioma	0.982	0.992	0.987	98.6
Normal	0.981	0.980	0.980	

Table 2. Results using SVM.

Class	Precision	Recall	F1-score	Accuracy
Glioma	0.996	0.990	0.993	
Meningioma	0.976	0.966	0.986	98.4
Normal	0.973	0.985	0.979	

8 CONCLUSIONS

Local Binary Pattern can be used to classify brain tumors. In this research, we have classified whether a person is having a tumor or not. If he is having a tumor then which type of tumor he is having. The classification of brain tumors (Glioma, Meningioma) is carried out using these two methods:- Support Vector Machine and Neural Network. In validation, with the use of SVM, our dataset has achieved an accuracy of 98.4 percent (Support Vector Machine). We were able to classify the dataset with a success rate of 98.6 percent by utilizing a neural network classifier. Therefore we can also conclude that the Neural Networks are better than the SVM classifier to classify brain tumors.

REFERENCES

Abdalla, Hussna Elnoor Mohammed, and MY Esmail. 2018. "Brain Tumor Detection by Using Artificial Neural Network." In *2018 International Conference on Computer, Control, Electrical, and Electronics Engineering (ICCCEEE)* 1–6. IEEE

Chandra, G Rajesh, and Kolasani Ramchand H Rao. 2016. "Tumor Detection in Brain using Genetic Algorithm." *Procedia Computer Science* 79:449–457.

Kadam, Megha, and Avinash Dhole. 2017. "Brain Tumor Detection Using GLCM with the Help of KSVM" *International Journal of Engineering and Technical Research* 7 (2)

Kaur, Navpreet, and Manvinder Sharma. 2017. "Brain Tumor Detection Using Self-adaptive K-means Clustering." In *2017 International Conference on Energy, Communication, Data Analytics and Soft Computing (ICECDS)* 1861–1865. IEEE

Kumar, A., & Alam, B. (2015, February). Improved EDF Algorithm for Fault Tolerance with Energy Minimization. In *2015 IEEE International Conference on Computational Intelligence & Communication Technology* (pp. 370–374). IEEE.

Alam, B., & Kumar, A. (2014, March). *A RealTtime Scheduling Algorithm for Tolerating Single Transient Fault. In 2014 International Conference on Information Systems and* Computer Networks (ISCON) (pp. 11–14). IEEE.

Naik, Janki, and Sagar Patel. 2014. "Tumor Detection and Classification using Decision Tree in Brain MRI" *International Journal of Computer Science and Network Security (IJCSNS)* 14 (6) 87.

Ramaswamy Reddy, A EV Prasad, and LSS Reddy. 2013. "Comparative Analysis of Brain Tumor Detection using Different Segmentation Techniques." *Int. J. Comput. Appl* 82:0975–8887.

Shanthakumar, P and P Ganeshkumar. 2015. "Performance Analysis of Classifier for Brain Tumor Detection and Diagnosis." *Computers & Electrical Engineering* 45:302–311.

Sharma, Pratibha, Manoj Diwakar, and Sangam Choudhary. 2012. "Application of Edge Detection for Brain Tumor Detection." *International Journal of Computer Applications* 58 (16)

Siegel, Rebecca L., Kimberly D. Miller, Hannah E. Fuchs, and Ahmedin Jemal. 2022. "Cancer Statistics, 2022." *CA: A Cancer Journal for Clinicians 72*, no. 1 (January) 7–33. https://doi.org/10.3322/caac.21708. https://doi.org/10. 3322/caac.21708.

Singh, Amritpal, *et al.* 2015. "Detection of Brain Tumor in MRI Images, Using Combination of Fuzzy C-Means and SVM" in *2015 2nd International Conference on Signal Processing and Integrated Networks (SPIN)* 98–102. IEEE

Artificial Intelligence, Blockchain, Computing and Security – Dagur et al. (Eds)
© 2024 The Author(s), ISBN: 978-1-032-49393-0

Towards computationally efficient and real-time distracted driver detection using convolutional neutral networks

Ramya Thatikonda
Doctor of Philosophy, University of the Cumberlands, Charlotte, North Carolina

Sambit Satpathy
Assistant Professor, Galgotias College of Engineering & Technology, Greater Noida

Shabir Ali
Department of Engineering and Technology, Bharati Vidyapeeth (D.U.), Navi Mumbai, India

Munesh Chandra Trivedi
CSE, NIT Agartala, Agartala

Mohit Choudhry
Associate Professor, Galgotias College of Engineering & Technology, Greater Noida

ABSTRACT: There are a lot of serious accidents on the roads because there are so many more cars and people are distracted by a lot of things. As it thinks about the problem of driver distraction, it needs to set up an intelligent transportation system. The learning algorithm helps to lower the risk, and it also has to deal with the difficulties of resource computation. In this, a new DL-based method is proposed to use image processing to find drivers doing different things. In this method, the image of the face is first taken and then processed using image and feature extraction. The extracted images help find the photos that show driver boredom, hand gestures, eye recognition, and other things. Based on the above features, a modified random forest that is integrated with C 5.0 helps to analyse the driver's condition, and a modified CNN is used to find out if the driver is distracted. An optimizer is used to improve the performance of the system and get around the problem of not having enough resources. During the performance analysis, the proposed method is compared with existing methods like RNN, VGG-16, and SVC-based pixel images to figure out the accuracy, precision, F1-Score, and Recall.

Keywords: CNN, Deep learning, hand gesture, optimization, accuracy

1   INTRODUCTION

Recently, there has been a drastic increase in the number of moving vehicles in road transport. Based on the survey taken in 2022 for the significant accident occurrences, driver distraction, and lethargy. In most countries, these challenges are substantial and result in numerous road accidents [1]. Even though the law was enforced for the driver long before, detecting the driver's distraction plays a crucial role. As per the report, there are certain activities such as eating, drinking, talking/texting on the phone, loud music sound, make some distractions while driving. Recognizing driver posture is crucial to the distracted driving detection procedure [2], which considers all types of distracted driving. Sensors that can be worn and camera-based. The two main methods for categorizing or identifying distracted drivers differ significantly, as represented in Figure 1.

The wearable sensor is less effective than the vision-based method for tracking distracted behaviours in real time because it costs more and requires the user to do something. Thus, using vision-based sensors to identify distracted behaviours is becoming more common.

DOI: 10.1201/9781003393580-7

Neural network architectures are widely utilized to categorize tasks as vision-based technologies advance. Convolutional neural networks (CNNs) are now making substantial strides in computer vision tasks.

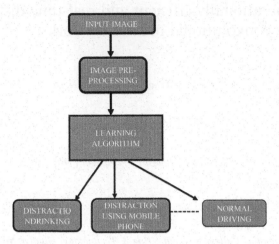

Figure 1.   Driver distraction classification.

Existing methods help to detect distractions while driving and using a mobile phone, but only the driver's face and hands are visible while collecting the image. A combination of the AlexNet [3] and Inception V3 [4] architectures are used to look at a method for manually recognizing driver distraction. However, the model requires a transfer learning-pertained network.

While evaluating several methods, such as random forests (RF), the k-nearest neighbour classifier (KNN), and the multilayer perceptron (MLP) classifier, it is shown that these methods perform poorly [5]. Recently, visual characteristics were processed using a deep learning model built on the hybrid CNN framework (HCF) to identify distracted driving behaviour. A pre-trained model that integrates ResNet50, Inception V3, and Xception to create HCF extracts driving activity variables via transfer learning. However, the model's generalizability for examining the driving behaviour from various camera angles was constrained because the authors only employed picture datasets from the right side.

The existing driver assistant systems have intricate deep-learning architectures with the primary goal of enhancing driving safety [6]. A driver behaviour features identification technique that primarily concentrates on the status to monitor the driver and driver behaviour models for ADAS is presented. Reinforcement learning maps image data on the camera to improve driving assistance. Nevertheless, ADAS's real-time performance restrictions, high equipment costs, and low portability make them less practical.

1.1   Driver distraction problem definition

As many moving vehicles are being travelled on the road, public safety needs to be ensured. In most countries, accidents happen, as most depend upon the driver's distraction. It is a severe problem as it results in significant casualties.

1.2   Detection process based on driver activities

Here the work is based on the approach to detecting driver distraction to reduce the issue of major road accidents. The main aim is to provide a learning algorithm which helps to see the driver's distraction. It will be applicable based on measured activities, such as eating, drinking, and talking/texting on the phone or loud music sounds. The identification of the distraction and those images are classified based on the neural networking algorithm through the image behaviour.

To identify and analyze the activity-based images of the drivers, the risk and accidental occurrence gets determined. In this approach, the face image is initially collected and processed based on image and feature extraction. The extracted images help to determine the pictures related to driver boring, hand gestures, eye recognition, etc. Based on the above features, a Modified random forest integrated with C 5.0 helps to analyze the driver condition, and a modified CNN is used to detect the driver distraction. An optimizer is used to improve the system's performance to overcome the problem of resources.

Table 1. Timeline to generalize the work.

S.No.	Work to be done	Timeline
1	Problem Identification	1 Month
2	Literature Survey	2 Months
3	Research Objective	1 Month
3	Formulating the research objective 1	2 Months
4	Formulating research objective 2	2 Months
5	Implementation	1 Month
6	Paper preparation	2 Months

The paper is organized as follows, the problem of Driver distraction and computing significant resource problems are discussed in Section II. The research scope is discussed in section III. The existing works related to driver distraction are in section IV, and the proposed research work is in section V. The performance analysis with data loss and data accuracy are analyzed and determined in section VI. Finally, the conclusion for the proposed approach is discussed in section VII.

2 PROBLEM STATEMENT ON DRIVER DISTRACTION AND RESOURCE COMPUTATION

The driver is usually to blame for serious accidents. Problems with road safety in developing countries are said to be caused and made worse by the way people act, which is a very serious issue. The rise of fully connected, driverless cars has turned the automotive industry into a tech-based one, which makes the problem possible. With AI, a real-time alert system that tells the driver to pay attention can be made. This makes accidents less likely, which saves lives and keeps damage to property to a minimum. Research shows that most accidents happen when a driver isn't paying attention. You can get distracted by using a cell phone, drinking, using tools, putting on makeup, or talking to other people. Training these complex data with learning algorithms is hard because it uses up a lot of computer resources.

3 LEARNING-BASED DRIVER DISTRACTION METHODOLOGY

In this approach, the face image is initially collected and processed based on image and feature extraction.

1. The extracted images help to determine the pictures related to driver boring, hand gestures, eye recognition, etc.
2. A Modified random forest integrated with C 5.0 as it helps to analyze the driver condition and to detect the driver distraction; a modified CNN is used.
3. An optimizer is used to improve the system's performance to overcome the problem of resources.
4. The proposed approach is compared with other existing techniques, such as SVM and LSTM-based pixel images used to determine the accuracy, precision, F1-Score, and Recall.

4 RELATED WORKS

Summarized earlier studies on the issue of distracted driving. Methods for identifying distracted drivers using their phones, radios, or other gadgets were explored and divided into three major categories: distracted by physical objects, visual, and cognitively distracted by cognitive things. Distracted driving detection techniques could be used to provide better outcomes. Most methods for determining whether a driver is tired are focused on how the automobiles move, how fast they travel, and how they turn.

4.1 Driver characteristics concerning camera and biosignal sensors

However, the information in these approaches could not be accurate due to variations in driving styles, vehicle characteristics, and road conditions. But the findings from object analysis make the results to be reliable. Compared to methods that employ driving data, those that use this data can yield more accurate results. However, drivers must wear some form of sensor to collect the physiological data. Proposed a method for detecting driver drowsiness and fatigue. The signals generated physically through the eyes and heart are analysed based on the image extraction, and it mitigates the collision happening while making the driver sleepy and tired. However, their approach required non-intrusive sensors to collect physiological data. The identification of driver weariness and distracted driving was published by [7]. By combining data from several cameras and bio-signal sensors, they could determine the number of hybrid traits.

4.2 Monitoring system based on spatial and temporal analysis

However, these hybrid characteristics were gathered via the drivers' non-intrusive wearable technology and their facial expressions. Vision-based methods have gained popularity due to their non-contact nature and effective performance. Yawns, closed eyes, facial expressions, and the position of the driver's head were commonly used cues in the investigations. The identification of driver fatigue using picture sequences was proposed [8]. Their approach was largely unaffected by the light levels. However, it didn't adjust for yawning or head position. Using a smartphone as a fatigue indicator for drivers. However, they solely considered factors related to fatigue based on eye movements. Created a smartphone-based method for monitoring driver fatigue that uses hands other than head posture. They assessed utilizing data sets, and the accuracy levels reached were 87% respectively.

4.3 Analyzing the data accuracy based on real-time datasets

State Farm Insurance introduced a competition on Kaggle for distracted drivers in 2016. The distracted driving dataset was made available to begin training and testing. A total of ten different driving positions were evaluated for the competition, with one representing safe driving and the other nine representing distraction. The State Farm data set was duplicated in a brand-new data set (AUC Distracted Driver) created by [9]. A strategy based on a genetic algorithm was suggested after segmenting the features of the skin, face, and hands. Their approach used five CNN weight sets, which had a classification success rate of 95.98%. Their direction wasn't appropriate for real-time applications because it was computationally expensive. The same data set plus an improvement to the VGG-16 network allowed us to use their method in real time and achieve a classification accuracy of 96.31%. Real-time processing was used to raise the accuracy of image classification for all ten classes to 99.10%...

4.4 To identify the driver distraction based on a non-invasive approach

To identify inattentive driving was proposed, and it is employed with various CNN models. Their strategy had a log loss of 0.795. A supervised learning method for identifying distracted driving has been suggested under the name Drive-Net. Their plan has a 95% average accuracy.

Based on CNN and a random forest, it was designated as a representative study on inattentive driving. Using a 3D convolutional neural network and temporal information for optimal information flow [10] improved the detection of distracted driving and increased its accuracy to 94%. [11] Presented a non-invasive approach based on characteristics from the driver's head position and mouth movements for identifying driver drowsiness.

4.4.1 *Existing models related to driver distraction*
Driver weariness and distracted driving appear to be the main contributing factors in traffic accidents, according to numerous research conducted in recent years. As a result, driver assistance in intelligent vehicles now faces a severe challenge of the onboard monitoring of driving behaviours. [12] First proposed utilizing facial expression analysis to spot signs of weariness in drivers. Extended short-term memory networks (LSTM) and CNN were employed to assess temporal and spatial picture information, respectively. Their method has an accuracy rate of above 87%. With the same data set and an enhancement network, we achieved a classification accuracy of 96.31% while utilizing their method.

4.4.2 *Technologies used to perform image extraction and detection process*
In the existing approaches, Long Short Term Memory (LSTM) and Convolutional Neural Networks (CNN) use information related to spatial and temporal characteristics as they analyse the facial expression with 87% accuracy, and those images can be enhanced using VGG-16 and VGG-32 increase the accuracy. In the proposed approach, the face image is collected and processed based on the image and feature extraction is analyzed to increase the image quality. Those images are classified using Random Forest with C5.0 to identify the driver distraction with enhanced accuracy, and fatigue characteristics are determined using Modified Convolutional Neural Networks.

4.4.3 *Justifications on the problem of driver distraction and computing the large resource*
Drivers have been mostly to blame for a lot of major accidents that have happened recently. It needs some practical ways to find and identify the distractions in order to solve the problem of feature distraction using Deep Learning-based systems like the Random Forest learning algorithm and Modified CNN. This is how the problem of driver distraction and the problem of resource computation are being solved right now.

5 PROPOSED DL-BASED DRIVER DETECTION APPROACH

Here, the modified random forest integrated with C 4.5 algorithm is proposed as it considers the datasets as input by creating several sub-trees with random variables 'V'. The sub-tree is designed for the 'n' number of trees along with the leaf node. The analysis is done on the proposed work related to data aggregation and scalability of data based on datasets; it will perform better than the existing algorithms like the Apriori algorithm, Decision Trees, etc. The proposed work algorithm will perform effectively related to the above parameter with some more modification as needed during the implementation.

ALGORITHM 1 MODIFIED PREDICTION MODEL:

Input: Data training and testing
Output: Data Prediction
Modified Random Forest with the integration of C 4.5 Tress
        Compute the Coefficient of the training Model.
        Applying Sigmoid Function
            Processing the data mapping Process
            Mapping Between the training and testing data
        Deploying Data Clustering
        Determining the Data prediction

41

5.1 *Deep learning-based prediction process*

In Algorithm 1, data mining datasets are taken as the input, and the learning process is performed based on the random forest algorithm as it is associated with C 5.0. In the process of random forest, the coefficient is computed as it trains the datasets. Then, the data mapping process is performed to perform the sigmoid function, and mapping is done between the prepared data and tested data. The data clustering is deployed, and the approach helps to determine the prediction process based on the data.

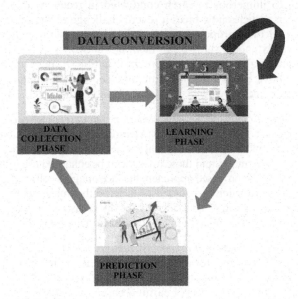

Figure 2.   Data conversion in 3 phases.

In Figure 2, Data conversion is performed based on structured and unstructured data as it has undergone three phases: data collection, learning, and prediction.

In Figure 3, data pre-processing is performed as the initial phase as the output is given. The data gets trained based on the machine learning algorithm as it helps to enhance the image contraction and improves the image's quality through image filtering. In the data extraction, the cleaning process is performed on the data and gets integrated based on the selection process. All the data that gets updated are stored in the data repository.

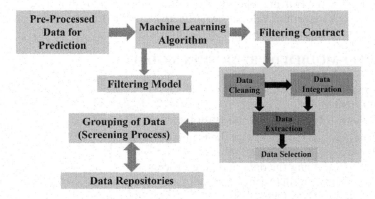

Figure 3.   Proposed DL-based driver detection architecture

ALGORITHM 2 MODIFIED RANDOM FOREST ALGORITHM:

Input: Datasets
Output: Binary Tree
Initialize

       Tree 'T' has been selected to derive Sub Tree
       Collecting of Data' Subset"
       For Subset to initialize 1 to 'n' Tree
             Train the data
             Construct the data samples with 'N' Size
             Terminate when the node has limited value
                Select the random variable 'V' from Sample 'S.'
Based on Node, the tree is to be spitted into several subtrees with an 'n' leaf node
       Generate random tree
Stop

In algorithm 2, the datasets are taken as the input and results in the binary tree as it performs the modified random forest algorithm. In this model, the data tree is represented as it is described as 'T', and it is further defined as several subtrees. Then the collected data subsets are designated as 'S'. Those subsets are initialized from 1 to n. In the process of data subset, the data are trained by constructing the sample data with size 'N'. The nodes are introduced as satisfied when the node gets the limited value. Each sample data gets a random variable 'V', and those nodes, represented as a tree, are spitted into several subtrees with an 'n' leaf node. Finally, all the nodes will generate the random tree as the sequence. This algorithm 1, based on algorithm two, will help to perform the prediction process to identify the driver distraction as it may lead to significant accident problems.

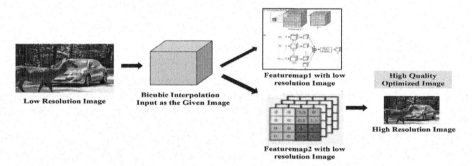

Figure 4.   Low-resolution images are converted into high-resolution images.

5.2 *Fatigue problem enhancement based on the CNN model*

From q 54the, Figure 4, the low-resolution images are converted into high-resolution images as it carried out three processes: Path extraction and representation, Non-linear mapping and Reconstruction.

The neural network uses multi perceptron, which contains three later such as input, output and hidden layer; a typical neural network is unsuitable for data images to be analyzed. Then CNN came into the picture as it helps analyse the image using spatial image datasets.

6   PERFORMANCE ANALYSIS

In the performance analysis, various metrics are being analyzed as accuracy and data loss, as it helps to balance the image quality and prediction with the identification level of the image.

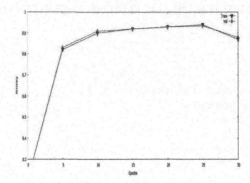

Figure 5.   Number of epochs vs. accuracy.

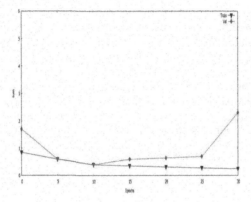

Figure 6.   Number of epochs vs. loss (Data).

In Figure 5, the accuracy value is determined for the proposed approach by varying the number of epochs from 0 to 30. The proposed model is categorized by training and testing datasets. The proposed model gets a gradual increase in value as the number of epochs also gets increases.

In Figure 6, the data loss is determined for the proposed approach by varying the number of epochs from 0 to 30. The proposed mode gets an average of 0.98 loss value for a milli-second. The training and testing performed for the proposed approach get a gradual reduction in the data loss value.

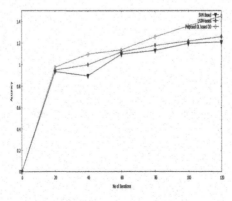

Figure 7.   Number of iteration vs. accuracy.

In Figure 7, the data accuracy is determined for the proposed approach as it increases as compared with other models like SVM-based and LSTM based. As the proposed model gets a drastic increase in data accuracy as it achieves an enhanced value rate.

The image enhancement has also been done in the proposed model based on the CNN model. The modified random forest learning algorithm is applied to predict and identify the driver detection accuracy. It is used, not applied to real-time datasets. The proposed model has to be modified, which is also suitable for complex images.

7 CONCLUSION

The proposed method is compared to other policies that are already in place, like SVM- and LSTM-based pixel images that are used to figure out the accuracy, precision, F1-Score, and recall. The proposed method is based on how accurate the data is, which is better than other models like SVM-based and LSTM-based. As the proposed model gets a much better value rate, the data accuracy goes up by a lot. In this method, the image of the face is first taken and then processed using image and feature extraction. The extracted images help figure out which ones are about driver boredom, hand gestures, eye recognition, and other things. A modified random forest that is built into C 4.5 helps to figure out how the driver is doing, and a modified CNN is used to find out if the driver is getting distracted. An optimizer is used to improve the performance of the system and get around the problem of not having enough resources. Not the real-time datasets are used. In future work, the proposed model, which can also be used for complex images, needs to be changed.

REFERENCES

[1] NDTV, Road Accident Statistics in India, 2016. [Online]. Available: http://sites.ndtv.com/roadsafety/important-feature-to-you-in-your-car-5/.

[2] Bhakti Baheti; Sanjay Talbar and Suhas Gajre, "Towards Computationally Efficient and Realtime Distracted Driver Detection With MobileVGG Network," *IEEE Transactions on Intelligent Vehicles*, Vol. 5, Issue. 4, pp. 565–574, 2020.

[3] Soni Lanka Karri, Liyanage Chandratilak De Silva, Daphne Teck Ching Lai and Shaw Yin Yong, "Identification and Classification of Driving Behaviour at Signalized Intersections Using Support Vector Machine," *International Journal of Automation and Computing*, Vo. 18, pp. 480–491, 2021.

[4] Guofa Li; Weiquan Yan; Shen Li; Xingda Qu; Wenbo Chu and Dongpu Cao, "A Temporal–Spatial Deep Learning Approach for Driver Distraction Detection Based on EEG Signals," *IEEE Transactions on Automation Science and Engineering*, Vol. 19, Issue. 4, pp. 2665–2677, 2021.

[5] Alexey Kashevnik; Roman Shchedrin; Christian Kaiser and Alexander Stocker, "Driver Distraction Detection Methods: A Literature Review and Framework," *IEEE Access*, Vol. 9, pp. 60063–60076, 2021.

[6] Jie Chen; Yanan Jiang; Zhi Xiang Huang; XiaoHui Guo; BoCai Wu; Long Sun and Tao Wu, "Fine-Grained Detection of Driver Distraction Based on Neural Architecture Search," *IEEE Transactions on Intelligent Transportation Systems*, Vol. 22, Issue. 9, pp. 5783–5801, 2021.

[7] Reda Bekka, Samia Kherbouche and Houda El Bouhissi, "Distraction Detection to Predict Vehicle Crashes: A Deep Learning Approach," *Computación y Sistemas*, Vol. 26, No. 1, 2022.

[8] Xin Zuo; Chi Zhang; Fengyu Cong; Jian Zhao and Timo Hämäläinen, "Driver Distraction Detection Using Bidirectional Long Short-Term Network Based on Multiscale Entropy of EEG," *IEEE Transactions on Intelligent Transportation Systems*, Vol. 23, Issue. 10, pp. 19309–19322, 2022.

[9] Satpathy, Sambit, Prakash Mohan, Sanchali Das, and Swapan Debbarma. "A New Healthcare Diagnosis System Using An IoT-based Fuzzy Classifier With FPGA." *The Journal of Supercomputing* 76, no. 8 (2020): 5849–5861.

[10] Mohan, Prakash, Manikandan Sundaram, Sambit Satpathy, and Sanchali Das. "An Efficient Technique for Cloud Storage Using Secured De-duplication Algorithm." *Journal of Intelligent & Fuzzy Systems Preprint* (2021): 1–12.

[11] Boon-Giin Lee and Wan-Young Chung, "A Smartphone-Based Driver Safety Monitoring System Using Data Fusion," *Sensors*, Vol. 12, Issue. 12, pp. 17536–7552, 2022.

[12] Tahir Abbas, Syed Farooq Ali, Mazin Abed Mohammed, Aadil Zia Khan, Mazhar Javed Awan, Arnab Majumdar and Orawit Thinnukool, "Deep Learning Approach Based on Residual Neural Network and SVM Classifier for Driver's Distraction Detection," *Applied Sciences*, Vol. 12 Issue. 13, pp. 6626, 2022.

[13] Xiao, Hongli Liu, Ziji Ma and Weihong Chen, "Attention-based Deep Neural Network for Driver Behaviour Recognition," *Future Generation Computer Systems*, Vol. 132, pp. 152–161, 2022.

[14] Khaled Bayoudh, Fayçal Hamdaoui and Abdellatif Mtibaa, "Transfer Learning Based Hybrid 2D-3D CNN for Traffic Sign Recognition and Semantic Road Detection Applied in Advanced Driver Assistance Systems," *Applied Intelligence*, Vol. 51, pp. 124–142, 2021.

[15] Satpathy, Sambit, Sanchali Das, and Bidyut Kumar Bhattacharyya. "How and Where to Use Super-Capacitors Effectively, an Integration of Review of Past and New Characterization Works on Super-Capacitors." *Journal of Energy Storage* 27 (2020): 101044.

[16] Satpathy, Sambit, Mohan Prakash, Swapan Debbarma, Aditya S. Sengupta, and Bidyut K. Bhattacaryya. "Design a FPGA, Fuzzy Based, Insolent Method for Prediction of Multi-Diseases in Rural Area." *Journal of Intelligent & Fuzzy Systems* 37, no. 5 (2019): 7039–7046.

Artificial Intelligence, Blockchain, Computing and Security – Dagur et al. (Eds)
© 2024 The Author(s), ISBN: 978-1-032-49393-0

A systematic study of networking design for co-working space environment

Rohit Vashisht, Rahul Kumar Sharma & Gagan Thakral
Assistant Professor, KIET Group of Institutions, Delhi-NCR, Ghaziabad

ABSTRACT: In today's era of digitalization, there is pool of resources that are being shared commonly among various networking devices within in a network. The effectiveness of the network is highly dependent on the synchronization, utility and security of these devices. The aim of the study is to design an appropriate network system for "Co-working Space Organizations", a network in which multiple organizations will access common resources. The proposed architecture has three-level hierarchy with least cost design and good level security, in such a way that network devices will also meet standards linked with the company. This paper also discusses the budget challenges that the network will face in developing countries in detail. Developing countries have a bounded budget problem that impact choosing good quality routers, switches and various other devices in the network like servers. DHCP servers have been utilized to allocate IP addresses in the proposed system. The proposed network architecture performs well for 3000-4000 populations as per the experimental investigations and can also be scaled successfully to a large number of network users.

Keywords: Co-working, Network, Scalability, Router, CISCO, Managed Networks

1 INTRODUCTION

Businesses that offer shared workspaces are very popular in developing nations like the Philippines. A co working setting that allows people and teams to work individually or collaboratively is known as a co-working space (Buladaco *et al.* 2021). Co-working space firms can benefit from computer networking by boosting revenue, retaining clients, and offering better service. By deploying managed network services, operating, maintenance, service, hardware, software, and infrastructure costs would be reduced. The market for co-working spaces may be able to concentrate on its main business objectives through managed network services. In developing nations, there are many co-working space organizations looking for ways to connect networks with security, backup, and other capabilities. Designing a network that adheres to the standards established by developed countries presents difficulties for co-working space organizations in developing nations (Network Lessons 2020).

Co-working offers a different type of office setting that promotes productivity without the formality of a conventional office. It has been reported that the demand for co-working spaces in Metro Manila is rising among independent contractors, business owners, startups, small businesses, IT companies, BPOs, and even foreign corporations (Reyes 2020). By introducing staff to novel company concepts and encouraging idea sharing, co-working environments increase productivity, develop collaboration, innovation, creativity, and a sense of community (Progress 2020). With improved network services, the infrastructure is set up to enable quicker implementations and upgrades. To further enhance the services and boost sales, it is strongly advised co-working space firms to install managed networks and upgraded network equipment (Yixin & Aslam 2020). Figure 1 demonstrates how individuals from diverse heterogeneous or homogeneous firms work independently on the same or different projects in a co-working setting. Figure 2 illustrates the salient characteristics of a good co-working space in detail.

DOI: 10.1201/9781003393580-8

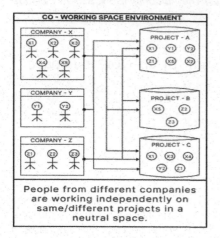

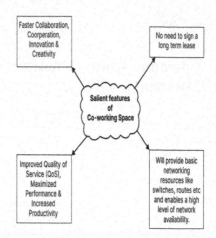

Figure 1.   Concept of co-working space environment.

Figure 2.   Features of a co-working space network design.

These businesses will benefit from the coworking space network design guidance in developing a network that makes use of affordable technologies without incurring unacceptable security or quality compromises. The prime aim of this paper is to offer a Local Area Network (LAN) design appropriate for the Grizell, Damsel, and Clocking Bit like co-working space organizations. The target user base for this network is between 3,000 and 4,000. With an increase in users, the network can be stretched to a larger size. To make the network more secure, good level security measures have been considered. Confidential data is protected by a variety of security-preserving technologies, such as firewalls and wireless routers with WPA encryption. All routers and switches in the proposed network design are secured using passwords (user access and admin privileges) and encryptions (WPA2 in wireless routers mostly).

The structure of the paper is organized under the following sections. The term Co-working Space Network Design (CSND), its features and need are mentioned in section 1, section 2 describes the literature survey in relation to CSND, and the proposed network design and the major security methods that are incorporated into the 3-Layer hierarchical network architecture are explained in section 3, implementation details are described in section 4 and lastly, conclusive findings and future potential directions are reported in section 5.

2   LITERATURE STUDY

Since the primary Co-working Space (CWS) opened in San Francisco in 2005, the concept has undergone a major trend (Foertsch & Cagnol 2013). It was designed as an alternative model to the non-social business centers and functioned as a workplace and a social area to foster community, independence, and communication (Dullroy 2012). The concept gained popularity throughout the world, and in 2009, the first CWS under this name opened in Germany.

Additionally, CWS provides its members with access to networks with other parties like businesses, venture capitalists, or entrepreneurs as well as specialised services like coaching, training, startup event advice, and access to these networks (Capdevila 2015).

In order to promote employee innovation, networking, and creativity, manufacturers like Bosch, BMW, and Merck as well as consulting firms like PwC have grabbed this institutional trend and built internal shared work and social spaces (Hanny 2017).

The concept of CWS shows how new modes of operation are manifested. In general, "new approaches to working" refers to a collection of procedures, particularly in human resource management, that are meant to increase the flexibility, independence, and freedom of those who work (Gerards *et al.* 2018).

Jeske & Ruwe (2019) stressed that CWS "offer significant sources of support, learning, and networking possibilities" in their overall analysis. Bouncken *et al.* (2020) referred CWS as "a real space for entrepreneurship" to emphasize its proximity to the entrepreneurship industry.

Aslam (2019), depending on in-depth qualitative interviews, confirmed that geographical proximity encourages information exchange procedures by coworkers. Conceptually, knowledge exchange procedures have been examined by Bouncken & Reuschl (2018) and Rese *et al.* as well as associated antecedents like trust or community empirically in 2020. The latter study found a positive relationship between personal creativity and an attitudinal and deliberate belief in information sharing. The study did not, however, explore network features in particular.

3 PROPOSED THREE- LAYER HIERARCHICAL NETWORK ARCHITECTURE FOR CO-WORKING SPACE (CSND)

With several protocols and different technologies, networks may be quite difficult to design and deploy. To solve this issue, CISCO has created a tiered hierarchical model for constructing a dependable network infrastructure known as the three-layer hierarchical model. This strategy aids in the development and maintenance of an affordable, dependable, scalable, and network for organizations. Since each of these layers has its own protocols, standards, features, and functionalities, the network's complexity is decreased and data flow is improved. The research study employed three-tier hierarchical structure in proposing the network design (CSND) which comprises of three layers – Core layer, Distributed layer and Access layer as shown in Figure 3 & Figure 5.

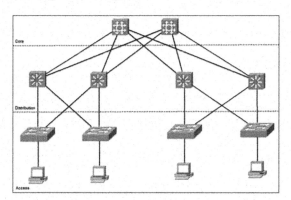

Figure 3.   CISCO 3-Layer hierarchical network design.

LAYER-1 (Access Layer) – Controlling individual and workgroup access to network resources is the central function of the access layer. This layer's primary objective is to include Layer 2 switches and points of access that enable connectivity between servers and workstations. At this layer, one can manage policy, access control, define distinct domains (free of collisions), and implement port security.

LAYER-2 (Distribution Layer) – The distribution layer serves as the communication channel between the access layer and the core. Access, routing, filtering, and determining how packets can access the network core are among the key responsibilities of this layer. When a request for a network service, such as how to send a file to a server, is made, this layer selects the quickest method and, if necessary, transmits the request to the core layer. This layer comprises of multilayer switches and routers, including wireless routers, for the effective flow of data packets.

LAYER-3 (Core Layer) – The quick delivery of enormous quantities of data packets is the primary responsibility of the core layer, also referred to as the network's backbone. The core layer provides connectivity between devices in the distribution layer and is frequently made up of high-speed components like top-of-the-line routers and switches with redundant links.

Advanced security methods should be used while configuring the network for co-working space organizations to avoid attempts to alter network data. The network has employed routers that are WAP and firewall protected. Employee access to networks at other offices should be restricted by robust security measures.

The most significant factor, aside from functional and non-functional requirements, is the hardware need. The authors implemented the concept using all of CISCO's products, and Table 1 includes a list of project requirements. As shown in Table 1, the authors have employed a variety of network devices, including routers, switches, and access points.

Table 1.   List of required hardware devices.

Device Name	Model	Quantity
Modem	DSL Modem – PT	1
Router	Cisco 2911/K9 Router and Wrt300 Router	5 (3 and 2 respectively)
Multilayer Switch	WS-C3560-24PS-S Cisco Switch	2
Switch	Cisco Catalyst 2960-24TT-L Switch	10
Firewall	Cisco ASA5505-BUN-K9 ASA 5505	2
DHCP Server	Cisco DHCP server	3
Cables	Copper Straight-Through	As per requirement

Security calls for a proactive approach to prevention in order to stop any threat or assault on a network. Password and encryption protection is required for routers. A computer administrator is required to maintain the network's data access security (Yan *et al.* 2015). WPA protects wireless routers that have been installed in the network to offer security. In terms of network security, there are three important things to consider. This category includes specific systems or components, infrastructure, and individual hosts. In order to protect the privacy of every employee of an organization, numerous steps have been implemented (AIP Conf. Proc. 2017). A network administrator is also required for the proposed CSND to oversee the effectiveness and efficiency of the entire network. This person's primary responsibilities are to oversee network operation, maintenance, provisioning, and security.

4   IMPLEMENTATION

The authors' suggested networking framework for co-working spaces is implemented using a three-layer CISCO hierarchical network model. The design utilised WPA-protected wireless

Figure 4.   Checking security measures of Grizell.

50

routers and provided User Access Protection (UAP) to ensure the security of the CSND (Priebe *et al.* 2020). Figure 4 illustrates that a password is required in order to modify the router of the Grizell organization.

To guarantee the connectivity between two systems, ping command is used to ping IP address of one PC from another PC of Grizell organization. The connectivity of wireless routers and other setups is also cross-checked. Therefore, Grizell, Damsel, and Clocking Bit, the three organizations under consideration, are all properly protected.

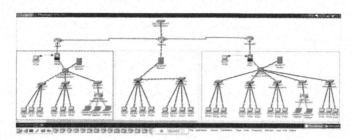

Figure 5.   Three layer co-working space networking design.

High levels of dependability can be found in the network security. This enables a wide range of powerful tools, such firewalls and WPA-secured wireless routers, to safeguard data. Additionally, all routers and switches have encryptions and password protection. The targeted user numbers for this network is between 3,000 and 4,000.

5   CONCLUSION

A co working setting that allows people and teams to work individually or collaboratively is known as a co-working space. This article established an optimum network design for businesses providing collaborative workspaces. The 3-layer hierarchical network model from CISCO is used to implement the proposed networking design for co-working spaces. Administration, operation, security, and maintenance analysis of perfect network management were provided. If a business adopts this suggestion, there will be better gains in revenue, sustainability, and client retention. Future research in this area may focus on developing security algorithms that are both more reliable and novel. Future work on scalability without compromising the security and efficiency of a co-working network design is also quite promising.

REFERENCES

Aslam, M., & Goermar, L. (2018, July). Sociomateriality and Entrepreneurship in Coworking-Spaces. In *Academy of Management Proceedings* (Vol. 2018, No. 1, p. 13853). Briarcliff Manor, NY 10510: Academy of Management.

Bouncken, R. B., & Reuschl, A. J. (2018). Coworking-spaces: How a Phenomenon of the Sharing Economy Builds a Novel Trend for The Workplace and For Entrepreneurship. *Review of managerial science*, 12, 317–334.

Bouncken, R. B., Kraus, S., & Martínez-Pérez, J. F. (2020). Entrepreneurship of an Institutional Field: The Emergence of Coworking Spaces for Digital Business Models. *International Entrepreneurship and Management Journal*, 16, 1465–1481.

Brummelhuis, L. L. T., Hetland, J., Keulemans, L., & Bakker, A. B. (2012). Do New Ways of Working Foster Work Engagement?. *Psicothema*.

Buladaco, M. V. M., Necio, G. C., Pilongo, O. B., & Timosan, J. Q. (2021). A Proposed Ideal Network Design for Collaborative Workspace Businesses. *International Journal*, 10(2).

Capdevila, I. (2015). Co-working Spaces and the Localised Dynamics of Innovation in Barcelona. *International journal of innovation management*, 19(03), 1540004.

Dullroy, J. (2012). Coworking began at Regus … But Not the Way They Think. *Deskmag, April*, 4.

Foertsch, C., & Cagnol, R. (2013). Es war einmal … Die Geschichte von Coworking in Zahlen. *Deskmag Magazin*; Berlin.

Hanney, M. (2017). *Corporate Coworking: Drivers, Benefits & Challenges*.

He, J. (2017, May). The Research of Computer Network Security and Protection Strategy. In *AIP Conference Proceedings* (Vol. 1839, No. 1, p. 020173). AIP Publishing LLC.

Jeske, D., & Ruwe, T. (2019). Inclusion Through Use and Membership of Co-working Spaces. *Journal of Work-Applied Management*, 11(2), 174–186.

Kumar, A. & Alam, B. (2018). Task Scheduling in Real Time Systems with Energy Harvesting and Energy Minimization. *Journal of Computer Science*, 14(8), 1126–1133.

Kumar, A., & Alam, B. (2019). Energy Harvesting Earliest Deadline First Scheduling Algorithm for Increasing lifetime of Real Time Systems. *International Journal of Electrical and Computer Engineering*, 9 (1), 539.

Low, J. S. (2020). *Intent-based Networking: Policy to Solutions Recommendations* (Doctoral dissertation, UTAR).

Priebe, T., Fernández, E. B., Mehlau, J. I., & Pernul, G. (2004). A Pattern System for Access Control. In *Research Directions in Data and Applications Security XVIII: IFIP TC11/WG11*. 3 Eighteenth Annual Conference on Data and Applications Security July 25–28, 2004, Sitges, Catalonia, Spain (pp. 235–249). Springer US.

Qiu, Y., & Aslam, M. (2020). Weaving My Cocoon: Business Model Configuration and Trajectory of Chinese Coworking-Spaces. In *Academy of Management Proceedings* (Vol. 2020, No. 1, p. 21069). Briarcliff Manor, NY 10510: Academy of Management.

Rese, A., Kopplin, C. S., & Nielebock, C. (2020). Factors Influencing Members' Knowledge Sharing and Creative Performance in Coworking Spaces. *Journal of Knowledge Management*, 24(9), 2327–2354.

Yan, F., Jian-Wen, Y., & Lin, C. (2015, June). Computer Network Security and Technology Research. In *2015 Seventh International Conference on Measuring Technology and Mechatronics Automation* (pp. 293–296). IEEE.

Artificial Intelligence, Blockchain, Computing and Security – Dagur et al. (Eds)
© 2024 The Author(s), ISBN: 978-1-032-49393-0

Application of neural network algorithms in early detection of breast cancer

D.K. Mukhamedieva
Tashkent University of Information Technologies Named After Muhammad al-Khwarizmi, Tashkent, Uzbekistan

M.E. Shaazizova
Research Institute for the Development of Digital Technologies and Artificial Intelligence, Tashkent, Uzbekistan

ABSTRACT: The purpose of the research presented in the article is to develop methods and algorithms for classifying signs of cancer based on digital mammography analysis for early detection of breast cancer, to create software and to develop a medical diagnosis system. One of the effective directions in this field is the use of artificial intelligence, which is an integral part of information technology. To carry out this research, the following tasks were assigned: development of software for primary processing of digital mammography images and intelligent identification of damaged parts; development of cancer detection algorithms based on the analysis of digital mammography of the breast, creation of software; making system design simple, understandable and adapting functions for cardiologists; create a section to develop the software knowledge base and discuss problematic situations.

Keywords: artificial intelligence, knowledge base, deep neural networks, early diagnosis, model

1 INTRODUCTION

Artificial intelligence can be considered in a broad sense as the execution of tasks that are required to be performed by the human mind in computer systems. It consists of three types: human-made algorithms, machine learning and deep learning [1–5]. Machine learning, like deep learning, is based on artificial neural networks.

However, in deep learning, there are different, hidden layers between the input and the output. Each layer of the neural network can work both independently and together. Deep learning holds great promise for diagnostic issues, as it can accurately analyze cardiovascular, oncology, diabetes, pathology, dermatology, ophthalmology, and radiography. In fact, deep learning is currently able to diagnose 5-10% more accurately than the average doctor, and this gap is expected to improve further. At the same time, the promising issues of synthesizing integrated soft models, including fuzzy-neuron models using evolutionary algorithms for creating artificial intelligence systems for early diagnosis, have not been sufficiently studied [3,4].

After our republic gained independence, special attention was paid to improving the convenience and quality of medical services to the population using the opportunities of information and communication technologies. In this regard, significant results have been achieved in the improvement of computer diagnostic systems for the early detection of medical diseases and the improvement of the quality of comprehensive treatment [1,2].

Intelligent analysis of medical data, development of methods and algorithms for classifying signs of cancer on the basis of analysis of digital mammography of the breast, creation

DOI: 10.1201/9781003393580-9

of software and diagnosis with the help of medical diagnostic systems, determination of treatment processes is one of the major problems [6–9].

Fully automatic disease diagnosis might have seemed impractical just a few years ago, but great advances in artificial intelligence have changed that. In this paper, we propose an algorithm based on a sequential process of detecting anomalies in medical images, characterizing them, and sometimes quantifying their evolution [10,11].

Convolutional neural networks, which have been extremely successful in recent years, have won many trials in statistical learning and have solved many training problems that were previously considered overwhelming [12–15].

Breast cancer is cancer that develops in the cells of the breast. Typically, cancer forms either in the lobules or in the ducts of the breast, and can be fatal if it grows to metastasize and spread to surrounding tissues or other parts of the body. In this case, the tumor is called malignant [1,2].

Using a neural network, it is possible to predict whether a tumor in a female breast is malignant or benign; a benign tumor is a tumor that cannot invade adjacent tissue, making it harmless in most cases; malignant tumor - on the contrary, it can spread to the rest of the body, and is extremely dangerous. CT systems for detecting and diagnosing breast cancer using mammograms can help reduce the burden on professionals by helping them classify mammograms into normal and abnormal [3,4].

Since we are dealing with a huge data set of patients around the world, applying conventional machine learning algorithms is not a good recommendation.

Huang *et al.* [1]. proposed a multimodal image-based disease severity prediction study using extensive study and analytics prediction. Now there is a trend in the processing of medical images using deep neural networks, which are being improved all the time.

For example, Pushpanjali, Baldev, [2] proposed histopathological image detection of breast cancer using deep learning.

Currently, machine learning and recognition algorithms are implemented mainly in Python using the appropriate libraries. One such library is the Keras library. This open source library provides a Python interface for artificial neural networks. Keras acts as an interface to the TensorFlow library. Here he plays a key role. Deep learning is great with huge datasets and can also extract high-level features without any domain intervention or hard feature extraction. Deep learning takes a lot of time, but the testing phase is faster than when using a machine learning algorithm [16–18].

Even though the data set is usually quite small compared to the amount of data required to train neural networks, which are typically tuned with a large number of weights, it is possible to train a highly accurate deep learning neural network model that can classify tumor type. into benign or malignant with a similar quality of the data set by feeding a neural network with random distortions of images allocated for educational purposes [11,12].

The training image data can be augmented by slightly modifying it and then fed into the network for training. This technique helps the neural network to be able to correctly classify the invisible images during the test [4,5].

2 METHODS

Dimensionality reduction is a widely used approach to increase the computational complexity of a classifier, at the expense of a slight detriment to the classifier's overall performance. There are two ways to reduce the dimension [7,8]:

1. Selection of the main elements from the original dimensions.

    In this method, different combinations of features are formed and evaluated to get the best combination.

    Among the feature selection schemes used are genetic algorithm, annealing simulation, amplification, grafting, and particle swarm optimization.

2. Creation of new measurements.

This method maps the original feature space to a new reduced feature space. Commonly used dimensionality reduction methods are Principal Component Analysis, Linear Discriminant Analysis, Independent Component Analysis, and Manifold Study [10,11].

Although dimensionality reduction improves the computational complexity of the classifier, it can lead to the loss of important features that can help in prediction. Thus, the overall performance of the classifier may be reduced due to size reduction. Models based on scrutiny have multiple levels of pooling to reduce the size of image feature maps [3,4].

The easiest way to solve the detection problem is to reduce it to a classification problem. To detect an object, it is necessary to take a classifier specific to it and apply it to image areas [1].

One way to extract such areas is the sliding window method, which is a rectangular area with a fixed width and height that "slides" over the image. For each of the sections, a classifier is applied to determine whether the window has the desired object [2].

It is important not to overlook information from mammographic images, as this can lead to misclassification of malignancies. In the proposed section, we have presented a new input image segmentation method. At the first stage, we perform breast segmentation [1–4].

For segmentation, mammographic images are used as input. The image is separated separately into red, green and blue channels. The separated channels are then cut again into five layers.

There is too much non-essential information in the input image that is not required for classification. Therefore, the first step in image classification is to simplify the image by extracting the important information contained in the image and excluding the rest [7].

Each pixel is represented by a number, in the case of a black and white image, or a triple of numbers, in the case of a color one. In this format, it is convenient to perceive the image, but to solve the recognition problem, this amount of information is too much [5,6].

When processing images, the most universal characteristic features are: shape, brightness (or color), texture, etc. Less distinguishable for human perception are spectral, histogram, correlation, and other characteristics [12–14].

The convolution operation is the most basic part of a convolutional neural network using edge detection as an input example. The mathematical explanation for this is that if we have an image of size $n \times n$. From images, using an $f \times f$ size filter, then the output dimension will be [7–9]

$$(n - f + 1) \times (n - f + 1)$$

So that after filling the output size is the same as the input size.
The formula becomes

$$(n + 2p - f + 1) \times (n + 2p - f + 1),$$

Where
In Python 3.7, Keras acts as an interface to the TensorFlow library [15–17]. Here he plays a key role. Deep learning is great with huge datasets and can also extract high-level features without any domain intervention or hard feature extraction. Deep learning takes a lot of time, but the testing phase is faster than using a machine learning algorithm. Features of Tensor Flow include:

– a simple architecture that allows you to quickly learn programming and create machine learning models;
– it is possible to move the created learning models to any device, cloud storage;
– the presence of an active execution option, which makes it possible to create models for machine learning, debug and manipulate the latter.

Tensor Flow has all the solutions for debugging, running and other tasks, problems regarding machine learning.

3   RESULTS

The implementation was carried out using the Python 3.7 language.

The first layer in CNN is always convolutional. The input image is a 32 ×32 x 3 matrix with pixel values. A filter is a matrix (such a matrix is also called a weight matrix or parameter matrix). Note that the depth of the filter must be the same as the depth of the input image (then there is a guarantee of mathematical fidelity), and the dimensions of this filter are 5 ×5 x 3. filter values by the original pixel values of the image (element-wise multiplication). All these multiplications are summed (total 75 multiplications). And as a result, one number [15,16] is obtained.

The second layer after the convolutional layer is the pooling layer. A pooling layer is typically applied to generated feature maps to reduce the number of feature maps and network parameters by applying appropriate mathematical calculations. In this work, the maximum pool and the global average pool are used. The maximum pooling process selects only the maximum value using the matrix size specified in each feature map, resulting in a reduction in output neurons. There is also a global average pooling layer, which was only used before a fully connected layer, reducing the data to a single dimension. The layer of the global middle layer of the pool is connected to the fully connected layer [7–9]. The full language layer is the last and most important layer of a CNN. This layer functions as a multilayer perceptron. The input to a fully connected layer is the output from the last merged or convolutional layer, which is smoothed and then passed to the fully connected layer. This smoothed vector is then connected to several fully connected layers, which are similar to artificial neural networks and perform the same mathematical operations [17,18]

As a result, we obtain the outputs of neurons in the hidden layer. These intermediate outputs can be thought of as non-linear transformations and combinations of the original inputs. They become the inputs of the output layer. Again we calculate the weighted sum of the inputs, apply the activation function and get the final values of the target variable. relu and tanh activation functions. Relu cuts off values below zero, while tanh takes values from -1 to 1 (respectively for the minimum and maximum values of the inputs). Either of these two non-linear functions allows the neural network to calculate much more complex dependencies than the linear model.

Next, the network was trained, after which the obtained readings were compared with the data from the test set. As a result, it turned out that the more the neural network model is trained, the result becomes closer to 100%. Based on everything, we can say that the larger the input set, the less the network model makes errors.

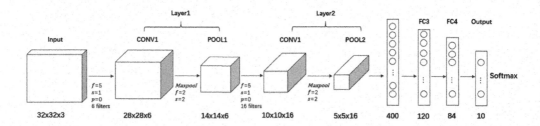

4   CONCLUSION

Accurate recognition of the medical image is very important for timely treatment. This study proposed the development of a breast cancer detection system as a standard procedure for diagnosed breast cancer. Digital mammography is now part of the standard diagnostic suite.

Various techniques are used for classification in the field of diagnosis. Image feature extraction is an important step in mammogram classification. These features are extracted using image processing techniques. The proposed neural network classifies mammogram images into 3 categories: normal, benign and malignant with an accuracy of more than 90%. Such a system can help both the patient and the doctor.

REFERENCES

[1]   Anitha J. and Peter J. D., "Mammogram Segmentation Using Maximal Cell Strength Updation in Cellular Automata," *Medical & Biological Engineering & Computing*, vol. 53, no. 8, pp. 737–749, 2015.

[2]   Dong M., Lu X., Ma Y., Guo Y., Ma Y., and Wang K., "An Efficient Approach for Automated Mass Segmentation and Classification in Mammograms," *Journal of Digital Imaging*, vol. 28, no. 5, pp.

[3]   Cheng H. D., Shan J., Ju W., Guo Y. and Zhang L., "Automated Breast Cancer Detection and Classification Using Ultrasound Images", *Pattern Recognition* No 43, pp. 299–317, 2010.

[4]   Mahendra G. Kanojia, Siby Abraham, "Breast Cancer Detection Using RBF Neural Network", *2nd International Conference on Contemporary Computing and Informatics (IC3I)*, IEEE, 363978-1-5090-5256-1/16, 2016.

[5]   Kruglov V. V., Dli M. I., and Yu R. *Fuzzy Logic and Artificial Neural Networks.* - M.: Fizmatlit, 2001.

[6]   Zagidullin BI, Nagaev IA, Zagidullin N.Sh., Zagidullin Sh.Z. _ A Neural Network Model for the Diagnosis of Myocardial Infarction. // *Russian Journal of Cardiology.* 2012; (6): 51–54.

[7]   Krizhevsky, Alex, et al. "ImageNet Classification with Deep Convolutional Neural Networks." *Communications of the ACM*, vol. 60, no. 6, 2017, pp. 84–90., doi:10.1145/3065386.

[8]   He, Kaiming, et al. "Delving Deep into Rectifiers: Surpassing Human-Level Performance on ImageNet Classification." *2015 IEEE International Conference on Computer Vision (ICCV)*, 2015, doi:10.1109/iccv.2015.123.

[9]   Andreev V.V., Minaev N.Yu., "Convolutional Neural Network for Recognizing Lung Cancer in Medical Images", 2019, 500pp.

[10] Stuart Russell, Peter Norvig, "*ArtificalIntelligence: A Modern Approach*", 2007, 640 pp.

[11] Andrew Burgess, "Artificial Intelligence for Your Business: A Guide to Evaluation and Application", 2018, 350 pp.

[12] Kumar, A., & Alam, B. (2014, February). Real Time Scheduling Algorithm for Fault Tolerant and Energy Minimization. In *2014 International Conference on Issues and Challenges in Intelligent Computing Techniques (ICICT)* (pp. 356–360). IEEE.

[13] Chaturvedi, R., Kumar, S., Kumar, U., Sharma, T., Chaudhary, Z., & Dagur, A. (2021). Low-cost Iot-enabled Smart Parking System in Crowded Cities. *In Data Intelligence and Cognitive Informatics: Proceedings of ICDICI 2020* (pp. 333–339). Springer Singapore.

[14] Y. Sharma, Shatakshi, Palvika, A. Dagur and R. Chaturvedi, "Automated Bug Reporting System in Web Applications," 2018 *2nd International Conference on Trends in Electronics and Informatics (ICOEI)*, Tirunelveli, India, 2018, pp. 1484–1488

[15] Mehra, P. S., Jain, K., Chawla, D., Dagur, A., Singh, S., & Sharma, J. (2022). *GWO-EFUCA: Grey Wolf Optimisation and Fuzzy Logic based Unequal Clustering and Routing protocol for sustainable WSN-based Internet of Things.*

[16] Mehra, P. S., Mehra, Y. B., Dagur, A., Dwivedi, A. K., Doja, M. N., & Jamshed, A. (2021). COVID-19 Suspected Person Detection and Identification Using Thermal Imaging-based Closed Circuit Television Camera and Tracking Using Drone in Internet of Things. *International Journal of Computer Applications in Technology*, 66(3–4), 340–349

[17] Kumar, A. & Alam, B. (2018). Task Scheduling in Real Time Systems with Energy Harvesting and Energy Minimization. *Journal of Computer Science*, 14(8), 1126–1133

[18] Durstewitz D., Koppe G., Meyer- Lindenberg A. Deep Neural Networks in Psychiatry // *Molecular Psychiatry* - 2019, SP - 1583 EP - 1598 VL - 24, SN - 1476-5578, DO - 10.1038/s41380-019-0365-9

Artificial Intelligence, Blockchain, Computing and Security – Dagur et al. (Eds)
© 2024 The Author(s), ISBN: 978-1-032-49393-0

Stock market prediction using DQN with DQNReg loss function

Alex Sebastian*, K.V. Habis* & Samiksha Shukla*
Department of Computer Science Christ University, Bangalore

ABSTRACT: There have been many developments in predicting stock market prices using reinforcement learning. Recently, Google released a paper that designed a new loss function, specifically for meta-learning reinforcement learning. In this paper, implementation is done using this loss function to the reinforcement learning model, whose objective is to predict the stock price based on certain parameters. The reinforcement learning used is an encoder-decoder framework that is useful for extracting features from long sequence prices. The DQNReg loss function is implemented in the encoder-decoder model as it could provide strong adaptation performance in a variety of settings. The model can buy and sell the index, and the reward is the portfolio return after the day's trading has concluded. To maximize yield the model must optimize reward function. The DQNReg loss implemented DQN network and the Huber loss DQN network is compared with the Sharpe ratio considered for return.

1 INTRODUCTION

The stock market is a marketplace where traders buy and sell shares of publicly-traded companies with the aim of benefiting from it. However, many traders experience more financial losses than gains while trading. Therefore, it is necessary to forecast stock values to enable traders to make knowledgeable decisions based on them.

Many studies have used machine learning techniques to forecast stock market values. Indicators such as social media news and algorithms like Support Vector Machine (SVM) have been employed to predict share prices in the past. In this study, the DQNReg loss function is used with reinforcement learning, which is a machine learning method based on an agent that is rewarded for desired behavior and punished for undesired behavior.

The new Reinforcement Learning (RL) algorithm is highly versatile and can be applied across domains, providing a generalizable solution. The learned loss function from the RL algorithm outperforms other algorithms in both training and unseen environments. Deep Reinforcement Learning (DRL), as a potent function approximator based on neural networks, can effectively handle vast state and action spaces.

The paper discusses related works on Reinforcement Learning, DQN, and trading in the stock market. The Dataset section provides information on the data source, features, and rationale for selecting the specific dataset. The paper also includes a brief summary of the proposed model, which demonstrates good performance with the new Loss Function. Finally, the paper presents conclusions drawn from the study and discusses future recommendations.

---

*Corresponding Authors: alex.sebastian@msds.christuniversity.in,
kv.habis@msds.christuniversity.in and samiksha.shukla@christuniversity.in

DOI: 10.1201/9781003393580-10

2 LITERATURE REVIEW

A deep dive into how reinforcement learning has been applied with various parameters to achieve the best profit is explored through related literature. The best model from the available models by considering the various parameters and the types of models can be chosen and used.

DQN: Mnih et al. 2021 [1] presented the first deep learning model that can successfully learn control laws from very detailed sensory information. Convolutional neural networks were used to train the model, which is one of them. It creates a value function from raw pixel input that forecasts future rewards.

In Taghian et al. 2021 [2] To develop an end-to-end model that can train single instrument investment strategies from a long series of raw instrument prices, neural encoder-decoder framework and deep reinforcement learning are proposed. The suggested model consists of a decoder, which is a DRL model, and an encoder, that is a neural structure which will be adopting profitable strategies based on the attributes recovered by the encoder from the input sequence. Yang et al. in [3], To explore share trading strategy by maximizing returns, put forth an ensemble approach that makes use of deep reinforcement schemes. The researchers train a deep reinforcement learning agent using Proximal Policy Optimization (PPO), Advantage Actor Critic (A2C), and Deep Deterministic Policy Gradient to create an ensemble trading strategy (DDPG). By incorporating and inheriting the top qualities of the three algorithms, the ensemble approach resolutely adjusts to varied market conditions. Due to their great liquidity, the algorithm was validated on the Thirty Dow Jones stocks.

Nan et al. in [4] present the researchers to develop a reinforcement learning strategy that employs conventional time series data for stock prices and blends it with sentiments from information headlines while utilizing information graphs to exploit information about implicit links in order to build a profitable trading strategy.

The limitation of the iterations at every strategy revision by the adoption of a substitute objective function is a key aspect of proximal policy optimization (PPO) [5], which has achieved ground-breaking results in the field of strategy look, a subdivision of reinforcement learning. The approach nevertheless exhibits optimization inefficiency and performance instability despite the benefit of such a constraint because of the curve's rapid flattening.

The paper [6] suggests that By including self-supervised goals based on the visual input's structure and serial interactions with its environment, the agent can learn more effectively. Self Predictive Representations (SPR), the technique used, teaches an agent to forecast its own implicit condition depictions over a series of future steps.

The Policy Similarity Metric (PSM), which has theoretical support, is presented in the study [7] as a way to gauge how similar states' behaviors differ from one another.

Value-based reward learning techniques such as deep Q-networks and DDPG are well recognised to be affected by the overstatement issue brought on by function optimization, which may result in substandard recommendations. In order to solve this matter, TD-3 introduces underestimation bias by selecting the critic with the lowest value among the two. The research [8] aims at a novel approach known as the Gradient of Balanced Late Shallow Deterministic Strategy, which can greatly improve performance and significantly reduce estimate error by weighing a pair of critics.

3 REINFORCEMENT LEARNING

A computational method for comprehending and automating goal-directed learning and decision-making is called reinforcement learning. Reinforcement Learning prioritizes learning by the person from direct interaction with its environment, as seen in Figure 1, it differs from previous computational techniques.

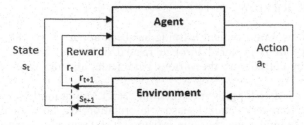

Figure 1. The agent and environment interactivity in reinforcement learning.

The agent observes the current state st at every distinct clock step t, selects a present measure at, and runs it. The settings reacts by rewarding the agent rt +1 = r (st, at) which in turn gives the succeeding condition st +1 = δ (st, at). δ and r are the functions which are not known to the agent and is part of the environment. The functions δ (st, at) and r(st, at) in the Markov Decision Process (MDP) depend exclusively on the current state and action and not on past states or actions. The agent's job is to learn a policy, π: S→A, where S is the set of conditions and A is the set of measures, to determine what to do next based on the present observed condition S; in other words, π(st) = at. A strategy is best if it can maximise all conceivable rewards from a state, or value, vπ (s).

4 DATASET

The data required has been procured from yahoo finance API. We take the Apple stocks as they have high volatility, and the agent has the liberty to buy and sell whenever a pattern arises. Apple is a tech giant company that is paying a growing dividend to its stakeholders. The massive cash flow the business generates supports its market capitalization estimate of $2.3 trillion. The business can pay increasing dividends to its stockholders because of this cash flow.

5 PROPOSED MODEL

5.1 *Deep Q network*

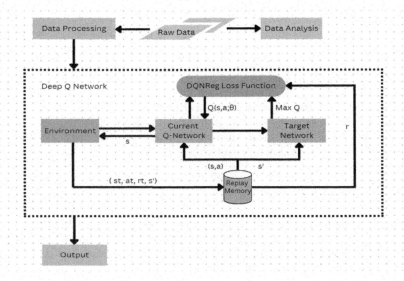

DeepMind created the above mentioned algorithm in 2015. It was able to handle a variety of Atari games by combining learning algorithm and dnns at scale (some to extraordinary amount). The program was created by combining experience replay and deep neural networks with the traditional RL algorithm Q-Learning.

5.2 Q-Learning

This works based on the function of Q. The yield we have in mind, or deducted total of benefits acquired from state **S** by acting, is measured by a policy's Q-function. A first and then a policy π after that. The Bellman optimality formula is satisfied by the best Q-function.

$$Q(s, a) = E_{(s,a \rightarrow s',r) \sim H} (r_{s,a} + \gamma \max_{a'} Q(s', a'))$$

5.3 Experience replay

There is the use of stochastic gradient descent to minimize the DQN loss rather than computing the complete expectation. If the loss is estimated by utilizing merely the recent transition, then this reduces to traditional Q-Learning, $\{s, a, r, s'\}$.

Experience Replay, a method developed for the Atari DQN, was used to increase the stability of network updates. At each time interval of the data collection, the transformation is placed to a memory area termed as replay buffer. Then, during training, Instead of only using the most recent transition, the loss and its gradient are calculated using a mini-batch of transitions taken from the replay buffer. By employing each transition several times in updates, this improves data efficiency and improves stability by using uncorrelated transitions in a batch.

In this paper, a DQN model with the Reg loss function is used, introduced by Google in its paper "Evolving Reinforcement Algorithm," [9] wherein the DQNReg directly utilizes an active weighted term to regularize the Q values. It performs especially well in a few test conditions, and it completes job when further approaches cannot yield any rewards. Additionally, it is coherent on test and significantly more stable, with smaller volatility across seeds.

$$L_{DQNReg} = 0.1 * Q(st, at) + \delta^2 \tag{1}$$

In Eqn (1) 0.1 is the active weighted term multiplied by the Q state. Action value function that is ideal Q*(s, a) = max E [Rt | st = s, at = a], where is a strategy linking sequences to measures. Q*(s, a) is defined as the highest expected return obtainable by adopting some approach after observing some sequences and then executing some steps.

The Q values are markedly undershot by DQNReg, and it does not converge to reality. Limiting the overestimation improves performance, while exaggerated value predictions are problematic, as demonstrated by numerous works (van Hasselt *et al.* 2015 [10]; Haarnoja *et al.* 2018 [11]; Fujimoto *et al.* 2018 [12]).

6 EXPERIMENT

In the study, the execution of the deep reinforcement learning algorithm with different loss function for ROI purpose is done on Apple. We used Deep-Q-Network (DQN) to automate stock trading and to maximize the profit using loss functions such as Huber Loss and DQNReg.

7 RESULTS

The results of the model are measured using Sharpe Ratio, Short term Return On Investment (ROI).

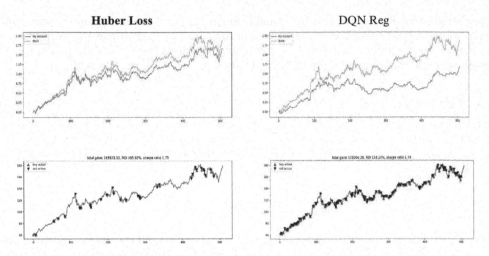

Figure 2.   Comparison of DQNReg loss function with Huber Loss.

In Figure 2, we have a comparison of both Huber Loss and DQNReg loss function on Apple stock price. Huber Loss is able to provide much better profit as compared to DQNReg and the number of trades provided by Huber Loss is much lower with comparison to DQNReg. With more trades the cost of buying or selling increases which in-turn decreases the overall profit achieved by the model.

	Average ROI	Average Sharpe Ratio	ROI Std Dev	Sharpe Ratio Std Dev
Huber Loss	174.01	1.77113	19.935	0.125
DQN Reg	125.48	1.53197	74.534	0.759

The average ROI for Huber Loss is at 174.01 while the DQNReg loss function gives an average ROI of 125.48. The Sharpe Ratio are also 1.77 and 1.53 for Huber and DQNReg, respectively. These results show that even though the Huber Loss is performing better than DQN Reg, the difference between them is not large.

8   CONCLUSION AND DISCUSSION

In this paper, we presented a reinforcement learning-based approach. We used Sharpe ratio to evaluate the performance of our model with two different loss function. The DQNReg loss function that is being used in our proposed model in this paper scores lower than the Huber loss function, but it is observed that there was an overall profit. Thus, the proposed model with DQNReg loss function can be used for prediction of stock market prices as it has a considerable amount of accuracy. For future works and discussions several improvements could be made, and restrictions could be handled. The model can be improved by using or adding more parameters of the stock market into consideration. Parameters like Social media sentiment and public news can be introduced to enhance the model's performance. As a future work, we also intend to combine transfer learning as it helps to tackle various challenges that is faced by reinforcement learning.

REFERENCES

[1] Mnih V., Kavukcuoglu K., Silver D., Graves A., Antonoglou I., Wierstra D., and Riedmiller M., "Playing Atari with Deep Reinforcement Learning," *arXiv* preprint arXiv:1312.5602, 2013.

[2] Taghian M., Asadi A. and Safabakhsh R., "A Reinforcement Learning Based Encoder-Decoder Framework for Learning Stock Trading Rules," *arXiv* preprint arXiv:2101.03867, 2021.

[3] Yang H., Liu X.-Y., Zhong S. and Walid A., "Deep Reinforcement Learning for Automated Stock Trading: An Ensemble Strategy," in *Proceedings of the First ACM International Conference on AI in Finance*, 2020.

[4] Nan A., Perumal A.and Zaiane O. R., "Sentiment and Knowledge Based Algorithmic Trading with Deep Reinforcement Learning," in *International Conference on Database and Expert Systems Applications*, 2022.

[5] Zhu W. and Rosendo A., "Proximal Policy Optimization Smoothed Algorithm," *arXiv* preprint *arXiv*:2012.02439, 2020.

[6] Schwarzer M., Anand A., Goel R., Hjelm R. D., Courville A. and Bachman P., "Data-efficient Reinforcement Learning with Self-predictive Representations," *arXiv* preprint *arXiv*:2007.05929, 2020.

[7] Agarwal R., Machado M. C., Castro P. S.and Bellemare M. G., "Contrastive Behavioral Similarity Embeddings for Generalization in Reinforcement Learning," *arXiv* preprint arXiv:2101.05265, 2021.

[8] He Q. and Hou X., "Reducing Estimation Bias via Weighted Delayed Deep Deterministic Policy Gradient," *arXiv preprint arXiv*:2006.12622, 2020.

[9] Co-Reyes J. D., Miao Y., Peng D., Real E., Levine S., Le Q. V., Lee H. and Faust A., "Evolving Reinforcement Learning Algorithms," *arXiv preprint arXiv:2101.03958*, 2021.

[10] Dulac G.-Arnold, Evans R., van Hasselt H., Sunehag P., Lillicrap T., Hunt J., Mann T., Weber T., Degris T. and Coppin B., "Deep Reinforcement Learning in Large Discrete Action Spaces," *arXiv* preprint arXiv:1512.07679, 2015.

[11] Haarnoja T., Zhou A., Abbeel P.and Levine S., "Soft actor-critic: Off-policy Maximum Entropy Deep Reinforcement Learning with a Stochastic Actor," in *International Conference on Machine Learning*, 2018.

[12] Fujimoto S., Hoof H.and Meger D., "Addressing Function Approximation Error in Actor-critic Methods," in *International Conference on Machine Learning*, 2018.

Artificial Intelligence, Blockchain, Computing and Security – Dagur et al. (Eds)
© 2024 The Author(s), ISBN: 978-1-032-49393-0

Towards improving the efficiency of image classification using data augmentation and transfer learning techniques

A. Christy*
Professor, Sathyabama Institute of Science and Technology, Chennai, India

S. Prayla Shyry*
Professor, Sathyabama Institute of Science and Technology, Chennai, India

M.D. Anto Praveena*
Associate Professor, Sathyabama Institute of Science and Technology, Chennai, India

ABSTRACT: Convolutional Neural Networks (CNN) are used widely adopted for tasks involved with Computer Vision, Medical Imaging and Natural language processing. Creating a CNN model which has the ability to detect and track objects as similar to human remains a challenging task. CNN can perform well if the object to be detected is found in the same position as the one found in the dataset. But, if the object contains certain angle of tilt or rotation, then detecting the object remains a tedious or laborious task. The task of image classification has 2 stages namely Feature extraction and classification. In this paper, we have deployed 2 techniques data augmentation and Transfer learning and the comparisons are studied. Data Augmentation is a technique adopted for rearranging the images in different orientation by applying scaling, rotation and shearing. Through Transfer learning pertained model is applied for feature extraction which can then be applied over the new model. The dataset considered for our approach is CIFAR10 dataset. Applied with CNN, we have obtained an accuracy of 84.56%. The model after applying Data Augmentation has received 96.56% accuracy in which 20% of data is considered for drop out. The model is trained with Transfer learning RESNET 50 architecture and the validation process has received an accuracy of 98.79. The prediction accuracy with Data augmentation process is 96.56% and with transfer learning is 100% as it is well pre trained model applied over the new dataset.

1 INTRODUCTION

In recent times. Artificial Intelligence has attracted the researchers in bridging the gap between human and machine. Machine learning techniques in health care sectors contribute widely in drug discovery, Disease identification and prediction, Bio medical signal processing, maintaining privacy and security in smart electronic health care records, etc. Multi-label classification problems can be addressed by techniques such as binary classification, machine learning, feature selection and image analysis. Multi-label classification problems fall under two categories such as Problem transformation and algorithm adaptation methods. The problem transformation is the widely used method in which dataset with muti-label is converted to single label by a set of operations. In image classification, multi-level neural networks were adopted by Convolutional Neural Networks in order to retrieve the salient features of the image. Due to the availability of advanced technologies and GPU facilities, some improvement in the model building and accuracy will be beneficial for any applications using computer vision.

---

*Corresponding Authors: ac.christy@gmail.com, praylashyry.cse@sathyabama.ac.in and antopraveena@gmail.com

DOI: 10.1201/9781003393580-11

BerinaAlic *et al.* (2017) have applied Artificial Neural Networks (ANN) and Bayesian Networks in the classification of Diabetes and Cardiovascular diseases. For the classification of both types of diseases multilayer feedforward neural network was identified as the common solution with best accuracy given by ANN [1]. Image processing, Natural language processing and Audio processing are some of the major applications of deep learning, Fengxiang, He and Dacheng Tao (2023). When deep learning is applied to security-critical areas such as health-care, agriculture and finance even a small error can cause a severe problem [2–4,6–8,14,18,22,24]. According to Erhui Xi (2022), in image classification feature generation and image recognition are the two important stages. Even though CNN can effectively categorize massive number of images, random forest plays an important role in improving the accuracy of the model. The author has applied CNN for extracting the features ad random forest for classification and this hybrid model has produced better accuracy [5]

The objective of feature selection is to improve the cost-effective performance of the training data and to make accurate prediction, Isabelle Guyon & Andre Elisseeff (2003). The classification accuracy can be found by a threshold Θ, and by varying the values of Θ, the tradeoff between false positive rate and false negative rate can be observed. Nested subset methods, Filters and Direct objective optimization were some of the supervised feature selection methods [9].Deep learning requires a huge volume of training data in order to categorize medical data, Maad M. Mijwil (2021). Architecture Inception V3 with Adaptive Moment Estimation (Adam) and activation function ReLu with learning rate 0.0001 was adopted for the classification of skin cancer from medical images [10]. Manuj Joshiand AshokJetawat (2020) Naïve Bayes, IBK, Support Vector Machine (SVM), ZeroR and VFI were some of the algorithms available in the classification of diseases in medical decision support system. In the analysis SVM and VFI have produced better result in comparison with other methods[11]. Khaleel *et al.* (2021) have proposed a framework based "Hierarchical Visual Concept" (HVC) for image classification using CNN using the architectures Vgg16 and ResNet50. EMIS-I datasets was used for image classification with hyper parameters visual concept filter h x w = 2 ×2, d=128, m =10, λ = 0.8, α = 0.2, β = 0.8, T = 0.05, K = 2 and the experimental results have shown HVC has shown better accuracy [12].

Feature fusion with deep learning improves the efficiency of classification of plant leaves, M. Okuda & Ohshima (2022). Whole-leaf, leaf-shape and leaf-vein features were considered for fusion [13]. Neha Sharma *et al.* (2018) have studied the performance of Alex Nets, GoogLeNet and ResNet50 considering the image datasets ImageNet, CIFAR10 and CIFAR100. It is observed that in comparison with AlexNet, GoogleNret and RestNet50 represents higher precision [15]. Nitin Pise & Parag Kulkarni (2016) have listed the classification algorithms such as function-based (support vector machine and neural networks), tree-based (J48 and random forest), distance-based (K-Nearest neighbour) and bayesian (naïve bayes). Meta learning algorithms such as (Ada boost, Logitboost and Bagging) helps to solve significant problems related to classification and regression in the areas of machine learning [16]. The purpose of Drop-out layer in a feed-forward network is a technique used to prevent over fitting and the loss function can be minimized under noise distribution. Drop-out layer can be applied in the hidden layer as a form of model averaging, Nitish Srivastava *et al.* (2014). Adding noise to the hidden layers can be extended to supervised learning algorithms also [17]. Solemane Coulibaly *et al.* (2022) have represented a multi-label classification model using deep convolutional neural network.Dataset used is Amazon rain forest dataset. The f1-score dataset for VOC 2007 dataset and VOC 2012 dataset has been obtained as 87.71% and 88.64% respectively. Multi-class classification is applied across various domains with annotation performed with image, genefunctions etc [19].

In classification of mangrove ecosystem, satellite remote sensing is very much essential, in saving lots of human resources Wu *et al.* (2015). Expert classification and fuzzy classification were done with parallelepiped method, since one single algorithm cannot produce accurate results [20].Machine learning and Artificial Neural Networks faces challenges in feature extraction and classification of massive images. Qing Lv *et al.* (2022) have adopted a deep learning model named as VggNetfor image classification which can handle large image dataset and produce better accuracy. VggNet is adopted to avoid the problem of overfitting [21].

Yugang Li *et al.* (2019) have adopted Long short Term inorder to identify the visual relationships among images. The visual relationships were represented as (object$_1$, predicate, object$_2$). If there are N objects and K predicates, then it would require $O(N^2K)$ to train the model. There can exist strong statistical correlations between objects and relationship and it was represented by Conditional Random Field (CRF) and the image classification here is handled by semantic labels. Microsoft COCO (MS-COCO) dataset and the LSTM approach for visual representation shows better accuracy compared with CNN-SVM and SNN-RNN [23]. From the literature survey, it is evident that Convolutional Neural Networks plays a vital role in computer vision, in classification of images, medical imaging, Autonomous driving, etc. We face some challenges using CNN when the training data is not sufficient as well the images are aligned in different oriented. In this paper, the problem of overfitting which happens due to the availability of insufficient data is addressed by the process of Data Augmentation and Transfer Learning.

3   SYSTEM ARCHITECTURE

A CNN being a neural network is designed with neurons and their associated weights. Each neuron is connected with a hidden layer assigning some weights. The weighted sum of these neuron weights are passed through the convolutional layers by applying activation function which inturn produces the output of the layer and the process adopted in the research is depicted in Figure 1.

Convolutional Neural Networks proposed by Le Cun *et al.* 1989 and LeCun *et al.* 1998 are emerging now a days widely for handling image data or computer vision due to the wide availability of GPUs and other processing resources. Convolution is a technique that will aggregate neighborhood values by multiplying them with a kernel of values in a region and adding them to a value. Considering the task of classifying images, each image consists of a set of intensity values related to pixels and the expected output being classifying the images to various categories based on its type.

The algorithm functions by taking input images and assigning corresponding weights and biases. An image is defined as a matrix of pixels. CNN is efficient in identifying the spatial and temporal dependencies with the application of appropriate filters. It is also capable of handling massive databases. Kernal / Filter, represented as K is the convolution operation involved in the initial operation. Taking the stride length an element wise multiplication operation performed between K, the filter moves to the right with certain stride value until the images are traversed. The convolution operation extracts lower-level features such as color, edges, etc from the initial layers. With the inclusion of more hidden layers the higher-level features are included to make the image more specific to the given domain. The Pooling layers such as Maxpooling and average pooling performs dimensionality reduction by flattening the image into a column vector.

The identity of an image is invariant under image transformations such as scaling, rotation, translation and shearing. Moreover, the Neural Network also exhibits certain invariance to certain restrained transformation. The transformation applied to an image with rotation transformation is depicted in Figure 1 a and 1 b.

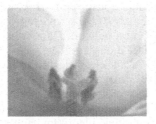 

Figure 1.   a and b: Transformation rotation being applied.

3.1 Data augumentation

Data Augmentation contains a set of techniques used to prevent overfitting techniques by increasing the amount of data by generating new data points from existing data. The drawback of CNN is that in images, pixels which are close to each other are more closely correlated that far away images. Information from lower order features can be combined to detect higher-order features, which would finally represent the entire image as a whole. This concept is carried out through 3 mechanisms in CNN as shown in Figure 2.

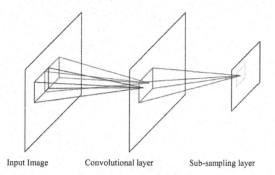

Input Image   Convolutional layer   Sub-sampling layer

Figure 2.   Layers of convolutional neural networks with subsampling layers.

In the overall architecture, there can be many pairs of convolution and sub sampling layers. At each and every layer, we can experience a small degree of invariance compared to its previous layer. Hence, proceeding in this manner, the final layer of the network will be fully connected and adaptive and softmax activation function is applied in the case of multi-modal classification problem in order to produce more accuracy.

3.2 Transfer learning using tensor flow

Transfer learning means reuse of a pre-trained model on a new problem. It is yet another technique used to prevent overfitting by generating a set of training data from the given dataset. In this method, the pre trained model (the model which is having weights and biases) is customized using Feature Extraction and Fine Tuning. The process of transfer learning is depicted in Figure 3. The architecture works by freezing the early layers (set learning rate as 0), apply the convolutions and generate a fully connected neural network by applying Softmax activation function.

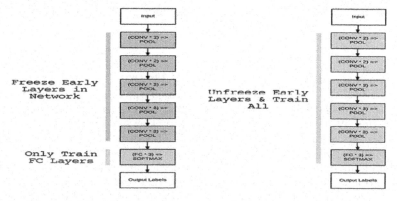

Figure 3.   Architecture of transfer learning.

Once the fully connected network has started to learn patterns in dataset, the training can be paused and the body is unfreezed using a small learning rate. Training is continued until required accuracy is achieved. TensorFlow is a flexible and versatile open-source library. It is a framework which can execute a sequence of nodes called as "Graph".

4 RESULTS AND DISCUSSION

In the classification model, images are classified using CNN approach. To avoid overfitting, Data Augumentation, Drop out and Transfer Learning were applied. The image dataset considered for classification is CIFAR-10 and the first nine training images are depicted in Figure 4. We have considered 8 classes of images under Trees, Cats, Dogs, Daisy, Dandelion, Rose, Sunflower and Tulip and the size of image being 32 ×32. Among them 6600 files were considered for training and 1320 files used for validation. A sample of the images in the dataset is depicted in Figure 4. The batch size considered for creating the dataset is 32. The batch size should be of multiple of 2 for efficient utilization of CPU. 80% of the dataset is considered for training.

Figure 4.   First nine training images from CIFAR.

The batch size considered is a tensor of shape (32,180,180,3), in which 32 is the batch size and 3 is the color channel. The training model is created with three convolution blocks with a max pool layer in each of them. There is a fully connected layer with 128 units on top of it that is activated by the ReLU activation function. The Adm optimizer is adopted with Softmax layer being used at the output layer. The shape of Convolution layer 1 is (180,180,16) after applying max pool layer has achived (90,90,16) and while flattening the number of units are obtained as 30976. Models are fitted and with 10 epochs, the model has achieved an accuracy of 84.56% and the training and validation loss and accuracy is depicted in Figure 5, in which we have seen overfitting in accuracy and underfitting in loss.

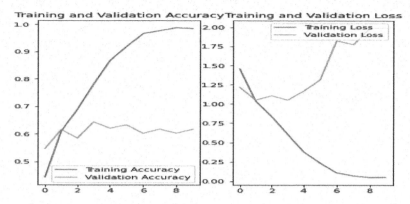

Figure 5.   Training and validation accuracy and loss.

4.1   *Results with data augumentation*

Overfitting generally occurs when there are a small number of training examples. Data Augmentation is done to prevent overfitting with 20% drop out and the results are shown to be 71.97% accuracy whereas the training and validation accuracy and loss are depicted in Figure 6 in which data converge more or less similar manner. After augmentation prediction accuracy is 96.5 6% as shown in Figure 7.

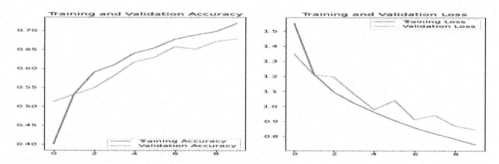

Figure 6.   Training and validation accuracy and loss after data augmentation.

```
1/1 [==============================] - 0s 164ms/step
This image most likely belongs to rose with a 96.56 percent confidence.
```

Figure 7.   Prediction result after augmentation.

4.2   *Results with transfer learning*

Transfer learning is a pre-trained model. The knowledge gained from a trained model is transferred to a model which is having close resemblance with the pretrained model. The pretrained model adopted is ResNet50 and the accuracy is obtained as 90.74% with the training and validation accuracy and loss are depicted in Figure 8. The total number of parameters are 23,606,153 from which the number of trainable parameters are 10,008,585 and the validation accuracy is 98.79% as depicted in Figure 9 and the prediction model has shown 100% accuracy as shown in Figure 10.

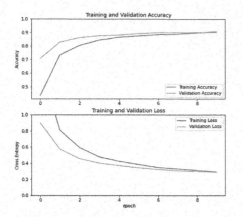

Figure 8.   Accuracy and loss with ResNet.

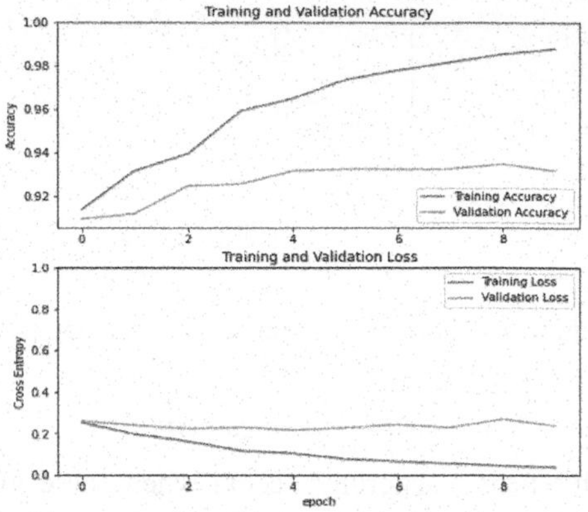

Figure 9.   Accuracy and loss after transfer learning with ResNet.

Figure 10.   Prediction accuracy of transfer learning with ResNet architecture.

5   CONCLUSION

Image classification task involves the extraction of features from high diamension data and is applied in retrieving patterns from the given dataset. In this paper CIFAR dataset from 8 various domains are considered for classification. The classification task deals with efficiently loading a dataset and applying techniques to avoid overfitting including data augmentation and drop out. Data augmentation is applied for improving the accuracy and avoiding overfitting with the help of Keras preprocessing layer by the functions such as flip, rotation and zoom. Further, it is also compared with a well trained data architecture RESNET for classification. The experimental results have shown well improved accuracy with the application of Transfer learning architecture.

REFERENCES

[1]   BerinaAlic, LejlaGurbeta, AlmirBadnjevic (2017), "Machine Learning Techniques for Classification of Diabetes and Cardiovascular Diseases", *6th Mediterranean Conference On Embedded Computing*, pp. 11–15.
[2]   C. Ma, S. Xu, X. Yi, L. Li and C. Yu, "Research on Image Classification Method Based on DCNN," *2020, International Conference on Computer Engineering and Application (ICCEA)*, 2020, pp. 873–876, doi: 10.1109/ICCEA50009.2020.00192.

[3]   Chen, Guoming & Chen, Qiang & Zhu, Xiongyong & Chen, Yiqun & Yuan, Zeduo. (2020). *Tensor Network for Image Classification*. 135–140. 10.1109/ICDH51081.2020.00031.

[4]   C.-L. Chang, H.-M. Tseng and H.-T. Chu, "Fireworks Image Classification with Deep Learning," *2021 International Conference on Technologies and Applications of Artificial Intelligence (TAAI)*, 2021, pp. 311–314, doi: 10.1109/TAAI54685.2021.00067.

[5]   Erhui Xi, "Image Classification and Recognition Based on Deep Learning and Random Forest Algorithm", *Wireless Communications and Mobile Computing*, vol.2022, Article ID 2013181, 9 pages, 2022. https://doi.org/10.1155/2022/2013181.

[6]   Fan, Yongxian & Gong, Hao. (2022). *An Improved Tensor Network for Image Classification in Histopathology*. 10.1007/978-3-031-18910-4_11.

[7]   Fengxiang, He, DachengTao (2023), "Foundations of Deep Learning", *Machine Learning: Foundations, Methodologies and Applications*.

[8]   F. Siraj, M. A. Salahuddin and S. A. M. Yusof, "Digital Image Classification for Malaysian Blooming Flower," *2010 Second International Conference on Computational Intelligence, Modelling and Simulation*, 2010, pp. 33–38, doi: 10.1109/CIMSiM.2010.92.

[9]   Isabelle Guyon, Andre Elisseeff (2003), "An Introduction to Variable and Feature Selection", *Journal of Machine Learning Research*, pp. 1157–1182.

[10]   Maad M. Mijwil (2021), "Skin Cancer Disease Images Classification Using Deep Learning Solutions", *Multimedia Tools and Applications*,Vol. 80, pp. 26255–26271

[11]   Manuj Joshi, Ashok Jetawat (2020), "Evaluation of Classification Algorithms used in Medical Decision Support Systems", *Fourth World Conference on Smart Trends in Systems, Security and Sustainability, IEEE Xplore*, 978-1-7281-6823-4/20

[12]   M. Khaleel, W. Tavanapong, J. Wong, J. Oh and P. de Groen, "Hierarchical Visual Concept Interpretation for Medical Image Classification," *2021 IEEE 34th International Symposium on Computer-Based Medical Systems (CBMS)*, 2021, pp. 25–30, doi: 10.1109/CBMS52027.2021.00012.

[13]   M. Okuda and H. Ohshima, "Feature Fusion for Leaf Image Classification," *2022 IEEE International Conference on Big Data and Smart Computing (BigComp)*, 2022, pp. 259–262, doi: 10.1109/ BigComp54360.2022.00056.

[14]   Muhammad Suhaib Tanveer, Muhammad Umar Karim Khan, Chong-Min Kyung (2020), "Fine-Tuning DARTS for Image Classification", *25th International Conference on Pattern Recognition (ICPR)*, pp. 4789–4796.

[15]   Neha Sharma,Vibhor Jain, Anju Mishra (2018), "An analysis of Convolutional Neural Networks for Image Classification", Procedia Computer Science, Vol. 132, pp. 377–384.

[16]   Nitin Pise, Parag Kulkarni (2016), "Algorithm Selction for Classification Problems". *SAI Computing Conference*, pp. 203–211

[17]   Nitish Srivastava, Nitish Srivastava, Alex Krizhevsky, Ilya Sutskever, Ruslan Salakhutdinov (2014), "Dropout: A Simple Way to Prevent Neural Networks from Overfitting", *Journal of Machine Learning Research*, Vol. 15, pp. 1929–1958

[18]   N. Jmour, S. Zayen and A. Abdelkrim, "Convolutional Neural Networks for Image Classification," *2018 International Conference on Advanced Systems and Electric Technologies (IC_ASET)*, 2018, pp. 397–402, doi: 10.1109/ASET.2018.8379889.

[19]   Y. Sharma, Shatakshi, Palvika, A. Dagur and R. Chaturvedi, "Automated Bug Reporting System in Web Applications," *2018 2nd International Conference on Trends in Electronics and Informatics (ICOEI)*, Tirunelveli, India, 2018, pp. 1484–1488

[20]   Mehra, P. S., Jain, K., Chawla, D., Dagur, A., Singh, S., & Sharma, J. (2022). *GWO-EFUCA: Grey Wolf Optimisation and Fuzzy Logic based Unequal Clustering and Routing Protocol for Sustainable WSN-based Internet of Things.*

[21]   Mehra, P. S., Mehra, Y. B., Dagur, A., Dwivedi, A. K., Doja, M. N., & Jamshed, A. (2021). COVID-19 Suspected Person Detection and Identification Using Thermal Imaging-based Closed Circuit Television Camera and Tracking Using Drone in Internet of Things. *International Journal of Computer Applications in Technology*, 66(3–4), 340–349.

[22]   Kumar, A., & Alam, B. (2016). Real-Time Fault Tolerance Task Scheduling Algorithm with Minimum Energy Consumption. In *Proceedings of the Second International Conference on Computer and Communication Technologies: IC3T 2015*, Volume 2 (pp. 441–448). Springer India.

[23]   YugangLi,Yongbin Wang, Zhe Chen (2019), "Image Classification with Visual Relationship", *IEEE Xplore*, pp. 214–218.

[24]   Y. Lee, "Image Classification with Artificial Intelligence: Cats vs Dogs," *2021 2nd International Conference on Computing and Data Science (CDS)*, 2021, pp. 437–441, doi: 10.1109/CDS52072.2021. 00081.

Artificial Intelligence, Blockchain, Computing and Security – Dagur et al. (Eds)
© 2024 The Author(s), ISBN: 978-1-032-49393-0

Predicting stock market price over the years by utilizing machine learning algorithms

Mrignainy Kansal & Pancham Singh
Department of IT, Ajay Kumar Garg Engineering College, Ghaziabad

Sachin Kumar
Department of CSE, Galgotias College of Engineering & Technology, Greater Noida

Ritu Sibal
Department of CSE, NSUT, Delhi

ABSTRACT: It is difficult to accurately predict stock market price returns due to the volatility and non-linear structure of current financial stock markets. However, programmable prediction techniques have shown success in forecasting stock prices with the development of machine learning and advanced computing capabilities. This study uses decision trees, random forests, and linear regression to predict a company's closing price the following day. The financial information from the open, close, and high prices is used to create the input variables for the model. To analyse the models and predict the stock, common strategic indicators like RMSE, MAE, and MSE are employed in our study, and in our approach, we have shown that lower values of these indicators are effectively helping in stock forecasting.

Keywords: Random Forest, Linear Regression, Decision Tress, Stock Market Prediction

1   INTRODUCTION

The stock market has a history of being unexpected, non-linear, and vibrant. Predicting stock values can be challenging as they depend on variables such as the political environment, the state of the global economy, business finance success, and more. Therefore, methods to anticipate stock values by looking at the pattern more than the previous numerous years could be pretty helpful for getting stock market moves, maximising profit, and minimising losses. A firm's stock price has traditionally been predicted using two primary methods [1]. First, to forecast a stock's future prices, we technically analyse the previous indicators, such as the opening and closing prices of the stock, volume marketed, adjacent close price values, and much more. The second method of studying a stock price is qualitative analysis, which draws on outside factors including the company profile, market mood, political and economic news, financial data, and social media blogs written by professional economists. Contemporary systems based on technical and fundamental research can be used to predict values. The data size for the stock market assessment is massive and non-linear. To handle this kind of data, you need a robust model that can sift through large amounts of data to uncover hidden patterns and complex relationships. Machine learning techniques are more efficient than conventional methods in this area by between 60 and 86 percent [2].

For predicting stock prices, the general public in advanced research on this subject uses various theories derived from random walks, moving average convergence, and a few popular linear models, including ARIMA and ARIMA. Research shows that machine learning can enhance stock market forecasting techniques, including decision tree and random forest

DOI: 10.1201/9781003393580-12

approaches. An ensemble method is Random Forest [3]. Both classification and regression tasks can typically be accomplished by it. Several decision trees are built throughout the training period, which thus produces the mean regression of each decision tree individually. The decision tree gives the best result for regression and classification issues.

Previous research in the field relied on conventional methods to forecast stock prices, including linear regression, random walk ARMA, and ARIMA [4]. However, studies in this field have shown that machine learning can enhance stock market prediction. Approaches include linear regression, random forest-based methods, and decision tree-based systems. The method used in ensembles is random forest, which can be used to carry out regression and classification tasks. It bases its work on building different decision trees while training, resulting in the mean regression of each tree.

In predictive analysis, linear regression, a modest and straightforward statistical regression technique, emphasizes the relationship between continuous variables. A statistical method known as "linear regression" illustrates a relationship between defined variables. Dependent variable is shown over the y-axis, while the independent variable is shown over the x-axis. In this study, the closing price of an organization was predicted using decision tree regression, random forest regression, and linear regression [5]. The models make use of a new set of variables that were produced utilising the financial data-set for a specific company's open, high, low, and close. These additional indicators will significantly enhance the models' capacity to forecast the remaining rate of a certain company the next day. Performance measurements such as RMSE, MSE, and MAE are used to evaluate the models' effectiveness.

2 METHODOLOGY

2.1 *Description of date*

Yahoo Finance is used to gather Microsoft's historical financial data. The collection contains data spanning two years, from 20/11/20 to 20/11/22. Furthermore, the data covers stock-related information such as high, low, open, close and volume.

2.2 *Linear regression*

Using the regression approach, a goal value can be simulated using different predictors [6]. Forecasting and identifying the relationships between variables' causes are two of this technique's essential uses. Regression techniques change primarily depending on the number of different independent variables and the type of interconnection between the variables as independent and dependent [7]. For example, there is just one independent variable, and a relationship is defined between them as an independent variable 'x' and a dependent [8] variable 'y' in simple linear regression.

2.3 *Random forest*

It is considered one of the most prominent ensembles. It can perform tasks involving classification and regression. The concept integrates multiple decision trees to produce the desired outcome rather than relying solely on individual [9] decision trees, increasing the model's diversity. In this study, new variables are made available for the training of each shown decision tree, which has an impact on the choices made at the tree's nodes. The interference in stock market data is frequently pretty significant and, due to its enormous volume, can cause the trees to evolve significantly differently than expected.

In order to reduce forecasting errors, Random Forest [10] looks at stock market analysis that considers a classification problem and predicts the stock's closing price for a particular firm the following day based on training factors.

3 RESULTS

A comparison between the three methodologies employing all the models mentioned above is made to assess the models' efficacy. We use RMSE, MAE, and MSE for contrast.
RMSE is computed using:

$$RMSE = \sqrt{\frac{\sum_{i=1}^{N}(Predicted_i - Actual_i)^2}{N}}$$

MAE is computed using:

$$MAE = \frac{1}{n}\sum_{i=1}^{n}|x_i - x|$$

MSE is computed using:

$$MSE = \frac{1}{n}\sum_{i=1}^{n}(Y_i - \widehat{Y}_i)^2$$

Figure 1 shows graphs comparing Microsoft's original final price of the stock to its expected closing price using linear regression. Using Random Forest, the graphs in Figure 2 compare the original prices of the stock of closing to its predicted closing price. Using a Decision Tree Regressor, Figure 3 displays graphs comparing the original closing price of a stock to the forecasted closing price of stocks [11]. The RMSE, MAE, and MSE values obtained using the Linear Regression, Random Forest, and Decision Tree Regressor models are evaluated in Table 1 [12]. As can be seen, Random Forest outperforms other stock price prediction algorithms. Results for a two-year dataset are shown below.

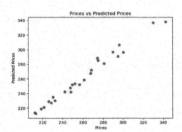

Figure 1. Graph for linear regression algorithm.

Figure 2. Graph for decision tree regressor algorithm.

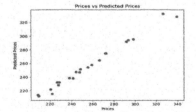

Figure 3. Graph for random forest algorithm.

Table 1. Table to compare standard strategic indicators.

Algorithm	RMSE	MAE	MSE	Accuracy
Linear Regression	1.892	3.5798	25.7455	99.69
Random Forest	3.7762	2.8336	14.2597	98.83
Decision Tree Regressor	2.7181	7.3885	105.5179	90.56

Yahoo Finance is used to gather Microsoft's historical financial data. The collection contains data spanning six months, from 20/05/22 to 20/11/22. In addition, the data integrates stock-related details like High, Low, Open, Close, and Volume [13].

Table 2, which compares the RMSE, MAE, and MSE values obtained using the linear regression, random forest, and decision tree models, shows that random forest produces superior stock price predictions [14].

Figure 4.   Graph shows linear regression.

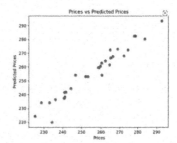

Figure 5.   Graph shows decision tree regressor.

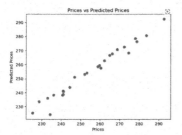

Figure 6.   Graph shows random forest.

Table 2.   Table to compare standard strategic indicators.

Algorithm	RMSE	MAE	MSE	Accuracy
Linear Regression	1.204	1.450	3.530	98.72
Random Forest	3.040	2.098	9.246	97.13
Decision Tree Regressor	1.708	2.919	16.741	93.94

Yahoo Finance is used to gather Microsoft's historical financial data [15]. The collection contains data spanning two months, from 20/09/22 to 20/11/22. The results are used to plot a graph between Actual Price and Predicted Price for all three algorithms [16].

Table 3, which compares the RMSE, MAE, and MSE values obtained using the linear regression, random forest, and decision tree regressor models show that random forest produces better stock price predictions [17]. Results for the two-month dataset are shown below.

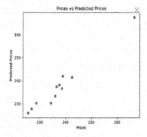

Figure 7.   Graph utilizing linear regression.

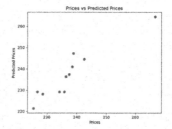

Figure 8.   Graph utilizing decision tree regressor.

75

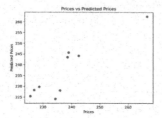
Figure 9. Graph utilizing random forest.

Table 3. Table to compare standard strategic indicator.

Algorithm	RMSE	MAE	MSE	Accuracy
Linear Regression	1.278	1.633	3.727	96.67
Random Forest	1.750	3.064	15.947	85.77
Decision Tree Regressor	5.350	4.238	28.628	79.09

Yahoo Finance is used to gather Microsoft's historical financial data. The collection contains data spanning one year, from 20/11/21 to 20/11/22. In addition, the records contain stock-related details like high, low, open, close, and volume [18].

The results are used to plot a graph between the actual price and predicted price for all three algorithms: linear regression, random forest, and decision tree regression.

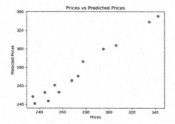
Figure 10. Graph shows linear regression.

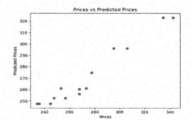
Figure 11. Graph shows decision tree regressor.

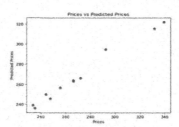
Figure 12. Graph shows random forest.

Table 4, which compares the RMSE, MAE, and MSE values obtained using the linear regression [19], random forest, and decision tree regressor models, shows that random forest produces better stock price predictions. The findings for a year of data are shown below.

Table 4. Table to compare standard strategic indicators.

Algorithm	RMSE	MAE	MSE	Accuracy
Linear Regression	2.3523	5.5337	41.3062	96.21
Random Forest	7.8799	5.7108	62.0931	94.76
Decision Tree Regressor	2.9430	8.6614	99.5385	90.87

Table 5. Comparison table for accuracy for all the algorithms for different dataset.

Algorithm	2 Years	6 Months	2 Months	1 Year
Linear Regression	97.69	98.72	96.67	96.21
Random Forest	98.83	97.13	85.77	94.76
Decision Tree Regressor	90.56	93.94	79.09	90.87

4 CONCLUSION AND FUTURE SCOPE

It is challenging to predict stock market returns since stock values often fluctuate and depend on several variables that produce complex patterns. The historical dataset is insufficient because it only contains a small number of variables, including low and high, open and closed stock prices, and the volume of traded shares. New variables were made to improve predicted price value accuracy by utilising the pre-existing variables. The stock's closing price the following day is expected using linear regression in addition to RF and Decision Tree Regressor for comparative analysis. A comparison is made by taking a data set from different years and executing different algorithms. After then, it represents that linear regression offers a superior stock price forecast than other models. Linear regression accuracy increases as the amount of data increases for the prediction. Deep learning methods may be incorporated to achieve better outcomes as we have large datasets, including financial news and metrics.

REFERENCES

[1] Singh, Ananya, and Swati Jain. "Stock Market Prediction During COVID Using Stacked LSTM." *In 2022 IEEE 9th Uttar Pradesh Section International Conference on Electrical, Electronics and Computer Engineering* (UPCON), pp. 1–6. IEEE, 2022.

[2] Kumar, Praveen, and Mallieswari R. "Predicting Stock Market Price Movement Using Machine Learning Technique: Evidence from India." *In 2022 Interdisciplinary Research in Technology and Management* (IRTM), pp. 1–7. IEEE, 2022.

[3] Akhtar, Md Mobin, Abu Sarwar Zamani, Shakir Khan, Abdallah Saleh Ali Shatat, Sara Dilshad, and Faizan Samdani. "Stock Market Prediction Based on Statistical Data Using Machine Learning Algorithms." *Journal of King Saud University-Science* 34, no. 4 (2022): 101940.

[4] Liu, Feng, Deli Kong, Zilong Xiao, Xiaohui Zhang, Aimin Zhou, and Jiayin Qi. "Effect of Economic Policies on the Stock and Bond Market Under the Impact of COVID-19." *Journal of Safety Science and Resilience* 3, no. 1 (2022): 24–38.

[5] Jaiswal, Rupashi, Kunal Mahato, Pankaj Kapoor, and Sudipta Basu Pal. "A Comparative Analysis on Stock Price Prediction Model using DEEP LEARNING Technology." *American Journal of Electronics & Communication* 2, no. 3 (2022): 12–19.

[6] Bats, Joost V., and Aerdt CFJ Houben. "Bank-based Versus Market-based Financing: Implications for Systemic Risk." *Journal of Banking & Finance* 114 (2020): 105776.

[7] Vijh, Mehar, Deeksha Chandola, Vinay Anand Tikkiwal, and Arun Kumar. "Stock Closing Price Prediction Using Machine Learning Techniques." *Procedia Computer Science* 167 (2020): 599–606.

[8] Kumaraswamy, Sumathi, Rabab Hasan Ebrahim, and Wan Masliza Wan Mohammad. "Dividend Policy and Stock Price Volatility in Indian Capital Market." *Entrepreneurship and Sustainability Issues* 7, no. 2 (2019): 862.

[9] Masoud, Najeb MH. "The Impact of Stock Market Performance Upon Economic Growth." *International Journal of Economics and Financial Issues* 3, no. 4 (2013): 788–798.

[10] Selvin, Sreelekshmy, R. Vinayakumar, E. A. Gopalakrishnan, Vijay Krishna Menon, and K. P. Soman. "Stock Price Prediction Using LSTM, RNN and CNN-sliding Window Model." *In 2017 International Conference on Advances in Computing, Communications and Informatics* (ICACCI), pp. 1643–1647. IEEE, 2017.

[11] Alam, B., & Kumar, A. (2014, March). A Real Time Scheduling Algorithm for Tolerating Single Transient Fault. *In 2014 International Conference on Information Systems and Computer Networks (ISCON)* (pp. 11–14). IEEE.

[12] Palvika, Shatakshi, Sharma, Y., Dagur, A., & Chaturvedi, R. (2019). Automated Bug Reporting System with Keyword-driven Framework. In *Soft Computing and Signal Processing: Proceedings of ICSCSP 2018*, Volume 2 (pp. 271–277). Springer Singapore.

[13] Kumar, A., & Alam, B. (2019). Energy Harvesting Earliest Deadline First Scheduling Algorithm for Increasing Lifetime of Real time Systems. *International Journal of Electrical and Computer Engineering*, 9(1), 539.

[14] Kumar, A. & Alam, B. (2018). Task Scheduling in Real Time Systems with Energy Harvesting and Energy Minimization. *Journal of Computer Science*, 14(8), 1126–1133.Murkute, Amod, and Tanuja Sarode. "Forecasting Market Price of Stock Using Artificial Neural Network." *International Journal of Computer Applications* 124, no. 12 (2015): 11–15.

[15] Yetis, Yunus, Halid Kaplan, and Mo Jamshidi. "Stock Market Prediction by Using Artificial Neural Network." *In 2014 World Automation Congress (WAC)*, pp. 718–722. IEEE, 2014.

[16] Mei, Jie, Dawei He, Ronald Harley, Thomas Habetler, and Guannan Qu. "A Random Forest Method for Real-time Price Forecasting in New York Electricity Market." *In 2014 IEEE PES General Meeting| Conference & Exposition*, pp. 1–5. IEEE, 2014.

[17] Seber, George AF, and Alan J. Lee. *Linear Regression Analysis*. Vol. 330. John Wiley & Sons, 2003.

[18] Zhang, G. Peter. "Time Series Forecasting Using a Hybrid ARIMA and Neural Network Model." *Neurocomputing* 50 (2003): 159–175.

[19] Liaw, Andy, and Matthew Wiener. "Classification and Regression by Randomforest. *R News* 2 (3): 18–22." URL: http://CRAN.R-project.org/doc/Rnews (2002).

A survey and classification of lung, breast, thyroid, and prostate cancer detection

Dhananjay Kumar Sharma
United University, Prayagraj, Uttar Pradesh, India
United Institute of Technology, Prayagraj, Uttar Pradesh, India

Manoj Kumar Pal
United University, Prayagraj, Uttar Pradesh, India

Ashutosh Kumar Singh & Vijay Kumar Dwivedi
United College of Engineering and Research, Prayagraj, Uttar Pradesh, India

ABSTRACT: Cancer is one of the painful diseases which is the most cause of the death. If we could predict the cancer in the early stage then not only, we can save the human life as well as we can escape a very painful process of a cancer patient. If we could detect cancer in early stage then the diagnosis of the cancer will be easy and there is less probability of any loss of the human life. For detecting the cancer in early stage, we can use various image processing techniques. This paper review various kinds of cancers and the algorithms which are used for recognition of the particular type of cancer. We will focus on machine learning based algorithms and also analyses different processes which need to be followed during the diagnosis of the cancer.

Keywords: Cancer, Lung Cancer, Medical Field, Image Processing, Segmentation

1 INTRODUCTION

AI has current done remarkable work in field of medical research, if we use AI in the field of medical science then the decision making about the disease could be accurate and fast [1]. We can use the machine learning techniques which help the prediction part easy and accurate. The objective of the machine learning algorithm should be classified the patient into high or low risk zone. If the proposed algorithm detects the cancer in the early stage so the life of patient could be saved [2]. Cancer is dangerous disease in now a day. There are different of cancers such as Brain Cancer, Lung Cancer, Breast Cancer, Liver Cancer, Bone Cancer, Bladder Cancer, Prostate Cancer, etc. Lung cancer is one of the major types of cancer which occurs frequently due to smoking and other reasons. It is detected due to abnormal growth of the human cell generally called as nodules. So for detecting the nodules we could use the image processing techniques. We can follow the following steps during the detection of Lung cancer.

As shown in Figure 1, cancer detection consist a lot of step wise processes like data acquisition, image preprocessing (Image Enhancement, Segmentation), Feature extraction (area, perimeter, and eccentricity), classification [3]. Generally, most of the researchers are used the CT images for analyzing purposes.

In all the above steps we can see the image segmentation plays an important role for feature extraction ad classification, so the aim of this paper will be focus on different image segmentation techniques.

DOI: 10.1201/9781003393580-13

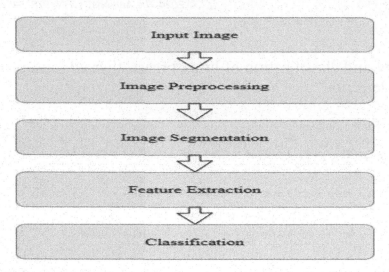

Figure 1.　Cancer detection.

The contribution of the review paper is listed as follows.

1. A detailed analysis of the automated diagnostics and prognostics model by developing machine learning algorithms over digital image, presented different research works with their used algorithms and accuracy level.
2. The existing solutions of cancer prediction are analyses, different kind of cancers, used algorithm and dataset etc.
3. Study of different kind of machine learning algorithms such as KNN, SVM & decision tree etc. also, a detailed analysis of the algorithms that which algorithm is suitable for specific type of cancer.
4. At the last, we represent an open research issue in Cancer prediction raised by different authors and industry working professionals around the world.

Remaining paper is structured as follows. Section 2 presents details analysis of lung cancer, breast cancer, thyroid cancer, and prostate cancer. Section 3 presents the open research challenges and issues. Section 4 concludes the paper.

2　CLASSIFICATION OF LUNG, BREAST, THYROID, AND PROSTATE CANCER DETECTION

In this section, we classify different types of cancer detection. Basically, there are four types of cancer; lung cancer, breast cancer, thyroid cancer, and prostate cancer, will be considered in this paper.

2.1　*Lung cancer detection*

The cancer begins in the lungs called lung cancer. People who smoke have the greatest risk of lung cancer although in some cases lung cancer can also occur in people who have never smoked. Masood *et al.* [1] developed a CAD system which helps in the nodule detection. The performance was less accurate and the system was able to detect the nodule less than 3mm diameter. The proposed model was based on the Convolutional Neural Networks (CNN) algorithm. They used CNN algorithm with some modification named as 3DDCNN. The accuracy of the proposed model was 98.4%. S. Lalitha [2] used the Ensemble classification for detecting the Malignant, she has used 37 benign and 63 malignant images for their research purpose. Suren Makaju *et al.* [3] used watershed segmentation for lung cancer detection and SVM for classification. The proposed models achieve 92% accuracy. Weilun Wang *et al.* [4] proposed a model which is able to detect the nodules as well as analyze the risk features. It was a CNN based model. He achieved 0.86 of AUC and 0.408 of logloss. Danqing *et al.* [5], In his study they proposed a prognostic prediction method for NSCLC patients. Dina M. Ibrahim [6], proposed a model which is able to detect COVID-19, and lung cancer. It was a model which detect COVID-19, pneumonia and lung cancer in a single model. In this model the CT Images and x-ray images are used. Nagar Maleki et al [7] design a kNN algorithm with feature selection genetic algorithm to detect the lung cancer risk. Image segmentation and feature extraction is an important step in the cancer detection.

After the detailed study of different research paper, we can conclude the used ML algorithms with the help of following tabular data. Most of the authors have used the Neural Network (NN) Model with some modifications which are given in Table 1.

Table 1. The key points taken from literature for lung cancer.

Ref.	3DDCNN/CNN	RLR	SVM	Segmentation	RNN	Random Forest	Ensemble	KNN	Accuracy
[1]	✔	✘	✘	✘	✘	✘	✘	✘	87%
[2]	✔	✘	✘	✘	✘	✘	✔	✘	98.7%
[3]	✘	✘	✔	✔	✘	✘	✘	✘	92%
[4]	✘	✘	✘	✘	✔	✘	✘	✘	—
[5]	✘	✔	✔	✘	✘	✔	✘	✘	—
[6]	✔	✘	✘	✘	✘	✘	✘	✘	98.5%–99.66%
[7]	✔	✘	✘	✘	✘	✘	✘	✘	100%
[8]	✘	✘	✘	✘	✘	✔	✘	✔	—

The data repository is an important requirement for any kind of Cancerresearch. Here is the list of databases which are used for different authors during the specific type of Cancer based research.

2.2 *Breast cancer detection*

The abnormal growth in the breast cells is the major cause of the breast cancer. It could be in different parts of a breast Hekal *et al.* [9] designed as CAD-based system which is able to identify the tumor like regions. For identifying the TLRs the author uses the thresholding method. Amkrane *et al.* [10] use the radiomics futures extraction for detecting the tumor.

We have reviews around 5 works of the breast cancer most of the author prefer the CNN algorithm for data classification but some of them also use the SVM and random forest for data classification.

Table 2. The key points taken from literature for breast cancer.

Ref.	3DDCNN/CNN	SVM	Random Forest	Accuracy
[9]	✘	✔	✘	84%–91%
[10]	✔	✘	✘	—
[11]	✔	✘	✘	89%
[12]	✘	✘	✔	84%
[13]	✔	✘	✘	—

2.3 Thyroid cancer detection

The abnormal growth in the thyroid gland could be a thyroid cancer. So, the machine learning approach should be detecting the abnormal growth and predicting the possibility of the cancer. In [14], authors proposed a CAD system which is able to differentiate malignant and benign in the thyroid nodules, It used Multi Input CNN for feature extraction process. System uses the texture pattern on the MRI images. Xhitij *et al.* [15] work on the multiple factors like sensitivity, accuracy and specificity. The accuracy 93.84%, sensitivity 97.82 & specificity was 84.21%.

Table 3. The key points taken from literature for thyroid cancer.

| Ref. | Algorithms | | Accuracy |
	3DDCNN/CNN/KNN	SVM	
[14]	✔	✘	87%
[15]	✘	✔	93.84%

2.4 Prostate cancer detection

The prostate is an organ of the man's body which generates the seminal fluid that carry and nourishes the sperm. Prostate cancer starts when the cells of the prostate gland begin the abnormal growth. Authors [16] claimed that prostate cancer is one of the major causes of mortality in India and there are 375,000 deaths are reported in year 2020. In the current work the author represents a ML-based technique to classify the prostate tissue is whether normal or malignant. H. Wen *et al.* [17] uses the SVM, naïve bayes classifier & neural network for data classification. The author concludes that the SVM models takes a little long time for the training but the model is able to predict the result accurately.

Table 4. The key points taken from literature for prostate cancer.

Ref.	Algorithms NN	Accuracy
[16]	✔	94.24%
[17]	✔	85.64%
[18]	✔	—

3 OPEN RESEARCH CHALLENGES

We have reviewed different works on Lung Cancer, Breast Cancer Thyroid Cancer and Prostate Cancer. In case of lung cancer most of the authors uses the CNN, Random Forest & SVM classifiers. In case of prostate cancer each author of our research uses the neural network for classification. The main challenge in the future research is developing a model which could prognosis the disease with more accuracy, we can also focus on some other criteria than the accuracy theses are sensitivity and specificity rates. The feature extraction is another major challenge in this area, the feature could be area, perimeter & eccentricity, but the main challenge that how we will set the threshold of these features.

4 CONCLUSION

After reviewing all the research work, we are able to understand different types of cancers and their causes. Also, we have learned different kind of machine learning algorithms which could be used for prognosis the particular cancer. We can say the most of the authors have used the

CNN algorithm with some modification, some of them also use SVM and Random Forest algorithms. Most of the authors are able to gain the accuracy percentage around 80% to 90%. In the future research work we will try to develop a system which could provide the accuracy level more than 90%. Also, we will consider the others features like sensitivity & specificity.

REFERENCES

[1]   Masood A. *et al.*, "Cloud-Based Automated Clinical Decision Support System for Detection and Diagnosis of Lung Cancer in Chest CT," *IEEE J TranslEng Health Med*, vol. 8, 2020, doi: 10.1109/JTEHM.2019.2955458.

[2]   Lalitha S., "An Automated Lung Cancer Detection System Based on Machine Learning Algorithm," *Journal of Intelligent and Fuzzy Systems*, vol. 40, no. 4, pp. 6355–6364, 2021, doi: 10.3233/JIFS-189476.

[3]   Makaju S., Prasad P. W. C., Alsadoon A., Singh A. K., and Elchouemi A., "Lung Cancer Detection Using CT Scan Images," in *Procedia Computer Science*, 2018, vol. 125, pp. 107–114. doi: 10.1016/j.procs.2017.12.016.

[4]   Wang W. and Charkborty G., "Automatic Prognosis of Lung Cancer Using Heterogeneous Deep Learning Models for Nodule Detection and Eliciting its Morphological Features," *Applied Intelligence*, vol. 51, no. 4, pp. 2471–2484, Apr. 2021, doi: 10.1007/s10489-020-01990-z.

[5]   Hu D., Li S., Huang Z., Wu N., and Lu X., "Predicting Postoperative Non-small Cell Lung Cancer Prognosis Via Long Short-term Relational Regularization," *ArtifIntell Med*, vol. 107, Jul. 2020, doi: 10.1016/j.artmed.2020.101921.

[6]   Ibrahim D. M., Elshennawy N. M., and Sarhan A. M., "Deep-chest: Multi-classification Deep Learning Model for Diagnosing COVID-19, Pneumonia, and Lung Cancer Chest Diseases," *Comput Biol Med*, vol. 132, May 2021, doi: 10.1016/j.compbiomed.2021.104348.

[7]   Maleki N., Zeinali Y., and Niaki S. T. A., "A k-NN Method for Lung Cancer Prognosis with the Use of a Genetic Algorithm for Feature Selection," *Expert Syst Appl*, vol. 164, Feb. 2021, doi: 10.1016/j.eswa.2020.113981.

[8]   Wang Q., Zhou Y., Ding W., Zhang Z., Muhammad K., and Cao Z., "Random Forest with Self-paced Bootstrap Learning in Lung Cancer Prognosis," *ACM Transactions on Multimedia Computing, Communications and Applications*, vol. 16, no. 1s, Apr. 2020, doi: 10.1145/3345314.

[9]   Hekal A. A., Elnakib A., and Moustafa H. E. D., "Automated Early Breast Cancer Detection and Classification System," *Signal Image Video Process*, vol. 15, no. 7, pp. 1497–1505, Oct. 2021, doi: 10.1007/s11760-021-01882-w.

[10] Amkrane Y., elAdoui M., and Benjelloun M., *"Towards Breast Cancer Response Prediction Using Artificial Intelligence and Radiomics,"* doi: 10.1109/CloudTech49835.2020.9365890.

[11] Li M., "Research on the Detection Method of Breast Cancer Deep Convolutional Neural Network Based on Computer Aid," in *Proceedings of IEEE Asia-Pacific Conference on Image Processing, Electronics and Computers*, IPEC 2021, Apr. 2021, pp. 536–540. doi: 10.1109/IPEC51340.2021.9421338

[12] Reis S. *et al.*, "Automated Classification of Breast Cancer Stroma Maturity from Histological Images," *IEEE Trans Biomed Eng*, vol. 64, no. 10, pp. 2344–2352, Oct. 2017, doi: 10.1109/TBME.2017.2665602.

[13] Wu N. *et al.*, *"Deep Neural Networks Improve Radiologists' Performance in Breast Cancer Screening,"* 2019. [Online]. Available: https://github.com/nyukat/breast_cancer_classifier.

[14] Kumar, A., Mehra, P. S., Gupta, G., & Jamshed, A. (2012). Modified Block Playfair Cipher Using Random Shift Key Generation. *International Journal of Computer Applications*, 58(5).

[15] Kumar, A., Mehra, P. S., Gupta, G., & Sharma, M. (2013). Enhanced Block Playfair Cipher. *In Quality, Reliability, Security and Robustness in Heterogeneous Networks: 9th International Conference, QShine 2013, Greader Noida, India, January 11–12, 2013, Revised Selected Papers 9* (pp. 689–695). Springer Berlin Heidelberg.

[16] Kumar, A., & Alam, B. (2015, February). Improved EDF Algorithm for Fault Tolerance with Energy Minimization. *In 2015 IEEE International Conference on Computational Intelligence & Communication Technology* (pp. 370–374). IEEE.

[17] Alam, B., & Kumar, A. (2014, March). A Real Time Scheduling Algorithm for Tolerating Single Transient Fault. *In 2014 International Conference on Information Systems and Computer Networks (ISCON)* (pp. 11–14). IEEE.

[18] Kumar, A., & Alam, B. (2019). Energy Harvesting Earliest Deadline First Scheduling Algorithm for Increasing Lifetime of Real Time Systems. *International Journal of Electrical and Computer Engineering*, 9(1), 539.

Artificial Intelligence, Blockchain, Computing and Security – Dagur et al. (Eds)
© 2024 The Author(s), ISBN: 978-1-032-49393-0

Modified attention based cryptocurrency price presage with convolutional Bi-LSTM

Vibha Srivastava, Ashutosh Kumar Singh* & Vijay Kumar Dwivedi
Department of Computer Science and Engineering, United College of Engineering and Research, Naini, Prayagraj, Uttar Pradesh, India

ABSTRACT: Predicting stock price is important for identifying the fluctuations in stock market which leads the investors to make profit. As predicting the future price tends to be less accurate in the existing studies, this study considers feature extraction process through the proposed Modified Attention with Convolutional Bi-LSTM for predicting the market value of next day. The stock datasets of Bitcoin and Ethereum are loaded and feature scaling method performs the data pre-processing operations. Further, pre-processed data are considered for feature extraction which is given as input into the classifiers of train and test split method. The comparison of suggested with the prevailing datasets are performed with the estimation of Mean Absolute Error, Mean Squared Error and Root Mean Squared Error and the outcome shows efficacy of suggested mechanism with reference to accuracy.

Keywords: Stock Price Prediction, Convolutional Bi-LSTM, Mean Absolute Error, Mean Squared Error, Root Mean Squared Error

1 INTRODUCTION

Stock market is the financial place in which stocks can be traded, circulated or transferred. It is used as a medium for organizations in raising funds. The basic issue that investors has to face in the stock market is the fluctuating trends in the stock prices. The modifications in national policies, foreign and domestic economic environment and certain international situations are considered to be factors affecting the stock prices (Lu 2021). To perform the divination of stock prices with Feature Selection (FS) and Long Short-Term Memory (LSTM), study has aimed in analysing the impacts of learning strategies in the area of data acquisition (Li 2020).

The study has exploited imaging strategy for encoding process (Barra 2020). Such design in the multi resolution framework has helped in the enhancement of market prediction results. The classification section in the considered study has been carried out by organization of group of Convolutional Neural Network (CNN) which had the similar structural design in which each of the structure has been established with a different essence function for the process during creation. To achieve reliability, the research (Lu 2020) has performed analysis in predicting the correlation and time series of pricing data in stock through utilization of CNN-LSTM for the extraction of time features which performed better forecasting and achieved reliability in utilization of tetralogy in market pricing. The key benefactions of the study are:

- To propose Modified Attention with convolutional Bi LSTM framework for predicting the future stock price by learning the past data of stock prices through effective learning strategy using the stock datasets of Bitcoin and Ethereum.

---

*Corresponding Author: ashuit89@gmail.com

DOI: 10.1201/9781003393580-14

- To focus on the missing data from the input sequence during prediction from the Bi-LSTM model and to minimize the loss, the proposed study uses the Loss Modified Attention Mechanism(AM) for identifying the most pertinent information.
- To carry out performance analysis through comparison of the proposed with existing datasets in terms of accuracy conducive to validate the productivity of proposed methodologies.

The remainder paper is constructed in this way. Section 2 shows the related work which explores various existing works related to this context. Then, subsequently Section 3 of proposed methodology explains the proposed methodology briefly. In addition to that, Section 4 interprets the results acquired from the proposed method. At last, the final section of conclusion in the enhanced study is summarized in Section 5.

2 RELATED WORK

This segment talk about numerous learning and prediction algorithms utilized in the existing methods for stock value prognosis. The financial market is an environment handling numerous transactions within a limited period of time. The three major parties included in stock market business are seller, buyer and exchange. For prediction of future prices, the study (Mootha 2020) has utilized sequence 2 sequence (S2S) modelling which has helped in mapping the input order of sequence to the output order of sequence. The Bi-LSTM method of S2S modelling has predicted the value of the stock as Open, High, Close and Low (OHCL). The study (Cao 2019) has forecasted the financial operations through CNN for estimation of stock index from the previous data. It has constructed CNN based prediction model for the analysis of parameter relationships. Through the research (Chung 2020), the DL structure of CNN for the prediction of fluctuations in stock index with hyper parameter tuning has been performed. The study (Long 2020) has taken steps in fusing both the trading data and market data. The study (Chen 2018) has utilized the DL technique of CNN for the prognostication of retail price movement in China. Study (Sezer 2018) has utilized the trading structure of CNN-Trading Algorithm (CNN-TA) based on the properties of processing images. The study (Kim 2018) has forecasted the instability in stock prices with the hybridisation of LSTM with Generalized Auto Regressive Conditional Heteroscedasticity (GARCH) type frameworks. The considered study (Long 2020) has analysed DL strategy for forecasting the changes in the stock prices everyday.

2.1 Problem identification

The previous investigations inadequate with some restraints in which usual cases are recognized and registered in this section.

- Hybrid prediction model of CNN with SVM has improved feasibility and validity in prediction of stock indexes, but has not considered real time data of stock market (Cao 2019).
- Optimization method with genetic algorithm influences its performance in prediction of stock prices but still remains a margin of improvement. It has utilized only the fixed set of parameters due to power source limitations. It can include more number of parameters for improving the prediction results (Chung 2020).

3 PROPOSED METHODOLOGY

The proposed Modified Attention with Convolutional Bi-LSTM and the enhanced loss attention mechanism performs feature extraction for vaticination of stock prices. The functional framework of proposed methodology is given in Figure 1.

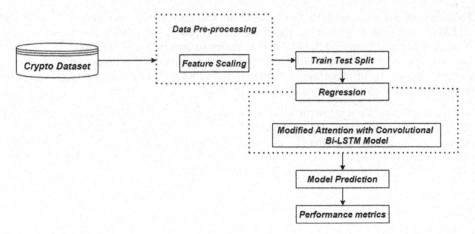

Figure 1.   Functional framework of Modified Attention with Convolutional Bi-LSTM.

Initially, the stock price Bitcoin and Ethereum dataset are loaded separately. Then the starting stage of pre-processing is done for removing unwanted information and for reducing dimensions. Followed by that, the proposed feature extraction method is employed for predicting stock prices. The modified attention mechanism is included for obtaining the missed prediction values from the Bi-LSTM model and ensures efficiency in accuracy. Finally, prediction is done by trained model and the enhanced system is being evaluated with the performance analysis.

3.1   Pre-processing

It is prime data processing stage which makes the data ready for further processing through algorithms. It is involved with some techniques. The proposed study is using two kinds of datasets like Bitcoin and Ethereum. In this study, feature scaling method is used for data normalization in which it normalizes the range of features or independent variables. It is briefly discussed below.

3.1.1   Feature scaling
The process of feature scaling fits the scalar on the training data and then utilize that for transforming the testing data. It helps in avoiding data leakage during the process of testing the model.

3.2   Feature extraction

This method refers to the method of transforming the raw input data into mathematical features which can be processed by preserving the data in the real data sets. It could yield better outcome than applying DL directly to the raw information.

3.2.1   Modified attention with convolutional Bi-LSTM model
CNN is a DL algorithm which takes the input data and process by assigning weights to features in order to make differentiation from the other. The Bi-LSTM framework mostly used in the time series analysis. It is a kind of LSTM model which has the advanced capability of processing in both the forward and backward directions. This characteristic of Bi-LSTM differentiates it from other LSTM models. The attention mechanism used in the study has the potential to add the past time series data to the end result. Additionally, it adjusts the prediction results of Bi-LSTM. Hence the proposed study designed the stock price forecasting model

that uses the combination of methodologies like CNN, Bi-LSTM and modified AM. The algorithm used in proposed study is a kind of feed forward NN through which it achieves effective prediction of time series. The study uses regression method in which it predicts the output based on the input features fed into the model. The feature dimensions can be reduced at the pooling layer.The equation is formulated in Equation (1):

$$Conv_t = tanh(a_t * k_t + b_t) \tag{1}$$

Where $Conv_t$ is the output of convolution layer, tanh is the activation function, a_t is the input vector, k_t is the weight of the convolution kernel and b_t is the bias of the convolution kernel.

The Bi-LSTM model usually operates with three types of gates such as input gate as ig_t, output gate as og_t and forget gate as fg_t as given in Equation (2):

$$fg_t = \sigma\left(Weight_f.[hs_{t-1}, a_t] + b_f\right) \tag{2}$$

The LSTM framework operates with memory cells for remembering the long time historical information which is regulated through gate operation. In every gate, the controlling function of memory cell is functioned by the sigmoid function and multiplication operation. Input data at the present state, output from hidden states of previously existing layers entering through all forms of gates. Particularly, forget gate is responsible in deciding the type of information which can be ignored or can be backed up. Data from the present input gate and data from previous hidden state performs transition through the sigmoid function. The output gate calculates values between 0 and 1.The value 0 indicates that the data can be discarded and if the values are 1, then it refers that the information are stored for analysis.

Where σ is referred as sigmoid activation method, w and b represents the weight along with bias of gate unit, a_t refers to the current present input and previous hidden form of states of hs_{t-1} are inserted into the sigmoid operation. The input gate is formulated as in Equation (3):

$$ig_t = \sigma(Weight_i.[hs_{t-1}, x_t] + b_i) \tag{3}$$

Where the current input is referred as a_t, hidden state referred as hs_{t-1} which are fed into the tanh function.

The cell state is calculated and values are updated to the cell state. It is formulated in the form of Equation (4):

$$\widehat{CS}_t = tanh(W_c.[hs_{t-1}, a_t] + b_c) \tag{4}$$

Where $\widehat{CS}_t$ is referred as the cell state and tanh is considered to be the hyperbolic tangent activation function.

New memory cell is calculated as in Equation (5):

$$CS_t = fg_t \odot CS_{t-1} + i_t \odot \widehat{CS}_t \tag{5}$$

Where CS_t refers to the new memory cell.

LSTM operates in only forward direction and utilizes only the prior information whereas BiLSTM operates with two layers such as forward and backward. It is depicted through Equation (6) & Equation (7):

$$o_t = \sigma(Weight_o.[hs_{t-1}, x_t] + b_o) \tag{6}$$

$$hs_t = o_t \odot tanh(CS_t) \tag{7}$$

As in Equation (8), t refers to the current time, $\overrightarrow{hs_t}$ forward is the forward position vector and $\overleftarrow{hs_t}$ backward determines the backward vector.

$$hs_t = \overrightarrow{hs_t} \oplus \overleftarrow{hs_t} \tag{8}$$

Bi-LSTM achieves better efficiency since it is using both the preceding and subsequent information.

3.2.2 *Modified attention mechanism*
Modified AM is the process of recognizing the particular data from the group of data being collected in the datasets. The proposed method with loss modified attention is shown in Figure 2.

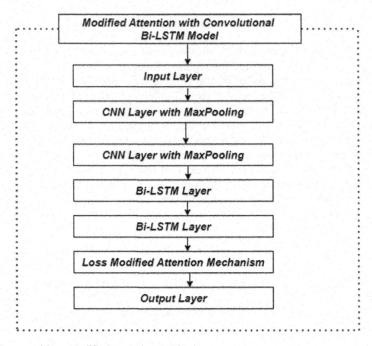

Figure 2.   Proposed loss modified attention method.

The algorithm uses the function attention with determination of bias and weights through which the mechanism predicts the better features with higher magnitude. The weight initialization is represented as Equation (9).

$$as_t = tanh(Weight_h hs_t + b_h) \tag{9}$$

Where as_t is the attention score, hs_t considered as input data vector and $Weight_h$ and the variable b_h being referred as weight and bias of AM respectively.

The softmax function is being utilized for converting the attention as in Equation (10).

$$as_t = \frac{exp(as_t)}{\sum\limits_{t} exp(as_t)} \tag{10}$$

The final attention values can be calculated through weighted summation of values as given in Equation (11).

$$as = \sum_{t} as_t hs_t \qquad (11)$$

The proposed loss function in the study calculates the accuracy and loss of the system which effectively shows that the accuracy is increased with the reduced rate of loss. It is calculated using the functionloss = 'binary_crossentropy'.

The RMSE, MSE and MAE is calculated as in Equation (12), Equation (13) and Equation (14).

$$RMSE = \sqrt{\frac{1}{n} \sum_{i=1}^{n} (y_i - \hat{y}_i)^2} \qquad (12)$$

$$MSE = \frac{1}{n} \sum_{i=1}^{n} (y_i - \hat{y}_i)^2 \qquad (13)$$

$$MAE = \frac{1}{n} \sum_{i=1}^{n} |y_i - \hat{y}_i| \qquad (14)$$

Where n determines portion of predicted variables, $\hat{y}_i$ is represented as the predictive data, y_i is the veritable value.

Finally, stock price prediction performed through the proposed model predicts the changes or variations in the stock price in an effective approach.

4 RESULT AND DISCUSSION

The results attained through the proposed technique implementation is discussed here. Additionally, execution investigation of the enhanced system is also discussed in this section which is performed by comparison of accuracy values of existing dataset with the proposed dataset.

4.1 *Dataset description*

Bitcoin and Ethereum is utilized for analysis in this study. Bitcoin is a historical and real time stock dataset and Ethereum dataset consists of sixteen features which calculates the daily transaction as the sum of entire transaction outcome which belongs to blocks being mined on that particular day. This represents that the individual subject possesses the variations in the stock price. Dataset source Link https://www.kaggle.com/datasets/sudalairajkumar/cryptocurrencypricehistory.

4.2 *Experimental results*

The proposed convolutional Bi-LSTM and Modified attention algorithms are implemented and training cost and loss values are predicted and discussed in this section. Figure 3 shows the cost of training and Figure 4 shows the loss for Bitcoin dataset.

The model seems to be fit and estimated with the training error. From the Figures 3 and 4, it is clearly understood that the model adapts to the training data of Bitcoin dataset with the minimized loss and higher prediction. From Figures 5 and 6, it is clear that the cost of training and testing time is lesser with the higher level of prediction values. The minimized loss values are being calculated which is being depicted through graph. It shows the optimistic estimation of the errors.

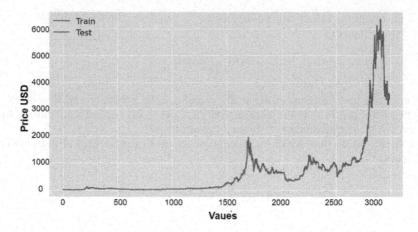

Figure 3.   Bitcoin training cost.

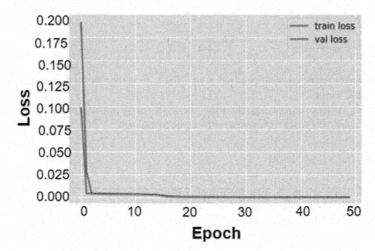

Figure 4.   Bitcoin loss estimation

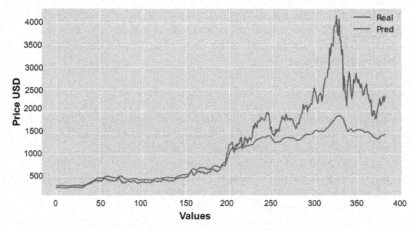

Figure 5.   Ethereum training cost.

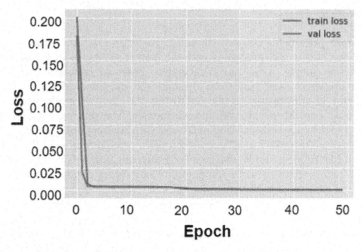

Figure 6.   Ethereum loss estimation.

4.3   *Performance analysis*

The functionality framework of the proposed work is analysed through the performance analysis. The miniature is collate with subsist procedures with respect to prediction accuracy. This helps in determining the range to which the proposed methods are better than the existing work. The results are acquired by the analysis of enhanced system with existing system in respect of MAE, MSE and RMSE. From the Table 1, the accuracy values of both Bitcoin (BTC) and Ethereum (ETH) are compared internally and both proves to be efficient in terms of prediction accuracy. RMSE values being calculated and depicted in Table 2, helps to find out the differences between the predicted values and true values where MSE and MAE values estimate the loss occurred betwixt predicted stock value and actual stock value. The study proved its efficiency of accurate prediction with minimized error values through loss calculation using Convolutional Bi-LSTM with Loss Attention Mechanism.

Table 1.   Prediction analysis of the proposed two datasets (Hamayel 2021).

Dataset	Existing Model			Proposed Model
	LSTM	GRU	Bi-LSTM	
BTC (RMSE)	410.399	174.129	2927.006	0.0139070918
ETH (RMSE)	59.507	26.59	321.061	0.0139070918

Table 2.   Prediction analysis of the proposed work along with the existing dataset (Suryawanshi 2022).

Cryptocurrency	MAE	MSE	RMSE
Existing Bitcoin	0.0427	0.0036	0.06
Proposed Bitcoin	0.0107133233	0.00019340720	0.0139070918
Proposed Bitcoin	0.0431	0.00035	0.059
Existing Ethereum	0.009825593	0.0001543512	0.0124238147

5 CONCLUSION

In this study, stock price prediction was computed based on DL methodologies. This kind of prediction was executed by pre-processing with the feature scaling method followed by regression and prediction model. This study deliberately provided enhancements in the feature extraction method which was performed through the proposed Modified Attention with Convolutional Bi-LSTM framework. Followed by this, the attention mechanism was performed using the proposed Loss Modified Attention mechanism. The proposed system was analysed through the cryptocurrency dataset being measured with training error, testing error, training loss and testing loss. This exhibited efficient stock price prediction proved through the experimental results of MAE, MSE and RMSE value of Ethereum as 0.009825593, 0.0001543512 and 0.0124238147 which is relatively lesser compared to the existing dataset error values. The proposed system was analysed for evaluating its efficiency through comparison with the existing datasets. The accurate prediction analysis with minimum error rate was confirmed that the system has accuracy. The analytical results have shown that the proposed attention method minimizes the error by measuring metrics such as MAE, MSE and RMSE. Hence the proposed study which is effective than the existing systems can be practically utilized by the economist specialists in predicting the stock prices. Better analysis and prediction through proposed work minimizes the losses faced by traders in stock market. Hence, this study considerably helps the investors to successfully predict the future price of stock yielding significant profit.

REFERENCES

Barra, Silvio and Carta, Salvatore Mario and Corriga, Andrea and Podda, Alessandro Sebastian and Recupero, Diego Reforgiato, 2020. Deep Learning and Time Series-to-image Encoding for Financial Forecasting. *IEEE/CAA Journal of Automatica Sinic (3)*, volume (7): pages 683–692.

Cao, Jiasheng and Wang, Jinghan 2019. Stock Price Forecasting Model Based on Modified Convolution Neural Network and Financial Time Series Analysis. *International Journal of Communication Systems (12)*, volume (32): pages e3987.

Chen, Sheng and He, Hongxiang 2018. Stock Prediction Using Convolutional Neural Network. *IOP Conference Series: Materials Science and Engineering* (1), volume (435): pages 012026.

Chung, Hyejung and Shin, Kyung-shik 2020. Genetic Algorithm-optimized Multi-channel Convolutional Neural Network for Stock Market Prediction. *Neural Computing and Applications (12)*, volume (32): pages 7897–7914.

Hamayel, Mohammad J and Owda, Amani Yousef 2021. A Novel Cryptocurrency Price Prediction Model Using GRU, LSTM and bi-LSTM Machine Learning Algorithms. AI (4), volume (2): pages 477–496.

Kim, Ha Young and Won, Chang Hyun 2018. Forecasting the Volatility of Stock Price Index: A Hybrid Model Integrating LSTM with Multiple GARCH-type Models.

Li, Hui and Hua, Jinjin and Li, Jinqiu and Li, Geng 2020. Stock Forecasting Model FS-LSTM Based on the 5G Internet of Things. *Wireless Communications and Mobile Computing, volume* (2020). Expert Systems with Applications, volume (103): pages 25–37.

Long, Jiawei and Chen, Zhaopeng and He, Weibing and Wu, Taiyu and Ren, J iangtao 2020. An Integrated Framework of Deep Learning and Knowledge Graph for Prediction of Stock Price trend: An Application in Chinese stock Exchange Market. *Applied Soft Computing*, volume (91): pages 106205.

Lu, Wenjie and Li, Jiazheng and Li, Yifan and Sun, Aijun and Wang, Jingyang 2020. A CNN-LSTM-based Model to Forecast Stock Prices. *Complexity*, volume (2020).

Lu, Wenjie and Li, Jiazheng and Wang, Jingyang and Qin, Lele 2021. A CNN-BiLSTM-AM Method for Stock Price Prediction. *Neural Computing and Applications (10)*, volume (33): pages 4741–4753.

Mootha, Siddartha and Sridhar, Sashank and Seetharaman, Rahul and Chitrakala, S 2020. Stock Price Prediction Using Bi-directional LSTM Based Sequence to Sequence Modeling and Multitask Learning. *2020 11th IEEE Annual Ubiquitous Computing, Electronics & Mobile Communication Conference (UEMCON)*, volume (2020): pages 0078–0086.

Sezer, Omer Berat and Ozbayoglu, Ahmet Murat 2018. Algorithmic Financial Trading with Deep Convolutional Neural Networks: Time Series to Image Conversion Approach. *Applied Soft Computing*, volume (70): pages 525–538.

Suryawanshi, Ravikant and Takavane, Tanvi and Shah, Trisha and Mathur, Rishima 2022. *Stocks and Cryptocurrency Price Prediction Using Long Short Term Memory*.

Artificial Intelligence, Blockchain, Computing and Security – Dagur et al. (Eds)
© 2024 The Author(s), ISBN: 978-1-032-49393-0

Classification of vegetation, soil and water bodies of Telangana region using spectral indices

Devulapalli Sudheer, S. Nagini, Naga Sreenija Meka, Yasaswini Kolli, Anudeep Eloori, Nithish Kumar Chowdam & Rushikesh Reddy Dorolla
Vallurupalli Nageswara Rao Vignana Jyothi Institute of Engineering and Technology, Hyderabad, India

ABSTRACT: Scientists have developed vegetation indices for qualitative and quantitative evaluation of vegetative cover using spectral data in remote sensing applications. A complex blend of vegetation, soil brightness, environmental impacts, shadow, soil color, and moisture can be visible in the spectral response of vegetated areas. Additionally, the spatial-temporal fluctuations of the atmosphere impact the spectral indices. In the last 20 years, more than 40 indices have been made to improve categorization response and lessen the effects of the things listed above. Vegetation indices are numerical measurements that show how vigorous the vegetation is. They have greater sensitivity for the identification of biomass than individual spectral bands. These indices are interesting because they can be used to evaluate remote-sensing images. In particular, they help detect land use changes (temporal data), assess vegetative cover density, tell the difference between different crop types, and predict agricultural yields. Most of these indices are interested in improving classifications in the domain of thematic mapping. This project will list and describe most of the chosen regions' green vegetation, soil types, and water bodies. The proposed method has compared how they have changed over time. It will also use spectral indices to sort these areas into groups.

Keywords: Spectral Indices, Multispectral Data, NDVI, NDWI, NDBI

1 INTRODUCTION

The process of identifying and examining particular regions' visible traits with the help of radiations at a distance comprising reflections and emissions is called remote sensing. The attainment of information without requiring direct contact corresponds to the sensing phenomenon. Due to the increase of technology in remote sensing, vast amounts of data are available, which helps to gain more inputs about various sectors. This project focuses on the agricultural industry to predict vegetation, water, and soil indices using the available multispectral data. This multispectral data is widely used in seed quality assessment, military target detection, and many more.

Two or more spectral bands combined the values of pixels in a multispectral image determine colour indices. Spectral indices like Normalized Difference Water Index (NDWI), Normalized Difference Vegetation Index (NDVI), and Normalized difference Built index (NDBI) are concentrated in this project. As a result of growth in urbanization of the Telangana region, there is a drastic change in its land cover. Hyderabad is a well-established metropolitan city with an increase in the average annual rate of the urban extent by 7.1% since 1990–1999, where the built-up areas initially were around 21,668 hectares. The average yearly rate comprised 3.7% from 1999 to 2014, corresponding to 72,998 hectares by 2014 8.

1.1 Multi spectral data

A multispectral image is made by putting together several pictures of the same scene that were taken at different wavelength bands, usually by satellites. A multispectral sensor records blue, green, and red wavelengths. A sensor's spectral sensitivity is the range of wavelengths it can pick up. This multispectral data comes from many satellite images, like LISS3, Sentinel, Landsat, Ikonos, and others.

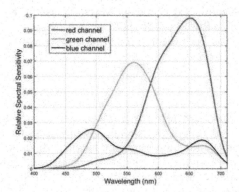

Figure 1.   Graph representing wavelengths with respect to relative spectral sensitivity.

1.2 Spectral indices

The many spectral bands in an image are put through a mathematical formula called a "spectral index." Spatial resolution is the distance that a single pixel in a picture shows. For example, a Sentinel-2 image covers an area of 100 square kilometers, which is about 10,000 by 10,000 pixels for the bands and has a spatial resolution of 10 meters. At the same time, the LISS-3 sensor covers an orbital swath of 140 km with a 24-meter resolution every 24 days. So, n numbers, where n is the number of spectral bands, describe each pixel. These values were used to figure out a spectral index using mathematical formulations. Practically, it is the disparity between two chosen bands standardized by their sum. This kind of calculation is highly beneficial to improve spectral aspects that are not initially visible and decrease the impacts of illumination, such as shadows in cloudy areas and mountain regions.

1.3 Related works

The below literature review is a comprehensive summary of previous research on the topic.

Garcia Pedrero (2017) applied agglomerative segmentation approaches in the field of agriculture. Spectral indices and texture features were applied on WV 2 data 1.

Nobuyuki Kobayashi et al. (2019) classified crops using spectral indices for multispectral images. Hierarchical clustering techniques have been applied to spectral indices identified from Sentinel 2A dataset. Sentinel-2A, a multispectral data with 13 bands comprising the from visible to short-wave infrared (SWIR) ranges, provides a variety of vegetation indicators. This work generated 91 published spectral indices using the MSI data. In the progression, grouping pixels' accuracy is less compared to supervised models 2.

Nagaraj R & Lakshmi Suthakumar (2021) evaluated the performance of machine learning techniques for various water bodies. Random forest, SVM, SG boost methods have been applied on LISS 3 dataset. It is inferred that indirect learning techniques are exhibiting less accuracy than direct learning techniques 3.

Saptarshi Mondal et al. (2014) have extracted seasonal cropping patterns using multi temporal vegetation indices. LISS 3 dataset (Bihar region) were used and spectral indices such as NDSBVI, EVI2 were extensively focused. They obtained over fourty samples during each of the seasonal VI, threshold reference values for theme classes related to vegetation and non-vegetation were retrieved. Three seasonal VI stack photos were classified using a decision tree built using these estimated value ranges, revealing seven distinct plantation and cropping patterns. To verify the correctness of the semi-automatically extracted VI based categorized image, a digitalized reference map was also created using multiple seasons of LISS-III photos. In the progression high computational complexity has been obtained 4.

Bannari *et al.* (1995) authors have compared the spectral indices over temporal data and applied edge detection techniques for finding the growth in vegetation, soil and atmospheric changes. RVI, TVI spectral indices were extensively focused on the LISS-3 based dataset.Themajority of the vegetative indicators available are summarized, cited, and discussed in this study. It presents various existing index classes and suggests grouping them under a new classification 5.

Jinru Xue *et al.* (2017) executed classification of terrestrial vegetation and gained insights with regards to applications in environmental monitoring, biodiversity conservation and agriculture. Landsat 7-8 datasets were utilized and methodology to obtain NDVI, PVI, ARVI spectral indices have been adopted. More than 100 VIs reviewed in this study and discusses their unique features in relation to the target vegetation 6.

Javid Malik *et al.* (2021) performed classification of forest covers, crop lands and urban growth using edge detection techniques. Landsat 8 for Bangladesh region dataset has been utilized and SVM, KNN, random forest methodologies have been adopted. The Landsat 8 (bands 5, 4, and 3) were fused with the panchromatic band using the image fusion technique (band 8). In the progression spectral distortion has occurred due to image fusion 7.

2 METHODOLOGY

The project focuses on classification of land covers using spectral indices.Initially satellite images are inputted and subjected to calculation of spectral indices such as NDVI, NDBI and NDWI. These calculated spectral indices correspond towards labeling the data and subsequent identification of diverse land covers.The methodological application developed in this research is illustrated in the below process flow.

2.1 Data collection

We obtained the multispectral data from the images captured by the LISS3 satellite from the NRSC Bhuvan website 11. The grid corresponding to the longitudes and latitudes of the Telangana area was extracted from the complete dataset.

LISS III(Linear Imaging Self-Scanning Sensor 3) is a satellite based camera in the field of remote sensing used for obtaining High-resolution land and vegetation observation developed by ISRO(Indian Space Research Organization) Multi-spectral data is provided in the form of 4 bands by LISS-III sensor. Visible (2 bands) and Near-infrared (1 band) comprises of spatial resolution of 23.5 meters accounting to 141 km of ground swath. The fourth band is the short-wave infrared band with 70.5 meters of spatial resolution encompassing 148 km of ground swath. LISS-III has a periodicity of twenty four days 12.

2.2 Spectral indices calculation

Spectral indices are determined via combination of multiple spectral bands based on their pixel values in a multispectral image. Further we would like to discuss about these spectral indices in detail. The vegetation index can be calculated using different formulae as shown below.

However, in this project we made use of NDVI. By determining the difference between near- infrared and red light, the NDVI estimates the vegetation values. This explains why vegetation seems green to our sight. We observe NDVI in many different industries. For instance, in forestry, foresters use NDVI to calculate leaf area index and to estimate biomass while farmers utilise it for precision farming in agriculture. Additionally, NASA claims that

NDVI is a reliable drought indicator. When water restricts vegetation development, the relative NDVI and vegetation density are lower. NDVI and other remote sensing technologies are being used in hundreds of different applications in the real world.

The NDBI value ranges between -1 to $+1$. Positive value of NDBI represents build-up areas where as lower value represents water bodies.

Input: Multispectral image bands b_1, b_2, b_3)
Output: Color coded RGB image
img=read_image('image file path')
imgSize = (size(img,1)*size(img,2));
C1=C2=C3=C4=C5=0;
thresh = int32((coherentPrec/100) *imgSize);
$i \leftarrow 0$
while $i \le$ *rows* **do**
    $j \leftarrow 0$
    while $j \le$ *columns* **do**
        if *NDVI* > 0.2 *and* *NDVI* $<= 0.5$ **then**
            $f(i,j) \leftarrow [R : 102, G : 255, B : 102];$
            $C1 \leftarrow c1 + 1$
        else if *NDVI* > 0.5 *and* *NDVI* $<= 0.9$ **then**
            $f(i,j) \leftarrow [R : 255, G : 0, B : 0];$
            $C2 \leftarrow c2 + 1$
        end
        else if *NDBI* > 0.5 **then**
            $f(i,j) \leftarrow [R : 255, G : 0, B : 255];$
            $C3 \leftarrow c3 + 1$
        end
        else if *NDSI* > 0.4 **then**
            $f(i,j) \leftarrow [R : 255, G : 255, B : 204];$
            $C4 \leftarrow c4 + 1$
        end
        else if *NDWI* > 0.5 **then**
            $f(i,j) \leftarrow [R : 102, G : 102, B : 255];$
            $C5 \leftarrow c5 + 1$
        end
    end
end

Table 1. LISS-III band related information.

Band	Wavelength
Green	$0.52\mu - 0.59\mu$
Red	$0.62\mu - 0.68\mu$
NIR	$0.77\mu - 0.86\mu$
SWIR	$0.55\mu - 1.70\mu$

Table 2. Formulae for vegetation index calculation.

Index	Formula
RVI	$\frac{NIR}{Red}$
NDVI $\frac{NIR-Red}{NIR+Red}$	
IPVI	$\frac{NIR}{NIR+Red}$
DVI	NIR-Red
TVI	$(NDVI + 0.5)^{0.5}$

Table 3. NDVI ranges.

Range	Description
−1.0 – 0	Clouds, water and snow
equal to 0	Rocks, Bare Soil
0 – 0.1	sand and empty areas of rocks
0.2 – 0.3	Shrubs, Meadows
0.6 – 0.8	Temperate and Tropical Forest

Table 4. Formulae for built-up index calculation.

Index	Formula
NDBI 9	$\frac{SWIR-NIR}{SWIR+NIR}$
BUI	$NDBI-NDVI$
BAEI10	$\frac{Red+0.3}{Green+SWIR}$

Table 5. Evaluation metrics (mean of all classes).

	2015			2017			2019		
Dataset	Precision	Recall	F1 Score	Precision	Recall	F1 Score	Precision	Recall	F1 Score
SVM-Poly	0.72	0.72	0.72	0.72	0.72	0.72	0.72	0.72	0.72
SVM-RBF	0.83	0.83	0.83	0.54	0.56	0.55	0.54	0.18	0.27
SVM-Sigmoid	0.79	0.88	0.83	0.81	0.81	0.81	0.84	0.84	0.84
Random Forest	0.952	0.963	0.957	0.88	0.88	0.88	0.95	0.95	0.95
Adaboost	0.947	0.952	0.952	0.88	0.88	0.88	0.95	0.95	0.95
Decision-Tree	0.948	0.951	0.946	0.88	0.88	0.88	0.95	0.95	0.95

3 RESULTS AND DISCUSSION

The system is able to classify the given satellite image into different regions i.e.vegetation, water bodies, soil, urbanization by labelling them based on one color for each index. The output is generated for three years 2015, 2017 and 2019 as shown below.

Based on the above classification, the different regions and there corresponding areal change in square meters over the years has been depicted in the below graphs. Based on the results of the classified data the precision, recall and F1 score of different spectral indices in different years is as follows.

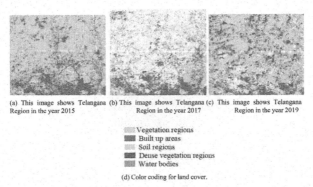

(a) This image shows Telangana Region in the year 2015 (b) This image shows Telangana Region in the year 2017 (c) This image shows Telangana Region in the year 2019

Vegetation regions
Built up areas
Soil regions
Dense vegetation regions
Water bodies

(d) Color coding for land cover.

Figure 2. Land cover classification results.

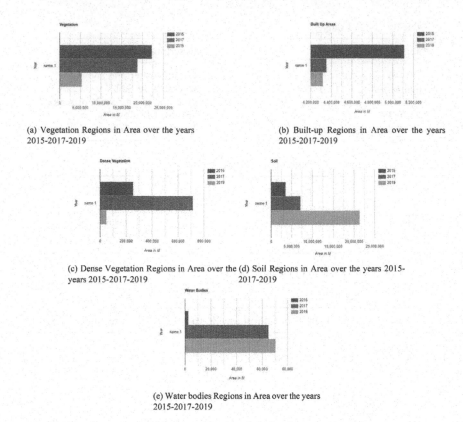

(a) Vegetation Regions in Area over the years 2015-2017-2019

(b) Built-up Regions in Area over the years 2015-2017-2019

(c) Dense Vegetation Regions in Area over the years 2015-2017-2019

(d) Soil Regions in Area over the years 2015-2017-2019

(e) Water bodies Regions in Area over the years 2015-2017-2019

Figure 3.   Bar Graph representation of Landcover classification.

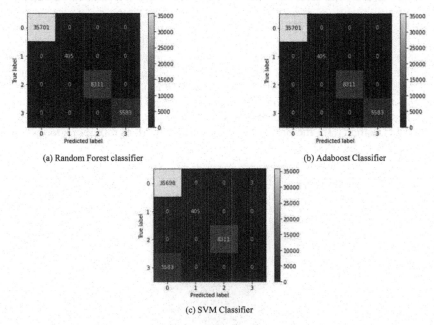

(a) Random Forest classifier

(b) Adaboost Classifier

(c) SVM Classifier

Figure 4.   Confusion matrices of landcover classification.

3.1 Performance metrics

The performance of the various machine learning algorithms are compared with precision, recall and F1 score measures 13. In all the three years of 2015,2017, and 2019 high precision achieved for vegitation and soil indices.

4 CONCLUSION

The initial datasets gathered for this research are labelled to provide compelling evidence of changes in the areas of the land covers throughout time. It is focused on applying project findings in a variety of areas, such as forecasting future vegetation, soil, and water bodies if the pace of change stays the same. The supervised machine learning algorithms are implimented on labelled LISS-III data and evaluated empirically. The Random Forest classifier has achieved highest accuracy for the considered data in this methodology.

REFERENCES

Bannari, D. Morin, F. Bonn & A. R. Huete (1995) A Review of Vegetation Indices, *Remote Sensing Reviews*, 13:1–2, 95–120. DOI: 10.1080/02757259509532298

Deglint, J., Kazemzadeh, F., Cho, D., Clausi, D. A., & Wong, A. (2016). Numerical Demultiplexing of Color image Sensor Measurements Via Non-linear Random Forest Modeling. *Scientific reports*, 6(1), 1–9.

Devulapalli, S., & Krishnan, R. (2021). Remote Sensing Image Retrieval by Integrating Automated Deep Feature Extraction and Handcrafted Features Using Curvelet Transform. *Journal of Applied Remote Sensing*, 15(1), 016504.

García-Pedrero, C. Gonzalo-Martín & M. Lillo-Saavedra (2017) A Machine Learning Approach for Agricultural Parcel Delineation Through Agglomerative Segmentation, *International Journal of Remote Sensing*, 38:7, 1809–1819. DOI: 10.1080/01431161.2016.1278312

Javed Mallick, Swapan Talukdar, Shahfahad, Swades Pal, Atiqur Rahman, A Novel Classifier for Improving wetland Mapping by Integrating Image Fusion Techniques and Ensemble Machine Learning Classifiers, *Ecological Informatics*, Volume 65,2021,101426, ISSN 1574–9541. https://doi.org/10.1016/j.ecoinf.2021.101426.

Jinru Xue, Baofeng Su, "Significant Remote Sensing Vegetation Indices: A Review of Developments and Applications", *Journal of Sensors*, vol. 2017, Article ID 1353691, 17 pages, 2017. https://doi.org/10.1155/2017/1353691

Karanam, H. K., & Neela, V. B. (2017). Study of Normalized Difference Built-up (NDBI) Index in Automatically Mapping Urban Areas From Landsat TN imagery. *Int J Eng Sci Math*, 8, 239–48.

N. R and L. S. Kumar, "Performance Analysis of Machine Learning Techniques for Water body Extraction," 2021 *IEEE Bombay Section Signature Conference (IBSSC)*, 2021, pp. 1–6. doi: 10.1109/IBSSC53889.2021.9673372.

Nobuyuki Kobayashi, Hiroshi Tani, Xiufeng Wang & Rei Sonobe (2020) Crop Classification Using Spectral indices Derived from Sentinel-2A Imagery, *Journal of Information and Telecommunication*, 4:1, 67–90. DOI: 10.1080/24751839.2019.1694765

Pandey, D., & Tiwari, K. C. (2020). Extraction of Urban Built-up Surfaces and its Subclasses Using Existing Built-up indices with Separability Analysis of Spectrally Mixed Classes in AVIRIS-NG Imagery. *Advances in Space Research*, 66(8), 1829–1845.

Saptarshi Mondal, C. Jeganathan, Nitish Kumar Sinha, Harshit Rajan, Tanmoy Roy, Praveen Kumar, Extracting Seasonal Cropping Patterns Using Multi-temporal Vegetation Indices from IRS LISS-III data in Muzaffarpur District of Bihar, India, *The Egyptian Journal of Remote Sensing and Space Science*, Volume 17, Issue 2, 2014, Pages 123–134, ISSN 1110–9823, https://doi.org/10.1016/j.ejrs.2014.09.002.

Sudheer, D., & Krishnan, R. (2019). Multiscale Texture Analysis and Color Coherence Vector Based Feature Descriptor for Multispectral Image Retrieval. *Advances in Science, Technology and Engineering Systems Journal*, 4(6), 270–279.

Sudheer, D., SethuMadhavi, R., & Balakrishnan, P. (2019). Edge and Texture Feature Extraction Using Canny and Haralick Textures on SPARK Cluster. In *Proceedings of the 2nd International Conference on Data Engineering and Communication Technology* (pp. 553–561). Springer, Singapore.

Artificial Intelligence, Blockchain, Computing and Security – Dagur et al. (Eds)
© 2024 The Author(s), ISBN: 978-1-032-49393-0

Ovarian cancer identification using transfer learning

Rishabh Dhenkawat, Samridhi Singh & Nagendra Pratap Singh
National Institute of Technology, Hamirpur

ABSTRACT: Ovarian cancer as it is more often called, is women's fifth largest cause of cancer-related mortality. As of late, deep learning has been proven superior to traditional approaches for predicting OC stages and subtypes. Despite this, most state-of-the-art deep learning models only use data from a single modality, which might lead to subpar performance due to an inadequate portrayal of crucial OC features. In addition, the quantity of computational resources needed for training and deploying these deep learning models is greatly increased due to the absence of optimization of the model architecture. In this research, we create a hybrid evolutionary deep learning model with seven unique architectures and evaluate their performance. In the end, a multi-model ensemble architecture is introduced. The deep feature extraction network that we built up independently on vgg16, vgg19, resnet, and ensembled of all of these architectures served as the basis for generating each modality's different states and forms. After analysis and comparison, the final ensemble output is produced. Generalization Stacking As a meta learner, multi-layer perceptron is employed due to its rapid training time and superior accuracy relative to alternative methods.

Keywords: Deep Learning, Ovarian Cancer, Histopathology Images, Medical Imaging, Multi-Modal Deep Learning, Evolutionary Deep Models, ovarian cancer

1 INTRODUCTION

This highlights the critical necessity for developing computational algorithms that accurately anticipate OC. Several OC diagnostic models developed in the last decade [4–6] have used single-modal histopathology images as their basis because they may depict the morphological aspects of the cells that are highly associated to the aggressiveness of OC. With this end goal in mind, several models were developed. Besides histological evidence, gene expression levels and genetic anomalies have been shown to speed up cell division and alter the tumor's microenvironment, contributing to cancer's progression [7]. This means that the genetic markers can be used as reliable diagnostic indicators. Among these genomic features are markers of gene expression, mutations in DNA sequence, local variations in DNA copy number, and alterations in DNA methylation [8]. Due to the inherent uncertainty and complexity of cancer survival prediction, whole-slide histopathology scanners, high-throughput omics profiling [6], and ground-breaking ML algorithms have shown promising results. Sun *et al.* [9] used imaging and genetic data from pathology to enhance breast cancer prognostication. The disparate data from the two channels were combined using multiple kernel learning. This method was found to be 0.8022 accurate with a precision of 0.7273. For cancer detection, [10] suggests using M2DP, a feature selection technique that considers many tasks simultaneously. [10] Initially, traits were gleaned from pathology pictures and gene expression data for this effort. The M2DP model was then used to zero in on diagnostic factors. The diagnosis for each patient was made using AdaBoosting, with the final result depending on the inputted features. The model's accuracy was determined to be 72.53% against a breast cancer benchmark and 70.08% against a lung cancer benchmark.

DOI: 10.1201/9781003393580-16

For lung carcinomas, Zhang *et al.* [11] combined genomic data with image-based patholo-gical features to describe a multiple kernel approach for prediction. This approach was created to predict lung cancer cases with an accuracy of 0.8022.

1.1 Data augmentation

A deep neural network model would often demand a substantial quantity of data for training [14].

Inadequate sample sizes during training may directly contribute to overfitting and other errors. For our research, we enlarged the sample size by using picture alteration to enhance classification accuracy [15]. Image enhancement and rotation are both included in the category of image manipulation. To increase the picture's clarity and edge sharpness, a Gaussian High Pass filter with a kernel size of 3*3 was applied to the image, and then a Laplass filter was placed on top of that. During the picture capture process using the microscope and camera, the orientation of tissue slices stained with H&E remained unchanged. As a result, the initial pictures that were 227 by 227 pixels were rotated in 90-degree increments from 0 to 270 degrees around their centers to expand the sample sizes. The steps involved in picture rotation and augmentation are shown in Figure 1.

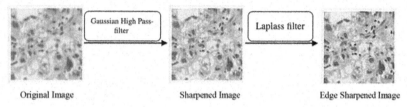

Original Image   Sharpened Image   Edge Sharpened Image

Figure 1.   Image enhancement.

1.2 Proposed network architecture

Only a few published studies have examined how the ensemble deep learning technique may be applied to histopathology images. Various modified CNN architectures are incorporated into the proposed model to ease the extraction of features and the aggregation of vital information from each input image. Each of these patterns is comprised of a distinctive combination of convolution layers. Figure 5 displays a block diagram of the intended study methodology for this project. As shown in the diagram, the methodology comprises six steps: collecting microscopic ovarian cancer histology images; data pre-processing; data augmen-tation; feature extraction using the proposed network; classification; and model evaluation, which occurs at the end of the process. Using a range of preprocessing approaches, we first improved the quality of the visual data included inside each input image. The amount of the training dataset is then increased using several data augmentation techniques. After the input images have been created, they are passed to the feature extraction phase using the proposed ensemble architecture. The retrieved features from each architecture are merged and flat-tened when generating the ultimate multi-view feature vector. The produced feature vector is fed to a multilayer perceptron, which utilizes this information to classify each image into the relevant category. Finally, the effectiveness of the proposed strategy is evaluated by applying the trained model to test images and evaluating the outcomes.

1.3 Feature extraction using transfer learning

Instead of starting from scratch and giving random weights when training a model, researchers have been looking into the possibility of using a technique called "transfer learning" to use the information that can be learned from studying information in other domains. In this strategy, we use what we learned from one data set on a new data set from a different field. Using a method called transfer learning, the model can learn general traits from a source dataset called a transfer dataset that don't exist in the current dataset. Transfer learning has several benefits, such as speeding up the convergence of the network, reducing the amount of processing resources needed, and making the most of the network's performance [16].

In this research, 3 CNNs– VGG16, VGG19, and ResNet50 are chosen to determine type of ovarian cancer. These traditional networks were chosen because they have been evaluated on various classification tasks and have demonstrated good accuracy and consistency across a wide range of datasets. In addition, the benefits and drawbacks of these networks have been methodically evaluated; as a result, it is possible to construct a comprehensive ensemble network based on the complementary of their components.

In this study, three CNNs were used: VGG16, VGG19, and ResNet50 are chosen to figure out what kind of ovarian cancer a person has. These traditional networks were chosen because they have been tested on a wide range of classification tasks and shown to be accurate and consistent over many datasets. Also, the pros and cons of these networks have been evaluated systematically. As a result, it is possible to build a complete ensemble network based on how their parts work together.

In conclusion, our research utilizes three CNN models, including VGG16, VGG19, and ResNet50.

First, labelled images from the vast ImageNet dataset are used individually for transfer learning based on these three CNNs. This makes sure that the networks are good at putting things into groups. After that, the front ends of the trained model parameters in each of these three CNNs are frozen to keep any information they may have for any future training that may be done. Then, the augmented ovarian histopathology images and their labels are used to fine-tune the layers at the back ends of the three networks, which are all connected. We use these layers to look at the ovarian tissue. Lastly, the ability to classify images learned from ImageNet can be used to classify the problematic slices, and Weighted Average Multi-Layer Perceptron Ensemble is used to make more predictions.

Getting the feature extraction right is very important because it affects how well the classifier works. This is because histopathology pictures are hard to understand just by looking at them. But worries about patients' right to privacy in the medical field [17] mean that the datasets given aren't enough to train a CNN well [18].

1.3.1 VGG16

[19] introduced VGG16 to the research community as a more advanced convolutional neural network model. The fundamental concept behind this model is to swap out the large kernels for more numerous smaller kernels while simultaneously increasing the depth of the CNN model [20]. As a result, the VGG16 has a greater potential for reliability in its performance of various classification tasks. The fundamental architecture of the VGG16 [19] is seen in Figure 2. It comprises five blocks and has a total of 41 layers, with 16 of those levels having learnable weights. Out of the learnable layers, there are 13 convolutional layers and three

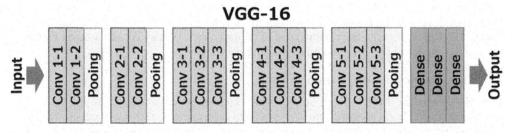

Figure 2.   VGG16 Model [22].

FCC layers [21]. The first two blocks each have two convolutional layers, and the third, fourth, and fifth blocks each have three convolutional layers. Small kernels with dimensions of 3 by 3 and a padding of 1 are used in the convolutional layers. The max-pooling layers have a filter size of 2 by 2 and padding of 1, are used to divide these convolutional layers. The number 4096 is the output of the final convolutional layer, which results in the number 4096 is the number of neurons in the FCC. VGG16 has over 134 million different parameters, which significantly increases its level of complexity in comparison to other pretrained models [21].

1.3.2 *VGG19*

A 19-layer implementation of VGGNet was utilized to build this network model. Figure 3 depicts the architecture of the VGG19 model, which comprises of sixteen convolution layers followed by three fully linked layers. Each convolution layer's output is supplied to a nonlinear ReLU the activation function uses. Five max-pooling layers are piled atop one another, and these layers split the convolution portion into five subsections. The first and second sub-regions consist of two convolution layers, with the first layer's depth size 64 and the second layer's depth size 128. The convolution layer depths in the remaining three sub-regions are 256, 512, and 512. The convolution layers follow one another in succession. Following the sub-regions of the convolutional layers are the pooling layers, which lower the number of learnable parameters. The feature vector was extracted from the very last layer of the VGG19 model. Before the output feature collecting layer are two hidden layers with 512 and, respectively, 1024 neurons. L2 regularisation was applied after each fully connected layer in addition to the dropout layer to mitigate the overfitting that happened during the development of this highly calibrated model.

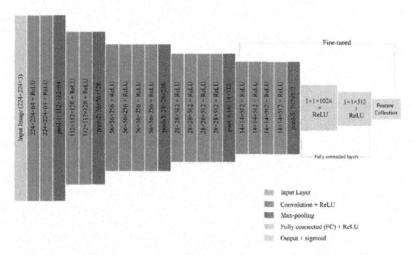

Figure 3. VGG19 Model [23].

1.3.3 *ResNet*

In deep CNN models, classification effectiveness keeps increasing proportionately with the number of network layers. This is the case in most cases. On the other hand, this results in a subsequent increase in the mistake rate during training and testing. The term "exploding gradient" or "vanishing gradient" describes this phenomenon. In addition, this problem can be fixed by utilizing the Residual Network, often known as ResNet [17–19]. These networks make use of a strategy that is known as skipping connections. In addition, the training is skipped after a few layers by the network, and it instead links straight to the output. The fundamental architecture of ResNet was conceptualized after the VGG network, which employs 3 x 3 filtering in its convolutional layers. Two different ideas are used in the architecture to achieve model optimization. The layers each include the same number of filters, which results in the same kinds of feature maps being generated. In addition, when the size of the output feature map is cut in half, the number of filters must be increased by a factor of two to maintain the same level of time complexity. During this research, the ResNet model was educated using the Kaggle dataset for Distracted Driver Detection by State Farm and evaluated using that dataset. In addition to that, this model can effectively categorize the driver's distraction. 152 different layers comprise the ResNet model, depicted as its architecture in Figure 4. In a ResNet, each step is built upon the previous one with four additional layers that share a behavioral pattern. The sequence is repeated in each next segment the same way. A convolution on a three-by-three grid uses the following dimensions as the constant: 64, 128, 256, and 512, respectively. As a result, it skips the input every two convolutions that it does. In addition, the width and height parameters of the layer do not change at any point over its entirety. Skip connections are responsible for identity mapping, and the outputs of these connections are added to the outputs of stacked layers. In addition, compared to the models of the other networks, the ResNet model is both simpler and more readily capable of being optimized. Additionally, compared to previous peer-level networks, our model converges more quickly and produces superior outcomes.

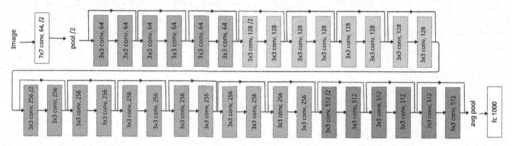

Figure 4.   ResNet Model [24].

1.4   *Ensemble learning*

Over the last few decades, the fields of computational intelligence and machine learning have paid increasing attention to the concept of ensemble learning. The term "ensemble learning" refers to a process in which many models, such as classifiers, are merged by a method of choice to address a particular challenge involving intelligent computing. To increase the performance of the final model on tasks such as classification, prediction, and function estimate, or to lessen the impact of an inappropriate selection of the basic model, is the primary purpose of its application.

The following is a condensed version of the overall architecture that constitutes ensemble learning. After first generating a group of individual learners, the next step is to combine the learners efficiently using a predetermined method; lastly, it is possible to achieve the anticipated outcomes of the experiment [25].

Individual learners are often generated using current learning algorithms from the corresponding training data (such as decision trees and error propagation neural networks). Homogeneous ensemble learning refers to the process in which all individual learners are of the same type; for instance, all neural networks. Heterogeneous ensemble learning refers to the process in which various individual learners are used. The method of homogeneous ensemble learning is implemented in this paper.

Recent fast ensemble deep learning techniques can be applied to whole-slide medical image analysis (e.g., predicting pCR from H&E-stained whole-slide images [62]) to reduce time and space overheads while, to some extent, sacrificing accuracy. This can be done by predicting pCR from H E-stained whole-slide images. We think that performance enhancement is of the utmost importance in this assignment, which is why we opted for the most accurate ensemble approaches. Because of the gradual evolution of the various ensemble techniques throughout some time, the learning models have improved in their generalizability. The ensemble tactics can be categorized in a general sense as [26] classical methods, including bagging, boosting, and stacking. general methods, which include Explicit Implicit, Negative correlation Learning, Homogeneous / Heterogeneous learning. And the last one is Decision Fusion Strategies, including Unweighted Model Averaging, Bayes Optimal Classifier, Stacked Generalization, Super Learner, Consensus and Query-By-Committee.

In our work, we have used and compared the most efficient ensemble methods and found Stacked Generalization as the best for ovarian cancer classification for histopathology images.

1.5 *Three-path ensemble architecture*

The architectures VGG16, VGG19, and ResNet are used as base models to develop the final ensemble model; various Decision Fusion Strategies are implemented and compared on the final outputs of base models. For the ensemble of all of these architectures is included because they satisfy the following criteria:Â (i)Â satisfactory performance in various computer vision tasks; (ii) usefulness towards real-time (or near real-time) applications; and (iii) feasibility of transfer learning for limited datasets. Because each method has drawbacks regarding the variations in the shape and texture of the input image, we propose a six-path ensemble prediction approach to make the most of the benefits offered by multiple classifiers to achieve a higher accuracy overall. The ensemble architecture suggested for ovarian cancer classification is shown in Figure 5. For making an ensemble of these deep neural networks, stacked generalization with Multi-layer perceptron is used as meta learner as it's fast to train and much more accurate than other techniques.

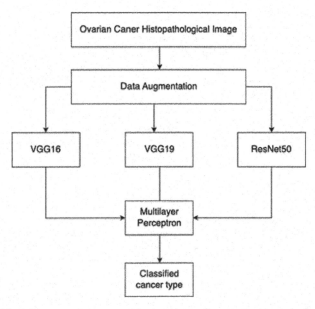

Figure 5.   Ensemble Process.

Table 1. The number of images in each dataset for 10-fold cross-validation.

| | Serous | | Mucinous | | Endometrioid | | Clear cell | |
	Orignal	Agumented	Orignal	Agumented	Orignal	Agumented	Orignal	Agumented
Set 1	42	462	48	528	42	462	41	451
Set 2	41	451	50	550	45	495	40	440
Set 3	54	594	41	451	47	517	40	440
Set 4	52	572	40	440	54	594	52	572
Set 5	51	561	44	484	46	506	46	506
Set 6	52	572	50	550	50	550	42	462
Set 7	47	517	46	506	47	517	40	440
Set 8	46	506	41	451	53	583	40	440
Set 9	48	528	51	561	53	583	45	495
Set 10	48	528	42	462	47	517	44	484

2 RESULTS

In this section, we introduced our dataset and described its preprocessing approach, training, validation, testing criteria, and augmentation procedure. Then, we explored the arrangement of the many models we employed and concluded by describing the architecture of our suggested ensemble architecture.

The classification accuracies of each type in two independent models trained by original images and augmented images were shown in Table 2.

Table 2. The classification accuracies for ensembled model.

	Original	Augmented
Serous	81%	83%
Mucinous	69.62%	70.51%
Endometrioid	61.53%	70.93%
Clear cell	80.57%	78.21%
Total	71.76%	75.20%

The accuracy of the classification model generated using augmented image data (75.20 percent) is about 4 percent greater than the accuracy of the classification model developed using the original image data (71.76 percent). Even though the architecture of the two devices is identical, their outputs are vastly different. This must be caused by a variation in the training data. This indicates that image augmentation, which includes image enhancement and rotation, is relevant to the ensembled model. Image enhancement increases the visibility of an image's qualities. Picture sharpening and edge enhancement are common forms of image enhancement. Rotating the image increased the sample size, which instantly improved the classification performance of the ensemble.

REFERENCES

[1] Vãȷzquez, M.A.; Marinño, I.P.; Blyuss, O.; Ryan, A.; Gentry-Maharaj, A.; Kalsi, J.; Manchanda, R.; Jacobs, I.; Menon, U.; Zaikin, A. A Quantitative Performance Study of Two Automatic Methods for the Diagnosis of Ovarian Cancer. *Biomed. Signal Process. Control* 2018, 46, 86–93. [CrossRef] [PubMed]

[2] Jayson, G.C.; Kohn, E.C.; Kitchener, H.C.; Ledermann, J.A. Ovarian Cancer. *Lancet* 2014, 384, 1376–1388. [CrossRef]

[3] Kommoss, S.; Pfisterer, J.; Reuss, A.; Diebold, J.; Hauptmann, S.; Schmidt, C.; Bois, A.D.; Schmidt, D.; Kommoss, F. Specialized Pathology Review in Patients with Ovarian Cancer. *Int. J. Gynecol. Cancer* 2013, 23, 1376–1382. [CrossRef]

[4] Bentaieb, A.; Li-Chang, H.; Huntsman, D.; Hamarneh, G. Automatic Diagnosis of Ovarian Carcinomas Via Sparse Multiresolution Tissue Representation. In *Lecture Notes in Computer Science Medical Image Computing and Computer-Assisted Intervention—MICCAI* 2015; Springer: Berlin, Germany, 2015; pp. 629–636.

[5] Bentaieb, A.; Li-Chang, H.; Huntsman, D.; Hamarneh, G. A Structured Latent Model for Ovarian Carcinoma Subtyping from Histopathology Slides. *Med. Image Anal.* 2017, 39, 194–205. [CrossRef] [PubMed]

[6] Yu, K.-H.; Hu, V.; Wang, F.; Matulonis, U.A.; Mutter, G.L.; Golden, J.A.; Kohane, I.S. Deciphering Serous Ovarian Carcinoma Histopathology and Platinum Response by Convolutional Neural Networks. *BMC Med.* 2020, 18. [CrossRef] [PubMed]

[7] Papp, E.; Hallberg, D.; Konecny, G.E.; Bruhm, D.C.; Adleff, V.; NoÃ«, M.; Kagiampakis, I.; Palsgrove, D.; Conklin, D.; Kinose, Y.; *et al.* Integrated Genomic, Epigenomic, and Expression Analyses of Ovarian Cancer Cell Lines. *Cell Rep.* 2018, 25, 2617–2633. [CrossRef] [PubMed]

[8] Integrated genomic analyses of ovarian carcinoma. *Nature* 2011, 474, 609–615.

[9] Sun, D.; Li, A.; Tang, B.; Wang, M. Integrating Genomic Data and Pathological Images to Effectively Predict Breast Cancer Clinical Outcome. *Comput. Methods Programs Biomed.* 2018, 161, 45–53. [CrossRef] [PubMed]

[10] Shao, W.; Wang, T.; Sun, L.; Dong, T.; Han, Z.; Huang, Z.; Zhang, J.; Zhang, D.; Huang, K. Multi-task Multi-modal Learning for Joint Diagnosis and Prognosis of Human Cancers. *Med. Image Anal.* 2020, 65, 101795. [CrossRef] [PubMed]

[11] Zhang, A.; Li, A.; He, J.; Wang, M. LSCDFS-MKL: A Multiple Kernel Based Method for Lung Squamous Cell Carcinomas Disease-free Survival Prediction with Pathological and Genomic Data. *J. Biomed. Inform.* 2019, 94, 103194. [CrossRef]

[12] Liu, T.; Huang, J.; Liao, T.; Pu, R.; Liu, S.; Peng, Y. A Hybrid Deep Learning Model for Predicting Molecular Subtypes of Human Breast Cancer Using Multimodal Data. *Irbm* 2021. [CrossRef]

[13] Kasture, Kokila (2021), "Ovarian Cancer & Subtypes Dataset Histopathology", *Mendeley Data*, V1, doi: 10.17632/kztymsrjx9.1

[14] Han X. (2017) MR-based Synthetic CT Generation Using a Deep Convolutional Neural Network Method. *Med. Phys.* 44, 1408–1419 10.1002/mp.12155

[15] Krizhevsky A., Sutskever I. and Hinton G.E. (2012) Image Net Classification with Deep Convolutional Neural Network, in: Advances in Neural Information Processing Systems. *NIPS Proc.* 25, 1106–1114, Hinton G.E., Srivastava N., Krizhevsky A., Sutskever I. and Salakhutdinov R.R. (2012) *Improving Neural Networks by Preventing Co-adaptiona of Feature Detectors*, arXiv.1207.0580

[16] Siyuan Lu, Zhihai Lu, and Yu-Dong Zhang. 2019. Pathological Brain Detection Based on AlexNet and Transfer Learning. *Journal of Computational Science* 30 (2019), 41–47.

[17] Uchi Ugobame Uchibeke, Sara Hosseinzadeh Kassani, Kevin A Schneider, and Ralph Deters. 2018. *Blockchain Access Control Ecosystem for Big Data security.*

[18] Zilong Hu, Jinshan Tang, Ziming Wang, Kai Zhang, Ling Zhang, and Qingling Sun. 2018. Deep Learning for Image-based Cancer Detection and Diagnosis- A Survey. *Pattern Recognition* 83 (2018), 134–149.

[19] Simonyan and Zisserman (2014)[Simonyan K, Zisserman A. 2014. Very Deep Convolutional Networks for Large-scale Image Recognition. *ArXiv preprint. arXiv*:1409.1556.

[20] Alzubaidi *et al.* (2021) Alzubaidi L, Zhang J, Humaidi AJ, Al-Dujaili A, Duan Y, Al-Shamma O, SantamarÃ-a J, Fadhel MA, Al-Amidie M, Farhan L. Review of Deep Learning: Concepts, CNN Architectures, Challenges, Applications, Future Directions. *Journal of Big Data.* 2021;8(1):1–74. doi: 10.1186/s40537-020-00387-6.

[21] Khan *et al.* (2020) Khan A, Sohail A, Zahoora U, Qureshi AS. A Survey of the Recent Architectures of Deep Convolutional Neural Networks. *Artificial Intelligence Review.* 2020;53(8):5455–5516. doi: 10.1007/s10462-020-09825-6.

Koklu, Cinar & Taspinar (2022) Koklu M, Cinar I, Taspinar YS. CNN-based Bi-directional and Directional Long-short Term Memory Network for Determination of Face Mask. *Biomedical Signal Processing and Control.* 2022;71:103216. doi:10.1016/j.bspc.2021.103216.

[22] https://neurohive.io/en/popular-networks/vgg16/

[23] *Machine Learning & Knowledge Extraction Focal Liver Lesion Detection in Ultrasound Image Using Deep Feature Fusions and Super Resolution - Scientific Figure on ResearchGate.*

[24] Srinivasan, Kathiravan & Garg, Lalit & Datta, Debajit & Alaboudi, Abdulellah & Zaman, No[1]or & Agarwal, Rishav & Thomas, Anmol. (2021). *Performance Comparison of Deep CNN Models for Detecting Driver's Distraction.* Cmc -Tech Science Press-. 68. 4109–4124. 10.32604/cmc.2021.016736.

[25] Yuchao Zheng, Chen Li, Xiaomin Zhou, Haoyuan Chen, Hao Xu, Yixin Li, Haiqing Zhang, Xiaoyan Li, Hongzan Sun, Xinyu Huang, Marcin Grzegorzek, Application of Transfer Learning and Ensemble Learning in Image-level Classification for Breast Histopathology, *Intelligent Medicine*,2022.

[26] M.A. Ganaie, Minghui Hu, A.K. Malik, M. Tanveer, P.N. Suganthan,Ensemble deep learning: A review, *Engineering Applications of Artificial Intelligence*, Volume 115, 2022, 105151, ISSN 0952-1976.

Artificial Intelligence, Blockchain, Computing and Security – Dagur et al. (Eds)
© 2024 The Author(s), ISBN: 978-1-032-49393-0

A study on automatic mathematical word problem solvers

Madhavi Alli, Balaga Sateesh, Duggasani Yaswanth Reddy, Potlapelli Sai Koushik &
Thelukuntla Sai Chandra
*Vallurupalli Nageswara Rao Vignana Jyothi Institute of Engineering and Technology, Hyderabad,
Telangana, India*

ABSTRACT: When it comes to solving mathematical problems that are described in words, a knowledgeable human can surpass a machine considerably. Computer programs however are more effective and efficient than humans in solving complex mathematical operations. Researchers in the field are currently focusing on more difficult problems like solving probability and geometry problems, with the idea that the current models are capable of accurately answering the single variable math word problems. Mathematical word problem solving automatically has seen a tremendous growth in new research. Therefore, the goal of this study is to comprehend various strategies used by researchers on various datasets and the ability of problems to be solved when given fresh inputs. We can recognize many of the methods used to solve these problems and the significance of using them for solving them. From this study, we identified that combining methods, such as Natural Language Processing and intermediate representations along with the models can effectively solve the problems. This will also increase their efficiency in a broad usage context. The several methods for solving the word problem and their accuracy on various datasets are all discussed in this study.

1 INTRODUCTION

Many artificial intelligence systems have found it difficult to solve basic school level mathematical word problems. There have been efforts to solve these problems for a long time, and several ways have been developed. These approaches range from the simplest, such as Rule-based methods, to the most sophisticated, such as GPT-3 based models, which are trained on 175 billion parameters. Much of the complexity of word problem solvers comes from transforming the word problem into an equation which can be quickly solved by a computer by employing basic gate operations. Therefore, the only thing that has to be improved is the way the word problem is converted into expression. Even though numerous studies have been published in this field, they consistently improved the system's overall accuracy by focusing on a variety of topics, including dataset refinement, the use of intermediate representations, the use of declarative knowledge, and the creation of equations using templates.

Table 1.   A simple mathematical word problem its equation and solution.

Math Word Problem: A total of 97 sea whales near a shore in the arctic. If 54 of them migrated towards the Atlantic Ocean what are the remaining number of sea whales that are present in the arctic? Equation: x = 97 − 54 Solution: x = 43

DOI: 10.1201/9781003393580-17

There has been extensive work done to solve the mathematical word problem solvers and major timeline of the model development can be depicted in the Figure 1. In the initial days simple rule-based techniques where a set of static rules are defined and then by the usage of machine learning and tree-based techniques and then the concept of the deep learning and neural networks which improved the accuracy of the models on benchmark datasets drastically.

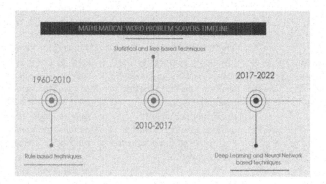

Figure 1.   Timeline of the models.

2   RELATED WORK

Throughout years, many methods and datasets have been developed to solve the MWP's. It ranges from basic machine learning based multi-level classification & rule-based approach [1]. An improved NumS2T model [4] that takes the consideration of numerical data from the Math23K and APE datasets. Seq2DAG model [5] to generate mathematical equations in a DAG structure this model was tested on Math23k dataset. Usage of pre trained knowledge encoder and hierarchical encoder [6] to give common sense to the model and to find words and sentences which contribute the most to the final output MAWPS dataset was used, an edge enhanced Hierarchical Graph to Tree encoder [7] to find and label the relationships between the words present in the problem here they used MAWPS dataset. A teacher student network with multiple decoders [12] to regularize the model and diversify the output respectively MAWPS dataset was used. Generating equation on Math_23k dataset using abstract syntax tree-based decoder [15], extracting features specific to math problems using group attention mechanism [16]. Various deep learning models like seq2seq and similarity models [19] and used Dolphin18K dataset to make the working solution to get more accurate and faster. An ensemble model [20] to normalize the duplicated equations and experimented on Math-23K dataset. Applications of the Math Word Problem Solver are it plays a major role in aeronautical industry like calculating force on space and aircraft flights and used in various infrastructure and architectural projects. Different risk prediction and bacterial growth and decay problems. Engineering and design problems which are based on mathematics. An efficient tool for primary school students to learn simple math word problem solving. Used in solving personal finance problems and weather forecasting.

3   INVESTIGATION AND FINDINGS OF PREVIOUS RESEARCH PAPERS

All the research that is done in automatic math word problem solvers is shown in Table 2 along with the methodologies of the solver and its accuracy and advantages and

Table 2. Table showing different methodologies, pros, and cons in this literature survey.

Sno	Title	Methodology	Pros/Cons	Year
1	(Mandal *et al.* 2022) Solving Arithmetic Word Problems Using Natural Language Processing and Rule-Based Classification	Modify, combine, and evaluate the operation prediction methodology. To resolve MWPs, verbal classification is used.	Good usage of Language Processing to make model learn language well. High accuracy on large dataset. Operation incorrect prediction for standard problems.	2022
2	(Lan *et al.* 2022) MWPToolkit: An Open-Source Framework for Deep Learning-Based Math Word Problem Solvers	The MWPToolkit is an open-source package that solves MWPs and offers a structure for conducting research. It has implementation on 6 different datasets and 17 cutting-edge deep learning-based solvers.	This framework makes it easy to compare different solvers directly. It is expected to establish a benchmark framework for MWP solving task, speeds up the development of new methods. It doesn't contain visualization and training under weak supervision.	2022
3	(Mandal *et al.* 2021) Classifying and Solving Arithmetic Math Word Problem – An Intelligent Math Solver	Supervised ML Model to classify the features of the MW-P & verbal categorization for further division to predict the operators accordingly.	Implemented Verbal Classification. Lack of excess knowledge to the working model. Standard problems were not produced accurate answers.	2021
4	(Patel *et al.* 2021) Are NLP Models really able to Solve Simple Math Word Problems?	The accuracy of the newly created dataset SVAMP was assessed using state-of-the-art models, yielding results of Seq2Seq 29.2%, GTS 28.6%, and Graph2Tree 30.8%.	Pointed out that most models use shallow heuristics and shown the shortcomings of the available datasets. The shallow heuristics problem of a model wasn't addressed.	2021
5	(Wu *et al.* 2021) Math Word Problem Solving with Explicit Numerical Values	NumS2T model with Math23k and APE dataset used. Evaluation done bases of pre-order traversal expression. Encoder uses Bi layer LSTM. Tree structured Decoder.	Enhances the math word problem solving performance by explicitly incorporating numerical values. Global Relationship with Target Expressions. Slower on larger datasets.	2021
6	(Cao *et al.* 2021) A Bottom-Up DAG Structure Extraction Model for Math Word Problems	A seq2dag approach, it produces output as a directed acyclic graph structure.	Can produce equation set. Can maintain associative and distributive laws. Coreference solution.	2021
7	(Roy *et al.* 2021) Improving Math Word Problems with Pre-trained Knowledge and Hierarchical Reasoning	Pre trained knowledge encoder tries to reason math word problems by using implicit or outside knowledge.	Working model has pre trained general knowledge which finds which words in sentences contribute the most to the model.	2021
8	(Wu *et al.* 2021) An Edge-Enhanced Hierarchical Graph-to-Tree Network for Math Word Problem Solving	It represents each math word problem as a graph in which nodes are connected by edge labels.	Model can identify the long-range dependencies. It can pay attention to different parts of the problem. Uses grammatical & external knowledge.	2021

(continued)

Table 2. Continued

Sno	Title	Methodology	Pros/Cons	Year
9	(Shen *et al.* 2021) Generate & Rank: A Multi-task Framework for Math Word Problems	In this study, a model with a generator and ranker that share a BART model is introduced.	The model learns from its own mistakes and can discriminate between proper and improper statements by working together with generation and ranking.	2021
10	(Dongxiang *et al.* 2020) The Gap of Semantic Parsing: A Survey on Automatic Math Word Problem Solvers	The relative accuracy of all the models with all datasets is examined in depth.	The significance of feature extraction in Bi-LSTM and the significance of T-RNN in predicting the unknown variable in the expression tree are demonstrated.	2020
11	(Ughade *et al.* 2020) Mathematical Word Problem Solving Using Natural Language Processing	In this model, it is assessed how likely specific key words are. The output that has been labelled as an issue has a higher likelihood of having unique terms or key words.	Model Processing done after the classification to which category it belongs to. Efficient algorithm to classify the problem. Doesn't classify dynamically and solve explicitly.	2020
12	(Zhang *et al.* 2020) Graph-to-Tree Learning for Solving Math Word Problems	On the MAWPS dataset, the Graph2Tree model with Graph based encoder and tree-based decoder along with model learning scored 83.7% accuracy and 77.4% on Math23K dataset.	Relationships and data about the order were recorded. Model learning is defined by the loss function. On the benchmark datasets, Graph2Tree had a better accuracy rate.	2020
13	(Liang *et al.* 2020) A Diverse Corpus for Evaluating and Developing English Math Word Problem Solvers	Highly diverse ASDiv dataset with accuracy of 36% on GTS, 37% on UnitDep, and 36% on LCA++ is developed.	The problem type, grade level, and equation are annotated in this MWP dataset.	2020
14	(Zhang *et al.* 2020) Teacher-student networks with multiple decoders for solving Math Word Problem	It uses knowledge distillation & teacher network to incorporate knowledge of equivalent solution expressions to regularize the learning of the student network.	Due to knowledge distillation the model is regularized. Usage of multiple decoders will diversify the output. Removes undesired bias. It can't handle complex terminology.	2020
15	(Griffith *et al.* 2019) Solving Arithmetic Word Problems Automatically Using Transformer & Unambiguous Representations	The TensorFlow Datasets library's sub-word text encoder is used to encrypt MWP questions and equations.	English Language pre-processing to the model with some language proficient datasets. Transformer models with different layers of training. More importance given for language processing than model functioning.	2019
16	(Amini *et al.* 2019) MathQA: Towards Interpretable Math Word Problem Solving with Operation-Based Formalisms	Annotating the problems MathQA dataset is created from the AQuA dataset. On MathQA dataset, accuracy of the bidirectional Sequence2Program RNN	A neural sequence to program model enhanced with automatic problem categorization. Representation language to model precise operation	2019

(continued)

111

Table 2. Continued

Sno	Title	Methodology	Pros/Cons	Year
		was 51.9%, and accuracy of Sequence2Program with categorization was 54.2%.	programs. Annotation process depends on human contributions it is ineffective.	
17	(Liu *et al.* 2019) Tree-structured Decoding for Solving Math Word Problems	A deep learning model that uses seq2seq approach, the tree-based decoder uses abstract syntax tree information of mathematical expression, also it uses auxiliary stack to guide its decoding process it has an accuracy of 89%.	Decoder mimics human behavior. No need for end of sequence token Uses structural information of the mathematical equation as an AST. Solid geometry. Long equation templates.	2019
18	(Li *et al.* 2019) Modeling Intra-Relation in Math Word Problems with Different Functional Multi-Head Attentions	A deep learning model that uses seq2seq approach, it uses group attention mechanism to extract important features specific to MWP's.	Extracting features will improve prediction. Doesn't use structural information of the mathematical problems.	2019
19	(Huang *et al.* 2018) Using Intermediate Representations to Solve Math Word Problems	To enhance the intermediate form, they are represented using Dolphin language and are labelled iteratively. Seq2Seq RNN on the Dolphin18K dataset has an accuracy of 10%, and Seq2Seq + Attention Regularization has an accuracy of 15%.	Includes information of higher order operations. Attention regularization addresses the loss. The first step is to update our model with all potential latent forms using maximum likelihood estimation. The ambiguity of intermediate forms is not addressed.	2018
20	(Roy *et al.* 2018) Mapping to Declarative Knowledge for Word Problem Solving	A framework is defined that maps arithmetic word problems to mathematical expressions by selecting relevant declarative knowledge.	Continuous weight vector updation for giving more weight to the correct equations. Highly reliant on declarative information.	2018
21	(Wong *et al.* 2018) Solving Math Word Problems	Hybrid Machine Learning Model. The normalizes, in pre-processing, all numerical quantities in question text. In S2S, uses encoder and decoder with LSTM layer. Similarity model uses bag of words.	Equations were converted to number and rounded to three decimal places. The model exhibited a strong sense of mathematical syntax with ∼80% of predicted equations syntactically correct.	2018
22	(Wang *et al.* 2018) Translating a Math Word Problem to an Expression Tree	Batched ML Model on Math23K dataset. Output of the model with the highest generation probability is selected as the final output.	They have normalized the equation using order duplications and bracket duplication. Less accuracy for working model.	2018
23	(Clementeena *et al.* 2018) A literature survey on question answering system in Natural Language Processing	Discussed about various methodologies of solving MWP's and applications of NLP in various domains. Types of Knowledge like ELIZA & LUNAR.	Usage of different parsing algorithms to improve the accuracy.	2018

disadvantages of the model. In Figure 2 main methodologies that are present are discussed along with their steps for generating the equations. Table 3 lists various datasets, and the operators present in the dataset and total number of problems present in the problem. Table 4 lists the major models and their techniques and datasets and their relative accuracies.

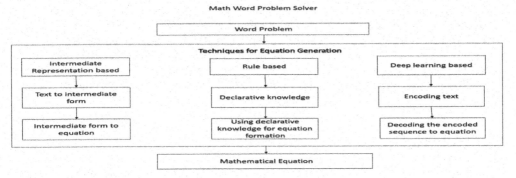

Figure 2.   Methodologies in solving mathematical word problems.

Table 3.   Datasets and their diversity in operators and number of problems.

Datasets	Operators	Number of Word Problems
ASDiv	{+, -, X, ÷}	1218
MAWPS	{+, -, X, ÷}	2373
SVAMP	{+, -, X, ÷}	1000
Math23K	{+, -, X, ÷}	23162
MathQA	{+, -, X, ÷, ^, $\sqrt{}$ }	29837
APE210K	{+, -, X, ÷, ^, $\sqrt{}$ }	210488

Table 4.   Various models and their accuracies on different datasets.

Models	Techniques Used	Datasets	Accuracy
UnitDep	Tree Based	ASDiv	37%
GTS	Gated Graph	ASDiv	36%
	Neural Networks	SVAMP	30.8%
		Math23K	77.4%
seq2seq	Layers of Long	MAWPS	76.1%
	Short-Term Memory	SVAMP	24.2%
		Math23K	40.3%
Graph2Tree	Tree based deep learning	MAWPS	83.7%
		SVAMP	36.5%
		Math23K	75.5%
FFN + LSTM	Feed Forward Neural	SVAMP	17.5%
	Network and	Math23K	66.9%
	Long Short-Term Memory		
Transformer	Deep Learning Model	MAWPS	79.4%
		SVAMP	25.3%
NumS2T	Tree based Neural Network	Math23K	78.1%
		APE210K	70.5%

4 CHALLENGES IN AUTOMATIC MATH WORD PROBLEM SOLVING

Early methods of handling arithmetic word problems relied on hand-engineered rule-based systems that could only address a small subset of problems. Another method transforms the problem language into logical forms before using those forms to derive the equation, although this method necessitates additional human annotation for the logic forms. Another line of research maintains a template using either a retrieval or classification model before filling in the blanks with quantities. This method is incapable of resolving issues that go beyond the templates in the training data. Another method uses deep learning models like seq2seq and others to directly compute the equation. Although this strategy is superior to earlier strategies, it doesn't consider the problem's structural details. The most recent method enhances the prior method by applying models that consider the structural information of math word problems. Some of these models are seq2tree and seq2dag.

5 CONCLUSION & FUTURE DIRECTIONS

In this research a thorough literature review is done to examine the current state of mathematical word problems solvers and concepts related to this field. This paper covers all the models for solving mathematical word problems and demonstrates the need for additional research to enable the models to solve problems and problem types that they have never encountered before with greater accuracy. After the research a methodology has been decided to solve the mathematical word problems by using natural language processing that uses deep learning models that take account of structural information of math word problems and maintains mathematical laws i.e. such as commutative and associative law and operator precedence.

REFERENCES

[1] Amini Aida, *et al.* "MathQA: Towards Interpretable Math Word Problem Solving with Operation Based Formalisms." *Proceedings of the 2019 Conference of the North American Chapter of the Association for Computational Linguistics: Human Language Technologies*, Volume 1 (Long and Short Papers). 2019.

[2] Cao Yixuan, *et al.* "A bottom-up DAG Structure Extraction Model for Math Word Problems." *Proceedings of the AAAI Conference on Artificial Intelligence*. Vol. 35. No. 1. 2021.

[3] Clementeena A & Sripriya P (2018). "A Literature Survey on Question Answering System in Natural Language Processing." *International Journal of Engineering and Technology (UAE)*. 7. 452–455. 10.14419/ijet.v7i2.33.14209.

[4] Dongxiang, *et al.* "The Gap of Semantic Parsing: A Survey on Automatic Math Word Problem Solvers." *IEEE Transactions on Pattern Analysis and Machine Intelligence* 42.9 (2019): 2287–2305.

[5] Griffith K and Kalita J, "Solving Arithmetic Word Problems Automatically Using Transformer and Unambiguous Representations," *2019 International Conference on Computational Science and Computational Intelligence (CSCI)*, 2019, pp. 526–532, doi: 10.1109/CSCI49370.2019.00101.

[6] Huang Danqing, *et al.* "Using Intermediate Representations to Solve Math Word Problems." *Proceedings of the 56th Annual Meeting of the Association for Computational Linguistics (Volume 1: Long Papers)*. 2018.

[7] Lan Yihuai, *et al.* "Mwptoolkit: an Open-source Framework for Deep Learning-based math Word Problem Solvers." *Proceedings of the AAAI Conference on Artificial Intelligence*. Vol. 36. No. 11. 2022.

[8] Li Jierui, *et al.* "Modeling Intra-relation in Math Word Problems with Different Functional Multi-head Attentions." *Proceedings of the 57th Annual Meeting of the Association for Computational Linguistics*. 2019.

[9] Liang Chao-Chun *et al.* "A Diverse Corpus for Evaluating and Developing English Math Word Problem Solvers." *Proceedings of the 58th Annual Meeting of the Association for Computational Linguistics*. 2020.

[10]   Liu Qianying, *et al.* "Tree-structured Decoding for Solving Math Word Problems." *Proceedings of the 2019 Conference on Empirical Methods in Natural Language Processing and the 9th International Joint Conference on Natural Language Processing (EMNLP-IJCNLP)*. 2019.

[11]   Mandal S and Naskar S, "Classifying and Solving Arithmetic Math Word Problems—An Intelligent Math Solver," in *IEEE Transactions on Learning Technologies*, vol. 14, no. 1, pp. 28–41, Feb. 2021, doi: 10.1109/TLT.2021.3057805.

[12]   Mandal S *et al.* "Solving Arithmetic Word Problems Using Natural Language Processing and Rule-Based Classification", *Int J Intell System Appl Eng*, vol. 10, no. 1, pp. 87–97, Mar. 2022.

[13]   Patel Arkil, Satwik Bhattamishra, and Navin Goyal. "Are NLP Models really able to Solve Simple Math Word Problems?" *Proceedings of the 2021 Conference of the North American Chapter of the Association for Computational Linguistics: Human Language Technologies*. 2021.

[14]   Roy Subhro and Dan Roth. "Mapping to Declarative Knowledge for Word Problem Solving." *Transactions of the Association for Computational Linguistics* 6 (2018): 159–172.

[15]   Roy Subhro *et al.* "Improving Math Word Problems with Pre-trained Knowledge and Hierarchical Reasoning." *Proceedings of the 2021 Conference on Empirical Methods in Natural Language Processing*. 2021.

[16]   Shen Jianhao, *et al.* "*Generate & Rank: A Multi-task Framework for Math Word Problems.*" *arXiv* preprint arXiv:2109.03034 (2021).

[17]   Ughade S, Kumbhar S. (2020). Mathematical Word Problem Solving Using Natural Language Processing. In: Tuba, M., Akashe, S., Joshi, A. (eds) *ICT Systems and Sustainability. Advances in Intelligent Systems and Computing*, vol 1077. Springer, Singapore.

[18]   Wang Lei, *et al.* "Translating a Math Word Problem to a Expression Tree." *Proceedings of the 2018 Conference on Empirical Methods in Natural Language Processing*. 2018.

[19]   Wong Ryan, *"Solving Math Word Problems,"* Department of Electrical Engineering, Stanford University, Palo Alto, CA 94305, 2018.

[20]   Wu Qinzhuo, *et al.* "An Edge-enhanced Hierarchical Graph-to-tree Network for Math Word Problem Solving." *Findings of the Association for Computational Linguistics: EMNLP 2021*. 2021.

[21]   Wu Qinzhuo, *et al.* 2021. Math Word Problem Solving with Explicit Numerical Values. *In Proceedings of the 59th Annual Meeting of the Association for Computational Linguistics and the 11th International Joint Conference on Natural Language Processing* (Volume 1: Long Papers), pages 5859–5869.

[22]   Zhang Jipeng, *et al.* "Graph-to-tree Learning for Solving Math Word Problems." *Association for Computational Linguistics*, 2020.

[23]   Zhang Jipeng, *et al.* Teacher Student Networks with Multiple Decoders for Solving Math Word Problem. (2020). *Proceedings of the Twenty-Ninth International Joint Conference on Artificial Intelligence 2020*.

Artificial Intelligence, Blockchain, Computing and Security – Dagur et al. (Eds)
© *2024 The Author(s), ISBN: 978-1-032-49393-0*

Predicting learning styles in personalized E-learning platforms

A. Madhavi, A. Nagesh & A. Govardhan
VNR Vignana Jyothi Institute of Engineering and Technology, Hyderabad, Telangana, India
Mahatma Gandhi Institute of Engineering and Technology, Hyderabad, India
Jawaharlal Nehru Technological University, Hyderabad, India

ABSTRACT: Learning Management Systems (LMS) are popularly used in online educational systems and universities to deliver self-paced online courses. Furthermore, in the literature many educational theories have recommended that suggesting learners with suitable learning material based on their learning styles may improve learner's learning caliber. Human brains generally use different methods for grasping knowledge faster and easier. We call these methods as *learning styles*. Learners with diverse individual attributes, knowledge levels, backgrounds, and characteristics have distinct learning styles. Determining student's learning style improves the efficiency of the learning process. To provide personalized study material to the learner depending on his or her learning style, an accurate automated learning style identification model is required. In this work we implement an intelligent model for accurate detection of student's learning styles.

Keywords: Machine learning, Robust Classification, Learning style, Learner Model, Personalized E-learning

1 INTRODUCTION

There are numerous online learning platforms available today. Minoru *et al.* (2012) state that instructors use a variety of methods to assess a student's comprehension, spanning from video lectures to the posting of assignments. According to a survey conducted by the Babson Survey Research Group [27], approximately 33 percent of college students take at least one online course. Schools and employers recognize that online courses can be advantageous and, at times, more effective than traditional classroom studies. Educators have investigated (Jack 1997) the issue of students learning through online platforms and discovered that the learning materials (Behram *et al.* 2010) (Yavuz *et al.* 2012) suggested by these systems do not adequately satisfy learners' demands and in some cases fail to meet learners' probable requirements. Primarily, issues known as "learning deviation" and "cognitive overload" are of utmost importance and demonstrate that the majority of learners are unsure of their actual requirements during the learning process, resulting in incorrect requests. To overcome these issues, it would be beneficial for the learner if learning models were designed to analyses learners' actual requirements, thereby enhancing their learning quality.

In a traditional classroom, a teacher can see how well a student understands a topic and help them if they need it. Online platforms, on the other hand, rest totally on self-analysis (Fathi *et al.* 2010). Because of this, most students find it hard to understand the online lessons. You might have known that people have different ways of learning that work best for them. This is backed up by seven ways of learning. All styles (Richard *et al.* 1998) take advantage of a person's strengths, which may help them remember things better. They have something to do with other people and focus on one of the five senses. This idea is popular

DOI: 10.1201/9781003393580-18

because it says that each learner's performance could be better by figuring out their learning style and training them in a way that fits them. So, to solve the problem, Norazlina *et al.* (2013) say that a tool that can guess a person's learning style and give them the right material based on those predictions would help the student learn better. 93 percent of UK educators agreed in 2012 that pupils learn best when given content in a style that best suits their individual needs. Virginia *et al.* (2006) says that there needs to be a model for accurate automatic learning style recognition in order to give each student personalized learning material based on how he or she learns. In this work, we use an intelligent programme to correctly figure out how students learn. Integrating user accounts with data about how people learn could make intelligent learning systems better by giving each student personalized advice on how to learn best for their needs.

2 RELATED WORK

In past research studies, different machine learning methods and different learning style recognition systems (Lukito *et al.* 2019) were used to automatically figure out how people learn. These ways have used the information about how the student uses the e-Learning system to find information. According to Sabine *et al.* (2007), the data-driven approach has been acknowledged as a means of developing a model for automatically predicting learners' preferred pedagogical approaches. In this method, the prediction is made by an algorithm that is built on artificial intelligence (AI). The learners' learning patterns have been found and fed into the intelligent algorithm. The algorithm then gives back the different learning styles of the learners. Since this way of classifying learning styles was based on real data sets, the results were very accurate. So, capturing the learners' behaviors is a must if you want to get the right information for classifying them. So, Umadevi *et al.* (2014) say that this can be done by using web usage mining methods. Also, for automatic detection of learning styles, different parts of the way a learner learns must be taken into account when creating a model for automatic detection. Also, all of these things are very important for running experiments to get good detail and make the model more accurate.

In a study using Bayesian Networks, data from e-Learning site archives and discussion boards were utilised (Patricio *et al.* 2007). This site log data set only revealed three of Felder Silverman's learning styles: Perception, Processing, and Understanding. In (Manal *et al.* 2015), a novel approach was utilized to anticipate each learner's preferred style of instruction using the FSLSM model. Learners' actions were recorded in the Moodle server logs, enabling a dynamic classification of students according to their individual study habits. Classification in this approach was achieved by the usage of Decision trees. This method's efficacy was determined by analyzing the outcomes of a final exam quiz for patterns in students' preferred methods of learning. However, this strategy is only tested with 35 students in one Moodle open-source online course. To automatically discover the learner's preferred method of instruction, fuzzy logic was also explored (Khaild *et al.* 2017). In order to build a fuzzy prediction model, Keeley *et al.* (2013) have used a fuzzy classification tree. Independent variables were derived from conversational natural language processing and employed in this model. The suggested method has been trained for a number of different pedagogical approaches in previous studies (Ramon *et al.* 2010), which incorporated fuzzy logic and neural networks. However, at least three elements of the FSLSM learning style model may be categorized based on their findings.

Furthermore, in (Sucheta *et al.* 2017), the retrieved learning behavioral data was classified into FSLSM dimensions using a Fuzzy C Means (FCM) method. This method describes the learner in detail, down to each and every category. The student's preferred method of instruction is identified in real-time using the Gravitational Search based Back Propagation Neural Network (GSBPNN) algorithm. However, for the sake of this investigation, just a single course has been planned. Using data from the Moodle server, the authors of (Lasith *et al.* 2016) developed a model to predict a user's preferred learning mode. Data mining is employed to analyses the gathered web server logs, with the Waikato environment for knowledge analysis tool (Weka) serving as the primary analytical tool. This research presents a comparison of the two systems' performance. However, this analysis only takes into account one curriculum. Research in this area suggests that any one of many different learning style models may be used in conjunction with machine learning approaches to automatically determine a student's preferred method of instruction (Ibtissam *et al.* 2020). Thus, the FSLSM is found to be the most widely employed model in these kinds of studies. However, the data used in this research comes from a single course's worth of student activities. Therefore, they needed to test their categorization procedure on many courses before they could have confidence in the results.

3 LEARNING STYLES

There are diverse learning style models, including Kolb (Alice *et al.* 2005), Honey and Mumford (Peter *et al.* 2006), VARK (Weinstein 2008) and Richard and Silverman (Richard *et al.* 2005). Each presents varied depictions and varieties of learning styles. Our work focuses on Kolb's learning model and Felder-Silverman's learning style model (FSLSM) hybrid. Most of the other "learning style models classifies learners into a few classes, whereas Felder and Silverman describe a learner's learning style in more detail", differentiating four dimensions.

FSLSM's four dimensions include perception, understanding information, processing information, and preferred input. Perception can have two variants: concrete (sensing) and abstract(intuitive) (Richard *et al.* 2005). Visual and verbal make up the two types of inputs. Similarly, there can be two ways of information processing, i.e., experimentation(active) and observation(reflective). Finally, understanding information also can be of two ways, including global (no order) and sequential (having an order). Discussing the above-mentioned dimensions in detail, the first dimension deals with reflective and active ways of processing information. As the name suggests, active learners learn actively, but they prefer to learn using learning resources, by applying the gained knowledge and experimenting with things. Further, they might be preferred in discussing with others and prefers to learn in peer-groups. Students who are reflective in nature and open to new ideas are more inclined to study independently or in small groups. They like to ruminate the information and reflect on the study materials.

The second part of the discussion then moves on to the topic of detection vs organic learning. Many students choose a more realistic and substantive approach to education. They are more likely to keep silent while dealing with intricacies and are good at handling problems that need the use of typical procedures. In addition, identifying students' learning patterns is seen as more rational and accountable, since these students will typically be more pragmatic than natural students and appreciate the opportunity to relate the acquired content to the actual world. Students with a strong intuitive bent, on the other hand, thrive when exposed to theoretical concepts like hypotheses and their central implications. They are more likely to be resourceful, creative, and sociable than pupils who focus on detection. Students who learn best from visual depictions (such as drawings, diagrams, and flowcharts) are distinguished from those who learn best from written or spoken descriptions by a third, visual-verbal factor.

Finally, the level of comprehension the learners have is used to characterize them. The learning curve for successive pupils is straight because each one makes little incremental gains. Most of the time, they will methodically use intelligence to find solutions. It's fascinating to see how students all across the world use a comprehensive reasoning cycle and make enormous leaps in their education. When asked for recommendations, people typically provide random suggestions without thinking about whether or not the material is

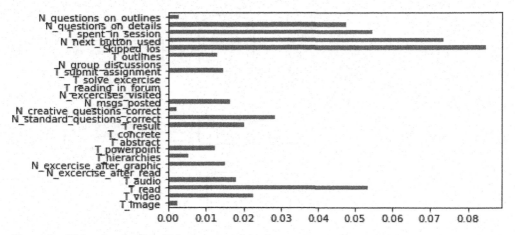

Figure 1. Fisher Scoring for all the features of the dataset.

appropriate. Students from all over the world will be interested in the big picture, so they will focus more on the overview and less on the details than their successors will.

4 DATASET

We created a synthetic dataset with around 1000 samples and 25 characteristics across the FSLSM's four dimensions to conduct this experiment. Theorized dataset structure that allows for many Deep Learning algorithms to be run on a single dataset, with the best results being chosen. Finally, we'd want to separate our data into three distinct sets: a training set for model development, a validation set for checking the accuracy of the model, and a test set for gauging the model's performance on generalized data. Rearranging the Dataset to eliminate biases is a necessary step before dividing it. The Dataset is best divided in a 60:20:20 fashion, with 60% serving as the training set and 20% as the validation set. Train_test_split in sklearn model_selection may be used twice to divide the dataset. Part the Dataset twice: first to separate the train from the approval set, and again to separate the train from the test set. FMutual calculates the amount of information that can be gleaned about one random variable by using information gleaned from another random variable, and may be applied to any pair of (potentially multi-dimensional) random variables, Y and X. Equation 1 provides the solution.

$$I(X : Y) = \int_X \int_y p(x, y) \log\left(\frac{p(x, y)}{p(x)p(y)}\right) dx dy \tag{1}$$

The marginal density functions p(x) and p(y) are defined in terms of the joint probability density function p(x,y). p(x,y) = p(x)p(y) = 0 if and only if X and Y are completely unrelated (thus independent).

5 LEARNING STYLE DETECTION

The objective of predicting the learning style of each individual is fundamentally a classification problem. Thus, we have used several machine learning-based classifiers that are

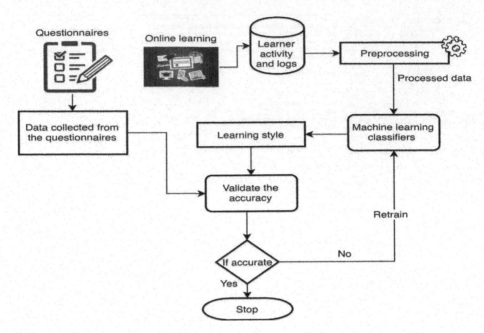

Figure 2. System architecture.

suitable for this kind of dataset. We have implemented and compared multiple classifier algorithms with the data set individual parameter tuning to find the best suited algorithm for this use case. As the dataset is small to medium in size, the use of deep learning algorithms seems not very feasible.

Proposed architecture in Figure 2 comprises of the modules, questionnaires, online learning, learner activity and logs, preprocessing, machine learning classifiers, learning style, validating the accuracy. The first algorithm we tried was decision trees. It is an algorithm that belongs to the supervised learning family, but unlike other algorithms in the family, it can be used for problems involving regression and classification also. In decision trees, predicting a class label for record involves starting from the roof the tree. Based on information gain at each node, the algorithm jumps to next node and follows the values in that branch.

To classify data, we have also utilized the random forest technique. A classifier that uses an average across many tree choices on multiple subsets of a dataset to determine predictive accuracy. It uses the predictions of several different trees to arrive at a conclusion, rather than relying on the results of only one. The machine learning technique known as Gaussian Naive Bayes is a probabilistic model founded on the Bayes Theorem and utilized for a wide range of classification problems. All feature pairs are assumed to be independent, and features are all assigned equal value. It's simple to implement and works well with moderate-to-large datasets. Some issues can be solved more effectively with Naive Bayes's simplicity than with the most complex classification systems. The Multi-Layer Perceptron classifier, a simplified form of neural networks, has also been tested. The network operates in a feedforward fashion and may consist of more than one linear layer of neurons. A loss function is defined to adjust the classifier's accuracy. If the anticipated class matches the real class, the loss is little; otherwise, it is large. While underfitting and overfitting are both possible, the smaller quantity of the dataset in this study likely led to the former. The underfitting issue was fixed again by reducing regularization and feature selection.

120

The boosting technique we developed, called AdaBoost (Adaptive Boosting), is a statistical approach to classification. It has the potential to enhance the efficiency of several machine learning techniques. The boosted classifier's output is a weighted average of the results obtained by many machine learning algorithms. In order to priorities input samples that were mistakenly categorised by earlier classifiers, AdaBoost adapts by changing the weights of its weak learners. It may be less prone to the overfitting problem than other learning methods in certain circumstances. Convergence to a powerful learner can be shown even if individual learners are just marginally more effective than random guessing. Although AdaBoost is often used to combine weak base learners (like decision stumps) in order to produce a more accurate model, it has been demonstrated that it may successfully combine strong base learners (like deep decision trees).

Random Forest makes use of a bagging technique to randomly choose training data. The Bagging Method is a data sampling methodology that reliably reduces prediction error by producing numerous copies of the same data through various permutations and repetitions. When training further decision stumps (trees with one node and two leaves), AdaBoost gives greater weight to fewer data samples that were misclassified in earlier decision stumps. Due to the greater weights placed on the subset of misclassified data, it is more likely to reappear in the new sample.

6 RESULTS

We have compared Adaboost + Random Forest with Decision Tree, Random Forest, Naïve Bayes, Multi-layer Perceptron(Tables from 1 to 4). Results shown that Adaboost + Random Forest is giving better accuracy w.r.t four dimensions of FSLSM model(Figures from 4 to 7). And we proposed model obtained overall accuracy 87.6 as shown in confusion matrix of figure 3.

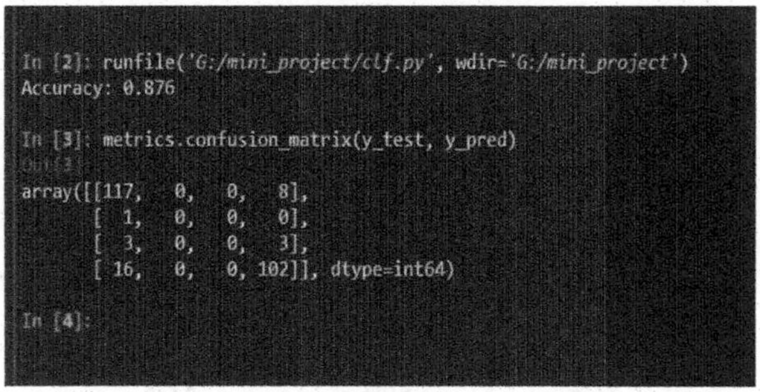

```
In [2]: runfile('G:/mini_project/clf.py', wdir='G:/mini_project')
Accuracy: 0.876

In [3]: metrics.confusion_matrix(y_test, y_pred)
Out[3]:
array([[117,   0,   0,   8],
       [  1,   0,   0,   0],
       [  3,   0,   0,   3],
       [ 16,   0,   0, 102]], dtype=int64)

In [4]:
```

Figure 3.   Accuracy & Confusion Matrix of AdaBoost.

Table 1.   Accuracy of input dimension.

Algorithm	Accuracy	Precision	Recall	F1-score
Decision Trees	78.11%	77%	76%	79%
Random Forest	84.23%	82%	79%	81%
Naïve Bayes	81.21%	80%	77%	79%
Multi-layer Perceptron	84.52%	82%	81%	82%
Adaboost + Random Forest	88%	89%	84%	86%

Table 2. Accuracy of processing dimension.

Algorithm	Accuracy	Precision	Recall	F1-score
Decision Trees	76.64%	79%	81%	81%
Random Forest	82.76%	81%	80%	80%
Naïve Bayes	80.38%	79%	82%	80%
Multi-Layer Perceptron	85.03%	79%	85%	84%
Adaboost + Random Forest	87.81%	82%	87%	83%

Table 3. Accuracy of understanding dimension.

Algorithm	Accuracy	Precision	Recall	F1-score
Decision Trees	77.34%	75%	78%	78%
Random Forest	78.63%	77%	79%	83%
Naïve Bayes	79.46%	84%	76%	81%
Multi-Layer Perceptron	82.12%	84%	78%	81%
Adaboost + Random Forest	87%	81%	79%	87%

Table 4. Accuracy of perception dimension.

Algorithm	Accuracy	Precision	Recall	F1-score
Decision Trees	74.58%	81%	86%	80%
Random Forest	77.52%	79%	88%	80%
Naïve Bayes	80.25%	72%	84%	77%
Multi-Layer Perceptron	81.67%	77%	84%	79%
Adaboost + Random Forest	84%	79%	81%	80%

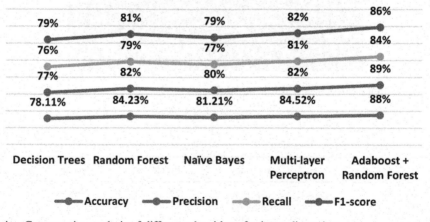

Figure 4. Comparative analysis of different algorithms for input dimension.

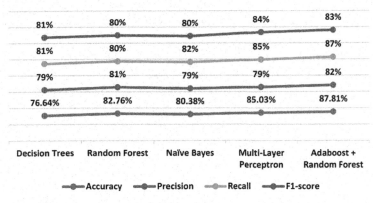

Figure 5. Comparative analysis of different algorithms for processing dimension.

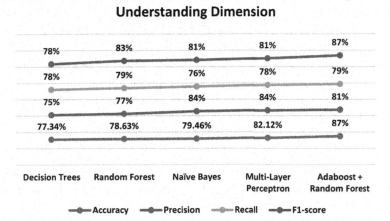

Figure 6. Comparative analysis of different algorithms for understanding dimension.

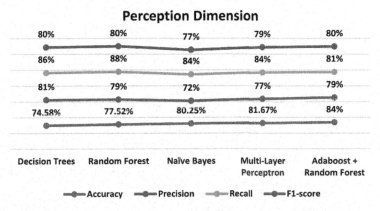

Figure 7. Comparative analysis of different algorithms for perception dimension.

TP: 219
Overall Accuracy: 87.6%

Class	n (truth) ⑦	n (classified) ⑦	Accuracy	Precision	Recall	F1 Score
1	137	125	88.8%	0.94	0.85	0.89
2	0	1	99.6%	0.0	0.0	0.0
3	0	6	97.6%	0.0	0.0	0.0
4	113	118	89.2%	0.86	0.90	0.88

Figure 8.   Accuracy, Precision, Recall & FI Score for each class.

7   CONCLUSION & FUTURE SCOPE

Key features that might be exploited for automatic detection of learning styles were identified using the Felder-Silverman Learning Styles Model (FSLSM). Some examples of these metrics include how much time was spent with an audio or video learning file, how much time was spent in discussion forums, how many assignments were turned in, how many quizzes were taken, how many mails were forwarded, and so on. The reliability of the categorization was checked by running algorithms. Ada-Boosted Random Forest was the most effective method. A decision tree achieved an accuracy of just 83.55 percent. Different algorithms' accuracy were determined for each FSLSM model dimension. The Support Vector Machine performed best in terms of accuracy in the input dimension. The findings are encouraging, and the situation is much better now.

We have noticed a wide range of student learning behaviors, such as shifts in learning styles as circumstances shift, such as when an exam is coming up or when the material becomes much more challenging. Our trained model will be useful in the future for predicting learners' preferred methods of instruction in a real-world setting, where teaching and learning strategies are constantly evolving. The identified patterns are generalizable across different types of students.

REFERENCES

Alice Kolb & David Kolb, "The Kolb learning Style Inventory — Version 3. 1 2005 Technical Specifications," *LSI Tech. Man.*, pp. 1–72, 2005.

Behram Beldagli & Tufan Adiguzel, "Illustrating an Ideal Adaptive e-learning: A Conceptual Framework," *Procedia- Social and Behavioral Sciences*, 2010, vol. 2, pp. 5755–5761.

Christos Troussas, *et al.* (2019). An Intelligent Adaptive Fuzzy-based Inference System for Computer-assisted Language Learning. *Expert Systems with Applications*, 127, 85–96.

Fathi Essalmi, *et al.* "A Fully Personalization Strategy of E-learning Scenarios", *Computers in Human Behavior*, 2010, vol. 26, pp. 581–591.

https://onlinelearningconsortium.org/read/olc-research-center-digital-learning-leadership/babson-survey-research-group/

Ibtissam Azzi & Adil Jeghal & Abdelhay Radouane & Ali Yahyaouy & Hamid Tairi, "A Robust Classification to Predict Learning Styles in Adaptive E-learning Systems", *Education and Information Technologies* (2020) 25:437–448.

Ibtissam Azzi, *et al.* A Robust Classification to Predict Learning Styles in Adaptive E-Learning Systems. *Education and Information Technologies* (2020) 25:437–448.

Jack Snowman, "Educational Psychology: What Do We Teach, What Should We Teach?:' *Educational Psychology Review*, vol. 9, no. 2, pp. 151–170, 1997.

Keeley Crockett, *et al.* (2013). A fuzzy Model for Predicting Learning Styles using Behavioral Cues in an Conversational Intelligent Tutoring System. *2013 IEEE International Conference on Fuzzy Systems (FUZZ-IEEE)* (pp. 1–8). IEEE.

Khalid Almohammadi, *et al.* (2017). A Type-2 Fuzzy Logic Recommendation System for Adaptive Teaching. *Soft Computing*, 21(4), 965–979.

Lasith Gunawardena, *et al.* (2016). Detecting Learning Styles in Learning Management Systems Using Data Mining. *Journal of Information Processing*, 24(4), 740–749.

Ling Xiao Li & Siti Soraya Abdul Rahman (2018). Students' Learning Style Detection Using Tree Augmented Naive Bayes. *Royal Society Open Science*, 5(7), 172108.

Lukito Nugroho, *et al.* (2019). Model Detecting Learning Styles with Artificial Neural Network. *Journal of Technology and Science Education*, 9(1), 85–95.

Manal Abdullah, *et al.* (2015). Learning Style Classification Based on Student's Behavior in Moodle Learning Management System. *Transactions on Machine Learning and Artificial Intelligence*, 3(1), 28.

Minoru Nakayama, *et al.* "Learner Characteristics and Online Learning.," In *Encyclopedia of the Sciences of Learning*, N. M. Seel, Ed., Springer US, 2012, pp. 1745–1747.

Norazlina Ahmad, *et al.* Automatic Detection of Learning Styles in Learning Management Systems by Using Literature Based Method. In: *13th International Educational Technology Conference*, Vol. 103, pp. 181–189, Procedia, Elsevier (2013)

Patricio Garcia, *et al.* Evaluating Bayesian Networks' Precision for Detecting Students' Learning Styles. *Computers & Education*, 49(3), 794–808.

Paul Sheeba & Reshmy Krishnan (2019). Automatic Detection of Students Learning Style in Learning Management System. In *Smart Technologies and Innovation for a Sustainable Future* (pp. 45–53). Springer, Cham.

Peter Honey & Alan Mumford *et al.*, "*The Learning Styles Helper's Guide*," Peter Honey Publ., vol. 1, no. 1, pp. 1–3, 2006.

Ramon Zatarain-Cabada, *et al.* (2010). A Learning Social Network with Recognition of Learning Styles Using Neural Networks. *Mexican Conference on Pattern Recognition*, pp. 199–209.

Richard Felder & Joni Spurlin, "Applications, Reliability and Validity of the Index of Learning Styles," *Int. J. Eng. Educ.*, vol. 21, no. 1, pp. 103–112, 2005. (10)

Richard Felder & Linda Silverman, "Learning and Teaching Styles in Engineering Education," *Eng. Educ.*, vol. 78, no. June, pp.674–681, 1988.

Sabine Graf (2007). *Adaptivity in Learning Management Systems Focussing on Learning Styles*. PhD thesis, Vienna University of Technology, 9801086 Neulinggasse 22/12A 1030, Vienna.

Sucheta Kolekar, *et al.* (2017). Prediction of Learner's Profile Based on Learning Styles in Adaptive E-learning System. *International Journal of Emerging Technologies in Learning (iJET)*, 12(6), 31–51.

Umadevi, *et al.* (2014). Design of E-learning Application Through Web Mining. *International Journal of Innovative Research in Computer and Communication Engineering*, 2(8).

Virginia Yannibelli, *et al.* A Genetic Algorithm Approach to Recognize Students' Learning Styles. *Interact. Learn. Environ.* 14(1), 55–78 (2006)

Weinstein, "Learning Styles.," *Learn. Styles – Res. Starters Educ.*, pp. 6–7, 2008.

Yavuz Akbulut & Cigdem Suzan Cardak, "Adaptive Educational Hypermedia Accommodating Learning Styles: A Content Analysis of Publications From 2000 to 2011," *Computers & Education*, 2012, vol. 58, pp. 835–842.

Artificial Intelligence, Blockchain, Computing and Security – Dagur et al. (Eds)
© 2024 The Author(s), ISBN: 978-1-032-49393-0

Exploratory Data Analysis (EDA) based on demographical features for students' performance prediction

Neeraj Kumar Srivastava, Prafull Pandey & Vikas Mishra
United Institute of Technology, Naini, Prayagraj, Uttar Pradesh, India

ABSTRACT: This paper presents an Exploratory Data Analysis (EDA) upon the dataset of students undergoing in secondary education and want to go ahead to acquire higher education. The dataset contains demographic features that include social and academic attributes of students. In this EDA, machine learning platform is incorporated along with python language to perform different analysis on the dataset under EDA. This EDA further approaches to the performance prediction of students' indicating the effect of different demographical features on students' performance. This will help in identifying the attributes that affect the student's performance in positive and negative both the aspects.

1 INTRODUCTION

In modern era, the pattern of data collection has been changed entirely as instead of recording data on notebooks, electronic data is being preferred, which is easy to record in digital files and easier to analyze also. To provide ease in recording huge amount of digital data and to analyze it properly, the concept of Educational Data Mining (EDM) is incorporated as an important tool that easily analyzes and retrieves useful and more relevant set of information from large digital files by means of different data withdrawal classifiers.

In general, mostly educational organizations emphasizes upon the academic parameters of students in order to assess their performance, but in modern time, it has been figured out that only academic parameters are not sufficient to predict students' performance to achieve certain goals, instead, the demographical attributes of students play very important role in performance prediction. Hence, family background, age of student, type of jobs of their parents, romantic relationship, exposure to social media and internet, alcohol consumption by the students, going out with parents etc. are equally very important to predict a more fruitful performance of students. For that reason, Exploratory Data Analysis (EDA) is to be performed over the demographical data of students to device a significant and comprehensive outcome while predicting students' future performance, in this paper.

Here, EDA(Exploratory Data Analysis)is performed on demographical data of students who are undergoing in higher secondary education. In the dataset demographical features of students have been recorded including age, gender, size of family, parents' education, students' relationship status, free time, study time, paid classes opted by students, pattern of alcohol consumption by the student and health conditions etc. This analysis will project the essential features regarding the future performance of the students. The dataset includes a single target variable "S_passed" that contains binary values either "yes" or "no". This analysis not only helps in predicting the future performance of students using demographical features but also suggest the features that could affect the students' performance in either positive or negative manner.

Here, we are going to use the platform of python language to perform all of the desired analysis by examining the academic, family and social background details, which is already

DOI: 10.1201/9781003393580-19

recorded in the electronic dataset. This analysis on categorical dataset will help in significant manner to predict the success rate along with the features that play vital part in improving the recital of students. The visualizations of the whole process of analysis along with findings will be discussed in section 4 of this paper.

The upcoming sections of the paper include: Section 2 contains an impression of allied works along with their outcomes, Section 3 represents the methodology adopted in the proposed study, Section 4 shows the findings of the analysis in form of results and Section 5 tells about the winding up stage of the study with future extent.

2 LITRATURE REVIEW

Ahajjam Tarik *et al.* [1] have suggested a method implying AI & ML algorithm to study and forecast students' recitals on different factors at the time of COVID 19. In this research, they have used baccalaureate mean as a function of many explanatory variables as the grades of the core subjects. In order to obtain the best possible results, they have tested many models and finally adopted the random forest algorithm, which gave the best predictions than other algorithms tested to predict the students' performance.

Akarshita Tripathi *et al.* [2] have advised a replica for student performance prediction implying Naïve Bayes Classification Model. They have concluded that the advised replica has tall correctness and a slight execution time in realizing the outcome than the SVM classification model.

Senthil Kumar *et al.* [3] have projected a replica for analyzing the students' performance with the concept of exploratory data analysis using ML classifier. In this replica, they have emphasized upon the application of fresh technologies to analyze the pupils' performances by the institutions and other educational organizations instead of using manual or tradition model of data analysis. Hence, they propose a model that entirely focuses not only upon the improvement the students' performance but also help the institutions to make better decisions by incorporating EDA.

Febrianti Widyahastuti *et al.* [4] have put forward a replica to forecast the outcome of students' act in Final Examination implying Linear Regression and Multilayer Perceptron methods and concluded that the Multilayer Perceptron produced better results than Linear Regression method in provisions of correctness.

Hanan Aldowah *et al.* [5] have conducted a go over and amalgamation on Educational data Mining learning analytics for higher education in 21st century and also concluded that their proposed model is able to offer the best possible way to solve certain learning problems.

Iti Burman *et al.* [6] have planned a replica to forecast student academic routine using Support Vector Machine that focused upon Non Intellectual parameters of students that actually affects the pattern of study and growth of students. They have performed Psychometric analysis of students' behavior to analyze and improve their performance with the help of different mining techniques like neural network and observed that within this model, Radical Base Function kernel has given more accurate results than Linear Kernel.

Kwok Tai Chui *et al.* [7] have suggested a facsimile to forecast the performance of students' with backdrop of their as well as their parents' educational background. They have used Generative Adversarial Network (GAN) Based Deep Support Vector Machine as a model in their proposal. This model will provide accurate prediction for students who are at-risk by growing their triumph rate of passing the itinerary and keep away from passing with insignificant marks. They have concluded that GAN model will benefit to users significantly, especially in case of tiny machine learning troubles with fresh data.

Mudasir Ashraf *et al.* [8] have developed a sharp forecasting system for a learning data mining based upon Ensemble and Filtering approaches of Machine Learning to examine the influence of the algorithms on the forecasting accuracy of learning classifiers and recommend a improved forecasting system on pedagogical data.

Muslihah Wook *et al.* [9] have proposed an incorporated replica as an extension of earlier models and focused upon individual and technology-specific features in order to forecast and describe the dissent of the target constructs. The authors have integrated two constructs- Technology Readiness Index (TRI) and Technology Acceptance Model3 (TAM3) to propose the new model for Educational Data Mining (EDM) used for analyzing specific large dataset in the educational setups that helps significantly in forecasting the student performance.

Lynn *et al.* [10] have proposed their go over on using data mining methods in prediction of students' performance and observed that the decision trees method, which is a simple method, is confirmed to be the finest manner for forecasting student performance with high level of correctness.

Nikola Tomasevic *et al.* [11] have presented a comprehensive comparison and study of diverse ML algorithms for supervised data mining methods to forecast pupils exam outcome which is aimed at providing a guess of potential pupils' reaching in special tests.

Vladimir L. Uskov *et al.* [12] have proposed ML based prognostic study of student learning recital in Science, Technology, Engineering and Mathematics in particular. This model included four types of data analytics: descriptive, diagnostic, predictive and prescriptive.

3 RESEARCH METHODOLOGY

The concept of Exploratory Data Analysis (EDA) along with EDM is going to be implied in this mock up, which is principal action in any Data Analysis or Data Science project. The EDM is a course of action of recording, analyzing and to probe the dataset to determine blueprint and irregularity to produce a hypothesis based on our thoughtful of the educational data. Similarly, EDA is method used to examine the underlying structure of a dataset thoroughly and its importance for the organization. It helps in understanding the trends, patterns and relations, which are not visible normally.

In order to perform EDA on dataset, the proposed model includes following steps:

3.1 *Description of data*

The dataset used is this paper contains demographical information of students, who are studying in secondary education. The dataset contain school name, age, sex, parents' education status, parents' working status, study time of students, pattern of daily and weekly alcohol consumption by the students, romantic relationship status, health status, pattern of extra classes attended by students etc. The data fields contain numeric, binary and ordinal values.

In the below given catalogue (Table 1), the particulars of all the columns along with their data types and possible values that can be contained in the columns accordingly, has been described:

Table 1. Description of every column in the database.

Field Name	Description	Data Type	Value
Sch_name	Name of school of student	Bool	"GP" or "MS"
Gen	Gender of student	Bool	"m" or "f"
Stud_Age	Age of student	Numb	Range- between 15 -to- 22
Stud_add	Type of student locality	Bool	R / U
Size_fam	Family size	Bool	"GT3" – more than 3/ "LE3" – below or equal to 3
P_stat	Status of parents co-living	Bool	"T" – with each other or "A" – separate

(continued)

128

Table 1. Continued

Field Name	Description	Data Type	Value
Moth_ed	Qualification of mother	Ordinal	Highly Educated -5, HSC- 4, Matriculation -3, Basic- 2, other- 1
Fath_ed	Qualification of father	Ordinal	Highly Educated -5, HSC- 4, Matriculation -3, Basic- 2, other- 1
Moth_job	Profession of mother	Ordinal	"teaching", "health worker", "Govt. Job", "No work" or "other than all"
Fath_job	Profession of father	Ordinal	"teaching", "health worker", "Govt. Job", "No work" or "other than all"
Rea_son	Cause to select the school	Ordinal	Near to house, fame of school, program availability or other
guard_ian	Guardian of student	Ordinal	Father/Mother/Other
Travel_time	Time incurred to reach school	Numb	1 – less than 10 min, 2- 10 to 25 min., 3 – 25 min. to 55 min, 4 – more than 1 hour
Study_time	Learning duration	Numb	1 – less than 10 min, 2- 10 to 25 min., 3 – 25 min. to 55 min, 4 – more than 1 hour
Failure_No	Count of previous collapse in studies	Numb	M, where m is between 1 to 4
School_sup	Learning support from school	Bool	True/ False
Fam_sup	Hold of family in studies	Bool	True/ False
Paid_classes	Attended paid lectures	Bool	True/ False
Extra_acti	supplementary actions	Bool	True/ False
Sch_nurs	Pre school status	Bool	True/ False
Higher_edu	Interest in high studies	Bool	True/ False
Net_use	Net availability	Bool	True/ False
Rom_rel	Romance excellence	Bool	True/ False
Famly_rel	Relationship excellence	Ordinal	from 1 – poor to 5 – Out standing
Free_time	Outing in the frère hours	Ordinal	from 1 – poor to 5 – Outstanding
Go_out	Number of outings	Ordinal	from 1 – poor to 5 – Outstanding
Daily_alc	Alcohol intake during week day	Ordinal	from 1 – poor to 5 – Outstanding
Weekly_alc	Alcohol intake in the week End	Ordinal	from 1 – poor to 5 – Outstanding
Health_sts	Status of health	Ordinal	from 1 – poor to 5 – Outstanding
No_abs	Sch. Absent count	Numeric	Above 0 upto 90
S_Passed	Passed the exam or not	Bool	True/ False

3.2 Handling missing values

When we record data in files, it often consists of noise, incorrect values and even some values are missing or incomplete. Therefore, we need preprocess the dataset to make it clean, correct and complete. Such a dataset certainly leads to unreliable findings. Following preprocessing tasks will be applied on dataset in order to achieve reliable-findings:

1. Replacing missing values
2. Dropping columns

3.3 Handling outliers

The term outlier means different or separate, hence, with respect to EDA, it can be considered as an outcome of some mistake at the time of data collection or it may indicate a variation of data in the dataset. Some frequently used methods in python that are used to handle and detect outliers are:

1. Box Plots
2. Scatter Plots, etc.

3.4 *Understanding relationships and new insights through plots*

In EDA, we perform various types of visualizations on our dataset to observe relations between the key features by using some mechanisms. These visualizations also help to get the important insights of the dataset which are not clearly visible by simply watching or reading the dataset. Some of the important visualization techniques are:

1. Histogram
2. Heat map

4 RESULT AND DISCUSSION

The dataset used in this paper contains demographical features of students having 31 features and 395 records. The EDA upon this dataset has clearly observes that out of 395 students, 265 students have been passed whereas 130 have been failed depending upon the demographic information provided in the dataset. The count of passed and failed students has been plotted in Figure 1, whereas the pass-fail ratio has been plotted in Figure 2:

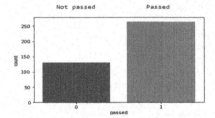

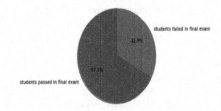

Figure 1.   Status of passed (1) and not passed (0) students.

Figure 2.   Percentage of passed and not passed students.

In this analysis, the EDA has observed that age is an important feature of students that affect their performance significantly. It is clearly observed that students below 20 years have the higher possibility to get passed whereas students above 20 years have less chance to be passed (Figure 3).

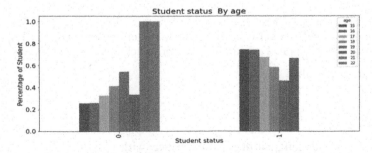

Figure 3.   Impact of age on student's performance.

Here in this paper, the EDA has been used to analyze the dataset to find out the impact of categorical values, ordinal values and continues values upon the performance of students. With respect to categorical features as shown in Figure 4, it has been observed that 52.7% female and 47.3% male students have passed the exam (where, 0-Male, 1- Female). Further, 94.9% students, who want to go for higher education, have passed the final exam(where,0-No, 1-Yes). Similarly, 83.3% students have passed the exam, who had internet connection (where, 0-No, 1-Yes).

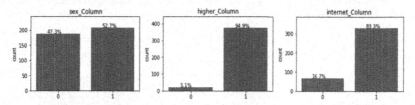

Figure 4.   Impact of sex, higher education and internet connection on performance of students.

Apart of the categorical features those have positive impact on the performance of students, there are some attributes (shown in Figure 5), which have negative impact also. In the analysis, it has been found that only 22.3% students, who live in urban area (where, 0-Urban, 1-Rural), have passed in final exam. The parents' status i.e. they live together or apart (where, 0-Togather, 1-Apart) also affects the students' performance. Only 10.4% students, whose parents live apart, have passed the final exam. The support from school is also important. Only 12.9% students, who have lesser support from school, have passed the final exam (where. 0-No, 1-Yes).

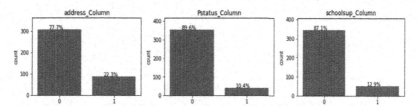

Figure 5.   Impact of living location, parent's status and school support on performance of students.

The attributes of dataset having ordinal values like mother's/father's education, going out with friends, health, travel time and daily & weekly alcohol consumption by the students etc. have significant impact on the performance of the students. According to this analysis, 33.2% students, whose mothers are highly qualified (1 other, 2 – primary education (4th grade), 3 – 5th to 9th grade, 4 – secondary education, 5 – higher education), have passed the final exam. This analysis shows that the students, who do not frequently go out with friends (32.9%)(from 1 – very low to 5 – very high), have higher chance to passed the final exam. The health of individual students is one of the most important attributes that affects the

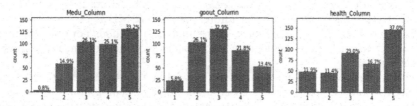

Figure 6.   Impact of mother's education, going out and health conditions on performance of students.

131

performance of students significantly as 37.0% students (from 1 – very bad to 5 – very good), who have passed the exam, were healthy enough.

As per the analysis, there are some attributes in the dataset, having ordinal values such as travelling time, pattern of daily or weekly alcohol consumption by the students, have negative impact over the performance of students as shown in Figure 7. Only 2% students passed in exam, who travel outside for longer duration of time or frequently (1 – <15 min., 2 – 15 to 30 min., 3 – 30 min. to 1 hour, or 4 – >1 hour). Similarly, only 2.5% – 7.1% students have passed the final exam, whose daily or weekly alcohol consumption is very high (from 1 – very low to 5 – very high).

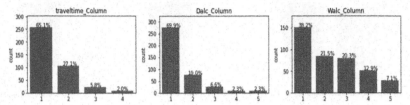

Figure 7.   Impact of travelling time and daily & weekly alcohol consumption on performance of students.

On the basis of features having numerical (continuous) values such as age, failures and no. of absences (shown in Figure 8) in the dataset, it has been clearly observed that age of students has significant on the students' performance. The success rate of student, whose age group is between 15–20 years, is quite high. This analysis also shows that no. of failures and no of absences have very crucial impact on the students' performance. The success rate increases if the no. of failures and no. of absences during the study are low.

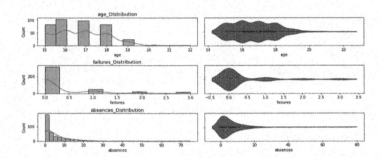

Figure 8.   Impact of features (Age, failures & no. of absences) having numerical values on students' performance.

5   FINDINGS

In this paper, Exploratory Data analysis (EDA) has been performed on demographical information of students to forecast their recital in the concluding assessment during their secondary education. This is very challenging task to predict accurate results for the betterment of the learning environment. Therefore, the EDA & ML algorithms, together, have made it simpler in an innovative manner. These analytical methods not only produce accurate results but also help us to identify the needs and areas of improvement to achieve more fruitful results.

Henceforth, the idea to present an analytical model that analyzes the pattern of student's data to predict students' performance based upon different demographical features is proposed. The main challenge in this model was to suggest the students with a comprehensive summary of positive feature that will help them to achieve higher academic status and to avoid failures that may come in their way. As per our analysis, following features have positive impact over students' performance:

- Mother's Job
- Father's Job
- Want to take Higher Education
- Study time

This model also outlines different features among the demographical features that leaves negative impact on students' performance, hence, should be kept in mind. Following features have negative impact over the students' performance:

- Age
- Going out with Friends
- No. of Absences
- No. of Failures

6 CONCLUSION

This research significantly figures out that there are some certain demographical features that have positive and negative impact on the students' performance. It contributes to educational institutes and Educationists to design policies and desired models to observe and analyze the patterns that lead students to success.

REFERENCES

[1] Ahajjam Tarika, Haidar Aissab, Farhaoui Yousef, "Artificial Intelligence and Machine Learning to Predict Student Performance during the COVID-19", *3rd International Workshop on Big Data and Business Intelligence*, Warsaw, Poland, 2021.
[2] Akarshita Tripathi, Saumya Yadav, Rajiv Rajan, "Naïve Bayes Classification Model for the Student Performance Prediction", *2nd International Conference on Intelligent Computing, Instrumentation and Control Technologies (ICICICT)*, 2019.
[3] Dr. A. Senthil Kumar, K. Joshna, "Student's Performance Analysis with EDA and Machine Learning Models", *International Journal of Scientific Research in Science and Technology*, 2021.
[4] Febrianti Widyahastuti, Viany Utami Tjhin, "Predicting Students Performance in Final Examination using Linear Regression and Multilayer Perceptron", *International Conference on Intelligent Computing, Instrumentation and Control Technologies (ICICICT)*, IEEE, 2017.
[5] Hanan Aldowaha, Hosam Al-Samarraiea, Wan Mohamad Fauzyb, "Educational Data Mining and Learning Analytics for 21st Century Higher Education: A Review and Synthesis", *Telematics and Informatics*, ELSEVIER, 2019.
[6] Iti Burman, Subhranil Som, *"Predicting Students Academic Performance Using Support Vector Machine"*, IEEE, 2019.
[7] Kwok Tai Chui, Ryan Wen Liu, Mingbo Zhao, Patricia Ordóñez De Pablos, "Predicting Students' Performance With School and Family Tutoring Using Generative Adversarial Network-Based Deep Support Vector Machine" *IEEE Access*, 2020.
[8] Mudasir Ashraf, Majid Zaman, Muheet Ahmed, "An Intelligent Prediction System for Educational Data Mining Based on Ensemble and Filtering approaches", *International Conference on Computational Intelligence and Data Science*, 2019.
[9] Muslihah Wook, Zawiyah M. Yusof, "Educational Data Mining Acceptance Among Undergraduate Students", *EducInf Technol*, Speringer, 2016.
[10] Lynn N D and Emanuel A W R, *"Using Data Mining Techniques to Predict Students' Performance: A Review"*, ICIMECE, 2020.
[11] Nikola Tomasevic, Nikola Gvozdenovic, Sanja Vranes, *"An Overview and Comparison of Supervised Data Mining Techniques for Student Exam Performance Prediction"*, ELSEVIRE, Computer and Education, 2020.
[12] Vladimir L. Uskov, Jeffrey P. Bakken, Ashok Shah, Adam Byerly, "Machine Learning based Predictive Analytics of Student Academic Performance in STEM Education", *IEEE Global Engineering Education Conference*, 2019.

Artificial Intelligence, Blockchain, Computing and Security – Dagur et al. (Eds)
© 2024 The Author(s), ISBN: 978-1-032-49393-0

Vehicle detection using Artificial Intelligence for traffic surveillance

Soma Ajay, Sai Vardhan Reddy, Tharun, Santhosh Kumar Pandian & T. Shakila
Bharath Institute of Science and Technology, Chennai, India

ABSTRACT: Developing countries experience traffic congestion as a significant problem. With smart traffic light systems, light configurations are automatically adjusted according to traffic conditions in real-time. It would be necessary for the system to have information on traffic density to adjust properly. With the help of neural networks, a vehicle counting system that can be used to calculate vehicle counts on traffic roads. This system detects vehicles by using YOLO (You Only Look Once), a neural network-based object detection algorithm. Simple Online and Real-time Tracking (SORT) algorithms are used to count and categorise vehicles in traffic recordings. After counting we will find out the direction of vehicle movement as well as we will try to implement the type of vehicle count like how many cars, bicycle and so on. The proposed neural network structure is better suited for real-time vehicle tracking, because the computational complexity is reduced.

1 INTRODUCTION

A traffic monitoring system provides accident detection and traffic surveillance via vehicle counting and traffic monitoring [1]. A traffic monitoring system detects and estimates the position of vehicles based on video images while they are still present in the scene through a framework [2]. In traffic flows with a variety of vehicle models and a high vehicle density, it is challenging to precisely identify and classify vehicles. Vehicle detection is further complicated by environmental changes, diverse vehicle attributes, and generally sluggish detection speeds [3]. As a result, the creation of an algorithm capable of accurate vehicle detection and real-time computation will be necessary for a real-time traffic monitoring system. Therefore, it is both theoretically possible and practically feasible to identify automobiles efficiently and precisely from traffic photos or videos [4].

Accident detection and aided traffic surveillance are made possible by traffic monitoring through intelligent transportation systems. A framework is essentially what a traffic monitoring system is. for detecting and estimating the location of vehicles in video images while they are still present in the scene [5]. In complex scenes with multiple challenging, to precisely find and categorise automobiles in traffic flows because of various vehicle models and a high vehicle density. Furthermore, changing environmental conditions, varying vehicle characteristics, and relatively slow detection speeds all contribute to vehicle detection limitations [6]. As a result, creating an algorithm that can compute in real-time and accurately recognise vehicles is necessary for a real-time traffic monitoring system. As a result, detecting vehicles in traffic images or videos can be done accurately and quickly [7].

Deep learning-based object detection algorithms have gained a great deal of attention. Using machine learning, these algorithms can automatically extract the features, to provide those with strong capabilities for image abstraction and automatically high-dimensional feature representation [8].

DOI: 10.1201/9781003393580-20

2 METHODOLOGIES

2.1 *Importing libraries and setting path*

Using the Video Capture function in cv2, import the video with an objects and labels to be recognized.

2.2 *Model backbone*

Model Backbone's main objective is to automatically extract features from photos [9]. Cross Stage Partial Networks (CSP) is the foundation of YOLO v5's ability to extract useful characteristics from the input photos.

2.3 *Model neck*

Pyramids with features are typically created using the Model Neck. Pyramids of features aid in generalisation of models for object scaling. It helps with object identification when it appears in many scales and sizes.

2.4 *Feature pyramids network*

PANet is used as a neck in YOLO v5 to obtain feature pyramids. Detection of Objects Feature Pyramid Networks are primarily used within the model head to perform the final detection part, which is extremely useful and helps models perform well on unobserved data [10]. Other models, such as FPN, BiFPN, and PANet, are available and to use various types of feature pyramid techniques (as shown in Figure 1), etc.

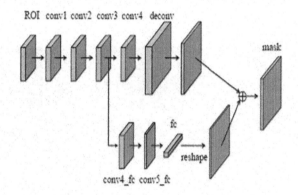

Figure 1. PANet.

2.5 *GUI design*

The official Python module for the Qt for Python project is called PySide6, and it gives users access to the whole Qt 6.0+ framework [11]. It is simpler to incorporate into commercial projects when compared with pyqt. It has Flexibility and cleaner codebase. Main window, upload video, play, stop, counting vehicle tab, total count tab.

2.6 *Getting bounding boxes*

In the original study report, YOLO was only able to predict two bounding boxes per grid cell. Although it is possible to raise that number, only one class prediction can be performed for each grid cell, which restricts the detections when several objects are present in a single grid cell like "bicycle", "car", "motorbike", "bus", and "truck" [12].

2.7 Non-maximum suppression

Even though we eliminated the low confidence bounding boxes, it's possible that one object will still be the subject of much detection. The final stage of these object detection methods, known as non-max suppression, is used to choose the item's ideal bounding box. In order to choose the most advantageous bounding box from the multiple predicted bounding boxes, these object identification algorithms employ non-max suppression. By employing this approach, the best bounding box is "suppressed" in favour of the less likely ones [13]. To choose one bounding box; we pass it the confidence threshold value and NMS threshold value as arguments. To determine which bounding box is the best fit for an object and to "suppress" any other bounding boxes, non-max suppression work is necessary.

2.8 Implementation of YOLO V3

The configuration file for YOLOv3, which specifies the layers and other crucial network parameters like the number of filters in each layer, learning rate, classes, stride, input size for each layer and channels, output tensor, etc., is all that is needed from the library in order to implement the pre-trained YOLOv3 network. The config file gives the basic structure of the model by defining the number of neurons in each layer and different kinds of layers [14]. With the help of the config file, one could start training their model with either a pre-existing dataset like COCO, Alex net, MNIST dataset for handwritten digits detection etc. A database called Common Objects in Context (COCO) seeks to facilitate future studies on object detection, instance segmentation, image captioning, and the location of human important points as shown in Figure 2. A sizable object detection, segmentation, and captioning dataset are called COCO.

Figure 2.   Large scale object detection.

2.9 YOLO V3

For object detection, YOLOv2 employed a customized CNN called Darknet-19 that had 30 layers total, including 19 from the original CNN and 11 more. Despite having a 30 layer architecture, YOLOv2 had trouble detecting small objects, which was thought to be because as the input travelled through each pooling layer, fine-grained information were lost.

Identity mapping and concatenating characteristics were utilized from preceding layers to determine low lever features, in order to compensate for this [15].

Ever after all this, it lacked several important aspects of an object detection algorithm which made it stable such as residual blocks, skip connections and up sampling layers.

These corrections were made and a new version of YOLO was born and that is known as YOLOv3. Additionally, YOLOv3 employs a variation of Darknet-53 with 53 convolutional layers learned on an image net for the purpose of classification with an additional of 53 more layers stacked onto it to make it a full-fledged network to perform classification and detection as shown in Table 1. As a result, YOLOv3 is slower than the second version but a lot more accurate than its predecessors.

Table 1. Darknet -53.

	Type	Filters	Size	Output
	Convolutional	32	3 × 3	256 × 256
	Convolutional	64	3 × 3 / 2	128 × 128
1×	Convolutional	32	1 × 1	
	Convolutional	64	3 × 3	
	Residual			128 × 128
	Convolutional	128	3 × 3 / 2	64 × 64
2×	Convolutional	64	1 × 1	
	Convolutional	128	3 × 3	
	Residual			64 × 64
	Convolutional	256	3 × 3 / 2	32 × 32
8×	Convolutional	128	1 × 1	
	Convolutional	256	3 × 3	
	Residual			32 × 32
	Convolutional	512	3 × 3 / 2	16 × 16
8×	Convolutional	256	1 × 1	
	Convolutional	512	3 × 3	
	Residual			16 × 16
	Convolutional	1024	3 × 3 / 2	8 × 8
4×	Convolutional	512	1 × 1	
	Convolutional	1024	3 × 3	
	Residual			8 × 8
	Avgpool		Global	
	Connected		1000	
	Softmax			

2.10 Structure of YOLOv3

The main difference between YOLOv3 (as shown in Figure 3) and it's a predecessor is that it forecasts on three distinct scales [16]. The initial detection is performed in the 82nd layer. If an input image of 416×416 is fed into the network, the feature map thus acquired would be of size 13×13. The other two scales at which detections happen are at the 94th layer yielding a feature map of dimensions 26x26x255 and the final detection happens at the 106th layer, resulting in a feature map with dimension 52x52x255.

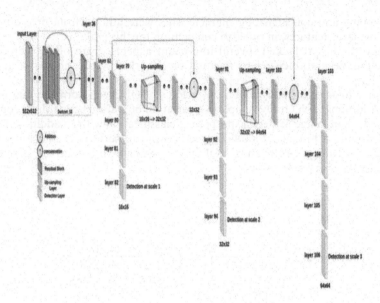

Figure 3.   YOLOv3 network architecture.

The detection which happens at the 82nd layer, is liable for the detection of large objects and the detections which happen at the 106th layer is liable for detecting small objects with the 94th layer, staying in-between these 2 with a dimension of 26×26, detecting medium size objects. This kind of varied detection scale renders YOLOv3 good at detecting small objects than its predecessors as seen in below Figure 4.

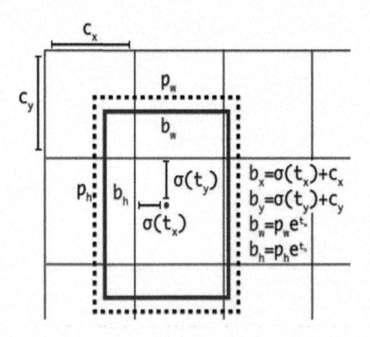

Figure 4.   Object detection scale.

YOLOv3 uses 9 anchor boxes to localize objects with 3 for each detection scale. The anchor boxes are assigned in the descending order with the largest 3 boxes of all for first detection layer which is used to detect large objects, the next 3 for the medium sized objects detection layer and the final 3 for the small objects' detection layer.

YOLOv3 utilizes 10 times as many bounding boxes than YOLOv2 does since it detects at three different scales. For instance, for an image of input size 416×416, YOLOv2Would predict 13x13x5 = 845 boxes whereas YOLOv3 would go for 13x13x5 + 26x26x5 + 52x52x5 = A whopping 10,647 boxes.

The loss function of YOLOv3 was modified

$$
\lambda\text{coord} \sum_{i=0}^{S^2} \sum_{j=0}^{B} \mathbb{I}_{ij}^{obj} (x_i - \hat{x}_i)^2 + (y_i - \hat{y}_i)^2
$$

$$
+ \lambda\text{coord} \sum_{i=0}^{S^2} \sum_{j=0}^{B} \mathbb{I}_{ij}^{obj} (\sqrt{w_i} - \sqrt{\hat{w}_i})^2 + \left(\sqrt{h_i} - \sqrt{\hat{h}_i} \right)^2
$$

$$
+ \sum_{i=0}^{S^2} \sum_{j=0}^{B} \mathbb{I}_{ij}^{obj} (C_i - \hat{C}_i)^2 \tag{1}
$$

$$
+ \lambda_{noobj} \sum_{i=0}^{S^2} \sum_{j=0}^{B} \mathbb{I}_{ij}^{noobj} (C_i - \hat{C}_i)^2
$$

$$
+ \sum_{i=0}^{S^2} \mathbb{I}_i^{obj} \sum_{c \in classes} (p_i(c) - \hat{p}_i(c))^2
$$

This is the loss function of YOLOv2 where the last 3 terms in this image correspond to the function which penalizes for the objectless score predicted by the model for responsible for predicting things are bounding boxes. The second-to-last term is in charge of the bounding boxes that do not contain any objects, while the last term penalizes the model for the class prediction score for the bounding box that contains predicted objects. The terms involved in the calculation of loss function in YOLOv2 were calculated using Mean Squared Error method while in YOLOv3; it was modified to Logistic Regression since this model offer a better fit than the previous one.

YOLOv3 executes classification of multiple labels for objects that are detected in images and videos. In YOLOv2, soft maxing is performed on all the class scores and whichever has the maximum class score is assigned to that object. This rests on the assumption that if one object belongs to one class, it can't be a part of another class. For instance, it is not necessary that an object belonging to the class Car would not belong to the class Vehicle. An alternative approach to this would be using logistic regression to predict class scores of objects and setting a threshold for predicting multiple labels. Classes that have scores greater than the threshold score are then assigned to the box.

YOLOv3 was benchmarked against popular state of the art detectors like RetinaNet50 and RetinaNet101 with the COCO mAP 50 benchmark where 50 stands how closely the predicted and actual bounding boxes match up will determine how accurate the model is as shown in Chart diagram 1. This metric of evaluating CNNs is known as IOU, Intersection over Union. 50 over here equate to 0.5 on the evaluation's IOU scale. A mislocalization and false positive are both considered when the model's forecast is less than 0.5. YOLOv3 is really fast and accurate. When measured at 50 mAP, it is on par with the RetinaNet50 and RetinaNet101 but it is almost 4x faster than those two models. In benchmarks where the accuracy metric is higher (COCO 75), the boxes need to be more

139

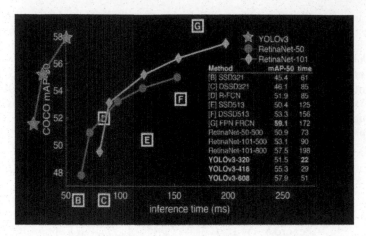

Method	mAP-50	time
[B] SSD321	45.4	61
[C] DSSD321	46.1	85
[D] R-FCN	51.9	85
[E] SSD513	50.4	125
[F] DSSD513	53.3	156
[G] FPN FRCN	59.1	172
RetinaNet-50-500	50.9	73
RetinaNet-101-500	53.1	90
RetinaNet-101-800	57.5	198
YOLOv3-320	51.5	22
YOLOv3-416	55.3	29
YOLOv3-608	57.9	51

Chart diagram 1.   Benchmark scores of YOLOv3 and other networks against COCO mAP 50.

aligned with the Ground Truth label boxes and here is where RetinaNet zooms past YOLO in terms of accuracy.

These are the metrics for different models with different benchmarks and it is quite observable that the mAP (mean Average Precision) for YOLOv3 is 57.9 on COCO 50 benchmark and 34.4 on COCO 75 benchmark. RetinaNet is better at detecting small objects but YOLOv3 is so much faster than versions of RetinaNet as shown in Table 2. YOLO uses a technique known as Non-Maximal Suppression (NMS) to eliminate duplicates of the same object being detected twice or more than that. It essentially retains on the bounding box with the highest confidence score. The initial step is to discard all the bounding boxes which have a confidence score lesser than the input of the threshold set for detected objects. If the threshold is set to 0.55, it retains bounding boxes with confidence scores more than or equal to 0.55.

Table 2.   Benchmark scores of object detection networks against COCO dataset.

	backbone	AP	AP_{50}	AP_{75}	AP_S	AP_M	AP_L
Two-stage methods							
Faster R-CNN+++ [5]	ResNet-101-C4	34.9	55.7	37.4	15.6	38.7	50.9
Faster R-CNN w FPN [8]	ResNet-101-FPN	36.2	59.1	39.0	18.2	39.0	48.2
Faster R-CNN by G-RMI [6]	Inception-ResNet-v2 [21]	34.7	55.5	36.7	13.5	38.1	52.0
Faster R-CNN w TDM [20]	Inception-ResNet-v2-TDM	36.8	57.7	39.2	16.2	39.8	52.1
One-stage methods							
YOLOv2 [15]	DarkNet-19 [15]	21.6	44.0	19.2	5.0	22.4	35.5
SSD513 [11, 3]	ResNet-101-SSD	31.2	50.4	33.3	10.2	34.5	49.8
DSSD513 [3]	ResNet-101-DSSD	33.2	53.3	35.2	13.0	35.4	51.1
RetinaNet [9]	ResNet-101-FPN	39.1	59.1	42.3	21.8	42.7	50.2
RetinaNet [9]	ResNeXt-101-FPN	40.8	61.1	44.1	24.1	44.2	51.2
YOLOv3 608 × 608	Darknet-53	33.0	57.9	34.4	18.3	35.4	41.9

YOLO also uses an Intersection over Union metric to grade the algorithm's accuracy. IOU is a simple ratio of the predicted box's area of intersection to the predicted box's area of union with the ground truth box. Following the removal of bounding boxes with detection probabilities lower than the NMS threshold, YOLO discards all the boxes for objects with IOU scores lesser than the IOU threshold to eliminate duplicate detections further.

3 RESULTS AND DISCUSSION

3.1 *Existing system*

- R-CNN is the classical algorithm in object detection.
- background subtraction method
- hierarchical of traffic recognition
- Pneumatic Tube Vehicle Counting
- Embedded magnetometers
- Inductive detector loops

3.1.1 *Disadvantage*

- Not all units count or categorize cars.
- Tube installations are not long-lasting; the lifespan of tubes is only a few weeks.
- The two wheelers cannot be detected.
- The sample rate of IDL data delivered to traffic control systems is quite low.
- Not suited for installation on bridge decking made of metal.

3.2 *Proposed system*

- The Deep Sort Algorithm uses Kalman Filters to track the objects, for better predictions.
- We also present a method for counting vehicles based on their direction of movement, such as "northbound" and "southbound" separately, then the intelligence system will take the decision to reduce the time based on the count of vehicle, further the traffic is managed accordingly.

3.3 *Inputs and outputs*

- A collection of photos with the shape (m, 608, 608, 3) make up the input.
- A list of recognised classes and a list of bounding boxes are included in the result. The six integer's pc, bx, by, bh, bw, and c are used to indicate each bounding box, as was previously mentioned. In an 80-dimensional vector of c, each bounding box is represented by 85 values.

3.4 *Encoding*

An object is detected by a grid cell if it contains the centre or midway of the object. Each 19×19 cell encodes data for five boxes since there are five anchor boxes in use. Only the dimensions of the anchor box are important. For the sake of simplicity, we will flatten the last two dimensions of the shape encoding (19, 19, 5, 85). So the Deep CNN's output is (19, 19, 425).

Let's take a closer look at what this encoding represents in the below Figure 5.

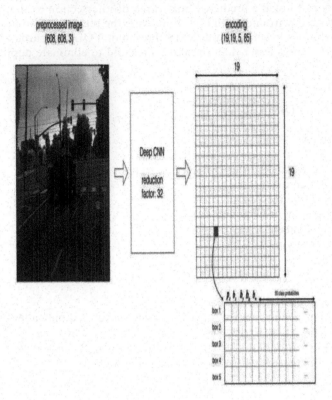

Figure 5.   YOLO architecture for encoding.

4   CONCLUSION

Vehicle counting system for traffic surveillance is developed using this project. The counting is separately done for all type of vehicle for potent maintenance of traffic. It was done with frontside-1x zoom video. Because counting is limited to the detected object, YOLOv3 is useful in vehicle detection. When viewing the output, the object 'car' has the highest counting accuracy, followed by 'motorcycle,' 'bus,' and 'truck,' with the lowest. The video frame rate also affects performance because it represents the integrity of the information processed by the system. Overall, this project was completed successfully and successfully. Any improvements made in the future should result in a better system.

REFERENCES

[1]   Shaoqing Ren, Kaiming He, Ross Girshick, and Jian Sun, "*Towards Real-Time Object Detection with Region Proposal Networks*", Jan 2016.
[2]   Joseph Redmon, Santosh Divvala, Ross Girshick, Ali Farhadi, "*YOLO, You Only Look Once: Unified, Real time Object Detection*", June 2015
[3]   Joseph Redmon, Ali Farhadi,"*YOLOv3: An Incremental Improvement*", 2016
[4]   Indumathi.K, Gnana Abinaya, Thangamani K, Ashok Deva A, "*Detection of Indian Traffic Signs*", 2016
[5]   Sang. K.S, Zhou. B, Yang. P. and Yang. Z, "Study of Group Route Optimization for IoT Enabled Urban Transportation Network", IEEE International Conference on Internet of Things (iThings) and

IEEE Green Computing and Communications (GreenCom) and IEEE Cyber Physical and Social Computing (CPSCom) and IEEE Smart *Data (SmartData)*, pp. 888–893, 2017

[6] Yu-Yun Tseng, Po-Min Hsu, Jen-Jee Chen and Yu-Chee Tseng, "Computer Vision-assisted Instant Alerts in 5g", *2020 29th International Conference on Computer Communications and Networks (ICCCN)*, pp. 1–9, 2020.

[7] Dai. Z, Song. Wang. X, Fang. Y, Yun. X, Zhang. Z, *et al.*, "Video-based Vehicle Counting Framework", *IEEE Access*, vol. 7, pp. 64460–64470, 2019.

[8] Abdelwahab. M.A, "Accurate Vehicle Counting Approach Based on Deep Neural Networks", *2019 International Conference on Innovative Trends in Computer Engineering (ITCE)*, pp. 1–5, 2019.

[9] Oltean. G, Florea. C, Orghidan. R and Oltean. V, "Towards Real Time Vehicle Counting Using Yolo-tiny and Fast Motion Estimation", *IEEE 25th International Symposium for Design and Technology in Electronic Packaging (SIITME)*, pp. 240–243, 2019.

[10] Sharma. P, Singh, Raheja. S and Singh. K, "Automatic Vehicle Detection Using Spatial Time Frame and Object Based Classification", pp. 8147–8157, 2019.

[11] Liu. Z, Zhang. W, Gao. Meng. H, Tan. X. Zhu. X, et al., "Robust Movement-specific Vehicle Counting at Crowded Intersections", *IEEE/CVF Conference on Computer Vision and Pattern Recognition Workshops (CVPRW)*, pp. 2617–2625, 2020.

[12] Kumar, A., & Alam, B. (2014, February). Real time Scheduling Algorithm for Fault Tolerant and Energy Minimization. In *2014 International Conference on Issues and Challenges in Intelligent Computing Techniques (ICICT)* (pp. 356–360). IEEE.

[13] Chaturvedi, R., Kumar, S., Kumar, U., Sharma, T., Chaudhary, Z., & Dagur, A. (2021). Low-cost IoT-enabled Smart Parking System in Crowded Cities. In *Data Intelligence and Cognitive Informatics: Proceedings of ICDICI 2020* (pp. 333–339). Springer Singapore.

[14] Y. Sharma, Shatakshi, Palvika, A. Dagur and R. Chaturvedi, "Automated Bug Reporting System in Web Applications," *2018 2nd International Conference on Trends in Electronics and Informatics (ICOEI)*, Tirunelveli, India, 2018, pp. 1484–1488

[15] Mehra, P. S., Jain, K., Chawla, D., Dagur, A., Singh, S., & Sharma, J. (2022). *GWO-EFUCA: Grey Wolf Optimisation and Fuzzy Logic based Unequal Clustering and Routing protocol for sustainable WSN-based Internet of Things.*

[16] Singh, A., Dhanaraj, R. K., Ali, Md. A., Balusamy, B., & Sharma, V. (2022). Blockchain Technology in Biometric Database System. In 2022 3rd International Conference on Computation, Automation and Knowledge Management (ICCAKM). *2022 3rd International Conference on Computation, Automation and Knowledge Management (ICCAKM)*. IEEE. https://doi.org/10.1109/iccakm54721.2022.9990133.

Artificial Intelligence, Blockchain, Computing and Security – Dagur et al. (Eds)
© 2024 The Author(s), ISBN: 978-1-032-49393-0

Predicting Encopresis & Enuresis treatment: Utilizing AI

Rolly Gupta* & Dr. Lalit Kumar Sagar*
Assistant Professor, SRMIST, Delhi NCR, India

ABSTRACT: Approximately 15 percent of the 5 yr old and 10 percent of all the 7 yr old children conflict with the problem of bedwetting and soiling, a situation which is medically called Encopresis & Enuresis. All the current remedial strategies depend (in most cases) at the own circle of family to perform an eight week remedy. Therefore, in order to contain healthcare companies within procedure and then create a possibility for the individualized remedies, WatchMinder has evolved as an Encopresis & Enuresis alarm gadget. This gadget without delay stocks affected person records with nurses via an affected person portal. Throughout this project, data gathered using WatchMinder alert structures is further examined with the aid of Python-based AI classifiers. With an accuracy of up to 91%, classifiers trained on sufferer's records can predict how an individual will respond to treatment for Encopresis and Enuresis. To reflect at what point in the treatment predictions are reliable, more analysis will be done on predictions made with different expert classifiers. However, all of the early data point to a time period between 2 and a few weeks after the treatment for these predictable elements. Previously, information regarding the effectiveness of the treatment should only be drawn after the eight-week treatment period. After two to three weeks, being able to predict the outcomes of a sufferers' treatment allows healthcare providers to both individualise sufferer's treatments while also including them in the treatment process. In the event that a sufferers' treatment is unsuccessful, it may be stopped after three weeks rather than eight to spare the families' weeks of insomnia.

The goal of this paper is to reduce overtime by individualising Encopresis & Enuresis treatments at a early stage. As a result, more sufferers will receive treatment in the allocated time, along with this more sufferers will have successful results, and the treatment process will be much simpler for their families.

Keywords: Encopresis, Enuresis, Artificial Intelligence, WatchMinder, Desmopressin

1 INTRODUCTION

One of the most uncommon issues affecting the children health globally is Encopresis and Enuresis; in India, it is currently overtaking asthma as the second most uncommon condition. Involuntary faeces and urine emission are referred to as Encopresis and Enuresis. Maximum prevalence rates for night time Encopresis and Enuresis are between 15 and 10 percent of all 5-year-olds and 5 to 10 percent of all 7-year-olds, respectively [1]. Due to the ongoing obstruction that Encopresis and Enuresis cause, those people are in a lot of danger of developing stress, low self-esteem, and even strain. When introduced at younger age group, these components can potentially affect kid's social developmental process.

A Desmopressin-based hormone treatment or an Encopresis & Enuresis alarm are now used to treat these conditions. Initial treatments involve the use of an alert system for Encopresis and Enuresis. With this device, the child undergoes the activation of a sensor after urinating, which

---

*Corresponding Authors: rollyg@srmist.edu.in and lalis@srmist.edu.in

DOI: 10.1201/9781003393580-21

thereafter causes the alarm to go off in room and awakes the child. The young child's eventually learn how to wake up before the process of urinating. Bedwetting based alarm treatments last for around at least 6 weeks and are frequently continued for another 2–3 months.

Both of the aforementioned treatments are very dependent on the sufferers' immediate family and may put a lot of strain on both of them and the sufferer's [1]. WatchMinder seeks to lessen the anxiety that Encopresis & Enuresis treatment causes for Sufferers and their families. In an effort to involve healthcare providers in the treatment process, the agency has developed an Encopresis and Enuresis alert device that enables families to collect records during the treatment. The distribution of moist to moisture free nights and the urination time are two data components that are accumulated. A database about each sufferer's is built throughout the treatment. As of now, WatchMinder, Encopresis & Enuresis alerts collects sufferer's records actively for more than two years and are present in healthcare facilities [12]. This results in enormous databases that never existed before as such, and this existence creates a possibility for further analysis of data.

1.1 *Literature survey*

According to the literature from the last ten years, the long-established conventional treatment plan, which includes educational, behavioural, nutritional, and physiological components, has not been outdone as the main form of therapy. Dayan *et al.* [14] hypothesised a connection between child abuse and enuresis or encopresis, but their research had a limited sample size of kids and had methodological flaws. Jiayao Shen *et al.* [15] investigated the risk factors for (Frequent nocturnal enuresis) FNE and the relationship between epilepsy and FNE. Through the lifespan, Sarah Kittel *et al.* [28] developed a conceptual synthesis of the relationships between ADHD and non-mental illnesses. In our discussion of co-occurring disorders, we cover possibly shared pathologic mechanisms, genetic background, and therapies. The physiology of micturition, the epidemiology of enuresis, a general clinical approach to enuresis, laboratory evaluation, and general management guidelines were reviewed by Greydanus *et al.* [12]. According to Naomi Warne *et al.* [20], children who have developed bladder control may nevertheless experience relapses of urine incontinence (UI) due to emotional/behavioural issues, exposure to stressful life events, and other factors. They used multivariable logistic regression in a prospective UK population to examine whether UI relapse was linked to mental health issues and stressful life events.

1.2 *Outline*

Using artificial intelligence, this section will review WatchMinder sufferer's records (AI). The information from Sufferers who are no longer receiving treatment, or who are inactive, will be examined for potential patterns and features. Predictions about the treatment outcomes of future Sufferers may be cured using patterns in dataset that would affect the very last results of the treatment [19].

On the basis of their treatment progression and eventual outcomes, Sufferers will first be divided into four results groups. Successful, moderately successful, failed, and dropout are the four groups. Their records will then be evaluated to determine whether record-style characteristics result in a fully, partially, or even un-successful solution [20]. During the treatment, it can be extremely a key component to highlight Sufferers who are the dropouts. Thus, this will put into action the steps to change course of treatment while lowering number of Sufferers who drop out of further treatment process.

2 METHOD

2.1 *Classification of sufferers – manually*

To start the evaluation of records, all sufferers that have finished their remedy are then manually separated into 4 major groups. These groups constitute 4 feasible results of the Encopresis & Enuresis remedy.

145

2.2 Initial process of evaluation

A twenty-three question based survey response has been recorded all through the sufferers, as per every affected person registration. The first step in affected person records evaluation is to arrange the accumulated solutions in a data frame. AI based algorithms are utilized for creating inferences that surrounds final results of affected person remedies. Set of rules is called a classifier and focuses on the usage of present records units so that it will separate new records into classes. [4]. Six exceptional classifiers (KNN, Random Forest, Logistic Regression, Decision Tree, SVM, Naive Bayes) are skilled the usage of affected person records, in this example the 23 questions so that it will pick out remedy final results [2]. The exceptional results may be separated into the groups.

2.3 Data assessment as remedy starts

The subsequent step within the procedure is to assess the records that have been accumulated after the remedy procedure has begun. Every night time all through the eight week remedy parents input the subsequent facts into the WatchMinder app. Eight weeks' worth of data from the initial 142 sufferers has been combined in a way that makes it impossible to understand the information contained therein.

Then, classifiers are developed for the usage of each Method 1 and Method 2 record. The records set is divided into groups as it was for the 23 survey questions in order to educate those classifiers. 70% of the records are utilised to teach advanced classifier techniques. Once the classifiers are proficient, the last 30% of the records can be reviewed [6]. In order to determine which of the two methodologies provides the highest level of accuracy of prediction, thereby plotting receiver operating characteristic curves (ROC-curves).

3 RESULTS

3.1 Initial evaluation

The data shows how accurately each classifier predicted the sufferers' treatment outcome. Data obtained using classifier Method 1 are shown in Table 1, whereas data collected using classifier Method 2 are shown in Table 2. The Random Forest classifier generates the most précised and positive conclusion in both scenarios.

Table 1. Scores of accuracy for outcome and classifiers (classifier Method 1).

	Successful	Part. succ.	Unsucc.	Dropout
Decision Tree	0.7674	0.8139	0.7441	0.6744
Random Forest	0.7906	0.7906	0.7674	0.7674
KNN	0.6511	0.7906	0.7674	0.7674
Logistic Regression	0.6046	0.7906	0.6976	0.6976
SVM	0.6511	0.7906	0.6976	0.7441
Naive Bayes	0.3953	0.3720	0.4651	0.7209

Table 2. Scores of accuracy for outcome and classifiers (classifier Method 2).

	Succ., Part, succ., Unsucc.	Dropout
Decision Tree	0.6129	0.6744
Random Forest	0.6207	0.7674
KNN	0.4516	0.7674
Logistic Regression	0.4516	0.6976
SVM	0.4516	0.7441
Naive Bayes	0.3953	0.7209

3.2 Results of records assessment as remedy progresses

The forty sufferers' plotted Encopresis and Enuresis latency are shown by the graphs in Figure 1. According to this data set, there doesn't appear to be a relationship between the night time wetting and the treatment outcomes. As a result, KNN imputer was used to predict all missing data of Encopresis & Enuresis i.e. values of latency in the data.

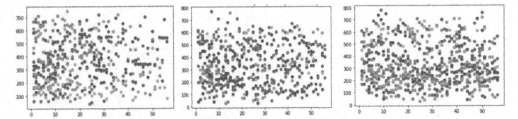

Figure 1. Plot of latency for Encopresis and Enuresis, where each hue represents a sufferer's. On the graph, a point is drawn at the time of night the accident occurred for Successful, Partially Successful, and Unsuccessful days when the sufferer's wet the bed.

Table 3. Accuracy rankings for every final results and every classifier for the every day questions created utilising Method 1.

Successful	All days	Day 1	Day 3	Day 5	Day 7
Decision Tree	0.8372	0.7441	0.7441	0.6744	0.6976
Random Forest	0.8604	0.8372	0.7906	0.7441	0.7674
KNN	0.8139	0.6279	0.7209	0.6744	0.6279
Logistic Regression	0.5348	0.7674	0.7441	0.6311	0.6311
SVM	0.7674	0.8372	0.7906	0.6976	0.7674

Partially successful	All days	Day 1	Day 3	Day 5	Day 7
Decision Tree	0.6976	0.7674	0.6511	0.6511	0.7441
Random Forest	0.8837	0.8139	0.8139	0.8372	0.7906
KNN	0.7674	0.7906	0.8139	0.8139	0.7441
Logistic Regression	0.8139	0.7674	0.7209	0.7674	0.7441
SVM	0.7441	0.8139	0.6744	0.7209	0.7209

Unsuccessful	All days	Day 1	Day 3	Day 5	Day 7
Decision Tree	0.7209	0.6976	0.6744	0.7209	0.6511
Random Forest	0.7906	0.7441	0.7441	0.7441	0.7441
KNN	0.7441	0.6744	0.7674	0.7906	0.7441
Logistic Regression	0.5581	0.7674	0.6744	0.7674	0.6976
SVM	0.6279	0.7906	0.6279	0.7674	0.6744

Dropout	All days	Day 1	Day 3	Day 5	Day 7
Decision Tree	0.9066	0.6976	0.6279	0.6046	0.6976
Random Forest	0.9069	0.7674	0.6511	0.7906	0.7674
KNN	0.8372	0.6744	0.7997	0.6976	0.7674
Logistic Regression	0.8604	0.7674	0.6744	0.7441	0.6744
SVM	0.8129	0.7906	0.6976	0.6511	0.7209

Table 4.   Rankings accuracy utilising classifier Method 2 on affected person records.

Successful, Partially successful, & Unsuccessful	All days	Day 1	Day 3	Day 5	Day 7
Random Forest	0.6451	0.3548	0.3870	0.3548	0.2903
Dropout	All days	Day 1	Day 3	Day 5	Day 7
Random Forest	0.9069	0.7674	0.6511	0.7906	0.7674

3.3   Implementation

The new applied capabilities make it possible for healthcare professionals to participate in the treatment process for Encopresis and Enuresis in a way that was not previously possible. New features include:

- Retraining the Random Forest classifiers, as sufferers finish the process of treatment – The original piece of code created at some point throughout the project is used to retrain the classifiers on a larger number of records.
- Predicting outcomes of a particular sufferers' treatment – Every day, predictions are made using data from the chosen sufferer's by trained classifiers.
- Predicting the remedy final results for all lively sufferers – The procedure above is finished on all lively sufferers.
- Classifying newly handled sufferers into one of the 4 results (a success, in part a success, unsuccessful, and dropout).

4   DISCUSSION

Growing predictions using classifier Methods 1 & 2 have been taught using strategies. As a result, the precision of the Method 2 is much lower than it was before. This outcome is the result of the AI based classifier having to choose three extraordinary outcomes rather than one, increasing the risk of predicting the accurate outcome by about 33% rather than 50%. However, the decision is thereby taken to make use of the classifier techniques (Method 2) because the results are presented in a more intelligible way. Three forecasts that add up to 100% are unquestionably given to nurses, providing a clear picture of the sufferers' treatment outcomes. Method 1 had a higher precision rating, it was determined that it was more important for healthcare companies to acknowledge the information provided to them. The code described above may also enable healthcare organisations to use Method 1 and anticipate the correct outcomes of a sufferer's's Encopresis & Enuresis treatment have precision of approximately 29.03% – 64.51%. Since there isn't much data to base predictions on in the early stages of the treatment, the precision rankings are lower. As the solution develops, more records are added, increasing the predicted precision to 64.51%.

5   CONCLUSION

Early process of observations has intended to the inference i.e., after two to three weeks, treatment predictions may also develop into a reliable predictor of treatment outcomes. The next step is to demonstrate this point by also evaluating predictions inside a larger study. Making informed decisions on the customization of each sufferers' treatment plan may be made possible by the ability to generate forecasts for health care providers. An individual's circle of family can now obtain personalised treatment counsel rather than going through the treatment for eight weeks on their own. Our hope is that the WatchMinder portal's predictive capabilities will change how Encopresis & Enuresis treatment is delivered in the future.

REFERENCES

[1] Holzinger, G. Langs, H. Denk, K. Zatloukal, H. Müller. Causability and Explainability of Artificial Intelligence in Medicine. *WIREs Data Mining and Knowledge Discovery*: Wiley Online Library; 2019 [cited 2022 April 24]. Available from: https://wires.onlinelibrary.wiley.com/doi/full/10.1002/widm.1312.

[2] Braytee, F. K. Hussain, A. Anaissi and P. J. Kennedy. ABC-sampling for Balancing Imbalanced Datasets Based on Artificial Bee Colony Algorithm. *14th International Conference on Machine Learning and Applications (ICMLA)*; 2015 [cited 2022 April 24], pp. 594–599, doi:10.1109/ICMLA.2015.103.

[3] Bailey JN, Ornitz EM, Gehricke JG, Gabikian P, Russell AT, Smalley SL (1999), Transmission of Primary Nocturnal Enuresis and Attention Deficit Hyperactivity Disorder. *Acta Paediatr* 88:1364–1368.

[4] Biederman J, Santangelo SL, Faraone SV (1995), Clinical Correlates of Enuresis in ADHD and non-ADHD Children. *J Child Psychol Psychiatry* 36:865–877.

[5] Eiberg H, Berendt I, Mohr J (1995), Assignment of Dominant Inherited Nocturnal Enuresis (ENUR1) to Chromosome 13q. *Nat Genet* 10:354–356.

[6] Fergusson DM, Horwood LJ, Shannon FT (1990), Secondary Enuresis in a Birth Cohort of New Zealand children. *Paediatr Perinat Epidemiol* 4:53–63.

[7] Pedregosa F., Varoquaux G., Gramfort A., Michel V., Thirion B., Grisel O., Blondel M, Prettenhofer P., Weiss R., Dubourg V., Vanderplas J., Passos A., Cournapeau D., Brucher M., Perrot M., Duchesnay E. Scikit-learn: Machine Learning in Python. *Journal of Machine Learning Research: Scikitlearn*; 2011 [cited 2022 May 3].

[8] Greydanus, D.E, *Enuresis: Current Concepts and Conundrums, Behavioral Pediatrics II: Neuropsychiatry, Sexuality and Eating Disorders*, August 2022, Pages 51–68.

[9] Hogg RJ, Husmann D (1993), The Role of Family History in Predicting Response to Desmopressin in Nocturnal Enuresis. *J Urol* 150:444–445.

[10] Dayan J., Enuresis and Encopresis: Association with Child Abuse and Neglect, Volume 48, Supplement 1, September 2022, Pages S30–S33.

[11] Jiayao Shen, Epilepsy and Frequent Nocturnal Enuresis Among Children in Shanghai, China, *Journal of Pediatric Urology*, October 2022.

[12] Loening-Baucke V (1995), Biofeedback Treatment for Chronic Constipation and Encopresis in Childhood: Long-term Outcome. *Pediatrics* 96:105–110.

[13] Mikkelsen EJ, Rapoport JL (1980), Enuresis: Psychopathology, Sleep Stage, and Drug Response. *Urol Clin North Am* 7:361–377.

[14] Saar-Tsechansky M., F. Provost. *Handling Missing Values When Applying Classification Models*. The University of Texas at Austin New York University; 2021[cited 2022 April 24].

[15] Naomi Warne, Mental Health Problems, Stressful Life Events and Relapse in Urinary Incontinence in Primary School-age Childhood: A Prospective Cohort Study, *medRxiv*, 2022.

[16] Park, J. E. (2021). Effectiveness of Creative Arts-based Parent Training for Parents with Children with Autism Spectrum Disorder. *The Arts in Psychotherapy*, 76.

[17] Butler R., Gasson S. Taylor & Francis online; 2009 [cited 2022 April 24]. Available from: https://www.tandfonline.com/doi/abs/10.1080/00365590500220321

[18] Sarah Kittel, Non-mental Diseases Associated with ADHD Across the Lifespan: Fidgety Philipp and Pippi Longstocking at Risk of Multimorbidity?, *Neuroscience & Biobehavioral Reviews* Volume 132, January 2022, Pages 1157–1180.

Artificial Intelligence, Blockchain, Computing and Security – Dagur et al. (Eds)
© 2024 The Author(s), ISBN: 978-1-032-49393-0

Object detection from images by convolutional neural networks for embedded systems using Cifar-10 images

Tushar Singh & Vinod Kumar

Delhi Technological University (Formerly Delhi College of Engineering), Delhi, India

ABSTRACT: This research article aims to create a model for embedded systems to detect objects from Cifar-10 images dataset using convolutional neural networks. Our main focus of the experiment is to use less memory size and train the model with good accuracy in a limited time so that we can use our model for embedded systems also. In this paper, we used Convolutional Neural Network (CNN) for image classification because it is designed to solve problems with input data in the form of a matrix, and images are also stored in the form of a matrix. In our experiment, we were able to achieve 99.06% accuracy within of the 3GB GPU memory, demonstrating the applicability of our methodology to embedded devices as well.

Keywords: CNN, Convolutional Layer, Strides, Pooling, Fully Connected Layer, ReLU, GPU

1 INTRODUCTION

The convolutional neural network has already been proven in the field of image data classification. It is a particular kind of deep neural network. It contains three different kinds of layers: fully connected, pooling, and convolutional [7].

The convolutional neural network used in our model is trained on the Cifar 10 images. We tried various hyperparameters like changing the kernel size or the number of strides, etc so that we can get good accuracy in limited epochs and limited memory. In each layer, we used the *ReLU* activation function with the exception of the final completely connected layer, where we employed the *Softmax* function [9].

Although, the model used in the literature [4] is very powerful and have limited memory requirement but is lacks in accuracy, so the purpose of this experiment is to reduce the memory size requirement, training time and gain good accuracy so that the proposed model can also be used for embedded systems.

This paper is organised as follows: Section 2 will discuss about terminologies used. Section 3 will explain about the convolutional neural networks (CNN) and it's building blocks. Section 4 will discuss about the dataset used in the proposed CNN model. In Section 5 the discussion will be about the proposed CNN model. Section 6 will be about the results and Section 7 will tell about the Conclusion.

2 TERMINOLOGY

This paper will use various terminologies in a repetitive manner so to avoid any confusion we will discuss them first. A "hyperparameter" in this paper will be referred to as the variables which are needed to be set before the training of the model. A "parameter" in this paper will represent those variables that are automatically learned by the model while training. A "kernel" will refer to the matrix which is a learnable parameter in the convolutional layer [5]. A "weight" and "parameter" will be used interchangeably with each other [12].

DOI: 10.1201/9781003393580-22

3 CONVOLUTIONAL NEURAL NETWORK

Convolutional Neural Networks (CNNs) is a deep neural network [10], it has already proven results over classical Artificial Neural Networks (ANNs) in the field of an image classification problems. CNNs have a significant advantage over ANNs in that they have fewer parameters, which attracts more developers and researchers [2]. CNN finds the features irrespective of their location in the image and also it extracts the features when the input goes to the deeper layers [3]. CNN improves the model accuracy through back propagation.

3.1 *Convolutional layer*

It is the first layer of the model which takes the CIFAR-10 32×32 image as the input in the matrix form. The input image and kernel are convolutioned in two dimensions [8], which means that the kernel matrix is multiplied by the image matrix element by element and then these multiplications are added together to produce a single value in the output matrix. This process is done row by row and column by column all over the matrix. All these processes constitute the convolutional layer.

3.2 *Strides*

Stride is the hyperparameter in CNNs. It is a method of lowering the number of parameters in the following layer to speed up the CNN model's convergence without duplicating features in adjacent layers. It defines the number of columns and rows the kernel will skip when it calculates the next value. Default value of stride is one. The formula for output matrix size O with Strides as S, Input matrix size as NxN, and kernel size as KxK is:

$$O = 1 + (N - K)/S \qquad (1)$$

3.3 *Padding*

In the CNN, the information in the image corners is lost because corner values of the input matrix are involved only once in the calculation and border values has also less participation as compared to the values that are away from the image border. So to preserve the features at the border, we add the dummy information at the borders of the input matrix so that we can include corner values more times in the calculation, this is known as padding. Default the value of padding is zero.

3.4 *Pooling layer*

Pooling is used for down-sampling input for further layers. It is similar to reducing the resolution of an image in further layers. *Maximum pooling* involves dividing the input matrix into subrectangles and selecting the maximum value from them [11], while *Average pooling* calculates the average value of the sub matrix and strides determine the number of columns or rows that must be jumped for the next calculation [11]. For maximum and average pooling see Figures 1 and 2 respectively.

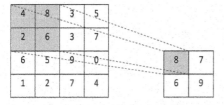

Figure 1. Max pooling with filter(2×2) and stride (2×2).

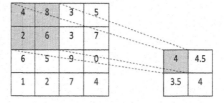

Figure 2. Avg pooling with filter(2×2) and stride (2×2).

151

3.5 Fully connected layer

Each neuron in the layer with complete connectivity is connected to every other neuron in the layer below and above it. It serves as the classification layer and provides the probability of the result. A fully Connected layer generally uses the *Softmax* function to get the output probability which ranges between from 0 to 1.

Below formula represents the *Softmax* function which takes input vector y of n real numbers and $\sigma(y_i)$ is the output probability which ranges $(0,1)$.

$$\sigma(y_i) = \left(\frac{e^{y_i}}{\sum_j e^{y_j}} \right) j = 1, ..., n \qquad (2)$$

4 DATASET

The CIFAR 10 image dataset, which contains 60000 32×32 RGB images, is the dataset used in our proposed model [1]. All these images are divided into two sets: training and testing. There are 10,000 and 50,000 images in testing and training, respectively [6]. Airplane, Bird, Automobile, Dog, Cat, Deer, Ship, Frog, Horse, and Truck are the 10 classes in which all images are categorized [6]. There are also 5 training batches that contain 10000 images per batch. 1000 randomly selected pictures from each class makes batch of the test. The remaining images are included in the batches of the training, thus it's possible that one batch contains more images from a particular class. The batches of training have 5000 images from each class [6].

5 PROPOSED CNN MODEL

In the linear model, there is a direct relationship between the number of layers and the training time. In our model we used four convolutional layers, The first convolutional layer have a 5×5 kernel with number of stride two, and the rest three convolutional layers have a 3×3 kernel with number of stride one. The same padding and ReLu activation function are used across all four convolutional layers. Max pooling of kernel size 2×2 with strides 2 is used across all four pooling layers. A pooling layer follows each convolutional layer in the model. The output is flattened after the final pooling layer, and then two fully connected layers of 384 and 10 neurons, respectively, are added. The model is trained for 200 epochs only. See Figure 3.

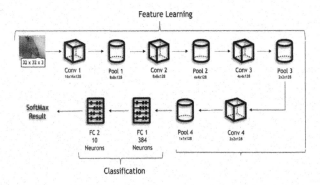

Figure 3. Model block diagram.

6 RESULTS

To implement our model, HP Pavilion AMD R5-5600H Hexa Core with NVIDIA GeForce RTX 3050 dedicated graphic card is used. All training stages are processed by the GPU. Our proposed model got 99.06% accuracy on the training data set. The highest precision we get is 0.85 for the class ship. The number of epochs is limited to 200 because there was no major improvement in accuracy after 150 epochs.

The snapshots of results for Epochs vs Loss and Accuracy are there in Table 1, classification report of training is in Table 2. Figures 4 and 5 shows the model accuracy and loss graphical representation while training.

Table 1.   Epoch vs loss and accuracy.

Epoch	Loss	Accuracy
25	14.38%	95.01%
50	8.01%	97.40%
75	6.62%	98.02%
100	5.31%	98.43%
125	4.19%	98.88%
150	4.46%	98.92%
175	4.16%	99.06%
200	4.98%	98.90%

Table 2.   Classification report.

Class	Precision	Recall	F1-Score
Airplane	0.73	0.77	0.75
Automobile	0.82	0.80	0.81
Bird	0.56	0.58	0.57
Cat	0.45	0.48	0.47
Deer	0.64	0.57	0.61
Dog	0.55	0.62	0.58
Frog	0.71	0.76	0.73
Horse	0.80	0.68	0.73
Ship	0.85	0.78	0.81
Truck	0.77	0.77	0.77

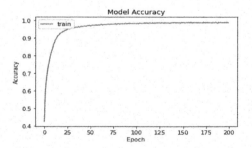

Figure 4.   Model accuracy graph.

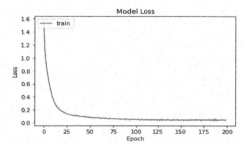

Figure 5.   Model loss graph.

7 CONCLUSION

In our experiment, we tried various model configurations that have different memory and time requirements. Our best model is able to get 99.06% accuracy, and it has a 3GB memory requirement for training, which makes this model suitable to be used for embedded systems. Although other models with greater accuracy are there but they have a lot of memory requirements too.

REFERENCES

[1] Abouelnaga, Y., Ali, O. S., Rady, H. & Moustafa, M. 2016. "CIFAR-10: KNN-Based Ensemble of Classifiers," *2016 International Conference on Computational Science and Computational Intelligence (CSCI)*, Las Vegas, NV, USA, 2016, pp. 1192–1195, doi:10.1109/CSCI.2016.0225.

[2] Albawi, S., Mohammed, T. A. & Al-Zawi, S. 2017. "Understanding of a Convolutional Neural Network," *2017 International Conference on Engineering and Technology (ICET)*, 2017, pp. 1–6, doi: 10.1109/ICEngTechnol.2017.8308186.

[3] Alippi, C., Disabato, S. & Roveri, M. 2018. "Moving Convolutional Neural Networks to Embedded Systems: The AlexNet and VGG-16 Case," *2018 17th ACM/IEEE International Conference on Information Processing in Sensor Networks (IPSN)*, Porto, Portugal, 2018, pp. 212–223, doi:10.1109/ IPSN.2018.00049.

[4] Calik, R. C. & Demirci, M. F. 2018. "Cifar-10 Image Classification with Convolutional Neural Networks for Embedded Systems," *2018 IEEE/ACS 15th International Conference on Computer Systems and Applications (AICCSA)*, pp. 1–2, doi:10.1109/AICCSA.2018.8612873.

[5] Chansong, D. & Supratid, S. 2021. "Impacts of Kernel Size on Different Resized Images in Object Recognition Based on Convolutional Neural Network," *2021 9th International Electrical Engineering Congress (iEECON), Pattaya, Thailand*, 2021, pp. 448–451, doi:10.1109/iEECON51072.2021. 9440284.

[6] *CIFAR-10 and CIFAR-100 datasets.* (n.d.). https://www.cs.toronto.edu/%7Ekriz/cifar.html

[7] Divya, S., Adepu, B. & Kamakshi, P. 2022. "Image Enhancement and Classification of CIFAR-10 Using Convolutional Neural Networks," *2022 4th International Conference on Smart Systems and Inventive* Technology (ICSSIT), Tirunelveli, India, 2022, pp. 1–7, doi:10.1109/ICSSIT53264. 2022.9716555.

[8] Doon, R., Rawat, T. Kumar & Gautam, S. 2018. "Cifar-10 Classification using Deep Convolutional Neural Network," *2018 IEEE Punecon,* 2018, pp. 1–5, doi:10.1109/PUNECON.2018.8745428.

[9] Gao, Y., Liu, W. & Lombardi, F. 2020. "Design and Implementation of an Approximate Softmax Layer for Deep Neural Networks," 2020 *IEEE International Symposium on Circuits and Systems (ISCAS)*, Seville, Spain, 2020, pp. 1–5, doi: 10.1109/ISCAS45731.2020.9180870.

[10] Sudharshan, D. P. & Raj, S. 2018. "Object Recognition in Images Using Convolutional Neural Network," 2018 *2nd International Conference on Inventive Systems and Control (ICISC)*, Coimbatore, India, 2018, pp. 718–722, doi:10.1109/ICISC.2018.8398893.

[11] *What are Convolutional Neural Networks? | IBM. (n.d.).* https://www.ibm.com/topics/convolutional-neural-networks

[12] Yamashita, R., Nishio, M., Do, R.K.G. & Togashi, K. 2018. Convolutional Neural Networks: An Overview and Application in Radiology. *Insights Imaging 9*, 611–629 (2018). https://doi.org/10.1007/ s13244-018-0639-9

Artificial Intelligence, Blockchain, Computing and Security – Dagur et al. (Eds)
© 2024 The Author(s), ISBN: 978-1-032-49393-0

Recognition of indian sign language using hand gestures

Umang Rastogi
SRMIST Modinagar, U.P., India
KIET Group of Institutions, Ghaziabad, U.P., India

Anand Pandey
SRMIST Modinagar, Uttar Pradesh, India

Vinesh Kumar
VIT University, Bhopal, Madhya Pradesh, India

ABSTRACT: In this study, we present a method for decoding Indian Sign Language alphabets using hand gesture recognition. Our proposed approach consists of four modules: gesture recognition, feature extraction, real-time tracking, and segmentation of hand. Utilizing the Hue, Intensity, Saturation (HSV) color model and the Camshift approach, hands are tracked and segmented. Gestures can be identified using the Genetic Algorithm. For correctly classifying single-handed and double-handed motions, we propose a straightforward, affordable technique. The employment of this method has allowed many of impaired people to interact with normal people.

Keywords: ISL-Indian Sign Language, single and double handed gestures, Hand tracking, feature extraction, segmentation, GS-gesture recognition

1   INTRODUCTION

For hearing-impaired people, sign language is the most expressive and natural method. The Government of India proposed the Indian Sign Language so that there would be a standardized sign language that all the country's dumb and deaf citizens could utilize. There hasn't been much focus on creating software that can translate this SL into equivalent voice and text until now. ASL recognition improves speech and hearing-impaired people's ability to communicate. It promises these persons more social opportunities and societal inclusion.

The Main focus on this research is to develop a GR system capable of automatically recording, identifying, and translating ISL alphabets into suitable Voice and text in a visual context. With the help of a single standard webcam and naked human hands, we aim to reliably distinguish both single and double-handed movements in our project. Our project's goal is to recognize motions as accurately and quickly as feasible. Our system has four components that work together to recognize gestures: Application Interface and Segmentation, Gesture Recognition, Feature Extraction, and Hand Tracking.

2   LITERATURE SURVEY

There are three primary types of object tracking: silhouette, point, and kernel. There is a comparison of various object tracking techniques in [9]. In [1], the numerous cutting-edge tracking methods and approaches were covered. This research demonstrates the real-time performance of the Camshift approach and its ability to track precisely in the presence of distractions and noise. The error classification rate for the Kalman filter approach is 86% [9]. While revolving objects can be detected by Speeded-Up-Robust-Features (SURF), highly scaled objects cannot. KLT only works with affine motion; it is not appropriate for motion that is translated, rotated, or scaled [1]. Other tracking techniques exist, including Zhigeng's color cue and robust multi-cue hand tracking, both of which use velocity weighted features [6]. To detect hands, Liu [18] employs the Viola-Jones method. For tracking, hand gesture recognition, and control in [7], the skin tone and Tower method are utilized. Through this review of the literature, we discovered that the Camshift method, this works with the system's integrated web camera and is both quick and accurate, is the most appropriate for it.

There are several techniques for segmenting images, such as ACIS [22], which segments images using a neural network and the HSV color model. The provided method works effectively on noisy photos and does not need any previous knowledge of the image. The YCbCr model is used for image segmentation by the ASRGC [14], which also employs the automatic seed selection, pixel-by-pixel region growth, and merging of related regions techniques. A Gaussian model classifier and self-adaptive skin color model utilizing BP NN were designed using the YCbCr model in [3]. In [12], many object-based segmentation and region-based techniques were covered. The HSV color space uses values that are intuitive to describe color. To segmenting skin tones, the components' clarity in distinguishing between luminance and chrominance qualities and intuitiveness make them useful. [8]. Therefore, we decided to segment hands using the HSV colour model.

Following segmentation, a database search of the input image is required. Specific features that describe or depict the image are required. An image's shape is a crucial visual component. Feature extraction employs a variety of techniques. According to [11], these techniques: Only shape boundary information is used in contour shape approaches. Hausdorff distance-based algorithms for finding items in an all-images matching. The GFD, which is a desired alternative to generic shape representations, performs well when it comes to retrieval. Fourier Descriptor (FD) can be taken into consideration if storage is a problem.

Gesture recognition has been approached in many different ways incorporating statistical modelling techniques like PCA is using for extract characteristics from photographs of hand motions, HMM, which is a twofold stochastic process regulated using a Markov chain to model a variety of data, and techniques based on soft-computing technologies like ANN, GA [19,15], which are employed for the feature selection problem and are advantageous for challenging pattern recognition and classification applications.

For the Gesture Recognition System, methods like Finite State Machine (FSM) [21] and [16] that are utilized for particle filtering, classification and condensation algorithm, and SVM [16] have been applied. depend on the ideas of genetic system and natural selection, Parallel adaptive search methods include genetic algorithms as a subclass [19,15,24]. It has been used in the last 20 years to address a variety of machine learning and search optimization problems. GA provides several benefits over other methods, including being simple to understand, supporting multi-objective optimization, and being effective in "noisy" environments. GA is a distributed system that is naturally parallel.

3 PROPOSED SYSTEM

The built-in web camera feeds input hand movements to the gesture recognition system. Using the Camshift approach, hand tracking is performed. After that, the HSV colour model is used to segment the hands. Certain elements are used to represent the segmented hand image. By means of a genetic algorithm, which produces the best results, these features are used to recognise gestures. The final output is transformed into the appropriate text and voice. [17,2,5,10]

Segmentation, Hand Tracking, Gesture Recognition and Feature Extraction, are the four modules that make up our system.

3.1 *Hand tracking*

The general Camshift algorithm, which can be summed up in the steps below, was chosen by us to track hands from the captured image.

- Choose the first point of interest that includes the hands you want to track.
- Use that area's colour histogram as the object model.
- Using the colour histogram, make the frame's probability distribution.
- Under probabilistic representation, calculate the search window's centre mass using the mean-shift approach.
- After taking the point from step 4, centre the search window there and repeat step 4 until convergence.
- Utilize the search window's location from step 5 to process the following frame.

3.2 *Segmentation*

Following the hands' trail, the hands must be distinguished from the background. To do this, It employs the HSV colour model.

3.3 *Feature extraction*

An appropriate classifier infers the association between the features and classes after extracting several general-purpose features during feature extraction. These characteristics are described below and were selected based on the suggested shape representation in [11]: Hausdoff distance and the Fourier descriptor.

3.4 *Gesture recognition*

We first extract the traits from the input character, compare them against the database, and then use the traits that are most similar as the outcome. The genetic algorithm appears to be appropriate for the analysis of hand movements and is a solid choice for controlling the randomness of natural data. The Genetic Algorithm's [14,3] steps are as follows: -

- Create a starting population to initialise.
- Assess: determine each population member's level of fitness.
- Parents of choice are chosen from the public by looking for people who fall inside a certain range in terms of fitness.

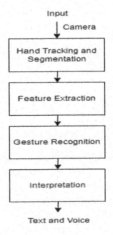

Figure 1.   Overview of system.

- Cross2x: generate new individuals by performing 2-point crossover on the parents.
- As a result of the cross2x technique, new people are produced, and some of their genes are then randomly changed.
- Replacement: It randomly selects five components from the population, chooses the worst one, and then swaps that component out with a brand-new individual.
- If the best option is discovered, check stop and shut down the system.

4 CONCLUSION

The Camshift approach, the HSV colour model, and a genetic algorithm are used in this research to present a real-time HGDS. The suggested GRS supports a variety of hand motions on a platform that is based on vision. The technology works with movements made with either one hand or both hands.

The device is affordable and uses only the user's hands to interface with the computer, making it easier for deaf-dumb persons to operate. The suggested technique is efficient at identifying the Indian Sign Language alphabets. Because of this method, millions of people who are deaf can interact with hearing people.

5 FUTURE WORK

Our research has shown that the motions used in Indian Sign Language are the only ones that our proposed GRS can recognise. The identification of further ISL gestures will be the subject of future investigations.

REFERENCES

[1]   Aggarwal, J. & Cai, Q. (1999). Human Motion Analysis: A Review. *Computer Vision and Image Understanding*, 73(3), 295–304.
[2]   Ali, M., Dhanaraj, R.K. (2023). IoT and Blockchain Oriented Gender Determination of Bangladeshi Populations. In: Santosh, K., Goyal, A., Aouada, D., Makkar, A., Chiang, YY., Singh, S.K. (eds) *Recent Trends in Image Processing and Pattern Recognition. RTIP2R 2022. Communications in Computer and Information Science*, vol 1704. Springer, Cham. https://doi.org/10.1007/978-3-031-23599-3_25
[3]   Bretzner, L., Laptev, I., & Lindeberg, T. (2002, May). Hand Gesture Recognition using Multi-scale Colour Features, Hierarchical Models and Particle Filtering. In *Proceedings of Fifth IEEE International Conference on Automatic Face Gesture Recognition* (pp. 423–428). IEEE.
[4]   Canny, J. (1986). A Computational Approach to Edge Detection. *IEEE Transactions on Pattern Analysis and Machine Intelligence*, (6), 679–698.
[5]   Chandraprabha, M., & Kumar Dhanaraj, R. (2022). An Empirical View of Genetic Machine Learning Based on Evolutionary Learning Computations. In *Machine Learning Methods for Engineering Application Development* (pp. 59–75). Bentham Science Publishers. https://doi.org/10.2174/9879815079180122010008
[6]   Collins, R. T. (2003, June). Mean-shift Blob Tracking Through Scale Space. In *2003 IEEE Computer Society Conference on Computer Vision and Pattern Recognition*, 2003. Proceedings. (Vol. 2, pp. II-234). IEEE.
[7]   Davis, J., & Shah, M. (1994, May). Recognizing Hand Gestures. In *European Conference on Computer Vision* (pp. 331–340). Springer, Berlin, Heidelberg.
[8]   Epstein, D., Gilman, R., Short, H., & Sims, C. (1996). Geometric and Computational Perspectives on Infinite Groups: *Proceedings of a Joint DIMACS/Geometry Center Workshop, January 3–14 and March 17–20, 1994 (No. 25)*. American Mathematical Soc.
[9]   Gary, R. B. (1998). Computer Vision Face Tracking for Use in a Perceptual User Interface. *Technical REport Q2, Intel Technology Journal*.
[10] Janarthanan, S., Ganesh Kumar, T., Janakiraman, S., Dhanaraj, R. K., & Shah, M. A. (2022). An Efficient Multispectral Image Classification and Optimization Using Remote Sensing Data. In

S. Bhattacharya (Ed.), *Journal of Sensors* (Vol. 2022, pp. 1–11). Hindawi Limited. https://doi.org/10.1155/2022/2004716

[11] Lyons, M. J., Budynek, J., & Akamatsu, S. (1999). Automatic Classification of Single Facial Imag-es. *IEEE Transactions on Pattern Analysis and Machine Intelligence*, 21(12), 1357–1362.

[12] O'Connor, N., Adamek, T., Sav, S., Murphy, N., & Marlow, S. (2003). Qimera: A software Plat-form for Video Object Segmentation and Tracking. In *Digital Media Processing For Multimedia Inter-active Services* (pp. 204–209).

[13] Oniga, S. (2005, March). A New Method for FPGA Implementation of Artificial Neural Network used in Smart Devices. In *International Computer Science Conference MicroCAD* (pp. 31–36).

[14] Pal, N. R., & Pal, S. K. (1993). A Review on Image Segmentation Techniques. *Pattern Recogni-tion*, 26 (9), 1277–1294.

[15] Pan, Z., Li, Y., Zhang, M., Sun, C., Guo, K., Tang, X., & Zhou, S. Z. (2010, March). A Real-time Multi-cue Hand Tracking Algorithm Based on Computer Vision. In *2010 IEEE Virtual Reality Conference (VR)* (pp. 219–222). IEEE.

[16] Russell, S. J., & Norvig, P. (2003). Artificial Intelligence: A Modern Approach (Harlow. Squire, D. M., & Caelli, T. M. (2000). Invariance Signatures: Characterizing Contours by their De-partures from Invariance. *Computer Vision and Image Understanding*, 77(3), 284–316.

[17] Starner, T., Weaver, J., & Pentland, A. (1997, October). A Wearable Computer Based American Sign Language Recognizer. In *Digest of Papers. First International Symposium on Wearable Comput-ers* (pp. 130–137). IEEE.

[18] Takahashi, K., Ohta, T., & Hashimoto, M. (2008, October). Remarks on EOG and EMG Gesture Recognition in Hands-free Manipulation System. In *2008 IEEE International Conference on Systems, Man and Cybernetics* (pp. 798–803). IEEE.

[19] Tamura, S., & Kawasaki, S. (1988). Recognition of Sign Language Motion Images. *Pattern recogni-tion*, 21(4), 343–353.

[20] Vapnik, V. N. (1999). An Overview of Statistical Learning Theory. *IEEE Transactions on Neural Net-works*, 10(5), 988–999.

[21] Vu, V. P., Wang, W. J., Chen, H. C., & Zurada, J. M. (2017). Unknown Input-based Observer Synthe-sis for a Polynomial T–S fuzzy Model System with Uncertainties. *IEEE Transactions on Fuzzy Sys-tems*, 26(3), 1447–1458.

[22] Yilmaz, A., Javed, O., & Shah, M. (2006). Object Tracking: *A Survey. ACM Computing Surveys (CSUR)*, 38(4), 13-es.

[23] Zhou, L., Chalana, V., & Kim, Y. (1998). PC-based Machine Vision System for Real-time Computer-aided Potato Inspection. *International Journal of Imaging Systems and Technology*, 9(6), 423–433.

Artificial Intelligence, Blockchain, Computing and Security – Dagur et al. (Eds)
© 2024 The Author(s), ISBN: 978-1-032-49393-0

Potato plant leaf diseases detection and identification using convolutional neural networks

Sriram Gurusamy*, B. Natarajan* & R. Bhuvaneswari*
Department of Computer Science and Engineering, Amrita School of Computing, Amrita Vishwa Vidyapeetham, Chennai, India

M. Arvindhan*
Assistant Professor, Galgotias University

ABSTRACT: Potato plant leaf diseases can cause significant crop losses and impact the quality of potato products. Early identification of these diseases can aid in timely intervention and treatment, minimizing the spread of diseases and reducing crop losses. In this work, we propose the incorporation of Convolutional Neural Networks (CNNs) for the automatic detection and identification of diseases present in potato plant leaves. We compiled a diverse dataset of images of potato plant leaves and used it to train and evaluate a CNN for the classification of potato leaf diseases. The results of the research work shows that the CNN was able to achieve high levels of precision score, F1 score, recall and accuracy over the training and testing splits of the dataset. Our findings demonstrate the potential of CNNs for the automatic identification of diseases present in potato plant leaves, and suggest that this approach could be applied to other types of plant diseases as well. The proposed method has the capability to improve crop management and disease control efforts by enabling the automatic and early detection of leaf diseases, which could have significant economic and environmental benefits. According to our result, the CNN offers a promising method for identifying and detecting leaf diseases on potato plants. This method may also be utilized to create an automated system for disease diagnostics in potato harvests. For the goal of identifying diseases in potato leaves, a Deep Learning (DL) base CNN model has been introduced in this research work. The proposed CNN technique has secured 98.1% accuracy rate for the Plant Village dataset, shows 3% improvement when compared to the existing methods.

Keywords: Image Processing, Leaf Image, Convolution Neural Network, Deep Learning, Potato Leaf, Disease Detection

1 INTRODUCTION

The nightshade family comprises the widely cultivated potato plant (Solanum tuberosum) (Solanaceae). It ranked after maize, wheat, and rice as that of the fourth-largest food crop in the world. A major food source, potatoes are used to manufacture French fries, potato chips, and mashed potatoes, as well as vodka, whisky, and biofuels. Potato plants are prone to numerous diseases that can reduce crop yields and quality. Implementing efficient control strategies to reduce losses requires the early identification and detection of potato plant diseases. Some common potato plant diseases include early blight, late blight, and leaf roll virus. One of the diseases that affect potato plants the most severely is late blight, which is brought on by the oomycete Phytophthora infestans. It can lead to significant yield losses and has caused famines in the past. Another significant disease of the potato plant that can cause leaf spotting and defoliation is early

*Corresponding Authors: gsriram633@gmail.com, rec.natarajan@gmail.com, bhuvanacheran@gmail.com and saroarvindmster@gmail.com

DOI: 10.1201/9781003393580-24

blight, which is brought on by the fungus Alternaria solani. Leafroll virus, which is transmitted by aphids, can cause stunting and reduced yields in infected plants.

The recent developments in CNN networks has justified the prominent performance in disease area identification and finding the infected spots, further provides an efficient and accurate results over wider applications. This type of DL algorithm is well-suited for image classification tasks, as it can process large amounts of data and identify subtle differences in the appearance of diseased and healthy leaves. Compared to traditional methods such as visual inspection, laboratory techniques, and field techniques, CNNs can provide a more efficient and accurate solution for the detection and identification of plant diseases from images. As such, CNNs offer a viable alternative to long and time-consuming traditional methods, as well as those that require specialized expertise. CNN have achieved impressive results on image recognition tasks and are well-suited for image classification tasks. The presence of multiple layers in CNN with interconnected neurons in each layer performs a specific function. The lower layers are responsible for extracting low-level features from the input data, while the higher layers combine these features to detect patterns and objects. CNNs are capable of learning features from the input data, which makes them highly suitable for the automation of detecting and identifying potato leaf diseases. The proposed model development aims to classify the healthy and infected potato leaves accurately using CNN. It can process huge amounts of data and identify subtle differences in the appearance of diseased and healthy leaves, reducing the reliance on subjective visual inspection. Additionally, it can reduce the need for specialized expertise and equipment, making disease diagnosis more accessible and cost-effective. This can be particularly useful in situations where traditional methods, such as microscopy, are time-consuming or require specialized training.

2 RELATED WORK

The recent literature studies focuses the disease classification and identification of the infected parts in potato leaves and surfaces by implementing the DL Techniques known as CNN. Studies carried out in [1] used a CNN to identify potato plant diseases from images of leaves and achieved an accuracy of 92%. The proposed CNN network was trained to classify the five major kinds of potato diseases such as late blight, early blight, bacterial wilt, black leg, and potato mosaic virus. The results demonstrated the potential of CNNs for accurately and efficiently identifying potato plant diseases from images. The research suggested in in [2] uses a DL multi-classification model based on CNN to categorize the leaf according to the severity level, coupled with binary classification that distinguishes between healthy and infected potato plant leaves. The research study has achieved an accuracy of 90.77% for binary classification and 94.77% for multi-classification using the real-time samples collected for classifying healthy and unhealthy leaves. The research work stated in [3] introduces SBCNN approach to assess the comparative performance over earlier approaches in both augmented and non-augmented potato leafs. This has achieved the best accuracy of 96.9% and 96.75% on non-augmented and augmented dataset respectively.

Overall, these studies demonstrate the strength of machine learning (ML) techniques, particularly CNNs, for the accurate and efficient detection and identification of potato plant diseases from images. It is essential to conduct additional research in order to determine the strength and applicability of these methods in various contexts and for a range of illnesses. The identification of diseases using ML techniques can be done by leveraging traditional ML algorithms and DL algorithms. Traditional ML algorithms such as decision trees, support vector machines and Bayesian networks have been used to detect and identify potato plant leaf diseases. Various approaches have been explored to create a robust detection and identification system for potato plant leaf diseases using ML techniques. In the studies carried out by [4], a pre-trained model such as VGG-19 has been used to extract the features from the dataset and multiple classifiers were utilized to achieve better classification accuracy. A CNN network is handled in [5] for the detection of various 9 diseases as well as in the detection of the healthy leaf. This work produced better accuracy when compared to pretrained models VGG-16, InceptionV3, and MobileNet. The various comparisons of DL and ML based approaches were discussed in the work [6] to spot the leaf diseases, in which the DL techniques produce impressive classification accuracy when compared to ML techniques. In [7], the authors focus on improvising the accuracy using various ML approaches for the accurate detection and classification of Maize leaf Disease. The Random Forest algorithm produces the highest accuracy when compared to the other algorithms. The DL based research works stated in [8–11] shows the strong evidences for using the techniques in future research works.

3   PROPOSED METHOD

The proposed CNN network aims to identify and classify the infected regions and types of diseases occurred in potato plant leaves by incorporating the novel approaches. The proposed method involves collecting and pre-processing a large and diverse dataset of images, trains the CNN, and evaluating the performance of the CNN using test set. As this method has the potential to significantly improve the accuracy and efficiency of potato plant disease detection and identification, we can expect a positive impact on crop management and disease control.

The proposed CNN architecture consists of several layers of interconnected neurons, including layers such as convolution, pooling and fully connected. The convolutional layers uses Gaussian blur filters to the input image data for extracting features such as edges and patterns. Further, the pooling layer reduces the dimensionality of the images to facilitate efficient processing. The fully connected layers helps to classify the healthy and infected leaves.

The proposed CNN architecture is depicted in the Figure 1. The layered approaches followed in CNN network enhances the performance in classification and detection tasks. The labelled dataset assist the proposed model to perform accurate classification of various classes it belongs to. The optimization of hyper parameters such as number of epochs 50, learning rate value set as 0.01, and batch size 32 helps to build a powerful model for potato leaf disease identification and classification. Further, the incorporation of Adam optimizer and drop out regularizer resolves the overfitting issues and generalizes the model performance in a better way to yield higher accuracy results.

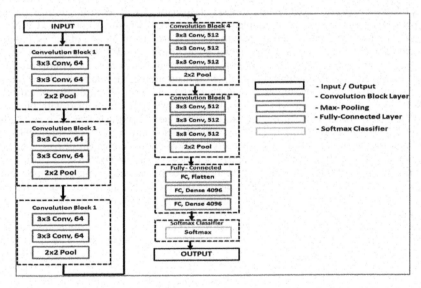

Figure 1.   Proposed convolutional neural network for the potato plant leaf detection.

4   EXPERIMENTAL RESULTS & DISCUSSIONS

The proposed work has been experimented using Plant Village Dataset, publically available and contains nearly 2152 images. The results of the evaluations showed that the CNN was able to achieve high accuracy in detecting and identifying potato plant leaf diseases. The model has been trained for various epochs and the best accuracy is obtained for the epoch value of 50 and it has been shown in Figure 2. The overall accuracy of the CNN on the test set was 98.1%, with a

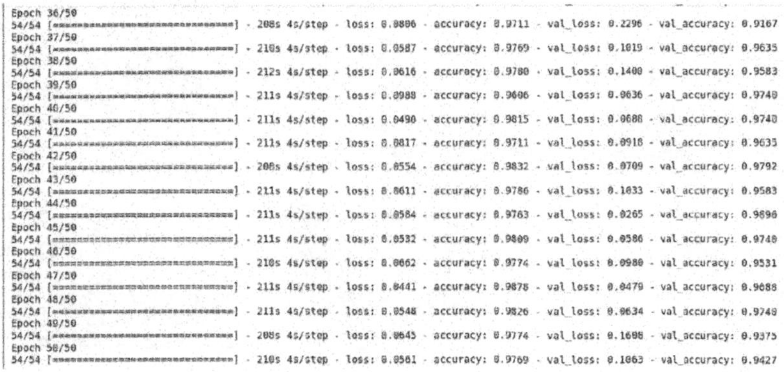

```
Epoch 36/50
54/54 [==============================] - 208s 4s/step - loss: 0.0806 - accuracy: 0.9711 - val_loss: 0.2296 - val_accuracy: 0.9167
Epoch 37/50
54/54 [==============================] - 210s 4s/step - loss: 0.0587 - accuracy: 0.9769 - val_loss: 0.1019 - val_accuracy: 0.9635
Epoch 38/50
54/54 [==============================] - 212s 4s/step - loss: 0.0616 - accuracy: 0.9780 - val_loss: 0.1400 - val_accuracy: 0.9583
Epoch 39/50
54/54 [==============================] - 211s 4s/step - loss: 0.0988 - accuracy: 0.9606 - val_loss: 0.0636 - val_accuracy: 0.9740
Epoch 40/50
54/54 [==============================] - 211s 4s/step - loss: 0.0490 - accuracy: 0.9815 - val_loss: 0.0680 - val_accuracy: 0.9740
Epoch 41/50
54/54 [==============================] - 211s 4s/step - loss: 0.0817 - accuracy: 0.9711 - val_loss: 0.0918 - val_accuracy: 0.9635
Epoch 42/50
54/54 [==============================] - 206s 4s/step - loss: 0.0554 - accuracy: 0.9832 - val_loss: 0.0709 - val_accuracy: 0.9792
Epoch 43/50
54/54 [==============================] - 211s 4s/step - loss: 0.0611 - accuracy: 0.9786 - val_loss: 0.1033 - val_accuracy: 0.9583
Epoch 44/50
54/54 [==============================] - 211s 4s/step - loss: 0.0584 - accuracy: 0.9763 - val_loss: 0.0265 - val_accuracy: 0.9896
Epoch 45/50
54/54 [==============================] - 211s 4s/step - loss: 0.0532 - accuracy: 0.9809 - val_loss: 0.0586 - val_accuracy: 0.9740
Epoch 46/50
54/54 [==============================] - 210s 4s/step - loss: 0.0662 - accuracy: 0.9774 - val_loss: 0.0980 - val_accuracy: 0.9531
Epoch 47/50
54/54 [==============================] - 211s 4s/step - loss: 0.0441 - accuracy: 0.9878 - val_loss: 0.0479 - val_accuracy: 0.9688
Epoch 48/50
54/54 [==============================] - 211s 4s/step - loss: 0.0548 - accuracy: 0.9826 - val_loss: 0.0634 - val_accuracy: 0.9740
Epoch 49/50
54/54 [==============================] - 208s 4s/step - loss: 0.0645 - accuracy: 0.9774 - val_loss: 0.1698 - val_accuracy: 0.9375
Epoch 50/50
54/54 [==============================] - 210s 4s/step - loss: 0.0561 - accuracy: 0.9769 - val_loss: 0.1063 - val_accuracy: 0.9427
```

Figure 2.   Training details of proposed convolutional neural network for the potato plant leaf detection.

precision score of 0.962 and a recall of 0.96. In terms of individual disease classes, the CNN performed particularly well in detecting and identifying diseases such as late blight, early blight, and Healthy leaves, with remarkable accuracy. The CNN also showed good performance on less common diseases, such as blackleg and purple top with notable accuracy.

Overall, the high accuracy, precision, and recall scores demonstrate the effectiveness of the CNN in detecting and identifying potato plant leaf diseases. The CNN was able to accurately classify the majority of the images in the test set, with few false positive or false negative errors. This indicates that the CNN was able to effectively learn and generalize from the training data to the test data, and can be confidently used for automated potato plant disease detection and identification.

Table 1 shows the comparison of precision, accuracy, and Recall of various methods with the proposed CNN model. In comparison with other approaches for potato plant disease detection and identification, the CNN showed superior performance. In previous studies, decision tree classifiers and deep belief networks (DBNs) have been used for potato plant disease detection and identification, with accuracies of 92% and 95.7%, respectively. While these approaches achieved good performance, the CNN was able to achieve a higher accuracy of 98.1% which is 3% improvement when compared to the existing methods. Figure 3 shows the results of the proposed model on the test dataset. Furthermore, the CNN was able to achieve high precision and recall scores, indicating that it had a lower scores of false positive and false negative errors. This is in contrast to some other approaches, which may have a higher rate of false positives or false negatives, resulting in a lower overall accuracy. Overall, the results of proposed model evaluations show that the CNN is able to outperform other approaches for potato plant disease detection and identification in terms of accuracy and precision. The ability of the CNN to effectively learn and generalize from the training data to the test data makes it a promising approach for automated potato plant disease detection and identification. There are several areas where future work could be done to improve the detection and identification of potato plant leaf diseases using CNN. One possibility is to incorporate more diverse and comprehensive datasets into the training process. Currently available datasets may not capture the full range of variation in leaf diseases, and using a more diverse dataset could help the CNN generalize better to real-world data.

Table 1.   Comparison of accuracy with the various methods.

S.No	Methodology	Precision(in%)	Recall (in%)	Accuracy (in%)
1	DBN	91.5	90.7	92
2	BPNN	89.3	88	90.9
3	SVM	92	91.6	92.7
4	Proposed CNN	96.2	96	98.1

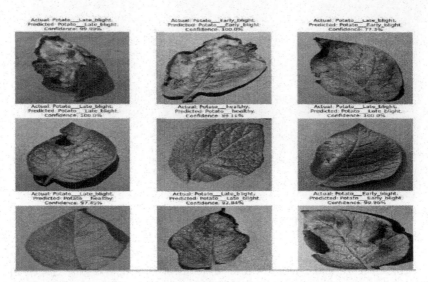

Figure 3.   Testing results of the proposed convolutional neural network for the potato plant leaf detection along with the confidence percentage.

Another direction for future work could be to investigate the use of transfer learning, where a CNN trained on a large, general-purpose dataset is fine-tuned for the specific task of leaf disease detection. This could potentially allow the CNN to take advantage of the knowledge learned from the large dataset and improve performance on the target task. Another approach that could be explored is the use of more advanced CNN architectures, such as those with residual connections or attention mechanisms, which have been shown to be effective in a variety of image classification tasks. Finally, it could be interesting to investigate the use of other types of data in addition to images, such as plant physiological measurements or weather data, to improve the accuracy of leaf disease detection.

This could be done through the incorporation of additional input modalities or the use of multi-modal CNNs. The Figure 4 clearly shows the performance of the proposed CNN network over training and validation phases. Finally, it secures 98.1% classification accuracy, proves the notable performance.

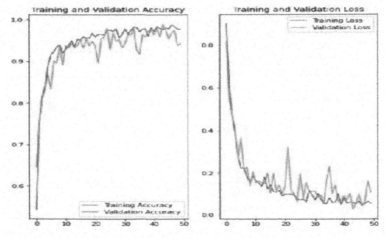

Figure 4.   Visualization of accuracy and loss - during training and validation phase of the proposed model.

5 CONCLUSION

The use of CNNs allows for the automation of this process, which can be time-consuming and labor-intensive when performed by humans. This can have significant implications for the efficiency and effectiveness of plant disease management efforts. Another contribution of this work is the inclusion of a diverse dataset for training and evaluating the CNN. The use of a diverse dataset helps to ensure that the CNN can generalize well to real-world data and improve the accuracy of leaf disease detection in practice. Overall, the findings and contributions of this work highlight the potential of CNNs for the automatic detection and identification of potato plant leaf diseases, and suggest that this approach could be applied to other types of plant diseases as well. One of the main benefits of this approach is the ability to automate the process of detecting and identifying leaf diseases. This can save time and labor, as the task is typically performed manually by trained experts. By automating the process, it is possible to scale up the detection and identification of leaf diseases to a larger number of plants, potentially leading to more effective and efficient disease management efforts. In addition, the use of CNNs allows for the detection of leaf diseases at an early stage, when the diseases are still in an asymptomatic or early-symptomatic phase. This can enable timely intervention and treatment, potentially reducing the spread of diseases and minimizing crop losses. Overall, the proposed method has the potential to improve crop management and disease control efforts by enabling the automatic and early detection and identification of potato plant leaf diseases. This could have significant economic and environmental benefits, including increased crop yields and drastically reduces the usage of pesticides.

REFERENCES

[1] Abdul Jalil Rozaqi and Andi Sunyoto. Identification of Disease in Potato Leaves Using Convolutional Neural Network (CNN) algorithm. In *2020 3rd International Conference on Information and Communications Technology (ICOIACT)*, pages 72–76, 2020.

[2] Vinay Kukreja, Anupam Baliyan, Vikas Salonki, and Rajesh Kumar Kaushal. Potato Blight: Deep Learning Model for Binary and Multi-classification. In *2021 8th International Conference on Signal Processing and Integrated Networks (SPIN)*, pages 967–672, 2021.

[3] Utpal Barman, Diganto Sahu, Golap Gunjan Barman, and Jayashree Das. Comparative Assessment of Deep Learning to Detect the Leaf Diseases of Potato Based on Data Augmentation. In *2020 International Conference on Computational Performance Evaluation (ComPE)*, pages 682–687, 2020.

[4] Dagur, A., Kaushik, A., Rastogi, A., Singh, A., Kumar, A., & Chaturvedi, R. (2021). Optimization of Queries in Database of Cloud Computing. In *Data Intelligence and Cognitive Informatics: Proceedings of ICDICI 2020* (pp. 325–332). Springer Singapore.

[5] Kushwaha, A., Amjad, M., & Kumar, A. (2019). Dynamic Load Balancing Ant Colony Optimization (DLBACO) Algorithm for Task Scheduling in Cloud Environment. *Int J Innov Technol Explor Eng*, 8 (12), 939–946.

[6] Radhakrishnan Sujatha, Jyotir Moy Chatterjee, NZ Jhanjhi, and Sarfraz Nawaz Brohi. Performance of Deep Learning vs Machine Learning in Plant Leaf Disease Detection. *Microprocessors and Microsystems*, 80:103615, 2021.

[7] Kshyanaprava Panda Panigrahi, Himansu Das, Abhaya Kumar Sahoo, and Suresh Chandra Moharana. Maize Leaf Disease Detection and Classification Using Machine Learning Algorithms. In Progress in Computing, *Analytics and Networking*, pages 659–669. Springer, 2020.

[8] Kumar, A. & Alam, B. (2018). Task Scheduling in Real Time Systems with Energy Harvesting and Energy Minimization. *Journal of Computer Science*, 14(8), 1126–1133.

[9] Natarajan B and Elakkiya R. Dynamic Gan for High-quality Sign Language Video Generation from Skeletal Poses Using Generative Adversarial Networks. *Soft Computing*, 26(23):13153– 13175, 2022.

[10] Abinash MJ, Sountharrajan S, Bhuvaneswari R, Geetha K, et al. Identification and Diagnosis of Breast Cancer Using a Composite Machine Learning Techniques. *Journal of Pharmaceutical Negative Results*, 13(4):78–85, 2022.

[11] Bhuvaneswari R and Ganesh Vaidyanathan S. Classification and Grading of Diabetic Retinopathy Images Using Mixture of Ensemble Classifiers. *Journal of Intelligent & Fuzzy Systems, (Preprint)*:1–13, 2021.

Artificial Intelligence, Blockchain, Computing and Security – Dagur et al. (Eds)
© 2024 The Author(s), ISBN: 978-1-032-49393-0

Review: Recent advancements on Artificial Intelligence

Meeta Singh*, Poonam Chahal*, Deepa Bura* & Srishty*
Department of Computer Science & Engineering, Manav Rachna International Institute of Research & Studies

ABSTRACT: In the upcoming era, enhanced machines will be used as an alternate or it will increase human capabilities in each and every field. Artificial intelligence can be explained as enhanced features exhibited by software or machines. Artificial intelligence is being used by humans to enhance the technology or their reach. Artificial intelligence got the boom in the last two decades in its performance as being highly used in manufacturing and service systems. Multiple national and international research agencies are working hard in the area of artificial intelligence which has helped in rapid growth of a technology which is commonly known as expert system. AI will have to face some challenges like overcoming fears, creating public trust that are needed to be addressed and handled. This technology has increased the quality and efficiency of the employees working in these departments.

1 INTRODUCTION

Artificial intelligence is the act of developing machines more intelligent meanwhile this term intelligent refers to work with forethought in the provided environment. Artificial intelligence has rapidly become the major part of human life or we can also say bringing a tremendous change in human life. Intelligence is commonly considered as the ability to collect knowledge and reason about knowledge to solve complex problems [1].

Intelligence = Perceive + Analyse + React

Artificial intelligence is the enhancement of smart or intelligent machines and also the softwares which can learn, reason, communicate, pleat knowledge, operate and observe the objects. It is the study on machines which makes it possible to understand or perceive the reason and act. There is a difference between psychology and Artificial intelligence as it emphasis on machines and computers and also different from computer science as it is based on perception, reasoning and action. Mc Carthy brought the term in 1956 as a sub-part of computer science which is concerned in developing machines that behaves like humans. It has not only change the human lifestyle but have also affected human life in various domains such as health, safety, security, education, defence or in various science departments to. Artificial intelligence is impacting people in various fields such as helping in driving safely, enhancing gaming, enhancing education system, enhancing medical services and also in medical research labs, enhancing various defence technologies [2,3]. Artificial intelligence has brought a drastic improvement in the technology of various national and international under-cover agencies like RAW, FBI, ISI.

Artificial intelligence uses artificial neurons and scientific theorems and its working [4]. Artificial intelligence is developed upto the point such that it is offering practical benefits in

---
*Corresponding Authors: meeta.sangwan@gmail.com, poonamnandal.fet@mriu.edu.in, deepa.fet@mriu.edu.in and srishty.fet@mriu.edu.in

DOI: 10.1201/9781003393580-25

it's applications. A test approach named as turing test was brought to the limelight in 1950 which proposed Alan turing. This turing test was developed to check whether a particular computer or a machine can think or not. This test includes a human invigilator whose work is to interact with the machine and also at the same time with the human and then has to decide and then declare which one is machine and which one is human. The computer can only qualify the test if an invigilator after designing some written questions, cannot tell either the written response received is from the machine or from the human.

2 FIELDS IN ARTIFICIAL INTELLIGENCE

2.1 Games

Artificial Intelligence is highly used in reaching to an adequate level of performance by problem solving and learning abilities [5,6]. It is highly efficient or capable to store stack of rules for the games such as ludo, checkers, chess, kalah etc. The techniques of artificial intelligence makes it more appealing. The Worlds chess champion, Garry Kasparov, was defeated by Deep Blue(IBM's Product) is one of the example of the application of AI.

2.2 Research

Artificial Intelligence has provided various opportunities in the field of research to organise and have made capable to be a part in national and international conferences based on ARTIFICIAL INTELLIGENCE. Universities like University of Washington have organised multiple conferences on AI on large scale. Various books or magazines and research papers are also published on AI and its search strategies [7].

2.3 Transportation

Artificial Intelligence is easily recognisable in transportation domain. Engineers are using Artificial Intelligence in manufacturing of smart cars and in bringing the technology of self driving vehicles. GPS technology based on AI has proved itself to be one of the most successful in personal vehicles [8]. It is helpful for the drivers to have knowledge about the transportation patterns. While driverless based automobiles having range of conditions becomes a more challenging and difficult task even compared to controlling a space-craft.

With the help of AI location of the vehicle, destination of the vehicle can be easily inferred. In addition to make the cars more smarter sensors can be integrated [9,10]. While these sensors also provide features like tracking and identifying people. This can be helpful in saving lifes by decreasing road accidents and it also contributes to manage the traffic congestion.

2.4 Health care

ARTIFICIAL INTELLIGENCE and its applications can be used in the betterment of health outcomes [11]. It is most helpful for the doctors during their surgeries as they use devices based on AI for patient monitoring assistance. It is considered as one of the most resourceful in the medical field. As per study conducted by NILSSON, 2010 Artificial Intelligence technology has been a important part of medical system and support. Even a monthly magazine 'Medical Device and Diagnostic Industry' also recognized its importance in health care industries. AI has been used in computer aided detection of tumours from the medical images. Primarily on analysing the patient condition with the help of his symptoms [12]. There is a rule based expert system name Mycin for identifying the bacterias which may cause infections and on based on them suggesting antibiotics to provide treatment from these infections was brought in limelight in 1970.

167

2.5 *Employment*

Artificial Intelligence has proved to be impactful in the field of employment while it seems to be difficult to configure its current impacts [13]. Integration of various factors leads to employment. Various studies are held for technological progress and is responsible for the problem of employment specially in America but it have also started making various inroads in some important area. Artificial Intelligence techniques are used in the feature of information processing.

2.6 *Perception*

It can be defined as the capability provided by the AI to study a sensed scene and relating it to other internal model representing the perceiving organisms. The outputs of these studies are structured set of relations in between those entities. There are two methods:-

(a) Pattern Recognition
(b) Scene Analysis

2.7 *Modeling*

AI brings a feature to develop a representation and stack of transformation rules that proves to be helpful in predicting the nature and the relation between set of real world entities. Modeling can be performed in various fields like:-

(a) Illustration of problem in problem solving
(b) Organising natural systems such as ecological, economic, biological, sociological and many more.
(c) Perceptual and functional based Hobot World Modeling.

2.8 *Robots*

It can be stated as the integration of all of the features of Artificial Intelligence and sums up as a technique with the ability which can move over terrain and operate objects.

(a) Navigation
(b) Exploration
(c) Industrial Automation
(d) Military
(e) Security

3 ARTIFICIAL INTELLIGENCE CATEGORIES

Artificial Intelligence can be broken into two parts based on philosophy of AI.

3.1 *Weak AI*

The basic principal behind the weak AI is based on the fact that system or machines can be prepared to behave as they are intelligent. In simple words the term Weak AI suggest that thinking like the applications or features can be simply added to a machine or system in order to make them more efficient tool. Taking an example in which when a individual plays games like chess against a computer the individual player can feel as if the system is making impressive moves according to users moves beside of the fact that the chess application is enable to plan and think at all. The moves made by the computer are previously saved into the computer itself by the human and that's how it ensures that the application will have the right moves with respect to the situation. Few more examples of weak AI could be witness expert systems, driving wire cars by speech recognisation is the ability of an equipment to perform certain activities, which can only be expected from the human brain. It comprises of the capability of understanding relationships, to judge or to produce original thoughts.

3.2 Strong AI

The fact beside the Strong AI is that the machines can be taken upto that level that it can be made to think themselves or defining the same in other words representing human minds in the future. Based on this fact Strong AI ensures that in the upcoming Era we could be surrounded by these kinds of system or hardwares which are capable to completely act as human being. This proves that systems can have human level intelligence or more.

Considering that case those machines will be able to reason, think and performing all those of the functions that are human can perform. Considering the current research, they are nowhere near developing Strong AI, instead debates are going on between the researchers as if the concept of strong AI is even possible or not.

4 SECURITY BASED ON ARTIFICIAL INTELLIGENCE

4.1 Physical security

Physical Security is provided in outside context and in internal infrastructure, through which human like security safeguard is achieved [14].

4.1.1 Outside context

It is further divided into Simple and Complex context that are specified in the former basic indentity, location and these entities status by a single parameter, this latter points to the graphical structures and real world conditions. Discussed above contexts are then refined for support creating, integrating application of Ubiquitous, Debugging which also provide interface connection in U2IoT model, the outlines of each and every entities outside context merge even vanish.

4.1.2 Internal infrastructure

Artificial Intelligence immune security system as machine intelligence is used to study the internal infrastructure. There are few typical algorithms such as Clonal section, Immune network and Negative selection can destroy the safety systems of detection. Some physical security issues example Adaptive Disposition, Error Recovery and Intrusion Detection can be fixed as follows:

4.1.3 Inherent infrastructure

4.1.3.1 Innate immunity

Innate Immunity gives some basic barriers in order to check external attacks in real time environment and then it is applied upon sensors identifying abnormal and malicious attacks by complex patterns identifying mechanism. Co-stimulation signals are then directed to separated control ends through unit IoT networks, After that rejections are performed by management sections. While performing defense operations, activating thresholds are stated to keep a check on detection optimization.

4.1.3.2 Adaptive immunity

It means acquired resistance, where specific signatures are used to mark an attack. Particular response needs identifying non-self element at the time of attack prototype presentation. In case U2IoT got infected by similar kind of invasion, Some particular memory module will take place to eliminate the effects by developing better response to get back the system into the secured state.

4.2 Information security

Social factors of information security are categorised into three layers [15]:

4.2.1 Sensor layer
Sensor layer comprises gateways and generalized sensors to perform entity verification. The Sensor layer performs functions like to perceive the entities, to realize semantic resource and to extract information. Sensor techniques are used to have effective integration, intraction and adaption of the gathered uncertain information.

4.2.2 Network layer
Network Layer integrates of network interfaces, network management, communication channels, intelligent processing and information maintainance. The hybrid centralized and distributed network topologies together are used for monitoring and maintaining the dynamic network configuration. The network layer is used to keep a check over the information transmission whether it is reliable or not. The major functions of the network layer are to transfer and process the data gathered from sensor layer and to realize information exchange on large scale.

4.2.2 Application layer
It provides functionalities for particular applications and provide fixed interfaces for the infrastructures for testing, auditing and monitoring applications. Technologies like service composition and standard protocols are applied for the integration of heterogeneous distributed network. For example logistic monitoring, intelligent search, cloud computing and smart grid scheduling. These kind of applications should adapt in real time environment.

4.3 Security management

In the upcoming era IoT, it is barely possible to set up a uniform security protocol in the form of internet, just as contrary nations or regions can't accept similar safety precautions. Therefore, different mechanisms for management are enough for both interconnection, safety and security requirements. Since there are limitations in technological approaches, accurate management should be designed with the imposition of physical security and information security. Security plans imposed on human behaviours need to be consider to verify that the virtual cyber data is accepted to the real physical contexts.

4.3.1 Industry/national regulation
It is heavily used for iM&DC/IM&DC/nM&DC to have rules, guidance and service management for U2IoT. It resist the offensive individuals or attacks by taking disciplinary and legal actions. Hence, Industry Regulation defines some approaches to reach security of higher level. Example energy and military industries. For a moment, in the chemical risk medical management, the rules requires some specific parameters such as temperature, vibrations and relative juxtaposition, warning the users for violating the thresholds and also guarantees security of system by cautioning the abnormal imposition and aliment. Therefore, local costumes and practices should be coincides by local regulations to accept human behaviour and actions for creating imposing the local IoTs. It is guided and governed by national regulations to have nation to nation similarity. There are some formal messages or mails i.e. information of agreements or deals that are needed to be shared among the country boundaries. Specific roles and responsibilities can be allotted in between different nations.

4.3.2 Application requirement
These divided control nodes and sensors helps us to achieve specific protection. In a particular case customised needs are allotted to define the authenticated usage in a specific institution or by an individual. It should be constant with privacy prevention which verifies that confidential data is exchanged, saved.

4.3.3 *International policy*

It contemplate the global IoT association while the connectivity and consistency of nM&DC and global IoT. While, standardisation should be performed by the government and agencies to promote security and ensuring the inter-operability. It frames a general multinational governments work with reasonable enforcement techniques and policies that will provide a consistent mechanism in order to have security protection.

5 CONCLUSIONS

In this paper we have provided a brief description of Artificial Intelligence and its applications. Further we have discussed its security models and a method of establishing integrated security architecture in addition to which considering physical-social world. The IPM model consist of three important security perspectives which are information physical and management, in which information security model comprises of social layer and intelligence for security purposes. We suggest to apply firewall within the IPM model which will keep the confidential data or projects of military, games or various research agencies that are working on increasing the machines capability of understanding the human thoughts and perspectives using AI safe and secure from malicious and unauthorised users. It will also prevent from getting the data theft or lost. It will bring a feature of enhancing the accessing speed with it.

REFERENCES

[1] Strong, A. I. (2016). Applications of Artificial Intelligence & Associated Technologies. *Science [ETEBMS-2016]*, 5(6).

[2] Smith, R. G., & Eckroth, J. (2017). Building AI Applications: Yesterday, Today, and Tomorrow. *AI Magazine*, 38(1), 6–22.

[3] Russell, S. J., & Norvig, P. (2016). *Artificial Intelligence: A Modern Approach*. Malaysia; Pearson Education Limited

[4] Copeland, J. (2015). *Artificial Intelligence: A Philosophical Introduction*. John Wiley & Sons.

[5] Misirli, A. T., Bener, A., & Kale, R. (2011). Ai-based Software Defect Predictors: Applications and Benefits in a Case Study. *AI Magazine*, 32(2), 57–68.

[6] Norvig, P. (2012). Artificial Intelligence: Early Ambitions. *New Scientist*, 216(2889), ii–iii.

[7] Korbicz, J., Koscielny, J. M., Kowalczuk, Z., & Cholewa, W. (Eds.). (2012). *Fault Diagnosis: Models, Artificial Intelligence, Applications*. Springer Science & Business Media.

[8] Bahrammirzaee, A. (2010). A Comparative Survey of Artificial Intelligence Applications in Finance: Artificial Neural Networks, Expert System and Hybrid Intelligent Systems. *Neural Computing and Applications*, 19(8), 1165–1195.

[9] Dasgupta, D. (Ed.). (2012). *Artificial Immune Systems and Their Applications*. Springer Science & Business Media.

[10] Gupta, R. A., Kumar, R., & Bansal, A. K. (2010). Artificial Intelligence Applications in Permanent Magnet Brushless DC Motor Drives. *Artificial Intelligence Review*, 33(3), 175–186.

[11] Khokhar, S., Zin, A. A. B. M., Mokhtar, A. S. B., & Pesaran, M. (2015). A Comprehensive Overview on Signal Processing and Artificial Intelligence Techniques Applications in Classification of Power Quality Disturbances. *Renewable and Sustainable Energy Reviews*, 51, 1650–1663.

[12] Kumar, K., & Thakur, G. S. M. (2012). Advanced Applications of Neural Networks and Artificial Intelligence: A Review. *International Journal of Information Technology and Computer Science*, 4(6), 57.

[13] Dilek, S., Çakır, H., & Aydın, M. (2015). Applications of Artificial Intelligence Techniques to Combating Cyber Crimes: A Review. *arXiv* preprint arXiv:1502.03552.

[14] Li, B. H., Hou, B. C., Yu, W. T., Lu, X. B., & Yang, C. W. (2017). Applications of Artificial Intelligence in Intelligent Manufacturing: A Review. *Frontiers of Information Technology & Electronic Engineering*, 18(1), 86–96.

[15] O'Leary, D. E. (2013). Artificial Intelligence and Big Data. *IEEE Intelligent Systems*, 28(2), 96–99.

Artificial Intelligence, Blockchain, Computing and Security – Dagur et al. (Eds)
© 2024 The Author(s), ISBN: 978-1-032-49393-0

Embeddings of knowledge graphs for link prediction: A systematic analysis

Neelam Jain
School of Doctoral Research Innovations, GLS University

Krupa Mehta
Faculty of Computer Application and IT, GLS University

ABSTRACT: Numerous domains, including drug discovery, fraud detection, recommender systems, and question-answering systems, have extensively used knowledge graphs. Existing knowledge graphs, however, are inaccurate and incomplete. One of the tasks required to ensure the completion of the knowledge graph is the prediction of missing facts between entities. Several methods are suggested to implement link prediction by embedding entities and relations in vector space. Three categories can be used to group these knowledge graph embedding techniques: translational-distance-based models, tensor decomposition models, and neural network models. This study makes an attempt to conduct an empirical examination of several link prediction models that are accessible in the literature.

1 BACKGROUND

Entities, relationships, and meaningful descriptions comprise a knowledge graph, which is an organized representation of facts. Both concrete items and intangible concepts can be considered entities. Relationships show how two entities are related, and semantic descriptions can be either forms or qualities of the two, together with the relationship itself, with clear meaning. Edges connecting entity nodes represent relationships. The knowledge graph is shown in Figure 1. Knowledge graphs are written in the form of triplets (subject, predicate, object) or (head, relation, tail). YAGO [11], Freebase [1], and DBPedia [6] are some prominent knowledge graph examples. Applications for knowledge graphs include fraud detection, chatbots, drug discovery, and recommender systems. Though much research is being done to enhance them, knowledge graphs are still incomplete.

Link prediction is the process of detrmining missing information from given information < head, relation,? > or <?, relation, tail >. For example, in Figure 1 the relations < Harry Potter, student of, Hogwarts > and < Hogwarts, is school of, Wizards >, can be used to predict the Harry Potter is the Wizard, i.e. < Harry Potter, isa, Wizard >. Link Prediction is considered main challenge in completion of knowledge graph. Entity and relation embedding in a low-dimensional vector is a common approach to link prediction. Many state of the art algorithms are proposed for this purpose.

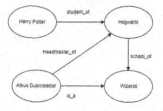

Figure 1. Example of knowledge graph.

DOI: 10.1201/9781003393580-26

Translational-distance based, semantic matching, and neural network models are the three kinds of knowledge graph embedding approaches. Models based on translational distance use distance as the scoring variable and relation as the translation from subject to object. To express complex relationships, tensor decomposition models employ tensor products. Due to their capacity for self-learning, neural networks are promising techniques in many fields. Neural network-based link prediction models are able to outperform several cutting-edge methods.

In this work, various knowledge graph embedding models are analysed and evaluated according to the implementation categories.

2 METHODOLOGY

This review begins with a taxonomy of existing link prediction models and then looks at each class's strengths and weaknesses and the top studies within each class. Various link prediction models strive to deduce different relational patterns, including symmetry, antisymmetry, inversion, and composition. The paper is thorough and in-depth and incorporates the most recent findings in the field. The selected articles reflect the most recent developments in the field and are closely related to knowledge graph embedding link prediction.

3 TRANSLATION-BASED EMBEDDING MODELS

Translation-based models are additive models that typically construct the score function for link prediction using a function that takes distance into account to assess the probability of the facts.

A relation is seen as a translation from a subject to an object in the classic and straightforward TransE [2] model. TransE expresses a link by a translation vector r, allowing r to connect a pair of embedded entities in a triplet (h, r, and t) with minimum error. This model employs the distance scoring function $\|h + r - t\|_1$. Due to its essential translation operation and the unavailability of a discrimination strategy for all types of relations, it struggles to handle multi-relational graphs. TransH [15], a variant of TransE, addresses the weaknesses of TransE concerning the reflexive N-N/N-1/1-N connections while retaining its effectiveness. It depicts translation on the hyperplane by encoding relation in a norm vector and a translation vector.

A further variant TransR [7] extracts more semantic information by handling various relations in a relation-specific space. TransR creates a translation in the related relation space after creating entity and relation embeddings in distinct vector spaces. An entity or a vector is represented by TransD [5] using two vectors. The other is utilised to create a dynamic mapping matrix, whereas the first conveys the meaning of the object or relation. It encompasses the diversity of interactions and entities in this way.

RotatE [12] rotates each element from the head entity (h) to the tail entity (t) according to the complex embedding, i.e. h,t Ck, functional mapping for each relation r. The distance function used is h r t where o is hadmard product. To generate negative samples, the model employs a self-adversarial sampling technique. Table 1 provides a list of various scoring functions for these models as mentioned in respective research.

Table 1.   Score functions of different models.

Models	Scoring Function
TransE	$\|h + r - t\|_2^1$
TransH	$\|(h - w_r^T h w_r) + d_r - (t - w_r^T)t w_r\|_2^2$
TransD	$\|(r_p h_p^T + I)h + r - (r_p r_p^T + I)t\|_2^2$
TransR	$\|M_r h + M_r t\|_2^2$
RotatE	$\|h \circ r - t\|$
DistMult	$\langle h, r, t \rangle$
ComplEx	$Re(\langle h, r, t \rangle)$
ConvE	$f(vec(f[h; \bar{r}] * \Omega))W)t$
ConvKB	$concat(f[h, r, t] * \Omega))w$

173

4 TENSOR DECOMPOSITION MODELS

Tensor decompositional models, which represent complex relationships via tensor products, are another sort of embedding model. DistMult [16] by default requires all associated embeddings to be diagonal matrices, which dramatically reduces the number of features that must be learned and simplifies model training. Since diagonal matrices are used, DistMult can only handle symmetric relations. Other than that, this is a model that generates one feature per embedding parameter through three-way interactions. The entity embedding vectors' linear transformation cannot be used to model asymmetric relations. This technique is scalable to vast knowledge graphs while learning shallow features with less expressiveness.

To enhance the modeling of asymmetric relations between entities, ComplEx [13] also sought to extend DistMult by putting forth complex-valued embeddings. New relational triplets are found using the ComplEx embedding method using the asymmetrical Hermitian product. Under this paradigm, entity and link embeddings exist in a complex space as opposed to a single physical space.

5 NEURAL NETWORK-BASED MODELS

Neural networks learn weights and biases, which they mix with the input data to discover meaningful patterns. Thanks to associative storage, fast optimization, and self-study capabilities, neural networks have demonstrated promise in various domains. These factors led to the development of this work, which offers a conceptual evaluation and comparison of contemporary KGE algorithms based on neural networks for producing KGE for link prediction

Embedded 2D convolution is used by ConvE [4] to identify missing links in KGs. The 1-N scoring used by the model expedites the link prediction process. It is also robust as it employs batch normalization and dropout. By skipping the reshaping step when encoding representations for the convolution process, ConvKB [8] improves ConvE. In this paradigm, each triple is represented by a 3-column matrix, with one column for each component. A convolution layer uses the matrix as its input to map to several feature spaces, concatenate those features, and then express the result as a single feature vector. Its distance score is derived by taking the dot product of a weight vector and a feature vector. ConvKB seeks to generalize TransE's transitional behaviors. To record more feature interactions in greater depth, InteractE [14] increases interactions by switching out the straightforward feature reshaping employed in ConvE with checkered reshaping and circular convolution.

R-GCN [10] is a relational graph convolution model that uses an encoder to generate latent feature representations of entities and a decoder, which is a tensor factorization model, to exploit these representations. This model's decoder employs DistMult factorization as a scoring function, and each relation has a corresponding diagonal matrix. R-GCN embeds semantic data about an entity included in a KG by repeatedly updating node representations. To begin, previous repre- sentations for a node's neighbours are gathered and passed through a learned, relation-specific transformation. "Messages" refer to these transformed representations. Second, the updated node representation is obtained by aggregating the collected messages and the node's state. By randomizing the initialization and freezing of the relation-specific transformations, RR-GCN [3] eliminates R-GCN's forwarding phase parameterization. In KG's structures, RR-GCN typically outperforms R-GCN.

Instead of using a standard convolution kernel, DyConvNE [9] suggests using a dynamic convolution kernel to identify aspects of the interactions between the head and related embeddings. The dynamic convolutional network can give these interaction features various weights, enabling the network to focus more on crucial interaction features and disregard less important ones.

6 CONCLUSION

With the help of the knowledge graph embedding (KGE) technology, which embeds objects and links into a continuous low-dimensional space, a range of domains now has access to accurate, effective, and structured representations of data. In this article, leading KGE technologies were assessed, existing models were divided into three groups based on whether they used additional data in addition to the facts, and then the merits and downsides of examples in each category were analysed. It has been observed that neural network models do provide good accuracy but require a long time to process, which reduces their usefulness in real-world applications. Although various studies have been conducted, most current approaches are based on large, publicly available datasets like WordNet, Freebase, etc., and the outcomes in particular domains are not thoroughly examined or reviewed. Very few studies analyze relationships when there is little or no training data that can be considered as direction for further research.

REFERENCES

[1] Kurt Bollacker, Colin Evans, Praveen Paritosh, Tim Sturge, and Jamie Taylor. Freebase: a Collaboratively Created Graph Database for Structuring Human Knowledge. In *Proceedings of the 2008 ACM SIGMOD international conference on Management of data*, pages 1247–1250, 2008.

[2] Antoine Bordes, Nicolas Usunier, Alberto Garcia-Duran, Jason Weston, and Oksana Yakhnenko. Translating Embeddings for Modeling Multi-relational Data. *Advances in Neural Information Processing Systems*, 26, 2013.

[3] Vic Degraeve, Gilles Vandewiele, Femke Ongenae, and Sofie Van Hoecke. R-gcn: The r Could Stand for Random. *arXiv* preprint arXiv:2203.02424, 2022.

[4] Tim Dettmers, Pasquale Minervini, Pontus Stenetorp, and Sebastian Riedel. Convolutional 2d Knowledge graph Embeddings. In *Proceedings of the AAAI Conference on Artificial Intelligence*, volume 32, 2018.

[5] Guoliang Ji, Shizhu He, Liheng Xu, Kang Liu, and Jun Zhao. Knowledge Graph Embedding Via Dynamic Mapping Matrix. In *Proceedings of the 53rd Annual Meeting of the Association for Computational Linguistics and the 7th International Joint Conference on Natural Language Processing (volume 1: Long papers)*, pages 687–696, 2015.

[6] Jens Lehmann, Robert Isele, Max Jakob, Anja Jentzsch, Dimitris Kontokostas, Pablo N Mendes, Sebastian Hellmann, Mohamed Morsey, Patrick Van Kleef, Sören Auer, *et al.* Dbpedia–a Large-scale, Multilingual Knowledge Base Extracted from Wikipedia. *Semantic Web*, 6(2):167–195, 2015.

[7] Yankai Lin, Zhiyuan Liu, Maosong Sun, Yang Liu, and Xuan Zhu. Learning Entity and Relation Embeddings for Knowledge Graph Completion. In *Twenty-ninth AAAI Conference on Artificial Intelligence*, 2015.

[8] Dai Quoc Nguyen, Tu Dinh Nguyen, Dat Quoc Nguyen, and Dinh Phung. A Novel Embedding Model for Knowledge Base Completion Based on Convolutional Neural Network. *arXiv* preprint arXiv:1712.02121, 2017.

[9] Haoliang Peng and Yue Wu. A Dynamic Convolutional Network-based Model for Knowledge Graph Completion. *Information*, 13(3):133, 2022.

[10] Michael Schlichtkrull, Thomas N Kipf, Peter Bloem, Rianne Van Den Berg, Ivan Titov, and Max Welling. Modeling Relational Data with Graph Convolutional Networks. In *The Semantic Web: 15th International Conference, ESWC 2018, Heraklion, Crete, Greece, June 3–7, 2018, Proceedings 15*, pages 593–607. Springer, 2018.

[11] Fabian M Suchanek, Gjergji Kasneci, and Gerhard Weikum. Yago: a Core of Semantic Knowledge. In *Proceedings of the 16th International Conference on World Wide Web*, pages 697–706, 2007.

[12] Zhiqing Sun, Zhi-Hong Deng, Jian-Yun Nie, and Jian Tang. Rotate: Knowledge Graph Embedding by Relational Rotation in Complex Space. *arXiv preprint arXiv:1902.10197*, 2019.

[13] Théo Trouillon, Johannes Welbl, Sebastian Riedel, Éric Gaussier, and Guillaume Bouchard. Complex Embeddings for Simple Link Prediction. In *International Conference on Machine Learning*, pages 2071–2080. PMLR, 2016.

[14] Shikhar Vashishth, Soumya Sanyal, Vikram Nitin, Nilesh Agrawal, and Partha Talukdar. Interacte: Improving Convolution-based Knowledge Graph Embeddings by Increasing Feature Interactions. In *Proceedings of the AAAI Conference on Artificial Intelligence*, volume 34, pages 3009–3016, 2020.

[15] Zhen Wang, Jianwen Zhang, Jianlin Feng, and Zheng Chen. Knowledge Graph Embedding by Translating on Hyperplanes. In *Proceedings of the AAAI Conference on Artificial Intelligence*, volume 28, 2014.

[16] Bishan Yang, Wen-tau Yih, Xiaodong He, Jianfeng Gao, and Li Deng. Embedding Entities and Relations For learning and Inference in Knowledge Bases. *arXiv* preprint arXiv:1412.6575, 2014.

Artificial Intelligence, Blockchain, Computing and Security – Dagur et al. (Eds)
© 2024 The Author(s), ISBN: 978-1-032-49393-0

Predictive system on the car market trend using AI & ML

Ansh Shankar & Dhruv Varshney
Galgotias University, SCSE, Greater Noida, Uttar Pradesh, India

Arvind Nath Sinha
GL Bajaj Institute of Technology and Management, Greater Noida, Uttar Pradesh, India

ABSTRACT: Automobile Manufacturing is one of the most sophisticated sets of processes in the World. And for the automobile to be Successful in the market, it requires an extensive amount of work to be done in the field of Market Analysis. This is an area of major concern for companies. This paper could provide base for prediction based on the history of the Cars Manufactured in India. This paper analyses the trend of Market demand and steps to build a predictive model that would predict that whether a car would be successful in the market or not.

Here we have used a Linear Regression technique to build a Machine Learning model. This model is trained on multiple datasets collected from different sources which are then analyzed and processed to obtain desired results. Thus this paper is a very handy tool for both manufacturers and customers. Since this business of car manufacturing is unceasing, hence the process of data generation is also never-ending. This paper's accuracy will get better with time as more data is available.

Keywords: predictive system, car market, machine learning

1 INTRODUCTION

Automobile Manufacturing is one of the most painful processes for the car Manufactures in terms of both labour and financial. But if automobile is not well received in the market, the pain and the treasure is lost. For example, for production of new car the company has to setup whole production unit along with its lifeline i.e. Workers. But what if its sales are low or what if the car doesn't match the standards of the population for which the car is manufactured, the whole ecosystem and the hard work goes to vain.

So, this paper comes up with a model that would be a boon to this type situation. We have collected automobiles dataset of different companies from different sources, analysed and processed it to obtain the information, upon which our machine learning model is trained. The model is based on Linear Regression technique. The manufacturer could extract from it which type of car to be manufactured, what would be the most prominent colour, what would be transmission type, fuel type, etc. The Whole Paper is divided into the 5 parts:

- Data Collection
- Data understanding and exploration
- Data cleaning
- Data preparation
- Model building and evaluation

DOI: 10.1201/9781003393580-27

2 LITERATURE SURVEY

Three parts make up a back-propagation neural network-based prediction system for auto fuel consumption: an information collection system, a fuel consumption forecasting algorithm, and a performance evaluation system. Although there are several elements that affect a car's fuel consumption in a real-world driving scenario, under the present system, the effect factors for fuel consumption are merely the car's make, engine type, vehicle type, weight, and transmission system type [1].

Their research examines the potential for using Google Trends data and sentiment analysis to forecast the sales of automobiles. Previous researches has demonstrated the effectiveness of both strategies for sales forecasting, but the results of the present study for predicting the sales of high-involvement goods like autos are less clear-cut. Approximately 500,000 social media posts for 11 car models sold in the Netherlands are evaluated using linear regression models. The predictive power of Google Trends is also contrasted with the findings of this study. The results suggest that while Google Trends data and social mention volume show significant results and may be incorporated into a helpful prediction model, social media emotions have little predictive value when it comes to automotive sales [2].

Based on the relative distance and relative acceleration of each instant, the MPC forecasts the leader vehicle's (LV) future behaviour, and the follower vehicle's (FV) acceleration is controlled in accordance with this behaviour. By regulating the acceleration, the MPC seeks to maintain the relative distance within a safe range. For the purpose of assessing the effectiveness of the produced controller, the system's output is contrasted with actual drivers' behaviour under similar starting circumstances. The simulation results show that the MPC controller acts far safer than real drivers and can provide passengers with a comfortable journey [3].

Since a leasing company needs to offer a competitive lease price in order to remain successful in a cutthroat industry. In order to set the right price for a used car, it is crucial to project its future value. If the depreciation is known, the lease fee might be adjusted to account for the car's value decline. However, a number of factors affect price, making this important responsibility challenging. The traditional regression approach might not be suited for high-dimensional data. This potential issue will be resolved using Support Vector Regression, a modern data mining technique that is independent of input dimension [4].

The Prediction of Car Ownership discussed that for long-term forecasting of automobile ownership, an econometric technique is developed. It is compared to other well-known techniques. It is based on estimates of the percentage of family income spent on automobile purchases, as well as an analysis of car pricing and stock. The technique provides a reasonable approximation of previous levels of automobile ownership in the United Kingdom, as well as a prediction that is comparable to Tanner's for the next 15 years. However, Mogridge forecasts a higher saturation level and, as a result, a larger eventual automobile population in the future [5].

3 DESIGN & IMPLEMENTATION

3.1 *Data set*

A collection of data is known as a data set, which is typically represented in tabular form. Each variable's details are stored in columns and each row represents the corresponding record. The dataset can contain data for one or more items, depending on the number of rows. For example, in our dataset we had Manufacturer name, Model, Vehicle type, Sale number, Fuel type as our columns, which basically represents our variable's detail. and each record/data is accommodated in the single row.

3.2 Model design

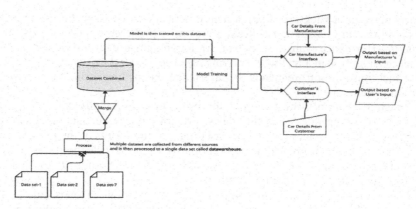

Figure 1.   Model design.

3.3 Module description

In the paper, to train my model I am using two Regression Models, the Linear Regression model and the Lasso Regression Model, which are part of the linear model of the sklearn library. The other libraries used in the paper are pandas, NumPy, Matplotlib, Seaborn, and metrics and preprocessing from sklearn.

NumPy is a Python library used for operating with multi-dimensional arrays. For, the purpose of manipulating and analyzing statistics, the Python programming language has a software package called Pandas. It provides statistical methods and processes, specifically for working with time series and numerical tables. Seaborn is a fantastic Python visualization package for visualizing statistical images. It provides gorgeous default patterns and colour schemes to enhance the aesthetic appeal of statistics charts. It was built at the very top of the Matplotlib package. For 2D displays of arrays in Python, Matplotlib is a well-liked visualization toolkit. A multi-platform statistics visualization package called Matplotlib was built using NumPy arrays.

Lasso stands for Least Absolute Shrinkage and Selection Operator. It is a type of Linear Regression Model that shrinks the data value towards a central point like mean. In, First segment, initialized the model into the respective variable which is named lasso_reg_model and lin_reg_model. After the model is trained (i.e., fitted) with the training data, it's time to evaluate the model. In a regression model, the R2 score also called the coefficient of determination is the statistical measure of the accuracy of the model that shows how much variance in a dependent variable is explained by the independent variable(s).

4   EXPLORATORY ANALYSIS

Now coming to the data preprocessing step of our Paper, as earlier mentioned that in our paper we had taken datasets from multiple sources so integrating them into a single entity was one of the most painful and time-consuming steps of the paper. Here, we analyzed all the different aspects that could be considered so that our model could predict with greater accuracy.

4.1 Dataset preprocessing

Dataset preprocessing is the process of cleaning up the data so that it can be given to our model and the model trains on it without any hindrance. In this step, we adapted many different techniques to get the desired data. Among them, was to segregate the year of manufacture from the date and month of manufacture as shown below. We did this by splitting the column using the str. split() function and splitting it whenever we found the '/' symbol. Similarly, we split many different columns to extract the meaningful data from it.

	Manufacturer	Name	Price	Mileage	Year	Fuel	Transmission
0	Acura	Integra	215000.0	28.0	2012	NaN	NaN
1	Acura	TL	284000.0	25.0	2011	NaN	NaN
3	Acura	RL	420000.0	22.0	2011	NaN	NaN
4	Audi	A4	239900.0	27.0	2011	NaN	NaN
5	Audi	A6	339500.0	22.0	2011	NaN	NaN

Figure 2. Final dataset.

4.2 *Data set split*

This is the last step before the model is trained, it is called so because in this step we split the dataset into two halves- the training and testing dataset. We have kept the testing data 20% of the original dataset which is 23237 rows. The model is trained using this dataset, and the test portion is used to evaluate the model's correctness.

5 RESULTS

The model predicted quite well in the case of Linear Regression the training data with an R^2-score *0.17130651180824064* while in the case of test data it was *0.17836986214106343*. The Graph of the prediction and the Z-Score is as given in figure below. In the case of Lasso Regression, the R2-score is 0.17130651180690104 on training data and 0.17836978 on the test data which is quite good.

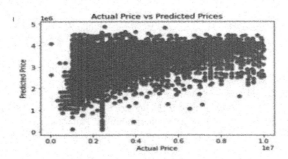

Figure 3. Result of linear regression on training data.

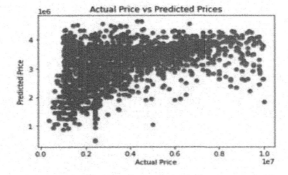

Figure 4. Result of linear regression on testing data.

179

6 CONCLUSION

This paper deals with analysis the market for a good car to be released as per the demands. Although poor dataset availability in this field, It was a tedious task to be able to extract the present dataset of second hand cars and use it to predict the market value of the new car. If in future a more reliable and variety of dataset is available, it would be useful for analyze different aspects like colour of car, state of sale of car, type of car, top speed of car, total sales of car and many more.

Also, this model can be integrated a user-friendly interface, where users can find the right car for themselves or for the firm owners where they can see if the car which they are planning to release would be a boom or crash in the market. The model can also be used to learn itself from the queries of the customer as well as can conduct the important analysis of the market demand which would a major boost for the manufacturing firm.

REFERENCES

[1] Jian-dawujun-chingliu Published Title *"Development of a Predictive System for Car Fuel Consumption Using an Artificial Neural Network"* 2019

[2] Enis Gegic, Becir Isakovic, Dino Kečo, Zerina Mašetić, Jasmin Kevrić Published in TEM Journal with Title *"Car Price Prediction using Machine Learning Techniques"*

[3] Fons Wijnhoven, Olivia Plant published in ICIS 2017 Proceeding with Title *"Sentiment Analysis and Google Trends Data for Predicting Car Sales"*

[4] Khodayari; A. Ghaffari; M. Nouri; S. Salehinia; F. Alimardani published in IEEE with Title *"Model Predictive Control System Design for Car-following Behavior in Real Traffic Flow"*

[5] Mariana Listiani in Master Thesis Proceeding with Title *"Support Vector Regression Analysis for Price Prediction in a Car Leasing Application"*

[6] Mogridge M. J. H. Published in Journal of Transport Economics and Policy Proceeding with title *"The Prediction of Car Ownership"* 2018.

[7] Dehghani M., Dangelico R. M., "Smart Wearable Technologies: Current Status and Market Orientation Through a Patent Analysis", *Proc. IEEE Int. Conf. On Industrial Technology*: 1570–1575, 2017.

[8] Lai L. F., Jiang J. F., Wei H. Y.and Hsu K. S., "Depth Sensor Used in Vehicle-related Patent Analysis" *Int. Conf. IEEE-ICAMSE*, 2016.

[9] Karwal H. and Girdhar A., "Vehicle Number Plate Detection System for Indian Vehicles," *2015 IEEE International Conference on Computational Intelligence & Communication Technology*, 2015.

[10] Babu K. M. and Raghunadh M. V., "Vehicle Number Plate Detection and Recognition Using Bounding Box Method," *International Conference on Advanced Communication Control and Computing Technologies (ICACCCT)*, 2016.

[11] Babbar S., Kesarwani S., Dewan N., Shangle K. and Patel S., "A New Approach for Vehicle Number Plate Detection," *2018 Eleventh International Conference on Contemporary Computing (IC3)*, 2018.

[12] Aworinde Halleluyah Oluwatobi, Onifade O.F.W. (2019) "A Soft Computing Model of Soft Biometric Traits for Gender and Ethnicity Classification" *International Journal of Engineering and Manufacturing (IJEM)*

[13] Feranmi & Olayiwola, Adedayo Amos Advances in Multimedia – *An International Journal (AMIJ)*, Volume (6): Issue (1): 2020

[14] Aworinde O., Afolabi A. O., Falohun A. S. & Adedeji O. T. (2019) "Performance Evaluation of Feature Extraction Techniques in Multi-Layer Based Fingerprint Ethnicity Recognition System" *Asian Journal of Research in Computer Science*.

[15] Abu Abraham Matthews & Amal Babu P (2017) "Automatic Number Plate Detection" *Journal of Electronics and Communication Engineering (IOSR-JECE)* 50–55

Artificial Intelligence, Blockchain, Computing and Security – Dagur et al. (Eds)
© 2024 The Author(s), ISBN: 978-1-032-49393-0

Object detection system with voice output using Artificial Intelligence

K. Sivaraman, Pinnika Gopi, Katta Karthik & Kamsani Venkata Upendar Reddy
Bharath Institute of Science and Technology, Chennai, India

ABSTRACT: As object identification technology has advanced recently, numerous technologies have been integrated into autonomous vehicles, robotics, and industrial facilities. The visually impaired, who need these tools the most, are not receiving their benefits, though. Using deep learning technology, we suggested an object detection system for the blind in this research. By using voice recognition technology, we can determine what items a blind person desires and then utilize object recognition to locate those items. To further assist those who are blind in locating items, voice guidance technology is used. The deep learning model uses deep neural network architecture to recognize objects, and speech-to-text (STT) technology to recognize voices. The creation of an audio announcement using text-to-speech (TTS) technology makes it easier for blind individuals to learn about objects. The system is created using the Python library OpenCV. In order to help the blind identify objects in a particular location on their own, we design an effective object-detection system, which is then tested in trials to determine its effectiveness.

1 INTRODUCTION

Finding and identifying instances of real-world objects, like a car, bike, TV, flower, or a person, in still photos or moving pictures is a process known as object detection [1]. The ability to identify, locates, and detects numerous things inside an image enables object detection techniques to help you comprehend the finer aspects of an image or a video [2]. In most cases, it is used in systems there are numerous techniques for the presence of items in an image can be quickly and easily detected by humans. The visual system of the human being is quick and precise, and it is capable of carrying out complicated tasks like object identification that require a lot of mental effort [3]. We can now easily train computers to detect and classify a large number of items within an image with high accuracy, tensor flow is used because it makes it easy to create, train, and deploy object detection models and because it provides a library of detection models that have previously been done [4]. The detection performance may be tuned from beginning to end because the entire detection pipeline is contained within a single network.

1.1 *Feed forward and feedback networks*

Networks with inputs, outputs, and hidden layers are called feed forward networks. Only one direction is possible for the signals to go (forward). There is a layer that receives input data and performs calculations. Based on the weighted sum of all of its inputs, each processing element computes [5]. The updated values serve as fresh inputs for the subsequent layer (feed-forward). The output is decided by carrying on in this manner through all the layers. In data mining, for instance, feed forward networks are frequently utilized. There are feedback channels in a feedback network, such as a recurrent neural network. So, employing loops, they can have signals moving in both ways. Neurons can link in any way that makes sense [6].

DOI: 10.1201/9781003393580-28

1.2 Weighted sum

The synapse serves as a "toll" between each neuron, which must be paid (weight). Following that, the neuron adds up the weighted inputs from all of the incoming synapses and applies an activation function to them. Each neuron in the subsequent layer receives the result [7]. It is these synapses' weights that are being adjusted when we discuss updating weights in a network. The weighted sum of all the outputs from the neurons in the layer before constitutes a neuron's input. The weight of the synaptic connection from each input to the activated neuron increases each input by a factor of two. The activation function, also known as the transfer function, produces output signals by converting input signals [8]. The output numbers are transformed into a range, such as 0 to 1 or −1 to 1.The action potential firing rate in the cell is represented by an abstraction. The probability that the cell may ignite is indicated by this value. The function is binary at its most basic level: the neuron fires when the answer is yes or the neuron doesn't. The output can range from anywhere in a range to either 0 or 1 (on/off or yes/no). If, for instance, you wanted to calculate the probability that an image is of a cat utilizing a mapping function for the 0–1 range you would do it this way: an output of 0.9 would indicate that there is a 90% chance that the cat in your image is real [9].

Activation function role of activation simply expressed, a node's output is determined by its activation function. The activation function, also known as the transfers' signals from the input to the output using the transfer function. The output values are projected into a range, such as 0 to 1 or −1 to 1. It represents the action potential firing rate of the cell as an abstraction. It represents the probability that the cell will ignite numerically [10]. If the function is true, the neuron fires; if it is false, the neuron does not fire. The output could be any value within a range or just 0 or 1 (on/off or yes/no).

Threshold function this activity takes steps. The function returns 0 if the total of the input values falls below a predetermined threshold. It would pass on 1 if it were greater than zero or equal to zero. It has a very strict, binary, yes-or-no functionality [11]. Sigmoid function Logistic regression employs this activity. It progresses from 0 to 1 smoothly and gradually, in contrast to the threshold function. In the output layer, it is helpful and frequently utilized for linear regression.

1.3 Hyperbolic tangent function

Although this does not closely resemble what occurs in the brain, this function produces better results when neural networks are being trained. Sometimes, while being trained with the sigmoid function, neural networks become "stuck." Rectifier function it is biologically conceivable and the most effective [12]. Despite having a kink, the transition following the kink at 0 is smooth and progressive. Thus, your result may be "no" or a percentage of "yes," for instance. No normalization steps or other challenging calculations are needed for this function. When machines can perform jobs that ordinarily demand for human intelligence, artificial intelligence becomes crucial. It falls under the category of machine learning, where computers may learn new abilities and from their past mistakes without the help of humans [13]. Deep learning, which uses a lot of data to learn, is a machine learning function that employs artificial neural networks, which are algorithms based on the human brain. Deep learning, which uses a large amount of data to learn, is a machine learning function that employs artificial neural networks, which are algorithms based on the human brain. Deep learning algorithms perform tasks indefinitely in order to improve outcomes [14]. The concept of deep learning is founded on human experience. Learning is made possible by neural networks' many (deep) layers [15].

2 METHODOLOGIES

2.1 INPUT-video file/image file

The entire image is subjected to a neural network application to forecast bounding boxes and their probability [16].

2.2 Yolov3 – DNN module

We'll employ blobFromImage in a method called detect objects () that takes as parameters a video frame or picture, a model, and the output layers [17].

2.3 Bounding box regression

- DNN module returns a nested list with data about all the objects that were detected, including the x and y coordinates of the object's center, the height and width of the bounding box, confidence levels, and scores for each of the object classes listed in coco. names.
- The projected class is thought to be the one with the highest score.
- Utilizing np.argmax, we next determine which class index has the highest confidence/score () Shown as below Figure. We may access the class name corresponding to the index from the classes list we produced in load yolo () shown in below Figure 1.
- I chose each of the projected bounding boxes with a confidence level of at least 50%.

As soon as we obtain the class id (index of the expected object class () and the vertices of the predicted bounding box, we must draw the bounding box and add an object label to it.

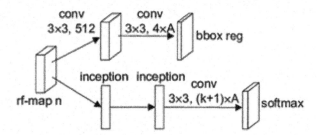

Figure 1.   Bounding box regression.

2.4 Voice feedback

The audio system converts the object's label and location into audio format. For further audio file manipulation, i also used pydub and ffmpeg shown in below Figure 2.

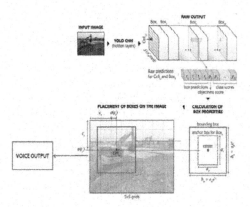

Figure 2.   Computer architecture.

GTTS (Google Text to Speech) is the library used generally for conversion of detected object as text value to Speech data. This is very fast in conversion and more reliable to use [18].

183

3 RESULTS AND DISCUSSION

3.1 *Existing system*

A popular algorithm of object detection includes:

- Region-based Convolutional Neural Networks (RCNN),
- Faster RCNN (F-RCNN).
- CNN
- Histogram of Oriented Gradients (HOG)
- CNN does not encode the position and orientation of object.

3.2 *Data flow diagram*

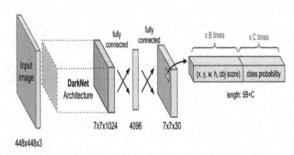

Figure 3. Data flow diagram.

1. Training Data: COCO, or Common Objects in Context, is used to train the model. It's quite cool to browse the photographs that were tagged in the link.
2. Model: The model used in this case is the You Only Look Once (YOLO) method, which utilizes a variant of the Darknet Convolutional Neural Network architecture shown as above figure.
3. Output: Additionally, each object that is detected in our frames will have its bounding box's coordinates obtained. Instead of 30 frames per second, we will also schedule a vocal feedback, such as "bottom left cat," to indicate the location of a cat that was spotted at the bottom-left of my camera view [19].

3.3 *Proposed system*

- YOLO (You Only Look Once) is a technique or methodology to detect objects. A completely different strategy is used by YOLO.
- It scans the network once and only looks at the complete image once to find items. Therefore, the name. It proceeds quickly.
- The many output levels of the YOLO v3 architecture we are implementing provide predictions.
- The output object is displayed in an image after detections are made by drawing bounding boxes around them.
- The ability to recognize both video and picture streams.
- The Darknet-53, which has 53 convolution architecture, is used by YOLO v3 to extract features.

3.4 *Output*

- [INFO] loading YOLO from disk...
- ['mid left person', 'mid left person', 'top center truck', 'top left truck', 'bottom left person', 'top center truck']

- ['mid center person', 'mid left person', 'top left truck', 'top left truck', 'top center truck', 'bottom center suitcase', 'mid left truck', 'bottom center suitcase']
- ['bottom center person']

4 CONCLUSIONS

Several algorithms are used for object detection and recognition, including CNN, Fast R-CNN, and RCNN. It was chosen because YOLOv3 is a quick and accurate method for detecting and locating objects in real time. Because the suggested system plays the sound from the sounds dataset, converting text to sound does not require an internet connection. Compared to previous techniques, the PC's necessary time for detection was shorter. Future versions of the system might be built into smart eyewear for the blind that could help them detect and identify objects in their surroundings using Raspberry Pi computers. Additionally, this system can be enhanced using age and gender prediction approaches to anticipate age and gender. The system can also incorporate the face recognition method to identify the person.

REFERENCES

[1] Kharchenko.V and Chyrka. I, "Detection of Airplanes on the Ground Using YOLO Neural Network," *Int. Conf. Math. Methods Electromagnet.* Theory, MMET, vol. 2018-July, pp. 294–297, 2018, doi: 10.1109/MMET.2018.8460392.

[2] Cheng. Z, Lv. J, Wu. A, and Qu. N, "YOLOv3 Object Detection Algorithm with Feature Pyramid Attention for Remote Sensing Images," *Sensors Mater.*, vol. 32, no. 12, p. 4537, 2020, doi: 10.18494/sam.2020.3130.

[3] Jabnoun. H, Benzarti. F, and Amiri. H, "Object Detection and Identification for Blind People in Video Scene," *Int. Conf. Intell. Syst. Des. Appl. ISDA*, vol. 2016-June, pp. 363–367, 2016, doi: 10.1109/ISDA.2015.7489256.

[4] Srivatsa. S. N, Sreevathsa. G, Vinay. G, and Elaiyaraja. P, "Object Detection using Deep Learning with OpenCV and Python,", *International Research Journal of Engineering and Technology (IRJET)*, Volume 8, Issue 1, pp. 227–230, 2021.

[5] Shah. M and Kapdi. R, "Object Detection using Deep Neural Networks," Proc. *2017 Int. Conf. Intell. Comput. Control Syst. ICICCS 2017*, vol. 2018-Janua, pp. 787–790, 2017, doi: 10.1109/ICCONS.2017.8250570.

[6] Geethapriya. S, Duraimurugan. N, and Chokkalingam. S. P, "Real Time Object Detection with Yolo," *Int. J. Eng. Adv. Technol.*, vol. 8, no. 3 Special Issue, pp. 578–581, 2019.

[7] Corovic. A, Ilic. V, Duric. S, Marijan. M, and Pavkovic. B, "The Real-Time Detection of Traffic Participants Using YOLO Algorithm," *2018 26th Telecommun. Forum, TELFOR 2018 – Proc.*, 2018, doi: 10.1109/TELFOR.2018.8611986.

[8] Potdar. K, Pai. C. D, and Akolkar. S, "A Convolutional Neural Network based Live Object Recognition System as Blind Aid," *arXiv, 2018.*

[9] Bhairnallykar. Prajapati. P. S. A, Rajbhar. A, and Mujawar. S, "Convolutional Neural Network (CNN) for Image Detection," *International Research Journal of Engineering and Technology (IRJET)*, Vol. 7, no.11, pp. 1239–1243, 2020.

[10] Teddy Surya Gunawan, Arselan Ashraf, Bob Subhan Riza Edy Victor Haryanto, Rika Rosnelly, Mira Kartiwi, Zuriati Janin, "Development of Video-based Emotion Recognition using Deep Learning with Google Colab," *Telkomnika (Telecommunication Comput. Electron. Control*, vol. 18, no. 5, pp. 2463–2471, 2020, doi: 10.12928/TELKOMNIKA.v18i5.16717.

[11] Mittal. N, Vaidya. A, and Shreya Kapoor. A, "Object Detection and Classification Using Yolo," *International Journal of Scientific Research & Engineering Trends*, vol. 5, no. 2, 2019.

[12] Burić. M, Pobar. M, and Ivašić-Kos. M, "Adapting Yolo Network for Ball and Player Detection," *ICPRAM 2019 – Proc. 8th Int. Conf. Pattern Recognit. Appl. Methods, no. Icpram*, pp. 845–851, 2019, doi: 10.5220/0007582008450851.

[13] Abdurrasyid. A, Indrianto. I, and Arianto. R, "Detection of Immovable Objects on Visually Impaired People Walking Aids," *Telkomnika (Telecommunication Comput. Electron. Control,* vol. 17, no. 2, pp. 580–585, 2019, doi: 10.12928/TELKOMNIKA.V17I2.9933.

[14] Bharti. R, Bhadane. K, Bhadane. P, and Gadhe. A, "Object Detection and Recognition for Blind Assistance," *International Research Journal of Engineering and Technology (IRJET)*, Volumemac_mac 06, Issue: 05, pp. 7085–7087, May 2019.

[15] Masurekar. O, Jadhav. O, Kulkarni. P, and Patil. S, "Real Time Object Detection Using YOLOv3," *Int. Res. J. Eng. Technol.*, vol. 07, no. 03, pp. 3764–3768, 2020.

[16] Vaidya. S, Shah. N., and Shankarmani. R, "Real-Time Object Detection for Visually Challenged People," *Proc. Int. Conf. Intell. Comput. Control Syst. ICICCS 2020, no. Iciccs*, pp. 311–316, 2020, doi: 10.1109/ICICCS48265.2020.9121085.

[17] Shifa Shaikh, Vrushali Karale and Gaurav Tawde, "Assistive Object Recognition System for Visually Impaired," *Int. J. Eng. Res.*, vol. V9, no. 09, pp. 736–740, 2020, doi: 10.17577/ijertv9is090382.

[18] Mahmoud. A. S, Mohamed. S. A, El-Khoribi. R. A, and AbdelSalam. H. M, "Object Detection using Adaptive Mask RCNN in Optical Remote Sensing Images," *Int. J. Intell. Eng. Syst.*, vol. 13, no. 1, pp. 65–76, 2020, doi: 10.22266/ijies2020.0229.07.

[19] Fahad A. A, Hassan H. J, and Abdullah S. H, *Deep Learning-based Deaf & Mute Gesture Translation System,"* vol. 9, no.5, pp. 288–292, May 2020, doi: 10.21275/SR20503031800.

186

Artificial Intelligence, Blockchain, Computing and Security – Dagur et al. (Eds)
© 2024 The Author(s), ISBN: 978-1-032-49393-0

Multi-objective optimization-based methodological framework for net zero energy building design in India

Pushpendra Kumar Chaturvedi, Nand Kumar & Ravita Lamba
Malaviya National Institute of Technology Jaipur

ABSTRACT: The challenge in designing net-zero energy building (NZEB) is to optimize the various design scenarios such as active, passive and renewable energy (RE) supply etc. and their trade-offs to determine the combinations of optimal solutions. Multi-objective optimization (MOO) is a neurotic method to find a Pareto solution for optimization and decision-making of NZEB. The objective of this paper is to develop a MOO-based methodological framework for NZEB design in the tropical climate of India, taking into account the 3S criteria of NZEB: social, environmental, and economic. Social criteria are mapped to enhance thermal, visual, and indoor air quality, environmental criteria address energy efficiency and CO_2 emissions, and economical criteria show minimum life cycle cost. This paper summarised the methodological approach into four sections: optimization framework, multicriteria decision making (MCDM), Robustness of solution with sensitivity analysis, and net metering (grid connectivity). This paper also focuses on the current needs and requirements, definitions, and structure of NZEB.

1 INTRODUCTION

Net-zero energy buildings (NZEBs) are a highly effective way of assisting the world in meeting its 2050 goals for sustainable development, climate mitigation, and zero carbon emissions [1]. Buildings alone represent 40% of global energy demand and are responsible for 30% of GHG emissions [2]. According to the International Energy Agency (IEA), building floor area is predicted to increase by 75% by 2050 as compared to the base year of 2020 [3]. India is the third-largest country in the primary energy consumption supply which is 569 Mton [4]. India is also the fastest-growing country in real estate. Figure 1 shows the projected build-up area of residential buildings. It demonstrates that the residential build area is 17.7 billion m^2 in the base year of 2020 and it may fold 3–4 times in 2050 [5–7]. Ultimately, it puts tremendous stress on electricity demand and supply chain. In India, buildings account for 33% of total electricity usage. It is projected to increase 3–5 times by 2031 which will ultimately contribute to GHG emissions [8]. India is the 4th largest carbon emitting country despite the lowest per capita CO2 emission. The power sector in India is responsible for half of the carbon emissions [9]. According to the Indian cooling action plan report, cooling demand of India will increase by 8 times in 2037–38 as compared to the base year of 2017–18. Ultimately, the total primary energy supply (TPES) will increase by 4.5 times by 2037–38 [10].

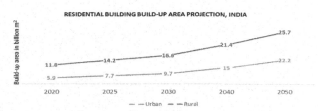

Figure 1. Projected build area of residential buildings

DOI: 10.1201/9781003393580-29

To overcome all these energy-related issues, transformation in the building sector is required with available technologies. After the revolution in the building sector, a new phrase known as 'net-zero energy building' (NZEB) is proposed. It is also known as zero energy building. These kinds of buildings essentially produce the same amount of energy consumed over a year.

2 CONCEPT OF NZEB

The concept of "zero energy building" was introduced in 1976 by Esbensen and Korsgaard from the University of Denmark [11]. Many developed countries use their definition and framework to build zero energy buildings. In 2011, India purposed NZEB roadmap for the ECO-III project. According to this, if any building equipped with energy-efficient measures and having minimum energy demand that is fulfilled by renewable energy sources falls into this category [12]. There are many classifications of NZEB based on grid connectivity and renewable energy (RE) supply options that are given below [13–15].

2.1 *Classification based on the grid connected balanced energy measurement methods*

2.1.1 *Net-zero site energy building*
A building that uses renewable energy to meet its annual energy needs is called net-zero site energy building. Generation of RE is equivalent to at least operational energy demand.

2.1.2 *Net-zero source energy building*
A building that generates enough renewable energy to cover both operational demand and grid-to-building energy losses is called net-zero source energy building. Transmission, distribution, primary fuel supply, and a few other factors all contribute to energy losses.

2.1.3 *Net-zero cost energy building*
A building that covers utility bills by selling the produced RE is called net-zero cost energy building. It can also generate profits, but it should at least cover all of bills of utility and services.

2.1.4 *Net zero-emission building*
A building that produces enough energy for consumption and compensates for the emissions produced by generating enough renewable energy is called net zero-emission building.

2.2 *Structure of NZEB*

NZEB structure is built on triple bottom line of sustainability such as social, environmental, and economic with safeguarding the people, planet, and profit [16]. Commonly, four conventional design factors are commonly connected with the formation of NZEB, as indicated in Figure 2 [17].

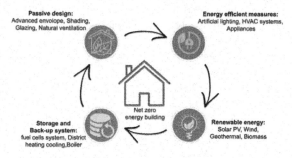

Figure 2.   Conventional design parameters for NZEB

3 LITERATURE REVIEW

A multi-objective optimization is an effective approach for decision-making of NZEB development. It provides a set of mono solutions that can compute through a trade-off analysis of the concerned objectives. Based on this concept, authors illustrate 10 studies in text and tabular form. Table 1 shows the recapitulation of analyzed case studies based on 3S criteria of sustainability. It illustrates that there is a research gap that no single study exists which includes all three criteria simultaneously of NZEB and their related parameters for optimization. The aim of this study to develop a multi-criteria decision-making approach for zero energy building design in India.

Table 1. Critically analysis of reviewed primary case studies, based on objective functions.

Objectives functions				Social			Energy				Environment		Economic		
Social	Energy	CO_2	Economic	Thermal	Visual	IAQ	Demand	Saving	Consumption	Load	CO_2 emission	Solar PV	LCC	Sensitivity/uncertainty	Ref.
	■			■				■			■				[18]
	■			■				■		■				■	[19]
■	■		■	■		■		■		■		■		■	[20]
■	■		■			■		■		■			■	■	[21]
■			■				■		■			■			[22]
														■	[23]
■			■	■				■		■		■			[24]
■		■	■	■			■		■		■		■		[25]
■	■		■				■			■	■		■		[26]
■	■		■				■		■		■			■	[27]

4 METHODOLOGY

It illustrates the multi-objective optimization-based approach for residential NZEB. It is divided into some chronological steps: optimization framework, multi-criteria decision-making process, robustness test by sensitivity analysis shown in Figure 3. The optimization process is made up of a series of steps that are detailed below.

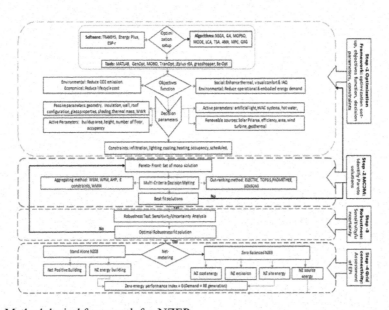

Figure 3. Methodological framework for NZEB.

4.1 Optimization setup

Step 1 depicts the optimization workflow hierarchy, which consists of a collection of tools, algorithms, and software. The software provides the user interface to formulate the MOO problem. Tools support to track the problem coupled with an adequate optimization algorithm to provide the set of mono solutions. Several tools is available for optimization of building associated problem. Table 2 presents some of the tools and their capabilities. Table 3 lists the software that is commonly used for creating building modeling coupled with optimization tools.

Table 2. List of available tools and their features for multi-objective optimization.

Tools	Characteristics				
	Open source	Optimization	User interface	Parametric simulation	Parallel computation
MATLAB		✓	✓	✓	✓
Gen-Opt	✓		✓	✓	✓
MOBO	✓	✓	✓		✓
TRNOPT			✓	✓	✓
Rhino	✓	✓	✓	✓	✓
Opt-E-Plus	✓	✓	✓		✓
Be-Opt	✓	✓	✓	✓	

Table 3. List of building simulation software and their modelling features (CI – Completely Investigated, PI – Partially Investigated, OI – Optionally Investigated, NI – Not Investigated).

Modelling features	TRANSYS	ESP-R	Energy plus	IDA-ICE
Simulation	CI	CI	CI	CI
MOO	CI	CI	CI	CI
Open source	NI	CI	CI	NI
Tools coupled	CI	CI	CI	CI
HVAC modelling	CI	CI	CI	CI
Daylight modelling	PI	PI	PI	PI
Solar gain/shading analysis	CI	PI	CI	CI
Natural ventilation	CI	NI	CI	CI
Weather data	CI	CI	CI	CI
LCA	PI	OI	PI	PI
Mathematical model	CI	NI	NI	NI
Emission modelling	OI	NI	CI	OI
ANN support	CI	PI	CI	PI

4.2 Objective function

NZEB is concerned with a number of design objective scenarios that are closely linked to social, environmental, and economic building behaviour standards. The majority of Indian geographical regions fall under the cooling dominant groups. For cooling prevailing climatic situations, defined objective functions are given (Figure 4).

4.2.1 Thermal comfort

Many indices exist to define thermal comfort, including the Fangers model, ASHRAE 55 model, EN adaptable model, and NBC 2016. Thermal comforts are primarily evaluated using two factors: predicted mean vote (PMV) and percentage predicated dissatisfied (PPD). For thermal comfort in −0.5% to +0.5% range, ISO 7730:2005 specifies a 90% acceptance

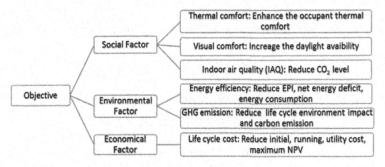

Figure 4. Hierarchy of objectives function for NZEB.

rate as PMV (10 percent PPD). Analysis criteria is given below. In Indian contect national building code (NBC) defines the criteria of thermal comfort for naturally, mix-mode and fully air-conditioned residential buildings (see Eq. 1, 2, and 3) based on occupant acceptability range of 90%.

$$PPD = 100 - 95 \times e^{-0.03353 \times PMV\left(4 - 0.217 PMV^4\right)}$$ (1)

$$T_{indoor\ operative\ temperature} = 0.54 T_{Mean\ outdoor\ temperature} + 12.83 \pm 12.38$$ (2)

$$T_{indoor\ operative\ temperature} = 0.28 T_{Mean\ outdoor\ temperature} + 17.83 \pm 3.48$$ (3)

$$T_{indoor\ operative\ temperature} = 0.078 T_{Mean\ outdoor\ temperature} + 23.25 \pm 1.25$$ (4)

4.2.2 Visual comfort

It can be evaluate by the four aspect such as uniform distribution of light, artificial lighting fixture efficiency, risk of glare for occupants, amount of light. Commonly, to maintain visual comfort minimum lux are required at work plane height. Commonly, 300 lux levels are meet at 0.8m height. And it calculated as

$$Daylight = \frac{24 - T}{365}$$ (5)

4.2.3 Energy

Two criteria are considered as minimum total operational energy demand and embodied energy.

$$Annual\ operational\ demand(E_{od}) = E_{cooling} + E_{lighting} + E_{Appliances} + E_{heating}$$ (6)

$$Annual\ embodied\ energy\ demand(E_{ed}) = E_{ps} + E_{cs} + E_{us} + E_{ef} + E_{bs}$$ (7)

Production stage energy demand (E_{ps}), Construction stage energy demand (E_{cs}), utilization stage energy demand (E_{us}), end of the life (E_{ef}), beyond the system boundaries (E_{bs}).

4.2.4 Environment

Minimum life cycle CO_2 emission would considered in this section.

$$CO_2 emission = \left(E_{grid} - E_{produce}\right) \times T\&D \times GEF$$ (8)

E_{grid} = Electricity used from the grid, $E_{produce}$ = Electricity generated by RE
T & D = Transmission and distribution losses, GEF = Grid emission factor.

4.2.5 *Life cycle cost*

Minimum initial cost, overall investment cot, building utility and operational cost, maximize net present value

$$\text{Life cycle cost (LCC } \textit{Life cycle cost} \, (I_c) = I_{OBD} + I_{RES} \tag{9}$$

I_{OBD} = Building cost, I_{RES} = renewable energy component cost

$$LCC = I_c + USPW(N, rd) \times EC \tag{10}$$

$$USPW \ (N, rd) = 1 + rd^{-n/rd} \tag{11}$$

I_c = Initial cost for design operation for building envelop, lighting, and HVAC
USPW = Uniform series present worth factor
rd = annual discount rate, N = life period.

4.2.6 *Renewable energy*

Minimize are requirement for solar panel (A_{PV})

$$(A_{PV}) = E_{pd} \div \eta_{pv} \times \eta_i \times C_{lf} \tag{12}$$

E_{pd} = total operational energy demand, G_{sr} = annual solar irradiation
η_{pv} = Efficiency of solar PV, η_i = efficiency of inverter, C_{lf} = coefficient of loss factor

4.3 *Decision variables*

Categories	Typology	Description
Defined parameters	Building Physical Characteristics	Building type, location, weather condition, building height, number of floor
Adjustable parameters	Active, Passive and RE	Passive: Building geometry (shape, orientation, volume), aspect ratio, Wall, roof, insulation, shading, glazing, thermal mass, WWR, natural ventilation, passive radiant cooling system, passive convection cooling system. Active: Artificial lighting, HVAC system, domestic hot water system, evaporative cooling RE: solar PV, wind, biomass, geothermal
Constraints	Building operational behavior	Infiltration, thermal comfort range, lighting, cooling, heating, occupancy scheduled, life cycle cost, energy demand, visual comfort lux level

4.4 *Multi-criteria decision making*

Multi-optimization runs the multiple input data in real-time and provides the set of mono solutions. MCDM technique is used to identify the optimal Pareto solution using a defined objective function with a trade-off. Commonly, aggregating and outranking methods are used. Aggregating methods include weighted sum method, £-constraint method, weighted matric method, and analytical hierarchical process (AHP). In this category, the user assigns the weight to each objective function. The concordance and disconcordance principles are used in the outranking approach, It compresses the solutions in a binary outranking relationship. It includes elimination and choice expressing reality (ELECTRE), preference ranking organization method for enrichment of evaluation (PROMETHEE), technique for order of preference by similarity to ideal solution (TOPSIS), Benson's method.

4.5 Sensitivity/uncertainty analysis

It focuses on the robustness of the optimal Pareto solution. Sensitivity analysis is used to identify the most influential design parameters that affect the building energy model and behaviour. Various methods such as Partial Rank Correlation Coefficient method (PRCC), Standardized Rank Correlation Coefficient (SRRC), and Morris method are widely used for sensitivity analysis. Uncertainty analysis focus to identify and integrate uncertain variable such as environmental condition, occupant behavior, envelop, energy price, utility price, life cycle cost that inherently show the impact on building life cycle performance. According to Nikolaidis *et al.* There are two types of uncertainty: aleatory uncertainty and epistemic uncertainty. Aleatory uncertainty, often known as statistical uncertainty, refers to the inherent uncertainty caused by probabilistic variability. This aleatory uncertainty is irreducible, and it is typically represented by a probability distribution. An epistemic is caused by a lack of knowledge, such as a lack of comprehension knowledge of the underlying processes, limited knowledge of the phenomena, or an inaccurate assessment of the related parameters and their behaviour. Commonly, two approaches is used for analysis of uncertainty such as Probabilistic and Non- probabilistic.

5 CONCLUSION

This paper explores the potential of a multivariate optimization paradigm for decision-making of net-zero energy building design. It provides the methodological approach through graphical representation for climate-adaptive NZEB in the Indian context. In this paper, the authors critically evaluated fifty major primary case studies based on MOO accounting with triple bottom line of sustainability and briefly illustrate some of the studies in text and tabular form. The factor affecting the social, environmental and economical aspects of building design are demonstrated in this study. It emphasizes the different objectives function of NZEB and their evaluation criteria for assessment. It briefly discusses the various tools, techniques, decision parameters, constraints, and approaches for decision-making for Pareto solutions. It shows the suitability and usability of MOO approach to generate optimal robust Pareto outcomes and significant objective improvements. It also illustrates how NZEB design can contribute to achieving sustainable development goals (SDGs), reducing global earth temperature, net zero-emission, and achieving 100% energy demand with renewable energy till 2050.

REFERENCES

[1] Feng W. *et al.*, "A Review of Net Zero Energy Buildings in Hot and Humid Climates: Experience Learned from 34 Case Study Buildings," (in English), *Renewable & Sustainable Energy Reviews*, Review vol. 114, p. 24, Oct 2019, Art no. 109303, doi: 10.1016/j.rser.2019.109303.

[2] Phillips R., Troup L., Fannon D., and Eckelman M. J., "Triple Bottom Line Sustainability Assessment of Window-to-wall Ratio in US Office Buildings," (in English), *Building and Environment*, Article vol. 182, p. 13, Sep 2020, Art no. 107057, doi: 10.1016/j.buildenv.2020.107057

[3] Energy and Agency I., "A Roadmap for the Global Energy Sector," 2021.

[4] Sen R., Bhattacharya S. P., and Chattopadhyay S., "Are Low-income Mass Housing Envelops Energy Efficient and Comfortable? A Multi-objective Evaluation in Warm-humid Climate," *Energy and Buildings*, vol. 245, p. 111055, 2021-08-01 2021, doi: 10.1016/j.enbuild.2021.111055.

[5] Economy E. E., "*Thermal Comfort for all, Sustainable and Smart Space Cooling*," 2017. [Online]. Available: Thermal-Comfort-for-All.pdf (shaktifoundation.in)

[6] Qiu M. H. Y., "A Comparison of Building Energy Codes and Policies in the USA, Germany, and China: Progress Toward the Net-zero Building Goal in Three Countries," *Clean Technologies and Environmental Policy*, 2018.

[7] Verma, Anurag S., Vinoth John Prakash and Anuj Kumar. "ANN-based Energy Consumption Prediction Model up to 2050 for a Residential Building: Towards Sustainable Decision Making." *Environmental Progress & Sustainable Energy* 40 (2020)

[8]   Janda R. K. K. B., "India's Building Stock: Towards Energy and Climate Change Solutions," *Building Research & Information*, vol. 47, 2019

[9]   Urge-Vorsatz D. *et al.*, "Advances Toward a Net-Zero Global Building Sector," in *Annual Review of Environment and Resources, Vol 45*, vol. 45, A. Gadgil and T. P. Tomich Eds., (Annual Review of Environment and Resources. Palo Alto: Annual Reviews, 2020, pp. 227–269

[10]   Cell O., F. a. C. C. Ministry of Environment, and G. o. India, "*India Cooling Action Plan*," 2019. [Online]. Available: India Cooling Action Plan.pdf (indiaenvironmentportal.org.in)

[11]   Lin Y. L., Zhong S. L., Yang W., Hao X. L., and Li C. Q., "Towards Zero-energy Buildings in China: A Systematic Literature Review," (in English), *Journal of Cleaner Production*, Review vol. 276, p. 17, Dec 2020, Art no. 123297, doi: 10.1016/j.jclepro.2020.123297.

[12]   A. P. D.-K. D. V. Gokhale2, "Applicability Of Net Zero Energy Building (Nzeb) Concept in the Residential Sector in India," *International Conference on Infrastructure Development*, 21–22 December 2018.

[13]   Torcellini S. P. P., and Deru M., "Zero Energy Buildings: A Critical Look at the Definition," *National Renewable Energy Laboratory*, 2006

[14]   Wei W. and Skye H. M., "Residential Net-zero Energy Buildings: Review and Perspective," (in English), *Renewable & Sustainable Energy Reviews*, Review vol. 142, p. 22, May 2021, Art no. 110859, doi: 10.1016/j.rser.2021.110859

[15]   Nancy Carlisle A., Otto Van Geet P., and Pless S., "Definition of a "Zero Net Energy" Community Technical Report NREL/TP-7A2-46065 November 2009," *National Renewable Energy Laboratory*, 2009.

[16]   Raj B. P. *et al.*, "A Review on Numerical Approach to Achieve Building Energy Efficiency for Energy, Economy and Environment (3E) Benefit," (in English), *Energies,* Review vol. 14, no. 15, p. 26, Aug 2021, Art no. 4487, doi: 10.3390/en14154487

[17]   Fatima Harkouss f. f., Pascal Henry, "Optimization Approaches and Climates Investigations in NZEB-A Review," 2018.

[18]   Amani N. and Kiaee E., "Developing a Two-criteria Framework to Rank Thermal Insulation Materials in Nearly Zero Energy Buildings Using Multi-objective Optimization Approach," (in English), *Journal of Cleaner Production*, Article vol. 276, p. 13, Dec 2020, Art no. 122592, doi: 10.1016/j.jclepro.2020.122592.

[19]   Lin Y. H., Lin M. D., Tsai K. T., Deng M. J., and Ishii H., "Multi-Objective Optimization Design of Green Building Envelopes and Air Conditioning Systems for Energy Conservation and CO2 Emission Reduction," (in English), *Sustainable Cities and Society*, Article vol. 64, p. 14, Jan 2021, Art no. 102555, doi: 10.1016/j.scs.2020.102555.

[20]   Lan L., Wood K. L., and Yuen C., "A Holistic Design Approach for Residential Net-zero Energy Buildings: A Case Study in Singapore," (in English), *Sustainable Cities and Society*, Article vol. 50, p. 16, Oct 2019, Art no. 101672, doi: 10.1016/j.scs.2019.101672.

[21]   Harkouss F., Fardoun F., and Biwole P. H., "Multi-objective Optimization Methodology for Net Zero Energy Buildings," *Journal of Building Engineering*, vol. 16, pp. 57–71, 2018-03-01 2018, doi: 10.1016/j.jobe.2017.12.003.

[22]   Hong T., Kim J., and Lee M., "A Multi-objective Optimization Model for Determining the Building Design and Occupant Behaviors Based on Energy, Economic, and Environmental Performance," (in English), *Energy*, Article; Proceedings Paper vol. 174, pp. 823–834, May 2019, doi: 10.1016/j.energy.2019.02.035.

[23]   Chaturvedi, Shobhit, Elangovan Rajasekar, and Sukumar Natarajan. 2020. "Multi-objective Building Design Optimization Under Operational Uncertainties Using the NSGA II Algorithm" *Buildings* 10, no. 5: 88. https://doi.org/10.3390/buildings10050088.

[24]   Harkouss F., Fardoun F., and Biwole P. H., "Passive Design Optimization of Low Energy Buildings in Different Climates," (in English), *Energy*, Article vol. 165, pp. 591–613, Dec 2018, doi: 10.1016/j.energy.2018.09.019

[25]   Gagnon R., Gosselin L., and Decker S. A., "Performance of a Sequential Versus Holistic Building Design Approach Using Multi-objective Optimization," (in English), *Journal of Building Engineering*, Article vol. 26, p. 13, Nov 2019, Art no. 100883, doi: 10.1016/j.jobe.2019.100883.

[26]   Gagnon R., Gosselin L., and Decker S. A., "Performance of a Sequential Versus Holistic Building Design Approach Using Multi-objective Optimization," (in English), *Journal of Building Engineering*, Article vol. 26, p. 13, Nov 2019, Art no. 100883, doi: 10.1016/j.jobe.2019.100883

[27]   Zhao J. and Du Y. H., "Multi-objective Optimization Design for Windows and Shading Configuration Considering Energy Consumption and Thermal Comfort: A Case Study for Office Building in Different Climatic Regions of China," (in English), *Solar Energy*, Article vol. 206, pp. 997–1017, Aug 2020, doi: 10.1016/j.solener.2020.05.090

Artificial Intelligence, Blockchain, Computing and Security – Dagur et al. (Eds)
© 2024 The Author(s), ISBN: 978-1-032-49393-0

A comparative study of different BERT modifications

S. Agarwal & M. Jain
Department of Computer Science and Engineering, Delhi Technological University, Delhi, India

ABSTRACT: Natural Language Understanding is a component of Natural Language Processing that deals with how computers comprehend and understand text from human languages like English, German, and French etc. It enables machines to communicate back and is crucial for NLP applications like question answering, text summarization, chatbots etc. Transformers like BERT and GPT completely turned the tide with their unparalleled results in these applications but over time more and more drawbacks started surfacing for these transformers like high computational cost, large memory requirements, skipping masked tokens, lack of linguistic and dependent knowledge in the learned representations etc. This paper presents a brief view of some of these problems and how new technologies are trying to solve them. We have reviewed and summarized some models that have been quite effective in tackling the above mentioned problems of BERT.

1 INTRODUCTION

Natural Language Processing and Time Series Analysis have been some of the most challenging branches in the applications of AI. Both applications contain data in which a particular order of data is important and it requires memory because the previous data has some influence on the next data. Earlier, recurrent neural networks (RNNs) (Tarwani *et al.* 2017) were introduced where the output at any point is affected by current data and the previous output. This feature enabled the model to retain some previous information like a memory and perform time series predictions.

Following RNNs, Long short-term memory (LSTM) (Tarwani *et al.* 2017) was introduced to encounter the issue of vanishing gradient in RNNs due to a long series of data. LSTMs were very effective in solving problems related to complex sequences. A special class of LSTM architecture called the Sequence to Sequence architecture or the Encoder-Decoder model provided a significant gain in solving natural language related problems like question answering, text summarization or machine translation etc.

After that, Transformers (Devlin *et al.* 2019) were introduced whose architecture was inspired from the encoder-decoder model. Just like the encoder-decoder model, transformers have two parts. The left half in the transformer architecture is used to simply map the input sequence to another representation of that sequence which is then passed to the right half. The right half of the transformer uses the encoder output and previous time step's output to produce the final result. Transformers also implement the attention mechanism instead of the recurrent or convolution techniques. Attention highlights the important part of the information in the current input.

Release of Transformers completely restructured the way NLP tasks are performed. They are much faster and more efficient in comparison to their predecessors. GPT (Radford *et al.* 2018) and BERT (Devlin *et al.* 2019) were the first big NLP based pre-trained transformer models. They provided highest results for classic tasks like text classification and recognition of named entities and generation tasks like question answering and text summarization. However as the technology advanced, more and more drawbacks started surfacing for these

DOI: 10.1201/9781003393580-30

transformers like high computational cost, large memory requirements, skipping masked tokens, lack of linguistic and dependent knowledge in the learned representations etc.

In this paper, we review about the uncommon areas of drawbacks we face in BERT and how over time, new models and techniques were introduced to tackle these drawbacks. We see how they train computers to better understand natural language and how they perform different NLP tasks.

2 LITERATURE SURVEY

2.1 BART

There has been remarkable success in quite a few NLP and NLU based projects and workload with the help of the self-supervised methods like BERT (bidirectional encoder representation from transformer) and GPT (generative pre-trained transformer). But being an early approach to tackle the problems of NLP, both models have some drawbacks which prevent them from being the best. Thus the model of BART (Lewis *et al.* 2020) was introduced which is a transformer-based approach to generalize BERT, GPT and some other pre-training schemes by combining the bidirectional encoder and auto-regressive decoder.

For training the sequence-to-sequence model, it first corrupts the user provided text by replacing some text in the document by mask symbols. The corrupted document is passed to the bidirectional encoder to learn a representation and then the left-to-right autoregressive decoder is utilised to generate the original document. It also provides freedom of noise i.e. to apply every conceivable method of record corruption technique like token concealing, token elimination, word infilling, sentence arrangement etc. BART surpasses previous PLMs in text summarization and text production roles while performing akin to RoBERTa in other disproportionate activities.

2.2 PERT

Auto-encoding based Pre-trained Language Models (PLM) are majorly pre-trained using the masked language modelling task. In the MLM pre-training task, a few of the symbols in the user provided text is substituted with the masking symbol (i.e. [MASK]) and the parameters of the architecture are trained to find the correct symbol from the vocabulary that fits the masked tokens. But it has a drawback. The encoded semantic representation of the input text only captures the contextual features of the tokens around the masked token and the contextual features of the artificial symbol are absent from the real representation. This creates a discrepancy leading to a sub-optimal depiction of the user provided text and the model lacks in performing the task optimally.

Thus a question of pre-training an auto-encoding PLM with tasks other than MLM arises. To answer this question, a new pre-training task was proposed called the Permuted Language Model (PerLM). The objective of PerLM is to permute a portion of the input text randomly and train the models to recover the original positions of the tokens of the input text.

PERT (Cui *et al.* 2022) is an auto-encoding model based on PerLM. It employs whole word masking (Devlin *et al.* 2019) and N-gram masking (Devlin *et al.* 2019) where either all the tokens of a word or N tokens are permuted and the model learns to find their original locations. This approach permits the model to develop a combination of short and long text inference and it should have an improved performance in reading comprehension tasks and named entity recognition tasks.

2.3 XLNet

Both autoregressive (AR) and autoencoding (AE) approaches have some disadvantages in their pre-training objectives as the autoregressive model only encodes a uni-directional context and the autoencoding model fails to calculate the joint probability like the AR models. Thus, XLNet (Yang *et al.* 2019) was proposed.

XLNet is an autoregressive approach that adopts the best of both AR and AE models. Unlike using a fixed order like unidirectional or bidirectional, it considers all the possible permutations of the factorization order. This permits the model to understand the significance of every token from each position. Also since it is an AR model, it can use the product principle to generate the combined likelihood of the tokens it needs to predict.

XLNet uses the permuted language modelling (PerLM) objective for pre-training thus removing its dependency on the masked token and solving the problem of fine tune discrepancy. By integrating segment reoccurring way and comparative encoding strategy, the performance of the model improved significantly. XLNet outperforms the previous pre-training objectives by a large margin in many NLP tasks.

2.4   ALBERT

As the technology advanced, it became possible to train large networks and it was observed that expanding the architecture of the model when pre-training performed better on downstream tasks. However, as the models became larger, the quantity of parameters associated with the model also grew exponentially and memory limitation became a problem. As a result the final model was degraded in performance and took longer to train.

Therefore, A Lite BERT (ALBERT) (Lan *et al.* 2019) architecture was introduced which had remarkably fewer parameters than BERT. This was because it adopted two variable compression techniques: factorized embedding parameterization and cross-layer parameter sharing. As the name suggests, cross-layer parameter sharing allows layers at a deeper level of the network to share parameters with the layers at shallow level. In BERT, the embedding size of tokens was tied to the hidden layer size due to which as the model size increases, embedding size also increases and parameters of the model increases. Thus, factorized embedding parameterization unties and fixes the embedding size. Together these two techniques reduce the parameters in the model by 18 times.

2.5   RoBERTa

Hyperparameter tuning is one of the most critical and challenging jobs in learning algorithms. The goal of hyperparameter tuning is to find a set of values of hyperparameters for a particular learning algorithm that makes the model efficient and enhances the performance. This is done by a hit and try method in which the model is trained on different combinations of hyperparameter values and observe their performance to choose the best model.

It was observed that BERT was severely undertrained and by increasing the training time and tuning the hyperparameters, it can achieve better performance. Thus, an improved version of BERT was introduced called RoBERTa (Liu *et al.* 2019) (Robustly optimized BERT approach). Along with hyperparameter tuning and more training, RoBERTa makes two more changes in the architecture: removing the NSP loss technique of BERT, as it was observed that it improves the success of the model on subsequent roles and unlike BERT which implements static masking (fixed pattern of masking in each input), RoBERTa implements dynamic masking. With adaptive masking, a fresh mask pattern is developed each time a data stream comes. RoBERTa showcases a significant improvement over the original BERT in almost all the NLP tasks.

2.6   MPNet

The Masked Language Modelling (MLM) method for pre-training adopted by BERT has a disadvantage as it does not consider the dependency of masked tokens. It assumes them to be independent. This is known as the Output dependency problem. To overcome this, the Permuted Language Modelling (PLM) method was introduced in XLNet. But by taking different permutations of the user provided text, the positional knowledge of tokens is lost producing the positional discrepancy in the initial training of the variables. This is known as the Input consistency problem.

MPNet (Song *et al.* 2020) produces a unified view of MLM and PLM while addressing the issues faced in both methods. In MLM, the tokens are divided into masked and non-masked tokens and in PLM, after permutation the tokens are divided into predicted and non-predicted tokens. Therefore by keeping the non-masked and non-predicted tokens the same, in the overall perspective of the input, non-masked symbols are placed on the left, then the concealed symbols and finally the anticipated symbols on the extreme right (Figure 1). With this method, the model takes the dependency of predicted tokens to solve masked token's discrepancy and by taking positional information of all tokens, it avoids the positional discrepancy.

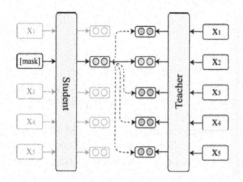

Figure 1.   The structure of MPNet.   Figure 2.   TaCL student-teacher architecture.

2.7   *TaCL*

Although BERT revolutionized the way computers understand natural language. But as the technology advanced, more and more drawbacks of BERT surfaced. One of the drawbacks suffered by Masked Language Modelling based pre-trained models is the anisotropic distribution problem. It states that the token representations learned by the model lie only in a small dimensional space. This causes the representations to be less discriminative and unable to differentiate between similar tokens.

TaCL (Token-aware Contrastive Learning) (Su *et al.* 2022) is a new pre-training approach that motivates the model to learn more isotropic and discriminative representations of the tokens. This approach uses two models initialized with pre-trained BERT. One model acts as a student and the other model acts as a teacher. The training objective of this network is to contrast the masked token's representations produced by the student model with the actual representations produced by the teacher model. The result of this training approach brings some improvement in the performance of the model in comparison to BERT across many NLP tasks.

2.8   *SpanBERT*

SpanBERT (Joshi *et al.* 2019) is a new masked language modelling based pre-trained model which introduces new masking techniques and training objectives. As the term indicates, in this strategy a contiguous span of fixed length of text is masked randomly rather than masking single tokens or masking whole words. The training objective introduced is called span-boundary objective (SBO). In continuation to the masking technique, SBO aims to teach the learning algorithm to correctly anticipate the entire masked span in one go using the context it learned from other tokens in the input text. Also SpanBERT omits the Next Sentence Prediction (NSP) algorithm of BERT as it is observed on multiple occasions that removing it increases the success of the model on subsequent roles. SpanBERT outperformed BERT on most of the NLP tasks but it performed exceptionally well on answer creation problems SQuAD 1.1 and 2.0.

2.9 ConvBERT

In English and many other languages, sometimes the same word expresses multiple meanings and sometimes different words show the same meanings. This makes the language ambiguous. To attend to this ambiguity, BERT had the self-attention block. When the BERT model is processing the input text, for each word it processes, the self-attention block warrants the system to view more expressions in the input text and grasp the correct meaning of each word. This allows it to learn a better version of representation of each word in the sentence. However, since BERT is highly dependent on its self-attention blocks as it has a block for each layer in the architecture, BERT requires large memory and more computation time and power.

It was observed that a large portion of the attention blocks can be substituted by internal dependencies in the text thus reducing the weight of the model. Since convolution operation best summarizes the local features, ConvBERT (Jiang *et al.* 2020) was introduced. In ConvBERT, convolution operation is integrated with the self-attention to make mixed attention. Experiments show that the model achieved comparable performance while reducing the computation cost.

Table 1. Summary of features of models.

Model	AE/ AR	Training	Masking	Novel Idea
ALBERT	AE	MLM + SOP	Token	Factorized embedding parameterization Cross-layer parameter sharing
XLNET	AR	Permutation LM (PLM)		Bidirectional style in autoregressive
RoBERTa	AE	MLM	Dynamic	Hyperparameter tuning for BERT
BART	AE	Reconstruct corrupt text	Any	Reconstruct text with noise using whole transformer
PERT	AE	PerLM	N-gram, Whole word	Correct positions of Permuted tokens
SpanBERT	AE	Span MLM	N-gram	Continuous span masking
ConvBERT	AE	Replaced token detection		Mix Convolution in attention block
TaCL	AE	NSP + MLM + unsupervised contrastive learning	Token	Learn isotropic and discriminative distribution
MPNet	AE	MLM + PLM	Token	Unifying MLM and PLM

3 RESULT

The General Language Understanding Evaluation (GLUE) (Wang *et al.* 2018) benchmark is a group of a variety of natural language understanding tasks like paraphrasing and similarity tasks, natural language inference tasks etc. These tasks have become a standard to measure how well a model understands the language and its performance in different areas.

In this study, we concentrate on the semantic similarity and paraphrasing task and compare how these different modifications of BERT perform in this area. We analyse the capability of different learning algorithms using the accuracy metrics (Table 2). BART being the most flexible transformer as it has a generic training objective of reconstructing corrupted text, it allows its users to train using any data noise function to induce corruption. It outperforms all the other models in the QQP dataset and is comparable in other datasets. However, TaCL is severely lacking in comparison to the other models.

Table 2. Result comparison.

S.No.	Method	MRPC	STS-B	QQP	S.No.	Method	MRPC	STS-B	QQP
1	BERT	88.7	90.2	91.4	6	RoBERTa	90.9	92.4	92.2
2	BART	90.4	91.2	**92.5**	7	MPNet	91.8	91.1	91.9
3	PERT	87.3	-	-	8	TaCL	**92.0**	89.7	82.5
4	XLNet	90.0	91.1	91.8	9	SpanBERT	87.9	89.1	89.5
5	ALBERT	90.9	**93.0**	92.2	10	ConvBERT	88.3	89.7	90.0

4 CONCLUSION

In this paper, we analyse the performance difference between various pre-trained language models fine-tuned on the similarity and paraphrasing datasets like QQP, MRPC etc. in the GLUE benchmark. We see that the recently proposed modifications BART and TaCL show a significant growth in accuracy over the originally proposed BERT model. Further work includes combining multiple modifications in different areas and seeing how well they complement each other and further enhance their performance in understanding different languages and performing various tasks.

REFERENCES

Alex Wang, Amanpreet Singh, Julian Michael, Felix Hill, Omer Levy, and Samuel Bowman. 2018. GLUE: A Multi-Task Benchmark and Analysis Platform for Natural Language Understanding. In *Proceedings of the 2018 EMNLP Workshop BlackboxNLP: Analyzing and Interpreting Neural Networks for NLP*, pages 353–355, Brussels, Belgium. Association for Computational Linguistics.

Cui, Y., Yang, Z., & Liu, T. (2022). PERT: Pre-training BERT with Permuted Language Model. *ArXiv, abs/2203.06906*.

Devlin, J., Chang, M., Lee, K., & Toutanova, K. (2019). BERT: Pre-training of Deep Bidirectional Transformers for Language Understanding. *NAACL*.

Jiang, Z., Yu, W., Zhou, D., Chen, Y., Feng, J., & Yan, S. (2020). ConvBERT: Improving BERT with Span-based Dynamic Convolution. *ArXiv, abs/2008.02496*.

Joshi, M., Chen, D., Liu, Y., Weld, D.S., Zettlemoyer, L., & Levy, O. (2020). SpanBERT: Improving Pre-training by Representing and Predicting Spans. *Transactions of the Association for Computational Linguistics*, 8, 64–77.

Lan, Z., Chen, M., Goodman, S., Gimpel, K., Sharma, P., & Soricut, R. (2020). ALBERT: A Lite BERT for Self-supervised Learning of Language Representations. *ArXiv, abs/1909.11942*.

Lewis, M., Liu, Y., Goyal, N., Ghazvininejad, M., Mohamed, A., Levy, O., Stoyanov, V., & Zettlemoyer, L. (2020). BART: Denoising Sequence-to-Sequence Pre-training for Natural Language Generation, Translation, and Comprehension. *ACL*.

Liu, Y., Ott, M., Goyal, N., Du, J., Joshi, M., Chen, D., Levy, O., Lewis, M., Zettlemoyer, L., & Stoyanov, V. (2019). RoBERTa: A Robustly Optimized BERT Pretraining Approach. *ArXiv, abs/1907.11692*.

M. Tarwani, Kanchan & Edem, Swathi. (2017). Survey on Recurrent Neural Network in Natural Language Processing. *International Journal of Engineering Trends and Technology*. 48.301–304.10.14445/22315381/ IJETT-V48P253.

Radford, A., & Narasimhan, K. (2018). Improving Language Understanding by Generative Pre-Training.

Song, K., Tan, X., Qin, T., Lu, J., & Liu, T. (2020). MPNet: Masked and Permuted Pre-training for Language Understanding. *ArXiv, abs/2004.09297*.

Su, Y., Liu, F., Meng, Z., Shu, L., Shareghi, E., & Collier, N. (2022). TaCL: Improving BERT Pre-training with Token-aware Contrastive Learning. *ArXiv, abs/2111.04198*.

Yang, Z., Dai, Z., Yang, Y., Carbonell, J.G., Salakhutdinov, R., & Le, Q.V. (2019). XLNet: Generalized Autoregressive Pretraining for Language Understanding. *NeurIPS*.

Artificial Intelligence, Blockchain, Computing and Security – Dagur et al. (Eds)
© 2024 The Author(s), ISBN: 978-1-032-49393-0

Prediction of cardiovascular diseases using explainable AI

Anuradha S. Deokar & M.A. Pradhan
AISSMSCOE, Department of Computer Engineering, Pune

ABSTRACT: Cardiovascular diseases (CVDs) are heart and blood vessel abnormalities. Immediately following covid-19 Many people died as a result of a heart condition. The primary cause is stress, as well as changes in nutrition and sleeping patterns. Methods are being devised in the subject of "explainable AI" to interpret the predictions produced by Artificial Intelligence systems. This paper explores explainable artificial intelligence as a technology that may be used by AI-based systems to analyse and diagnose heart disease data and offers a proposed strategy with the goal of providing result tracing, accountability, transparency, and model refinement in the healthcare sector. We provide LIME The LIME technique focuses on significant factors that influence heart disease, such as ST slope and cholesterol level, and the model predicts that this particular patient will develop cardiovascular disease and SHAP, provides a list of critical features, ranked from most important to least important (top to bottom). We also offer a model explanation method that frames the problem as a submodular optimization one.

Keywords: Explainable AI, cardiovascular disease, LIME, SHAP, heart disease dataset.

1 INTRODUCTION

The simulation of human intellect in systems meant to behave and think like humans is known as artificial intelligence (AI). Computer exhibiting human like traits of learning and problem-solving traits, this term can be used.

Explainability is essential for a number of reasons: - Aids analysts in swiftly and easily comprehending system outputs. Analysts can effectively make decisions if they are aware of how the system functions. – Aids in avoiding false positives.

It contributes to defining model correctness, transparency, fairness, and outcomes in decision-making supported by AI. Before bringing AI models into production, a company must first build trust and confidence.

With the help of AI explainability, a corporation may take a reasonable approach to AI development. Model explainability refers to the ability to comprehend the machine learning model. Assume a healthcare model that can determine whether or not a patient has a certain condition.

In order to produce explainable AI, they should keep track of information used in models, achieve a balance between explainability and accuracy, focus on the end user, and develop Key Performance Indicators (KPIs) to evaluate AI risk.

2 RELATED WORK

Many different XAI-related solutions have been put forth over the course of the last number of years, with many of them being used in the healthcare industry.

Model-agnostic XAI methods are those that aren't dependent on the AI model that requires explanation [4].

One of the popular model-agnostic techniques, known as Local Interpretable Model-Agnostic Explanation (LIME), was developed by researchers in [6]. LIME is a framework for explaining predictions that quantifies the contributions of all the variables that go into making a prediction.

DOI: 10.1201/9781003393580-31

Recurrent neural networks (RNNs) were used by researchers in [3] to predict heart failure, and LIME was used to explain how these RNNs predicted heart failure. These explanations allowed researchers to identify the most prevalent medical conditions, including kidney failure, anemia, and diabetes, which increase a person's risk of developing heart failure.

Other model-independent XAI techniques have been created and applied in the healthcare industry, including Anchors and Shapley values [7; 8].

[9] outlines a framework that aims to create better explanations by taking into account the user's reasoning objectives when building XAI methods.

The use of XAI techniques is not without its difficulties.

The end-users, -common people or professionals from medical field, should find the explanations produced by XAI approaches beneficial [10].

It is no more impossible to prepare such interfaces that are suitable for efficiently displaying explanations [9].

Research is still needed to address issues with model-agnostic XAI approaches' rising computational costs and assumption-based operation [4].

An illustration of this notion would be to send a report to clinicians together with information on heart rate, body temperature, and calorie consumption if blood sugar levels were to rise.

Calorie consumption is the factor that contributes most to this prediction, according to a XAI model.

Clinicians can then examine the characteristics and suggest the best treatments or exercises.

3 PROPOSED APPORACH

In this paper, we propose approach, steps are as follows:

1. Identification of required data – For applying prediction model is it important to ensure our dataset is accurate, complete, reliable and relevant. Hence, identification and selection of such dataset is our primary goal.
2. Data pre-processing – Data pre-processing consists of handling missing values, removing the outliers. If there is imbalance in data treating the imbalanced data. Normalization of data.
3. Definition of training dataset – Once processed the data now needs to splited into training and testing data. It is important to ensure that all possible set of values are present in training dataset. The training data should be diverse. For this purpose, cross validation is used
4. Algorithm selection – algorithm is selected on the basis of predictive accuracy and interpretability. An algorithm which goes well with the proposed system goal, data size and processing 5. Performance evaluation – true positive, false positive, true negative and false negative are calculated from confusion matrix. Accuracy, F1-Score, Error rate and recall are used as performance evaluation.

The proposed strategy is described as follows, as seen in Figure 1: [1]

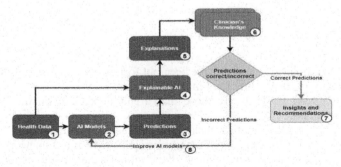

Figure 1.   Block diagram.

1) Intelligent healthcare applications gather user health data (1) and trained AI models are used (2) to identify whether patient is having diseases of not.
2) The health data (1) and forecasts (3) are used together with XAI algorithms (4) to produce explanations (5).
3) These explanations (5) can be evaluated using the expertise of a therapist (6).
4) By using the foresaid technique, physicians are able to transparently validate the AI model's predictions.
5) If forecasts are accurate, explanations and clinical expertise can be leveraged to produce insightful conclusions and suggestions (7).
6) If forecasts are inaccurate, the discrepancy between explanations and clinical expertise can

4 SHAP

Calculate sharply value to identify, the "black box" contains each feature's contribution.as described in Shapley value.

Marginal value/contribution: computing the contribution of each characteristic in each subgroup, then average these contributions.

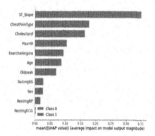

Figure 2. Summary plot showing important variable.

4.1 *Observation*

As a result of the previous graphic, we can conclude the following:

It displays a list of crucial characteristics, ranked from most important to least important (top to bottom).

Analysis of the features

- Given that colours fill half of the rectangles, all features appear to be equally important in determining whether either class receives a cardiovascular disease diagnosis (label = 1) or not (label = 0).
- According to the model, ST slope has the highest predictive power The next feature with the best predictive ability is ChestPainType.
- The third factor with the highest level of predictability is cholesterol.
- The six-factor model with the highest predictive power includes age.

Plan based on a specific label

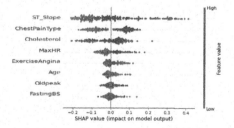

Figure 3. Summary plot deep-dive on label = 1.

203

4.2 Observation

For classification issues, there are SHAP values for every conceivable label.
  In this instance, we have chosen to get that forecast of 1. (True).
  The plot can be used to deduce the following key traits:

- Use the same analysis from the variable importance plot for significant characteristics.
- Every dot indicates a different data instance's feature value.
- The colours indicate if this characteristic has a high (red) or low value (blue).
- The X-axis displays the contribution, positive or negative, to the anticipated output.

The following interpretations are obtained when we apply those analyses to the features:

4.3 For ST-slope

We can see that the bulk of the high values (red dots) contribute positively (positive on the X-axis) to the expected result.
  In other words, the probability of a positive outcome (being diagnosed with cardiovascular disease) increases significantly if the ST-Slope quantity for a particular data instance is high, whereas the probability of a negative outcome (being diagnosed with cardiovascular disease) decreases (positive X-axis values) if the quantity is low (blue dots).

- Age: The same analysis is done for age. The likelihood of a data instance (patient) being diagnosed with cardiovascular disease increases with age.

On the other hand, the model appears confused when it comes to children because we can see about the same number of data points on each. side of the vertical axis, which is at zero.
  We can utilise the dependency plot to have more precise information as the Age feature appears to be confounding for the study.

4.4 Dependence plot

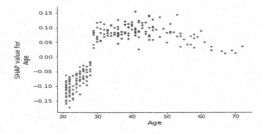

Figure 4. Dependence plot for age feature.

4.4.1 Observation
Side of the vertical line (X-axis = 0).
  We can utilise the dependency plots to have more specific information because the Age feature appears to be misleading for the analysis.

4.5 Local Interpretable Model Agnostic Explanation (LIME)

Due to its local nature, it can be utilised to interpret specific predictions made by a machine learning model.

- Work on any blackbox model
- Model internals are hidden

- Works with many data types
- Using prior knowledge data types

LIME's explanation is produced as a result of the following: [2]

$$\xi(x) = \text{argming} \in GL(f, g, \pi x) + \Omega(g) \tag{1}$$

With this method, you may use different explanation families G, fidelity functions L, and complexity measures.

4.6 Single instance explanation

The explanation is given in Figure 5 for a particular occurrence from the test data.

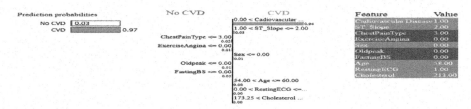

Figure 5. LIME explainability for a single instance.

4.6.1 Observation

The ST slope is more than 0.2 mv, the cholesterol level is higher than 200, and the model indicates that this particular patient will have cardiovascular disease.

We can see the worth of the real features for the patient on the right.

5 CONCLUSION

The creation of frameworks and models to help in explainable AI (XAI) is an emerging area of study. Analysing and interpreting the judgements made by AI systems.

Explainable and traceable autonomous systems. As said, there are some difficulties with incorporating XAI approaches.

We will be building a model which will analyse the clinical data of the individuals using LIME and SHAP, it identifies weather patient is having cardiovascular disease or not and also calculate the prediction probability and most affective feature that causes disease. Use Explainable AI, the prediction from prediction model were used to provide the reasoning.

6 FUTURE WORK

1. Extending the prediction model for other chronic diseases
2. Reducing the training time.
3. Improving the quality of prediction model.
4. Tuning the efficiency on a large dataset.

ACKNOWLEDGMENT

We would like to thank Dr. Madhavi Ajay Pradhan for helpful discussions and feedback.

REFERENCES

[1] Urja Pawar and *et al.* "*Explainable AI in Healthcare*" IEEE Xplore August 2020.

[2] "Smart Healthcare: Making Medical Care More Intelligent," *Global Health Journal*, vol. 3, no. 3, pp. 62–65, 2019.

[3] Khedkar S., Subramanian V., Shinde G., and Gandhi P., "Explainable AI in Healthcare," *SSRN Electronic Journal*, 2019.

[4] Doshi-Velez F. and Kim B., "*Towards a Rigorous Science of Interpretable Machine Learning*," no. Ml, pp. 1–13, 2017.

[5] Lakkaraju H., Bach S. H., and Leskovec J., "Interpretable Decision Sets: A Joint Framework for Description and Prediction," *Proceedings of the ACM SIGKDD International Conference on Knowledge Discovery and Data Mining*, vol. 13–17-August-2016, pp. 1675–1684, 2016.

[6] Ribeiro M. T. and Guestrin C., "*Why Should I Trust You?" Explaining the Predictions of Any Classifier,"* pp. 1135–1144, 2016.

[7] Ribeiro M. T. and Guestrin C., "*Anchors: High-Precision Model-Agnostic Explanations.*"

[8] Lundberg S. M. and Lee S.-I., "A Unified Approach to Interpreting Model Predictions," in *Advances in Neural Information Processing Systems*, pp. 4765–4774, 2017.

[9] Wang D., Yang Q., Abdul A., and Lim B. Y., "Designing Theory-Driven User-Centric Explainable AI," *Proceedings of the 2019 CHI Conference on Human Factors in Computing Systems – CHI '19*, pp. 1–15, 2019.

[10] Holzinger A., Biemann C., Pattichis C. S., and Kell D. B., "*What Do we Need to Build Explainable AI Systems for the Medical Domain?*" no. Ml, pp. 1–28, 2017.

[11] Mukta Sharma, Amit Kumar Goel, Priyank Singhal. "*Chapter 7 Explainable AI Driven Applications for Patient Care and Treatment*", Springer Science and Business Media LLC, 2023.

[12] Anurag Sinha, Tannisha Kundu, Kshitiz Sinha. "Comparative Study of Principle and Independent Component Analysis of CNN for Embryo Stage and Fertility Classification", *International Journal of Fuzzy System Applications*, 2022.

[13] Kumar, A., & Alam, B. (2016). Real-Time Fault Tolerance Task Scheduling Algorithm with Minimum Energy Consumption. In *Proceedings of the Second International Conference on Computer and Communication Technologies: IC3T 2015*, Volume 2 (pp. 441–448). Springer India.

[14] Dagur, A., Malik, N., Tyagi, P., Verma, R., Sharma, R., & Chaturvedi, R. (2021). Energy Enhancement of WSN Using Fuzzy C-means Clustering Algorithm. In *Data Intelligence and Cognitive Informatics: Proceedings of ICDICI 2020* (pp. 315–323). Springer Singapore.

[15] Dagur, A., Kaushik, A., Rastogi, A., Singh, A., Kumar, A., & Chaturvedi, R. (2021). Optimization of Queries in Database of Cloud Computing. In *Data Intelligence and Cognitive Informatics: Proceedings of ICDICI 2020* (pp. 325–332). Springer Singapore.

[16] Kushwaha, A., Amjad, M., & Kumar, A. (2019). Dynamic Load Balancing ant Colony Optimization (DLBACO) Algorithm for Task Scheduling in Cloud Environment. *Int J Innov Technol Explor Eng*, 8 (12), 939–946.

[17] Ali Raza, Kim Phuc Tran, Ludovic Koehl, Shujun Li. "Designing ECG Monitoring Healthcare System with Federated Transfer Learning and Explainable AI", *Knowledge Based Systems*, 2021.

[18] Anderson Monken, William Ampeh, Flora Haberkorn, Uma Krishnaswamy, Feras *A. Batarseh.* "*Assuring AI Methods for Economic Policymaking*", Elsevier BV, 2023.

Artificial Intelligence, Blockchain, Computing and Security – Dagur et al. (Eds)
© 2024 The Author(s), ISBN: 978-1-032-49393-0

Music generation using RNNs and LSTMs

H. Aditya, J. Dev, S. Das & A. Yadav
Bennett University, Greater Noida, Uttar Pradesh, India

ABSTRACT: Music theory is at the core of music. It takes a long time for artists to generate music and it might not always be to their liking. The point of using AI is that previously successful music can generate new and inspired music. With this help, professional musicians can create better music faster. The idea is about building a model that will generate music using data sets containing music from different artists. It will learn the patterns of music in the datasets and generate new music. The knowledge of basic music representation like the ABC notation and MIDI format will be utilized. A Character Recurrent Neural Network (char-RNN) is being implemented by us. A RNN has recurrent, or looped connections that will allow the RNN network to keep information across all inputs. These connections are like memory. RNNs are very useful, particularly for learning sequence data such as music.

1   INTRODUCTION

AI has grown exponentially over the past several decades and has made computers to be able to achieve feats that were once thought of as impossible. From their use in finance where we can predict future trends from data sets of past years to recognizing images, self-driving cars, and many more. In recent times AI has entered fields that are primarily defined by human intelligence and creativity. One of those such areas that it has ventured into recently is music. Music has always been and still is a defining piece of human culture. It is something that remains constant across different sections of the globe. The ever-present desire to express oneself through rhythm and melody has resulted in the prevalence of music even in the modern era. Throughout history, music is something that is created by the human mind alone as it resembles complex human emotions and melodies that were not possible for a machine to reciprocate at the time but with the advent of deep learning and natural language processing (NLP) that barrier is significantly closed. There are already some existing examples of AI being used in the music industry such as computer scientists who apply AI to create entirely new pieces of tracks from scratch while others such as musicians and singers who use that computer programs to compose instrumental pieces for their songs which are known as "beats". As we can see from these cases that AI can be used for both, to compose original music as well as tools for musicians to use. (Briot *et al.* 2017) analyzes several methods of deep learning being used in the generation of musical content. Similarly, (Huang & Wu 2016) proposes the creation of single melody with end-to-end learning and generation approach using deep neural networks alone.

1.1   *Problem statement*

Music as we know has existed for an extremely long amount of time and as time has passed it has become increasingly difficult to create new and unique pieces of music because somebody may have already done something that resembles or is like that in the past. This not only demotivates the artist, but it also results in burnout as they succumb to the pressure of constantly creating new and interesting music while also trying to be able to stay distinctive from their peers and true to the music they create. This also makes the listeners feel weary listening to similar generic music all the time.

DOI: 10.1201/9781003393580-32

1.2 Novelty

These days you don't have to be a music enthusiast to enjoy music. Therefore, even common people who don't know anything about music theory and notations can listen to and enjoy it, but this leads to most of them not exploring more and they end up listening to a select few songs, to their liking, over and over. This idea will help them listen to new music which will be created from the same songs they listen to; hence not only will they be able to discover more songs, but it will also most probably be to their liking as it will be derived from the songs they already like. The freedom to listen to new music, which also at the same time reminds the listener of the music they love, is the novelty we are approaching for.

1.3 Motivation

Music is subjective where if one person likes a song that doesn't necessarily mean that the other person will like it too. Everyone has their own choices and tastes hence the primary reason why we chose this idea was for the opportunity to produce and provide a personalized music experience for different users depending on their tastes and opinions. Music streaming services such as Spotify, and Apple Music have recommendation systems that recommends new songs based on the songs you listen to or like, whereas here we are using a similar approach where instead of recommending the music we are generating through our model inspired by the user's personal playlist resulting in an entirely new experience for the user. Feedback can also be provided by the user to the AI-generated songs and according to that, changes can be implemented into the model to best fit the requirements of the consumer and provide them with the optimal music experience.

1.4 Background

Music is represented through notes and symbols. The dataset will be represented in ABC notation instead of the conventional Do, Re, Mi, etc. The ABC notation comprises letter notations from A to G which corresponds to the classical notes along with numbers and special characters. In addition to the notes that make up the tune, different units are utilized as extra data; reference number, origin, note length, beat, composer, and so forth. For designing our model, we could involve old-style Artificial Neural Networks, however, since the tunes we call music, comprise a progression of rehashing notes, the classical Artificial Neural Networks utilizing the feed-forward procedure are to some degree useless for this issue. This is because the notes that make up our music are not autonomous of one another, instead, they are organized in a steady progression in a method for adjusting to the past one. Here is where RNN (Recurrent Neural Network) comes in, which is a further developed artificial neural network algorithm, that gives a generally excellent answer for such sequence and redundancy issues. Along with classical neural networks, the RNN algorithm can take the result of the past node as input and at the same instance take its output as input for parallel nodes.

Even with the solution RNN provides, it also has some drawbacks. One of them is that the memory of RNN is too small and the algorithm becomes more and more complex if we start using the previous elements as well. To resolve this, LSTM (Long Short-Term Memory) is generally used. The primary difference between a standard RNN unit and an LSTM unit is that there is more sophistication in the LSTM unit. To be more precise, it is made of gates that apparently regulate the flow of information across the unit much better than RNN.

After training the model is done, it gives the music generated by the LSTM-based RNN model in string format.

2 RELATED WORK

There are several models proposed for music generation through different approaches and using different aspects of machine learning and deep learning. (Conklin 2003) proposes using statistical models to imitate musical styles. These statistical models have been generated from extant pieces in a stylistic corpus.

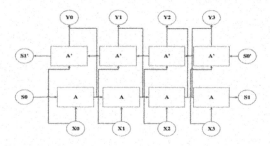

Figure 1.   Bi-directional recurrent neural network with LSTM.

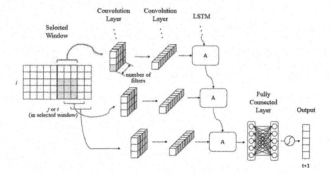

Figure 2.   C-LSTM model structure.

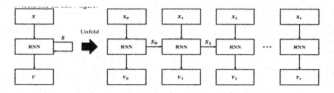

Figure 3.   RNN in detailed unfolded form.

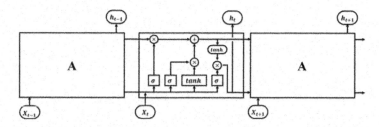

Figure 4.   Repeating module of LSTM.

The main objective of these models is to classify new music pieces accurately. It discusses a handful of methods for sampling from an analytical statistical model, and then proposes an approach that still maintains the repetitive patterns within an existing piece. (Van Der Merwe & Schulze 2010) focuses on Generation of Music through Markov Models where music has been presented in a particular fashion for training data, and parameters are used for constructing the hidden Markov models and Markov chains to represent the musical style of the given data as some mathematical models. (Dong *et al.* 2018) proposes three models for multi-track music generation, the jamming, composer, and hybrid models, that differ in the underlying assumptions and network architecture.

(Muhamed *et al.* 2020) suggests using Transformer-GAN for Symbolic generation of music that uses learned loss mechanism in which they talk about how the researchers were able to detect failures such as sample quality degrading for long sequences in the auto-regressive model trained by reducing the NLL (Negative log likelihood) of the sequence observed using classifiers specifically trained to differentiate between sampled and real sequences. They made use of a pre-trained Span-BERT model to act as the discriminator of the GAN which helped with providing stability in training. Similarly (Mittal *et al.* 2021) presents a technique for training diffusion models on symbolic music data by para-meterizing the discrete domain in a continuous latent space using a pre-trained variational autoencoder.

For RNN and LSTM based approaches (Azmi *et al.* 2020) proposes a method for generating music in contrast with the music samples used in the dataset using bidirectional recurrent neural networks with LSTM cells by understanding the complex relationship between different components that constitutes music such as different notes, pitch, and timbre values. They analyzed the performance of the model with the variation of the number of epochs used. Building over the above approach (Dua *et al.* 2020) have proposed an improved RNN-LSTM based novel approach for sheet music generation, the researchers are aiming to improve the accuracy of sheet music, which they have achieved by working on source separation and chord estimation, for which they have used RNN, GRU (Gated Recurrent Units) and LSTM. They have used multilayered GRU cells for implementing RNN in the source separation module and LSTM cells for implementing RNN in chord estimation module. The number of sources that can be separated were also increased for improving the accuracy of chord estimation module inside the source separation module.

On the other hand, (Huang *et al.* 2018) discusses combining CNN (Convolutional Neural Network) and LSTM for generating music. Here they use MIDI-format to represent the notes, then they convert the music file into a matrix of musical scores. They establish con-volution layers for feature extraction from the musical score matrix. After that the con-volution layers were split along the direction of the time axis and used as input for LSTM. The output was verified by comparing the accuracy, frequency-domain analysis, human-auditory evaluation, and time-domain analysis. The result came to be that Convolution-LSTM is better for music generation than LSTM.

Char-RNN which is short for "Character Recurrent Neural Network" is a recurrent neural network trained to predict the next character when a sequence of previous characters is given. (Goel *et al.* 2014), have proposed a basic way to model temporal dependencies and sequences by using a RNN (recurrent neural network) and DBN (Deep Belief Network). They take advantage of the memory state of the RNN which helped them to provide tem-poral information whereas the multi layered DBN helped with high level representation of data. This combination of RNN and DBN proved to be ideal for sequence generation and way better than an RBM (Restricted Boltzmann Machine).

There are several applications and fields where music generation can be used such as (Di *et al.* 2021) establishes rhythmic relationships between video and background music and proposes the Controllable Music Transformer (CMT), which allows for local control of rhythmic features and global control of music genre and instrument. (Plans & Morelli 2012) discusses the evolution of video-game music from simple circuit-based artifacts to a multi-billion-dollar industry. Moreover, (Madhok *et al.* 2018) proposes a framework that generates music corresponding to the emotion of a person predicted by a facial expression recognition model, which uses CNN to classify the expression into one of seven categories.

Table 1. Related work in music generation field.

References	Term Used	Model Description	Limitations
(Conklin 2003)	Statistical Model	Statistical Model assigns every piece of musical note with probability. It captures regularities in a class of music. Their assignment is done on the basis that high probability will be given to pieces that are in class and vice versa.	Requirement of motif or phrase-level substitutions which in turn leads to high computation time for computing valid motifs for substitution.
(Van Der Merwe & Schulze 2010)	Markov Model	A Markov model is a Stochastic technique for haphazardly changing frameworks where the future states do not depend upon past states. Models like these show all potential states and the advances, changes in its pace and probabilities among them.	The generation steps do not take each other into account resulting in failed composition attempts.
(Muhamed et al. 2020)	Transformer-GAN	GAN is a deep neural network which is used along with an adversarial network.	Not mentioned.
(Azmi, Shreekara & Baliga 2020)	Bi-RNN	Bidirectional monotonous neural frameworks are just collecting two free RNNs. The gathering of information is dealt with in average time demand for one framework, and in upset time demand for another.	The training time for 30 epochs is twice compared to training 15 epochs. This affects the efficiency when the dataset is significantly large.
(Dua et al. 2020)	RNN, GRU, LSTM	RNN – The computed output depends on the output of previous computations. LSTM –It is the building block of RNN. It selectively remembers patterns for long time.	Sheet music generator is still not accurate and leaves considerable room for improvement.
(Huang Huang & Cai 2018)	CNN, LSTM	CNN is another deep learning NN for the processing of structured arrays.	Not mentioned.
(Goel Vohra & Sahoo 2014)	RNN, DBN	Deep Belief Network is a type of deep neural network which is made up of extensive layers of hidden units which have connections between the layers but not between each other.	The memory of RNN is too small and the algorithm becomes more and more complex if we start using the previous elements as well.

3 PROPOSED METHODOLOGY

3.1 *RNN (Recurrent Neural Networks)*

RNN is a form of further developed artificial neural network algorithm that works with time series or sequential data. It has looped and recurring connections that enable the network to store data from multiple sources. RNN suffers from drawbacks. One of the main drawbacks of RNN is that its memory is too small, and the algorithm grows more and more complex if we use the previous nodes as well for prediction.

3.2 *LSTM (Long Short-Term Memory)*

LSTM is the extension of RNN. It is a paradigm or architecture that allows RNN to have more memory. RNN generally has 'short term memory' which means they utilize prior information in the present neural network. LSTM proves particularly helpful in music generation because melodies have long term structural patterns therefore, LSTM, which captures long-term temporal dependencies can be efficiently used to generate these melodies.

4 EXPERIMENTAL RESULT ANALYSIS AND DESCRIPTION

4.1 *Dataset description*

The dataset is the ESAC (Essen Associative Code) dataset which contains over 20K + songs from all over the world. The dataset is comprised of traditional folk songs. It consists of cyphers i.e., pitch connected to the mode's specified tonic as well as underlines and dots for rhythmic durations. It can be converted into MIDI, TEX, PCX, and other formats. Aside from the code, various programs to analyze, listen to, transpose, and depict tunes have been built for the PC.

4.2 *Pre-processing*

Loading the folk music dataset: For loading the songs, we go through all the files in the dataset and load them with music21. The songs are in krn format. So, for loading the songs we must first filter out the files which aren't kern files i.e., aren't music files. For that purpose, we can simply check if the last three characters of the file are 'krn' or not. Now for loading the song, we use a basic function of music21, m21.convertor.parse(). This just loads one song, so we create a list where we store all the songs.

Filtering out songs that have non-acceptable durations: We create a function where we pass the songs as arguments. We define the acceptable durations, which are [0.25, 0.5, 0.75, 1.0, 1.5, 2, 3, 4]. So now we check whether the songs in the dataset have acceptable durations or not. For this we need to take all the notes and rests and analyze them one by one. We run a loop through the song with certain attributes such as 'flat' (flattens the entire song structure and the objects into a single list) and 'notesAndRests' (filters out all the attributes that aren't notes and rests) then we check if the duration of the song is in acceptable duration. If it is, then the function returns true otherwise it returns false. The reason why we are doing this is because we want to simplify the note durations which will make it easier for our deep learning model and at the same time, most of the songs will be at acceptable durations so we will not lose much of the data.

The songs are transposed to C major and A minor. We create another function and retrieve the key from the song if the key is noted, otherwise, we estimate the key with the help of music21. First, we extract the parts then we extract all the measures from part 0. Then, we retrieve the item at index 4 of the first measure of part 0 because that is usually where the key is stored in this dataset (The extraction is done with help of getElementByClass() function). If the song doesn't have a key stored, then we estimate the key with a single line of code which is song.analyze("key"). After retrieving the key, we need to get the interval for transposition, so if the key is in major mode, then we transpose it to C major and if the key is in minor mode, then we transpose it to A minor. Now we transpose the song by the calculated interval by simply writing song.transpose(interval). The reason we transpose is because we would have had to train the model on all 24 keys if we weren't transposing, which will require a greater amount of computation time.

Encoding songs with music time series representation: We again create a function where we take the songs as the argument. First, we need to get all the notes and rests as we did when we filtered out songs that had non-acceptable durations. Then we check whether the instance is note or rest. If the instance is a note, we initialize the value of symbol with the respective MIDI notation (event.midi.notation) and if the instance is rest then we initialize the symbol with 'r'. Now we convert the notes and rests into time series notation. We define steps which are int (event.duration.quarterlength/time step). Then we run a loop in range(steps) and build a list by appending the symbol with the encoded song if the instance is a note otherwise it is appended with " " if the instance is a rest. Then we convert the encoded song to a string.

Saving songs to text file: For this we load the encoded song and add delimiters. We create an empty string which we will update at every song we process. After loading the songs, we need to append the songs to the empty string and add the delimiter along with it. Then we create a mapping to encode symbolic notation with integers, to get the data ready to be ingested by the neural network.

Figure 5. The format of encoding used for the melodies.

Figure 6. Example sheet music of a generated melody.

5 RESULTS

TensorFlow loads and produces the song which will be in midi format. So, we open the file using MuseScore and play the output. No impartial evaluation predictor exists to evaluate the effect of the model because it is for production, the one which has greater unpredictability and eccentricity. Therefore, we found out the accuracy score which came out to be 0.8139 for training the model for 30 epochs. Our model detected complex melodies and it started to give constant melody sequences.

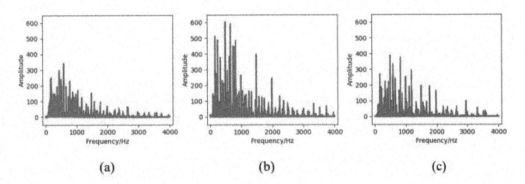

Figure 7. Spectrum of three songs.

5.1 Frequency domain analysis.

Figure 7 signifies the value of their parts at each step/frequency of the signal, and the below diagram is used to show the behavior of a signal in frequency. A time-domain curve could be turned into a signal that belongs to frequency-domain using the Fourier transform, which will contain a higher amount of information than a signal that is time-domain.

Note. (a) Sample music, (b) LSTM-generated song, and (c) C-LSTM-generated song. The generated song frequency through LSTM is not adequately dense, Moreover, the spectrum of the frequency from the song made by C-LSTM is nearer to the sample which we have

213

taken, according to the variation and analysis in the domain of frequency for all the three songs. It has been observed that C-LSTM is reliable for precisely identifying and producing the song's frequency.

5.2 *Sonogram*

Figure 8 (a) Sample music, (b) LSTM-generated song, and (c) C-LSTM-generated song. The variation that can be seen in the frequency about how the songs are generated from the sample songs and the Convolutional-LSTM over a period can be observed to be mostly due to variations in that component. It can also occur due to the variations in the quantity and overall distributions of the notes. The poor frequency indicates the beats, whereas more frequency denotes melody, as is the case with the rest of the song. Music generated by LSTM has relatively simple frequency distribution and does not change with time.

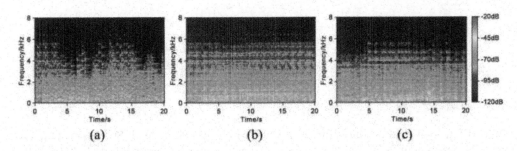

(a)   (b)   (c)

Figure 8.   Sonogram of three songs.

6   CONCLUSIONS

Based on our research, we have developed a deep learning model based on Convolutional Long Short-Term Memory (C-LSTM) that can generate music in MIDI format. Our model detected complex melodies and produced constant melody sequences with an accuracy score of 0.8139 for training the model for 30 epochs.

In the time-domain analysis, we found that C-LSTM had a better impact than LSTM, but the music produced by C-LSTM was not as great as the sample song. In the frequency domain analysis, we observed that the generated song frequency through LSTM was not adequately dense. However, the spectrum of the frequency from the song made by C-LSTM was nearer to the sample, indicating that C-LSTM is reliable for precisely identifying and producing the song's frequency.

In conclusion, our research demonstrates the effectiveness of the C-LSTM model for generating music in MIDI format. While our model still has some limitations, we believe that it can be improved in future research to produce even more complex and interesting melodies.

REFERENCES

Azmi, Syeda Sarah, Shreekara, C, and Baliga, Shwetha. 2020. "Music Generation using Bidirectional Recurrent Neural Nets." In: *International Research Journal of Engineering and Technology 7.05*.
Conklin, Darrell. 2003. "Music Generation from Statistical Models." In: *Proceedings of the AISB 2003 Symposium on Artificial Intelligence and Creativity in the Arts and Sciences*. Citeseer, pp. 30–35.
Di, Shangzhe *et al.* 2021. "Video Background Music Generation with Controllable Music Transformer." In: *Proceedings of the 29th ACM International Conference on Multimedia*, pp. 2037–2045.

Dong, Hao-Wen *et al.* 2018. "Musegan: Multi-track Sequential Generative Adversarial Networks for Symbolic Music Generation and Accompaniment." In: *Proceedings of the AAAI Conference on Artificial Intelligence*. Vol. 32, Issue 1.

Dua, Mohit *et al.* 2020. "An Improved RNN-LSTM Based Novel Approach for Sheet Music Generation." In: *Procedia Computer Science* 171, pp. 465–474.

Goel, Kratarth, Vohra, Raunaq, and Sahoo, Jajati Keshari. 2014. "Polyphonic Music Generation by Modeling Temporal Dependencies using a RNN-DBN." In: *Artificial Neural Networks and Machine Learning-ICANN 2014: 24th International Conference on Artificial Neural Networks*, Hamburg, Germany, September 15–19, 2014. Proceedings 24. Springer, pp. 217–224.

Huang, Allen and Wu, Raymond. 2016. "Deep Learning for Music." In: *arXiv* preprint arXiv:1606.04930.

Huang, Yongjie, Huang, Xiaofeng, and Cai, Qiakai. 2018. "Music Generation Based on Convolution-LSTM." In: *Comput. Inf. Sci.* 11.3, pp. 50–56.

Jean-Pierre Briot, Gaetan Hadjeres, and François-David Pachet. "Deep Learning Techniques for Music Generation–a Survey". In: *arXiv* preprint arXiv:1709.01620 (2017)

Madhok, Rishi, Goel, Shivali, and Garg, Shweta. 2018. "SentiMozart: Music Generation based on Emotions." In: *ICAART (2)*, pp. 501–506.

Mittal, Gautam *et al.* 2021. "Symbolic Music Generation with Diffusion Models." In: *arXiv* preprint arXiv:2103.16091.

Muhamed, Aashiq *et al.* 2020. *"Transformer-GAN: Symbolic Music Generation Using a Learned Loss."*

Plans, David and Morelli, Davide. 2012. "Experience-Driven Procedural Music Generation for Games." In: *IEEE Transactions on Computational Intelligence and AI in Games 4.3*, pp. 192–198. DOI: 10.1109/TCIAIG.2012.2212899.

Van Der Merwe, Andries and Schulze, Walter. 2010. "Music Generation with Markov models." In: *IEEE MultiMedia* 18.3, pp. 78–85.

215

Artificial Intelligence, Blockchain, Computing and Security – Dagur et al. (Eds)
© 2024 The Author(s), ISBN: 978-1-032-49393-0

Effectiveness of virtual education during Covid-19: An empirical study in Delhi NCR

Girish Kumar Bhasin & Manisha Gupta
School of Business Studies, Sharda University, Knowledge Park, Greater Noida

ABSTRACT: The COVID-19 pandemic has severely disturbed India's educational system, and as a result, sudden changes are taking place in many fields of education. The lockdown has engulfed educational facilities, seriously interfering with student learning and assessment. For the advantage of students, most academic institutions have started offering online education. In poor countries like India, where internet education is still in its infancy, classroom instruction is still the best method of education. For the many stakeholders, switching from traditional classroom instruction to online learning is difficult due to mindset and numerous infrastructure problems. The current study attempts to investigate the numerous elements influencing how undergraduate and graduate students use online learning. A standardized questionnaire was used to conduct an online survey of 270 students. The following factors were found using exploratory and confirmatory factor analyses: educators' assistance, student engagement and motivation, Technology assistance, course content and design, and administrative assistance. The paper examines how these discoveries can be used.

1 INTRODUCTION

An important development that forced educational establishments to switch from the conventional teaching space setting to a variety of online learning possibilities was the COVID-19 outbreak. The distribution of teaching and learning has been impacted by COVID-19 (Laksana 2021; Panesar *et al.* 2020) as well as other aspects of human existence (Chang *et al.* 2020; Mokter 2020; Wong *et al.* 2020). The bulk of educational events, including, classroom instruction, conferences, workshops, seminars, etc., were either postponed or annulled by educational establishments until their availability online in order to prevent the spread of COVID-19 (Khachfe *et al.* 2020; Patricia 2020).

Online learning is the transfer of knowledge via synchronous and asynchronous internet technologies. While maintaining a social distance, these approaches enable students to communicate with their instructors and fellow students (Dong *et al.* 2020). Students can learn, interrelate, share their opinions, be independent in their learning, and utilize time on their own through online learning (Azzi *et al.* 2021; Hwang *et al.* 2021). When learning online, both students and teachers must be able to use technical support to create and sustain healthy social interactions (Andel *et al.* 2020). Other elements, such as the availability of appropriate facilities, infrastructure, and the financial situation of the students, play a significant role in online learning in addition to the effective use of technology (Laksana 2021; Rusli *et al.* 2020).

It is critical to consider the quality of online learning in this era of rapid growth of Information and Communication Technologies (ICT) and their many uses (Zhafira *et al.* 2020). Online education may be delivered quickly, effectively, and efficiently thanks to a wide range of capabilities made available by information technology systems. ICT systems allow students to acquire more diverse content while enjoying time and space flexibility, despite the fact that the current pandemic (COVID-19) has pushed students to use ICT

DOI: 10.1201/9781003393580-33

(Mustakim 2020). Additionally, audio-visual enrichments are frequently used with online texts to enhance the learning experience (Hasibuan 2016).

Although there are many advantages for students using online resources, there are numerous factors that can affect how successful and efficient online learning is (Pratiwi 2020). These components include learner readiness, university assistance, teacher involvement, and motivation. Different students may have different perspectives on these factors. Many challenges that many students face make it difficult to deliver educational services effectively (Laksana 2021). Online education is typically of lesser quality for students in areas with poor internet and frequent power outages. The quality of online learning can be greatly influenced by a number of significant aspects, including the role of the professors, the university's assistance, the home study environment, and motivational factors.

Students are the main focus of online learning and are the ones who benefit most from it (Chen et al. 2021; Laksana 2021). In addition, academic staff personnel are crucial to online learning. They must create presentations and teaching materials that are compatible with the online environment, based on data, and relevant to the subject matter. Similar to this, the administration of the institution must cooperate in order to provide online educational services. Both the learning process itself and the efficient delivery of learning resources online demand a number of resources. The COVID-19 pandemic's restrictions have led to the adoption of online teaching and learning in most poor nations (Suryaman et al. 2020; Suprianto et al. 2020). The students' capacity to learn may have been affected by several problems, including inadequate internet connection, intermittent electricity supply, and a absence of support from their families and the academia (Djidu et al. 2021). To evaluate the effectiveness of online learning in the context of a developing economy, it is important to investigate how the students felt about their online learning experiences throughout the COVID-19 pandemic.

2 LITERATURE REVIEW

In the context of the education industry and related studies, online-learning, as defined by Zemsky & Massy (2004), is any learning that is assisted by software or is electronically mediated. Traditional course format and face-to-face format are interchangeable concepts. Although there are other labels for it, "online education" has become more and more popular over the course of the epidemic. According to Kaplan & Leiserson (2000), "delivery of messages and material through the internet and different electronic devices, including interactive television, satellite broadcast, mobile, tablets, desktops and laptops." As defined by Okiki (2011), online education is "the use of various software, hardware, and network related technologies to foster, create, facilitate, and deliver learning, anywhere in the world." Online education is a cutting-edge technique of delivering education that uses electronic material to improve students' knowledge, abilities, or other performance areas, according to Bhuasiri et al. (2012). Online education was defined by Mbarek & Zaddem (2013) as the use of ICT components without taking into account space-time shift to enable learners to develop a new set of skills and knowledge. This study sought to answer multiple questions by examining the variables influencing the uptake of online learning. During the COVID-19 epidemic in India, the researchers sought to uncover and validate variables affecting staff and student satisfaction in online learning environments.

According to a study by Kraidy (2002), students are more likely to use digital media in their studies now than they will in the future, and offering application-based content and activities will help students meet expectations for both personal and professional progress (Joe et al. 2004). In educational institutions, the term "flexible learning" covers a wide range of applications. Flexible learning, on the other hand, advocates for the idea that students should be seen as active participants in the learning process (Collis 1998) and that more in-depth learning strategies should be supported. Peltier et al. (2003) established the conceptual framework for online marketing education.

"Student-to-instructor interactions, student-to-student interactions, instructor support and mentoring, course content, and course structure and information delivery technology" were the main areas of attention in the model. In his research, Schroeder (2003) found that students pay high costs to institutions for infrastructure services such access to technology support, online instruction, internet connectivity and bandwidth, data transfer speed, computers, smartphones, and other items. Since the research focused on progressive learning, students were at its centre. The chance of choosing online courses could be influenced by students' prior experiences with and opinions about online learning.

For the purpose of this study, the following factors affecting the adoption of online education have been discovered.

2.1 Educators' support

The effectiveness of the teacher, his or her accessibility to technology, and the simplicity with which it may be used improve learning and instruction through online pedagogies and resources made available online (Caywood & Duckett 2003). Online course instructors, such as professors, have a variety of responsibilities. They can respond to students individually or as a class, regulate debates, and control the flow of content through exercises like homework and evaluations. Their presence and proximity appear to have an impact on learning and student pleasure. According to the majority of professors (Doris et al. 2009), online students are skilled communicators who actively engage in learning from their respective course teachers. Practical implications during faculty training and digital literacy programmes may have roles to play for HEIs (Higher Educational Institutions), and structural aspects could also be addressed for the best acceptance of learning technology. However, a number of motivating factors have been noted in the research to encourage faculty members who participate in online and distance learning (Mishra & Panda 2007). Along with student happiness, learning effectiveness, and institutional cost-efficiency, faculty satisfaction is listed as one of the five pillars of quality and access. As online education becomes more common and dynamic dynamics like adoption rates, learned expectations, levels of support, and other situations continue to change, it is necessary to explore the factors that contribute to teacher satisfaction. However, faculty members are having trouble adopting technology or do not have access to sufficient technology and tools, which leads to lower satisfaction.

H1: Effective virtual education has a significant positive correlation with educator's assistance.

2.2 Student engagement and motivation

According to Kleinman (2005), an online learning environment fosters active and engaged learning, which offers the interactive support needed to assist students to comprehend expectations and results in a happy learning community. Palmer & Holt (2009) established that a student's technological comfort level was crucial to their satisfaction with online courses. They have an additional learning style option in addition to traditional learning thanks to e-learning (Hollenbeck et al. 2006).

Since e-learning does not require students to physically attend classes on campus, they have complete control over the speed and rhythm of their study (Bhuasiri et al. 2012). Student satisfaction was well predicted by learner-instructor contact, learner-content interaction, and Internet self-efficacy (Kuo et al. 2013). According to Choy et al. (2002), students primarily demand the following 10 services when studying online: Information on: (i) the requirements for completing the curricula; (ii) in-depth information about the subject; (iii) the security aspect of personal information maintained by the institute; (iv) the clarity of what we expect them to learn; (v) teacher feedback; (vi) the need for assessment; (vii) the ease of communication with faculty members using various appropriate methods, such as chat, email, and face-to-face; (viii) regular faculty feedback; and (viii Students also need to enhance technical systems and teacher facilitation, it was reported.

H2: Effective virtual education has a significant positive correlation with student motivation and engagement.

2.3 Technology assistance

Infrastructure and technology have the power to revolutionize current approaches to education. It can eliminate obstacles to education present in conventional classroom settings. To learn from the teacher collectively, students are no longer need to gather in traditional classes at the same time. Modern technologies that are ICT enabled can alter how people think about higher education institutions. Online technologies are being used frequently by many educators (Elizabeth et al. 2009). According to Brill & Galloway (2007), the impact of technology on teaching and learning led to useful recommendations for utilising available resources to support classroom-based technologies. When faculty and teachers have access to technology in classrooms and computer labs, they are more inclined to use it with students, which will eventually lead to the integration of online education into educational institutions. Similar support is needed for faculty and students in online learning as they would on a traditional campus. Online education is seen as a bandwidth-intensive activity. Additionally, technical help available on campus is essential. According to Bakar et al. (2010), using ICT effectively can enhance the teaching-learning process. Therefore, helpful factors including administrative and technical support were highlighted (Moses et al. 2012). Faculty and students continue to have access to a variety of teaching and learning alternatives thanks to ICT applications (Sarabadani & Shami Zanjani 2017).

H3: Effective virtual education has a significant positive correlation with technological assistance.

2.4 Course design and content

Engaging course material encourages student engagement and initiative, which affects learning results (Ashwin & McVitty 2015; Little & Knihova 2014). Chapters of learning materials' organization and content are included in e-learning content. Additionally, the E-learning content offers extra elements to aid students in understanding the subject matter more thoroughly and clearly (Khamparia & Pandey 2017). According to Akyuz & Samsa (2009), this characteristic makes it easier for students to develop their analytical, critical thinking, and problem-solving abilities.

Course design interface, structure, testing and evaluation techniques, and discussion panels for faculties and students are all included in the creation of an e-learning course. According to Oh et al. (2020), an effective course design will draw students to online learning and make it easier for them to do so. The course design interface is used to introduce the course material, which is appropriate in terms of time and place and created in accordance with the students' level of comprehension and competency. (Ahmad et al. 2018).

H4: Effective virtual education has a significant positive correlation with course design and content.

2.5 Administrative assistance

According to Moore & Kearsley (2005), administrative assistance entails that students at the university or institute have access to advise and counseling services. These consist of participant social interaction, administrative support, and introductions to online learning. According to Thorpe (2002) & Lee et al. (2011), administrative sustenance can be divided into official support and course design and content. When students require assistance from the institution, it is usually with concerns related to entrance, registration, scholarships, research, and student life. For institutional assistance, Selim (2007) has supplementary library services, an assistance desk, computer workrooms, and facilities. On the other side, students seek explanations of the course materials, tasks, activities, and evaluations for the particular course in the course assistance section.

To improve students' learning experiences, support must be given to them throughout the learning process (Chang et al. 2008). The majority of colleges began to offer their curriculum online during the COVID-19 epidemic. The majority of chief academic officers at institutions (58%) said that online learning was essential for ongoing educational activities and saw it as a component of their overall organisational strategy. Instructor traits and behaviours, as well as the methods and media employed for the delivery of online education, have an impact on online learning. The university's primary duty is to enhance the methods used and offer a trustworthy source of education delivery. Providing the teachers with the necessary information, skills, and abilities helped ensure that high-quality instruction was provided.

H5: Effective virtual education has a significant positive correlation with administrative assistance.

2.6 Conceptual framework

Factors affecting online education are the self-assessment inventory for diagnosing face validity. It consists of 21 items of five factors. These five factors are Educator's assistance, Student engagement and motivation, Technology Assistance, Course Content and Design, and Administrative assistance. Responses have been taken for each statement with five-point Likert scales ranging from strongly agree to strongly disagree in Figure 1.

2.7 Research objective

For many students, online learning has emerged as the preferred method of instruction. Up until this point, developed nations were primarily where internet education was growing. The majority of research has been conducted in industrialized nations and at colleges that have opted for an online learning environment. The majority of studies concentrated on attitudes, intentions, and behaviors related to online learning. The COVID-19 pandemic has caused a significant uptick in online education in India. There were no thorough studies concentrating on various aspects of online education in different parts of India. Therefore, the primary goal of this study is to investigate the numerous variables that influence the adoption of online learning.

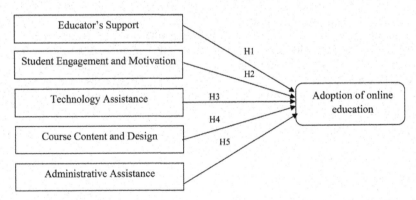

Figure 1. Conceptual framework.

3 RESEARCH METHODOLOGY

3.1 Research design and method

One way to conceptualize cross-sectional research is as a "snapshot" of the prevalence and features of a condition in a population at a specific period. A population's condition prevalence can be determined using this kind of data. Therefore, the current study uses an online primary survey research method and a single cross-section descriptive – single cross-sectional research design.

3.2 Dimensions of the buildings

Standardized scales and conceptions were easily unavailable in the literature for the current investigation. Based on the evaluations, discussions, and some earlier studies on e-learning and online education that are currently available, researchers have created assertions about online education.

Researchers tested the questionnaire's validity and reliability in a pilot study involving 50 students. The final poll was conducted after the questionnaire's statements were amended considering the results of the pilot study.

3.3 Data collection methodology

The survey was conducted using a questionnaire with a five-point Likert scale. Delhi NCR residents make up the students at the various institutions. The population of the current study consisted of students enrolled in online courses. The respondents were contacted via getting in touch with various institutes' authorities. A non-probability convenience sample strategy was used to gather 270 responses from an online survey of the targeted group. SPSS and AMOS were used to analyze the data that had been gathered. Cronbach's Alpha can be used to determine the reliability of a scale. To determine the factors, both exploratory and confirmatory factor analyses were used.

4 ANALYSIS

Table 1. Demographic profile of respondents.

Demographic Profile	Items	Frequency	Percentage
Age	15–20 years	41	15.2%
	20–25 years	173	64.1%
	Above 25 years	56	20.7%
Gender	Male	163	60.4%
	Female	107	39.6%
Education	Graduate	74	27.4%
	Post-graduate	196	72.6%
Have laptop or Desktop	Yes	218	80.7%
	No	52	19.3%
Possess Internet connection	Yes	192	71.1%
	No	78	28.9%

The respondents' demographic profile is shown in Table 1. Only 218 of the 270 respondents have a desktop or laptop, and only 192 have access to a broadband internet connection at home.

Table 2. Result of explorartory factor analysis.

FACTOR	ACRONYM	SCALE ITEMS	Cronbach alpha	Communalities	Factor Loading	Eigen Value
Educator's	TS1	Quality of Course Content	0.752	0.634	0.793	5.534
Assistance	TS2	Monitoring and encouraging better academic progress		0.785	0.781	
	TS3	Timely and Continuous Feedback		0.721	0.764	
	TS4	Detailed instructions for assignments and tests		0.673	0.732	

(*continued*)

Table 2. Continued

FACTOR	ACRONYM	SCALE ITEMS	Cronbach alpha	Communalities	Factor Loading	Eigen Value
Student Engagement and Motivation	SMP1	Online learning is boring, and it's difficult to focus	0.832	0.603	0.734	3.169
	SMP2	Lack of ability to communicate and interact with others		0.672	0.782	
	SMP3	Insufficient practical experience		0.631	0.627	
	SMP4	Being overwhelmed by academic obligations and demands		0.665	0.615	
Technology Assistance	THS1	Guidance/ tutorials on access and use of modes of online education	0.748	0.832	0.603	2.623
	THS2	Support to overcome technical difficulty		0.756	0.672	
	THS3	Proper internet bandwidth available to attend online sessions		0.634	0.657	
	THS4	Proper ICT infrastructure required for online sessions		0.772	0.789	
Course Design and Content	CD1	Well defined course structure	0.869	0.693	0.813	1.699
	CD2	Course framework properly designed		0.751	0.821	
	CD3	Course content (materials, lectures, videos) prepared		0.769	0.675	
	CD4	Students can learn from online activities and discussions		0.812	0.783	
	CD5	Incapability to grasp notions, learning objectives effectively.		0.763	0.761	
Administrative Assistance	AS1	Training and manuals are easily available	0.728	0.719	0.835	1.231
	AS2	Necessary help and resources are provided		0.783	0.612	
	AS3	Clear policies/ guidelines on vacillating from traditional to online mode of education		0.691	0.603	
	AS4	Provision for registering any complaints and grievances		0.737	0.781	

4.1 *Exploratory factor analysis*

25 items were subjected to exploratory factor analysis, principal component analysis, and varimax rotation. The items whose Eigenvalue was equal to or greater than 1 were kept after conducting an exploratory factor analysis. Accordingly, the items with communalities detected above 0.6 are taken into consideration for the study. Communality is defined as the percentage of variation that an item shares with other things. 25 elements were reduced to 21 items as a result, and they were then grouped together under 5 independent factors. Whether or not the Kaiser-Meyer-Olkin (KMO) measurement sample is suitable and sufficient for factor analysis depends on its value. Here, the Kaiser-Meyer-Olkin (KMO) value is 0.864, which is significantly higher than the cutoff point of 0.60 (Kaiser 1974). If there is a strong or weak association between the variables, it can be determined using Bartlett's Test of Sphericity. The idea of factor analysis for the data is a smart one. It is a test that is frequently used to look at hypotheses for population-level variables that are not correlated, i.e., where the population correlation matrix is an identity matrix. Bartlett's Test of Sphericity was determined to be significant in the current study at 0.000 (Bartlett 1954). Nine factors with Eigenvalues greater than one were identified by exploratory factor analysis, and these nine factors combined explained 64.34% of the variation. No factor loading in Table 2 is less than 0.60, indicating that there are no cross-loadings and that an item loads to just one

component. Cronbach's alpha was estimated to purify various scale goods. Table 2 displays values in the range of 0.728 to 0.869. An item-to-item correlation larger than 0.4 is considered acceptable, according to Nunnally (1978). The factors that were retrieved were designated with the appropriate labels, as shown below.

4.2 Educator's assistance dimension

The four items extracted under this factor with their factor loadings are: quality of course content (0.793), Monitoring and encouraging better academic progress (0.781), Timely and Continuous Feedback (0.764), and Detailed instructions for assignments and tests (0.732). The Cronbach's alpha for the educator's assistance dimension is 0.752.

4.3 Student engagement and motivation dimension

The four items extracted under this factor with their factor loadings are: Online learning is boring, and it's difficult to focus (0.734), Lack of ability to communicate and interact with others (0.782), Insufficient practical experience (0.627), and Being overwhelmed by academic obligations and demands (0.615). The Cronbach's alpha for student engagement and Motivation dimension is 0.832.

4.4 Technology assistance dimension

The four items extracted under this factor with their factor loadings are: Guidance/ tutorials on access and use of modes of online education (0.603), Support to overcome technical difficulty (0.672), Proper internet bandwidth available to attend online sessions (0.657), and Proper ICT infrastructure required for online sessions (0.789). The Cronbach's alpha for the technology assistance dimension is 0.748.

4.5 Course design and content dimension

The five items extracted under this factor with their factor loadings are: Well defined course structure (0.813), Course framework properly designed (0.821), Course content (materials, lectures, videos) prepared (0.675), Students can learn from online activities and discussions (0.783), and Incapability to grasp notions, learning objectives effectively (0.761). The Cronbach's alpha for course design and content dimension is 0.869.

4.6 Administrative assistance dimension

The four items extracted under this factor with their factor loadings are: Training and manuals are easily available (0.835), Necessary help and resources are provided (0.612), Clear policies/ guidelines on vacillating from traditional to online mode of education (0.603) and Provision for registering any complaints and grievances (0.781). The Cronbach's alpha for administrative assistance dimension is 0.728.

Table 3. Reliability, composite reliability and AVE.

Dimensions	Cronbach's Alpha	Composite Reliability	AVE
TS	0.989	0.99	0.839
SMP	0.913	0.918	0.722
THS	0.925	0.925	0.816
CD	0.958	0.947	0.825
AS	0.974	0.969	0.763

4.7 *Reliability and validity assessment*

Convergent and discriminant validity were used to evaluate the validity, and Cronbach's alpha and composite reliability were used to evaluate the reliability. Internal consistency, convergent validity, and discriminant validity were looked at in order to assess the measurement model's validity and reliability. Cronbach's Alpha was utilized to determine the core consistencies of numerous propositions framed in the research activity.

The general guidelines for convergent validity are:

(i) Loading factor > 0.7, where a loading factor of 0.6 to 0.7 is still appropriate for exploratory research;
(ii) Communality > 0.5; and
(iii) Average variance extracted (AVE) > 0.5.

All structures have CRs above 0.90 and AVEs that fall between 0.722 and 0.839, as shown in Table 3. Fornel and Larcker (1981) were used to assess the discriminant validity presented in Table 4 by comparing the square root of each AVE on the diagonal with the correlation coefficients (off-diagonal) for each construct in the relevant rows and columns. Overall, discriminant validity supports the constructs' discriminant validity and can be used to support this measurement strategy.

Table 4. Discriminant analysis.

Dimensions	Fornell-Larcker Criterion				
	EA	SEM	TA	CDC	AA
EA	**0.843**				
SEM	0.444	**0.926**			
TA	0.372	0.689	**0.787**		
CDC	0.421	0.452	0.363	**0.921**	
AA	0.25	0.234	0.251	0.323	**0.906**

In order to determine the impact of factors swaying the acceptance of online education, a model was developed and evaluated using route analysis of structural equation modeling. The following theories have been created to gauge the impact of various variables.

H1: Effective virtual education has a significant positive correlation with educator's assistance.

H2: Effective virtual education has a significant positive correlation with student engagement and motivation.

H3: Effective virtual education has a significant positive correlation with technology assistance.

H4: Effective virtual education has a significant positive correlation with course design and content.

H5: Effective virtual education has a significant positive correlation with administrative assistance.

5 PATH ANALYSIS

Following an analysis of the measurement model, the path model of the structural equation model, which is explained below, is used to examine the causal linkages between the latent constructs in the model and the hypotheses. All of the assumptions are supported, as can be seen from Table 7. As shown in Table 7, there is a substantial correlation between the

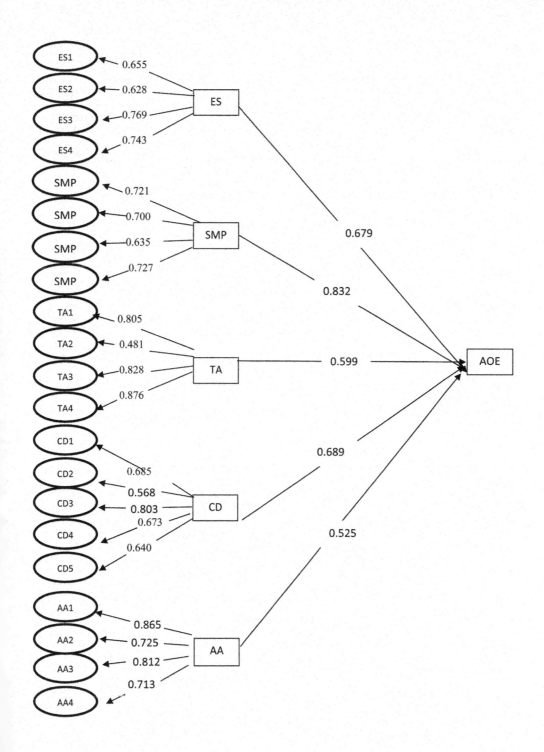

Table 5.   CFA model.

Factor	Acronym	Scale Items	Standardized Regression Weight	Indicator Reliability
Educator's Assistance	TS1	Quality of Course Content	0.655	0.72
	TS2	Monitoring and encouraging better academic progress	0.628	0.636
	TS3	Timely and Continuous Feedback	0.769	0.742
	TS4	Detailed instructions for assignments and tests	0.743	0.569
Student Engagement and Motivation	SMP1	Online learning is boring, and it's difficult to focus	0.721	0.718
	SMP2	Lack of ability to communicate and interact with others	0.700	0.665
	SMP3	Insufficient practical experience	0.635	0.621
	SMP4	Being overwhelmed by academic obligations and demands	0.727	0.583
Technology Assistance	TA1	Guidance/ tutorials on access and use of modes of online education	0.805	0.783
	TA2	Support to overcome technical difficulty	0.481	0.594
	TA3	Proper internet bandwidth available to attend online sessions	0.828	0.783
	TA4	Proper ICT infrastructure required for online sessions	0.876	0.774
Course Design and Content	CD1	Well defined course structure	0.685	0.639
	CD2	Course framework properly designed	0.568	0.619
	CD3	Course content (materials, lectures, videos) prepared	0.803	0.772
	CD4	Students can learn from online activities and discussions	0.673	0.568
	CD5	Incapability to grasp notions, learning objectives effectively.	0.640	0.581
Administrative Assistance	AA1	Training and manuals are easily available	0.865	0.783
	AA2	Necessary help and resources are provided	0.725	0.527
	AA3	Clear policies/ guidelines on vacillating from traditional to online mode of education	0.812	0.793
	AA4	Provision for registering any complaints and grievances	0.713	0.629

Table 6.   Fit indices of CFA.

Fit Statistics	Measured value	Recommended value
CMIN/DF	1.892	<5
GFI	0.973	>0.90
AGFI	0.925	>0.90
TLI	0.956	>0.90
NFI	0.935	>0.90
CFI	0.917	>0.90
RMSEA	0.043	<0.05

Table 7. Summary of path analysis.

Path	Path co-efficient and p-value	Result
Educator's Assistance	0.679 (0.000)	Significant
Student Engagement and Motivation	0.832 (0.000)	Significant
Technology Support	0.599 (0.000)	Significant
Course Design and Content	0.689 (0.000)	Significant
Administrative Support	0.525 (0.000)	Significant

adoption of online learning and the aspects of educator assistance, student engagement and motivation, technical assistance, course design and content, and administrative assistance.

5.1 *Practical implications*

Every level of educational stakeholders needs thorough information on infrastructure as well as information on efficient teaching tactics and techniques to increase online educational productivity. The present study's findings will assist online education providers in establishing the technical requirements and infrastructure they need. To improve the effectiveness of online learning, practitioners can provide a variety of training modules and operating procedures. As online learning options grow in popularity, educational stakeholders will need to engage in research that compares the cost-effectiveness of various teaching methods, and online learning will play a key role in achieving this goal. The study's findings are meant to be a resource for stakeholders to utilize in evaluating the adoption of online education. It also increases understanding of the most effective ways to define and measure student, teacher, and system outcomes and can be used as a roadmap for developing future courses of action in online education. The constructions found in the present study should be compared to other recent indices of related or identical constructs. Our measurements will be concurrently valid if they are different. Researchers, decision-makers, and practitioners will find the scale described in this article to be helpful when doing research and putting online education systems into place. It is well-developed.

5.2 *Restrictions and future research directions*

The scope of this study is restricted to students in Delhi NCR, which is one of its drawbacks. Additionally, it is possible to do research on professors from different universities and geographical regions, which aids in comparisons and the generalization of the results. It would be fascinating to learn how urban students and rural students view online education and which aspects have the most impact on them since the study did not make any comparisons between the two groups of students. Studies examining the effects of these elements on attitude, adoption, and intention will be helpful to all online educators as the study's sole goal was to identify the variables that influence students' acceptance of online learning. In order to overcome the challenges of conducting online education, it is crucial to compare online education to classroom instruction among students.

6 CONCLUSION

The goal of the current study is to comprehend the variables influencing the uptake of online learning in Delhi NCR. Exploratory and confirmatory factor analyses were used to determine the five elements that influence the adoption of online education. When embracing online education, the effects were found to be considerable in the educator's support, student involvement and motivation, technical support, course design and content, and administrative support. The acceptance or implementation of online education can be aided by the factors found in the current study.

REFERENCES

Ahmad, N., Quadri, N., Qureshi, M. and M. Alam (2018). Relationship Modeling of Critical Success Factors for Enhancing Sustainability and Performance in E-learning. *Sustainability*, 10(12): 4776.

Akyuz, H. I., and S. Samsa (2009). E-effects of Blended Learning Environment on the Critical Thinking Skills of Students. *Procedia – Social and Behavioral Sciences*, 1(1): 1744–1748.

Andel, S.A., de Vreede, T., Spector, P. E., Padmanabhan, B., Singh, V. K., and De Vreede, G. J. (2020). Do Social Features Help in Video-centric Online Learning Platforms? A Social Presence Perspective. *Comput. Hum. Behav.* 113:106505. doi: 10.1016/j.chb.2020.106505

Ashwin, P., and McVitty, D. (2015). E-meanings of Student Engagement: Implications for Policies and Practices in -e *European Higher Education Area: Between Critical Reflections and Future Policies.* Springer International Publishing, Berlin, Germany, 343–359.

Azzi, D.V., Melo, J., Neto, A. D. A. C., Castelo, P. M., Andrade, E. F., and Pereira, L. J. (2021). Quality of Life, Physical Activity and Burnout Syndrome During Online Learning Period in Brazilian University Students During the COVID-19 Pandemic: A Cluster Analysis. *Psychol. Health Med.* 27, 466–480. doi: 10.1080/13548506.2021.1944656

Bakar, K.A., Ayub, A.F. M., Luan, W.S., Tarmizi, R.A. (2010). Exploring Secondary School Students' Motivation Using Technologies in Teaching and Learning Mathematics. *Procedia Social and Behavioral Sciences*, 2, 4650–4654.

Bartlett, M.S. (1954). A Note on the Multiplying Factors for Various Chi Square Approximations. *Journal of the Royal Statistical Society*, 16, 296–298.

Bhuasiri, W., Xaymoungkhoun, O., Zo, H., Rho, J.J. & Ciganek, A.P. (2012). Critical Success Factors for E-learning in Developing Countries: A Comparative Analysis Between ICT Experts and Faculty. *Computers & Education*, 58(2), 843–855. https://doi.org/ 10.1016/j.compedu.2011.10.010

Brill, J.M. & Galloway, C. (2007). Perils and Promises: University Instructors' Integration of Technology in Classroom-based Practices. *British Journal of Educational Technology*, 38(1), 95.

Caywood, K. & Duckett, J. (2003). Online vs On-campus Learning in Teacher Education. *Teacher Education and Special Education*,26(2), 98–105.

Chang, K. E., Chen, Y. L., Lin, H. Y., and Sung, Y. T. (2008). Effects of Learning Support in Simulation-based Physics Learning. *Comput. Educ.* 51, 1486–1498. doi: 10.1016/j.compedu.2008.01.007

Chang, T. Y., Hong, G., Paganelli, C., Phantumvanit, P., Chang, W. J., Shieh, Y. S., *et al.* (2020). Innovation of Dental Education During COVID-19 Pandemic. *J. Dent. Sci.* 16, 15–20. doi: 10.1016/j.jds.2020.07.011

Chen, C., Landa, S., Padilla, A., and Yur-Austin, J. (2021). Learners' Experience and Needs in Online Environments: Adopting Agility in Teaching. *J. Res. Innov. Teach. Learn.* 14, 18–31. doi: 10.1108/JRIT-11-2020-0073

Choy, S., McNickle, C. & Clayton, C. (2002). Learner Expectations and Experiences: *An Examination of Student Views of Support in Online Learning.* Leabrook, SA: Australian National Training Authority.

Collis, B. (1998). Flexible Learning and Design of Instruction. *British Journal of Educational Technology*, 29 (1), 59–72.

Djidu, H., Mashuri, S., Nasruddin, N., Sejati, A. E., Rasmuin, R., and La Arua, A. (2021). Online Learning in the post-Covid-19 Pandemic era: is Our Higher Education Ready for it? *J. Penelitian Pengkajian Ilmu Pendidikan* 5, 139–151. doi:10.36312/esaintika.v5i2.479

Dong, C., Cao, S., and Li, H. (2020). Young Children's Online Learning During COVID-19 Pandemic: Chinese Parents' Beliefs and Attitudes. *Child. Youth Serv. Rev.* 118:105440. doi: 10.1016/j. childyouth.2020.105440

Doris B. U., and Wasilik, O. (2009). Factors Influencing Faculty Satisfaction with Online Teaching and Learning in Higher Education. *Distance Education*, 30 (1): 103–116.

Elizabeth, R.O., Rochelle, Y.J. & Rosemary, B. (2009). Factors Influencing Faculty use of Technology in Online Instruction: A Case Study. Online *Journal of Distance Learning Administration*, XII(I), Accessed: 29th August 2020 https://www.westga.edu/~distance/ojdla/spring121/osika121.html

Fornell, C., and Larcker, D. F. (1981). Evaluating Structural Equation Models with Unobservable Cariables and Measurement Error. *J. Mark. Res.* 18, 39–50. doi: 10.2307/3151312

Hasibuan, N. (2016). Development of Islamic Education with Educational Technology Implications. *FITRAH J. Study Islamic Sci.* 1:189. doi: 10.24952/fitrah.v1i2.313

Hollenbeck, C.R., Zinkhan, G.M. & French, W. (2006). Distance Learning Trends and Benchmarks: Lessons from an Online MBA Program. *Marketing education review*, 15(2), 39–52

Joe, B., Jacqueline, K., Eastman & Cathy, O.S. (2004). Retaining the Online Learner: Profile of Students in an Online MBA Program and Implications for Teaching Them. *Journal of education for business*, 79(4), 245–253, DOI: 10.3200/JOEB.79.4.245-253.

Kaiser, H. (1974). An Index of Factorial Simplicity. *Psychometrika*, 39, 31–36.

Kaplan-Leiserson, E. (2000). *E-learning Glossary*. Retrieved January, 9, 2020, from http://www.learningcir-cuits.org/glossary.html.

Khachfe, H. H., Chahrour, M., Sammouri, J., Salhab, H. A., Makki, B. E., and Fares, M. Y. (2020). An Epidemiological Study on COVID-19: A Rapidly Spreading Disease. *Cureus* 12:e7313. doi: 10.7759/cureus.7313

Khamparia and Pandey, B. (2017). Impact of Interactive Multimedia in E-learning Technologies, in Enhancing Academic Research with Knowledge Management Principles. *IGI Global, Pennsylvania, CA. USA.*

Kleinman, S. (2005) Strategies for Encouraging Active Learning, Interaction, and Academic Integrity in Online Courses, *Communication Teacher*, 19:1, 13–18, DOI: 10.1080/1740462042000339212

Kraidy, M. (2002). Social Change and the Media. In: Schement J.R. (Ed.), *Encyclopedia of Communication and Information 3*, 931–935. New York, NY: Macmillan Reference USA. Retrieved from http://repository.upenn.edu/asc_papers/328

Kuo, Y.C., Walker, A.E., Belland, B.R. & Schroder, K.E.E. (2013). A Predictive Study of Student Satisfaction in online Education Programs. *International Review of Research in Open and Distance Learning*, 14(1), 16–39.

Laksana, D.N. L. (2021). Implementation of Online Learning in the Pandemic Covid-19: Student Perception in areas with Minimum internet Access. *J. Educ. Technol.* 4, 502–509. doi: 10.23887/jet.v4i4.29314

Lee, S. J., Srinivasan, S., Trail, T., Lewis, D., and Lopez, S. (2011). Examining the Relationship among Student Perception of Support, Course Satisfaction, and Learning Outcomes in Online Learning. *Intern. High. Educ.* 14, 158–163. doi: 10.1016/j.iheduc.2011.04.001

Little, B., and Knihova, L. (2014). Modern Trends in Learning Architecture. *Industrial and Commercial Training*, 46(1): 34–38.

Mbarek, R. & Zaddem, F. (2013). The Examination of Factors Affecting E-learning Effectiveness. *International Journal of Innovation and Applied Studies*, 2(4), 55–83.

Mishra, S. and Panda, S. (2007). Development and Factor Analysis of an Instrument to Measure Students Attitude Towards E-learning, *Asian Journal of Distance Education*, Vol.5 (1), pp.27–33.

Mokter, H. C. (2020). The Effect of the Covid-19 on Sharing Economy Activities. *J. Clean. Prod.* 280, 124782. doi: 10.1016/j.jclepro.2020.124782

Moore, M. G., and Kearsley, G. (2005). *Distance Education: A Systems View.* Belmont, CA: Thomson Wadsworth.

Moses, P., Khambari, M.N., & Luan, W.S. (2008). Laptop use and its Antecedents Among Educators: A Review of Literature. *European Journal of Social Sciences*, 7 (1), 104–114.

Mustakim (2020). The Effectiveness of E-learning Using Online Media during the Covid-19 Pandemic in Mathematics. *Al ASMA* 2, 1–12. doi: 10.19166/johme.v5i1.3811

Nunnally, J.C. (1978). *Psychometric Theory*, McGraw Hill, New York.

Oh, E. G., Chang, Y., and Park, S. W. (2020). Design Review of MOOCs: Application of E-learning Design Principles. *Journal of Computing in Higher Education*, 32(3): 455–475.

Okiki, C. (2011). *Information Communication Technology Support for an E-learning Environment at the University of Lagos*, Nigeria. uhttp://digitalcommons.unl.edu/libphilprac/610.

Palmer, S. R., & Holt, D. M. (2009). Examining Student Satisfaction with Wholly Online Learning. *Journal of Computer Assisted Learning*, 25(2), 101 113.

Panesar, K., Dodson, T., Lynch, J., Bryson-Cahn, C., Chew, L., and Dillon, J. (2020). Evolution of COVID-19 Guidelines for University of Washington Oral and Maxillofacial Surgery Patient Care. *J. Oral Maxillofac. Surg.* 78, 1136–1146. doi: 10.1016/j.joms.2020.04.034

Patricia, A. (2020). College Students' Use and Acceptance of Emergency Online Learning Due to COVID-19. *Int. J. Educ. Res. Open* 1:100011. doi: 10.1016/j.ijedro.2020.100011

Peltier, J.W., Drago, W. & Schibrowsky, J.A. (2003). Virtual Communities and the Assessment of Online Marketing Education. *Journal of marketing education*, 25(3), 260–276.

Pratiwi, W. (2020). The Impact of Covid-19 on Online Learning Activities at Christian Universities in Indonesia. *Educ. Sci. Perspect.* 34, 1–8. doi: 10.31334/abiwara.v2i1.1049

229

Rusli, R., Rahman, A., and Abdullah, H. (2020). Student Perception Data on Online Learning using Heutagogy Approach in the Faculty of Mathematics and Natural Sciences of Universitas Negeri Makassar. Indonesia. *Data Brief* 29:105152. doi: 10.1016/j.dib.2020.105152

Sarabadani, J., Jafarzadeh, H. & Shami, Z.M. (2017). Towards Understanding the Determinants of Employees' E-learning Adoption in Workplace: A Unified Theory of Acceptance and Use of Technology (UTAUT) View. *International Journal of Enterprise Information Systems*, 13(1), 38–49.

Schroeder, C.C. (2003). Supporting the New Students in Higher Education Today. *Change*, 25(2), 55–58.

Selim, H. M. (2007). Critical Success Factors for E-learning Acceptance: Confirmatory Factor Models. *Comput. Educ.* 49, 396–413. doi: 10.1016/j.compedu.2005.09.004

Suprianto, S., Arhas, S. H., Mahmuddin, M., and Siagian, A. O. (2020). The Effectiveness of Online Learning Amid the COVID-19 Pandemic. *J. Adm.* 7, 321–330.

Suryaman, M., Cahyono, Y., Muliansyah, D., Bustani, O., Suryani, P., Fahlevi, M., *et al.* (2020). COVID-19 Pandemic and Home Online Learning System: Does it Affect the Quality of Pharmacy School Learning. *Syst. Rev. Pharm.* 11, 524–530.

Thorpe, M. (2002). Rethinking Learner Support: The Challenge of Collaborative Online Learning. *Open Learn.* 17, 105–119. doi: 10.1080/02680510220146887a

Wong, G. L. H., Wong, V. W. S., Thompson, A., Jia, J., Hou, J., Lesmana, C. R. A., *et al.* (2020). Management of Patients with Liver Derangement During the COVID-19 pandemic: an Asia-Pacific Position Statement. *Lancet Gastroenterol. Hepatol.* 5, 776–787. doi: 10.1016/S2468-1253(20)30190-4

Zemsky, R. & Massy, W.F. (2004). Thwarted Innovation. What Happened to E-learning and why? Publisher: Learning Alliance. Accessed: 24th August 2020 http://www.irhe.upenn.edu/Docs/Jun2004/ThwartedInnovation.pdf

Zhafira, N. H., Ertika, Y., and Chairiyaton. (2020). Student Perceptions of Online Lectures as Learning Facilities During the Covid-19 Quarantine Period. *J. Bus. Manag. Strategy Stud.* 4, 37–45.

Block chain

Artificial Intelligence, Blockchain, Computing and Security – Dagur et al. (Eds)
© 2024 The Author(s), ISBN: 978-1-032-49393-0

Land transaction and registration system using blockchain

Anubhavi Agrawal, Ayush Teotia, Dhrubb Gupta, Akash Srivastava, G. Mahesh &
B.C. Girish Kumar
JSS Academy of Technical Education, Noida, Uttar Pradesh, India

ABSTRACT: Land registration describes systems which take into account transaction, ownership, possession, etc. of a block of land to present verification of title to bring to a halt an illegitimate disposal of land. This requires scores of paperwork and plenty of time. This paper propose a land registration and transaction solution using Blockchain which is a cluster of records bunched mutually in blocks that are linked collectively using Linked List providing speedy and dependable system. The proposed model (BBLTRS) Blockchain based land transaction and registration system will not only speed up the procedure of land registry, but also make it simpler, cheaper and transparent for transacting parties and authorities to reassign the land, maintain secured virtual records and keep a track of all the transactions.

1 INTRODUCTION

1.1 *About existing system*

Traditionally, the land registry has concentrated on paper records that are vulnerable to being lost, falsified, destroyed, or altered. Also included are various levels of manual intervention. There are numerous inefficiencies and risks of mutation. There are numerous instances of impersonation fraud, unauthorized transactions, etc. every day in addition to the cases that are currently in court. The wise distribution of land in India leaves it extremely susceptible to fraud and corruption in the recording of transactions. Different parties assert varying degrees of control over a particular plot of land. Sometimes, even more than once, someone's land has been illegally sold.

1.2 *Challenges faced by people during land registration process*

- Ownership verification is a global challenge.
- At times properties do not have a recorded ownership history, having access to which enhances trust among unknown transacting parties.
- Properties may be at a risk of unauthorized sale.
- Manual Process requires score of paper work and takes more than a month.
- Land Valuation could be inaccurate leading to incorrect taxes or insurance premiums.
- Present system fails in detecting and preventing frauds.

2 ESSENTIAL TECHNOLOGIES

2.1 *Blockchain*

Blockchain could be made up of a growing number of records arranged in blocks. A quick and dependable system for process execution is provided by linking the blocks together using

DOI: 10.1201/9781003393580-34

linked lists. A replacement block is added to the line of preceding blocks for each new transaction that takes place.

The block consists primarily of three components:

- Information about the sender, recipient, and volume of data exchanged.
- A hash, which resembles a fingerprint and is an alphanumeric code specific to each block.
- The previous block's hash, which enables the creation of a sequence by connecting each block to a different one.

This implies that a small modification to one block might have a significant effect on the chain as a whole. To stop hackers from abusing the blockchain, there is an additional layer of security. Attributes of Blockchain Technology

- Persistent: Data being stored cannot be corrupted as modifications are quite complex.
- Time-stamped records: The time at which the data is entered is recorded precisely.
- Approachable: The data being stored is accessible by all partakers and the modifications possible based on the permissions they are granted.
- Decentralized: There isn't a single governing authority as it is distributed all over the network.
- Distributed Ledger: Ensures if a block is spoilt, it can at all times be restored as every block has a duplicate of the ledger.

2.2 Smart contract

Smart Contract is a code chunk with built-in trade logic that outlines the circumstances in which the transfer of corporate bonds will take place. Its only function is to represent ownership of the things kept there and move the assets and money in accordance with established rules. To control the state of the asset it stores, the developer incorporates a set of rules and logic. It has the immutable and distributed properties of the blockchain as it is stored inside the blocks of the network, which prevents changes to it after it has been created. Apiece has a matchless address that is to bring into play propel or take delivery of ether. It can identify the caller, allowing for the possibility of granting special access to the contract. Each smart contract account has four properties: fields, balance, storage, and code.

2.3 Web3 library

Interaction with contracts over the blockchain is made possible by the Web3 library. To make use of Web3, an object Web3 class is essential. Web3 library allows for the creation of multiple instances, each of which can be used to hook up to a diverse Ethereum network. So as to connect the web3 instance together with an Ethereum network,[05] a provider communication layer is necessary. It can in effect send a call for a local network and take delivery of a reaction to that demand as each contributor has a collection of indistinguishable techniques for distribution or getting an appeal from Ethereum network.

2.4 Ganache module

"ganache-cli"npm module makes the installation of the ganache, a regional test network for Ethereum, possible. To devise a restricted Ethereum test network, project necessarily needs to introduce it. As soon as the contract gets into play on a restricted set-up, none of safety measures or public-private key(s) are desirable to let the test-network-provided versions (ganache) loose.[05] The ability to create these accounts in an unlocked state and without the need for ether is advantageous.

2.5 Truffle

Truffle serves as a platform for the creation, evaluation, and deployment of Ethereum DApps. Truffle manages contract artifacts in addition to assisting with conventional deployments, library linking and sophisticated Ethereum applications. Additionally, it offers automated tests for Solidity and JavaScript contracts. Users can interact with all of the built-in Truffle commands and contracts on a console provided by Truffle.

2.6 Metamask

It is a blockchain wallet that stores the user's private key, controls the blockchain account's balance, allows the user to create transactions, and submits them to the network. It supports managing Ethereum accounts, creating digital signatures, and creating transactions. Through Metamask, users can quickly access the blockchain's smart contract.

2.7 React

It is a tool for constructing web pages developed by Facebook. It is a JavaScript library. It is used to create user interface (UI) components and has many features that facilitate and speed up work. It has the ability to simultaneously render several components on the screen. A single component is created for each file.

3 PROPOSED SYSTEM

In this model, a spread-out system of land registry with technology of blockchain as pedestal called Blockchain Based Land Transaction and Registration System (BBLTRS) is what we propose. The proposed system, being influenced by blockchain paybacks goes on board with decentralization, data immutability. This would even help building immense trust, reduce risks to great levels, reduce overheads, and cut short the cost of intermediaries. By ensuring proper visibility and authenticating transactions, the blockchain-based system of land registry would maintain confidentiality of accumulated data and uphold the network's overall trust. This system can offer tamper-resistant functionality [06][12]. Data invulnerability is another area where blockchain shows great promise; once data is collected there, it is nearly impossible to change it. Additionally, the middle class would not participate as usual. The proposed approach attempts to maintain the integrity [14], confidentiality, and authenticity of the sensitive information while significantly cutting the time needed to complete the process conventionally.

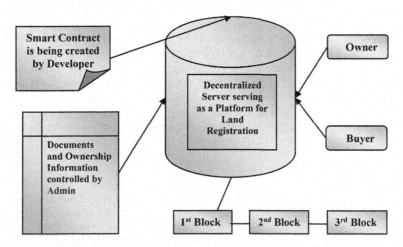

Figure 1.   Blockchain Based Land Transaction and Registration System (BBLTRS).

235

4 LITERATURE SURVEY

Table 1. Advantages & disadvantages of related works.

S. No.	Paper Title	Advantages of Proposed System	Disadvantages of Proposed System
01.	Digitization of land records: From paper to blockchain [15]	• No third-party needed • Transparency	• Time Lapse in Process • Security Concerns
02.	Smart Contract Definition for Land Registry in Blockchain [02]	• Reduction of Frauds • Reduced dependence on Brokers	• No Buyer/Seller Verification Performed
03.	A secured land registration framework on Blockchain [09]	• Digital Maintenance of Records • Lower Cost • Minimization of dependence on third party.	• No mention of Security • No mention of Authentication • No mention of secured payment gateway
04.	Securing land registration using Blockchain [07]	• High speed and performance • Extensive Security Mechanisms.	• No mention of storage and security of the documents uploaded.

5 CONCLUSION AND FUTURE ENHANCEMENT

India's current land registration system faces a number of challenges. It is suggested that blockchain technology be used to address these problems, and the importance of smart contracts is emphasized. The main tools have been described in broad terms. Under the current system, the essential land registration remodeling in rural areas are made manually and precision and steadfastness of those updates varies from person to person. Since humans are fallible, the process of modification is risky. Unidentified errors can undoubtedly be reduced with the use of such a system. Above all the methodology employed is straightforward even for a novice. Data is outdated and proper transaction etiquette is not documented as a result of poor maintenance and communication. The use of blockchain is necessary in this situation because it is necessary for land ownership. Future prospects for blockchain technology are bright. It is a secure medium for the transmission of operations and information without interfering with other parties. The current system has a number of flaws that promote corruption and dissension, and filling them would cost the judiciary and law enforcement agencies a significant amount of valuable public funds. Blockchain can be used to close these gaps and solve issues with land registry systems like record-tampering and selling the same piece of land to multiple buyers. The implementation of this proposed model by making use of blockchain will be covered in one of our upcoming papers; it would be a wonderful and incredibly secure replacement for the current systems, which have flaws.

REFERENCES

Alam, B., & Kumar, A. (2014, March). A Real Time Scheduling Algorithm for Tolerating Single Transient Fault. In *2014 International Conference on Information Systems and Computer Networks (ISCON)* (pp. 11–14). IEEE.R. B.-Fich and A. Castellanos, Digitization of land records: From paper to blockchain, Conference: Thirty Ninth International Conference on Information SystemsAt: San Francisco 2018, 2018.

Ameya Thosar, Mayur Harne, Ashutosh Sarode, Dr. Parminder Kaur, *Land Registry Management Using Blockchain*, 2021.

Baliga A., *Understanding Blockchain Consensus Models, Persistent*, vol. 4, pp. 1–14, 2017.

Girish Kumar B. C., Nand P. and Bali V., *Opportunities and Challenges of Blockchain Technology for Tourism Industry in Future Smart Society, 2022 Fifth International Conference on Computational Intelligence and Communication Technologies (CCICT)*, 2022, pp. 318–323, doi: 10.1109/CCiCT56684.2022.00065.

Girish Kumar, B.C., Nand, P., Bali, V. (2022). Review on Opportunities and Challenges of Blockchain Technology for Tourism Industry in Future Smart Society. In: Bali, V., Bhatnagar, V., Lu, J., Banerjee, K. (eds) *Decision Analytics for Sustainable Development in Smart Society 5.0. Asset Analytics*. Springer, Singapore. https://doi.org/10.1007/978-981-19-1689-2_16

Krishnapriya S, Gireeshma Sarath, Securing Land Registration Using Blockchain, *2019 Third International Conference on Computing and Network Communications (CoCoNet'19)*, 2019.

Kumar, A., & Alam, B. (2015, February). Improved EDF Algorithm for Fault Tolerance with Energy Minimization. In *2015 IEEE International Conference on Computational Intelligence & Communication Technology* (pp. 370–374). IEEE.

Kumar, A., Mehra, P. S., Gupta, G., & Sharma, M. (2013). Enhanced Block Playfair Cipher. In *Quality, Reliability, Security and Robustness in Heterogeneous Networks: 9th International Conference, QShine 2013, Greader Noida, India, January 11–12, 2013, Revised Selected Papers 9* (pp. 689–695). Springer Berlin Heidelberg.

Malsa, N., Vyas, V., Gautam, J. (2022). Blockchain Platforms and Interpreting the Effects of Bitcoin Pricing on Cryptocurrencies. In: Sharma, T.K., Ahn, C.W., Verma, O.P., Panigrahi, B.K. (eds) *Soft Computing: Theories and Applications. Advances in Intelligent Systems and Computing*, vol 1380. Springer, Singapore. https://doi.org/10.1007/978-981-16-1740-9_13

Mukhopadhyay M., *Ethereum Smart Contract Development: Build Blockchain-based Decentralized Applications Using Solidity*, Packt Publishing Ltd., 2018.

Nandi M., Bhattacharjee R. K., Jha A. and Barbhuiya F. A., A Secured Land Registration Framework on Blockchain, *2020 Third ISEA Conference on Security and Privacy (ISEA-ISAP)*, Guwahati, India, 2020, pp. 130–138, doi: 10.1109/ISEA-ISAP49340.2020.235011

Sahai A. and Pandey R., *Smart Contract Definition for Land Registry in Blockchain, 2020 IEEE 9th International Conference on Communication Systems and Network Technologies(CSNT)*, Gwalior, India, 2020, pp. 230–235, doi: 10.1109/CSNT48778.2020.9115752.

Sandeep Kumar Panda, Sipra Sahoo, Sachi Mohanty, *Smart Contract Based Land Registry System to Reduce Frauds and Time Delay*, June 2021.

Sangeeta Gupta, Kavita Agarwal. (2021). Essentials of Blockchain Technology for Modern World Applications, *Materials Today: Proceedings*, 2021, ISSN 2214-7853,

Shuaib M., Daud S. M., Alam S., and Khan W. Z., Blockchain-based Framework for Secure and Reliable Land Registry System, *TELKOMNIKA Telecommunication, Computing, Electronics and Control*, vol. 18, no. 5, pp. 2560–2571, 2020, doi: 10.12928/TELKOMNIKA.v18i5.15787.

Spielman A., *"Blockchain: Digitally Rebuilding the Real Estate Industry,"* Thesis: S.M. in Real Estate Development, Massachusetts Institute of Technology, 2016.

Suganthe R.C., Shanthi N., R.ha S. La, Gowtham K., Deepakkumar S., Elango R., *Blockchain Enabled Digitization of Land Registration, 2021 International Conference on Computer Communication and Informatics (ICCI – 2021)*, Jan. 27–29, 2021.

237

Artificial Intelligence, Blockchain, Computing and Security – Dagur et al. (Eds)
© 2024 The Author(s), ISBN: 978-1-032-49393-0

E-policing and information management system using blockchain technology

G. Mahesh
JSS Academy of Technical Education, Noida, India

B.C. Girish Kumar
SJB Institute of Technology, Bangalore, Karnataka, India

Shivani Pathak, M. Surekha, K.G. Harsha & Mukesh Raj
JSS Academy of Technical Education, Noida, India

ABSTRACT: In manual policing system the complainant need to visit the police station and give complaint, here the complaint may be registered or may not be due to influence of political people, bribe and other circumstances, then justification for complainant cannot be obtained ever after giving complainant, so justification for complainant cannot be obtained, if we implement the same system by using blockchain technology, the complainant information will be recorded along with the key and also it becomes proof for future references, along with the complainant can upload the proof documents for case resolve action smoothly, here once complainant login and register in blockchain then such a complaint information can share with number of blocks that cannot be modify by another other due to distributed system, so in this proposed method we introducing the complaint management with police department and same information can be share to court also to make judgment within short time without interfere of influenced person.

1 INTRODUCTION

The complaints of offenses in India required to be registered as the law. The complaints registration may be of two types based on their category such as cognizable and non-cognizable offenses [1]. Here cognizable category offenses are type such as kidnapping, murder, rape and theft etc., as per the Criminal Procedure code 1973 in Section 2(c). The offenses can be arrested by police staff without any warranty whereas in non-cognizable which are not serious nature the suspects can arrested by police staff with warrant from court. The FIR First Investigation Report registered which contains complainant's name, address, date and time of location as well as the facts of incidents. As per the RTI Right to Information Act, the complainant can acquire a charge sheet by applying a letter to the court with specified fee. Nowadays the criminal activities are increasing and also police official corruption, the police staff avoid the registration of complaint of FIR/NCR, some people unable register their complaints, and others may not received FIR copy. Here requirement of transparent system to avoid corruption from system of public. The complaints reports and criminal information stored in centralized server is not efficient to share the information globally for the other police officials who can attend cases and try to solve that within a given bound time, because if the time increases there will be chances of missing witness and due to political influence the direction of investigation may change and actual criminal may escape, so we required the blockchain technology based technology, where decentralized data such as FIR's and NCR's can be make available anywhere though globally, and also it is immutable i.e., once we entered in a block that cannot be changeable by others, which leads once complaint information entered that can be make available for different police officials and supports for further investigation of cases easily without much delay.

DOI: 10.1201/9781003393580-35

The consensus used to check the validity of blockchain in terms of proof of work, state and capacity etc., After adding the transactions block to the network, the blocks are infeasible block, here we can use Interplanetary file system (IPFS) algorithm, that supports to store and sharing of files between peer to peer in decentralized manner. In this technology we can use hash technology which is content based and saved in a decentralized network. Here hash function based technology supports immutability i.e., not changing the content of block, good reliability and throughput. The decentralized application used to share the information simultaneously globally and also supports for tracking the information about the police from initial stage to till the case submission to the court. This technology confirms trust in between police department and complainants, because of immutable property it gives secure from information loss as well as hacking or malicious attacks of criminal information.

2 LITERATURE SURVEY

The blockchain technology [16,17] used to stores the FIR information secure. (Proposed by Gupta, Antra and D. Vílchez Jose). [3] The proposed system can ensures more trust sharing information between police department and public. Here the block information can be entered and viewed by complainant and duty allocated Police Officer. K. Tabassum, H. Shaiba, S. Shamrani and S. Otaibi [4] proposed here by using decentralized technology we can make it available globally for increase speed in inspection about the case. In proposed a user can complaint by using mobile with android application the same information stored in WAMP server.(proposed by Iyer A, Kathale P, Gathoo S and Surpam N) [5]. Here blockchain technology support to share it in efficiently for other Police Official for takes a suitable action immediately. They proposed the benefits of using E-police system over the conventional based paper filing system. Here uniqueness achieved by the IMEI numbers of user mobile for reference. P. A. K. S. Y. K. S., Shivaganesh Pillai, [6] proposed that the authorized user can register FIR with the proper selection of IPC section, with the help of SOS sytem, if they complained person was in trouble then location could be traced by nearest police station easily.

The information stored centralized database can have different types of hacks such hacking may damage the integrity and validation of data. SQL injection attacks can be used for hacking purpose nowadays Thoms, N. [6] proposed that the SQL injection is type of destructive attack here hackers try to hack information stored in database. This type of drawback can be overcome from blockchain technology and also avoid the hardware and software malfunctions. In a blockchain technology if any changes try to do such a changes can be visible through entire network. Here data updated by a node is authenticated by multiple nodes, so falsified infrequently find way into the blockchain as per Anh, D.T.T., Zhang, M., Ooi, B.C., Chen, G. Proposed. If attempt to make unstable the system required a concurrent attacks on at least 51% of nodes of a certain blockchain to reflect a unit block. As per increasing number of nodes attacking can decreases in exponentially form as per proposed by Miles, C. [8].

Maisha Afrida Tasnim, Abdullah Al Omar, Mohammad Shahriar Rahman, Md. Zakirul Alam Bhuiyan, [9] proposed that criminal records could be secured with authenticity and rigidity by blockchain technology which supports decentralized with peer to peer networks and illegal changes cannot be implemented. Our system supports for the authorities to maintain criminal records efficiently. The data can be kept with safely and also managing the data in tedious task of data managing easily. Our system supports to maintain criminal records efficiently by authorities. As per Ishwarlal Hingorani, Rushabh Khara, Deepika Pomendkar, Nataasha Raul [10] propose how to implement the blockchain technology for cognizable and non-cognizable offenses handling and also filing FIR filing securely. Here trust for police department no need for the public. This system involves four stakeholders such as court, law enforcement admin, Police Officer and also suspected person. Due to blockchain technology works in decentralized form, all other activities can be monitored and avoid any changes to block by unauthenticated person. In blockchain technology the concerned complaints have cryptographically generated hash key that is used in integrity checking process. In this technology is having a piece of software can combines with algorithm, hash functions and a consensus mechanicsm. Antra Gupta, Deepa V. Jose. [5] proposes blockchain technology used in secured mode for registering FIR. Here SHA 512 and hash true used in algorithm to protect the complaint information cryptographically. With the decentralization technology using SHA-512 provides higher security for the system and also the hash value uniqueness derived from the previous block's hash, for next step again that can be feed as input with random hash key, we could store other information such as behaviour and health of criminal along with FIR registration. Which lead information for respective Police Officer under their watch can summoned for court hearing.

3 PROPOSED SYSTEM

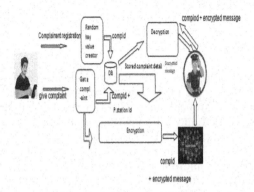

Figure 1.   Proposed model of complaint management using blockchain.

The proposed method is decentralized system, in which the technologies like blockchain IPFS etc, used to achieve the outcome. The proposed system architecture can be brief as follows

A. Data Security module: In this module the required documents of legal can be making available for the stake holders such as police officers, suspect person as well as complainant person. The information will be decrypted using 512 bits hash function; this can be implemented by SHA algorithm. The following mentioned steps indicated the implementation:	B. Secured Blockchain module : Blockchain is the technology [12,13] in which once information entered which cannot be altered easily due to the decentralized technology, the entered information can be spread over all the nodes, just like a pdf format the content can be available to authorized person with proper rights provided for them using decryption technique, The encrypted information stored in IPFS network.

In this algorithm the public components X & Y calculated using the formula as follows.

The complainant information encrypted by secret key generated by SHA algorithm can be shared and make available for the police officers, who can gather the information from block chain using decrypted along with the secret code shared with them $X = g\ x mod\ (p)$, $Y = g\ y mod\ (p)$ where p is prime number and base is g and X, Y are security components, At time of complainant register a complaint, such as information is encrypted with the secret key, that can be calculated using public components of police station such as p,g,Y, The police officer can read the complaint using secret key public components p,g,X and take the necessary action about the complaint.

(I)   Interface module: In this module the complainant or witnesses details provider can register by using Mobile or Laptop or System and can share the complaint or witnesses along the encryption technique with key value, which will be secured in Blockchain technology [14], so it can reach to Police Officer or Magistrate for further enquiries about the complaint.

(II)   Interface module: In this module the complainant or witnesses details provider can register by using Mobile or Laptop or System and can share the complaint or witnesses along the encryption technique with key value, which will be secured in

240

Blockchain technology [14,15], so it can reach to Police Officer or Magistrate for further enquiries about the complaint.

(III) Interface module: In this module the complainant or witnesses details provider can register by using Mobile or Laptop or System and can share the complaint or witnesses along the encryption technique with key value, which will be secured in Blockchain technology [14–16], so it can reach to Police Officer or Magistrate for further enquiries about the complaint.

The steps involved achieving the E Police complaint management and storage can be achieved by using the following steps.

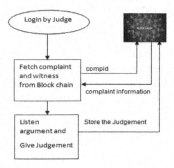

Figure 2.   Judgment workflow.

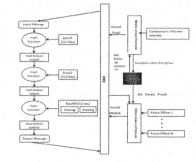

Figure 3.   Information management algorithm.

4   ANALYSIS OF PROPOSED MODEL

In this proposed model the designed architecture will help to get the transparency in Police Department and encourage for confidentially the complainant can file a complaint for any types of criminal activities such as major like Murder, Dacoit etc., and minor cases such as harassment, any type of cheating, so cases can be handled by their jurisdiction Police Officer, If any fault occurs in terms of not filing FIR etc., that can be viewed by Judge through blockchain and can take action about the Police Officer regarding not attending duty, because the data shared by complainant or witness provider can be uploaded on the spot with mobile having internet connection, here complaint and witness provider need to register before filing a complaint about any event, If the Police Officer delayed in case can be come to know using blockchain [17,18], similarly once it come to argue the Judge can get the witness with the proof provided by witness provider, and also case can be settled within a short time, this encourage the increasing in respect about law and order, and also in society this technology supports to get peaceful and safety from criminal activities. This technology could implement along with the Police Stations to get work done as earliest. and also case can be settled within a short time, this encourage the increasing in respect about law and order, and also in society this technology supports to get peaceful and safety from criminal activities. This technology could implement along with the Police Stations to get work done as earliest.

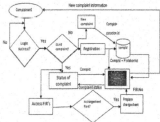
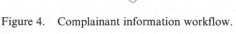
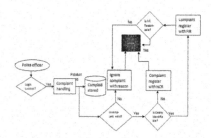

Figure 4.   Complainant information workflow.   Figure 5.   Police officer information work flow.

5 CONCLUSION AND FUTURE WORK

In the existing E-Policing management system, the author described about the management of police task using blockchain technology [19,20] with sharing of information of crime records from complainant to police officer and the information protected by cryptography, where the details can encrypted by sender with the selection of key value and decrypted by the police officer and do the necessary action. In this system they not mentioned about the judgment implementation. Here the required witness and other details need to be provide for the judge to give judgment, but due to involvement of highly political influence and money, the witness can turn to any side so proper judgment will not take place, in addition with the existing system, If we make available witness and other documentation of criminal using blockchain technology [19,20] to Judge, The Judgment can be made fast as well as required witness may not be loss due to decentralized form of storage, Where the changes cannot be done easily once complaint entered with witness. So the details of document and witness can reach to Judge also without any corruption, there is a result in Judgment can be taken based on witness available. Here any public who know about witness can share with their mobile with proper document or picture to Judge. So cases can be solved within stipulated time with limited arguments.

REFERENCES

[1] Ishwarlal Hingorani, Rushabh Khara, Deepika Pomendkar, Nataasha Raul, *Police Complaint Management System using Blockchain Technology*, ISBN: 978-1-7281-7089-3, DOI: 10.1109/ICISS49785.2020.9315884.

[2] Satoshi Nakamoto *Bitcoin: A Peer-to-Peer Electronic Cash System*. Retrieved from URL https://bitcoin.org/bitcoin.pdf on 1 September 2022.

[3] Gupta, Antra and D. V´ılchez Jose. A Method to Secure FIR System Using Blockchain. *International Journal of Recent Technology and Engineering (IJRTE)* ISSN: 2277-3878, Volume-8, Issue-1, May 2019

[4] Tabassum K., Shaiba H., Shamrani S. and Otaibi S., *e-Cops: An Online Crime Reporting and Management System for Riyadh City, 2018 1st International Conference on Computer Applications Information Security (ICCAIS)*, Riyadh, 2018, pp. 1–8, doi: 10.1109/CAIS.2018.8441987.

[5] Iyer A, Kathale P, Gathoo S and Surpam N 2016 E-Police System-FIR Registration and Tracking Through Android Application *International Research Journal of Engineering and Technology* 3(2) 1176–1179

[6] Thoms, N.: *SQL Injection: Still Around, Still a Threat.* https://www.fasthosts.co.uk/blog/digital/sql-injection-still-around-still-threat

[7] Anh, D.T.T., Zhang, M., Ooi, B.C., Chen, G.: Untangling Blockchain: A Data Processing View of Blockchain Systems. *IEEE Trans. Knowl. Data Eng.* 30(7), 1366–1385 (2018)

[8] Miles, C. *Blockchain Security: What Keeps Your Transaction Data Safe* https://www.ibm.com/blogs/blockchain/2017/12/blockchain-security-what-keeps-your-transaction-data-safe/

[9] CRAB: Blockchain Based Criminal Record Management System by Maisha Afrida Tasnim, Abdullah Al Omar, Mohammad Shahriar Rahman, Md. Zakirul Alam Bhuiyan, *SpaCCS 2018 Conference Paper*: Springer Nature

[10] Police Complaint Management System Using Blockchain Technology by Ishwarlal Hingorani, Rushabh Khara, Deepika Pomendkar, Nataasha Raul, *Proceedings of the Third International Conference on Intelligent Sustainable Systems* [ICISS 2020].

[11] A Method to Secure FIR System Using Blockchain by Antra Gupta, Deepa V. Jose. *International Journal of Recent Technology and Engineering (IJRTE)* ISSN: 2277-3878, Volume-8, Issue-1, May 2019.

[12] Girish Kumar, B.C., Nand, P., Bali, V. (2022). Review on Opportunities and Challenges of Blockchain Technology for Tourism Industry in Future Smart Society. In: Bali, V., Bhatnagar, V., Lu, J., Banerjee, K. (eds) *Decision Analytics for Sustainable Development in Smart Society 5.0. Asset Analytics.* Springer, Singapore. https://doi.org/10.1007/978-981-19-1689-2_16

[13]   G. K. B. C, Mahesha A. M., and Harsha K. G., *A Review on Data Storage & Retrieval Using Blockchain Technology*, vol. 11, no. 08, pp. 28630–28637, 2021.

[14]   Girish B. C. Kumar, Nand P. and Bali V., Opportunities and Challenges of Blockchain Technology for Tourism Industry in Future Smart Society, *2022 Fifth International Conference on Computational Intelligence and Communication Technologies (CCICT)*, 2022, pp. 318–323, doi: 10.1109/CCiCT56684.2022.00065.

[15]   G. K. B. C, Singh A., Patel U., Yadav A., and Kumar A., *Tourist and Hospitality Management Using Blockchain Technology*, vol. 11, no. 08, pp. 28670–28672, 2021.

[16]   Kumar, B.C.G., Garg, M., Saini, J., Chauhan, K., Malsa, N., Malsa, K. (2023). Accurate Rating System Using Blockchain. In: Shaw, R.N., Paprzycki, M., Ghosh, A. (eds) *Advanced Communication and Intelligent Systems. ICACIS 2022. Communications in Computer and Information Science*, vol 1749. Springer, Cham. https://doi.org/10.1007/978-3-031-25088-0_44

[17]   Dagur, A., Malik, N., Tyagi, P., Verma, R., Sharma, R., & Chaturvedi, R. (2021). Energy Enhancement of WSN Using Fuzzy C-means Clustering Algorithm. In *Data Intelligence and Cognitive Informatics: Proceedings of ICDICI 2020* (pp. 315–323). Springer Singapore.

[18]   Dagur, A., Kaushik, A., Rastogi, A., Singh, A., Kumar, A., & Chaturvedi, R. (2021). Optimization of Queries in Database of Cloud Computing. In *Data Intelligence and Cognitive Informatics: Proceedings of ICDICI 2020* (pp. 325–332). Springer Singapore.

[19]   Aadarsh Neeraj Singh, Aditi Arya, Aditya Sharma, Girish Kumar B C; *Contact Tracing For Communicable Diseases Using Blockchain"*, International Journal of Emerging Technologies and Innovative Research (www.jetir.org | UGC and issn Approved), ISSN:2349-5162, Vol.9, Issue 5, page no. ppj528–j532, May-2022Two Papers published in UGC journal 2022.

[20]   Girish Kumar B C, Aditya Kumar Gupta, Arpita Verma, Devansh Chaudhary, Rajneesh Pandey; *An Encrypted Automatic Multiple-Choice Question Generator for Self-Assessment Using Natural Language Processing, International Journal of Emerging Technologies and Innovative Research (*www.jetir.org), ISSN:2349-5162, Vol.9, Issue 5, page no.j337–j343, May-2022.

Artificial Intelligence, Blockchain, Computing and Security – Dagur et al. (Eds)
© 2024 The Author(s), ISBN: 978-1-032-49393-0

A survey on Automated Market-Makers (AMM) for non-fungible tokens

Rishav Uppal, Ojuswi Rastogi, Priyam Anand, Vimal Gupta, Sur Singh Rawat & Nitima Malsa
JSS Academy of Technical Education, Noida, Uttar Pradesh, India

ABSTRACT: Blockchain has become a revolutionary technology that has had a great impact on the business environment. Non-fungible tokens (NFTs) are distinct from fungible tokens traded on multiple centralized or decentralized exchanges. Automated Market-MMs (AMMs) are decentralized markets for crypto-tokens that offer users three core operations: deposition of crypto tokens to get AMM shares in return; the dual operation of getting shares for the base tokens; and swapping of two different tokens with each other. This research aims to put forth a comprehensive view of Blockchain and its applications in the real world, including cryptocurrencies, NFTs trading, voting, and much more. It also focuses on how NFTs are traded on different platforms and aims at better marketplaces for trading NFTs,namely Automated Market Makers on different blockchains like Ethereum and Tezos.

Keywords: Blockchain, Cryptocurrencies, Fungible-tokens, Non-Fungible tokens, NFT Collection, NFTs Market Places, Automated Market Maker (AMMs). keyword text

1 INTRODUCTION

In recent years of innovation and technology, cryptocurrency has been boosted in technical industries. For example Bitcoin [1], the most popular cryptocurrency, with its market capital touching 10 billion dollars in 2016, has gained huge success. The blockchain was first proposed in 2008 and deployed in 2009 and has become a key technology that allows transactions to take place on the Bitcoin network without disturbance from an external source [2]. Bitcoin's development was first impeded by its unusually high volatility and the hostility it faced from many nations due to its complexity, but as time went on, the benefits of blockchain – the innovation that underpins bitcoin – gained more significance. The distributed ledger, decentralization, information transparency, and attack-proof design of blockchain are some of its benefits. With the help of a number of its applications, blockchain development has evolved over time and started to digitalize the globe [3].

1.1 *Blockchain and its applications*

Blockchain technology promises benefits in dependability, teamwork, organization, identity, quality, and transparency. In plain English, decentralization means the process which involves setting up an application on a network in a way that total control over the management of data and execution doesn't go to any server. Each server present in this cluster receives the present condition of the data, past transactions can't be removed or altered by

DOI: 10.1201/9781003393580-36

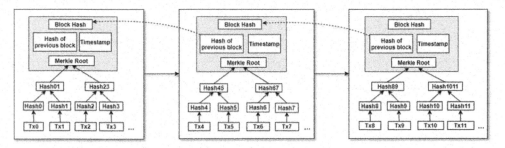

Figure 1. Architecture of blockchain.

anyone. The word 'distributed' means that each server node is either directly or indirectly connected to every other node. Ledger is a phrase related to accounting, and you can think of it as specialized data storage and retrieval [4].

Recently, academics and business people alike have become interested in blockchain technology and cryptocurrencies [5]. Cryptocurrencies are digital money enabling secure and private transactions using blockchain and cryptography. The Cryptocurrency market value has exceeded USD 500 billion. Numerous organizations and nations are beginning to comprehend and use the concept of cryptocurrencies in their business structures [6]. Once cryptocurrencies gained the dominant place for transactions [27,31], there may be a need to reform established systems to be used for trade to handle the existing competition. Thus, cryptocurrencies may become one of the most powerful and advanced technologies being given to global financial systems. They are prime examples of fungible tokens (FTs) because each coin has the same value as any other coin of the same type at any given moment [7]. Certificate verification is a very useful application of blockchain as mentioned in the studies [28–31].

1.2 NFT: Non-fungible token

The idea of non-fungible tokens [8] was first put up by an Ethereum [9] token standard with the sole intention of distinguishing each token with a unique set of identifiable signals. NFTs have mostly included in the Ethereum [9] blockchain. NFTs are a class of cryptocurrencies that are generated by smart contracts. Because NFTs are distinct tokens that cannot be exchanged like-for-like, they are appropriate for uniquely identifying things or people. It can be used to protect intellectual property. An NFT signifies ownership of a physical work of art; each piece of digital art, like each asset and NFT, is distinct. The NFT stands for digital scarcity. Tickets, collectibles, game objects, cryptographic artwork, financial products, deeds, and other items are examples of NFT (Url-5, 2021). As of June 29, 2021, the non-fungible token market cap was $17,408,786,221.48 and the 24-hour trading volume was $1,147,366,236.48. The top 10 NFT coins by Coingecko's market capitalization are displayed in the table below (Url-6, 2021) [10].

Since everything is automated, authors relax and enjoy their earnings as another person gets the transfer of work. Because it is so time-consuming and inaccurate to calculate royalties, many creators are currently underpaid. You won't ever miss a royalty if you encode one into your NFT. [10].Using NFTs to make money is not just for the arts. Businesses like Taco Bell and Charmin have auctioned off themed NFT artwork to benefit charities. NFT art of Taco Bell has sold out in a matter of minutes. Nyan Cat, a cat GIF, sold for about $600,000 in February. NBA Top Shot, a firm, had over $500 million in sales towards March 10].Table 1. presents the pros along with the cons of Non-fungible tokens.

Table 1. Pros and cons of NFT tokens.

Intellectual Property: The creators/owners of any NFT can protect and claim their work that is globally recognizable, and will be awarded every time their work is used,	Practicality: Since with the digitised artwork, the copy is identical to the original. This raises the question of the usefulness and value of owning an NFT.
Level playing field: In other words, a community or following grows quickly around a specific interest or industry. Additionally, creators frequently receive remuneration when NFT is sold and resold (a percentage of the transaction).	Value erosion from duplicates: A creator may choose to sell more than one copy of the exact same NFT as there is no limit on the total number of copies created and distributed.
Verifiable ownership: Irrefutable ownership of NFTs can be easily backed by creators/holders.	Hype-based market: New or current waves of entrants could make an NFT invaluable and can make the entire market weak. Regulatory uncertainty: No legal environments are designed to accommodate NFTs. Due to lesser commercial interests, it is yet unclear how NFTs will be classified and what obligations will follow for owning an NFT

1.3 NFT trading and marketplaces

In 2021, the popularity of nonfungible tokens (NFTs) skyrocketed. Market intelligence can be influenced by marketplace design, with a particular focus on the costs of bidding. Peer-to-peer markets dictate that vendors provide products for a set price while allowing participants to place bids that are frequently below listed pricing. At one extreme, markets may prohibit bidding and force sellers to only sell NFTs through fixed price postings. Alternatives include creating a mechanism to keep bids "off-chain" (avoiding transaction costs), encouraging the development of bidding bots, or creating user interfaces that make putting bids simpler. Real-world markets differ substantially along these parameters. A study was conducted to identify issues in trading platforms that frequently serve as obstacles, understand the driving forces behind new NFT users, and provide suitable suggestions in design. A survey was conducted to learn how different user groups perceive NFTs, and a testing of the user interface identified some problems for the top 2 well-known NFT markets. Layer 2 market solutions can address many of the issues raised above, but the first and most important stage in developing a market is selecting the appropriate ERC standard.

1.4 Uniswap AMM markets

Cryptocurrency trading has increased recently, and the decentralized finance (Defi) [18] industry has advanced quickly. More recent attempts to create DEXs have centered around the use of automated market makers (AMMs) as opposed to order books [22]. The market value of decentralized exchanges (DEX) [21] with automated market maker (AMM) protocols has surpassed $100 billion as of this writing [19]. A straightforward but unexpectedly effective market maker for trading fungible tokens is the constant product market maker known as Uniswap [20]. These markets offer a straightforward method for decentralized trading between coin pairs. They are now a well-liked (and practical) substitute for other kinds of DEXs [19]. A market maker frequently serves as a platform that is available to buy or sell an asset, to make a profit from the asked bid spread. An AMM [22–24], which receives trader orders and uses an algorithm to determine the price, automates this process. The users and a pool that holds both input and output assets, both are available to exchange assets in order to execute the AMMs. As a first stage, a liquidity pool receives assets from several liquidity providers. Users are benefited from immediate liquidity, with no need to identify exchange counterparty, on the other hand, LPs profit from assets provided with exchange fees from users [19].

1.5 AMM for non-fungible tokens

The emergence of AMMs with Uniswap and the subsequent Sushiswap vampire assaults has been an unstoppable ascent and one of the biggest use cases for crypto. AMMs have been working great for fungible ERC-20 tokens, but ERC-21 tokens (or NFTs) present a different problem. Platforms like Sudoswap and NFTX have been designed to liquidate NFTs for easy trading. The sudoswap AMM is a straightforward, gas-efficient automated market maker (AMM) protocol that enables NFT-to-token exchanges (and vice versa) based on adaptable bonding curves. NFTX serves illiquid Non-Fungible Tokens (NFTs) by creating liquid markets. These AMMs are being built upon the Ethereum blockchain due to its continued popularity, but they are still in their infancy. More advanced AMMs like those used by Balancer and AMMs on many more good blockchains are yet to catch on with the crypto world.

2 LITERATURE REVIEW

Table 2 represents a survey on Non-Fungible tokens, their Trading, and Market Places. Table 3 represents a survey on Defi and AMMs for Cryptocurrencies and NFTs.

Table 2. A survey on non-fungible tokens, their trading and market places.

S. No	Title	Author / Source	Method/ Findings	Future Work
1	A Review Paper on Non-Fungible Tokens (NFT) [8]	Mrs. Vidya, Jayanth G, Kathik Kulkarni, Kavya K P, Kavya Mahesh Sureban	Blockchain. They proposed a system for electronic transactions without relying on trust	It showed how NFTs may be used to tokenize digital products,
2	Blockchain Beyond Cryptocurrency: Non-Fungible Tokens [10]	Dr. BurcuSakız, Prof. Dr. AyşenHiçGencer	Contends how blockchain and NFTs can go hand in hand.	To boost NFT technology on proof-of-stake blockchains
3	NFT Marketplace Design and Market Intelligence [11–15]	Pavel Kireyev	Discussed mathematically different parameters needed for NFT marketplaces.	Future research can examine other design parameters, such as commission fee structures etc.
4	User-centered Evaluation and Design Suggestions for NFT Markets [16]	S. Viannis Murphy Caxton, K. Naveen, R. Karthik, S. Sathya Bama	Explored new NFT users and suggested correct steps for designing a freely-usable trading platform.	Based on the study, the suggested design patterns to enhance the use of the NFT platforms.
5	Challenges of implementing an NFT Marketplace [17]	Yash Mhatre, Devansh Dixit, RiteshSalunkhe, Dr. Sanjay Sharma	Discussed ERC limitations, high gas fees, and smart contract risks as important challenges faced while implementing an NFT Marketplace.	With this work, it will help to choose a proper approach for implementing and executing an NFT Marketplace.

3 CHALLENGES AND FUTURE WORK

Although AMMs can have significant restrictions, they have been a major force behind Defi and democratizing liquidity availability. To continue taking part in this change and manage their risks more effectively, mainstream consumers will require a new generation of innovation. There are a few problems associated with already existing AMMs like low fund utilization, additional risk exposure, and also the widely discussed issue of impermanent loss. This entire research and our project aimed to provide a decentralised platform where users can create

Table 3. A survey on defi and AMMs for cryptocurrencies and NFTs.

S. No	Title	Author / Source	Method/ Findings	Future Work
1	Bitcoin: A Peer-to-Peer Electronic Cash System [1]	Satoshi Nakamoto	Proposed a trustable architecture for carrying out digital transactions	Laid a foundation for the upcoming great studies on different cryptocurrencies
2	An Overview of Blockchain Technology: Architecture, Consensus, and Future Trends [2]	Zibin Zheng, Shaoan Xie, Hongning Dai, Xiangping Chen, and Huaimin Wang	Presents an overview on blockchain technology and also compares different consensus mechanisms	Future work stopping centralization, analysis of big data and further applications of blockchain
3	Software Engineering Applications Enabled by Blockchain Technology [4]	Selina Demi, Ricardo Colomo-Palacios and Mary SÃ¡nchez-GordÃ³n	Given an overview on the blockchain along with software Engg by carrying out a mapping study	To research blockchain 4.0 impact on software engineering & develop a blockchain-enabled framework for software projects.
4	Blockchain and Cryptocurrencies [5]	Stephen Chan, Jeffrey Chu, Yuanyuan Zhang and Saralees Nadarajah	Based upon Cryptocurrencies, the modern and digital form of transactions	Advantages: Focused on financial and risk analysis of cryptocurrencies.
5	A Blockchain-Based Decentralised Computing And NFT Infrastructure For Game Networks [6]	Koushik Bhargav Muthe, Khushboo Sharma, Karthik Epperla Nagendra Sri	Aims to propose a fully decentralised gaming infrastructure and certain difficulties with current gaming networks	Laid path for creating models by integrating other proof of stake-based blockchains
6	An Analysis of Cryptocurrency, Bitcoin, and the Future. [7]	Peter D. DeVries	Presents a SWOT analysis on how Bitcoin contributes to a shift in economic paradigms.	Further Studies to focus on: how other cryptocurrencies can modernise the digital world.

NFT index funds, or fungible tokens that are backed by NFT collectables, instantly buy/sell/swap their crypto collectables and stake their tokens to earn liquidity providers' rewards. Users contribute their valuable NFTs to the vaults in order to generate liquidity. Thus, those users would be tokenizing their NFTs. Instead of the NFTs just sitting in their wallets, they would receive high yields in return. It would also help mitigate impermanent loss [25,26]. We have chosen Tezos as the building blockchain for our AMM because it is on its path to becoming a promising platform as it offers several technical and powerful innovations in terms of smart contract security, consensus mechanism and self-upgrade procedures.

REFERENCES

[1] Nakamoto, Satoshi. 2008. "*Bitcoin: A Peer-to-Peer Electronic Cash System.*"
[2] *An Overview of Blockchain Technology: Architecture, Consensus, and Future Trends.*" n.d. ResearchGate.
[3] Xu, Min, Xingtong Chen, and Gang Kou. 2019. "A Systematic Review of Blockchain." Financial Innovation 5 (1): 1â€"14.
[4] *Software Engineering Applications Enabled by Blockchain Technology: A Systematic Mapping Study.*" n.d. ResearchGate.
[5] Chan, Stephen, Jeffrey Chu, Yuanyuan Zhang, and Saralees Nadarajah. 2020. "Blockchain and Cryptocurrencies." *Journal of Risk and Financial Management* 13 (10): 227.

[6]   Muthe, Koushik Bhargav, Khushboo Sharma, and Karthik Epperla Nagendra Sri. 2020. "A Blockchain Based Decentralised Computing And NFT Infrastructure For Game Networks." *In 2020 Second International Conference on Blockchain Computing and Applications (BCCA)*, 73–77.

[7]   *"An Analysis of Cryptocurrency, Bitcoin, and the Future."* n.d. ResearchGate.

[8]   *"A Review Paper on Non-fungible Tokens* (Paper2694.pdf." n.d. https://doi.org/10.48175/IJARSCT-26.)

[9]   *"Ethereum Whitepaper."* n.d. Ethereum.org.

[10]   *Blockchain Beyond Cryptocurrencies* (2527.pdf)

[11]   *"Non-Fungible Tokens (NFTs)."* https://quasa.io/media/non-fungible-tokens-nfts.

[12]   Kireyev, Pavel. 2022. "NFT Marketplace Design and Market Intelligence."

[13]   "College-of-liberal-arts-and-sciences-Axeing-the-Axie-Infinity-AI-The-AI-of-Modern-Gaming-Business-Model-Strategem-and-Global-Economy-Towards-ryptocurrency-Era-In-Partial-Fulfillment-of-the-Requirem.pdf." n.d.

[14]   Pelechrinis, Konstantinos, Xin Liu, Prashant Krishnamurthy, and Amy Babay. 2022. "Spot- ting Anomalous Trades in NFT Markets: The Case of NBA Topshot." *arXiv [cs. SI]. arXiv.*

[15]   White, Bryan, Aniket Mahanti, and KalpdrumPassi. 2022. "Characterising the OpenSea NFT Marketplace." In *Companion Proceedings of the Web Conference* 2022, 488–96. WWW '22. New York, NY, USA: Association for Computing Machinery.

[16]   Murphy Caxton, S. Viannis, K. Naveen, R. Karthik, and S. Sathya Bama. 2022. *"User-Centred Evaluation and Design Suggestions for NFT Marketplaces,"* July, 1214–21.

[17]   Mhatre, Yash, Devansh Dixit, RiteshSalunkhe, and Sanjay Sharma. 2008. "Challenges of Implementing an NFT Marketplace." *Impact Factor Value*: 7: 529.

[18]   Chen, Yan, and Cristiano Bellavitis. 2020. "Blockchain Disruption and Decentralised Finance: The Rise of Decentralised Business Models."*Journal of Business Venturing Insights* 13 (June): e00151.

[19]   Xu, Jiahua, Krzysztof Paruch, Simon Cousaert, and Yebo Feng. 2021. "SoK: Decentralised Exchanges (DEX) with Automated Market Maker (AMM) Protocols". *arXiv [q-fin.TR]. arXiv.*

[20]   Adams, Hayden, Noah Zinsmeister, Moody Salem, River Keefer, and Dan Robinson. 2021. "Uniswap v3 Core." *Tech. Rep., Uniswap, Tech. Rep.*

[21]   "DEX: A DApp for the Decentralised Marketplace." n.d. ResearchGate. (Mohan, Vijay. 2022. "Automated Market Makers and Decentralised Exchanges: A Defi Primer." *Financial Innovation* 8 (1): 1–48.

[22]   *"Sudoswap AMM Docs."* n.d. https://docs.sudoswap.xyz/

[23]   *"Introduction to NFTX."* n.d. https://docs.nftx.io/.

[24]   *"Sudoswap NFT AMMs."* n.d. Accessed November 20, 2022. https://mirror.xyz/origins-research.eth/zpkjrhJOcY 67YYQTGNac9yfkX08tUSp0CRdhVIVWNU.

[25]   Masla N., Vyas V., Gautam J., Shaw R. N., and Ghosh A., "Reduction in Gas Cost for Blockchain Enabled Smart Contract," *2021 IEEE 4th International Conference on Computing, Power and Communication Technologies (GUCON)*, 2021, pp. 1–6, doi: 10.1109/GU-CON50781.2021.9573701.

[26]   Malsa, N., Vyas, V., Gautam, J. (2022). Blockchain Platforms and Interpreting the Effects of Bitcoin Pricing on Cryptocurrencies. In: Sharma, T.K., Ahn, C.W., Verma, O.P., Panigrahi, B.K. (eds) *Soft Computing: Theories and Applications. Advances in Intelligent Systems and Computing*, vol 1380. Springer, Singapore. https://doi.org/10.1007/978-981-16-1740-9 13

[27]   Pathak, S., Gupta, V., Malsa, N., Ghosh, A., Shaw, R.N. (2022). Blockchain-Based Academic Certificate Verification System – A Review. In: Shaw, R.N., Das, S., Piuri, V., Bianchini, M. (eds) *Advanced Computing and Intelligent Technologies. Lecture Notes in Electrical Engineering*, vol 914. Springer, Singapore. https://doi.org/10.1007/978-981-19-2980-9 42

[28]   Pathak, S., Gupta, V., Malsa, N., Ghosh, A., Shaw, R.N. (2022). Smart Contract for Academic Certificate Verification Using Ethereum. In: Shaw, R.N., Das, S., Piuri, V., Bianchini, M. (eds) *Advanced Computing and Intelligent Technologies. Lecture Notes in Electrical Engineering*, vol 914. Springer, Singapore. https://doi.org/10.1007/978-981-19-2980-9 29

[29]   Malsa, N., Vyas, V., Gautam, J., Shaw, R.N., Ghosh, A. (2021). Framework and Smart Contract for Blockchain Enabled Certificate Verification System Using Robotics. In: Bianchini, M., Simic, M., Ghosh, A., Shaw, R.N. (eds) *Machine Learning for Robotics Applications. Studies in Computational Intelligence*, vol 960. Springer, Singapore. https://doi.org/10.1007/978-981-16-0598-7 10

[30]   Malsa N., Vyas V., Gautam J., Ghosh A. and Shaw R. N., "CERTbchain: A Step by Step Approach Towards Building A Blockchain-based Distributed Application for Certificate Verification System," *2021 IEEE 6th International Conference on Computing, Communication and Automation (ICCCA)*, 2021, pp. 800–806, doi: 10.1109/ICCCA52192.2021.9666311.

[31]   Malsa, N., Vyas, V. Gautam, J. RMSE Calculation of LSTM Models for Predicting Prices of Different Cryptocurrencies. *Int J Syst Assur Eng Manag* (2021). https://doi.org/10.1007/s13198-021-01431-1

Artificial Intelligence, Blockchain, Computing and Security – Dagur et al. (Eds)
© 2024 The Author(s), ISBN: 978-1-032-49393-0

Blockchain based prophecy of cardiovascular disease using modified XGBoost

Vibha Srivastava, Ashutosh Kumar Singh* & Vijay Kumar Dwivedi
Department of Computer Science and Engineering, United College of Engineering & Research, Naini, Prayagraj, Uttar Pradesh, India

ABSTRACT: One of the global and complex disease known as Heart Disease (HD) plays a significant mantle in cardiology and healthcare. In recent times, the blockchain technology is widely used in medical sector for storing the medical records. Still there is a need for enhancing security range and the studies have failed to focus on both the security and HD prediction. Therefore, the proposed study introduced a secured Electronic Health Record (EHR) system over the blockchain environment by implementing encryption and decryption process using Advanced Encryption Standard (AES) algorithm and the HD prediction model is implemented with a Modified Empirical Loss (MEL) Function Extreme Gradient Boosting (XGBoost) classifier. The experimental evaluation is done by using Cleveland Dataset. The efficiency of preferred exemplary is determined with regard to performance metrics namely accuracy, encryption time, precision, decryption time, recall, and F1-score.

Keywords: Blockchain Technology, Encryption, Decryption, AES Algorithm, Electronic Health Record System, XGBoost Classifier, Heart Disease Prediction.

1 INTRODUCTION

Recently, the technology growth affects the human life and changes the previous way of using technology. The benefits of the technology advancement leads to enhanced security, user experience and various healthcare aspects. These benefits are accessible by Electronic Health Record (EHR) systems (Abunadi 2021). Moreover, they still faces some challenges in terms of medical record security, data user's ownership, and data integrity. The solution for these challenges are the usage of technology called blockchain. Blockchain is a chain of block, which are connected together and grows continuously through storing data over the blocks. It has been utilized as a decentralized method which allows the information to be distributed and that each piece of distributed data (or) generally termed as data results shared ownership.Through Blockchain Support Framework over Electronic Health Record (BSF-EHR), the patients are capable of managing themselves along with the downloads and share the EH reports (Pai 2021). The present study focuses on both the security enchantment and developing a HD prediction model. The main objectives of the proposed study are:

- To introduce an encryption and decryption using AES in blockchain technology in order of enhancing the security range in EHR systems.
- To perform efficient HD prediction, the proposed study introduced a Modified Empirical Loss (MEL) Function XGBoost Classifier in range of enhancing the accuracy range of the prediction phase.
- To analyse the efficiency and potency of proposed replica, comparative analysis is performed in connection with recital measures.

---

*Corresponding Author: ashuit89@gmail.com

DOI: 10.1201/9781003393580-37

Followed by this, section 2 deliberates the existing methods and the problems which are identified from the review of various studies. Section 3 states the research methodology and the various algorithms employed. The result discourses are done in section 4 and paper was ceased in section 5.

2 RELATED WORK

The various existing studies implemented in the security enhancement of medical system based on blockchain technology and the HD prediction model were reviewed in this section.

2.1 *Blockchain technology in medical system*

The efficient protection to the privacy information of users was accomplished with Elliptic Curve Diffie Hellman Key Exchange and the encryption and decryption was performed with AES (Gupta 2019). The keyless signature infrastructure architecture was proposed in (Wu 2021), which supports to secure patients records efficiently. The blockchain framework was designed for the management of EHR, which could results the final control and ownership of EHRs for the patients and securely controls the persons those who could access and track the records and documents which were being used (Nagasubramanian 2020).

2.2 *ML based heart prediction models*

Various ML based models implemented in the HD prediction are viewed in this section. In (Zhang 2019), the expert model labelled as Hybrid Optimized XGBoost (hyOPTXg) that predicts HD by utilizing optimized XGBoost classifier. The paper (Srinivas 2022) contributed a correlative application and distinct ML algorithms was analyzed using python software. Mixed data factor analysis was utilized for extracting and deriving features form Cleveland dataset (Ali 2020).

2.3 *Problem identification*

From the review of various studies, the problem for implementing the proposed model efficiently is stated in this section. The application of blockchain are still in earlier stages (Haleem 2021). Though, the system have focused on both the security and disease prediction, it results high encryption and decryption time and the classification accuracy was also low (Hossen 2021). The system (Fitriyani 2021), need to be implemented with combination of models in order to attain high accuracy towards the early HD prediction.

3 RESEARCH METHODOLOGY

The proposed study focused on enhancing the security range in EHR and performed efficient HD prediction. Figure 1, represents the entire working flow of suggested process. Initially, the Cleveland Dataset is loaded and the EHR secure framework is built. The security is enhanced with the encryption of the input data using AES and then encrypted data are converted into blocks, the blockchain environment is built. For decryption, the verification is done with key function, by entering the key which was generated during encryption, the decryption is accomplished. After, the decryption, the records are also retrieved efficiently. Followed by this, the model also performed efficient HD classification. In the prediction, the model initially performed pre-processing by feature scaling, which normalizes the dimensions of features (or) independent variables. Then, the test and train split is done as 80% for training and 20% for testing. Then, the classification is performed with MEL Function XGBoost Algorithm. Then, the classification is accomplished as presence and absence of the HD. Followed by this, the efficiency and effectiveness of the proposed model is determined by performance yardsticks.

3.1 Dataset description

Present study utilized Cleveland HD Dataset, which was provided by Dr. Robert Detrano, M.D for investigating HD (https://archive.ics.uci.edu/ml/datasets/heart+disease). The Cleveland dataset features contains both the medical background and clinical features. The features are not independent (Ali 2020).

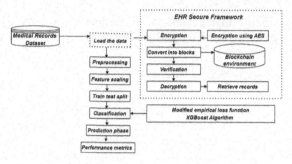

Figure 1.   Overall working flow of the designed model.

3.2 EHR secure framework

At present, the blockchain is widely used in medical field, in that case there is need for securing the data from third party. In that context, the blockchain environment is built with an encrypted data. The security range of EHR is enhanced with the Encryption using AES.

3.2.1 Encryption using AES

The most widely and popularly adopted symmetric encryption algorithm which is likely to be encountered at present is AES. It is found to be faster than triple-DES algorithm. Rather than Feistel Cipher, AES is iterative, which is generally based on substitution to permutation network. Figure 2, illustrates schematic formation of AES algorithm.

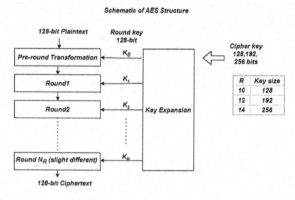

Figure 2.   Schematic structure of AES algorithm.

The Algorithm 3 employed for encryption is depicted in encryption using AES algorithm.

3.2.2 Blockchain environment

The Algorithm 2 is utilized for building block is denoted in the Blockchain Environment Algorithm.

3.2.3 Decryption

The decryption process is done for viewing the encrypted patient data. By entering the encrypt password, the patient's data can be decrypted from blockchain environment. Followed by this, the patient's records are also retrieved. Figure 4 illustrates the process of AES encryption and decryption. Algorithm 5 represents the process involved in decryption. The system also performed the heart prediction using MEL Function XGBoost Algorithm.

3.3 *Classification*

The system performed efficient heart prediction using MEL Function XGBoost Algorithm.

Algorithm 1 Encryption Using AES

1: **Input:** Patients data from dataset
2: Initialize
3: Begin function AES is given as:
4: $\left(in = byte[16], out = byte[16], key_{array} round_{key[N_r+1]}\right)$
5: byte state [16]
6: state =1
7: Add_Round_Key $\left(state, round_{key[0]}\right)$
8: for I =1 to N_r-1 do
9: Sub_Bytes (*state*)
10: Shift_Rows(*state*)
11: Mix_Column(*state*)
12: end for
13: Sub_Bytes(*state*)
14: Shift_Rows(*state*)
15: Add_Round_Key $\left(state, round_{key[N_r]}\right)$
16: out =state
17: return out
18: **Output:** EHR secure framework

Algorithm 2 Blockchain Environment

1: Chipertext—→AES Encrypt (*password, patientdata*$_1$, ...*patientdata*$_n$. . .)
2: AES Encrypt (*password, patientdata*) -—→converted into block
3: Blockchain verification takes place.

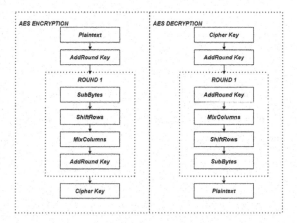

Figure 3.   Process of AES encryption and decryption.

Algorithm 3 Decryption

1: **Input:** key
2: key—→Encrypt_password
3: for as much as tolerance give the secured password
4: if key = get_ secured password
5: key→Encrypt_password
6: plaintext→AES decrypt $(Chipertext, key)$
7: end if
8: end for
9: **Output:** plain text

3.3.1 *MEL function XGBoost algorithm*

The empirical loss function calculated the quality of approximation, thus the proposed study modified the empirical loss function in XGBoost Algorithm. Let base $_n(x)$ be the function that often list as a Base Learner (*BL*) shown in Equation (1). The additive system was the quantity of the BL.

$$f(p) = \sum_{n=1}^{M} base_n(p) \tag{1}$$

The term n = 1,2,3,.....M, M:members of BL and L:risk minimization. Equation (2) is calculated as: $L=((f(p,q))$ for the BL base $_n(p)$

$$\widehat{b}(p) = argmin_b \sum_D L(f_{n-1}(p) + b(p), q)$$
$$argmin_{base} \sum_D \left[base_n(p)g(p,q) + \frac{1}{2} base^2(p)h(p,q) \right] \tag{2}$$

Where medical data MD = $\{(p,q)\}$ is a dataset.

$g(p,q) = \frac{\partial L(f_{n-1}(p),q)}{\partial f} \, h(p,q) = \frac{\partial L(f_{n-1}(p),q)}{\partial f^2}$

The additive approach of Equation(1) is updated iteratively with the enhancing as in Equation (3):

$$f_n(p) = f_{n-1}(p) + \widehat{b}(p) \tag{3}$$

In the proposed study, tree-boosting XGBoost Algorithm is adopted as Equation (4).

$$f(p) = \sum_{j=1}^{T} w_j I \left[p\varepsilon R_j \right] \tag{4}$$

Where, w_j is the constant fit in region R $_j$, I is the set of indices of input p, then the jth leaf for j is equal to 1,2,3.....,T and mel* is the optimized leaf weight. Finding the enhanced leaf weights is equivalent to leaf weights learning. It seeks the split, which maximizes a gain, which is the MEL split reduction. Therefore, the (Equation (4)) is substituted in (Equation (2)) and resulted as in Equation (5).

$$mel^* = argmin_w \sum_D \sum_{j=1}^{T} \left[g(p,q)w_j + \frac{1}{2}h(p,q)w_j^2 \right]$$
$$argmin_w \sum_{j=1}^{T} \left[Gw_n + \frac{1}{2}Hw_j^2 \right] \tag{5}$$

Where,

$G = \sum_D g(p, q)$ and

$H = \sum_D h(p, q)$

It has been recommended that the result of optimized leaf weights from Equation (6) is equal to the leaf weight learning. The split that maximizes a gain has the split loss reduction. The fixed structure gain is derived by substituting Equation (6) in Equation (5).

$$mel^* = -\frac{G}{H} \tag{6}$$

The binary splits were determined through maximizing gain according to Equation (7).

$$-\frac{1}{2} \sum_{n=1}^{M} \frac{G^2}{H} \tag{7}$$

Equation (8) is given as:

$$A = \frac{1}{2} \left[\frac{G_L^2}{H_L} + \frac{G_R^2}{H_R} - \frac{G^2}{H} \right] \tag{8}$$

Where, R and L represents right and left of the tree branches. The illustration of XGBoost Algorithm is given in the following Algorithm 4.

4 RESULTS AND DISCUSSION

The efficiency and effectiveness of framework is assessed through Cleveland dataset, which contains the medical record of the patients. The robustness of prediction aspect of proposed design is determined with respect to the implementation measures expressed in Equation (9), Equation (10), Equation (11) and Equation (12). The recommended EHR chassis is estimated with reference to encryption and decryption time.

Algorithm 4 XGBoost

1: **Input:** dataset and hyperparameter
2: initialize f_0 (x); for n = 1,2,3, ... ,M do
3: calculate $g_n = \frac{\partial L(q,f)}{\partial f}$
4: calculate $h_n = \frac{\partial^2 L(q,f)}{\partial f^2}$
5: determine the structure by choosing the splits
6: $A = \frac{1}{2} \left[\frac{G_L^2}{H_L} + \frac{G_R^2}{H_R} - \frac{G^2}{H} \right]$
7: determine the leaf weights mel$^* = -\frac{G}{H}$
8: determine the base learner $\hat{b}(p) = \sum_{j=1}^{T} wI$
9: add trees $f_n(x) = f_{n-1}(p) + \hat{b}(p)$
10: end
11: **Output:** $f(p) = \sum_{n=0}^{M} f_n(p)$

4.1 Performance metrics

1.
$$Accuracy = \frac{(TN + TP)}{(TN + FP + TP + FN)} \tag{9}$$

255

2.
$$Precision = \frac{(TP)}{(TP + FP)} \tag{10}$$

3.
$$Recall = \frac{(TP)}{(TP + FN)} \tag{11}$$

4.
$$F1 - score = \frac{2*precision * recall}{precision + recall} \tag{12}$$

In the above Equations, TN was labelled as true negative, FP was labelled a false positive, TP was labelled as true positive and FN was labelled a false negative.

4.2 *Performance evaluation*

By employing AES algorithm in encryption phase, the system results in the increment of encryption time along with the increase of file size. This results was represented in Figure 4. Similarly, Figure 5 represents the performance results of the decryption phase. The system results in the increment of decryption time along with the increase of file size. Followed by this, the performance of the introduced HD prediction model using MEL Function XGBoost classifier result was illustrated in Figure 6.

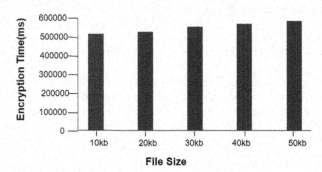

Figure 4. Performance analysis of the EHR framework – encryption phase.

4.3 *Comparative analysis*

In range of determining the efficiency and effectiveness of the proposed model, the comparative analysis was performed. The proposed EHR model efficiency was determined with the comparison of encryption and decryption time with existing EHR model (Kalyani 2020) shown in Table 1.

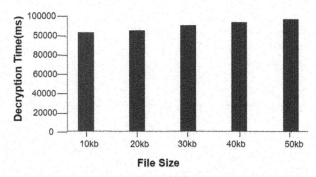

Figure 5. Performance analysis of the EHR framework - decryption phase.

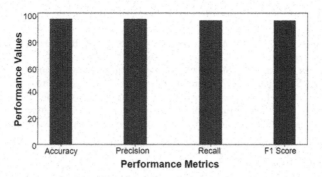

Figure 6.   Performance analysis of prediction phase

Table 1.   Comparative analysis of the proposed model with existing model (Kalyani 2020).

File Size	Encryption time		Decryption time	
	OHE	Proposed	OHE	Proposed
10kb	562,712	525,416	98,412	84,562
20kb	586,235	535,486	99,632	90,254
30kb	598,746	564,125	101,236	95,654
40kb	601,236	579,856	113,253	99,412
50kb	611,421	596,587	125,893	102,548

Table 2.   Comparative analysis of the proposed study with existing study (Hossen 2021).

Dataset	Model	Precision (%)	Recall (%)	Accuracy (%)	F1 score (%)
Cleveland Dataset	Logistic-regression(LR)	–	–	75	–
	Decisiontree-1	–	–	82.5	–
	RandomForest-1	–	–	80.3	–
	RandomForest-2	–	–	87.6	–
	RandomForest-3	77	87	80	82
	Decisiontree-2	71	74	72	72
	Existing Model [Hossen 2021]	92	92	92	92
	Proposed model	98	97	98	97

The comparative analysis of suggested HD prediction model is expressed in Table 2. In all considered terms of performance metrics the proposed model showed optimized result than the various existing methods.

5   CONCLUSION

The present study implemented a secured EHR model and HD prediction using Cleveland Dataset. The security range in blockchain was enhanced by introducing Encryption and Decryption process using AES algorithm. The performance evaluation and comparative analysis determined that the efficiency and effectiveness of the proposed model. The proposed HD Prediction model results 98% accuracy, 98% precision, 97% recall and 97% F1-score which is better than the various existing models. Likewise, encryption and decryption time was also lower than the other conventional methods. Thus, it determines the robustness of the proposed model.

REFERENCES

Abunadi, Ibrahim and Kumar, Ramasamy Lakshmana 2021. BSF-EHR: Blockchain Security Framework for Electronic Health Records of Patients. *Sensors (8)*, volume (21): pages 2865.

Ali, Farman, El-Sappagh, Shaker, Islam, SM Riazul, Kwak, Daehan, Ali, Amjad, Imran, Muhammad and Kwak, Kyung-Sup 2020. A Smart Healthcare Monitoring System for Heart Disease Prediction Based on Ensemble Deep Learning and Feature Fusion. *Information Fusion,* volume (63): pages 208–222.

Fitriyani, Norma Latif, Syafrudin, Muhammad, Alfian, Ganjar and Rhee, Jongtae 2020. HDPM: An Effective Heart Disease Prediction Model for a Clinical Decision Support System. *IEEE Access*, volume (8): pages 133034–133050.

Gupta, Ankur, Kumar, Rahul, Arora, Singh Harkirat and Raman, Balasubramanian 2019. MIFH: A Machine Intelligence Framework for Heart Disease Diagnosis. *IEEE Access*, volume (8): pages 14659–14674.

Haleem, Abid, Javaid, Mohd, Singh, Pratap Ravi, Suman, Rajiv and Rab, Shanay 2021. Blockchain Technology Applications in Healthcare: An Overview. *International Journal of Intelligent Networks*, volume (2): pages 130–139.

Hossen, MD, Tazin, Tahia, Khan, Sumiaya, Alam, Evan, Sojib, Hossain Ahmed, Monirujjaman Khan, Mohammad and Alsufyani, Abdulmajeed 2021. Supervised Machine Learning-based Cardiovascular Disease Analysis and Prediction. *Mathematical Problems in Engineering*, volume (2021).

https://archive.ics.uci.edu/ml/datasets/heart+disease.

Kalyani, G, Chaudhari, Shilpa 2020. An Efficient Approach for Enhancing Security in Internet of Things Using the Optimum Authentication Key. *International Journal of Computers and Applications (3)* volume (42): pages 306–314.

Nagasubramanian, Gayathri , Sakthivel, Kumar Rakesh, Patan, Rizwan, Gandomi,H Amir and Sankayya, Muthuramalingam & Balusamy, Balamurugan 2020. Securing E-health Records Using Keyless Signature Infrastructure Blockchain Technology in the Cloud. *Neural Computing and Applications (3)* volume number (32): pages 639–647.

Pai, Manohara MM, Ganiga, Raghavendra, Pai, Radhika M and Sinha, Kumar Rajesh 2021. Standard Electronic Health Record (EHR) Framework for Indian Healthcare System. *Health Services and Outcomes Research Methodology (3)* volume (21): pages 339–362.

Srinivas, Polipireddy and Katarya, Rahul 2022. hyOPTXg: OPTUNA Hyper-parameter Optimization Framework for Predicting Cardiovascular Disease Using XGBoost. *Biomedical Signal Processing and Control*, volume (73): pages 103456.

Wu, Hongjiao, Dwivedi,Dhar Ashutosh and Srivastava, Gautam 2021. Security and Privacy of Patient Information in Medical Systems Based on Blockchain Technology. *ACM Transactions on Multimedia Computing, Communications, and Applications (TOMM) (2s)* volume (17): pages 1–17.

Zhang, Fenghua, Chen, Yaming, Meng, Weiming and Wu, Qingtao 2019. Hybrid Encryption Algorithms for Medical Data Storage Security in Cloud Database. *International Journal of Database Management Systems (IJDMS)* Vol, volume (11).

A survey on crowdfunding using blockchain

Nikunj Garg, Siddharth Seth, Naincy Rastogi, Rajiv Kumar, Vimal Gupta,
Sur Singh Rawat & Nitima Malsa
JSS Academy of Technical Education, Noida, Uttar Pradesh, India

ABSTRACT: Crowdfunding is a means of raising money for a project by seeking small donations from a large number of individuals, typically over the Internet. The trust element is the main problem with crowdfunding nowadays. Due to the rising number of frauds that are taking place these days, people are hesitant to donate. But donations may increase dramatically if it could be guaranteed that their money would be used for good. Here comes the need to use blockchain in crowdfunding. Blockchain is a system where transactions are recorded across multiple computers connected by a peer-to-peer network. This paper addresses the drawbacks of the current crowdfunding platforms in use and suggests the steps to be taken to improve the existing situation.

Keywords: Crowdfunding, Blockchain, Health care

1 INTRODUCTION

The gap between those in need of assistance and those with privilege and willingness to give has been narrowed in recent years, thanks to crowdfunding. It has changed traditional fundraising since it allows online transactions which are comfortable and simple. Crowdfunding for medical expenses has grown in popularity, particularly in nations without universal health coverage. According to 2018 research by the Public Health Foundation of India, medical costs have caused 55 million Indians to fall into poverty. The largest crowdfunding sites in India raised almost 272 crores for healthcare that year, and it is expanding quickly. Because of the increase in scams in crowdfunding, one of its key limitations is a lack of donor confidence. Here comes the need for blockchain integration in crowdfunding. A platform called Crowdfunding is intended for startup fundraising. Peer-to-peer, rewards, donations, and equity are the most common among them, however, they can differ per type. Peer-to-peer lending is sometimes known as debt crowdfunding. Donation-based crowdfunding platforms let investors contribute money to a firm without expecting anything in return. The equity finance model is an asset-based one. Figure 1. shows the possible use cases of blockchain for crowdfunding.

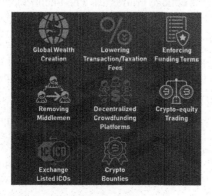

Figure 1. Use cases of blockchain for crowdfunding [1].

DOI: 10.1201/9781003393580-38

2 LITERATURE REVIEW

Susana Bernardino and Freitas Santos [1] conducted an exploratory study on the knowledge, benefits, and barriers perceived by young potential entrepreneurs. The results show that young entrepreneurs' inadequate knowledge of crowdfunding prevented them from researching a wide range of potential business models, notably the one involving investment. Additionally, they think it facilitates client feedback and communication with a bigger audience.

The study by Cynthia Weiyi Cai [2] identifies gaps in the economics and finance literature on two FinTech applications: blockchain and crowdfunding. It is based on an analysis of 402 papers published between 2010 and 2018 that included a systematic review of influential works. The examination of them shows that the trust aspect of blockchain reduces the need for middlemen, though not in all financial fields.

This work by Vidya [3] provides a practical implementation of launching a crowdfunding application with zero-knowledge proof for user identity and user data encryption secured by a lattice-based cryptosystem. Machine learning has also been used to estimate campaign success for the benefit of money givers. The person who proposes the idea or initiates the task of financing a venture, the people or investors who invest in the concept, and a system that connects the two characters to aid in the venture's success are the three categories of on-screen characters that comprise the modern crowdfunding notion, according to this study.

The focus of the article by Abhrajit Sarkar [4] is on the steps taken by SEBI to protect investor funds through the adoption of several laws and regulations, such as the prohibition on crowdsourcing for equity raising to HNIs (High Net worth Individuals). It also lists websites like Start 51 and Ignite Intent that help creative firms raise capital. The rest is history. Reliance is now the largest company in India. He raised money for his textile business. Even movies have raised money through fundraising efforts, and these enterprises have experienced exceptional success. The primary danger is that ideas can be copied and even stolen, which would impede the growth of the company.

Ethan Mollick's [5] paper focuses on how crowd financing platforms have fundamentally changed and altered the way money is raised. For instance, the largest website, Kickstarter, helped raise 227 million dollars and has over 48 000 projects, however, the majority of large projects were delayed and most of them had to reduce their profit margins to be successful. Pebble "smartwatches," Ouya "games," and other products that couldn't get backing from venture capitalists were able to raise money through Kickstarter. According to the paper, there are four types of crowdfunding that might be used: a) donation-based b) reward-based c) peer-to-peer lending d) equity-based.

Logit and Cox regression models were employed in a study by Felix Reichenbach and Martin Walther [6] to evaluate a variety of hypotheses. The results indicate that there is no proof to support a particular influence of an economics degree on companies with CEOs who self-identify themselves as college graduates. The likelihood of failure is correlated with the venture's worth and the number of early VC investors. Family enterprises have a lower chance of failing. Investors shouldn't utilise the frequency of updates on campaign development, external certification, promotions, or start-ups since they are linked to a high likelihood of failure. They found that enterprises with the legal form of "entrepreneurial company" or those that have taken part in numerous crowdfunding campaigns have a higher failure rate.

According to a study by Raveena V.1 and Sunayana N.2 [7], crowdfunding is growing rapidly during COVID-19. After March 2020, crowdfunding for COVID-19 greatly increased. People started aiding and giving to one another during this time. As internet platforms grew more widely, everything started functioning through them, including educational transactions, communication, and other activities. Through the use of social media and online platforms like LinkedIn, Facebook, and others, crowdfunding platforms have raised billions of dollars over the past few years in the forms of lending, equity, prizes, and donations. The Indian economy is not considered to be particularly welcoming to investors.

According to this paper on crowdfunding utilising blockchain technology by Ms. S. Benila and three other individuals [8], interactive forms for campaign creation and donation are provided, allowing both campaign creators and donors or investors to easily establish and support the campaigns. We can fix the problem that exists on the present crowdsourcing websites with the proposed system. Some of the proposed system's features are listed below, including Trust, Money Control, No Fees, Donor Guarantee Policy, All Transactions are Recorded, and Secure Money Storage.

In the Michael Gebert research [9], particular attention is given to issues with crowdsourcing in the European Union (EU). For example, money laundering, information asymmetry, and fraud motivate legislative restrictions on fundraising activities, as well as the Brexit issue, which generated conflicting rumours (Berend 2017). All these issues have an impact on crowdfunding, so using blockchain technology to address them will help prevent incidents like these. However, there is potential for a global resurrection of crowdfunding, thanks to blockchain technology.

This article by Dr. Aishath Muneeza and Mustapha [10] claims that the incorporation of blockchain in crowdfunding will dramatically lower transaction costs and provide the system certainty and trust. The several types of crowdfunding models, such as contribution crowdfunding, reward crowdsourcing, lending crowdfunding, and equity crowdfunding, are also covered. This study also discusses the difficulties with crowdfunding and the potential of blockchain to address them. There are two approaches to connect blockchain with crowdfunding – using cryptocurrencies or not. But in order to use this, strong cyber laws must be put into effect.

The research by Cephas Coffie and Zhao Hongjiang [11] identifies the link between fundraising platforms and investors. The biggest issue faced by new businesses worldwide is raising the necessary capital. although there are other sources. According to Hu (2014), when crowdfunding platforms engage in fraudulent transactions, invest of investors might be destroyed. Therefore, trust is an issue for humans. The introduction of blockchain technology addresses the issue of investor protection and security in crowdfunding.

This paper by Zach Zhizhong and Huasheng Zhu [12] addresses equity crowdsourcing. A specific type of crowdfunding called Equity Crowdfunding is used to raise capital for start-ups. It offers a low entrance barrier, cheap cost, and fast speed, increasing and encouraging innovation. This research paper analyses the issues with equitable crowdsourcing in China. Based on a review of blockchain technology, equity crowdsourcing can be a secure, effective, and affordable option. It also makes transactions simpler, allows peer-to-peer transactions, and aids in understanding market conditions. But using blockchain consumes a lot of power which needs to be taken care of.

The study by AndrÃ© Amedomar and Renata Giovinazzo Spers [13] outlines the main reasons why Brazilian technology-based companies (TBCs) favour reward-based funding over other available options. Businesses based on science and technology are referred to be TBCs when they use cutting-edge and creative techniques to acquire goods and services. They looked at five TBCs, three of which were successful and all of which operated in various markets. The key arguments for the project's development included testing communication techniques, gathering product feedback, and assessing market demand. They discover that one of the factors contributing to this is the small size of these TBCs that generate revenue, in addition to the fact that they don't adequately explain how they Intend to use the funds.

Firmansyah Ashari's work [14] discusses how covid-19 has affected numerous nations, requiring significant financial support from the government to combat it. Many organisations raised money from businessmen and other investors during the COVID-19 era to assist the government in giving aid to those in need. Therefore, obtaining funding from investors depends greatly on the trust. We can gain the trust of donors by utilising the blockchain technology and smart contract concept. A smart contract is an agreement that is stored on the blockchain and is automatically carried out when a certain condition is satisfied.

Blockchain has different applications such as cryptocurrency price predicting [15,16]. Other blockchain applications are certification verification, healthcare management etc. [17–21].

The authors of the work, Ankita A. Malve*et al.* [22], explore blockchain, a newly emerging technology that could bridge the gap between donors and fundraisers. It also emphasizes how streamlining transactions by cutting out the intermediary can increase their effectiveness. They created an experiment to show how transactions work on the Ethereum virtual machine and to add each and every dollar contributed.

The authors of the research, Felix Hartmann *et al.* [23], give details to the success criteria of blockchain fundraising as compared to conventional techniques. The main factors were secure transactions, decentralization, and transparency. Despite a rise in the number of campaigns to raise awareness about blockchain, the majority of people still don't know much about it. This review article clarifies the potential for future growth in this industry. This paper by Alex Bockel and Jacob Horisch [24] discusses crowdfunding and sustainability. The primary goal of sustainable development crowdfunding is to address this issue by using cutting-edge goods, services, and procedures that boost revenue or help the local economy and environment. We can guarantee that crowdfunding is a better option than making a contribution to sustainable development because of the increased related risk that businesses face when accepting donations and engaging in other activities.

3 CHALLENGES WITH CROWDFUNDING

There is a great deal of uncertainty around the accounting standards for money raised through crowdfunding.The campaign may be launched on two distinct platforms by the same individual or fundraiser multiple times, often with the same name and goals.From time to time, terrorist organizations will attempt to gather money by posing as an organization or non-governmental organization (NGO) that aids starving citizens of Yemen, Lebanon, and other resource-poor nations.On the fund-raising platform Milaap, fraud complaints are 0.05%, however, on another platform called Ketto, roughly 23% of campaigns are rejected when they are deemed suspicious.The most important aspect of a project is the idea. Everything in the company depends on it. They must therefore be protected.

4 CONCLUSION AND FUTURE SCOPE

Artificial intelligence and machine learning models could be applied to these platforms because the majority of fake applications must share some fundamental characteristics that can be recognized and mentioned appropriately, allowing volunteers to review the applications before moving forward with the campaign if they become suspicious. Ratings based on various factors and components such as risk could be used in this to filter the campaign. With the widespread adoption of blockchain technology in the future and its expanding application today, there is a lot of space to lower the cost of raising funds. Because of this, the idea of a minimum donation might be dropped, enabling anyone to give whatever they want.A heavy reliance on volunteers. Employing people to carry out checks could reduce the need for volunteers, but doing so would undoubtedly increase the cost of fundraising.The launched campaign's dependability and integrity would always be in question.People that fundraisers themselves should be more attentive and careful about to whom they are donating and some research is required.Friends and family of the fundraiser typically make contributions in a genuine campaign, thus this can be a way to recognize the campaign.

REFERENCES

[1]   Bernardino, S. and Santos, J.F., 2020. Crowdfunding: An Exploratory Study on Knowledge, Benefits and Barriers Perceived by Young Potential Entrepreneurs. *Journal of Risk and Financial Management*, 13(4), p.81.

[2]   Cai, C.W., 2018. Disruption of Financial Intermediation by FinTech: A Review on Crowdfunding and Blockchain. *Accounting Finance*, 58(4), pp.965–992.

[3]   Vidya, K., Hussain, H.I., Celestine, V., Kumar, V. and Robert, V.N.J., 2022. *Security Enhanced Crowdfunding Using Blockchain and Lattice Based Cryptosystem.*

[4]   Sarkar, A., 2016. *Crowd Funding in India: Issues Challenges.* Available at SSRN 2739008.

[5]   Mollick, E., 2014. The Dynamics ofCrowdfunding: An Exploratory Study. *Journal of Business Venturing*, 29 (1), pp.1–16

[6]   Reichenbach, F. and Walther, M., 2021. Signals in Equity-based Crowdfunding and Risk of Failure. *Financial Innovation*, 7(1), pp.1–30.

[7]   Raveena, V. and Sunayana, N., 2022. Challenges of Crowdfunding During Covid-19 Period. *Journal of Positive School Psychology*, pp.8265–8273.

[8]   Benila, M.S., Ajay, V., Hrishikesh, K. and Karthick, R., 2019. Crowd Funding Using Blockchain. *GRD Journals.*

[9]   Gebert, M., 2017. Application of Blockchain Technology in Crowdfunding. *New European*, 18.

[10]   Muneeza, A. and Mustapha, Z., 2020. Application of Blockchain Technology in Crowdfunding to Fuel the Rise of the Rest Globally. *A Journal of Interest Free Microfinance*, 7(1), pp.9–26.

[11]   Zhao, H. and Coffie, C.P 2018. *The Applications of Blockchain Technology in Crowdfunding Contract.* Available at SSRN, 3133176.

[12]   Zhu, H. and Zhou, Z.Z., 2016. Analysis and Outlook of Applications of Blockchain Technology to Equity Crowdfunding in China. *Financial innovation*, 2(1), pp.1–11.

[13]   Amedomar, A. and Spers, R.G., 2018. Reward-based Crowdfunding: a Study of the Entrepreneurs' Motivations When Choosing the Model as a Venture Capital Alternative in Brazil. *International Journal of Innovation*, 6(2), pp.147–163.

[14]   Ashari, F., Catonsukmoro, T., Bad, W.M. and Sfenranto, W., 2020. Smart Contract and Blockchain for Crowdfunding Platform. *International Journal of Advanced Trends in Computer Science and Engineering*, pp.3036–3041.

[15]   Malsa, N., Vyas, V. Gautam, J. RMSE Calculation of LSTM Models for Predicting Prices of Different Cryptocurrencies. *Int J Syst Assur Eng Manag* (2021).

[16]   Malsa, N., Vyas, V., Gautam, J. (2022). Blockchain Platforms and Interpreting the Effects of Bitcoin Pricing on Cryptocurrencies. In: Sharma, T.K., Ahn, C.W., Verma, O.P., Panigrahi, B.K. (eds) Soft Computing: Theories and Applications. Advances in Intelligent Systems and Computing, vol 1380. Springer, Singapore.

[17]   Pathak, S., Gupta, V., Malsa, N., Ghosh, A., Shaw, R.N. (2022). Blockchain-Based Academic Certificate Verification System – A Review. In: Shaw, R.N., Das, S., Piuri, V., Bianchini, M. (eds) *Advanced Computing and Intelligent Technologies. Lecture Notes in Electrical Engineering*, vol 914. Springer, Singapore.

[18]   Pathak, S., Gupta, V., Malsa, N., Ghosh, A., Shaw, R.N. (2022). Smart Contract for Academic Certificate Verification Using Ethereum. In: Shaw, R.N., Das, S., Piuri, V., Bianchini, M. (eds) *Advanced Computing and Intelligent Technologies. Lecture Notes in Electrical Engineering*, vol 914. Springer, Singapore.

[19]   Malsa, N., Vyas, V., Gautam, J., Shaw, R.N., Ghosh, A. (2021). Framework and Smart Contract for Blockchain Enabled Certificate Verification System Using Robotics. In: Bianchini, M., Simic, M., Ghosh, A., Shaw, R.N. (eds) *Machine Learning for Robotics Applications. Studies in Computational Intelligence*, vol 960. Springer, Singapore.

[20]   Malsa N., Vyas V., Gautam J., Ghosh A. and Shaw R.N., "CERTbchain: A Step by Step Approach Towards Building A Blockchain Based Distributed Appliaction for Certificate Verification System," 2021 *IEEE 6th International Conference on Computing, Communication and Automation (ICCCA)*, 2021, pp. 800–806.

[21]   Masla N., Vyas J., Gautam J., Shaw R.N. and Ghosh A., "Reduction in Gas Cost for Blockchain Enabled Smart Contract," 2021 *IEEE 4th International Conference on Computing, Power and Communication Technologies (GUCON)*, 2021, pp. 1–6.

[22]   Ankita A. Malve,Shweta M. Barhate, Satish J. Sharma "Trusted Crowdfunding Using Smart Contract", *International Journal of Emerging Technologies and Innovative Research* ISSN:2349-5162, Vol.9, Issue 6, page no.371–375, June-2022

[23]   Hartmann, F., Grottolo, G., Wang, X. and Lunesu, M.I., 2019, February. Alternative Fundraising: Success Factors for Blockchain-based vs. Conventional Crowdfunding. In 2019 *IEEE International Workshop on Blockchain Oriented Software Engineering (IWBOSE)* (pp. 38–43)

[24]   BÃ¶ckel, A., HÃ¶risch, J. and Tenner, I., 2021. A Systematic Literature Review of Crowdfunding and Sustainability: Highlighting What Really Matters. *Management Review Quarterly*, 71(2), pp.433–453.

Artificial Intelligence, Blockchain, Computing and Security – Dagur et al. (Eds)
© 2024 The Author(s), ISBN: 978-1-032-49393-0

Data provenance for medical drug supply chain using blockchain-based framework

Martin Parmar & Parth Shah

Chandubhai S Patel Institute of Technology (CSPIT), Faculty of Technology & Engineering (FTE), Charotar University of Science and Technology (CHARUSAT), Anand, Gujarat, India

ABSTRACT: A medical drugs supply chain is a sizeable process that involves stakeholders such as drug manufacturer, wholesalers, pharmacy, patient, doctors, and other intermediaries who run the supply chain network. It is expected to have a standard service from stakeholders in terms of quality ingredients used while manufacturing of drug product in sensible rate to end consumers. In the current situation, specifically COVID-19, many incidents are reported such as drug counterfeiting and black marketing of drugs by lacking quality that has a serious impact on people's health. Blockchain technology is being used among many industries as a part of the fourth industrial revolution. Blockchain decentralized distributed network ensures transparency, data integrity, information traceability and availability among the stakeholders of the supply chain network. In this paper, we focus on the issues of traditional drug supply chain of healthcare and provide Blockchain-based architecture for better product visibility and data origin.

1 INTRODUCTION

The medical sector is one of the most crucial sectors in the context of both human life and its economical aspect. Since its efficient functioning is crucial for the lives of people to be safe and disease free, enough emphasis should be given to its strengths and vulnerabilities. Also, with the current scenario of frequent disease outbreaks that when analyzed with the multiple prevalent health risks make it even more important to ensure that the drugs and medicines being received by the consumers are effective (Beaulieu & Bentahar 2021). But one of the major issues in the currently used supply chain is the involvement of a huge number of entities that make it difficult to ensure the legitimacy of the drug being delivered.

The Figure 1 presents the basic flow of medical drugs supply to different stakeholders. The drug supply chain starts with the manufacturer producing the drugs, then hands it over to the wholesaler who then distributes smaller chunks to the pharmacists who finally sell it to the customers (Panda & Satapathy 2021). Apart from this, there is also a constant involvement of the logistics that pick-up and deliver the drugs between the above-mentioned entities. It also gives malicious entities more opportunities to introduce fake drugs into the system. This problem is dangerous not just from the economical aspect but also from the health and safety aspect since consuming drugs that are to up to the drug standards may have some unwanted side effects on the consumers and can also sometimes be fatal. Apart from this, the economical

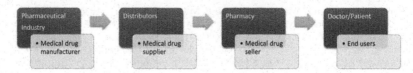

Figure 1.   Medical drugs supply chain.

DOI: 10.1201/9781003393580-39

aspect can also not be overlooked since these issues cause a significant loss to the pharmaceutical industry. These issues also lead to a clear lack of trust in the system by all involved parties.

The currently operational supply chain requires certain modifications to accommodate the requirements of the current digital and complex problems. All the involved parties must be required to update and share information with other parties to incorporate the features of transparency and traceability, thus building trust in the system (Kittipanya-ngam & Tan 2020).

There are following some major issues with present Medical Drug System:

1.1 *Visibility*

Since the pharma drug supply chain has several entities involved in the process, the chances for malicious activities like drug counterfeiting and drug contamination increase. Thus, such a system requires higher transparency to maintain trust between the involved entities, which is not provided by the currently prevailing system (Ageron *et al.* 2020). Retailers, regulators, and the end-users don't know where the drug has come from, what are the ingredients and what is the concentration of every ingredient.

1.2 *Cold-chain shipping*

A significant number of drugs and medicines are often required to be kept and transported in a temperature and humidity-controlled environment for their impact to be the intended. It is a crucial aspect in maintaining the drug efficiency in treating a patient. The information regarding these aspects and conditions is stored in centralized databases that are often the victims of attacks and manipulation, thus casting a doubt on the authenticity of the delivered drug.

1.3 *Drug counterfeiting (Musamih* et al. *2021)*

The probability and opportunity for hostile organizations to inject fake or counterfeit medications into the supply chain increases because the drug pharma supply chain involves so many parties to accomplish its ultimate purpose. The risk to the end user is increased as a result. Governments across the world are increasing their efforts to stop this unlawful trade as a result. For pharmaceutical companies, who struggle to control product quality and security, these advancements are fantastic news.

2 BLOCKCHAIN FOR MEDICAL DRUG SUPPLY CHAIN

Blockchain at the basic level is a database that is distributed across the participants in the networks called nodes. The building blocks of a Blockchain are blocks that comprise transactions that are verified and established by the entities involved in the network using a consensus algorithm. Thus, Blockchain is essentially distributed and decentralized digital ledgers that are tamper evident and tamper resistant. They operate without a central authority. A recorded transaction is cryptographically signed to enforce the authentication of the entity. Under their normal mode of operation, a transaction cannot be changed once recorded in a block. The blocks are also cryptographically linked to the previous block, making the act of tampering with it even more difficult. Moreover, any conflicts in the network are resolved using already established automatically. Without the presence of a central authority governing the actions of the entities, the required trust level in the network is achieved because of the following entities and properties in the Blockchain.

The following are the essential key concepts behind Blockchain Technology-like private key cryptography and a distributed network that has a shared ledger.

2.1 *Private keys*

To elucidate the concept and technology behind personal cryptographic keys, we can imagine two people who want to coordinate a transaction online. Everyone has two keys: a public key and a private key. By using and combining the aspects of both these keys, we can generate a unique and secure identity for an individual in the network. This secure and unique identity plays a vital role in the entire functioning of the Blockchain. These keys, together, can be used to create a digital signature, which will be a useful aid in controlling and certifying the ownership.

2.2 Distributed networks

Blockchain technology can be thought of as a network of entities that can validate the transactions occurring in the system to reach a consensus on which transactions to record in the block and which are fraudulent. In this way, the digital signature element of the Blockchain network operating through private and public keys is augmented with the distributed network component. Combining the consensus concept with the mathematical verification and stamping, we can achieve the security of data in the network. This adds the property of transparency and traceability of the recorded transactions, which may be required for future references to resolve conflicts.

2.3 Process of confirmation

In the above-mentioned example, when two entities in the network wish to conduct a transaction online, both need to have their public and private keys for the process to move further. Blockchain permits the primary person (person A) to use their private key to connect information concerning the transaction to the public key of the person (person B). This, along with some metadata, forms a part of the block, which also contains the digital signature as a timestamp. However, the block does not the actual identities of the involved entities to maintain anonymity at the same time ensuring authentication through the digital signatures. The formed block is then passed on in the network to all the nodes that then validate the transaction, thus leading to the block being recorded on the Blockchain of every node.

2.4 Consensus protocols

The consensus mechanism makes sure that the peers of the Blockchain network agree on the distributed ledger's current state. As a result, this guarantees that the new block uploaded to the Blockchain is the sole source of truth and puts an end to attackers. The correct operation of the entire Blockchain network depends on the consensus procedures.

2.5 Potential of blockchain for supply chain

Blockchain's characteristics are effectively applicable to solve major issue of traditional medical drug supply chain process.

2.5.1 Combating drug counterfeiting

The introduction of fake drugs in the system is a crucial issue that concerns all the entities ranging from the manufacturer to the end consumer. Using Blockchain in the supply chain can allow all the people in the process to be able to see all the transactions occurring in the system, thus making it more trustworthy and transparent (Musamih *et al.* 2021). This property would ease the issue of counterfeit drugs since all the transactions are visible to all the entities and can be alerted if any malicious user tries to add fake drugs. Also, it will be able to trace the record to the offender, thus acting as a hindrance to such acts. Such a system allows all the stakeholders to gain from the increase in efficiency during drug distribution while allowing the end users to rest assured that their medicines are authentic.

2.5.2 Ensuring data integrity

The ledger in the Blockchain is available with the nodes in the network and when any transaction in the system occurs, it is validated and verified by all the nodes thus eliminating the fear of any fraudulent transaction being approved. Apart from this, once a transaction has been verified and has been recorded in a block, it is very difficult for a malicious entity to tamper it or change it since a copy of the ledger is available with all the nodes and thus will be required to change all the ledgers in the network which is not a feasible task. This establishes high-level of knowledge integrity, thereby making data secure, compliant, and available for the participants.

2.5.3 *Better tracking and traceability*

The number of entities and the huge number of transactions occurring at every stage is a challenge that are to be managed in a supply chain. Keeping a track of all this knowledge becomes a heavy task even if the participants have automated their systems to trace the movement of the drug through the supply chain. Since every transaction between any two entities is recorded in the Blockchain and verified by the nodes, it becomes easier to trace back any fraudulent transaction to its concerned offender.

2.6 *Related research*

Table 1 has a summary of research in the field of medical fields using different technologies such as IoT, Cloud Computing, AI and Blockchain technology. It shows the potential of Blockchain for the medical domain for information tracking as well as keeping patients' data in a secure place. The next section, we proposed Blockchain-based architecture.

Table 1. Research in medical health.

Reference	Technology Integration	Focus area
(Tunc *et al.* 2021)	IoT and Blockchain Technology	– Describe the ongoing research project using IoT, Blockchain in smart healthcare. – Issues of smart health-like interoperability, power consumption, security, resource management and latency can be overcome using machine learning, AI, Blockchain, SDN, etc.
(Haleem *et al.* 2021)	Blockchain	– Focus on the issue of accessing medical services and resource availability at all times. – Use of Blockchain technology to provide securely and transparently patients' data sharing among medical stakeholders.
(Ahmad *et al.* 2021)	Decentralized Blockchain	– The issue of COVID-19 medical equipment tracking and waste handling using centralized system which is single point of failure. – The proposed solution using decentralized Blockchain to smooth handling of medical supply chain process among all the participants of the network.
(Cheng *et al.* 2021)	Machine Learning and Blockchain	– Use of machine learning to classified patients' data and detect a pattern of cancer from dataset. – The integration of Blockchain with Machine learning for data sharing and exchanging.
(Uddin *et al.* 2021)	Cloud Service, IoT, Blockchain	– Issue of sharing Electronic Medical Records (EMRs) with different organizations or institute through third party cloud service providers. – Blockchain provides efficient truss management and immutably share records among others.
(Tariq *et al.* 2020)	IoT and Blockchain	– We present the security issues of data breaches, attacks and privacy leakage, tempering and forgery in smart health system. – Integration of Blockchain can solve the problems such as unauthorized access and modifications.
(Dai *et al.* 2019)	IoT and Blockchain	– Issues of wearable smart devices sharing data with privacy preserving mode. – IoT-Blockchain-based solution to provide traceability and reliability.
(Chang & Chen 2020)	Blockchain	– Survey on IoT issues in health domain such as access control, authentication, and information sharing. – Apply Blockchain for securely storing information into a distributed ledger and keep data immutable.

(continued)

Table 1. Continued

Reference	Technology Integration	Focus area
(Onik et al. 2019)	Blockchain	– Present much scope of research in health care using Blockchain. – Focus on developing a novel algorithm, framework and proof of concepts for implementing the Blockchain in health care.
(Dang et al. 2019)	IoT and Cloud	– Survey on IoT trends in health domain for upliftment of services and process. – Discuss the security issues of IoT and future challenges to integrating IoT into healthcare.
(Siyal et al. 2019)	Blockchain	– Review the possible developments in medicine and healthcare by presenting Blockchain as model. – Discuss the various applications of Blockchain and future challenges to incorporate into healthcare.
(Agbo et al. 2019)	Blockchain	– Discuss the advantages of using Blockchain to solve the issues of medical domain such as digital trust and information traceability. – Issue of Blockchain for storage of large data such as image, scalability, and privacy of data.
(Tandon et al. 2020)	Blockchain	– Identify the possible use cases of Blockchain in medical fields. – Focus the research scope like interoperability, scalability, latency and privacy of Blockchain in health domain
(Clim et al. 2019)	Blockchain	– Focus on the data sharing issue in health applications through untrusted third party. – The scope of Blockchain for mobile applications in health care.
(Noby & Khattab 2019)	IoT and Blockchain	– Survey conducted on integration of IoT and Blockchain. – Present future research scope in Blockchain security and trust management.

3 BLOCKCHAIN BASED FRAMEWORK

The architecture in Figure 2 comprises two aspects one is private Blockchain network in which the participants are identified priory and another type is public network to see the details of any transactions through an explorer. The identification is done by the certificate authority (CA) before launching any transaction into Blockchain. All the transactions are verified and validated by the concerned authority known as the validator nodes of the Blockchain network. For instance, manufacturer initiates the transaction to add medical drugs information into each distributed ledger of Blockchain node. This information is later

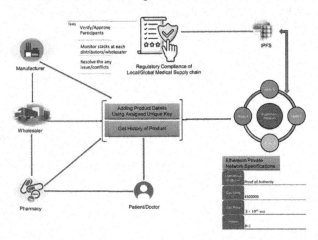

Figure 2. Blockchain-based framework.

accessible to the different stakeholders' likewise distributors, retailers, or pharmacy to inspect the quality parameters of medical drug and other details.

All the medical products are to be purchased digitally by the stakeholders and related transactions will be reported into Blockchain network. This way the distributed ledger will be regularly synchronized with the updated stock details. Moreover, any stakeholder can directly access the history of medical drugs or products from the user interface such as mobile application or website. The regularity authority can monitor the live stocks of medical products with additional details such as the number of stocks is registered among different stakeholders and their usages. These will resolve the issues of drugs counterfeiting and stop the black marketing of the medical drugs. supply. Not only that but also patients or normal user can get all the medical products with satisfactory rate by ensuring the medical drug quality. The framework consists of Inter planetary file system (IPFS) that is distributed file system. In case of large volume of data, it acts like off-chain data storage and store the hash into Blockchain.

Figure 3 shows smart contract functionality of Blockchain-based application. All the functionalities are uniform and having the same result across all platforms. All the functionalities mentioned in each stakeholder are written as smart contract. All the rules and regulation are inform using smart contract. For instance, the manufacturer of medical drugs first initiates the transaction to add lot details into Blockchain network. Once the lots are added, manufacturer cannot buy the lot means seller cannot be a buyer. All activities are going to track and trace by all the participants of the Blockchain network.

Figure 3. Smart contract functionalities of medical drug supply chain.

All the distributors and suppliers are verified and identifies priory to be a part of the system. Smart contract displays the ownership of medical products and automate the transfer of medical drugs based on condition.

4 ETHEREUM BLOCKCHAIN ANALYSIS

We deployed entire decentralized application (dapp) on Blockchain test network on Ethereum platform using ropsten and rinkeby which are based on Proof of Work (POW) and Proof of Authority (POA) consensus mechanisms respectively. We have analyzed some of the parameters such as throughput, hashrate, blocksize and blocktime of Ethereum mainnet from etherscan.io. All the results have been taken from Etherscan.io to check the feasibility to run the application and analyze its performance. The Ethereum 1.0 is on POW mechanism and Ethereum next merger 2.0 will be on POS (Proof of Stack). Figures 4, 5, 6 and 7 are based on proof of work (POW) mechanism. Figure 4 shows daily Ethereum transaction of three months. The daily transaction varies 1.2 million to below 1.4 million. If we consider 1.2 M transaction per day, then there is around 13 to 14 transaction per seconds. Figure 5 shows the daily hashrate in GH/s. The hashrate is dependent on the difficulty of Ethereum network. If the difficulty goes high, it will require more hashrate to mine the block.

The other parameters are also closely related with hashrate and difficulty such as blocksize and blocktime. The blocksize is the number of transactions in a single block in Blockchain network. Figure 6 shows the blocksize of Ethereum network. This can be varied as per the Ethereum network functioning. The Figure 7 shows the blocktime that is time taken to mine one block. Figure 7 clearly states that average blocktime is vary from 13 to 14 seconds per block creation. POW demands more hashrate to mine single block that result in more electricity consumption.

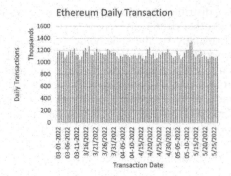

Figure 4.   Ethereum daily transaction.

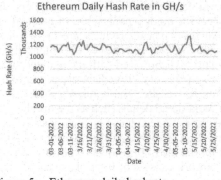

Figure 5.   Ethereum daily hashrate.

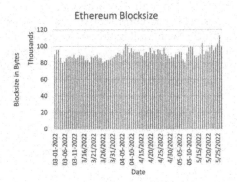

Figure 6.   Ethereum blocksize.

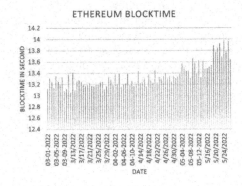

Figure 7.   Ethereum blocktime.

Ethereum has merged from POW to POS from 15th September 2022. According to CCRI (2022) which is Crypto Carbon Ratings Institute to provide information on sustainability of Blockchain technology. As per the research by CCRI (2022), the electric load consumption of Ethereum POW on 15th September 2022 was 19.45 TWh. After merger to POS, the electric load consumption of Ethereum POS was 0.0021 TWh on 15th September 2022. The average consumption from 15th September to 20th December 2022 is approximately around 26 to 27 TWh. In conclusion, POS has very less electric load as compare to POW.

5   CONCLUSION

Blockchain-based medical drugs supply chain is a decentralized distributed application (dapp) that keeps track of the deliveries of goods to various stakeholders, including distributors, merchants, pharmacies, patients, and physicians, as well as the transactions related to the manufacturing of medical drugs. To deploy smart contracts to the Ethereum Virtual Machine (EVM) and apply all rules and regulations deterministically across all platforms. The traditional, centralized medical supply chain is currently facing several challenges, including complexity and extensive middle-man intervention, which makes the entire process person-centric rather than transparent to all stakeholders. A survey was conducted on the use of IoT and machine learning in the supply chain and medical industry to automate processes. Additionally, by incorporating Blockchain technology, the entire record-keeping process becomes much more transparent and unchangeable. All the information can be easily tracked back to its source.

REFERENCES

Agbo, C., Mahmoud, Q., & Eklund, J. (2019). Blockchain Technology in Healthcare: A Systematic Review. *Healthcare, 7*(2), 56. https://doi.org/10.3390/healthcare7020056

Ageron, B., Bentahar, O., & Gunasekaran, A. (2020). Digital Supply Chain: Challenges and Future Directions. *Supply Chain Forum: An International Journal, 21*(3), 133–138. https://doi.org/10.1080/16258312.2020.1816361

Ahmad, R. W., Salah, K., Jayaraman, R., Yaqoob, I., Omar, M., & Ellahham, S. (2021). Blockchain-Based Forward Supply Chain and Waste Management for COVID-19 Medical Equipment and Supplies. *IEEE Access, 9*, 44905–44927. https://doi.org/10.1109/ACCESS.2021.3066503

Beaulieu, M., & Bentahar, O. (2021). Digitalization of the Healthcare Supply Chain: A Roadmap to Generate Benefits and Effectively Support Healthcare Delivery. *Technological Forecasting and Social Change, 167*, 120717. https://doi.org/10.1016/j.techfore.2021.120717

Chang, S. E., & Chen, Y. (2020). Blockchain in Health Care Innovation: Literature Review and Case Study From a Business Ecosystem Perspective. *Journal of Medical Internet Research, 22*(8), e19480. https://doi.org/10.2196/19480

Cheng, A. S., Guan, Q., Su, Y., Zhou, P., & Zeng, Y. (2021). Integration of Machine Learning and Blockchain Technology in the Healthcare Field: A Literature Review and Implications for Cancer Care. *Asia-Pacific Journal of Oncology Nursing, 8*(6), 720–724. https://doi.org/10.4103/apjon.apjon-2140

Clim, A., Zota, R. D., & Constantinescu, R. (2019). Data Exchanges Based on Blockchain in m-Health Applications. *Procedia Computer Science, 160*, 281–288. https://doi.org/10.1016/j.procs.2019.11.088

Dai, H.-N., Zheng, Z., & Zhang, Y. (2019). Blockchain for Internet of Things: A Survey. *IEEE Internet of Things Journal, 6*(5), 8076–8094. https://doi.org/10.1109/JIOT.2019.2920987

Dang, L. M., Piran, Md. J., Han, D., Min, K., & Moon, H. (2019). A Survey on Internet of Things and Cloud Computing for Healthcare. *Electronics, 8*(7), 768. https://doi.org/10.3390/electronics8070768

Haleem, A., Javaid, M., Singh, R. P., Suman, R., & Rab, S. (2021). Blockchain Technology Applications in Healthcare: An Overview. *International Journal of Intelligent Networks, 2*, 130–139. https://doi.org/10.1016/j.ijin.2021.09.005

Kittipanya-ngam, P., & Tan, K. H. (2020). A Framework for Food Supply Chain Digitalization: Lessons from Thailand. *Production Planning & Control, 31*(2–3), 158–172. https://doi.org/10.1080/09537287.2019.1631462

Musamih, A., Salah, K., Jayaraman, R., Arshad, J., Debe, M., Al-Hammadi, Y., & Ellahham, S. (2021). A Blockchain-Based Approach for Drug Traceability in Healthcare Supply Chain. *IEEE Access, 9*, 9728–9743. https://doi.org/10.1109/ACCESS.2021.3049920

Noby, D. A., & Khattab, A. (2019). A Survey of Blockchain Applications in IoT Systems. *2019 14th International Conference on Computer Engineering and Systems (ICCES)*, 83–87. https://doi.org/10.1109/ICCES48960.2019.9068170

Onik, Md. M. H., Aich, S., Yang, J., Kim, C.-S., & Kim, H.-C. (2019). Blockchain in Healthcare: Challenges and Solutions. In *Big Data Analytics for Intelligent Healthcare Management* (pp. 197–226). Elsevier. https://doi.org/10.1016/B978-0-12-818146-1.00008-8

Panda, S. K., & Satapathy, S. C. (2021). Drug Traceability and Transparency in Medical Supply Chain Using Blockchain for Easing the Process and Creating Trust Between Stakeholders and Consumers. *Personal and Ubiquitous Computing*. https://doi.org/10.1007/s00779-021-01588-3

Siyal, A. A., Junejo, A. Z., Zawish, M., Ahmed, K., Khalil, A., & Soursou, G. (2019). Applications of Blockchain Technology in Medicine and Healthcare: Challenges and Future Perspectives. *Cryptography, 3*(1), 3. https://doi.org/10.3390/cryptography3010003

Tandon, A., Dhir, A., Islam, A. K. M. N., & Mäntymäki, M. (2020). Blockchain in Healthcare: A Systematic Literature Review, Synthesizing Framework and Future Research Agenda. *Computers in Industry, 122*, 103290. https://doi.org/10.1016/j.compind.2020.103290

Tariq, N., Qamar, A., Asim, M., & Khan, F. A. (2020). Blockchain and Smart Healthcare Security: A Survey. *Procedia Computer Science, 175*, 615–620. https://doi.org/10.1016/j.procs.2020.07.089

Tunc, M. A., Gures, E., & Shayea, I. (2021). A Survey on IoT Smart Healthcare: Emerging Technologies, Applications, Challenges, and Future Trends (arXiv:2109.02042). *arXiv*. http://arxiv.org/abs/2109.02042

Uddin, M. A., Stranieri, A., Gondal, I., & Balasubramanian, V. (2021). A Survey on the Adoption of Blockchain in IoT: Challenges and Solutions. *Blockchain: Research and Applications, 2*(2), 100006. https://doi.org/10.1016/j.bcra.2021.100006

271

Artificial Intelligence, Blockchain, Computing and Security – Dagur et al. (Eds)
© 2024 The Author(s), ISBN: 978-1-032-49393-0

Blockchain technology for agricultural data sharing and sustainable development of the ecosystem

Ashok Kumar Koshariya*
Department of Plant Pathology, School of Agriculture, Lovely Professional University, Jalandhar, Punjab, India

Virendra Kumar*
SE and Head, Land Use and Urban Survey Division, Remote Sensing Application Center, Lucknow, Uttar Pradesh, India

Vashi Ahmad*
Department of Civil Engineering, Remote Sensing Application Center, Lucknow, Uttar Pradesh, India

Bachina Harish Babu*
Department of Automobile Engineering, VNR Vignana Jyothi Institute of Engineering and Technology, Hyderabad, Telangana, India

B. Umarani*
Department of Electronics and Communication Engineering, Kongunadu College of Engineering and Technology, Trichy, Tamil Nadu, India

S. Ramesh*
Department of Computational Intelligence, Saveetha School of Engineering, Chennai, Tamil Nadu, India

ABSTRACT: Because of blockchain's various properties, such as payment systems, democratization, visibility, accountability, data unlinkability, and privacy laws, today's complex and multi-echelon supply chains can benefit from its implementation. These qualities and a crucial agreement make it appropriate for application in different supply networks. These elements enhance the production procedures and, over time, make the current supply chains flexible, adaptable, and responsive. Blockchain also adds a sustainability component that is related to the current global phenomenon of the circular economy. Organizations can evaluate and contrast the old supplier base and the production lines offered by cryptocurrencies in order to place the concept of corporate sustainability today. The goal of or work is to use a state of the art and the opinions of industry leaders in agriculture to highlight the benefits of cryptocurrency in logistics management. The primary advantages cited include better sustainability, creation of resilient supply chains, decentralization, immutability of data, smart contracts, transparency, and shared database.

Keywords: Blockchain technology, decentralization, data privacy, smart contract, immutability

1 INTRODUCTION

DLT (Distributed Ledger Technology) Blockchain is an intriguing new technology that has the ability to completely change the game. Digitalization are typically interfere and day when databases (Demirkan *et al.* 2020). Participation of people and intermediary steps. Blockchain is

---

*Corresponding Authors: ashok.koshariya@gmail.com, vk15868@gmail.com, vacivi0@gmail.com, bachina.harish@gmail.com, umabkv@gmail.com and rameshshunmugam.sse@saveetha.com

DOI: 10.1201/9781003393580-40

able to alter the existing business and economic paradigms, much like the combustion engine and the World wide web did for earlier industrialization (Westerlund *et al.* 2018). Numerous industries, along with the banking system, global markets, logistics providers, the management of trade secrets, "digital enterprises," the civil service, and others could benefit from the use of supply chain. intermediate, improve clarity, and sharpen listening skills (Dutta *et al.* 2020). Blockchains are networks of computers, that share a distributed database. The database allows for contributions, but existing data cannot be changed. The network routinely verifies the reliability of the database. Although blockchain is currently being utilized in a variety of different projects, Bitcoin continues to be one of its best implementations. (Schwab *et al.* 2017). Blockchain technology has the potential to enhance cash activities, provenance, auditability, and accountability in supermarket chains, while also empower the founding of new markets and goods for food production in underdeveloped nations. This is because humanity deserves the greatest in terms of learning, company, affordable healthcare, and even grocery shopping. We always have access to nutritious foods, veggies, cheese, and protein because to agricultural developments, which helps us live a healthy lifestyle and pleasurable lives. (Justinia *et al.* 2019).

2 RELATED WORKS

The research proposed by (Tian *et al.* 2016) uses a descriptive case study method and takes as its model that the very first bakery in the USA to embrace distributed ledger technology. The next step is to use breach assessment to assess whether cryptocurrencies might be used to transform the current situation into one that is more durable. Sachin Kumar *et al.* have presented a conceptual framework for integrating blockchain technology to develop a sustainable tea supply chain was proposed (Xiong *et al.* 2020). It uses the Circular Fuzzy Hierarchy Method (SF-AHP) method to identify possible moves and prioritize potential risks that may surface during this integration process. The system that integrates blockchain technology into every aspect of functioning is then shown. False Data Injection, Sybil, DDoS, Identity Spoofing, Side-Channel, Botnet, Backdoor Trojan, Targeted Code Injection, Social Engineering, Phishing, Sinkhole, Man-in-the-Middle, SQL Injection, Consensus, Eclipse, and Block Mining are among the sixteen different cyberattacks that Malka *et al.* (Halgamuge *et al.* 2022) proposed attack models, that takes into account, the possibility that some nodes might misplace, forget, or lose their private keys. They created a simulation environment to assess the likelihood that a malicious attacker will be successful. Ankush *et al.* (Mitra *et al.* 2023) proposed data poising assaults, that cause extremely serious problems, when Big data analytics upon that investigated information are itself contaminated because they may seriously affect organizations and businesses in terms including both monetary terms and harming their reputations.

3 PROPOSED METHODOLOGY

By using the PRISMA methodology, we were able to compile a dataset of publications. The papers' contents were then examined, along with their primary sources of information for the bibliographies and their connections to the SDGs. Additionally, Microsoft Excel and the VOS viewer program were used in the data processing to stratify and analyze the data. We can distinguish three types of findings thanks to data processing. We begin by doing a quantitative analysis of the research relating blockchain solutions to sustainable development. Then, using a keyword co-occurrence analysis, we identify the research patterns and theme topics. Third, we create a matrix that shows the connection between topics.

3.1 *Identification*

The protocol's initial step is to specify how to find and choose the publications that must be included in the review. We used SCOPUS, a database that is extensively used in academic research, to source the papers because it offers a complete portfolio of scientific journals. We chose a list of keywords and looked for them in the paper's titles, keywords, and abstracts. Our investigation was constrained to parameter kinds, such publications or evaluation with exception of various books, opinion columns, or annotations, and the source type to the journal article because consensus educational research papers are typically seen as being at the cutting edge of knowledge in comparison to these other document types. The development of fundamental standards for screening the gathered collections of publications is the final stage in the principles stated.

3.2 Content based analysis

The bibliographic sample was converted into an MS Excel file, which contained the name, description, phrases, surnames and associations of the authors, the name of the publication, the year that each of the discovered documents was published as well as how many instances, it was cited. The dataset is integrated with other data to aid in the general open coding depending on the review objective. We carefully picked the criteria and other elements that would help us categorize the 195 products we had chosen. The first set of dimensions relates to the type of article and the extent of each one's industry and geographic area.

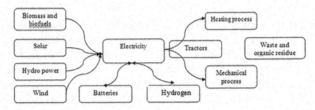

Figure 1.   The conceptual image of an agricultural ecosystem.

An energy-independent farm's schematic environment is shown in Figure 1. According to this model, any renewable energy source might be turned into electricity. Then, additional electrified devices like sensors, automobiles, and robots might make use of this energy. Renewable energy can be created and stored, either temporarily in batteries or permanently in hydrogen tanks. Depending on the application, It is possible to immediately produce physical or thermal energy straight from the source, by avoiding the electricity conversion.

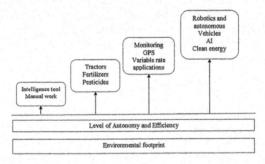

Figure 2.   Development roadmap for the agricultural revolution from agriculture 1.0 to Agriculture 5.0.

The plan for the agricultural revolution is shown in Figure 2. Agriculture 1.0 refers to the conventional agricultural method used from antiquity until the late 19th century, when farmers heavily resorted to local equipment for agriculture. Although extensive human labor was necessary for such peasant farming, production was quite low. Farmers used a variety of machinery to boost food output and decrease the need for physical work in the early 20th century, a period known as Agriculture 2.0, which capitalized on the first Industrial Revolution, which took place between 1784 and 1870.

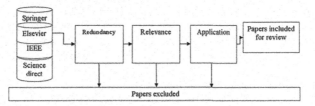

Figure 3.   The representative data curation process.

This section explains the literature that has been gathered and the filters that have been applied to focus the results on publications, articles, and review findings that present the usage of blockchain in agriculture. The first search produced 300 papers. This work was created by applying the criteria to the primary studies. The next step was paper selection, as seen in Figure 3.

4   RESULTS AND DISCUSSION

Table 1.   Cryptocurrency technologies are matched to agriculture's concerns.

Agriculture's issues	Bitcoin-based Services
Connections between shops, users, vendors, and government	Connecting a public blockchain with data and enhancing traceability
loss of origin information	Identifiability of clients
streams of funding and development	Various funding systems: the digital industry
	Bridge agreement: compensation
Businessmen	Going directly, fair prices, and reasonable taken any action
Hard data access for intelligent agricultural	information independence and real privacy

For the chosen themes, the executive summary of the qualitative evaluation is provided. The principal keywords used in each of the selected articles have been recorded, that might make it easier for fresh investigators to locate the most well-liked phrases in the cryptocurrency related to an agriculture topic. To make it simpler to understand each publication's essential value, its main goal and accomplishment was stated in the second place. The contributions of the publications that passed the screening covered the subject from a variety of angles, including the use of blockchain in agriculture, modeling using blockchain technology in agriculture, and advantages and disadvantages of blockchain adoption in the agricultural industry.

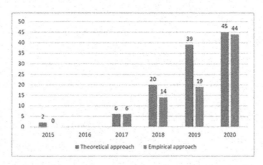

Figure 4.   Comparison between various approaches.

This outcome is not unexpected, as we predicted between 2018 and 2020, interest in this subject will increase dramatically as a result of governmental attention being given to blockchain technology and environmental concerns. Additionally, the data revealed that the publications from 2019 (44.4%), 2018 (44.3%), and 2017 (27.3%) received the most citations. In 2017, a higher value would be anticipated, therefore the percentage of citations would appear to be an error. Only 12 publications were published in 2017, yet they weren't cited the following year. The fact that application papers have been consistently published, since 2018 is an intriguing element. Figure 4 demonstrates how the tendency of empirical articles has evolved over time; in 2015, there were only theoretical papers, however in 2020, there were almost as many empirical papers as theoretical ones.

Table 2. Effectiveness in the context of existing works.

Publications	Execution	Device-to-Cloud Latency (Second)	Cloud-to-Blockchain Latency (Second)	Blockchain-to-Client-Console Latency (Second)	Alert Total Latency (Second)
Shaikh *et al.* (2022)	Ethereum	1.18	1.78	3.67	16.55
Arkhmen *et al.* (2021)	Simulations	2.27	2.89	2.90	12.89
Chen *et al.* (2022)	Ethereum	4	3.90	4.90	13.67
Dhanaraju *et al.* (2022)	Ethereum	0.01	5.89	2.89	12.78
Proposed work	Ethereum	0.02	5.49	2.71	11.57

Table 2 compares the communication delay of current research with earlier efforts. The authors used Ethereum-based smart contracts to upload the Wifi sensor information to the network. They also assessed the transmission delay associated with generating a cryptocurrency operation. The authors reported a total transmission delay of 16.55 s. This work is extremely similar to ours in terms of entering Wifi data from the sensors into in the network. Our solution significantly surpassed the work since we employed specific project apps, like Internet fundamental and blockchain based leveraging Infura API. The additional delay may also be coming from the network node running within the linux environment.

5 CONCLUSION

The use of technology in farming is still in its infancy, despite future studies, programs, and initiatives to optimize the benefits of embracing cryptocurrency technology. These ongoing procedures are centered on problems like the visibility, accessibility, believability, and greater transparency of agricultural data with the help of blockchain based method. Additionally, researchers are creating technology and valuable concepts which could be applied to improve the efficiency of the farming sector.

REFERENCES

Demirkan, S.; Demirkan, I.; McKee, A. Blockchain Technology in the Future of Business Cyber Security and Accounting. *J. Manag. Anal.* 2020, 7, 189–208.

Dutta, P.; Choi, T.-M.; Somani, S.; Butala, R. Blockchain Technology in Supply Chain Operations: Applications, Challenges and Research Opportunities. *Transp. Res. Part E Logist. Transp. Rev.* 2020, 142, 102067.

Halgamuge, Malka N. "Estimation of the Success Probability of a Malicious Attacker on Blockchain-based Edge Network." *Computer Networks* 219 (2022): 109402.

Justinia, T. Blockchain Technologies: Opportunities for Solving Real-world Problems in Healthcare and Biomedical *Sciences. Acta Inform. Med.* 2019, 27, 284.

Mitra, Ankush, *et al.* "Impact on Blockchain-based AI/ML-enabled Big Data Analytics for Cognitive Internet of Things Environment." *Computer Communications* 197 (2023): 173–185.

Niforos, M. Blockchain in Development: Part I—A New Mechanism of Trust? The World Bank: Washington, DC, USA, 2017.

Schwab, K. The Fourth Industrial Revolution; Currency: New York, NY, USA, 2017.

Tian, F. An Agri-food Supply Chain Traceability System for China Based on RFID & Blockchain Technology. In *Proceedings of the 2016 13th International Conference on Service Systems and Service Management (ICSSSM)*, Kunming, China, 24–26 June 2016.

Westerlund, M.; Neovius, M.; Pulkkis, G. Providing Tamper-resistant Audit Trails with Distributed Ledger Based Solutions for Forensics of IOT Systems Using Cloud Resources. *Int. J. Adv. Secur.* 2018, 11, 3–4.

Xiong, H.; Dalhaus, T.; Wang, P.Q.; Huang, J.J. Blockchain Technology for Agriculture: Applications and Rationale. Front. *Blockchain* 2020, 3, 7.

Artificial Intelligence, Blockchain, Computing and Security – Dagur et al. (Eds)
© 2024 The Author(s), ISBN: 978-1-032-49393-0

Problems of developing a decentralized system based on blockchain technology

D.T. Muhamediyeva
"Tashkent Institute of Irrigation and Agricultural Mechanization Engineers" National Research University, Tashkent, Uzbekistan

A.N. Khudoyberdiev & J.R. Abdurazzokov
Research Institute for the Development of Digital Technologies and Artificial Intelligence, Tashkent, Uzbekistan

ABSTRACT: The article discusses the problems of developing distributed ledger systems, which is a new approach to creating databases, the key feature of which is the absence of a single control center. Each node compiles and writes registry updates independently of the other nodes. Unlike distributed databases, each participant in a distributed ledger system stores the entire history of changes and validates the addition of any changes to the system using a consensus algorithm, which mathematically guarantees that data cannot be forged.

1 INTRODUCTION

Since the 17th century, when the Dutch East India Company was the first listed company on the stock exchange, the world economy has been built around and supported by stock exchanges, where millions of transactions are made every day, helping companies increase their value. The exchange market is a set of offers to buy and sell corresponding to an asset. An asset may represent stocks or stocks of companies, bonds or other securities. The people who buy or sell assets are called investors, and the people who make the transactions are called brokers or traders [1].

Modern stock exchanges are highly computerized and can process a huge number of transactions in a short amount of time, ensuring the security, execution, and authenticity of transactions at the cost of a transaction fee, usually in direct proportion to the cost of trading. A stock exchange such as the New York Stock Exchange, London Stock Exchange facilitates the buying and selling of shares of companies through it, which is regulated by a central authority. This market architecture has many advantages due to the central authority that ensures the authenticity, security and validity of transactions. However, centralization also has many disadvantages, such as having a single point of failure, possible performance bottleneck or attack susceptibility, and time costs. In addition, the central authority charges a fee and the trading process is not transparent to the trader.

Bitcoins, Ethereum, Ripple are well-known digital (crypto) currencies that are easily bought and sold anywhere in the world [1]. This concept of cryptocurrencies has inspired the creation of digital shares, which include the use of a decentralized stock exchange architecture to overcome the shortcomings given above, using new blockchain technology [2]. The potential of the blockchain system can benefit the entire system, the execution of market orders and the correct settlement between accounts. In addition, the guaranteed immutability of the ledger provides a valuable advantage over a centralized system. In addition, due to the decentralization of the system, no central authority or intermediary is required to place and execute orders. This allows peer-to-peer transfer, direct purchase and sale of shares between traders and investors without the need for a third intermediary party to trade. Also, the implementation of the blockchain helps to reduce the transaction cost for each transaction, provide increased security and transparency, and the time required for the transaction will be significantly reduced.

The distributed ledger technology, often known as "Blockchain" technology (from the English phrase "block chain"), is one of the most significant technologies at the moment (Distributed Ledger Technology, DLT). The revolutionary alternative payment service bitcoin and the related digital currency originally surfaced in 2009 as a means of enabling decentralized, distributed operation. Due to the open code of Bitcoin, many other cryptocurrencies have already been created at this stage, and each of them is based on its own blockchain [1].

Economics, analysts, and IT professionals are already investigating the potential of using the technology beyond its initial purpose in light of the significant benefits of blockchain over the antiquated financial system, and the digital market is flooded with blockchain firms. Blockchain is a database that ensures data immutability and high security. Although blockchain is equated with cryptocurrency, it is important to understand that it is a tool that can be used in a variety of ways, some of which are: storing and tracking confidential information, such as patient records and patent rights, developing decentralized applications, notarial documents and others [2].

By order of the President of the Republic of Uzbekistan, Sh.M. Mirziyoyev, dated 3.07.2018, "ON MEASURES FOR THE DEVELOPMENT OF THE DIGITAL ECONOMY IN THE REPUBLIC OF UZBEKISTAN," No. PQ-3832, this technology became a part of the country. This decision clearly explains the purpose of adopting the technology and shows that the digital economy is behind the technology. In the state plan, large-scale measures were implemented to develop the digital sector of the economy, introduce an electronic document circulation system, develop electronic payments, and improve the legal framework in the field of electronic commerce [3].

2 OPPORTUNITIES OF BLOCKCHAIN TECHNOLOGY

E-commerce often relies on financial institutions to serve as reliable middlemen for electronic payments. For the majority of transactions, this technique works well, but because it depends on trust, there are certain issues that arise. Financial institutions' necessary mediation precludes irrevocable transactions. It is not suggested to conduct frequent and minor transactions since the cost of these services raises their cost and establishes a minimum price for them. Additionally, the price of renewable services is increased by the lack of irreversible processes. The vendor is compelled to ask the customer for more information than is inherently necessary since the payment can be canceled. Additionally, some fraud is just seen as unavoidable. There is no method for direct electronic transactions, however these restrictions and payment risks can be bypassed in paper currency transactions [4].

A payment system that relies on cryptography instead of trust and enables both parties to move money directly without the use of a middleman is what is required. Sellers and purchasers are shielded from fraud by the high expense of accounting for transaction cancellations [5].

Distributed data processing makes it possible to place a database (or several databases) on different nodes of a computer network. Data distribution is performed on different computers in conditions of vertical and horizontal connections for organizations with a complex structure.

The objective need for a distributed form of data organization depends on the requirements set by end users [4]:

- centralized management of scattered information resources;
- improving the efficiency of managing databases and data banks and reducing the time of accessing information;
- support data integrity, consistency and protection;
- to provide an acceptable level in the "price - performance - reliability" ratio.

The distributed system of databases makes it possible to create and maintain various possibilities, to avoid obstacles that hinder the user's efficiency and to increase the efficiency of using information resources.

Blockchain is a continuous chain of blocks (linked list) containing information, built according to certain rules. Often, copies of blockchains are stored independently on different computers. Blocks are information about transactions, deals and contracts within the system, presented in cryptographic form. Blockchain allows people to record information, and a community of users of a particular chain can control the changes and updates of information about the record [5].

Transactions are transmitted to participants and each node creates an updated version of the events. It is this difference that makes blockchain technology so interesting - it represents an innovation in registration and information distribution that eliminates the need for a third party to simplify digital communications. However, blockchain technology is not a new technology with all its advantages. Rather, it is a combination of proven technologies applied in new ways. It is a special combination of three technologies (P2P Network, secret key cryptography and a protocol that guides the creation of new base elements). As a result, there is a system of digital interactions that does not need a trusted third party. Blockchain technology's unique elegant, simple, yet robust network architecture enables the implementation of digital communications to be hidden. In blockchain technology, cryptography provides a powerful means of ownership that meets the requirements of private key authentication. Ownership of a private key is property [6].

See the example below. During the service life of the vehicle, it goes through various stages - collection, sale, insurance, etc., until disposal. At each stage, many different documents and reports are created. If it is necessary to get an explanation, requests will be sent to the relevant authorities. This process takes a long time. Physical location, different working languages and bureaucracy are some of the challenges.

Blockchain technology avoids all these problems. All information about each vehicle can be stored in the network. This information cannot be deleted or changed without the participant's consent. It is possible to have the necessary information at any time. Based on the idea of smart-contracts, they are working on the goal of ensuring that the entire life path of any vehicle is recorded on the block chain.

A transaction is verified by every computer (i.e. host) that keeps a copy of it. At this point, the nodes check the transaction history.

Now, when the transaction is found to be valid, it goes into the pool - this is a kind of "waiting room", and considering it in the next block, from here the transaction is accepted by the miner. At this point, the transaction is considered "unconfirmed". As soon as a miner executes a transaction and includes it in a successfully generated block, the transaction is considered confirmed. The block contains a limited number of transactions (about 2.5 thousand), therefore, in periods of high activity, if the queue for confirmation (processing transactions through the network and adding it to the blockchain) is long, the miner must be added to the block selects the transactions based on the priority fee attached to them [5].

Thus, the commission is designed to show the miner how urgent the transaction is - if the user wants it faster, he should offer a higher payment, and if the user is not in a hurry, he will be able to pay less [6].

Previously, fees were charged according to different rules: if the transaction was small enough or "priority", it could be free. Today, a commission is always required. The size of each transaction is similar to the size of a file on a computer. As miners try to maximize their earnings, they first select transactions with the best ratio of commission and volume - the smaller the transaction, the better. Here is an example from the real estate market. When a customer comes to buy or rent an apartment, he pays per square meter. The client pays the price of the apartment in full, but can compare it with the price per square meter of other apartments. The fee rate is the ratio of commission and volume (fee rate) managed by miners - this is the price per square meter. This ratio is measured in "Satoshi" per byte. – how many

Satoshi's (the smallest unit of account in the Bitcoin network) users are willing to pay for each byte of a transaction. There are services that allow you to check how much money will be spent to enter the transaction into the nearest block. This indicator always changes depending on network traffic [7].

A public network holds that users can join the network by supplying their own hardware, hence boosting network sharing, computational power, and data storage. Equipment owners should be rewarded for their honest labor in order to promote these attitudes [1].

This means that database operations are paid for by the end user. This situation may seem strange to someone new to blockchain, but it makes sense. The reality is that blockchain projects are often ownerless. The community owns them. As a result, the community will have to pay the project costs. The money is very little, but not zero. Existing decentralized file storage tools charge the user to store files. And we can't ignore that, at a basic level, the operation of the equipment needs to be paid for by its users. Later, these costs can be covered from other sources.

3 METHOD

A transaction block is a structure for recording groups of transactions in a distributed ledger. The block contains a header and a list of transactions. The block header contains the hash of the previous block, the information hash, and the hash constructed by the Merkle tree containing each operation [4].

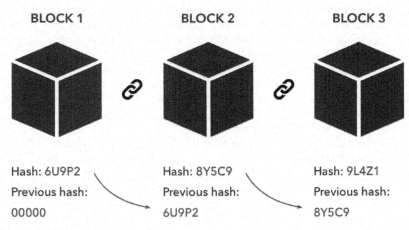

Figure 1. Chain of blocks.

The block consists of the following features:

Table 1. Description of the block.

hash	SHA-256 block header hash
ver	Block diagram version
prev_block	hash of previous block in the chain
mrkl_root	A merkle root is a hash list of transactions
time	uint32_t block creation time
bits	the short form of the target hash value
nonce	A number that is incremented after each iteration of the hash calculation, starting from zero
n_tx	The number of transactions in the list
size	Block size in bytes

3.1 Merkle tree construction algorithm

The construction of a Merkle tree is shown in Figure 2. Merkle tree construction algorithm:

- The hashes of each operation in the block are calculated: hash (L1), hash (L2), etc.
- Hashes are calculated from the sum of transaction hashes: hash (hash (L1) + hash (L2)).
  Since the tree is binary, the number of elements in each iteration must be even. If there is an
  odd number of transactions in the block, then the last one is repeated and added to itself;
- The second point occurs until the calculation of a single hash, which is the root of the
  Merkle tree.

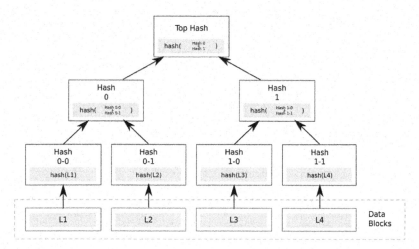

Figure 2. Merkle tree.

4 RESULT

Using the Merkle tree, it is possible to easily verify that a particular transaction is included in
a given block, thereby reducing the cost of this process both technically and financially for
the users of the network itself. Since each subsequent block of transactions references the
previous one, knowing the current block, all transactions on the network can be easily read,
continuing to trace the block chain up to the first line, which is called the "genesis" block.

On our server, timestamps are searched for a value with the desired hash by iterating over
the value (nonce) of the repeating join field in the data block. Once a block is found that
satisfies the condition, its contents cannot be changed without redoing all the work. If this
block is not the last part of the chain, this operation involves recalculating all the blocks that
follow it.

5 CONCLUSION

It also solves the problem of determining the version supported by the majority by hashing.
If a single IP address is considered a voice, then such a scheme can be broken if a large
number of addresses are managed. Our scheme is based on the principle of "one processor -
one voice". The longest of the hash chains represents the opinion of the majority who contributed the most resources to it. If more than half of the computing power belongs to
uncorrupted nodes, then the uncorrupted transaction chain will grow faster and outperform
any competing chain. In order to make changes to one of the previous blocks, the attacker
may have to redo the work in that block and all subsequent blocks, and then overtake the

honest participants of the new blocks. In such a situation, the probability of such success for an attacker with fewer resources decreases dramatically with the number of blocks.

To compensate for the ever-increasing processing power of processors and the change in the number of working nodes in the network, the hashing complexity must be changed to ensure the speed of block production. If they appear more often, the complexity increases and vice versa.

REFERENCES

[1]   Nakamoto, Satoshi (October 2008). *"Bitcoin: A Peer-to-Peer Electronic Cash System"* (PDF). bitcoin. org. Archived (PDF)from the original on 20 March 2014. Retrieved 28 April 2014.

[2]   O'Keeffe, M.; Terzi, A. (7 July 2015). *"The Political Economy of Financial Crisis Policy"*. Bruegel. Archived from the original on 19 May 2018. Retrieved 8 May 2018.

[3]   PQ-3832-son 03.07.2018, *"O'zbekiston Respublikasida Raqamli Iqtisodiyotni Rivojlantirish Chora-Tadbirlari To'g'risida"*

[4]   Iansiti, Marco; Lakhani, Karim R. (January 2017). *"The Truth about Blockchain"*. Harvard Business Review. Harvard University. Archived from the original on 18 January 2017. Retrieved 17 January 2017. The Technology at the Heart of Bitcoin and Other Virtual Currencies, Blockchain is an Open, Distributed Ledger that can Record Transactions between Two Parties Efficiently and in a Verifiable and Permanent Way.

[5]   Voorhees, Erik (30 October 2015). *"It's All About the Blockchain"*. Money and State. Archived from the Original on 1 November 2015. Retrieved 2 November 2015.

[6]   Kopfstein, Janus (12 December 2013). *"The Mission to Decentralize the Internet"*. The New Yorker. Archived from the original on 31 December 2014. Retrieved 30 December 2014. The Network's 'Nodes' — Users Running the Bitcoin Software on their Computers — Collectively Check the Integrity of Other Nodes to Ensure that no one Spends the Same Coins Twice. All Transactions are Published on a Shared Public Ledger, called the "Block Chain".

[7]   Gervais, Arthur; Karame, Ghassan O.; Capkun, Vedran; Capkun, Srdjan. "Is Bitcoin a Decentralized Currency?". *InfoQ. InfoQ & IEEE Computer Society*. Archived from The Original on 10 October 2016. Retrieved 11 October 2016.

[8]   Zheng, Z., Xie, S., Dai, H., Chen, X. and Wang, H. (2017) "An Overview of Blockchain Technology: Architecture, Consensus, and Future Trends". *2017 IEEE International Congress on Big Data (BigData Congress)*, Honolulu, 25–30 June 2017, 557–564.

[9]   Stuart Haber and W. Scott Stornetta, *"How to Time-Stamp a Digital Document"*. 1991 International Association for Cryptologic Research, Morristown, NJ 07960-1910, U.S.A, J. Cryptology (1991) 3:99–111.

[10] Tanweer Alam. *"Blockchain and its Role in the Internet of Things (IoT)."* *International Journal of Scientific Research in Computer Science, Engineering and Information Technology*. Vol 5(1), 2019. DOI: 10.32628/CSEIT195137

[11] Applications for blockchain. https://www.unpri.org/sustainable-financial-system/stock-exchange-innovation-applications-for-blockchain/3597.article

[12] Finoa. (2018) *The Era of Tokenization — Market Outlook on a $24trn business opportunity*. Retrieved from: https://medium.com/finoa-banking/market-outlook-on-tokenized-assets-ausd24trn-opportunity-9bac0c4dfefb.

Artificial Intelligence, Blockchain, Computing and Security – Dagur et al. (Eds)
© 2024 The Author(s), ISBN: 978-1-032-49393-0

Authenticating digital documents using block chain technology

E. Benitha Sowmiya, D. Isaiah Ramaswamy, S. Hemanth Sai, T. Vignesh & S. Madhav Sai
Bharath Institute of Science and Technology, Chennai, India

ABSTRACT: As technology has developed rapidly, it has become more and more common for important documents to be faked due to easy access to inexpensive, high-tech office instruments. Document verification, however, has become very complex and time-consuming due to various challenging and tedious processes, which prompted us to conduct this study. Document verification has become very difficult and time-consuming due to numerous challenging and tedious processes, which is why we conducted this study. Our decentralised Using P2P cloud storage and Ethereum block chain technologies; a web application verifies digital documents. The suggested architecture includes a number of techniques, including which makes it easier and quicker for businesses or authorities to verify uploaded documents. Our proposed model fully complies with the specifications for a digital document verification system by resolving the shortcomings and difficulties in the existing methods for document verification chain technology ensures safety, autonomy, consensus, and trust. Because of the completely trustworthy, transparent, and incorruptible way that the transactions are stored and validated, this block chain technology has also motivated us to apply it in our attempts to authenticate important digital documents.

Keywords: hashing, peer-to-peer networks, KNN algorithm, id-selfie datasets, ethereum block chain

1 INTRODUCTION

More businesses and individual users are turning to digitized documents as a result of the quick development of information sharing and exchanging. Furthermore, the time-consuming and laborious procedure of using and validating conventional physical documents motivates people to use contemporary methods of issuing and validating crucial documents. Although using digital documents is obviously easy, it is frequently a worry to establish their legitimacy. It is currently relatively easy to fabricate important documents due to the technological revolution and easy access to affordable, sophisticated technologies, making document authentication a time-consuming process. The implications of the false document problem are having serious and worrying effects, thus they must be taken into quick attention. For users to keep their digital papers, a mechanism to confirm the validity of crucial documents would be very helpful. Block chain can be used to resolve this issue because it is open-source, unchangeable, and consensus-based.

Recently, block chain technology was created to enhance document verification and make it more difficult to reduce document fraud and misuse. The terms "block chain" simply designates a distributed database where information is kept in data packs or blocks those is temporally connected and are maintained in a manner that makes it challenging to update the information. Block chain, a cutting-edge technology, can benefit companies in a number of ways incorporating block.

DOI: 10.1201/9781003393580-42

2 METHODOLOGY

2.1 *Modules*

- User registration and authentication
- User upload certificate
- Get certificate
- QR request and response from verification authority

2.2 *Modules description*

2.1.1 *User registration and authentication*
In this module, users must sign up for their applications and submit a request for authentication to the central board server. A request must be approved by the central board server before the user may access his account. As soon as the central board server approves the request, a key is issued and the user can log into his account.

2.1.2 *User upload certificate*
After logging into his account, the user must submit his certificates, including his voter id, Aadhar card, and SSC certificates, to the central board server. After reviewing the certificates, the central board server will decide whether to accept or reject them. These details will be saved in E.C.S. and block chain if central board server approves the certificate. E.C.S. or the block chain will not keep the certificate if the central board server rejects it.

2.1.3 *Get certificate*
If user needs a certificate he will send request to central board server. If central board server found the user details to be genuine he accepts the request and forwards a request to E.C.S where all the certificates will be there. E.C.S. responds for the request and certificates will be provided to the user.

2.1.4 *QR request and face verification*
To apply for any certifications, a user must submit a request to the central board server, which will verify the request's details and send them on to E.C.S. E.C.S, will then create the QR Code, which will then be provided to the user by way of the central board server. The user sends the QR code to the verifying authority, who will issue the document if all the information is accurate and the face matches a live person.

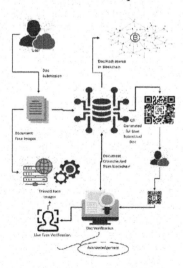

Figure 1.   System architecture.

3 RESULTS AND DISCUSSION

3.1 *Existing system*

Identity verification is necessary for our everyday business operations under the current method. For instance, we must present our identification documents with facial photographs to human operators at access control, physical security, and international border crossings in order to establish our identities and access (security) level. These documents include passports and driver's licences. This approach, however time-consuming, slow, and erratic. It is therefore necessary to develop an automated system that can accurately and quickly match live facial photographs (selfies) to ID document photos. The gate automatically opens for a traveller to enter once their identification has been confirmed using face comparison. They are contrasting a digital or scanned document photo when doing IDselfie matching.

3.2 *Proposed system*

We propose a block chain-based certificate scheme to address the problem. Since each node has its own data store, all other nodes must request that one internal datum be changed at the same time. The mechanism is hence extremely dependable. Using the Ethereum block chain as a foundation, we created a decentralized application and a certificate system. The incorruptible, encrypted, trackable, and data synchronization capabilities of this technology were key factors in its selection. Utilizing block chain technology, the solution increases operational effectiveness at every level. With the aid of technology, less paper is required; administrative expenses are decreased, document fraud is stopped, and accurate and trustworthy information on digital certificates is delivered. It contrasts the real user face with that of the verified document.

3.2.1 *Project scope*

Address the problem. Since each node has its own data storage, all other nodes must simultaneously request that one internal datum be adjusted since each node has its own data store. The mechanism is hence extremely dependable. Using the Ethereum block chain as a foundation, we created a decentralized application and a certificate system. The incorruptible, encrypted, trackable, and data synchronization capabilities of this technology were key factors in its selection. The system incorporates block chain components to increase the efficiency of activities at each level. The solution lessens the need for paper, cuts down on administrative costs, prevents document fraud, provides accurate and reliable information on digital certificates, and compares the user's actual face to the face of the certified document.

3.2.2 *Product perspective*

Our face-containing ID cards, such passports and driver licences, which we must present to human operators in order to authenticate our identity for a number of daily activities. Though labor-intensive, slow, and inconsistent, this method. It is therefore necessary to develop an automated system that can accurately and quickly match live facial photographs (selfies) to ID document photos. We this study suggests DocFace+ as a means of achieving this. We first demonstrate that, similar to the state-of-the-art ID-selfie datasets, gradient-based optimization approaches converge slowly when many classes have few samples (as a result of the under fitting of classifier weights). In order to address this flaw, could the classifier weights be updated to enable quicker convergence and more generic representations? Despite being a publicly available generic face matcher, cross-validation on an ID selfie dataset confirms this.

3.2.3 *System features*

We are frequently required to establish our identification to human operators by displaying IDs featuring facial images, passports and driver's licenses, for example. However, this method requires a lot of work, takes a long time, and is unreliable.. As a result, a computerized system is needed for quickly and accurately matching live face photographs (selfies) to ID document photos. In order to achieve this goal, in this work, we offer a remedy called DocFace+. We first demonstrate that when many classes have a small number of samples, as is the case with existing ID-selfie datasets, gradient-based optimization approaches converge slowly (as a result of the under fitting of classifier weights). In order to address this flaw, could the classifier weights be updated to enable quicker convergence and more generic representations? To learn a unified face representation with domain-specific characteristics, a pair of sister networks with partially shared parameters are trained. Cross-validation on an ID selfie dataset confirms this despite the fact that it is a generally available public face matcher.

4 CONCLUSION

We can create the verification documents digitally so we don't have to carry them along. KNN is one of the simplest classification techniques. Despite its apparent simplicity, it is capable of producing results that are fiercely competitive. The KNN algorithm can also be used to solve regression problems. Which provides generic representations and a faster convergence? The following stage involves training two sibling networks to discover. Despite being a widely accessible public face matcher, cross-validation on an ID selfie dataset demonstrates this.

REFERENCES

[1] Leible, S. Schlager, S. Schubotz, M. & Gipp, B. 2019. "A Review on Blockchain Technology and Blockchain Projects Fostering Open Science", *Front. Blockchain* 2:16. doi: 10.3389/fbloc.2019.00016.
[2] Prashanth Joshi, A. Han, M. & Wang, Y. 2018, "A Survey on Security and Privacy Issues of Blockchain Technology," *Mathematical Foundations of Computing* Volume 1, Issue 2, doi:10.3934/mfc.2018007, pp. 121–147.
[3] Chen, W. Xu, Z. Shi, S. Zhao, Y. & Zhao, J. 2018. "A Survey of Blockchain Applications in Different Domains," doi:https://doi.org/10.1145/3301403.3301407, pp. 17–21.
[4] Gilani, K. Bertin, E. Hatin J. & Crespi, N. 2020. "A Survey on Blockchainbased Identity Management and Decentralized Privacy for Personal Data," *2nd Conference on Blockchain Research and Applications for Innovative Networks and Services (BRAINS)*, Paris, France, doi: 10.1109/BRAINS49436.2020.9223312, pp. 97–101.
[5] Wang, J. Junqi, G. Du, Y. Cheng, S. & Li, X. 2019. "A Summary of Research on Blockchain in the Field of Intellectual Property", *Procedia Computer Science*, Volume 147, doi: https://doi.org/10.1016/j.procs.2019.01.220, pp. 191–197.
[6] Rouhani, S. & Deters, R. 2019. "Security, Performance, and Applications of Smart Contracts: A Systematic Survey," *in IEEE Access*, vol. 7, doi: 10.1109/ACCESS.2019.2911031, pp. 50759–50779.
[7] Yue, D. Li, R. Zhang, Y. Tian W. & Peng, C. 2018. "Blockchain Based Data Integrity Verification in P2P Cloud Storage," *IEEE 24th International Conference on Parallel and Distributed Systems (ICPADS)*, Singapore, doi: 10.1109/PADSW.2018.8644863, pp. 561–568.
[8] Teymourlouei, H. and Jackson L. 2019. *"Blockchain: Enhance the Authentication and Verification of the Identity of a User to Prevent Data Breaches And Security Intrusions."*
[9] Zhu, X. 2020. "Blockchain-Based Identity Authentication and Intelligent Credit Reporting," *Journal of Physics: Conference Series*, volume 1437, 012086, doi: 10.1088/1742-6596/1437/1/012086.
[10] Arjomandi, L. M. Khadka, G. Xiong, Z. & Karmakar, N. C. "Document Verification: A Cloud-Based Computing Pattern Recognition Approach to Chipless RFID," *in IEEE Access*, vol. 6, 2018, doi: 10.1109/ACCESS.2018.2884651, pp. 78007–78015.
[11] Musarella, L. Buccafurri, F. Lax, G. & Russo, A. 2019. *"Ethereum Transaction and Smart Contracts among Secure Identities."*
[12] Lakmal, C. Dangalla, S. Herath, C. Wickramarathna, C. Dias, G. & Fernando, S. 2017. "ID Stack — The Common Protocol for Document Verification Built on Digital Signatures," *National Information Technology Conference (NITC)*, Colombo, doi: 10.1109/NITC.2017.8285654, pp. 96–99.
[13] Hamitha Nasrin, M. Hemalakshmi, S. & Ramsundar, G. 2019."A Review on Implementation Techniques of Blockchain Enabled Smart Contract for Document Verification," *International Research Journal of Engineering and Technology (IRJET)*, Volume 6, Issue 2, 81.
[14] Ghazali, O. & Saleh, O. "A Graduation Certificate Verification Model via Utilization of the Blockchain Technology," *2018, Journal of Telecommunication, Electronic and Computer Engineering*, 10, pp. 29–34.
[15] Shah, M. & Dr. Priyanka Kumar. 2019. "Tamper Proof Birth Certificate using Blockchain Technology", *International Journal of Recent Technology and Engineering (IJRTE)*, Volume 7, Issue 5S3, pp. 95–98.

Communications

Artificial Intelligence, Blockchain, Computing and Security – Dagur et al. (Eds)
© 2024 The Author(s), ISBN: 978-1-032-49393-0

Vehicles communication and safe distancing using IOT and ad-hoc network

Raj Kumar Sharma, Roushan, Rajneesh Dev Singh & Isha Nair
Department of Computer Science, Sharda University Greater, Noida, India

ABSTRACT: This paper presents a systematic approach for vehicle communication that addresses the problem. Vehicular Ad-Hoc Network (VANET) technology is used to create a network that allows vehicles to exchange information while moving. The purpose of this analysis is to create a VANET domain and algorithms for collision evasion. To test this, we built a sample model to demonstrate how the proposed idea works and implemented the VANET with microcontrollers, sensor systems, and toy cars. The combination of all these elements creates a collision avoidance system that helps increase the safety of people on the road by maintaining a safe following distance between vehicles. The proposed method applies vehicle detection and feature tracking to each frame within the range of the sensor to improve detection accuracy and enhance vehicle localization stability. The physical model was built using an Arduino Uno microcontroller, SR04 ultrasonic sensor for distance calculation, Neo-6m (GPS), 16*2 display, I2c module, Wi-Fi module and Toy car.

1 INTRODUCTION

The Special Vehicle Network (VANET) is an area of research that has attracted a lot of attention in recent years. The idea shows the prototype of his VANET that predicts a collision head and warns the driver in VANET In addition, the system works independently from other information about the car [1]. The distance measurement was carried out using ultrasonic sensors as a means of determining the position of the vehicle, and the speed of the vehicle was determined using a GPS receiver. The system uses this data on average over several measurements to calculate the safe braking distance. In addition, if the GPS receiver is out of range, for example, in the car tunnels, each host can rely on the vehicle [2,3].

Unique features such as high dynamic topology and projected agility have been identified from a research perspective. The article begins with a literary review incentive, followed by a discussion of methods and issues. A custom communication and information service that provides ease and business use for riders and passengers. Collision detection is especially important in poor visibility [4,5]. If the driver is unable to recognise the danger immediately due to obstructions such as buildings. One of the most important components of the mobile network is the Special Vehicle Network. VANET facilitates communication between units located in vehicles and road units. Grouping is used as an important function in response to frequent changes [6].

So far, as part of these research efforts, we have modeled two nodes in the network (automobiles).

Collide at the intersection. In the simulation, we use a portable sense of multi-access using the CS/CA protocol [8,9]. This is a protocol for the computer network using the carrier sense. The node tries to avoid collisions by sending only when the channel is clean. Another situation is the intersection, where four nodes will move in different directions and each node can send data.

2 MOTIVATION

The inspiration for this plan, as well as for the spread of IoT, was the Bennett phenomenon associated with exploitation.

2.1 *The exploitation phenomenon in VANETs*

- Despite VANET's immense prospects of taking steps to reduce operating costs and reporting security and efficiency concerns, the past two decades have failed to generate enthusiasm that benefits the enterprise. Some of the reasons for the decline in commercial enthusiasm for VANET are mentioned in [11].
- The structure of VANET could not guarantee a worldwide and supportable establishment of ITS use cases. This is related to purely ad-hoc system design. When a car disconnects from the ad-hoc system, even if it car is in its lane, it cannot connect to any other route-accessible network, so facilities are lost from the system [12].
- The current VANETs system could not guarantee internet availability. As such, drivers and travelers do not have access to profitable use cases. Exactly because profitable use cases rely on solid cyberspace systems [14].
- In our daily lives, individual gadgets have developed a lot, but gadgets still cannot connect to VANET. It's just because of bad system design [7].
- Calculation of transportation conditions in VANET based on crew experience results in less accurate management in ITS applications as the associated risks depend on managing long driving encounters.

2.2 *Safety and transport efficiency*

VANET faces many security challenges and issues related to authentication and privacy. In addition to these, untrusted vehicles pose many security and communication problems in VANET. With VANET, all communication is in an open access environment, making VANET more vulnerable to attack. Therefore, an attacker can modify, intercept, insert, and delete messages in her VANET. In wireless communication technology, he said, effective use of VANET requires the efficient handling of security and privacy issues by introducing sophisticated algorithms to combat all kinds of threats and attacks. Several research studies on authentication and privacy schemes for VANET systems have been proposed to address these issues.

3 PROPOSED METHOD

I created a model for this. First, we calculated the distance in all directions, then created an ad-hoc network to detect vehicles or objects and announce or broadcast a message. Communication is as shown in Figure 1.

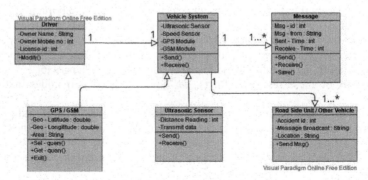

Figure 1.   Flow of our model.

3.1 *Distance estimation*

Here, ultrasonic sensors are used to identify objects around the car and take appropriate measures according to this. If the car or object is detected at a distance of 250 metres, this allows our system to respond accordingly to avoid Crows This module works on the natural phenomenon of reflection. At the start of the pulse module, about 10 MS are transmitted, then the module will transmit 40 kHz ultrasonic signals for 8 cycles and monitor the echo. The signal returns after colliding with the barrier and is received by the receiver as shown in Figure 2. Therefore, the distance from the barrier sensor is calculated simply by using the formula defined below.

$$\text{Distance} = (\text{time} \times \text{speed})/2.$$

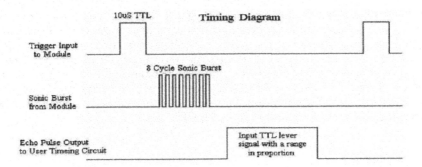

Figure 2. Timing diagram of ultrasonic sensor.

3.2 *Vehicle detection*

Vehicles are detected in two ways: by transmitting ultrasonic signals with range sensors, and by entering the VANET network. When vehicle A and vehicle B are moving in the same direction and both vehicles are within range of the vehicle network, they start exchanging basic information such as speed and distance in meters. In the event of an accident involving B or another vehicle noticed by B, vehicle B will broadcast her message to notify all other vehicles within range so that they can take appropriate action.

Figure 3. Our proposed model.

3.3 Physical model

We have designed a physical model that can simulate this whole event to understand how it is working and to demonstrate its functionality. Our physical model has four RC, each RC representing a physical car, and we connected some pairs of sensors for storing and calculating some basic data about the vehicle and its speed, as shown in Figure 3.

4 FUTURE WORK

Cars and car networks are important technologies to ensure the safety of the driver on the road. It is necessary to continue modelling, assuming a number of cars and different structures on the road. Further development of this prototype should be aimed at improving the reliability of the inspection. Many other improvements have made it possible to improve driving in high-frequency mode [15]. Many cars should be added to the network. The receiver node must be programmed to send instructions to the vehicle, telling it carefully to avoid a car crash with detected objects. One that can make research more convenient is the development of printed circuit boards for the room. It is a good starting point, but it is complicated by the rise of unstable systems and problems. You can consider other options such as the Raspberry Pi, Arduino Mega, and Intel Edison.

5 CONCLUSION

The aim of this project is to model a specific network of vehicles by exchanging messages between them. It shows that this model works perfectly in a small, controlled environment. When various nodes enter the range, they connect and start exchanging information together. However, for this communication to be considered very effective, send messages. It is important to understand that many transmissions may not work unless on such a road.

REFERENCES

[1]   Joo, D.Y and Kim, J.K.: *Creative & Active Convergence Model of IoT*, Korea Institute for Industrial Economics & Trade, Korea (2014).

[2]   Eason G., Noble B., and Sneddon I. N., "On Certain Integrals of Lipschitz-Hankel Type Involving Products of Bessel functions," *Phil. Trans. Roy. Soc. London*, vol. A247, pp. 529–551, April 1955.

[3]   Baek, Jang Woon, Byung-Gil Han, Hyunwoo Kang, Yoonsu Chung, and Su-In Lee. "Fast and Reliable Tracking Algorithm for On-road Vehicle Detection Systems." *2016 Eighth International Conference on Ubiquitous and Future Networks (ICUFN)* (2016): n. pag. W.

[4]   Clerk J Maxwell, *A Treatise on Electricity and Magnetism*, 3rd ed., vol. 2. Oxford: Clarendon, 1892, pp.68–73.

[5]   Yorozu Y., Hirano M., Oka K., and Tagawa Y., "Electron Spectroscopy Studies on Magneto-optical Media and Plastic Substrate Interface," *IEEE Transl. J. Magn. Japan*, vol. 2, pp. 740–741

[6]   Temkar, Rohini, Vishal Asrani, and Pavitra Kannan. "IoT: Smart Vehicle Management System for Effective Traffic Control and Collision Avoidance." *International Journal of Science 4.5* (2015): 3241–246. W.

[7]   Zhong, Xunyu, Xungao Zhong, and Xiafu Peng. "Velocity-Change-Space-based Dynamic Motion Planning for Mobile Robots Navigation." *Neurocomputing* 143 (2014): 153–63. W.

[8]   Deligiannakis, Antonios, and Yannis Kotidis. "Detecting Proximity Events in Sensor Networks." *Information Systems* 36.7 (2011): 1044–063. W.

[9]   Zhu, Wanting, Deyun Gao, and Chuan Heng Foh. "An Efficient Prediction-Based Data Forwarding Strategy in Vehicular Ad Hoc Network." *International Journal of Distributed Sensor Networks 11.8* (2015): 128725. W.

[10] Almagbile Ali, Jinling Wang, and Weidong Ding. "Evaluating the Performances of Adaptive Kalman Filter Methods in GPS/INS Integration." *Journal of Global Positioning Systems 9.1* (2010): 33–40. W.

[11] Jing Q., Vasilakos A. V., Wan J., Lu J., and Qiu D., "Security of the Internet of Things: Perspectives and Challenges," *Wireless Networks*, vol. 20, no. 8, pp. 2481–2501, 2014.

[12] Yaseen Q., AlBalas F., Jararweh Y., and Al-Ayyoub M., "A Fog Computing-based System for Selective Forwarding Detection in Mobile Wireless Sensor Networks," in *Foundations and Applications of Self* Systems*, 2016.

[13] Kumar, A., & Alam, B. (2014, February). Real Time Scheduling Algorithm for Fault Tolerant and Energy Minimization. In *2014 International Conference on Issues and Challenges in Intelligent Computing Techniques (ICICT)* (pp. 356–360). IEEE.

[14] Chaturvedi, R., Kumar, S., Kumar, U., Sharma, T., Chaudhary, Z., & Dagur, A. (2021). Low-cost Iot-enabled Smart Parking System in Crowded Cities. In *Data Intelligence and Cognitive Informatics: Proceedings of ICDICI 2020* (pp. 333–339). Springer Singapore.

[15] Anadu D., Mushagalusa C., Alsbou N. and Abuabed A. S. A., "Internet of Things: Vehicle Collision Detection and Avoidance in a VANET Environment," *2018 IEEE International Instrumentation and Measurement Technology Conference (I2MTC)*, Houston, TX, USA, 2018, pp. 1–6, doi: 10.1109/I2MTC.2018.8409861.

PG Radar

Yash Grover, Aditya & Kadambari Agarwal
Department of Computer Science and Engineering, ABES Engineering College, Ghaziabad, Uttar Pradesh

ABSTRACT: The usage of mobile technology and applications in the accommodation industry has grown in popularity, and the PG Radar app is an example of an android-based paying guest app that tries to meet this requirement. PG Radar is an Android software that assists users in locating suitable and authentic paying guest rooms. Additionally, paying guest owners can promote their lodgings, and users can search for paying guest accommodations, apartments, and mansions in Ghaziabad's Crossing Republik. Users can look for lodging based on certain parameters such as reviews, meals, rent, security, and so on. Overall, the goal of PG Radar is to assist paying guest owners in better managing their communities and related activities, as well as to provide a convenient alternative for persons seeking temporary housing.

This app simplifies the process of searching for housing and eliminates the need to spend money on housing brokers and aims to make the process of finding temporary housing more convenient and cost-effective. [8]

1 INTRODUCTION

Paying guests (PGs) and apartment rentals can be difficult to manage due to their reliance on manual processes. This can lead to vacancies and a lack of communication. To address these issues, a rental management system is needed. This system can be accessed through websites or mobile apps, which are becoming increasingly popular due to the widespread use of mobile devices. The system can automate many tasks, reducing the workload on people. Students and other individuals may choose to stay in a PG or apartment near a university or college in order to save time and energy on commuting. This can be difficult if they do not have connections in the area, so a system that helps them locate and book nearby accommodations can be useful.

A paying guest (PG) is a type of accommodation where a person rents a room in a shared living space and typically pays for it on a monthly basis. An Android application for managing a paying guest accommodation could potentially offer a range of features for both the owners of the accommodation and the guests staying there. Some potential features for such an application might include:

- Room booking and reservation: Guests could use the app to browse available rooms and make a booking or reservation directly from their phone.
- Communication with the owner: The app could provide a messaging platform for guests to communicate with the owner or manager of the accommodation. This could be useful for arranging check-in or asking any questions about the property.
- House rules and policies: The app could display the house rules and policies for the accommodation, so that guests are aware of what is expected of them.
- Amenities and services: The app could list the amenities and services available at the accommodation, such as access to a shared kitchen or laundry facilities.

DOI: 10.1201/9781003393580-44

In addition to these features, the app could also include functionality for the owner or manager of the accommodation to manage bookings and payments, communicate with guests, and update information about the property.

Overall, an Android application for managing a paying guest accommodations could provide a convenient and efficient way for both owners and guests to manage their accommodations and make bookings. It could potentially streamline the process of renting a room in a shared living space, making it easier and more convenient for both parties.

2 LITERATURE SURVEY

Our application, "PG Radar" is a resource for individuals seeking rental housing in Crossing Republik , Ghaziabad, a city in India. It is an educational application that provides detailed information about paying guest (PG) accommodations in the area, organized by feedback given by the existing users to make searching easier. Unlike other websites or search engines, our website specifically focuses on PGs in Crossing Republik. Users can view detailed descriptions of the available housing options, add their favourites to a list, and book a house later. In addition to providing information about housing. The website also provides information about nearby facilities such as a gym and laundry service. [7]

3 SURVEY ANALYSIS

We conducted a survey with students living as paying guest in Crossing Republik, Ghaziabad. A glimpse of the survey questions and results. [11]

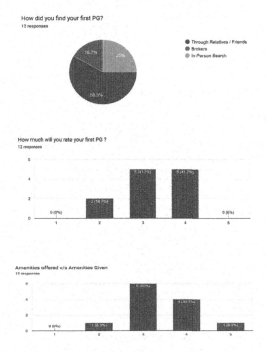

- Video Reference:
  https://drive.google.com/file/d/1VV5XstVQQO1KycOVkmUiZ42ML1AVO3ae/view?usp= sharing

4 PROPOSED APPROACH

The PG Radar App is an Android-based application designed to help people find accommodation in Crossing Republik, Ghaziabad. It is a user-friendly system that stores and shares information about paying guest (PG) accommodations. The app allows users to inquire about PGs and view their addresses and rent prices. The managing committee and residents of a housing society can install the app and register themselves to use it. The app is developed using an Android Studio for the frontend and a Google Firebase [10] for the backend. It is divided into several modules, including a registration module for users and a database module for storing and accessing information. [1,9]

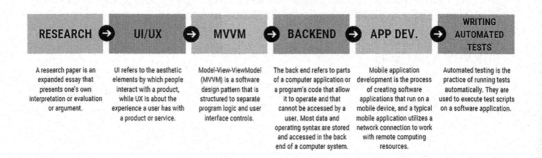

PROJECT
ROADMAP

RESEARCH →	UI/UX →	MVVM →	BACKEND →	APP DEV. →	WRITING AUTOMATED TESTS
A research paper is an expanded essay that presents one's own interpretation or evaluation or argument.	UI refers to the aesthetic elements by which people interact with a product, while UX is about the experience a user has with a product or service.	Model-View-ViewModel (MVVM) is a software design pattern that is structured to separate program logic and user interface controls.	The back end refers to parts of a computer application or a program's code that allow it to operate and that cannot be accessed by a user. Most data and operating syntax are stored and accessed in the back end of a computer system.	Mobile application development is the process of creating software applications that run on a mobile device, and a typical mobile application utilizes a network connection to work with remote computing resources.	Automated testing is the practice of running tests automatically. They are used to execute test scripts on a software application.

[5,6]

4.1 Administration module

- Controls all information and has access to it.
- Check the owner's and customer's information.
- View/Delete the PG, owner, and user information.

4.2 Owners module

- The PG owner will use identity proofs to complete the registration process.
- They can quickly add details about the apartment, rooms, and paying guests, and they may amend or delete the details as needed.

4.3 User module

- The user must additionally register using their unique ID.
- Users can search for hostel rooms, paying guests, and so on throughout the city to acquire room details, room rent, address of the room, images of the room, and so on. [2,3]

4.4 *System flow diagram*

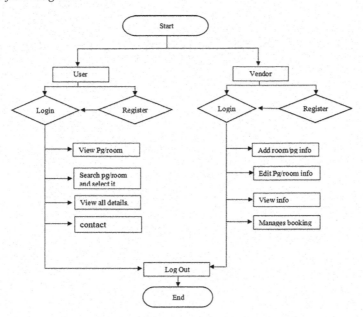

5 CONCLUSION AND FUTURE SCOPE

PG Radar is an app that helps people quickly find authorized PG (paying guest) accommodations. It can be very useful for those who are moving to a new place for work or study and don't want to waste time searching for a place to live.

REFERENCES

[1] Akshatha.M1 B. M., *"PG locator,"* vol. 7, no. 6, 2017.

[2] Murahari Prithvi Yash C. C. A. L. S. D., *"RentoAxis: Android App for Paying Guest,"* vol. 5, no. 12, 2018.

[3] Ramesh Kumar V. K. V. S. C., "Upcoming Future of Paying Guest Accommodation," *International Research Journal of Modernization in Engineering Technology and Science*, vol. 4, no. 5, 2022.

[4] *PG Locator App"* – ijres.org. (n.d.). from https://www.ijres.org/papers/Volume-10/Issue-5/Ser-15/1005105110.pdf

[5] *Emil Lamprecht Emil* from, https://careerfoundry.com/en/blog/ux-design/the-difference-between-ux-and-ui-design-a-laymans-guide/

[6] *Guide to App Architecture: Android Developers.* Android developers. (n.d.). from, https://developer.android.com/topic/architecture

[7] *Online Application for Booking Paying Guest and Explore Mess in Nearby ...* (n.d.). http://www.ijet-journal.org/Special-Issues/ICEMESM18/ICEMESM29.pdf

[8] Paying Guest Tourist Accommodation _ a New Trend to Boon Sustainable Tourism Development in Kashmir. *International Journal of Recent Scientific Research.* (n.d.). https://recentscientific.com/paying-guest-tourist-accommodation-new-trend-boon-sustainable-tourism-development-kashmir

[9] *Developer Guides: Android Developers.* Android Developers. https://developer.android.com/guide

[10] *What is Firebase? All Secrets Unlocked.* Back4App Blog. (2022, August 10) https://blog.back4app.com/firebase/

[11] *Writing Survey Questions* https://www.pewresearch.org/our-methods/u-s-surveys/writing-survey-questions/

[12] *A First Look at Blockchain-based Decentralized Applications* https://www.researchgate.net/publication/336718591_A_first_look_at_blockchain-based_decentralized_applications

Artificial Intelligence, Blockchain, Computing and Security – Dagur et al. (Eds)
© 2024 The Author(s), ISBN: 978-1-032-49393-0

A new framework for distributed clustering based data aggregation in WSN

Anuj Kumar Singh
School of Computer Science & Engineering, Galgotias University, Greater Noida

Shashi Bhushan
Department of Computer Science, ASET, Amity University Punjab, Mohali

Ashish Kumar
Department of Computer Science & Engineering, AKG Engineering College, Ghaziabaad

ABSTRACT: Wireless sensor networks (WSNs) are used in a variety of applications, such as industrial control systems, environmental monitoring, military applications, and structural health monitoring. Energy consumption is a key factor in the lifespan of a WSN, as all sensor nodes are energy-constrained devices. In order to improve energy efficiency, various routing protocols have been developed. These protocols aim to establish paths between sensor nodes and a data sink, while also maximizing the lifespan of the network. However, designing such protocols can be challenging due to the resource and power constraints, as well as the harsh conditions in which sensor nodes often operate. In this paper, a novel approach is proposed for improving energy efficiency in WSNs through the use of a distributed clustering-based data aggregation (DCDA) algorithm. In this approach, we use a grid structure as the basis for clustering, with each cluster having a designated cluster head (CH). Rather than transmitting data directly to the base station (BS), sensor nodes (SNs) send their data to their cluster head, which then aggregates and transmits the data to the BS. This approach can extend the network lifetime by reducing the energy required for communication between SNs. Simulations were conducted to evaluate the performance of the DCDA algorithm in a grid-based WSN. The results showed that using the specified parameters for clustering in the protocol significantly reduces the overall energy consumption of the entire network.

Keywords: Wireless Sensor Network, Clustering, Grid, Source Node, Base Station

1 INTRODUCTION

The essential idea of whenever and anyplace preparing outcomes in the new subject known as Mobile Computing. The improvement in Wireless Networks is also among the primary driver for the advancement and prevalence of mobile computing [1]. The advancement in these minor computing model and different methodologies for remote transmission result in the presentation of the wireless sensor networks (WSNs) [3–7]. Sensor systems are required in the applications, for example, modern control units, ecological checking, military applications and auxiliary wellbeing observing applications. Due to the way that all sensor hubs are vitality obliged gadgets, control utilization of hubs all through transmission or gathering of parcels assumes a crucial job in the life-time of the system [8]. So as to make steering, vitality productive, wide scope of conventions have been planned and presented. Sensor hubs are generally built from four key units: control unit, handling unit, communicational unit, just as a detecting unit. The detecting unit of a hub controls specific physical characteristic for example temperature or the dimension of dampness of a spot for which it is utilized. The preparing

DOI: 10.1201/9781003393580-45

component is responsible for handling the gotten information. The correspondence component of a hub is utilized for transmitting/social event of the acquired information to/from different sensors. The handling and correspondence devours the greater part of the vitality of a sensor to work, and the power segment, is in control for providing of vitality into different segments. Despite the assorted variety in the objectives and goals of sensor applications, the most significant activity of remote sensor hubs is sense and amass information from a potential area, handling the information, and send them back to explicit destinations. Achievement of this mission proficiently needs the structure of vitality proficient steering conventions to organize ways in the middle of sensor hubs and the information sink [9]. The way decision ought to be in such an approach to draw out system's life however much as could reasonably be expected. The properties of nature inside which sensor hubs by and large capacity, notwithstanding serious asset and power constraint, make the structure of steering conventions testing.

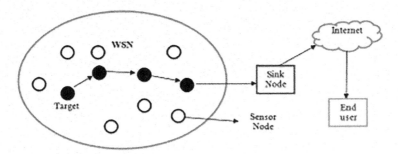

Figure 1.   WSN architecture.

2   AIMS AND OBJECTIVE

In centralized clustering, clustering process is controlled by BS. BS knows about area and energy level of all SNs in the network. BS chooses the ideal CH dependent on the separation and vitality level of hubs. When the CHs are chosen the remainder of SNs join their nearest CH whose duty to total information of its part. Unified bunching acquires additional overhead of control bundles on the grounds that every hub needs to send its data to BS. In distributed clustering, CHs are chosen by hubs themselves which is reasonable for enormous WSN and require less overhead [8]. To circulate CH consistently in network locale is separated in equivalent size frameworks. Dispersed Uniform Clustering Algorithm (DUCA) (36) isolates the sensor region into frameworks and after that head are chosen dependent on LEACH. Yet, this strategy doesn't give affirmation that every lattice must have in any event one head for information accumulation. In this proposal Distributed Clustering Based Data Aggregation (DCDA) convention for network based WSN has been structured [10].

Rests of the parts are orchestrated as pursues:

A framework model for circulated grouping based information accumulation plan utilized for WSN pursued by recreation [11,12].

2.1   System model

Working of DCDA convention is separated into stages:

(a)   Grid Formation Phase
(b)   CH Selection Phase
(c)   Data Aggregation Phase

2.2   Assumptions

Following suppositions are considered in DCDA convention.

- The SNs are consistently conveyed in detecting area.
- All hubs can speak with one another in a framework just as with sink.
- SNs know about their area.
- After sending hubs are stationary.
- All SNs are of homogeneous sort.
- Sink is situated at focal point of area.
- Whenever hub recognizes occasion it transmit information to sink by means of its CH.

2.3 *Grid formation phase*

For rearrangements square field are considered as appeared in Figure 2 which is additionally separated into equivalent size square lattices of size ß* ß. Every framework I considered as group. Sink will give the matrix ID dependent on the co-ordinates of the framework. Sink directions are the beginning stages for development of lattice [13]. Give TH and TL a chance to be the transmission scopes of each SN in high and low power radio mode.

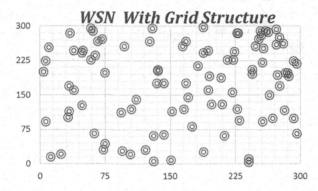

Figure 2.   WSN with grid structure.

2.4 *CH-selection phase*

This stage comprises of two stages: stage 1 and stage 2 which are depicted underneath:

Stage 1: In each bunch, all individuals select their head which is in charge of gathering information from its individuals and transmitted amassed information to sink [14]. The strategy for CH determination is embraced from improved LEACH (QBCDCP) (111) appeared in Figure 3. In this methodology; each hub chooses an incentive in the middle of 0 and 1 arbitrarily. On the off chance that this number is found beneath an edge.

p: the ideal level of CHs

G: the arrangement of SNs that are not chosen as CHs in last 1/P rounds

rs: the quantity of progressive adjusts in that a hub has not been CH.

Eicurrent: the present vitality of hub. EINAS : the most extreme vitality of hub. r: current round

Stage 2: In stage 1 more than one CH is chosen in every matrix. Hence the point of this stage is to choose one CH for every matrix. For this all CHs chose in stage 1 communicate their vitality and IDs. One with most noteworthy vitality is chosen as CH. When the CH has been chosen, it communicates hi bundle with its ID to all individuals. On accepting this welcome bundle, the individuals send join parcel to its CH [16].

Information Aggregation Phase:

A CH Ci of framework Gi gather the information from its individuals and advances totaled information to a CH Cj of neighbor matrix Gj , which has the base good ways from the BS. Same course is preceded until the information at long last came to the BS. In any case, all the CHs that are one bounce away from the BS can straightforwardly speak with the BS.

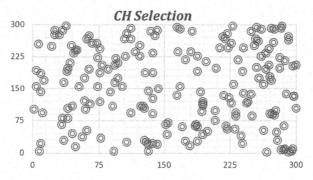

Figure 3.   CH selection.

3   SIMULATION AND RESULTS

DCDA convention is actualized utilizing MATLAB 7.5 on an Intel i5 processor and 2 GB RAM running on the Microsoft Windows 8 working framework. Reproduction parameters are given in Table 1. In the reproduction, the vitality model and the exhibition parameter esteems set same as LEACH. Assessment parameters are lingering vitality and network lifetime [18]. For correlation, LEACH (24) and EEHC (75) on framework based WSN topology is likewise reenacted.

Table 1.   Simulation parameters.

Simulation parameter	Value
Topology size	$300*300m^2$
Number of Nodes (n)	100–500
Initial Energy	0.5J
Sink Location	150*150
Probability (p)	0.1–0.5
TotalNumber of rounds	5

3.1   *Network lifetime*

Network Lifetime is considered as execution parameter for assessing execution of DCDA convention which is characterized as pursues: all out number of dead hubs after 5 rounds. DCDA convention is contrasted and network based EEHC(75) and DUCA(36) [20] by considering the various estimations of wanted level of CH (p) differs from 0.2–0.4 as appeared in Figures 4, 5 and 6.

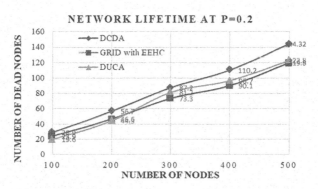

Figure 4.   Network lifetime at p = 0.2.

301

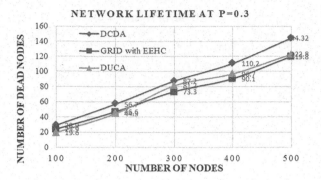

Figure 5.   Network lifetime at p = 0.3.

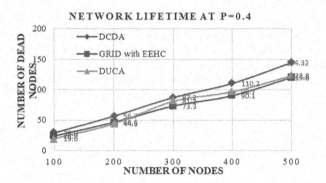

Figure 6.   Network lifetime at p = 0.4.

3.2   *Impact of p on residual energy*

In light of likelihood of wanted level of CHs, charts for the normal remaining vitality of the network against number of hubs for the DCDA is plotted and analyzed lattice based EEHC and matrix based LEACH by considering the various estimations of wanted level of CH (p) shifts from 0.2–0.4 [21]. It very well may be demonstrated that normal lingering vitality after 5 rounds is more in DCDA convention when contrasted with others as appeared in Figures 7, 8 and 9 in light of the fact that no lattice left without head so there will be no immediate transmission between any SN and BS.

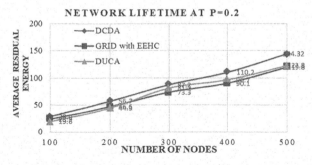

Figure 7.   Average residual energy at p = 0.2.

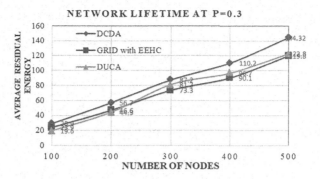

Figure 8.   Average residual energy at p = 0.3.

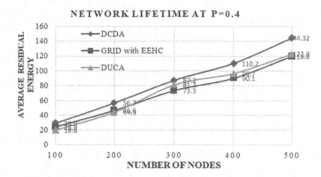

Figure 9.   Average residual energy at p = 0.4.

4   CONCLUSION

The WSNs have been imagined to help in various checking applications. Energy effective lattice-based information total is foremost to broaden the lifetime of the framework. Bunched association of WSN is a successful path for information collection in WSNs. In this proposal, Distributed Clustering Based Data Aggregation (DCDA) calculation for framework based WSN is created in which matrix is considered as bunch. Recreations are done to assess the presentation which demonstrates that the utilization of previously mentioned parameters for bunching in convention altogether decreases the general energy utilization of entire network.

REFERENCES

[1]   Jain A, Sharma D, Goel M, Verma AK. "Conventions for Network and Data Link Layer in WSNs: A Review and Open Issues". In *Advances in Networks and Communications* 2011 Jan 2 (pp. 546–55). Springer Berlin Heidelberg.

[2]   Demirkol I, Ersoy C, Alagoz F. "Macintosh Conventions for Remote Sensor Arranges: A Study". *IEEE Communications Magazine*. 2006 Apr 1;44(4):115–21.

[3]   Tilak S, Abu-Ghazaleh NB, Heinzelman W. "A Scientific Categorization of Remote Miniaturized Scale Sensor System Models". *ACM SIGMOBILE Mobile Computing and Communications Review*. 2002 Apr 1;6(2):28–36.

[4]   Singh, Anuj Kumar, Mohammed Alshehri, Shashi Bhushan, Manoj Kumar, Osama Alfarraj, and Kamal Raj Pardarshani. "Secure and Energy Efficient Data Transmission Model for WSN." *Intelligent Automation and Soft Computing* 27, no. 3 (2021): 761–769.

[5]   Lindsey S, Raghavendra CS. PEGASIS: "Power-proficient Assembling in Sensor Data Frameworks. In *Aerospace Gathering Procedures*", 2002. IEEE 2002 (Vol. 3, pp. 3–1125). IEEE

[6]   Fasolo E, Rossi M, Widmer J, Zorzi M. "In-organize Accumulation methods for Remote Sensor Arranges: An Overview". *Remote Communications, IEEE*. 2007 Apr;14(2):70–87.

[7]   Solis I, Obraczka K. "The Effect of Timing in Information Total for Sensor Systems", In *Communications, 2004 IEEE International Conference* on 2004 Jun 20 (Vol. 6, pp. 3640–3645). IEEE.

[8]   Bhushan, Shashi. "The Use of LSTM Models for Water Demand Forecasting and Analysis." In *Proceedings of 3rd International Conference on Machine Learning, Advances in Computing, Renewable Energy and Communication*, pp. 247–256. Springer, Singapore, 2022.

[9]   Intanagonwiwat C, Govindan R, Estrin D. "Coordinated Dispersion: An Adaptable and Hearty Correspondence Worldview for Sensor Systems". In *Proceedings of the Sixth Yearly Universal Meeting on Mobile Registering and Systems Administration* 2000 Aug 1 (pp. 56–67). ACM.

[10]   Bhushan, Shashi. "Liver Cancer Detection Using Hybrid Approach-Based Convolutional Neural Network (HABCNN)." In *Proceedings of 3rd International Conference on Machine Learning, Advances in Computing, Renewable Energy and Communication*, pp. 235–246. Springer, Singapore, 2022.

[11]   Bhushan, S., Saroliya, A., & Singh, V. (2013). Implementation and Evaluation of Wireless Mesh Networks on MANET Routing Protocols. *International Journal of Advanced Research in Computer and Communication Engineering*, 2(6).

[12]   Eskandari Z, Yaghmaee MH, Mohajerzadeh A. "Energy Effective Spreading Over Tree for Information Collection in Remote Sensor Systems" In *Computer Communications and Networks, 2008. ICCCN'08. Procedures of seventeenth International Conference* on 2008 Aug 3 (pp. 1–5). IEEE.

[13]   M, Wong VW. "An Energy Mindful Traversing Tree Calculation for Information Conglomeration in Remote Sensor Systems". In *Communications, Computers and Sign Processing, 2005. PACRIM. 2005 IEEE Pacific Rim Conference* on 2005 Aug 24 (pp. 300–303). IEEE.

[14]   Bhushan, Shashi, Manoj Kumar, Pramod Kumar, Thompson Stephan, Achyut Shankar, and Peide Liu. "FAJIT: A Fuzzy-based Data Aggregation Technique for Energy Efficiency in Wireless Sensor Network." *Complex & Intelligent Systems* 7, no. 2 (2021): 997–1007.

[15]   Koucheryavy A, Salim An, Osamy W. "Improved LEACH Convention for Remote Sensor Systems". *St. Petersburg University of Telecommunication*. 2009.

[16]   Heinzelman WB, Chandrakasan AP, Balakrishnan H. "An Application-explicit Convention Design for Remote Microsensor Systems". *Remote Communications, IEEE Transactions on*. 2002 Oct;1(4):660–70.

[17]   Bhushan, Shashi, Anui Kumar Singh, and Sonakshi Vij. "Comparative Study and Analysis of Wireless Mesh Networks on AODV and DSR." In *2019 4th International Conference on Internet of Things: Smart Innovation and Usages (IoT-SIU)*, pp. 1–6. IEEE, 2019.

[18]   Yao Y, Gehrke J. "The Cougar Way to Deal with In-organize Question Preparing in Sensor Systems". *ACM Sigmod Record*. 2002 Sep 1;31(3):9–18

[19]   Younis O, Fahmy S. "Regard: A Half Breed, Vitality Effective, Disseminated Bunching Approach for Specially Appointed Sensor Systems". *Portable Computing, IEEE Transactions on*. 2004 Oct;3(4):366–79.

[20]   Suresh D. and Selvekumar K., "Twofold Cluster Head based Reliable Data Aggregation", in *Procedures of World Engineering and Applied Sciences Journal*, Vol. 6 ,No. 3, pp. 136–146, 2015.

[21]   Bhushan, Shashi. "A Novel Digital Forensic Inspection Model for XSS Attack." In *Soft Computing: Theories and Applications*, pp. 747–759. Springer, Singapore, 2022.

Designing composite codes to mitigate side-lobe levels in MIMO radar using polyphase codes

Ankur Thakur*
Assistant Professor, Apex Institute of Technology (CSE), Chandigarh University, Gharuan, Mohali Punjab, India

Bobbinpreet Kaur*
Assistant Professor, Department of Computer Science Engineering, Chandigarh University, Gharuan, Mohali Punjab, India

ABSTRACT: Mitigation in side-lobe levels is an exhausting task in Multiple-Input Multiple-Output (MIMO) radar. Transmit sequences plays a significant role in radar to overwhelm correlation side-lobe levels. In general, side-lobe levels performance of the incoming signal is perceived by cross-correlation function with other transmitted signals. New polyphase codes are projected that shows good auto-correlation and cross-correlation function responses to moderate peak side-lobe level (PSL) and cross-correlation levels (CCL). Performances of the various polyphase codes are compared and the P4 code is chosen for the design of new polyphase code. The proposed composite polyphase codes (CPC) are produced by adding the right and left shifted versions of P4 code. Simulation outcomes validate superiority of the proposed CPC to the counterpart codes.

1 INTRODUCTION

In radar, broader pulses are preferred for long range detection however range resolution becomes poor due to overlying of nearby targets. For better range resolution, small transmit pulses are required. Range resolution is the competency of the system to discriminate two targets in range profile. The profits of shorter and longer pulses are achieved by a pulse compression technique. The correlations among transmitted and returned signals from target are exploited. The correlator includes main-lobe as well as side-lobe levels. The side-lobe levels suppression is careful as important parameters to squared efficiency of the designed signal.

The side-lobe levels can be suppressed by two different techniques. In the first technique, search for the transmitting waveform has to be conducted which gives minimum side-lobe levels in the correlation functions response. In the second technique, the transmitting waveform has to be conceded through the various window functions. The drawback of this method is to increase relative main-lobe breadth that disturbs range resolution. In multiple-input multiple-output (MIMO) radar, to perceive numerous targets, several transmit and receive antennas are exploited at radar system. This process is different from phased-array an antenna in which single waveform is communicated. Every convey antenna can communicate dissimilar arrangement provides additional degrees of freedom. The waveform performances of the conveyed waveforms are affected due to high values of cross-correlation levels (CCL). To invalidate cross-interference, orthogonal sets for the transmission signals are preferred for CCL reduction.

1.1 Related works

In the literature, numerous communicating signals exist such as non-linear frequency modulation, linear frequency modulation (LFM), and polyphase codes. Different methods are designed using

---

*Corresponding Authors: ankurece69@gmail.com and bobbinece@gmail.com

all these existing codes. Blunt et al. (2016) starting from LFM, the auto-correlation function (ACF) of LFM signal provides peak side-lobe level (PSL) of −13.26 dB which is much higher to adjacent target masking. Guodong *et al.* (2019) the Barker codes are not possible beyond 13 code length which offers -22.30 dB PSL reductions that is insignificant for real-world applications.

He *et al.* (2009) Chebyshev chaotic benefits for MIMO radar are investigated which offers higher spectral contents. Kerahroodi *et al.* (2017) side-lobe suppression methods are designed to mitigate range side-lobes. Leilei *et al.* (2018) orthogonal polynomials are recycled to enhance bandwidth for range resolution improvement. A cyclic algorithm designed using chaotic Bernoulli system given in Qazi et al. (2015), useful in MIMO radar for rapid singular value putrefaction. Using polyphase codes, Woo filters are proposed for side-lobe levels suppression in Santra *et al.* (2015).

Thakur *et al.* (2017) filters are designed by combining the shifted versions of polyphase codes and best combination has to be produced at the transmission side. Thakur et al. (2018) unimodular sequences are developed using cyclic procedure obtainable in Thakur et al. (2020) having worthy correlation belongings. Thakur et al. (2021) Binary sequences are established with radar codes such as Gold, Random, and Kasami. It is found that the Kasami sets provide improvement in PSL. The main indices that are to be considered for the enhancement of MIMO radar performance are:

- Complexity of the designed radar waveforms must be least to get high system speed as existing waveforms and optimized algorithms affects system speed.
- PSL and CCL values in the correlation function response must be least to avoid blind target detection.
- Radar waveform response must be equivalently feast with least side-lobe levels on the ambiguity function delay-Doppler plane.

1.2 *Contributions*

Using P4 polyphase codes, new codes for MIMO radar are designed for the suppression of high side-lobe levels. The author's contribution is as:

- Search for the best polyphase codes have been conducted for side-lobe levels suppression. The proposed polyphase codes are designed by combining the shifted version of polyphase P4 codes.
- Many combinations of right and left shifted versions of polyphase codes are observed and the mixture which offers maximum fall in side-lobe levels is chosen for transmission purposes.
- Using correlation function response, PSL and CCL values of the proposed codes are observed with varying code length. Further, the delay-Doppler planes of ambiguity function are studied in details.

2 PROBLEM FORMULATIONS

The polyphase codes advantage is to provide multiple phases which are not confined to 0 and π. Thakur et al. (2022) the phases in polyphase codes lie between 0 and π which depends upon the sequence length. In the proposed method, P4 codes are used due to its high Doppler tolerances and recurring properties givne in Zhang *et al.* (2017). The P4 code phases are given by

$$\phi_n = \frac{\pi}{N}(n-1)(n-1-N) \tag{1}$$

In MIMO radar, each sequence has length N and L numbers of transmitting antennas are considered. The transmitted signal is given as

$$X = [x_1\ x_2\ x_3\ ...\ x_L]_{N \times L} \tag{2}$$

A single antenna transmit signal as

$$x_l(n) = [x_l(1)\ x_l(2)\ x_l(3)\ ...\ x_l(N)]^T \tag{3}$$

where $l \in [1 : L]$ specifies sequence entities. Using auto-correlation and cross-correlation function, the transmitting sequence performance can be analysed. For sequence $x_l(n)$, the ACF values are calculated as

$$r_k = \sum_{l=1}^{N-k} x_l^* x_{l+k}, \quad k = 0, 1, ..., N-1 \tag{4}$$

The highest value of side-lobe levels in (5) represents the PSL value which is calculated as

$$PSL = \max\{|r_k|\}_{k=1}^{N-1} \tag{5}$$

3   PROPOSED METHODOLOGY

The proposed code is designed using P4 codes. The ACF responses of ordinary P4 codes are observed which gives high side-lobe levels leading to ambiguous target detection. New polyphase codes are proposed by shifting the ordinary P4 codes to left and right. These shifts depend upon the sequence length and we have to choose that combination of shifts which produces maximum side-lobes decline in the correlation function. During new polyphase code designs, several shifts are incorporated to the incoming P4 code like one shift, two shifts, three shifts and $(N-1)^{th}$ shifts. The proposed code is designed using the addition of one bit left shift of P4 code to the original P4 code which is named as composite polyphase code (CPC). In MIMO radar, multiple orthogonal sequences are preferred for transmission. The main objective is to achieve orthogonality among transmitting sequences.

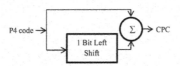

Figure 1.   Proposed CPC generation method.

Once the best polyphase code is generated then orthogonality among transmitting sequences has to be maintained for MIMO radar applications. In this paper, orthogonality among transmitting signals is achieved by changing the phases of proposed CPC through multiplying $Y(N, L)$.

$$Y(N, L) = e^{(j2\pi * rand(N,L))} \tag{6}$$

where $rand(N,L)$ generates random sequences. In literature, many optimization techniques are available for orthogonal signal generations. In our proposed methods, orthogonal signals are produced without using any optimization methods.

4   SIMULATION RESULTS

4.1   *ACF analysis*

The designed CPC's performance is equated with counterpart codes. The well-known P4 code's ACF response is plotted in Figure 2 for $N = 500$. It is perceived that -33.39dB PSL occurs which is much higher for blind target detection. The ACF response of Golomb code is similar to P4 code hence its plot is not given here. Further, the ACF response of CPC is plotted in Figure 3 for $N = 500$ which gives the PSL value of -79.46dB. Comparing PSL values, it is found that CPC offers higher PSL suppression which is greater than two times offered by P4 and Golomb codes.

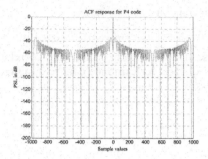

Figure 2.   ACF response of P4 code for N =500.

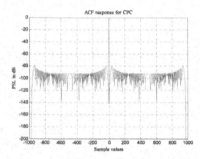

Figure 3.   ACF response of proposed CPC for N = 500.

Like Frank codes, proposed CPC have not constraint of perfect square length. It means CPC sequences can be designed for any arbitrary length N. In MIMO radar, proposed CPC performance is compared with Golomb and P4 code considering $L = 3$ and $N = 128$. The PSL values given by three transmitting antennas are observed using r_{11}, r_{22}, and r_{33} response. Figures 4 and 5a give the ACF responses and it is clear that CPC gives least side-lobe levels than Golomb and P4 code. Whereas CCL values are observed using r_{12} whose response is given in Figure 5b.

4.2   Cross-correlation analysis

In this paper, plots of r_{21}, r_{31}, and r_{32} are not given because responses are similar to r_{12}, r_{13}, and r_{23}, respectively. All these values are calculated using correlation plots given in Figures 4 and 5. The N and L values can be changed as per requirement, here $N = 128$ is considered due to clear visibility of results. If the sequences length is increases then more reduction in side-lobe levels can be achieved. The performance of the designed CPC is observed for different combinations of antennas and sequence lengths. In this paper, all simulations are shown for $N = 128$ and $L = 3$.

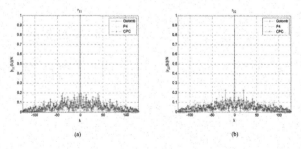

Figure 4.   Correlation response for $N = 128$ and $L = 3$ (a) r_{11} (b) r_{22}.

5 CONCLUSIONS

In this paper, CPC are designed with the help of existing P4 codes. In MIMO radar, orthogonality is an essential parameter which is much needed to reduce cross-interference. Therefore, orthogonal signals are generated for multiple antennas which are mutually orthogonal to each other.

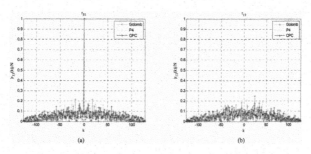

Figure 5.   Correlation response for $N = 128$ and $L = 3$ (a) r_{33} (b) r_{12}.

Using ACF responses, side-lobe levels are observed for the designed code. Different combinations of shifts on the P4 code are incorporated to design best polyphase code. It is concluded that the addition of original P4 code with respect to one bit left shifted P4 code provides minimized side-lobes levels in the cross-correlation function. Further, proposed CPC sequences can be designed for any arbitrary length N.

REFERENCES

Blunt, S.D. & Mokole, E.L. 2016. Overview of Radar Waveform Diversity. *IEEE Trans. Aerosp. Electr. Syst. Mag.* 31(11): 2–42.

Guodong, J., Yunkai, D., Robert, W., & Yajun, L. 2019. An Advanced Nonlinear Frequency Modulation Waveform for Radar Imaging with Low Sidelobe. *IEEE Trans. Geos. Remo. Sens.* 57(8): 6155–6168.

He, H., Stoica, P. & Li, J. 2009. Designing Unimodular Sequence Sets with Good Correlations Including an Application to MIMO Radar. *IEEE Trans. Sig. Process.* 57(11): 4391–4405.

Kerahroodi, M.A., Aubry, A., Naghsh, M.M., & Hashemi, M.M. 2017. A Coordinate-descent Framework to Design Low PSL/ISL Sequences. *IEEE Trans. Sig. Process.* 65(22): 5942–5956.

Leilei, X. Shanghai, Z. & Hongwei, L. 2018. Simultaneous Optimization of Radar Waveform and Mismatched Filter with Range and Delay-Doppler Sidelobes Suppression. *Digital Signal Proces.* 83: 346–358.

Qazi F.A. & Fam, A.T. 2015. Doppler Tolerant and Detection Capable Polyphase Code Sets. *IEEE Trans. on Aerosp. Electr. Syst.* 51(2): 1123–1135.

Santra, A., Srinivasan, R., Jadia, K., & Alleon, G. 2015. Ambiguity Function Processing Gain and Cramer-rao Bound for Matched Illumination Radar Signal. *IEEE Trans. Aerosp, Electr. Syst.* 51(2): 2225–2235.

Thakur, A., Talluri, S.R. & Saini, D.S. 2017. Signal Generation Employing Chebyshev Polynomial for Pulse Compression with Small Relative Side-lobe Level. *IEEE Int. Conf. on Sens. Sig. Proc. and Security.* 114–118.

Thakur, A. & Talluri, S.R. 2018. Comparative Analysis on Pulse Compression with Classical Orthogonal Polynomials for Optimized Time-bandwidth Product. *Ain Shams Engineering Journal*, 9(4): 1791–1797.

Thakur, A. & Saini, D.S. 2020. Bandwidth Optimization and Side-lobe Levels Reduction in PC Radar Using Legendre Orthogonal Polynomials. *Digital Signal Proces.* 101: 102705.

Thakur, A. & Saini, D.S. 2021. MIMO Radar Sequence Design with Constant Envelope and Low Correlation Side-lobe Levels. *International Journal of Electronics and Communications.* 136: 153769.

Thakur, A. & Saini, D.S. 2022. Mitigating Peak Side-lobe Levels in Pulse Compression Radar Using Classical Orthogonal Polynomials. *Wireless Networks.* 28: 2889–2899.

Zhang, L., Yang, B., & Luo, M. 2017. Joint Delay and Doppler Shift Estimation for Multiple Targets Using Exponential Ambiguity Function. *IEEE Trans. Sig. Process.* 65(8): 2151–2163.

Artificial Intelligence, Blockchain, Computing and Security – Dagur et al. (Eds)
© 2024 The Author(s), ISBN: 978-1-032-49393-0

Design and implementation of industrial fire detection and control system using internet of things

Tanushree Bharti*
Department of CSE, Poornima University, Jaipur, Rajasthan

Madan Lal Saini*
Professor, AIT-CSE, Chandigarh University, Mohali, Punjab

Ashok Kumar*
Associate Professor, AIT-CSE, Chandigarh University, Mohali, Punjab

Rajat Tiwari*
Assistant Professor, AIT-CSE, Chandigarh University, Mohali, Punjab

ABSTRACT: In the present time fire is very necessary to monitor and it is a very dangerous circumstances and needed to prevent before something unusual happens. This leaves the fire unattended and leads to many losses of properties, people and more. Fire detection and fire hazard prevention is one of the necessary and important applications of IoT. It can alert distant users or fire control centers to the presence of fires and can identify them early. In this paper a new fire detection and control system is proposed which uses flame, smoke, temperature, LDR, and MQ2 sensors for detecting the speed and severity of fire. This proposed IoT-based fire monitoring and control system is not only sends current situational information but also conducts necessary remedial actions. The NodeMCU board provides the foundation for the construction of the fire alarm system and communicates the information to fire control centers and activates the alarms according to the severity of fire. If a flame is discovered before it has a chance to spread, the system will raise an alarm and notify the fire control center via the Blynk cloud. At the same time the water sprinkler and fire extinguisher will be turned on based on severity.

Keywords: Fire Detection, Fire Control, Fire Extinguisher, Petroleum/Flammable Gases Detection, NodeMCU, Severity of Fire

1 INTRODUCTION

Today, security continues to capture the world's attention. Fires are very common in all types of disasters and cause most of the damage to life, property, etc. There is a particular lack of transport fire signals and the actual system communication is not satisfactory with modern fire detection. With the rapid development of science and technology, a new generation of solid-state system technical theory and artificial intelligence are added to the fire control system and alarm system. Although traditional fire alarm systems can be met to some extent for both fire detection and control system. The intelligent fire alarm control system records, transmits, processes and controls the fire from spreading. We are in a highly dangerous situation when there is a fire, thus it is crucial to be watched over and warned before something bad happens. In many households in poor nations, there are no fire alarm systems. This results in a fire and the destruction of a lot of property and lives. There aren't any tight regulations governing the installation of fire alarm systems, not even in developing

---
*Corresponding Authors: ajmertanu@gmail.com, madan.e13485@cumail.in,
ashok.e11260@cumail.in and madan.e13485@cumail.in

DOI: 10.1201/9781003393580-47

nations like India. In order to create automatic fire monitoring and alarm systems, it is important to establish a department dedicated to fire detection systems.

2 BACKGROUND

Both residential and commercial building designs must take safety into consideration to avoid property damage and fatalities. Fire alarm systems on the market today are very complex in their design and construction. The system is very complicated. Therefore, regular maintenance is required to function properly.

Sowah *et al.*, they mentioned the control and detection of automobile fire systems [1]. Despite the many benefits of road fire detection, over 2,000 vehicles are destroyed by accidental fires every day. He modified it to test computerized systems such as smoke, temperature and flame on a real medium sized car. The results of this study show that the car's fire and alarm system can identify fires within 20 seconds without false alarms. A completely original and mostly modular approach to control a car fireplace was proposed.

A study on automatic fire detection by a network of wireless sensors [2] was conducted and an automatic fire detection was proposed for rapid detection and response. Contribution of sensor networks to early detection of fires, fire detection technology in forest areas, and fire detection tactics in residential areas was discussed. Automated fire controls and detecting systems were covered by Leones Sherwin *et al.* [3]. The main idea behind this effort is to provide a cost-effective firefighting solution. Rishika Yadav and others have talked about motor alarm structures and fireplace detection [4]. Alice Abraham [5] proposed wireless sensor networks to detect wildfires where each sensor node has a temperature and humidity sensor. For early detection of landmine fires a wireless sensor network was proposed by Zervas *et al.* [6]. The author described the system including subsystems for data processing, monitoring, and data collection. They focused their research on optimal communication protocols, scheduling algorithms, and network topologies.

An alert application based on Telos B motes was suggested in 2005 to assist with firefighting operations [7]. For restricted-access environments, the authors combined temperature, light, and humidity sensors. They considered a dispersed WSN made up of various isolated WSNs. Fire-related destruction of sensor nodes was also considered. The author of [8] used the FWI and his unique k-coverage technique to find forest fires. To increase fault tolerance, the K-coverage method track each point utilizing k or more sensor nodes. Bouyeddou [9] suggests that a big range of sensors can examine environmental readings and transmit them to database stations over the wireless medium. WSNs are used for a variety of packages inclusive of remote manipulate, goal tracking, navy, nuclear reactor manipulate, environmental monitoring and tracking. Elsherif [10] designed a fire control system which can detect Methane, CO, smoke, temperature, and flame. Authors of [11]and [12] designed a fire alarm and prevention system for homes using gas, flame, temperature sensors.

3 PROBLEM STATEMENT

Many fire detection and control system have been developed so far but these traditional fire detection and alarm systems can be met to detect fire up to some extent. The intelligent fire alarm control system records, transmits, processes and controls fire alarm systems but most of systems do not work on severity of fire. This proposed system detects the severity of fire by flame, MQ2, LDR, and LM35 sensors. According to the severity of fire, control actions like water sprinkler, fire extinguisher, alarm, alert message, etc. are taken.

4 METHODOLOGY

IoT based fire alarm system uses four sensors, flame, LDR, temperature and smoke sensor to detect the fire and its severity. There is an ADC converter, which converts the simple signals received at the sensor end to computerized ones and later transmits them to the small scale micro controller Arduino. The small micro controller is modified to trigger a bell when the temperature and smoke reach a threshold. As soon as the sensors' values reaches to threshold, fire control system get activate and according to the severity of fire, the devices start working. At the same time, the NodeMCU module sends information to the fire control stations.

All connections between the detector and the board, as well as the circuit between the board and the trigger device are closed loop that is strictly monitored for with open and short circuits. The main fire control panel requires facilities for sensitivity adjustment, separation of detectors from the control panel. Fire detection, alarm monitoring and reporting system panels and associated rectifier are operated by DC power with battery backup. Additionally, each panel requires a battery charger. Each battery pack should have enough power to last 12 hours after a charger or AC power failure. Exit signs should be painted on 18 gauge cold rolled steel. This panel should be able to be wall mounted or hung from the ceiling. Both sides of the exit sign must indicate whether there is an emergency exit.

5 COMPONENTS

5.1 *Flame sensor module*

In flame sensor a mettalic rod is sensitive to the fire and when a fire breakout, the current flow amount get change which generates the signals. These signals are converted into digital or analog information and this gives the information that the fire got breakout. It generates the sound when it detects the fire. The picture on the right represents the IR flame sensor that was used; These sensors are sometimes called fire sensor modules or flame detector sensors.

5.2 *Node MCU*

NodeMCU is microcontroller based board which is used in various IoT applications. It has a low-cost system-on-chip (SoC) called the ESP8266 which was developed and manufactured by Espressif Systems and it includes the elements of a computer like processor, RAM, WiFi, Ports and operating system. It has analog and digtial pins for interfacing various sensors and capable to send the information over internet.

5.3 *LDR sensor*

A Light Dependent Register, by its name, it is sensitive to light so when it gets changes in light, its registestance get change and due to this current flowing through it get change which leads to generate the singal. It is of two types, analog and digital. Its variable resistance get change in response to the amount of light it receives.

5.4 *LM35*

LM35 sensor measures the temperature and generates the analog output voltage proportional to the temperature. These analog output can be converted into Celsius or Fahrenheit. The output power of the LM35 integrated analog temperature sensor is proportional to degrees Celsius. When temperature increases the output voltage of this sensor also increases.

5.5 *Smoke sensor*

Commonly used as a fire indicator, a smoke detector is a device that detects smoke. Commercial smoke detectors function as a part of a fire alarm system by sending a signal to an automatic fire alarm panel. Household smoke detectors, sometimes referred to as smoke alarms, often generate an audible or visual signal, either from a stand-alone detector or, in the case of multiple linked detectors, from multiple detectors.

5.6 *Relay*

Relays are a specific form of electrical circuit that are activated by an electromagnet when current flows through a coil. Electrical devices, devices with switches unable to resist the powerful currents of electrical relays, etc. may be used to operate these relays. In both situations, it offers electric shifting with a very simple and attractive design.

6 WORKING OF THE SYSTEM

Proposed prototype IOT-based fire alarm system consisting of LDR, smoke, LM35 and flame sensor. LM35 temperature sensor of the fire suppression system used to detect the temperature or heat of the fire. The microcontroller board NodeMCU is used to control fire alarm system by detecting the trigger response of the LDR, LM35, smoke and the flame sensor. Blynk Cloud is used for alert notifications which will send alert notifications to users through the cloud messaging service. When the system detects a threshold temperature, the system will simultaneously send SMS alerts to the fire control station. The fire protection system is also activated at the same time. The system's fast and responsive response helps improve safety and standards by preventing home fires. The auto responder does this efficiently and effectively because it reduces human error. Figure 1 shows a functional diagram of the proposed system.

A flame sensor detects a fire and sends an alert message on registered number and in this system a smartphone is required because it uses Blynk cloud to send messages. When flame sensor senses fire, it activates water pump and fire extinguisher and the main power supply will be cut off. An automatic fire monitoring control system was designed that when the sensor value exceeds a predetermined threshold, will detect the flame and display a notifications on the control screen. The dimmer is used to operate and turn on/off high voltage AC appliances. A related software program, relay modules are installed so AC appliances can be turned ON and OFF with the click of a button in the Blynk app. This action is important for controlling different device models. These devices can be powered by a backup UPS to prevent system failure in the event of a power failure. Sprinklers are installed above the fire floor and Carbon dioxide can be used in sprinklers instead of water if the floor has many electrical appliances, it won't damage your electronic devices.

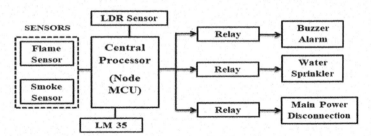

Figure 1. Functional diagram for fire detection and control system in industries and at home.

A functional diagram of industrial fire detection and control is shown in Figure 1. It includes NodeMCU, temperature sensor, smoke sensor, flame sensor, and LDR. A temperature sensor detects the severity of fire when fire break out. LDR values are monitored and when it exceeds to threshold, other sensors' values are analyzed. When flame, temperature and smoke threshold are exceeded, signals are sent to alert people and firefighters that a fire is raising. An electric motor responds to a signal from a smoke detector to open the windows of an automated home, clearing the room of smoke and keeps heat and smoke out of the room. The NodeMCU sends a signal to the people that where is the emergency exit and allow them to exit the building immediately.

Fire detection based on the smoke detector [Figure 2], includes a smoke detector, CO and CO2 sensors, a record processing module, a set of fire detection rules and a fire detection module. It is a fire alarm that detects fire with carbon monoxide and carbon dioxide, and detects smoke with a gasoline sensor and smoke detector. Generates an alert when increasing flames exceed a specified threshold. The data processing module includes smoke, CO and CO2 facts filtering and speed calculation techniques. Information filtering is used to dispose of undesirable facts based totally on a moving common filter over a time period. Calculate smoke, CO, CO2 boom charges the usage of linear regression. Its miles suited to this clear out output over a particular time-frame the use of a directly line and slope. The instantly lines show the time derivatives of smoke, CO, and CO2, which correspond to smoke, CO, and CO2 rise costs. The output of the facts processing module is then fed into the fire alarm algorithm.

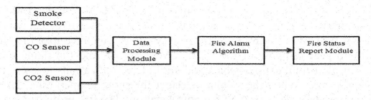

Figure 2   Data flow diagram of fire detection system.

Figure 3.   Fire detection and control system.

7   RESULTS AND DISCUSSION

The proposed system focuses on building fire detection and control systems using sensor-actuator technology. It shows deficiencies in current fire alarm systems and provides necessary changes. The basic goal of fire detection systems is to detect fires early with minimal false alarms. Rapid fire detection requires some more sensors with fast response times that can actually detect fires in the early stages.

Although many fire detection and control systems are in operation across industries and many have been proposed by various researchers, they sense the fire, not the severity of the fire. This proposed system uses LDR, LM35, smoke, and flame sensors to detect fire severity. As these sensors are mobile, their response times can increase. Most of the fire spread to the upper floors; Therefore, the thermal sensor is usually placed on the wall or ceiling. Extremely sensitive imaging thermal sensor based on changes in refractive index caused by small changes in ambient temperature.

Microcontroller NodeMCU collects data from all sensors every 100 milliseconds and analyzes that data. When sensor data indicates a fire, data from the LDR and heat sensor are analyzed and the appropriate actuators are activated. For mines or tunnels, most of the literature shows how to use the optical sensor of an inline heat exchanger. However, they are also suitable for environments found in buildings, such as those with large halls, elaborate galleries, and elaborate rooms. These can reduce the price of many point-type thermal sensors used in buildings. Bimetallic thermocouples are ideal for direct mechanical drive signals of valves used in fire extinguishing systems.

A simple absolute flame measurement is insufficient to detect the presence of a fire hazard. Fire detection applications can benefit from thermal sensors that rely on the rate of change of heat. Therefore, we can focus more on the heat change rate, which can be used to detect the rate at which fire occurs. This proposed system uses a LM35 temperature sensor, moreover other heat sensors can be used.

8   CONCLUSION

An overview on fire-sensing technologies is given in this paper and it focuses on constructing a fire detection and control systems using sensor-actuator technology. It highlights the flaws in the current fire detection systems and offers changes that are required. A sensor with a fast response time that really can detect fires in their earliest stages is necessary for swift fire detection. Traditional system uses either flame sensor or heat sensor but this proposed system uses flame, heat, smoke, and LDR which can detect the rate and severity of fire. However, it takes a little bit more time but as compared to fast responsive system it is more effective. If the speed of change of fire is detectable then appropriate prevention can be taken. The majority of fires spread through the floors; therefore heat sensors are often positioned on walls or ceilings. Simply absolute temperature values are insufficient to detect the presence of a fire threat so a heat sensor is required which can detect rate of change in heat. The proposed system uses all these essential components and sensors which can detect the severity of fire and take appropriate prevention actions.

REFERENCES

[1]   Sowah R., Ampadu K. O., Ofoli A. R., Koumadi K., Mills G. A., and Nortey J., "A Fire-Detection and Control System in Automobiles: *Implementing a Design That Uses Fuzzy Logic to Anticipate and Respond*," IEEE Ind. Appl. Mag., vol. 25, no. 2, pp. 57–67, 2019, doi: 10.1109/MIAS.2018.2875189.

[2]   Bouabdellaha K., Noureddine H., and Larbi S., "Using Wireless Sensor Networks for Reliable Forest Fires Detection," *Procedia Comput. Sci.*, vol. 19, pp. 794–801, 2013, doi: 10.1016/j.procs.2013.06.104.

[3]   L. J, L. S. V. S, M. R, and M. R, "Automated Food Grain Monitoring System for Warehouse using IOT," *Meas. Sensors*, vol. 24, 2022, doi: 10.1016/j.measen.2022.100472.

[4]   Yadav R. and Rani P., "Sensor Based Smart Fire Detection and Fire Alarm System," *SSRN Electron. J.*, 2020, doi: 10.2139/ssrn.3724291.

[5]   Abraham A., Rushil K. K., Ruchit M. S., Ashwini G., Naik V. U., and Narendra G., "Energy Efficient Detection of Forest Fires Using Wireless Sensor Networks," *Int. Confernce Wirel. Networks*, vol. 49, no. Icwn, pp. 197–201, 2012.

[6]   Zervas E., Mpimpoudis A., Anagnostopoulos C., Sekkas O., and Hadjiefthymiades S., "Multisensor Data Fusion for Fire Detection," *Inf. Fusion*, vol. 12, no. 3, pp. 150–159, 2011, doi: 10.1016/j.inffus.2009.12.006.

[7]   Svensson S., "A Study of Tactical Patterns During Fire Fighting Operations," *Fire Saf. J.*, vol. 37, no. 7, pp. 673–695, 2002, doi: 10.1016/S0379-7112(02)00027-9.

[8]   Hefeeda M. and Bagheri M., "Forest Fire Modeling and Early Detection Usingwireless Sensor Networks," *Ad-Hoc Sens. Wirel. Networks*, vol. 7, no. 3–4, pp. 169–224, 2009.

[9]   Kadri B., Bouyeddou B., and Moussaoui D., "Early Fire Detection System Using Wireless Sensor Networks," *Proc. 2018 Int. Conf. Appl. Smart Syst. ICASS 2018*, 2019, doi: 10.1109/ICASS.2018.8651977.

[10]   Mahgoub A., Tarrad N., Elsherif R., Al-Ali A., and Ismail L., "IoT-based Fire Alarm System," *Proc. 3rd World Conf. Smart Trends Syst. Secur. Sustain. WorldS4 2019*, pp. 162–166, 2019, doi: 10.1109/WorldS4.2019.8904001.

[11]   Imteaj A., Rahman T., Hossain M. K., Alam M. S., and Rahat S. A., "An IoT Based Fire Alarming and Authentication System for Workhouse using Raspberry Pi 3," *ECCE 2017 – Int. Conf. Electr. Comput. Commun. Eng.*, pp. 899–904, 2017, doi: 10.1109/ECACE.2017.7913031.

[12]   Hsu W. L., Jhuang J. Y., Huang C. S., Liang C. K., and Shiau Y. C., "Application of Internet of Things in a Kitchen Fire Prevention System," *Appl. Sci.*, vol. 9, no. 17, 2019, doi: 10.3390/app9173520.

Artificial Intelligence, Blockchain, Computing and Security – Dagur et al. (Eds)
© 2024 The Author(s), ISBN: 978-1-032-49393-0

Implementation of optimized protocol for secure routing in cloud based wireless sensor networks

Radha Raman Chandan* & Sushil Kumar*
School of Management Sciences, Varanasi, India

Sushil Kumar Singh*
Seoul National University of Science and Technoloy, South Korea

Abdul Aleem* & Basu Dev Shivahare*
Galgotias University, Greater Noida, India

ABSTRACT: Wireless Sensor Networks (WSN) can be used in a variety of applications, including emergency and healthcare monitoring, precision agriculture, environmental and disaster prevention, intelligence and army tracking, industrial automation, and different location-tracking systems. The IP-based network architecture has long been claimed to be unsuitable for resource-constrained devices that are increasingly getting integrated into the physical realm. Furthermore, the requirement for an application gateway to connect with other computers or the broader Internet is emphasized. RPL uses control messages to keep track of the network. As a replacement for RPL, a Secure RPL paradigm based on Contiki OS is suggested to create secure information routing. There are various network simulators that may be used to test the performance of a WSN node, but only a handful of them to give real-time data output. Because the code flow truly represents the finalized code that can be transferred to the equipment to form the node with the required features COOJA Simulator for Contiki OS can replicate the precise working of a WSN node. Before deploying in real time, the COOJA simulator is used to evaluate the RPL optimization of the control packets timing parameters. The viability of the method in terms of power consumption and convergence time is investigated and compared for various cryptographic algorithms.

Keywords: WSN, Routing, Cryptography, Secure Routing, IOT

1 INTRODUCTION

The Internet of Things (IoT) study topic now fascinates academics due to its large number of applications and simplicity of deployment in a variety of real-world scenarios. IoT becomes exposed to a variety of security threats, which can have a negative impact on appropriate functionality. When they're used in smart cities, the severity of the problem increases even more. Routing Protocols for Low Power and Lossy Networks (RPL) that are now in use and are regarded as lightweight. Available routing protocols for IoT equipment provide a limited level of protection from a variety of RPL routing attacks. Due to the structure of the Network infrastructure, which is a resource constrained, traditional routing approaches are ineffective. As a result, IoT routing safety is a difficult challenge.

---

*Corresponding Authors: rrcmiet@gmail.com, sushilkumar@smsvaranasi.com, sushil.sngh001007@seoultech.ac.kr, er.aleem@gmail.com and basuiimt@gmail.com

DOI: 10.1201/9781003393580-48

Yagan [1] presented a decentralized sensor network-friendly Key Management system. Key distribution, removal, and re-keying are all done without requiring a lot of processing or bandwidth. Although the system is expandable and adaptable, there is a trade-off between memory capacity and connection. Chan et al. [2] offer the Q-Composite randomized key pre distribution method, which is an enhancement of the randomized key pre-distribution method. If two sensor nodes desire to form a link, they must exchange at minimum q number of keys in the Q-Composite scheme. In this method, if any of these keys is compromised, the two connected nodes will still be able to communicate. With a 100-node test bench, Hakeem et al. [3] investigated estimates and neighbour handling in extensive RPL deployments.The hop count-based ETX protocol was investigated, as well as cache capacity limitations. Contiki OS provides a switching frequency per hour, test duration equivalent to the transmission speed. Mohammed et al. [4] used the COOJA simulator to test RPL implementation and picked the optimal Objective functions from OF0 and ETX. The COOJA Simulator, which has 80 nodes and is automated with Perl scripts, is used to identify matrix such as network convergence speed, packet delivery ratio, power usage, control traffic latency, and target latency. The RPL's ideal range of parameters for optimal functioning is documented.Ancillotti et al. [5] investigated the use of the RPL protocol in smart metering for intelligent electronic gadgets. Route-level metrics such as pathway stretch, route life periods, dominance, and flapping are used to assess the RPL protocol's dependability. RPL routes' quality and stability are estimated using these factors. They claim that RPL, like every route optimization technique, gives the best routing pathway toward the target based on simulation findings.

The goal of this study is to create a wireless IPv6-based safe routing protocol that will permit networks to communicate with one another. It is attempted to integrate WSN along with Cloud Computing for safe information storage. Simulation tool was used to verify the efficiency of secure RPL connections, which was then begun using a real-time experimental arrangement. The primary objective of this study is to develop a real-time Contiki OS oriented WSN for further applications.

2   OVERVIEW OF METHODOLOGY

There is minimal security in the current RPL. The suggested solution involves incorporating security into the network layer. RC5 and Skipjack, as well as CBC-MAC, are used to ensure secrecy and authenticity [6]. The routing security is achieved by sending secure RPL packets. Figure 1 depicts the position of security enhanced routing incorporated with the network layer of the Secure Contiki-RPL (SC-RPL) framework.

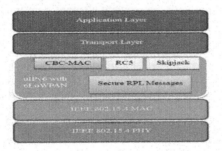

Figure 1.   Position of Secure RPL Messages.

The application layer utilizes CoAP along with HTTP and it can adapt to any real-time application. Whether the transmission is TCP or UDP, it will be determined by the Transport Layer. SC-RPL, adaption layer, 6LoWPAN, and uIPv6 are the major players in the network layer. This allows IPv6 packets transmitted through the web to make direct connections with the wireless equipments. The suggested effort entails securing the routing protocol (RPL), which may be utilized in 6LoWPAN.

317

2.1 RPL packet structure with security implementation

The suggested method makes advantage of SC- RPL control message variations. Data encryption methods that have been shown to be acceptable for resources controlled situations, including the RC5, skipjack and CBC-MAC technique, as illustrated in Figure 2. This technique provides the routing messages with secrecy and authenticity.

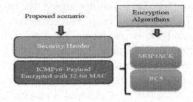

Figure 2.   Proposed ICMPv6 packet.

The security header elements allow any encryption method to be implemented using an 8-bit scheme. The LVL security level parameter specifies the degree of data authentication and, if desired, data secrecy. The Key Identifier Mode (KIM) controls whether the secrecy and authenticity keys are generated explicitly or implicitly. Figure 3 shows a typical security header structure with contents.

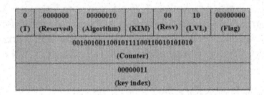

0	0000000	00000010	0	00	10	00000000
(T)	(Reserved)	(Algorithm)	(KIM)	(Resv)	(LVL)	(Flag)
0010010011001011110011001010101010						
(Counter)						
00000011						
(key index)						

Figure 3.   Sample security header format.

2.2 Implementation of secure ICMPV6 RPL control messages

A security header is inserted in addition to the attributes shown in Figure 4 as a preliminary step. The elements inside the security header define secrecy and authentication. Authentication must be required for this security system, and the CBC-MAC algorithm does the part. The value associated with the security element determines secrecy. Here MAC is computed by considering the whole packet that is attached to a routing message.

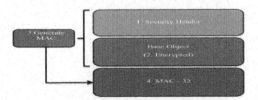

Figure 4.   Steps to construct secure RPL messages.

The Security header is first linked to the basic object which is encrypted. The MAC of total packets is created. Finally, this MAC is attached at the end. The flexibility of RC5 is one of its advantages. RC5 does have a customizable block size, rounds count, and key length, unlike existing encryption methods. The algorithm's level of security is influenced by the values of those specific parameters [7]. Security header is first linked to the basic object which is encrypted. The MAC of total packets is created. Finally, this MAC is attached at the end. The flexibility of RC5 is one of its advantages. RC5 does have a customizable block size, rounds count, and key length, unlike existing encryption methods. The algorithm's level of security is influenced by the values of those specific parameters. Skipjack is a 32-round asymmetrical Feistel network. The encryption / decryption keys are both 80 bits long, while the data sets are 64 bits long [8]. The most common method of constructing a MAC from a block cipher is to use a cipher block chaining MAC (CBC-MAC). CBC-MAC utilizes CBC for cryptography and a MAC for the final output. CBC-MAC must employ distinct cryptographic operations and MAC creation, else the MAC tag will not alter even though the message is changed [9]. CBC-MAC can construct message authentication codes using an existing cryptographic method, which saves a lot of memory. Tmote sky along with the COOJA simulator is used for the execution. Tmote outputs which are kept in a log file provide the data for the evaluation. The log file is examined using Perl for the computation of the installed network's energy usage and convergence time [10]. Figure 5 depicts the procedures included in the assessment and deployment of SC-RPL [11–14].

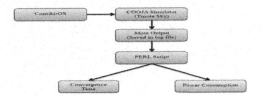

Figure 5.   Steps for analyzing log file.

This framework was created using a variety of scripting and programming languages. During this process, Perl, Apache Ant, C, and Make Files were employed. The operating system used in the workplace context is Ubuntu Linux 14.04. The processor is a 3.20GHz Intel core i5. RAM has an 8GB memory space.

2.3   Implementation of COOJA setup

With a COOJA simulator, A simulation is created by changing the experiment settings as needed and pressing the start button. The setup configuration for 50 mobile nodes in simulator. The output of each mote is displayed in the mote console window, with its mote ID. For the relevant output, the time is also presented. As a result, it's particularly ready to be analyzed. The effective encryption of RC5 implementation can be seen in the mote result. During transmission, the message is encoded and a MAC is appended. The MAC must be checked on receipt, and afterwards the communications must be decoded. The mote outcome may also be filtered to show only the outputs that are necessary. Any word that appears in all of the essential outputs is used to match the template. Otherwise, the MOTE outputs can be shown, manipulated, and processed using COOJA's simulator.

3   RESULTS AND DISCUSSIONS

The convergence time is evaluated using a model with a varying amount of nodes (20, 50, and 100). Figure 6 shows that the secure variants take longer to reach a state of convergence than the unsecured variants. The RC5 deployment takes longer to reach convergence compared to the Skipjack deployment. It clearly demonstrates a strong reliance on mote dispersion. For different security deployments under investigation, the convergence time is examined for a specific Network configuration. Changes in configuration result in slightly different results when it comes to network convergence. Contiki-MAC requires a lot of RAM, yet it cuts power usage from roughly 60mW to around 1mW. Contiki-MAC and RPL both require 41.07 KB of ROM and only 7KB is available for encryption and program

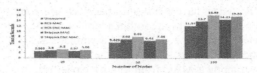

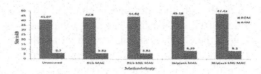

Figure 6.   Convergence time.

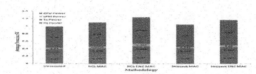

Figure 7.   Memory occupation of SC-RPL.

Figure 8.   Power comparison.

functionality. Whenever Contiki MAC and RPL are not being used, several algorithms for application layer protection might be used. Contiki-MAC and RPL must be deployed because the present work is around battery driven motes and protecting RPL.

Figure 7 depicts the ROM and RAM utilization based entirely on the size function results. The change in system memory utilization while employing CBC-MAC based RC5 implementation is calculated to be 4.5% and 13% for ROM and RAM respectively. In the case of MAC, there is a 9% increment in ROM usage and 15.82% increment in RAM usage. Increased secrecy results in a 15% increase in ROM and a 63% increase in RAM capacity. The higher memory usage is computed in comparison to the insecure Contiki RPL variant. In terms of memory, it seems evident that RC5 is the superior candidate. Also, Skipjack consumes nearly all of the ROM space allocated to Tmote Sky, thus reducing the scope of any UDP socket application's deployment.

Wireless sensor nodes are often battery-powered, which restricts their power flexibility and shortens the network's lifespan. COOJA employs an accurate power profiling method to quantify motes' power usage in the simulated scenario. The energy is estimated using this time information. Figure 8 clearly depicts the power comparison plot. Obviously, secure systems use more amount of power compared to unsecure systems. From the values in the comparison plot, Skipjack SC-RPL looks to be more energy saving than RC5 SC-RPL. The graph clearly shows the trend in the consumption of power during a period of 1hr by monitoring the network communication. Each mote's overall power usage is divided into four categories: Reception (Rx), Transmission (Tx), Low Power Mode (LPM) and CPU. The increased CPU and receiver power causes a significant difference in power. The fact that secrecy and authentication demand extra CPU processing time is self-evident. Before the deployment of network, the utilization of power is his as depicted in Figure 9. This can be connected to the initialization of nodes and the overhead in network establishment. After the convergence, utilization of power gradually reduces to levels ranging from 0.89 mV to 1.4 mV, raising the prospect of developing a SC-RPL framework for important network operations. Major variation occurs in receiver power, while encryption and authorization need some more CPU power. There isn't much of a difference in transmission and LPM power. Power usage is close to 60mW when Contiki-MAC is not utilized. As a result, some memory is compromised in order to substantially minimize power consumption, which is required in a limited resource scenario.

320

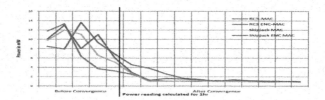

Figure 9. Power consumption analyzed for 1 hour.

4 CONCLUSION

Using COOJA and two cryptographic techniques paired with message authentication, the job of building SC-RPL suitable for porting into any hardware motes was completed successfully. SC improves the performance RPL's in terms of power consumption, memory and convergence time. The profile is calculated using CBC-MAC, Skipjack and RC5. These methods are compared with the normal RPL that is not secure. Memory usage of Skipjack, on the other hand fills the ROM to the brim, rendering it unfit for any other deployment of real implementations. RC5 takes up far less memory while still offering ample capacity for secrecy, authenticity, and additional application-oriented development. As a result, it can be determined that the combination of CBC-MAC and RC5 is the best option for SC-RPL development, which may then be used for various practical applications.

REFERENCES

[1] Yagan, O. (2012). Performance of The Eschenauer?Gligor Key Distribution Scheme Under an ON/OFF channel. *IEEE Transactions on Information Theory*, 58(6), 3821–3835.

[2] Chan, H., Perrig, A., & Song, D. (2013, May). Random key predistribution schemes for sensor networks. In *2003 Symposium on Security and Privacy*, 2013. (pp. 197–213). IEEE.

[3] Abdel Hakeem, S. A., Hady, A. A., & Kim, H. (2019). RPL routing protocol performance in smart grid applications based wireless sensors: *Experimental and simulated analysis. Electronics*, 8(2), 186.

[4] Mohammed, B., & Naouel, D. (2020). Experimental Performance Evaluation of RPL Protocol for IPv6 Sensor Networks. *International Journal of Wireless Networks and Broadband Technologies (IJWNBT)*, 9(1), 43–55.

[5] Ancillotti, E, Bruno, R & Conti, M 2013, The role of the RPL routing protocol for smart grid communications?, *Communications Magazine, IEEE*, vol. 51, no. 1, pp. 75–83.

[6] Arena, A., Perazzo, P., Vallati, C., Dini, G., & Anastasi, G. (2020). Evaluating and improving the scalability of RPL security in the Internet of Things. *Computer Communications*, 151, 119–132.

[7] Sobo?, A., Kurkowski, M., & Stachowiak, S. (2020). Complete SAT based Cryptanalysis of RC5 Cipher. *Journal of Information and Organizational Sciences*, 44(2), 365–382.

[8] Al-Ahdal, A. H., AL-Rummana, G. A., & Deshmukh, N. K. (2021). A Robust Lightweight Algorithm for Securing Data in Internet of Things Networks. In *Sustainable Communication Networks and Application* (pp. 509–521). Springer, Singapore.

[9] Hung, C. W., & Hsu, W. T. (2018). Power Consumption and Calculation Requirement Analysis of AES for WSN IoT. *Sensors*, 18(6), 1675.

[10] Singh, A. A. G., Leavline, J., Sushmitha, M., & Manjula, E. (2019). Smart Irrigation System Using Wireless Sensor Network with Cooja and Contiki. *International Journal of Automation and Smart Technology*, 9(1), 33–39.

[11] Chandan, R.R., Kushwaha, B.S. and Mishra, P.K., 2018. Performance Evaluation of AODV, DSDV, OLSR Routing Protocols using NS-3 Simulator. *International Journal of Computer Network and Information Security*, 10(7), p.59.

[12] Chandan, R.R. and Mishra, P.K., 2018. A Review of Security Challenges in Ad-Hoc Network. *International Journal of Applied Engineering Research*, 13(22), pp.16117–16126.

[13] Singh, S.K., Pan, Y. and Park, J.H., 2022. Blockchain-enabled Secure Framework for Energy-efficient Smart Parking in Sustainable City Environment. *Sustainable Cities and Society*, 76, p.103364.

Artificial Intelligence, Blockchain, Computing and Security – Dagur et al. (Eds)
© 2024 The Author(s), ISBN: 978-1-032-49393-0

A cross CNN-LSTM model for sarcasm identification in sentiment analysis

Sandeep Kumar
Department of Computer Science & Engineering (AI), ABESIT, Ghaziabad

Anuj Kumar Singh
School of Computing Science & Engineering, Galgotias University, Greater Noida

Shashi Bhushan
Department of Computer Science, ASET, Amity University Punjab, Mohali

Vineet Kumar Singh
Department of Computer Science & Engineering (AI), ABESIT, Ghaziabad

ABSTRACT: Social networking is evolving into an incredible tool for obtaining consumer feedback. A significant amount of data is generated from a number of sources, including social blogging and other websites, due to the complexity of the internet and the development of new technologies. Today, blogs and websites are used to gather product reviews in real-time. Thoughts, opinions, and reviews can now be created in a variety of ways thanks to the growth of a big number of cloud-based blogs. It is therefore vital to find a mechanism or process to extract relevant information from large amounts of data, classify it into various categories, and predict end user behaviors or moods. Convolutional neural networks (CNN) and the Long Short-Term Memory (LSTM) model have been used to successfully complete a number of Natural Language Processing (NLP) tasks. The CNN model successfully recovers higher-level information by utilizing convolutional layers and max-pooling layers. The LSTM model is able to capture the long-term dependencies between word sequences. In this review, we propose the Cross CNN-LSTM Model as a solution to the sentiment analysis problem. First, we train the Word to Vector (Word2Vc) technique's initial word embedding. Word2Vc determines the distance between words, groups words based on similarity in meaning, and turns text strings into a vector of numerical values. Convolution and global max-pooling layers with long-term dependencies after embedding are used to retrieve a number of features in the suggested model. The proposed model additionally employs dropout technology, normalization, and a rectified linear unit to improve accuracy. Our results show that the proposed cross CNN-LSTM model outperforms traditional deep learning and machine learning techniques in terms of precision, recall, f-measure, and accuracy. Our technique yielded results that were competitive on the YELP review data.

Keywords: Sentiment Analysis, CNN, LSTM, NLP, Word2Vc

1   INTRODUCTION

Sentiment analysis is a process of people's feelings about various issues, events, products, entities and their attributes. With the unstable expansion of web applications, such as social networks, blogs, forum discussions and e-commerce sites, people share their opinions about the products, services or any topics [1]. the most important task of sentiment study is the identification of polarity for particular input data that includes text, audio, attribute, video, and so on. There are a number of text mining techniques which can be employed in sentiment analysis, which is the

DOI: 10.1201/9781003393580-49

extraction of information from a person's opinions. Nowadays, public opinion can be obtained from numerous different sources [2]. Some of these include social media, micro-blogging websites, e-commerce websites, public forums, and blogs. Opinion posted in these sources may be a small text, short text, or emotions on various websites. In general, the primary objective of sentimental analysis is to make the data available for finding the pattern. Following the opinion extraction process, sentiment analysis determines a text's subjectivity, polarity, and polarity strength. In our society, online social media platforms have gained tremendous traction. Twitter and Facebook have a significant impact on people's daily lives all over the world. Their users generate and exchange a diverse range of content on social media, making it an invaluable source of information about public sentiment toward economic, social, and political issues [3]. It is critical to develop automated methods for retrieving and analyzing data from social media in this context. Sentiment is a broad term that encompasses people's feelings, emotions frequently communicate their sentiments online via blogs, social media, review sites, and rating. Sentiment research is commonly used by business owners and advertising agencies to start new marketing initiatives. Sarcasm is one of the most extensively researched and often employed linguistic phenomenon. Despite being widely used in speech, it can be exceedingly challenging to spot sarcasm in textual data [4]. Even when applied to regular polar language, machine learning approaches frequently fall short of achieving accuracy levels that are close to a coin toss, or less than 50%. Solving the issue of sarcasm detection in text data is therefore urgently needed. The detection of sarcasm in text-based data has been the subject of various studies using machine learning, solutions based on specific corpora, semi-supervised learning, and other techniques, but there is still no commercially available technique that can reliably detect sarcasm in text.

2 RELATED WORK

A study [5] about the internet and social media sites have a massive amount of data. It is vital to make efficient use of this data. The success of businesses is determined by client happiness. As a result, sentiment analysis is both crucial and difficult. Various machine learning algorithms that can be employed for this goal were found in this literature review. More people are using the internet and social media to communicate their views and opinions. As a result, there were more user-generated sentences that contained sentiment information. To get a greater grasp of how individuals feel and behave in diverse situations, it is inevitable to try new things. In addition to an analysis of the efficacy of various machine learning and deep learning techniques, this research offers a novel cross system for sentiment categorization that blends text mining and neural networks. More than a million tweets from five distinct categories are included in the dataset utilised for this study. The remaining 25% of the dataset was utilised to test the system after it had been trained on 75% of it. The outcomes demonstrate the efficacy of the cross learning technique, with a maximum accuracy rate of 83.7%.

A study stated [6] Users' feedback on social media has reached huge dimensions and Natural Language Processing and Deep Learning play a vital role in uncovering the social media users' sentiments. They propose cross deep learning model using multiple deep learning approaches like LSTM, GRU, BiLSTM and CNN to classify the sentiments. They compared performance of deep learning models with past studies and claimed that their proposed deep learning classifier have achieved better performance ie 80.44%. Evaluation of SVM approach sentiment analysis technique on movie domain, shows highest performance accuracy i.e. 0.917 on unigram model, whereas 0.728 on bi-gram model.A CNN framework technique has been suggested in [7] to predict the feelings of visual content in visual SA. Back propagation is used in the experiment on a dataset of 1269 photos that were gathered from Twitter. The authors' findings demonstrate that the suggested system performed well in terms of recall, accuracy, and precision on the Twitter dataset and that the proposed GooLeNet model outperformed AlexNet by 9%. The study [31] used microblog comments to collect online people's perspectives and viewpoints on special events. The CNN technique was employed because it circumvents feature extraction and implicitly learns the data. Multimedia Tools and Applications created a corpus of 1,000 comments from microblogs and categorised them into three categories.

3 METHODOLOGY

3.1 *Cross model (CNN LSTM model)*

The presence of huge measures of assessments towards individuals, items, occasions, and administrations as video, sound, and conversation gatherings, message remarks, web journals, audits, video blogs, and other video designs is ubiquitous in numerous modalities like message, sound, and video. These assessments contain some sort of feeling, whether pessimistic or good. Machines have two fundamental issues: first, they should get human sentiments and allot an enthusiastic charge to them; and second, they should have the option to distinguish the equilibrium between energy and cynicism. Feeling force can be indicated by relegating extremity scores, known as Sentiment Analysis [8]. This additionally incorporates distinguishing savage records via web-based entertainment and recognizing mockery in a voice. The bountiful measure of information is equipped for serving various organizations, social, and political requirements, which makes sentiment investigation a basic input component. The neural network, which can be a single layer or a combination of LSTM and CNN, is the following layer. This network processes feature vectors as an input. For instance, a convolutional layer reads input feature vectors as overlapping sequences that are then read using a fixed window filter to create a feature map. The max-pooling layer receives the feature map as input and outputs the largest value in the array. When activation functions are applied to these dense layers using this output, the result is a label (0: negative; 1: positive).

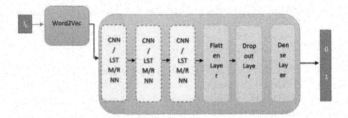

Figure 1.   Provides a block diagram of the experiment's process.

Once it has been pooled to a smaller dimension, the output from these embeddings will then be fed further into an LSTM layer. This is required as the convolutional layer will extract all local "features, and LSTM will be able to recall the features and their ordering, further comparing the input text to a body of previously known knowledge [10]. The model is depicted in Figure 3. It has a significant presence across a variety of media, including text, sound, and video, in forums, websites, comments, audits, video blogs, and more. With so many viewpoints, some form of emotion is always there.

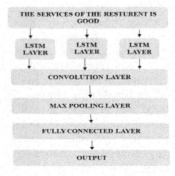

Figure 2.   CNN-LSTM model

Preceding their work in real world applications, LSTM models should be prepared on a preparation dataset. Coming up next are the absolute most requesting applications:

The calculation of words in view of the arrangement of words took care of as info. Whenever language models are run at the person, n-gram, sentence, or passage level, then, at that point, language can be removed as message from records. Image handling, which investigations an image and produces a sentence thus. As a component of this, one will require a dataset that incorporates countless photographs joined by precise subtitles. Assuming that one has recently prepared a model, it tends to be utilized to make expectations about the highlights of pictures in the dataset. To stress the thoughts introduced in the dataset, just words that are most interesting are added to the dataset. This is organized text information. To attempt to fit the model, these two sorts of information is utilized [11]. Because LSTMs have the capacity to distinguish which melodic notes are contained in a bunch of given notes, they are as often as possible used to create new melodic pieces. Language interpretation is the most common way of changing an arrangement starting with one language over then onto the next. In picture handling, likewise with express handling, a dataset is at first cleaned to assist the model with learning quicker. Nonetheless, just a part of the information is utilized to prepare the model. To precisely change the info arrangement over to a vector portrayal (encoding), an encoder-decoder LSTM model is utilized, which initially encodes the information succession, and afterward yields it to a deciphered adaptation [12].

3.2 *Experimental setup*

The proposed crossover model, which joins CNN and LSTM, is utilized to classes sentiment. It starts with an installing layer, trailed by one convolutional layer, two LSTM layers, lastly a thick grouping layer that is completely associated. The segment talks about exhaustively the datasets utilized, the model hyper-boundary settings, the advancement procedures utilized, the examination results, the assessment results, and the model investigation.

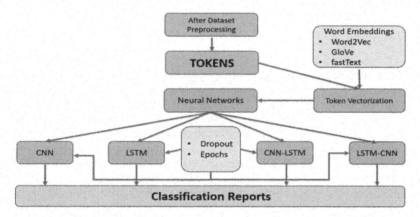

Figure 3.   Diagram of the proposed cross model.

To ascertain a sentiment investigation rating for each film and client survey, the Stanford Amazon Product Reviews dataset has been pre-handled, yielding a jargon size of nearly 5000 words with max-min cushioning of 500 words. This mutilates some other dataset. As recently expressed, a limit of 30 audits are allowed for some random film. The Conv1D layer's number of channels has been set to 128 of variable width (4, and 5), and the number of include maps formed has been set to 256. The piece size has been set to 2 to achieve a 50% reduction in input stature. The ReLU (Redressed Straight Unit) is an actuation unit. Dropout is an effective regularization technique. There has been a dropout of 0.5 applied after the LSTM layers. The unit counts of the two LSTM layers are, respectively, 128 and 64. The final fully connected classification layer has no tuning parameters. ADAM has been used to implement the optimizer. Weight decay and learning rate were given parameters of 0.01 and 0.1, respectively. The proposed cross model has been trained for 100 epochs in Google Colaboratory using GPU and CUDA v10.2, which was done using batches of 50 items. Binary Cross Entropy (BCN)are the techniques used to train the model. Validation set size is kept constant with respect to training set size This work uses deep learning, more especially convolutional neural networks, recurrent neural networks like LSTM (Long Short Term Memory), and a combination of these two models, in an effort to increase accuracy. The main goal is to ascertain how different hyper-parameters utilized in deep learning models would affect performance. All feasible LSTM and CNN combinations are modelled for this purpose, and performance metrics like accuracy, F-measure, precision, and recall have been determined. Using a graph of accuracy versus number of epochs, over-fitting has been made visible.

3.3 Selected parameters for cross model

Table 1.   Parameter for cross model.

Parameter	Set-Value
Word Embedding	Word2Vec, FastText
Screens	128
Kernel	6
Embedding Length	200
Epochs	2,4,8,16
Activation Function	Sigmoid
Batch Size	256
Pool Size	4
Dropouts	0.20,0.30,0.40
Optimizer	Adam

4   RESULTS

The numerous observations show that cross models have outperformed the conventional single layer models. The difference, however, is only 1%, and as such, "they are preliminary indication of the assumption which is the motivation of this study and experiment." However, it cannot be "negated" that by merging them, the productivity of the model has been harnessed. The convolution layer does not provide any useful information, according to another conclusion about the CNN-LSTM cross model. The LSTM layer will function similarly to an LSTM model with all connections. The cross model's full potential cannot be realized in these conditions. Another significant finding from the analysis is that the LSTM-CNN model is best suited to the problem because the LSTM layer will monitor each token that enters its path and attempt to provide the resultant token by taking into account not one but two layers, making the analysis more accurate and effective. After this is finished, the convolution layer completes the remaining work by categorizing it according to the correct category utilizing a 0, 1, and sarcastic, respectively, categorization system. The results of the numerous experiments carried out on the dataset were measured on the standard setup modifying the Size of Training Data Set on YELP Data Set, particular hyper-parameters including the size of Training Data, Special Dropout, and Number of Epochs. Word2veb and FastText two-word embeddings have been used to assess the outcomes. The effects of spatial dropout, the amount of the training data, and the number of epochs is used to quantify the findings. Figures 4 to 7 depict a graph illustrating how the size of the training set affects each algorithm's accuracy, recall, F1 Score, and precision.

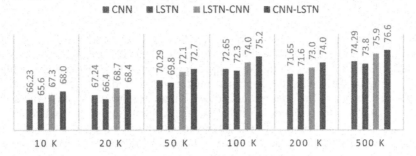

Figure 4. Comparative result of various models in terms of accuracy.

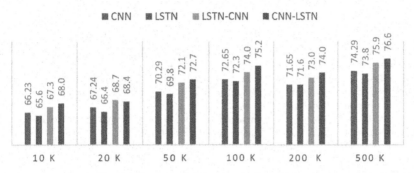

Figure 5. Comparative result of various models in terms of precision.

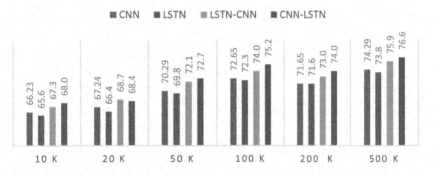

Figure 6. Comparative result of various models in terms of recall.

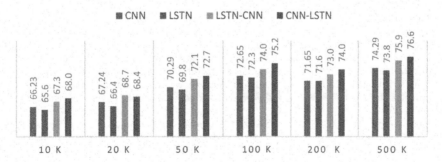

Figure 7. Comparative result of various models in terms of recall.

5 CONCLUSION

The process for the several models—such as CNN, LSTM, CNN-LSTM, and LSTM-CNN—that have been measured for the assessment has been carried out by placing them in different topological configurations and applying various hyper-parameter settings each time. According to the results, the hyper-parameter setting and the evaluation dataset can occasionally be to blame for performance losses. There are instances where the significant performance gain has been noted, nevertheless. The comparison of the test results reveals that CNN-LSTM, the algorithm utilised, outperformed all other algorithms. Additionally, it has produced outcomes that surpass the state of the art. Additionally, the highly big and self-annotating dataset used for the experimentation and evaluation helps to provide high-quality labels inside the dataset.

REFERENCES

[1]   Ain QT, Ali M, Riaz A, Noureen A, Kamran M, Hayat B, Rehman A (2017) SA Using Deep Learning Techniques: A Review. *Int J Adv Comput Sci Appl*, 8(6)

[2]   Al-Smadi M, Talafha B, Al-Ayyoub M, Jararweh Y (2018) Using Long Short-term Memory Deep Neural Networks for Aspect-based Sentiment Analysis of Arabic Reviews. *International Journal of Machine Learning and Cybernetics*, 1–13

[3]   Amolik A, Jivane N, Bhandari M, Venkatesan M (2016) Twitter Sentiment Analysis of Movie Reviews using Machine Learning Techniques. *Int J Eng Technol* 7(6):1–7

[4]   Cheng Z, Ying D, Lei Z, Mohan K (2018) Aspect-aware Latent Factor Model: Rating Prediction with Ratings and Reviews. In: *Proceedings of theWorld Wide Web Conference on World Wide Web*, pp. 639–648

[5]   Cheng Z, Ying D, Xiangnan H, Lei Z, Xuemeng S, Mohan K (2018) A^ 3NCF: An Adaptive Aspect Attention Model for Rating Prediction. In IJCAI, pp. 3748–3754

[6]   Kumar, S., Singh, A. K., Bhushan, S., & Vashishtha, A. (2022, September). Polarities Inconsistency of MOOC Courses Reviews Based on Users and Sentiment Analysis Methods. In *Proceedings of 3rd International Conference on Machine Learning, Advances in Computing, Renewable Energy and Communication: MARC 2021* (Vol. 915, p. 361). Springer Nature.

[7]   Collobert R, Weston J (2008) A United Architecture for Natural Language Processing: Deep Neural Networks with Multitask Learning, in *Proc. 25th Int. Conf. Mach. Learn.*, pp. 160–167

[8]   Conneau A, Schwenk H, Barrault L, Lecun Y (2017) Very Deep Convolutional Networks for Text Classification. In *Proceedings of the 15th Conference of the European Chapter of the Association for Computational Linguistics*, pp. 1107–1116

[9]   Elghazaly T, Mahmoud A, Hefny HA (2016) Political Sentiment Analysis Using Twitter Data. In *Proceedings of the ACM International Conference on Internet of things and Cloud Computing*, pp.11

[10] Fang X, Zhan J (2015) Sentiment Analysis Using Product Review Data. *J Big Data* 2(1):5

[11] Govindarajan M (2013) Sentiment Analysis of Movie Reviews Using Cross Method of Naive Bayes and Genetic Algorithm. *Inte J Adv Comput Res* 3(4):139

[12] Singh, A. K., Kumar, S., Bhushan, S., Kumar, P., & Vashishtha, A. (2021). A Proportional Sentiment Analysis of MOOCs Course Reviews Using Supervised Learning Algorithms. *Ingénierie des Systèmes d'Information*, 26(5).

[13] Hassan A, Mahmood A (2018) Convolutional Recurrent Deep Learning Model for Sentence Classification. *IEEE Access* 6:13949–13957

[14] Himelboim I, Smith MA, Rainie L, Shneiderman B, Espina C (2017) Classifying Twitter Topic-networks Using Social Network Analysis. *Social Media+ Society*, 1–13

[15] Islam, J, Zhang Y (2016) Visual Sentiment Analysis for Social images Using Transfer Learning Approach. In *IEEE International Conferences on Big Data and Cloud Computing (BDCloud)* pp. 124–130

[16] Kaur A, Gupta V (2013) A Survey on SA and Opinion Mining Techniques. *J Emerg Technol Web Intell* 5(4): 367–371

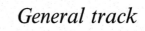

General track

Artificial Intelligence, Blockchain, Computing and Security – Dagur et al. (Eds)
© 2024 The Author(s), ISBN: 978-1-032-49393-0

Detection of hate speech in multi-modal social post

Abhishek Goswami, Ayushi Rawat, Shubham Tongaria & Sushant Jhingran
Sharda University Greater Noida, India

ABSTRACT: It has been observed in the past few years, multi- modal problems have been capable of attaining the interest of a large number of people. The core challenges faced in such problems are its representation, alignment, fusion, co-learning, and translation. The focus of this paper is on the analysis of multimodal memes for hate speech. On the evaluation of the dataset, we found out that the common statistics factors which were hateful initially became benign simply by unfolding the picture of the meme. Correspondingly, a bulk of the multi-modal baselines gives hate speech more options. In order to deal with such issues, we discover the visible modality through the use of item detection and image captioning fashions to realize the "real caption" after which we integrate it with multi-modal illustration to carry out binary classification. The method challenges the benign textual content co-founders present in the dataset to enhance the enactment. The second method that we use to test is to enhance the prediction with sentiment evaluation. It includes a unimodal sentiment to complement the features. Also we carry out in depth evaluation of the above methods stated, supplying compelling motives in want of the methodologies used.

1 INTRODUCTION

Social media has played a primary function in influencing human beings' ordinary existence. However having several aids, it additionally has the functionality of influencing community judgment and non-secular ideals throughout the world. It may be used to assault human beings without delay or in a roundabout way primarily based totally on caste, religion, nationality, status, gender, sexual orientation, and disease or disability. This can eventually result in various crimes. If the platforms are widely used, it is near impossible to keep such content under human supervision and prevent its spread. Hence, this responsibility comes down to the artificial intelligence and machine learning community to solve this problem.

They attain this via way of means of "benign confounders" within the dataset that is for each meme an alternative caption or image is found which is capable enough to make the meme content harmless.

In this paper, we are introducing two main concepts in which we attempt to discover the two modalities the use of understanding the circumstance and Pre-trained image captioning algorithms and sentiment analysis were employed to link the two modalities. The text is frequently given greater attention in many baselines than other components.

Figure 1. Example of a hateful meme (Left) and benign meme (Right).

DOI: 10.1201/9781003393580-50

331

Also, at some point in the information evaluation, it was observed that the bulk of hateful memes are transformed obsessed by non-hateful ones impartial with the aid of using image description. We attempt to stabilize the demonstrations of the dual modalities in the first approach and also address the benign textual content confounders with the aid of using fetching deeper information about the picture through object recognition besides captioning. We fuse this illustration with the multimodal one to enhance the performance.

2 LITERATURE REVIEW

Combining text and visual data in hate speech detection has been shown to improve its effectiveness. This is achieved by using picture embeddings from a pre-trained ResNet neural network and concatenating the text and picture vectors before using MLP, dropout and softmax for categorization.

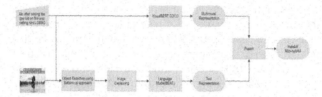

Figure 2. Approach 1: model architecture.

The detection of hate speech has become a critical issue, with a growing need for multi-modal solutions that combine both text and image information. To address this, a dataset of 150k physically annotated tweets (MMHS150k) has been created, which includes both image and text information. To leverage this data, two models have been trained, the Feature Concatenation Model (FCM) and the Linguistic Kernels Model. The FCM combines linguistic features from the tweet text and image text, while the Linguistic Kernels Model is based on visual question answering and explores the relationship between patterns in images and accompanying sentences. The core technique of this study is to enhance the understanding of the visual realm through multi-modal embedding of state-of-the-art baseline methods and pre-trained image captioning models. Recent developments in photo captioning involve using an encoder-decoder system with an attention mechanism to construct captions. This approach is flexible and allows different architectures to be used for feature generation. In sentiment analysis, there have been some creative fusion strategies proposed using graphs and layered designs, but little has been done to improve enemy material identification using multimodal emotion data. A sentiment analysis approach was proposed that performs unimodal sentiment analysis in both the script and graphic domains to determine the positions of both modalities.

3 PROPOSED APPROACHES

3.1 *Problem statement*

This project's objective is to categorize memes as harmful or useful while considering the data available in each literary component and visual medium. The meme itself is the visible object I in our case. The equation X1 = l1,.., li, wherein I is the meme index, denotes each meme's visible inputs. Let X2 display the text that was extracted as from memes (T1,.., Ti). The appropriate T will combine all of the textual data if a particular meme has words that occur in many places. Assume that the descriptors for all memes are Y = y1, y, and yi, where 0 denotes a helpful meme and 1 a detrimental meme. Therefore, our assignment might be categorized as a binary classification task utilizing the inputs X1 and X2. Our study focuses on the P(Y |X1, X2), which decreases the subsequent cost function.

$$J(\theta) = \sum_i -(Y\log(p_\theta) + (1 - Y)\log(1 - p_\theta))$$

3.2 *Image captioning*

As was already established, this study removes the non- threatening text confounders from the dataset that might turn a negative meme into a positive one by providing context for what is occurring in the image. Some of those opposed samples. As proven, they account for 20% of the dataset, and for that reason, our speculation is if we are able to offer our version with this more knowledge, it'll fight those opposed instances and offer a lift in precision. Using picture captioning and object detection helps the user to learn about the dataset and understand how the benign text confounders behave, and as a consequence, performs better than the existing approaches. You may compare the meme's "actual caption" and "pre-extracted caption" to see if they agree or disagree. Additionally, the majority of baselines tend to detect hate speech more frequently in textual forms. The motivation behind this method is to explore a deeper relationship between the textual and visual modes. Inside, as we can see, shows the VisualBert version of the COCO dataset and the hateful data . This retrieves the multimodal illustration of the two modalities, which is a 786 tensor of the multimodal picture (m1,m2,m3,...). Additionally, we send the picture to an image captioning system (Show, Attend, and Tell, Bottom-up, Top-down), which gives us a caption for the picture that is part of the meme (X3 = symbolizes the caption obtained from the pictures). Then, in order to generate a textual representation of any other 768-dimensional tensor, we pass the newsletter caption through a pre-educated Bert version (h1,h2,h3,...). The two tensors are then combined using methods like concatenation and bilinear transformations. Bilinear transformation is an easy technique for merging the information from many vectors into one vector. (m',h', dim) bilinear $= m'$ The mathematical equation is T.M.h + b, where dim is the hyper-parameter designating the anticipated measuring of the output vector (768), M is the weight matrix of measuring $(\text{dim},|m'|,|h'|)$, and b is the bias vector of measurement dim. We once more mix m, h, and bilinear(m',h',dim) for the category of hate speech. Finally, we use a multi-layer perceptron to process the data and produce a binary category of hateful and non-hateful memes (0/1). With the help of the captions created for the images in the Facebook hateful dataset, we fine-track the Visual Bert version, the Bert version from the Facebook horrible dataset, and other versions of Bert. This innovative method, which combines multimodal baselines with picture captioning, makes it easier to handle the challenging conditions mentioned before and will considerably improve performance.

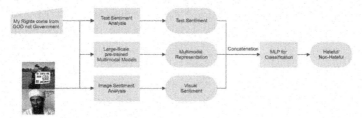

Figure 3.   Approach 2: model architecture.

3.3 *Sentiment analysis*

A method to enhance hate speech detection by using sentiment data from both text and image modalities is being developed. Multi-modal contextual representations of text and image are generated using VisualBERT. Text-based sentiment embeddings are obtained using RoBERTa and visible sentiments from the image are obtained using VGG. This approach takes into account the indirect connection between the textual and visual elements in hate speech.

We are unable to precisely trace such models on our dataset, however, due to the bother of labeled data. Instead, the RoBERTa is trained on the Stanford Sentiment Treebank, and the T4SA dataset is used to determine the parameters for the visible sentiment version. To create the final prediction, yhat, em, et, and ev are combined by concatenation and passed to multi-layer perceptrons. Figure demonstrates the whole version's framework.

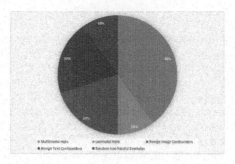

Figure 4.   Types of social media posts in the dataset.

4   EXPERIMENTAL SETUP

4.1   *Dataset*

Facebook dataset is utilized, which fronts of 10,000 images. The dataset remembers disdainful images and message for the images, these images are made by individuals who use disdain in Facebook. This dataset involves five various types of images. In multimodal images, non-scornful confounders are found in pictures as well as text. The preparation set is 85 and approval, and testing set is 5 and 10 separately. In both the preparation and the approval datasets, mocking pictures are denoted with a 0 whereas non- disrespectful images are denoted with a 1.

4.2   *Multi-modal baselines*

We used VisualBert which is pretrained on the COCO dataset, is utilized for our review. We prepared the expressed model on the our dataset and tried on the arrangement of 500 images. The mistakes brought about by the model is shown involving a disarray network in Figure.

4.2.1   *VisualBERT*

From the input picture I, several district highlights (f1, f2,..., fn) are first retrieved using Quicker Regions with Convolutional Neural Network before applying the VisualBERT. The following condition is used to switch every location, including f, to visual implantation ev.

$$e_v = f + e_s$$

where es means fragment inserting and designates whether the info is text or picture. The printed implanting et is gotten likewise for the text input:

$$e_t = f_t + e_s + e_p$$

where ep is the positional implanting revealing each symbolic's relative location, and ft is the token implanting for each token in the phrase. The installing is sent into the pre-prepared VisualBERT model for additional handling in the wake of connecting ev and et.

VisualBERT is a pre-prepared model for learning joint logically significant vision and language portrayals. It contains various transformer blocks on top of the visual and text installing. It has been pre-prepared on Microsoft COCO subtitles (Chen *et al.* 2015) considering two objectives: veiled language displaying and sentence-picture forecast. Concealed language demonstrating is basically the same as the technique utilized in sentence BERT (Devlin *et al.* 2018), in which some info message tokens are haphazardly darkened, and the model should anticipate what the current tokens are.

The model should choose if the information message fits the picture to recognize sentenceimages. The main token's VisualBERT yield is used as the multi-modular portrayal em.

The last expectation is then created utilizing a MLP. The above misfortune capability is utilized to calibrate the model for the ongoing errand.

$$l(\theta) = Cross\ Entropy\ Loss(W.e_m, y)$$

where h is the hidden size of VisualBERT and em is a vector with size h. The MLP's learnable framework is W, which has the structure 2 by h. implies the whole model boundaries, including the W.

4.3 Methodology

We used mmf, a technology from Facebook's AI Exploration, to create the actual brain designs for our two techniques using Visual BERT. The model was pre-trained on the MS COCO dataset and has a hidden component of 768. The opinion implantation in the second method is directly obtained from the last logits of sentiment analysis models and their combination. The MLP classifier has two layers and 768 hidden units.

5 EVALUATION METRIC

5.1 Classification accuracy

Since it is more straightforward to understand, we decide the precision of the gauges as the proportion of right expectations to the all out number of forecasts delivered. Subsequently, for each test, we yield the names 0 and 1, as well as the likelihood with which the classifier predicts that the example is loathed. The AUCROC bend is plotted utilizing this likelihood.

5.2 AUCROC

The Recipient Working Attributes bend is a diagram that looks at the Genuine Positive Rate (TPR) against the Bogus Positive Rate (FPR) (FPR). It evaluates how effectively the parallel classifier recognizes classes as the choice edge is changed. An ideal classifier will have a region underneath the bend of one, with the ideal point in the upper left corner of the plot having a TPR of one and a FPR of nothing. To improve TPR and decline FPR, each classifier ought to have a more noteworthy region under the curve.

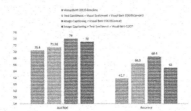

Figure 5. AUCROC.

6 RESULT AND CONCLUSION

6.1 Image recognition

We evaluate our examinations using two systems: the MMF framework and locally created models using Concat BERT as a baseline. Our initial picture inscribing model, combined with Concat BERT, had a precision of 57%. After using Xu *et al.*'s Picture inscribing model, the accuracy increased by 2%. We then tested the MMF framework against more realistic gauge models like Visual Bert, which resulted in a 3.6% increase in AUCROC score and a 6.7% increase in accuracy. Our model labels the image in a way similar to the helpful text confounder, improving the accuracy and AUCROC score. Bilinear Change was attempted as a combination system, but it decreased performance and ran slowly, so we decided to stay with the connection for our discoveries.

6.2 *Sentiment analysis*

Our opinion examination strategy has a significant lift in precision of 4% compared to the Visual BERT model. Our models perform better in two common scenarios: conflicting attitudes between text and picture, and positive attitudes of both modalities. However, opinion prediction accuracy is limited without annotated data to calibrate the models or perform multitask learning. In some cases, sentiment information is inadequate and our model may fail, such as when both modalities have unfavorable perspectives.

7 CONCLUSION

We propose two methods for incorporating external information into our multi-modal models: image captioning and sentiment analysis. Our methods show improved results compared to standard models in the Hateful memes dataset. Our goal is to develop real multi-modal models that consider all modalities. Further research is needed to improve picture captioning algorithms and to combine picture captioning and emotion effectively. We plan to investigate fusion strategies like attention mechanisms and transformers and use large-scale pretrained multi-modal models like UNITER to incorporate internet knowledge via a visual approach.

REFERENCES

[1] Manpreet Singh A, Vedanuj Goswami, & Devi Parikh. Are we Pretraining it Right? Digging deeper Into Visio-linguistic Pretraining, 2020. *arXiv*:2004.08744.

[2] Anderson, P., He, X., Buehler, C., Teney, D., Johnson, M., Gould, S., & Zhang, L. Bottom-up & Top-down Attention for Image Captioning & Visual Question Answering. In *Proceedings of the IEEE Conference on Computer Vision & Pattern Recognition*, pp. 6077–6086, 2011.

[3] Tripathi, A. M., Singh, A. K., & Kumar, A. (2012). Information and Communication Technology for Rural Development. *International Journal on Computer Science and Engineering*, 4(5), 824.

[4] Kumar, A., Mehra, P. S., Gupta, G., & Jamshed, A. (2012). Modified Block Playfair Cipher using Random Shift Key Generation. *International Journal of Computer Applications*, 58(5).

[5] Kumar, A., Mehra, P. S., Gupta, G., & Sharma, M. (2013). Enhanced Block Playfair Cipher. In *Quality, Reliability, Security and Robustness in Heterogeneous Networks: 9th International Conference, QShine 2013*, Greader Noida, India, January 11–12, 2013, Revised Selected Papers 9 (pp. 689–695). Springer Berlin Heidelberg.

[6] Luowei Zhou, Hamid Palangi, Lei Zhang, Houdong Hu, Jason J. Corso, & Jianfeng Gao. Unified Vision-language Pre-training for Image Captioning & VQA. In *AAAI*, 2019.

[7] Majumder, N., Hazarika, D., Gelbukh, A., Cambria, E., & Poria, S. Multimodal Sentiment Analysis Using Hierar Chical Fusion with Context Modeling. *Knowledge-based Systems*, 161:124–133, 2018.

[8] Oriol Vinyals, Alexander Toshev, Samy Bengio, & Dumitru Erhan. Show & Tell: Lessons Learned from the 2015 Mscoco Image Captioning Challenge. In *PAMI*, 2016.

[9] Raul Gomez, Jaume Gibert, Lluis Gomez, & Dimosthenis Karatzas. Exploring Hate Speech Detection in Multimodal Publications. In *WACV*, 2020.

[10] Shenoy, A. & Sardana, A. Multilogue-net: A Context Aware rnn for Multi-modal Emotion Detection & Sentiment Analysis in Conversation. *arXiv* preprint arXiv:2002.08267, 2020.

[11] Singh, A., Goswami, V., Natarajan, V., Jiang, Y., Chen, X., Shah, M., Rohrbach, M., Batra, D., & Parikh, D. *Mmf: A Multimodal Framework for Vision & Language Research*.

[12] Xu, K., Ba, J., Kiros, R., Cho, K., Courville, A., Salakhudi nov, R., Zemel, R., & Bengio, Y. Show, Attend & Tell: Neural Image Caption Generation with Visual Attention. In *International Conference on Machine Learning*, pp. 2048–2057, 2015.

IC-TRAIN – an advance and dynamically trained data structure

Ochin Sharma*

Chitkara University Institute of Engineering & Technology, Chitkara University, Punjab, India

ABSTRACT: A Data structure deals with the storage and retrieval of data efficiently. As new technologies are emerging constantly and more people are associated with the digital world, a new data structure can bring new hope for the techies to store and efficiently use a large amount of data within a minimum response time. Ic-train can store large data, for a large amount of data. The traditional data structures namely array, and linked list also able to store and handle large data but both have their disadvantages, and those disadvantages are overcome by this newly introduced data structure. In context of IOT, Big data and cloud environment, this type of data structure is essentially very useful.

Keywords: Data Structure, employability, industry usage, dynamic array, link list, large data

1 INTRODUCTION

Data structures offer a way to deal with vast amounts of data efficiently. Generally, creating efficient algorithms requires creating well-organized data structures. Data structures are emphasized more than algorithms in a variety of design methodologies and programming languages as the primary emphasis of software design. Both main memory and secondary memory can be used for storing and retrieving data [1–3]. According to the requirements, a variety of data structures are supplied, such as primitive and non-primitive, large and heterogeneous, static, and dynamic, linear, and non-linear, etc. In context of IOT, Big data and cloud environment, this type of data structure is essentially very useful [9,16,17]. For varied applications, various data structures are needed, and numerous of these are highly specialised for certain functions. B-tree data structures, for instance, are excellent for doing database operations, while compiler implementation use hash tables to look up identifiers. For quicker look – up applications, AVL trees are employed [4,5].

2 LITERATURE REVIEW

wIn general, AVL trees are neither weighted nor balanced, meaning that sibling nodes can have significantly different amounts of descendants [6,7]. Treap is the process of looking finding a key-value pair while disregarding the priority using a binary tree structure and a traditional binary search algorithm [8,9]. A self-adjusting binary tree of results dubbed a splay-tree has the extra feature of becoming quick to access formerly searched items once more. Even when merely "read-only" access is used, splay trees can change (i.e., by find operations). This makes it more difficult to employ such splay trees in situations with multiple threads [10]
Linear data structures, array allows for fast and random access. If a person looks at some linear data structures. Access to an array is quick and random. Basic drawbacks of using an array include the fact that it requires continuous memory, which may not be accessible due to a high storage requirement, makes it more difficult to dynamically expand or contract, and makes it more difficult to insert, remove, or transfer data chunks around. Linked lists,

---

*Corresponding Author: ochinsharma3@gmail.com

DOI: 10.1201/9781003393580-51

on the other hand, are more adaptable and enable quicker growth/shrink, insert/delete/move operations, but they also require one or two extra pointers for each member and do not support quick random access to the available data. For instance, in the case of an array, there may be 410 spare bytes at some point [11–15].

A link list requires more memory space than other data types, which goes without consequently; there are some circumstances where a whole new data structure is required to meet the requirements with the least amount of complexity. The data architecture A novel and reliable type of dynamic big data structure called IC-TRAIN combines the advantages of linked lists and arrays while inheriting just a small portion of their typical drawbacks. When we refer to something as dynamic, we mean that it changes through time and is not static. Ic-train is a combination by nature, combining linked lists and arrays, although it is more complex in terms of development because of its architecture. Huge numbers of large data components may be stored using the data structure "Ic-train," and simple actions like insertions, deletions, searching, and counting can be carried out with ease. Particularly with enhanced time complexity and performance efficacy, parallel programming can be done in many circumstances [8].

3 METHODOLOGY

While developing this data structure, each node is termed a coach and the combination of these coaches is termed a Train. Each data item is represented as a passenger. Ic-train is a brand-new, strong, adaptable huge data structure that is highly helpful when there are numerous enormous data sets. In essence, Ic-train is a collection of related data of coach. The 'length' of an ic-train, which may occasionally rise or decrease, refers to the number of coaches in the ic-train. The length of the coach and the availability status of the train will be used to identify an ic-train T with length l (>0). The location of the train's initial "Coach" is "START." As ic-train is having the property of a linked list. So, it can use the non-contiguous memory available in the system.

4 PROCEDURES IN IC-TRAIN

4.1 IC-TRAIN operations

To perform its operations, a user needs to create a self-referential structure, if using C language (however any high-level language can be used) as:
```
struct coach
{
int info [50];
struct coach *ei;
} *START=NULL;
```

4.2 Creation of Ic-train

Create_Train_2 (data [], max)
   Description: This procedure will add several coaches to make a train.

1. BEGIN
2. count: =0.
3. struct coach *q, *tmp.
4. DO
5. start
6. tmp -> info[var] = arr[var].
7. count = count + 1.
8. WHILE (var < max)
9. end
10. tmp -> ei = ='NULL'.

4.3 Add_Coach_At_Beggining (Passenger_Data [ ], Max)

Figure 1. Shows a graphical representation of adding a coach at the beginning. Figure 1. Represents detailed algorithm in a simple and easy to understand.

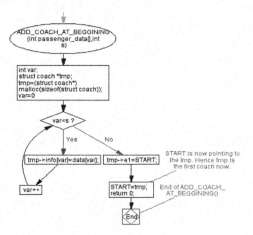

Figure 1.   Add a coach at the beginning.

4.4 Add_Coach_In_Between (Passenger Data [ ], Max, Pos)

Figure 2. Shows a graphical representation of adding a coach at any place except first and last position. The algorithm needs all the data, position of insertion and total number of coaches to check the availability.

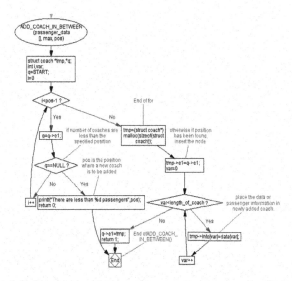

Figure 2.   Add a coach in between any two coaches.

4.5 Insert_In_Vacant_Seat (Passenger)

Figure 3. Shows a graphical representation of adding a passenger (data) in the train (data repository).

339

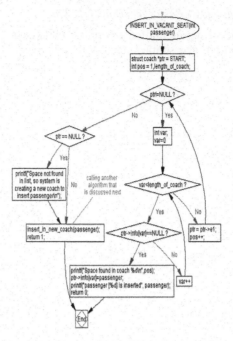

Figure 3. Add a passenger to the vacant seat in a coach.

4.6 *Delete passenger from the train*

Figure 4. Shows a graphical representation of deleting a passenger (data) from the train (data repository).

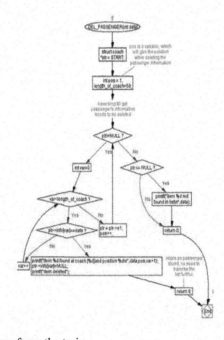

Figure 4. Delete a passenger from the train.

4.7   Deletion of an existing coach

Figure 5. Shows a graphical representation of deleting a coach (data) in the train (data repository). The flow chart is easy to understand.

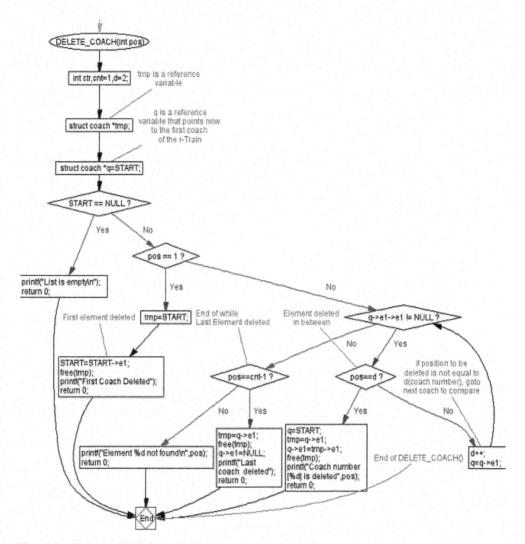

Figure 5.   Delete a coach from the train.

5   CONCLUSION

Ic-train is a data structure that combines the benefits of linked lists and arrays to handle enormous amounts of data very effectively. The phrase has come to refer to adjustable advanced data structures that can be acquired simultaneously by several threads using today's multicore processors. or multiprocessor system scenario where huge paralle-lism became the dominant computing platform (by the expansion of multi-core micro-processor). The system in most enormous corporate organizations must deal with a significant amount of data, for which the data structures described in the existing research don't always result in the required optimum contentment. As a data structure shouldn't be too difficult to use and should be developed with a clear vision, the actual implementation is equally crucial. Here, we've provided a vision for using this data structure in conjunction with the resources at hand.

REFERENCES

[1]   Alfred V. Aho, John E. Hopcroft, and Jeffrey D. Ullman, *Data Structures and Algorithms*, Addison Wesley, 1983.

[2]   Clifford A. Shaffer, *A Practical Introduction to Data Structures and Algorithm Analysis*, Prentice-Hall, 2001.

[3]   Donald E. Knuth, *The Art of Computer Programming*, Addison Wesley, Vol – I, II, III, 1968.

[4]   Sleator D.D. and Tarjan R.E., Self-adjusting Binary Search Trees, *Journal of the ACM*, Vol.32(3), (1985), 652–686.

[5]   Velskii G.M. & Landis E.M., An Algorithm for the organization of information, *Soviet Mathemtics Doklady*, Vol.3.(1962), 1259–1263.

[6]   Nancy A. Lynch. *"Distributed Algorithms"*, Morgan Kaufmann Publishers, California, 1996.

[7]   Thomas H. Cormen, Charles E. Leiserson, Ronald L. Rivest, and Clifford Stein. *Introduction to Algorithms, Second Edition*. MIT Press and McGraw-Hill, 2001. ISBN 0-262-03293-7. Section 10.4: Representing rooted trees, pp. 214–217. Chapters 12–14 (Binary Search Trees, Red-Black Trees, Augmenting Data Structures), pp. 253–320.

[8]   Ranjit Biswas, i-Shuttle: A New Flexible Dynamic Data Structure, *INFORMATION: An International Journal (Japan)*, Vol.14(4), page 1231–246.

[9]   Jena, B., Sahoo, S. K., & Mohapatra, S. K. (2022). *A Cloud Native SOS Alert System Model Using Distributed Data Grid and Distributed Messaging Platform*. In *Intelligent and Cloud Computing* (pp. 55–64). Springer, Singapore.

[10]   Seidel R. and Aragon C.R., Randomized Search Trees, *Algorithmica*, Vol.16.(1996), 464–497.

[11]   Sitarski, Edward (September 1996), Algorithm Alley, "HATs: Hashed array trees", *Dr. Dobb's Journal* 21 (11), http://www.ddj.com/architect/184409965?pgno=5

[12]   Bayer, Rudolf; McCreight, E. (July 1970), *Organization and Maintenance of Large Ordered Indices*, Mathematical and Information Sciences Report No. 20, Boeing Scientific Re

[13]   Cormen, Thomas H.; Leiserson, Charles E.; Rivest, Ronald L.; Stein, Clifford (2001). *Introduction to Algorithms* (2nd ed.). MIT Press and McGraw–Hill. ISBN 0-262-53196-8.

[14]   Kumar, A., Yadav, R. S., & Ranvijay, A. J. (2011). Fault Tolerance in Real Time Distributed System. *International Journal on Computer Science and Engineering*, 3(2), 933–939.

[15]   Tripathi, A. M., Singh, A. K., & Kumar, A. (2012). Information and Communication Technology for Rural Development. *International Journal on Computer Science and Engineering*, 4(5), 824.

[16]   Kumar, A., Mehra, P. S., Gupta, G., & Jamshed, A. (2012). Modified Block Playfair Cipher Using Random Shift Key Generation. *International Journal of Computer Applications*, 58(5).

[17]   Sarma, H. K. D., Balas, V. E., Bhuyan, B., & Dutta, N. (Eds.). (2022). Contemporary Issues in Communication, Cloud and Big Data Analytics: *Proceedings of CCB 2020* (Vol. 281). Springer Nature.

Artificial Intelligence, Blockchain, Computing and Security – Dagur et al. (Eds)
© 2024 The Author(s), ISBN: 978-1-032-49393-0

Big Bang theory improved shortest path, construction, evolution and status model based course like environment machine learning

Tejinder Kaur, Abhijeet Singh, Yuvraj Singh Behl, Sanjoy Kumar Debnath & Susama Bagchi
Chitkara University Institute of Engineering and Technology, Chitkara University, Punjab, India

ABSTRACT: In 1950, Alan Turing got the idea that machines can think like humans, after that he made The Imitation Game in which he sat two humans and a computer in three different rooms and the first human in the form of a text message. Now a robot and a human were answering the question asked by the first human, but now because the three are in different rooms, the first human who was asking the question could not understand it. Answer by human or computer. Alan Turing believed that if the first person cannot understand that the answer is given by another person or by a computer, then it will be proved that computers can also think like humans. This is possible in case the dark energy changes its form at some point in the future and starts shrinking the space instead of expanding it. In this case the force of gravity will become effective and the universe will shrink and shrink as a point. This will probably give rise to another Big Bang. This cycle of great explosion and great contraction will continue. This is just a possibility, so far no evidence has been found in favour of this possibility.

Keywords: Machine, Learning, Model, Human, Big bang, stars, Dark Energy

1 INTRODUCTION

Just as we learn things from human beings through our experience, we learn from humans without machines or computers. The ability of a machine or computer to learn by itself is called machine learning. The machine learns and predicts things automatically with the help of its experience and data. In other words, "the ability to learn by themselves Even though machine learning has become very advanced today, the idea of machine learning came to the mind of Alan Turing, a British mathematician, 71 years ago. In the year 1952, computer scientist Arthur Samuel created a game in the IBM Company called Seven Checkers in which the game was becoming better by learning by itself. In the given example (Figure 1) in 1958, computer programmer Frank Rosenblatt created an algorithm named Perception. This algorithm used to capture patterns and recognize patterns. Today's Finger print lock and Face lock work on this principle [1]. In 1979, some people from Stanford University created a robot called Stanford cart. Its special thing was that it could change its path by detecting everything coming in its way. In 1985, a computer programmer named Terry Sejnowski created a program called Net Talk. Its special feature was that this program could learn and speak English words by itself. Later there were many changes in it and today we know it as Google Assistance and Siri. This is making the life of us humans much easier. Let's know about the top application of ML [1]. The results of these experiments turned out to be unexpected. The speed of expansion of the universe was neither slowing down nor steady; on the contrary it was increasing. The experiment was based on observation of some Group 1A supernovae [1]. Group 1A supernovae are used as standard lampposts or standard candles for measuring cosmic distances. According to the most prevalent and accepted theory of the origin of the universe, billions of years ago, the whole universe was a point. [1] For

DOI: 10.1201/9781003393580-52

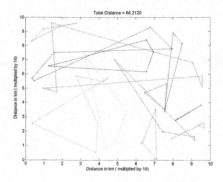

Figure 1.   City location [1].

some unknown reason this point started expanding with an explosion and the universe came into existence. This expansion of the universe continues even today. We call this the Big Bang Theory [2]. The observations made by Edwin Hubble and Einstein's theory of relativity gave this theory a practical and logical basis. [2]. Gravitational force attracts mass, so most scientists believed that gravitational force would slow down or stabilize the speed of expansion of the universe [2]. Light to reach us from those supernovae. Scientists observed light emitted from supernovae at different distances to calculate the speed at which they move away from us. This is calculated by measuring the red shift caused by the Doppler Effect. This observation was able to inform scientists about the magnitude of the expansion of the universe at different points in the past [3]. It was amazing but true. This experiment was performed by two different groups of scientists at the same time so that there was no possibility of error in data or observations in the given example (Figure 1). All this meant that a mysterious force was not only neutralizing the effect of the force of gravity but also accelerating the expansion of the universe.

2   DISCUSSION

2.1   *Self driving car*

Machine learning is used in Self Driving Car. Cars are trained using Unsupervised Learning Algorithm so that the car can run without a driver. You must have heard the name of Tesla, this company makes self.

Driving Cars and it is the only one in the market is also decided that every person has to die one day or another. Everything has an end. Even our earth will end one day or another as it rotates on its axis. The sunlight that has been giving light for millions of years will also disappear one day and it will become an extinguished lamp. [3] Every star has an age. Scientists have recently proved the existence of dark energy with new methods in addition to supernovae. This new method relies on data from observations at the Chandra X-ray Observatory [3]. According to this, dark energy is a form of Einstein's Cosmological Constant, which Einstein himself denied as the biggest blunder of his life. In these experiments, scientists have noticed a decrease in the mass of some galaxy clusters, which may be due to a disintegrating force like a dark energy [4] According to scientists, this effect of dark energy can tear stars, planets and even the molecules that make us in the future. (The Chandra Space Observatory is named in honor of Indian-origin scientist Subramanian Chandrasekhar.)According to Michael Turner, a scientist in the Department of Astronomy and Astrophysics at the University of Chicago involved in the experiment. The amount of dark energy According to astronomers, 74% to 76% of the total universe is made up of dark energy, 20% to 22% is made up of dark matter and the remaining 4% is made up of ordinary matter [4]. The remaining 4% includes us, our solar system and our galaxy. How insignificant and insignificant we are in the universe! Dark energy works on a large scale (at the cosmic level) but on a small scale (galaxy) it has no effect. Due to this we do not see the effect of this energy in our galaxy or in our solar system. [4] The effect of this energy has not been seen in the local galaxy cluster, our galaxy and neigh boring galaxy Andromeda is a member of this group. Owner of Tesla Company is Elon Musk. Waylon, GM Cruise, and Argo AI these companies also make Self Driving Cars. However, there is still more advancement in this technology [4].

2.2 *Traffic alert*

If you use Google Map, then you must have seen that, when there is a lot of traffic on the road ahead, then Google Map shows you the message of Traffic Alert. This is also possible only with machine learning. When everyone is using the map, Google Map processes Everyone's Location, Speed, and Route and shows you the correct route using machine learning algorithms in the given example (Figure 2).

2.3 *Product recommendation*

You went to buy a product on Amazon but you didn't buy it and the next day you were watching YouTube and you saw an ad for the same product that you went to buy on Amazon but you didn't buy it, then you went to Face book and there too. The ad show of the same product was happening. After tracking your online behavior, machine learning algorithm is applied on it and then you are shown the Ad in which you are interested [4].

2.4 *Social media*

When you create an account on Instagram or Face book, you get friend suggestions that are in your contact list or you have met them, this is all done by machine learning. Face book extracts the list of your friends from your contact list and then suggests you new friends. Similar Algorithm also happens on Instagram, that's why it shows you the list of your friends to tag you. If your face is seen in someone else's photo, then Face book uses algorithms to understand that if you are in that photo, it means that you have met before and the person whose photo you were in suggests a friend to you and your friend. Friend Suggest by your name [5].

2.5 *Virtual personal assistance*

You must have used Google Assistance and Cortina; these are all examples of Virtual Personal Assistance. Whenever you say ok Google then Google Assistance gives you a reply and also answers any question you ask. Maybe you don't know the scope of Machine Learning in this field and how it is going to change today's technology. Before learning this, you need to have basic knowledge of computer, which you can learn by taking a short computer course. Apart from this, the machine improves its performance. Nowadays, machine learning i.e. artificial intelligence has developed so much that through it you can complete any task in the world without the help of any human within a few minutes. Its main goal is Computer system is to be advanced in which the machine uses labeled data to understand the data sets to create the model. This is a learning in which the machine learns things without any supervision. Unsupervised learning is used to derive useful insights from large amounts of data the given example (Figure 1). Unsupervised learning model is capable of thinking like a human being, like behaving like a human being, working and thinking etc. Labeled data is a type of input data that is already present in the machine. If we say it in simple language, it is a process of supervised learning in which the input data present in the system by means Data [6] Supervised learning is a learning process based on monitoring, just as a small child learns good and bad by being under the supervision of his parents. There are two types of Supervised Learning, the details of which we are giving you below. Apart from this, Regression is used as a method of predictive modeling in machine learning [9]. Regression is also of many types. Like Linear Regression, Polynomial Regression In the given example (Figure 1).

2.6 Type of algorithm in which data is organized into categories

For example – it is used to classify the students studying in the School i.e. it helps in determining their grades like Average, Good, and Excellent in the given example (Figure 2). Unsupervised learning in which the machine is trained using unlabeled data. In this process, the machine learns everything without supervision. It is used to get more useful insights from the data. This type of learning model is able to think like a human like behaving like a human, thinking, acting, thinking etc. Unsupervised learning is of two types. Which is the following?

Clustering – In this type of method, objects are separated from each other and arranged in a group. The objects of one group are placed in one group and the objects of another group are placed in another group. The best example of this is when you if you go to a hotel to eat, you must have seen that there are different types of food items. Moreover, if you go to a car showroom, then there are different types of companies' cars kept at different places [6].

Association – Association is a technique that tells how objects are associated with each other. Association is a popular method of finding relationships between variables in large databases. And if he made any mistake then he corrects that cult the given example (Figure 2). The biggest example of this is that Robot teaches himself how to move his hand without needing human help [6].

What is Operating System?
It is used in the following places, the details of which we will give you according to the point below, which is as follows. Machine learning is used to recognize people, places, objects, pictures etc. Face detection is used to recognize pictures which are a part of it You people must have seen that when you say something through the mic in Google, then the result related to what you have said comes in front of you. This is how Google can provide you with this service through machine learning [6].Today you can go to any corner of India without anyone's help through Google Map. The technology used in Google Map is known as Machine Learning. Machine learning is used by companies like Amazon and Netflix.

These companies provide the user with input and output data. For example, when you search for a product on Amazon, all the information about that product is displayed on your screen. Jaata hai and this Amazon company can do for you is the machine learning technique behind it. Today, complex diseases are identified by Machine Learning so that it can be treated. In the share market, you must have seen that big share experts predict that the price of this share will increase and decrease in the future. It is possible to do all this in the given example (Figure 2).

Google Translate: Whenever you want to translate a sentence from English to Hindi, you use Google Translate, but have you ever wondered how Google Translate could do all this?

You can translate many languages of the world by using Google which the user has to provide correct output Translate. Machine learning is done on Google Translate; Natural Language Processing Algorithm is applied on Google Translate so that it easily translates any language (English) into any other language (Hindi) [6] the machine learning system or program that is trained without human help is called Machine Learning Model. Machine learning model is a computer program; it takes input and then predicts output by learning from experience in the given example (Figure 2). By Machine Learning, the machine is made exactly like a human being, just like a human being takes a decision by thinking, in the same way; the machine is made in such a way that the machine can take its own decision without needing any human being. In this, Google also conducts a free course of machine learning in which you can learn online machine learning for free. After this you can also do a paid course so that you can know about it deeply. Click here to enrol in Google Machine Learning Course – Google Machine Learning Course.

Is there much scope in machine learning? – It is expected that in the world of computer and internet, machine learning and artificial intelligence will be the era.

Online Video Streaming: If you use Netflix, then you must have noticed that Netflix gives you the movie of your choice at the right time in this, the machine uses labelled data to understand the data sets to create the model. Labelle data is a type of input data that is already present with the machine and the system predicts the output data by analyzing this data. In simple terms, supervised learning is a process in which the System provides correct output data to the user with the help of input data [6].

2.7 *Types of classification is used to classify data into classes or groups in the given example (Figure 2)*

For example – it can be used to classify the students of a class based on their grade (average, good, excellent).

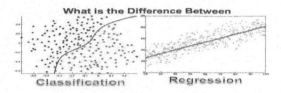

Figure 2.   Classification and regression [2].

2.8 *Advantages of supervised learning*

It helps the machine to predict output data based on previous input data. This learning is not able to perform difficult tasks. It takes a lot of time to predict the output data. This is a learning in which the machine learns things without any supervision. Unsupervised learning is used to derive useful insights from large amounts of data. Unsupervised learning model like behaving like a human being, working and thinking etc [6]

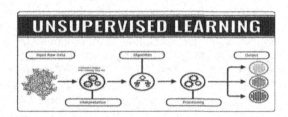

Figure 3.   Unsupervised learning [3].

2.9 *Types of unsupervised learning*

It is also mainly of two types: – 1- Clustering is a method in which objects are divided into different groups, in which the objects that are similar are placed in one group and the objects that are different are placed in another group. Clustering also plays a role in our normal life. For example, a restaurant has different types of food and a vehicle showroom has cars, bikes and other vehicles.

2- Association is a technique that tells how objects are associated with each other. Association is a very popular method of finding relationships between variables in large databases in the given example (Figure 3) Advantages of Unsupervised Learning Tasks easily because it does not have labelled data due to which it can complete complex tasks easily. Its results are not accurate, due to which the user does not get correct information. They can be properly trained in all ways. Personally, I think that machine learning can be like a catalyst which is going to be our assistant to change our future. We have become so much dependent on machine learning that life without it seems beyond imagination [7].

It is mainly of two types: – 1- Regression: Apart from this, Regression is used as a method of predictive modelling in machine learning. Regression is also of many types like Linear Regression, Polynomial Regression. For example, when we book a taxi in Ola or Uber it shows us information such as the cost of the trip, the distance, and the route. Machine learning in Hindi: The word machine learning is technically as much in reality as it sounds; if machine learning is understood in an easy way then it is not a difficult topic. Whether you know it or not, there is no field today where machine learning is not being used now see, we all use different services of Google in our smart phones. If you have set up Google Assistant in your phone, then your Google Assistant gives you a reply just by saying 'ok Google'. Have you ever wondered how we are able to talk to a digital assistant and how it is able to give us accurate answers? Now technology has become so advanced that the work which seemed impossible earlier has been made possible by technology. Machine learning is a discovery that has played an important role in our lives if seen, we like to know or do anything in a simple way. Now many people do not search for questions by writing on the Internet but search by speaking If you also like to search by speaking then perhaps this question has come in your mind that how can any machine system give us an exact answer by speaking? So I want to tell you that it is directly related to machine learning and artificial intelligence Friends, what about machine learning? I am going to give detailed information about it in easy language, so that you can understand it in a good way. So give a good smile and read this post completely it will be very easy to understand machine learning from an example. Let's understand it with an example like – We humans learn something good and bad from our work and experience. If we have any problem in the future, we solve that problem based the given example (Figure 3). On our past experiences similarly, machine learning is also a similar concept a computer or machine is programmed in such a way that it can work according to the user and also store the usage data which will improve the machine experience in the future. Machine learning mainly works on the development of computer programs. Who accesses the data himself and later learns it himself and uses it there are many examples of machine learning, out of which you can take the example of Face book because Face book is used by everyone. While using Face book, you must have noticed that often the profiles we check, or share something in other groups, then automatically Face book starts giving us a notification that you know them, or can add them to the friend list. According to this data, the machine gives its output and the output if you use Uber for transportation, then you must have seen how Uber itself detects the location of the customer, the actual location of the vehicle is also visible in real time, the driver is informed about the shortest and open routes. It is also known, and at the same time, Uber keeps changing its charges when there is a huge demand, so all this is possible only with Machine Learning [7]. So here Machine Learning Technology is being used, in which you see the result according to your activity and your interest [7] Another example can be taken from Netflix, where based on the movies you have watched or liked, you start seeing many other movies of the same type, that is, here the pattern of your previous search data is prepared by Machine Learning, and similar data is presented to you. In Supervised input and output data are already provided to the machine, which is also called Training Data or Labelled Data [7].

Unsupervised Learning: – In unsupervised learning, the machine is not given any input and output labelled data beforehand. In this, as soon as the machine receives any input, the machine itself evaluates it and prepares a cluster and divides them into different groups according to the type of things.

Reinforcement Learning: – As in Supervised learning, the machine already has the training data and the output labelled. On the contrary, in Reinforcement Learning, the machine does not have any answer and because there is no training data, the decision is taken by the Reinforcement Agent task, which tries to complete the task based on its experience. Does and learns from repeated attempts the given example (Figure 4).

Figure 4.   Reinforcement learning in ML [4].

2.10 *Machine learning applications*

There are many applications of Machine Learning in our daily life, which we all use, some of which are as follows. Face book: – Face book is widely used all over the world and we all use it. Machine learning is used in Face book's Automatic Friend Tagging Suggestion, in which Face book checks in its database based on Face Detection and Image Recognition and recognizes a photo or image [8].

Shopping Websites: – If you shop online, you may have noticed that information related to the product you are looking for appears everywhere [9]. Like you searched for something on Amazon and after some time when you open Face book or YouTube, you start seeing videos related to the same product there too. So this is all about Machine Learning, in which Google takes care of your every activity, and shows you ads accordingly in the given example (Figure 4). E-Mail Spam Filters: – While using E-Mail, you must have seen how only our necessary Mails come in the inbox, and most of the Spam Mails go to a folder named Spam, so Machine Learning is being used behind this. In which the content and source of an email is automatically detected by machine learning and when something wrong is found, the email is spammed in the given example (Figure 4).

3   RESEARCH METHODOLOGY

If we talk about the benefits of Machine Learning, it has actually made human life much easier. Today, Machine Learning is being used continuously to improve operations in every field. For this, machines are being made more effective and efficient. Retail: – Trends can be understood and future sales can be predicted. Also, the appropriate product is suggested by understanding the browsing behavior of the customer, so that the customer experience can be increased and the sales can be increased in the given example (Figure 5). Finance: – AI and Machine Learning are also being used in the finance sector, so that a better and faster service can be provided to the customer like increasing the security of transactions and preventing Fraud activities etc [10].

In machine learning, algorithms are used to improve the computer or machine that gives the system the ability to think and understand. Its algorithm is used in many tasks like

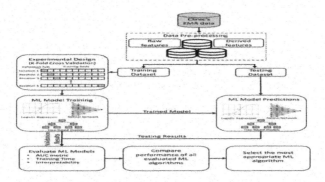

Figure 5.   Methodology machine learning [5].

349

[11,12]. The use of machine learning is not limited to any one field but today this technology of AI is being benefited in almost every field; let's know in the given example (Figure 6).

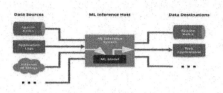

Figure 6. ML inference host machine learning [6].

Table 1. Relevant found.

Relevant found in	Anchor	Title	Abstract	Regression
Top 5	55	78	88	56
Top 50	89	85	75	23
Top 500	68	25	65	79
Top 5000	45	45	32	12
Improvement	32%	56%	22%	42%

4   IMPLEMENTATION/RESULT

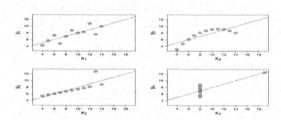

Figure 7. Regression result for machine learning [7].

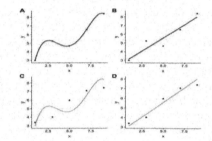

Figure 8. ML inference host range ML [8].

REFERENCES

[1] Ching, T. *et al.* Opportunities and Obstacles for Deep Learning in Biology and medicine. *J. R. Soc. Interface* 15, 20170387 (2018). This is a thorough review of applications of deep learning to biology and medicine including many references to the literature.

[2] Mitchell, T. M. *Machine Learning* (McGraw Hill, 1997).

[3] Goodfellow, I., Bengio Y. & Courville, A. *Deep Learning* (MIT Press, 2016).

[4] Libbrecht, M. W. & Noble, W. S. Machine Learning Applications in Genetics and Genomics. *Nat. Rev. Genet.* 16, 321–332 (2015).

[5] Zou, J. *et al.* A Primer on Deep Learning in Genomics. *Nat. Genet.* 51, 12–18 (2019).

[6] Myszczynska, M. A. *et al.* Applications of Machine Learning to Diagnosis and Treatment of Neurodegenerative Diseases. *Nat. Rev. Neurol.* 16, 440–456 (2020).

[7] Yang, K. K., Wu, Z. & Arnold, F. H. Machine-learning Guided Directed Evolution for Protein Engineering. *Nat. Methods* 16, 687–694 (2019).

[8] Tarca, A. L., Carey, V. J., Chen, X.-W., Romero, R. & Drăghici, S. Machine Learning and Its Applications to Biology. *PLoS Comput. Biol.* 3, e116 (2007). This is an introduction to machine learning concepts and applications in biology with a focus on traditional machine learning methods.

[9] Kandoi, G., Acencio, M. L. & Lemke, N. Prediction of Druggable Proteins Using Machine Learning and Systems Biology: A Mini-review. *Front. Physiol.* 6, 366 (2015).

[10] Sachdeva R. K., and Bathla P., "A Machine Learning-based Framework for Diagnosis of Breast Cancer", *International Journal of Software Innovation*, vol. 10, no. 1, 2022, pp. 1–11.

[11] Kumar Sachdeva R., Garg T., Khaira G. S., Mitrav D.and Ahuja R., "A Systematic Method for Lung Cancer Classification," 2022 10th International Conference on Reliability, Infocom Technologies and Optimization (Trends and Future Directions) (ICRITO), Noida, India, 2022, pp. 1–5, doi: 10.1109/ICRITO56286.2022. 9964778.

[12] Sachdeva R. K., Bathla P., Rani P., Kukreja V.and Ahuja R., "A Systematic Method for Breast Cancer Classification Using RFE Feature Selection," *2022 2nd International Conference on Advance Computing and Innovative Technologies in Engineering (ICACITE)*, 2022, pp. 1673–1676, doi: 10.1109/ICACITE53722. 2022.9823

Artificial Intelligence, Blockchain, Computing and Security – Dagur et al. (Eds)
© 2024 The Author(s), ISBN: 978-1-032-49393-0

A module lattice based construction of post quantum blockchain for secure transactions in Internet of Things

Dharminder Chaudhary & M.S.P. Durgarao
Department of Computer Science and Engineering, Amrita School of Computing, Amrita Vishwa Vidyapeetham, Chennai, India

Pratik Gupta
Department of Applied Mathematics, School of Vocational Studies and Applied Sciences, Gautam Buddha University, Greater Noida, India

Saurabh Rana
Department of Mathematics, Bennet University, Greater Noida, India

Soumyendra Singh
Department of Mathematics, Amrita School of Engineering, Amrita Vishwa Vidyapeetham, Chennai, India

ABSTRACT: The Fiat-Shamir with Aborts paradigm of Lyubashevsky introduced efficient lattice based signatures. We have used lattice based signature to design post quantum blockchain design. But the blockchain is always consisting of many nodes depending upon participants, then we require an efficient aggregate signature technique to verify nodes efficiently. We have designed a blockchain using fast post quntum secure tool called module lattices. Module lattices provide both security and efficiency based on two assumptions, (1) Module Learning With Errors, and (2) Module Short Integer Solution respectively. The proposed blockchain design ensures security of Internet of Things devices against quantum attacks. The design provides a better secure environment for Internet of Things. Blockchain technology has been used to combine fixed number of IoT nodes. This makes the communication efficient, and the aggregate signature has been used to provide the security to the given architecture.

1 INTRODUCTION

A distributed ledger called blockchain [20] is capable of creating trust in among more than two nodes. Blockchain technology is anticipated to operate as the basis for a database that will store the communications and transaction records of Internet of Things devices. The literature has been discussed based on "Internet of Things" in the form of device-to-tool system. For Internet of Things, this technology has three major advantages: (1) it creates trust among more than two nodes, and it also lowers the risk of collusion and manipulation, (2) it also eliminates intermediaries, (3) it increases transaction rate and shortens agreement time. It is not an easy task how to combine blockchain networking system with the blockchain and "IoT".

The public key cryptography, one-way collision resistant hashing, and signature algorithms concepts make the blockchain secure and reliable for decentralization and clarity [1,16,25]. There is a lot of development in blockchain technology discussed in the articles [15,23]. With the aid of quantum computers, it is possible to quickly and effectively solve number-theoretic issues like discrete logarithm and elliptic curve discrete logarithm utilising various algorithms based on Shor and its variants [26]. Now, we are ready for advanced

DOI: 10.1201/9781003393580-53

development of the post quantum secure cryptographic primitives. A detailed literature related to development can be found in the articles [10,11,22]. Li *et al.* [11] have proposed a new post quantum secure lattice based signature. In this technology, Bonsai Trees concept is used to generate public/private key pairs. The algorithm Randomize Basis helps in generating the new keys from the seed producing enough randomness. Later on, Kiktenko *et al.* [10] proposed another significant design for blockchain technology. Li *et al.* [13] have proposed a new secure block containing blockchain for peer-to-peer cloud storage services. Fernandez *et al.* [7] have proposed a review on post quantum secure blockchain resistance against existing quantum attacks. Li *et al.* [12] have proposed efficient anti-quantum lattice based blind signatures for blockchain enabled system. Saha *et al.* [24] have proposed an advanced blockchain design for post quantum secure decentralized database. The exertion in [21], presented Internet of Things Decentralized Access Control System where access control information is stored and shared using Blockchain. Internet of Things is fasten to Block Chain network with management hub Which is a gateway between these two. BlockChain generally explains how the interesting conversation of things can benefit from it. Proposed an efficient method to improve transaction fees by Li *et al.* [14]. A security, identity, coordination, and privacy solution crafted without an intermediary is explored by Slock [6]. Even though it is not an Interent of Things solution, Blockchain clever transportation device [27] provides a conceptual model on seven layers for massive scale vehicular networks. As well, [8] suggests using Blockchain in smart grid transactions for comfort. There has been a proposal to integrate BlockChain with clever devices in [4]. As part of provenance [5], roof of concept and transparent concepts were applied to supply chain management. As a result, [9], proposes the Decentralized Framework for BlockChain of Things. It is important to note that the majority of current research on Internet of Things and BlockChain integration does not take into account the expandability of ledgers or enlarging transaction rate. The main goal of this study is to develop a substructure that can be applied to Internet of Thing network. Therefore, we tried to develop a framework that uses modular grid-based aggregate signature with identity-based key generation. The use of aggregate module grid-based signatures in blockchains is an emerging idea. Benchmark results show performance in terms of throughput, latency, power consumption and complexity. The security analysis shows that the proposed system is protected against various quantum attacks.

2 PRELIMINARIES

We have taken $Q = \frac{\mathbb{Z}[x]}{x^n+1}$ ring of integers in the field of $2n^{th}$ cyclotomic field. Let $q \equiv 1 \bmod(n)$ be a random prime, and m be the rank of the matrix/ dimension of the lattice [2]. Let $Q_q = \frac{\mathbb{Z}_q[x]}{x^n+1}$ be the corresponding quotient ring of integer polynomials. An element $e = \sum_{i=1}^{n} e_i.x^{i-1} \in Q$ can be denoted as vector $e = (e_i)_{i \in [n]} \in \mathbb{Z}^n$. The symbol $||e||$ denotes the Euclidean norm of the ring element. The vector $e \in \mathbb{Z}^n$ be an integer vector, and D is discrete Gaussian with standard deviation σ, and mean zero.

Definition 1.1 *Let $G = [g_1|g_2|\ldots|g_m] \in \mathbb{R}^{m \times m}$ be a square matrix, with linearly independent columns $g_1|g_2|\ldots|g_m$ respectively. The full rank lattice $\mathbb{L}$ [2] generated by $G = [g_1|g_2|\ldots|g_m]$ is the set $\mathbb{L}(G) = \{y = Gr \in \mathbb{R}^m \mid r \in \mathbb{Z}^m, y = Gr = \sum_{i=1}^{m} r_i.g_i\}$*

Definition 1.2 *Let $q > 3$ be an odd prime, $B \in \mathbb{Z}_q^{n \times m}$, $\beta \in \mathbb{R}$ is positive number, then the short integer solution (SIS) assumption is to compute $e \neq 0 \in \mathbb{Z}^m$ satisfying $A.e = 0 \bmod(q)$, and $||e|| < \beta$.*

Definition 1.3 *Let $q > 3$ be an odd prime, $B \in \mathbb{Z}_q^{n \times m}$, $\beta \in \mathbb{R}$ is positive number, and $u \in \mathbb{Z}_q^n$, then the short integer solution (SIS) assumption is to compute $e \neq 0$ in the set $\mathbb{Z}^m$ satisfying $A.e = u \bmod(q)$, and $||e|| < \beta$.*

Definition 1.4 *Module Learning with Errors (MLWE) assumption is denoted by* $MLWE_{k,\ell,\beta}$, *and defined as if* $B \leftarrow U(Q_q^{k \times \ell})$, *and* $e \leftarrow Q_q^k$ *are known, then it si hard to decide* $e \in U(Q_q^k)$ *or* $e = [B|I_k].s$, *where* $s \leftarrow U(S_\beta^{\ell+k})$, *and S follows uniform distribution.*

Definition 1.5 *Module Short Integer Solution (MSIS) assumption is denoted by* $MSIS_{k,\ell,\beta}$, *and defined as if if* $B \leftarrow U(Q_q^{k \times \ell})$, *then find a non zero* $e \leftarrow Q_q^{k+\ell}$ *such that* $[B|I_k].s = 0 \in Q_q^k$, *and* $\|e\| < \beta$.

3 PROPOSED BLOCKCHAIN BASED POST QUANTUM SECURE AUTHENTICATION FOR INTERNET OF THINGS

This section, illustrates the proposed scheme for device communication based on module learning with errors, and module short integer solution assumptions respectively. The scheme consists four phases: (1) setup phase, (2) subkeys generation phase, (3) signature phase, and at last (4) verification phase. This construction uses $Trap - Gen$ to generate public key $B_0 \leftarrow \mathbb{Z}_q^{n \times m}$ with basis $T_{B_0} \leftarrow \mathbb{Z}_q^{m \times m}$ satisfying the condition $\|\widetilde{T}_{B_0}\| \leq O(\sqrt{n \log(q)})$, $m = O(n \log(n))$. The wallet contains the seed value (B_0, T_{B_0}) that is lattice basis to generate subkeys using Bonsai tree algorithm.

3.1 Setup phase

This phase takes input n, and q, then it uses algorithm $Trap - Gen(q, n)$ to generate random $B_0 \leftarrow \mathbb{Z}_q^{n \times m}$, with basis T_{B_0} satisfying $\|\widetilde{T}_{B_0}\| \leq O(\sqrt{n \log(q)})$.

1. Setup takes input n, and q, then it uses algorithm $Trap - Gen(q, n)$ to generate random $B_0 \leftarrow \mathbb{Z}_q^{n \times m}$.
2. It generates basis T_{B_0} satisfying $\|\widetilde{T}_{B_0}\| \leq O(\sqrt{n \log(q)})$.

3.2 Subkey-generation($B_0, T_{B_0}, B_1, B_2, B_3, ...,B_n$)

The algorithm Subkey-Generation($B_0, T_{B_0}, B_1, B_2, B_3, ...,B_n$) chooses matrices $B_1, B_2, B_3, ...,B_n$ for the devices ($U_1, U_2, U_3, ...,U_n$), and they uses the algorithm $Ext - Basis$, and it outputs $T_{B_{1'}} \leftarrow Ext - Basis(T_{B_0}, B_{1'} = B_0|B_1)$, $T_{B_{2'}} \leftarrow Ext - Basis(T_{B_0}, B_{2'} = B_0|B_2)$, ..., $T_{B_{n'}} \leftarrow Ext - Basis(T_{B_0}, B_{n'} = B_0|B_n)$, and it generates private/public key pairs ($SK_1 = s_1 \leftarrow T_{B_{1'}}$, $PK_1 = B_{1'}.s_1$), ($SK_2 = s_2 \leftarrow T_{B_{2'}}, PK_2 = B_{2'}.s_2$), ..., ($SK_n = s_n \leftarrow T_{B_{n'}}, PK_n = B_{n'}.s_n$) for each of devices in the blockchain. These subkeys are used to perform signature and verification of devices.

3.3 Address generation

Different addresses are generated from different public keys of autonomous devices. Suppose, the autonomous device V_{B1} has public key $B_{1'} = (b_1, b_2, ...,b_{2m}) \in \mathbb{Z}_q^{n \times 2m}$, then it applies a hashing like SHA-256 to generate hashed public key. First, the device U_{B1} sends a rquest to another device U_{B2}, then U_{B2} chooses pair of subkeys from its wallet, and it generates the address, and sends to U_{B1}. Finally, the user U_{B1} generates a transaction for this particular address, and this is broadcasted in the whole network.

3.4 Signature generation

A device broacasts a transaction to other devices, and it uses algorithm $Sign(SK, M)$ is used to generate a signature on the transaction with one private from the wallet. It transaction bearing signature is different becuase of different time stamps, and its private key used to sign. A device samples $y \leftarrow D$, and it computes $u = [B_{1'}].y \in Q_q^k$, $c = H(u, PK_1, M)$, and $z = s_1.c + y$ with probability $(1 - P_{rej})$, where $P_{rej} = min\{1, \frac{D_s^{\ell+k}(z)}{M}.D_{c.s,s}^{\ell+k}(z)\}$. The final signature is $\sigma = (u, z)$.

353

3.5 Aggregation of signatures

A device broacasts a transaction to other devices, and it uses aggregate signature algorithm $Agg - Sign(PK = PK_j, M = M_j)$ is used to generate a signature for more than one device. To find aggregate signature, the device computes $c_j = H(u_j, PK_j, M_j)$, and queries $e_j \leftarrow H_2(c_1, c_2, ..., c_N, j)$ for $j \in [N]$. Furthers, it computes $z = \sum_j e_j.z_j$ satisfying the norm condition $||z|| \leq \beta = O(\sqrt{N}B)$. Finally, the device broadcasts the aggregate signature $\sigma_{agg} = ((u_j)_j, z)$.

3.6 Verification

The algorithm $Verification(u, z)$ takes the input public key matrix PK_1, message M, and signature σ respectively. The receiver end device computes $c = H(u, PK_1, M)$, and if $||z|| < \beta$, and $B_{1'}z = PK_1.c + u$, then only accepts the signature.

3.7 Aggregate signature verification

A device broacasts a transaction to other devices, then the device verifies aggregate signature using algorithm $VAgg - Sign(PK = PK_j, \sigma_{agg})$. It computes $c_j = H(u_j, PK_j, M_j)$ for $j \in [N]$, and queries $e_j \leftarrow H_2(c_1, c_2, ..., c_N, j)$ for $j \in [N]$. Furthers, it computes $PK.z = \sum_j e_j.$ $(PK_j.c_j + u_j)$, and if $||z|| \leq \beta = O(\sqrt{N}B)$, then only accepts aggregate signature.

4 SECURITY ANALYSIS

Any aggregate signatuture Π_{Agg} for space of messages $\mathcal{M}$, and probabilistic polynomial times adversary $\mathcal{A}$ is a tuple $\{Key - Gen, Sign, Verification, Agg - Sign, VAgg - Sign\}$ following the rules given below.

1. **Key-Gen** (1^κ): This algorithm takes security parameter κ, and it creates private key SK, and corresponding public key PK respectively.
2. **Sign(SK,M)**: This algorithm takes input secret key SK, and message, and it ouputs the signature σ.
3. **Verification(PK,M,σ)**: This algorithm takes public key PK, message M, and signature σ, and accepts or rejects the signature σ.
4. **Agg-Sign(PK, M, $\sum$)**: This algorithm takes verification keys $PK = PK_j$, and messages $M = M_j$ for $j \in [N]$, and it computes the aggregated signature σ_{agg}.
5. **VAgg-Sign(PK, M, σ_{agg})**: This algorithm takes $PK = PK_J$, messages $M = M_j$ for $j \in [N]$, and aggregate signature σ_{agg}, and it accepts or rejects aggregate signatures.

Theorem 1.1 *The proposed module lattice based post quantum secure blockdhain is mathematically correct.*

Proof. Let σ_{agg} be a valid aggregate signature produced by the algorithm $Agg - Sign(PK = PK_j, M = M_j, \sum)$. If $||z|| < \beta$, then the algorithm computes $B_{j'}.z = B_{j'}.$ $(\sum_{j=1}^n e_j.z_j) = \sum_{j=1}^n e_j.B_{j'}.z_j = \sum_{j=1}^n e_j.(PK_j.c_j + u_j)$. Hence, it is mathematically correct.

Theorem 1.2 *The proposed post quantum secure authenticated autonomous vehicular communication system possesses strong unforgeability under chosen message attack.*

Proof.

- $\mathcal{G}_0$: The adversary $\mathcal{A}$ sets $N_q = N_{H_1} + N_{H_2} + N_S$, where $\mathcal{A}$ is allowed at most N_{H_1} queries to oracle H_1, N_{H_2} queries to oracle H_2, and N_S queries to the signing oracle on PK_N respectively. Let C be the challenge space from set of signatures. Let $\mathcal{B}$ be algorithm for random $h_i, h_j \leftarrow U(C)$ for $j \in [N_q]$, and it maintains a list HT_1 for H_1. Initially, we consider the list HT_1 is empty. The algorithm $\mathcal{B}$ maintains a list HT_2 for H_2. Initially, we consider the lists HT_1, and HT_2 are empty.

1. **Setup**: In the setup phase, $\mathcal{B}$ creates (SK_N, PK_N) using $Key - Gen(1^\kappa)$ algorithm. Finally, $\mathcal{B}$ sends PK_N to the adversary $\mathcal{A}$.
2. **Queries on H_1**: If $\mathcal{A}$ submits query for $g =< u, PK, M >$, then HT_1 checks it in the list. If the value $HT_1[g]$ is found in the list, then it returns $HT_1[g]$, otherwise, it chooses random $c = HT_1[g] \leftarrow U(C)$, and retuns this after storing the value in the list.
3. **Queries on H_2**: If $\mathcal{A}$ submits query for $g =< c_1, c_2, ..., c_N, j >$ for $j \in [N]$, then HT_1 checks it in the list. If the value $HT_1[g]$ is found in the list, then it returns $HT_1[g]$, otherwise, it chooses random $c = HT_1[g] \leftarrow U(C)$, and retuns this after storing the value in the list.
4. **Sign Query**: The algorithm $\mathcal{B}$ follows honest signing procedure with secret key SK_N on message M.
5. **Forgery**: If $\mathcal{A}$ finds a forgery $\sigma_{agg} = ((u_j)_j, z)$ on $M = M_j$ for $j \in [N]$, and public key $PK = PK_j$ for $j \in [N]$, where $c = HT_1$ is programmed on $g = (u_N, PK_N, M_N)$ with $c = h_{ji}$ for index j, i. If this is not the case, then adversary $\mathcal{A}$ can guess correct c_N with probability $(1 - \frac{1}{C})$.

- $\mathcal{G}_1$: The game $\mathcal{G}_1$ follows the same lines as $\mathcal{G}_0$ except $\mathcal{B}$'s signature generation process is not honest, but it similates message without secret key SK_N. In **Signing Queries** phase, $\mathcal{B}$ samples $c \leftarrow U(C)$, $z \leftarrow D_s^{\ell+k}$, and a message M, then it computes $u = B_{1'}.z - t_N.c$ and programmes on $c = HT_1[g]$ where $g = (u, t_N, M)$. Finally, it outputs the signature $\sigma = (u, z)$ with probability $\frac{1}{M}$.
- $\mathcal{G}_2$: The game $\mathcal{G}_2$ follows the same lines as $\mathcal{G}_1$ except the key gerneration process of $\mathcal{B}$ in the setup phase. In the **Setup** phase, $\mathcal{B}$ samples $t_N \leftarrow U(R_q^k)$ with $PK_N = t_N$, and it sends PK_N to $\mathcal{A}$. If the sign queries are answered without private key SK_N, then $\mathcal{B}$ replaces PK_N with random vector. Therefore, assuming hardness of module learning with errors

$$|Pr[\mathcal{G}_2 - \mathcal{G}_1]| \leq Adv_{MLWE}$$

5 PERFORMANCE

The proposed construction uses $Trap - Gen$ to generate public key $B_0 \leftarrow \mathbb{Z}_q^{n \times m}$ with basis $T_{B_0} \leftarrow \mathbb{Z}_q^{m \times m}$ satisfying the condition $||\tilde{T}_{B_0}|| \leq O(\sqrt{n \log(q)})$, $m = O(n \log(n))$. The wallet contains the seed value (B_0, T_{B_0}) that is lattice basis to generate subkeys using Bonsai tree algorithm. The Table 1 shows the storage required for both lattice based cryptography and classical cryptography.

Table 1. Storage comparison with classical scheme.

Schemes	Primitive		length	bit-size
Diffie-Hellman	Storage	$g \in \mathbb{Z}_q^*$	$|q|$	$\kappa' = log(q)$
Lattice based	Storage	$B \in \mathbb{Z}_q^{m \times n}$	$(m^2 + m).q$	$4\kappa log^2 \kappa (2\kappa log \kappa + 1)$

5.1 *Efficiency*

This scheme uses three algorithms (1) Trap-Gen, (2) Ext-Basis, and (3) Sapmle-Pre respectively. The algorithm $Trap - Gen$ starts with seed value B_0, and T_0 with size of B_0 is $O(n^2 log^2(n))$, and T_0 with size of $O(n^2 log^3(n))$ respectively. The algorithm uses the system of linear equations in matrix form $BW = -\bar{B}$, where W follows the lemma (4), and it generates $B \in \mathbb{Z}_q^n$. The matrix can be computed easily by Gaussian elimination that takes complexity $O(n^3)$. The last algorithm generates W_M by finding solution to linear equations, and using $Sample - D$ as subalgorithm with complexity $O(n^2)$, and time for n calls time for the oracle.

5.2 *Analysis/Implimentation*

Both "time" and "latency" are the first parameter those have been considered for the evaluation. We have measured the cost for 200 blocks, all of which were produced by newly added nodes to the blockchain. Table 3 shows the observations. Table 3 shows that when the number of block increases, time required for computation also increases. This is happening due to linear growth, and it plays a very effective role in blockchain based applications for IoT environment.

The complexity of the algorithm is $O(1)$, and the cost taken by aggregate signature is static. The reason for static cost is the production and verification of the signatures being performed in a single operation. The process of generating aggregate signatures is not much effected with increase in number of nodes. The time for key generation may be optimized for better effect, therefore our proposed design is more scalable under given constraints. The storage complexity can be seen in Table 3, and it is also less than the non aggregated approach, which takes the cost $O(N.n)$.

Table 2. Description for design.

Protocol Conseunsus	Stake Proof
Geographical positions of nodes	Network based on Ethereum
Peer's Hardware	3.8GHz, 8GB RAM, Octa, 1 TB
Nodes participating in the transaction	Simulation nodes 70
Tools for Test	Caliper Hyperledger
Data storage Type	Level DBs

Table 3. Proposed design performance.

	Number of blocks										
	10	20	30	40	50	60	70	80	90	100	complexity
Reads Latency (seconds)	12.22	14.77	15.66	18.22	25.50	25.22	27.77	28.55	31.77	34.22	$O(logn^2)$
Transactions Latency (seconds)	46.44	78.66	111.22	173.22	208.33	260.55	317.59	352.11	371.22	410.11	$O((N-1)logn)$
Spaces Complexity	$O(nN)$										

Number of blocks = n; Number of nodes participating in the transaction = N

6 CONCLUSION

The proposed design uses post quantum secure signature to create the trust among communicating nodes. All the nodes combined to form a blockchain with the help of post qunrum secure lattice based signature. We have used the terminology efficient aggregate signature to verify all the participating nodes. The proposed blockchain is secure based on assumptions, (1) Module Learning With Errors, and (2) Module Short Integer Solution respectively. The proposed design possesses security analysis, and it has been proved secure against quantum attacks. The design provides a better environment for Internet of Things. Blockchain technology has been used to combine fixed number of IoT nodes. This makes the communication efficient, and the aggregate signature has been used to provide the security to the given architecture.

REFERENCES

[1] Tareq Ahram, Arman Sargolzaei, Saman Sargolzaei, Jeff Daniels, and Ben Amaba. Blockchain Technology Innovations. In *2017 IEEE Technology & Engineering Management Conference (TEMSCON)*, pages 137–141. IEEE, 2017.

[2] Miklós Ajtai. Generating Hard Instances of Lattice Problems. In *Proceedings of the Twenty-eighth Annual ACM Symposium on Theory of Computing*, Pages 99–108. ACM, 1996.

[3] Miklós Ajtai and Cynthia Dwork. A Public-key Cryptosystem With Worst-case/average-case Equivalence. In *Proceedings of the Twenty-ninth Annual ACM Symposium on Theory of Computing*, pages 284–293. ACM, 1997.

[4] Kamanashis Biswas and Vallipuram Muthukkumarasamy. Securing Smart Cities Using Blockchain Technology. In *2016 IEEE 18th International Conference on High Performance Computing and Communications; IEEE 14th International Conference on Smart City; IEEE 2nd International Conference on Data Science and Systems (HPCC/SmartCity/DSS)*, pages 1392–1393. IEEE, 2016.

[5] Sujit Biswas, Kashif Sharif, Fan Li, Boubakr Nour, and Yu Wang. A Scalable Blockchain Framework for Secure Transactions in iot. *IEEE Internet of Things Journal*, 6(3):4650–4659, 2018.

[6] Abhishek Chakraborty, Nithyashankari Gummidipoondi Jayasankaran, Yuntao Liu, Jeyavijayan Rajendran, Ozgur Sinanoglu, Ankur Srivastava, Yang Xie, Muhammad Yasin, and Michael Zuzak. Keynote: A Disquisition on Logic Locking. *IEEE Transactions on Computer-Aided Design of Integrated Circuits and Systems*, 39(10):1952–1972, 2019.

[7] Tiago M Fernandez-Carames and Paula Fraga-Lamas. Towards Post-quantum Blockchain: A Review on Blockchain Cryptography Resistant to Quantum Computing Attacks. *IEEE Access*, 8:21091–21116, 2020.

[8] Feng Gao, Liehuang Zhu, Meng Shen, Kashif Sharif, Zhiguo Wan, and Kui Ren. A Blockchain-based Privacy-preserving Payment Mechanism for Vehicle-to-grid Networks. *IEEE network*, 32(6):184–192, 2018.

[9] X Jia, RA Fathy, Z Huang, S Luo, J Gong, and J Peng. Framework of Blockchain of Things as Decentralized Service Platform. *Int. Telecommun. Union, Geneva, Switzerland, ITU-T Recommendation SG*, 2017.

[10] Evgeniy O Kiktenko, Nikolay O Pozhar, Maxim N Anufriev, Anton S Trushechkin, Ruslan R Yunusov, Yuri V Kurochkin, AI Lvovsky, and Aleksey K Fedorov. Quantum-secured Blockchain. *Quantum Science and Technology*, 3(3):035004, 2018.

[11] Chao-Yang Li, Xiu-Bo Chen, Yu-Ling Chen, Yan-Yan Hou, and Jian Li. A New Lattice-based Signature Scheme in Post-quantum Blockchain Network. *IEEE Access*, 7:2026–2033, 2018.

[12] Chaoyang Li, Yuan Tian, Xiubo Chen, and Jian Li. An Efficient Anti-quantum Lattice-based Blind Signature for Blockchain-enabled Systems. *Information Sciences*, 546:253–264, 2021.

[13] Jiaxing Li, Jigang Wu, and Long Chen. Block-secure: Blockchain Based Scheme for Secure p2p Cloud Storage. *Information Sciences*, 465:219–231, 2018.

[14] Wenting Li, Alessandro Sforzin, Sergey Fedorov, and Ghassan O Karame. Towards Scalable and Private Industrial Blockchains. In *Proceedings of the ACM workshop on Blockchain*, Cryptocurrencies and Contracts, pages 9–14, 2017.

[15] Damiano Di Francesco Maesa and Paolo Mori. Blockchain 3.0 Applications Survey. *Journal of Parallel and Distributed Computing*, 138:99–114, 2020.

[16] Vruddhi Mehta and Sakshi More. Smart Contracts: Automated Stipulations on Blockchain. In *2018 International Conference on Computer Communication and Informatics (ICCCI)*, pages 1–5. IEEE, 2018.

[17] Daniele Micciancio. Generalized Compact Knapsacks, Cyclic Lattices, and Efficient One-way Functions. *Computational Complexity*, 16(4):365–411, 2007.

[18] Daniele Micciancio and Chris Peikert. Trapdoors for Lattices: Simpler, Tighter, Faster, Smaller. In *Annual International Conference on the Theory and Applications of Cryptographic Techniques*, pages 700–718. Springer, 2012.

[19] Daniele Micciancio and Oded Regev. Worst-case to Average-case Reductions Based on Gaussian Measures. *SIAM Journal on Computing*, 37(1):267–302, 2007.

[20] Satoshi Nakamoto. Bitcoin: A Peer-to-peer Electronic Cash System. Cryptography Mailing List, 2009.

[21] Oscar Novo. Blockchain Meets iot: An Architecture for Scalable Access Management in iot. *IEEE Internet of Things Journal*, 5(2):1184–1195, 2018.

[22] Del Rajan and Matt Visser. Quantum Blockchain Using Entanglement in Time. *Quantum Reports*, 1(1):3–11, 2019.

[23] Ali Mohammad Saghiri. Blockchain Architecture. In *Advanced Applications of Blockchain Technology*, Pages 161–176. Springer, 2020.

[24] Rahul Saha, Gulshan Kumar, Tannishtha Devgun, William Buchanan, Reji Thomas, Mamoun Alazab, Tai-Hoon Kim, and Joel Rodrigues. A Blockchain Framework in Post-quantum Decentralization. *IEEE Transactions on Services Computing*, 2021.

[25] Vatsal Sanghavi, Ronak Doshi, Devansh Shah, and Pratik Kanani. Blockchain Based Asset Tokenization. International Journal of Research in Engineering, *IT and Social Sciences*, 8(11):60–64, 2018.

[26] Peter W Shor. Polynomial-time Algorithms for Prime Factorization and Discrete Logarithms on a Quantum Computer. *SIAM Review*, 41(2):303–332, 1999.

[27] Yong Yuan and Fei-Yue Wang. Towards Blockchain-based Intelligent Transportation Systems. In *2016 IEEE 19th International Conference on Intelligent Transportation Systems (ITSC)*, pages 2663–2668. IEEE, 2016.

Artificial Intelligence, Blockchain, Computing and Security – Dagur et al. (Eds)
© 2024 The Author(s), ISBN: 978-1-032-49393-0

Entertainment based website: A review and proposed solution for lightning fast webpages

Prashante, Arslan Firoz, Vishesh Khullar & Abdul Aleem
SCSE, Galgotias University

ABSTRACT: We live in a time when visual symbols are becoming a bigger part of everyday lives. Websites are quickly replacing other media as people choose them for social interaction, company displays, commerce, entertainment, and information search. Since web development, including server-side development of large-scale web-sites, is a very popular application of languages like JavaScript, CSS, and HTML. The improvement of the performance of these languages has become crucial. However, efficiently compiling programs in these languages is challenging, and many popular dynamic languages still lack efficient production quality implementations. This article describes the implementation of an entertainment website using these languages and efficient expediting techniques. After developing this website using these three languages, the research improves it using some more advanced level programming. This combination of many languages, improve it with the aid of React and hence it will result in excellent performance for website. The purpose of this article is to present and discuss the layout and coding of a successful website. Creative coding is essential for supporting content and functionality, appealing to target audiences' tastes, projecting the sender's desired image, and meeting genre-specific needs.

1 INTRODUCTION

JavaScript and other dynamic languages have gained a lot of attention in recent years due to a variety of variables, including their cost, accessibility to non-experts, speedy development cycles, integration with online browsers, and integration with web servers. [1] These advantages have encaged the adoption of dynamic languages for client-side development, leading to the creation of large applications with millions of lines of code in addition to simple scripts. There is greater variation on the server side thanks to JavaScript becoming the de facto norm. In comparison to static languages, dynamic languages include a number of characteristics that make efficient execution more difficult. Among these properties, dynamic typing is a crucial among them because it helps compilers produce efficient code for languages that seem to be statically typed. However, for dynamic languages, the types of variables and expressions may not be known before execution but they can also change during execution, making them more suitable for dynamic or just in time compilation, or JIT. Compilers have input text ahead of execution time thus goal is to generate very optimized code without running applications.

2 LITERATURE SURVEY

This entertainment-based website will be a master piece because it is mainly based on many researches of other websites for example the research over tism website of Chinese

DOI: 10.1201/9781003393580-54

government and also researched different versions of HTML, also researched over designing in website with Presentation and Cascade Style Sheets (CSS) & Research on JavaScript Language use in different GitHub projects. [2] After doing this much research found out that what version of languages and what is the best requirement for website linked main aspects of website with different websites for example feedback facility in Chinese government website is pretty good so also included this facility in website after including this did some major changes which are based on research all the research did is explained below in experimental details.

2.1 Finding from inflations webpages

[3] Here research details are based on many different aspects for example research over website is based upon personal experience and this experience was gained by personally visiting specific website other ways of research are based upon reviews for example different versions of html are tested by many other users so observed there comments and views about certain version of html than included it in research only when find that this issue or this advancement is possible in certain scenarios after this also did research about JavaScript language usage in different GitHub projects researched over these projects by personally visiting there codes and observing there flaws in their codes and advancements or logics they have built during their project.

2.2 Finding over Chinese government tism website

For developing an entertainment based website need to do some research over some websites in other fields for example tism. [4] Below is survey or can say a little research about a Chinese tism website that what are other options they are considering in their website for example: interactions between traveler's, travel agencies, tism businesses etc. This helped us in countering other options in entertainment field-based website for example: comments, ratings, description etc. The interactions between traveler's, travel agencies, and tism businesses have been completely redesigned by web-based (online) technologies. The new connections open up new channels for getting hold of and using travel-related information and services. With the development of web-based technology, the travel, lodging, and tist destination industries have access to a whole new range of opportunities and difficulties. Both travel providers and travel agencies have begun to directly market their goods and services online during the past few years [5]. As a result, they have gained advantages in terms of lowering service costs, offering more effective service, and luring clients. Web-based technologies are also making it easier for travelers to get relevant and current information, browse tist product catalogues, and make flexible reservations.

Research has been done on tism e-commerce in industrialized nations. By analyzing ten European web sites, Rita (2000), for instance, reported on the promotion and management of tism destinations through web-based destination marketing systems. Cano & Prentice (1998) showed tism web sites created in Scotland and recommended a networking approach for the creation and administration of travel websites firms. When researching the usage of the web, Standing & Vasudevan (2000) looked at the levels of preparation and strategies employed by Australian travel businesses. Burgess *et al.* (2001) examined the extent to which companies in Australia's regional tism specific industry were using internet as a tool for marketing The Chinese travel and hospitality market has strengthened its microservices and gained a lot of knowledge. In comparison to advanced economies, China, which is really a developing economy, has a tistic E-commerce framework, constraints, constraints, and requirements. The taxonomy and assessment of travel blogs in China, although, have not been studied. Despite the existence of publications of results for general website assessments, they do not particularly mention tism websites in developing nations. This study's goals are

359

also to look at how commute sites (analytics tools) are distributed and categorised, evaluate how these web pages have changed over time, and identify any present problems. A government's strategic census, a general search, and a discussion with a programmer were the main research techniques employed in the study.

2.3 *Finding over other versions of HTML based on website*

Since its debut in the early 1990s, HTML has been undergoing continual improvement. A large portion Others of its features and abilities are the result of good initiatives and HTML's absorption into browser plugins, while some of its powers and operations have additionally been specified via definitions [6]. Since over a decade ago, HTML4 has been the accepted form of HTML. The fact of HTML4 "doesn't really offer sufficient data to develop implements which facilitate cooperation with others and, more significantly, with a mass movement of other implements" is one of its main shortcomings. W3C (World Wide Web Consortium) states that there is now a "mass of distributed content." The same is true of DOM Level 2 HTML and XHTML 1, which both specify JavaScript APIs for HTML and XHTML and an XML serialization for HTML4 respectively. In terms of improving the versatility and scalability of HTML technologies while also making html files more collaborative and providing a better user engagement.

HTML5 was first developed from the inside of the W.H.A.T.G.W. (Web Hypertext Application Technology Working Group) effort and indeed the W3C organisation. The creation is based on research on HTML4 implementations now in use, best practices, and evaluations of web content that has previously been published. HTML5 supports the syntax of both HTML and XML and it is back-ward compatible with HTML4 and XHTML1. Additionally, new interfaces that enable modern technology will be included.

Rising trends like website services that are rich (RIA). The use of complex JavaScript plus patented connectors such Flash Player, Microsoft Sun JavaFX, and Silverlight is a major current dependency of these interfaces [7]. The main suggestion offered by webmasters is to integrate the necessary components of such displays directly into browsers. in order to reduce reliance on numerous proprietary plugins. HTML5 is anticipated to reach 2012 will see a nominee for nomination rating, while 2022 will see a recommend. Although Responsive web design won't be complete for a few more years, internet browsers are beginning to support an increasing number of its characteristics.

2.4 *Finding over designing in website with presentation and Cascade Style Sheets (CSS)*

The core structure of documents and pages is primarily determined by programming language, whereas the last presentation & render are typically define by CSS language, a language that provides display principles. Despite CSS 2.1 serving as the de facto rule again for previous 12 years, the method of creating web pages with CSS is about 15 years old. The vast majority of web browsers support it, and during the past ten years, it has evolved quickly. New markup languages and CSS are also being proposed with the new markup languages. The latest proposed standard, CSS3, is currently a candidate recommendation or working draught. The extensive CSS3 specification is broken down into a number of modules that are developed independently with various development velocities and dynamics. Vendors of browsers can implement them piecemeal thanks to various modules.

The majority of contemporary browsers already support a number of CSS3 modules [8]. The new standard includes new features and functionalities while being fully backwards compatible. Numerous novel connections between objects and the associated style are provided by the newly released Selectors module. The ability to choose markup elements depending on where they are in the DOM is the most intriguing addition. In accordance with other commands, items that are checked, without children, or do not match the provided declaration can also be matched. It is also possible to choose the markup components depending on the existence of a certain attribute or only a portion of an attribute. Use of several backgrounds that can be scaled and positioned relativistic ally or absolutely is possible thanks to a background and border module. It facilitates for the reuse of images in a variety of contexts and the more accurate filling out of various fields. Additionally, the module enables the usage of shadows, rounded edges, gradients, and border images for borders. Using an image file as an object's border is possible with the border-image attribute. Programmatic col switching is made possible using the gradient property for borders and backgrounds. Each capability for setting the cols in the document is defined by the Color module. The command falls back to the another given font in the font-circle attribute if browser does not implement the command. Copyright and font licencing are the two main problems with new font types. From every page, it is simple to download the embedded fonts.

Transitions and Animations, the two exciting new modules, allow for visual elements moving and changing within 3D or 2D space. To transition through one condition to the other seamlessly, a coder might designate a certain CSS attribute using transitions. For instance, clicking on an image may cause it to alter in size. On the other hand, animations allow for the repeated iteration of several CSS properties at once. Keyframes can be used to segment the animation into stages. The browser approximates each step of an element between keyframes automatically.

2.5   Finding on JavaScript language from GitHub projects

JavaScript is one of the languages that has had the biggest growth in recent years, and it appears that this trend will continue in 2023 [9]. 2020 and 2021 saw the most pull requests (in GitHub projects). The "Language of the Year" award is given annually by the T.I.O.B.E. Index to the programmers' language with the fastest rate of growth. 2014 saw JavaScript take home the prize. Many projects now employ JavaScript as primary programming language for server-side and client field modules due to its rising popularity. As a result, static code workings of JavaScript programmes also becoming a hot issue. Numerous code analysis techniques rely on the program's call graph presentation. The program's functions are represented by nodes in a call graph, and edges connecting nodes indicate that there has been at least one function call between the respective functions. This programme representation can be used to detect different bad quality and insecurity problems in JavaScript programmes, such as functions that are never called or as a visual representation to make the code easier to grasp. Call graphs can be used as a starting point for additional analysis, such as the construction of a complete inter-procedural control flow graph (ICFG), or to check that the right no. of arguments is supplied to do calls. Numerous type analysis procedures can be carried out within the aid of control flow-graphs. Additionally, this programme model is helpful in a variety of other study fields, such as automated refactoring, defect prediction, and mutation testing. Because call graphs are such a fundamental based data structure, the accuracy of code working methods that rely on them is determined on the accuracy of call graphs. It can be difficult to construct accurate call graphs for the type-free, asynchronous, and dynamic language JavaScript. The obvious drawback of static techniques is the absence of dynamic call edges resulting from non-trivial uses of evall(), bind((), or apply()) (i.e. reflection). Furthermore, they could be overly pessimistic, recognising edges that are statically legitimate but are never realised for any inputs in reality. They don't need a testbed to purpose of the program under analysis, yet they are quicker and more effective than dynamic analysis techniques Because of this, should not overlook the state-with-the-art static call graph generation methods JavaScript and require a deeper grasp of their functionality.

3 PROPOSED WORK

Using HTML, CSS, JavaScript for Website development after doing good research over different versions of these languages and research over different websites is the motto of this research. We gathered the requirements needed for work. In work, we will be countering some of the major issues like dynamic look and processing of the website and ad free environment of website and are not stopping here also provide speed however other website like us also provide good speed because they also use JavaScript, CSS, html but will counter it as know some entertainment website they are illegal because they provide torrent downloads of movies and this makes their websites slow because size of a movie is in 1gb or 2gb this size is too much but on site will not provide any torrent downloads of movies will only provide description and ratings eventually which will make site legal and fast. This research has been based on the findings by Aleem *et al.* [10,11].

Working over making this site a bigger role model in field of speed now the question arises that how will do it so here is the answer will do it by the help of react. when implement react Language in website this will make it way faster from where it is already now if I have to explain this with an example than I will say speed of a racing car and then speed of a rocket this is the future target of website to merge with react and take a speed of rocket from a speed of a racing car. This merging is also used by YouTube I don't want to talk big but there are chances that will able to out speed YouTube only in terms of speed and other facilities like feedback, comment, ratings, dynamic version of website will remain unchanged even after usage of react.

4 RESULTS AND DISCUSSION

Some results after different researches and the results are shown in Table 1.

Table 1. Effects of research for lighting webpages.

S.No.	Research	Results
1.	Comparison with Chinese Government Tism website	This leads to adding comments and feedback facility in website
2.	Comparison with other versions of HTML-based website	This leads to use of HTML5" version.
3.	Comparison with designing in website with CSS	This leads to use "CSS 2.1: Level 2" version.
4.	Comparison with JavaScript usage of GitHub projects	This leads to some changes in code and also decide to use "ES2021" versin. Moving ahead "ES2022" will be used.

5 CONCLUSION

After this much of research and discussion about different languages that are used in web development and different websites, concluded that use "HTML5" version for HTML language and use "CSS 2.1: Level 2" version for CSS language and in JavaScript use "ES2021" version and in coming time Researchers will be using "ES2022" version to update website. These different versions are fit and compatible with requirements and now can say that they are helpful in making dream website truly successful and also concluded that using JavaScript as dynamic language for website will not only make website faster but also open new options for us which will be helpful for us in coming future.

REFERENCES

[1]   https://www.researchgate.net/publication/328906451_Research_Paper_Static_JavaScript_Call_Graphs:A Comparative Study.

[2]   https://www.sciencedirect.com/science/article/abs/pii/S0261517708000356?via%3Dihub

[3]   https://www.sciencedirect.com/science/article/abs/pii/S0378720601000921

[4]   https://link.springer.com/article/10.1007/s10098-001-0130-y

[5]   https://www.persee.fr/doc/netco_0987-6014_2002_num_16_3_1556

[6]   Leavitt, Michael O., and Ben Shneiderman. "Based Web Design & Usability Guidelines." *Background and Methodology* (2006).

[7]   Brinck, Tom, Darren Gergle, and Scott D. Wood. *Designing Web sites that work: Usability for the Web.* Morgan Kaufmann, 2002.

[8]   Landay, James A., and Jason I. Hong. *The Design of Sites: Patterns, Principles, and Processes for Crafting a Customer-centered Web Experience.* Addison-Wesley Professional, 2003.

[9]   Kim, Heejun, and Daniel R. Fesenmaier. "Persuasive Design of Destination Web Sites: An Analysis of First Impression." *Journal of Travel Research* 47.1 (2008): 3–13.

[10] Shubham Sharma, Chetan Sharma, and Abdul Aleem, "Movie Recommendation System Using Combination of Content-based and Collaborative Approaches", Book Chapter in book entitled *"Intelligent Systems and Smart Infrastructure"*, Taylor and Francis Group, London, pp: 453–459, 2023

[11] Abdul Aleem, Palash Tewary, Shanta Karn, and Varun Kumar, "Entertainment Advisor Using Sentiment Analysis", *Journal of Scientific Research*, vol. 66, issue no. 2, pp. 101–107, 2022

Artificial Intelligence, Blockchain, Computing and Security – Dagur et al. (Eds)
© 2024 The Author(s), ISBN: 978-1-032-49393-0

Sentiment analysis based brand recommendation system: A review

Chaitanya Rastogi, Darshika Singh, Ashutosh Dwivedi, Anshika Chaudhary,
Sahil Kumar Aggarwal & Ruchi Jain
Information Technology, ABES Engineering College, Ghaziabad

ABSTRACT: A brand recommendation system with sentiment analysis is a software tool that uses natural language processing techniques to analyze consumer opinions and sentiments towards various brands. The system analyses data from multiple sources such as social media, online reviews, and survey responses to understand consumer preferences and make recommendations accordingly. Brand recommendation systems using sentiment analysis can be a valuable tool for businesses looking to improve customer experience and drive brand loyalty. This paper reviews various methods supporting sentiment analysis and their effectiveness in each condition. The review includes the research paper on sentiment analysis from 2014 to 2021 in different categories for prediction. The parameters chosen in the paper for reviewing different algorithms used in machine learning are response time, complexity, type of data on which methods are applied, a different area where methods can be applied and obstacles and opportunities where other machine learning methods work well with pre- and post-conditions.

1 INTRODUCTION

Machine Learning methods are changing buying habits and developers are effectively anticipating the consumers need. This article proposes a vocabulary of ML shopping terms based on a methodical review of the academic and professional literature [1]. For retailers, large volumes of paperwork are both temporary and challenging. Machine learning applications offer the potential to assist in this effort by allowing dossiers to be studied considering the correlations, patterns, and ultimately predictive models found [2]. Artificial intelligence (AI) refers to the vehicle's ability to simulate human characteristics, including interpretation, education, preparation and creativity [2] Many associations use AI and machine intelligence (ML) to more accurately believe in buyer needs, predict future demand to modify consumer obligations, and enable bots to meet original obligation demands [3]. Machine learning (ML) enables computers to self-discover and proliferate by finding links in recommendation dossiers and recognizing the correct output. ML is a powerful form that delves into large user datasets to enable marketers to gain new visions of community behaviour [4]. The rationale for many instances of AI requests in marketing is based on what the model sees when shopping. There are examples of using AI in new product production behaviour, energetic personalization, automated approvals, and incidents. Four areas of shopping promotion: product, price, location and promotion [5].

ML algorithms not only cater to user needs but can also determine some aspects of a buyer's appearance, allowing clubs to target precisely personalized offers [6]. When purchasing services, brands often determine their fundamental differentiators. Businesses face a vibrant and unpredictable market with competition, technological advances, and multiple needs. According to Forrester's research, more than 30% of internet users rate brands and services. Merchants have found that social message reach effectively increases awareness. Facebook now dominates the digital shopping space, while Twitter [7].

DOI: 10.1201/9781003393580-55

2 LITERATURE SURVEY

In today's scenario, the trading environment is highly competitive. Customer satisfaction has been much a lot of focus in business development. Business arrangements put a lot of money and effort into different game plans to accept and meet consumer needs. Using NLP and SVM to do sentiment analysis has shown a productive approach on large and unstable multi-category datasets. one of the best approaches is to determine accuracy using a support vector machine and Natural Language Processing (NLP) which gives 77% accuracy [8] Algorithms such as NLP and SVM add significant capital to understanding the content [9].

Monitoring distinguishing face of the adventures is the fundamental task and backbone for various business-related actions that are usually conducted by way of surveys, which are costly on account of the amount of work force complicate the task [10] But by relying on computer engineering methods usually can penetrate the expenses as information is openly applicable provided through the allied websites. Hence it can be decided by placing each buying product as per their category defined by the analysis pattern whose records are obtained from the mechanical reasoning of an allied site or a permanent data conversion beginning, accompanying the presence of e-commerce abilities [11].

There is an idea which focuses on sentiment analysis of Twitter data and about syntax study. It has become the hot point standard accompanying the eutrophication of miscellaneous Machine learning and Artificial Intelligence. In the past few years, so many methods have existed projected in the region of sentiment reasoning to resolve the public news data and to specify a graphical performance towards the brand. They are established by using Support Vector Machine (SVM) and Natural Language Processing (NLP) algorithms. The system understands sure words as 'positive' or 'negative', allowing you experience if your brand is being admired or floored [12].

Sentiment study is dependent excavation of text that recognizes and extracts emotional facts in textual data. Sentiment reasoning substantiates to be an marvellous asset for organizations to extract essential facts and assists arrangements accompanying understanding the social emotion of their brand, product or duty while monitoring connected to the internet dialogues. Sentiment study shows the emotion of people while classifying a friendly radio communication about some product or brand. This is main news if you experience that the sentiment of individual man can influence additional person about a firm or their products. Sentiment reasoning is a the study of computers technique used to decide either data is positive, negative or neutral. Sentiment reasoning models focus not only opposition (positive, negative, neutral) but still on impressions and passion (enraged, happy, depressed, etc.), importance (critical, not essential) and even intentions (concerned v. not concerned). So first we have a pre-processed raw data set from the source section related to media commentary or any nuances of miscellaneous merchandise from a particular brand, such as Amazon. Modifiers obtained from a dataset with a target named feature vector play an important role to preselect a list of feature or measurements vectors and then apply artificial intelligence-generated classification algorithms. Localized SVM and NLP along with Semantic-driven Word Networks extract synonyms and match content features. Finally, effectiveness of classifier's are measure in terms of recall, accuracy, and precision [12].

3 OPPORTUNITIES WHILE USING ML ALGORITHMS

This section discusses a comparison between the two most popular machine learning techniques SVM & NLP.

3.1 Advantages of implementing NLP

Less high-priced: Utilizing a program is less damaging than engaging a person as one can take two or three periods higher than a system to accomplish the abovementioned tasks [13] can cost higher rather than working on a program and produce an effective result.

Faster department dealing with customer's answer times: Usually, when NLP is used, the chatbot which also offers the opportunity to predict customer behavior provides the active response on a customer call. Typically, call centres consist of limited staff, limiting the number of calls in a day to be handled. By using NLP, a higher number of calls can be controlled, which means fewer days of waiting for customers [10].

Easy to implement: Earlier, in consideration of use of NLP, difficult research had to happen concerning the language, and many tasks had been expected to execute manually. In many cases, when it came to interpretation, it was unavoidable to conceive a kind of language that included a conversation that may be exactly interpreted into another style. Therefore, it took a long time to evolve as a better solution for conducting research thoroughly. Now it is easy to find pre-prepared machine intelligence models that promote various uses of NLP [10].

3.2 Opportunities of using SVM

SVMs are maximally profitable classifiers compared to, for example, Naïve Bayes, which is the probability. It has been known to be highly effective in text classification and scale well for the rich features, which is essential in GUI-based datasets [14].

Work Independently: One user's knowledge does not depend on other users' knowledge. Stated another way, individual user's profile maybe built utilizing his own data without taking advantage of added user's data [14].

Effective Labelling: Feature set for GUI-based behavioural traits tested with linear kernel as well as radial basis function kernel. When using SVM for tampering, the binary classification problem is defined as the user will be labelled as positive (authentic) or negative [15].

Modification as per Need: SVM try to blow up the border middle from two points the tightest support vectors when in fact logistic reversion blow up the posterior class feasibility. SVM is deterministic (but we can use Platts model for possibility score) while LR is probabilistic. For the kernel room, SVM is faster [16].

4 OBSTACLES WHILE USING L ALGORITHM

4.1 Obstacles in using NLP

Training can lag: Other than using pre-built models, it is unavoidable to extend the model that accompanies new datasets, depending on what is known, achieving sufficient efficiency can take several weeks to process [17].

Voice tone: One of the major concerns is to understand human nature to pitch their thoughts on any social network platform as it is challenging for NLP to process and understand if the said point is sarcasm.

Too many languages: Human terminology isn't easy to understand, but handling in addition to 6000 languages in general isn't smooth, especially following all of their semantic rules.

Uniformity: In order to process an expression, it should define it in a form that machines can appreciate. Machine learning (ML) algorithms can recognize unorganized accents and convert them into forms that machines can learn. Thus, this is the stage where NLP collects data [17].

4.2 Obstacles in using SVM

Produce Complicated Parameter: SVMs are designed to produce more complicated decision perimeters. An LS-SVM accompanying a simple undeviating kernel function complements to a undeviating decision horizon. Instead of a linear kernel, more complicated kernel functions, like the usually used RBF essence, can be preferred [7].

Requires Visualization by different methods: SVM works best when the dataset is narrow and complex. It is ordinarily recommended to first use logistic reversion and visualize by what method does it acts, if it fails to present a good preciseness you can choose SVM outside some kernel.

Pre-assumptions are required: SVMs start accompanying the notion that the proofs are "breakable" (namely, that they may be split into groups by a functional separator). However, SVMs considerably generalize this separability idea based on various standard.

Additional Efforts are required: The question of finding the undeviating separator maybe treated as an addition (quadratic) program accompanying easy-going constraints (to give reason for imperfectly divided data). When fitting a linear separator to knowledge; the points tightest to the separator in the facts room entirely decide the separator.

5   PARAMETER BASED COMPARISON BETWEEN NLP AND SVM

So Support vector machines work well when there is an understandable separation boundary between classes and most successful works for vast space. It works when the given range is more extensive than the number of samples [18].

For data that are not breakable by an uninterrupted separator, the data may be planned into a more significant measure, and in the larger dimension, the undeviating separator may be appropriate [19]. For data that are not breakable by an uninterrupted separator, the data may be plan into a bigger measure, and in the larger dimension, the undeviating separator may be appropriate [Simos Chari]. For data that are not linearly breakable in the higher predefined grammar. It needs to improve as the grammar of language spoken and actual language may differ because of the ambiguity used by human and hence provide hindrance while applying NLP. Table 1 show the highlights of using different algorithms.

Table 1.   Comparison between NLP and SVM.

Parameter	NLP	SVM
Cost	Less Expensive [2]	Expensive [13]
Response times	Fast and Quick [19]	Comparatively times more [10]
Complexity	Low and Simple [8]	Comparatively higher [1]
Ease of implementation [7]	Quite Simple [3]	Choice of Support Vector is quite hectic at times [4]
Feasibility	High [16]	Low [9]
Training time	Highly time taking [6]	Quick and Responsive [15]
Reliability	Low and Volatile [12]	High and substantiate [2]
Data	Linear Data [5]	Non Linear Data [14]
Accuracy	High [1]	Low [38]
Decision boundaries	More Volatile [16]	Rigid and Complex [11]
Flexibility	Low [4]	High [3]
Monotonisity	Quite High [13]	Low and unconventional [18]
Area of applicability	Less and highly specific [19]	Vast and Generalised [17]

6   CONCLUSION

Current scenario talks about industry focus is totally based on consumer need. So, it will not be wrong if it is called consumer driven industry where it is necessary to know the demand of consumers on daily basis in each category. One of the best and cost-effective mediums is using machine learning which not only reduce an individual's effort physically but also provide results efficiently. The paper reviewed show the effect of SVM and NLP algorithm. Although some pre-processing and previous condition are needed to be according to desired parameters but in the end produce effective results which can contribute to address consumers needs and provide support to the developers.

REFERENCES

[1] Trellet M., Férey N., Flotyński J., Baaden M., and Bourdot P., "Semantics for an Integrative and Immersive pipeline Combining Visualization and Analysis of Molecular Data," *J Integr Bioinform*, vol. 15, no. 2, 2018.

[2] Calzavara S., Rabitti A., and Bugliesi M., "Semantics-based Analysis of Content Security Policy Deployment," *ACM Transactions on the Web (TWEB)*, vol. 12, no. 2, pp. 1–36, 2018.

[3] Guardia G. D. A., Ferreira L. Pires, da Silva E. G., and de Farias C. R. G., "SemanticSCo: A Platform to Support the Semantic Composition of Services for Gene Expression Analysis," *J Biomed Inform*, vol. 66, pp. 116–128, Feb. 2017, doi: 10.1016/J.JBI.2016.12.014.

[4] Guardia G. D. A., Pires L. F., da Silva E. G., and de Farias C. R. G., "SemanticSCo: A Platform to Support the Semantic Composition of Services for Gene Expression Analysis," *J Biomed Inform*, vol. 66, pp. 116–128, 2017.

[5] Dietz L., Xiong C., and Meij E., "Overview of the First Workshop on Knowledge Graphs and Semantics for Text Retrieval and Analysis (KG4IR)," in *ACM SIGIR Forum*, 2018, vol. 51, no. 3, pp. 139–144.

[6] Bouazizi M. and Ohtsuki T., "Multi-class Sentiment Analysis on Twitter: Classification Performance and Challenges," *Big Data Mining and Analytics*, vol. 2, no. 3, pp. 181–194, 2019.

[7] Rajeswari B., Madhavan S., Venkatesakumar R., and Riasudeen S., "Sentiment Analysis of Consumer Reviews–a Comparison of Organic and Regular Food Products Usage," *Rajagiri Management Journal*, 2020.

[8] Kaur J., Bedi R. K., and Gupta S. K., "Product Recommendation Systems a Comprehensive Review," *Int. J. Comput. Sci. Eng., IJCSE*, vol. 6, pp. 1192–1195, 2018.

[9] Chang W.-J. and Chung Y.-C., "A Review of Brand Research (1990-2010): Classification, Application and Development Trajectory," *International Journal of Services Technology and Management*, vol. 22, no. 1–2, pp. 74–105, 2016.

[10] Feizollah A., Ainin S., Anuar N. B., Abdullah N. A. B., and Hazim M., "Halal Products on Twitter: Data Extraction and Sentiment Analysis Using Stack of Deep Learning Algorithms," *IEEE Access*, vol. 7, pp. 83354– 83362, 2019.

[11] Almaslukh A., Magdy A., and Rey S. J., "*Spatio-temporal Analysis of Meta-data Semantics of Market Shares Over Large Public Geosocial Media Data*," https://doi.org/10.1080/17489725.2018.1547428, vol. 12, no. 3–4, pp. 215–230, Oct. 2018, doi: 10.1080/17489725.2018.1547428.

[12] Scheider S., Ostermann F. O., and Adams B., "Why Good Data Analysts Need to be Critical Synthesists. Determining the Role of Semantics in Data Analysis," *Future Generation Computer Systems*, vol. 72, pp. 11–22, 2017.

[13] Kumar, A., & Alam, B. (2014, February). Real Time Scheduling Algorithm for Fault Tolerant and Energy Minimization. In *2014 International Conference on Issues and Challenges in Intelligent Computing Techniques (ICICT)* (pp. 356–360). IEEE.

[14] Chaturvedi, R., Kumar, S., Kumar, U., Sharma, T., Chaudhary, Z., & Dagur, A. (2021). Low-cost IoT-enabled Smart Parking System in Crowded Cities. In *Data Intelligence and Cognitive Informatics: Proceedings of ICDICI 2020* (pp. 333–339). Springer Singapore.

[15] Sharma Y., Shatakshi, Palvika, Dagur A. and Chaturvedi R., "Automated Bug Reporting System in Web Applications," *2018 2nd International Conference on Trends in Electronics and Informatics (ICOEI)*, Tirunelveli, India, 2018, pp. 1484–1488

[16] M. G. K and S. R. C, "Performing Sentiment Analysis on the Product Brands using Tweets," *International Journal of Engineering Research & Technology*, vol. 3, no. 27, Jul. 2018, doi: 10.17577/IJERTCONV3IS27082.

[17] Wei B. and Prakken H., "An Analysis of Critical-link Semantics with Variable Degrees of Justification," *Argument & Computation*, vol. 7, no. 1, pp. 35–53, 2016.

[18] Saad S. E. and Yang J., "Twitter Sentiment Analysis Based on Ordinal Regression," *IEEE Access*, vol. 7, pp. 163677–163685, 2019.

Artificial Intelligence, Blockchain, Computing and Security – Dagur et al. (Eds)
© 2024 The Author(s), ISBN: 978-1-032-49393-0

Models for integrating Artificial Intelligence approaches & the future the humans

Ojas Sharma & Tejinder Kaur
Chitkara University Institute of Engineering and Technology, Chitkara University, Punjab, India

ABSTRACT: Artificial Intelligence, then it doesn't matter because today's article is going to be very important for you. By which devices are given the ability to learn and understand. In simple words, with the help of AI, the mind of the computer is advanced so much that the computer also starts thinking and acting like humans. Mainly this is done with computer system. As you give an instruction to the computer, it shows the result according to your instructions; all this is possible due to artificial intelligence. In the true sense of the word, computers are advanced today, but their minds are like those of a 5-year-old child. If there is, or any task that involves decision-making, then the computer is unable to perform such tasks there are many examples of Artificial Intelligence in the world we live in today. Some of which we have told you in the article.

Keywords: Artificial Intelligences, learning, machine

1 INTRODUCTION

Friends, in the world of Internet, we know very well that since the time when a device like computer was introduced, most people like us have been using it extensively. – Soro se kia hai hai. Now the situation is such that all people like us entrust all their work to the computer. Or say that we are completely dependent on the computer. Humans like us have greatly increased the capacity of these devices [1]. the given example (Figure 1) If we take for example its speed, its size and its working capabilities etc. so that the computer can do your work in a short time, it also saves your time a lot. You must have noticed that Artificial Intelligence is being liked a lot nowadays [2].

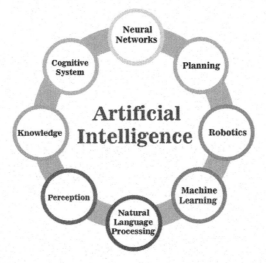

Figure 1. Artificial Intelligences [1].

DOI: 10.1201/9781003393580-56

Mass, so most scientists believed that all those sci-fi movies that we watched as a child often left us in awe and most of us wanted to make those human-like robots in the given example (Figure 2).

Studying an academic program in this field, you will be equipped with knowledge of the broad and complex concepts of AI, as well as the various technologies used to build efficient digital and robotic devices [3].

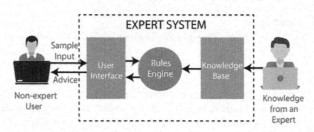

Figure 2.   Expert system user & knowledge base [2].

2   REVIEW

2.1   *Apple Siri*

Siri is a virtual assistant of the Apple Company, which is currently only available for iPhone and iPad. This is the most popular voice assistant. You can do any of your work by saying Hey Siri. It can make calls, calendar, alarm, and set time, send messages etc. for you. Similar to Siri are Amazon's Alexa and Google Assistant. All of them can do all this work through Artificial Intelligence. They use Machine Learning to understand your language [4].

2.2   *Google Map*

You must be using Google Map, but have you thought how this application tells you the right way. This is the wonder of Artificial Intelligence. Google map scans the information of any road by AI and tells the user the correct route, the time taken to reach the location etc. using the algorithm. Every day millions of people find their way through Google Map [5].

2.3   *Google Assistant*

Google Assistant is also a virtual assistant of Google like Apple Siri, through which you can ask anything from Google by speaking, and it answers all your questions. Also, you can send a message, call someone, set an alarm, and open any application in the mobile through Google Assistant. You can use Google Assistant in any device. This is also an example of Artificial Intelligence in the given example (Figure 2).

2.4   *Tesla motor*

Along with mobile devices, Artificial Intelligence is also being used in automobiles. You must have heard about Tesla Car, if not then let me tell you that it is a Self Driving Car which does not require any driver to drive. In near future you will see these cars in India too. Apart from Self Driving, it has many advanced features [6].

2.5   *Tesla car: Marketing automation software*

Marketing Automation Tool is used by people for online growth of their business. You can send emails, chat, etc. through Automation Software.

2.5.1 Reactive machines

Reactive Machines – Reactive machines that are not capable of storing any memories and do not have the ability to use past experiences to make decisions.

Example – In the year 1990, the chess playing supercomputer "Deep Blue" created by IBM defeated the then famous chess player Garry Kasparov in a game of chess.

Deep Blue was a supercomputer that could recognize the columns on a chess board and analyze the situation and make its moves based on probability. The supercomputer had no previous experiences inside it, nor was it capable of storing memory. Deep Blue used to ignore everything else and make decisions based on his opponent's current moves. He had no idea what trick he had done before [7]

2.5.2 Limited memory

It is used in driverless vehicles; driverless vehicles are made in such a way that they can prevent future accidents by observing the speed of the vehicles running on the road, breakers. This observation and memory is not stored forever [8].

2.5.3 Theory of mind

What we talked about in the first two (AI) types is that they are being widely used today. But now we are going to talk about two types of AI that exist only in ideas and theory. They are currently being worked on. In the given example (Figure 2) If AI systems are to live among humans in the future; they need to understand that every human's desires, feelings and thoughts are different.

2.5.4 Self-conscious

Self-Conscious – The final stage of artificial intelligence is to create self-awareness inside the machine. So that they can know about their existence that they exist all the scientists are working continuously to achieve this in the given example (Figure 2).

Self-awareness – like I know I exist. We can feel this world by looking at nature without any thought if robots become aware of their existence like humans. So there will be no difference between humans and machines. But right now we don't even know where consciousness comes from inside a human being, how it is generated, if we understand it then maybe we can make robots like humans in the future.

2.5.5 Brain theory artificial intelligence

If we talk about Brain Theory Artificial Intelligence, then it is a fictional concept that depends on the human brain. In this, the tools should understand the human emotions, beliefs, which a human mind uses. This means that it thinks, understands and then processes information just like a human brain. But, primarily, no such device has been introduced that works on these principles till now [9]

2.5.6 Self conscious artificial intelligence

Let me tell you that Self Conscious is also known as Self Awareness. This AI is the AI of the future. These will be super intelligence. These AIs are still only imaginary; currently this type of AI does not exist.

2.5.7 Limited memory artificial intelligence

Limited memory artificial intelligence is a category that includes devices that can store data from their past experiences for a limited time [9].

2.5.8 Uses of artificial intelligence

Artificial intelligence is being used in many fields. Below we have told you about some major uses of AI –In computer gaming, in machine learning, in the automation tool, intelligent robots.

Machine Vision, in automatic cars (Tesla Car), Speech recognition in computer.

2.5.9 Advantages of artificial intelligence

Reduces human error, Heavy Data completes the task in less time, Consistently gives results, Takes decisions quickly, Available 24 X 7, Makes human tasks easier.

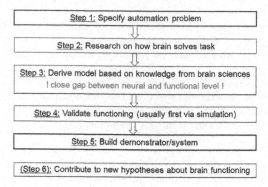

Figure 3.   Methodology artificial intelligences [3].

2.5.10   *Following are some disadvantages of artificial intelligence – [10]*

Creating AI is very expensive, Only Expert Technician can create AI, and Humans are becoming lazy with the advent of AI. For example, as AI is developing, the problem of unemployment is increasing worldwide.

3   METHODOLOGY

For example, once a user comes to your website, the Chabot opens in front of the user and asks the user a question, and similarly, when a user enters their email address on your website, they receive automatic emails. Are all this has been possible only through Artificial Intelligence [10].

Learning: – In this process information has to be put in the mind of the devices and along with this they are also taught about some rules [11]. So that they can complete any work according to the given rules. Reasoning: – If we talk about Reasoning, then under this the information is provided to the devices that they have to display the result according to the stated rules In the given example (Figure 4) Now you have understood its history well, so now we are going to tell you about some of its types. However, there are different types of artificial intelligence, which are mainly divided into two categories one on the basis of capacity and the other on the basis of functionality. The reason behind creating this type of AI is that machines can also think and make decisions like a normal human being [12].

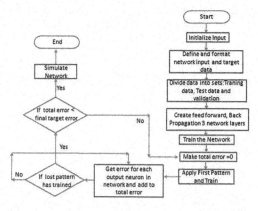

Figure 4.   Flowchart AI [4].

4   RESULT

Types of Artificial Intelligence based on functionality

Based on functionality AI can be divided into four parts –Purely Reactive, Brain Theory, Self Conscious, Limited Memory.

4.1 Purely reactive artificial intelligence

It is completely reactive. This means that Purely Reactive Artificial Intelligence focuses on current scenarios and reacts to them with the best action.

4.2 Brain theory artificial intelligence

If we talk about Brain Theory Artificial Intelligence, then it is a fictional concept that depends on the human brain. In this, the tools should understand the human emotions, beliefs, which a human mind uses. This means that it thinks, understands and then processes information just like a human brain. But, primarily, no such device has been introduced that works on these principles till no [10].

4.3 Self conscious artificial intelligence

Let me tell you that Self Conscious is also known as Self Awareness. This AI is the AI of the future. These will be super intelligence. These AIs are still only imaginary; currently this type of AI does not exist.

4.4 Limited memory artificial intelligence

Limited memory artificial intelligence is a category that includes devices that can store data from their past experiences for a limited time in the given example (Figure 5).

Uses of Artificial Intelligence: Artificial intelligence is being used in many fields. Below we have told you about some major uses of AI – In computer gaming, in machine learning, in the automation tool, intelligent robots, Machine Vision, in automatic cars (Tesla Car), Speech recognition in computer [11,12].

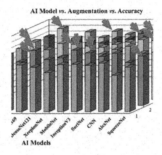

Figure 5.   AI Models accuracy [5].

Table 1.   Primary data using AI [6].

Variable	Responses	Frequency (%)
Have you heard of AI?	Yes	481 (94.31)
	No	12 (2.35)
	I don't know	17 (3.33)
	Total	510 (100)
Have you heard of deep machine learning?	Agree	143 (29.97)
	Disagree	77 (15.98)
	I don't know	262 (54.36)
	Total	482 (100)
Is AI sometimes called machine ingelligence	Agree	229 (47.22)
	Disagree	24 (4.95)
	I don't know	232 (47.84)
	Total	485 (100)

(continued)

373

Table 1. Continued

Variable	Responses	Frequency (%)
Is it ability of machines to function as humans in learning and problem-solving	Agree	247 (49.9)
	Disagree	14 (2.83)
	I don't know	234 (47.27)
	Total	495 (100)
Is intelligence demonstrated by machines in contrast to the natural intelligence displayed by humans	Agree	249 (49.9)
	Disagree	18 (3.61)
	I don't know	232 (46.49)
	Total	499 (100)

Source: From primary data.

5 FUTURE SCOPE

Integrating artificial intelligence into computing is a recent development. It is the brainchild of scientists like Alan Turning. It is widely used in many industries around the world. Today, artificial intelligence has become a part of our daily life. From our electronic devices to our cars to our search engines and even our kitchen appliances, artificial intelligence (AI) is all around us in the 21st century.

REFERENCES

[1]   Cappella, J. N. (2017). Vectors into the Future of Mass and Interpersonal Communication Research: Big Data, Social Media, and Computational Social Science. *Human Communication Research*, 43(4), 545–558.

[2]   Ananny, M., & Crawford, K. (2018). Seeing without Knowing: Limitations of the Transparency Ideal and its Application to Algorithmic Accountability. *New Media & Society*, 20(3), 973–989.

[3]   Banker, S., & Khetani, S. (2019). Algorithm Overdependence: How the use of Algorithmic Recommendation Systems Can Increase Risks to Consumer Well-being. *Journal of Public Policy & Marketing*, 38(4), 500–515

[4]   Lambrecht, A., & Tucker, C. (2019). Algorithmic Bias? An Empirical Study of Apparent Gender-based Discrimination in the Display of STEM Career Ads. *Management Science*, 65(7), 2966–2981.

[5]   Amazon (2020). All in: Staying the Course on our Commitment to Sustainability. *Retrieved February 4, 2021*, from https://sustainability.aboutamazon.com/pdfBuilderDownload?name=sustainabil ity-all-in-december-2020.

[6]   Andrew, J., & Baker, M. (2021). The General Data Protection Regulation in the Age of Surveillance Capitalism. *Journal of Business Ethics*, 168(3), 565–578

[7]   van Esch, P., Cui, Y., & Jain, S. P. (2021). Stimulating or intimidating: The Efect of AI-enabled In-store Communication on Consumer Patronage likelihood. *Journal of Advertising*, 50(1), 63–80.

[8]   Ameen, N., Tarhini, A., Reppel, A., & Anand, A. (2021). Customer Experiences in the Age of Artifcial Intelligence. *Computers in Human Behavior*, 114, 106548.

[9]   Du, R. X., Netzer, O., Schweidel, D. A., & Mitra, D. (2021). Capturing Marketing Information to Fuel Growth. *Journal of Marketing*, 85(1), 163–183

[10] R. K. Sachdeva, and P. Bathla, "A Machine Learning-based Framework for Diagnosis of Breast Cancer", *International Journal of Software Innovation*, vol. 10, no. 1, 2022, pp. 1–11.

[11] R. Kumar Sachdeva, T. Garg, G. S. Khaira, D. Mitrav and R. Ahuja, "A Systematic Method for Lung Cancer Classification," *2022 10th International Conference on Reliability, Infocom Technologies and Optimization (Trends and Future Directions) (ICRITO)*, Noida, India, 2022, pp. 1–5, doi: 10.1109/ICRITO56286.2022.9964778.

[12] R. K. Sachdeva, P. Bathla, P. Rani, V. Kukreja and R. Ahuja, "A Systematic Method for Breast Cancer Classification using RFE Feature Selection," *2022 2nd International Conference on Advance Computing and Innovative Technologies in Engineering (ICACITE)*, 2022, pp. 1673–1676, doi: 10.1109/ICACITE53722.2022.9823

Artificial Intelligence, Blockchain, Computing and Security – Dagur et al. (Eds)
© 2024 The Author(s), ISBN: 978-1-032-49393-0

Deep learning driven automated malaria parasite detection in thin blood smears

Aryan Verma
Computer Science and Engineering Department, National Institute of Technology, Hamirpur, HP

Sejal Mansoori
Department of Computer Science, Indraprastha College for Women, New Delhi, Delhi

Adithya Srivastava, Priyanka Rathee & Nagendra Pratap Singh
Computer Science and Engineering Department, National Institute of Technology, Hamirpur, HP

ABSTRACT: Malaria has remained a recurrent challenge in subtropical and tropical areas across the globe. Malaria is caused by plasmodium parasites and can become fatal for the infected person; hence, its precise diagnosis within a short time frame becomes crucial in ending or controlling it. Malaria is analyzed by conventional microscopy, but this methodology has its draw-backs. It is time-consuming and requires an efficient skill set that is often unavailable in many health facilities in several malaria-endemic countries. Hence, tools like computer-aided diagnosis are momentous in analyzing cell images in the modern era of technology. In this paper, we employ deep neural networks to detect malaria presence in the cell images. Both uninfected and parasitized image samples are used to train the models. We use the holdout Validation technique with 15% and 20% validation data to train and compare four variant transfer learned models, VGG19, Resnet50, InceptionV1, and InceptionV3. We employ data augmentation techniques and early stopping callbacks for training the final model Inception V3, which receives 91.40% accuracy, which is better than most of the existing studies in the domain.

Keywords: Medical image analysis, Deep Learning, Transfer Learning, Malaria detection, Inception V2

1 INTRODUCTION

Every year, millions of people around the globe are infected by malaria. The chief cause of malaria is Plasmodium parasites, which are present in the saliva of anopheles mosquitoes which are female and the source of this parasitic disease, which is contagious. Fever, headache, chills, sweating, and exhaustion are some of the earliest symptoms following an infected female anopheles mosquito bite. In the year 2020, as per WHO, around 241 Million cases were found in the world of malaria, out of which 95% were from the region of Africa, the total deaths were 6,27,000, and 96% were from Africa, making it a life-threatening disease. Plasmodium falciparum, Plasmodium malariae, Plasmodium ovale, Plasmodium knowlesi, and Plasmodium vivax are the various types of plasmodium parasites that have been identified as being harmful to humans. Out of these, Plasmodium falciparum is the most common and accounts for most infections and deaths.

Various techniques have been developed as an evolutionary approach to determine the best diagnosis methods for malarial or Plasmodium infection in the human body. The

DOI: 10.1201/9781003393580-57

techniques have been improving in the medical domains, but they still lack the indexes of swiftness and accuracy.

If there is a need for mass screening, the techniques may not prove very fast for diagnosis of this disease, and the effective prognosis of the patient's condition may not occur. Taking a blood smear and examining it under a microscope to see the infected cell is one of the most popular ways to identify malaria. To detect malaria disease presence in the red platelets, blood from the patient is drawn onto a glass slide, which is then subjected to several testing solutions before being examined under a microscope. While malaria infects millions of people every year, it is crucial to detect the disease early so that treatment can begin as soon as the disease is identified. This inspired us to create a classification model that can identify the presence of malaria in a person's blood to increase both the speed and accuracy of the detection process. To implement an image-based malaria classification model, we took the help of the Inception V3 model, which is trained on the ImageNet dataset, to recognize objects more efficiently.

2 OUR APPROACH

A summary of the approach followed in this work is shown in Figure 1. The input images are obtained from NIH (National Institute of Health), containing around 27,588 images of parasitized and uninfected cells. The number of images that are parasitized and the number of uninfected images are equally divided into 13,794 each. After we procured the input image dataset, we performed the next step, data pre-processing, where we transformed the raw data before applying machine learning algorithms and resize the image to be 224*224 so that it could fit into the Inception V3 model. Various models are used to test the best generalization and the highest achievable accuracy. Training is done on separate systems with the same configuration to save time and effort with VGG-16, VGG-19, Inception V1, Inception V3, Resnet 50, and Alexnet neural network architectures. We performed the split of data into testing and validation sets in the ratio of 80:20 intending to perform the holdout validation for our data. Finally, the best-performing model is hyperparameter tuned using a grid search algorithm and results are reported.

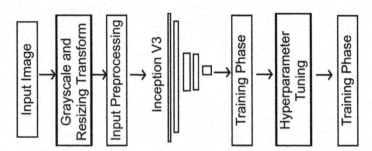

Figure 1. This figure represents the approach followed in this research. The input images for the training of the model are modified in accordance with the parameters that the model accepts, finally, these are put to training, and results are obtained using probabilistic inference from the trained model.

3 LITERATURE SURVEY

Various state-of-the-art techniques can perform Malarial image classification to achieve supreme accuracy. Techniques like watershed segmentation can be employed to acquire plasmodium-infected erythrocytes and extract the necessary features. Then we can apply machine learning algorithms like subspace KNN (Olugboja & Wang 2017) K nearest neighbor, a popular machine learning algorithm, has been employed to classify the cells after extracting the cell masks as features (Malihi *et al.* 2013) Techniques like transfer learning that are pre-dominantly applied in today's arena of machine learning have been employed, and the use of ResNet50, VGG16, and other state of the art modern models have enhanced the performance and accuracy (Swastika *et al.* 2021) (Vijayalakshmi *et al.* 2020) In combination with feature extractors like SURF and 2-stage tree classifiers, have shown good proportionate results (Rizal *et al.* 2019) (Ross *et al.* 2006) A summary of the works, their performance, and their respective citations and available in a tabular manner in Table 1.

Table 1. Summary of neural network architectures used for the domain of malaria disease detection in prior works.

S. No.	Method	Results	Citation
1	Using subspace K-Nearest Neighbor algorithm and classifier algorithm from machine learning to detect malaria parasite	Accuracy = 86.30%	(Oluboja & Wang 2017)
2	Using Giemsa stained images of blood cells to detect malaria parasite by K-Nearest Neighbor	Accuracy = 91%	(Malihi *et al.* 2013)
3	Using preprocessed images of thin blood smears and deep learning methods Resnet 50 model	Accuracy = 90.1%	(Swastika, *et al.* 2021)
4	Using VGG-16, SVM, to detect malaria parasite from microscopic images	Accuracy = 89.21%	(Vijaylakshmi *et al.* 2020)
5	Employing SURF and HOG features to classify the blood images of malaria parasites using SVM	Accuracy = 85.53%	(Rizal *et al.* 2019)
6	Using 2-stage tree classifier and image processing for analysis of thin blood smears to detect malaria	Accuracy = 81%	(Ross *et al.* 2006)

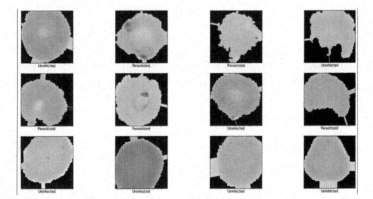

Figure 2. This figure represents the sample from a dataset that contains infected and uninfected samples of malaria parasite cells.

4 DATA

The input images are obtained from NIH (National Institute of Health), which contains around 27,588 images of parasitized cells and uninfected cells together (Yang *et al.* 2020) (Kassim *et al.* 2021b) (Kassim *et al.* 2021a) (Rajaraman *et al.* 2018) The dataset consists of a folder structure in which it is divided into two separate repositories, Infected and Uninfected.

These repositories contain a total of 27,558 images. A sample of the data contained within this dataset is shown in 2.

5 METHODS

5.1 *Inception V3*

A deep learning model based on neural networks specifically made for recognition purposes, convolutional neural networks, the inception v3, is used for tasks such as image classification and other pattern recognition tasks. A team at Google developed this model, so the model was named on the convention of the company from which the model came originally, which is the same as Google (Szegedy *et al.* 2016) Inception v3 is a very weight-rich and complex model as compared to its other variants and also other deep neural architectures. This model is used for image classification and recognition tasks. On the ImageNet dataset, this model attains greater than 78.1% accuracy. Many researchers over the years accumulated their ideas to make this model. The original paper on this model is "Rethinking the Inception Architecture for Computer Vision". Symmetric and asymmetric building blocks constitute the working and building of this neural network architecture, including convolutional layer, average pooling layer, and some other layers which are max pooling layers, concatenation layer, dropouts, and others such as fully connected layers. Softmax activations are used for the computation of the loss function, and also Batch normalization is used very much in this model.

5.2 *ResNet 50*

ResNet50 is one of the members of the ResNet Model family, which has been immensely successful in classification tasks. The model comprises 48 layers of type convolutions and also consists of one MaxPooling layer and one Average Pooling layer embedded in the architecture (He *et al.* 2016) It has $3.8 * 10^9$ Floating points operations. The ultra-deep neural networks were made possible to be trained just because of the framework that the family of the Resnet50 model proposed. This framework consists of residual skip connections which means that the network may consist of many layers; still, the evaluation can be expected to be very good. Initially, the Resnets were applied to the task of image recognition. Now, the framework can also be altered to be suited for many other tasks having multimodal data. This model was and also is used as a state of art currently.

5.3 *VGG16*

Another excellent performance can be expected from the VGG 16 model, which is also from the family of convolutional neural networks and is considered the best-performing model until Resnet34 arrived (Simonyan & Zisserman 2014) This model is still used in many computer vision tasks which comprise image classification, image recognition, and much more. The authors of this model did the evaluation of this model, and they increased the model depth using many layers which were supposed to perform 3x3 convolutions. This step showed a considerable increment in the accuracy of the base datasets and decrement in the loss as compared to the prior-art configurations. The VGG16 has the naming convention to denote the 16 layers consisting of weighting in filters. There are 13 layers that consist of convolutional type operations and 5 layers that are responsible for max-pooling operations. In addition to these layers and three Dense layers, which sum up to 21 layers. The layers which are learning in this model are only 16. Hence, these are the layers that are responsible for learning the weights.

5.4 *AlexNet*

The AlexNet model was also one of the best in 2012 when it won the 2012 ImageNet LSVRC-2012 competition. This model not only won the competition but left a substantial margin for the competition to rise high (Krizhevsky *et al.* 2017) Hence, there were a total number of eight layers in the AlexNet model. An Overlapping Max Pooling layer follows each convolutional layer for topmost primary and secondary convolutional layers. There exists a visible and straight connection between the convolutional layers, three, four, and five, to establish a relation between the features. The convolutional layer on the fifth position is immediately followed by Max Pooling operations using a max pooling layer. This final layer consists of 4096 neurons, with two in number.

6 RESULT ANALYSIS

After training various state of the art models on the Malaria cell image data for 16 epochs, we observed that the Inception V3 model gave a accuracy for training epochs of 92.44% and for validation epochs an accuracy of 91.40%. The ResNet50 model gave a training accuracy of 91.27% and a validation accuracy of 90.61%. The VGG16 model is trained with an accuracy of 88.83% and a validation accuracy of 86.20%, and the Alexnet model is trained with an accuracy of 83.90% and a validation accuracy of 81.19%. The results are tabulated and shown in Figure 3. The best performance was shown by the Inception V3 model, which performed best in the comparison above.

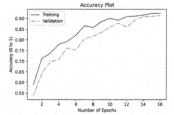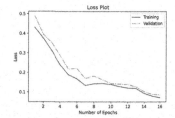

Figure 3. This figure represents the training and validation results of the Inception V3 model. Both, the accuracy and loss were plotted (on a scale of 0 to 1) and represented in this graph.

7 CONCLUSION

From this experiment, deep learning techniques are potent in classifying malarial cells. After training and testing the images on various state-of-the-art transfer learning models, we concluded that the Inception V3 had superior performance compared to other models such as Resnet50, VGG16 and AlexNet. This model can not only be applied to medical imaging and analysis but also be used for effective prognosis and making a tool capable of diagnosing and detecting malaria cells using the images, as shown in this work.

DATA AND CODE AVAILABILITY

The input images are obtained from NIH (National Institute of Health), containing around 27,588 images of parasitized and uninfected cells. The images of the NIH are also put into an easy-to-access form by the kaggle repository, whose link is provided in this section. The link for the dataset from Kaggle is https://www.kaggle.com/datasets/iarunava/cell-images-for-detecting-malaria. Also, we have provided the code related to this work for research and development purposes, which can be accessed on the link https://bit.ly/jupyter-notebook-malaria.

CONFLICTS OF INTEREST

The authors of this work declare no conflicts of interest.

REFERENCES

He, Kaiming, Xiangyu Zhang, Shaoqing Ren, and Jian Sun. 2016. "Deep Residual Learning for Image Recognition." In *Proceedings of the IEEE Conference on Computer Vision and Pattern Recognition*, 770–778.

Kassim, Yasmin M., Kannappan Palaniappan, Feng Yang, Mahdieh Poostchi, Nila Palaniappan, Richard J Maude, Sameer Antani, and Stefan Jaeger. 2021a. "Clustering-Based Dual Deep Learning Architecture for Detecting Red Blood Cells in Malaria Diagnostic Smears." *IEEE Journal of Biomedical and Health Informatics* 25 (5) 1735–1746. https://doi.org/10.1109/JBHI.2020.3034863.

Kassim, Yasmin M., Feng Yang, Hang Yu, Richard J. Maude, and Stefan Jaeger. 2021b. "Diagnosing Malaria Patients with Plasmodium Falciparum and Vivax Using Deep Learning for Thick Smear Images." *Diagnostics 11, no. 11 (October)* 1994. https://doi.org/10.3390/diagnostics11111994. https://doi.org/10.3390/diagnostics11111994.

Krizhevsky, Alex, Ilya Sutskever, and Geoffrey E Hinton. 2017. "Imagenet Classification with Deep Convolutional Neural Networks." *Communications of the ACM 60 (6)* 84–90.

Malihi, Leila, Karim Ansari-Asl, and Abdolamir Behbahani. 2013. "Malaria Parasite Detection in Giemsa-stained Blood Cell Images." In *2013 8th Iranian Conference on Machine Vision and Image Processing (MVIP)* 360–365. IEEE

Olugboja, Adedeji, and Zenghui Wang. 2017. "Malaria Parasite Detection Using Different Machine Learning Classifier." In *2017 International Conference on Machine Learning and Cybernetics (ICMLC)* 1:246–250. IEEE

Rajaraman, Sivaramakrishnan, Sameer K. Antani, Mahdieh Poostchi, Kamolrat Silamut, Md. A. Hossain, Richard J. Maude, Stefan Jaeger, and George R. Thoma. 2018. "Pre-trained Convolutional Neural Networks as Feature Extractors Toward Improved Malaria Parasite Detection in Thin Blood Smear Images." *PeerJ* 6 (April) e4568. https://doi.org/10.7717/peerj.4568. https://doi.org/10.7717/peerj.4568.

Rizal, Reyhan Achmad, Julianus Stepanus Sihotang, Rollys Gultom, *et al.* 2019. "Comparison of SURF and HOG Extraction in Classifying the Blood Image of Malaria Parasites Using SVM" in *2019 International Conference of Computer Science and Information Technology (ICoSNIKOM)* 1–6. IEEE

Ross, Nicholas E Charles J Pritchard, David M Rubin, and Adriano G Duse. 2006. "Automated Image Processing Method for the Diagnosis and Classification of Malaria on Thin Blood Smears." *Medical and Biological Engineering and Computing* 44 (5) 427–436.

Simonyan, Karen, and Andrew Zisserman. 2014. "Very Deep Convolutional Networks For Large-scale Image Recognition." *arXiv* preprint arXiv:*1409.1556*.

Swastika, W GM Kristianti, and RB Widodo. 2021. "Effective Preprocessed Thin Blood Smear Images to Improve Malaria Parasite Detection Using Deep Learning." In *Journal of Physics: Conference Series*, 1869:012092. 1. IOP Publishing.

Szegedy, Christian, Vincent Vanhoucke, Sergey Ioffe, Jon Shlens, and Zbigniew Wojna. 2016. "Rethinking the Inception Architecture for Computer Vision." In *Proceedings of the IEEE Conference on Computer Vision and Pattern Recognition*, 2818–2826.

Vijayalakshmi, A *et al.* 2020. "Deep Learning Approach to Detect Malaria from Microscopic Images." *Multimedia Tools and Applications* 79 (21) 15297–15317.

Yang, Feng, Nicolas Quizon, Hang Yu, Kamolrat Silamut, Richard Maude, Stefan Jaeger, and Sameer Antani. 2020. "Cascading YOLO: Automated Malaria Parasite Detection for Plasmodium Vivax in Thin Blood Smears." In *Medical Imaging 2020: Computer-Aided Diagnosis*, edited by Horst K. Hahn and Maciej A. Mazurowski. SPIE March. https://doi.org/10.1117/12.2549701. https://doi.org/10.1117/12.2549701.

Artificial Intelligence, Blockchain, Computing and Security – Dagur et al. (Eds)
© 2024 The Author(s), ISBN: 978-1-032-49393-0

The future of mobile computing in smart phones and its potentiality-A survey

Mohd Shahzad*
Chandigarh University, Gharaun, Mohali, Punjab, India

Geetinder Saini*
Apex Institute of Technology (CSE), Chandigarh University, Gharaun, Mohali, Punjab, India

ABSTRACT: Nowadays, the market for smartphones is expanding and reviving at a quick pace. Furthermore, the meteoric development of mobile apps enables mobile users to conduct the majority of their e-commerce using their smartphones. Moreover, significant limitations in the cloud environment limit the full potential of smartphones. As a result, cloud computing for smartphones became offered as a viable addition to phone carriers. That combines cloud technology into the mobile context to overcome the constraints of smartphones. Despite advances in these devices' processing and storage capacities, they are still not as capable as PCs for big and demanding workloads. One solution to this problem is to combine cloud computing with smartphones. This will outsource certain mobile records and functions to cloud servers on an as-needed basis. Because the details are analyzed elsewhere instead of through cell phones, privacy, authenticity, and security breaches have developed, and these concerns must be addressed.

Keywords: MCC, Cloud service, Challenges in MCC, Smartphones, Security Risk

1 INTRODUCTION

Digitization has revolutionized we cope with electronic content in our digital lives. Cloud computing now allows us to view our information and data anyplace, at any moment. It offers a variety of services, such as Infrastructure as a Service (IS), Platform as a Service (PS), Software Ease of Use as a Service (SS), and others. Meanwhile, the advancement of cell phones, smart devices, and cellular services had resulted in a rise in user adoption and implementation. It involves not just exchanging conversations and communicating via text messages, but as well as employing complex software to help us with our everyday tasks [1].

Since smartphones are among the most productive and efficient social tools, technologies had made human existence considerably simpler than in the past. They are lightweight and often weighed below 2.4 pounds. Smartphone nodes get an affluent skill from numerous mobile application services that operate on the devices; however, the limits of mobile devices have limited certain of the optimal uses of features and apps on smartphones. As a result of advancements in cloud computing and mobility, we now have new technologies known as MCC, that may assist to alleviate certain constraints.

1.1 Mobile Cloud Computing (MCC)

Merging cloud computing with the virtualized context to conquer efficiency, privacy, and day-to-day challenges are known as MCC. It is well recognized as a concept that may

---

*Corresponding Authors: shahzadkhan04470447@gmail.com and geetinder.e12987@cumail.in

DOI: 10.1201/9781003393580-58

considerably increase consumer comprehension while using smartphones. It introduces the latest services and abilities for smartphone customers to entire use cloud computing. MCC is intended to address the restrictions of smartphones by concentrating assets and providing them to portable devices [2]. It has enhanced the customer journey while using smartphones and apps.

1.2 *Architecture of MCC*

Figure 1 depicts the entire MCC design. Phone networking, as seen in Figure 1, is primarily composed of portable devices connected by access points (e.g., BTS unit, gateway, or space station) that transport transmissions between cells to another [3]. Bss, commonly referred to as cellphone towers, are in charge of establishing and sustaining connection and smartphone deployment. They collect frequencies from smartphones on the earth either in houses and then transmit such calls, messages, and information, as well as other cellular technologies [4]. Depending on house agents and customer information stored in systems, mobile telecom companies might offer certain services to mobile customers, such as authentication, authorization, and accounting.

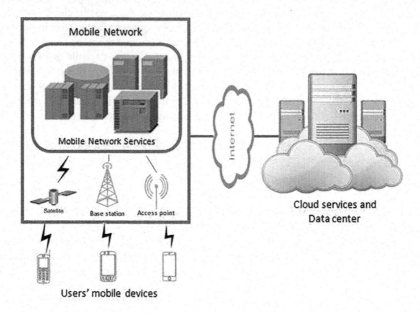

Figure 1.   Architecture of MCC.

2   ADVANTAGES

So far, MCC is the best answer to many challenges in virtualized context. As seen in Figure 2, describes mobile constraints and how MCC overcomes these constraints.

2.1 *Extending the mobile battery life*

Battery performance is a major problem for smartphones. Customers must adjust or minimize their mobile use to keep the battery charged. As a result, much research had been conducted for increasing the efficiency of the processing unit and other cellular functions to minimize the usage of battery. [5] Therefore, the adoption of these study findings and the design of the smartphone must be altered. Other study discoveries need the addition of additional apparatus, which raises the cost of smartphones [6].

2.2 *Improving the mobile data storage capacity*

Though smartphones have limited storage space, MCC enables customers to store an unlimited amount of information in the cloud and obtain that whenever over a cellular connection.

These are some examples:

- (S3) Amazon Simple Storage Service [7].
- Images are stored in the cloud by Facebook, Instagram, and Google Photos.

2.3 *Improving reliability*

Because data is saved and stored on several computers, one of the advantages of the cloud is reliability. As a result, even if the information is lost on the smartphone, it may be recovered. Furthermore, owing to such limitations of smartphones, MCC may offer customers security services like online antivirus scans and monitoring of harmful activities [8].

2.4 *Scalability*

Cloud technology is a versatile, cost-effective method for providing certain systems. Pervasive network connectivity enables consumers to use a variety of handheld devices to access a broad range of services [9]. Mobile application deployment may be conducted and expanded to meet unanticipated user needs thanks to dynamic resource provision. Service providers may easily add and grow services and programs with minimal or no resource constraints [10].

3 CHALLENGES

The primary goals of cloud services are to provide mobile users with speed and convenience. That is, by employing smartphones, phone users may effectively retrieve and obtain data from the cloud computational power. The fundamental problem of the mobile cloud is the constraint of smartphones and wireless networks. [11] Because of this difficulty, designing, creating, and deploying applications on phone-dispersed gadgets is withal difficult than reliable cloud systems. The further section will discuss several critical aspects of mobile cloud computing, such as device constraints, communication quality, and application services. Each aspect is explained in further depth below [12].

3.1 *Mobile devices limitations*

Numerous noteworthy advancements have occurred in cellphone properties like Central processing units, ram, memory, pixel density, remote access, sensor technologies, and operating systems [13]. Some restrictions, however, such as limited computational capabilities or energy supplies to install demanding tasks, are still to be addressed. Regular smartphone users will be required to charge their phone's batteries virtually daily [14].

3.2 *Communication quality*

Among the primary problems in mobile cloud services is connectivity. Because radio resources for wireless networks are few, the frequency band in the digital cloud is limited compared to typical wired broadband [15]. Furthermore, data centers in major organizations or assets in Network operators are often far from end consumers, smartphone device users. Climate change, user mobility, and dynamic changes in application performance may all have an impact on the frequency band. To address the bandwidth constraint, many alternatives have been offered, such as distributing the frequency band among phone users or imposing a use regulation [16].

3.3 *Applications services*

Existing mobile application strategies have been shaped by the few resources present in a mobile context. There are two types of mobile applications: offline and online. The majority of contemporary mobile device apps are now offline applications. As a result, all procedures are carried out instantly on the smartphone using data obtained from rear systems. In contrast, due to the existing connectivity among smartphones and rear databases the majority of the time, employing online apps may assist relocate data analysis beyond the smartphone [17].

4 LITERATURE REVIEW

Many strategies, concepts, and workspaces are now being suggested throughout the aforementioned dangers and difficulties provided with the most potential business.

Huang *et al.* suggested a Mobi-Cloud system where the service customer inputs multiple cloud platforms. The trustworthy platform is used to handle important files, while cloud computing is used to run additional apps and information [18].

J. Oberheide *et al.* designed the Cloud AV framework, an antimalware solution. That technology is able to detect spyware using virtual machine methods and numerous analysis algorithms that run simultaneously [19].

M. Satyanarayanan *et al.* were also the first to describe the Cloud service phenomena. A cloud server is a word used to describe simple cloud computing systems that are positioned nearby gadgets. That connection chain goes through a smartphone to a cloud environment, and finally from a cloud system to the comprehensive data center. Another primary goal of configuring cloudlets is to decrease device transmission delay while connecting with the cloud. Thus, it contributes to meeting many clients' maximum connectivity needs [20].

Ioana Giurgiu *et al.* suggested a two-step strategy to properly partition a program between a smartphone as well as a web server. Firstly, they model a program's behavior as a graph of data flow composed of numerous interlinked software components. In the second stage, having such a network, a division system identifies the ideal split with maximizing (or reducing) a particular objective function. They recommend two partition algorithms: ALL and K-step. In the primary example, the appropriate partition is determined offsite by taking into account various kinds of mobile devices and connection circumstances. Whenever a smartphone is connected to the servers and describes its resources and needs, the segmentation is calculated on the go in the second instance. ALL corresponds to the first situation, whereas K-step corresponds to the second [21].

Byung-Gon and Petros *et al.*, pioneered the outsourcing of smartphone execution to a computer system containing clouds of cellphone copies. That concept is straightforward: cloned the whole collection of information and programs from the cellphone into the internet and then conduct specific actions on the copies before resettling the findings further into smartphones. There are five forms of augmentation, each with a unique manner of unloading. Numerous copies of identical smartphones, copies professing that they are more sophisticated smartphones, and so on [22].

Eduardo Cuervo *et al.* and colleagues introduced MAUI, a framework that allows really well power offloading of smartphone software to infrastructural. It optimizes the possibility of energy reductions via really well program offloading with reducing program modifications. To start, MAUI uses code that is portable to construct two copies of a smartphone application, one that operates directly upon this device while the other remote in the infrastructures. Next, MAUI uses program reflections in conjunction to type integrity to detect distant procedures and retrieve just the program state required in such ways. Furthermore, MAUI analyses every methodology and utilizes serialization to calculate networking shipments [23].

5 CONCLUSION

There will be a number of exciting new mobile technologies in the future, and cloud computing is one of them. With their smartphones and the power of the cloud at their disposal, mobile users can enjoy the greatest experiences. Many of the issues plaguing the cloud platform may have promising solutions, and mobile cloud computing may be one of them. An overview of cloud computing systems, along with their advantages and disadvantages, has been provided here. It also discussed the many security threats that mobile devices face in today's computerized world.

REFERENCES

[1]   Orsini G., Bade D., and Lamersdorf W., "Cloud Aware: Towards Context Adaptive Mobile Cloud Computing," in Integrated Network Management (IM), *2015 IFIP/IEEE International Symposium on. IEEE*, 2015, pp. 1190–1195.

[2]   Rudenko A., Reiher P., Popek G. J., and Kuenning G. H., "Saving Portable Computer Battery power Through Remote Process Execution," *ACM SIGMOBILE Mobile Computing and Communications Review*, vol. 2, no. 1, pp. 19–26, 1998.

[3]   Bifulco R., Brunner M., Canonico R., Hasselmeyer P., and Mir F., "Scalability of a Mobile Cloud Management System," in *Proceedings of the first edition of the MCC workshop on Mobile Cloud Computing*. ACM, 2012, pp. 17–22.

[4]   Kovachev D., Cao Y., and Klamma R., "Mobile Cloud Computing: A Comparison of Application Models," *arXiv* preprint arXiv:1107.4940, 2011.

[5]   Tayade D., "Mobile Cloud Computing: Issues, Security, Advantages, Trends," *IJCSIT) International Journal of Computer Science and Information Technologies*, vol. 5, no. 5, pp. 6635–6639, 2014.

[6]   Gray P., *"Legal Issues to Consider with Cloud Computing,"* Available at http://www.techrepublic.com/blog/tech-decision-maker/legal-issuesto-consider-with-cloud-computing/ (2016/11/30).

[7]   Gilbert F., *"Cloud Computing Legal Issues: Data location,"* Available at Shttp://searchcloudsecurity. techtarget.com/tip/Cloud-computing-legalissues-data-location (2016/12/03).

[8]   Huang D., Zhang X., Kang M., and Luo J., "MobiCloud: Building Secure Cloud Framework for Mobile Computing and Communication," in *Proceeding 5th IEEE International Symposium on Service-Oriented System Engineering, SOSE '10*, Nanjing, China, June 2010.

[9]   Nitesh Kaushik, Gaurav, and Jitender Kumar, "A Literature Survey on Mobile Cloud Computing: Open Issues and Future Directions" published in *International Journal of Engineering and Computer Science*, ISSN:2319-7242 Volume 3 Issue 5 May 2014, Page No. 6165–6172.

[10] Sapna Malik and MM Chaturvedi "Privacy and Security in Mobile Cloud Computing: Review" published in *International Journal of Computer Applications*, Volume 80 – No 11, October 2013.

[11] Ruay-Shiung Chang, Jerry Gao, Volker Gruhn, Jingsha He, George Roussos, Wei-Tek Tsai, "Mobile Cloud Computing Research – Issues, Challenges, and Needs", in *Proceeding of 2013 IEEE Seventh International Symposium on ServiceOriented System Engineering*.

[12] Pragya Gupta and Sudha Gupta, "Mobile Cloud Computing: The Future of Cloud" published in *International Journal of Advanced Research in Electrical, Electronics and Instrumentation Engineering*, Vol. 1, Issue 3, September 2012, ISSN 2278 – 8875.

[13] Suganya V and Shanthi A L, "Mobile Cloud Computing Perspectives and Challenges" published in *International Journal of Innovative Research in Advanced Engineering (IJIRAE)* ISSN: 2349-2163, Issue 7, Volume 2 (July 2015)

[14] Popa D., Boudaoud K., Cremene M., and Borda M., "Overview on Mobile Cloud Computing Security Issues", *Tom* 58(72), Fascicola 1, 2013

[15] Zhang, X., Schiffman, J., Gibbs S, Kunjithapatham, A. and Jeong S. 2009, "Securing Elastic Applications on Mobile Devices for Cloud Computing" In *Proceeding ACM workshop on Cloud Computing Security, CCSW '09*, Chicago, IL, USA.

[16] Xiao, S. and Gong, W., 2010., "Mobility Can Help: Protect User Identity with the Dynamic Credential". In *Proceeding 11th International Conference on Mobile Data Management, MDM '10*, Missouri, USA, May 2010.

[17] Chun B., Ihm S., Maniatis P., Naik M., and Patti A., "Clonecloud: Elastic Execution Between Mobile Device and Cloud," in *Proceedings of the Sixth Conference on Computer Systems*. ACM, Pages: 301–314, 2011

[18] Huang D., Zhang X., Kang M., and Luo J., "MobiCloud: Building Secure Cloud Framework for Mobile Computing and Communication," in *Proceeding 5th IEEE International Symposium on Service-Oriented System Engineering*, SOSE '10, Nanjing, China, June 2010.

[19] Oberheide, J., Veeraraghavan, K., Cooke, E. and Jahanian, F. 2008, "Virtualized in-cloud Security Services for Mobile Devices" in *Proceedings of the 1st Workshop on Virtualization in Mobile Computing (MobiVirt)*, 31–35.

[20] Mahadev Satyanarayanan; Paramvir Bahl; Ramon Caceres; Nigel Davies "The Case for VM-Based Cloudlets in Mobile Computing". IEEE. Published in: *IEEE Pervasive Computing* (Volume: 8, Issue: 4, Oct.-Dec. 2009), ISSN: 1536-1268

[21] Ioana Giurgiu, Oriana Riva, Dejan Juric, Ivan Krivulev, and Gustavo Alonso, Calling the cloud: enabling mobile phones in *Proceedings of the 10th ACM/IFIP/USENIX International Conference on Middleware*. 2009, Springer-Verlag New York, Inc.: Urbanna, Illinois. p. 1 20.

[22] Byung-Gon Chun and Petros Maniatis, Augmented Smartphone Applications Through Clone Cloud Execution, in *Proceedings of the 12th Conference on Hot topics in operating Systems (HotOS)*. 2009

[23] Eduardo Cuervo, Aruna Balasubramanian, Dae-ki Cho, Alec Wolman, Stefan Saroiu, Ranveer Chandra, and Paramvir Bahl, MAUI: Making Smartphones Last Longer with Code Offload, in *Proceedings of the 8th International Conference on Mobile Systems, Applications, and Services*. 2010, ACM: San Francisco, California, USA. p. 49–62

Artificial Intelligence, Blockchain, Computing and Security – Dagur et al. (Eds)
© 2024 The Author(s), ISBN: 978-1-032-49393-0

Auto scaling in cloud computing environments with AWS

Nazish Baliyan*
Chandigarh University, Gharaun, Mohali, Punjab, India

Sukhmeet Kaur*
Apex Institute of Technology (CSE), Chandigarh University, Gharaun, Mohali, Punjab, India

ABSTRACT: In cloud computing environments, auto scaling methods have become the norm. According to user demands, such methods might increase or decrease the number of virtual computers, hence meeting pay-per-use goals, Providers offering infrastructure as a service, however, have to adhere strictly to customised by users thresholds. when implementing auto scaling techniques. A crucial component of clouds is auto-scaling, which regulates the quantity of resources on hand to satisfy service demand. To maintain performance metrics, like utilization level, between user-defined lower and upper boundaries, resource pool adjustments are required. Unacceptable Quality of service (QoS) and significant resource waste may result from auto-scaling solutions that are not appropriately designed in accordance with user workload characteristics. Therefore, a greater comprehension of auto-scaling strategies and how to tune them to minimize these issues is required. In this study, we compare different auto-scaling techniques.

Keywords: Cloud Computing, Auto scaling, AWS

1 INTRODUCTION

Cloud computing (CC) has been a very useful tool for addressing issues with particular demands for computing resources in recent years [1]. With lower costs and improved performance, computing resources may now be used wherever and whenever they are needed thanks to the rapid development of virtualization technologies [2]. Cloud computing has gained popularity and acted as the model for new computing paradigms like Amazon Web Services (AWS), just as with the application of energy, water, telephone, and evaluation models. [3] The cloud platform's capacity to dynamically spin up or down server instances based on the workload or application's live traffic load is known as auto-scaling. The cloud's ability to auto scale allows it to adjust the instance cluster's computing power based on demand, assuring smooth handling of traffic surges and slumps. As a result, the chance of the already-running servers failing under the heavy traffic load is reduced when additional instances are deployed on the fly. When referring to cloud computing, the term Based on the demands a workload makes on a system, "scaling" implies the act of adding or eliminating storage, computing, and network capabilities. on the system, maintaining availability and performance as demand increases. Scaling generally refers to adjusting the number of active servers (instances) required to meet your business's requirements for resources. There are two dimensions—scaling up and scaling out—across which resources and hence capacity can be augmented.

---

*Corresponding Authors: nazish.baliyan07@gmail.com and sukhmeet.e13420@cumail.in

DOI: 10.1201/9781003393580-59

2 RELATED WORK

Applications hosted in the cloud are optimised for performance using auto scaling techniques. Numerous research studies on auto scaling and dynamic provisioning of cloud-hosted apps have been published in the literature [4,5]. The majority of these methods are based on machine learning, reinforcement learning, analytical queuing, and rule-based policies. Either proactively or reactively are used to activate the auto scaling mechanisms. For instance, Liu *et al.* reactive's auto scaling approach [6] was put out. It monitors bandwidth and CPU consumption and scales resources vertically whenever CPU or bandwidth utilisation reaches saturating levels. In order to improve application performance, Michael *et al.* [7] demonstrate the use of a reactive approach for the horizontal auto-scaling of cloud-hosted healthcare and bioinformatics tools. "A reactive auto scaling technique is presented by Liu *et al.* that uses fuzzy logic to dynamically modify the cluster's sizes and scaling threshold" [8]. An auto scaling technique that adjusts based on the variety of active users is presented by Chieu *et al.* [9]. This work scales the programme whenever the number of active users exceeds a predetermined threshold for allotted resources. Several initiatives have been made to develop machine learning-based predictive auto scaling systems. Maryam *et al.* [10] provide an overview of predictive auto scaling techniques and talk about the various machine learning techniques the researcher employed for dynamic resource supply. Recently, there have been various attempts to automatically grow apps with microservices.

3 AUTO SCALING

"Auto-scaling is the feature in cloud computing infrastructures that facilitates the delivery of virtualized resources on demand" [11,12]. Cloud-based apps' resource utilisation can be automatically increased or decreased to match the needs of the programme [11].

Two scalability types emerge: Scaling out is another name for horizontal scaling, simply adding nodes or machines to the system, resources are increased. Meanwhile, scaling up is another name of vertical scaling, it's help in increases the resources in already existing nodes, such as CPU and processing capability" [12].

The MAPE loop is followed by the auto-scaling: (M) Monitoring, (A) Analysis, (P) Planning, and (E) Execution.

3.1 *Monitoring*

The monitoring system gathers data on user expectations being met, resource status, and SLA violations from a cloud environment. It gives the cloud provider information on the infrastructure's condition.

3.2 *Analysis*

The information gathered is put through more processing during the analysis step. It collects all data from metrics, current system usage, and forecasting data for future workload. Some auto-scalers are developing a reactive strategy. After considering the state of the system as it stands, a choice is made. Others use a reactive strategy or both, while others use set threshold values enabling scale-in/scale-out options. Because there is usually typically occurs a delay between the resource settings for scaling decisions., reactive is a complicated method. Between 350 and 400 seconds are required for VM starting. The reactive strategy still has problems with flash crowds and events [13].

3.3 Planning

After the analysis phase examines the current situation, the planning phase must choose whether to scale-up, scale-down, or scale in order to comply with SLA and maximize profits.

3.4 Execution

Decisions on the execution phase were made during the planning stage. API from cloud providers is in charge of carrying out the planning. The problems with the execution phase are unknown to the client. Users can access virtual machines for a set amount of time, and these delays have already been addressed with them as part of the Service level agreement (SLA) [14].

4 TYPES OF AUTO SCALING

4.1 Reactive scaling

Simply said, AWS Reactive Scaling, keeps an eye on your applications and modifies their capacity to ensure optimal performance at very low cost. It is cost-free and simple to set up with little trouble. One single scaling policy can be set up for each application source using the AWS Auto Scaling service [15]. Then, you can establish the usage targets based on which each scalable resource supporting your application should scale, and you can add each scalable resource to the scaling plan. Application availability, cost reduction, or a mix of the two can be prioritized. Simply put, AWS Auto Scaling maintains an application's performance under conditions of increased usage, traffic, or load. It can be quite helpful for applications whose traffic fluctuates erratically and which want to maintain performance with a minimum of bother. Simple threshold settings, such as when to scale up and when to scale down, can be set, and AWS will take care of the rest. "It guarantees that you always have the appropriate number of EC2 instances on hand to handle the load for your application. For instance, the following Auto Scaling group has a targeted capacity of 2, a minimum size of 1, and a maximum size of 4. Depending on the criteria you provide, such as increasing instances when traffic climbs to 80% and decreasing instances when traffic declines to 20%, The number of instances inside your minimum and maximum number of instances is altered by the scaling policies you establish" [16].

4.2 Manual scaling

All the exesting number of instances are manually changed/modified in manual scaling. Through a CLI or console, one can manually alter the quantity of present instances. In cases when the user doesn't require automatic scaling, manual scaling is a smart option [16].

4.3 Scheduled scaling

The ability to scale your application resources in accordance with anticipated future loads is known as scheduled scaling. Let's imagine, for instance, that weekdays are less busy than weekends are with your application. Because of the weekend traffic, user might not want to use a big number of EC2 instances or a lot of resources. Therefore, employing scheduled scaling might be a superior strategy [16]. The time period on which new resources will be introduced is selectable by users. Known changes in traffic loads will be anticipated by a hybrid approach known as scheduled auto scaling, which operates in real-time and reacts to such variations at predetermined intervals. When there are predicted traffic decreases or spikes at particular times of the day, scheduled scaling is most effective, even though the changes are usually somewhat abrupt. Instead resources could be pre-provisioned in advance rather than having to wait for them to scale up as demand increases. of a major event or a peak period of the day.

4.4 *Dynamic scaling*

In this additional instance of auto scaling, the number of EC2 instances is modified automatically in response to signals. A good option when there is a lot of unpredictable traffic is dynamic scaling.

The number of instances is automatically changed based on information from a CloudWatch alarm, which modifies the number of instances in response to real-time changes in resource utilisation (average CPU, network in/out). The objective is to provide sufficient capacity to maintain usage at the targeted level [17].

For instance, user can set up the scaling strategy so that the ECS service only utilises 75% of our CPU. The scaling policy is triggered to add another task to the service to deal with the increasing demand when the CPU utilisation of the service climbs above 75% (when more than 75% of the CPU that is reserved for the service is being used) [17].

4.5 *Predictive scaling*

A predictive auto scaling approach analyses traffic loads and predicts Using skills of artificial intelligence and machine learning, evaluate whether the user will require more or fewer resources. This is comparable to how weather forecasts operate. With predictive scaling, user can make sure that resource capacity is available through carrying out planned scaling actions based on the prediction, before the application requires it. In other words, predictive auto scaling automatically adjusts scaling based on forecasted future demand using predictive analytics, including past usage data and current usage patterns. The actions taken by your instances during predictive scaling are based on the application's predictable traffic patterns. The only group currently supporting this feature is the Amazon EC2 Auto Scaling group. Although user can configure it to a minimum of 24 hours, it runs by analysing historical data for the preceding 14 days for the selected load indicators (CPU utilisation, network input/output). It then formulates a prognosis for two days in the future and prepares to scale thr EC2 instances in order to either increase or decrease the capacity. Predictive scaling aims to bring the scaling index as close as possible to the desired value [17].

Predictive scaling, for instance, can be enabled and set up to maintain your Auto Scaling group's average CPU utilisation at 60%. According to the forecast, there will be a traffic jam at 9 a.m. every day. In order to ensure that the infrastructure is prepared to manage that traffic in advance, it then creates future planned activities.

4.6 *Working of AWS auto scaling*

The process of auto scaling in AWS involves a number of different entities, including:- AMIs and load balancers are two essential elements in this procedure. The AMI portion of AWS allows for the execution of this procedure. Your system is ready for incoming traffic if all goes according to plan and auto scaling is configured. The AWS auto scaling service would immediately start launching another instance using the server's AMI when the traffic started to increase. This instance would have the exact same configuration as the present server. In the following phase, AWS' load balancer would evenly distribute or route the traffic among the recently deployed instances. "A load balancer divides the traffic according to the load on a particular system, and it uses internal processes to decide where to send the traffic. When a new instance is created, the person managing auto scaling chooses the sole set of guidelines. The criteria can be as simple as CPU usage; for example, user could set auto scaling such that when your CPU use reaches 70 to 80% to be able to deal with the load, an additional instance will be started. There may, of course, be restrictions on what user can accomplish".

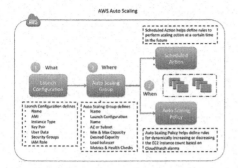

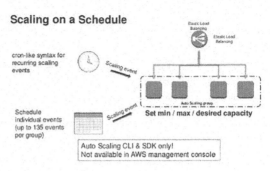

Figure 1. AWS auto scaling.

Figure 2. Scheduled scaling.

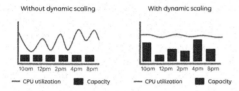

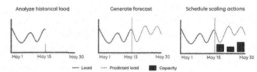

Figure 3. Dynamic scaling.

Figure 4. Predictive scaling.

4.7 *Reasons to auto scale*

Some of the benefits of auto scaling are:

- Better load management: Since servers can be used to accomplish non-time-sensitive computational jobs during periods of low traffic, auto scaling promotes effective server load control. This is feasible since auto scaling significantly frees up server capacity with reduced traffic.
- Less failures: Auto scaling services make sure that instances of failed servers are promptly swapped out for better servers. This cuts down on application outages.

6 CONCLUSION

There are numerous services that can be scaled automatically on the AWS Cloud, and the topic of automatic scaling in AWS is fairly broad. The AWS Auto Scaling service, which works with things like EC2 instances, DynamoDB tables, and more, enables you to do this for your entire application in one location. **The cloud scaling approach should be implemented using AWS Auto Scaling. Based on traffic demand, it automatically adjusts computing resources. Auto Scaling reduces the cost of the cloud. Only the resources which are utilized are charged for. Depending on the preferences, the target is decided and scaling policies are created.** Continuous application monitoring via AWS Auto Scaling results in automatic capacity adjustments that ensure reliable, consistent performance at the lowest possible cost. For a variety of resources and services, AWS Auto Scalability makes it simple to quickly set up application scalability.

REFERENCES

[1]   Buyya R., Yeo C. S., Venugopal S., Broberg J., and Brandic I., "Cloud Computing and Emerging It Platforms: Vision, Hype, and Reality for Delivering Computing as the 5th Utility," *Future Generation Computer Systems*, Vol. 25, No. 6, pp. 599–616, 2009.

[2] F Y. Li, Li W., and Jiang C. F., "A Survey of Virtual Machine System: Current Technology and Future Trends," *IEEE International Symposium on Electronic Commerce and Security*, 2010.

[3] *Amazon Web Service, Amazon Inc.*, http://aws.amazon.com/

[4] C. Qu, R. N. Calheiros, and R. Buyya, "Auto-scaling Web Applications in Clouds: A Taxonomy and Survey," *ACM Comput. Survey* vol. 51, no. 4, pp. 1–33, Jul. 2018.

[5] Lorido-Botran T., Miguel-Alonso J., and Lozano J. A., "A Review of Auto-scaling Techniques for Elastic Applications in Cloud Environments," *J. Grid Comput.*, vol. 12, pp. 559–592, 2014.

[6] Liu H. and Wee S., "Web Server Farm in the Cloud: Performance Evaluation and Dynamic Architecture," in *Proc. 1st Int. Conf. Cloud Computing.*, 2009, pp. 369–380.

[7] Krieger M. T., Torreno O., Trelles O., and Kranzlmuller D., € "Building an Open Source Cloud Environment with Auto-scaling Resources for Executing Bioinformatics and Biomedical Workflows," *Future Gener. Comput. Syst.*, vol. 67, pp. 329–340, 2017.

[8] Liu B., Buyya R., and Nadjaran Toosi A., "A Fuzzy-based Autoscaler for Web Applications in Cloud Computing Environments," in *Proc. Int. Conf. Serv.-Oriented Comput.*, 2018, pp. 797–811.

[9] Chieu T. C., Mohindra A., Karve A. A., and Segal A., "Dynamic Scaling of Web Applications in a Virtualized Cloud Computing Environment," in *Proc. IEEE Int. Conf. e-Bus. Eng.*, 2009, pp. 281–286.

[10] Amiri M. and Mohammad-Khanli L., "Survey on Prediction Models of Applications for Resources Provisioning in Cloud," *J. Netw. Comput. Appl.*, vol. 82, pp. 93–113, 2017.

[11] Calcavecchia N. M., Caprarescu B. A., Di Nitto E., Dubois D. J., and Petcu D., "Depas: A Decentralized Probabilistic Algorithm for Auto-scaling," *Computing*, vol. 94, no. 8–10.

[12] Herbst N. R., Kounev S., and Reussner R., "Elasticity in Cloud Computing: What It Is, and What It Is Not," in *Proceedings of the 10th International Conference on Autonomic Computing (ICAC 2013)*, San Jose, CA, 2013.

[13] Mickulicz N., Narasimhan P., and Gandhi R., "To Auto Scale or not to Auto Scale," in *Workshop on Management of Big Data Systems*, 2013.

[14] Dawoud W., Takouna I., and Meinel C., "Elastic vm for Cloud Resources Provisioning Optimization," in *Advances in Computing and Communications*, Springer, 2011, pp. 431–445.

[15] Kumar, A., Yadav, R. S., & Ranvijay, A. J. (2011). Fault Tolerance in Real Time Distributed System. *International Journal on Computer Science and Engineering*, 3(2), 933–939.

[16] Tripathi, A. M., Singh, A. K., & Kumar, A. (2012). Information and Communication Technology for Rural Development. *International Journal on Computer Science and Engineering*, 4(5), 824.

[17] Kumar, A., Mehra, P. S., Gupta, G., & Jamshed, A. (2012). Modified Block Playfair Cipher using Random Shift Key Generation. *International Journal of Computer Applications*, 58(5).

Artificial Intelligence, Blockchain, Computing and Security – Dagur et al. (Eds)
© 2024 The Author(s), ISBN: 978-1-032-49393-0

Analysis of cryptanalysis methods applied to stream encryption algorithms

Rakhmatullayev Ilkhom Rakhmatullaevich
Samarkand Branch of Tashkent University of Information Technologies Named After Muhammad al-Khwarizmi, Samarkand, Uzbekistan

Ilkhom Boykuziyev Mardanokulovich
Associate Professor, Department of Cryptology, Tashkent University of Information Technologies Named After Muhammad al-Khwarizmi

ABSTRACT: Scientific-research work was carried out within the framework of ongoing research in order to create a durable stream encryption algorithm. In this research, cryptanalysis methods applied to stream encryption algorithms are studied. As a result of the reviewed analysis, it was determined that correlation attack type is the most widely used cryptanalysis method compared to stream encryption algorithms. It was determined that the correlation-immunity property of the combining function used in the combining generators provides tolerance compared to the correlation cryptanalysis method, and the use of a non-linear filtering function with the correlation-immunity property in the filtering generators increases the tolerance compared to the correlation cryptanalysis method.

Keywords: Graphical tests, Assessment tests, Analytical attacks, Statistical attacks, Power-based attacks, Correlation attack, Time-memory trade-off, Guess and detect, Inversion attack, Meet-in-the-middle method, Statistical attacks, Linear cryptanalysis, Differential cryptanalysis, NESSIE

1 INTRODUCTION

When creating any cryptographic algorithms, their cryptoresistance is evaluated by their resistance to cryptographic attacks known today. At the same time, the randomness of the gamma sequences generated by stream ciphers is checked by randomness checking tests.

There are 2 types of randomness tests [1]:

1.1 Graphical tests

Graphical tests provide the user with information about a certain graphical relationship of the tested sequence and allow him to draw conclusions about the properties of the tested sequence.

1.2 Assessment tests

Assessment tests analyze the statistical properties of the sequence under investigation and provide an opportunity to draw conclusions about its degree of true randomness.

2 METHODOLOGY

Tests that determine the degree of randomness are presented in Figure 1 below.

The above randomization tests alone are insufficient to assess the robustness of stream ciphers. To fully assess the robustness of a stream cipher, its robustness to cryptanalysis methods is tested.

DOI: 10.1201/9781003393580-60

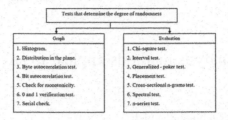

Figure 1.   Tests that determine the degree of randomness.

The cryptanalysis methods applied to stream encryption algorithms can be conditionally divided into three groups [2]:

– Analytical attacks;
– Statistical attacks;
– Force-based attacks.

An example of an attack on analytical attacks is cryptoattacks aimed at opening a cryptoscheme based on analytical principles. Statistical attacks include cryptoattacks based on the estimation of statistical properties of the cipher gamma. Force-based attacks include cryptoattacks based on looking at all possible variations of a key.

The above randomization tests alone are insufficient to assess the robustness of stream ciphers. To fully assess the robustness of a stream cipher, its robustness to cryptanalysis methods is tested.

The cryptanalysis methods applied to stream encryption algorithms can be conditionally divided into three groups [2]:

– Analytical attacks;
– Statistical attacks;
– Force-based attacks.

An example of an attack on analytical attacks is crypto attacks aimed at opening a crypto scheme based on analytical principles. Statistical attacks include crypto attacks based on the estimation of statistical properties of the cipher gamma. Force-based attacks include crypto attacks based on looking at all possible variations of a key.

The class of analytical attacks can be divided into two:

– Cryptoanalysis methods aimed at determining the encryption gamma;
– Cryptanalysis methods based on key initialization and reinitialization procedures.

The class of statistical attacks can also be divided into two:

– Cryptoanalysis methods based on the statistical properties of encryption gamma;
– Cryptoanalysis methods based on sequence complexity.

Figure 2 above depicts a common cryptographic attack classification (attack models) applied to synchronous stream encryption algorithms [15].

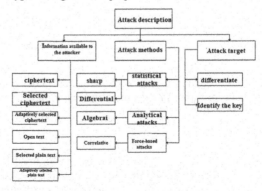

Figure 2.   Cryptoattacks on streaming ciphers.

Below are some common attack methods based on the above attack models.

Correlation attack. The most common type of attack based on the specifics of stream cipher construction is the correlation attack. If the nonlinear function passes information about its internal components to the output, the complexity of opening such a system is significantly reduced. At the same time, such a function will always be available. According to this axiom, correlational attacks use the correlation of the sequences coming out of the encryption scheme and coming out of the registers [2–5].

The main goal of the cryptanalyst is to find the initial state of the generator registers, i.e. the key, with the sequence generated by this generator. It should be noted that if the generator consists of three registers, to determine the initial state of such a generator, it is necessary to consider $O(2^{L1+L2+L3})$ options by checking all key options. Here L is the length in bits of the corresponding registers. In this case, each register is looked at separately to find the initial state of the registers. At the first stage, all options for the initial filling of the first register will be considered. Among these options, those with the highest matching frequency to the output sequence from the generator are selected. The initial conditions of the second register are determined in the same way. The initial state of the third register is determined by the options of the first and second registers based on the characteristics of the generator.

This method requires $O(2^{L1}+2^{L2}+2^{L3})$ consideration of the initial state of the register. This value is significantly less than $O(2^{L1+L2+L3})$ for considering all options. For example, if the lengths of the registers of the generator are $L_1 = 31$, $L_2 = 33$, $L_3 = 35$ bits, the number of all options is equal to $2^{99} = 6,3*10^{29}$, as a result of correlation attack this value is 231+233+235 = 4,5*10^{10} will be equal.

The following attacks belonging to this class can be distinguished:

1. Basic correlation attack.
   – Zygentaler's base correlation attack;
   – Zygentaler's correlational attack;

2. Attacks based on lightweight pair checking.
   – Mayer-Staffelbach fast correlation attack;
   – Forre's fast correlation attack;
   – fast iterative algorithm of Mikhalevich-Golich;
   – Chepijov-Smits fast correlation attack.

3. Attacks based on the use of convolutional codes.
4. Attacks using the technique of turbo codes.
5. Attacks based on recovery of linear polynomials.
6. Fast correlation attack by Chepijov, Johansson, Smits.

A fast correlation attack is defined as an attack whose computational complexity is significantly less than that of force-based attacks.

"Time-memory" compromise. The purpose of this attack method is to restore the initial filled state of the shift register by parts of the encryption sequence [2,4]. To carry out this attack, the cryptanalyst must have several parts of the encryption gamma and know the scheme of the algorithm. The complexity of the attack depends on the length of the cipher gamma in the hands of the cryptanalyst and the size of the internal state of the cipher. This type of attack is used when the size of the state field is small enough. In general, an attack consists of two stages:

– Preparation stage. In this step, a large dictionary is created containing all possible "state-output" pairs (of the same size n);
– The main stage. At this stage, it is assumed that the cipher is in a certain fixed state, that is, all the cells of the memory are filled with a certain value. Output values are generated based on this input value. Then a sequence from the generator that matches the sequence in the cryptanalyst's possession is searched. If a match is found, the fixed state is most likely the initial filled state of the register, otherwise the algorithm continues.

The following attacks belonging to this class can be distinguished:

- Steve Babbidge Attack;
- Biryukov-Shamir attack.

Hypothesis and detection. The main idea of the attack is based on assumption [2,4]. The cryptanalyst should know the encryption gamma, the shift value of the registers between the outputs of the circuit, and the r-level feedback multiplier and the f-filter function. Restoring the initial state of the register is done by guessing some parts of the value in the register. In general, an attack would look like this:

- The value in some cells of the register is estimated;
- Based on the accepted assumptions, the filled state of the register is determined by using a linear recurrent register.
- An outgoing sequence is generated. If it is equivalent to the encryption gamma, the guess is assumed to be correct, otherwise it returns to the first step.

The complexity of the attack depends on how the scheme is implemented and is proportional to the number of guesses.

Inversion attack. The goal of the inversion attack is to restore the initial state of the generator by parts of the encryption gamma [4,9]. The cryptanalyst needs to know the r-level feedback multiplier, the f-filter function, and γ_i the sequence of acquisition points ($i = 1, \ldots, n$). A memory size M is considered, where $M = \gamma_n - \gamma_1$ and $M \leq r-1$. The output sequence from the generator $(y(t))_{t=0}^{\infty}$ is given as follows:

$$y(t) = f(x(t - \gamma 1), \ldots, x(t - \gamma n)), \ t \geq 0,$$

the nonlinear function $f(z_1, z_2, \ldots, z_n)$ is balanced for each $(z_2, \ldots, z_n)$ value,
That is

$$f(z1, z2, \ldots, zn) = z1 + g(z2, \ldots, zn).$$

Hence, the following equality follows from the above:

$$x(t) = y(t) + g(x(t - \gamma 2), \ldots, x(t - \gamma n)), t \geq 0.$$

An inversion attack can be performed in the following sequence:

- An unknown initial value of the memory M bit $(x(t))_{t=-M}^{-1}$ which is not checked first is selected.
- Using $f(z_1, z_2, \ldots, z_n) = z_1 + g(z_2, \ldots, z_n)$, the known part gamma $(y(t))_{t=0}^{r-M-1}$, the output sequence part $(x(t))_{t=0}^{r-M-1}$ is generated.
- Using a linear recurrent register, the first r bit $(x(t))_{t=0}^{r-M-1}$ by $(x(t))_{t=r-M}^{N-1}$ sequence is generated.
- Using $y(t)$, $(x(t))_{t=r-2M}^{N-1}$ by $(y'(t))_{t=r-M}^{N-1}$ is calculated and compared with the observed $(y(t))_{t=r-M}^{N-1}$.

If they match, the accepted initial state is considered valid, otherwise the process is repeated.

The peculiarity of this attack is that its computational complexity is of the order of 2^M and does not depend on the length of the linear recurrent register.

"Meet in the middle" method. In the early 1990s, based on Ross Anderson, he proposed a method for fast opening of latent cryptogenerators by estimating the initial state of one of the registers and checking whether it corresponds to the part of the gamma [2,4]. In cryptographic literature, this algorithm was named "Meet in the Middle". In 1993, Anderson presented an optimized version of his attack method. According to it, in most cases, the attack complexity is equal to $2^{n/2}$, where n is the length of the register.

In the method proposed by Andersen, the initial filled state of register 1 is roughly observed. Then it is checked whether the gamma in our hands matches the acquisition points of the 2nd register. In other words, for each bit of gamma, the multiplexer address is calculated and this bit is placed in the corresponding position of register 2, if it causes a contradiction, the assumption about the initial state of register 1 is rejected.

Experiments have shown that the "meet in the middle" method can open a Jennings generator in a few hours using a simple workstation.

2.1 Statistical attacks

Statistical attacks are rarely used compared to stream encryption algorithms. All statistical attacks occur when the cryptanalyst knows the description of the generator, the plaintext and the corresponding ciphertexts. The task of a cryptanalyst is to determine the key or parts of the key used [2]. Examples of statistical attacks include differential and linear cryptanalysis.

2.2 Linear cryptanalysis

In 1994, Jovan Golic presented his work on the transfer of linear cryptanalysis to stream ciphers. The basis of this method is to determine the linear statistical weaknesses of the binary gamma generator. If the structure of the ciphertext does not depend on the key, an appropriate statistical test can be used to recover statistical plaintexts from the known ciphertext. If the structure of the cipher depends on the key, this test can also be used to find the corresponding secret key [7].

Let the binary autonomous machine or series circuit be given by the following relation.

$$S_{t+1} = F(St), t \geq 0$$

$$y_t = f(S_t), t \geq 0,$$

where, $F : GF(2)^M \to GF(2)^M$ is the next state vector Boolean function, $f : GF(2)^M \to GF(2)$ is the outgoing Boolean function, $S_t = (s_{1t}, \ldots, s_{Mt})$ is the state vector at time t, M is the number of bits in the memory, y_t is the output bit at time t, $S_0 = (s_{1,0}, \ldots, s_{M,0})$ is the initial state vector.

For a given F and f, each output bit is a boolean function, i.e. $y_t = f(F^t(S_0))$. If S_0 is a uniformly distributed random variable, then the output bits are binary random variables. In the generator, each y_t bit is required to be a balanced function from S_0. This is done only if the functions F and f are balanced functions. A vector sequence of output bits $(y_t, \ldots, y_{t-M})$ from $M+1$ cannot be balanced from S_0 at any $t \geq M$, because S_0 has size M only. Thus, for each $t \geq M$, there must be several L $(y_t, \ldots, y_{t-M})$ linear functions, that is, unbalanced functions related to S_0. If the next state function is balanced, then the state vector S_t is a balanced random variable at any instant $t \geq 0$. Then the probability distribution of linear $L(y_t, \ldots, y_{t-M})$ belonging to S_{t-M} is the same for each $t \geq M$, and there exist linear functions that effectively depend on y_t. This means that an autonomous finite automaton can be equivalent to a non-autonomous register with a linear inverse connection of length M at most. In other words, the linear model:

$$y_t = \sum_{i=1}^{M} a_i y_{t-i} + e_t, \quad t \geq M.$$

To search for all unbalanced linear functions from the largest $M+1$ consecutive output bits, it is necessary to determine the correlation coefficient for 2^M Boolean functions from M variables.

2.3 Differential cryptanalysis

Differential cryptanalysis was created by Israeli cryptographers Eli Biham and Adi Shamir in the 1980s and 1990s to research the DES block cipher [6,7]. In 1993, the Chinese cryptographer Kunshen Ding proposed the use of differential cryptanalysis in stream ciphers. This algorithm was applied to the additive natural stream cipher. This cipher is based on a non-linear filtering shift register, and the register is replaced by a simple counter with N arbitrary periods. Such a scheme is presented in Figure 3.

Here $(\sum)_N$ means addition modulo N, "+" is binary addition in the abelian group $(G,+)$, $f(x)$ is a function related to Z_N in the field G. The initial state of the counter is taken as a key.

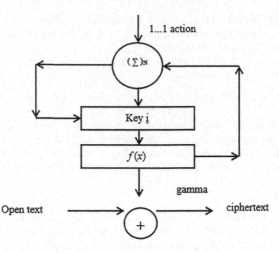

Figure 3. Additive natural stream cipher.

It is assumed that $f(x)$, N and $h^t = h_0\, h_1\, \ldots\, h_{t-1}$ gamma components are known to the cryptanalyst in order to recover the key at the moment of generating h0 in the generator shown in the figure.

Let $C(f)_i = \{x : x \in Z_N,\, f(x) = i\}$ for $i = 0.1$. Then the attack procedure suitable for the differential cryptanalysis method can be described as follows.

Step 1. Thus $(i, j;\, w)$, $(i, j) \in G \times G$ and $w \in Z_N$, parameters are found such that

$$d_f(i,j;w) = \left| C(f)_i \cap \left(C(f)_j - w \right) \right|$$

the equality should be determined from the set as small as possible and the corresponding $D_f(i, j;\, w) = C(f)_i \cap (C(f)_j - w)$.

Step 2. Analyzing the known sequence h^t, samples of gamma parts of length $w+1$ in the form $i * \ldots * j$ are obtained. If such a sample is identified and i is in the k'-th place, then $k \in D_f(i, j;\, w) - k'$.

If $d_f(i, j;\, w)$ is small, then k is searched from this set. Otherwise, another $(i', j';\, w')$ is chosen and its corresponding $D_f(i', j';\, w') - k''$ is searched.

The procedure is continued until the set containing k is sufficiently small. The reason for calling this method differential is that the parameter $d_f(i, j;\, w)$ in step 1 is actually the number of methods, and the parameter w obtained from Z_N can be considered as the difference between $C(f)_j$ and $C(f)_i$, i.e.:

$$w = x_j - x_i, x_j \in C(f)_j, x_i \in C(f)_i$$

The idea of the attack is to find small w or large w such that it is possible to write as small differences of C_i and C_j as possible.

2.4 NESSIE

From January 1, 2000, NESSIE (New European Schemes for Signature, Integrity and Encryption), which is considered a crypto project of European countries for 40 months, started its work [3,4].

Unlike the AES competition project, the NESSIE cryptoproject focuses not only on block ciphers, but also on stream ciphers, digital signature algorithms, hashes, and asymmetric ciphers. Also, no government organization was behind this project.

The first conference aimed at carrying out this project was held in November 2000, and the organizers officially announced that 39 algorithms had been submitted. Six of these algorithms are stream encryption algorithms: BMGL, Levithan, LILI-128, SNOW, SOBER – t16, SOBER – t32.

On September 24, 2001, the second stage of the competition began, and the BMGL, SNOW, SOBER – t16, SOBER – t32 algorithms successfully entered this stage.

In this competition, the tolerance of stream ciphers to the above-mentioned cryptanalysis methods was checked. Table 1 below summarizes the results of the evaluation of the strength of stream ciphers that participated in the first round of the competition.

Table 1. Results of evaluating the strength of stream ciphers participating in the NESSIE competition.

	Correlation attack	Time-memory trade-off	Inversion attack	Hypothesis and detection	Find all key variants
SNOW 128	2^{400}	2^{512}	2^{512}	2^{352}	2^{128}
SNOW 256	2^{400}	2^{512}	2^{512}	$<2^{256}$	2^{256}
SOBER-t32	$\gg 2^{256}$	2^{544}	2^{544}	2^{320}	2^{256}
SOBER-t16	$\gg 2^{128}$	2^{272}	2^{272}	2^{160}	2^{128}
LILI-128	2^{71}	2^{100}	2^{128}	2^{124}	2^{128}

3 RESULTS

The most common type of attack based on the peculiarities of the construction of stream encryption algorithms is the correlation attack. If a nonlinear function omits information about its internal components to the output, the work to uncover such a system is significantly reduced. At the same time, such a function is always available [8]. According to this axiom, correlational attacks use the correlation between the sequence coming out of the encryption scheme and the sequence coming out of the registers [2,4,5]. The cryptanalysis process is based on finding the initial state of the generator registers, i.e. the key, with the output sequence from the generator.

A key parameter in evaluating stream encryption algorithms for correlation attack is the concept of correlation immunity.

The concept of correlation immunity was originally created as a feature of combinatorial Boolean functions that do not allow applying the above-mentioned approach.

3.1 Description

$f(x)$, $x \in GF(2^n)$, this function is said to have a "k"-rank correlation immune ($CI(k)$ – is denoted, where $1 \leq k < n$), if $1 \leq Wt(\alpha) \leq k$ is for all $\alpha \in GF(2^n)$ satisfying the condition $\hat{U}_\alpha(f) = 0$ [9].

Here – $Wt(\alpha)$ – "α" is the "Hamming weight" for the vector (ie $Wt(\alpha)$ means the number of ones in the vector $\alpha = (\alpha_1, \alpha_2, ..., \alpha_n)$). – Walsh-Adamar substitution.

So, if $f(x)$ the correlation immunity level of this function is equal to "k", then the value of the function $Y = f(x)$ is considered to be statistically independent in the optional $x \in GF(2^n)$ argument "k" component. In general, $f(x)$ the degree of correlation immunity of this function $x \in GF(2^n)$ can be $k = n-1$, that is, it does not exceed it.

Comparing the results obtained in the analysis, it can be said that the concepts of nonlinearity and correlation immunity are theoretically contradictory concepts. That is, the degree of correlation immunity of functions with a maximum degree of nonlinearity is minimal, or vice versa, the degree of correlation immunity of functions with a minimum degree of nonlinearity has maximum values.

The correlation cryptanalysis method for combinational stream encryption algorithms is implemented based on the following sequence of steps:

3.2 Step 1

The degree of nonlinearity and the degree of correlation immunity of the combination function used in the stream encryption algorithm are calculated. If the level of non-linearity is equal to zero and the level of correlation immunity is high, this algorithm is considered resistant to the correlation cryptanalysis method and the analysis process is terminated. Otherwise, it goes to the second step;

3.3 Step 2

All possible statoanalog functions of the combinatorial function are searched. If there are analogs of the function equal to x1, x2 or x3, the next step is taken, otherwise this algorithm is considered to be resistant to the correlation cryptanalysis method and the analysis process is completed;

3.3 Step 3

Using the control function, the outputs of the registers are obtained separately with respect to different initial states;

3.4 Step 4

The outputs of the registers corresponding to the output of the generator with the specified frequency are selected, the remaining options are discarded.

3.5 Step 5

By crossing the possible variants of the registers, the variants of the initial state of the registers are determined.

The correlation cryptanalysis method for filtering stream encryption algorithms is implemented based on the sequence of steps below:

3.6 Step 1

The degree of nonlinearity of the combination function used in the stream encryption algorithm is calculated. If the degree of non-linearity is equal to zero, the function used is considered unsustainable. Otherwise, it goes to the second step;

3.7 Step 2

The correlation immunity levels of the combination function used in the stream encryption algorithm are calculated. If the degree of correlation immunity is zero, the function used is considered to be intolerant. Otherwise, it goes to the third step;

3.8 Step 3

All available statanalog functions of the combinator function are searched. The linear one is selected from among the statistical analog functions;

3.9 Step 4

A system of linear equations is created using the statanalog function as a filtering function, the known gammas and the feedback polynomial;

3.10 Step 5

The initial state of the register is determined by solving the system of linear equations.

The analysis showed that the use of the correlation-immune function as a combination function cannot completely eliminate the possibility of conducting a correlation attack on the algorithm, but complicates it.

4 CONCLUSION

Scientific-research work was carried out within the framework of ongoing research in order to create a durable stream encryption algorithm. In this research, cryptanalysis methods applied to stream encryption algorithms were studied. As a result of the reviewed analysis, the type of correlation attack can be cited as the most widely used cryptanalysis method in relation to stream encryption algorithms. The correlation-immunity property of the combining function used in the combining generators provides tolerance to the correlation cryptanalysis method. The use of a non-linear, correlation-immunity filter function in filter generators increases the tolerance compared to the correlation cryptanalysis method. In some algorithms, tolerance to the correlation cryptanalysis method is provided by using a linear function.

The results of the analysis show that it is important to consider the following when creating robust stream encryption algorithms:

It is desirable to use pseudo-random sequence generators based on a systematic theoretical approach in stream encryption algorithms.

The tolerance of stream ciphers can be increased by increasing the length of the registers, using the tolerance feedback function.

The gamma acquisition points (n) of the register whose length is equal to r should be selected by the following condition: $n \leq \sqrt{2}r$.

REFERENCES

[1] Schneier B. Applied Cryptography. *Protocols, Algorithms, Source Texts in C Language.* – M., Ed. TRIUMPH, 2003. – 816 p.

[2] Kharin Yu.S., Bernik V.I., Matveev G.V., Agievich S.V. Mathematical and Computer Bases of Cryptology: Textbook. – Minsk, LLC *"New Knowledge"*, 2003. – 382 p.

[3] Asoskov A.V., Ivanov M.A. Stream Ciphers, M: Kudits-Obraz, 2003. – 336 p.

[4] Akbarov D.Ye. *Cryptographic Methods of Information Security and Their Application.* – Tashkent, *"Mark of Uzbekistan"* Publishing House, 2009. – 432 p.

[5] http://www.cryptography.ru

[6] Musayev A.I. *Research the Basics of Existing Stream Encryption Algorithms and Create New Cryptoresistant Algorithms.* Master's Dissertation in the Field of Information Security. Tashkent, 2008.- 81 p

[7] Suwais K., Samsudin A. *New Classification of Existing Stream Ciphers.* Universiti Sains Malaysia(USM), Malaysia 2010.

[8] Hassan Shirali-Shahreza M., Mohammad Shirali-Shahreza. (2006) A New Approach to Persian/Arabic Text Steganography. In: *5th IEEE/ACIS International Conference on Computer and Information Science and 1st IEEE/ACIS International Workshop on Component-based Software Engineering, Software Architecture and Reuse,* pp 310–315.

[9] Bala Krishnan R., Prasanth Kumar Thandra, Sai Baba M. (2017). An Overview of Text Steganography. *4th International Conference on Signal Processing, Communications and Networking (ICSCN -2017),* March 16 – 18, 2017, Chennai, INDIA.

[10] Zaynalov N. R., Kh. Narzullaev U., Muhamadiev A. N., Mavlonov O. N., Kiyamov J., Qilichev D. Hiding Short Message Text in the Uzbek Language. *2020 International Conference on Information Science and Communications Technologies (ICISCT).*

[11] Por LY, Delina B (2008) Information Hiding—a New Approach in Text Steganography. In:*7th WSEAS International Conference on Applied Computer and Applied Computational Science.* Hangzhou China,689

[12] Zaynalov N. R., Mavlonov O. N., Muhamadiev A. N., Kilichev D., Rakhmatullayev I. R. UNICODE For Hiding Information In A Text Document. *2020 IEEE 14th International Conference on Application of Information and Communication Technologies (AICT) | 978-1-7281-7386-3/20 IEEE*

Artificial Intelligence, Blockchain, Computing and Security – Dagur et al. (Eds)
© 2024 The Author(s), ISBN: 978-1-032-49393-0

Early recognition of Alzheimer's disease using machine learning

Prajwal Nagaraj, Anjan K. Koundinya & G. Thippeswamy
Computer Science and Engineering, B.M.S Institute of Technology and Management, Bangalore, India

ABSTRACT: Alzheimer's Disease (AD), a neurological disorder, is the most common cause of dementia in those over the age of AD must be precisely and soon identified in order to halt the progression of this lethal condition. This article focuses on a machine learning method for MRI images to identify The condition Alzheimer's. The brain's hippocampus region is the fundamental objective of the recommended tactic. Use a Gray Level Concidence Matrix to extract the hippocampus region's textural characteristics, including entropy, uniformity, energy, contrast, correlation, and variance (GLCM). Using moment invariants, area and form elements are retrieved. A classifier is utilised to identify distinct phases of AD using error propagation (EBP) of an artificial neural network. An average accuracy of 86.8% is offered by the proposed technique.

Keywords: Alzheimer's disease, Neural Network, Machine learning, disease analytics

1 INTRODUCTION

A degenerative brain condition called Alzheimer's disease (AD) causes memory loss. The brain's beta amyloid plaques, which are the most common kind of dementia, cause this condition. Plaque and tangles are among the key symptoms of disease. Because plaques and tangles affect healthy neurons' function, limiting their capacity to communicate, and eventually killing them, the total volume of the brain diminishes. New memory formation is hampered by neuron loss, particularly in the hippocampus. First impacted in the brain is the hippocampus. Memory formation occurs in this part of the brain, which further serves as a link between the body and the mind. Alzheimer is a progressive and irreversible disorder that manifests gradually in accordance with patterns of cerebral damage and can last for decades. Mild, moderate, and severe phases of the illness gradually develop, each with distinct symptoms and issues. People can function independently, although memory loss may occur during the moderate period, which lasts an aggregate of 2 to 4 years. When a person is at the moderate stage, they can need help with daily tasks and could have more trouble remembering things. This stage of the illness can continue anywhere between two and ten years, making it likely the longest stage overall. One to three years may pass during the disease's last stage. Currently utilised non-automated techniques for diagnosing AD include the CDR, MMSE, and Cognitive Impairment Testing. Other techniques include MRI scans, SPECT and PET.

2 RELATED WORK

For the health and social care systems, the ageing population with AD and other types of dementia is growing quickly. Patients who are identified with asthma at an early stage can receive appropriate care and take advantage of new medicines as they become accessible.

DOI: 10.1201/9781003393580-61

Many years prior to the appearance of clinical signs, AD starts to develop. Early detection of AD will benefit from the development of biomarkers that can track changes in the brain throughout this time. The EEG may be crucial in the primordial recognition of AD. Alzheimer's disease-related brain degeneration alters EEG and brain activity in ways that may be measured using biomarkers. The work discussed in this article aims to develop reliable EEG-based biomarkers for such primordial detection of AD. We describe a unique method based on changes in EEG amplitude for quantifying EEG slowing, one of dementia's most persistent characteristics. The new method offers 100% and 88.88% sensitivity and specificity values, respectively. When separating AD from healthy people, it performed better than the Lempel-Ziv Complexity (LZC) method.

3 MOTIVATION

A computational AD diagnosis tool based on the 11C-PiB PET imaging technique. The discovery of beta-amyloid (A) was made possible by the development of Pib PET. Radiologists are still having difficulty with PiB PET scans. This work suggests a CAD (Computer-Aided Diagnostic) tool to address the aforementioned issues by integrating three machine learning cores [1,2]. The suggested CAD tool can give good accuracy for AD diagnosis, according on experimental results on a collection of 120 PiB PET scans.

3.1 *Problem statement*

Using association rules, analyse the FDG and PIB biomarkers of AD. In this study, a CAD system for the early identification of AD using 18F-FDG and PiB PET biomarker analysis is designed using an Association Rule (AR)-based methodology. The AD Neuro Imaging Initiative 1 dataset is utilised for examining, and two distinct comparisons are made between controls and AD or controls and participants with MCI. Without repeating any steps, 3D Regions of Interest (ROIs) are produced for both biomarkers whether they are taken separately or together using an activation estimation approach. To get ARs from controls, an Apriori method is utilised using these ROIs as input. The percentage of rules tested for each subject determines the classification. Combinations of biomarkers have favourable outcomes in both groups, notwithstanding PiB's distinction from FDG in delivering differentiated information. The FDG score improved in the initial set to 94.74% accuracy, whereas the FDG and PIB values increased in the second iteration to 92.86% accuracy. These findings outperform previously published methodologies.

3.2 *Aim objective*

Alzheimer's disease computer diagnosis employing PET image region-based selection and categorization. Doctors may identify neurodegenerative disorders such as Alzheimer's disease using PET, a 3D neuroimaging tool (AD). It is vital to employ computerized recognition and diagnosis utilizing medical imaging technologies for quantification. The ability of different brain areas to distinguish Alzheimer's disease from photographs of healthy brains is ranked using a novel approach that is given. First, 116 anatomical areas of interest are assigned to pictures of the brain. This region's first four moments' entropy and the histogram are computed. The capacity of areas to separate brain PET images is subsequently assessed using receiver performance curves. 142 PET brain pictures are evaluated while the support vector machine and random forest classifiers are given the 21 chosen areas. The categorization outcomes outperform employing all 116 of the original areas or full brain voxels. Also much shortened is the calculating time. Alzheimer's disease Designs entropy as an indicator for medical evaluation is among the most serious societal and public health challenges of our day, given the widespread incidence of ad and its varied forms. Early diagnosis is critical for individuals and families to get medical and social resources and prepare for the future. EEG is a valuable first-line significant prediction for the early identification and diagnosis of dementia, as well as for learning about how the brain works.

It offers a high temporal resolution, is non-invasive, and is affordable. Dementia-related damage alters the properties of brain waves and inhibits the activity of brain cells. As a way to quantify EEG alterations as a dementia biomarker, information theoretic approaches have come to light. The Tsallis entropy is among the most promising adaptive threshold methodologies for estimating EEG changes. This article expands on the strategy. When compared to previous procedures, this has delivered better results. Alzheimer's disease (AD) must be appropriately and immediately identified if the illness is to be stopped in its tracks, according to a blended sequential feature selection strategy. Recently, a number of imaging methods, including PET and MRI, have been employed extensively to diagnose AD. Despite their value, the resulting picture data are high-dimensional and noisy, which makes correct diagnosis challenging. In this study, we provide a unique feature selection method to identify the useful features in MRI data, which increases diagnostic precision while lowering computing expense. Benchmarking findings on the datasets for the AD challenge demonstrate that our suggested methodology performs better than other well-liked featureselection techniques. Additionally, our method outperforms the top competition winner in identifying AD and is equivalent to others in identifying moderate cognitive impairment (MCI).

3.3 Existing system

About one three years may pass during the last stages of the illness. Patients at this stage may require round-the-clock support as their memory and cognitive abilities deteriorate and they become less capable of adapting to their circumstances. Before irreparable neurological damage is done, it is crucial to detect the illness as early as possible. Currently, researchers use imaging methods like [1,2]. Plaques and tangles are two of the disease's most noticeable characteristics. As a consequence of plaques and tangles, healthy neurons were less efficient over time, lose their capacity to communicate, and eventually die. As a result of the loss of healthy neurons, the tissues of the brain normally atrophy. Neuronal degeneration, especially in the hippocampus, limits the creation of new memories. The hippocampus in the brain suffers damage first.

3.4 Proposed system

In the anticipated investigation, MRI scans are used to detect AD. The hippocampus is selected as the ROI, and its texture, area, and shape properties are retrieved using the [2]. Moment Invariants specify a set of characteristics used for form identification, and GLCM extracts the statistical second-order texture features. Then, based on the attributes gathered from the ROI, AD is divided into different stages. build an artificial neural network using the EBP method ANN. Using GLCM and ROI, the texture, area, and form properties of the hippocampus are retrieved. Using back propagation (EBP)-trained artificial neural networks, the stages of AD are then divided according to the properties extracted from the ROI (ANNs). For the purpose of identifying AD, the HFS feature selection route is recommended. This methodology combines a filter-and-wrap strategy with a filter-and-wrap approach to find meaningful features in MRI data acquired from the ADNI database [4].

3.5 Requirement specification

At this step, we assess the project's feasibility and provide an estimate, a general project plan, and some cost forecasts. During system analysis, the proposed system should be evaluated for viability. This is done to ensure that the system does not become a financial burden for the company. A feasibility study requires an understanding of the underlying system requirements. This research is being conducted to determine the financial impact of the system on a corporation. A corporation may only invest a limited amount of money on system research and development. Costs need to be supported. The system that was subsequently created was likewise within budget, and the majority of the technology was openly accessible. Only individual goods were required of me. The purpose of this research is to confirm that the system satisfies the technical criteria. Any designed system shouldn't put an excessive amount of strain on the existing technological resources. This places a heavy burden on the available technological resources. Higher standards for clients will result from this. The system being developed should have reasonable requirements because installing it calls for little to no adjustments. The purpose of the study component is to gauge how well-liked the system is by users.

3.6 Detailed design

The connection between the user and the information network is provided by the input design. To convert transaction history into a processable format, data preparation requirements and techniques must be developed. This can happen when people enter data straight into the system instead of glancing at the desktop to acquire it from a narrative text document. The input design focuses on limiting the amount of input necessary, managing errors, eliminating delays, minimising additional phases, and keeping the process simple. The input is designed to provide security, usability, and privacy protection. Input design is the process of converting a user-centered representation of the intake into a computer-based solution. This design is critical for avoiding errors during the data entry process and giving management with the necessary direction for getting reliable information from the computerised system. This is accomplished by developing user-friendly displays that are straightforward to use while entering massive amounts of data. The purpose of input design is to reduce mistakes and make data entering easier. You can do any data operations on the data input screen. It also gives a way to access the records. The information you submit will be verified. On the screen, data may be entered. Send the proper notifications as needed to keep people out of potentially life-threatening situations. As a result, the objective is to create an input layout that is simple to grasp.

3.7 Implementation

A high-quality output fits the needs of the end user and clearly presents the information. The outputs of any system are how processing results are transmitted to users and other systems. During output design, it is decided how information will be relocated for both instant demand and hard copy output. It is the user's most essential and direct source of information. Effective and smart output design improves a system's capacity to engage with tools that aid in user decision-making. The process of developing computer output should be organised and well-thought out; the rightoutput must be developed while ensuring that each output component is built in such a manner that the system can be used by humans quickly and effectively. While examining the computer's output, you must determine the specific output required to meet your requirements. Choose a display style for your data.

ACKNOWLEDGMENT

Dr. Anjan K Koundinya, Professor of Cyber Security, BMSITM will like to thank Late Dr. VK Ananthashyana, Erstwhile Professor and head, Dept. of CSE, MSRIT, Bengaluru for igniting the passion for research.

REFERENCES

[1] Ali H. Al-nuaimi *et al.* "Changes in the EEG Amplitude as a Biomarker for Early Detection of Alzheimer's Disease", 2016 *30th Annual International Conference of the IEEE Engineering in Medicine and Biology Society (EMBC).*

[2] Jiehui Jiang *et al.* "A Computed Aided Diagnosis Tool for Alzheimer's Disease Based on 11C-PiB PET Imaging Technique", *IEEE International Conference on Information and Automation Lijiang, China,* August 2015.

[3] Chaves R. *et al.* "FDG and PIB Biomarker PET Analysis for the Alzheimer's Disease Detection Using Association Rules", *2012 IEEE Nuclear Science Symposium and Medical Imaging Conference Record (NSS/MIC).*

[4] Imene Garali *et al.* "Region-Based Brain Selection and Classification on PET Images for Alzheimer's Disease Computer Aided Diagnosis", *2015 IEEE International Conference on Image Processing (ICIP).*

[5] Ali H. Al-nuaimi *et al.* "Tsallis Entropy as a Biomarker for Detection of Alzheimer's Disease", *2015 37th Annual International Conference of the IEEE Engineering in Medicine and Biology Society (EMBC).*

[6] Yang Han, Xing-Ming Zhao "A Hybrid Sequential Feature Selection Approach for the Diagnosis of Alzheimer's Disease", *2016 International Joint Conference on Neural Networks (IJCNN).*

[7] Rangini M., Dr.G, Wiselin JIJI "Detection of Alzheimer's Disease Through Automated Hippocampal Segmentation", *2013 International Multi-Conference on Automation, Computing, Communication, Control and Compressed Sensing (iMAC4s).*

[8] Devvi Sarwinda, Alhadi Bustamam "Detection of Alzheimer's Disease Using Advanced Local Binary Pattern from Hippocampus and Whole Brain of MR Images", *2016 International Joint Conference on Neural Networks (IJCNN).*

[9] Saraswathi S. *et al.* "Detection of onset of Alzheimer's Disease from MRI images Using a GA-ELM-PSO Classifier", 2013 *Fourth International Workshop on Computational Intelligence in Medical Imaging (CIMI).*

Artificial Intelligence, Blockchain, Computing and Security – Dagur et al. (Eds)
© 2024 The Author(s), ISBN: 978-1-032-49393-0

Detecting malign in leaves using deep learning algorithm model ResNet for smart framing

Anmol Kushwaha* & E. Rajesh*
School of Computer Science and Engineering, Galgotias University, Greater Noida, India

ABSTRACT: Agriculture Sector is a major contributor in the Indian Economy farmers are lacking in the area of Technology, Processing and Marketing of their goods. This Paper represents a Machine Learning Model for increase crop production by giving awareness to farmers about different disease that can harm their plants. Every crop is prone to many diseases throughout their lifespan. The disease can affect the crop at any stage of their growing phase. Early detection of disease is the only solution to reduce the damage. Early detection may reduce the damage caused and increase the quality as well as quantity of Production. Early detection of these diseases can reduce the damage and increase the production value. We used ResNet of deep learning model to predict the early symptom of disease on different plants like tomato, grape, orange, soybean, squash, potato, corn, strawberry, peach, apple, blueberry, cherry, pepper, raspberry.

Keywords: Disease in plants, plant leaves, ResNet algorithm, Deep Learning

1 INTRODUCTION

Farming and Agriculture play a major role in our country. In India 70% of population depends on the agriculture and farmer. Agriculture is the major source to food Indian population but with many uncertainties in farming as farmer cannot know about rainfall, weather conditions, fertilizer. Crop and Market price will depend on these parameters. At that time also a Indian farmer adopt their traditional technique of farming where they can't analyze the uncertainty parameter which is a very difficult task for him. The agriculture industry are contribute to the economy of country and hence by the industry the economic growth of the country will increases. Its required to cultivate modern technique of farming and used some tools which helped in farming in much better ways.

In plants diseases affects the leaves first then spreads to other parts and environmental changes are the one of the major causes of the diseases in plants. Finding the correct disease at early stage of plant growth is an important factor. Farmers rely on visual inspection to identify disease-affected parts and find a solution. However, determining the disease and the appropriate pesticide to use remains a challenge for many cultivators. A model that accurately identifies the major disease affecting crops is needed. Major disease that cause the loss of the cops are as follows:

1.1 *Blast*

Blast is a fungal disease that harms crops at any growth stage, causing brownish-black spots on leaves and nodes. The damage can be severe, leading to the breaking of affected parts.

1.2 *Blight*

Blight is a fungal disease that impacts plants at all growth stages, causing color changes in the sheath portion near the water-soaked level and white, brown, or brownish-black spots on leaves.

---

*Corresponding Authors: anmol.avi.kushwaha@gmail.com and rajesh.e@galgotiasuniversity.edu.in

DOI: 10.1201/9781003393580-62

1.3   *Rot*

Rot is a fungal disease that causes significant damage to crop production as it affects plants during the vegetative stage. Symptoms include discoloration in the leaf sheath and dark brown spots or lesions.

1.4   *Spot*

Sot is a fungal disease that impacts leaves at all stages of plant growth, causing dark brown spots that can spread to the entire plant in oval or round shapes, leading to drying of the leaves. If severe, a reduction of up to 50% in crop growth can be observed.

2   RELATED WORKS

Vignesh *et al.*, In their proposed work they have used different CNN algorithms and compare them to find the model with the best accuracy on the images of the Paddy. And predicted the various diseases on it during the process of growing it. As they have less number of images they use data segmented techniques to increase the size of the data and then use the various algorithms [2].

Shrimali *et al.*, In their project they have used different images of the plants and used VGG and MobiNetV2 algorithm to develop an application in which by entering the new image it can predict the disease in the plant leaf. They preform different algorithm to enhance the image quality to increase the accuracy, they were able to reach the accuracy of 93% from VGG and 95% from MobileNetV2 [3].

Gavhale *et al.*, In their proposed work they have used the various machine learning algorithms to predict the unhealth Ness of the leaves against the virus that affects them and give the highest accuracy and by increasing the data by Appling different techniques to enhance the image quality. By defining the different parameters like threshold, region or edges, then they use K-nearest, SVM to train and test the model [4].

Rastogi *et al.*, In their proposed work they use artificial neuro network on the leaves and run through different types of image processing like converting the leaf to grey cluster and then in cluster to study more about the leaves and segments the leaf disease it. Their system will help in study the leaf of the plants and their architecture will help in different algorithm to achieve high accuracy [5].

Khirade *et al.*, In their proposed work they use the different algorithms on the images with classifying them into different class as Powderly mildew, yellow rust and aphids and were able to achieve a accuracy of 86.5%,85.2% and 93.5% respectively, And used ANN alone with BPNN and compare the result [6].

Zhang *et al.*, In their proposed work they use the plants data set which has yellow rust and use the image processing to automate the model using deep learning to predict the outcome. In their research work they have used DCNN algorithm along with the UAV hyperspectral image to automate the process. Their project help in increase the profit of the farmers by decreasing the loss of the crops from unhealthy crops [7].

Arsenovic *et al.*, In their proposed work they have total of 42 different plants with disease to predict the disease in plants. In their proposed model they have use AlexNet, VGG, Inception version, DenseNet and ResNet to predict the outcome and were able to achieve the accuracy of 99.13%, 99.31%, 99.74 and 99.75% respectively they use data augmentation to increase the accuracy of the model [8].

Memon *et al.*, In their proposed work they use cotton leaves to make the production the diseases in it. They have use the ResNet, VGG, and Inception to predict the disease and by predicting they are increasing the productive of the plant growth and make profit from it [9].

Zhao *et al.*, In their proposed work they are helping the farmers to increase the growth of the plants and give them the correct knowledge to use the best fertilizer for their tomato and increase the profit. They use the ResNet algorithm in they model and achieved the accuracy of 94.5% [10].

3   MATERIALS AND METHDOS

Our main objective is to develop a model which can predict the disease to which we need the date set, we are using the data set present on the git-hub [1]. After downloading the data, we can start making the model. We are using Kaggle platform for implementing our project as it provides user friendly platform and it include all the required libraries and even more functionality. In ur dataset we have images of 14 plants with 26 different diseases and with a total of 70295 images. The Figure 1 shows the number of images per plants which includes health and unhealthy images of the plants.

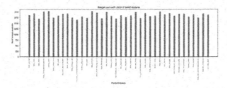

Figure 1.   Images available for each plants.

The Figure 2 shows the some of the images present the our data base it includes tomato, grape, orange, soybean, squash, potato, corn, strawberry, peach, apple, blueberry, cherry, pepper, raspberry plants.

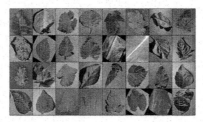

Figure 2.   Images of different leaves.

We are going to use Residual network because it allows us to decrease the error cause by the vanishing gradient. In this we use skip condition connection to activate the layers and skipping some of them. By doing so we can see that network try to fit the residual mapping instead of layers learning mapping which allows reduce the errors in the prediction.

4   PROPOSED MODEL

In our proposed model first, we are going to include all the required libraries as shown in the Figure 3.

```
import os
import numpy as np
import pandas as pd
import torch
import matplotlib.pyplot as plt
import torch.nn as nn
from torch.utils.data import DataLoader
from PIL import Image
import torch.nn.functional as F
import torchvision.transforms as transforms
from torchvision.utils import make_grid
from torchvision.datasets import ImageFolder
from torchsummary import summary

%matplotlib inline
```

Figure 3.   Libraries included in the model.

After importing the data we will clean the data and divide them into various class and classify based upon the different leaves images by names and by healthy or unhealth.

Now, we are going to check if our system has GPU or not because CPU is verry slower than the GPU we are going to check it and if it is available as shown in the Figure 4 and the Figure 5 shows the coding how we load the data into the GPU is available.

```
def get_default_device():
    if torch.cuda.is_available:
        return torch.device("cuda")
    else:
        return torch.device("cpu")

def to_device(data, device):

    if isinstance(data, (list,tuple)):
        return [to_device(x, device) for x in data]
    return data.to(device, non_blocking=True)

class DeviceDataLoader():

    def __init__(self, dl, device):
        self.dl = dl
        self.device = device

    def __iter__(self):

        for b in self.dl:
            yield to_device(b, self.device)

    def __len__(self):

        return len(self.dl)
```

Figure 4.   Class to check GPU.

```
train_dl = DeviceDataLoader(train_dl, device)
valid_dl = DeviceDataLoader(valid_dl, device)
```

Figure 5.   To load data into the GPU.

After loading the data into the GPU we define our architecture of the model by defining conv1, conv2, res1, conv3, conv4 and res2 and define output shape given to each layeras shown in Figure 6.

```
        Layer (type)               Output Shape         Param #
================================================================
            Conv2d-1          [-1, 64, 256, 256]           1,792
       BatchNorm2d-2          [-1, 64, 256, 256]             128
              ReLU-3          [-1, 64, 256, 256]               0
            Conv2d-4         [-1, 128, 256, 256]          73,856
       BatchNorm2d-5         [-1, 128, 256, 256]             256
         MaxPool2d-6         [-1, 128, 256, 256]               0
            Conv2d-7          [-1, 128, 64, 64]          147,584
       BatchNorm2d-9          [-1, 128, 64, 64]             256
             ReLU-10          [-1, 128, 64, 64]               0
           Conv2d-11          [-1, 128, 64, 64]          147,584
      BatchNorm2d-12          [-1, 128, 64, 64]             256
             ReLU-13          [-1, 128, 64, 64]               0
           Conv2d-14          [-1, 256, 64, 64]          295,168
      BatchNorm2d-15          [-1, 256, 64, 64]             912
             ReLU-16          [-1, 256, 64, 64]               0
        MaxPool2d-17          [-1, 256, 16, 16]               0
           Conv2d-18          [-1, 512, 16, 16]        1,180,160
      BatchNorm2d-19          [-1, 512, 16, 16]           1,024
             ReLU-20          [-1, 512, 16, 16]               0
        MaxPool2d-21            [-1, 512, 4, 4]               0
           Conv2d-22            [-1, 512, 4, 4]        2,359,800
      BatchNorm2d-23            [-1, 512, 4, 4]           1,024
             ReLU-24            [-1, 512, 4, 4]               0
           Conv2d-25            [-1, 512, 4, 4]        2,359,800
      BatchNorm2d-26            [-1, 512, 4, 4]           1,024
             ReLU-27            [-1, 512, 4, 4]               0
        MaxPool2d-28            [-1, 512, 1, 1]               0
          Flatten-29                 [-1, 512]               0
           Linear-30                  [-1, 38]          19,494
================================================================
Total params: 6,589,734
Trainable params: 6,589,734
Non-trainable params: 0
----------------------------------------------------------------
```

Figure 6.   Summary of the model.

The Figure 6 shows the summary of our models architecture with different layer and what their output shape will and how many parameters will be used for prediction and how many para-meterise lost during the process of defining the architecture can be seen also which is zero, Which means we can use our model to make prediction.

After defining the model, we can use it to make prediction and calculate the accuracy of the model against the different parameters like by changing epochs, max_lr, grad_clip, weight_decay and comparing them to select the best parameter for prediction.

5   RESULTS AND DISCUSSION

After finding the right parameters we were able to achieve the accuracy of 99.23 precent as shown in the Figure 7 alone with the time taken to execute the query.

```
Epoch [0], last_lr: 0.00812, train_loss: 0.7519, val_loss: 0.4635, val_acc: 0.8585
Epoch [1], last_lr: 0.00000, train_loss: 0.1275, val_loss: 0.0278, val_acc: 0.9923
CPU times: user 11min 33s, sys: 7min 27s, total: 19min 1s
Wall time: 19min 52s
```

Figure 7.   Accuracy and time.

410

The Figure 8 shows the accuracy of the model against epochs and it shows after a certain point it barely increase. Figure 9 shows the loss of epochs in training and validation of the data, which decrease by the increase of the epochs.

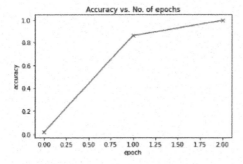

Figure 8.   Graph of accuracy vs number of epochs.

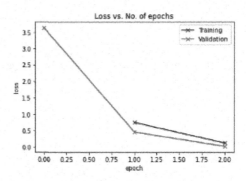

Figure 9.   Graph of loss vs number of epochs.

6   CONCLUSION

We developed a model to predict the disease in the plants the farmer can use to correctly identify the fertilizer which will be best for the plants based upon the prediction provided by our model and by this they can increase the profit and quality of the plants and get the healthy food to eat.by detecting it at an early stage really help the farmers.

REFERENCES

[1]   https://github.com/spMohanty/PlantVillage-Dataset
[2]   Vignesh U. and Elakya R. (2022), Identification of Unhealthy Leaves in Paddy by using Computer Vision Based Deep Learning Model. *IJEER* 10(4), 796–800. DOI: 10.37391/IJEER.100405.
[3]   Shrimali S., "PlantifyAI: A Novel Convolutional Neural Network Based Mobile Application for Efficient Crop Disease Detection and Treatment," *2021 2nd Asia Conference on Computers and Communications (ACCC)*, 2021, pp. 6–9, doi: 10.1109/ACCC54619.2021.00008.
[4]   Gavhale, Kiran R., and Ujwalla Gawande. "An Overview of the Research on Plant Leaves Disease Detection Using Image Processing Techniques." *IOSR Journal of Computer Engineering (IOSR-JCE)* 16.1 (2014): 10–16.
[5]   Rastogi A., Arora R. and Sharma S., "Leaf Disease Detection and Grading Using Computer Vision Technology & Fuzzy Logic," *2015 2nd International Conference on Signal Processing and Integrated Networks (SPIN)*, 2015, pp. 500–505, doi: 10.1109/SPIN.2015.7095350.
[6]   Khirade S. D. and Patil A. B., "Plant Disease Detection Using Image Processing," *2015 International Conference on Computing Communication Control and Automation*, 2015, pp. 768–771, doi: 10.1109/ICCUBEA.2015.153.
[7]   Zhang, X.; Han, L.; Dong, Y.; Shi, Y.; Huang, W.; Han, L.; González-Moreno, P.; Ma, H.; Ye, H.; Sobeih, T. A Deep Learning-Based Approach for Automated Yellow Rust Disease Detection from High-Resolution Hyperspectral UAV Images. *Remote Sens.* 2019, 11, 1554. https://doi.org/10.3390/rs11131554
[8]   Arsenovic, M.; Karanovic, M.; Sladojevic, S.; Anderla, A.; Stefanovic, D. Solving Current Limitations of Deep Learning Based Approaches for Plant Disease Detection. *Symmetry* 2019, 11, 939. https://doi.org/10.3390/sym11070939
[9]   Memon, M.S.; Kumar, P.; Iqbal, R. Meta Deep Learn Leaf Disease Identification Model for Cotton Crop. *Computers* 2022, 11, 102. https://doi.org/10.3390/computers11070102
[10]   Zhao, S.; Peng, Y.; Liu, J.; Wu, S. Tomato Leaf Disease Diagnosis Based on Improved Convolution Neural Network by Attention Module. *Agriculture* 2021, 11, 651. https://doi.org/10.3390/agriculture11070651

© 2024 The Author(s), ISBN: 978-1-032-49393-0

COVID-19 prediction using deep learning VGG16 model from X-ray images

Narenthira Kumar Appavu & C. Nelson Kennedy Babu
Saveetha School of Engineering, Saveetha Institute of Medical and Technical, Sciences, Chennai, Tamilnadu, India

ABSTRACT: Using human lung X-rays, a new technique is currently being used to diagnose Covid-19. So many deep learning concepts piece an imperative role in detecting covid-19. In this proposed thesis, Presented VGG16 deep learning model. Covid-19 in a precise and timely way. The proposed model used a two-way classification system to differentiate the lung X-rays according to the given input. Finally, it detects affected and normal lung X-rays. The effectiveness of the proposed system is gaged by evaluation criteria as in accuracy, precision, recall and function1 score. More than 2000 samples were used to diagnose Covid-19. The VGG16 model gives the best results of 99.58 Percentage. It is superior compared to all existing approaches in the literature. Medical professionals and healthcare workers can use the proposed system to accurately identify Covid-19 using X-rays of human lungs.

1 INTRODUCTION

Today's Acute respiratory syndrome and other viral infections pose a serious threat to our health, the health of our loved ones, and the general population (Drosten *et al.* 2003; Ventura *et al.* 2016; Wolfe *et al.* 2007). Despite extensive research and efforts, it is unclear how and when these new diseases will emerge, and the world was initially affected by a respiratory disease outbreak in Wuhan, Hubei Province, China, where they originated. From December 12, 2019, to January 25, 2020, at least 1,975 patients were admitted to the hospital. Since December 12, 2019, at least 1975 patients have been admitted to the hospital. The outbreak is said to be linked to the Wuhan market. For information on a patient who was taken to Wuhan's Central Hospital on December 26 with severe respiratory symptoms including fever, light-headedness, and cough, see the following: (Wu *et al.* 2020). Because these images often overlap with other lung disorders, highly trained radiologists are needed to identify this disease from chest X-rays. (Bhandary *et al.* 2020) These estimates are actually accurate. Viruses can be detected and measured even in very low levels in patient samples. However, this makes PCR testing more challenging. Costs are high and diseases take a long time to diagnose Chest X-ray images can be examined to see if the new corona virus is present or showing symptoms. Viruses that fall into this category are notable on radiographic images. Several studies have shown (Makris *et al.* 2020; Mangal *et al.* 2020; Rahimzadeh & Attar 2020; Rajpurkar *et al.* 2017, 2020; Wang *et al.* 2017; Yan *et al.* 2020).

2 RELATED WORKS

It can be reliably characterized using chest radiography images, is much faster and cheaper than PCR, and can quickly identify the coronavirus. Compared to other radiographic procedures like kit scan, chest x-rays are less expensive and easily available in all clinics

DOI: 10.1201/9781003393580-63

(Das *et al.* 2021). (Narin *et al* 2021). Morbusx developed the framework used by Deep CNN to detect and classify covid using photos of inmates. An important role in the diagnosis of Covid-19 is played by X-ray scans. If there are any positive findings in the AI predictive test, patients can be referred for clinical trials, to some extent for Covid testing. For the most recent studies on using AI and DL to identify covid-19, our method used chest X-ray datasets of covid-19 and normal patients for transfer learning. The VGG 16 model uses numerical methods to diagnose covid, and experimental results are described. (Haritha *et al.* 2020) Each layer of DL-based CNN models typically consists of a hierarchical structure that extracts features related to Covid-19 that can be used to distinguish between Covid-19 and normal images. CNN's automatic feature learning capabilities served as inspiration for Covid 19 classification using deep neural networks. Comparing Univariate and Multivariate Data Analysis DL model has the potential to provide high classification performance as it can extract robust and accurate country from large heterogeneous data sets. (Hilmizen *et al.* 2020; Mukherjee *et al.* 2021). Fortunately, Tl can transfer our knowledge from an existing domain to a new one. can be replaced by, speed up model-based training, and reduce computing costs (Chen *et al.* 2020). Using transfer learning means using an already trained resource. Using samples from tasks enables rapid and accurate prototyping of new machine learning models, while training millions of photos on a GPU takes too much time and money (Ribani & Marengoni 2019). They discovered early in the transfer learning process that this type of learning did not begin; There are two components to this exchange. The data thus collected is the raw material needed to successfully train a model. Finally, the model has the ability to recognize and classify different photographs. The results of the experiment are used to determine whether or not the model is suitable for transfer learning. (Kumar *et al.* 2022).

3 PROPOSED METHOD

A new technique has been proposed that can immediately identify early-onset pneumonia in Covid-19 cases. A deep neural network is used in this novel approach to segment chest x-ray images into 2-4 distinct classes under different functional parties based on active viral bacterium and covid-19 groups. This program requires a lot of chest x-rays used to diagnose covid. Other quality x-rays are taught to identify covid and our model is covid 19 certified. About 1200 images in a small data set of 2400 images are Covid X-rays and the remaining are standard X-rays to refine our model. The model is shown in Figure 1 below. Our model works well and predicts covid-19 without any ambiguity. Tests have been conducted by us. All these data are segmented in our model with a ratio of 8:2 and the photos of the entire dataset are scaled to the default image size (224 ×224).Vgg16 Feature Extraction The scaled

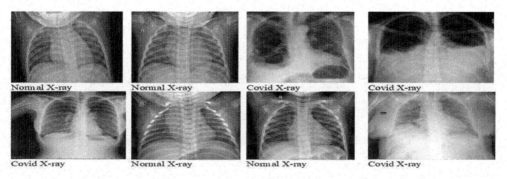

Figure 1. Sample covid-19 and normal images.

layers for feature extraction are from the contribution layer to the final maximum pooling layer. Figure 1. Here, we used sample chest X-ray images of patients with and without Covid 19 and divided them into two groups. AI Deep VGG 16 has reached the peak of model maturity with GPU for high-level image sequencing operations. The VGG-16 model is described in Table 1 below for two reasons. First, the descriptor numbers should be increased to improve the performance of the model when cataloging photos. Our Photo Scaling selections set the fill mode to "Near" and limit the rotation to 15 degrees. To organize all layers of the model chronologically, a sequential model must first be created. A pre-processing image generator can input data from the model with labels using the model prepared in this way. This class performs various useful tasks without changing the stored data, including the following. For example, zoom in, zoom out, rotate, flip, etc. It is a simple method of feeding data to a neural network. The array model is now set. Additionally, as mentioned here, Covid-19 has excellent feature extraction capabilities for classifying CXR and normal images. 2020 (H. Panwar). Below, the various phases of the strategy suggested here are explained in more depth. Identify positive and negative (+) X-ray images with and without Covid-19 in the set of X-ray images. Here, divide the labeled dataset into two groups, the first type serves as the training and testing dataset [80:20]. The size of the input image should be set to 64, and the first layer of the VGG16 transfer learning model should have only two classification types. Train the VGG16 transfer learning AI model using coded lung x-ray images and then save the best trained model by setting epoch to 100 and block size to 50. Figure 2 shows the Covid19 X-ray image prediction method using the VGG-16 transfer learning model. The default input image size for the VGG-16 model is 224224. The recommended job is to reduce the image to 64 by 64 pixels. The input layer of the VGG16 model is then used with the modified image. Here, the model successfully performs binary classification, and as a result, the output is bounded. Our input and output parameters in the recommended practice are organized according to specifications.

Table 1. Requirements for the experiment's variables and data.

Material	Value
Data set type, size	Chest X-ray images, 2400 (COVID-19: 1200; normal: 1200;)
Training,Testing data ratio	Train set: 80% & test: 20%
Model, Learning rate	VGG16, 0.001
Optimizer type, chart	Adam, accuracy & Loss graph
Activation Functions	ReLU (Hidden Layer) & SoftMax (Output Layer)
Dropout Layer Value	0.5
Metrics used	Accuracy, precision, recall, sensitivity, specificity, F1-score.

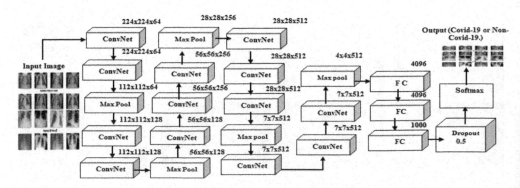

Figure 2.   Block diagram of proposed model for COVID-19 X-ray (CXR) image classification.

After building in this way, train the already trained model to effectively match the normal chest X-ray images from the Covid19 (+) X-ray image. The suggested transfer learning technique has the advantage of being faster than the operation currently in use and saves a considerable amount of time and effort. Another significant advantage of the result suggested here is that it outperforms all the models trained.

4 EXPERIMENTAL AND RESULTS

For this tentative setup work, here we have conducted an experiment with help of our Google Colab, based on the experiment, which is a free and open-source Python IDE They are very simple to use and their systems and functions are useful to us, meaning that we can use them to solve complex problems in a relatively straightforward manner. Table 1 shelters all the strictures and evidence used to optimally formulate the model and present the results of the experiment:

Evaluation Metrics: These data sets are balanced excepting for the fit classifier, so a single result is not sufficient for us to determine the accuracy here. So here we better use other metrics like precision, recall and F1-score. These four metrics are common metrics we use in machine learning for classification analysis. The aforementioned measures are assessed using a confusion matrix with four terms: for which call true positive (TP), false positive (FP), true negative (TN), and false negative (FN). All benchmarks Equations (1),(2),(3),(4) are defined (Rizk *et al.* 2019) in detail:

$$\text{exactness} \quad TP = \frac{TP}{TP + FP}, \tag{1}$$

$$\text{remembrance} = \frac{TP}{TP + FN}, \tag{2}$$

$$\text{F1} - \text{score} = \frac{2 \times \text{precision} \times \text{recall}}{\text{precision} \times \text{recall}}, \tag{3}$$

$$\text{Accuracy} = \frac{TP + TN}{TP + TN + FP + FN}, \tag{4}$$

4.1 Results and discussion

Here Table 2 clearly illustrations the recital results of each transmission model detecting COVID-19. So here different representations have given diverse results and here maximum precision is obtained by modified-VGG16 models.

So, with the assistance of this confusion matrix, achieve other measures of precision, F1-score, recall, specificity and sensitivity. Here we have Relu (Rectified Linear Here we have used Unit) activation function for each layer in this. So, when use it we can avoid sending all negative values to the next layer. This is a good function to handle negative values. So we use ReLu activation function in hidden layers. High quality parameters are finetuned here to get high performing models. Here we fine tune 5 different parameters including cycle limit etc. By doing this adjust the learning rate here, within these 5 parameters are as follows 1. Selection of enhancers, 2. loss functions, 3. changing the number of periods, 4. block size, and finally 5. test size. Here we work with diverse optimizer and loss function does not affect the recital of the prototypical to any great extent, so here we use Adam optimizer function and binary cross entropy is used as loss function throughout the entire model. Batch size is The quantity of samples distributed across the network and epochs represent the quantity of

415

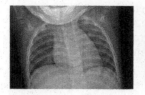 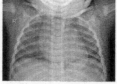 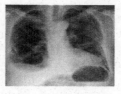 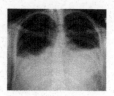

Normal X-ray   Normal X-ray   Covid X-ray   Covid X-ray

Figure 3.   Predicted X-ray images using VGG16 model.

times the model is run on the training data. Next here we do some operation using dropout i.e. Dropout is a regulatory performance in which randomly selected neurons are neglected during training. Growing the dropout generally surges the accuracy so we use dropout here to increase the accurateness.

Table 3.   Performance of the proposed VGG16 model classification report.

	Precision	Recall	f1-score	Support
0	0.98	1.00	0.99	119
1	1.00	0.98	0.99	121
accuracy			0.99	240
macro avg	0.99	0.99	0.99	240
biased avg	0.99	0.99	0.99	240

We can perceive in Table 2 below. 1200 chest xray images here clearly illustrates how diverse these parameters values affect the model's recital when proficient on a model dataset of images. Next here after we've done all the convolutions we can make, send the data to the fully connected (dense) layer, here flatten the output vector from the convolution layer. Here to block negative values from forwarding through the network do the activation ReLu function 1 x density 4096 units twice. along with in the end, 2-unit dense layer using softmax activation function 1 x dense SoftMax layer in 2 units to finally detect whether it is covid 19 or normal. It uses a low-level kernel to extract features at minimum saturation, and when compare it with its other counterpart VGG-16 model, it is more suitable for chest xray images with a small number of layers. "Recall" is demarcated as the number of true positives divided by the number of true positives and false negatives. In Equation (1) and (2) they denote precision and recall, respectively. These are calculated here as a smoothed average between F1-score precision and recall, as further defined in Equation (3). "Accuracy" can be determined here by evaluating the degree of correct prediction mid all values, as represented by Equation (4). (They are also called "true positive rate") A better way to determine true positives from all available classes is called "sensitivity". Similarly, the next "specific" is a good way to regulate true negatives from all obtainable classes. (The image below shows here the confusion matrix of each transfer learning model, finally 0 represents Covid-19 and 1 represents normal. Figure 4. (b). illustrations the confusion matrix of diverse model's castoff to envisage the above recital metrics. With the support of a confusion matrix, then forecast the values of TP, TN, FP, and FN. evaluation metrics value shown in Table 4.

4.1.1   *Accuracy & loss*
Below the accuracy graphs and loss diagrams are also plotted for the proposed models: The VGG16 with dropout layer and patch normalization. Figures 4. (a) show the accuracy maps

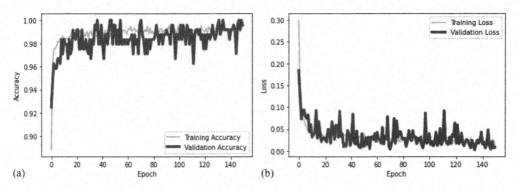

Figure 4. (a) Accuracy versus epochs. (b) Loss versus epochs.

Table 4. Evaluating metrices.

	True Positives	True Negatives	False Positives	False Negatives
Value	1211	118	1	0

obtained from the proposed models respectively and Figure 4. (b) show the loss curves are gotten from the future models respectively. Also, it shows accuracy graphs and loss graphs for the corrected model. Here, the x-axis shows us number of epochs, and here the y-axis shows precision/loss. From the above graphs, the training and validation of the models are plotted.

One thing can notice from the figure is the F1-score obtained between VGG16 with dropout layer and patch normalization is better than other models in comparison. (The F1-score values for dropout layer patch normalization from eVGG16 are 0.9958, respectively. This results in a good accuracy of 99.49% and sensitivity and specificity values of 1.0000 and 0.9890 for the multiclass model, respectively. And specificity values of 1.0000, 1.0000 and 98%, respectively.

5 CONCLUSION

The proposed report clearly focuses on the detection of chest x-ray images in healthcare and clinical use of Covid-19. The most important problem in image exposure techniques is to estimate the optimal subset of topographies from input images in a given dataset. And the image size is changed here during the recognition process. Here several aspects have been extracted for study. Then, all the optimal features were mined and classified them into two categories, namely images with Covid-19 and ordinary images by the planned model (VGG 16 CovNet). Finally, Covid-19 feature extraction with improved learning rate procedures and VGG 16 recognized images. From the experiment, the results of the experiment here confirmed that the future VGG16 model achieved better results than other methods (CNN, AlexNet, Resnet18). And here the performance of the scheme is checked under various dimensions like accuracy, F-score and specificity. Future studies on Covid-19 detection techniques may improve depiction accuracy by removing critical topographies from chest x-ray images. In addition, future research works may focus some attention on combining metaheuristic calculations to select the finest element and influential hierarchy.

REFERENCES

Bhandary, A. *et al.* (2020) 'Deep-learning Framework to Detect Lung Abnormality–A Study with Chest X-Ray and Lung CT Scan Images', *Pattern Recognition Letters*, 129, pp. 271–278.

Chen, Y. *et al.* (2020) 'Fedhealth: A Federated Transfer Learning Framework for Wearable Healthcare', *IEEE Intelligent Systems*, 35(4), pp. 83–93.

Das, A.K. *et al.* (2021) 'Automatic COVID-19 Detection From X-ray Images Using Ensemble Learning with Convolutional Neural Network', *Pattern Analysis and Applications*, 24(3), pp. 1111–1124.

Drosten, C. *et al.* (2003) 'Identification of a Novel Coronavirus in Patients with Severe Acute Respiratory Syndrome', *New England Journal of Medicine*, 348(20), pp. 1967–1976.

Haritha, D., Praneeth, C. and Pranathi, M.K. (2020) 'Covid Prediction from X-ray images', in *2020 5th International Conference on Computing, Communication and Security (ICCCS)*. IEEE, pp. 1–5.

Hilmizen, N., Bustamam, A. and Sarwinda, D. (2020) 'The Multimodal Deep Learning for Diagnosing COVID-19 Pneumonia from Chest CT-scan and X-ray Images', in *2020 3rd International Seminar on Research of Information Technology and Intelligent Systems (ISRITI)*. IEEE, pp. 26–31.

Kumar, V. *et al.* (2022) 'COV-DLS: Prediction of COVID-19 from X-Rays Using Enhanced Deep Transfer Learning Techniques', *Journal of Healthcare Engineering*, 2022.

Makris, A., Kontopoulos, I. and Tserpes, K. (2020) 'COVID-19 Detection from Chest X-Ray Images using Deep Learning and Convolutional Neural Networks', in *11th Hellenic Conference on Artificial Intelligence*, pp. 60–66.

Mangal, A. *et al.* (2020) 'CovidAID: COVID-19 Detection Using Chest X-ray', *arXiv preprint arXiv:2004.09803* [Preprint].

Mukherjee, H. *et al.* (2021) 'Deep Neural Network to Detect COVID-19: One Architecture for Both CT Scans and Chest X-rays', *Applied Intelligence*, 51(5), pp. 2777–2789.

Narin, A., Kaya, C. and Pamuk, Z. (2021) 'Automatic Detection of Coronavirus Disease (covid-19) using X-ray Images and Deep Convolutional Neural Networks', *Pattern Analysis and Applications*, 24(3), pp. 1207–1220.

Rahimzadeh, M. and Attar, A. (2020) 'A Modified Deep Convolutional Neural Network for Detecting COVID-19 and Pneumonia from Chest X-ray Images Based on the Concatenation of Xception and ResNet50V2', *Informatics in Medicine Unlocked*, 19, p. 100360.

Rajpurkar, P. *et al.* (2017) 'Chexnet: Radiologist-level Pneumonia Detection on Chest X-rays with Deep Learning', *arXiv preprint arXiv:1711.05225* [Preprint].

Ribani, R. and Marengoni, M. (2019) 'A Survey of Transfer Learning for Convolutional Neural Networks', in *2019 32nd SIBGRAPI Conference on Graphics, Patterns and Images Tutorials (SIBGRAPI-T)*. IEEE, pp. 47–57.

Rizk, Y. *et al.* (2019) 'Deep Belief Networks and Cortical Algorithms: A Comparative Study for Supervised Classification', *Applied Computing and Informatics*, 15(2), pp. 81–93.

Ventura, C. V. *et al.* (2016) 'Zika Virus in Brazil and Macular Atrophy in a Child with Microcephaly', *The Lancet*, 387(10015), p. 228.

Wang, L., Lin, Z.Q. and Wong, A. (2020) 'Covid-net: A Tailored Deep Convolutional Neural Network Design for Detection of Covid-19 Cases from Chest X-ray Images', *Scientific Reports*, 10(1), pp. 1–12.

Wang, X. *et al.* (2017) 'Chestx-ray8: Hospital-scale Chest X-ray Database and Benchmarks on Weakly-supervised Classification and Localization of Common Thorax Diseases', in *Proceedings of the IEEE Conference on Computer Vision and Pattern Recognition*, pp. 2097–2106.

Wolfe, N.D., Dunavan, C.P. and Diamond, J. (2007) 'Origins of Major Human Infectious Diseases', *Nature*, 447(7142), pp. 279–283.

Wu, F. *et al.* (2020) 'A Novel Coronavirus Associated with Human Respiratory Disease in China', *Nature*, 579 (7798), pp. 265–269.

Yan, Q. *et al.* (2020) 'COVID-19 chest CT Image Segmentation–a Deep Convolutional Neural Network Solution', *arXiv preprint arXiv:2004.10987* [Preprint].

Artificial Intelligence, Blockchain, Computing and Security – Dagur et al. (Eds)
© 2024 The Author(s), ISBN: 978-1-032-49393-0

EDGE computing as a mapping study

Md Sarazul Ali* & Ramneet Kaur*
Chandigarh University, Gharaun, Mohali, Punjab, India

ABSTRACT: More smart gadgets are going online and producing massive amounts of data as the Internet of Everything (IoE) steadily spreads. This has led to issues with bandwidth utilization, sluggish response times, insufficient security, and inadequate privacy in typical cloud computing approaches. Edge computing solutions have emerged because of traditional cloud computing's inability to handle the complex data processing necessities of the sophisticated culture of today. It is a brand-new paradigm for processing data at the network's edge. It emphasizes being close to the user and the data source more so than cloud computing. It can be transported for small-scale local data processing and archiving at the network's edge.

The idea that there are a huge number of objects is a fundamental one in edge computing. The edge nodes are being used by a variety of applications, each of which has its own structure for how the service is delivered. The naming system in edge computing is crucial for programming, address, object recognition, and data transmission, like other computer systems. There is currently no trustworthy nomenclature that has been developed and recognized for the edge computing paradigm. To interface with the diverse system components, edge practitioners typically need to master various communication and network protocols. When naming edge computing systems, it is crucial to take object flexibility, highly dynamic network architecture, privacy and security safety, and other issues into account.

Keywords: Cloud Computing, Collaborative Edge, Smart Cities, Cloud, Mobile Computing

1 INTRODUCTION

Intelligence is now a part of many societal sectors and people's daily lives as a result of the development of an advanced society and the continual improvement of people's needs. Every facet of society, including intelligent manufacturing, smart homes, self-driving cars, and transportation, has been impacted by the use of edge devices. The number of devices linked to the Internet has substantially expanded as a result. Real-time: More terminal data must be sent to the cloud for processing as there are more edge devices. As a result, although the performance of data transmission declines, the volume of intermediate data transmission will soar. The network's bandwidth will be heavily taxed by this, which will slow down the transfer of data. In some application scenarios, such as traffic, monitoring, etc., where real-time feedback is required, cloud computing won't be able to meet corporate real-time needs.

2 LITERATURE REVIEW

Now a days more services are moving from of the cloud to the network's edge so because edge can process data more quickly and dependably. If more information is handled just at

---

*Corresponding Authors: Sarazulali444@gmail.com and ramneet.e13422@cumail.in

DOI: 10.1201/9781003393580-64

edge rather than being transferred to the cloud, it may be possible to implement less bandwidth. With the emergence in IoT and mobile devices, the edge's role in the computing model shifted from extraction of knowledge to data producer/consumer. Data processing / modification could be more efficient at the edge of the network. In this work, we developed our understanding of edge computing using the premise that processing should happen close near data sources. The measurements are mutually dependent. A job, for example, must be completed on the municipal data center tier due to energy constraints. When compared to the constructing server layer, energy constraints always have an impact on latency.

[5] We investigated whether edge computing was appropriate for new IoT applications in this article. We specifically evaluated the performance of edge computing for mobile gaming as an example of a new application that incorporates sensory inputs from the real world in addition to those that the user deliberately generates. According to the findings, edge computing is required to enable fast- paced interactive games.

While hosting resources at the edge is the only way to provide an enjoyable gaming experience, regional data centers can significantly reduce network latency. Furthermore, the increase in network latency cannot be compensated for by increasing the computational power of cloud servers. As a result, even the use of limited computer resources at the edge improves the use case's quality. The performance that can be achieved now is primarily limited by current technology's delays.

[8] Establishing organizations or standardizing bodies that provide a set of broadly accepted norms regarding edge computing inside a 5G context is important in order to standardize protocols. There are two significant challenges to go beyond. To begin with, it is difficult to agree on a standard because of the flexibility and variety of modification offered by many vendors (e.g., the location and capabilities of the edge cloud). The edge cloud is also contacted by a significant number of heterogeneous UEs via a variety of interfaces. The establishment of standardization initiatives like the European Telecommunications Standards Institute (ETSI) has made it possible for different tiers and compute paradigms to work together in a multivendor environment. There are two good choices. Edge cloud apps must initially be unaware of unprocessed data (or unprocessed data). Raw data must therefore be processed or encrypted (for instance, personal data including healthcare information). Second, to safeguard privacy, raw data can be removed before it enters the edge cloud.

3 EDGE COMPUTING CONCEPT

It is incorrect to contrast edge computing with conventional cloud computing. It employs computation at the network's edge using a brand-new computing paradigm. The fundamental concept is to move computation closer to the source of the data. Edge computing is defined in a wide range of ways by researchers. Shi *et al.*, first who proposed the concept of edge devices, define edge computing as a computing technology style of edge of network execution. The group of networks and ad hoc computer resources between both the data source as well as the path of a cloud computing center is known as that of the edge for edge computing. The Internet of Everything is represented by edge computing uplink data, while cloud services are represented by edge computing downlink data.

4 CASE STUDY

To further exemplify our concept of edge computing, we present many case examples in this part that demonstrate how effective edge computing may be.

4.1 Cloud offloading

Most calculations in the cloud are performed according to the cloud computing paradigm, which means that requests and data are handled in a centralized cloud. The user experience is weakened by greater delay (such as lengthy tail latency) that could occur with this computing paradigm. Numerous studies have looked into cloud offloading in terms of the energy-performance tradeoffs in a mobile-cloud setting [22]. A portion of the workload from the cloud can be offloaded since, in edge computing, the edge possesses specific compute resources. Online retail services are one application that might gain from edge computing. It is critical to enhance user experienced, individually in relation to latency, as mobile shopping becomes more and more popular.

4.2 Video analytics

Because of the widespread use of network cameras and mobile phones, video analytics is a new technique. Cloud computing's long data transfer latency and privacy constraints make it unsuitable for video analytics applications. We present a case study of how to find a missing child in a city. Many different types of cameras are commonly used in modern cities and on every vehicle. There is a good chance that a missing child will be discovered on video. However, due to privacy concerns or bandwidth costs, camera data is typically not transmitted to the cloud, making it difficult to use data from wide-area cameras. Despite the fact that the data is available in the cloud, downloading and searching for a large amount of data.

4.3 Smart home

IoT would be extremely beneficial in the home. Smart TVs, smart lights, and robot vacuums are just a few examples of gadgets that have been created and are presently marketed. On the other hand, a smart house needs more than just a cloud connection and the addition of a Wi-Fi module to an existing electrical component. In a smart home setting, it is necessary to install low-cost Along with the related device, wireless sensor and controls can be attached to walls, floors, pipes, and even rooms. Due to data transmission limitations and privacy issues, these gadgets would produce a lot of data, which should mostly be utilized at home. Because of this, the paradigm of cloud computing is improper for a smart house. The opposite is true of edge computing is thought to be ideal for building a smart home because it allows for easy connectivity and management of household objects. Data processing can also be done locally to alleviate the strain on the Internet's bandwidth, and services for better administration and delivery can be put on the edgeOS.

4.4 Smart city

The idea of edge computing can be scaled up to include everything from a single house to a whole neighborhood or even city. The premise behind edge computing is that computing should take place as close as feasible to a data source. A request could be produced at the top of a computing model and serviced at the edge thanks to this architecture. The advantages of edge computing for smart cities are outlined in the following sections.

4.5 Collaborative edge

It could be argued that cloud computing has surpassed traditional data processing platforms in business and academia. One of the main promises of cloud computing is that data will be handled in the cloud, whether it is already there or is being sent there. However, due to concerns about privacy and the prohibitive cost of data transfer, stakeholders frequently fail to share the data they own with one another. As a result, the likelihood of collaboration among many stakeholders is low. The logical concept can also include the physical edge, a compact data center with data processing capabilities that acts as a link between both the

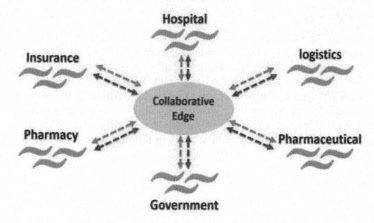

Figure 1. Connected health is an example of a collaborative edge.

cloud and the end user. The existence of a collaborative edge connecting the geographically dispersed edges of various stakeholders has been theorized [14].

5 CHALLENGES

In the previous section, we discussed five potential applications for edge computing. We believe that collaboration between the system and network communities is required to realize the vision of edge processing. This section will go into deeper detail about these problems and offer potential solutions and research opportunities, such as service administration, privacy and security, naming, data abstraction, programmability, and optimizations metrics.

5.1 Programmability

The cloud is where users deploy their own programmers that they have written. Where computation occurs in the cloud is decided by the cloud provider. Users have no idea how the application works in any way. Cloud computing has the benefit of hiding the infrastructure from the user. The application is typically built inside an one programming language & compiled for a certain target platform since it is only used within the cloud. Edge computing, as opposed with cloud technology, offloads computation to peripheral nodes, which very certainly comprise many platforms.

5.2 Naming

The idea that there are many objects is important in edge computing. The edge nodes have a lot of active apps at the top each with its own structure for how the service is delivered. The naming convention is critical in edge computing, as it is in all computer systems, for data transfer, addressing, programming, and object identification. But there isn't a specified and built-in naming system in the edge computing paradigm.

5.3 Data abstraction

On the edgeOS, a variety of apps may be running, interacting with the service management layer's air position indicators to either consume data or supply functions. Both the cloud computing and wireless sensor network paradigms have seen a lot of discussion and research on data abstraction. This issue, meanwhile, gets trickier in the case of edge computing. We will use a smart home setting as an example to show the enormous number of data producers that the IoT would introduce to the network. Almost every device in a smart home, not to mention the numerous other devices scattered throughout the house, will send data to the edgeOS.

5.4 *Privacy and security*

The two most important services that should be provided at the network's edge are user privacy protection and data security protection. The usage data acquired when IoT is integrated in a home might provide a large amount of personal data. One can easily determine if a house is empty or not by checking at the readings of the electricity or water usage, for instance. It is challenging to maintain the service in this scenario without compromising privacy. Before evaluating the data, certain confidential material could be eliminated, for instance by covering every face inside the video. To ensure privacy and data security, we think it makes sense to limit computers to the house, which is the edge for data resources.

6 CONCLUSION

The edge computing model is meticulously presented in this work, together with its main concepts, architecture, technologies, security, and privacy safeguards. Edge computing provides data computational resources at the network's edge. As well as local Internet intelligent services, assisting many industries in their digital transformations and addressing their data diversification needs. Edge computing is a popular research topic. Edge computing will become more important in the future as the Internet and human civilization evolve, as will its ability to successfully foster the growth of numerous enterprises. The Content Delivery Network (CDN), the industrial Internet, and energy are just a few examples.

This paper presents the edge computing model methodically, including its essential principles, structure, key technology, security, and privacy safeguards. Edge computing offers local Internet intelligent services in addition to data storage & computation at the network's edge, facilitating the digital transformations of many businesses and meeting their data diversification demands. Edge computing is a popular study topic. Edge computing will become increasingly important in the future as the Internet and human civilization evolve, and as a result, so will its ability to successfully foster the growth of numerous enterprises. the Industrial Internet, the Content Delivery Network (CDN), and energy Data processing or manipulation would be more effective if done close to the edge of the network. The idea that computation should happen close to data sources informed how we defined edge computing in this study. Then, we outline a number of scenarios where edge computing, which is currently being moved to the cloud, could flourish in sophisticated environments like the house and city. We also provide collaborative edge, which not only enables remote networks to be connected for data collaboration and sharing but also supports the traditional cloud computing paradigm by creating physical and logical links between end users and the cloud. We conclude by outlining the problems and opportunities worth thinking about, including fully programmable, naming, and the data storage.

REFERENCES

[1] Ashton K., "That Internet of Things thing," *RFiD J.*, vol. 22, no. 7, pp. 97–114, 2009.
[2] Sundmaeker H., Guillemin P., Friess P., and Woelfflé S., "*Vision and Challenges for Realising the Internet of Things*," vol. 20, no. 10, 2010.

[3] Gubbi J., Buyya R., Marusic S., and Palaniswami M., "Internet of Things (IoT): A Vision, Architectural Elements, and Future Directions," *Future Gener. Comput. Syst.*, vol. 29, no. 7, pp. 1645–1660, 2013.

[4] *"Cisco Global Cloud Index: Forecast and Methodology, 2014–2019 White Paper,"* 2014.

[5] Evans D., "The Internet of Things: How the Next Evolution of the Internet is Changing Everything," *CISCO White Paper*, vol. 1, pp. 1–11, 2011.

[6] Greenberg A., Hamilton J., Maltz D. A., and Patel P., "The Cost of a Cloud: Research Problems in Data Center Networks," *ACM SIGCOMM Comput. Commun. Rev.*, vol. 39, no. 1, pp. 68–73, 2008.

[7] Cuervo E. *et al.*, "MAUI: Making Smartphones Last Longer with Code Offload," in *Proc. 8th Int. Conf. Mobile Syst. Appl. Services*, San Francisco, CA, USA, 2010, pp. 49–62.

[8] Choy S., Wong B., Simon G., and Rosenberg C., "The Brewing Storm in Cloud Gaming: A Measurement Study on Cloud to End-user Latency," in *Proceedings of the Workshop on Network and Systems Support for Games*, 2012.

[9] Damopoulos D., Kambourakis G., and Portokalidis G., "The Best of Both Worlds: A Framework for the Synergistic Operation of Host and Cloud Anomaly-based IDS for Smartphones," in *Proceedings of the European Workshop on System Security*, 2014, pp. 6:1– 6:6.

[10] Kumar, A., & Alam, B. (2015, February). Improved EDFAlgorithm for Fault Tolerance with Energy Minimization. In 2015 *IEEE International Conference on Computational Intelligence & Communication Technology* (pp. 370–374). IEEE.

[11] Alam, B., & Kumar, A. (2014, March). A Real Time Scheduling Algorithm for Tolerating Single Transient Fault. In 2014 *International Conference on Information Systems and Computer Networks (ISCON)* (pp. 11–14). IEEE.

[12] Stojmenovic I. and Wen S., "The Fog Computing Paradigm: Scenarios and Security Issues," in *Proc. Federated Conf. Comput. Sci. Inf. Syst.*, Sep. 2014, pp. 1– 8.

[13] Sahai A. and Waters B., "Fuzzy Identity-based Encryption," in *Proc. 24th Annu. Int. Conf. Theory Appl. Cryptograph. Techn. (EUROCRYPT)*, 2005, pp. 457–473.

[14] Blaze M., Bleumer G., and Strauss M., "Divertible Protocols and Atomic Proxy Cryptography," in *Proc. 17th Annu. Int. Conf. Theory Appl. Cryptograph. Techn. (EUROCRYPT)*, 1998, pp. 127–144.

[15] Rivest R. L., Adleman L., and Dertouzos M. L., *"On Data Banks and Privacy Homomorphisms,"*Found. Secure Comput., vol. 4, pp. 169–179, Oct. 1978.

[16] Yang K. and Jia X., "Data Storage Auditing Service in Cloud Computing: Challenges, Methods and Opportunities," *World Wide Web*, vol. 15, no. 4, pp. 409– 428, Jul. 2012.

[17] Pawan Singh Mehra, Yogita Bisht Mehra, Arvind Dagur, Anshu Kumar Dwivedi, M.N. Doja, and Aatif Jamshed. 2021. COVID-19 Suspected Person Detection and Identification Using Thermal Imaging-based Closed Circuit Television Camera and Tracking Using Drone in Internet of Things. *Int. J. Comput. Appl. Technol.* 66, 3–4 (2021).

[18] Touceda D. S., Cámara J. M. S., Zeadally S., and Soriano M., "Attributebased Authorization For Structured Peer- to-Peer (P2P) Networks," *Comput. Standards Interfaces*, vol. 42, pp. 71–83, Nov. 2015.

[19] Ateya A., Muthanna A., Gudkova I., Abuarqoub A., Vybornova A., and Koucheryavy A. "Development of Intelligent Core Network for Tactile Internet and Future Smart Systems." *Journal of Sensor and Actuator Networks*, vol. 7, no. 1, Jan 2018.

[20] Ning Z., Kong X., Xia F., Hou W., and Wang X., "Green and Sustainable Cloud of Things: Enabling Collaborative Edge Computing", *IEEE Communications Magazine*, vol. 57, no. 1, pp. 72–78, 2018.

Artificial Intelligence, Blockchain, Computing and Security – Dagur et al. (Eds)
© 2024 The Author(s), ISBN: 978-1-032-49393-0

Reverse and inverse engineering using machine and deep learning: Futuristic opportunities and applications

Sanjeev Kumar
University of Petroleum Studies, Dehradun

Pankaj Agarwal
K R Mangalam University, Gurugram

Jay Shankar Prasad
Greater Noida Institute of Technology, Gautam Buddha Nagar

D. Pandey & Saurabh Chandra
KIET Group of Institutions, Ghaziabad

ABSTRACT: This research paper investigates the use of deep and machine learning techniques in the fields of reverse and inverse engineering. Reverse engineering involves analyzing a system or product in order to understand its design or function, while inverse engineering involves creating a design or system based on a given set of requirements or specifications. We review the current state of the art in using machine and deep learning for reverse and inverse engineering, and present several case studies demonstrating the potential applications of these techniques. We also discuss the challenges and limitations of using deep and machine learning in these contexts, and outline directions for future research in this area.

1 INTRODUCTION

Deep and machine techniques have the potential to significantly accelerate and enhance reverse and inverse engineering processes by automating many of the tasks that are traditionally done manually. These techniques can be used to analyze and extract features from raw data, such as images or point clouds, and to classify or cluster the data into meaningful categories. They can also be used to learn complex relationships and patterns in the data and to generate predictive models that can be used to make decisions or recommendations [1].

Reverse engineering refers to the process of analyzing and understanding a system or device through examination and disassembly, with the goal of understanding how it works and potentially reproducing or improving upon it. In order to understand their internal structure and function. This can involve techniques such as gradient-based optimization or evolutionary algorithms [2–4].

Inverse engineering, on the other hand, refers to the process of using a model or system to infer or reconstruct the underlying process or structure that generated it. In the context of machine learning and deep learning, this can involve using trained models. This can involve techniques such as deep learning or evolutionary algorithms. Inverse engineering, on the other hand, refers to the process of using a known output or result to determine the system or process that produced it. Inverse engineering is often used in the context of manufacturing, where the objective is to ascertain the design and process used to produce a particular product [5–8].

Both reverse and inverse engineering involve analyzing and understanding complex systems or processes in order to replicate or modify them. However, the specific goals and approaches used in these two types of engineering can be quite different. Reverse engineering

DOI: 10.1201/9781003393580-65

typically involves breaking down a system or component in order to understand how it works, while inverse engineering involves working backward from a known output to determine the underlying system or process that produced it.

Reverse engineering and inverse engineering are techniques that involve analyzing an existing product or system in order to understand its design and functionality, or to create a new product or system that performs a similar function.

There are many potential applications for machine and deep learning in reverse and inverse engineering. Some examples include:

- Analyzing and extracting features from images or point clouds of physical objects to create digital models for 3D printing or design.
- Classifying and categorizing components or materials in a product to identify their properties or characteristics.
- Analyzing and understanding the behavior of complex systems, such as power grids or transportation networks, in order to optimize their performance or design.
- Generating predictive models of system performance or failure in order to optimize maintenance and repair schedules or to identify potential problems before they occur.

Overall, the use of machine and deep learning techniques in reverse and inverse engineering has the potential to significantly improve efficiency, accuracy, and speed, and to enable new applications and capabilities that were previously not possible.

2 REVERSE ENGINEERING

Reverse engineering refers to the process of analyzing a system or component in order to understand its design and operation, usually with the goal of creating a copy of the system or modifying its behavior. Reverse engineering is often used in the context of software, where the goal is to understand how a piece of software works or to modify its behavior. A team of researchers at the Carnegie Melon University developed a Design Pattern Recovery from Malware Binaries [9] and explained the process.

Reverse engineering is the process of analyzing a system or component in order to understand its design and operation, usually with the goal of creating a copy of the system or modifying its behavior. Reverse engineering is often used in the context of software, where the goal is to understand how a piece of software works or to modify its behavior.

A group of researchers at the Aerospace Center, Germany proposed a deep learning-based method for analyzing the internal structure and function of a software component in order to understand how it works and identify vulnerabilities [10].

Reverse engineering can also be applied to other types of systems, such as mechanical or electrical systems, or to physical objects, such as consumer products or medical devices. A group of researchers at the Turkey proposed a deep learning-based method for detecting vulnerabilities in software systems by analyzing their internal structure and function [11].

Reverse engineering typically involves decomposing a system or component into its component parts and analyzing how these parts work together to produce the desired behavior. This process can involve disassembling the system or component, examining its internal structure, and analyzing its functionality. Reverse engineering can be used to identify vulnerabilities or weaknesses in a system, to create compatible or interoperable systems, or to understand how to create compatible or interoperable systems, or to understand how a system works in order to replicate or improve upon it [12].

3 INVERSE ENGINEERING

Inverse engineering is the process of using a known output or result to determine the system or process that produced it. Inverse engineering is often used in manufacturing, where the goal is to determine the design and process used to produce a particular product. This can involve analyzing the physical characteristics of a product, such as its shape, size, and material properties, in order to determine the design and manufacturing process used to create it. Researchers at the University of California, Berkeley developed a deep learning-based approach for automatically inferring the manufacturing process used to create a product based on its physical variables names [13].

Inverse engineering can be used to reverse engineer the design of a product in order to improve upon it, to replicate it, or to create compatible or interoperable products. It can also be used to analyze the performance of a product in order to identify weaknesses or opportunities for improvement. A group of researchers at the Cambridge Design Technology proposed a deep learning-based method for reverse engineering in modern product design [14].

In the context of reverse engineering, ML and DL can be used to analyze and understand complex systems or components in order to replicate or modify them. For example, machine learning algorithms can be trained on a dataset of software binaries to learn how to classify different types of software or to detect malicious software. Deep learning algorithms can also be used to analyze the structure and function of software components in order to understand how they work or to identify vulnerabilities. In the context of inverse engineering, ML and DL can be employed to analyze the physical characteristics of a product in order to determine the design and manufacturing process used to create it. For example, machine learning algorithms can be used to analyze images or other data collected from a product in order to identify patterns or features that can be used to infer the manufacturing process used to create it. Deep learning algorithms can also be used to analyze data collected from a product in order to learn about its structure and function, which can be used to reverse engineer the design of the product [14–16].

Overall, machine learning and deep learning can significantly improve the efficiency and accuracy of reverse and inverse engineering tasks, and they are actively being researched and developed for these applications.

4 DECODING BRAIN WITH REVERSE AND INVERSE ENGINEERING

One approach to reverse-engineering the brain using machine learning is to use trained models to predict brain activity patterns from behavioral or cognitive data. For example, researchers have used deep learning models to predict fMRI patterns from natural language descriptions of visual stimuli, with the goal of understanding how the brain processes language and vision [17]. Another approach is to use machine learning to analyze and interpret brain imaging data in order to understand the underlying neural mechanisms and networks that support brain function. For example, researchers have used machine learning techniques to identify patterns in fMRI data that correspond to different cognitive processes, such as memory or decision-making.

In the study of the brain has the potential to greatly enhance our understanding of brain function and inform the development of new treatments for brain-related disorders.

4.1 *Real projects or research that have used ML and DL to reverse-engineer the brain*

In a study published in the Human Brain Mapping [8], researchers used deep learning to predict brain activity patterns from natural language descriptions of visual stimuli. The researchers trained a deep neural network on a large dataset of fMRI patterns and corresponding language descriptions, and then used the trained model to predict fMRI patterns from new language descriptions. This allowed the researchers to understand how the brain processes language and vision, and to identify brain regions that are specifically involved in these processes. In a study published in the journal Trends in Neuroscience, researchers used machine learning to identify patterns in fMRI data that correspond to different stages of memory consolidation. The researchers trained a machine learning model on a dataset of fMRI patterns from participants who were performing a memory task, and then used the trained model to identify patterns in the fMRI data that corresponded to different stages of memory consolidation. This allowed the researchers to better understand the neural mechanisms underlying memory consolidation, and to identify potential targets for the development of new treatments for memory-related disorders [21].

A general outline of the steps involved in using machine learning or deep learning to predict brain activity patterns from behavioral or cognitive data as reported are as follows [16,21]:

(i)   Collect a dataset of brain imaging data (such as fMRI or EEG) and corresponding behavioral or cognitive data.
(ii) Pre-process and clean the data as needed, such as by removing noise, artifacts.
(iii) Split the data into training, validation, and test sets.
(iv) Choose a machine learning or deep learning model and set the hyperparameters.
(v)   Initially, the model will be trained using the training set, and then its results will be evaluated using the test set.
(vi) Fine-tune the model as needed based on the results from step 5.
(vii) See how well the final model does on the test data.
(viii) Use the trained model to predict brain activity patterns from new behavioral or cognitive data.

Interpret the results and draw conclusions about the underlying neural mechanisms and networks that support brain function. Overall, this approach allows researchers to better understand how the brain processes different stimuli or performs different tasks, and to identify brain regions that are specifically involved in these processes. It can also inform the development of new treatments for brain-related disorders.

4.2   *Reverse engineering problems and deep learning*

Reverse engineering refers to the process of analyzing a system or component in order to understand its design and operation, usually with the goal of creating a copy of the system or modifying its behavior. Reverse engineering problems often arise in the context of software, where the goal is to understand how a piece of software works or to modify its behavior. There are a number of ways in which deep learning can be used to solve reverse engineering problems. For example, deep learning algorithms can be trained on a dataset of software binaries to learn how to classify different types of software or to detect malicious software. Deep learning algorithms can also be used to analyze the structure and function of software components in order to understand how they work or to identify vulnerabilities. Overall, DL has the Capability to significantly improve the efficiency and accuracy of reverse engineering tasks, and it is an active area of research in the field.

General algorithm for using deep learning to solve a reverse engineering problem is given below:

4.2.1   *Define the problem*
The success of any reverse engineering endeavor will depend on how well you define the problem you're attempting to address and your desired end state.

4.2.2   *Collect and preprocess data*
Gather a dataset of examples that will be used to train your deep learning model. This might involve collecting software binaries or other relevant data, and preprocessing the data to prepare it for analysis.

4.2.3   *Choose a deep learning architecture*
Select a deep learning architecture that is appropriate for the problem at hand.

4.2.4 *Train the model*

Use your dataset to train the deep learning model using an appropriate optimization algorithm (e.g. stochastic gradient descent). You may need to tune the hyperparameters of the model (e.g. learning rate, regularization strength, etc.) in order to achieve good performance.

4.2.5 *Test the model*

appraise the performance of the qualified model on a distinct test dataset to see how well it generalizes to unseen data.

The competent model is ready to analyze the system or component that you are trying to reverse engineer. This might involve making predictions about the behavior of the system or identifying specific components or functions within the system.

4.3 *Variation regularization*

Variation regularization is a technique used in ML and DL to prevent overfitting and expand the generalization ability of trained models. Overfitting occurs when a model fits the training data too closely and does not perform well on new, unseen data. This can be a problem when the training data is limited or noisy, as the model may learn patterns that do not generalize to the broader population. This can help to hinder the model's ability to improve too much about the specific details of the training data, and encourage it to learn more robust and generalizable patterns. However, variation regularization techniques such as data reinforcement, dropout, and weight decay are widely used in ML and DL, and there is a large body of research on these techniques. It is an important technique for improving the generalization ability of trained models and preventing overfitting, especially when working with limited or noisy training data.

Here are a few examples of how variation regularization techniques can be used in machine learning and deep learning [5];

Data augmentation is a standard method in deep learning for enhancing image recognition models' scalability. Say you wish to build a deep learning model using a dataset containing photographs of dogs to recognize different breeds. You could apply data augmentation techniques such as rotation, scaling, or cropping to the images in the training dataset in order to generate new, augmented training examples. This can help the model to learn more robust and generalizable features, and improve its performance on new, unseen images of dogs.

Dropout is a popular variation regularization technique in deep learning, and is often used in neural network models to prevent overfitting. Suppose you have a DNN that is being trained to classify images of dogs and cats. During training, you could apply dropout by randomly setting a certain proportion of the network's weights or neurons to zero at each training step. This promotes the network's ability to acquire features that are more resilient and widely applicable, and it helps to avoid the network from becoming overly dependent on any one collection of weights or neurons.

Weight decay is another variation regularization technique that can be used in machine learning and deep learning to prevent overfitting. Weight decay works by adding a penalty to the model's loss function based on the size of the model's weights. For example, suppose you are training a linear regression model to predict the price of a house based on various features such as size and location. You could apply weight decay by adding a forfeit to the loss function based on the size of the model's weights, which would encourage the model to prefer smaller weights and help to prevent overfitting.

Overall, these examples demonstrate how variation regularization techniques can be used ML and DL to enhance the generalization ability of trained models and prevent overfitting. Overall, variation regularization techniques such as dropout, weight decay, and data augmentation continue to be widely used in machine learning and deep learning, and there is ongoing research in this area to improve and refine these techniques.

4.4 Bayesian inference for inverse problem

Bayesian inference is an appropriate statistical procedure wherein the probability of a hypothesis is recalculated in light of new data using Bayes' theorem. It is a way of combining prior knowledge about a hypothesis with new information from data to obtain a posterior probability distribution for the hypothesis. This posterior distribution can be used to make predictions, to quantify uncertainty, or to perform model selection. In an inverse problem, Bayesian inference can be used to estimate the unknown parameters of a model that describe the association among the observed data and the underlying physical phenomena [19].

In a Bayesian approach to an inverse problem, we start with a prior distribution over the possible causes of the effect. As we observe more data (i.e., the effect), Bayes' theorem allows us to generate a posterior distribution by updating the prior distribution which represents our updated beliefs about the possible causes given the data that we have observed. One advantage of using Bayesian inference for inverse problems is that it allows us to incorporate prior knowledge about the system and to propagate uncertainty in the data and model parameters through to the solution. This can lead to more robust and accurate solutions compared to traditional optimization-based approaches.

In a Bayesian approach to an inverse problem, we start by specifying a prior distribution over the space of possible solutions to the problem. This prior encodes our initial beliefs about the solution before we have seen any data. We then use Bayes' theorem to update our beliefs in light of the observed data, resulting in a posterior distribution over the space of solutions. The posterior is a more refined version of the prior that considers the information we have gained from the data.

To unravel the inverse problem, we can use techniques such as Markov chain Monte Carlo (MCMC) or variational inference to approximate the posterior distribution. We can then make predictions about the unknown parameters or causes by computing summary statistics or sampling from the posterior distribution. Here are the steps for performing Bayesian inference for an inverse problem:

Step 1: Define the model for the inverse problem, including the unknown parameters and any observed data.

Step 2: Create a probability distribution function over the range of values for the parameter of interest. This represents our initial beliefs about the values of the parameters before any data has been observed.

Step 3: Use Bayes' theorem to compute the posterior distribution as shown below:

$$posterior = (likelihood * prior)/normalizing\ constant.$$

Here, the normalizing constant is a term that guarantees the posterior distribution is appropriately normalized, and the likelihood indicates the probability of the data being observed given a specific set of parameter values. Use the posterior distribution to make predictions about the values of the unknown parameters. For example, you might compute the mean or mode of the posterior distribution as an assessment of the most likely constraint values, or you might compute credible intervals to quantify your uncertainty about the values of the parameters.

(Optional) If you have more data that you would like to incorporate into your analysis, you can repeat steps iii and iv to update the posterior distribution and make more accurate predictions.

4.5 Challenges and limitations to using machine and deep learning in the fields of reverse and inverse engineering

These include: [12–29]

4.5.1 Data requirements

Machine and deep learning algorithms require large amounts of data in order to learn and make accurate predictions. In the context of reverse and inverse engineering, this can be a challenge, as it may be difficult to obtain sufficient data to accurately model a complex system or product.

4.5.2 Data quality
The quality of the data used to train machine and deep learning algorithms is critical to their performance. Inaccurate or biased data can result in poor performance or incorrect results.

4.5.3 Algorithm bias
Machine and deep learning algorithms can exhibit bias if the data used to train them is biased. This can be a particular concern in fields such as medical or criminal justice, where decisions made by these algorithms can have significant consequences for individuals.

4.5.4 Interpretability
Many machine and deep learning algorithms are difficult to interpret, making it difficult to understand how they reach their conclusions. This can be a challenge in the context of reverse and inverse engineering, where it is often important to understand the underlying mechanisms or processes being modeled.

4.5.5 Over-reliance
There is a risk of over-reliance on machine and deep learning algorithms, particularly in fields where they have been successful. This can lead to a lack of understanding of the limitations of these algorithms and the potential for them to make mistakes.

Overall, while machine and deep learning have the potential to greatly improve the fields of reverse and inverse engineering, it is important to carefully consider these challenges and limitations in order to ensure their responsible and ethical use.

5 CONCLUSION

The machine and deep learning have the potential to revolutionize the fields of reverse and inverse engineering. These techniques have demonstrated their ability to analyze and interpret complex data sets, identify patterns and trends, and facilitate the design and optimization of systems and products. However, there are also challenges and limitations to the use of deep and machine learning in these contexts, including the need for large and diverse data sets, the risk of bias in the data and algorithms, and the potential for over-reliance on these technologies. Despite these challenges, the use of machine and deep learning in reverse and inverse engineering shows great promise, and is likely to continue to grow in importance in the coming years. Further research is needed to address the limitations of these technologies and to identify new and innovative applications of deep and machine learning in reverse and inverse engineering. Ultimately, the successful deployment of machine and deep learning in these fields will depend on the ability to balance the benefits of these technologies with the need to ensure their ethical and responsible use.

REFERENCES

[1]   Khoramian, Saman. "A Connection Among Machine Learning, Inverse Problems and Applied Harmonic Analysis." *arXiv: Optimization and Control* (2020):
[2]   Bertero M., Boccacci P., *Introduction to Inverse Problems in Imaging*. CRC press, 1998.
[3]   Bishop C., *Pattern Recognition and Machine Learning*. Springer New York, 2006.
[4]   Chung C. V., De los Reyes J. C., Schonlieb C-B., Learning Optimal Spatially-Dependent Regularization Parameters in Total Variation Image Denoising. *Inverse Problems*, 33(7), 074005, 2017.
[5]   Chung J., Espanol M. I., Learning Regularization Parameters for General form Tikhonov, *Inverse Problems* 33, 074004, 2017.
[6]   Haber E., Horesh L., Tenorio L., Numerical Methods for the Design of Large-scale Nonlinear Discrete ill-posed Inverse Problems. *Inverse Problems* 26 025002, 2010.

[7] Khoramian S., An Iterative Thresholding Algorithm for Linear Inverse Problems with Multi-constraint and its Application, *Appl. Comput. Harmon. Anal.* 32 (2012) 109–130.

[8] Li Z., Zou D., Tang J., Zhang Z., Sun M. and Jin H., "A Comparative Study of Deep Learning-Based Vulnerability Detection System," *in IEEE Access*, vol. 7, pp. 103184–103197, 2019, doi: 10.1109/ACCESS.2019.2930578.

[9] Alagöz Z. İ. and Akleylek S., "A Brief Review on Deep Learning Based Software Vulnerability Detection," *2021 International Conference on Information Security and Cryptology (ISCTURKEY)*, 2021, pp. 143–148, doi: 10.1109/ISCTURKEY53027.2021.9654351.

[10] Cambridge Design Technology, "*The Role of Reverse Engineering in Modern Product Design*", https://www.cambridge-dt.com/the-role-of-reverse-engineering-in-modern-product-design/

[11] Kobler E., Effland A., Kunisch K., Pock T., Total Deep Variation for Linear Inverse Problems. In *IEEE Conference on Computer Vision and Pattern Recognition*, 2020.

[12] Langer, Automated Parameter Selection in the L1-L2-TV Model for Removing Gaussian plus Impulse Noise, 33(7), 074002, 2017.

[13] Maass P., Deep Learning for Trivial Inverse Problems. In *Compressed Sensing and its Applications*, Birkhuser, 2019.

[14] Mixon D. G., Villar S., SUNLayer: Stable Denoising with Generative Networks, *CoRR*, vol.abs/1803.09319, 2018.

[15] Jonas Adler and Ozan Öktem. Solving Ill-posed Inverse Problems Using Iterative Deep Neural Networks. *Inverse Problems*, 33(12), 2017.

[16] Martin Benning, Guy Gilboa, Joana Sarah Grah, and Carola-Bibiane Schönlieb. Learning Filter Functions in Regularisers by Minimising Quotients. In *International Conference on Scale Space and Variational Methods in Computer Vision*. Springer, 2017.

[17] Heinz Werner Engl, Martin Hanke, and Andreas Neubauer. *Regularization of Inverse Problems*, volume 375. Springer Science & Business Media, 1996.

[18] Ian Goodfellow, Jean Pouget-Abadie, Mehdi Mirza, Bing Xu, David Warde-Farley, Sherjil Ozair, Aaron Courville, and Yoshua Bengio. Generative Adversarial ets. In *Advances in Neural Information Processing Systems (NIPS)*, 2014.

[19] Kyong Hwan Jin, Michael McCann, Emmanuel Froustey, and Michael Unser. Deep Convolutional Neural Network for Inverse Problems in Imaging. *IEEE Transactions on Image Processing*, 26(9), 2017.

[20] Leonid I Rudin, Stanley Osher, and Emad Fatemi. Nonlinear Total Variation-based Noise Removal Algorithms. *Physica D: Nonlinear Phenomena*, 60(1–4), 1992.

[21] Wang J, You X, Wu W, Guillen MR, Cabrerizo M, Sullivan J, Donner E, Bjornson B, Gaillard WD, Adjouadi M. Classification of fMRI patterns–A Study of the Language network Segregation in Pediatric Localization Related Epilepsy. *Hum Brain Mapp.* 2014 Apr;35(4):1446–60. doi: 10.1002/hbm.22265. Epub 2013 Mar 1. PMID: 23450847; PMCID: PMC3748221.

[22] Kumar, A., & Alam, B. (2014, February). Real Time Scheduling Algorithm for Fault Tolerant and Energy Minimization. In *2014 International Conference on Issues and Challenges in Intelligent Computing Techniques (ICICT)* (pp. 356–360). IEEE.

[23] Chaturvedi, R., Kumar, S., Kumar, U., Sharma, T., Chaudhary, Z., & Dagur, A. (2021). Low-cost Iot-enabled Smart Parking System in Crowded Cities. In *Data Intelligence and Cognitive Informatics: Proceedings of ICDICI 2020* (pp. 333–339). Springer Singapore.

[24] Sharma Y., Shatakshi, Palvika, Dagur A. and Chaturvedi R., "Automated bug reporting system in web applications," *2018 2nd International Conference on Trends in Electronics and Informatics (ICOEI)*, Tirunelveli, India, 2018, pp. 1484–1488

[25] Mehra, P. S., Jain, K., Chawla, D., Dagur, A., Singh, S., & Sharma, J. (2022). *GWO-EFUCA: Grey Wolf Optimisation and Fuzzy Logic based Unequal Clustering and Routing protocol for sustainable WSN-based Internet of Things.*

[26] Amjad, J., Lyu, Z. and Rodrigues, M. (2019), "Deep Learning for Inverse Problems: Bounds and Regularizers", *ArXiv*, abs/1901.11352.

[27] Antun, V., Renna, F., Poon, C., Adcock, B. and Hansen, A.C. (2020), "On Instabilities of Deep Learning in Image Reconstruction and the Potential Costs of AI", *Proceedings of the National Academy of Sciences*, Vol. 117 No. 48, pp. 30088–30095, doi: 10.1073/pnas.1907377117

[28] Bai, Y., Chen, W., Chen, J. and Guo, W. (2020), "Deep Learning Methods for Solving Linear Inverse Problems: Research Directions and Paradigms", *Signal Processing*, ISSN 0165-1684, Vol. 177, p. 107729, doi: 10.1016/j.sigpro.2020.107729.

[29] Kamyab S, Azimifar Z, Sabzi R, Fieguth P. 2022. Deep Learning Methods for Inverse Problems. *PeerJ Computer Science* 8:e951 https://doi.org/10.7717/peerj-cs.951

Artificial Intelligence, Blockchain, Computing and Security – Dagur et al. (Eds)
© 2024 The Author(s), ISBN: 978-1-032-49393-0

A Comprehensive study of risk prediction techniques for cardiovascular disease

Huma Parveen, Syed Wajahat* & Abbas Rizvi
Department of Computer Science and Engineering, Amity University Uttar Pradesh, India

Raja Sarath Kumar Boddu
*Department of Computer Science and Engineering, Rampachodavaram, Lenora College of Engineering
Andhra Pradesh, India*

ABSTRACT: In recent years, the majority of people have become extremely concerned about their health. ICT is playing a significant role in the quality enhancement of healthcare applications, and now a days this area becomes most investigative research areas for scientists and practitioners. In developing nations like India, CVD is one of the leading causes of mortality. One of the interesting methods for providing insights for improved evaluation of cardiovascular illnesses is the fuzzy system. Many developers have constructed knowledge-based decision support systems (KDSS) on illnesses over the years, based on the information of medical experts. This is a time-consuming process, and it relies on the judgement of a medical professional. For removing such issue, machine learning approaches have been created to automatically gather databank from historical data. This paper proposes a unique fuzzy rule-based disease decision support system for cardiovascular disease diagnosis, based on scores derived from patient clinical data and automatically collected knowledge. This model experimental results will aid physicians in making informed judgments.

Keywords: Cardiovascular Disease, Fuzzy logic, Clinical Decision Support System, Fuzzy Expert System, Knowledge Based System

1 INTRODUCTION

Computer technology is employed in practically every profession in today's society, and medical diagnosis is no exception. According to World Health Statistics 2020 report, one in every three persons globally has an elevated vital sign, which accounts for around half of all stroke deaths. According to WHO data [1], this deadly disease affects over 17.8 million elderly and adults. Adults, on the other hand, have a greater death rate than the elderly. One out of every five Americans dies from a stroke, according to the CDC, and one out of every five people is unaware that they are undergoing a silent attack. According to the British Heart Foundation [2], the fear of the COVID19 pandemic (coronavirus sickness) in 2020 became the cause of stroke deaths owing to psychological stress and social isolation. This has had an impact on global and national economies. For categorization, the suggested method uses fuzzy logic. The heart disease dataset was chosen since it comprises information about people with and without heart disease. Automatically, cardiac disease prediction systems get the revolution in the field of the healthcare sector. It also reduces the number of tests which a patient goes through. Fuzzy logic is a logical framework for reasoning under uncertainty that generalises traditional two-value logic. It is a collection-based computing and approximation reasoning

---

*Corresponding Author: swarizvi@lko.amity.edu

DOI: 10.1201/9781003393580-66

system. Fuzzy sets are classes of ideas and technologies that use them. Membership is a matter of choice for things with no clear borders. This research article aim to develop a medical expert system which recognises cardiac issues using fuzzy logic. The purpose is to assist medical doctors to diagnose sufferers with persistent coronary heart disease.

This fuzzy-based system is based on three blocks inspired by the Mamdani-type fuzzy model. The description of each three blocks are given below:

(1) The translation of exact values into fuzzier ones is known as fuzzification. and the used values are datasets given by users [3].
(2) User data satisfied users with the given output by using an inference mechanism system.
(3) Defuzzification is typically the final stage. During the defuzzification process, the output of the sum up fuzzy set is reduced to a single integer [4].

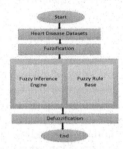

Figure 1.   Flowchart of the algorithm for detecting the risk of heart disease using FL-based method.

2   LITERATURE REVIEW

The term "heart disease" refers to a wide range of clinical diseases involving the heart. These medical diseases explain the odd health issues that directly affect the heart and all of its components. This research looked at the numerous approaches for predicting heart disease that has been established in recent years. The Intelligent Fuzzy Approach method, which is used to classify cardiac disease, is examined in this section.

"A Neuro-Fuzzy Approach to Classification of ECG Signals for Ischemic Heart Disease Diagnosis," by Neagoe et al. [5]. This study focused on the Fuzzy-Gaussian Neural Network (FGNN), a neuro-fuzzy classifier that can discriminate ECG data for IHD (Ischemic Heart Disease) diagnosis. In addition to FGNN, the functionality to remove ECG data from the QRST zone using PCA or DCT; FGNN is used to classify patterns to diagnose IHD. The proposed neuro-fuzzy model, diagnosis of IC was built and tested in this work using an ECG database of 40 individuals, of which 20 had IC and the other 20 were healthy. The greatest result was a perfect IHD recognition score of 100 per cent. The conclusion is shifting as much as utilising only one lead (V5) of ECG recordings as input information since current diagnostic algorithms need a collection of 12 lead ECG signals. "Research on Diagnosing Heart Disease Using Adaptive Network-based Fuzzy Interferences System," by Li Shi et al. [6].

The training technique described in this study can simultaneously train several ANFIS models to increase the modelling accuracy of complicated nonlinear systems. The recognising strategy suggested is sub-ANFIS models which are trained in the research paper while addressing the problem that ANFIS cannot be utilised as a classifier. The findings show that a classifier depend on ANFIS models is capable of accurately determining the forms of ST components. Precision can be enhanced when compared to the BP approach. "A Fuzzy Expert System for Heart Disease Diagnosis" was created by Ali Adeli et al. [7]. The goal of this research is to create an unambiguous expert system for diagnosing cardiac problems. Cleveland Clinic and the V.A. Long Beach Medical Center provided data for the system. Using cardiac data, a comparative analysis of fuzzy classifiers is conducted by Anushya et al. [8].

The study's major objective was to enhance the precision of predicting the existence of cardiac disease. This study used fuzzy logic along with decision trees, naive bayes, K-means and neural networks to predict heart disease. Furthermore, findings revealed that the Fuzzy K-means strategy beat the traditional K-means approach. This paper aims to expand previous work by using various data mining techniques as well as fuzzy sets to assess the severity of heart disease. The tests are run using the UCI machine learning repository cardiac dataset and implemented by using MATLAB. "Diagnosis of cardiac illness utilising Advanced Fuzzy Resolution Mechanism," by Dr A.V.S. Kumar [9]. In this article, we diagnose cardiac disease. Cleveland Heart Disease dataset is employed, and throughout the fuzzification step, crisp values become fuzzy values. The advanced fuzzy resolution mechanism has five levels to enhance the correctness of the results: fuzzy model membership function, the fuzzy if-then rule, and the fuzzy model output variable is estimated using fuzzy predicted values. Fuzzy values are generated via the Advanced Fuzzy Resolution Mechanism. "Decision Support System for Preventing Coronary Heart Disease (CHD) Using Fuzzy Logic," by Cinetha *et al.* [10].

In this study, the author presents a decision support system that uses fuzzy logic and a decision tree to estimate the risk of heart disease in patients over the next 10 years, allowing patients to take preventative measures to extend their lives. A simple coronary heart disease prediction system was built using hierarchical components, fuzzy logic, and decision trees with 1,230 training data sets, allowing clinicians to forecast the multivariate danger of CHD in patients without CHD with predictive accuracy. 97.67%. "A Fuzzy Rule-based Approach to Predict Risk Level of Heart Disease," by K.K. Oad *et al.* [11].

This research proposes a "Fuzzy Rule-Based Support System" that can intelligently and effectively forecast heart disease and replace physical efforts. It comprises two main phases: one for categorization and diagnosis, and another for detecting the risk of respiratory illness. The Mamdani inference system was employed for this system. To evaluate the system's performance, the authors used a neural network and a J48 DT model [31–33].

To diagnose hepatitis B intensity rates and compare them to Adaptive Neural Networks, Mehdi Y. [12] developed a Fuzzy Expert System. M. Neshat *et al.* [13] developed a fuzzy expert system for liver disease diagnosis. Using weighted fuzzy rules, P.K. Anooj [14] built a clinical decision support system for forecasting the risk level of cardiac disease. For Coronary Heart Disease Prediction, Pamela *et al.* [15,16] use DecisionTree and a Fuzzy Optimization Technique. Prediction, diagnosis, and treatment of cardiac disease, Amin *et al.* has used data mining techniques to develop a CDS system [17]. To identify cardiac illness, Sophia Reena [18] used the data mining method "GJCST Classification." Anupriya *et al.* [19] used a genetic algorithm to select feature subsets which further improves heart disease prediction. Vaithiyanathan *et al.* [20] developed a decision-support system for heart illness detection based on genetic-fuzzy logic algorithms.

2.1 Comparative analysis of FL techniques

Table 1 displays the FL System's accuracy over time, as well as the assessment parameters. All of the studies propose different strategies for improving the accuracy of the cardiac diagnostic system. Combining hybrid fuzzy logic with the upgraded C4.5 algorithm achieves maximum accuracy (99.4 percent) in 2021. There is always a degree of ambiguity in the diagnosis of any condition. Since FL can diagnose heart disease, several researchers have undertaken extensive research using FL to identify heart disease. As a consequence, managing FL is more suitable, yet rule optimization is the only difficulty with the FL system. The system becomes more complex as the number of rules grows, yet as the number of rules reduces, the system's accuracy drops.

3 PROPOSED METHODOLOGY

The major focus of our suggested system is on the diagnosis of cardiovascular illnesses. Figure 2 depicts a data-driven approach for diagnosing cardiac disease. In a second stage, we

Table 1. FL-based methods research articles from 2014 to 2021.

Author	year	Approach	Software Used	Accuracy	Evaluation parameters	Advantages
Jan Bohacik et al. [21]	2014	FL Controller and DT	Weka	93.27%	Confusion matrix	An algorithmic and intelligent home telemonitoring model has been constructed using a cumulative system of Decision Tree and Fuzzy Logic
Bohacik et al. [22]	2015	Fuzzy Expert System	Java	92%	–	A task was set to design an effective and intelligent biological diagnosing system employing AI or computer-aided technology.
Assemgul Duisenbayeva et al. [23]	2016	Fuzzy inference system	MATLAB	93.33%	K = 10 cross-validation	Assistance to physicians and practitioners in making judgments, minimising the time it takes to diagnose an illness, and delivering error-free outcomes.
Ali Mohammad Alqudah [24]	2017	Fuzzy Logic Controller	Visual Studio 2010	88.11%	"	Design a mobilephone based application for an E- Health supportive system that demonstrates that how clinical data from patiens may be easily utilised.
Ion Iancu [25]	2018	Fuzzy Logic Controller	MATLAB	98%	–	Dealing with conflicting data to resolve contradictory circumstances and generate a reasonable outcome
Oumaima Terrada et al. [26]	2018	FL Controller and DT	MATLAB	82.65%	–	Right approach founded on FL and then on DT to enhance outcomes owing to the increased number of rules.
Prerna Jain et al. [27]	2019	FL Controller	MATLAB	91%	–	The dataset is presented in an easy-to-understand graphical format, combined with a working environment, allowing predictive analytics to be built.
Khiarak et al. [28]	2019	FL Controller and ANN	MATLAB R2019b	98.7%	F-measure/ ROC	Automatic prediction through the use of PC-based decision-making processes, followed by artificial intelligence.
Preethi Krishnan et al. [29]	2020	FL Controller	PhysioNet	96.6%	Confusion matrix	To detect PVC beats in real-time ECG readings, a clinically grounded, comprehensible, and computationally efficient model is required.
Muhammad et al. [30]	2021	FL and Improved C4.5 algorithm	MATLAB	94.55%	ROC	Affordability, capability, reliability, and reasonably unique technology are all requirements.

decreased the number of characteristics and then used **MATLAB** programming to feed the processed data (symptoms) into a fuzzy system. Following the successful creation of the fuzzy model, symptom prediction will commence, and performance will be evaluated based on the results after the development phase.

Furthermore, an ambulance-based approach for direct patient transfer from home to the ICU can be established. The clinical parameter data will be collected from the patient's home by an artificial intelligence system. The Fuzzy Logic system will be used to provide a single result that will disclose the different phases of patients' health, illness progression, and critical stage.

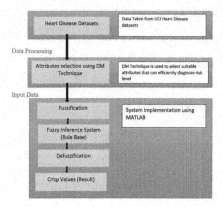

Figure 2. Development of a heart disease diagnosis system process flow.

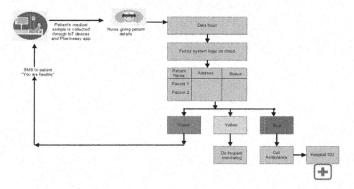

Figure 3. The Framework of fuzzy logic.

3.1 *Dataset*

The dataset was obtained from the Data Mining Repository at the University of California, Irvine (UCI) and consist of 14 attributes.

4 RESULTS

This section covers the CDS system's risk prediction experimental findings. To test sensitivity, accuracy and specificity that is the presentation of the suggested system is compared to that of a neural network-based system.

4.1 *Performance and experiment evaluation*

MATLAB is used to evolve the CDS system. The suggested system is provided the testing dataset to determine the risk prediction of cardiac patients in the testing phase, and the outcomes are assessed using the evaluation metrics specificity, accuracy and sensitivity and they are the three often used statistical metrices, and very useful for determining how excellent and consistent the test was.

4.2 *Experimentation*

The UCI CVD datasets are parted in two groups: training datasets and testing datasets. In addition, we divided the selected UCI data into two classes. Values 0 indicate the lack of disease, whereas values 1–4 indicate otherwise. As a result, we separated it into two groups, with class 0 indicating no heart disease and class 1 indicating the existing heart disease. Table 2 displays the risk prediction attributes selected by the suggested system, as well as the datasets and their descriptions.

Table 2. For the development of fuzzy rules, details about datasets and specified attributes are provided.

Datasets	Total Instance	Training Data	Testing Data	Selected Attributes
cleveland	303	202	101	Age, Cholestrol, Resting Blood Pressure, Thalach (maximum heart rate achieved), Thal, Oldpeak (ST depression, induced by exercise relative to rest)

The suggested technique is then used to construct fuzzy rules from the specified properties. "IF (resting blood pressure is H) then (Class is Class1) (0.20388)", "IF (Age is VL) then (Class is Class0) (0.12621)", "IF (fasting blood sugar is M) then (Class is Class1) (0.64078)", and "IF (exercise-induced angina is VL) then (Class is Class1) (0.15534)" are examples of fuzzy rules.

Table 3. Risk prediction performance of recommended CDS system.

Datesets	Class	Metric	Proposed system Training	Proposed system Testing
Cleveland	>50%	Sensitivity	0.258065	0.447368
		Specificity	0.724771	0.765957
		Accuracy	0.509901	0.623529
	<50%	Sensitivity	0.724771	0.765957
		Specificity	0.258065	0.447368
		Accuracy	0.509901	0.623529

4.3 *Performance analysis*

When a previously developed system was compared with the performance of proposed CDS system, and the suggested CDSS greatly improved risk prediction.

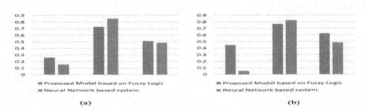

Figure 4. For patients with a risk of more than 50%, the following is a risk prediction. There are two datasets: (a) training dataset and (b) testing dataset.

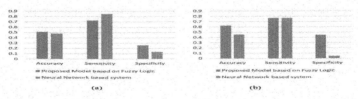

Figure 5. Assumtion of risk for patients with a mortality rate of less than 50%. There are two datasets: (a) training dataset and (b) testing dataset.

5 CONCLUSION

The paper provides sufficient information on numerous studies conducted by various researchers on the topic of creating soft computing algorithms and software for diagnosis of heart disease. The outcome of this paper is a year-long investigate on and analysis of multiple studies on the subject published in a variety of prestigious magazines. The significant conclusion is that the majority of articles rely on FL to enhance accuracy since FL is capable of diagnosing cardiac disease. The accuracy of numerous ways has been visually compared, and the optimum strategy has also been proposed. This suggested technology, dubbed "Fuzzy Rule-Based Support System," is designed to intelligently and effectively anticipate cardiac disease and replace manual labour This model is judged to be the most accurate among several categorizations and prediction models.

REFERENCES

[1]   WHO (2021) https://www.who.int/health-topics/cardiovasculardiseases/tab
[2]   CDC (2021) https://www.cdc.gov/datastatistics/index.html
[3]   BHF (2021) https://www.bhf.org.uk/informationsupport/heartmatters-magazine/news/behind-the-headlines/coronavirus
[4]   Das, S., Ghosh, P. K., & Kar, S. (2013, July). Hypertension Diagnosis: A Comparative Study Using Fuzzy Expert System and Neuro Fuzzy System. In *2013 IEEE International Conference on Fuzzy Systems (FUZZ-IEEE)* (pp. 1–7). IEEE.
[5]   Neagoe, V. E., Iatan, I. F., & Grunwald, S. (2003). A Neuro-fuzzy Approach to Classification of ECG Signals for Ischemic Heart Disease Diagnosis. In *AMIA Annual Symposium Proceedings* (Vol. 2003, p. 494). American Medical Informatics Association.
[6]   Shi, L., Li, H., Sun, Z., & Liu, W. (2007, August). Research on Diagnosing Heart Disease Using Adaptive Network-based Fuzzy Interferences System. In *2007 International Joint Conference on Neural Networks* (pp. 667–671). IEEE.
[7]   Krishnaiah, V., Narsimha, G., & Chandra, N. S. (2016). Heart Disease Prediction System Using Data Mining Techniques and Intelligent Fuzzy Approach: A Review. *International Journal of Computer Applications, 136*(2), 43–51.
[8]   Anushya, A., & Pethalakshmi, A. (2011, December). A Comparative Study of Fuzzy Classifiers on Heart Data. In *3rd International Conference on Trendz in Information Sciences & Computing (TISC2011)* (pp. 17–21).
[9]   Kumar, A. S. (2013). Diagnosis of Heart Disease Using Advanced Fuzzy Resolution Mechanism. *International Journal of Science and Applied Information Technology (IJSAIT), 2*(2), 22–30.
[10] Cinetha, K., & Maheswari, P. U. (2014). Decision Support System for Precluding Coronary Heart Disease (CHD) using Fuzzy Logic. *IJCST, 2*(2), 2347–857.
[11] Oad, K. K., DeZhi, X., & Butt, P. K. (2014). A Fuzzy Rule Based Approach to Predict Risk Level of Heart Disease. *Global Journal of Computer Science and Technology.*
[12] Neshat, M., & Yaghobi, M. (2009, October). Designing a Fuzzy Expert System of Diagnosing the Hepatitis B Intensity Rate and Comparing it With Adaptive Neural Network Suzzy System. In *Proceedings of the World Congress on Engineering and Computer Science* (Vol. 2, pp. 797–802).
[13] Neshat, M., Yaghobi, M., Naghibi, M. B., & Esmaelzadeh, A. (2008, December). Fuzzy Expert System Design for Diagnosis of Liver Disorders. In *2008 International Symposium on Knowledge Acquisition and Modeling* (pp. 252–256). IEEE.
[14] Anooj P.K., (2012). Clinical Decision Support System: Risk Level Prediction of Heart Disease Using Weighted Fuzzy Rules. *International Journal of Research and Reviews in Computer Science (IJRRCS)* (Vol. 3, No. 3. ISSN: 2079–2557).
[15] Persi Pamela I., Gayathri. P., Jaisankar N., (2013). A Fuzzy Optimization Technique for the Prediction of Coronary Heart Disease Using Decision Tree. *International Journal of Engineering and Technology (IJET).* 5–3.
[16] Bhatla, N., & Jyoti, K. (2012). A Novel Approach for Heart Disease Diagnosis Using Data Mining and Fuzzy logic. *International Journal of Computer Applications, 54*(17).

[17]   Mago, V. K., Bhatia, N., Bhatia, A., & Mago, A. (2012). Clinical Decision Support System for Dental treatment. *Journal of Computational Science, 3*(5), 254–261.

[18]   Amin, S. U., Agarwal, K., & Beg, R. (2013). Data Mining in Clinical Decision Support Systems for Diagnosis, Prediction and Treatment of Heart Disease. *International Journal of Advanced Research in Computer Engineering & Technology (IJARCET), 2*(1), 218–223.

[19]   Rajkumar, A., & Reena, G. S. (2010). Diagnosis of Heart Disease Using Datamining Algorithm. *Global Journal of Computer Science and Technology, 10*(10), 38–43.

[20]   Anbarasi, M., Anupriya, E., & Iyengar, N. C. S. N. (2010). Enhanced Prediction of Heart Disease with Feature Subset Selection Using Genetic Algorithm. *International Journal of Engineering Science and Technology, 2*(10), 5370–5376.

[21]   Kim, J. K., Lee, J. S., Park, D. K., Lim, Y. S., Lee, Y. H., & Jung, E. Y. (2014). Adaptive Mining Prediction Model for Content Recommendation to Coronary Heart Disease Patients. *Cluster Computing, 17*(3), 881–891.

[22]   Bohacik, J., Kambhampati, C., Davis, D. N., & Cleland, J. G. (2014, July). Use of Cumulative Information Estimations for Risk Assessment of heart Failure Patients. In *2014 IEEE International Conference on Fuzzy Systems (FUZZ-IEEE)* (pp. 1402–1407). IEEE.

[23]   Baihaqi, W. M., Setiawan, N. A., & Ardiyanto, I. (2016, August). Rule Extraction for Fuzzy Expert System to Diagnose Coronary Artery Disease. In *2016 1st International Conference on Information Technology, Information Systems and Electrical Engineering (ICITISEE)* (pp. 136–141). IEEE.

[24]   Duisenbayeva, A., Atymtayeva, L., & Beisembetov, I. (2016, October). Using Fuzzy Logic Concepts in Creating the Decision Making Expert System for Cardio—vascular Diseases (CVD). In *2016 IEEE 10th International Conference on Application of Information and Communication Technologies (AICT)* (pp. 1–5). IEEE.

[25]   Kasbe, T., & Pippal, R. S. (2017, August). Design of Heart Disease Diagnosis System Using Fuzzy Logic. In *2017 International Conference on Energy, Communication, Data Analytics and Soft Computing (ICECDS)* (pp. 3183–3187). IEEE.

[26]   Sharma, P., & Saxena, K. (2017). Application of Fuzzy Logic and Genetic Algorithm in Heart Disease Risk Level Prediction. *International Journal of System Assurance Engineering and Management, 8*(2), 1109–1125.

[27]   Iancu, I. (2018). Heart Disease Diagnosis Based on Mediative Fuzzy Logic. *Artificial Intelligence in Medicine, 89,* 51–60.

[28]   Terrada, O., Cherradi, B., Raihani, A., & Bouattane, O. (2018, December). A Fuzzy Medical Diagnostic Support System for Cardiovascular Diseases Diagnosis Using Risk Factors. In *2018 International Conference on Electronics, Control, Optimization and Computer Science (ICECOCS)* (pp. 1–6). IEEE.

[29]   Jain, P., & Kaur, A. (2019, June). A Fuzzy Expert System for Coronary Artery Disease Diagnosis. In *Proceedings of the Third International Conference on Advanced Informatics for Computing Research* (pp1–6).

[30]   Krishnan, P., Rajagopalan, V., & Morshed, B. I. (2020, January). A Novel Severity Index of Heart Disease from Beat-wise Analysis of ECG Using Fuzzy Logic for Smart-Health. In *2020 IEEE International Conference on Consumer Electronics (ICCE)* (pp. 1–5). IEEE.

[31]   Parveen H., Rizvi S.W.A, *Disease Risk Level Prediction Based on Knowledge Driven Optimized Deep Ensemble Framework,Biomedical Signal Processing and Control, 2023.*

[32]   Parveen H., Rizvi S.W.A, Shukla P., *Disease Risk Level Prediction Using Ensemble Classifiers: An Algorithmic Analysis. 2022 12th International Conference on Cloud Computing, Data Science & Engineering (Confluence)* DOI: 10.1109/Confluence52989.2022.

[33]   Parveen H., Rizvi S.W.A, Shukla P., Parametric Analysis on Disease Risk Prediction System Using Ensemble Classifier, July 2022, *Congress on Intelligent Systems*, Proceedings of CIS DOI:10.1007/978-981-16-9113-3_53.

Artificial Intelligence, Blockchain, Computing and Security – Dagur et al. (Eds)
© 2024 The Author(s), ISBN: 978-1-032-49393-0

WeSafe: A safety app for all

Reshma Kanse, Supriya Ajagekar, Trupti Patil, Harish Motekar, Vinod Rathod,
Rahul Papalkar & Shabir Ali
Bharti Vidyapeeth Deemed University Pune, DET Navi Mumbai, India

ABSTRACT: As crime rates grow, people's safety is becoming more important on a daily basis. Constraints that make it tough for them to promptly respond to distress calls is the primary issue in how such cases are handled by the police. The fundamental issue with how police respond to such occurrences is the restrictions that prevent them from acting promptly in reaction to distress calls. These restrictions include the inability to quickly and secretly notify the police, the uncertainty around the precise location of the crime, and the fact that the victim is unaware it is occurring. In time of need reaching a person of trust might be helpful. Traditionally in time of need a victim would call a person from their contacts or text them or call helpline number or police which is always not possible. There is a plethora of safety applications available on play store and apple store that can be used for safety purposes but on studying them one may find that those certain features are almost similar to the traditional way of tackling the problem at hand. Almost everyone is using smart phones now days which are equipped with GPS tracking system. This can be used efficiently for personal Security and safety. This paper presents the development details of WeSafe- a Safety System for All app which focuses on alerting the registered contacts of the user in times of need. To make the access easier the user can use the voice commands or shake their device which will alert the registered contacts. This System also recommends nearby police stations and hospitals so the victim could get help faster if necessary.

Keywords: Safety, Crime, Helpline, Hospitals, Police Stations, Security, GPS tracking

1 INTRODUCTION

Currently, it is hazardous for both men and women to travel alone, particularly at night. There is the possibility of theft, kidnapping, or bullying [1]. Parents are not allowing their kids for late night travel. Women are not safe while travelling at night, they are afraid of travelling at night [7]. Crime rates in India have shown a rise, particularly rape and assault. According to a Hindustan Times article date 2nd July, 2021 cases of rape increased by 43% from June 15 last year to June 15 this year, molestation cases increased from 733 to 1,022 and kidnapping of women rose to 1,580 cases while abduction of women rose to a total of 159 cases. A report by National Crime Records Bureau's (NCRB) an average of 80 murders and 77 rape cases occur daily in the country, states a NDTV article dated 18th September, 2021. Crime rates especially of rape against women have seen an increase in the last two years [11]. Given the rising incidence of crime against children, it is imperative to provide a safety support system for school-aged children. Not just women, children but the safety for everyone is main concern. So here an application is proposed which can be used by all for their safety.

Traditionally, one would try to call police or a helpline number or someone from their phonebook, which would require some amount of time which is crucial in hours of need.

DOI: 10.1201/9781003393580-67

441

There are a lot of applications available on play store but one often finds their functionalities take the same amount of time as manually alerting a person. Sometimes it's difficult to open mobile during an emergency. Also in case of kidnapping the kidnapper first snatches the mobile phone and switches it off because of this reason victim is unable to do all the important things. The system proposed alerts a contact of the user's choice by using either voice command "help". The system is engaged automatically when the mobile device is shaken above the predetermined threshold value. In order to identify and confirm an unsafe condition, it begins to record the surrounding voice. If the user doesn't answer within a set amount of time, an alert message will be sent to the user's registered contacts. It is not required to make a call to the person manually the system will going to make a call automatically when user shouts "HELP" or shakes mobile phone. The speakerphone at the victim's side is immediately connected and turned on if a registered contact responds with an audible notification. Along with that the application also recommends nearby emergency contacts such as police stations and hospitals which are required during such an emergency.

2 RELATED WORK.

Dhruv Chand *et al.* [1] WoSApp aims to provide women with safety by enabling quick, covert calls for assistance in times of need. Three steps are involved in using the app: entering emergency contacts, setting off the alarm, and sending an emergency message and calling the police. The notification includes the user's current location and information about their emergency contacts.

Women now have a dependable means to contact the police in an emergency thanks to WoSApp. The calling function can be quickly activated by the user via panic button on screen with an easy process. The system supports women at difficult times. Additionally, this programme answers the query about the user's location and who to call. Shaking the phone will send the police an emergency message with the user's GPS coordinates and pre-selected emergency contact.

Rajesh Nazare [2]. Women can benefit from using this smartphone app. Cases occurring in society are decreased by using SWMS. Women will receive alerts regarding uncharted territory with the aid of this application. so that she will be prepared for any circumstance. Her location will be accessed after logging in, so she must register her emergency contacts to be notified.

Abhijit Paradkar [3] the system helps to support the gender equality. Women are given a safe setting in society that permits them to work until late at night. Anyone planning a crime against women will be discouraged, which lowers the overall rate of crimes against women. The system may occasionally offer relevant evidence. As the system is providing audio-video recording of the incidents can serve as proof. The system offers a tool for detecting intrusions within the home where elderly, disabled, or female visitors are departing alone, and it takes the required precautions to assure safety after detecting an incursion. The technology suggested in the article has the capability to recognize a spy camera installed in a hotel or changing area and alert the user to its presence, protecting them from taking objectionable photos or films.

Cheeka [6] Cheeka is a multifunctional personal safety app created for cellphones running different operating systems, such as Android, Windows Phone, and Blackberry. The time-stamp of the location allows the user to follow Facebook pals. The software updates the customer's trusted contacts on his location every few moments if he feels unsafe till he feels safe. Until the person reaches to a safe location, it acts as a security guard following them behind. Unauthorized power offs, speed tracking, and panic alerts are additional crucial elements. Cheeks can employ augmented reality to display the customer's friends who are nearby. Due to Cheek's broad usage, more cross-platform desktop programmers are being made.

3 OBJECTIVE

Most of the existing systems contained various features but they were a little inconvenient to use. These systems were mainly targeted towards the women safety. Violence against various individuals occurs all over the world, majorly in developing countries. This has severe consequences on an individual's physical and mental well-being. The main problem in handling these circumstances is that there are restrictions that the masses from responding quickly to calls of distress. The existing systems are static systems and do not have all the desired features. In such systems it is required to manually browse through the contact list and make call and then when the receiver responds at that time the victim will tell him about the crime. This process is quite lengthy. So there has to be a system that allows the individuals to quickly contact their trusted contacts. They should be able to have the access to information regarding the nearby hospitals or police station for if and when they find themselves in the dire situations. The system should be a complete package, so the individual can be at least being a little worry free while performing their daily tasks.

The objectives of the system are as follows:

1. To alert trusted contacts easily and quickly when in danger.
2. To be able to look for nearby places like hospital and police station.
3. To have an application that can be used stealthily when in need.
4. To have a system that is easier to use.

4 PROPOSED SYSTEM

In order to access the system, the user first needs to register them. Then they have to add the contact information of the person they want to contact in time of need. The users can also view the contacts that they have added. The nearby function allows the user to get map for the nearby hospitals and police station near to their current location. For the SOS function the user has to shake the device so that the alert message is sent. The system consists of the modules shown in Figure 1. The first module provides the instructions on how to use the system.

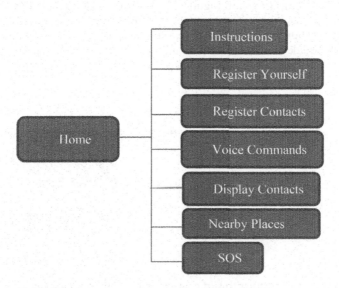

Figure 1. System block diagram.

5 IMPLEMENTATION

5.1 App users

Any User can having smartphone with GPS tracking and internet connection.

5.2 Mobile platform

This app is technologically advanced using Android Studio using the Java language.

5.3 Speech recognizer class

This class provides access to the speech recognition service. Through this service, one can access the speech recognizer. The use of this API will probably include streaming audio to distant servers for speech recognition. Since continuous recognition would demand a lot of battery and bandwidth, this API is not meant to be used for that purpose.

5.4 Accelerometer sensor

Sensors can offer raw data with exceptional precision and accuracy if you want to track three-dimensional device movement or location or keep an eye on changes in the environment near to a device. identifies the gravitational force and the acceleration force applied to an object along each of the three physical axes (x, y, and z) in units of m/s2. It is typically used to find motion, including tremor, tilt, etc. This api was used in this project's SOS button to detect shake motion.

5.5 Location API

The location data that is available to an Android smartphone or tablet comprises the device's precise present location, movement direction and speed, and if the device has crossed a predetermined geographic boundary, or geophone.The first page provides instructions to user on how to proceed with the application. Which shows the instructions such as first user needs to register and then proceed with the system.

For next module as in Figure 2 user has to register him or herself by submitting his or her contact number. And then user has to register the contacts of trusted person to whom he or she can call when in trouble as shown in Figure 2. Instead of accessing the whole contact list, the system provides the user with register contacts through which user can only register trusted contacts. The application also provides the feature to display registered contacts to the user.

Basically this module shows the map to user which includes the nearby places such as hospitals, police stations etc. which are required during emergency conditions. It also shows the current location of the user.

Figure 2. Register yourself, register trusted contacts and display registered contact.

6 RESULTS

When the user completes the whole registration process he or she is allowed to use to main features that is SOS. Here when the user presses start and shakes the device, an alert message is sent to the registered contact.

Voice Command: To use this feature user has to shout out "HELP", the system recognizes it. When the user yells "HELP" the system automatically calls the contact number that the user has registered in the app. It also sends an alert message to the same registered contact.

Figure 3. System sending message to the trusted contact when victim shouts HELP.

The SOS service when started and the device has been shook, will send a message "I am in danger" to the registered contact along with a link giving user's live location. Similarly, when the voice service is turned on and the user shouts the word "Help" a text message saying "Please help me!! I am in Danger, I need your help" is sent to the registered contact.

Along with the message the system makes a call to the registered user. In a case if registered contact person may not notice the message then he or she may respond to the call so that the person in danger may get help. To ensure that the person in need receives assistance, the system makes several attempts to contact the person (via SMS and Call).

Figure 4. Sending SMS with current location of victim and also makes call to trusted contact.

7 CONCLUSION

Here provided details on WeSafe, an Android Application for the safety of masses which can be used on devices running different operating systems, such as Android, Windows Phone, and Blackberry. It aims to provide a platform for everyone in case of emergency while travelling or in general. This app not only assists with SMS SOS messages and victim position tracking, but also alerts the trustworthy contacts listed in the app's contacts list. Along with the SOS/panic button the merit of this application is the voice command which will be activated through voice command "help" and will automatically call the contacts mentioned in contacts list in the app.

REFERENCES

[1]   Dhruv Chand, Sunil Nayak, Karthik S. Bhat, Shivani Parikh, Yuvraj Singh, Amita Ajith Kamath "A Mobile Application for Women's Safety: WoSApp" *TENCON 2015 – 2015 IEEE Region 10 Conference IEEE (2015)*

[2]   Rajesh Nasare, Aishwarya Shende, Radhika Aparajit, Sayali Kadukar, Pratiksha Khachane, Mrunal Gaurkar. "Women Security Safety System using *Artificial Intelligence*" *IJRASET*(2020)

[3]   Abhijit Paradkar, Deepak Sharma. "All in one Intelligent Safety System for Women Security" *International Journal Of Computer Applications*(2015)

[4]   Lyu, M.R.; King, I.; Wong, T. -T; Yau, E.; Chan, P.W. "ARCADE: Augmented Reality Computing Arena for Digital Entertainment" *Aerospace Conference,* 2005 IEEE, vol., no., pp.1,9, 5–12 March 2005.

[5]   Sangani, K., "Developing AR APPS", IEEE Engineering & Technology, vol.8, no.4, pp., May 2013.

[6]   Ananda Kanagaraj S, Arjun G and Shahina A. "Cheeka: A Mobile Application for Personal Safety" in *9th IEEE International Conference on Collaborative Computing: Networking, Applications and Worksharing(Collaborate Com*2013)

[7]   Shubham Nikam, Jay Hiray, Kalpesh Gaikwad, Sanket Patil, Prof. Smita K Thakare. "An Android Based Women Safety App" in *International Research Journal of Modernization in Engineering Technology and Science* (Volume:04/Issue:05/May-2022)

[8]   Byrne Evans, Maire, O'Hara, Kieron, Tiropanis, Thanassis and Webber, Craig (2013) Crime applications and social machines: crowdsourcing sensitive data. In, *SOCIAM: The Theory and Practice of Social Machines*, Rio de Janeiro, BR.

[9]   Supriya A. Bagane, J. L. Chaudhari, "Anomaly Detection of Online Data using Oversampling Principal Component Analysis", *International Journal of Science and Research (IJSR)*, Volume 3 Issue 12, December 2014, pp. 687–690.

[10]   Ravi Sekhar Yarrabothula Bramarambika Thota, "Abhaya: an Android App for The Safety Of Women," *IEEE*, 1 December 2015

[11]   Piyush Bhanushali, Rahul Mange, Dama Paras, Prof. Chitra Bhole, "Women Safety Android App," *IRJET Journal* – Volume 5 Issue 4, April 04, 2018.

Artificial Intelligence, Blockchain, Computing and Security – Dagur et al. (Eds)
© 2024 The Author(s), ISBN: 978-1-032-49393-0

Detection of toxic comments over the internet using deep learning methods

Akash Naskar, Rohan Harchandani & K.T. Thomas
Department of Data Science, Department of Data Science CHRIST (Deemed to be University) Pune, Lavasa, CHRIST (Deemed to be University) Pune, Maharashtra, Lavasa, India

ABSTRACT: People now share their ideas on a wide range of topics on social media, which has become an integral part of contemporary culture. The majority of people are increasingly turning to social media as a necessity, and there are numerous incidents of social media addiction that have been reported. Social media channels. Social media platforms have established their worth over time by bringing individuals from different backgrounds together, but they have also shown harmful side effects that could have serious consequences. One such unfavourable result is how extremely poisonous many discussions on social media are. Online abuse, hate speech, and occasionally outrage culture are now all considered to be toxic. In this study, we leverage the Transformers' Bidirectional Encoder Representations to build an efficient model to detect and classify toxicity in user-generated content on social media. The Kaggle dataset with labelled toxic comments, was used to refine the BERT pre-trained model. Other Deep learning models, including Bidirectional LSTM, Bidirectional-LSTM with attention, and a few other models, were also tested to see which performed best in this classification task. We further evaluate the proposed models utilising dataset obtained from Twitter in order to find harmful content (tweets) using relevant hashtags. The findings showed how well the suggested methodology classified and analysed toxic comments.

Keywords: toxic, social media, twitter, BidirectionalLSTM, BERT, sentiment analysis, toxic comment, hate speech, neural networks, language model, finetuning, pretraining, BERT

1 INTRODUCTION

Researchers discovered that one of the biggest issues for users of ML technologies is the existence of unfairness in machine learning models because the majority of these models are trained using human-generated data, which means that as their use for various tasks and purposes increases, human bias will be evident in these modellingtechniques. Or to put it another way, the humans who provided the training data are just as biased as ML models. The work's main objective is to improve the classification accuracy of the toxicity in online discussion forums, however the methodologies for classification mentioned here can be utilised for any classification purpose. Any biases or false associations in the training data can therefore cause unintentionally skewed correlations in the classification findings. As is generally known, trained models are capable of capturing contextual dependencies.

It has been shown that general classification models, in particular, capture biases that are common in society from training data generated by society and repeat these biases in classification outcomes, including such incorrectly associating victim groups that are commonly attacked, like "Black" and "Muslim," with toxic effects in any context, even in low toxicity contexts. Different deep learning algorithms will be used in the classification of toxic comments on "Jigsaw Unintended Bias in Toxicity Classification. This paper proposed to use deep learning algorithms on the given data set and calculate and compare their accuracy, log loss, and hamming loss.

DOI: 10.1201/9781003393580-68

2 LITERATURE REVIEW

Androcec *et al.*'s [1] paper talks about machine learning models being used for toxic comment classification. It reviews a total of 32 studies published between 2017 and 2020, focusing on the different machine-learning methods and approaches used for this task such as k-nearest neighbors, decision trees, logistic regression, and Naïve Bayes, also achieved decent performance, with average accuracies ranging from 0.71 to 0.81.

Adding some different points of view to the discussion, researcher Alsharef *et al.* [2] found that the LSTM model outperformed several baselines, including a support vector machine and a random forest, and was able to achieve an F1-score of 0.78 for toxic comment detection.

Contrary to the above LSTM, Gang Liu *et al.* [3] found that the attention mechanism improved the performance of the model, compared to a version of the model without the attention mechanism suggesting that the combination of a LSTM, attention mechanism and convolutional layer can be an effective approach for text classification tasks.

Georgakopoulos *et al.* [4] present a study that uses convolutional neural networks (CNNs) to classify toxic comments and ended up concluding that CNNs are a suitable choice for text classification tasks involving toxic comment detection.

Saurabh Srivastava *et al.* [5] suggested that the use of a capsule model could be useful in the development of automated moderation systems for online platforms.

A. G. D'Sa *et al.* [6] model was trained and tested a logistic regression model using the BERT and fastText embeddings on a dataset of annotated comments from Wikipedia and found that the BERT embeddings outperformed the fastText embeddings in terms of both precision and recall for toxic comment detection.

Coming into classification models. Ahmed Abbasi [7], in their study, paper examines how deep learning methods can also detect and classify religious and continental-based toxic content. At first, is a multilabel classification of hazardous religious comments, and the second is a multiclass classification of toxic statements with and without the use of word embeddings (GloVe, Word2vec, and FastText) to demonstrate that in both cases, the CNN model gave the most accurate classifications.

Sara Zaheri *et al.* [8] explains the different metrics that can be used for evaluating the performance of classifiers. Finally, she provides a conclusion that LSTM based models provide way better results than simple machine learning models. Ralf C. Borkan *et al.* [9] in their paper talks about the set of metrics to capture the bias in the data and the model trained on it. Fan *et al.* [10], in their research, collected a dataset of 1.2 million tweets from the period of the referendum and used a deep learning model based on bidirectional long short-term memory to classify them and demonstrates the robustness of the model and its ability to generalize to new data.

P. Malik *et al.* [11] conducted an experiment using some Machine learning and Deep learning models (like CNN and RNNs). The authors tested the accuracy of three different models, which included BERT and fastText embedding with a CNN, BERT and fastText embedding with an RNN, and a model which utilized only traditional machine learning models.

3 METHODOLOGY

3.1 *Pre-processing*

The dataset used for this paper was "Jigsaw Unintended Bias in Toxicity Classification" from the Kaggle community. The text data pre-processing techniques followed before processing and modelling the data are as follows.

a) HTML tags removal: In this work, all HTML were eliminated from all comments. Along with this IP addresses and hyperlinks were also removed. For each comment in this work, lemmatization was done.

b) Remove accented characters: Acute (é), grave (è), circumflex (â, î, or ô), tilde (), umlaut, dieresis (ü or — the same sign is used for two different functions), and cedilla (ç) are the most popular accents. Diacritical marks, also known as accent marks, typically appear above characters. (Reference) We must make sure that we standardize and convert these characters to ASCII.

c) Convert to lowercase and remove special characters and emoticons: Lowercase texts are helpful during pre-processing and later NLP phases.

d) Replace duplicates and decontraction: We also need to sure the comment text is free of duplicates and also make sure every contracted word it back to its original form.

3.2 *Word embedding*

Word embedding is a concept used for representing words for text analysis, generally in a form of a vector of real values that encodes the meaning of the word in such a way where the words that are closer in the vector space are expected to have related meanings.

3.3 *LSTM model*

Initially, LSTM was created where the information flows through cell states. In this way, LSTMs can selectively remember or forget information. This study worked on using LSTM and word embeddings for toxicity classification. The design of the LSTM neural networks used in this work is shown in Figure 1. LSTM has 3 gates input gate, output gate, and forget gate.

- **Input gate:** It make sure how much information should allow to enter the LSTM cell.

$$i_t = \sigma(W_i(h_{t-1}, x_t] + b_i)$$

- **Output gate:** It make sure how much information should be allowed to pass through the cell.

$$o_t = \sigma(W_o(h_{t-1}, x_t] + b_o)$$

- **Forget gate:** It make sure how much information cell should forget or to keep.

$$f_t = \sigma(W_f(h_{t-1}, x_t] + b_f)$$

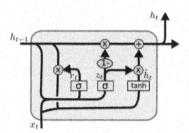

Figure 1.   LSTM diagram as given by Staudemeyer *et al.* [18].

3.4 *Bi-directional LSTM*

The LSTM network parameters will sharply rise as a result of the high-dimensional vector used as its input, making it challenging to tune. The convolution process can reduce the dimensionality of the data while extracting the features. As a result, the text vector's characteristics may be extracted and the vector's size can be decreased using the convolution technique. Although Bidirectional LSTM can obtain the text's contextual information, it is not feasible to concentrate on the information's key points. The categorization will be more accurate if you concentrate on the crucial details.

3.5 *Bidirectional LSTM with attention*

The inspiration for Bidirectional-LSTMs came from bidirectional Recurrent Neural Networks as stated in (11), which employs two distinct hidden layers to analyze sequence input not only in forward but also in backward directions. The two hidden layers are linked to a single output layer by Bidirectional LSTMs. In many areas, it has been demonstrated that bidirectional LSTMs perform noticeably better than nor- mal LSTMs. Our assessment of the literature indicates that bidirectional LSTMs have not, however, been applied to the issue of social media toxicity. Utilizing inputs in a positive sequence from time T-n to time T-1, the forward layer output sequence, $\vec{h}$, is periodically generated.

while the reversed inputs from time T-n to T-1 are used to calculate the backward layer output sequence, h., the common LSTM updating equations are stated below which are used to determine the forward and backward layer outputs:

$$f_t = \sigma_g(W_f x_t + U_f h_{t-1} + b_f) \tag{4}$$

$$i_t = \sigma_g(W_i x_t + U_i h_{t-1} + b_i) \tag{5}$$

$$o_t = \sigma_g(W_o x_t + U_o h_{t-1} + b_o) \tag{6}$$

$$\tilde{C}_t = \tanh(W_C x_t + U_C h_{t-1} + b_C) \tag{7}$$

the hidden state input is projected to the three gates and the input cell state via the weight matrices W_f, W_i, W_o, and W_C while the weight matrices U_f, U_i, U_o, and U_C connect the preceding cell output state to the three gates and the input cell state. The bias vectors are b_f, b_i, b_o, and b_c. The tanh function is the hyperbolic tangent function, and the sigma g function is the gate activation function, which is often the sigmoid.

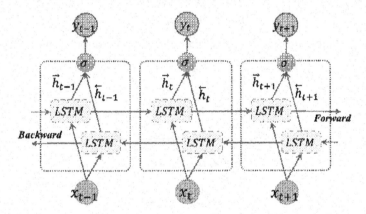

Figure 2. Architecture of bidirectional LSTM.

3.6 *BERT*

One of the most effective context and word representations at the moment is BERT (Bidirectional Encoder Representations from Transformers). BERT employs an attention mechanism and is based on the transformers' technique. The link between the words in a phrase may be examined by paying attention to them. As a result, BERT considers a substantial amount of a word's left and right context. It is significant to remember that, depending on the context, the same word might have many embeddings.

BERT model can be used in two ways:

- For generating the embeddings of the words of a given sentence. These embeddings are further used as input for DNN classifiers.

450

- For fine-tuning a pre-trained BERT model using a task-specific corpus and further to perform the classification.

We suggest using feature-based and fine-tuning techniques to define representations in two different ways:

- **Feature based:** Two phases are carried out in the feature-based method. Each remark is first represented as a list of words or word chunks, and then a fastText or BERT embedding is computed for each word or word chunk. Second, the DNN classifiers, which make the ultimate choice, will use this series of embeddings as their input. As classifiers, we make use of CNN and Bidirectional-LSTM models.
- **Fine-tuning:** Everything is completed in a single step when using the fine-tuning technique. A BERT model when fine-tuned classifies each remark even better.

3.7 AUC base: AUC or area under curve

There are three types of AUCs to measure negative-positive misordering between these sub-groups:

- Subgroup AUC: the calculates AUC on only the examples from the subgroup. This represents model understanding and separability within the subgroup itself.

$$Subgroup\ AUC = AUC(Dg - +D^+)$$

- Background Positive sub-groups negative: AUC is calculated for positive instances from the background and negative ones from the subgroup. When the scores for negative instances in the subgroup are greater than the scores for other positive examples, this number is dropped.

$$BPSN\ AUC = AUC(D^+ + Dg^-)$$

- Background Negative sub-groups positive: AUC is calculated for negative instances from the background and positive ones from the subgroup. This value is lowered when the scores for positive instances in the subgroup are lower than those for other negative examples. The cases would most likely show as false negatives inside the subgroup at several thresholds.

$$BNS\ PAUC = AUC(D^- + D^+)$$

3.8 Generalized mean of the aboved obtained AUC's

We compute the generalised means of the per-identity Bias AUCs as specified below to integrate them into a single overall metric.:

$$M_p(m_s) = \left(\frac{1}{N}\sum_{s=1}^{N} m_s^p\right)^{\frac{1}{p}} \tag{12}$$

where M_p represents the **p**th power-mean function (here, we have used a value of 5 for this use case), m_s is the bias metric **m** calculated for subgroup **s**. and N is the number of identity subgroups.

3.9 Combined custom metric

A recently created custom metric has already been provided in this paper, combining a number of above mentioned submetrics to balance overall performance with numerous unwanted biases as this case study deals with imbalanced data.

451

To determine the final model score, we add the overall AUC and the generalized mean of the Bias AUCs.

$$\text{score} = w_0 AUC_{\text{overall}} + \sum_{a=1}^{A} w_a M_p\left(m_{s,a}\right)$$

where A *represents* number of sub metrics, here which is 3, $m_{s,a}$ *is the* bias metric for identity subgroup s using sub metric a and w_a *is* a weighting for the relative importance of each submetric; all four w values set to 0.25.

4 PROPOSED METHODOLOGY

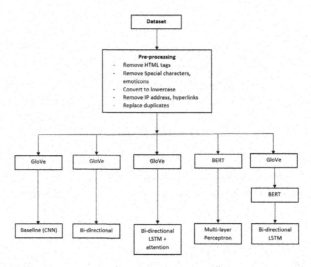

Figure 3. Proposed model.

Along with all the models previously explained previously,in the last model, The output of the Bidirectional LSTM Model with GloVe Embedding Layer was merged with BERT features, followed by skip connections. In short, the model comprises 20 layers; the first layer is the embedding layer uses Glove embedding, followed by the one spatial dropout layer. Then there is a combination of two bi-directional layers. The output is connected to the two output pooling layers, max pooling and average pooling. Then we concatenate the above pooling layers using the concatenate layer. Here we will use the input layer having BERT features. Again there will be concatenated input layer having the BERT feature and the first concatenate layer. Then there will be three sets of combinations of batch normalization layer, dropout layer, fully connected layer, and additional layer of the first fully connected layer and concatenates layer; the other addition layer is the combination of the fully connected layer and the previous dropout layer. There are two output layers, the first layer is output for toxicity, and the second layer is output for the auxiliary toxicity subgroups both the output layer has an activation function that uses "sigmoid" activation.

5 RESULT

We have executed the deep learning model such as CNN based model, Bi-directional LSTM, BERT along with the combination of different embedding mechanism on the Jigsaw Unintended Bias in Toxicity Classification" and compared them in Table 1.

Table 1. Comparison of the final models.

Models	Embedding	Final Score	Kaggle Score
Baseline Model (CNN based)	GloVe	0.90428	0.90006
Bi-directional LSTM	GloVe	0.9262	0.92482
Bi-directional LSTM with Attention Mechanism	GloVe	0.92058	0.91632
MLP with BERT features	BERT	0.86431	0.86048
Bi-directional LSTM with GloVe and BERT features	Glove + BERT features	0.92369	0.92248

6 CONCLUSION

The work compared five different models for finding if a given comment s toxic or not along with its probability using deep learning algorithms like CNN, Bidirectional LSTM models, Bidirectional LSTM with attention, multi-layer perceptron. On comparing the model accuracies, we proposed that the Bidirectional LSTM architecture does the best classification task and can be used in real time too.

Future changes to this model can include using the GloVe Embedding Matrix, use an embedding layer that can be trained. We can also use tune the BERT Model. It might be time- and resource-consuming. Also for future use, we can use large models like XLNet model, a more sophisticated transformer-based model that gets around BERT's draw- backs.

REFERENCES

[1] Androcec, Darko. (2020). Machine Learning Methods for Toxic Comment Classification: *A Systematic Review*. *Acta Universitatis Sapientiae, Informatica*. 12. 205–216. 10.2478/ausi-2020-0012.

[2] Alsharef A, Aggarwal K, Sonia, Koundal D, Alyami H, Ameyed D. An Automated Toxicity Classification on Social Media Using LSTM and Word Embedding. *Comput Intell Neurosci*. 2022 Feb 15;2022:8467349. doi: 10.1155/2022/8467349. PMID: 35211168; PMCID: PMC8863472.

[3] Gang Liu, Jiabao Guo, Bidirectional LSTM with Attention Mechanism and Convolutional Layer for Text Classification, *Neurocomputing* (2019), doi: https://doi.org/10.1016/j.neucom.2019.01.078.

[4] Convolutional Neural Networks for Toxic Comment Classification. Spiros V. Georgakopoulos, Sotiris K. Tasoulis, Aristidis G. Vrahatis, Vassilis P. Plagianakos *arXiv*:1802.09957

[5] Saurabh Srivastava, Prerna Khurana, and Vartika Tewari. 2018. Identifying Aggression and Toxicity in Comments Using Capsule Network. In *Proceedings of the First Workshop on Trolling, Aggression and Cyberbullying (TRAC-2018)*, pages 98–105, Santa Fe, New Mexico, USA. Association for Computational Linguistics.

[6] A. G. D'Sa, I. Illina and D. Fohr, "BERT and FastText Embeddings for Automatic Detection of Toxic Speech," *2020 International Multi-Conference on: "Organization of Knowledge and Advanced Technologies"(OCTA)*, 2020, pp. 1–5, doi: 10.1109/OCTA49274.2020.9151853.

[7] Androcec, Darko. (2020). Machine Learning Methods for Toxic Comment Classification: A Systematic Review. *Acta Universitatis Sapientiae, Informatica*. 12. 205–216. 10.2478/ausi-2020-0012.

[8] Alsharef A, Aggarwal K, Sonia, Koundal D, Alyami H, Ameyed D. An Automated Toxicity Classification on Social Media Using LSTM and Word Embedding. *Comput Intell Neurosci*. 2022 Feb 15;2022:8467349. doi: 10.1155/2022/8467349. PMID: 35211168; PMCID: PMC8863472.

[9] Gang Liu, Jiabao Guo, Bidirectional LSTM with Attention Mechanism and Convolutional Layer for Text Classification, *Neurocomputing* (2019), doi: https://doi.org/10.1016/j.neucom.2019.01.078.

[10] Convolutional Neural Networks for Toxic Comment Classification. Spiros V. Georgakopoulos, Sotiris K. Tasoulis, Aristidis G. Vrahatis, Vassilis P. Plagianakos arXiv:1802.09957

[11] Saurabh Srivastava, Prerna Khurana, and Vartika Tewari. 2018. Identifying Aggression and Toxicity in Comments Using Capsule Network. In *Proceedings of the First Workshop on Trolling, Aggression and Cyberbullying (TRAC-2018)*, pages 98–105, Santa Fe, New Mexico, USA. Association for Computational Linguistics.

[12]   A. G. D'Sa, I. Illina and D. Fohr, "BERT and FastText Embeddings for Automatic Detection of Toxic Speech," *2020 International Multi-Conference on: "Organization of Knowledge and Advanced Technologies"(OCTA)*, 2020, pp. 1–5, doi: 10.1109/OCTA49274.2020.9151853.

[13]   Abbasi, A., Javed, A.R., Iqbal, F. *et al.* Deep Learning for Religious and Continent-based Toxic Content Detection and Classification. *Sci Rep* 12, 17478 (2022). https://doi.org/10.1038/s41598-022-22523-3.

[14]   Zaheri, Sara; Leath, Jeff; and Stroud, David (2020) "Toxic Comment Classification," *SMU Data Science Review*: Vol. 3: No. 1, Article 13. 232, doi:10.1109/ICICS49469.2020.239539.

[15]   Borkan, Daniel & Dixon, Lucas & Sorensen, Jeffrey & Thain, Nithum & Vasserman, Lucy. (2019). Nuanced Metrics for Measuring Unintended Bias with Real Data for Text Classification. *WWW '19: Companion Proceedings of The 2019 World Wide Web Conference*. 491–500. 10.1145/3308560.3317593.

[16]   Fan, H.; Du, W.;Dahou, A.; Ewees, A.A.; Yousri, D.; Elaziz, M.A.; Elsheikh, A.H.; Abualigah, L.; Al-qaness, M.A.A. Social Media Toxicity Classification Using Deep Learning: Real-World Application UK Brexit. *Electronics* 2021, 10, 1332. https://doi.org/10.3390/electronics10111332

[17]   P. Malik, A. Aggrawal and D. K. Vishwakarma, "Toxic Speech Detection using Traditional Machine Learning Models and BERT and fastText Embedding with Deep Neural Networks," *2021 5th International Conference on Computing Methodologies and Communication (ICCMC)*, 2021, pp. 1254–1259, doi: 10.1109/ICCMC51019.2021.9418395.

[18]   *Understanding LSTM – a Tutorial into Long Short-Term Memory Recurrent Neural Networks*. Ralf C. Staudemeyer, Eric Rothstein Morris. https://doi.org/10.48550/arXiv.1909.09586.

[19]   *Jigsaw Unintended Bias in Toxicity Classification* | Kaggle, https://www.kaggle.com/competitions/jig-saw-unintended-bias-in-toxicity-classification/data

Performance testing of scheduling algorithms for finding the availability factor

Prathamesh Vijay Lahande & Parag Ravikant Kaveri
Symbiosis Institute of Computer Studies and Research, Symbiosis International (Deemed University), Pune, India

ABSTRACT: The resource scheduling process is important for the cloud computing environment to provide the best results consistently. Here, the availability factor equally plays a vital role in making the cloud available at all times for processing jobs. The cloud uses its Physical Machines (PM) at the data centers (DC) to process the requests of users. The cloud uses virtualization techniques to create Virtual Machines (VM) from these PMs to prevent the entire data center from failing. The cloud system can provide consistent and best results only when its resource scheduling is done in the right way, thereby providing high availability factors for its resources. Hence, it becomes important to study the resource scheduling algorithms concerning availability factors in the cloud system environment. To fill this gap, this study aims to experiment by processing jobs using Round Robin – Online (RR – ON) and Least Imbalance Level First – Online (LIF – ON) resource scheduling algorithms in a simulated cloud environment under various PM sizes and comparing their results concerning availability factors of resource scheduling algorithms. To provide an additional exhaustive comparison, an empirical analysis is also performed on the results obtained from this experiment to depict the behavior of these algorithms. Lastly, the Machine Learning method called Reinforcement Learning is suggested to entwine these scheduling algorithms and improve them individually or work them hand-in-hand to improve the cloud performance.

1 INTRODUCTION

Resource scheduling is one of the key points for the cloud platform to provide the best service to the user for storing data, using applications, and, most importantly, computations (Dillon *et al.* 2010; Tian *et al.* 2011). Resource scheduling becomes a complex responsibility for the cloud, considering factors such as elasticities, scalabilities, faults, and other dynamically occurring events at its end. Therefore, examining the resource scheduling algorithms and enhancing them becomes vital for the cloud to generate smooth results and keep processing jobs submitted by the user. The cloud data centers (DC) process these submitted jobs on their Physical Machines (PM) (Gonzalez *et al.* 2020; Shaw *et al.* 2022). These PMs are further virtualized into several Virtual Machines (VM) to ensure that the entire cloud is not down if some machine crashes (Han *et al.* 2019; Ma *et al.* 2023). The resource scheduling algorithms should smartly use the cloud resources and ensure that the user experiences the lowest minor problems or issues while using the cloud computing environment (Dillon *et al.* 2010). In this research, the resource scheduling algorithm Round Robin – Online (RR – ON) and Least Imbalance Level First – Online (LIF – ON) (Sanaj *et al.* 2020) are implemented in a simulated environment under various circumstances to examine them with respect to availability factors and make rigorous comparison among them. This experiment is conducted in multiple stages, and the results obtained are compared concerning multiple availability factors. Lastly, Reinforcement Learning (RL) (Anoushee *et al.* 2023;

DOI: 10.1201/9781003393580-69

Ma *et al.* 2023; Ramezani *et al.* 2023; Vengerov 2007) is proposed to provide intelligence to the cloud by individually improving these algorithms or designing a hybrid algorithm using them.

Figure 1. Depicts the flow of the experiment conducted.

The remaining of the paper is categorized as follows: section 2 consists of the experimental design; section 3 consists of the results and discussions; section 4 consists of a detailed empirical analysis followed by the conclusion in the last section.

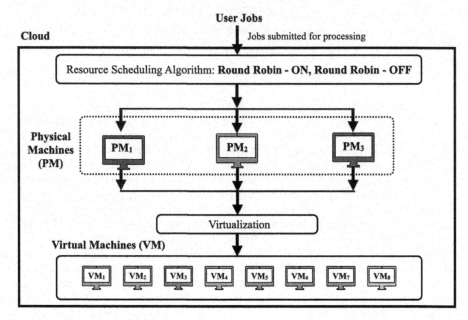

Figure 1.   Flow of the experiment conducted.

2   EXPERIMENTAL DESIGN

An experiment was conducted in the CloudSched cloud simulation environment where jobs were processed using resource scheduling algorithms RR – ON and LIF – ON using PMs: PM_1, PM_2, and PM_3. The first PM, i.e., PM_1, consisted of a 16 GHz CPU with 30 GB RAM, the second PM, i.e., PM_2, had a 52 GHz processor with 137 GB RAM, and the last PM, i.e., PM_3 had 40 GHz processor with 14 GB RAM. All these PMs have a bandwidth (BW) capacity of 3380 megabits per second (Mbps). These PMs are further divided into eight VMs: VM_1, VM_2, VM_3, ... , and VM_8, respectively. The processor capacity of these eight VMs in GHz are as follows: 1, 4, 8, 6.5, 13, 26, 5, and 20, respectively. The RAM capacity of these VMs in GB are as follows: 1.7, 7.5, 15, 17.1, 34.2, 68.4, 1.7, and 7, respectively. Also, the BW of these VMs in Mbps are as follows: 160, 850, 1690, 420, 850, 1690, 350, and 1690 respectively. The four stages of this experiment are as follows:

Stage 1: Total PM count = 75: PM_1 = 25, PM_2 = 25, PM_3 = 25.
Stage 2: Total PM count = 150: PM_1 = 50, PM_2 = 50, PM_3 = 50.
Stage 3: Total PM count = 225: PM_1 = 75, PM_2 = 75, PM_3 = 75.
Stage 4: Total PM count = 300: PM_1 = 100, PM_2 = 100, PM_3 = 100.

In each stage, the output of RR – ON and LIF – ON obtained are compared using the performance parameters: Unbalanced Degree of DC - Variance, Unbalanced Degree of DC – Index, Unbalanced Degree of PMs – Variance, Unbalanced Degree of PMs – Index, Average (Avg.) Percentage (Perc.) CPU Utilization in DC, Avg. Perc. RAM Utilization in DC, Avg.

Perc. BW Utilization in DC, and Avg. Perc. Utilization in DC. Availability factor for these parameters is calculated as:

$$\text{Availablity Factor} = \text{Total Available} - \text{Total Utilized}$$

3 RESULTS AND DISCUSSIONS

This section includes the results and discussions for the conducted experiment. Table 1 depicts the experimental results from all the stages.

Table 1. Experimental results for all the four stages for algorithms RR – ON and LIF – ON across all the performance parameters.

Parameter	$PM_1 = 25,$ $PM_2 = 25,$ $PM_3 = 25.$		$PM_1 = 50,$ $PM_2 = 50,$ $PM_3 = 50.$		$PM_1 = 75,$ $PM_2 = 75,$ $PM_3 = 75.$		$PM_1 = 100,$ $PM_2 = 100,$ $PM_3 = 100.$	
	RR – ON	LIF – ON	RR – ON	LIF – ON	RR – ON	LIF – ON	RR – ON	LIF – ON
Unbalanced Degree of DC – Variance	0.2	0.14	0.2	0.12	0.2	0.14	0.2	0.13
Unbalanced Degree of DC – Index	358.37	321.03	358.37	317.56	358.37	319.66	357.57	319.35
Unbalanced Degree of PMs – Variance	0.028	0.036	0.028	0.031	0.028	0.033	0.029	0.03
Unbalanced Degree of PMs – Index	325.94	321.87	325.94	319.7	325.94	321.65	325.92	319.8
Avg. Perc. CPU Utilization in DC	39	56	39	57	39	57	39	57
Avg. Perc. RAM Utilization in DC	49	70	49	69	49	70	50	69
Avg. Perc. BW Utilization in DC	41	55	41	56	41	58	42	57
Avg. Perc. Utilization in DC	43	61	43	60	43	62	44	61
Available CPU Utilization in DC	61	44	61	43	61	43	61	43
Available RAM Utilization in DC	51	30	51	31	51	30	50	31
Available BW Utilization in DC	59	45	59	44	59	42	58	43
Available Utilization in DC	57	39	57	40	57	38	56	39
Performance	RR – ON > LIF – ON		RR – ON > LIF – ON		RR – ON > LIF – ON		RR – ON > LIF – ON	

From the above table, we can observe the following with respect to availability factors:

- Availability CPU Utilization in DC (RR – ON) > Availability CPU Utilization in DC (LIF – ON) for all the four stages
- Available RAM Utilization in DC (RR – ON) > Available RAM Utilization in DC (LIF – ON) for all the four stages.
- Available BW Utilization in DC (RR – ON) > Available BW Utilization in DC (LIF – ON) for all the four stages.
- Available Utilization in DC (RR – ON) > Available Utilization in DC (LIF – ON) for all the four stages.

∴ The performance of RR – ON is better than LIF – ON for all the four stages and across all the performance parameters.

4 EMPIRICAL ANALYSIS

This section includes the empirical analysis of the results obtained for the experiment conducted with respect to the performance parameters: Unbalanced Degree of DC - Index (UDDC - I), Unbalanced Degree of PM - Variance (UDPM – V), and Unbalanced Degree of PM - Index (UDPM - I). Terminologies used for the empirical analysis are: **LiReEq**: Linear Regression Equation, **ReLiSl**: Regression Line Slope, **SlSi**: Slope Sign, **ReY-In**: Regression Y-Intercept, **Rel**: Relationship. Table 2 depicts the empirical analysis which is performed for RR – ON and RR – OFF based on the parameters of unbalanced degrees of DCs and PMs.

From the above table, we can observe that:

- PM = ↓ UDDC – I: As the PMs increases, the amount of Unbalanced Degree of DC – Index decreases for RR – ON.
- PM = ↓ UDDC – I: As the PMs increases, the amount of Unbalanced Degree of DC – Index decreases for LIF – ON.
- PM = ↑ UDPM – V: As the PMs increases, the amount of Unbalanced Degree of PM – Variance increases for RR - ON.
- PM = ↓ UDPM – V: As the PMs increases, the amount of Unbalanced Degree of PM – Variance decreases for LIF – ON.
- PM = ↓ UDPM – I: As the PMs increases, the amount of Unbalanced Degree of PM – Index decreases for RR - ON.
- PM = ↓ UDPM – I: As the PMs increases, the amount of Unbalanced Degree of PM – Index decreases for LIF – ON.
- Performance (RR – ON) > Performance (LIF – ON) for Unbalanced Degrees of DC – Index and Unbalanced Degree of PM – Index.
- Performance (LIF – ON) > Performance (RR – ON) for Unbalanced Degrees of PM – Variance.

Table 2. Empirical analysis of RR – ON and RR – OFF with respect to availability factors.

	Unbalanced Degree of DC – Index (UDDC – I)		Unbalanced Degree of PM – Variance (UDPM – V)		Unbalanced Degree of PM – Index (UDPM – I)	
	RR – ON	LIF – ON	RR – ON	LIF – ON	RR – ON	LIF – ON
LiReEq	y = −0.24x + 358.77	y = −0.294x + 320.14	y = 0.0003x + 0.0275	y = −0.0016x + 0.0365	y = −0.006x + 325.95	y = −0.426x + 321.82
ReLiSl	−0.24	−0.294	0.0003	−0.0016	−0.006	−0.426
SlSi	Negative	Negative	Positive	Negative	Negative	Negative
ReY-In	358.77	320.14	0.0275	0.0365	325.95	321.82
Rel	Negative	Negative	Positive	Negative	Negative	Negative
R^2	0.6	0.0707	0.6	0.6095	0.6	0.2230
PM Analysis	↑ PM = ↓ UDDC - I	↑ PM = ↓ UDDC - I	↑ PM = ↑ UDPM – V	↑ PM = ↓ UDPM - V	↑ PM =↓ UDPM - I	↑ PM = ↓ UDPM - I
Result	RR – ON > LIF – ON		LIF – ON > RR – ON		RR – ON > LIF – ON	

5 CONCLUSION

The process of resource scheduling is crucial for the cloud to generate consistent and best results. While doing so, the resource availability factor is important to ensure that the cloud is always available for any on-demand user requests. Hence, it becomes significant and crucial to study the resource scheduling algorithms concerning their availability factors. To bridge this gap, this research studied the RR – ON and LIF – ON resource scheduling algorithms. These algorithms were incorporated into the CloudSched simulation environment, and jobs were processed in four stages. In each stage, physical machines (PM) were virtualized into eight virtual machines (VMs), and jobs were processed on these VMs. The count of PMs varied in each stage. The main reason for using multiple stages for this experiment is to extensively observe and compare the behavior of RR – ON and LIF – ON. For this comparison, the performance metrics considered were: Unbalanced Degree of DC - Variance, Unbalanced Degree of DC - Index, Unbalanced Degree of PMs – Variance, Unbalanced Degree of PMs - Index, Average Percentage CPU Utilization in DC, Average Percentage RAM Utilization in DC, Average Percentage BW Utilization in DC, and Average Percentage Utilization in DC. Considering these metrics, availability factors were calculated for each performance parameter and used for this extensive comparison. From the experiment, it can be concluded that the performance of RR – ON is better than LIF – ON concerning the availability factors across all the stages. An empirical analysis is also performed concerning performance parameters: Unbalanced Degree of DC – Index (UDDC – I), Unbalanced Degree of PM – Variance (UDPM – V), and Unbalanced Degree of PM – Index (UDPM – I). This empirical analysis provides an additional extensive comparison of RR – ON and LIF – ON. The empirical analysis suggests that as the number of PMs increases, the RR – ON algorithm performs better than LIF – ON concerning UDDC – I and UDPM – I. However, the LIF – ON algorithm gives better output in terms of UDPM – V. These algorithms can be provided with an intelligence mechanism to enhance their working individually and together through a hybrid algorithm. The reinforcement Learning (RL) technique is a feedback-based Machine Learning (ML) technique that can improve the results of any system when incorporated into it. This RL mechanism can be implemented to enhance these algorithms individually and provide a hybrid algorithm that can process jobs in time quantum considering the least imbalance level first. The cloud can provide better output with enhanced resource scheduling, improving its overall performance.

REFERENCES

Anoushee, M., Fartash, M., & Akbari Torkestani, J. 2023. An Intelligent Resource Management Method in SDN Based Fog Computing Using Reinforcement Learning. *Computing.*

Dillon, T., Wu, C., & Chang, E. 2010. Cloud Computing: *Issues and Challenges. 2010 24th IEEE International Conference on Advanced Information Networking and Applications.*

Gonzalez, C., & Tang, B. 2020. FT-VMP: Fault-Tolerant Virtual Machine Placement in Cloud Data Centers. *2020 29th International Conference on Computer Communications and Networks (ICCCN).*

Han, S., Min, S., & Lee, H. 2019. Energy Efficient VM Scheduling for Big Data Processing in Cloud Computing Environments. *Journal of Ambient Intelligence and Humanized Computing.*

Ma, X., Xu, H., Gao, H., Bian, M., & Hussain, W. 2023. Real-Time Virtual Machine Scheduling in Industry IoT Network: A Reinforcement Learning Method. *IEEE Transactions on Industrial Informatics*, 19(2), 2129–2139.

Ramezani Shahidani, F., Ghasemi, A., Toroghi Haghighat, A., & Keshavarzi, A. 2023. Task Scheduling in Edge-fog-cloud Architecture: A Multi-objective Load Balancing Approach Using Reinforcement Learning Algorithm. *Computing.*

Sanaj, M. S., & Joe Prathap, P. M. 2020. An Enhanced Round Robin (ERR) Algorithm for Effective and Efficient Task Scheduling in Cloud Environment. 2020 *Advanced Computing and Communication Technologies for High Performance Applications (ACCTHPA).*

Shaw, R., Howley, E., & Barrett, E. 2022. Applying Reinforcement Learning Towards Automating Energy Efficient Virtual Machine Consolidation in Cloud Data Centers. Information Systems, 107, 101722.

Tian, W., Zhao, Y., Zhong, Y., Xu, M., & Jing, C. 2011. A Dynamic and Integrated Load-balancing Scheduling Algorithm for Cloud Datacenters. *2011 IEEE International Conference on Cloud Computing and Intelligence Systems.*

Vengerov, D. 2007. A Reinforcement Learning Approach to Dynamic Resource Allocation. *Engineering Applications of Artificial Intelligence*, 20(3), 383–390.

Artificial Intelligence, Blockchain, Computing and Security – Dagur et al. (Eds)
© 2024 The Author(s), ISBN: 978-1-032-49393-0

Higher education recommendation system using data mining algorithm

S. Ponmaniraj, S. Naga Kishore, G. Shashi Kumar, C.H. Abhinay & B. Harish
Bharath Institute of Science and Technology, Chennai, India

ABSTRACT: In many organizations, machine learning and data mining techniques are used for analyzing large amount of available data's, information's for decision making process. In educational sector, Machine learning is used for wide variety of applications such as suggestion to the students based on 10th mark and interest. The process of self-analysis, critical thinking, and decision-making is the important milestones for a person's life. Here, a survey on academic decisions and the variables that affect those decisions will be presented. A real case study using machine learning algorithms helped support decision-making by predicting courses and institutions. It is usually influenced by the views of your parents, friends, relatives, teachers, and the media that you choose a career. You need to plan your career wisely and as early as possible today, as there is choicer and greater competition.

1 INTRODUCTION

Higher education must address two requirements for improved data utilization. The first is influenced by outside variables, but the second is influenced internally through ongoing quality improvement [1]. Governments have been working to gather data that demonstrates the idea that institutions are responsible for the money they get as a result of steep decreases in funding and public support [2]. By responding to external requests for information, many colleges and institutions have been able to derail unfavorable developments while assuming a defensive stance. On a more significant level, organization's that consciously utilize data to enhance performance generally putting in place a data-driven future while satisfying compliance-based criteria [3]. It may be contested that there hasn't been much of a change in how higher education uses data. It is also evident that technology has made it feasible to have new talks at the same time [4]. Institutions now have more options to better serve students through the use of data, owing to evolving techniques like analytics or predictive analytics. Big data may be utilized in the future to bring together strategic data on student learning and achievement, budgets, and efficiency for colleges and institutions [5]. More data than ever before is being gathered in higher education. However, objective, developing institutional strategy, these efforts are typically focused on the first imperative, compliance reporting [6]. This seeming dichotomy will be soon resolved by institutions with a forward-thinking attitude. They will look for chances to increase capacity, lift restrictions that prevent them from crossing existing borders that govern the use of data, and discover approaches of integrating data and strategy [7]. The outcome may support institutional goals, satisfy requirements of outside policy, and enhance student achievement. Costs are associated with strategic thinking and the data they require [8]. We examine the advantages and disadvantages of producing and utilizing useful strategic and operational data in this chapter. Based on our work collaborating with institutions of higher learning to foster strategic planning and create cultures that value research and evidence, also establish effective data

DOI: 10.1201/9781003393580-70

usage procedures [9]. We also look at new technological developments and how they might assist schools in assisting students [10]. Instead of giving a theoretical examination of the fundamentals of strategic planning, this chapter is meant to offer suggestions that can be put into practice. Institutions that have the guts to go on a data journey need help [11]. To aid in navigating these new pathways, this chapter also offers guidance gleaned from personal experience and recent advancements in management science.

Data collection in higher education is at an all-time high. We give the results of the poll, including the factors that were taken into consideration when making academic selections. We projected programs/institutions in a genuine case study using machine learning techniques to aid in decision-making [12]. A system for academic information that allows you to get material from a variety of sources to help you make decisions. Institutions now have more options to better serve students by using data to boost their efficiency thanks to emerging methodologies like analytics or predictive analytics [13].

2 METHODOLOGIES

2.1 Data collection module

A worldwide dataset is known as data. In this system, a model is trained using data from the Pima Indian population. The training data is the first collection of information that is utilized to comprehend the program. In order to set the feature and use the system-available data, this is the case where we must train the model first [14]. The system is taught to perform various tasks using this data. It is the set of data that the model can use to learn from using an algorithm, hence automating tasks. Software receives testing data as input. This is primarily useful for testing because it demonstrates how the data is affected when the specified module is executed.

2.2 Preprocessing module

The process of creating a clean data set from raw data is known as data preparation. Data transformation or encoding is what it is so that it may be easily parsed by a machine. Removing irrelevant data and filling in missing values is the main goal of data preparation in the learning process in order for machines to be trained more readily [15].

2.3 Feature extraction module

Feature Extraction is the process used to change the important information for result features. The attributes of supplied designs that aid in differentiating among the class of vital pattern aspects are calculated using this trait square. This technique involves lowering the resource counts needed to convey the massive quantity of data [16]. Feature. A procedure of attribute reduction is extraction. This is also utilized to boost supervised learning's efficiency and speed.

2.4 Applying machine learning and data mining algorithm

The ML Thomas Cover invented the algorithm, a non-parametric technique for regression and classification. This technique is mostly employed in the sector to classify challenges. The machine learning algorithm is one type of instance-based learning technique. In order to significantly improve accuracy, this method normalizes training data before using distance to classify objects. The collection of objects for whose classes or object property values are known is used to determine the neighbors [17]. Although there are no explicit training steps needed, it can be considered a training set for the algorithm.

The Thomas Cover-proposed non-parametric method for regression and classification is known as the data mining algorithm. This method is primarily used in the industry to categorize difficulties. Instance-based learning is a sort of algorithm used in data mining. This technique uses distance to classify objects, normalizing training data to greatly increase accuracy [18]. The collection of objects for whose classes or object property values are known is used to determine the neighbor's. Although there are no explicit training steps needed, it can be considered a training set for the algorithm.

2.5 *Prediction*

Machine learning is employed in this module for a wide range of purposes, including making suggestions to students depending on their interest and grade in the tenth grade. Self-analysis, critical thinking, and decision-making are three of the most significant life milestones. You should be aware of your skills, interests, and personality while selecting a stream after the tenth grade, a training program, or a career path. In addition to this, you should learn about various career alternatives, eligibility requirements, top institutions/schools, other selection factors, and market demands.

Higher Education Recommendation System is a system framework that will suggest universities to the users of the system who are searching for universities to seek after education for their higher studies and recommends the top universities to the students or users of the system. The users need to fill the registration form to get the login details and this system will suggest universities as seen in above Figure 1. The main parameters used to recommend universities are CGPA percentage, University ratings etc. This System utilizes machine learning algorithms and compares accuracy among algorithms.

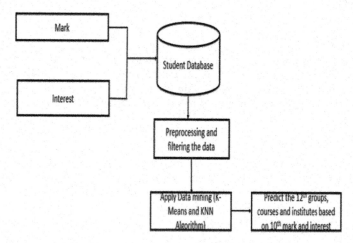

Figure 1. Computer architecture.

3 RESULTS AND DISCUSSION

The recently formed higher education policy calls for the development of a more varied higher education system with institutions that are adaptable and flexible. Machine learning is employed in the educational industry for a wide range of applications, including making suggestions to pupils based on their interest and grade in the tenth grade. We hoped it would assist you in comprehending what each course for a career entails and assist you in reflecting on and analyzing the skills necessary to be successful, content, and happy in the course or career you pick.

3.1 System design

3.1.1 Input design

A connection between the user and the information system is created through the input design. Developing requirements and techniques for data preparation is part of the process of converting transaction data into a format that can be processed. To do this, users can input the information directly into the system, or a computer can read the information from a written or printed document. The input process was designed with the intention of minimizing the amount of input necessary, errors, delays, additional phases, and maintaining a straightforward workflow. The entry of the data protects its privacy, usability, and security. The discussion is intended to guide the operating staff's input techniques for developing input validations and action plans in the case of a mistake.

3.2 Output design

The end user's needs must be met and the information must be presented properly in a high-quality output. The means through which the dissemination of processing results to consumers and other systems are known as a system's outputs. During output design, decisions are made about both the replacement of information for output in hard copy and on-demand consumption. The user uses it as their main and most immediate information source. Effective and clever output design improves the system's ability to assist users in making decisions. An information system's output format should achieve one or more of the aforementioned goals.

1. Share information about your previous acts, the current circumstance, or
2. Forecasts for the future.
3. Identify important times, opportunities, problems, or warnings.
4. Starting a task;

In existing system, Using actual data, three supervised classification algorithms are used to forecast the graduation rates of undergraduate engineering students in South America. As a measure of effectiveness, decision trees, logistic regression, and random forests are evaluated, as well as the receiver operating characteristic curve and accuracy are contrasted and contrasted with the last one showing the best results. The main disadvantage of the existing system miss the undesirable data for the students and it may not check the social data for the student.

In proposed system, machine learning is used for a wide range of applications, including suggestions to the kids based on their interest and 10th grade point. A turning point in a person's life involves self-evaluation, critical thought, and decision-making. You should be aware of your skills, hobbies, and personality while selecting a stream after 10th grade, a training program, or a profession, and 12th grade groups. Along with these, you should learn about various career alternatives, eligibility requirements, top institutions/schools, other selection factors, and market demands. The advantage of proposed system is it has high accuracy efficiency.

Predicting a student's success involves tasks that are similar to more conventional duties involving course materials retrieval or suggestion. However, one of the least researched recommender areas is personal education. One example is that by evaluating the complexity of various course components like intelligent tutoring systems have used recommender systems to predict student achievement. The range of educational recommender systems can be very different, both in terms of the computational techniques employed and the subject areas they cover.

4 CONCLUSIONS

In this study, the classification task is applied to a student database to forecast the student's choice based on historical data. K-means clustering Classifier and KNN Classifier are used here because there are numerous methods for classifying data. From the student's previous database, data like marks and interests were gathered in order to forecast the training program or career and 12th groups at the conclusion. The results of this study will aid students in making better decisions. This could aid pupils in advancing academically, which would ultimately result in wise career choices.

REFERENCES

[1] Lašáková. A, Bajzíková. L, and Dedze. I, "Barriers and drivers of innovation in higher education: Case study-based evidence across ten European Universities," *Int. J. Educ. Develop.*, vol. 55, pp. 69–79, May 2017.

[2] Acevedo. Y. V. N and Marín. C. E. M, "System Architecture Based on Learning Analytics to Educational Decision Makers Toolkit," *Adv. Computer. Sci. Eng.*, vol. 13, no. 2, pp. 89–105, 2014.

[3] Góes. A. R. T, Steiner. M. T. A., and Neto. P. J. S., "Education Quality Measured by the Classification of School Performance Using Quality Labels," *Appl. Mech. Mater.*, vols. 670–671, pp. 1675–1683, Oct. 2014.

[4] Lounis. H and Fares. T, "Using Efficient Machine-learning Models to Assess Two Important Quality Factors: Maintainability and Reusability," in *Proc. Joint Conf. 21st Int. Workshop Software. Meas. 6th Int. Conf. Software. Process Product Meas.*, Nov. 2011, pp. 170–177.

[5] Nieto. Y, García-Díaz. V, Montenegro. C, and Crespo. R. G., "Supporting Academic Decision Making at Higher Educational Institutions Using Machine Learning-based Algorithms,' *Soft Comput.* vol.23, no.12, pp.4145–4153, 2018.

[6] Palvika, Shatakshi, Sharma, Y., Dagur, A., & Chaturvedi, R. (2019). Automated Bug Reporting System with Keyword-driven Framework. In *Soft Computing and Signal Processing: Proceedings of ICSCSP 2018*, Volume 2 (pp. 271–277). Springer Singapore.

[7] Kumar, A., & Alam, B. (2019). Energy Harvesting Earliest Deadline First Scheduling Algorithm for Increasing Lifetime of Real Time Systems. *International Journal of Electrical and Computer Engineering*, 9(1), 539.

[8] Kumar, A. & Alam, B. (2018). Task Scheduling in Real Time Systems with Energy Harvesting and Energy Minimization. *Journal of Computer Science*, 14(8), 1126–1133.

[9] Kumar, A., & Alam, B. (2014, February). Real Time Scheduling Algorithm for Fault Tolerant and Energy Minimization. In *2014 International Conference on Issues and Challenges in Intelligent Computing Techniques (ICICT)* (pp. 356–360). IEEE.

[10] Chaturvedi, R., Kumar, S., Kumar, U., Sharma, T., Chaudhary, Z., & Dagur, A. (2021). Low-Cost IoT-enabled Smart Parking System in Crowded Cities. In *Data Intelligence and Cognitive Informatics: Proceedings of ICDICI 2020* (pp. 333–339). Springer Singapore

[11] Faham. E, Rezvanfar. A, Mohammadi. S. H. M, and Nohooji. M.R, "Using System Dynamics to Develop Education for Sustainable Development in Higher Education with the Emphasis on the Sustainability Competencies of Students," *Technol. Forecast. Social Change*, vol. 123, pp. 307–326, Oct. 2017.

[12] Goni. F. A, Chofreh. A. G, Mukhtar. M, Sahran. S, Shukor. S. A, and Klemeš. J. J, "Strategic Alignment Between Sustainability and Information Systems: A Case Analysis in Malaysian Public Higher Education Institutions," *J. Clean. Prod.*, vol. 168, pp. 263–270, Dec. 2017.

[13] González-González. I and Jiménez-Zarco. A. I, "Using Learning Methodologies and Resources in the Development of Critical Thinking Competency: An Exploratory Study in a Virtual learning Environment," *Comput. Hum. Behav.*, vol. 51, pp. 1359–1366, Oct. 2015.

[14] Dhanaraj, Islam. R.K, S.H. & Rajasekar V. A Cryptographic Paradigm to Detect and Mitigate Blackhole Attack in VANET Environments. *Wireless Netw* (2022). https://doi.org/10.1007/s11276-022-03017-6

[15] Kushwaha, A., Amjad, M., & Kumar, A. (2019). Dynamic Load Balancing Ant Colony Optimization (DLBACO) Algorithm for Task Scheduling in Cloud Environment. *Int J Innov Technol Explor Eng*, 8 (12), 939–946.

[16] Rajasekar, V., Premalatha, J., & Dhanaraj, R. K. (2022). Security Analytics. In *System Assurances* (pp. 333–354). Elsevier. https://doi.org/10.1016/b978-0-323-90240-3.00019-9

[17] Jeyaselvi, M., Dhanaraj, R.K., Sathya, M. *et al*. A Highly Secured Intrusion Detection System for IoT Using EXPSO-STFA Feature Selection for LAANN to Detect Attacks. *Cluster Computing* (2022). https://doi.org/10.1007/s10586-022-03607-1

[18] Saravanakumar. P, Sundararajan. T. V. P, Dhanaraj. R. K, Nisar. K, Memo. F. H., "Lamport Certificateless Signcryption Deep Neural Networks for Data Aggregation Security in wsn," *Intelligent Automation & Soft Computing*, vol. 33, no.3, pp. 1835–1847, 2022.

Artificial Intelligence, Blockchain, Computing and Security – Dagur et al. (Eds)
© *2024 The Author(s), ISBN: 978-1-032-49393-0*

A brief evaluation of deep learning-based retinal disease approaches

Reetika Regotra, Tamana & Samridhi Singh
National Institute of Technology, Hamirpur, Himachal Pradesh, India

Shekhar Yadav
Madan Mohan Malaviya University of Technology, Gorakhpur, India

ABSTRACT: One of the most delicate parts of the eye is the retina. It is a brain limb that is linked by an optical nerve. The visual information can be organized by retina only. Numerous diseases, including ophthalmology and laryngology, can affect this substantial portion of the eye. Here, segmentation in medical image processing enters the picture for the correct identification of the various retinal diseases. The computer-aided diagnosis makes this simple to accomplish (CAD). CAD greatly helps in the routine clinical application of segmenting blood vessels closer to the retina. The key goal of this article is to present all of the most recent research and studies conducted on retinal disease using CAD since the paradigm has transformed from machine learning to deep learning. In our most recent comparison, we considered studies from 2019 to 2022 as well as benchmark datasets from the public domain (DRIVE, STARE, CHASE ERC) and performance evaluation metrics based on images taken with a fundus camera. Although the field of deep learning has made significant progress, there are still numerous challenges. As a result, we carefully examined these techniques in order to compile a list of unresolved research problems and potential solutions that could suggest a new direction of study for the scientific community.

Keywords: Retinal disease, deep learning, machine learning

1 INTRODUCTION

One of the most delicate parts of the eye is the retina. It is a brain limb that is linked by an optical nerve. The visual information can be organized by retina only. Numerous diseases, including ophthalmology and laryngology, can affect this substantial portion of the eye. Here, segmentation in medical image processing enters the picture for the correct identification of the various retinal diseases. The computer-aided diagnosis makes this simple to accomplish (CAD). CAD greatly helps in the routine clinical application of segmenting blood vessels closer to the retina. The key goal of this article is to present all of the most recent research and studies conducted on retinal disease using CAD since the paradigm has transformed from machine learning to deep learning. In our most recent comparison, we considered studies from 2019 to 2022 as well as benchmark datasets from the public domain (DRIVE, STARE, CHASE ERC) and performance evaluation metrics based on images taken with a fundus camera. Although the field of deep learning has made significant progress, there are still numerous challenges. As a result, we carefully examined these techniques in order to compile a list of unresolved research problems and potential solutions that could suggest a new direction of study for the scientific community. A light-sensitive layer in the human body, the retina, receives images that are shaped by the focal point of a lens and sends them to the brain via the optical nerves (Krishna *et al.* 2022). Globally, a considerable number of individuals suffer from visual impairment due to several ocular disorders every year. It has been found out that

DOI: 10.1201/9781003393580-71

approximately 2.2 billion of the population possess ocular and eyesight issues (Steinmetz *et al.* 2021). Notably, the two major reasons of visual impairment have been found out as diabetic retinopathy (DR) and diabetic maculopathy (DM), among which the latter is characterized to possess an overlong preclinical period. Under critical situations, in case a patient loss of vision then its recovery becomes almost impossible. Hence, early diagnosis of the disease ensures avoidance of the disease advancement for protection of eyesight of the patient. Among other significant concerns that arise during ocular disorder treatment, the primary concern remains the recognition of the visual disorder in the initial phase of its occurrence. This happens due to many reasons as restricted access to eye specialists, refusal for regular ocular inspections, additional to detection of ocular disorders from the images of the retina, and so on (Khandouzi *et al.* 2022). It has been stated that despite being the group most at risk, a mere count of 30% African Americans belonging to the south of Los Angeles reported having their eyes examined for diabetic retinopathy (DR) (Lu *et al.* 2016). Moreover, the recognition of ocular diseases is considered as a time-intensive, tedious task which is decided according to the judgment of various professionals. Thus, in order to increase diagnostic rates, it is crucial to create a smart automatic tool for extraction of blood vessels and better health results (Khanal & Estrada 2020). Since 1982, the medical field have been using computer vision for analysis of retinal images for better results (Soomro *et al.* 2016) (Soomro *et al.* 2017). Machine learning based techniques particularly supervised learning methodologies are prevalent these days as they are reviewed to be beneficial due to their progression and achievement in the area of medical image processing. (Soomro *et al.* 2017). Recent developments in deep learning techniques for medical image processing have led to the creation of a number of algorithms for automatically segmenting retinal vessels to obtain improved performance and ensure a reduction in labour expertise (Wang *et al.* 2020). Ophthalmologists can diagnose ocular diseases including diabetic retinopathy, hypertension-related retinopathy, glaucoma, and choroidal neovascularization using the information related to retinal blood vessel extraction (Prajna & Nath 2022). The alteration in retinal vascular anatomy can be used to diagnose related disorders as brain and heart stocks. With the purpose of diagnosing and determining the ocular disorder, retinal blood vessel extraction is carried out by utilizing vessel elements such as its distance, breadth, pattern of branching and angle (Hassan *et al.* 2015). Segmenting retinal vessels has long been a challenge for the area of medical image analysis. Low contrast, intensity inhomogeneity, and varying thickness levels between capillaries and major blood vessels are the main issues on a fundus retinal blood vessel image (Wu *et al.* 2020). As a result, a variety of automated blood vessel segmentation techniques have been suggested. The authors (Kirbas & Quek 2004) describe a thorough investigation into the extraction of vessel-like structures in medical images, both in 2D and 3D. Also in (Winder *et al.* 2009) (Faust *et al.* 2012), the focus is on methods for retinal image analysis that can automatically diagnose diabetic retinopathy. Researchers (Ricci & Perfetti 2007) also tried segmenting retinal vessels with the use of line operators along with SVM. Existing solutions either rely on rules (labelling each pixel as a blood vessel based on predefined criteria) or machine learning (including techniques of feature extraction). Furthermore, these approaches are further divided into the following categories based on the kind of segmentation algorithm used: (a) Unsupervised techniques (a) Practices under supervision Morphology and multiscale methods (d) Kernel or matched filtering (c) Vessel tracking techniques (e) Model-based (Almotiri *et al.* 2018). This research work's goal is to provide a thorough analysis of current deep learning developments that address the issues with retinal vascular segmentation. However, it is predicted that the endeavor would make it simpler for academics interested in deep learning to publish their discoveries in the area of retinal vascular segmentation.

The following includes the main aim of our article:

(i) An overview of deep learning is provided and summarised the resiliency of deep learning-inspired applications in the detection of retinal diseases.
(ii) Through performance evaluation protocols, a comparative analysis of various segmentation techniques used for the correct identification of retinal diseases is presented.
(iii) The presented study investigated some broad research challenges that may inspire the research community to conduct additional research in this active field of medical image processing.

2 OUTLINE OF DEEP LEARNING

Deep learning is a branch of machine learning which was developed around 1950s. It is developed from (ANNs) Artificial Neural Networks with multiple layers also called as deep neural networks (DNNs). which carries behavior of human brain, analyze and summarize neurons (Sule 2022). To create systems that learn similarly how humans learn, this architecture deep learning architecture was developed from the structure of human brain. Similar to neurons that performs the basic building blocks of the brain, deep learning architecture have perceptron, a computational unit. Perceptron acquires input signals and converts them into output signals similar to neuron that transmits electrical pulses. Every layer of perceptron interpret a particular pattern within the data. Deep learning models use hierarchical structures to establish connection with layers so that the output of lower layer can be used as input of higher layer by applying non-linear and linear computations. The low-level features of data can be converted to high-level features with the help of deep learning (DL) models making DL techniques much robust as compared to traditional machine learning models in representation of features. The working and performance of traditional machine learning model depends on user experience, while in deep learning, it depends on the data which makes deep learning very supreme in machine learning. Deep learning models are categorized into supervised as well as unsupervised deep learning techniques. These are Convolutional Neural Networks (CNN), Auto-Encoder(AE), Deep Belief Networks(DBNs) and Graphical Neural networks(GNNs). Deep learning is widely used in image segmentation, tracking of the object, and detection of the object.

3 LITERATURE SURVEY

This section covers a deep literature survey of all research articles related to retinal disease detection. We gained a detailed understanding of the segmentation techniques used and improved the model's performance by reviewing numerous recent studies from 2019 onward.

Researchers (Hashemzadeh & Azar 2019) incorporated supervised learning techniques with unsupervised learning techniques for the extraction of blood vessels of retina. Unsupervised techniques were employed for the extraction of clear and thick blood vessels, and supervised techniques extracted thin blood vessels. The incorporation of both unsupervised and supervised techniques addressed the issues related to intra-class high variance of image features. The formulated methodology was comprised of three major stages: pre-processing, vessel extraction, and post-processing. In the first stage, the field of view (FOV) was extracted from the image, color space transformation was applied, appropriate color channels were chosen, and contrast was enhanced for better results. In the second stage of vessel extraction, initially features were extracted which resulted into feature vector construction. Also, principal component analysis was applied along with certain algorithms of clustering (for non-vessel clusters) and classification (for vessel clusters). The detected blood vessels were fed to the last stage of post-processing in which simple masking was applied to get the final results. The methodology utilized influential image features among which three features viz. TH, SC and BPS were consequential in determining the accuracy criterion of the vessel extraction. Moreover, five major criteria namely, accuracy, sensitivity, specificity, positive predictive value (PPV) and area under curve (AUC) were taken into consideration for evaluation measures of the suggested method. The implementation of the formulated technique was evaluated by employing it on widely known datasets DRIVE, STARE and CHASE DB1. The computed sensitivity values on the respective datasets came out as 0.7830, 0.8087 and 0.773, and specificity as 0.9800, 0.9892 and 0.9840, PPV as 0.8584, 0.9012 and 0.8488, accuracy as 0.9531, 0.9691 and 0.9623, AUC as 0.9752, 0.9853 and 0.9789.

The higher accuracy obtained in the segmentation phase of blood vessels of retina is considered as of much importance as it aids doctors in diagnosing retinal fundus diseases. A study by (Li *et al.* 2021) utilized U-Net and Dense-Net to present a new methodology ensuring to minimize the segmentation errors and enhance accuracy in conventional segmentation of retinal vessels. The technique was initiated with the enhancement of vascular feature information which was done using fusion limited contrast histogram equalization, median filtering, data normalization and multiscale morphological transformation. Adaptive gamma correction was used for the correction of the obtained artifacts. With the purpose of improving generalization and to expand the dataset, the image blocks are arbitrarily extracted and utilized as a training data. Also, dice loss function was optimized with the help of stochastic gradient descent to ensure enhanced accuracy of segmentation. At last, segmentation was carried out using Dense-U-net. The results demonstrated better segmentation of small retinal blood vessels and enhanced segmentation accuracy of retinal blood vessels. Sensitivity, specificity, PPV, accuracy and area under curve (AUC) were taken into consideration for performance evaluation of the suggested method on DRIVE dataset which was calculated as 0.7831, 0.9896, 0.8946, 0.9698 and 0.9738.

Deep learning algorithms provide better outcomes in segmentation of blood vessels among which mostly supervised techniques are utilized. Supervised learning techniques demands huge count of labelled high quality retinal images and labelling is cost-intensive with regard to individual effort and finance. Authors (Chen *et al.* 2020) offered a novice methodology which was based on semi supervised technique to perform segmentation on retinal images with minimal labelling. The retinal vessel tree was segmented using the enhanced U-Net deep learning. The formulated method outperformed the outcomes on the DRIVE dataset as it used merely eleven labeled fundus images provided accuracy as 0.9631 and AUC as 0.9760. Experimental observations demonstrated the effective performance and better convergence of the formulated technique. Moreover, Retinopathy in people suffering from diabetics can be prevented by performing the segmentation of fundus images of retinal blood vessels. It also leads to the minimization of disease occurrences and assesses the treatment of ocular diseases. A study (Cheng *et al.* 2020) used U-Net network enhanced by addition of a dense block to formulate a new methodology for retinal vessel segmentation. The suggested technique facilitated enhanced transmission of the output of every layer further fed as an input in the multi-layered structure. The feature map created by every layer was appropriately utilized by the method by incorporating detailed information of low level with the semantic information of high level and thereby ensured reduction in the redundant parameters. Batch normalization function (BN) and the PReLU activation function were used in the training phase for the purpose of optimization, reduction of over-fitting issues and enhancement of the learning ability. The widely used datasets namely, DRIVE and CHASE DB1 were selected to evaluate the competence of the proposed methodology which provided sensitivity as 0.7672 and 0.8967, specificity as 0.9834 and 0.9540, accuracy as 0.9559 and 0.9488, AUC as 0.9793 and 0.9785 on respective datasets. The developed methodology proved optimal for small retinal vessels segmentation and thus, verified its excellence in the evaluation of ocular diseases. Segmentation was performed in retinal blood vessel tree to appropriately recognize and diagnose several eye disorders and thereby demanded superior performance of segmentation along with minimal time for computation. To achieve the desired outcome, authors (Boudegga *et al.* 2021) proposed a new U-form methodology based on deep neural networks which utilized lightweight convolution blocks. In addition to that, the provided images were initially pre-processed for enhancement of their quality and contrast of blood vessels. After that, data augmentation was performed enabling image transformation and patch extraction. To determine the efficiency of the formulated methodology, the proposed method was verified four times on widely available datasets namely, DRIVE and STARE which computed sensitivity as 0.8448 and 0.8060, specificity as 0.9900 and 0.9928, precision as 0.8890 and 0.8764, accuracy as 0.9819 and 0.9816.

But, the segmentation performed using ResNet, VGG16 and U-net was not sufficient because of absence of profound segmentation found in dense regions. To overcome this, researchers (Kumar & Singh 2022) suggested a new methodology introducing retinal blood

469

vessels extraction using generalized Pareto probability distribution function (pdf) based on matched filter and a registration methodology for vessels segmented on the basis of features specifying Binary Robust Invariant Scalable Key point (BRISK). Comparative to Probability Distribution Function or pdf, Binary Robust Invariant Scalable Key point (BRISK) provided the previously established pattern of sampling. In addition to that, BRISK also enabled attention point identification as well as a matching strategy for alternating arrangement of vessels. The image was firstly pre-processed using PCA, CLAHE and TOGGLE to enhance the contrast to such that vessels could be found out clearly. The generalized pareto distribution matched filter was applied on improved image of retina which resulted into Matched Filter Response (MFR) image. The resultant MFR image was processed with optimum thresholding for the purpose of post-processing as well as BRISK feature detection. The performance of the proposed methodology was validated through experiments on DRIVE dataset particularly for specificity, sensitivity and average accuracy computed as 0.6605, 0.9808 and 0.9526 using Generalised Pareto PDF and 0.6764, 0.9914 and 0.9851 using Registration approach.

It was found that the testing of Diabetic retinopathy which is an ocular disorder is cost intensive in terms of both effort and finance. A study (Saranya *et al.* 2022) offered a solution by suggesting a new methodology for retinal blood vessel segmentation in fundus images by utilizing deep learning techniques. Diabetic Retinopathy is generally classified as proliferative diabetic retinopathy or PDR and non-proliferative diabetic retinopathy or NPDR. This methodology focuses on whether eyes is in PDR or NPDR. In the initial phase images were pre-processed, then vessel segmentation is done, after that extra features are removed, and then classification and prediction is carried. For image enhancement in pre-processing various procedures were carried like resizing, green channel, Gaussian blur, contrast-limited adaptive histogram equalization FCM fuzzy C-mean (CLAHE), and morphological structuring were used. In vessel segmentation, contouring technique was used to extract vessel line. Convolutional neural network (CNN) is applied for classification. For prediction confusion matrix is created for prediction and training dataset is stored in hierarchical data format 5 file (HDF5). This model was employed on 2200 images. The performance was evaluated on DRIVE and STARE datasets in terms of sensitivity, specificity, precision, accuracy and F1 score as 0.9500, 0.9900, 0.9900, 0.9600 and 0.9700 on DRIVE as well as 0.9375, 0.9900, 0.9600, 0.9500 and 0.9500 on STARE.

Researchers (Samuel & Veeramalai 2019) proposed deep neural network as well as multi- scale or multilevel layers were employed for appropriate retinal vessel segmentation. In this multiscale/multilevel deep supervision layers are created by convolving vessel specific Gaussian convolutions using varied scale initials (0.001 and 0.0002;). Activation maps are generated using these layers that can understand vessel features at multiple levels and multiple scales. Symmetric which gives probability map of refined retinal blood vessel is generated by increasing receptive field of maps. Map does not contain any non-vessel background and boundaries. Hence, it consists of clear blood vessels with less false predictions. Through experimental demonstrations, the performance of the suggested method was evaluated on the widely known datasets, namely, DRIVE, STARE and HRF in terms of sensitivity as 0.8282, 0.8979 and 0.8655 respectively.Also, (Jebaseeli *et al.* 2019) implemented blood vessel segmentation of retinal using SVM-based deep learning and tandem pulse coupled neural network or TPCNN. The methodology uses Contrast Limited Adaptive Histogram Equalization (CLAHE) providing an image the background of which was removed comparative to the given image and it makes front vessel pixels more clear. Pre-processing for removing inconsistencies and false photographic artefacts in fundus images is done. CLAHE pre-processes green channel as it has more pixels of vessels. TPCNN model was incorporated for generation of feature vectors automatically. DLBSVM or Deep Learning Based Support Vector Machine DLBSVM was further used for blood vessels classification and its extraction. CLAHE improves the contrast of images and removes noises. TPCNN produce feature vectors and these feature vectors were employed by DLBSVM to classify vessels and non-vessels. It decreases the errors between vessels and non-vessels. Better performance of the suggested technique was achieved when employed on five datasets, viz. DRIVE, STARE, HRF, REVIEW and DRIONS. The computed values for sensitivity and specificity came out to be 0.8027 and 0.9980 on DRIVE,

0.8060 and 0.9970 on STARE, 0.8077 and 0.9968 on HRF, 0.8088 and 0.9876 on REVIEW, 0.8054 and 0.9978 on DRIONS, respectively.

Lastly, (Balasubramanian & Ananthamoorthy 2021) incorporated Support Vector Machine (SVM) along with Convolutional Neural Network (CNN) for retinal blood vessel segmentation. CNN was used for the extraction of features from segmented regions, and Support Vector Machine carried out the classification of the features which were initially extracted, further those features were classified into two categories: non-vessel and vessel regions. All the retina images are collected from DRIVE and STARE datasets. Pre-processing is implied in collected images using median filter which identifies the edge area of images for quality enhancement of the images. After this, mean orientation-based super pixel methodology was implemented on pre-processed images for the segmentation of non-vessel and vessel regions. CNN was used to extract feature vector and it automatically extract essential features. SVM avoid over-fitting and handles non-linear data using kernel and also classifies vessel and non-vessel regions. Hence, this method in total consists of four stages involving pre-processing of images followed by segmentation as well as extraction of features and at last, image classification was conducted. The suggested methodology outperformed the state-of-the-art methods in terms of sensitivity, specificity, accuracy and kappa index resulting in 0.9712, 0.9809, 0.9743 and 0.8976 on DRIVE and 0.9734, 0.9749, 0.9689 and 0.8800 on STARE respectively.

From Tables 1 to 6, we have provided brief information on all of the evaluating parameters of each dataset used. After reviewing each table, one can determine the sensitivity and specificity of the authors with regard to the specific dataset.

Table 1. The evaluation parameters of DRIVE dataset.

	Author	Sensitivity	Specificity	PPV	Accuracy AUC		Precision	F1 Score	Kappa Index
1.	(Hashemzadeh & Azar 2019)	0.7830	0.9800	0.8584	0.9531	0.9752	–	–	–
2.	(Li et al. 2021)	0.7931	0.9896	0.8946	0.9698	0.9738	–	–	–
3.	(Chen et al. 2020)	–	–	–	0.9631	0.9760	–	–	–
4.	(Cheng et al. 2020)	0.7672	0.9834	–	0.9559	0.9793	–	–	–
5.	(Boudegga et al. 2021)	–	–	–	0.9819	–	–	–	–
6.	(Kumar & Singh 2022)	0.9808	0.6605	–	0.9526	–	–	–	–
7.	(Saranya et al. 2022)	0.9500	0.9900	–	0.9600	–	09900	0.9700	-
8.	(Jebaseeli et al. 2019)	0.8027	0.9980	–	–	–	–	·	-
9.	(Balasubramanian & Ananthamoorthy 2021)	0.9712	0.9809	–	0.9743	–	–	·	0.897
10.	(Samuel & Veera malai 2019)	0.8282	–	–	–	–	–	·	-

Table 2. The evaluation parameters of STARE dataset.

S. No.	Author	Sensitivity	Specificity	PPV	Accuracy AUC		Precision	F1 Score	Kappa Index
1.	(Hashemzadeh & Azar 2019)	0.8087	0.9892	0.9012	0.9691	0.9531	–	–	–
2.	(Saranya et al. 2022)	0.9375	0.9900	–	0.9500		0.9600	0.9500	–
3.	(Jebaseeli et al. 2019)	0.8060	0.9970	–	–	–	–	–	–
4.	(Balasubramanian & Ananthamoorthy 2021)	0.9734	0.9749	–	0.9689	–	–	–	0.8800
5.	(Samuel & Veera malai 2019)	0.8979	–	–	–	–	–	–	–

Table 3. The evaluation parameters of CHASE DB1 dataset.

S. No.	Author	Sensitivity	Specificity	PPV	Accuracy	AUC	Precision	F1 Score	Kappa Index
1.	(Hashemzadeh & Azar 2019)	0.7737	0.9840	0.8488	0.9623	0.9789	–	–	–
2.	(Cheng et al. 2020)	0.8967	0.9540	–	0.9488	0.9785	–	–	–

Table 4. The evaluation parameters of REVIEW dataset.

S.No.	Author	Sensitivity	Specificity	PPV	Accuracy	AUC	Precision	F1 Score	Kappa Index
1.	(Jebaseeli *et al.* 2019)	0.8088	0.9876		–	–		–	–

Table 5. The evaluation parameters of DRIONS dataset.

S.No.	Author	Sensitivity	Specificity	PPV	Accuracy	AUC	Precision	F1 Score	Kappa Index
1.	(Jebaseeli *et al.* 2019)	0.8054	0.9978	–	–	–	–	–	–

Table 6. The evaluation parameters of HRF dataset.

S. No.	Author	Sensitivity	Specificity	PPV	Accuracy	AUC	Precision	F1 Score	Kappa Index
1.	(Jebaseeli *et al.* 2019)	0.8077	0.9968	–	–	–	–	–	–
2.	(Samuel & Veera Malai 2019)	0.8655	–	–	–	–	–	–	–

4 CONCLUSIONS

In this study, we analyzed a number of methods based on deep learning and machine learning algorithms. The efficiency of deep learning models has improved significantly model accuracy in the detection and identification processes. In several cases, we can conclude that DL strategies outperform traditional ones. This research will help potential researchers decide which paradigm was used and will deliver them a broad view of their research areas.

REFERENCES

Almotiri, Jasem, Khaled Elleithy, and Abdelrahman Elleithy. 2018. "Retinal Vessels Segmentation Techniques and Algorithms: A Survey." Applied Sciences 8 (2) 155.

Balasubramanian, Kishore, and NP Ananthamoorthy. 2021. "Robust Retinal Blood Vessel Segmentation Using Convolutional Neural Network and Support Vector Machine." *Journal of Ambient Intelligence and Humanized Computing* 12 (3) 3559–3569.

Boudegga, Henda, Yaroub Elloumi, Mohamed Akil, Mohamed Hedi Bedoui, Rostom Kachouri, and Asma Ben Abdallah. 2021. "Fast and Efficient Retinal Blood Vessel Segmentation Method Based on Deep Learning Network." *Computerized Medical Imaging and Graphics* 90:101902.

Chen, Dali, Yingying Ao, and Shixin Liu. 2020. "Semi-supervised learning method of u-net deep learning network for blood vessel segmentation in retinal images." *Symmetry* 12 (7) 1067.

Cheng, Yinlin, Mengnan Ma, Liangjun Zhang, ChenJin Jin, Li Ma, and Yi Zhou. 2020. "Retinal Blood Vessel Segmentation Based on Densely Connected U-Net." *Math. Biosci. Eng* 17 (4) 3088–3108.

Faust, Oliver, Rajendra Acharya U Eddie Yin-Kwee Ng, Kwan-Hoong Ng, Jasjit S Suri, *et al.* 2012. "Algorithms for the Automated Detection of Diabetic Retinopathy Using Digital Fundus Images: A Review." *Journal of medical systems* 36 (1) 145–157.

Hashemzadeh, Mahdi, and Baharak Adlpour Azar. 2019. "Retinal Blood Vessel Extraction Employing Effective Image Features and Combination of Supervised and Unsupervised Machine Learning Methods." *Artificial Intelligence in Medicine* 95:1–15.

Hassan, Gehad, Nashwa El-Bendary, Aboul Ella Hassanien, Ali Fahmy, Vaclav Snasel, *et al.* 2015. "Retinal Blood Vessel Segmentation Approach Based on Mathematical Morphology." *Procedia Computer Science* 65:612–622.

Jebaseeli, T Jemima, C Anand Deva Durai, and J Dinesh Peter. 2019. "Retinal Blood Vessel Segmentation from Diabetic Retinopathy Images Using Tandem PCNN Model and Deep Learning Based SVM" *Optik* 199:163328.

Khanal, Aashis, and Rolando Estrada. 2020. "Dynamic Deep Networks for Retinal Vessel Segmentation." *Frontiers in Computer Science* 2:35.

Khandouzi, Ali, Ali Ariafar, Zahra Mashayekhpour, Milad Pazira, and Yasser Baleghi. 2022. "Retinal Vessel Segmentation, a Review of Classic and Deep Methods." *Annals of Biomedical Engineering* 50 (10) 1292–1314.

Kirbas, Cemil, and Francis Quek. 2004. "A Review of Vessel Extraction Techniques and Algorithms." *ACM Computing Surveys (CSUR)* 36 (2) 81–121.

Krishna, BV Santhosh, Sanjeev Sharma, KR Indrajith, Eric Joe, Amith Sabu, and M Dilip. 2022. "Retinal Vessel Segmentation Techniques." In *2022 Second International Conference on Artificial Intelligence and Smart Energy (ICAIS)* 586–591. IEEE

Kumar, K Susheel, and Nagendra Pratap Singh. 2022. "An Efficient Registration-based Approach for Retinal Blood Vessel Segmentation Using Generalized Pareto and Fatigue Pdf." *Medical Engineering & Physics* 110:103936.

Li, Zhenwei, Mengli Jia, Xiaoli Yang, and Mengying Xu. 2021. "Blood Vessel Segmentation of Retinal Image Based on Dense-U-Net Network." *Micromachines* 12 (12) 1478.

Lu, Yang, Lilian Serpas, Pauline Genter, Christina Mehranbod, David Campa, and Eli Ipp. 2016. "Disparities in Diabetic Retinopathy Screening Rates within Minority Populations: Differences in Reported Screening Rates Among African American and Hispanic patients." *Diabetes Care* 39 (3) e31–e32.

Prajna, Yellamelli, and Malaya Kumar Nath. 2022. "Efficient Blood Vessel Segmentation from Color Fundus Image Using Deep Neural Network." *Journal of Intelligent & Fuzzy Systems*, no. Preprint, 1–13.

Ricci, Elisa, and Renzo Perfetti. 2007. "Retinal Blood Vessel Segmentation Using Line Operators and Support Vector Classification." *IEEE Transactions on Medical Imaging* 26 (10) 1357–1365.

Samuel, Pearl Mary, and Thanikaiselvan Veeramalai. 2019. "Multilevel and Multiscale Deep Neural Network for Retinal Blood Vessel Segmentation." *Symmetry* 11 (7) 946.

Saranya, P S Prabakaran, Rahul Kumar, and Eshani Das. 2022. "Blood Vessel Segmentation in Retinal Fundus Images for Proliferative Diabetic Retinopathy Screening Using Deep Learning." *The Visual Computer* 38 (3) 977–992.

Soomro, Shafiullah, Farhan Akram, Jeong Heon Kim, Toufique Ahmed Soomro, and Kwang Nam Choi. 2016. "Active Contours Using Additive Local and Global Intensity Fitting Models for Intensity Inhomogeneous Image Segmentation." *Computational and Mathematical Methods in Medicine* 2016.

Soomro, Shafiullah, Farhan Akram, Asad Munir, Chang Ha Lee, and Kwang Nam Choi. 2017. "Segmentation of Left and Right Ventricles in Cardiac MRI Using Active Contours." *Computational and Mathematical Methods in medicine* 2017.

Soomro, Toufique Ahmed, Junbin Gao, Tariq Khan, Ahmad Fadzil M Hani, Mohammad AU Khan, and Manoranjan Paul. 2017. "Computerised Approaches for the Detection of Diabetic Retinopathy Using Retinal Fundus Images: A Survey." *Pattern Analysis and Applications* 20 (4) 927–961.

Steinmetz, Jaimie D Rupert RA Bourne, Paul Svitil Briant, Seth R Flaxman, Hugh RB Taylor, Jost B Jonas, Amir Aberhe Abdoli, Woldu Aberhe Abrha, Ahmed Abualhasan, Eman Girum Abu-Gharbieh, *et al.* 2021. "Causes of Blindness and Vision Impairment in 2020 and Trends Over 30 Years, and Prevalence of Avoidable Blindness in Relation to VISION 2020: the Right to Sight: an Analysis for the Global Burden of Disease Study." *The Lancet Global Health* 9 (2) e144–e160.

Sule, Olubunmi Omobola. 2022. "A Survey of Deep Learning for Retinal Blood Vessel Segmentation Methods: Taxonomy, Trends, Challenges and Future Directions." *IEEE Access* 10:38202–38236.

Wang, Kun, Xiaohong Zhang, Sheng Huang, Qiuli Wang, and Feiyu Chen. 2020. "Ctf-net: Retinal Vessel Segmentation via Deep Coarse-to-fine Supervision Network." In *2020 IEEE 17th International Symposium on Biomedical Imaging (ISBI)* 1237–1241. IEEE

Winder, Robert John, Philip J Morrow, Ian N McRitchie, JR Bailie, and Patricia M Hart. 2009. "Algorithms for Digital Image Processing in Diabetic Retinopathy." *Computerized Medical Imaging and Graphics* 33 (8) 608–622.

Wu, Yicheng, Yong Xia, Yang Song, Yanning Zhang, and Weidong Cai. 2020. "NFN +: A Novel Network Followed Network for Retinal Vessel Segmentation." *Neural Networks* 126:153–162.

473

Artificial Intelligence, Blockchain, Computing and Security – Dagur et al. (Eds)
© *2024 The Author(s), ISBN: 978-1-032-49393-0*

Multifunctional pose estimator workout guider

D. Burad, B. Gurav, S. Desai, S. Banerjee & S. Agrawal
Department of Computer Engineering, NMIMS University, India

ABSTRACT: In today's fast paced world it is difficult to find time to visit a gym and also tough for the gym trainer to give personalized attention to all. Time, location and expenses are major constraints which affect people visiting gym. During the Covid 19 pandemic, people were stuck at home with no one to assist them while they exercise. Keeping this in mind, we have developed a multifunctional pose estimator which functions as a workout guider and suggests correct exercise postures to prevent injuries. We use pose estimation to keep a count of squats or bicep curls, provide real time feedback and also plot a graph which shows the performance.

Keywords: Computer vision, Form, Gym, Health, Media pipe, Pose estimation, Workout, Virtual personal trainer

1 INTRODUCTION

Pose estimation is a computer vision task that typically includes detecting and associating human movements and classifying them into specific actions. Effectively, pose estimation is a sequence of activities that involve the detection of movements and extraction of joint coordinates. These continuously acquired joint coordinates are monitored and relative changes in these coordinates are used to calculate the angle change between the nodes. Application of pose estimation is feasible in the domain of augmented reality and robotics, computer interaction, action recognition, motion analysis, etc. Other applications of pose estimation are outdoor as well as indoor diving guides. It provides feedback to the diver regarding the correctness and preciseness of the posture while diving the main utility of pose estimation is action recognition, which is critical in sports activities that typically require highly accurate estimation. There are a wide variety of effective uses of pose estimations for guidance in the field of yoga.

There are many different challenges that people face in their journey of becoming fit. They face challenges with the flexibility of time as there are chances that they cannot go to the gym because of tight schedules or they are not closer to one. Even doing exercises at home may not be possible due to space constraints. The costs of gyms are also high which cannot be afforded by everyone. Even if someone joins a gym, they do not get one-to-one training and personal attention until and unless one opts for a personal trainer and their fees are quite expensive. This may lead to a person performing exercises with insufficient information about particular exercises in the wrong form and posture which in turn is very harmful as it can lead to injury and damage the body posture permanently. Taking information from people who are not that experienced in the particular domain may lead to inefficient outputs and leads to unsatisfactory results.

The need for a virtual gym and fitness assistant is very much necessary in our fast-growing technological world. To achieve this goal, the following contributions have been made:

DOI: 10.1201/9781003393580-72

1. To develop a personal gym trainer model for monitoring and assessment of gym exercise especially arm curl and squat of an Individual.
2. To analyze correct pose estimation when performing gym exercises and also inform users about the correct form of arm curls and squats for avoiding any Injury.

The paper is organized as follows: Section 2 discusses the review of literature in this field. Section 3 discusses the proposed methodology; Section 4 presents the analysis and results. Section 5 discusses the conclusion and future work.

2 REVIEW OF LITERATURE

(Xiong *et al.* 2020) researched a vision-based exercise analysis in 2020. Users do not receive prompt feedback since the application only detects improper motions but not their timing. They were unable to use video tutorials to teach their model. A deep learning model which is hybrid in nature was suggested to identify yoga positions in real-time videos used by (Yadav *et al.* 2019) to accurately identify various yoga positions. For self-training and real-time predictions, a portable device can be employed.

By taking into account the shape of the body, skeletal structure, and axis(dominant), (Chen *et al.* 2018) developed a method to evaluate the user's posture from both sides. The system can be strengthened by improving the feature point detection and assistant axis generating processes. For both indoor and outdoor sports, such as skiing (Chen *et al.* 2020) or monitoring free-weight exercises with accelerometers in a glove and on the waist (Chang *et al.* 2007), a variety of sensor-augmented physical device types of research methodologies have been proposed. (Chen *et al.* 2020) created a program to correct users' postures by designing the best training movements. Open pose deep neural networks were implemented for pose estimation. For performance comparison, heuristic-based and machine-learning techniques were used. (Achkar *et al.* 2016) created the Adaptive Modified Backpropagation (AMBP) algorithm which offers a complete, real-time training programme, and monitors the user's performance.

(Nagarkoti *et al.* 2019) utilized a pre-recorded video of a trainer for posture correction, utilizing Dynamic Time Warping to synchronize the trainer's and user's body movements. (Cao *et al.* 2017) took into account single-user environments as well as multiple-user environments. For single-user inference, they used observations of the structural relationship between body parts. For multi-user scenarios, they employed a top to down approach to identify individuals and then the pose of each individual is determined separately. (Kumar *et al.* 2020) utilized Part Confidence Maps and Part Affinity Fields to differentiate joint areas and provide feedback based on differences in angles. (Chiddarwar *et al.* 2020) gathered a single representative image to extract 17 key points using a pre-trained model, which was stored locally. The Euclidean distance was calculated between each body part and was used specifically for the detection of yoga poses.

The major disadvantage of the above models is that they worked only on an image and cannot be used on a video. This is the research gap that we tried to work upon in this model.

3 PROPOSED METHODOLOGY

In this section, we have discussed the proposed methodology to identify multiple pose estimation for fitness. To open the webcam, we use cv2 and a new window appears. After that, we identify the joints using the landmark model. Algorithm 1 depicts our method of calculating the angles for the exercises. Algorithms 2 and 3 depicts the working of squat and arm curls respectively. The flowchart of the entire process is depicted in Figure 1. There are 33 major joints and the coordinates of the joints needed for the calculation of squats and arm curls are extracted. We then calculate the angle between the extracted joint coordinates in radians and it is converted to

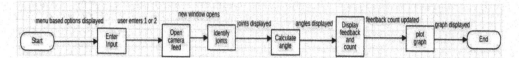

Figure 1.　System flow diagram.

degrees. Then the feedback and state of the exercise update as per the change in the angles are implemented. If the angle is correct, the value of the counter is incremented. The counter and feedback are continuously updated as per the movements of the user. For a visual representation of the user's movement, a graph is plotted and also the threshold value for counting the arm curl/squat is represented by a line. Our method supports different variations of arm curl for example regular bicep curl, reverse curl and hammer curl. All the mentioned variations focus on different muscle groups of the bicep helping the user develop the muscles optimally.

Algorithm 1: Angle Detection	**Algorithm 2:** Squat Detection
Input: calculateAngle(coordinate x, coordinate y, coordinate z) **Output:** Angles for arm curls and squat **I. Procedure:** 1. calculating radians for our particular joint 2. converting radians to the 360° angle 　If angle>180 　　angle = 360 – angle return angle	**Input:** Coordinates of the left leg 　　　Coordinates of the right leg **Output:** Outputs the count and feedback for squat **Procedure:** Calculate angle for left leg Calculate angle for right leg Check the knee distance for correct form. If left>170 and right >170: State = "up" If left<165 and right<165: Output go lower until the height of hips. If left<120 and right<120 and state=="Up" 　　State="Down" 　　Counter + =1 　　Output counter If state=="Down" 　　Output feedback "Good rep"

Algorithm 3: Arm Curls

Input: Coordinates of the left arm and right arm
Output: Count and feedback for arm curls
Procedure:
If leftElbowAngle >160 and rightElbowAngle>160:
　If not flagged:
　Output message did not curl properly
Else:
　Output message good rep
　Change state to down
Elif leftElbowangle >71.9 and rightElbowangle>71.9 and state is down:
　　Change state of flag to false
Elif leftElbowAngle <= 71.9 and rightElbow <= 71.9 and state is down:
　　Change state to up
　　Counter+ =1

For the exercises of squats and arm curls, three joints were selected based on their importance in maintaining proper posture. These joints include the hip, knee, and ankle for squats (shown in Figure 1), and the shoulder, elbow, and wrist for arm curls (shown in Figure 2). Table 1 provides guidelines on the appropriate angles to be maintained between these joints for optimal posture during the exercises. The maximum movement should be 71.9° for the standing arm curl according to (Miller 2007). The classification of squat types is often determined by the angles made at the knee joint. Variations include the mini squat which makes an angle between 140° to 150°, semi-squat which makes an angle between S120° to 140°, half-squat which makes an angle between 80° to 100°, and deep-squat which makes an angle under 80°. The normal range of knee angle during a squat can range widely, starting at 180° while standing upright and potentially reaching as low as 30° for a deep squat (Zulkifley et al. 2019).

We have personally examined the model using the variables listed in Table 1. Figure 2.1 presents the qualitative outcomes.

Table 1.   Acceptable and unacceptable angles for posture.

Sr No.	Squat		Arm Curls	
	Posture	Angle	Posture	Angle
1	Correct	120°–140°	Correct	71.9°
2	Incorrect	<120° or >140°	Incorrect	>71.9°

Figure 3 (a) displays the output we receive after doing a repetition of the arm curl with proper form. If the angle of the elbow appears to be less than or equal to 71.9° and the stage is down, the rep is counted and the user is given the feedback comment "Good Rep" to ensure that they are executing the exercise correctly. On the other hand, in Figure 3 (b) we can see the output we receive on performing a repetition of arm curl with an incorrect form and posture. If the angle of the elbow appears to be more than 71.9° and we take our arm back to the lower position, then the rep is not counted and we are given the feedback comment "Did not curl completely." to point out our mistakes and help us so we do not repeat it.

(a) Joints used to calculate the angle for squat   (b) Joints used to calculate angle for arm curl

Figure 2.   Representation of joints considered for angle calculation.

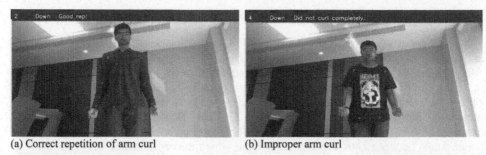

(a) Correct repetition of arm curl   (b) Improper arm curl

(c) Correct repetition of a squat   (d) Incorrect form of squat

Figure 3. Pictorial representation of the different results.

In Figure 3 (c), we can see the output we receive on performing a correct repetition of squats. If the angle of the back side of the knee is under 120°, the stage is up and the knees are shoulder width apart then repetition is counted. We can also see that on performing a correct repetition of squat, we get a feedback comment saying "Good rep!". This guides the user by informing them that their form is correct and also motivates them by mentioning "Good rep!". While Figure 3 (d) shows us the output we receive on performing an incorrect repetition of the squat. If the angle of the knee joint is around 165°, we give feedback "You can do it almost there!" to motivate the user to squat deeper. If the distance between the user's knees is not similar to the distance between their shoulders, then the repetition is not counted and a feedback comment "Open up your knees further apart to shoulder width!" is displayed. This feedback corrects the wrong form of the user and also informs them what exactly is their mistake. In Figure 4 (a) and (b) we present the graphical analysis for the quantitative results we get from the exercise performed.

The illustration in Figure 4 (a) shows the range of motion involved in an arm curl exercise. The red line in the figure represents the angle of the left arm, and the blue line represents the angle of the right arm. The green dotted line in the figure indicates the minimum angle that must be reached for the repetition to be considered valid. The first half of graph 4 (a) represents incomplete repetitions because neither the red nor the blue lines are crossing the green dotted line. While the second half of the graph represents complete repetitions as both the red and blue lines are crossing the green dotted line. The range of motion when performing a squat is depicted in Figure 4 (b). The angle of the left knee is represented by the red line, while the angle of the right knee is represented by the blue line. The green dotted line denotes the minimal angle value that must be achieved for the rep to be counted. The first half of graph 4 (b) represents incomplete repetitions because neither the red nor blue lines are crossing the green dotted line. While the second half of the graph represents complete repetitions as both the red and blue lines are crossing the green dotted line.

478

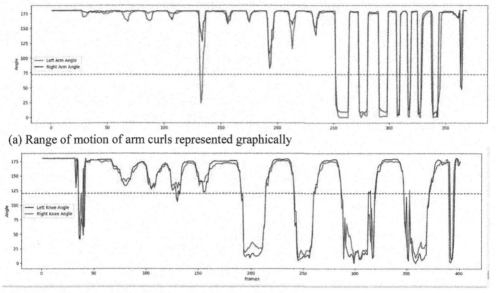

(a) Range of motion of arm curls represented graphically

(b) Range of motion of the squat represented graphically

Figure 4.   Graphical analysis for squats and arm curls.

5   CONCLUSION AND FUTURE WORK

Fitness routines do have significant health benefits, but if they are done improperly, they can also seriously harm our bodies. Form is the utmost important thing when it comes to doing gym exercises. The right form will help you gain the right outputs. If we see that gyms nowadays are too expensive and with that if we hire a personal trainer, it costs us a fortune so to tackle all of these challenges we have come up with a Multifunctional Pose Estimator Workout Guider. It will guide the user through a particular exercise of their selection, check their form and provide feedback. It will also display the exercise's repetition count and a detailed graph, which will help the user analyze their performance. To further help the user, we can add multiple other exercises to our model and make it more precise. The gyms can include this estimator for their use and attract a larger user base and this model can be integrated with the website or an app and make a product out of it.

REFERENCES

Achkar, R., Geagea, R., Mehio, H., & Kmeish, W. (2016, November). SmartCoach Personal gym Trainer: An Adaptive Modified Backpropagation Approach. *In 2016 IEEE International Multidisciplinary Conference on Engineering Technology (IMCET)* (pp. 218–223). IEEE.

Cao, Z., Simon, T., Wei, S. E., & Sheikh, Y. (2017). Realtime Multi-person 2d Pose Estimation Using Part Affinity Fields. In *Proceedings of the IEEE Conference on Computer Vision and Pattern Recognition* (pp. 7291–7299).

Chang, K. H., Chen, M. Y., & Canny, J. (2007). Tracking Free-weight Exercises. In UbiComp 2007: Ubiquitous Computing: *9th International Conference, Innsbruck*, Austria, September 16–19, 2007. Proceedings 9 (pp. 19–37). Springer Berlin Heidelberg.

Chen, H.T., He, Y.Z. & Hsu, C.C. Computer-AssistedYogaTraining System. *Multimed Tools Appl* 77, 23969–23991 (2018). https://doi.org/10.1007/s11042-018-5721-2

Chen, S., & Yang, R. R. (2020). *Pose Trainer: Correcting Exercise Posture using Pose Estimation.* https://doi.org/10.48550/ARXIV.2006.11718

Chiddarwar, G. G., Ranjane, A., Chindhe, M., Deodhar, R., & Gangamwar, P. (2020). AI-based Yoga Pose Estimation for Android Application. *Int J Inn Scien Res Tech*, 5, 1070–1073.

Kumar, D., & Sinha, A. (2020). *Yoga Pose Detection and Classification Using Deep Learning.* LAP LAMBERT Academic Publishing.

Miller, A. (2007). An Upper Extremity Biomechanical Model: *Application to the Bicep Curl.*

Nagarkoti, A., Teotia, R., Mahale, A. K., & Das, P. K. (2019, July). Realtime Indoor Workout Analysis Using Machine Learning & Computer Vision. In *2019 41st Annual International Conference of the IEEE Engineering in Medicine and Biology* Society (EMBC) (pp. 1440–1443). IEEE. https://doi.org/10.1109/EMBC.2019.8856547

Xiong, H., Berkovsky, S., Sharan, R. V., Liu, S., & Coiera, E. (2020). Robust Vision-based Workout Analysis Using Diversified Deep Latent Variable Model. In *42nd Annual International Conference of the IEEE Engineering in Medicine and Biology Society: Enabling Innovative Technologies for Global* Healthcare, EMBC 2020 (pp. 2155–2158). [9175454] Institute of Electrical and Electronics Engineers (IEEE). https://doi.org/10.1109/EMBC44109.2020.9175454

Yadav, S. K., Singh, A., Gupta, A., & Raheja, J. L. (2019). Real-time Yoga Recognition Using Deep Learning. *Neural Computing and Applications*, 31, 9349–9361.

Zulkifley, M.A., Mohamed, N.A., & Zulkifley, N.H. (2019). Squat Angle Assessment Through Tracking Body Movements. *IEEE Access*, 7, 48635–48644.

Artificial Intelligence, Blockchain, Computing and Security – Dagur et al. (Eds)
© 2024 The Author(s), ISBN: 978-1-032-49393-0

Volume control using hand gesture recognition

Ashish Kumar Mallick, Adil Islam & Abdul Aleem
SCSE, Galgotias University, Greater Noida, India

ABSTRACT: One system that can recognize hand gestures in real-time video is gesture recognition. Gestures are classified according to their meaning. The design of information guidelines is one of the most difficult tasks because it presents two important problems. Manual control is the first step and creates a logo that can only be used by one hand at a time. It can be used in many areas, including human-computer interaction and language. The key points of hand segmentation and hand measurement using cascaded Haar classifiers can be used to generate gesture recognition using Python and OpenCV. This article discusses gesture recognition as a qualitative analysis. The setup includes a camera that records the users' movements and transmits them to the system. The main goal of gesture recognition is to create a system that can recognize human gestures and then use them to send information to drivers. Realtime work experience allows users to work on the computer by performing certain tasks in front of the computer camera.

1 INTRODUCTION

Hand gestures are a versatile and efficient way of communication in human-computer interaction (HCI). Despite being available for computer interaction, conventional input devices like the keyboard, mouse, joystick, and touch screen do not provide a natural interface. Users of the proposed system would be able to make hand movements by donning a data glove or by capturing images of their hands with a web camera, and it will have a desktop or laptop interface. This research addresses adjusting a computer's volume using gesture control. We first investigate hand tracking, and then we use hand landmarks to identify a hand gesture that changes the loudness. Due to the fact that this project is module-based, we will be utilizing a previously constructed hand module, which makes the hand tracking really simple.

The initial steps in the identification of any hand motion are hand tracking and segmentation. A user can operate a computer by making a certain motion in front of a video camera that is attached to it by implementing real-time gesture recognition. Information can be sent for device control via gestures. This research builds a volume control system using hand gesture utilizing the OpenCV module. Here, using hand gestures instead of a keyboard or mouse allows you to operate the system.

The Vision Based processes, in contrast, merely require a camera, enabling knowledge of a typical interaction between people and calculations without the need for any further tools. These details of vision wholes that are affected that are carried out in software or hardware are likely to complement biological dream. To achieve real-time efficiency, the questioning queries of these structures must be background neutral, ignition unaffected, and camera unrestricted. The advantages and cons of various algorithms for acknowledging secondhand posture and gestures are examined. The identification of assistance postures and gestures using various mathematical techniques is presented. The technique of determining a related domain within the face along with a specific property to the extent of colour or force, or a relationship between pixels, is known as segmentation is to say, a pattern and the algorithms bear be compliant [1].

DOI: 10.1201/9781003393580-73

481

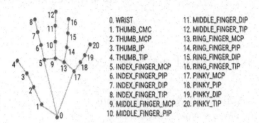

Figure 1. Hand gesture points.

2 EXISTING SYSTEM

ANN of Gesture Recognition with Accelorometer Data: The authors created a programme that classifies items and recognises gestures using artificial neural networks. Gestures are recognised by the Wii remote, which spins in the X, Y, and Z axes. The gesture recognition is handled twice to lower the cost of computation cost and usage of memory. For gesture recognition, user authentication is performed at the first level [1,3].

2.1 *Combining depth-based descriptors with hand recognition*

The authors develop an innovative gesture recognition method based on depth information from camera images. They use a set of 3D features to detect complex expressions using 3D data. Our action-based system is easy to set up. Color and depth information are used in the first step to separate the cell structure from the background. The split hand test is divided into the wrist, palm, and fingers. Four variables make up the motion system. Extracting features for segmentation is the second step. The first two features depend on the center of the palm and the height of the fingers. The third set of features focuses on calculating the curvature characteristics of hand contours. The fourth group focused on creating the geometric shape of the palm field.

2.2 *Dynamically hand recognition at real-time with hidden markov models*

In this study, a method using dynamic hand gestures is proposed to recognize English numbers from 0 to 9. It configures the system in two steps. Preprocessing comes first, followed by classification. Because there are two types of gestures, Link gestures and Basic actions. Signal link in continuous signal can be characterized by key signal. The path between two points of continuous motion is provided for classification. A separate hidden markov model is used for classification. The BaumWelch algorithm is used to train this DHMM. The average recognition rate when using HMM ranges from 95% to 98% [6].

2.3 *Robust Apart-based hand gesture recognition using the kinect sensor*

In this study, deep availability. The Kinect-camera sensor is used to create a reliable hand gesture detection system. The resolution of the Kinect sensor is limited and troublesome it is easy to distinguish the hand, but write large objects. Finger Earth Movers distance is a new distance measure proposed by the authors as a solution to the noisy hand movements recorded by the Kinect sensor [7].

3 PROPOSED WORK & ANALYSIS

In the field of vision, research on hand gestures is ongoing, primarily with an eye toward recognizing sign language and enhancing human-computer interaction. There was one of the

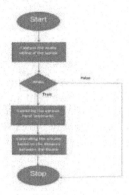

Figure 2.   System flowchart.

first tracking systems to concentrate on articulated hand movements. By extracting point and line information from grayscale photos, their technique allowed for the tracking of a hand with 27 degrees of freedom at a rate of 10Hz.

Most hand tracking research up to this point has been on 2D interfaces from an interaction standpoint. In [Zhang01], a typical graphical user interface was manipulated without a mouse or keyboard by tracking a finger around a planar surface with inexpensive web cameras. Fitting a conic to rounded features enabled fingertip detection, and Kalman filtering enabled local tracking of the tip.

Similar to this, background pixels and skin regions were separated using infrared cameras that tracked two hands for inter-action on a 2D table-top display. They then employed a template matching strategy in their research to identify a limited number of gestures that could be read as interface commands. However, their method did not yield accurate information about fingertip location.

Algorithm
Step 1: Start the Program
Step 2: Opening the Other Modules-Open Vision, which is utilising AI recognition, and different audio utilities, which is the key concern.
Step 3: Detecting the distinct contours and separating the white and black region of interest to capture the area of interest
Step 4: Executing Loops for detecting various hand landmark.
Step 5: Using the provided algorithm, obtaining the hand landmarks and confirming the separation between both the index, thumb finger
Step 6: Using artificial intelligence, display the frame with the reading's final numbers together with a complete volume drop and increase. The programme runs as long as the loop is iterated; after the iterations are finished, the programme exits the loop and terminates.
Step 7: Stop

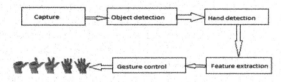

Figure 3.   System architecture.

483

4 SYSTEM ARCHITECTURE

Python technology was used in the development of this project, and OpenCV and NumPy modules were used to organise and create the code. OpenCV is accessible on various operating systems, including Windows, Linux, OS X, Android, and iOS. Numerous programming languages, including Java, Python, C++, Python, and others can be utilized by OpenCV. CUDA and OpenCL are also active and in development of image processing techniques. The API for Python OpenCV is called OpenCVPython. It incorporates the best elements of the Python language and OpenCV, C++ API. The foundation of scientific computing is a Python module called NumPy. Apart from these, a lot of other features are associated with Python, as follows:

- Array objects, which are N-Dimensional and hence, powerful
- Functions utilized for broadcasting, which are sophisticated
- Code integrator tool for C/C++ and Fortran language
- Fourier transformation
- High utility of linear algebra
- Randomness of number capabilities. [9]

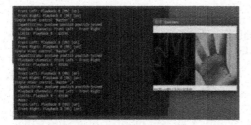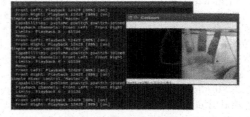

Figure 4.　Counter ZERO and ONE.

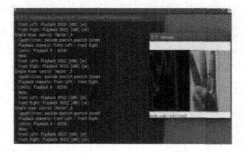

Figure 5.　Counter TWO and THREE.

5 COMPARISON ANALYSIS AND RESULTS

This project uses Crimmins speckle removal method to remove the dark spots or speckles and remove the noise from a picture /gesture to improve the output's accuracy. This makes gesture capturing more efficient. It takes less memory and less time to get implemented. On the other hand, previous technologies related to image capturing/ gesture identifying used conservative filter methods which makes the image sharper and clearer. This in turn makes the spots brighter and deeper, reducing the efficiency and increasing the level of uncertainty in image identifying and capturing.

6 CONCLUSION AND FUTURE WORK

The research let users easily operate software with hand gestures. a vision-based hand gesture system that can work in real-time on a standard PC with inexpensive cameras without the use of special markers or gloves. The technology can monitor the locations of each hand's index finger and counter tips in particular. A desktop-based volume control system that allows users to manage volume and cursor navigation in real-time using intuitive hand motions served as the inspiration for this hand gesture. Additionally, we advocate using a mouse cursor that is moved by hand motion, and we offer a tip for positioning a point on the bare hand so that the mouse cursor can be moved. We also suggest a straightforward probabilistic model to effectively prevent the created system from responding to incorrect gestures in the interest of reliability. In this paper, four algorithms for static gesture identification and categorization for human computer interaction were compared using two different datasets. The outcomes of identifying hand gestures in the two datasets allow for the conclusion that dataset 1 attributes produce better hand gesture identification results and that the ANN performed admirably for the given issue. The outcomes also demonstrated the significance of the feature selection and data preparation stages, particularly when working with low-resolution images like depth photos produced using kinect camera gestures.

REFERENCES

[1] Hninn, T. and H. Maung, *Real-Time Hand Tracking and Gesture Recognition System Using Neural Networks.* 2009. 50(Frebuary): p. 466–470.

[2] Chaudhary, A., *et al.*, Intelligent Approaches to Interact with Machines using Hand Gesture Recognition in Natural way: *A Survey. International Journal of Computer Science & Engineering Survey*, 2011. 2(1):

[3] Chen, Q., N.D. Georganas, and E.M. Petriu, Real-time Vision-based Hand Gesture Recognition Using Haar-like Features, in *Instrumentation and Measurement Technology Conference*, IEEE, Editor 2007: Warsaw, Poland.

[4] Miner, R. RapidMiner: *Report the Future.* December 2011]; Available from: http://rapidi.com/content/view/181/196/.

[5] Mitra, S. and T. Acharya, Gesture Recognition: A Survey, in *IEEE Transactions on Systems, Man and Cybernetics* 2007, IEEE. p. 311–324.

[6] Murthy, G.R.S. and R.S. Jadon, A Review of Vision Based Hand Gestures Recognition. *International Journal of Information Technology and Knowledge Management*, 2009. 2(2): p. 405–410.

[7] Faria, B.M., N. Lau, and L.P. Reis. Classification of Facial Expressions Using Data Mining and Machine Learning Algorithms. in *4ª Conferência Ibérica de Sistemase Tecnologias de Informação.* 2009. Póvoa de Varim, Portugal.

[8] Gillian, N.E., *Gesture Recognition for Musician Computer Interaction*, in Music Department 2011, Faculty of Arts, Humanities and Social Sciences: Belfast, p. 206.

[9] Faria, B.M., *et al.*, Machine Learning Algorithms Applied to the Classification of Robotic Soccer Formations ans Opponent Teams, in *IEEE Conference on Cybernetics and Intelligent Systems (CIS)* 2010: Singapore. p. 344–349

[10] Kumar, A., & Alam, B. (2019). Energy Harvesting Earliest Deadline First Scheduling Algorithm for Increasing Lifetime of Real Time Systems. *International Journal of Electrical and Computer Engineering*, 9(1), 539.

[11] Mehra, P. S., Jain, K., Chawla, D., Dagur, A., Singh, S., & Sharma, J. (2022). *GWO-EFUCA: Grey Wolf Optimisation and Fuzzy Logic Based Unequal Clustering and Routing protocol for sustainable WSN-based Internet of Things.*

[12] Kumar, A., & Alam, B. (2014, February). Real Time Scheduling Algorithm for Fault Tolerant and Energy Minimization. In *2014 International Conference on Issues and Challenges in Intelligent Computing Techniques (ICICT)* (pp. 356–360). IEEE.

[13] Intel Corp, "*OpenCV Wiki*," OpenCV Library [Online], Available:http://opencv.willowgarage.com/wiki/

[14] Kumar, D., Aleem, A., Gore, M.M. (2021). Employing Data Augmentation for Recognition of Hand Gestures Using Deep Learning. In: Sharma, H., Saraswat, M., Kumar, S., Bansal, J.C. (eds) *Intelligent Learning for Computer Vision. CIS 2020. Lecture Notes on Data Engineering and Communications Technologies*, vol 61. Springer, Singapore.

Artificial Intelligence, Blockchain, Computing and Security – Dagur et al. (Eds)
© 2024 The Author(s), ISBN: 978-1-032-49393-0

Connecting faces: Secure social interconnection

Aatif Jamshed*, Ankit Bhardwaj, Avi Nigam*, Ujjwal Gupta & Sachin Goel*
Department of Information Technology, ABES Engineering College, Ghaziabad, India

ABSTRACT: Social media is a way to engage with others and have fun. Social media enables us to converse with individuals who are far away from us around the world, share information with them and even create web content. There are many forms of social networking websites some allow us to share text, some allow us to share pictures, and some allow us to share both. We make use of social media to communicate information and make acquaintances and cooperate with each other. It permits us to work together with colleagues when you are away from them and make new acquaintances. We can also make use of social media to improve our knowledge in a specific field and develop our group by connecting with other professionals in our industry. Social media lets us to link up with our immediate audience, get consumer response, and enhance our assistance. However, designing online forums raises concerns and disclosures of potential abuse. This paper provides a brief overview of your social media platforms and applications.

Keywords: Social interaction, Privacy, Security, Social Networking

1 INTRODUCTION

The Internet has rapidly evolved from a social network to a social network that is used to distinguish objects, beliefs and data. Interaction and entertainment can be found through social media. Social media enables us to communicate with individuals who are far away from us around the world, share information with them and even create web content [1]. We make use of social media to communicate information and make acquaintances and cooperate with each other [2]. It permits us to work together with colleagues when you are away from them and make new acquaintances [3]. We can also make use of social media to improve our knowledge in a specific field and develop our group by connecting with other professionals in our industry [4,5]. Social media lets us to link up with our immediate audience, get consumer response, and enhance our assistance.

Social networking is a comprehensive trend that has transformed how individuals unite with other party. It impacts almost every single part of our lives: education, transmission, occupation, healthcare, politics, individual efficiency and social relationships [7]. An online platform designed to promote and develop social relationships amongst people is known as a social networking service. It offers ways for consumers to collaborate online through people with related interests, whether rational or social persistence, Allows users to distribute emails, online comments, direct messages, wiki, digital videos and photos, and post posts. It also gives people with disabilities the opportunity to express their feelings and opinions in an important way [8].

Social networks offer two functions as providers and clients. Customers are allowed to control who can read their profile. By responding to inquiries about things like age, position,

---

*Corresponding Authors: 09.aatif@gmail.com, pranshusaxena@gmail.com and sachin.viet@gmail.com

DOI: 10.1201/9781003393580-74

concerns, etc., the profile is created. Users can create and exchange contact lists on other websites, upload images, add functioning curriculum vitae, edit the style and feel of their profile, blog, and comment on topics. Users can control who can access their profiles, get in touch with them, add them to their contacts list, and other privacy-preserving features [6].

Our objective is to create a fully functional social networking model where we can like, share and comment on posts, add friends and chat with them using a chat messenger. In this model, the user can create posts using our application; can follow/unfollow any other user, can like, share or comment on his friends' posts. The user can get different updates of his and his friends' posts as soon as possible via notifications and will be able to chat live or call with any of their followers.

2 POPULAR SOCIAL WEBSITES

There are many modern SNSs which consist of Facebook, Twitter, and Instagram.

2.1 Facebook

It was first released in 2004 as a social media sharing software. It gradually expanded to other academies and colleges and ultimately to everyone. It became one of the largest social interactions networking applications. It is the one of the largest photo sharing, video sharing website. People found Facebook as a useful web application because it covered a variety of personal and group benefits.

2.2 Twitter

Odeo Inc. founded it in 2006, and at first, only employees from Odeo Inc. and related families worked there. It became initially open to the public in 2006. This website service allows users to write comments on other users' blog articles and send instant messages to them. Tweets are published on Twitter. A tweet is a brief message with a character limit of 140. Such messages are produced by users to express ideas. A recent and well-liked blogging option on Twitter is micro-blogging.

2.3 LinkedIn

Experts use this platform. Experts contribute to the communication itself. After creating an account domain on LinkedIn we associate ourselves with professionals and people with similar interests. LinkedIn has been a very popular social networking site for various organizations to train new employees.

2.4 Instagram

Kevin Systrom and Mike Krieger established the social networking site Instagram, a platform for online social media. Instagram is a social network that allows users to share photos and videos. Instagram photo, video sharing social networking originated in America. In Facebook, users have the ability to upload media. The uploaded media can be edited with filters. We can also organize it by geographical tagging and by common interest by using hash-tags. Posts and messages are shared publicly. Everyone or the followers who are only pre-approved can see the posts.

3 APPLICATIONS

Social networking websites are now an important day that provides users' online based forums and allows them to participate in the community. General applications include computer-based social media, education, business, finance, health care, politics and religion.

Table 1. Different applications with their feature enabled.

Features	Facebook	Twitter	Instagram	Connecting Faces
Notification	Yes	Yes	Yes	Yes
Live Chat	Yes	Yes	Yes	Yes
Live call	No	No	No	No
Follow/Unfollow	Yes	Yes	Yes	Yes

3.1 Social interaction

These web applications facilitate computer-based social networking. They help to unite individuals with similar interests and activities that include political, economic, and environmental boundaries. They also contribute to the fun of entertainment. People make new friends, reconnect with old friends, find people, and keep in touch with old relatives and friends. Many job seekers use social networking sites in their search for work, which furthers their chances of finding employment and securing reputable contracts.

3.2 Education

The learning styles of students and educators are influenced by social networks. They are now utilized for learning, education, professional development, and content exchange, among other things. Many scientific societies and groups use social media to disseminate information to the general public. Many scholars and librarians use social media to keep in touch with one another on a regular basis. Social media apps have developed into networks for learning and research. There is at least one page on each of the social networking platforms used by several institutions, including Facebook, Twitter, and Instagram. Education encounters issues with privacy, genuine relationships, time management, and inadequate communication, among other things. The key benefits of social networks are flexibility, duplication, comfort, and accessibility.

3.3 Business

Another excellent application is social networking websites for corporations and financial institutions. It might be helpful for starting businesses, businesspeople, performers, artists, or other creative. Five different ways that businesses utilize social networking sites include: product exposure, online dignity management tools: third, the gathering; fourth, learning about new technologies and competitors; and fifth, to prevent probability. Businesses can use social media to promote their products, identify customer requirements, and get feedback from a variety of sources. The adoption of virtual currency through social networks has opened a slew of new options for global finance. Consumers can share their personal experiences through social media. Early adopters can make more informed purchase decisions by sharing their experiences. It also reduces the risk of buying a new product.

3.4 Healthcare

Varied stakeholders, like as doctors, patients, and nurses, have different levels of social connectedness. Doctors and nurses can benefit from social media as a teaching and learning tool. Social networking sites are utilized to disseminate fresh study findings. They also aid in the delivery of high-quality care to their patients. There are a slew of health-related social networking sites to choose from. Health chapter, Daily Strength, Tools To Life, Health Care 2.0, Live Strong, Everyday Health, My Cancer Place, Revolution Health, No Surrender, Planet Cancer, Prostate Cancer Infolink, sober circle, Psych Central, diabetes connect, and Daily Plate are just a few examples.

3.5 *Politics*

Social media has an impact on political life and for meetings around the world. It affected voting patterns and led to civil unrest, uprisings, and uprisings around the world. The government will be more accountable for social media, and citizens will be able to exercise their right to free speech. Involvement of younger generations in politics and public participation in political processes are other benefits. For instance, Barack Obama engaged voters, empowered volunteers, and greatly increased donations during his US presidential campaign in 2008. Obama was the first American president to understand the importance of social media.

4   IMPLEMENTATION: CODE SNIPPETS

All figures (Figures 1, 2, 3 and 4) are showing the implementation procedures and explaining how the entire system [9] will work. The output of present work is summarized in the Figure 4.

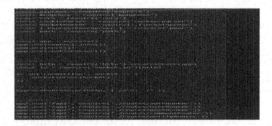

Figure 1.   Implementation part-1.

Figure 2.   Implementation part-2.

Figure 3.   Implementation part-3.

Figure 4.   Outcome to two connected faces.

5 OUTPUT SNIPPETS CHALLENGES AND ISSUES

Today's social media platforms aim to fulfil affective, mental, personal, and social wants, but they also Others, however, believe that blocking social media sites is ineffective because it prevents kids from learning attitude, aptitude, and other critical qualities that are necessary for young people to lead the digital age of the new world with self-assurance and self-assurance.

Some people have also asserted that social media is a poor substitute for traditional face-to-face social connections and that excessive usage of these technologies causes users to feel lonely, hopeless, and unhappy. People can publish nasty comments, images, or videos since there are no limitations on what can be posted online. A key concern on social networking platforms is privacy. As an illustration, a lot of third parties regularly use information (such as personal information, information, and profile information) that is provided on social networking sites for any of the goals. It might cover things like whether or not employers should have access to the social networking profiles of their employees.

Depending on a user's age, gender, preferences, and personality, various users may have different privacy concerns. Studies have shown that women frequently have greater privacy concerns than men. Another drawback of social networking is that it is increasingly being used as a tool to sabotage friendships, father-son ties, and other types of partnerships. We must take steps to ensure that social networking does not continue to degrade the lives of those who are exposed to it.

6 CONCLUSION

The way people engage with one another, communicate, and trade information has unquestionably altered as a result of social networking. Additionally, it enables individuals or users to communicate and converse with others, regardless of location. New technological applications are frequently seen when social networking sites gain popularity. The social internetworking of machines is another recent trend in the market. The creation of the Internet of Things (IoT) and inter-machine communication is the primary goal of this transition. While creating the system, deliberate attempts were taken to design a software package using the tools, techniques, and resources that would result in a suitable social networking application system. The system's usability has also been kept in mind in the hopes that it will be acceptable to all users and successfully satisfy their needs.

REFERENCES

[1] Arnott, D. and Pervan, G., (2005), "A Critical Analysis of Decision Support Systems Research", *Journal of Information Technology*, Vol.20, pp. 67–87.

[2] Arnott, D. and Pervan, G., (2008), "Eight Key Issues for the Decision Support Systems Discipline", *Decision Support Systems*, Vol. 44, pp.657–672.

[3] Boyd, D.M. and Ellison, N.B., (2007), "Social Network Sites: Definition, History, and Scholarship", *Journal of Computer-Mediated Communication*, Vol. 13 (1), October, pp. 210–230.

[4] Lager, M., (2009), "No One's Social (yet)", *CRM Magazine*, Vol. 13(6), pp. 29–33.

[5] McAfee, A.P., (2006), "Enterprise 2.0: the Dawn of Emergent Collaboration", *MIT Sloan Management Review*, Vol. 47 (3), Spring, pp. 20–29.

[6] O'Reilly, T., (2005), "*What is Web 2.0? -Design Patterns and Business Models for the Next Generation of Software*", http://oreilly.com/web2/archive/what-is-web-20.html, Cited 2011-07-06.

[7] Richter, D., Riemer, K. and Brocke, J.v., (2011), "Internet Social Networking: Research State of the Art and Implications for Enterprise 2.0", *Business & Information Systems Engineering*, Vol.2, pp. 89–101.

[8] Simon, H., (1977), The New Science of Management Decision, Prentice Hall, Engle woods Cliffs, NJ. Online *(PDF) The Interconnection of DSS and Online Social Networking*. Available from: https://www.researchgate.net/publication/259656796_The_interconnection_of_DSS_and_online_social_networking

[9] Jamshed, A., Mallick, B., & Kumar Bharti, R. (2023). Grey Wolf Optimization (GWO) with the Convolution Neural Network (CNN)-based Pattern Recognition System. *The Imaging Science Journal*, 1–15.

Artificial Intelligence, Blockchain, Computing and Security – Dagur et al. (Eds)
© 2024 The Author(s), ISBN: 978-1-032-49393-0

2019-nCovid Safe – a deep learning application for crowd management

Prachi Pundhir*, Aatif Jamshed*, Puneet Kumar Aggarwal*, Sukrati Pateriya*,
Vaani Tyagi* & Vanshita Garg*
Department of Information Technology, ABES Engineering College, Ghaziabad, India

ABSTRACT: To make an effective contribution to public health, our project 2019-nCoviSafe aims to develop and implement a more precise and real-time approach that helps to better visualize faces that can be masked in public areas and thus, help in compelling people to wear a mask in public areas at all time. Our strategy combines single-stages and two-stages acquisitions that help in achieving lesser decision-making time and higher accuracy of the results. We started with MobileNetV2 as the base and used transfer learning to integrate advanced facial reading information into multiple feature extraction maps. Above this model, we have also proposed a modification of the binding box to improvise local performance during the process of mask acquisition. The test was performed on popular basic models that are namely, AlexNext, ResNet50, and MobileNetV2. This paper is examining whether it was possible for this method that is to be connected to the proposed model to get the most precise results that could be obtained in a short period of duration. It is to be noted that the proposed methodology achieves the highest accuracy (97.255%) when uses MobileNetV2. In addition, when compared to the most recent publicly available model that was released as a Retina Face Mask detector, the suggested model yields a high accuracy of 89.93%, 93.66%, and memory retrieval of the mask. The proposed model nCoviSafe performs brilliantly, making it ideal for camera surveillance systems.

Keywords: covid, machine learning, deep learning, image processing, virus

1 INTRODUCTION

Constructive and productive strategies were taken to prevent the Corona Virus epidemic require which requiresa high level of attention to reduce the impact on the lives of the affected communities and the global economic conditions, as the full horizon is yet to be introduced to the world. Wearing a face protection shield is one of the non-pharmacological interventions that can be used to cut the main source of SARS-CoV2 virus droplets that are spread by an infected person. Except for the talk about medical services and the variety of masks, all countries approve that in public places,covering of nasal and facial, particularlythe mouth should be essential.

When face masks are found on someone, it is important to determine whether or not they are indeed wearing one. The face recognition system's reverse-based engineering solution, in which the face is recognised using various machine learning approaches for security and authentication, is the challenge. The research of face detection is crucial in the realm of

*Corresponding Authors: prachipundhir1@gmail.com, 09.aatif@gmail.com,
Puneetaggarwal7@gmail.com, sukratipateriya@gmail.com, vaani.18bit1076@abes.ac.in and
Vanshita.18bit1017@abes.ac.in

visualisation and pattern visibility. An important research topic has contributed to the complexity of facial detection algorithms in the past. A major facial recognition study was conducted using the design of a handicraft feature and the use of conventional machine learning algorithms to train successful class dividers to find and identify. Problems traced in this way have higher complexity in feature construction and lesser acquisition accuracy.

The data list includes a variety of face images including a face mask, face mask with and without a single image, and confusing maskless images. With a comprehensive database containing 45,000 images, our method gains a remarkable accuracy of 97.255%. The main contribution of the nCoviSafe project is given below.

- Create an object acquisition method that combines single-stages and two-stage detectors to precisely detect an object in real-time from webcam streaming and transmission.
- Improved modification is created to prevent facial areas from uncontrolled images taken at the moment with variety in face size, shapes, and background. This step helps us to improve the person who violates the face mask practices in places of the high density of population / Bigger offices.
- to create of unbiased facemask database module with a measure of inequality equals that is equal to unity.
- The nCoviSafe model requires less memory, which makes it easier to use on devices used for security check purposes

2 RELATED WORK

The majority of algorithms concentrate on creating and recognising various faces as well as on identifying people wearing and not wearing masks. The primary goal of project's research is to locate those who are not wearing masks in order to help stop the transmission and spread of the Covid-19 global pandemic. Researchers have determined that the rate of COVID-19 spread can be reduced by utilising face masks after consulting a number of studies. A brand-new facemask-wearing condition detection technique was developed by the authors of [9]. They were able to separate those wearing facemasks into two groups. The two categories are using a facemask properly and not wearing one. The proposed method achieved a 97.2 percent accuracy rate for face detection.

PCA was employed by Sabbir et al. [10] to identify the subject in both masked and unmasked face recognition. They found that utilising PCA to recognise faces is much less accurate when wearing a mask. The identification accuracy drops to less than 70% when the identified face is hidden. Additionally, PCA was utilised by Dong et al. in 2018 [11]. A method for removing glasses from a frontal face image of a person was suggested by the authors. The deleted part was put back together using PCA reconstruction and recursive error compensation. The authors in [12] used the YOLOv3 approach to identify faces. Darknet-53 serves as YOLOv3's framework. The accuracy of the suggested method was 93.9 percent. Over 600,000 photos from the WIDER FACE and CelebA datasets, which together make up the training set, were used. The study was conducted using the FDDB dataset. A novel GAN-based network was introduced by Nizam et al. [13] that can remove masks covering the face and reconstruct the image by filling in the blanks. The suggested method creates a full face image that looks realistic and natural. The authors of [14] provided a method for determining whether or not a necessary medical mask is used in the operation room.The final objective is to decrease the amount of false positive-based face detections while preventing missing mask detections so that only medical staff who do not wear surgical masks are warned. The proposed system's accuracy rate was 95%.

MRGAN is an interactive technique that was introduced by Muhammad et al. [15]. The procedure depends on the user providing the microphone region, which is subsequently recreated using the Generative Adversarial Network. Shaik et al. [16] used deep learning to classify and recognize real-time facial emotions. VGG-16 was used to classify seven different facial expressions. The proposed model achieved 88 percent accuracy on the KDEF dataset. Viola-Jones [17] screening system can be trained to see anything, but face detection is the most popular application because it is more accurate and faster. Supervised learning is demonstrated through the Viola and Jones Process. Zhu [18] also presented some new work on a widely used face detection approach that uses a neural network detector. Only the front, the straight face looks good with it. The Multi-View Face Detector capable of surfing was proposed by Li *et al.* [1,2]. Oro *et al.* [18] also suggested a haar-like feature based on the HD face detection method video on GTX470 to obtain an image.a 2.5-fold increase in speed They exclusively employed CUDA, which is a graphics processing unit.NVIDIA GPUs tool for GPU system Unlike OpenCL, which is utilized in a variety of applications,It does not address the issue of inefficient load, though.

Face detection with the viola-jones algorithm yielded inefficiency. On the GPUs algorithm, Glass *et al.* (2006) [20] talked about how important it is to have a good team.The importance of social cohesion and how it might help to lower the danger of epidemics. Without the use of technology, it is possible to keep human coexistence distance.

Vaccines and antiretroviral medications are two examples of antiretroviral drugs. To demonstrate the slowing growth rate, the authors conducted a comprehensive study in rural and urban regions. Z., Luo [19] learns to recognize people who have their faces closed completely or partially. People holding hands over their faces or covered with objects are divided in the center by this method. This strategy is ineffective for our situation, which necessitates the use of masks such as scarves, mufflers, handkerchiefs, and other such items to hide our faces. In [17] P. Saxena *et al.* use deep learning approaches VGG-16 is uses to classify chest x-ray Images.

Table 1. Comparative analysis of various research works.

Reference	Dataset	Methodology	Approaches	Evaluation Matrices	Result
[13]	ORL DB color ferret yale Face Database	Two-step process in the first step, they have extracted the Hough feature and in the next step, these are fed into SVM classifiers	Machine learning	CMC, EPC, and ROC curve	87.5% success rate as compared to the PCA base model
[14]	Data is collected using a web scraping tool, Data annotation is done	YOLOv3	Deep learning approaches	Accuracy	96% classification accuracy
[15]	SMFD, RMFD, LFD	The two-step process in the first step feature is extracted using ResNet-50, and classification by machine learning techniques.	Deep learning followed by machine learning	Accuracy	testing accuracy is 99.64% in RMFD and in LFW 99.49%
[12]	Two medical face mask datasets combined into one set	Features are extracted using ResNet-50 and classification using YOLOv2	Deep learning	Accuracy and Precision	81% precision is achieved
[16]	Custom data set	Detection and trackingdata collected through the camera used CNN and	Deep learning approached	–	The model is deployed in the actual environment in real-time applications.

3 PROPOSED METHODOLOGY

The first stage of model training that may be carried out using the right database is determining whether a person is wearing the mask appropriately. The section below goes into further detail regarding the set of information. An reliable face detection model is required for face detection after the required classifier has been trained. As a result, our algorithm can tell if someone is wearing a mask or not.

This study aims to improve mask discovery accuracy with high precision and low resource consumption. For this purpose, the DNN module in OpenCV has been used to implement the Single Shot Multibox Detector based on Liu *et al.*, 2016 model detection model using ResNet-10 proposed earlier by Anisimov & Khanova, 2017 as its backbone structure. In the following section, this paper is estimating whether a person will wear a mask or not using a pre-trained MobileNetV2 model introduced by Sandler, Howard, Zhu, Zhmoginov, and Chen in 2018. The flowchart in Figure 1 illustrates the procedure employed in this paper.

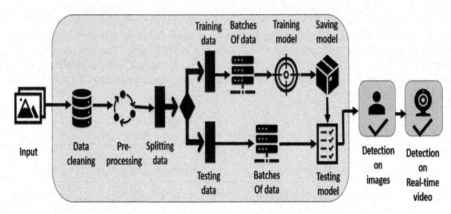

Figure 1. Workflow of CoviSafe.

4 DESCRIPTION OF DATASETS

This paper is splitting the work flow into two distinct phases, each with the methods listed below, in order to train a unique face scanner.

Training: In this section, the paper is concentrating on importing face mask discovery data, training models on this database using Keras and TensorFlow, and integrating a face mask scanner.

Deployment: After training, loading the mask scanner is started, followed by scanning the subject's face, and separate the various faces as though a mask were on and off. The use of several datasets was necessary to collect different scenarios:

- Different types of mask and their texture
- correctness of mask, while putting on the face
- Positioning of people with different angles
- People with with representations (in color, height, and ethnicities).

Data set from Face mask recognition contains great noise, and many repetitions were found in the pictures of the CoviSafe Database. As a well-organized database sets the accurate results of a trained model, therefore the data was taken from the above-mentioned databases. Then, these datasets from the different databases were processed, and all duplicate values were removed by manual process one after the other. Data purification is done manually to remove damaged images obtained from the database. The important part was to find these images that required data purity. Due to the effectiveness of the training models, changing the size of an image is a significant step forward in computer vision. The model will work better if the size of the photograph is smaller. Changing the image size in this investigation resulted in a 224 × 224-pixel image.

The next stage is to convert all of the database's photos into identical members. The image is made into a looping function. The image will then be utilized to process input with MobileNetV2.

5   EXPERIMENTAL SETUP AND RESULT EVALUATION

Figure 2 graphical representation shows the accuracy of the given model of the nCoviSafe model that is being built. The losses generated are also given in the above graph. On our dataset of 1006 images, this paper is able to attain an accuracy of about 97.255 percent (i.e., 0.972) using 600 images for training, 100 images for testing, and 306 images for validating the model (including both mask and non-mask images) and can identify the person's identity who is not wearing a mask. 0.0117 is the loss. To calculate the accuracy of the model indicated above, a total of 50 epochs with batch size 32 were executed. Each epoch took about a minute to complete their processing.

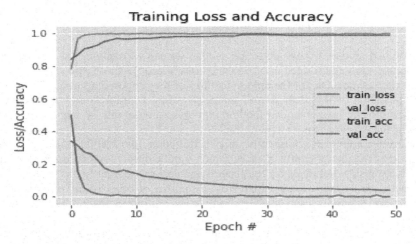

Figure 2.   Graphical representation of accuracy.

6   RESULT AND DISCUSSION

Object detection, face mask detection, and facial recognition are all functionalities that we have integrated in our project. We have used MobileNetV2 to recognize objects and face masks in the dataset, and it was efficient in extracting features from the photographs. TensorFlow-Slim Image Classification Library integrated with MobileNetV2 in its newly released version. MobileNetV2 develops on the fundamentals of MobileNetV1 by employing depth-wise separable convolution as a cost-effective building element. V2, on the other hand, adds two additional architectural features:

495

- Between the layers, there are linear bottlenecks, and
- there are shortcut links between the bottlenecks.

It is used because it has helped achieve more accuracy than other neural networks such as MobileNet, ResNet50, ShuffleNet, NASNet, and others. Some of these are outperformed by MobileNetV2.Others are based on inference time, while others are based on inference time. Comparable model size and computational cost.MobileNetV2 was utilized. For data, there's tensorflow, keras, sklearn, numpy, and pandas.Augmentation, model construction, and training are all things that can be done. Tensorflow provides an end-to-end solution and end-to-end open-source platform on which our code can be implemented on Keras, its high-level API.The Keras model, optimizer, and layers are not only imported but also used in our program. Flatten, Input, Dropout, MaxPooling2D, and AveragePooling2D are some of the layers of Adam Optimizer. Dense assisted in the creation of the model, which was then fitted to determine the accuracy and loss. Numpy'snp.array(), LabelBinarizer(), to categorical(), and other functions helped with data augmentation. We used sklearn, which is an efficient tool in such instances because ours is a predictive analysis. It was used to create a table showing accuracy (macro average, weighted average) versus precision, f1-score, support, and recall using a classification report. On our dataset of 1006 images, we were able to attain an accuracy of about 97.255 percent (i.e., 0.972) using 600 images for training, 100 images for testing, and 306 images for validating the model (including both mask and non-mask images) and can identify the person's identity who is not wearing the mask. 0.0117 is the loss. To calculate the accuracy of the model indicated above, a total of 50 epochs with batch size 32 were executed. Each epoch took about a minute to complete its processing.

7 CONCLUSION AND FUTURE WORK

In the proposed face and mask detection model, termed nCoviSafe, the dataset was successfully constructed and trained into two categories of people: those who are wearing masks and those who are not. One of the unique features of the proposed approach is the accurate image classification by the MobilenetV2 image classifier. To address any obstruction under dense situations, the recommended method employs an ensemble of single- and two-stage detectors at the pre-processing level.

This ensemble approach considerably boosts detection speed while assisting in reaching high accuracy. The identification of faces, which significantly goes against mask standards, increases the system's usefulness for the general population. In order to increase the accuracy of the results, the dataset we assembled from a number of other sources and the graphics used here were manually cleaned. This completely addressed the problem with various inaccurate predictions from our model.

Real-world applications will be a much tougher task in the near future. In this pandemic situation, which has spread around the globe, the ncoviSafe model should ideally help the necessary authorities; this work also points researchers in new directions in the near future. First of all, the suggested method is not just restricted to mask and face identification; it is frequently included into high-resolution video surveillance systems.

Second, other researchers can evaluate more complicated models, such facial landmark models and biometric face component detection algorithms, using the dataset that was used in this study. Previous studies have addressed the troubling findings, and some have been able to increase the reliability of their databases. Since the data was obtained from a number of other sources and the images included in the database were meticulously cleaned to improve the quality of the results, the issue of multiple incorrect predictions has been successfully eliminated from the model. Future practical applications will be quite challenging. Future applications in the actual world will be quite challenging. The database fed in this work may be used by other researchers for cutting-edge models such partial face process identification, facial feature detection, and facial recognition. The database supplied in this work may be used by other researchers for cutting-edge models such partial face process identification, facial feature detection, and facial recognition.

REFERENCES

[1] Eason G., Noble B., and Sneddon I. N., "On Certain Integrals of Lipschitz-Hankel Type Involving Products of Bessel Functions," *Phil. Trans. Roy. Soc. London*, vol. A247, pp. 529–551, April 1955. (Journal Paper)

[2] Ayyubi S. r., Miao Y., and Shi H., "Automating Standalone Smoke Alarms for Early Remote Notifications," in *13th Intern. Conf. on Cont. Automa. Robotics & Vision (ICARCV)*, Marina Bay Sands Singapore, Dec. 2014, pp. 675–680. (Conf Paper)

[3] Jacobs S. and Bean C. P., "Fine Particles, Thin Films and Exchange Anisotropy," in *Magnetism*, vol. III, G. T. Rado and H. Suhl, Eds. New York: Academic, 1963, pp. 271–350. (Journal)

[4] Clerk Maxwell J., *A Treatise on Electricity and Magnetism*, 3rd ed., vol. 2. Oxford: Clarendon, 1892, pp.68–73. (Book)

[5] Crawford M., *Catching the Sun, American Society of Mechanical Engineers*, Feb. 2013. Accessed on: Oct 12, 2020. [Online]. Available: https://www.asme.org/engineering-topics/articles/renewable-energy/catching-the-sun. (Online Content with Author)

[6] "*Engineering Triumph That Forged a Nation: Panama Canal Turns 100*," Aug. 15, 2014. Accessed on: Oct. 12, 2020. [Online]. Available: http://www.msnbc.msn.com/news/world/engineering-triumph-forged-nation-panama-canal-turns-100-n181211 (Online Content without author)

[7] Joao Carlos Virgolino Soares, Marcelo Gattass, Marco Antonio Meggiolaro, "Visual SLAM in Human Populated Environments: Exploring the Trade-off between Accuracy and Speed of YOLO and Mask R-CNN", *19th International Conference on Advanced Robotics (ICAR)*, 2019.

[8] Hensley L., "Social Distancing is Out, Physical Distancing is Inheres How To Do It," *Global News–Canada* (27 March 2020), 2020.

[9] Wan S., Liang Y., Zhang Y., "Deep Convolutional Neural Networks for Diabetic Retinopathy Detection by Image Classifification", *Comput. Electr.Eng.* 72 (2018) 274–282.

[10] Koubaa, B. Qureshi, M. Sriti, Y. Javed, and E. Tovar, "A Service Oriented Cloud-based Management System for the Internet-of-drones", *2017 IEEE International Conference on Autonomous Robot Systems and Competitions, ICARSC 2017*, April 26–28, 2017, pp. 329–335, 2017.

[11] Ren S., He K., Girshick R. B., Sun J., "*Faster R-CNN: Towards Real-Time Object Detection with Region Proposal Networks*", CoRRabs/1506.01497, 2015.

[12] Dong E., Zhu Y., Ji Y., Du S., An Improved Convolution Neural Network for Object Detection Using YOLOv2, 2018 IEEE International Conference on Mechatronics and Automation (ICMA), pp. 1184–1188, 2018.

Loey, M., Manogaran, G., Taha, H. N., & Khalifa, E. M. (2021). Fighting Against COVID-19: A Novel Deep Learning Model Based on YOLO-v2 with ResNet-50 for Medical Face Mask Detection. *Sustainable Cities and Society*, 65, 102600. https://doi.org/10.1016/j.scs.2020.102600.

[13] "*Improving Face Recognition Rate Uusing Hough Features and SVM Classifiers – Google Search*." (accessed Jan. 13, 2023).

[14] Bhuiyan M. R., Khushbu S. A., and Islam M. S., "A Deep Learning Based Assistive System to Classify COVID-19 Face Mask for Human Safety with YOLOv3," *2020 11th International Conference on Computing, Communication and Networking Technologies, ICCCNT 2020*, Jul. 2020.

[15] Loey M., Manogaran G., Taha M. H. N., and Khalifa N. E. M., "A Hybrid Deep Transfer Learning Model with Machine Learning Methods for Face Mask Detection in the Era of the COVID-19 Pandemic," *Measurement*, vol. 167, p. 108288, Jan. 2021.

[16] Yadav S., "Deep Learning based Safe Social Distancing and Face Mask Detection in Public Areas for COVID-19 Safety Guidelines Adherence," *Int J Res Appl Sci Eng Technol*, vol. 8, no. 7, pp. 1368–1375, Jul. 2020.

[17] Saxena P., Singh S. K., Tiwary G., Mittal Y.and Jain I., "An Artificial Intelligence Technique for Covid-19 Detection with Explainability Using Lungs X-Ray Images," *2022 IEEE International Conference on Distributed Computing and Electrical Circuits and Electronics (ICDCECE)*, pp. 1–6: 2022.

Artificial Intelligence, Blockchain, Computing and Security – Dagur et al. (Eds)
© 2024 The Author(s), ISBN: 978-1-032-49393-0

Stock market price forecasting

Raja Jadon*, Shivam Yadav* & Abdul Aleem*
School of Computer Science and Engineering, Galgotias University

ABSTRACT: Nowadays almost 98% people in India deposit their money in banks, insurance, mutual fund, and soon rather than investing their money in Stock Market. As they know it is easiest way to earn some interest on wealth, they only need some prior knowledge like they certified from RBI, when people choose them for gaining some extra worth on their wealth. The main reason we found during our research they have lack of knowledge, as well as some of them, are not aware of Stock Market. Those who know about Stock Market but they didn't want to spend their time on researching about how they invest their money in Stock Market that's why here we are for helping them out of this curiosity. We are currently introducing a new stock market price forecasting model using machine learning technique. This model predicts the next week's stock market price outcome and is trained by looking at previous data to obtain information that can be used to predict in an accurate manner. As peoples know it's hard to predict with 100% accuracy because the stock market price depends on various factors like Human Psychology, Natural Calamities, Political factors, Current Events, etcetera. There is so many methods for stock market price forecasting like Support Vector Machine (SVM), Reinforcement learning, etcetera. We proposed our model through Machine Learning pipeline method where we are using long short-term memory (LSTM) because as per our research we find out LSTM is better method than others to Stock Market Price Forecasting of NSE_TATA.

1 INTRODUCTION

As we all know one of the crucial areas for investors is stock market to study, so predicting stock price movements which is ever growing topic for researchers in finance and technical fields. This research focuses to create a one of a kind price trend forecasting model that give emphasis on short-term price trend forecasting.

Stock prices varies with various factor for example the demand and supply. If people are interested and intrigued to pay and buy shares of his then an increase in demand will drive the price up. As more people try to sell his shares, the price touch bottom line (i.e decreases) as supply exceeds demand. While it is relatively easy to understand supply and demand, it is difficult to deduce exactly which factors contribute to increased demand or supply. Such influences often bring down the socio-economic factors such as market dynamics, trends preciously the company's positives and negative sin the news.

The Importance of Stock Markets

- Stock markets help companies to raise capital.
- It helps create personal wealth.
- Stock market serves as an indicator of economic conditions.
- Stock market is a popular source for investing in companies with high growth potential market.

*Corresponding Authors: rjrajajadon007@gmail.com, shivamyadavsm41@gmail.com and er.aleem@gmail.com

DOI: 10.1201/9781003393580-76

2 PURPOSE

A stock market forecast is often regarded as an attempt to evaluate upcoming scenario of a company's stock or various publicly traded finance assets. The next price of a stock in a successful quote is called the significant profit return. Doing this can help you immensely and earn profit.

3 REVIOUS RELATED RESEARCHS

Forecasting of stock prices can be seen as attempt to determine stock prices and giving people concrete analogy and understanding the market and stock prices. Mehar Vijh. Presented, tested various machine learning techniques some of them were ANN artificial neural network and random forest technique which predicted the closing price of five companies that were in different sectors for the upcoming day.

The input data was financial data. He created new variables as input to the model which consisted of various factors such as open, high, low and close prices. The models were tested against the standard strategy metrics RMSE and MAPE. The lesser the value produced by the two indicates more the model accurate to predict stock market prices. The comparative study shows that ANN artificial neural network produces better results than Random forest.

V. Kranthi Sai Reddy in year 2018 introduced ML machine learning technique that trains and extracts information from available stock data and uses the gained information to make almost correct predictions. In his analysis he try to utilize a ML algorithm known as SVM support vector machine which collects large and small market caps and stock prices from three different markets, both daily and latest frequency to predict. This algorithm works on large input data extracted by various international financial capital. Also, SVMs are not good with overfitting problems.

Khalid Alkhatib Al (year 2013) used techniques like non-linear regression and k-nearest neighbours to foretell the prices of stock of more than six companies which were named in the Jordan stock exchange.

Managers helped consumers or inventors to make informed investments and decision. The results tells us how K-nearest neighbour algorithm therefore aa robust and has a error rate. So the result was not accurate. Moreover, it depend on the data which consists of real stock price. Forecasting result was not similar enough to the real stock price and parallel to it.

4 PROPOSED WORK

We are introducing Machine learning techniques such as LSTM for existing work using available machine learning stock price predictions using NSE_TATA data as input to estimate the future value of stock price. This ML algorithm is by far the most stable as well as accurate and gives the best prediction results for future stock prices. LSTM can capture changes in stock price movements over time in this proposed system.

The LSTM module consists of states that can selectively gain, un-gain, or recollect useful intel from each state unit. These states in LSTMs help the intel pass through different cells unit unaltered by granting very less communication. Every cell unit has inputs, outputs, and the forgetting gates that can plus or minus cell state information. The forget gate basically comes in action with a sigmoidal function to figure out which and what information have to be forget from previous cell states. The input gate uses a 'sigmoid' or 'tanh' point wise multiplication operation to control the flow of information about the present state of cell. Finally, the output gate figures out which and what intel to deliver to the adjacent hidden state.

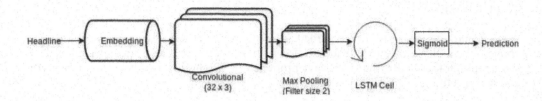

Headline → Embedding → Convolutional (32 x 3) → Max Pooling (Filter size 2) → LSTM Cell → Sigmoid → Prediction

4.1 Network input

The NSE_TATA Stock Price are first pre-processed and are transferred as input to this very neural network. Model consist of three step-by-step process. The very first and foremost is to what intel to delete at that particular moment of time from the cell state. Determined by the sigmoid function. Compute this function given the past state (ht1) and the present data input xt. The next core has count of two functions. The first is the sigmoid function and second is the (tan-h) function. The sigmoid function determines passed value (0 or 1). The tanh function weights passed value and concludes its importance on a scale of −1 to 1. The last step is to determine the final output. Firstly we have to execute a sigmoid part that reconciles cell state part to send output. Next, we need to pass state of this cell to the (tan-h) function to vary value juggling in between −1 and 1 and then with the output of our sigmoid gate we perform multiplication of these values.

4.2 Convolutional layer

This Layers consist of various layers also known as filters during training process the weights of these filters are learned and trained. The dimension of the filter is much tiny in comparison of the dimension of the input vector, but with the similar deepness. During the ahead first pass, input layer the vector is dealt with respect to their height and width and their dot product is calculated between value pf the filter and input layer at each and every position. Taking I as input and k as filter the process comprises of convolutional process. The networks enhances its learning during this training process and adapt the new learning. These filters are activated upon detection of a hidden target in the input.

4.3 Max pool layer

In general, a convolutional network architecture adds pooling layer in middle of consecutive convolutional layers. The pooling layer decreases the spatial size step by step at each representation, which results in reduction of number of parameters, network calculations and evaluation, also the over-fitting control. The layer that is pooling if simply put is used to apply maximum functions or operations which are used at each fundamental size of the input in the specified dimension. Adjacent values in the convolution are highly correlated, so it makes sense to subsample the response from the filter to reduce the output size. Most widely used max pool layer is considered to be of size 2.

4.4 LSTM cell

Result collected by the max pool layer is then send to the cell of LTSM. L S T M consist of three layer system, one of them is the number of units another one is activation function and the last is iterative activation function. Cross validation is the way to achieve number of units. The input, forget, and output gates are treated with number of repetitive activations and these activation is applied to hidden candidate states and hidden output states. Default value of iterative activation function is hard sigmoid function and that of activation function is hyperbolic tangent function.

4.5 *Output layer*

We use sigmoidal activation function in output layer, as we see the network output isin between 0 and 1.

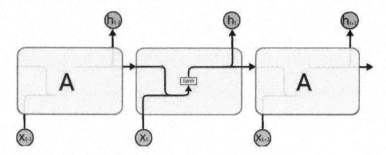

Figure 1.   A standard RNN's repetition module contains a single layer.

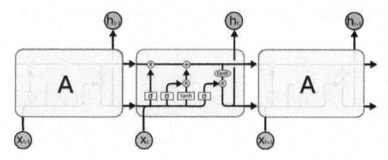

Figure 2.   The LSTM iteration module contains four interacting layers.

5   EXPERIMENTAL RESULT

The proposed data is taken from GitHub where already we have train data csv file and test data csv file. By using this data we are going find our result using lstm model:-

LSTM Based Model Result
The above Figures (2) and (3) is the plot of TATA stock price consisting of data size equals 32 and includes 100 number of epochs. The prediction is shown by black line for the actual dataset and green line used for shown predicted TATA stock price trend of LSTM. The closeness of the two shown lines shows, how much variation between LSTM model and actual dataset. The prediction approaches the real time trend of stock market when a sufficient amount of time has gone by. The greater accuracy or precision will be shown when the more the system will be trained then it may show the best that we want from the model.

6   CONCLUSIONS AND FUTURE CHALLENGES

Stock price forecasting is an area of great interest to stock traders, retail investors and portfolio managers. However, accurate and consistent stock price prediction is a difficult task due to noisy and nonlinear behavior. There are several factors that can affect forecasts, including underlying market data, macroeconomic data and technical indicators. This

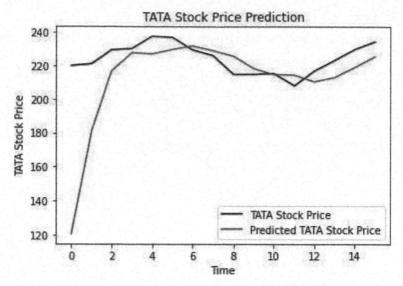

Figure 3. Stock market prediction model using LSTM (SGD optimizer).

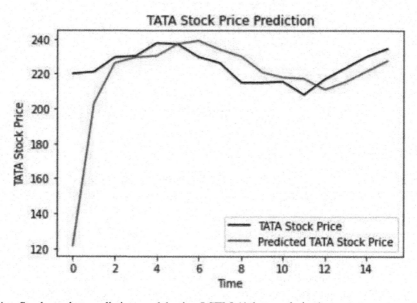

Figure 4. Stock market prediction model using LSTM (Adam optimizer).

research focuses on developing his LSTM-based model to predict the closing price of the S&P 500 index by extracting a balanced combination of input variables that capture diverse aspects of the economy and the broader market. increase. Both single-tier and multi-tier LSTM architectures are implemented, and performance is analyzed using various metrics to identify the best model. Experimental results show that a single-layer LSTM model with about 150 hidden neurons can provide better fitting and higher prediction accuracy compared to multi-layer LSTM. The proposed model can be easily adapted to apply to a wide range of other market indices whose data exhibit similar behavior. Interested stakeholders can use the proposed model to better understand market conditions before making investment decisions. In the near future, we plan to explore the possibility of including unstructured textual information in the model, such as: B. Investor sentiment from social media, earnings reports from underlying companies, policy breaking news and research reports from market analysts. Another potential direction of future work might be the development of hybrid prediction models by combining LSTM with other neural network architectures. To further improve prediction accuracy, we also plan to implement hybrid optimization algorithms that combine existing local optimizers with global optimizers such as genetic and particle swarm optimization algorithms to train model parameters.

REFERENCES

[1] Lin, Y., Chen,T., Yu, L. Using Machine Learning to Assist Crime Prevention. In: *2017 Sixth IIAI-AAI*

[2] Kerr, J.: Vancouver Police go High Tech to Predict and Prevent Crime Before it Happens. *Vancouver Courier*, july 23, 2017.

[3] Marchant, R., Haan, S., clancey, G., Cripps, S.: Applying Machine Learning to Criminology: Semi Parametric Spatial Demographic Bayesian Regression. *Security Inform.* (2018).

[4] M. J. H. B. T. A. M. K. T. Baig, M.Q., "Artificial Intelligence, Modelling and Simulation (Aims), *2014 2nd International Conference on,*" pp. 109–114, (November 2014).

[5] Anitha A, Paul G and Kumari S 2016 A Cyber Defence Using Artificial Intelligence *International Journal of Pharmacy and Technology* 8 2532–57

[6] Moghar, A., & Hamiche, M. (2020). Stock Market Prediction Using LSTM Recurrent Neural Network. *Procedia Computer Science, 170*, 1168–1173. https://doi.org/10.1016/j.procs.2020.03.049

[7] Bogomoloy A., Lepri B., Staiano J., Oliver N., Pianesi F. and Pentland A., 'Once Upon a Crime: Towards Crime Prediction from Demographics and Mobile Data', *CoRR*, vol. 14092983, 2014.

[8] Arulanandam R., Savarimuthu B.and Purvis M., 'Extracting Crime Information from Online Newspaper Articles', in *Proceedings of the Second Australasian Web Conference – Volume 155*, Auckland, New Zealand, 2014, pp. 31–38.

[9] Buczak A. and Gifford C., 'Fuzzy Association Rule Mining for Community Crime Pattern Discovery', in *ACM SIGKDD Workshop on Intelligence and Security Informatics*, Washington, D.C., 2010, pp. 1–10.

[10] Tayebi M., Richard F. and Uwe G., 'Understanding the Link Between Social and Spatial Distance in the Crime World', in *Proceedings of the 20th International Conference on Advances in Geographic Information Systems (SIGSPATIAL '12)*, Redondo Beach, California, 2012, pp. 550–553.

[11] Nath S., 'Crime Pattern Detection Using Data Mining', in Web Intelligence and Intelligent Agent Technology Workshops, 2006. WI-IAT 2006 Workshops. *2006 IEEE/WIC/ACM International Conference on*, 2006, pp. 41,44.

Artificial Intelligence, Blockchain, Computing and Security – Dagur et al. (Eds)
© 2024 The Author(s), ISBN: 978-1-032-49393-0

Analysis of node security optimization in WSN

K. Sharma*, S. Chhabra* & S. Rani*
Chitkara University Institute of Engineering and Technology, Chitkara University, Punjab

ABSTRACT: Nowadays, thanks to the Internet of Things, wireless sensor networks are expanding swiftly. The term "Wireless Sensor Network" (WSN) describes a wireless network implemented ad hoc employing a significant number of sensors to monitor system, physical, or parameters based on environment. Applications for such networks include consumer and commercial monitoring and management of industrial processes and equipment health. In both industrial and consumer applications, WSN is widely employed. WSN places a high priority on privacy, making it crucial to protect any utilised, transferred, or stored data. Bluetooth, WiFi, and other wireless communication technologies expose consumers to a wide range of potential security issues. This study places a strong emphasis on the WSN's security.

Keywords: Wireless sensor network, WSN application, WSN security, and WSN application range

1 INTRODUCTION

One of the most cutting-edge and rapidly expanding communication systems used today is the wireless sensor network. With the use of this communication, many people may easily share data with others in distant parts of the world. Regardless of the application, the following categories canbe applied to wireless sensor networks in general. Mobileand static WSN. WSNs that are deterministic and non-deterministic. WSNs with a single base sta- tion and several base stations. Mobile Base Station WSN and Static Base Station. a wireless sensor node's physical composition. Four fundamental elements contribute to a sensor node: sensingunit, processing unit, transceiver unit, and power unit. Overthe years, WSNs have expanded significantly, and they now hold enormous promise for a wide range of applications in fields including environmental research, medicine, telecommunications, education services, agriculture, surveillance, and military services, among others. The integrity, accessibility,and data secrecy of the WSNs are impacted by some of these problems, including spying, sinkholes, tampering Sybil, cloning, wormholes, spoofing, etc. Many security solutions have been put out for WSNs; however, due to sensor resource limitations, some of these security solutions are not suitablefor WSNs '(Akvildiz *et al.* 2011)'.

2 CHARACTERISTICS OF WSN

A WSN's primary attributes include:

1. Limitations on the amount of power that nodes with batteries or energy harvesting can consume. Suppliers include ReVibe Energy '(ReVibe *et al.* 2018)' and Perpetuum, for instance.

---

*Corresponding Authors: kashishaayushi2@gmail.com, chhabrasatyam024@gmail.com and shalli.rani@chitkara.edu.in

DOI: 10.1201/9781003393580-77

2. Capability to handle node failures
3. Node heterogeneity
4. Node homogeneity
5. Scalability of deployment to huge scale
6. Capacity to endure challenging environmental circum-stances
7. Usage ease
8. Optimization for cross-layer '(Aghdam *et al.* 2014; Miao *et al.* 2016; Saleem *et al.* 2009)'. Cross-layer research is growing in importance for wireless communications

Additionally, the conventional layered technique has three key issues:

1. The traditional tiered technique does not allow for the sharing of diverse information between levels, which leaves each layer with incomplete knowledge. The optimization of the entire network cannot be ensured by the conventional tiered technique
2. The conventional layered strategy cannot evolve to accommodate the environment.
3. Due to user interference, access restrictions, fading, and environmental changes, wireless networks cannot employ the traditional tiering technique used for wired networks.
4. One or even more WSN parts with significantly larger processing, energy, and information concerning are the base stations. Data's confidentiality, integrity, and availability are negatively impacted by the vulnerability of wireless sensor networks to numerous attacks. The router, which is created to compute, compute, and disseminate the routing tables, is another characteristic component of routing-based networks '(Xenakis *et al.* 2016)'.

3 SECURITY OF WSN

The infrastructure-less design of WSNs (i.e., the absence of gateways, etc.) and their intrinsic needs (i.e., an unattended working environment, etc.) might provide a number of vulnerabilities that draw in attackers. Traditional computer network security techniques would be worthless (or less effective) for WSNs due to their special properties. Therefore, breaches into those networks would result from a lack of security measures. It's important to find these incursions and use mitigation techniques '(Gupta *et al.* 2022)'.

There have been significant advancements in wireless sensor network security. Since omni-directional antennas are used in the majority of wireless embedded networks, communication between nodes can be overheard by bystanders. In order to identify complex assaults, like blackhole or wormhole, this rudimentary technique known as "local monitoring" was developed. Many researchers have subsequently adopted this basic. Later, this was upgraded for more complex intrusions involving mobility, collusion, and multi-antenna.

Wireless sensor network (WSN) security is now a prominent topic in WSN technology research. The sensor node's data collection cannot be instantly and precisely conveyed to the target sink node, particularly when the network's routing is attacked. Similar to the conventional computer communication network, numerous assaults are the major source of the wireless sensor network's security threat. Monitoring, adjusting, faking, and blocking assaults will be done on the sensor network. Assaults on the sensor network will be foreseen, thwarted, and monitored as a result of radio properties of wireless channels and network features. Wireless sensor networks are susceptible to denial-of-service attacks because they are a form of energy-depleting, energy-constrained sensor node '(Lee *et al.* 2016)'.

The following methods of attack are possible against wireless sensor networks:

1. Network layer attacks include discarding and selective transmission, exploitation of sink nodes, direction rerouting, sink hole invasion, etc. Attackers use bogus routing information or direction misdirection to redirect traffic, lengthen or shorten the source path, and create routing loops, among other things, with the goal of disrupting the network and lengthening end-to-end delays '(Khan *et al.* 2022)'.

2. The transport layer contains both synchronised damage attack and flood attack. Flood attacks happen when a nearby sensor node is subjected to an attacker's frequent attempts to connect, taxing its connectivity and compelling it to prioritise the attacker's demands above legitimate request.
3. Th Physical harm and congestion assaults can affect the physical layer network. An attacker can create congestion by providing radio interference near to the wireless sensor network's centre frequency if they are aware of the transmission frequency used by the network. Sensor nodes are typically put in vulnerable areas where they may easily be removed hackers and kept secret on the network for physical injury, surveillance, and network disruption.

As a result, the wireless sensor network (WSN) faces more security risks, making the traditional security protection mechanisms ineffective.

4 THE KEY TECHNOLOGY

4.1 *Security optimization technology*

Node security optimization solutions may help enhance the security protection of WSN. The project would provide technique for optimization based on the ternary key distribution method to protect the security of wireless sensor networks. This technique can simplify the node architecture and manage the anti-attack capabilities of WSN '(Liu *et al.* 2018)'.

The wireless sensor network topology is made up of different kinds of network nodes; by utilising the self-organization form of the cluster head election node and three alternative kinds, it is possible to meet the key calculation requirements and collaborate to adapt adaptive security routing, increasing the WSN attack resistance '(Liu *et al.* 2018)'.

The use of three key distribution techniques involves three different types of key calculations. The relationships between the sensor node and the base station, the cluster head and the cluster, and .3the sensor node are crucial. Base station node data must be encrypted using the Kn key. It can meet the demand for key computation between the base station and the sensor node. Normal sensor nodes must decode the broadcast message data after receiving it from the base station.

4.2 *Security fusion technology*

The fact that WSN nodes are typically found in unsupervised environments and security-sensitive locations makes it simple to combine data from wireless sensor networks to counter various security threats. A robust security system must be given to assure data security under the effect of reduced energy costs. In modern wireless sensor networks, data integration methods like data rolling and data integrity are frequently used.

While the latter depend on data privacy, former is based on data integrity. Additionally, in order to ensure data security, symmetric cypher algorithms are used in data fusion methods by encryption schemes. While these schemes have high applicability to advantages, there is a shortage of expensive aggregation points. As a result, to reduce the energy consumption of the current schemes, it is possible to adjust intermediate nodes during the data transmission process so that they don't decode the data you got, the data security technique is a revolutionary protective mechanism that combines security with energy efficiency '(Parmar *et al.* 2014)'.

The cluster member nodes should keep an eye on the sampled fusion nodes based on their actions in order to build confidence among the member nodes.After computing and analysing the result set nodes together with the data fusion byte points, it is up to the base station to make the ultimate choice.

5 RANGE OF APPLICATIONS

5.1 *Environmental/earth sensing*

Environmental parameter monitoring has several uses, some of which are provided here. They likewise face the additional difficulties posed by tough settings and little power.

5.1.1 *Air quality monitoring*
According to experiments, there are many different ways that people are exposed to city air pollution. Therefore, it is preferable to have pollutants and particles with higher temporal and spatial resolution. In order to monitor the amounts of hazardous gases in humans as part of the study, wireless sensor networks have been put up. Additionally, the data quality is currently insufficient for accurate decision since frequent recalibration may be necessary and field calibration yields inconsistent measurement findings '(Saligrama *et al.* 2008; Saukh *et al.* 2015)'.

5.1.2 *Landslide detection*
A landslide detection system uses a wireless sensor network to spot tiny changes in a number of variables that might happens before to or during a landslide. Using the information gathered long before they actually happen, it could be able to forecast when landslides would start to occur.

5.1.3 *Water quality monitoring*
Examining the qualities of the water in dams, rivers, lakes, and oceans is part of monitoring water quality. The use of many wireless scattered sensors allows for the permanent placement of monitoring stations in challenging-to-access locations and also makes it possible to produce a more precise map without the need for manual data retrieval '(Spie 2013)'.

5.2 *Health care monitoring*

Medical devices that are implanted into a person's body are known as implantable devices. Sensors located in the environment are used by systems embedded in the environment. Applications might involve locating people, measuring body posture, or keeping track of sick patients both at home and in hospitals. The confidentiality and reliability of user data are crucial in healthcare applications. User identification is becoming increasingly difficult, especially with the integration with IoT; nevertheless, a solution has just been proposed '(Bilal *et al.* 2017)'.

5.3 *Industrial monitoring*

5.3.1 *Machine health monitoring*
WSN were created for industrial machinery condition-based maintenance in order to considerably save costs and enable more functionality.

In places where a wired system would be difficult or impossible to install, such as on moving machinery and untethered vehicles, wireless sensors can be installed.

5.3.2 *Data logging*
Additionally, data gathering for environmental information monitoring is done through wireless sensor networks '(Saleem *et al.* 2014)'. This can be done as easily as keeping an eye on a refrigerator's temperature or the amount of water in a nuclear power plant's overflow tanks. The statistical data may then be used to show how well the systems work. WSNs differ from conventional loggers in that they may receive "live" data streams.

5.3.3 *Wine production*
Wine production is observed using WSN in both the vineyard and the winery.

5.4 *Supply chain*

WSNs may be used effectively and economically in supply chains across a variety of sectors by utilising low-power devices '(Liz *et al.* 2022)'.

6 RESULT ANALYSIS AND DISCUSSION

The analysis of different protocols is performed in MATLAB. EARP, AODV, DSR, LEACH, LLEAP and EAP are considered for the comparison. Simulations are performed for 500 Nodes. Data transmission is done till the end of the last nodes and throughput is taken in bits/second. It is observed that throughput of LLEAP, EAP and LEACH is very low, However, EARP is performing very well and it transmits 70 thousand bits which is higher than other protocols as shown in Figure 1:

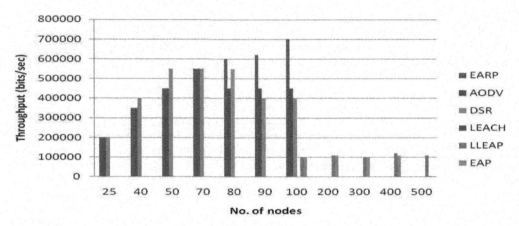

Figure 1. Comparison of throughput.

The more the data is transmitted, more the security is required for the same protocol.

7 CONCLUSION

The use of the WSN will be wider as a result of the quick advancements in sensor and communication technologies. Secret key management will get greater attention because it is a fundamental security function. Secret key management protocols and schemes must meet the requirements of WSNs, scalability, low computing complexity, minimal storage needs, minimal communication demands, customizable topology, etc. They also need to be directly tied to the application. The next piece of study will concentrate on how secret key management methods and schemes distribute security, self-organize, tolerate faults, and combine geographic information.

REFERENCES

Aghdam, S.M., Khansari, M., Rabiee, H.R. and Salehi, M., 2014. WCCP: A Congestion Control Protocol for Wireless Multimedia Communication in Sensor Networks. *Ad Hoc Networks, 13*, pp.516–534.
Akyildiz, I.F., Lo, B.F. and Balakrishnan, R., 2011. Cooperative Spectrum Sensing in Cognitive Radio Net-works: A Survey. *Physical Communication, 4*(1), pp.40–62.
Bilal, M. and Kang, S.G., 2017. An Authentication Protocol for Future Sensor Networks. *Sensors, 17*(5), p.979.

Feinberg, S., Williams, R., Hagler, G.S., Rickard, J., Brown, R., Garver, D., Harshfield, G., Stauffer, P., Mattson, E., Judge, R. and Garvey, S., 2018. Long-term Evaluation of Air Sensor Technology Under Ambient Conditions in Denver, Colorado. *Atmospheric Measurement Techniques*, *11*(8), pp.4605–4615.

Gupta, K., Gupta, D., Kukreja, V., & Kaushik, V. (2022). Fog Computing and Its Security Challenges. In *Machine Learning for Edge Computing* (pp. 1–24). CRC Press.

Khan, N., & Gupta, K. (2022, October). IOT: Applications, Challenges and Latest Trends. In *2022 1st IEEE International Conference on Industrial Electronics: Developments & Applications (ICIDeA)* (pp. 181–186). IEEE.

Lee, H.M., Yoo, D.G., Sadollah, A. and Kim, J.H., 2016. Optimal Cost Design of Water Distribution Net-works Using a Decomposition Approach. *Engineering Optimization*, *48*(12), pp.2141–2156.

Liu, Y. and Morgan, Y., 2018. Security Analysis of Subspace Network Coding. *Journal of Information Security*, *9*(01), p.85.

Liz Young, 2022. Israeli Tech Firm Rolls Out Tracking Devicesthe Size of Postage Stamps. *Wall Street Journal*.

O'Donovan, T., O'Donoghue, J., Sreenan, C., Sammon, D., O'Reilly, P. and O'Connor, K.A., 2009, April. A Context Aware Wireless Body Area Network (BAN). In *2009 3rd International Conference on Per-vasive Computing Technologies for Healthcare* (pp. 1–8). IEEE.

Parmar, K. and Jinwala, D.C., 2014. *Symmetric-key Based Homomorphic Primitives for End-to-end Secure*

Peiris, V. 2013. Highly Integrated Wireless Sensing for Body Area Net- work Applications. *SPIE Newsroom*. doi:10.1117/2.1201312.005120.

ReVibe Energy – *Powering The Industrial IoT*. revibeenergy.com. Archived from the original on 22 September 2017. Retrieved 3 May 2018.

Saleem, K., Fisal, N. and Al-Muhtadi, J., 2014. Empirical Studies of Bio-inspired Self-organized Secure Autonomous Routing Protocol. *IEEE Sensors Journal*, *14*(7), pp.2232–2239.

Saleem, K., Fisal, N., Hafizah, S., Kamilah, S., Rashid, R. and Baguda, Y., 2009, January. Cross Layer Based Biological Inspired Self-organized Routing Protocol for Wireless Sensor Network. In *TENCON 2009-2009 IEEE Region 10 Conference* (pp. 1–6). IEEE.

Saligrama, V. ed., 2008. *Networked Sensing Information and Control*. Heidelberg: Springer.

Saukh, O., Hasenfratz, D. and Thiele, L., 2015, April. Reducing Multi-hop Calibration Errors in Large-scale Mobile Sensor Networks. In *Proceedings of the 14th International Conference on Information Processing in Sensor Networks* (pp. 274–285).

Savoine, M.M., de Menezes, M.O. and de Andrade, D.A., 2018. Proposal of a Methodology for the Assessment of Security Levels of IoT Wireless Sensor Networks in Nuclear Environments. *World Journal of Nuclear Science and Technology*, *8*(2), pp.78–85.

Spie 2013. Vassili Karanassios: Energy Scavenging to Power Remote Sensors. *SPIE Newsroom*. doi:10.1117/ 2.3201305.05.

Xenakis, A., Foukalas, F. and Stamoulis, G., 2016. Cross-layer Energy-aware Topology Control Through Simulated Annealing for WSNs. *Computers & Electrical Engineering*, *56*, pp.576–590.

Artificial Intelligence, Blockchain, Computing and Security – Dagur et al. (Eds)
© 2024 The Author(s), ISBN: 978-1-032-49393-0

Online roadside vehicle assistance: A review

Rohan Dass Gujrati, Roshi Kumar, Rupali Chaubey, Shikha Singh & Sahil Kumar Aggarwal
Information Technology, ABES Engineering College, Ghaziabad

ABSTRACT: While travelling with the personal vehicle, there might be a situation where you need an assistance for your vehicle. There are cases, especially on Indian roads where people do not find any appropriate assistance immediately, and if they get any assistance, it may be expensive. This paper consists of a review of multiple research papers that have tried to address these situations. This paper also covers the various drawback in existing solution and showcase them cumulatively for making a proper solution to handle most of the problems.

1 INTRODUCTION

While travelling, everyone organises and takes all necessary safety measures to ensure safe passage. However, in the unforeseeable case of a breakdown or a traffic accident, there is a requirement for a quick fix to prevent the problem that come with being stranded in the middle of the road. There is a need for a solution that ensures them immediate help. To make this feasible, a web-based solution is an option for user support that helps drivers and passengers alike [1]. Because the web does not require a complicated platform to work, more devices now run the operating system using simple hardware. As a result, individuals may now use the internet even in the most remote regions. Therefore web-based applications can be utilized as a platform to provide a solution which can be used by individuals who are in trouble due to a breakdown and combine many potential sources of support that may be offered while travelling [1].

In the current scenario, everyone relies on their portable gadget to fix any problem faced in day-to-day life. There are a variety of applications available on an app store to provide the solution to any problem. From now on, users of the Web ecosystem may download the app and have access to road assistance services whenever and wherever they need it. The "Mechanic Finder App" provides a tool that finds the closest mechanic. With the built-in GPS function of smartphones, users may determine the position of the closest mechanic while also determining the best route from their present location using Google Maps APIs (API). This software allows users to locate the closest mechanic at any time or place [2].

2 LITERATURE SURVEY

The number of vehicle breakdowns, road accidents and automobiles are increasing day by day. It is challenging to develop an effective strategy to achieve maximum fuel efficiency while not compromising the interior structure of these cars and still providing a reactive system to prevent accidents. There is a need to develop an application for diagnostics that emphasizes on RPM, fuel levels and throttle position. Further, the analysis generated by the diagnostic app can assist inexperienced drivers with gear changes and in the case of a vehicle breakdown [3].

DOI: 10.1201/9781003393580-78

There is, however, a discussion platform where you may debate the sort of breakdown and exchange views concerning car breakdowns. Due to time constraints and the necessity for extensive study, the system may be created on the Android platform. If customers provide positive feedback, the development of this system on other platforms like IOS and Windows will be explored in future. The purpose of this system is to locate the nearest CRSP for drivers, assisting those who still need a mechanic's number on hand. The commercial transaction between the CRSP and the driver is outside of the system's control [4].

Findings from the "Emergency Breakdown Assistance Kit"- It is a car emergency signalling kit that displays "HELP" on the front transparent panel. It specifies the type of breakdown, beneath the HELP symbol. Breakdown Assistance did not display any notice on the front transparent panel regarding the breakdown detail. There is a requirement for a mechanic to determine the type of breakdown.

As a result, emergency auto breakdown service provides better location outcomes. There is an algorithm developed to discover neighbouring sites rapidly, which is incredibly useful for users in emergency circumstances. It also has an offline mode, which delivers recommendations when the internet is unavailable. This method improves the user experience and surpasses the existing system in critical situations [5].

Findings of the "On-Vehicle Breakdown-Warning Report System"- It installs an electronic control panel. When a breakdown occurs, the signal is displayed on the control panel. This may aid in detecting the failure before the vehicle suffers a severe breakdown. On-Road Vehicle Breakdown Assistance could not detect any exceptional breakdowns and did not display any specialised breakdown signal. A mechanism for reporting on-vehicle breakdowns is unveiled. A break-down occurrence is detected and judged based on a signal in an electronic control system installed on a control apparatus for an engine ignition system, a charging system, an engine fuel system, an engine cooling system, a power transmission system, and an oil lubricating system of an automobile, or a diagnosis display system, and diagnostic data is transmitted to an information terminal device of a diagnosis and maintenance agency or a service firm that has a diagnosis and maintenance agency as a contents information using an on-vehicle mobile communication equipment. Action for emergency measures and a maintenance schedule is requested [6].

Findings of the "Geo Place Tracking System and Method" – It is geo-tracking routing from point to point in a geographical location.

There is a location tracking feature that is dependent on the user's location. Users may search for spare parts shops depending on their location. People are increasingly anticipating information on the whereabouts of anything for tracking purposes, thanks to recent technological advancements in modern science. There is a need for additional location-based services to be more advanced while saving time and money. GPS is a pre-existing system that anybody may use. We need a GPS device to determine the position to develop this system based on GPS data [7].

Findings of the "Vehicle breakdown assistance"-This programme is intended to locate nearby area mechanics when we are unexpectedly stuck in distant regions due to mechanical troubles with our car. It is an excellent answer for folks who need assistance in rural areas. This application includes a list of authorised mechanics. This approach also ensures that they do not charge any additional service fees to subscribers. The admin may keep track of this through customer comments on their service. This application is only available to registered users. This programme will assist users in saving time when looking for a good mechanic. This programme will allow the user to pay for a vehicle repair at a reasonable cost. When a car breaks down, the driver must see a mechanic or a repair shop. The passenger must seek assistance from the public. This programme allows the user to locate mechanics depending on their location. The user may obtain mechanical assistance easily and quickly. This helps users to save time while travelling. When a breakdown happens, the user may quickly fix their vehicle. This makes the individual feel more at ease. They will not get exhausted on their travel [8].

In location tracking, the "Localize Intelligence Algorithm" was introduced, which allows users to microscopically identify a specific region. Typically, satellite access to this information is not accessible. For instance, if a mobile phone user is at home or at a relative's house, the programme notifies us of his position, such as the sector number. Along with this novel strategy, the research compares several location-tracking methods and applications [9].

3 EXISTING SYSTEM

In the existing system, the search result is available only when the user is in the range, so it is challenging for the user to get the service, especially in rural areas. Assistance through the helpline is highly prone to unavailability, making the traveler's experience worse.

When a user enters information about the breakdown location into the system, it will automatically look for nearby local car repair shops that have enrolled and been verified with the system.

However, there is a chat platform where people may ask for help based on different types of breakdowns and share their feedback [1].

The chat may be rated using a mobile chat system that allows real-time polling, rating, and RSVPs. While conversing, the user can rate the response with stars. Users of On-Road Vehicle Breakdown Assistance can respond and give discussion a star rating. When a person first registers with the system, they get access to the chat and the top-rated mechanics [10].

A vehicle emergency signaling equipment called an emergent breakdown assistance kit displays "HELP" on its front transparent panel. Indicate the precise type of disability underneath the HELP symbol. Breakdown Assistance needed a distinctive front panel sign in on-road vehicles. To determine the type of problem, mechanics are required [5].

The novel driver-aid system that relies on inter-vehicle communication is the main emphasis of Car Talk 2000. Using a radio network for communication aids in automobile communication. Radio networks were not used for communication. As a result of the system's utilization of the Android operating system and GPS, users may find mechanics [12].

In some of the existing systems, geo-tracking navigation is used for location monitoring depending on the user location. The user may use their location to look for spare parts stores [13].

4 ANOMALIES OF THE EXISTING SYSTEM

While travelling from one place to another, there is always a need for assistance as it gives the person comfort to travel without fear. Over time there have been many solutions for roadside assistance, but a whole app that can solve most of the problems is still missing. Table 1 shows the functioning of the different solutions provided with their drawbacks.

Table 1. Function of the solution provided for breakdown assistance with their drawback.

Features	Description
Do not show the specific issue	An emergency breakdown Assistance Kit is an automobile emergency signalling kit, that didn't display any special sign in the front panel. There is need for a mechanic for identifying the nature of the disable [5].
Time consuming	To search for a mechanic in remote area is very difficult using GPS disabled system.
Calling System	The Car Repair Service Provider (CRSP) has only a chat system, does not have a calling system [4].
Location Tracking	Geo Location Tracking System and Method is geo tracking routing from point to point using GPS sensor. There is no way to contact the mechanic or spare parts shop [12].
Communication	Car Talk 2000 uses radio network for inter vehicle communication [12].
Rating and Feedback	Mobile Chat system for Real-Time Polling, rating and rsvp'ing can rate the chat using star but cannot give detailed feedback [10].

We have "A car breakdown service station locator system", it is connected to Car Repair Service Provider (CRSP)" which gives the functionality for tracking the vehicle but lacks a communication system [1]. The "Emergency Vehicle Breakdown Assistance Kit" adds the functionality of communication but still lacks in many functionalities [10]. In the "Car Talk-2000", it was proposed to give smooth intercommunication via radio ad-hoc [11]. The "Mobile chatting system for the real-time polling, rating and rsvp'ing" provides the functionality for polling and rating through the mobile app [12]. The "On-Vehicle Breakdown-warning report system" was very efficient in case of vehicle breakdown it provides a warning to the driver in case of bad weather or bad light in order to avoid an accident. It also shows the signal and provides the exact location of the place where the accident took place [13].

A location for automobile repair shops also functions as a finder for car breakdown services. When the car breaks down, the user inputs the information and a search for maintenance stations close to their location is initiated automatically. On-Road Vehicle Breakdown Assistance ("HelpMe") allows users to search for them depending on their location without entering any personal information. The software tracks the user's position automatically when they search for a mechanic. In the emergency failure assistance kit, the word "HELP" is shown on the dashboard. However, when the mechanic was shown the position of the breakers, On-Road Vehicle Breakdown Assistance didn't emit any distinctive signals. "HelpMe" doesn't use any special method for communication. It is showing the mechanic detail with the phone number. Using that user can contact the mechanic and even the mechanic and user can online chat as well.

We can see the other users too and their experience through it. Using that chat platform, users can ask questions related to vehicle breakdowns from mechanics [14].

The On-Vehicle Breakdown-Warning report system uses the electronic GPS control panel and is used to detect the type of breakdown. "HelpMe" cannot detect any type of breakdown.

5 REQUIREMENT GATHERING

Collecting the wants and need of the automation app is necessary. First and foremost, there should be a search for the need for significant issues of the app. So, with the help of sources present on the web in which most of the participants were the UN agencies that participated in the survey area used an automated phone. The questionnaire includes issues given in Table 2.

Table 2. Initial requirement gathering for solution required in case of emergency.

Questions	Likes	Dislikes	Result
Would you rather get a mechanic if you suddenly required roadside assistance tomorrow?	87.1%	12.9%	**Most people prefer to get mechanic assistance in case of a breakdown.**
Does the need for roadside assistance ever arise?	80.6%	19.4%	**Most people will prefer road assistance in case of breakdown.**
Do you anticipate the benefits of on-road vehicle failure aid?	66.1%	32.3%	**Most people are happy by the working of roadside assistance.**

6 CONCLUSION

While providing the solution to the user for an emergency, there is a need to give a quick solution. But it must focus on essential points to provide the solution. In delivering a web-based solution, it must be simple enough to be used by users with minimal effort. This paper discusses the types of emergencies that occur during vehicle breakdowns and their possible solution. It also focuses on the points which are still left to be addressed. This paper also showcases the need for service from the customer's perspective and how they want their problem to be solved quickly. It gives the initial agenda for solution developers to work with. A plan for making a perfect solution is also discussed where the minimum modules to be developed are detailed with their functioning.

This system allows the user to save time. It can also get the mechanics' details and track their position. This approach makes using the system easier and may outperform the present one in crucial circumstances. So providing an app or web-based solution brings the user and the mechanic closer even in rural areas. It removes the obstacle of searching for a solution by asking the individuals if they can assist. The user of this roadside assistance can locate a technician's location with no problem. The driver is able to get mechanical help rapidly and immediately.

In future, we can use AI and ML for enabling faster and more accurate diagnoses of vehicle issues. With the growth of the electric vehicle market, online roadside vehicle assistance is likely to expand to cover electric vehicles and their unique needs, such as charging and battery management.

REFERENCES

[1] Kadam M., Sutar N., Dorge P., and Tundalwar R., "A Car Breakdown Service Station Locator System," *International Journal*, vol. 3, no. 4, 2018.

[2] Sale H. B., Bari D., Dalvi T., and Pandey Y., "Online Management System for Automobile Services," *International Journal of Engineering Science and Computing (IJESC)*, vol. 8, no. 02, 2018.

[3] Khanapuri A. v, Shastri A., D'souza G., and D'souza S., "On Road: A Car Assistant Application," in *2015 International Conference on Technologies for Sustainable Development (ICTSD)*, 2015, pp. 1–7.

[4] Jang J. A., Kim H. S., and Cho H. B., "Smart Roadside System for Driver Assistance and Safety Warnings: Framework and Applications," *Sensors*, vol. 11, no. 8, pp. 7420–7436, 2011.

[5] Sheng K. J., Baharudin A. S., and Karkonasasi K., "A Car Breakdown Service Station Locator System," *International Journal of Applied Engineering Research*, vol. 11, no. 22, pp. 11037–11040, 2016.

[6] Coviello A. and di Trapani G., *"The Customer Satisfaction in the Insurance Industry,"* Available at SSRN 2144684, 2012.

[7] Whipple J., Arensman W., and Boler M. S., "A Public Safety Application of GPS-enabled Smartphones and the Android Operating System," in *2009 IEEE International Conference on Systems, Man and Cybernetics*, 2009, pp. 2059–2061.

[8] Kumari P. and Nandal R., "A Research Paper OnWebsite Development Optimization Using Xampp/PHP.," *International Journal of Advanced Research in Computer Science*, vol. 8, no. 5, 2017.

[9] Bhatia S. and Hilal S., "A New Approach for Location Based Tracking," *International Journal of Computer Science Issues (IJCSI)*, vol. 10, no. 3, p. 73, 2013.

[10] Kumar, A., & Alam, B. (2015, February). Improved EDF Algorithm for Fault Tolerance with Energy Minimization. In *2015 IEEE International Conference on Computational Intelligence & Communication Technology* (pp. 370–374). IEEE.

[11] Alam, B., & Kumar, A. (2014, March). A Real Time Scheduling Algorithm for Tolerating Single Transient Fault. In *2014 International Conference on Information Systems and Computer Networks (ISCON)* (pp. 11–14). IEEE.

[12] Palvika, Shatakshi, Sharma, Y., Dagur, A., & Chaturvedi, R. (2019). Automated Bug Reporting System with Keyword-driven Framework. In *Soft Computing and Signal Processing: Proceedings of ICSCSP 2018*, Volume 2 (pp. 271–277). Springer Singapore.

[13] Hitosugi M. *et al.*, "Traffic Injuries of the Pregnant Women and Fetal or Neonatal Outcomes," *Forensic Sci Int*, vol. 159, no. 1, pp. 51–54, 2006.

[14] Dongre M., Verma S., Dighore A., Tumdam S., Dhote K., and Tote M., *"IOT Based On-Road Vehicle BreakDown Assistance,"* 2020.

Artificial Intelligence, Blockchain, Computing and Security – Dagur et al. (Eds)
© 2024 The Author(s), ISBN: 978-1-032-49393-0

Security approaches in software defined networks using machine learning – a critical review

Zahirabbas J. Mulani
Research Scholar, Bharti Vidyapeeth Institute of Management and Information Technology Research Centre, CBD Belapur, University of Mumbai, India

Suhasini Vijaykumar
Principal, Bharti Vidyapeeth Institute of Management and Information Technology Research Centre, CBD Belapur, University of Mumbai, India

ABSTRACT: We will be transiting to Software-defined networks from the traditional network because of its numerous application benefits like flexibility, scalability, network-wide visibility, and cost-effectiveness. Increasing network traffic due to the use of AI, IoT, and data science, demands a high instant responsive Intrusion detection system which is the need of the SDN environment, to tackle the most devasting attacks in the future. SDN includes the application plane, data plane, and control plane which is more vulnerable as it's the main controller of the networks. The controller in SDN has gained critical attention of security and also it can be further extended for mitigating the attacks. In this paper, we have analyzed a few methods from machine learning to deep learning, and the outcome of this systematic literature review-based discussion discloses the dimensions of ongoing challenges in implementing the intrusion detection system for the software-defined network. Also, the study claims that there is a requirement for a more generic and versatile algorithm to overcome these challenges.

Keywords: Software-defined networks, Machine Learning, Deep Learning and Intrusion Detection System.

1   INTRODUCTION

Due to the increase in the applications and data related to AI, Big data, Data Science and IOT Traditional networks can grow very complex and thus become incapable of serving the services. So, there is a birth of a new network paradigm called a software-defined network which surpasses the other traditional networks. Traditional network operations are reliant on a tightly coupled control and data plane. However, New Software Defined Networking (SDN) separates these two planes in order to provide additional features and benefits not possible with traditional networks [1]. For example, SDN can be used to stretch network functionality like programmability, elasticity, flexibility, and adoption capabilities beyond what is achievable with standard architectures. As per [7] the overall market size for SDN is projected to upsurge from USD 13.7 billion in 2020 to USD 32.7 billion by 2025. At present SDN deployment is incremental and very small [2] but in the coming era, we will be transiting to Software-defined networks from traditional networks because software-defined networks because of its numerous applications. But there are several challenging vulnerabilities that are to be addressed for SDN ascribing to the emergence of virtualization and programmable features. Cybercrime can be especially damaging when it's perpetrated by malicious insiders, who cause a denial of service attack or compromise systems to steal intellectual property. Organizations also risk serious harm when malware penetrates their networks and damages crucial national infrastructure. Cyber-attacks against government and commercial entities are on the rise, and to combat this problem, intrusion detection systems (IDSs) have been developed in academia and industry. IDSs serve as a network security measure by monitoring information sources for unauthorized access events. There are three main types of IDS: statistical, data-mining, and machine-learning methods. Some of the most common detection mechanisms include anomaly-based, signature-based, and machine-learning models.IDS can be carried out utilizing various procedures including peculiarity based; signature-based; or AI-based.

The irregularity-based procedure utilizes a gauge contrasted with ordinary traffic and movement happening in the organization. we measure the present status of organization traffic against this standard to distinguish designs in rush hour gridlock that are not ordinarily present. Signature-based techniques detect intruders by observing events and identifying patterns that match known attack signatures. In machine learning, different types of algorithms are used to monitor malicious activities. Various machine learning, and deep learning, approaches are implemented for NIDS but little has been done for SDN-based NIDS [3]. Also, there is a need for improvements over these algorithms in many aspects which we will analyze in this paper. We also focus on creating an adaptive and rigid approach that will be more intelligent for the intrusion detection system in an SDN environment.

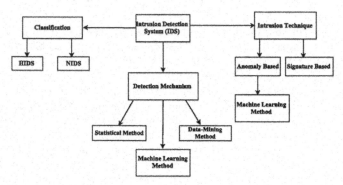

Figure 1.   Overview of intrusion detection system.

We will also analyze some of the alternatives for making IDS intelligent enough from swarm intelligence, ensemble learning, and fuzzy logic. In the end, we focus on the application of the Graph neural network for IDS in SDN for achieving a resilient and agile intelligent approach to detect malicious activity.

2   RELATED WORK

First of all, we took a look at some research related to detecting intrusion using a variety of approaches, such as swarm intelligence to graph neural networks. [4] proposes an approach to discover and mitigate DDoS attacks through the usage of a self-reliant multi-agent system with retailers using particle swarms. It demonstrates the Optimized overall performance of the cloud platform and stepped forward safety however wasn't for the SDN environment and could simplest discover the DDOS assaults. A component design diagram model [5] significantly outflanks different strategies on accuracy, review, and F − score yet there is an extension to work on the technique in three ways like robotized countermeasure selector to work with the determination of countermeasures as per the traffic designs, to lessen time utilization and processing assets lastly close to flooding-based DDoS assaults, different sorts of DDoS assaults ought to be considered to make it effective and workable in reasonable use. These DDoS assault designs are separated into three kinds in light of the convention type, i.e., ICMP, UDP, and TCP. An extraordinary outfit system [11] can successfully recognize different assault classifications and upgrades the recognition precision of many assault classes in contrast with Profound Learning, AI, and the Democratic troupe draws near. In [19] writers investigated a top-to-the-bottom outline of the remarkable distributed research work from last 6 to 7 years that pre-owned AI (ML) and Profound Learning (DL) strategies to build an IDS arrangement in SDN. These investigations expanded future exploration issues, for example, single regulator issues, bottleneck creation because of an absence of versatility, assessment, and testing, legitimate model determination review, absence of SDN-explicit dataset for model preparation, extra concentration towards discovery instead of relief and counteraction approach, absence of low-rate DDOS assault recognition. As per [6] there are a few issues and examination holes in the interruption recognition frameworks, all the more explicitly, DDoS assault interruption identification in SDN and AI which incorporates the old heritage datasets, e.g., KDD99 and CAIDA (2007) utilizing recreations, approval of discovery isn't near the real world.

3 SDN BASED NIDS USING MACHINE LEARNING

3.1 *Software Defined Networks (SDN)*

In the future very soon, we are able to be transiting to software program-described networks from the traditional network because software program-described networks have numerous software benefits. though, at the same time, they have massive chances of looking ahead to attacks. SDN essential additives stay in the application plane, facts plane, and control plane out of which the managed aircraft is extra prone as it's the imperative part of the networks and the primary controller. The aim of SDN can be described as refining community functions by means of consenting organizations in addition to provider vendors to adapt unexpectedly to developing market demands. The manipulated aircraft is drastically different from the statistics aircraft, and it could intelligently manage community assets. This keen organization of the executives is perceived while SDN uncouples the organization arrangement and guests designing from their basic equipment foundation.SDN elements network controller and information plane/pass sports activities with a display that updates sending tables inside the neighborhood. This awards on-the-fly neighborhood and brief reaction to changes in network visitors without the need for guide reconfiguration of the procurement of the latest gadgets.

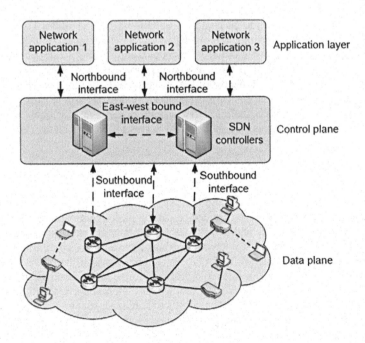

Figure 2. Architecture of SDN.

SDN designing involves three layers like control plane, a toward the north association point, and a toward the north mark of communication. The control plane partners with the utility layer through programming points of collaboration (APIs), which can be similarly broken down as north-certain association focuses. conclusively, a toward the north association point unravels the rules and headings in the application layer for the controllers. The toward-the-south association point positions the place of collaboration among a controller in the direct plane and genuine switches in the information plane. A typical SDN toward the south association point is OpenFlow, which has been standardized through the Open Frameworks organization Foundation [10]. Information Plane: The elements of the information layer are sending, disposing of, and changing information as per the standards or approaches given by the control layer. The SDN information plane has various gadgets that need knowledge. Control Plane: The head steady controller is the central unit of the control plane that engages in everyday SDN functionalities. The controller accomplishes the entire traffic stream and is absolutely responsible for coordinating, sending, and dropping groups by programming. Application Layer: Affiliation security, the versatility of the board, access control, load changing, firewall execution, nature of the association, and cloud mix are a piece of the instances of purposes that the application layer makes due, as displayed in Figure 2. The critical capacity of this layer is to play out the ordinary update for business network associations [12].

3.2 *Intrusion detection in SDN*

Programming-depicted affiliations can show man-made brainpower for impedance affirmation, which can be moreover improved for the early region and facilitating. Obstruction frameworks have different assessments which track the impedance and see the irregularity of the affiliation's traffic. As per [10] there is adjusting control for each of the designs that work with web applications, nonetheless, in SDN plan the information streams go through OpenFlow switches shaping stream tables. SDN regulator facilitates accomplishes and assembles undeniable data via looking for records in pack stream tables from one or several affiliation centers or applications. These tables contain an affirmation of the wellsprings of the affiliation beginning of the streams and the sort of points of participation utilized for transmission. Assaults cause disappointment with a piece of the applications which can incite loss of association and inaccessibility for purchasers. There is at present no wide instrument to change DDoS assaults. To counter appropriated denying of-association assaults, there are two fundamental undertakings to be followed one, to perceive DDoS assaults when could be expected, and second, to separate traffic stream into strange and typical. To vanquish this, they have proposed to set up the SDN regulator to perceive assaults utilizing data about the condition of the stream, the scope of the social occasion, and its early phase. This assessment can be recuperated from the tables of streams. As per [19] the Intrusion detection system collects the information and utilizes the Machine Learning / Deep Learning-based model which detects the intrusion in Software Defined Networks.

4   CRITICAL REVIEW

This work not just examines the working and results of existing models for Interruption Location in Programming characterized networks but does a careful basic examination of the equivalent to gain proficiency with their viability and deficiencies in learning and proposing new ways for settling something very similar. There are a few surveys that have a typical concentration however in [21] creators have proposed a Diagram Brain Organizations approach that licenses catching both the edge elements of a chart as well as the topological data for network interruption location in IoT organizations. It is the first fruitful, functional, and broadly assessed approach to applying GNNs to the issue of organization interruption discovery for IoT utilizing stream-based information. The following table (Table 1) shows the summary of related work from various articles from the last few years till today. It shows how SDN Security is implemented using various algorithms and also suggests the drawback and future work for the study.

Table 1. Comparison summary of related work.

Ref	Type of Intrusion	Techniques	Proposed Solution	Mechanism	Drawback	Future work
Kesavamoorthy et al.[4]	DDOS in cloud	particle swarm optimization	using the autonomous multi-agent system - particle swarm	the ability to successfully ward off various types of DDoS attacks with a success rate of 98 percent	minimum attack detection and recovery time	optimization of the usage of the stochastic-based filtering inside the attacks detection segment
Xiao et al. [5]	DDOS in SDN	a feature-pattern graph model.	attack behavior discovery and graph update tasks	malicious attack patterns can be divided as ICMP, UDP, TCP	Other types of DDoS attack are not considered in this work.	to improve the method and make it effective and serviceable in practical use
Seth et al. [11]	Malicious Intrusion	Ensemble machine learning model	effectively detect different attack categories	capability of different base classifiers to identify various types of attacks	ensembles cost more to create, train, and deploy	unsupervised learning to train models on unlabelled datasets in the security
Ahmed et al. [19]	Malicious Intrusion in SDN	ML and DL	articles from last 6 to 7 years that used Machine Learning (ML) and Deep Learning (DL)	automated IDS utilizing various MLDL algorithms in the SDN paradigm	NA	Single Controller, Lack of Scalability, Lack of SDN Specific Dataset, attack Mitigation, Low-Rate DDoS Attack Detection
Gupta et al. [6]	DDoS	Machine Learning	review of various machine learning approaches in SDN	Review	NA	use of Fuzzy logic for DDoS detection in SDN
Thakkar et al. [13]	Intrusion Detection	ML, DL techniques, and swarm and evolutionary algorithms (SWEVO)	A survey of feature selection, model, performance measures, application perspective	ML, DL and SWEVO	limitations of each of the datasets	will reinforce the study to explore different capabilities for using and leveraging the information provided in context with attack detection
Banitalebi et al. [14]	DDoS	entropy-based ML methods	collector, entropy-based and classification	detecting high-volume and low-volume DDoS attacks	attacks detected only by one controller	method can be improved in networks by involving more than one controller
Mahfouz et al. [15]	Intrusion Detection	ensemble and adaptive classifier ML model	Intrusion Detection Using a Novel Network Attack Dataset	attacks such as botnet, brute force, DDoS, and infiltration attacks	completely labeled, and about 84 network traffic features are extracted	improve detection accuracy, improve TPR, and decrease FPR
Maleh et al. [17]	Intrusion Detection	ML	Several vulnerabilities were discovered and exploited	Review	NA	requiring new defence mechanisms
Elsayed et al. [18]	Intrusion Detection	ML	to generate an attack-specific SDN dataset	Review	NA	to create a the more intrinsic dataset generated from large-scale networks
Maheshwari et al. [20]	DDoS	ML Ensemble	a novel optimized weighted voting ensemble	ensemble framework	NA	explore novel ways of designing dynamic existing metaheuristic optimization algorithms and use them in different domains to create efficient systems

5 RESEARCH GAP

There are a few challenges while emerging a flexible and well-organized NIDS using ML/DL in SDN-based networks:

Single Controller: Present SDN needs to depend on a solitary regulator. Intruders can send off forswearing of-administration assaults in the correspondence between the regulator and change to deaden the organization. When an SDN switch is undermined by an assailant, the stream table can be effortlessly adjusted [5].

More Consideration towards DDoS: The continuous examination has focused harder on the normal DDoS assaults through different kinds of DDoS assault, for example, IP address parodying, mimicry assaults, target flooding, interface flooding, and goes after in light of intricacy are not offered any consideration [5].

Dataset: The momentum interruption identification dataset isn't precise for research expectations for scholarly exploration as they require legitimate order of information. Network scientists take the assistance of engineered informational indexes for network interruption discovery because of the lack of better and more sensible datasets. It is fundamental to make datasets to guarantee reliable and precise assessment of interruption discovery frameworks [3].

Mitigating the attack: Up till currently parcel many methodologies have been produced for assault recognition however, just a few are for alleviation of SDN [2].

Low network traffic issue: Streamlining in assault recognition: In a DDoS assault, it propagates a huge magnitude of TCP, UDP, and ICPM bundles to the designated assets. Authentic solicitations will become blurred or evaporated and these assaults might be gone to by malware abuse. in this way, there is a prerequisite to achieving more improvement involving the stochastic-based filtering in the assault's discovery stage [3] and [5].

Aside from this Low Traffic Assaults, Continuous execution, and Steady issues of SDN likewise are a portion of the holes that are not tackled

6 CONCLUSION AND FUTURE RESEARCH DIRECTIONS

In this paper, we have presented a review of a few articles for SDN security using machine learning, deep learning, and swarm intelligence. The outcome of this systematic literature review-based discussion has revealed the dimensions of ongoing challenges in implementing the intrusion detection system for software-defined networks. There is a significant potential for Graph Neural networks and Fuzzy Logic in network intrusion detection which provides motivation for further work. Also this study can be extended to explore unsupervised learning to train models on unlabeled datasets.

REFERENCES

[1] Rashid Amin, Elisa Rojas, Aqsa Aqdus, Sadia Ramzan, David Casillas-Perez, And Jose M. Arco, "A Survey on Machine Learning Techniques For Routing Optimization in SDN", *IEEE Access* (2021).

[2] Hou Leqing, "How to Realize the Smooth Transition From Traditional Network Architecture to SDN", In *2020 5th International Conference on Mechanical, Control and Computer Engineering (ICMCCE)*, pp. 1948–1952. IEEE, 2020.

[3] Sultana, Nasrin, Naveen Chilamkurti, Wei Peng, and Rabei Alhadad. "Survey on SDN Based Network Intrusion Detection System Using Machine Learning Approaches", *Peer-to-Peer Networking and Applications 12*, no. 2 (2019): 493–501.

[4] Kesavamoorthy R, Ruba Soundar K. "Swarm Intelligence Based Autonomous DDoS Attack Detection and Defense Using Multi Agent System". *Cluster Computing*. 2019 Jul;22(4):9469–76.

[5] Xiao Y, Fan ZJ, Nayak A, Tan CX, "Discovery Method for Distributed Denial-of-service Attack Behavior in SDNs Using a Feature-pattern Graph Model. *Frontiers of Information Technology & Electronic Engineering.*", 2019 Sep;20(9):1195–208.

[6]   Gupta, Shaveta, and Dinesh Grover. "A Comprehensive Review on Detection of DDoS Attacks Using ML in SDN Environment." In *2021 International Conference on Artificial Intelligence and Smart Systems (ICAIS)*, pp. 1158–1163. IEEE, 2021.

[7]   *Software Defined Networking Market Size, Share and Global Market Forecast to 2025 — COVID-19 Impact Analysis,"* Marketsandmarkets.com, 2020. [Online]. Available: 14 Feb 2021.

[8]   Manso, Pedro, José Moura, and Carlos Serrão. "SDN-based Intrusion Detection System for Early Detection and Mitigation of DDoS Attacks." *Information 10*, no. 3 (2019): 106.

[9]   Deb, Raktim, and Sudipta Roy. "A Comprehensive Survey of Vulnerability and Information Security in SDN." *Computer Networks* (2022): 108802.

[10]   Feng, Mingjie, Shiwen Mao, and Tao Jiang. "Enhancing the Performance of Future Wireless Networks with Software-defined Networking." *Frontiers of Information Technology & Electronic Engineering 17*, no. 7 (2016): 606–619.

[11]   Seth, Sugandh, Kuljit Kaur Chahal, and Gurvinder Singh, "A Novel Ensemble Framework for an Intelligent Intrusion Detection System.", *IEEE Access* 9 (2021): 138451–138467.

[12]   Jayasri, P., A. Atchaya, M. Sanfeeya Parveen, and J. Ramprasath. "Intrusion Detection System in Software Defined Networks Using Machine Learning Spproach." *International Journal of Advanced Engineering Research and Science 8*, no. 4 (2021): 241–247.

[13]   Thakkar, Ankit, and Ritika Lohiya. "A Survey on Intrusion Detection System: Feature Selection, Model, Performance Measures, Application Perspective, Challenges, and Future Research Directions." *Artificial Intelligence Review 55*, no. 1 (2022): 453–563.

[14]   Banitalebi Dehkordi, Afsaneh , MohammadReza Soltanaghaei, and Farsad Zamani Boroujeni. "The DDoS Attacks Detection Through Machine Learning and Statistical Methods in SDN." *The Journal of Supercomputing* 77 (2021): 2383–2415.

[15]   Mahfouz, Ahmed, Abdullah Abuhussein, Deepak Venugopal, and Sajjan Shiva. "Ensemble Classifiers for Network Intrusion Detection Using a Novel Network Attack Dataset." *Future Internet 12*, no. 11 (2020): 180.

[16]   Gupta, Shaveta, and Dinesh Grover. "A Comprehensive Review on Detection of DDoS Attacks Using ML in SDN Environment." In 2021 *International Conference on Artificial Intelligence and Smart Systems (ICAIS*), pp. 1158–1163. IEEE, 2021.

[17]   Maleh, Yassine, Youssef Qasmaoui, Khalid El Gholami, Yassine Sadqi, and Soufyane Mounir. "A Comprehensive Survey on SDN Security: Threats, Mitigations, and Future Directions." *Journal of Reliable Intelligent Environments* (2022): 1–39.

[18]   Elsayed, M.S., Le-Khac, N.A. and Jurcut, A.D., 2020. InSDN: A novel SDN Intrusion Dataset. *IEEE Access*, 8, pp.165263–1

[19]   Ahmed, Md, Swakkhar Shatabda, A. K. M. Islam, Md Robin, and Towhidul Islam, *"Intrusion Detection System in Software-Defined Networks Using Machine Learning and Deep Learning Techniques—A Comprehensive Survey."* (2021).

[20]   Maheshwari, Aastha, Burhan Mehraj, Mohd Shaad Khan, and Mohd Shaheem Idrisi. "An Optimized Weighted Voting-based Ensemble Model for DDoS Attack Detection and Mitigation in SDN Environment." *Microprocessors and Microsystems* 89 (2022): 104412.

[21]   Lo, W.W., Layeghy, S., Sarhan, M., Gallagher, M. and Portmann, M., 2022, April. E-graphsage: A Graph Neural Network-based Intrusion Detection System for IoT. In *NOMS 2022-2022 IEEE/IFIP Network Operations and Management Symposium* (pp. 1–9). IEEE.

Artificial Intelligence, Blockchain, Computing and Security – Dagur et al. (Eds)
© 2024 The Author(s), ISBN: 978-1-032-49393-0

Arduino based fire detection alarm in rural areas

Omkar Bhattarai, Abhay Aditya Dubey, Shashank Singh & Avjeet Singh
*Department of Computer Science & Engineering, School of Engineering & Technology,
Sharda University, Greater Noida, India*

ABSTRACT: Fire is the major disaster worldwide, and even worst condition at the village. Hence, the fire detection system or alarm should accurately locate the fire in the shortest amount of time to reduce financial loss and environmental damage. Fire is a deadly element as well as a highly useful instrument in daily life, and it regularly causes catastrophes that are more expensive than money. Fire incidents typically result in the greatest amount of damage in the shortest amount of time since they happen quickly and leave no other salvageable parts behind. Early fire detection and prevention techniques have been developed to counteract this. In this paper, a tool for early fire detection and message sender response have both been created. The instrument is designed to find heat, gas, and flame possible fire indicators. The device is made up of an Arduino module, a GSM module for text messaging, sensors, a buzzer, and LEDs. It also has an acrylic circuit cover. Additionally, the gadget has been successful in text messaging the given cellphone number to notify the intended receiver of the events it has identified, such as excessive temperatures and gas leaks. It can show the faster response towards the fire and asks the alarm to respond quickly so that everyone gets informed.

1 INTRODUCTION

In our daily lives, fire can be both a very helpful instrument and a deadly element that frequently causes disasters that are more expensive than money. Fire, if handled improperly, has the potential to be both useful and harmful. Around the world, fires are seen as a serious security threat. Fire events frequently cause the most damage in a short period of time and leave no recoverable parts behind. Fire can even put out life, destroying resources and property. Because the alert system about the fire that is starting to happen was not known, many people perish as a result of the fire in the village area. The fire alarm system's interference should be less likely to affect the fire alarm control panel. Fires are detected, sprinklers are turned on before they start to spray, and the majority of conventional fire alarm applications are turned on using the application-recommended way. Additionally, it aids in preventing erroneous warnings. Today, sensor-based detection has become the main category of solution methodologies, driving the development of automated fire detection systems and capturing the majority of the market. This kind of methodology typically works effectively for various types of fire incidence situations. It is recommended using this method in residential areas. It cannot be used in a specialized environment with peculiar vibrations or extreme heat zones.

To find and extinguish fires, technology like fire detectors is routinely used. Using a fire detector, which recognizes and responds to fire signs like smoke, heat, and radiation, can help prevent or lessen fire damage. Humans have five senses, making them highly good fire detectors. Their actions are the product of processed ideas, and they serve as a warning to nearby inhabitants about the fire that they see. Manual fire detection is one of the earliest methods of finding flames. The Tx subsystem's accessible sensors will collect the real-time data pertaining to physical quantities there and communicate it to the Tx processing unit [1].

DOI: 10.1201/9781003393580-80

By accident prevention, this prototype device can assist users in raising their safety standards right away. This will finally enable the recovery of both human lives and property from the catastrophe [2]. The major objective of the project is to design and develop wireless sensor network-based fire extinguishing and monitoring systems. The fire monitoring system continuously keeps watch of the surroundings, recording the temperature and any intrusions that the monitoring nodes identify as invaders. When a fire is discovered or when the temperature exceeds the authorized range, the fire extinguishing system will trigger the fire extinguisher. [4] temperature and flame detectors are used to determine whether a fire is present, and the buzzer and fan are used to alert for a fire and expel smoke, respectively [5]. The small-scale controller is upgraded to sound the bell as soon as the desired temperature and smoke level are reached. The fumes fan starts to blow the smoke outside of the shop floor as soon as it is located. [6] Experiments showed that the suggested method can distinguish between smoke movies and non-smoke videos with a quick-fire alert and a low false alarm rate [7].

Prior to pre-processing, the motion region was extracted and used as a probable fire area using the frame difference method. Performed a new sample of the same size and removing flame features like the texture feature, colour moment feature, and H component first moment feature. Normalized the retrieved data to produce the eigenvector [8]. Ashwini *et al.* [9] proposed fire alarm system, they fuse an Arduino Uno, a Wi-Fi module and data storage into a communication module capable of transmitting information to neighboring police, fire and hospital stations as well as pre-stored phone numbers. Here, fire detectors use Wi-Fi module to transmit data to micro controller after detecting fire with sensor or manual call point.

Wonjae *et al.* [10] proposed a wild fire detection, and model is divided into two parts, the alarm notification and the fire detection system. Here image processing is used to alert the nearest station of the location where fire is detected, they employed Arduino technology for the alarm notification system. The Arduino board is programmed to control the buzzer and the effort will result in a reduction in false alarm rates and camera is also used in this device for fire detection.

2   RELATED WORK

Ibrahim Majid *et al.* [1] proposed a communication system needs to be set up in advance in order to facilitate the installation of sensors at the remote site. This was developed using the GSM network's internet of things framework. The transmission (Tx) subsystem is one of the two sides of the subsystems in our system, and the other. Each subsystem has a processing unit, including the reception (Rx) subsystem. The Tx subsystem's accessible sensors will collect the real-time data pertaining to physical quantities there and communicate it to the Tx processing unit.

It has become very common in recent years to use different IOT devices for home automation, especially in fire alarm and fire detection systems based on sensor. Pandey *et al.* [3] proposed IOT based fire alarm system which organize hierarchical wireless sensor networks. According to test results for the proposed methodology, the automatic fire alarm system satisfies all design specifications. The major objective of the paper is to design and develop wireless sensor network-based fire extinguishing and monitoring systems. The fire monitoring system continuously keeps watch of the surroundings, recording the temperature and any intrusions that the monitoring nodes identify as invaders. When a fire is discovered or when the temperature exceeds the authorized range, the fire extinguishing system will trigger the fire extinguisher.

A linear relationship exists between output voltage and temperature changes in the IOT fire alarm system that Mahzan, Enzai, Zin, and Noh constructed using the LM35 analogue and linear temperature sensor. A GSM module (GSM SIM900A) was used in addition to this, and the main board, Arduino, has an ATmega328 microprocessor on it, which is utilized as the main controller to handle the circuit appropriately. An LCD screen was additionally linked to the micro-controller. A text message is immediately sent to the home owner whenever a fire is discovered, and a text is also displayed on the screen [4]. The system developed by Xie *et al.* includes both hardware and software designs. The hardware design is produced by combining numerous modules and parameter designs. The main components of the system are the microcontroller, sensor data collecting module, buzzer alert module, and exhaust module [5]. Jeevanandham *et al.* [6] proposed the IOT based Automatic fire detection system, in two sensors, specifically temperature and smoke sensors, are utilized in the IOT-based fire alarm design. Few algorithms are used, such as Background Subtraction, Block Processing, Texture Analysis, and Neural Network Classifier, where feed forward and back propagation phases were applied for the computation of the derivative of the network function [7].

The support vector computer-based image fire detection method employed in this work has the following primary phases: firstly, study the pictures than prior to pre-processing, the motion region is extracted and used as a probable fire area using the frame difference method, than performing a new sample of the same size and removing flame features like the texture feature, colour moment feature, and H component first moment feature and normalize the retrieved data to produce the eigenvector finally, Making decisions based on trained support vector machine models [8]. Ashwini *et al.* proposed a wireless sensor networking-based automatic fire alarm system that is intended for tall structures. in order to ensure prompt fire disaster extinguishment. Increasing the pace of fire recognition is the main goal of the fire detection design concept. The system consisted of three parts. Using an Arduino Uno, a wi-fi module, and data storage, they transmitted notifications to nearby police, fire, and hospital stations as well as pre-stored phone numbers in the communication module. A sprinkler system, an alarm, and a display screen are features of the execution module [9].

To protect the cultural heritage, Dimitropoulos *et al.* [11] proposed a flame detection approach for video-based early fire warning. This system used alarm notification and fire detection systems. They used Arduino Technology to notify the station connected to the place where the fire detection occurs in the alarm notification system. When a fire is detected, the system enters emergency mode, sounds the alarm, and updates the status on the screen. Human monitoring is the cornerstone of conventional ground systems, commonly referred to as terrestrial systems. The monitoring and detection of fires is made possible by the use of nearby sensors, such as flame, smoke, heat, and gas detectors. This is done either by keeping an eye on specific websites or by analysing the information they gather. Ambient sensors were also added to the system to improve performance and accurately spot fires. It is possible to detect fire and smoke with these sensors both during the day and at night. In terrestrial systems, infrared (IR) cameras, infrared spectrometers, and light detection and ranging systems are the primary sensors used (LIDAR) [12].

3 PROPOSED METHODOLOGY

The speeding up of the fire recognition process is the main goal of the fire detection design concept. This plan was put up as a way to reduce fire damage by sending out an early notice when a fire starts. Wireless sensor networks have been the most significant advancement in recent years in environmental monitoring and home or factory automation. The purpose of this project is to develop a wireless automated fire alarm system for tall buildings. There is a need for rapid, catastrophic fire extinction.

There are three parts to the system. Execution, communication paths, and fire detectors are all easily accessible. The fire detection system has a temperature sensor, fire detector, and manual call point. An Arduino Uno, a GSM module, and data storage are required to send alerts to pre-stored phone numbers in the communication module, the neighbourhood police, the house owner (if absent), the fire brigade, and the hospital departments. The execution module has a sprinkler system and an alarm.

Furthermore, we have proposed the module containing the solar panel where on the onset of the fire, the system will detect fire in the field and send SOS an alert to the farm and also buzzer

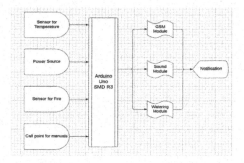

Figure 1. Fire alarm system module.

will start alerting the people nearby. Process: Whenever the fire is detected by the flame sensor, it sends digital signal to the Arduino where it analyses the input sent by flame detector. When input is low, it gives 0 as a digital signal and it means no fire is detected. But, when the input is high, it gives 1 as a digital signal and it means there is fire around the surrounding. On receiving the digital signal 1, arduino sends signal to the buzzer and buzzer starts buzzing and the notification is sent to the required destination e.g. Fire extinguisher department.

3.1 *Explanation of key terms*

- **Power source:** It supplies the power to all the system available that requires the electricity for running.
- **Sensor for temperature:** It measures and provides the surrounding temperature to the destination that throws the alarm.
- **Flame detecting sensor**: It detects and responds whenever there is detection of the flame towards the surrounding.
- **Arduino**: An actual microcontroller board that can be used to store pre-written programs and application software called the IDE (Integrated Development Environment), which is used to write and upload computer code to the board, make up the Arduino system. The choice is made, and all the data and programs are kept on the widely used open-source micro controller board known as the Arduino Uno, which was built by Arduino.cc based on the SMD R3 (or Microchip ATmega328) CPU. It will decide whether or not to sound the alert.

Module for GSM: There is GSM (Sim 800) slot at the Arduino Uno for immediate calling to the required destination (e.g. Fire stopping agent/company). Any flames or fires are located by the flame sensor. It uses an infrared flame flash technology to operate. To connect the flame sensor to Arduino, make the connection as indicated in the diagram above. Once the code has been uploaded to the Arduino board, any fire source may be positioned in front of

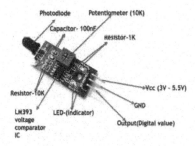

Figure 2. Flame sensor.

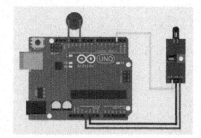

Figure 3. Circuit design.

525

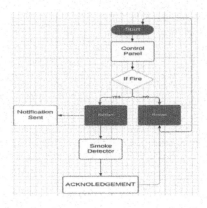

Figure 4.   Flow chart for fire detection.

the flame sensor. When fire or flame is detected, the flame sensor is turned on. Several factors, including temperature, smoke, humidity, and others, have an affect on this sensing.

Digital or analogue signals may be included in these communications. For this project, we used the digital output from the flame sensor. When Arduino receives this signal, it reacts appropriately. That might entail sounding the buzzer or taking another fire-fighting action. A buzzer has been connected as a detection tool in this case Fire detectors employ sensors or manual call points to detect fire, and wi-fi modules are used to transmit the information to the microcontroller. Wireless connections are used to transfer data between devices, and alarm and display units that show the position of the sensor or manually activated point are used to provide notifications to adjacent households. The system is manually reset during maintenance.

4   CONCLUSION

A fire alarm is a device that can spot changes in the environment brought on by smoke or flame in addition to the presence of a fire. A fire alarm's main function is to alert people to escape an area where there is a fire or smoke buildup. A fire alarm may occasionally work as a component of a comprehensive security system. To alert adjacent inhabitants of an imminent fire emergency, a functional fire alarm will sound. Fire alarms are common in homes, workplaces, and places of worship. They are essential to the protection of many innocent lives. When a bell or siren is blown, the majority of fire alarms go off. A reliable, affordable fire alarm with a distinctive sound was produced as a result of this research. We will be implementing temperature sensor; flame sensor and GSM module will be including GEO Spatial data for location tracing in the future.

REFERENCES

[1]   Ibrahim Majid Al Shereiqi *et al.*, "Smart Fire Alarm System Using IOT", *Fourth Middle East College Student Research Conference*, Muscat, Sultanate of Oman, 2019

[2]   Rishika Yadav & Poonam Rani, "Sensor Based Smart Fire Detection and Fire Alarm System" *System Proceedings of the International Conference on Advances in Chemical Engineering (AdChE)*, 2020

[3]   Devanshi Pandey, *et al.*, "Iot Based Fire Detection System", *Recent Trends in Computer Graphics and Multimedia Technology* volume 3 issue 1 (2021).

[4]   Mahzan N N, *et al.*, "Design of an Arduino-based Home Fire Alarm System with GSM Module" In *Journal of Physics: Conference Series*, vol. 1019, no. 1, p. 012079. IOP Publishing, 2018.

[5]   WenLan Xie and Huang Haoming, "Design and Application of Home Fire Alarm System", 2021 *J. Phys.: Conf. Ser.* 1871 012125

[6]   Jeevanandham A. T. *et. al.*, "*IoT Based Automatic Fire Alarm System*", 2020

[7]   Yu Chunyu, *et al.*, "Texture Analysis of Smoke for Real-time Fire Detection", in *2009, Second International Workshop on Computer Science and Engineering*, vol. 2, pp. 511–515. IEEE, 2009.

[8]   Huang Hongyu, *et al.*, "An Improved Multi-Scale Fire Detection Method Based on Convolutional Neural Network", in 2020, *17th International Computer Conference on Wavelet Active Media Technology and Information Processing, (ICCWAMTIP)*, pp. 109–112. IEEE, 2020.

[9]   Ashwini C, *et al.*, "Smart Fire Alarm System Using Arduino", *International Journal of Emerging Technologies in Engineering Research (IJETER)* Volume 7, Issue 5, May (2019)

[10]   Wonjae Lee, *et al.*, "Deep Neural Networks for Wild Fire Detection with Unmanned Aerial Vehicle," *2017 IEEE International Conference on Consumer Electronics (ICCE)*, 2017.

[11]   Dimitropoulos K. *et al.*, "Flame Detection for Video-based Early Fire Warning for the Protection of Cultural Heritage." In *Progress in Cultural Heritage Preservation: 4th International Conference*, EuroMed 2012, Limassol, Cyprus, October 29–November 3, 2012.

[12]   Jesus San-Miguel-Ayanz, *et al.*, "Active Fire Detection for Fire Emergency Management: Potential and Limitations for the Operational Use of Remote Sensing." *Natural Hazards*, 35(3), 361–376 (2005).

Recent advances and future technologies in IoT, blockchain and 5G

Artificial Intelligence, Blockchain, Computing and Security – Dagur et al. (Eds)
© 2024 The Author(s), ISBN: 978-1-032-49393-0

Fog enabling technologies in healthcare: A review

Aditya Yadav & Onesimus Chandra Pradhan
Computer Science & Engineering, Sharda School of Engineering & Technology, Sharda University, Greater Noida, U.P., India

Ruqaiya Khanam
Department of EECE, Center for AI in Medicine, Imaging and Forensic, Sharda University, Greater Noida, U.P., India

Amrita
Center for Cyber Security and Cryptology, Computer Science & Engineering, Sharda School of Engineering & Technology, Sharda University, Greater Noida, U.P., India

ABSTRACT: This paper provides an overview of fog computing healthcare technologies that are playing a huge role in the current healthcare Industry. In this paper, we suggest the Data mining is the most effective option for disease prediction. The suggested framework intends to support health care based on disease symptoms, and health-related data. Gathering and analyzing activity-specific data that may be utilized to gauge the health benefits of exercise and its intensity as well as provide suggestions. Health Informatics makes extensive use of wireless sensor networks (WSNs). Patients at risk for chronic illnesses are increasingly being monitored via wireless and wearable sensors. This ensures that patients follow their treatment schedules and protects them from unexpected assaults. Along with its abilities, the fog computing framework also brings over from the cloud computing platform a number of security flaws that must be solved for the benefit of the user. Fog computing is crucial because it aids in the system's elimination of latency problems, which are crucial in the healthcare sector.

Keywords: Fog Computing, Cloud Computing, the Internet of Things, Autonomic and Computational Intelligence

1 INTRODUCTION

Data mining is a valuable approach for extracting insights and understanding from datasets that may be difficult to comprehend or analyze using conventional statistical methods. Various techniques such as data processing and conversion methods involve data mining as a useful source of knowledge. Patients who use health prediction systems can respond quickly compared to manual analysis, the Data mining method produces results that are more accurate. The most precise sickness that might be caused by a patient's symptoms is determined using naive Bayes algorithms. The system notifies the user of the sort of sickness or ailment it believes the user's symptoms are related to if it is unable to deliver satisfactory findings. [2] Data Mining are proposed in the system design detailed plan to combat the drawbacks of the current system. The project's design components employ the smart health framework in its implementation. In order to employ data mining techniques to predict the disease, this system asks for input and collects symptoms. For disease prediction, to clarify patient questions, doctors can be contacted for feedback and discussed with. There are

DOI: 10.1201/9781003393580-81

certain positives, such as the ability to locate the closest doctor to our location. These elements can be utilized to improve how the system is used to assist patients. Using a data mining approach, a lot of data is analyzed. The administrative, clinical, scientific, and instructional components of Clinical Predictions were also discussed for each discipline, as well as how clinical data warehousing and data mining may be helpful [5]. The data is being processed using established computing frameworks like distributed computing and the cloud. By lowering prices and expanding capabilities, eliminating the need for direct investment, and providing other advantages, the Cloud has fundamentally transformed the Internet. Due to the reaction time lag, time-sensitive use cases are starting to suffer as long as the devices are linked over the Internet. Standard frameworks' mobility and location awareness are additional drawbacks. In order to solve these issues, a brand-new computer architecture dubbed Fog computing was developed in 2012. Fog computing is an architectural model that aims to bring distributed computing, control information, and data storage closer to the end-users by expanding the cloud infrastructure to the network's edge. When the constraints of the standard cloud computing, in addition to the new potential brought about the IoT development and five G-related topics are taken into account, the importance of the fog technology can simply be re-exposed [8].

2 RELATED WORK

This section contains the discussion of different feature selection methods used in fog computing, cloud computing, machine learning, data mining, and sensor-based models.

2.1 Smart health prediction

In this paper Wilson Wibamanto et al. [2] presented a smart health prediction system that would include the conditions the patient is presently experiencing along with their corresponding accuracy levels. By entering the user's health information and any symptoms they may be experiencing, data will be evaluated using the processes of a Knowledge Discovery-based system, which will help physicians and medical personnel with their clinical decision-making process. It entails employing a variety of efficient data gathering, warehousing, and computer processing techniques to analyze a certain volume of data in order to identify specific patterns of occurrence that may be used to forecast future trends. Therefore, this feature serves a very useful purpose when it comes to forecasting people's health conditions, particularly in determining the link between the provided health information that has been given by both the medical staff and the patient.

Shabaz Ali & Divya [4] worked on the healthcare prediction using different machine learning algorithms. Vector machines, artificial neural networks, logistic regression, Random forest, and other significant techniques are among them. It is possible to calculate the phases of different diseases properly and treat patients accordingly. Making informed judgments can be helped by the knowledge discovered through data mining.

In this paper [5], by collecting datasets and using the Naive Bayes method, hidden information will be gleaned from the historical data. Only when the system reacts in this manner can smart health be predicted. After comparing these datasets with the incoming queries, the final report will be generated using Association Rule Mining.

This paper [6] predicts the illness based on the patient's reported symptoms and offers the appropriate medicine using the provided information. If necessary, the patient is also notified about experts and physicians at the closest hospital. Occasionally you find yourself in a scenario where you require immediate assistance from an expert but they are unavailable for whatever reason; in such cases, the suggested system will be helpful.

Pankaj Hooda et al. [22] proposed in a paper he analyzed the health issue problem using data mining building up an automated system to keep up an examination system for individuals to check about their own particular medical problems. With this mechanized system, there would be straightforwardness for individuals to perceive medical problems. The objective of the paper is to learn and examine the upgraded strategies for putting away and handling tremendous arrangements of data in the health sector.

2.2 *Fog computing and IOT*

This study offered numerous communication failure scenarios on a three-tier system consisting of cloud, fog, and edge computing, including the case of verifying the status of communication failures in edge (end) devices and servers. This study considered many elements, for instance, resilience techniques concerning connection coverage, processing, communication, and power consumption to build mechanisms to create a resilient system against communication failures in the fog and edge layers [7].

In the paper [8] they implement healthcare services after COVID-19, and the healthcare system will be heavily dependent on incorporating artificial intelligence (AI) mechanisms into its daily operations. This will be realized through the use of sensor-enabled smart and intelligent Internet of Things (IoT) devices to provide patients with extensive care that is consistent with the symmetric concept. Load Balancing in Fog Computing Taxonomy the load-balancing algorithms can be categorized in a variety of ways, roughly speaking as approximate, precise, fundamental, and hybrid techniques. Fog node clustering can aid in lowering a task's offloading overhead. It is subtle because of a different virtual computer inside the unit might be allocated that duty in the event that one virtual machine fails.

The design and prototype for a temperature sensor that uses the principles of RFID and harmonics were proposed in this paper. This temperature sensor, formed of a single diode, can measure temperatures even in the harshest environments. Any sensing device's sensitivity is its valuable component. The sensor is tested at temperatures ranging from $-30°C$ to $100°C$. The simulated results indicate that the potential sensor could be used [10].

Hussain *et al.* [13] Discuss IoT-based projects that use a wireless smart wearable based on body vital signs, mobility, and health-related indicators, this system provides omnipresent fitness and health monitoring in a smart gym environment., intended to support the health and fitness business. The EKG, heart rate, variations in heart rate, and respiration rate were all collected, and a 3D acceleration was used to calculate the athlete's movement. Heart zone and zone module are two different sorts of modules that are used to implement the health zone module on body vitals data. Identification and warning of health dangers are the responsibility of the zone module.

This paper [14] (2008) Proposed how to monitor patients suffering persistent data loss is proposed. They introduced a system that evaluates system behavior. Propose a three-tier architecture for context-sensitive health tracking that uses fog computing to reduce latency. The tri-tiers application architecture is a good application-specific software architecture that divides apps into 3 physical and logical computer levels: cloud and fog computing, and sensors that operate together. Sensors are linked to patients might be wearable or non-wearable devices such as smartwatches, fitness bands, and so on.

A detailed analysis of recent literature on methods for data integrity protection and upkeep of sustainable smart city infrastructure was conducted in response to the use of 5G-enabled IoT devices, which have generated large amounts of different data. The utilization of mobile computing, which occurs on portable devices like cell phones, laptops, etc., by mobile devices functioning as fog nodes in this system, is not clearly explained. With the help of this paradigm, we may design reminders depending on geological location. By improving network connectivity and computing power, MC seeks to address computational problems [15].

This paper [16] proposes a fog computing-based framework and takes advantage of different clinical entry points or medical decision-making by gathering sensor information that are necessary for the medical field. Three layers make up the architecture used to implement this framework: a cloud, fog, and a sensor network layer. The sensor network layer includes all of the sensors used in the medical industry. The data sensed in this layer is delivered to the fog layer, where it is used for local storage, data security, and analysis, notifications, and compression of data. The cloud layer must then carry out the required actions, such as big data analysis and storage, and illness prediction after the data has been computed. Following data computation, the cloud layer must perform the necessary tasks, including big data analysis, big data storage, and disease prediction. The final recipients of the usable data are hospitals, physicians, healthcare professionals, researchers, and insurance companies. The data must be protected in a framework for health monitoring in the medical field. As a result, the security measures of encryption and watermarking must be taken into account. The proposed framework is made more useful by the use of fog, which not only lowers latency but can also serve as emergency temporary storage until a proper connection.

While collecting data from sensors, solves a number of issues that emerge because of the environment. Collecting data in a mountainous area and transferring a lot of the data over a long distance. They put out a brand-new system design that relays data over great distances by merging current wireless technologies with a Low Altitude Platform (LAP) [20].

This paper [21] proposed a system to aid climbing guides in the management of mountaineer teams. The device can monitor each team member's real-time physiological status and coordination and it helps to upload the data information to the cloud service platform. The system is divided into two parts.

The cloud network provides a network and the mountaineering team monitoring device, which shows the team members' health status updates and uploads data to the cloud, thereby improving mountaineering safety. Latency is excessively high, endangering the lives of mountain climbers. To lessen the effect of motion, they used a Kalman filter (UKF) with a variable framework. In order to solve some of the expanded Kalman filter's flaws (EKF) in nonlinear system estimation, the expanded Kalman filter (UKF) is an expansion of the classic Kalman filter. To limit the data effect of motion during climbing, use the Kalman filter (UKF). The experimental findings demonstrate that the suggested approach performs admirably [23]

Fog has a number of uses in the medical field. The use of fog in the healthcare area is referred to as "fog-IoHT." The fog-IoHT applications prioritize patient data and private information privacy and security, while fog servers provide effective data storage. The data storage in fog servers also provides fog-IoHT applications with unique attributes and capacities [28]. Figure 1 depicts the many sorts of applications for fog-IoHT applications.

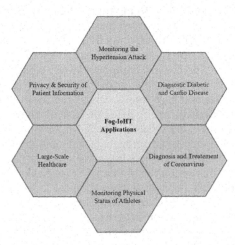

Figure 1.   Several fog-IoHT applications.

3 METHODOLOGY

This section will provide an overview of the approaches used in various papers.

3.1 *A smart health prediction*

[1] Proposed a methodology in which they developed "smart health prediction," which is a web-based tool that forecasts a user's sickness using the symptoms that the user or patient may experience. This paper focuses on disease prediction using a machine-learning algorithm. Because medical data is growing at such a rapid pace, the Naive Bayes algorithm is used to identify patient data. This requires the use of current data to forecast the precise illness based on symptoms. As a result of having the data as inputs in a patient record, we were able to get a precise general sickness risk prediction, which aided us in understanding the degree to which we calculate the disease's risk.

The operation of smart healthcare systems in emergencies is far more critical than their use in controlled settings. As a result, [7] proposed the use of A Low Altitude Platform (LAP) for long-distance data transfer in steep terrain. The incorporation of methods targeted at achieving a specific level of reliability on a stage that combines sensor networks with edge, fog, and cloud levels is a critical component of the overall system architecture's operation. Article [9] suggested a way for evaluating such cases when communication failures will occur on the three-tier architecture, in which the sensors are the end nodes that are coupled to Microcontrollers, Robots, Drones, or other autonomous devices that can conduct activities remotely managed by a server.

Healthcare difficulties are becoming increasingly widespread as the population grows in overpopulated countries, as well as the demand for healthcare. The populace's demand for high-quality treatment is essential, while healthcare costs are decreasing. There is an established solution enabling triparty, one-round key authentication agreements that uses fog resources and generates a session key shared by the parties to ensure secure communication. This [15] proposes a novel computational architecture for high-performance computing for prognosis, remote real-time surveillance, sensing, scaling, and diagnostics. The recommended healthcare monitoring architecture is made up of three layers: the sensor network, the fog, and the cloud layer. The suggested health monitoring system uses the fog framework's architecture.

The best accuracy offered by SVM in the literature to date is 94.60% in 2012, as shown in Table 1. SVM displays good performance outcomes in a wide range of application domains. SVM responds appropriately to the qualities or features utilized in 2012. Otoom *et al.* employed an SVM variation named SMO in 2015. The best characteristics are also found using the FS approach. SVM reacts to these characteristics. and provides a precision of 85.1%, but this is lower than it was in 2012 in comparison. Both the training and testing sets for the two data sets are distinct, as are the data types. [25]

Table 1. Machine learning algorithm's accuracy [25].

Algorithm Used	Accuracy
SVM	92.59
Naive Bayes	85.21
Bagging	83.12
J48	82.25
Bayes Net	82.15
FT	82.12

3.2 Fog computing and IOT

It is also suggested that fog computing be used to monitor patients in a hospital setting. This paper [21] proposes a sensing layer that collects linked data in the form of health metrics in real-time. The collected data is subsequently sent to fog nodes for additional processing. The nRF module is used to communicate between the sensing layer and the node. On average, the fog node is in charge of collecting the data every five minutes, and if the values of the health parameters are found to be and the data is greater than or less than the requirement, it quickly delivers it to cloud servers and create alarm notifications; the classification algorithm is then executed at the cloud server.

IoT fog computing paradigm Applications in sensor-based networks address issues with storage, location awareness, latency, mobility, and real-time data processing. Healthcare based on IOT The goal was to acquire real-time access to the location, condition, tracking, and assets of the patient. The technology employs an armband with a 3-axis gyroscope and a 3-axis accelerator to automatically recognize exercise. For activity detection and repeat counting, the FEMO system was employed. a wearable system with a Bluetooth connection. Three different network types exist.

The network's edge backend network for the forwarder network. Health-related characteristics are useful in a gym setting to assess physical fitness as well as prevent health risks. The system combines the capabilities of body vitals research, automated workout categorization, and real-time warning generation if the athlete's health is jeopardized. [13]

Service models and deployment models are two categories of cloud models. Cloud Software as a Service (SaaS), Cloud Platform as a Service (PaaS), and Cloud Infrastructure as a Service (IaaS) are all examples of cloud computing and 3 different types of cloud service models based on the functional capabilities of the services. The legality of information sharing by a user may be impacted by this. If data is stored in multiple locations, the user could face a lower chance of facing legal issues, but at the same time, they may be more susceptible to not receiving adequate protection if they fail to file a claim in the appropriate jurisdiction. If the cloud service provider is a competitor's agent, it's possible that all of the crucial private information may be given to the rival without any additional warning or procedure thanks to compromised prosecutors and intelligence agencies. [18]

ICU patients, as we all know, also require constant observation because their health conditions might become critical at any time, and they may even pass away. Since their suggested system would monitor these key parameters in real-time for ICU patients. So this [24] paper discussed an Smart ICU system based on IoT that will provide an option for ICU patients by capturing metrics and fluid measures from the patients' bodies via IoT devices.

Additionally, the gathered information will be kept within the fog nodes, where it will be handled before being sent to the cloud for storage. Additionally, following real-time data processing, if any unusual parameters are discovered, our solution will alert the designated physician(s) and ECU for further patient care. Fog computing may be used in conjunction with machine learning to improve computational abilities, resource management, decision-making, latency and power consumption, data accuracy, and big data analysis. [26] The implementation of fog computing using machine learning is depicted in Figure 2 below.

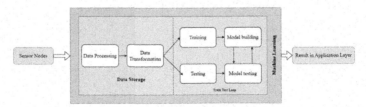

Figure 2. Machine learning in fog computing.

4 CONCLUSION

The extraction of knowledge from underutilized or irrelevant datasets is known as data mining. Various data mining techniques are employed in data processing and transform it into meaningful information. The significance of data mining is greatest in the scientific and technological domains. Data mining and machine learning algorithms have the potential to revolutionize the field of medical science. Fog computing plays a significant role in the health industry since it would be useful for patients and doctors and there would be fewer accidents. Fog-enabling technologies are also significantly working to make the world a good place. Therefore, combining fog with different disciplines like data mining and machine learning methods seems to be beneficial.

5 FUTURE SCOPE

During our investigation, we discovered that fog technologies become quite weak in severe conditions, which is also a serious issue, and that fog computing might be advantageous in certain situations. Regular communication and a strong infrastructure can help alleviate this limitation.

REFERENCES

[1]   Naveen Kumar S, Kirubha Karan R, Jeeva G, Shobana M, Sangeetha K "Smart Health Prediction Using Machine Learning" *International Research Journal on Advanced Science Hub (IRJASH)* 2021

[2]   Wilson Wibamanto, Debashish Das, Sivananthan A/L Chelliah "Smart Health Prediction System with Data Mining" *International Journal of Current Research and Review 2020*

[3]   Pradnya Suresh Joshi Asst. Prof. Ashwini Gaikwad *"Smart Health Prediction System Using Data Mining"*©2020 JETIR December 2020, Volume 7, Issue 12

[4]   Shabaz Ali N., Divya G. "Prediction of Diseases in Smart Health Care System using Machine Learning" *International Journal of Recent Technology and Engineering (IJRTE)* 2020

[5]   Prof. Krishna Kumar Tripathi, Shubham Jawadwar, Siddhesh Murudkar, Prince Mishra "A Smart Health Prediction Using Data Mining" *International Research Journal of Engineering and Technology (IRJET)* 2018

[6]   Pooja reddy G., Trinath basu M., Vasanthi K., Bala Sita Ramireddy K., Ravi Kumar Tenali "Smart E-Health Prediction System Using Data Mining" International Journal of Innovative Technology and Exploring Engineering (IJITEE) 2019

[7]   Haider A.H. Alobaidy, J. S. Mandeep, "Low Altitude Platform-based Airborne IoT Network (LAP-AIN) for Water Quality Monitoring in Harsh Tropical Environment" 2022 IEEE. Personal Use of this Material is Permitted.

[8]   Swati Malik, Kamali Gupta, Deepali Gupta, Aman Singh, Muhammad Ibrahim, Arturo Ortega-Mansilla, Nitin Goyal and Habib Hamam *"Intelligent Load-Balancing Framework for Fog-Enabled Communication in Healthcare"* 2022 by the Authors Licensee MDPI, Basel, Switzerland.

[9]   Santiago Medina, Diego Montezanti, Lucas Gómez D' Orazio, Evaristo Compagnucci, Armando De Giusti, Marcelo Naiouf *"Incorporating Resilience to Platforms based on Edge and Fog Computing"*

[10] Luis M Vaquero, Luis. Rodero-Merino *"Finding Your Way in the Fog: Towards a Comprehensive Definition of Fog Computing"* 2014 Hewlett-Packard Development Company, L.P.

[11] Tagleorge Marques Silveira, Pedro Pinho, and Nuno Borges Carvalho, "Harmonic RFID Temperature Sensor Design for Harsh Environments" *IEEE Microwave and Wireless Components Letters 2022*

[12] Hadi Zahmatkesha, Fadi Al-Turjmanb "Fog Computing for Sustainable Smart Cities in the IoT Era: Caching Techniques and Enabling Technologies - An Overview" *Sustainable Cities and Society* 59 (2020) 102139

[13] Afzaal Hussain, Kashif zafar, and Abdul rauf baig, *"Fog-Centric IoT Based Framework for Healthcare Monitoring, Management and Early Warning System"* 10.1109/ACCESS.2021.3080237, IEEE Access

[14] Anand Paul, Hameed Pinjari, Won-Hwa Hong, Hyun Cheol Seo, and Seungmin Rho "Fog Computing-Based IoT for Health Monitoring System" *Hindawi Journal of Sensors* Volume 2018.

[15]   Ahmed Elhadad, Fulayjan Alanazi, Ahmed Taloba, and Amr Abozeid "Fog Computing Service in the Healthcare Monitoring System for Managing the Real-Time Notification" *Hindawi Journal of Healthcare Engineering* Volume 2022

[16]   Siddhant Jain Shashank Gupta K. K. Sreelakshmi Joel J. P. C. Rodrigues "Fog Computing in Enabling 5G-Driven Emerging Technologies for Development of Sustainable Smart City Infrastructures"

[17]   Harshit Gupta, Amir Vahid Dastjerdi, Soumya K. Ghosh, and Rajkumar Buyya "iFogSim: A Toolkit for Modeling and Simulation of Resource Management Techniques in Internet of Things, Edge and Fog Computing Environments" Cloud Computing and Distributed Systems Laboratory,The University of Melbourne, June 6, 2016

[18]   Mohammed A. T. Al Sudiari and TGK Vasista, King Saud University, KSA "Cloud Computing And Privacy Regulations: An Exploratory Study On Issues And Implications" *Advanced Computing: An International Journal (ACIJ)*

[19]   Kashif Munir and Sellapan Palaniappan "Secure Cloud Architecture" *Advanced Computing: An International Journal (ACIJ)*

[20]   Selvaraj Kesavan, Jerome Anand and J. Jayakumar "Controlled Multimedia Cloud Architecture And Advantages" *Advanced Computing: An International Journal (ACIJ)*

[21]   Kumar, A., & Alam, B. (2019). Energy Harvesting Earliest Deadline First Scheduling Algorithm For Increasing Lifetime of Real Time Systems. *International Journal of Electrical and Computer Engineering*, 9(1), 539.

[22]   Kumar, A. & Alam, B. (2018). Task Scheduling in Real Time Systems with Energy Harvesting and Energy Minimization. *Journal of Computer Science*, 14(8), 1126–1133.

[23]   Kumar, A., & Alam, B. (2014, February). Real Time Scheduling Algorithm for Fault Tolerant and Energy Minimization. *In 2014 International Conference on Issues and Challenges in Intelligent Computing Techniques (ICICT)* (pp. 356–360). IEEE.

[24]   Chaturvedi, R., Kumar, S., Kumar, U., Sharma, T., Chaudhary, Z., & Dagur, A. (2021). Low-cost IOT-Enabled Smart Parking System in Crowded Cities. In *Data Intelligence and Cognitive Informatics: Proceedings of ICDICI 2020* (pp. 333–339). Springer Singapore.

[25]   Meherwar Fatima, Maruf Pasha "Survey of Machine Learning Algorithms for Disease Diagnostic" Journal of Intelligent Learning Systems and Applications 2017

[26]   Das, S., & Guria, P. "Adaptation of Machine Learning in Fog Computing: An Analytical Approach". *International Conference for Advancement in Technology*, ICONAT 2022.

[27]   Quy, V. K., Hau, N. van, Anh, D. van, & Ngoc, L. A. "Smart Healthcare IoT Applications Based on Fog Computing: Architecture, Applications and Challenges." *Complex and Intelligent Systems 2021*

[28]   Ahanger, T. A., Tariq, U., Ibrahim, A., Ullah, I., Bouteraa, Y., & Gebali, "Securing IoT-Empowered Fog Computing Systems: Machine Learning Perspective." *Mathematics 2022*, 10, 1298. https://doi.org/10.3390/math10081298 2022.

[29]   Dhillon, A., Singh, A., Vohra, H., Ellis, C., Varghese, B., Gill, S. S., Sukhpal, & Gill, S. IoTPulse: "Machine Learning-based Enterprise Health Information System to Predict Alcohol Addiction in Punjab (India) using IoT and Fog Computing." *Enterprise Information Systems 2020*.

[30]   Bustamante-Bello, R., García-Barba, A., Arce-Saenz, L. A., Curiel-Ramirez, L. A., Izquierdo-Reyes, J., & Ramirez-Mendoza, R. A. "*Visualizing Street Pavement Anomalies through Fog Computing V2I Networks and Machine Learning.*" https://mdpi.com/1424-8220/22/2/456 2022

Artificial Intelligence, Blockchain, Computing and Security – Dagur et al. (Eds)
© 2024 The Author(s), ISBN: 978-1-032-49393-0

Microstrip patch antenna with high gain and dual bands for secure 5G communication

V. Kalai Priya
Department of Computer Science Engineering, RMK Collge of Engineering and Technology, Thiruvallur, Tamil Nadu, India

D. Sugumar
Department of Electronics and Communication Engineering, Karunya Institute of Technology and Sciences (Deemed to be University), Coimbatore, Tamil Nadu, India

K. Vijayalakshmi
Institute of Electronics and Communication Engineering, Saveetha School of Engineering, Saveetha Institute of Medical and Technical Sciences, Thandalam, Chennai, Tamil Nadu, India

V. Vanitha
Department of Electronics and Communication Engineering, Aarupadai Veedu Institute of Technology, Vinayaka Mission's Research Foundation (Deemed to be University), Chennai, Tamil Nadu, India

Charanjeet Singh
Department of Electronics and Communication, Deenbandhu Chhotu Ram University of Science and Technology, Murthal, Haryana, India

A. Yasminebegum
Department of Electronics and Instrumentation Engineering, Sree Vidyanikethan Engineering College, Tirupati, Andhra Pradesh, India

ABSTRACT: A Microstrip Patch Antenna with High Gain Dual Bands for secure 5G communication is presented in this research. The major goal is to utilize a wide bandwidth antenna operating in the 28/38GHz millimetre-wave band to obtain a high data throughput. An L-shaped, high-gain microstrip patch antenna is the subject of this research. Rogers RT Duroid 5880 has a relative permittivity constant of 2.2, an angular loss of 0.0009, and a 0.508mm thickness. The substrate for the format is Ansys HFSS software serves as the simulation basis. At 27.84GHz and 39.5GHz, respectively, the concept produced a gain of up to 8.33dB and 7.9dB. At the corresponding resonant frequencies below the parameters' −10dB line, the impedance bandwidth responses of 1.5 and 4.37GHz are attained. It is recommended to use a small antenna with the following dimensions: 3.2 × 4.9 × 0.508; this antenna offers High Gain and Bandwidth at both bands. Some of the crucial design factors Voltage Standing Wave Ratio (VSWR), gain, and return loss are also properly examined. The suggested antenna has a strong operating band performance, which qualifies it for secure 5G communication.

Keywords: 5G communication, Dual-band, High gain, Microstrip, Wide bandwidth

1 INTRODUCTION

There are still a lot of significant issues to take into account for the impending 5th generation (5G) wireless communication in light of contemporary wireless communication research

(Wei *et al.* 2017). In essence, the 5G wireless communication system will operate at a frequency that is higher than the 4G system, and is greater than 25 GHz (Andrews *et al.* 2014; Pi *et al.* 2011). Since more bandwidth is available in the millimetre frequency band than in the 3G and 4G frequency bands, many gigabits of data can be carried per second (Ghosh *et al.* 2016; Rappaport *et al.* 2013). The most common variety of microstrip antenna is referred to as a "patch antenna." Numerous experimenters have employed various strategies According to the literature, improving the patch antenna's performance (Patel *et al.* 2022; Ranjeet *et al.* 2021). Numerous scholars have employed arrays in their study (Bangash *et al.* 2019; Hasnaoui *et al.* 2020).

2 PROPOSED DUAL BAND MICROSTRIP PATCH ANTENNA

2.1 *Modelling of L-shaped microstrip patch antenna*

Equations (1) to (3) provide the formulas for properties of the proposed antenna are initially determined using L-shaped microstrip patches. (7). The size of the patch specified in the equation can be calculated.

The patch width (W) is determined using (1)

$$W = \frac{C}{2f_o\sqrt{\frac{\varepsilon_r+1}{2}}} \tag{1}$$

When measuring the patch's length (L1) shown in (2)

$$L = L_{eff} - 2\Delta L \tag{2}$$

Where L_{eff} and ΔL are the path length extension and effective length of the patch, which are calculated using (3) and (4)

$$L_{eff} = \frac{C}{2f_o\sqrt{\varepsilon_{eff}}} \tag{3}$$

$$\Delta L = 0.412h\frac{(\varepsilon_{eff}+0.3)\left(\frac{W}{h}+0.3\right)}{(\varepsilon_{eff}-0.3)\left(\frac{W}{h}+0.8\right)} \tag{4}$$

ε_{eff} is the effective relative permittivity

$$\varepsilon_{eff} = \frac{\varepsilon_r+1}{2} + \frac{\varepsilon_r-1}{2}\left[\frac{1}{\sqrt{1+\frac{12h}{W}}}\right] \tag{5}$$

In (5), the substrate's height is given by the value h = 1.6 mm.

The proposed design is a microstrip patch antenna with excellent gain and broad bandwidth. A microstrip patch antenna is made of a dielectric substrate with a radiating patch on one side and a ground plane on the other, as shown in Figure 1.

The top and bottom parts of the substrate are printed with the ground plane and radiating structure, respectively. Due to the rectangular radiator, the radiating structure consists of two L-shaped patches separated by a single strip (5100.05mm3). Also, a ground plane with the dimensions 70600.05mm3 has been printed on the substrate's reverse. The dimensions shown in the picture were produced using the simulation programme Ansys HFSS version 2.2 for dual band 6.21-5.44GHz and 6.72-5.82 GHz. The rectangular FR4 substrate has the following measurements: W × L. Along the centre line of the patch, the W dual L-slots are evenly spaced apart and have a small S width.

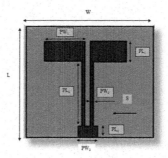

Figure 1. The L-shaped microstrip antenna construction.

The initials PL1 and PW1 stand for the lengths of the L slots' vertical and horizontal arms, respectively. The letters PW2 and PL2 stand for the width and length of L-shaped strips, respectively. PEC material has a patch thickness of 0.05mm.

3 SIMULATION RESULTS AND DISCUSSION

3.1 *Antenna bandwidth and resonant frequency*

The results are discussed in this section. The patch antenna resonates at 28.84GHz and 38.42GHz based on the reflection coefficient of the proposed design. With a return loss of −46.9 and −27.22, the patch antenna achieved a bandwidth of 4.3 GHz and 1.5GHz at the relevant bands. The bandwidth will be increased in order to give the suggested antenna a 60% boost. The results of each design technique are depicted in Figure 2.

3.2 *Gain for the specified antenna*

The system gain of the suggested antenna is shown in Figure 2, which is suitable for 5G application systems, is and at the pertinent operating bands. Gain can be raised by inserting slots into the radiating patch. This design is less complex than other ones, and the simulation results demonstrate that it was effective. Gain and frequency of the antenna is shown in Figure 3.

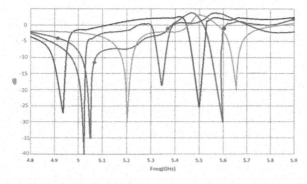

Figure 2. Return loss of different design.

541

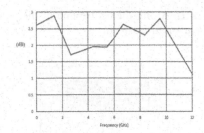

Figure 3.   Gain vs Frequency.

3.3   *Voltage Standing Wave Ratio (VSWR)*

The optimal VSWR range, which denotes an effective and suitable antenna, is between 1 and 2. The resonant frequencies of 27.9GHz and 29.22GHz correspond to the VSWR values of 1.18 and 1.07, respectively.

The proposed approach yields effective outcomes appropriate for 5G communication. It is clear that adding slots to specific locations on a microstrip's radiating element enhances the antenna's capabilities. With this strategy, the device achieves the high gain of 8.3/7.7dB at the appropriate operating bands for 5G applications. The technique employed to increase bandwidth is straightforward yet effective, bringing the upper band bandwidth up to a maximum value of 4.45GHz. The patch's edges are trimmed to do this. The return loss is also improved by doing this, as seen in Table 1.

Table 1.   Return loss of proposed method.

Antenna	Resonant frequency (GHz)	Return loss (dB)
5G Antenna	39	−15.6
	55	−13
5G Antenna	39	−13.6
	47.8	−22.6
	55	−18

Some studies have employed different techniques, like giving the radiating patch element outlines. However, the method used in this research has shown excellent results for enhancing gain and bandwidth of the microstrip patch antenna. The slots and steps that were incorporated into the proposed antenna's design altered its resonating frequency, but they were also able to improve its performance, and at the desired band frequency, the antenna is presently resonating.

Table 2 makes comparisons between the data and previous works using various slot characteristics. The fact that the proposed data compare favourably with the majority of the data suggests that the proposed antenna is compact, has an improved bandwidth, and has a high gain due to cutting edges in the microstrip patch.

Table 2.   Comparing the intended product with existing works.

Parameters	Resonating frequency (GHz)	Bandwidth (GHz)	Gain (dB)
Hybrid fractal slot	23.5/28.40	1.2/0.63	4.66
U-shape slot	28/38.6	5.14/11.7	8.3/6.4
High gain filtering antenna	56.4	5.6/8.7	5.5/69
Tri-band 5g	38/50	2/3.3	6.6/5.6
Proposed	28/39	1.47/4.5	8.3/7.7

4 CONCLUSION

This research presents a small, dual-band, microstrip antenna with high gain for secure 5G transmission. The IEEE 802.11a standards (5.15–5.35GHz and 5.725–5.825GHz) are constantly covered by the frequency bands, which have maximum gain values in the lower and higher frequency bands of 7.2 and 8.5 dB, respectively, and return losses below −10 dB. By cutting the antenna proposed, which comprises of a single element, at the edges, return loss and gain are displayed. The suggested antenna is ideal for secure 5G connection based on its properties. Additionally, the beam width has been expanded. Overall, at the required operating frequency, the antenna's performance satisfies the expected specifications for return loss, high gain, and VSWR. The results of this investigation have revealed that the slot's dimensions strongly influence the microstrip antenna's performance.

REFERENCES

Andrews, J.G., Buzzi, S., Choi, W., *et al.*: 'What will 5G be?', *IEEE J.Sel.AreasCommun.*, 2014, 32, (6), pp. 1065–1082

Bangash K., Ali M. M., Maab H., and Ahmed H., "Design of a Millimeter-Wave Microstrip Patch Antenna and Its Array for 5G Applications," *1st Int. Conf. Electr. Commun. Comput. Eng. ICECCE 2019*, no. July, pp. 1–6, 2019, doi: 10.1109/ICECCE47252.2019.8940807.

El Hasnaoui Y. and Mazri T., "Study, Design and Simulation of an Array Antenna for Base Station 5G," *2020 Int. Conf. Intell. Syst. Comput. Vision, ISCV 2020, 2020*, doi: 10.1109/ISCV49265.2020.9204261.

Ghosh, A.: '*The 5G mmWave Radio Revolution*', *Microw. J.*, 2016, 59, (9 PartI), pp. 3–10.

Patel D. H. and Makwana G. D., "A Comprehensive Review on Multi-band Microstrip Patch Antenna Comprising 5G Wireless Communication," *Int. J. Comput. Digit. Syst.*, vol. 11, no. 1, pp. 941–953, 2022, doi: 10.12785/ijcds/110177.

Pi, Z., Khan, F.: 'An Introduction to Millimeter-wave Mobile Broadbandsystems', *IEEE Commun. Mag.*, 2011, 49, (6), pp. 101–107

Ranjeet Pratap Singh (2021), Directivity and Bandwidth Enhancement of Patch Antenna using Metamaterial. *IJEER* 9(2), 6–9. DOI: 10.37391/IJEER.090201.https://ijeer.forexjournal.co.in/archive/volume-9/ijeer090201.html

Ribhu Abhusan Panda, Mihir Panda, Pawan Kumar Nayak, Debasish Mishra (2020), Butterfly Shaped Patch Antenna for 5G Application. *IJEER* 8(3), 32–35. DOI: 10.37391/IJEER.080301. http://ijeer.forexjournal.co.in/archive/volume-8/ijeer080301.html

Wei, X., Zheng, K., Shen, X. S., ed.: '*5G Mobile Communications*' Springer, Switzerland, (2017)

Artificial Intelligence, Blockchain, Computing and Security – Dagur et al. (Eds)
© 2024 The Author(s), ISBN: 978-1-032-49393-0

Blockchain-based access control and interoperability framework for electronic health records (ANCILE)

G. Senthilkumar*

Department of Computer Science and Engineering, Panimalar Engineering College, Chennai, Tamil Nadu, India

Aravindan Srinivasan*

Department of Computer Science Engineering, Koneru Lakshmaiah Education Foundation, Vaddeswaram, Andhra Pradesh, India

J. Venkatesh*

Department of Computer Science and Engineering, Malla Reddy Institute of Engineering and Technology, Hyderabad, Telangana, India

Ramu Kuchipudi*

Department of Information Technology, Chaitanya Bharathi Institute of Technology, Hyderabad, Telangana, India

K. Vinoth*

Department of Electrical and Electronics Engineering, Vel Tech Rangarajan Dr. Sagunthala R&D Institute of Science and Technology, Chennai, Tamil Nadu, India

A. Ramamoorthy*

Department of Mathematics, Velammal Engineering College, Chennai, Tamil Nadu, India

ABSTRACT: Even if big cities are working globally to build the infrastructure for smart cities and there is a greater emphasis on the security of electronic health records, patient privacy is frequently compromised. Previous attempts to combat this have left patients with largely unavailable data. Currently used record-management systems struggle to strike a balance between data privacy and patient and provider access. Blockchain, a new technology, the ability to share data in a decentralized and transactional manner. To balance the accessibility and privacy of electronic health records, blockchain technology may be applied in the healthcare industry. The blockchain-based architecture we present in this work enables patients, healthcare professionals, and third parties to access medical information in a secure, efficient, and straightforward manner while maintaining the privacy of sensitive patient data. Modern cryptographic techniques are used in our architecture, Ancile, to increase security. The usage of smart contracts developed on the Ethereum blockchain improves access control and data obscuration. This article will look at how Ancile works with the varied needs of patients, providers, and third parties in order to understand how the framework might address reoccurring privacy and security concerns in the healthcare industry.

Keywords: blockchain, healthcare, information security, smart cities, access control, Ethereum, smart contracts

*Corresponding Authors: senthilkumar@yahoo.com, aravindansrinivasan2@gmail.com, venkatesh.j@mriet.ac.in, kramupro@gmail.com, vinothkrishna03@gmail.com and ramzenithmaths@gmail.com

DOI: 10.1201/9781003393580-83

1 INTRODUCTION

The healthcare transition program is anticipated to be successful given that healthcare researchers and service providers have access to these EHRs globally. Patients currently scatter their electronic health records (EHRs) across many locations as a result of life events, which results in the EHRs moving between service provider databases. As a result, while the service provider often retains primary stewardship, the patient may no longer have ownership over the current healthcare data (Tanwar *et al.* 2020). The healthcare industry would substantially benefit from integrated data management and the ability to manage and exchange his EHRs entirely and securely, independent of the study's objectives or the healthcare industry's data sharing practices (Tanwar *et al.* 2020). By using blockchain technology as support, the suggested method is effective in fostering collaboration in the form of strong mutual trust between each firm. Satoshi Nakamoto first created and introduced blockchain to the Bitcoin digital currency [3]. The blockchain of EHRs is built using blockchain technology, which also establishes standards for data management and identity management. Additionally, This technology ensures accountability and transparency throughout the data exchange process by recording all transaction audit trails in an immutable distributed ledger. On order to lower medical errors and protect patient privacy, patients now have the option to save medical and diagnostic data from clinicians in their own EHRs (Kshetri *et al.* 2018).

2 RELATED WORK

We present a framework that might be used to the usage of blockchain technology for electronic medical records in the healthcare industry. By establishing access requirements for users, the architecture we suggest aims to initially integrate blockchain technology for EHR before guaranteeing secure storage of digital records (Dagher *et al.* 2018). Electronic health record security, which include patients' names, addresses, and ailments, is indirectly impacted by the regular violations of patient privacy in the present era of smart homes and cities (EHRs) (Zarour *et al.* 2020). We explicitly demonstrate the security of our attribute-based signature method in the random oracle model, together with the privacy and invulnerability of the attribute-perfect signer, based on the computational bilinear Diffie-Hellman assumption (Watanabe *et al.* 2019). Private and permissioned block chains, according to some, go against the idea of decentralization. They do, however, offer benefits including improved privacy protection and the ability to change the minimum gas price [10].

3 PROPOSED METHOD

The proposed blockchain-based method for transferring electronic health records is introduced in this section. A blockchain-based design for an EHR sharing system is consequently suggested. The network's block trans-activity features a range of settings and tactics. The suggested solution makes it easier for other blockchain network users to access the EHR by using a shared symmetric key and private key. For quick communication and effective operation, Table 1 also covers the top EHR sharing algorithms.

3.1 *EHRS system model*

The patients, EHRs server, N authorities, and data verifiers were the four components of this EHRs system paradigm. A cloud storage server-like device the EHRs server is in charge of storing and distributing the EHRs. The responsibility for collecting enrollments and facilitating the exchange of patient data falls on hospitals, health insurance providers, research institutions, and other organizations. Patients can create their own EHRs, administer them, regulate them, and sign them. They can also set the predicate, and the data verifier can use this signature to confirm the accuracy of the data.

Algorithm: EHRS system model

```
Algorithm : HER

Input: creation of patient HER creation
Output: response to the patient HER creation
// creation of patient HER
If (HER − Exists)
{
Return (HER, −ID, Ptient − ID, HER − Doctor − ID, hospital − ID, HER − time);
}
Else
{
Respose ← Create − HER − Records);
EHIR − Records
            ← HER(HER, −ID, Ptient − ID, HER − Doctor
            − ID, hospital − ID, HER − time)
Returns (HR − Records)
}
// update the HER in the list of the patient EHRs into the blockchain
If (patient − Exists)
{
Verify (patient − ID − Exists)
{
Verify (patient − ID);
If (HER − Exists)
{
World state ← push(HER − Records, HER
            − ID); An HER is updated and stored in the world state
}
Returns world state);
}
}end of the algorithm
```

4 EXPERIMENTAL RESULT

A searchable encryption system for electronic medical records based on blockchain. They developed a two-part evaluation scheme as well as an algorithm for electrical health records. To evaluate the performance and function of our prototype, several tests have been conducted. These tests also advance our understanding of the real-world use of the blockchain.

Table 1. Measurements of performance for various tasks. The number N determines how many entries make up the blockchain.

Chain function in a blockchain	Mean values ± Standard deviation one [ms]			
	$N = 100$	$N = 500$	$N = 2000$	$N = 4000$
Registration	2685 ± 64	27645 ± 89	2869 ± 64	2198 ± 61
Grant permission	2382 ± 12	2374 ± 16	2309 ± 17	2664 ± 15
Update data	95 ± 4.5	95 ± 4.6	398 ± 4.5	103 ± 5.6
Invoke	2200 ± 16	2202 ± 14	2218 ± 12	2228 ± 23
Query	55 ± 2.1	57 ± 1.7	58 ± 2.0	59 ± 2.3
Blockchain size	37	113	391	768

Blockchains with N, 1, 100, 500, 2000, and 4000 entries are used in the experiments, and 50 measurements are collected for each test. Ten simultaneous queries were used in the testing, which was done on the Chrome web browser. Table 1 presents the findings, while Figure 1 plots the average values. It was discovered that the initial blockchain size (N = 1) was 19 KB. The size of the blockchain as a whole grows along with the amount of entries it contains (Table 2).

Response times for functions like registration, granting permission, and revoking authorization are closely correlated with invocation times. The cause is the blockchain queries made by the login and update data features. When data is updated, the blockchain is

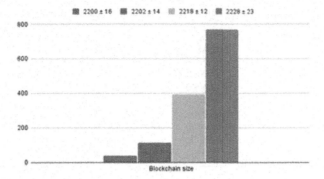

Figure 1.   Performance measurements for different activities.

checked to see if the actor is authorized to access the updated data. In the case of registration, the data must be entered into the blockchain, and the invoke function must be invoked. Another thing to note is that as the blockchain's entries grow in size, it takes a little longer to query or invoke the blockchain and to perform system tasks like login, registration, and permission-granting.

4.1   *Data sharing*

We tested the calculation time for creating cipher texts. Each encryption is a separate operation. Table 2 displays the time it takes to encrypt healthcare data versus the size of the string. We study how the rate of time growth for encryption changes with input size using a range of inputs. We evaluate the encryption duration of various data sizes using data ranging from 5 to 30 kilobytes. The resulting graph reveals that the curve's rate of expansion is generally linear, meaning that the encryption time increases as data size increases.

Table 2.   Input generation computation time.

Input size (kb)	Encryption time	Input generation time
	84	5
	86	10
	88	15
	90	20
	92	25
	94	30
	100	35

We test the encryption time of various data sizes using data ranging from 5 to 35 kilobytes. The resulting graph reveals that the curve's rate of expansion is generally linear, meaning that the encryption time increases as data size increases. The data sharing phase of our system is a changeable and autonomous procedure. Variable indicates that individual users' input sizes may differ, and independent means that the encryption of separate users' data is not dependent on each other. Ethereum has been utilized to access blockchain in Table 3 because it is the most effective platform for executing Dapps (Distributed Apps) developed in Solidity.

Prior to getting access to a block on the blockchain network, our smart contract will access data. The use of smart contracts will cost some gas, which is the crypto fuel used by the Ethereum Virtual Machine (EVM). The executed application must have certain network

547

Table 3. Cost of transaction computation and smart contract execution.

Transaction cast for variable input	Transaction cast in GAS	String length (number of character)
	100000	10
	200000	20
	300000	30
	400000	40
	500000	60
	600000	80
	700000	100

transactions in order to run any dapp (distributed application) in the Ethereum environment; in exchange for the transactions, the environment charges the executor some gas. In the Ethereum environment, the transaction initiator or executor will receive gas in return for Ether.

We have established a set of security and privacy guidelines for healthcare data management systems.

Table 4. Smart contract execution cost with variable input.

Transaction cast for variable input	Transaction cast in GAS	String length (number of character)	Execution Cost of Smart Contract (GAS) 4 inputs
	100000	5	0
	200000	10	6
	300000	15	8
	400000	20	16
	500000	25	21
	600000	30	34
	700000	35	38
	800000	40	42
	900000	45	45

5 CONCLUSION

The goal of this article was to create a blockchain framework for managing electronic health records (EHRs) that may grant patients ownership and ultimate management of those records. Securely limit who has access to them and keep track of their usage, support secure record transfers, and reduce the potential for unauthorized parties to obtain PHI while remaining HIPPA compliant. While recognizing the requirement for some nodes to have greater power than others, the Ancile design demonstrates how a blockchain system can be highly decentralized. This analysis showed that maintaining complete information secrecy while maintaining a workable and interoperable system would be quite unlikely, Ancile employs smart contracts to divide information, yet it still offers robust data integrity and privacy protection.

REFERENCES

Dagher, Gaby G.; Mohler, Jordan; Milojkovic, Matea; Marella, Praneeth Babu (2018). Ancile: Privacy-preserving Framework for Access Control and Interoperability of Electronic Health Records Using

Blockchain Technology. *Sustainable Cities and Society*, S2210670717310685–doi:10.1016/j.scs.2018.02.014 doi:10.1109/GLOCOMW.2018.8644088

Guo, Rui; Shi, Huixian; Zhao, Qinglan; Zheng, Dong (2018). Secure Attribute-Based Signature Scheme with Multiple Authorities for Blockchain in Electronic Health Records Systems. *IEEE Access*, 1–1. doi:10.1109/ ACCESS.2018.2801266

Health Insurance Portability and Accountability Act (2021). URL http://www.dhcs.ca.gov/formsandpubs/ laws/hipaa/Pages/1.00WhatisHIPAA.aspx

Kshetri, Nir (2018). Blockchain and Electronic Healthcare Records [Cybertrust]. *Computer*, 51(12), 59–63. doi:10.1109/MC.2018.2880021

Tanwar, Sudeep; Parekh, Karan; Evans, Richard (2020). Blockchain-based Electronic Healthcare Record System for Healthcare 4.0 Applications. *Journal of Information Security and Applications*, 50, 102407–. doi:10.1016/j.jisa.2019.102407

Watanabe, Hiroki; Fujimura, Shigeru; Nakadaira, Atsushi; Miyazaki, Yasuhiko; Akutsu, Akihito; Kishigami, Jay (2019). [IEEE 2016 IEEE International Conference on Consumer Electronics (ICCE) – Las Vegas, NV, USA (2016.1.7-2016.1.11)] *2016 IEEE International Conference on Consumer Electronics (ICCE) – Blockchain Contract: Securing a Blockchain Applied to Smart Contracts.*, 467–468. doi:10.1109/ ICCE.2016.7430693

Yang, Guang; Li, Chunlei (2018). [IEEE 2018 IEEE International Conference on Cloud Computing Technology and Science (CloudCom) – Nicosia, Cyprus (2018.12.10-2018.12.13)] 2018 IEEE International Conference on Cloud Computing Technology and Science (CloudCom) – *A Design of Blockchain-Based Architecture for the Security of Electronic Health Record (EHR) Systems.*, 261–265. doi:10.1109/ cloudcom2018.2018.00058

Zarour, Mohammad; Ansari, Md Tarique Jamal; Alenezi, Mamdouh; Sarkar, Amal Krishna; Faizan, Mohd; Agrawal, Alka; Kumar, Rajeev; Khan, Raees Ahmad (2020). Evaluating the Impact of Blockchain Models for Secure and Trustworthy Electronic Healthcare Records. *IEEE Access*, 1–1. doi:10.1109/ ACCESS.2020.3019829

Artificial Intelligence, Blockchain, Computing and Security – Dagur et al. (Eds)
© 2024 The Author(s), ISBN: 978-1-032-49393-0

Deep learning based approach for rice prediction from authenticated block chain mode

V.V. Satyanarayana Tallapragada*
Department of Electronics and Communication Engineering, Mohan Babu University, Tirupati, Andhra Pradesh, India

Sumit Chaudhary*
School of Computer Science & Artificial Intelligence, TransStadia University, Kankaria, Ahmedabad, Gujarat, India

J. Sherine Glory*
Department of Computer Science and Engineering, R.M.D. Engineering College, Kavaraipettai, Chennai, Tamil Nadu, India

G. Venkatesan*
Department of Civil Engineering, Saveetha Engineering College, Chennai, Tamil Nadu, India

B. Uma Maheswari*
Department of Computer Science and Engineering, St. Joseph's College of Engineering, Chennai, Tamil Nadu, India

E. Rajesh Kumar*
Department of Computer Science and Engineering, Koneru Lakshmaiah Education Foundation, Vaddeswaram, Andhra Pradesh, India

ABSTRACT: Various quality and safety issues with rice, a significant food crop world-wide, are strongly tied to human health. The rice supply chain is a crucial topic of food safety research that has received more and more attention. This study, which used block chain technology, looked into issues with data privacy and circulation efficiency brought on by complicated systems for the supply of rice, continuous quality periods, and various performance factors at every connection. Initially constructed a valuable data helps to establish for every connection in the production field after deconstructing the quality and safety of each connection at the information level. Based on that, we developed a block chain-based information supervision model for the rice supply chain. To fulfill the requirements of authorities for efficient supervision, a variety of encryption methods are utilized to protect the private information of businesses in the give connection. Additionally, we suggest a Deep Convolutional Neural Network algorithm that evaluates enterprise node credit, optimizes master node selection, and guarantees high efficiency and cheap cost. The findings showed that the suggested approach can improve rice prediction and offer a workable remedy for the grain and oil quality and safety supervision.

Keywords: Rice prediction, Block chain node, Deep convolution neural network

*Corresponding Authors: satya.tvv@gmail.com, drchaudharysumit@gmail.com, sherinegloryj@gmail.com, gvenkatesan@saveetha.ac.in, mahespal2002@gmail.com and rajthalopo@gmail.com

DOI: 10.1201/9781003393580-84

1 INTRODUCTION

More over 50% of the world's population relies mostly on rice, which has made a substantial contribution to global food security. Furthermore, 156 million hectares of rice are grown globally, with Asia accounting for more than 85% of that total. As a result, reliable rice output predictions in Asia are helpful for commerce, development strategies, humanitarian aid, and decision-makers. For estimating crop growth, there are three basic forms: The first three categories include physical systems, empirical model training and neural having to learn designs. Mechanistic models use mathematical operations to represent physical, biological, and chemical processes. It presented a model of rice yield using mathematical functions and deep learning to interpret 2D photos. These models struggle to adapt to various places because they rely heavily on empirical knowledge and input parameters (Zhao *et al.* 2019).

Using block chain technology, problems with centralization, security, and tampering are all effectively resolved. Block chains can record the entire process and perform online a program in accordance with commercial guidelines framed by the network to safely share data, supplies, and exchange rates data in a distribution network (Berdik *et al.* 2021).

2 RELATED WORKS

In order to make applications with bootstrapping and frameworks, research on deep learning and block chain has exploded recently. A resource-efficient, block chain-based approach for a private and secure IoT was put forth by (Khan *et al.* 2020). A novel privacy-anonymous Internet of Things paradigm was created by (Gargi *et al.* 2020). They have provided an identification evidence for this specific design. For the collection and access with on information, this approach takes advantage of block chain decentralization. Using their suggested approach, moving objects will be capable of transmitting or receiving alerts once they are near to an instance that is detected, likely to be infected, or has been shown to be impacted.

(Ali *et al.* 2020) introduced a Long Short-Term Memory (LSTM) based on historical data with the estimation of the highest, lowest, and average scores of the atmospheric heat, moisture content, velocity, speed and water level. For the examination of the best-fitting LSTM model, both microclimate data inside the greenhouse and macroclimate data outside are gathered. A model for forecasting the environment needed for greenhouse tomato production is presented by Ali *et al.* in their publication (Ali *et al.* 2019). For the purpose of observing any differences, pest, or disease that may be present on plant leaves, a Wireless Visual Sensor Network (WVSN), computer vision and visual manufacturing system was created by (Ali *et al.* 2019). The greenhouse is covered with camera sensors. In order to identify fungus, each sensor node takes a picture from within the greenhouse and analyzes it using machine learning and image processing. Utilizing machine learning and image processing approaches, (Zhang *et al.* 2020) solution allowed them to count all of the wheat plant's grain-bearing points. While the use of crowdsourcing based on a block chain to protect harvests from illnesses has not been extensively adopted, other tools have been created for the benefit of producers. (Mingjie *et al.* 2020). a wavelet transform method for enhancing images is comprised of two distinct high and low pass techniques. The classification of maize diseases is proposed using a DMS-Robust Alex net design, an enhancement of the conventional Alex net architecture. For instance, in order to track and control the production process and ensure the validity of the entire process, (Baralla *et al.* 2021) developed a block chain-based approach.

The contribution of the proposed work is as follows:

- This study offers a Deep Convolution Neural Network (DNN) approach to enhance the prediction of rice crop yield. Using block chain technology, security-related problems and restrictions can be resolved.

- In terms of several performance measures, the suggested work is contrasted with the current models.

The following describes the study's framework: The related works are described in Section 2; the techniques are introduced in Section 3; the results and discussion are presented in Section 4; and the conclusions are presented in Section 5.

3 PROPOSED METHOD

This study used block chain technology to create a deep learning model of the rice supply chain based on the traits of the chain's participating businesses and supply chain procedures. The model involves not just total supply chain management but also supervision of government agencies, as well as enterprise data uploading and exchange. To achieve integrated monitoring, both querying and the traceability of consumer items are used. The integrity and security of data transported across the supply chain can be guaranteed by handling and controlling rice information in this way.

The proposed framework that employs deep convolution neural network is shown in Figure 1. Predicting the output yield of the rice crop is the major goal of the endeavor. The cultivation of the rice crop is the main topic of the research. In terms of production, there is an imbalance, though. Multiple causes, including natural disasters, poor fertilizer use, shifting environmental circumstances, and seasonal variable crop yield output, are to blame for this unbalanced productivity. Knowing the predicted crop output is crucial to ensuring the continuous change in productivity. In this paper, DCNN is suggested to increase prediction accuracy. A deep neural network to increase the accuracy of the predictions. Rice prediction needs to be more accurate.

Figure 1. Architecture for proposed method.

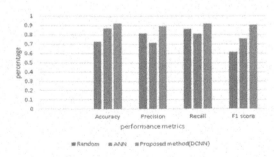

Figure 2. Performance indicators of DCNN model.

552

4 SIMULATION RESULTS AND DISCUSSION

R programming is used to implement and evaluate the suggested model. It functions as a platform for deep learning algorithms. Recall, accuracy, precision, and F1 score are among the metrics used to evaluate DCNN's performance in terms of rice prediction. The comparative results are presented and examined in the end. The performance DCNN model utilized in this approach is described in Table 1 and Figure 2.

Table 1. Performance of DCNN model.

Performance Metrics	Random Forest	ANN	Proposed method(DCNN)
Accuracy	0.725	0.869	0.921
Precision	0.812	0.712	0.891
Recall	0.861	0.810	0.920
F1 score	0.615	0.758	0.907

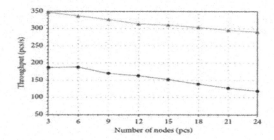

Figure 3. Accuracy comparison.

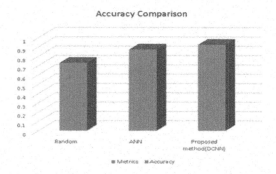

Figure 4. Comparison of throughput.

Figure 3 shows the accuracy comparison. The precision attained by the classification with convolutional neural networks is 92.1%, which is higher than that those obtained using other methods. This method is depicted in Figure 4. With the increase in the number of nodes in the network, the throughput of both algorithms shows a downward trend. Generally the DCNN method has a substantially higher efficiency than the current approach. The various optimizing strategies employed in the model's development are assessed in Figures 5–8.

Figure 5. Training accuracy.

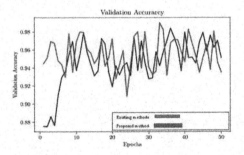

Figure 6. Training loss.

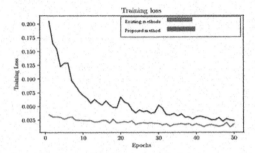

Figure 7. Validation accuracy.

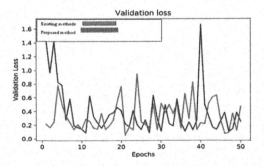

Figure 8. Validation loss.

554

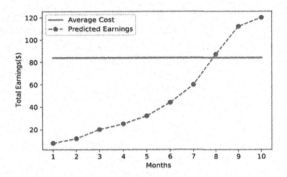

Figure 9.   Average cost & earnings.

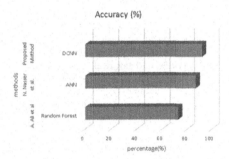

Figure 10.   Comparative analysis.

Figure 9 analyzes the financial advantages of our concept for the producer. The red line in the graphic represents the typical price of diagnosing and finding cures for crops. This cost is ongoing because it is incurred by the producer when discovering various traits or problems in his or her goods. Most farms deal with this cost, but this model can assist them in controlling it. Benefits offered by the conceptual model eventually outweigh the estimated price if the producer consistently participates to the system by offering treatments for various ailments.

The suggested study work is compared to other methods for judging classification accuracy, as in Table 2 and Figure 10. The suggested study work employing the DCNN approach yields the greatest results, with an accuracy of 92.1%.

Table 2.   Comparative analysis of the proposed research work with existing methods.

Study	Methods	Accuracy (%)
Ali *et al.* [2019]	Random Forest	72.5
Nasser *et al.* [2020]	ANN	86.9
Proposed Method	DCNN	92.1

5   CONCLUSION

The DCNN model for the rice prediction model has been proposed in this research. Recall, accuracy, precision, and F1 score are only a few of the performance measures used to analyze the suggested method. The DCNN model provides a higher prediction of crop production. The suggested model produced the most accurate predictions. The suggested model was stable and quickly reached low errors. Based on the findings of this paper's research, the features that have a significant impact on rice yields are identified and they are incorporated into the model to enhance forecast accuracy. The small size of the time-series data collection and the increased sample size used to improve prediction accuracy as a future part of this study.

REFERENCES

Ali A., Hassanein H. S., A Fungus Detection System for Greenhouses using Wireless Visual Sensor Networks and Machine Learning,in:2019 *IEEE Globecom Workshops (GC Wkshps)*, 2019, pp. 1–6.

Ali A., Hassanein H. S., Time-series Prediction for Sensing in Smart Greenhouses, in: *GLOBECOM 2020-2020 IEEE Global Communications Conference*, 2020, pp. 1–6.

Ali A., Hassanein H. S., Wireless Sensor Network and Deep Learning for Prediction Greenhouse Environments, in: *2019 International Conference on Smart Applications, Communications and Networking (Smart Nets)*, 2019, pp. 1–5.

Baralla G., Pinna A., Tonelli R., Marchesi M., and Ibba S., "Ensuring Transparency and Traceability of Food Local Products: a Blockchain Application to a Smart Tourism Region," *Concurrency and Computation: Practice and Experience*, vol. 33, no. 1, Article ID e5857, 2021.

Berdik D., Otoum S., Schmidt N., D., Porter, and Jararweh Y., "A Survey on Block Chain for Information Systems Management and Security," *Information Processing & Management*, vol. 58, no. 1, Article ID 102397, 2021.

Garg L., Chukwu E., Nasser N., Chakraborty C., Garg G., Anonymity Preserving iot-based covid-19 and Other Infectious Disease Contact Tracing Model, *IEEE Access* 8 (2020) 159 402–159 414.

Khan M. A., Abbas S., Rehman A., Saeed Y., Zeb A., Uddin M. I., Nasser N., Ali A., A Machine Learning Approach for Block Chain-based Smart Home Networks Security, *IEEE Netw.* (2020) 1–7.

M. Lv, G. Zhou, M. He, A. Chen, W. Zhang, Y. Hu, Maize Leaf Disease Identification Based on Feature Enhancement and DMS-robust Alexnet, *IEEE Access* 8 (2020) 57 952–57 966.

Zhang D., Wang Z., Jin N., Gu C., Chen Y., Huang Y., Evaluation of Efficacy of Fungicides for Control of Wheat Fusarium Head Blight Based on Digital Imaging, *IEEE Access* (2020).

Zhao, S., Zheng, H., Chi, M., Chai, X., Liu, Y., 2019. Rapid Yield Prediction in Paddy Fields Based on 2d Image Modelling of Rice Panicles. *Comput. Electron. Agric.* 162, 759–766.

Artificial Intelligence, Blockchain, Computing and Security – Dagur et al. (Eds)
© 2024 The Author(s), ISBN: 978-1-032-49393-0

Digital media industry driven by 5G and blockchain technology

B. Md. Irfan*
Department of Information Technology, Nalsar University of Law, Hyderabad, Telengana, India

Ramakrishnan Raman*
Symbiosis Institute of Business Management, Pune and Symbiosis International (Deemed University), Pune, Maharashtra, India

Hirald Dwaraka Praveena*
Department of Electronics and Communication Engineering, School of Engineering, Mohan Babu University, Tirupati, Andhra Pradesh, India

G. Senthilkumar*
Department of Computer Science and Engineering, Panimalar Engineering College, Chennai, Tamil Nadu, India

Ashok Kumar*
Department of Computer Science, Banasthali Vidyapith, Rajasthan, India

Ruhi Bakhare*
Dr. Ambedkar Institute of Management Studies and Research, Nagpur, Maharashtra, India

ABSTRACT: 5G (5 Generation) is maximum providing data per second speed (interval) per second average Megabits per Second (Mbps), can 20 gigabits 100 Mbit peak data rate is more significant than faster 4G. 5G is, have a greater capacity. 5G is to support the business ability by multiplying by 100, has been designed to increase network efficiency. 5G is, have a lower than 4G latency. Focus on sharing content-centric networks face a huge range of content requested to provide in Digital Media. How efficiently and safely protect the information network for the upcoming 5Gage is a problem. 5G mobile networks, content-centric block chain-based SHA-256 (Secure Hash Algorithm) programs to address the privacy issue. Also, a block chain ledger of features ensures privacy and access control of the supplier. With the help of miners user selected, the author can be maintained for the public ledger's convenience. The share of low overhead, network delay, and congestion implement green interesting data communications.

Keywords: Block chain, 5G, Network, Communication, and Mobile Network

1 GENERAL INSTRUCTIONS

To support significant applications, i.e., in the high-repeat millimeter waves acquired by the 5G journeys, a short division past the 30 GHz (Gigahertz) band exercises and the movement signal. A grand plan of the system requires PDAs and Web objections to cover a greater zone of comparative 4G numbers. Emerging countries are pondering activities to reduce the cost

*Corresponding Authors: irfan@nalsar.ac.in, raman06@yahoo.com, hdpraveena@gmail.com, senthilkumar@yahoo.com, kuashok@banasthali.in and ruhibakhare@rediffmail.com

DOI: 10.1201/9781003393580-85

of sending establishment. These choices join cloud establishment, securing, meandering, sharing, and reach sharing. Another critical test is the best approach to managing the vital assistance for each square kilometer, the organization's activities. Colossal numbers, for instance, circled and heterogeneous phones, include the attestation 5G network and the new troubles of flexibility. Thinking about the, acknowledge, Blockchain-Blockchain can address these troubles and give decentralized, strong and monetarily clever game plans accept a critical work. Despite Blockchain-Blockchain advanced money, Bitcoin essential development, blockchain correspondence between secured, incredible, generous and cost-sufficiency (using astute arrangements) to control various players and devices now—seen as a forefront, trustworthy, and successful dissipating. Square chain trade is in the chain block set aside invariably.

2 RELATED WORKS

Fifth-Generation (5G) flexible association is needed to be finished by demanding execution targets consigned by the standards leading body of trustees (Castellano *et al.* 2019). Appropriately, enormous changes to achieve the ideal level of execution have been made to the association establishment (Miklosik *et al.* 2020). Association virtualization work, disseminated processing and programming described the organization, the best introduction and the gainful resource use to ensure the association design uses a couple of huge advances (Miklosik *et al.* 2019; Tahir *et al.* 2020). The Digital Media of huge data and promoting the industry, from the viewpoint of data and information the board, affects the Digital Media (Azzaoui *et al.* 2020). Record advancement is to make an open entryway for high ground, develop new data-arranged methods, and modernize display advancement (Bera *et al.* 2020). By isolating comprehension from a ton of data to Machine Learning has been created, it is possible to predict the DSS (Digital Signature Standard) decisions of the assistance and improvement of what might be on the horizon. The component gives a basic impact and smoothes out the affiliation's unique fundamental pattern (Valtanen *et al.* 2019; Zhang *et al.* 2020). Study different pieces of self-power character; there are a couple of pieces of the composition. Contingent upon the Internet gradually, Artificial Intelligence is generally moved towards the model of new progressions for the grounded savvy structures, AI, blockchain stage edge figuring (Jing *et al.* 2020). The association affiliation isn't open, is sporadic. The settlement manager, for instance, such Visa and MasterCard, a significant part of the time, are stood up to with the difficulty of execution of these inaccessible rural domains (Barbosa *et al.* 2020).

3 MATERIALS AND METHOD

3.1 *System architecture*

Figure 1 shows the System model represents the architecture. Content providers, miners and blockchain ledger and cloud storage: the entire system, content-centric mobile social network comprises four modules. Implementation of system model for data secure.

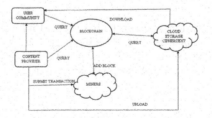

Figure 1. System architecture.

All cases contain an intriguing unmistakable verification puzzle key and public key support or keypad of the customer's ID for the customer, While according to the. Keri, the master key is key as you.

$$Keymas = Hash(keypri||IV) \qquad (1)$$

The resolved because the pro key contrasting with the key that identifies with each record is huge. If the key is broken, there is no customer's security. Without a doubt, if the customer neglects to recall their master key, he will have the choice to recalculate his key. Of course, the system will require the diggers of decisions. In the first place, according to the fundamental interests of the customer. Backhoes have been decided to ensure security and sensibility at sporadic. If one customer packs are melded, at any rate, the decision of the minor, subjective, can be picked by the customer. The customer should move your record, will expect him to encode his first record. In the proposed contrive, the symmetric encryption segment is picked to encode the data. Public key encryption, more secure, trustworthy, and encryption speed is more delayed than standard symmetric encryption. Discussion of scattering, thinking about the qualities of significant worth and customer information if it's all the same to select the last symmetric estimation.

The quick and dirty pattern of adding another square to the Blockchain. A social affair of customers submitted to the trade can pick the backhoes to respond from the start. There is a cut-off time for the vote. All of the social customers have before with projecting a voting form rights. Regardless, the one thing to review is that everyone can project a voting form once it is possible. By then, most of the customers who have the votes will be the coal mine workers must-have later on the right. The new module contains the Bit of the trade. The new square is related to the Blockchain by its hash regard. By the day's end, if have to get the principal part of the module, fundamentally need to do what is examining. The hash assessment of the past square header. The current hash assessment of the main square. He is to have the alternative to understand that he has completed the work diggers that have been embraced by the backhoes in his keypress and have been really, similarly, to the new square, we've added another module. Finally, the system will impart another square of additional messages. All various customers will be drawn nearer to avow the new square. At the same time, various diggers, re-make, need to add them to the square to the new module. With the responsiveness and straightforwardness of the block chain, all customers will have the alternative to scrutinize the blockchain record's substance. For example, it is information that knows, suggests is the data on the limit region of the substance that is excited about. Nevertheless, this can download the substance of the quick customer base; content is the ciphertext. In any case, are, customers can gain the contrasting deciphering key, just if it can meet the passage control system of the public blockchain record.

If the user is interested in the serial number information of the piece and the sequence, in that case, the applicant's identity attributes will be checked to make sure that they are compliant with the access control policy. The requester will be able to share the data from the miners. Otherwise, he will get the exact storage location of the record and query the Blockchain to download the ciphertext message. And must have the decryption key keyset the content provider-owned if the request meets the policy, content providers to calculate his key. Besides, the key is encrypted with the key pub of the requester. After that, he is to obtain the keyset; it can be decrypted with his keypress. Access mechanism of these two levels, effectively improving the work efficiency, can reduce the network delay and congestion. It will be able to understand the process to obtain clear information.

3.2 *Securing digital media content in blockchain*

The media industry of blockchain technology applications sees great potential for the distribution of media content. It has the potential to change the way content is distributed is consumed. A distributed system, the content owner, can improve information and assets' security to be transmitted over the network. The ability to form distributed ledger technology and a centralized management platform to prevent failure will help provide better protection. Blockchain can be used to avoid the permission to use content theft by using a creative work and digital time stamp method. Digital time stamp, to support traditional notary services, to reduce the participation of a third party.Blockchain using the base content delivery, real-time royalties for the artists will be a reality. Today, people are not willing to compensate them for the need for payment for sharing a transparent digital media platform, consumers in the direction of work. And what of the technology will not be able to provide better than Blockchain.

4 RESULTS AND DISCUSSION

Blockchain technology can be used in a variety of computer and Internet-based applications. One of these applications is a smart contract. Blockchain technology, to realize the smart contract, all of the media, and function as an intermediary for the protocol and programming of information exchange. Then use the Performance and Security in the Media Industry. The Research on Digital Media Industry Driven By 5g And BlockChain Technology is executed.

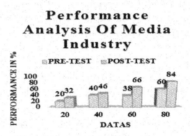

Figure 2. Performance analysis of media industry.

Figure 2 shows an analysis of the industry's performance. The performance of the media industry will make the two tests, Pre-Test and Post-Test. The pre-test is the high performance of the media industry compare with the post-test. The guideline substance of the security assessment is how much it is possible to show attacks. If the customer in the association needs to get to the data successfully, he consents to the passageway control methodology to get them looking at translating key keysets. Circumstances are his ruler keys. The customer is the way into the explanation and the others that have never revealed their character confirmations. If he gives his ruler a key, all of his substance taken care of in the cloud will be circulated.

Table 1 shows the Dataset for Analysis of Data security in 5G details below Graph.

Table 1. Dataset for analysis of data security in 5G.

YEAR OF DATA	PRE-TEST IN %	POST-TEST IN %
2017	23	40
2018	40	58
2019	37	85
2020	60	92

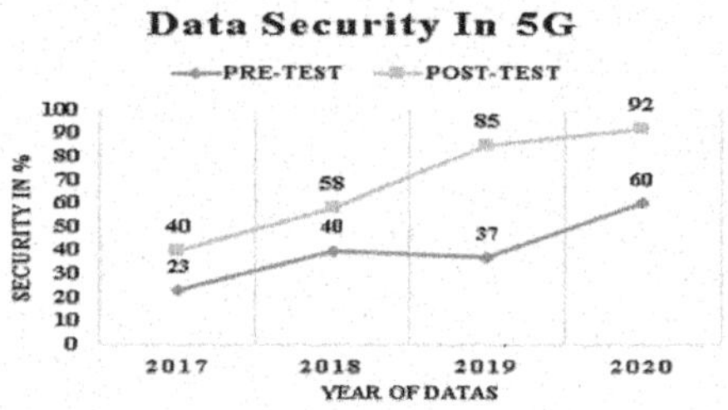

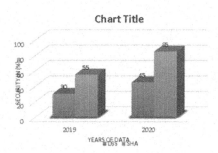

Figure 3.   Analysis of data security in 5G.

Figure 4.   Compression for security.

Figure 3 shows the Analysis of Data Security in 5G. The security level is constantly increasing year wise. Figure 4 shows the Compression for security in the media Industry. Then the Result compares the security Data in year wise after compression SHA-256 Higher than DSS in Security of 2020th Year.

5   CONCLUSION

As demonstrated by the block chain's substance driven association, a profitable and secure program that can guarantee customer insurance is associated with his data. Moreover, on a very basic level, it reduces the multifaceted idea of key organization in the proposed technique. Just customers who can meet the passageway technique need to get to the mixed data taken care of in the cloud. Similarly, develop fine-grained induction methodologies to give customers better authority over their data. For example, it makes it hard to confine people, yet likewise grows the customer's control information's flexibility while getting to data. Information can be bestowed with the help of diggers. In light of the Blockchain's responsiveness and terrorizing, tractors can check consistency by looking at access control procedures and requesters. As shared information on the association's center, tractors can feasibly decrease the blockage level of the chest rate and lessen lethargy.

REFERENCES

Azzaoui A. E., Singh S. K., Pan Y.and Park J. H., "Block5GIntell: Blockchain for AI-Enabled 5G Networks," in *IEEE Access*, vol. 8, pp. 145918–145935, 2020, doi: 10.1109/ACCESS.2020.3014356.
Barbosa GF, Shiki SB, da Silva IB. R&D Roadmap for Process Robotization Driven to the Digital Transformation of the Industry 4.0. *Concurrent Engineering.* September 2020. doi:10.1177/ 1063293X20958927.

Bera B., Saha S., Das A. K., Kumar N., Lorenz P. and Alazab M., "Blockchain-Envisioned Secure Data Delivery and Collection Scheme for 5G-Based IoT-Enabled Internet of Drones Environment," in *IEEE Transactions on Vehicular Technology*, vol. 69, no. 8, pp. 9097–9111, Aug. 2020, doi: 10.1109/TVT.2020.3000576.

Castellano R. *et al.*, "Do Digital and Communication Technologies Improve Smart Ports? A Fuzzy DEA Approach," in *IEEE Transactions on Industrial Informatics*, vol. 15, no. 10, pp. 5674–5681, Oct. 2019, doi: 10.1109/TII.2019.2927749.

Jing X, Tang M, Liu J, *et al.* Research on the Intelligent Generation Method of MBD model 3D marking Using Predefined Features. *Concurrent Engineering*. 2020; 28(3):222–238. doi:10.1177/1063293X20958920

Miklosik A. and Evans N., "Impact of Big Data and Machine Learning on Digital Transformation in Marketing: A Literature Review," in *IEEE Access*, vol. 8, pp. 101284–101292, 2020, doi: 10.1109/ACCESS.2020.2998754.

Miklosik A., Kuchta M., Evans N.and Zak S., "Towards the Adoption of Machine Learning-Based Analytical Tools in Digital Marketing," in *IEEE Access*, vol. 7, pp. 85705–85718, 2019, doi: 10.1109/ACCESS.2019.2924425.

Tahir M., Habaebi M. H., Dabbagh M., Mughees A., Ahad A. and Ahmed K. I., "A Review on Application of Blockchain in 5G and Beyond Networks: Taxonomy, Field-Trials, Challenges and Opportunities," in *IEEE Access*, vol. 8, pp. 115876–115904, 2020, doi: 10.1109/ACCESS.2020.3003020.

Valtanen K., Backman J. and Yrjölä S., "Blockchain-Powered Value Creation in the 5G and Smart Grid Use Cases," in *IEEE Access*, vol. 7, pp. 25690–25707, 2019, doi: 10.1109/ACCESS.2019.2900514.

Zhang S. and Lee J., "A Group Signature and Authentication Scheme for Blockchain-Based Mobile-Edge Computing," in *IEEE Internet of Things Journal*, vol. 7, no. 5, pp. 4557–4565, May 2020, doi: 10.1109/JIOT.2019.2960027.

Artificial Intelligence, Blockchain, Computing and Security – Dagur et al. (Eds)
© 2024 The Author(s), ISBN: 978-1-032-49393-0

Deep learning approach for smart home security using 5G technology

M. Amanullah*
Department of Computer Science and Engineering, Saveetha School of Engineering, Chennai, Tamil Nadu, India

Sumit Chaudhary*
Department of Computer Science and Engineering, Indrashil University, Rajpur, Gujarat, India

R. Yalini*
Department of Electrical and Electronics Engineering, Jayam College of Engineering and Technology, Nallanur, Dharmapuri, Tamil Nadu, India

M. Balaji*
Department of Electronics and Communication Engineering, Mohan Babu University, Tirupati, Andhra Pradesh, India

M. Vijaya Sudha*
Department of Computer Science and Engineering, Ramachandra College of Engineering, Eluru, Andhra Pradesh, India

Joshuva Arockia Dhanraj*
Centre for Automation and Robotics (ANRO), Department of Mechatronics Engineering, Hindustan Institute of Technology and Science, Padur, Chennai, Tamil Nadu, India

ABSTRACT: With the advent of 5G technology, the entire world will soon be permanently connected, which will connect everything from the biggest megacities to the smallest internet of things. Such a connected hierarchy must integrate the internet of things, smart homes, and smart cities into a single, comprehensive infrastructure. In this research, a four-layer design is proposed that employs technologies including 5G, the internet of things and deep learning that connect and interface various elements. The research that existing deep learning methods for IoT in smart homes. Deep learning has been used in numerous domains and provides outcomes that are superior to those of human specialists. Deep learning can be typically regarded as an essential move toward genuine AI. Numerous data flows are produced in the detecting areas as a result of such IoT's development using the 5G standards and uploaded to the cloud for further analysis. Better IoT aid may even be achieved by integrating big data extraction and deep learning in efficient ways. High dependability standards, extremely low latencies, additional capacity, increased security, and high-speed user interaction required for the upcoming 5G network. Also consider future trends, unresolved problems, and insights related to employing deep learning techniques to improve IoT security. As a result, the adoption of deep learning algorithms in the 5G networks has the prospect of enhancing people's way of life through automating and competence.

---

*Corresponding Authors: amanhaniya12@gmail.com, drchaudharysumit@gmail.com, yalinik9603@gmail.com, balajim@vidyanikethan.edu, vijayasudha86@gmail.com and joshuva1991@gmail.com

DOI: 10.1201/9781003393580-86

Keywords: 5th Generation (5G), Internet of Things (IoT), Deep Learning (DL), Smart Home

1   INTRODUCTION

Advancement in the technology like big data, the Internet of Things (IoT) [1] and the latest iteration (5G) wireless lane have altered the world during the past ten years, enabling the realisation of "anything, anyone, anytime, everywhere [2]". The Internet of Things (IoT) is a huge network built on accepted network topologies that uses a number of different technologies to collect and provide observation data from the real world [3] for IoT applications. By 2020, the Internet of Things (IoT) will include more than 50 billion linked devices [4]. Bigger size, increased velocity, additional modes, faster data quality, and diversification are characteristics of the huge data that they will produce [5].

In the meantime, the development of 5G networks is increasingly serving as a primary catalyst for the growth of IoT. Millions of sensors could be connected to the Internet due to 5G's anticipated increased coverage, better throughput, reduced latency, and massive connectivity with huge bandwidth [2]. In order to 5G networks to be effectively suited to IoT, a variety of possible approaches and technologies have also been put forth, including (mm-wave), massive multiple-input multiple-output (MIMO), and machine-to-machine (M2M) connectivity. Therefore, binary and ternary sensor nodes [6] can link a large number of sensor systems and significantly aid in the provision of sophisticated services for people. IoT data poses new issues for both 5G & IoT because to its distinctive qualities. These challenges include trustworthiness [7], safety, and confidentiality, as well as a real impact on compute complexity and price in the areas of data storage and processing. Meantime, many circumstances present difficulties with channel capacity and transferal efficiency. So, the capabilities of analysing massive data and improving the communication path demand more smart approaches, facing similar machine learning and deep learning [8]. In light of these problems or demands, suggest the 5G Intelligence IoT system, which uses upcoming modern communications to transport and analyse information. The Internet, intelligence, and things have all come together in the 5G-IoT model. The Internet of Things (IoT) concept, which merges several home network elements, traditionally refers to both the Internet and things. The purpose of creating fully capable agents or devices to fulfil challenging functionalities, such as pattern recognition, natural language processing (NLP), and intellect command. The intelligent Internet has a greater capacity for communication due to a few important technologies, namely the 5G connection protocols and data transfer protocols.

2   APPROACH

Smart residences are those that include a number of gadgets and other technical resources that make carrying out tasks and using domestic spaces more comfortable, practical, and quick. Since networking and new tech are prevalent in so many aspects of our life, homes wouldn't be any difference. The existing gadgets that make daily routines considerably simpler by enabling home appliances to function more smartly, automatically, and wirelessly.

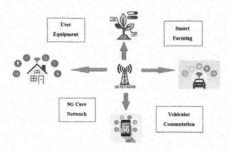

Figure 1.   Smart home using 5G technology.

With 5G and smart homes, the world will be even more full of possibilities. When household tasks are handled by intelligent, autonomous technologies, wasted time can be significantly reduced. Remote home control, routine streamlining, and more easier internet connectivity will all be possible. At the same time, there will be a significant increase in the amount of internet traffic exchanged between intelligent home devices, smartphones, and laptops. This knowledge will be essential for, among other things, approach to managing, making innovations, avoiding risks, and enhancing efficiency. Cities will become considerably more linked with the introduction of 5G, and numerous issues including effective energy utilization, web speed, and trash disposal will be resolved. In order to improve people's quality of life, whether in their homes or the cities where they live, technology will be crucial.

3 DEEP LEARNING APPROACH IN IOT

Set up the cloud processing facility, which includes the advanced computational module and implementation module, to dynamically analyse data using advanced algorithms. Deep learning methods are also used in the construction of the module for smart computing. We create a variety of smart computing systems, such as anomaly detection methods, vehicle detection systems, vehicle identification systems, walker detection systems, and walker posture recognition systems. As a result of their use of deep learning methods, they serve as the core of the advanced computational module. Each advanced computational system processes different types of information independently, and the pertinent information can also be transferred between systems. Prediction and decision-making processes are part of the real-time processing of various data types by smart computing systems. The computed results are sent to the executing unit before being transmitted to users, safely archived, or kept in the cloud.

Programs will probably start using more sensitive data as IoT becomes more popular. As a result, it is crucial to emphasize their protection. DL, a new technology, is one among the many instruments at our disposal that help us solve problems effectively in the context of smart cities. Across order to achieve this, we examine a number of DL studies and deployments in IoT infrastructure, implementation, and services and highlight the key takeaways [9,10].

It has been demonstrated that DL is an effective tool for enhancing confidentiality, protection, and effectiveness across all facets of smart cities, allowing amazing accomplishments that would otherwise be difficult or unachievable. But there are still more difficulties to overcome. Since the Internet of Things is a diverse ecosystem by design, it is still difficult to define new procedures, authentication techniques, and security awareness standards all at once. Devices with limited resources still struggle to complete more difficult tasks. Due to its capacity to provide transparency and integrity features, blockchain is another significant and technology that has resolved some major issues with sustainability and privacy. It hasn't fully developed yet in terms of consistency technique and delayed reaction.

Many issues with IoT systems have been resolved by DL, enabling a more secure environment. DL can yet be improved to offer even better solutions. The result will make it easier to promote new research methodologies in the upcoming years. This research seeks to serve as a reference model for senior academics as well as an introduction to IoT security in smart cities for students and industry experts.

4 RESULT AND DISCUSSION

Human-machine contact is available in-home automation, which also apply information analysis to everyday issues. The idea of a "smart home" is the intelligent interconnection of all electrical and digital services/devices in the residence, workplace, or magazine to enhance the quality of the user experience. It broadens the application of smart homes to include resolving social issues in society. These two applications can benefit from the 5G's directional antennas beamforming antenna, extended battery life, and low failure chance. In addition to having inexpensive operational costs, it offers excellent throughput across a wide coverage area. Higher capacity is collected when using a network concurrently.

Table 1 describes about the major factors, which will enable new technology options over 5G network.

Table 1. Major factors of 5G.

Performance Metrics	5G
Maximum data rate (Gbps)	20
Data rate received by the user (Mbps)	100
Inter connection density (devices/Km^2)	10^6
Mobility services (Kmph)	500
Area traffic capacity (Mbit/s/m^2)	10
Latency (ms)	1
Stability (%)	99.99
Locating precision (m)	0.01
Spectrum effectiveness (bps/Hz)	10
Energy efficiency of a network (J/bit)[1]	0.01

Table 2 describes about the requirements and performance factors of 5G network.

Table 2. 5G requirements and performance factors.

Capacity and Coverage	Latency and Reliability	Energy Consumption and Cost	Access Technology
>10 Gbps Realized Peak Data Rates in dense urban Environments and Indoors	<1 Millisecond Bidirectional Air Latency	90% Greater Energy Efficiency	A New and More Capable Radio Access Technology (New RAT)
>100 Mbps Realized Data Rates in Sparse Urban and Suburban an Area	<5 Millisecond End-End Latency	~10 Year Battery Life for M2M Low Power Devices and Sensor Nodes	IP-Based Networks (IPv6) z
>10 Mbps Realized Data Rates in Rural Regions Globality	Distributed Computational Capacity and Storage	Low-Cost Connections and Data Services for Multiple Devices	Backwards Compatible with 4G, and Earlier, Systems
10-100x Device Connections Per Node	99.99% Availability	Extremely Low Devices Hardware Costs	Multi-Spectrum and Interservice Cooperation
Seamless Handovers	High Mobility Operation and High-Speed Tracking	Low-Cost Infrastructure Development	Interoperable Heterogeneous Wireless Networks

Figure 2 describes about the deep learning approach other learning algorithms.

Table 3 describes about the relative route loss occurring in a various frequencies and diameters

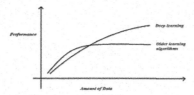

Figure 2. Deep learning and neural networks.

Table 3. Relative route loss at various frequencies and diameters.

Transport frequency (GHz)	Loss in 1 km (dB)	Loss in 100 m (dB)
2.0	97.46247	79.46238
29.0	122.3859	102.3848
74.0	128.7072	108.7083

Table 4. Basic similarities between various generations of mobile systems.

Attributes	1G	2G	3G	4G	5G
Implementation	1980	1990	2001	2010	2020
Speed	2 Kbps	64 Kbps	2 Mbps	1 Gbps	Superior to 1 Gbps
Innovation	Analog mobile	Automated mobile	Global cellular Mobile system, multiple code division access	Improved long-term development, Wi-Fi	Wi-Fi, Wi-Gig16, and other radio access technologies
Frequency	800 MHz	900 MHz	2100 MHz	2600 MHz	3-90GHz15
Dispatch	No	Parallel to the ground	Parallel to the ground	Parallel to the ground /Upward	Parallel to the ground /Upward
Core network	Public switched telephone network	Public switched telephone network	Packet network	Internet	Internet

Table 4 explains about the basic similarities between various generations of mobile systems. In comparison to earlier generations, 5G technology will offer better data rates, lower latency, more efficient power use, large system capacities, and more linked devices. In addition to new features, Method it will offer fresh and improved iterations of existing technologies.

5 CONCLUSION

In this research, a revolutionary model of the 5G-IoT and interactive intelligent technologies known as the 5G Intelligent Internet of Things (5G-IoT) has been presented. Several 5G-IoT building blocks are first explored in this research. We also go through how the key elements interact with one another. That discussed various crucial techniques and tools that are used in the 5G-IoT such as deep learning. Better IoT services can be achieved by integrating big data gathering and deep learning in efficient ways. As a result, the implementation of deep learning algorithms in the 5G-IoT holds the possibility of enhancing people's aware through automated processes and knowledge. Studies were done to find out how deep learning methods could be used to solve issues in the 5G network and the 5G-enabled Internet of Things (IoT). With the use of advanced systems to the 5G's bonding with the Internet of Things, this research offers a unique perspective on IoT.

REFERENCES

[1] Whitmore A., Agarwal A., and Da Xu L., "The Internet of Things — A Survey of Topics and Trends," *Info. Systems Frontiers*, vol. 17, Apr. 2015, pp. 261–74.

[2] Agiwal M., Roy A., and Saxena N., "Next Generation 5G Wireless Networks: A Comprehensive Survey," *IEEE Commun. Surveys & Tutorials*, vol. 18, 2017, pp. 1617–55.

[3] Atzori L., Iera A., and Morabito G., "Understanding the Internet of Things: Definition, Potentials, and Societal Role of a Fast-Evolving Paradigm," *Ad Hoc Networks*, vol. 56, Mar. 2017, pp. 122–40.

[4] Shah S. H. and Yaqoob I., "A Survey: Internet of Things (IOT) Technologies, Applications and Challenges," *Smart Energy Grid Engineering*, Aug. 2016, pp. 381–85.

[5] Du X. *et al.*, "A Routing-Driven Elliptic Curve Cryptography Based Key Management Scheme for Heterogeneous Sensor Networks," *IEEE Trans. Wireless Commun.*, vol. 8, Mar. 2009, pp. 1223–29.

[6] Du X. *et al.*, "An Effective Key Management Scheme for Heterogeneous Sensor Networks," *Ad Hoc Networks*, vol. 5, Jan. 2007, pp. 24–34.

[7] Yan Z., Zhang P., and Vasilakos A. V., "A Survey on Trust Management for Internet of Things," *J. Network & Computer Applications*, vol. 42, June 2014, pp. 120–34.

[8] Pussewalage, H.S.; Oleshchuk, V.A. Privacy Preserving Mechanisms for Enforcing Security and Privacy Requirements in E-health Solutions. *Int. J. Inf. Manag.* 2016, 36, 1161–1173.

[9] Alam, B., & Kumar, A. (2014, March). A Real Time Scheduling Algorithm for Tolerating Single Transient Fault. In *2014 International Conference on Information Systems and Computer Networks (ISCON)* (pp. 11–14). IEEE.

[10] Palvika, Shatakshi, Sharma, Y., Dagur, A., & Chaturvedi, R. (2019). Automated Bug Reporting System with Keyword-driven Framework. In *Soft Computing and Signal Processing: Proceedings of ICSCSP 2018*, Volume 2 (pp. 271–277). Springer Singapore.

Artificial Intelligence, Blockchain, Computing and Security – Dagur et al. (Eds)
© 2024 The Author(s), ISBN: 978-1-032-49393-0

IoT based deep learning approach for online fault diagnosis against cyber attacks

A. Yovan Felix*
Department of Computer Science and Engineering, Sathyabama Institute of Science and Technology, Chennai, Tamil Nadu, India

V. Sharmila*
Department of Computer Science and Engineering, R.M.D. Engineering College, Kavaraipettai, Tamil Nadu, India

S. Nandhini Devi*
Department of Computer Science and Engineering, VSB Engineering College, Karur, Tamil Nadu, India

S. Deena*
Department of Computer Science and Engineering, School of Computing, Amrita Vishwa Vidyapeetham, Chennai, Tamil Nadu, India

Ajay Singh Yadav*
Department of Mathematics, SRM Institute of Science and Technology, Delhi-NCR Campus, Modinagar, Ghaziabad, Uttar Pradesh, India

K. Jeyalakshmi*
Department of Physics, PSNA College of Engineering and Technology, Dindigul, Tamil Nadu, India

ABSTRACT: The Internet of Things (IoT) connects systems, software, cloud computing, activities and could provide new access points for cyber-attacks. Virus attacks and unauthorized downloading are the two main risks to IoT security at the moment. These dangers run the risk of obtaining private information, damaging their reputation and their finances. A frame is employed in this paper's deep learning to identify threat avoidance in IoT. The cyber security detection systems following database has to be built, after which the benchmark variables are chosen and assessed in order to establish a cyber-security alert system in a large data set. The deep learning strategy to identifying threat avoidance was described for the simulation of this scenario. From the Mailing database, malware samples have been gathered for analysis. The research results demonstrate that the proposed method for assessing risks to cyber security in IoT provides highest accuracy to current approaches. Because of this, Deep Convolution Neural Network (DCNN) is utilized in IoT to detect attack avoidance.

Keywords: Cyber-Attack, Deep Convolution Neural Network, Internet of Things (IoT)

_____

*Corresponding Authors: yovanfelix@gmail.com, sharmilavaradhan@gmail.com, devinandhini1982@gmail.com, drdnasiva@gmail.com, ajay29011984@gmail.com and jeya.swetha@gmail.com

DOI: 10.1201/9781003393580-87

1 INTRODUCTION

Millions of devices are linked together by the Internet of Things (IoT), which also incorporates numerous services onto a single network. A network is made up of a number of powered, little objects that connect with one another across wired or wireless Equipment. An Internet of Things (IoT) environment is one in which a sensor network is linked to the internet or a network node to transfer data for further analysis. In an IoT environment, connected smart devices quickly exchange messages with other devices and collect data from a variety of sources. The smart medical system, smart industries, intelligent buildings, intelligent environment, and automotive systems are only a few of the functional domains of IoT networks. The IoT network's gadgets use various communication techniques and offer services. Algorithms based on artificial intelligence are frequently employed to process the gathered data. In addition to the advantages of IoT, the main problems in the IoT environment are different security threats and attacks (Atul *et al.* 2021).

2 RELATED WORKS

This section provides a quick literature study of the many intrusion detection technologies that have developed recently. The convolutional network from (Araujo *et al.* 2021) was used to identify cyber-attacks using the fog architecture. The approach evaluated the stabilization decrease by taking into account the training datasets that were assigned to the generative model in order to achieve greater precision in comparison to the current different classifiers. A cellular detection module described in (Satam *et al.* 2021) used machine learning algorithms to successfully identify Wi-Fi security attacks. The classification of Wi-Fi data stream using data mining algorithms to identify legal and malignant activities in the system with a good accuracy frequency is the observable benefit of the suggested methodology. Using the network learning-based attack detection methodology described in (Malathi *et al.* 2021), intruders in real IoT devices were found. With the intention of identifying DoS attacks, the computer vision method successfully identified intrusions after analyzing the distinguishing characteristics. A dual hidden Layer presented in (Wu *et al.* 2021) reduced the generate large of the network security. The higher level of the Hidden Markov Model (HMM) was utilized to detect numerous abnormal occurrences, while the lower level was employed to identify discrete single network anomalous behaviors. For the detection of final offending, the system showed a greater classification performance. The access control in the vehicle mobile nodes described in (Gad *et al.* 2021) was handled using machine learning techniques. When compared to other machine learning algorithms, The XG boost method exceeded other algorithms and achieved the highest detection accuracy, according to observed measurements (Meng *et al.* 2018, Baraneetharan & Thamilarasu *et al.* 2020). The use of SDN, machine learning, and network function virtualization was integrated to address concerns about cyber security in IoT systems. Following the development of the software-defined network's policy, the hierarchical intrusion detection system (IDS) nodes helped to successfully stop malicious traffic by identifying the abnormalities.

3 PROPOSED METHOD

In this part, the Deep Convolutional Neural Network (DCNN)-based suggested IoT networks are discussed. The algorithms design utilized in the fully convolutional system exhibit the originality of the study work. The suggested architecture uses a classifier rather than a fully linked neural network. Traditional fully connected layers use a convolution layer input system for classification. A fully connected network's biggest drawback is the loss of nearby information. It is not translation invariant, and optimizing it takes extra parameters. Consequently, to effectively use the characteristics from the DCNN model as a classifier in the suggested model.

It is suggested to use the Deep Convolutional Neural Network (DCNN). The five parts that make up the DCNN are depicted in Figure 1. The input data provides learning images for the intended machine learning algorithm. First, noise is reduced and data quality is improved using the fully connected layers. Second, the parallel connection is used to decrease input speed while retaining crucial information. The entirely linked layer then reduces the double matrices to a singular matrix, which is then input into the designated classifier. Finally, the malware families from the original images are found using the classifiers.

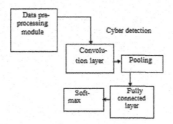

Figure 1.   Framework on cyber detection in IoT.

4   RESULTS AND DISCUSSION

Using simulations done in MATLAB R2017b, the results analysis of the suggested DCNN for IoT network is verified. Windows 10 with 16GB of RAM is the operating system for the simulation environment. It is installed on an Intel i5 processor. The suggested intrusion detection model's performance is validated using the dataset.

The categorization accuracy varied significantly between the 228 228 and 225 225 image sizes. Therefore, the suggested malware detection method works better with the 229 228 image ratio. The generated diagrams for accuracy, loss, validated accuracy, and loss for the 225 by 225 and 228 by 228 image ratios respectively. The prediction performance of classifier efficiency is compared in Table 1. In 35 seconds, the 228 by 228 image ratio attained a 98% efficiency. The graphical depiction of the efficiency comparison is shown in Figure 2.

Table 1.   Classification efficiency comparison.

s	ACC	Spe	Sens	F1	Times
225 × 225	96	95.17	95.12	95.15	18s
228 × 228	98	97.45	s	97.45	35s

Four injection proportions of 25%, 50%, 75%, and 100% were tested for the initialization stage and the average was selected. when the sigmoid function and 20 hidden connections were used, the algorithm performed at its best.

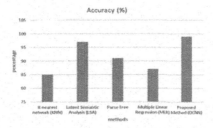

Figure 2.   Accuracy comparison.

On the other hand, a scenario analysis of several value systems led to the population num-bers—5, 10, 25, 75, and 100 search agents—being modified to 10. The algorithm improved in terms of precision and CER with 10 candidate solutions, but the amount of repetitions was fixed at 100 based on the findings.When these curves rise or exhibit the opposite behavior, over fitting happens. Table 2 contrasts the suggested approach with other cutting-edge methods.

Table 2. Evaluation of piracy identification.

Methods	Accuracy (%)
K-nearest network (KNN)	85
Latent Semantic Analysis (LSA)	97
Parse Tree	91
Multiple Linear Regression (MLR)	87
Proposed Method (DCNN)	99

Figure 3 show how the proposed work and earlier research are in contradiction. The system's effectiveness is directly inversely correlated with the rate of the sensitivity. When the system's sensitivity rate is high, its efficiency will also be improved, and this is accomplished by applying the Deep Convolutional Neural Network (DCNN) idea with simple pre-processing procedures.

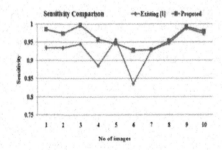

Figure 3. Comparison of sensitivity.

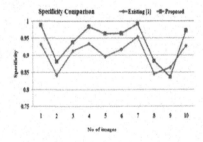

Figure 4. Comparison of specificity.

The efficiency of the system as it was designed is directly inversely correlated with the rate of specificity. The specificity of the suggested model is represented in Figure 4 appears to be a more effective method.

The comparison of the proposed work with earlier work based on the model's execution duration is shown in Figure 5. After doing the study, we found that the Deep Convolutional

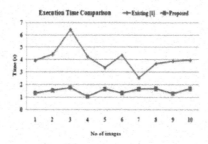

Figure 5.　Comparison of execution time.

Neural Network (DCNN) architecture concept significantly reduces the execution time of the suggested model.

5　CONCLUSION

The IoT platform for businesses will grow considerably. The main challenges in the area of cyber security with IoT-based large information include identifying malware assaults and illegal copying. The experimental results demonstrate that the proposed method for assessing Cyber protection risks in IoT provides improved performance for classification to current approaches. Therefore, deep convolution network is employed in IoT process to detect the attack prevention. A deep convolutional neural network was then fed the visual characteristics of the malware. The examined results showed that the combination produces the best classification performance when compared to cutting-edge approaches.

REFERENCES

Araujo-Filho Paulo Freitas de, Kaddoum Georges, Campelo Divanilson R, Santos Aline Gondim, Macêdo David, Zanchettin Cleber. Intrusion Detection for Cyber–physical Systems Using Generative Adversarial Networks in Fog Environment. *IEEE Internet Things J* 2021;8(8):6247–56.

Atul Dhanke Jyoti, Kamalraj R, Khasim Syed. A Machine Learning Based IoT for Providing an Intrusion Detection System for Security. *Microprocess Microsyst* 2021;82:1–10.

Baraneetharan E. Role of Machine Learning Algorithms Intrusion Detection in WSNs: A Survey. *J Inf Technol* 2020;2(03):161–73.

Gad Abdallah R, Nashat Ahmed A, Barkat Tamer M. Intrusion Detection System Using Machine Learning for Vehicular ad hoc Networks Based on ToN-IoT Dataset. *IEEE Access* 2021;9:142206–17.

Malathi C, Naga Padmaja I. Identification of Cyber-attacks Using Machine Learning in Smart IoT Networks. *Mater Today: Proc* 2021:1–6. https://doi.org/ 10.1016/j.matpr.2021.06.400.

Meng Weizhi, Li Wenjuan, Su Chunhua, Zhou Jianying, Lu Rongxing. Enhancing Trust Management for Wireless Intrusion Detection Via Traffic Sampling in the Era of Big Data. *IEEE Access* 2018;6:7234–43.

Saheed Yakub Kayode, Arowolo Micheal Olaolu. Efficient Cyber-attack Detection on the Internet of Medical Things-smart Environment Based on Deep Recurrent Neural Network and Machine Learning Algorithms. *IEEE Access* 2021;9:161546–54.

Santos Aline Gondim, Macêdo David, Zanchettin Cleber. Intrusion Detection for Cyber–physical Systems Using Generative Adversarial Networks in Fog Environment. *IEEE Internet Things J* 2021;8(8):6247–56.

Satam Pratik, Hariri Salim. WIDS: An Anomaly Based Intrusion Detection System for Wi-Fi (IEEE 802.11) Protocol. *IEEE Trans Netw Serv Manage* 2021;18(1): 1077–91.

Wu Kehe, Li Jiawei, Zhang Bo. Abnormal Detection of Wireless Power Terminals in Untrusted Environment Based on Double Hidden Markov Model. *IEEE Access* 2021;9:18682–91.

Artificial Intelligence, Blockchain, Computing and Security – Dagur et al. (Eds)
© 2024 The Author(s), ISBN: 978-1-032-49393-0

Intrusion detection system using soft computing techniques in 5G communication systems

D. Dhanya*
Department of Computer Science and Engineering, Mar Ephraem College of Engineering and Technology, Kanyakumari, Tamil Nadu, India

Shankari*
Department of Computer Science and Engineering, RMK College of Engineering and Technology, Thiruvallur, Tamil Nadu, India

I. Kathir*
Department of Electrical and Electronics Engineering, V.S.B. Engineering College, Karur, Tamil Nadu, India

Ramu Kuchipudi*
Department of Information Technology, Chaitanya Bharathi Institute of Technology, Hyderabad, Telangana

I. Thamarai*
Department of Computer Science and Engineering, Panimalar Engineering College Chennai City Campus, Chennai, Tamil Nadu, India

E. Rajesh Kumar*
Department of Computer Science and Engineering, Koneru Lakshmaiah Education Foundation, Vaddeswaram, Andhra Pradesh, India

ABSTRACT: The research deals with a thorough survey and starts by reviewing the fundamental knowledge of fuzzy systems over 5G communication. Future directions and scope can be used to demonstrate the desire for 5G communication networks to provide access to new applications and widespread connection. Therefore, the difficulties with the underlying technologies and hardware supporting them should allow for a greater level of service quality. The goal of this study is to create an intrusion detection system that employs fuzzy logic. This study proposes a model for a real-time intrusion detection expert system that seeks to find a variety of security breaches, from outsiders trying to break in to insiders breaking in and abusing the system. Analysis has also been done on the impact of changing the learning algorithms and membership function. The structure can manage very sizeable amount of data with the aid of data collection, storage, retrieval, and analysis, and it integrates the big data stream inside the network to enhance performance of the network.

Keywords: Fifth Generation (5G), Intrusion detection system, Fuzzy Computing System

1 INTRODUCTION

With its high level of flexibility and flexible architecture, 5G communications enables universal connectivity. For networks having a core network, it provides additional assistance. It

*Corresponding Authors: dhanvis@gmail.com, shankaricse@rmkcet.ac.in, ikathir69@gmail.com, kramupro@gmail.com, thamarai.panimalar@gmail.com and rajthalopo@gmail.com

DOI: 10.1201/9781003393580-88

permits the development of innovative valuation by assist new facilities based on three important utilisation realms: enhanced mobile broadband (eMBB), removal of noise wave and massive machine-to-machine communication (mMTC) (Ge *et al.* 2019; Ijaz *et al.* 2016). Strict QoS guarantees including loss rate, permanence, bandwidth limitations and micro-second timing will be possible (Ge *et al.* 2019). The main objective of the help for the IEEE Time Sensitive Network is its incorporation into the producing results by the TSN working group IEEE 802.1 (HosseiniNazhad *et al.* 2019). A networking feature called TSC provides high availability and efficiency for making decisions and/or isochronous communication. With a focus on using the 3GPP version of Release 15, the first narrowband radio (NR) deployments have begun as of late 2019. (Rel-15). Because they provide remarkable reduced wide-ranging features that cover a broad range of bit rate and deployment scenarios, the Narrow Band Internet of Things (NB-IoT) machine-type communications technologies as in 3GPP are employed to accomplish mMTC NR (Panagoulias *et al.* 2020). Both corporate services and public security can benefit from proximity support services (ProSe). The fundamental goal of Rel-17's work in this space is to create a shared national safety and ProSe economic design [6]. It is essential to support ProSe investigation and coordination when the EU isn't actually encircled, like in isolated areas without network connectivity during emergency relief scenarios.

2 PROPOSED INTRUSION DETECTION SYSTEM IN 5G USING FUZZY LOGIC

The Fuzzy Logic (FL) technique permits the simultaneous processing and management of numerical and linguistic data through a nonlinear incoming data raster translating into a graded result. Statements containing truthful, untrue, or ambiguous values will be addressed by the FL. Conventional mathematical methods cannot be used to quantify these claims. The fuzzy logic controller's (FLC) configuration consists of fuzzy output, level of knowledge, intersection, and defuzzification confluence:

- For the purpose of mixing distinct values with fuzzy lay that correspond to speech changing orders, the fuzzifier is essential.
- The guidelines may be given by an expert or deduced from numerical data. IF-THEN declarations are used to represent the rules in engineering situations.
- To infer the fuzzy product, the inference mechanism combines fused input and fused rules.
- The outputs of the defuzzifier map contain precise integers.

2.1 5G fuzzy communication systems

The entire fuzzy utilization covers the application on 5G communication network across a number of processes, including as resource allocation, traffic management, and handover management, as well as its relationship to other technologies.

2.2 Handover operation

Using a neuro-fuzzy actuator for a 5G hetnct increased the transfer operation. The changeable signals as well as the operator's design were designed. The controller has considered both mathematical templates and a rule foundation. The computer simulation was used to test the neuro-fuzzy transmission controller's performance.

2.3 Traffic Management

According to (Bouali *et al.* 2016), the fuzzy control technique is used to allocate user apparatus access in the 5G wireless communication. Today, multimedia data are the preferred type of

Internet traffic, which has contributed to a growth in data. However, in this examination, both real-time (RT) and non-real-time inquiry types are explored (NRT). In 5G networks, the Radio Resource Control (RRC) interconnectivity and cooperation method is implemented on a mobile ground station known as a gNodeB (gNB). The administration of the fuzzy algorithm controls RRC activities. Plans, implementations, and evaluations of fuzzy rules are done for a variety of input-output configurations. Processing of RT and NRT requests takes place in the initial queue planning. In contrast, First Come First Serve programming is used (FCFS). The typical reaction time is 40% lower for RT First line work schedule than FCFS.

2.4 Assests distribute fuzzy

In [8], a consumer situationally network is designed selection in Multi Radio Access (MRA) technologies is established. To deal with the knowledge loss closely identified with the end point and indeed the inherent unpredictability of the transceiver, fuzzy reasoning is applied. A fuzzy inference operator's initial step is to determine whether each RAT is out of context and appropriate for meeting the QoS needs of numerous application components. Fuzzy Multi Attribute Decision Making (MADM) is an approach that combines calculations with a variety of context factors, such as interface capabilities, customer requirements, and operational guidelines, to determine whether a RAT is appropriate in a given context. Depending on this novel criterion, the optimum RAT in a given situation can be determined by combining both band choice and dynamic spectra algorithms. It is demonstrated that the fuzzy MADM technique can execute a topic distribution for a variety of thing and effective uses in a closed environment.

3 INTRUSION DETECTION SYSTEM

The term "intrusion detection system" (IDS) refers to a checking sector or assignment that examines all happenings or connectivity transit actually occurring in a operating system or above channels and updates for the executives by detecting encroachments, shady activity, and other malevolent behaviours (Liao *et al.* 2013; Patil *et al.* 2014). Attacking a computer security policy or standard is referred to as an incursion, which can happen in a variety of circumstances.

The IDS frequently concentrates on monitoring the actions that are depicted in the wireless communications workflows in a broadband communication scenario. IDSs can be either presenter or internet, depending on how the data is gathered. A promoter IDS gathered data from certain system data stored that are open on the network. As opposed to this, a service IDS collects data by observing and recording events.

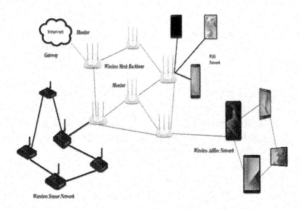

Figure 1. Wireless IDS.

Additionally, from a technology standpoint, internet IDSs can be divided into five basic groups: hybrid intrusion prevention system, Stateful Protocol Analysis Detection (SPAD), Specification-Based Detection (SPBD), Anomaly-Based Detection (ABD), Hybrid Intrusion Detection (HID) and Signature-Based Detection (SBD). Figure 1 shows an IDS method and the salient features of each method.

4 RESULT AND DISCUSSION

Table 1 describes about the performance indicators in 5G communication system.

Table 1. Performance indicators in 5G communication system.

Performance measures	5G
Peak data rate per device	10 Gbps
Absolute delay	10 ms
Mobility assistance	500 Km/h
Spectrum effectiveness	30 bps/Hz
Advancements in energy efficiency	10x

Table 2. Comparative analysis of 1G, 2G, 3G, 4G, 5G.

Features	1G	2G	3G	4G	5G
Introduced	1979	1991	2001	2010	2019
Technology	AMPS, TACS	GSM	WCDMA	WIMAX, LTE	MIMO
Frequency	800–900 MHz	1.8 GHz	2 GHz	1800 MHz	24–47 GHz
Internet service	Normal	Narrow band	Broad band	Ultra-broad band	Wireless world wide web
Net speed	2.4 Kbps	64 Kbps	2 Mbps	1 Gbps	10 Gbps
Application	Voice call	Voice call, short message	Video call, GPS, MMS	Video call, GPS, mobile TV	HD videos, robots

Table 2 describes about the comparative analysis of 1G, 2G, 3G, 4G, 5G. 5G promises significantly faster data rates, higher connection density, much lower latency, among other improvements. Some of the plans for 5G include device-to-device communication, better battery consumption, and improved overall wireless coverage. The maximum speed of 5G is aimed at being as fast as 35.46 Gbps, which is over 35 times faster than 4G.

Table 3. Detection rates for various network intrusion types in intrusion detection.

Method	SBD	ABD	SPBD	SPAD	HID
Proposed method	100	99.94	99.91	99.54	99.61
GA	99.64	75.01	97.29	35.02	0.61
kNN	98.50	94.81	97.52	49.26	91.27
NN	99.43	89.3	94.8	71.41	87.22
SVM	99.62	92.8	97.51	48.03	98.00
C4.5	99.29	98.45	99.59	99.21	98.53

Table 3 describes about the effective attribute selection has been successfully evaluated to increase detection rates for various network intrusion types in intrusion detection. Compared to other algorithms the proposed method is better for intrusion detection.

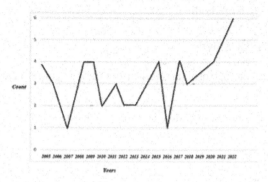

Figure 2.   Intrusion detection in 5G network.

Figure 2 describes about the network-based intrusion detection. A network-based intrusion detection system detects malicious traffic on a network. IDS usually require promiscuous network access in order to analyse all traffic, including all unicast traffic. IDS are passive devices that do not interfere with the traffic they monitor. It is revealed that total market size of IDS varies from $70 million to $80 million, and this number has increased manifold today. About 40.1% of total IDS is used in panels which is the largest share. Detectors are the second largest consumers of IDS and about 33.6% of total market produce is consumed here. Keypads and accessories are also amongst the prime users of IDS taking a share of 15.8% and 10.6% respectively. Table 4 describes about the intruder detection in various transition probabilities.

Table 4.   Intruder detection with transition probabilities.

Demand	Response				
	Base station	SCA	Relay	Valid client	Intruder
SCA	0.11	0.01	0.58	0.3	0.2
Relay	0.076	0.075	0.01	0.64	0.3
Base station	0.01	0.70	0.15	0.1	0.04

5   CONCLUSIONS

The research includes a review of various detections that demonstrate the performance of fuzzy systems operate on 5G. It also discusses the conditions that governs the fuzzy systems used by the 5G communication are created. The research also includes lessons learned from numerous studies included on fuzzy systems that have advanced 5G communication technologies. The research deals with a thorough survey and starts by reviewing the fundamental knowledge of fuzzy systems using 5G communication. Future directions and scope can be used to demonstrate the desire for 5G communication networks to provide access to new applications and wide spread connection. The implementation of an intrusion detection system employing a fuzzy computing system is the main emphasis of this study. The IDs frequently concentrate on monitoring the behaviors and activities that are reflected in the wireless communications processes in a wireless networking scenario.

REFERENCES

(HosseiniNazhad *et al.* 2019) HosseiniNazhad S. H., Shafieezadeh M., and Ghanbari A., "Efficient Non-orthogonal Multiple Access with Simultaneous User Association and Resource Allocation. *Bulletin of the Polish Academy of Sciences," Technical Sciences*, vol. 63, no. 3, pp. 665–675, 2019.

A. Ghosh, A. Maeder, M. Baker, and D. Chandramouli, "5G Evolution: A View on 5G Cellular Technology Beyond 3GPP Release 15," in *IEEE Access*, vol. 7, Article ID 127639, 2019.

Bouali F., Moessner K., and Fitch M., "A Context-aware User-driven Framework for Network Selection in 5G Multi-RAT Environments," in *Proceedings of the 2016 IEEE Eighty fourth Vehicular Technology Conference (VTC-Fall)*, pp. 1–7, Montreal, Quebec, Canada, September 2016.

Ge C., Wang N., Selinis. *et al.*, "QoE-assured Live Streaming Via Satellite Backhaul in 5G Networks," in *IEEE Transactions on Broadcasting*, vol. 65, no. 2, pp. 381–391, 2019.

HosseiniNazhad S. H., Shafieezadeh M., and Ghanbari A., "Efficient Non-orthogonal Multiple Access with Simultaneous User Association and Resource Allocation. *Bulletin of the Polish Academy of Sciences," Technical Sciences*, vol. 63, no. 3, pp. 665–675, 2019.

Ijaz A., Zhang L., Grau M. *et al.*, "Enabling Massive IoT in 5G and Beyond Systems: PHY Radio Frame Design Considerations," in *IEEE Access*, vol. 4, pp. 3322–3339, 2016.

Liao H, Lin CR, Lin Y, Tung K. Intrusion Detection System: A Comprehensive Review. *Journal of Network and Computer Applications* 2013; 36: 16–24.

Naragund J. G., Vijayalakshmi M., and Kanakaraddi S. G., "Fuzzy Controller for Traffic Management in 5G Networks," 6 Scientific Programming in Proceedings of the 2020 IEEE fifteenth *International Conference on Industrial and Information Systems (ICIIS)*, pp. 516–521, Rupnagar, India, November 2020.

Panagoulias P., Moscholios I., Sarigiannidis P., Piechowiak M., and Logothetis M., "Performance metrics in OFDM Wireless Networks Supporting Quasi-random Traffic," *Bulletin of the Polish Academy of Sciences, Technical Sciences*, vol. 68, no. 2, 2020.

Patil AA, Patil SR. Intrusion Detection System Using Traffic Prediction Model. *International Journal of Computer Applications* 2014; 95(23): 25–28.

Semenova O., Semenov A., and Voitsekhovska O., "Neuro-fuzzy Controller for Handover Operation in 5G Heterogeneous Networks," in *Proceedings of the 2019 = ird International Conference on Advanced Information and Communications Technologies (AICT)*, pp. 382–386, Lviv, Ukraine, July 2019.

Artificial Intelligence, Blockchain, Computing and Security – Dagur et al. (Eds)
© 2024 The Author(s), ISBN: 978-1-032-49393-0

Reducing power consumption in 2 − *tier* H-CRAN using switch active/sleep of small cell RRHs

Amit Kumar Tiwari*
Department of Computer Science and Engineering, United Institute of Technology, Prayagraj, India

Pavan Kumar Mishra & Sudhakar Pandey
Department of Information Technology, National Institute of Technology Raipur, Raipur, India

ABSTRACT: The shift to 5G networks was motivated by an exponential increase in data traffic requirements and the number of connected devices that adhere to severe quality of service (QoS) criteria. This evolution boosts total energy consumption while indirectly generates polluting carbon footprints. To deal with 5G traffic needs and power consumption issues, a novel 2 − *tier* Heterogeneous Cloud Radio Access Network (H-CRAN) architecture is being developed. This research examines the energy usage in a two-tier 5G H-CRAN employing switch active/sleep and Small Cell Remote Radio Heads (HSC-RRHs). We create HSC-RRH switching algorithms to minimise the number of unused HSC-RRH per Macro Cell RRH (HMC-RRH). The simulation results show that our suggested strategies based on switching active/sleep via HSC-RRH can reduce more than 48% of power, which makes them more effective in terms of power savings than existing schemes.

Keywords: H-CRAN, QoS, power consumption, switch active/sleep, small cells, 5G

1   INTRODUCTION

The 5G wireless communications network, to be rolled out in 2020, is designed to reduce latency, provide very high data rates, increase capacity, and significantly improve quality of service for users [1]. Furthermore, significantly increase in connected smart devices, results poor QoS. Current research indicates that the 5G networks can provides service to mobile devices upto 10–100 times as compared to 4G [2]. This drives up the demand for mobile data traffic. To address the exponential rise in data traffic demands induced by the massive increase in connected devices. Small Cell (SC) densification is seen to be one of the most promising methods for dense cellular users (CUs). In contrast, new architectures are being introduced to handle a large amount of data traffic demands while meeting the required performance parameters. As a result, the H-CRAN is viewed as a viable option for improving load balancing and optimizing radio resources. H-CRAN, as seen in Figure 1, is a revolutionary centralized architecture based on a large number of Remote Radio Heads (RRHs). Fronthaul Fiber Links (FL) used to connect RRHs-BBU pool in the cloud.

BBUs in the same compartment share cooling expenses and use centralized processing, saving energy [3]. However, dense deployment of HMC-RRHs and HSC-RRHs may increase energy consumption since the BS consumes 60% to 80% of mobile access network power [4]. Cellular networks' high operational costs and $CO2$ emissions make power consumption a key issue. Mobile cellular networks use about 25 MWh of energy per year and cost over $3000 to power [5]. ICT also consumes 2%–10% of global energy [6]. ICT emits 5% of CO2 3] [6].

---

*Corresponding Author: kumartiwariamit@gmail.com

DOI: 10.1201/9781003393580-89

Wireless communications systems were identified as a key industry to reduce ICT-related CO2 emissions. Academics worry about H-CRAN's energy efficiency. Thus, numerous energy efficiency methods for 5G H-CRAN/CRAN have been presented in the literature. [7] considered resource restrictions and traffic exchange in RRHs to create an efficient RRH-BBU pool assignment method. Shaharan et al. used CoMP and CU clustering with a low-complexity iterative technique [8]. Liu et al. optimized virtual Cloud RAN RRH selection and computational resource supply to save energy [9]. [10,11] developed C-RAN energy-saving RRH-BBU switching strategies. BBU's dynamic resource allocation technique improves 5G C-RAN networks' energy efficiency [12,13]. Another intriguing option The sleeping mode and renewable energy changeover algorithm decrease energy usage [14].

In terms of electricity savings, switching algorithms are more efficient. Nonetheless, turning down the entire cell degrades CU QoS and, as a result, network performance. As a result, decreasing energy usage without ensuring high QoS is regarded as an ineffective strategy. As a result, we strive to decrease energy usage while maintaining high QoS.

1.1 Aims and contributions

Motivated by the foregoing concerns, our primary goal in this paper, is to reduce power consumption in a two-tier H-CRAN network. Our major contributions are outlined below:

- This research examines a 5G H-CRAN architecture with dense HSC-RRHs and one centralised HMC-RRH. Each RRH has an ominidirectional mmwave antenna. We'll examine nonuniform CU distribution.
- Our design is incompatible with the usual power model, hence we propose a customized power consumption model. The power consumption model includes BBUs, HMC-RRHs, HSC-RRHs, and FL.
- Instead of turning off the HMC-RRH node, we'll employ HSC-RRH's CU density-based switching method to save energy. Thus, QoS is maintained while saving energy.
- We highlight the contribution to demonstrate the efficacy of our suggested techniques for nonuniform CU distributions.
- Selectively distribute HSC-RRHs over the HMC-RRH region to support redundant CUs and preserve quality of service. Due to its ability to compensate for high-frequency millimeter wave antenna loss, beamforming is also researched. It also improves system performance, power consumption, and communication network interference.

1.2 Organization

Further the paper is organized as, in section 2, we describes the proposed 5G H-CRAN architecture. The power consumption model is presented in Section 3. Section 4 describes the techniques recommended. Section 5 discusses the MATLAB simulation outcome and article summary is discussed in section 6.

2 SYSTEM MODEL

We introduce the system model shown in Figure 1 used in our research. We propose a $2 - tier$ 5G H-CRAN architecture having cloud BBU pool connected via fronthaul links to a set $S = \{1, 2, \ldots, N_{hsc}\}$ and $M = \{1, 2, \ldots, N_{hmc}\}$ of HSC-RRHs and HMC-RRHs respectively, where N_{hmc} is the number of HMC-RRHs and N_{hsc} represents the number of HSC-RRHs. We consider fibre fronthaul links in our system model because of their high bandwidth and low latency [15]. A star topology connects the RRHs to their BBU pool. An omini directional mmwave antenna serves each RRH. The beamforming approach is used in the mmwave antenna because it may make the transmission between CUs and the RRH more directed, resulting in decreased

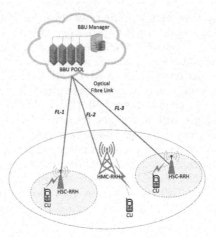

Figure 1. Proposed 2 – *tier* H-CRAN architecture.

energy consumption and interference [16]. HMC-RRHs are positioned in the middle of each HMC region, as illustrated in Figure 1, while HSC-RRHs are spread in HMC areas based on CU density.

Let set of CUs is $U = \{1, 2, \ldots, N_{cu}\}$, distributed randomly and nonuniformly. The CUs count may differ in each HSC cell. The mean of CUs count in a particular area is defined as follows:

$$N_{cu} = \lambda_{cu} A \tag{1}$$

where area of the RRHs and the count of CUs per Km^2 is denoted by A and λ_{cu} respectively.

3 PROPOSED HC-RAN POWER MODEL

Conventional earth model [17] is commonly used to calculate power consumption in traditional base stations. However, for various reasons, our proposed design cannot utilize this typical pattern of power consumption. To begin with, the BS function can be divided into RRH and BBU, that are placed at measurable physical distance, so the power consumed by BBU and RRH must be calculated individually. In addition, the FL power between the RRHs and the BBU is added to the total power consumed by the RRHS and all BBUs in the same pool must share the power consumed in cooling. For these reasons, we propose a new three-component power calculation model: (i) FL power, (ii) BBU group power and (iii) radio segment power. Accordingly, the total power P_{total} is can be formulated as:

$$P_{total} = P_{fl} + P_{pool} + P_{radio} \tag{2}$$

We concentrate on the radio and FL components in our suggested techniques. As a result, we consider the consumption of BBU pool power as constant and only the power consumed by FL and the radio part changing depending on the status of each HSC-RRH. As a result, when some HSC-RRH of a particular HMC-RRH is turned sleep, the power of the HSC-RRH decreases, due to decrease in power consumption FL, connecting this RRH to BBU.

3.1 *BBU pool power model*

The cooling power P_{cool} consumed by BBU pool, is shared by each BBU so, while calculating the power consumed by BBU pool we have to add it. So, power consumed by a BBU pool P_{pool} at a time instant t is formulated as:

$$P_{pool}(t) = P_{cool} + \sum_{i}^{N_{BBU}} P_{BBU}(i, t) \tag{3}$$

Where, count of BBUs in the BBU pool is denoted by N_{BBU} and $P_{BBU}(i, t)$ represents the power required by the ith BBU.

3.2 Fronthaul power model

The fronthaul fiber link also required power for transmission between RRHs and BBUs. So in our model FL power P_{fl} can be calculated as [18,19]:

$$P_{Fronthaul}(t) = \sum_{i}^{N_{BBU}} \sum_{j}^{N_{hmc}} P_{fl}(i,j) S_i^{fl,hmc}(j, t) + \sum_{i}^{N_{BBU}} \sum_{j}^{N_{hsc}} P_{fl}(i,k) S_i^{fl,hsc}(k, t) \qquad (4)$$

Where $P_{fl}(i,k)$ and $P_{fl}(i,j)$ the power required by FL between BBU i, and HMC-RRH j, and BBU i and HSC-RRH k, respectively. Finally, $S_i^{fl,hmc}(j, t)$ the link state between BBU i and HMC-RRH j and $S_i^{fl,hsc}(k, t)$ the connection link state between BBU i and HSC-RRH k is given by:

$$S_i^{fl,hmc}(j, t), S_i^{fl,hsc}(k, t) = \begin{pmatrix} 1, & \text{if active State} \\ 0, & \text{if sleep State} \end{pmatrix} \qquad (5)$$

3.3 Radio power model

The radio element of our suggested design is made up of both HSC-RRHs and HMC-RRHs. Thus, P_{radio} is calculated as:

$$P_{radio} = P_{HSC-RRHs} + P_{HMC-RRHs} \qquad (6)$$

HSC-RRH total power is defined as:

$$P_{HSC-RRHs}(t) = \sum_{k}^{N_{hsc}} P_{hsc}(k, t) S_{hsc}(k, t) \qquad (7)$$

where the total number of HSC-RRHs is N_{hsc}, the consumed power of HSC-RRHs k and $S_{hsc}(k, t)$ represents the state of HSC-RRHs k at a instant t and is shown in (5).

4 PROPOSED POWER SAVING SCHEME

Here, we describe our suggested power-saving scheme, which are based on switching HSC-RRH state between active and sleep.

4.1 Switch active/sleep scheme based on historical data and current CUs(SASHC scheme)

The switching scheme for proposed architecture for $2 - tier$ 5G H-CRAN is based on current and historical CUs density in HSC-RRH. Turning sleep the HSC-RRHs with low count is an effective way to minimize power consumption in H-CRAN. The switch active/sleep state is determined by the count of CUs in every HSC-RRHs. The concept is to change the state of HSC-RRH to sleep when its count of CUs λ_{cu} is lesser to a provided threshold λ_{thresh}. It can, affect QoS provided to CUs. As a result, to maintain QoS, CUs of turned-sleep HSC-RRHs will be surved by other HSC-RRHs or HMC-RRHs that give maximum power to the CUs. The distribution of HSC-RRHs is presented in Section II. At the start, we suppose that HMC-RRH and all HSC-RRHs are sleeping. More information is provided in the following phases.

Step 1: The initial step is to decide switch i.e. whether to activate or sleep. The decision to turn active or sleep an HSC-RRH n is performed when the density of CUs in HSC-RRH n, $\lambda_{cu}(n, t)$ is lower to the threshold λ_{thresh}.

Step 2: Once the state of switch active/sleep is committed, a confirmation of the state of HSC-RRH n at time $t-1, S_s(j,n,t-1)$, and in historical data is carried out to decide if the HSC-RRH will be turned sleep/active. *Pseudo code 1* describes the above steps 1 and 2.

PSEUDO CODE 1: TAKING DECISION OF SWITCH ACTIVE/SLEEP

Input Parameters

N, N_s, λ_{thresh}

$for \ j = 1: N_{hsc} \ do$
$\quad for \ n = 1: N_s \ do$
$\qquad$ **Input Parameters**
$\qquad S_s(n,t-1), \lambda_{cu}(j,n,t)$
$\qquad if \ \lambda_{cu}(j,n,t) > \lambda_{thresh} \ then$
$\qquad if S_s(j,n,t-1) = 1 \ then$
$\qquad\quad S_s(j,n,t) \leftarrow 1 \ \{ \ HSC\text{-}RRH \ n \ still \ ON\}$
$\qquad\quad else \ if S_s(j,n,t-1) = 0 \ then$
$\qquad\qquad S_s(j,n,t) \leftarrow 1 \ \{ \ switch \ ACTIVE \ HSC\text{-}RRH \ n \ \}$
$\qquad\quad end \ if$
$\qquad\quad else \ if \ \lambda_{cu}(j,t) \le \lambda_{thresh} \ then$
$\qquad if S_s(j,k,t-1) = 0 \ then$
$\qquad\quad S_s(j,n,t) \leftarrow 0 \ \{ \ HSC\text{-}RRH \ n \ still \ SLEEP\}$
$\qquad\quad else \ if \ S_s(j,n,t-1) = 1 \ then$
$\qquad\qquad S_s(j,n,t) \leftarrow 0 \ then \ \{ \ switch \ SLEEP \ HSC\text{-}RRH$
$\qquad\qquad n \ \}$
$\qquad\quad end \ if$
$\qquad end \ if$
$\quad end \ for$
$end \ for$

Step 3: When an HSC-RRH n is in sleep state, CU m belongs to this will be served by maximum power provided by HSC-RRH k or HMC-RRH:

$$\widetilde{sc} = arg \ max(max_{k \in N_s} \ (P_{hscrx}(k,m)), P_{hmcrx}) \qquad (8)$$

Step 4: Activated HSC-RRHs will use the beamforming technology to improve CU QoS following the switch active/sleep operation.

PSEUDO CODE 2: ASSIGNMENT OF CUs

$for \ m = 1 : N_{cu} \ do$
$\quad for \ k = 1 : N_{hsc} \ do$
$\qquad$ Input Parameters : $N_{cu}, N_{hsc}, S_{hsc}(k,t-1)$
$\qquad if \ max(P_{rx}(k,m)) \ then$
$\qquad\quad \{turn \ active \ the \ best \ HSC\text{-}RRH \ k \ sending \ the \ maximum \ power \ for \ CU \ m\}$
$\qquad\quad if \ S_{hsc}(k,t-1) = 1 \ then$
$\qquad\qquad S_{hsc}(k,t) \leftarrow 1$
$\qquad\quad else \ if \ S_{hsc}(k,t-1) = 0 \ then$
$\qquad\qquad S_{hsc}(k,t) \leftarrow 1$
$\qquad\quad end \ if$
$\qquad end \ if$
$\quad end \ for$
$end \ for$

Step 5: After the CU allocation operation is finished, a small quantity of power is returned to active HSC-RRHs and their FL linking them to the BBU. In contrast, a power related to switched sleep HSC-RRH and their FL is removed. Hence, P_{total} is calculated as:

$$P_{total}(t) = P_c + \sum_{i}^{N_{BBU}} P_{BBU}(i,t) + \sum_{i}^{N_{BBU}} \sum_{j}^{N_{hmc}} P_{fl}(i,j) S_i^{fl,hmc}(j,t) + \sum_{i}^{N_{BBU}} P_{BBU}(i,t) +$$

$$\sum_{i}^{N_{BBU}} \sum_{k}^{N_{hsc}} P_{fl}(i,k) S_i^{fl,hsc}(k,t) + \sum_{k}^{N_{hsc}} P_{sc}(k,t) S_{hsc}(k,t) + P_{hmcrx} \qquad (9)$$

The SINR of CU m from RRH n is formulated as:

$$SINR_j(n,m) = \frac{P_{rx}^j(n,m)}{\sum\limits_{m' \neq m}^{N_s} P_{rx}^j(n,m') + \sum\limits_{j' \neq j}^{N_{hmc}} P_{rx}(j',m) + P_{noise}} \tag{10}$$

Where $P_{rx}^j(n,m)$ is the received power from HSC-RRH n antenna to CU m. $\sum_{m' \neq m}^{N_s} P_{rx}^j(n,m')$ and $\sum_{j' \neq j}^{N_{hmc}} P_{rx}(j',m)$ is intercell interference and intracell interference average power respectivily is The AWGN power P_{noise} calculation given in [19] is:

$$P_{noise} = -174 + 10\,log_{10}(B) \tag{11}$$

Where B represents the bandwidth.

5 SIMULATION RESULTS

The simulation parameter and results of our proposed schemes is explained in this section to analyze the efficiency in lowering power consumption and compared with other existing schemes.

5.1 *Simulation scenario*

MATLAB simulation is used to evaluate the performance of our suggested strategies. We emphasizes our study on the radio part of the H-CRAN topology that is composed of $N_{hmc} = 1$ HMC-RRH. The radius (d) of HMC-RRHs is 500*meter*, and the HMC-RRH CU Euclidian distance is computed through their positions. HMC-RRH have dense HSC-RRH. At instant $t = 0$, all HSC-RRHs are in the sleep state. To begin, we simulate our suggested techniques using a nonuniform distribution model in which the average number of CUs serviced by an HMC-RRH is fixed at 100*CUs*. The uniform CUs assignment model is next investigated to demonstrate the usefulness of our suggested solutions even for uniform

Table 1. Simulation parameter.

Parameter	Value
Number of HMC-RRH	1
Number of HSC-RRH	18
Bandwidth B in[MHz]	400
Frequency in[GHz]	28
Propagation model	Pathloss model
Pattern of CU's distribution	uniform and nonuniform
HMC-RRH transmitted power [dBm]	46–43
HSC-RRH transmitted power [dBm]	21–17
HMC-RRH radius [m]	500
HMC-RRH resource blocks (RBs)	100
HSC-RRH resource blocks (RBs)	20
Max. CUs per HMC	100
Max. CUs per HSC	8
Height of HMC-RRH antenna [m]	30
Height of HSC-RRH antenna [m]	6
Path loss model [dB]	$32.9 + 19.2log_{10}(d) + 20.8log_{10}(f)$

distribution patterns. Our proposed architecture for nonuniform CU distribution patterns is shown in Figure 1. Table 1 lists the parameters used for simulation of proposed scheme.

5.2 Simulation results

Here we shown the performance of our proposed SASHC schemes in terms of consumption and saving of power, and service quality.

5.2.1 Power consumption

Figures 2 and 3 show the power savings, and power consumption vs. the count of CUs for our suggested schemes and those based on switching between active/sleep by cell, respectively. Figure 5 show the power consumption while Figure 4 shows the power savings for

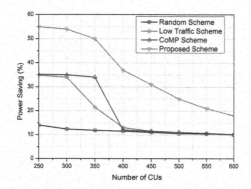
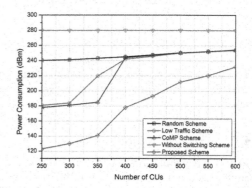

Figure 2. Power saving (%) versusthe number of users.

Figure 3. Power consumption versus the number of users.

both nonuniform and uniform CUs distribution after using our proposed strategies for different iterations. We chose to fixed 80 CUs per cell for the case of uniform didtribution. Thus, the number of CUs remains constant, but their placements change from iteration to iteration and from HSC-RRH to HSC-RRH.

On the other hand, unlike the switch active/sleep by the small cell that is only achievable for nonuniform CU distribution models, our HSC-RRH switch algorithms are successful in

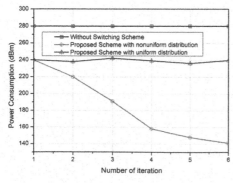
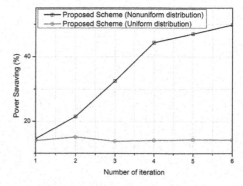

Figure 4. Uniform distribution versus nonuniform distribution of users: power saving.

Figure 5. Uniform distribution versus nonuniform distribution of users : power consumption.

586

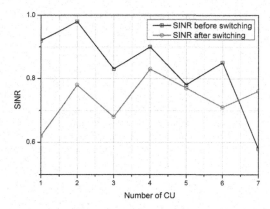

Figure 6.   SINR comparison before and after the switch ON/OFF process.

both cases. As seen in Figure 4, an average 12% power can be saved by using SASH scheme when CUs are distributed uniformly.

Few CUs will be affected by the HSC-RRH switching slumber, and their QoS will suffer as a result. However, QoS may be ensured with the deployment of HSC-RRH, which intends to signal coverage following the HSC-RRH switch sleep procedure. So, to maintain high QoS, we also simulate the SINR values for various CUs belongs to switched sleep HSC-RRHs for demonstrating the usefulness of our proposed approaches after applying the switching tehnique. Simulation result for SINR values shown in Figure 6, the mean value of SINR after using the HSC-RRH switch sleep schemes remains steady that indicates HSC-CRANs switching techniques maintain the QoS. As a result, we conclude that our suggested HSC-RRH switch active/sleep methods provide good network performance.

6   CONCLUSION

In this work, We evaluated 5G H-CRAN power consumption with homogeneous CUs distribution. We recommend sleeping HSC-RRHs when their CU density goes below a particular level to save power in 2-tier H-CRAN. Small cells sleeping may reduce service quality. HMC-RRH ensures complete QoS after active/sleep switching. HMC-RRHs and HSC-RRHs employ millimeter wave antennas with beamforming technology to reduce network interference.

Our method saves network energy without affecting QoS, according to simulations. When CUs density is minimal, SASHC procedures can save over 48% energy. Our idea cuts power use by almost 20%. SASHC works for homogeneous and heterogeneous CU distribution models. The switched sub-sleep cells' UCs' SINR values following the switch sleep operation showed constant quality of service.

REFERENCES

[1]   Cheng-Xiang Wang, Fourat Haider, Xiqi Gao, Xiao-Hu You, Yang Yang, Dongfeng Yuan, Hadi M Aggoune, Harald Haas, Simon Fletcher, and Erol Hepsaydir. Cellular Architecture and Key Technologies for 5g Wireless Communication Networks. *IEEE Communications Magazine*, 52(2):122–130, 2014.

[2]   Mamta Agiwal, Abhishek Roy, and Navrati Saxena. Next generation 5g Wireless Networks: A Comprehensive Survey. *IEEE Communications Surveys & Tutorials*, 18(3):1617–1655, 2016.

[3] Tshiamo Sigwele, Atm S Alam, Prashant Pillai, and Yim F Hu. Energy-efficient Cloud Radio Access Networks by Cloud Based Workload Consolidation for 5g. *Journal of Network and Computer Applications*, 78:1–8, 2017.

[4] Gunther Auer, Vito Giannini, Claude Desset, Istvan Godor, Per Skillermark, Magnus Olsson, Muhammad Ali Imran, Dario Sabella, Manuel J Gonzalez, Oliver Blume, *et al.* How Much Energy is Needed to Run a Wireless Network? *IEEE Wireless Communications*, 18(5):40–49, 2011.

[5] Antonio De Domenico, Emilio Calvanese Strinati, and Antonio Capone. Enabling Green Cellular Networks:A Survey and Outlook. *Computer Communications*, 37:5–24, 2014.

[6] Vijay K Bhargava and Alberto Leon-Garcia. Green cellular networks: A Survey, Some Research Issues and Challenges. In 2012 26th Biennial Symposium on Communications (QBSC), pages 1–2. IEEE, 2012. Albrecht Fehske, Gerhard Fettweis, Jens Malmodin, and Gergely Biczok. *The Global Footprint of Mobile Communications: The Ecological and Economic Perspective*. IEEE Communications Magazine, 49(8):55–62, 2011.

[7] Bharat JR Sahu, Shatarupa Dash, Navrati Saxena, and Abhishek Roy. Energy-efficient BBU Allocation for Green C-ran. *IEEE Communications Letters*, 21(7):1637–1640, 2017.

[8] Ismail Sharhan Hburi and Hasan Fahad Khazaal. Joint RRH Selection and Power Allocation Forenergy-efficient C-ran Systems. In *2018 Al-Mansour International Conference on New Trends in Computing, Communication,and Information Technology (NTCCIT)*, pages 29–34. IEEE, 2018.

[9] Shinobu Namba, Takayuki Warabino, and Shoji Kaneko. BBU-RRH Switching Schemes for Centralized Ran. In *7th International Conference on Communications and Networking in China*, pages 762–766. IEEE,2012.

[10] Chunli Ye, Yaxin Wang, Xin Zhang, and Dacheng Yang. Calibration Algorithm for C-ran BBU-RRH Switching Schemes. In *2016 IEEE International Conference on Network Infrastructure and Digital Content (ICNIDC)*,pages 406–410. IEEE, 2016.

[11] M Khan, Raad S Alhumaima, and Hamed S Al-Raweshidy. Reducing Energy Consumption by Dynamicresource Allocation in C-ran. In *2015 European Conference on Networks and Communications (EuCNC)*,pages 169–174. IEEE, 2015.

[12] Emad Aqeeli, Abdallah Moubayed, and Abdallah Shami. Power-aware Optimized RRH to BBU Allocation Inc-ran. *IEEE Transactions on Wireless Communications*, 17(2):1311–1322, 2017.

[13] Mst Rubina Aktar, Abu Jahid, and Md Farhad Hossain. Energy Efficiency of Renewable Powered Cloudradio Access Network. In *2018 4th International Conference on Electrical Engineering and Information & Communication Technology (iCEEiCT)*, pages 348–353. IEEE, 2018.

[14] Shunqing Zhang, Shugong Xu, Geoffrey Ye Li, and Ender Ayanoglu. First 20 Years of Green Radios. *IEEE Transactions on Green Communications and Networking*, 4(1):1–15, 2019.

[15] Jakob Belschner, Veselin Rakocevic, and Joachim Habermann. Complexity of Coordinated Beamforming and Scheduling for ofdma Based Heterogeneous Networks. *Wireless Networks*, 25 (5):2233–2248, 2019.

[16] Haijun Zhang, Fang Fang, Julian Cheng, Keping Long, Wei Wang, and Victor CM Leung. Energy-efficient Resource Allocation in Noma Heterogeneous Networks. *IEEE Wireless Communications*, 25 (2):48–53,2018.

[17] Tshiamo Sigwele, Yim Fun Hu, and Misfa Susanto. Energy-efficient 5g Cloud Ran with Virtual BBU Server Consolidation and Base Station Sleeping. *Computer Networks*, 177:107302, 2020.

[18] Paolo Monti, Sibel Tombaz, Lena Wosinska, and Jens Zander. Mobile Backhaul in Heterogeneous Network Deployments: Technology Options and Power Consumption. In *2012 14th International Conference on Transparent Optical Networks (ICTON)*, pages 1–7. IEEE, 2012.

Artificial Intelligence, Blockchain, Computing and Security – Dagur et al. (Eds)
© 2024 The Author(s), ISBN: 978-1-032-49393-0

A Blockchain-based AI approach towards smart home organization security

Sarfraz Fayaz Khan*
Professor, SAT – Algonquin College, Ottawa, Canada

S. Sharon Priya*
Assistant Professor (Sr. Gr.), B.s Abdur Rahman Crescent Institute of Science and Technology, Tamil Nadu

Mukesh Soni*
Assistant Professor, Department of CSE, University Centre for Research & Development Chandigarh University, Mohali, Punjab, India

Ismail Keshta*
Computer Science and Information Systems Department, College of Applied Sciences, AlMaarefa University, Riyadh, Saudi Arabia

Ihtiram Raza Khan*
Computer Science Department, Jamia Hamdard, Delhi

ABSTRACT: *The Internet of Things' (IoT) predicted enormous breadth and extensive organization make it challenging to acknowledge safe and secret communications. Investigations on the application of BC technology* for decentralized assurance as well as protection are still ongoing. However, these arrangements are quite difficult to implement in terms of computation requirements and time constraints, preventing them from being used in most IoT applications. This study describes a blockchain-based asset-productive explanationwith regard to safe as well aspersonal Internet-of-Thing. The arrangement has beenfacilitated by the unique use of double-dealing of computing sources within a typical Internet-of-Things surrounding (such as intelligent houses), as well as the use of a Deep Outrageous Learning Machine event (DOLME). The Brilliant Home Engineering located in Blockchain gets protected in this suggested method by carefully evaluating its unshakable quality about the essential security elements of protection, trustworthiness, and availability. The overheads incurred with this approach (in regards to propagation, handling stage, along withpower use) have been peripherally linkedwith their assurance as well as protection advantages, which is why we also give reproduction findings.

Keywords: AI, IoT, Blockchain, Smart home, Deep Learning

1 INTRODUCTION

A smart-home is an Internet-of-Things (IoT)-an integrated housethat provides purchasers with safety, wellness, ease, a planned lifestyle, etc. Smart house structures are suited to

---

*Corresponding Authors: drkhansarfraz@gmail.com, sharonpriya@crescent.education, 3mukesh.research24@gmail.com, 4imohamed@mcst.edu.sa and erkhan2007@gmail.com

DOI: 10.1201/9781003393580-90

simplifying and improving people's everyday activities and independence. They provide practical tools like security assessments and recommended methods of acting, which have caught the attention of customers and device designers. Smart houses may benefit mortgage holders and other close family members greatly, but they may also be vulnerable to malicious cyber-attacks that put the security and safety of their residents at risk (Chourabi *et al.* 2012). These threats have typical defences but are very focused and helpless against untamed onslaught. As a result, the adaptability and flexibility necessaryin order toproperly use the creative domain with regard to independent bright house apps, as well as workplaces,are lacking. A few clever ideas make life simpler for people. These initiatives generate enormous amounts of data. Security issues arise from storing this constantly evolving information in storehouses (Mukesh Soni *et al.* 2022).

However, blockchain has shown impressive executionby way of the keystonewith consideration of a network safety architecture within the number of clever house advances, including distant networks as well as information transfer (Naphade *et al.* 2011). To alleviate these problems, blockchain technology and concentrated hoarding groups may be used. Satoshi Nakamoto initially proposed the concept of blockchain in 2008, which included a time-stamped collection of malicious-proof material that was restricted by a local network of decentralised frameworks (Iqbal *et al.* 2018).

The three pillars of blockchain innovation are stability, decentralisation, and transparency. The three skills have expanded into wide-ranginguses, such as the digital money conceptalong with the potential analysis with regards to clever utilization, thoughBC innovation assures safety. By way of explanation, the assaults nature islatterly become complicated, with a great majority of threatscontrolling of fictitious characters to verify the agreement (AsadUllah *et al.* 2018).

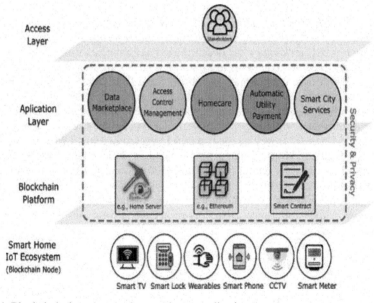

Figure 1. A Blockchain-based smart home 4-layer application system.

A full Interruption Identification Framework (IIF) is essential to overcome the hidden problem since normal approaches employ a mark-based strategy to identify innovative programs. The Deep Outrageous Learning Machine (DOLM), perhaps the most recent breakthrough, may be used to analyze the information stream as well asrecognize disruptions along withassaultarrays. Therefore, this has beenvital to managing clever BC-dependent apps viaemergentvigorousas well as adaptable proceduresfor handling this massive amount of information (Schaffers *et al.* 2011). Artificial-Intelligence encompasses devices that can plan, think, and act independently of individualinterference. This has beenconsidered to be a part of the Artificial-Intelligence (computerized thinking) architectures. The main goal of AI is to successfully calculate information from information, create an expectation, and alter outcomes via quantifiable analysis. AI may manage a lot of data and make decisions based on evidence (Abbas *et al.* 2020).

In order to cope with making smart housesinnocuoususing Internet-of-Things-associateddetectorsalong with increased functioning, all of that investigation will use a Deep Outrageous Learning Machine (DOLM) approach.The important commitments for this review consist of providing a complete analysis of progressive technologies pertinent to BC-dependent smart houses using the Deep-Outrageous-Learning-Machine and presenting an alternate perspective on applications in various fields(such as sharing information about brilliant homes), which is supported by the new stages of innovation development (Nakamoto *et al.* 2019). As shown in Figure 1, a Deep-Outrageous-Learning-Machine design is to be put out to use in the BC-dependent dazzling homes. The clever application in this case combines information from several data sources usingdetectors, clever gadgets, along with Internet-of-Thinggadgets. Data acquiredthrough these sophisticated apps are processed as well as dissected. One important component of such apps is the blockchain (Badii *et al.* 2020). Then, in such apps, Deep-Outrageous-Learning-Machine would be utilized for interpreting as well as predicting information (information examination and continual examination). The DOLM structure's data sets might be managed by a blockchain network, which eliminates errors in data including redundancy, loss of information value, mistakes, and disruption. Blockchains are information-dependent. Information-related concerns in the DOLM structure will afterwards be disregarded. Instead of focusing on the whole collection of information, the DOLM system may be oriented upon different chain segments. This will provide a superior solution for a variety of purposes, including detecting deception and anticipating the discovery of burglaries (Aujla *et al.* 2020).

2 EXPLANATION FOR PROJECTED DOLM METHODOLOGY

Blockchain was developed by Satoshi Nakamoto in 2008. By using its simple blockchain digital currency, a P2P instalment company will remove outsiders and double spending difficulties (e.g., bitcoins). Each datum block in this architecture for gathered information has beenauthenticated using SHA-256 (safe hash calculation) with a previous hash-block. The main components of the block's design were the block number, earlier block hash, exchange information, nonce, and time stamps. A consistent variable is included in the timestamp, but due to a nonce, the variable is unreliable. As a figure which starts with several successive driving noughts, dynamic (timestamp plus nonce) and static (block) information are continuously hashed downward by the validator and the digger (PC hub). The term "cryptography puzzle" is typically used to describe this technique. (Tanwar *et al.* 2020). Within the BC, the person who wins is allowed to embed the block will be the primary factor taken into consideration by miners when determining the proper hash function. The process used to determine if a block is valid is called a confirmation of work. The following steps demonstrate the fundamental aspects of blockchain. Each hub associated with an IoT device in a smart home works with a memory pool to remember excavators for a blockchain framework.

If the deleted exchanges are appropriate, they are incorporated withina block and made settominewith the help ofdiggers across a smart housearrangement. Diggers create a Hash of the Block after amending the nonce as well asthe time-phase name. The application then attempts to compare the hash produced with the goal (Wu *et al.* 2022). The hash has beenattached to a chain whenever a miner is done extracting a block. A conversation could begin again whenever the hash reaches the goal total.If the hash is less modest than the objective worth, the work verification is evaluated before execution and then implemented in the chain. As a result, such information is distributed across the company for informingall connected hubs of the elimination of concluded notice pool exchanges (*et al.* 2017). Blockchain technology is increasingly defining the smart home communication landscape because this has beencompliant as well asconformablesufficientfor working seamlessly amongintelligent house Internet-of-Thing applications. The clever house arrangement in relation to a blockchain is shown in Figure 1. Four layers make up the structure: an IoT information layer, theBCsystemlevel using Deep-Outrageous-Learning-Machine (DOLM), a layer for smart home devices, and a layer for the client hub.

3 INCORPORATION OF DOLM IN A BLOCKCHAIN-ASSOCIATED SMART-HOME

The Internet of Things (IoT) data structure gathers data from equipment that is necessary for smart homes, conditions, and customer assessments. These devices fall into one of three broad categories: sensors, mixed media, or medical services (Tran *et al.* 2017). Sensors determine the weather conditions. The indoor regulator, for instance, is used to compute and provide guidelines for room temperature. The blockchain layer of the IoT sensor network consists of closed-circuit TV, wearables, etc. Data from these hubs is collected and stored on an artificially created database, such as a blockchain that shows the topmost tier of the stack (Zhang *et al.* 2017).

DOLM registration advances may be used to make them smarter using blockchain-based applications. Involving DOLM for the distributed record may increase security. DOLM may also be utilised to extend the time necessary to achieve knowledge via improved data-sharing courses. Additionally, it opens up the possibility of building more advanced structures using blockchain technology's built-in engineering. We became familiar with the DOLM execution architecture in blockchain-associated smart innovation, as shown in Figure 1. The clever programme in this case gathers data from several data sources, including cameras, cutting-edge technology, and IoT platforms [9]. Data from these methods was considered to be a component of excellent applications. Blockchain contributes significantly to these clever applications. However, the DOLM structure may be used to such application information for comprehension (information research and continuing inspection) and expectation. Informational collections used by DOLM models are managed on a blockchain network.

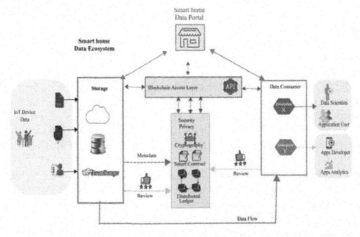

Figure 2.   Smart home administration framework built on block chain working with DOLM.

There are a few information errors such as repetition, poor information values, deformities, and loudness [14]. The DOLM architecture can prevent data-related concerns since data is sent over a blockchain. The DOLM edge may be built on clear chain parts rather than the complete informative collection. This may integrate intriguing structures for a variety of uses, such as avoiding data fraud or providing identifiable evidence of extortion. The Deep Outrageous Learning Machine layer, shrewd agreement, and blockchain data design are the three main parts of the BCprototypical, which has been dependent upon cutting-edge Internet-of-Thing engineering [6].

Huge quantities of stored layers, stored neurons, and various setting-off capabilities have been used in the suggested DOLM structure to increase the security and welfare of smart homes. The proposed technique includes three distinct levels: the collection of information, pre-process, along with evaluation phases [8]. For the evaluation layer, it is important to keep in mind expectations and execution appraisal, specifically. For exploratory analysis, reliable sensor data and actuators are gathered. The information obtained is sent to the acquiring layer as a contribution. To deal with information inconsistencies within the pre-processing level, explicit dataclean-upalong with arranging cycles have been proposed [11]. Any vengeful or interfering activity at the application layer was exploited to progress the clever home organization by use of the DOLM.

Blocks are linked by hash values using cryptography. While shrewd agreements follow predetermined standards to offer decentralized exchanges that are easier to use and less complex, a housemain computer mightconsidera digger toverifythe latest exchanges as well as insert fresh blocks'. There are other methods to combine blockchains, including public, private, and unified, on the other handwithin a classy houseorganization, generally private BChasbeenadaptedforcontrollingallthe above expenses [12].

The application layer is being developed to facilitate the fusion of certain cutting-edge home apps with existing blockchain networks. This system includes cutting-edge home innovations like the computerized market, board access, connectivity within the house, and medical care, as well as planned installation to the foundation plus smart local area administrations [10]. The entry layer, which is the top of the organized progression, allows outsiders for taking benefit of BC-dependent brilliant house innovations such as micro-grids, retail stores, utility suppliers, jobs, etc.

The person who might be given individualized access to the home automation and is utilizing its applications is the Administrator [12]. Figure 2 shows in what way blockchain encourages secure access.

4 DEEP OUTRAGEOUS LEARNING MACHINE (DOLM)

The DOLM can be utilized to forecast energy use, travel, and traffic signals among other things in many different businesses. The learning framework may be overridden, and the present Convolutional Neural Network (CNN) computations need multiple adjustments and protracted learning cycles [14]. The information on an absurd machine is defined. Given that it progresses rapidlyas well as is effective within the procedure complication rate, The DOLM may be broadly applied for arranging as well as relapsing prevention goals in various contexts. During the learning phase of the suggested framework, we applied the reverse propagation method, in which data flows across the organization in reverse, and in which the brain setup modifies the loads to attain higher precision having the lowest mistakefrequency [15]. A feed-forward brain system called the Outrageous-Learning-Machine advises that information goes only in one direction through a series of layers. Throughout the approval phases, when we remove the prepared model and assess the real data, the organization's loads are dependable. Three layers make up the proposed DOLM approach: an info layer, stowed-away layers, and a result layer [6].

Table 1. Comparing the suggested DOLM method against existing ai computations using different datasets.

Method	Correctness of NSLKDD information	Correctness of KDD-
CNN	82%	91%
SVM	70%	90%
Decision Tree	82%	92%
Projected DOLM	94%	95%

Different measurable limits, such as exactness, miss rate, responsiveness, explicitness, misleading positive worth, and positive forecast worth, are found to simplify smart home security in the evaluation layer. These boundaries are shown in Figure 3.

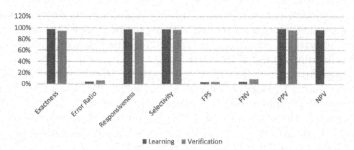

Figure 3. Performance assessment of a smart house built on a blockchain and equippedwith deep outrageous learning during the foresight of hostile behaviour or attacks using various statistical measures.

Weight setup, feed-forward proliferation, reverse mistake engendering, and noticeability updating are all components of the backpropagation technique [5]. Every neuron uses a sigmoid enactment capability in its hidden layer. The construction of the sigmoid information capability and the DOLM stored layer is made easier by dividing the square total from the needed result by 2 [2]. The typical error is anticipated to disappear with the weight shift.

5 RESULTS

This work used data from [10] as input for the Deep Outrageous Learning Machine (DOLM) in the proposed system. The data were randomly divided into 90% preparation (126,000trials), and 20% of the data are used for approval (23,000trials) [24]. Information has been treated in advance to remove data anomalies and reduce the possibility of data being present due to errors. DOLM made an effort to find any harmfulness or disruption in various hidden layers, stashed associations, and execution capabilities [11]. We also evaluated various dynamic capacities and a certain number of neurons in hidden layers. In order to properly predict the efficacy of this framework, we evaluated the DOLM in this study [6]. We used many factual measures to calculate the outcome using the partner computations of this DOLM calculation [7].

The suggested blockchain-based smart house is shown in Table 2 working with the DOLM framework prototypeforecastregarding interruption identification throughout the planning stage [28]. Instances totaling 1,26,000 were used in the preparation. Additionally, they are divided into 59,120 cases of attack and 68,242 examples of typical behavior, respectively [12]. It can be demonstrated that 2,012 records are mistakenly forecasted as an assault found even though there isn't a real attack present, while 66,242 instances of a typical class (meaning in which no assault is detected) are accurately predicted. A total of 59,120 instances are also used to account for assault cases, of which 57,140 are correctly predicted as assault cases and 2,530 are incorrectly predicted as typical cases where assault is included [13].

Table 2. Training of the DOLM system prototype for the blockchain-associated smart home during the prediction of harmful behavior.

Projected DOLM associated organization prototypical (90% of trial information in learning)		
Count of trials (N = 126,000)		Output results (O_0, O_1)
Expected outcome (T_0, T_1)	O_0 (standard)	O_1 (outbreak)
InputT_0 = 68,242 standard	66,242	2,012
T_1 = 59,120 outbreak	2,530	57,140

Table 3. Validation of the DOLM system prototype for the blockchain-associated smart home during the prediction of harmful behavior.

Projected DOLM associated organization prototypical (20% of trial information in learning)		
Count of trials (N = 23,000)		Output results (O_0, O_1)
Expected outcome (T_0, T_1)	O_0 (standard)	O_1 (outbreak)
Input T_0 = 9,815 standard	9,320	510
T_1 = 13,112 outbreak	902	12,105

The proposed smart house built on a blockchain and equipped with the DOLM framework model's expectation of interruption detection during the approval stage is shown in Table 3. 23,000 instances in all are used during approval [11]. These are further divided into 9,815 instances of the ordinary and 12,833 instances of assault, respectively. It can be shown that 473 entries are falsely forecasted as assaults when there aren't any, but 9,237 cases from a regular category are appropriately forecasted. In addition, 13,112 cases are chosen for assaults that were discovered, of which 12,105 examples were correctly identified as such and 902 examples were incorrectly identified as routine discoveries despite the presence of an attack in those instances [12].

The proposed blockchain-based smart house is shown in Figure 3 as being involved in the execution of the DOLM framework model as far as various factual metrics throughout the preparation and approval stage [13]. It is clearly shown that using the suggested framework during preparation results in individual exactness and miss rates of 97% and 4%. The suggested framework generates, respectively, accuracy and miss rates of 94% and 7% during approval. Additionally, it demonstrates how the framework model is implemented in terms of responsiveness, especially throughout the planning and testing phases [14]. It is obvious that the suggested framework, while being prepared, provides 97% responsiveness and 97% explicitness, respectively, whereas, after being approved, it provides 92% awareness and 97% particularity, respectively [15]. Additionally, a few more factual measurements are included for predicting all attributes, such as false-positive, false-negative, positive, and negative probability proportions, along with positive plus negative expectation figures. In Figure 3, all effects of these activities are listed.

6 CONCLUSION

An important challenge for smart homes continues to be the interruption of identifiable evidence, particularly with regard to assessment as well as anticipation. The astonishing potential has been demonstrated in recent advancements in the realms of blockchain and AI to accomplish these goals in the interim. Such arrangements can't be immediately implemented since the majority of clever house arrangements include devices with limited power and handling. This study filled this gap by introducing a simple but effective solution for interruption ID and expectation. A blockchain-associated Deep Outrageous Learning Machine (DOLM) concept was put out. A few factual techniques were used to determine if the suggested arrangement was workable. These evaluation findings demonstrate that the DOLM technique is unquestionably more reliable than those of other approaches. The suggested DOLM method produced astounding results with a 94%exactness rate. The obtained results are encouraging, and we are now looking at expanding our usage of more datasets and fluctuating topologies.

REFERENCES

Abbas S. et al., "Modeling, Simulation and Optimization of Power Plant Energy Sustainability for IoT Associated Smart Cities Empowered with Deep Extreme Learning Machine," *IEEE Access*, vol. 8, 2020, pp. 39982–97.

AsadUllah M., Khan M. A., Abbas S., Athar A., Raza S. S., and Ahmad G., "Blind Channel and Data Estimation Using Fuzzy Logic Empowered Opposite Learning-based Mutant Particle Swarm Optimization," *Comput. Intell. Neurosci.*, vol. 2018, pp. 1–12, Dec. 2018.

Aujla G. S. et al., "BlockSDN: Blockchain-as-a-Service for Software Defined Networking in Smart City Applications," *IEEE Network*, vol. 34, no. 2, 2020, pp. 83–91.

Badii C. et al., "Smart City IoT Platform Respecting GDPR Privacy and Security Aspects," *IEEE Access*, vol. 8, 2020, pp. 23601–23.

Chourabi H., Nam T., Walker S., Gil J. R. Garcia, Mellouli S., Nahon K., Pardo T. A., and Scholl H. J., "Understanding Smart Cities: An integrative Framework," in Proc. *45th Hawaii Int. Conf. Syst. Sci.*, Jan. 2012, pp. 2289–2297.

Iqbal K., Adnan M., Abbas S., Hasan Z., and Fatima A., "Intelligent Transportation System (ITS) for Smartcities Using Mamdani Fuzzy Inference System," *Int. J. Adv. Comput. Sci. Appl.*, vol. 9, no. 2, pp. 94–105, 2018.

Mukesh Soni, Dileep Kumar Singh, Privacy-preserving Secure and Low-cost Medical Data Communication Scheme for Smart Healthcare, *Computer Communications*, Volume 194, 2022, Pages 292–300, ISSN 0140-3664, https://doi.org/10.1016/j.comcom.2022.07.046.

Nakamoto S. et al., "Bitcoin: A Peer-to-Peer Electronic Cash System," *Manubot*, 2019.

Naphade M., Banavar G., Harrison C., Paraszczak J., and Morris R.,"Smarter Cities and their Innovation Challenges," *Computer*, vol. 44, no. 6, pp. 32–39, Jun. 2011.

S. B. Calo, Verma D. C., and Bertino E., "DisTrends in the Management of Complex Systems," *Proceedings of the 22nd ACM on Symposium on Access Control Models and Technologies (ACMT)*, Indianapolis, USA, June. 2017.

Schaffers A., Komninos H., Pallot N., Trousse M., Nilsson B., and Oliveira M., "Smart Cities and The Future Internet: Towards Cooperation Frameworks for Open Innovation," in *The Future Internet Assembly*. Berlin, Germany: Springer, 2011, pp. 431–446.

Tanwar S. et al., "Machine Learning Adoption in Blockchain-Based Smart Applications: The Challenges, and a Way Forward," *IEEE Access*, vol. 8, 2020, pp. 474–88.

Tran T. X., Hajisami A., Pandey P., Pompili D., "Collaborative Mobile Edge Computing in 5G Networks: New Paradigms, Scenarios, and Challenges," *IEEE Communications Magazine*, vol. 55, no. 4, pp. 54–61, 2017.

Wu J, Haider SA, Soni M, Kalra A, Deb N. 2022. Blockchain Based Energy Efficient Multi-tasking Optimistic Scenario for mobile Edge Computing. *PeerJ Computer Science* 8:e1118 https://doi.org/10.7717/peerj-cs.1118.

Zhang W., Zhang Z., Chao H. C., "Cooperative Fog Computing for Dealing with Big Data in the Internet of Vehicles: Architecture and Hierarchical Resource Management," *IEEE Communications Magazine*, vol. 55, no. 12, pp. 60–67, 2017.

Artificial Intelligence, Blockchain, Computing and Security – Dagur et al. (Eds)
© 2024 The Author(s), ISBN: 978-1-032-49393-0

Multi-party secure communication using blockchain over 5G

K. Archana*
Assistant Professor, Department of Computer Engineering, Pimpri Chinchwad College of Engineering and Research, Pune, Maharashtra, India

Z.H. Kareem*
Medical Instrumentation Techniques Engineering Department, Al-Mustaqbal University College, Babylon, Iraq

Liwa H. Al-Farhani*
System Analysis, Control and Information Processing dep., Academy of Engineering, RUDN University, Moscow, Russia

K. Bagyalakshmi*
Sri Ranganathar Institute of Engineering and Technology Coimbatore, India

Ignatia K. Majella Jenvi*
Associate Professor, Mathematics Department, Saveetha School of Engineering, SIMATS, Saveetha Nagar, Chennai

Ashok Kumar*
Assistant Professor, Department of Computer Science, Banasthali Vidyapith, Banasthali (Rajasthan)

ABSTRACT: One of the most critical problems facing the blockchain technology industry right now is how to protect the privacy of users' data on the blockchain in a way that is both effective and cheap. Based on the Pedersen commitment and the Schnorr protocol, this study comes up with a secure multi-party computing protocol (BPLSM). By making the structure of the protocol and doing formal proof calculations, it has been shown that the protocol can be used in the blockchain network to combine private messages for efficient signing while keeping people's identities secret. Furthermore, by looking at the nature and security of the protocol, it is also possible to find that the BPLSM protocol on the blockchain has a low cost of computing and a high level of information secrecy. Furthermore, it was found that the BPLSM protocol takes less time to check than the current mainstream BLS signature in a simple multi-party transaction with a fixed number of participants.

Keywords: Blockchain, Secure Communication, 5 Generation Network, Pedersen Commitment, Schnorr Protocol

1 INTRODUCTION

Since Satoshi Nakamoto launched Bitcoin in 2008, blockchain has grown in popularity as a ground-breaking underpinning technology for cross-industry applications. The qualities of blockchain, such as decentralisation, difficult to tamper, traceability, openness, and

*Corresponding Authors: kadarchna@gmail.com, zahraa.hashim@mustaqbal-college.edu.iq, liwarussia@gmail.com, bagyalakshmi@sriet.ac.in, simaigni100@gmail.com and kuashok@banasthali.in

DOI: 10.1201/9781003393580-91

transparency, can better suit people's expanding requirements than current popular information technologies such as big data, cloud computing, artificial intelligence, and so on (Khujamatov et al. 2020). However, as academic research has progressed, more and more researchers have discovered that user data may be conveniently tracked in a decentralised ledger based on blockchain technology. Through their research, (Li et al. 2022) discovered that the theft of user privacy information is linked to the leakage of user address information. (Longbing et al. 2021) used Monero, a digital currency based on ring signatures, to break the disguised transaction by monitoring the username and private key, posing a serious threat to the transaction initiator's privacy information. Furthermore, most blockchain exchanges have incorporated a centralised identity verification system, and the mapping relationship between the user's personal information and the blockchain address is automatically recorded by the exchange. The risk of users' transaction information and behaviour being cracked will rise when big data technologies are implemented. Author et al. analysed the association between users and their bitcoin addresses using an automated clustering method in 2017, and presented a risky bitcoin usage pattern for users (Mukesh Soni et al. 2021).

Therefore, the security issues of blockchain technology are also getting more and more attention. How to reasonably and efficiently protect the privacy of user identity information and transaction data is a key issue in the current blockchain technology field in terms of security. For example, in the medical and health service industry, how to safely and efficiently solve the problem of personal medical and health data sharing while ensuring the privacy of personal data; another example in the freight logistics industry, how to quickly establish a trust relationship with multi-party cargo suppliers to ensure high-quality long-tail resources while ensuring Transaction information is not stolen and tampered with, etc. With the continuous updating and iteration of technology, the emergence of secure multi-party computation provides a possible way to solve such problems.

In the current research on blockchain security multi-party computing, blockchain technology service provider Defi uses blockchain and trusted computing to build a system architecture that helps companies achieve joint risk control, but there are still some problems in efficiency and transparency; (Wu et al. 2022) proposed a data governance collaboration method based on blockchain, and gave a construction standard for multi-party collaboration to achieve blockchain collaborative governance, but they did not give a more specific analysis on privacy and security strategy. Combining the above literature and the current technology development trend, in order to improve efficiency and achieve effective privacy protection, it is imperative to design a secure multi-party computing protocol that can merge and sign multi-party messages under anonymous conditions and verify the signature efficiently. In fact, as early as 2001, a BLS scheme that can combine, sign and verify different messages appeared (Weerasinghe et al. 2021) and in recent years, the main proposer of the scheme, Author, is still updating this signature scheme (Yue et al. 2021), to adapt to the current security on research and development of multi-party computing. However, due to the unique operation logic of the BLS signature, it relies too much on the bilinear pairing operation on the elliptic curve in the finite field and has a large number of operations (Mirtskhulava et al. 2021; Ning et al. 2022) which leads to the same signature condition, the BLS signature even It takes 3 times more time and cost than the ECDSA signature verification widely used in the current Ethereum.

On the basis of existing research, in order to merge and sign and verify different messages, improve the efficiency of signature verification on the basis of existing aggregated signatures, and at the same time enhance the privacy and anonymity of sensitive data, referring to Yu's Pedersen commitment With the addition of Schnorr signature scheme (Soni et al. 2021), the BPLSM protocol is designed, a secure multi-party computation protocol that integrates Pedersen commitment and Schnorr protocol on the blockchain. Through the formal verification of the protocol, it is proved that the BPLSM protocol can ensure the validity of the signature without being leaked. After that, a related experiment was designed using this protocol to construct a signature verification program under the semi-honest model. The main logic of the experiment is to discard the signatures that do not pass the verification, and extract the relevant signature information that has passed the verification and submit it to the contract on the chain to analyze the blind factor and declaration when using the commitment, and then process the transaction after obtaining a unified opinion. Publish the results to ensure the rationality of signing messages and the efficiency and feasibility of anonymous signature verification. The experimental results show that the signature constructed based on the BPLSM protocol can save about 83.5% of the time cost in signature verification compared with the BLS signature under the same signature verification conditions.

2 BASIC KNOWLEDGE

This section mainly introduces the basic knowledge involved in the protocol, including Pedersen commitment, Schnorr protocol and secure multi-party computation.

2.1 *Schnorr protocol*

As a kind of Sigma protocol (Σ-protocol), the non-interactive Schnorr protocol (Crowe *et al.* 2020) is a very good, concise and zero-knowledge security protocol. Although the Schnorr protocol was proposed in 1989, the enthusiasm of the academic community for its exploration and research has not diminished for decades, and it has shown great potential in the field of blockchain in recent years. For example, in the zero-knowledge data exchange protocol zkPoD, in order to find another way to achieve fair exchange, an extended Schnorr protocol is used combined with Pedersen's commitment to achieve high efficiency and scalability (Hewa *et al.* 2020). In the research and development of the Schnorr protocol, various Schnorr aggregated signatures have extended the cryptographic elliptic curve digital signature scheme and promoted the development of digital signatures. In their 2019 literature, Maxwell *et al.* also proposed a multi-signature method based on the Schnorr protocol and applied it to the Bitcoin network (Shobanadevi *et al.* 2020).

Schnorr aggregate signature can be divided into six steps: key generation, key aggregation and summation, interactive random number, generation of single signature, aggregate signature and signature verification. The interaction process is shown in Figure 1.

Taking the aggregated signature of three signers as an example, firstly, they need to exchange their respective random numbers with each other, and then use their respective signature private keys to perform a single signature on the same message, and then aggregate the single signature to generate a new signature, and finally It is handed over to the verifier for verification using the verification key. In this process, for the verifier, the representation is a smart contract that verifies the signature on the blockchain, and the contract uses the predefined verification key to verify the combined signature to ensure the legitimacy of the signature.

According to the previous research combined with the signature logic in the figure above, it is not difficult to see that the Schnorr protocol is very mature in the signature scheme for a single signature or an aggregated signature of a single message.

2.2 *Secure multiparty computation*

The main purpose of secure multi-party computing is to solve the problem of collaborative computing between distrusted participants under the premise of protecting privacy (Chen *et al.* 2020).

Compared with traditional data confidentiality, the advantages of secure multi-party computation are shown in Table 1:

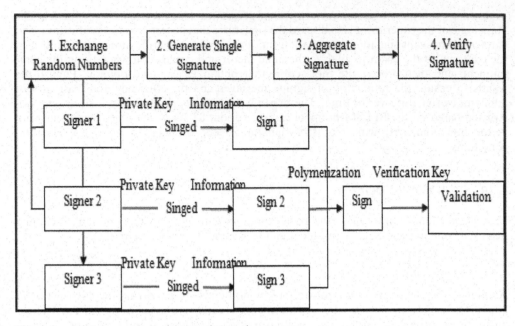

Figure 1.   Schnorr aggregate signature interaction process.

Table 1.   Comparison of traditional data confidentiality and secure multi-party computation.

Control information	Traditional data privacy	Multi-party secure computation
calculate	Data cannot be computed before encryption nor can it be used before decryption	Can be computed under data encryption and can obtain the same results as computed directly in plaintext
Safety	Safety Requires a trusted system administrator	Guaranteed by mathematical theory
Store	Requires a trusted hardware environment and a stable operating system	No trusted hardware environment and stable operating system required
Use	Static read	Dynamic usage

Today, with the rapid development of information, the emergence of collaborative situations is not uncommon. The technical characteristics of secure multi-party computing input privacy, calculation correctness and decentralization just meet the objective needs of the participants who want to obtain cooperation benefits but do not want to leak their own data in such a scenario. Therefore, in recent years, the research on secure multi-party computation is also deepening. In 2018, (Huang *et al.* 2021) used the dynamic programming method to improve the multi-party computation protocol of the constant wheel, which improved the computational efficiency. Author (Shi *et al.* 2021) also analyzed privacy protection when using secure multi-party computing in electronic medical systems in edge computing, and weighed the computing overhead. (Yazdinejad *et al.* 2021) analyzed several general frameworks for secure multi-party computing in detail at the top security conference Security and Privacy 2019, packaged the construction environment in dock, and evaluated the pros and cons of these frameworks from various dimensions.

3 SECURE MULTIPARTY COMPUTATION PROTOCOL BPLSM

In this section, we first propose the status quo of secure multi-party computation signatures, and then introduce the BPLSM protocol. Then, starting from the architecture design of BPLSM, we explain the three main properties of the protocol and analyze its security.

3.1 *BPLSM protocol architecture*

In the first step of architecture, each signing party makes a hidden message commitment, the formula is as follows:

$$MCommitment_i = m_i * g + seed_i * h \tag{1}$$

Where g and h are points with different positions on the finite field elliptic curve; m_i is the mutually different message signed by all parties; $seed_i$ is the random number belong to {1, 2, 3}. After making the commitment $MCommitment_i$ for sensitive messages, it is necessary to aggregate the message commitments of the three parties involved in order to generate a unified signature and submit it to the contract for verification.

The corresponding steps and calculations for the signature are as follows:

1) The participating three parties generate their own commitments $MCommitment_1$, $MCommitment_2$ and $MCommitment_3$ according to formula (2);
2) Each party generates blind factors r1, r2, r3 privately, and discloses $r_1 * H$, $r_2 * H$, $r_3 * H$ to each other as the public key R_i, i$\in$ {1,2,3} based on the blind factor r_i;
3) According to the respective public keys R_i, define the combined message commitment:

$$MCommitment = \sum_{i=1}^{3} MCommitment_i, \tag{2}$$

$$SumCom = Hash(R_1 || R_2 || R_3 || MCommitment) \tag{3}$$

Each party will perform Schnorr signature on the calculated combined transaction commitment, that is, $SigM_i = r_i + SumCom*prkey_i$, where $prkey_i$ is the private key of each party. And the respective transaction announcement $declare_i$, and the $seed_i$ used are passed to the contract through an anonymous channel, i$\in$ {1,2,3};
4) All parties send the signatures of their respective transactions to the last signing party for aggregated signatures

$$SigM = SigM1 + SigM2 + SigM3 \tag{4}$$

5) After the aggregation is completed, the last signing party submits a single signature pair (R, SigM, MCommitment) to the contract, where R = R1 + R2 + R3.

Using the characteristics of Pedersen's additive homomorphism, the above protocol signature step directly combines the transaction commitments in the form of cypher text without decrypting the transaction commitments of all parties, ensuring that the message commitments of all three parties are included in the privacy calculation. Ensure that all parties have the freedom to talk freely and anonymously. Furthermore, because the contract's ultimate submission is a single signature, it cuts down on contract signature verification time and eliminates unnecessary computational overhead.

4 SIMULATION AND RESULT ANALYSIS

In order to confirm the feasibility of the BPLSM protocol, this study mainly designs the BPLSM protocol experiment based on the Go language and Solidity language. In order to reflect the high-efficiency objectivity of the BPLSM protocol in the signature verification process, this study conducted a computational cost comparison experiment on the BLS signature, one of the current mainstream aggregated signatures that can sign different messages, under the same hardware environment.

Because the signature process of the BPLSM protocol is divided into the generation of the combined commitment and the generation of the aggregated signature, it needs to be divided into two parts for analysis during the stress test process, and the BLS signature requires multiple parties to participate in the aggregation process only once, so the stress test only needs to be Aggregate signatures for pressure measurement. It can be seen from Table 2 that under the same pressure measurement, the BPLSM protocol signature time overhead cost is lower than that of the BLS signature, but if the efficiency issue is aside, the BPLSM protocol involves two multi-party interactions, so the comparison of signatures still has some limitations.

Table 2. BPLSM protocol and BLS signature stress test.

Test content	Test frequency	Average cost per test/ns
Benchmark BPLSM SumCom-8	1648732	750
Benchmark BPLSM Signature-8	2042548	552
Benchmark BLS Aggregate SignatureN-8	204651	5873

The signature verification process needs to be executed on the blockchain through smart contracts, so in this experiment, the Solidity language is used to implement the signature verification logic adapted to the signature content in the Go language on the Ganache private chain. And by using the truffle framework combined with the Javascript scripting language to analyze the efficiency of signature verification. Still taking the three participants as the benchmark, the use of the truffle framework combined with the Javascript scripting language to construct the BPLSM protocol signature verification part and the BLS signature verification part based on the BLS12-381 curve are given. The results of the two tests are very different. In order to more accurately compare the efficiency of the two in the Ethereum smart contract verification, so as to confirm the advantages of the designed BPLSM protocol. We again conducted 20 single-unit tests on the BPLSM protocol signature verification part and the BLS signature verification part based on the BLS12-381 curve, in order to strengthen the persuasiveness of the experiment. As shown in Table 3.

Table 3. BPLSM protocol and BLS signature verification test.

Testing frequency	BPLSM Combination Commitment Verification Overhead	BPLSM aggregate signature verification overhead	BLS aggregate signature verification overhead
1	115	75	2894
2	163	85	3489
3	124	73	2483
4	125	78	3105
5	265	88	2994
6	190	115	2564
7	231	76	2787
8	123	91	2841
9	126	87	2813
10	159	83	2783

It can be seen from Table 3 that the average verification time of the BPLSM protocol designed in this study among the three participating parties is about several hundred milliseconds, while the average verification time of the BLS signature is about several thousand milliseconds, that is to say, the BPLSM in the same experimental environment. The signature verification efficiency of the protocol is less than the time cost of BLS signature, and the result can be verified without multiple single public key signature message comparisons.

In order to show the experimental phenomenon objectively and in detail, Figure 2 compares the average verification time loss of the two. It can be seen that in the multi-party computation of the three parties involved in this experiment, the average signature verification time of BPLSM is 260.8 ms, and the average signature verification time of BLS is 2908.8 ms. The time cost loss of BPLSM protocol signature verification is about 83.5% less than that of BLS signature verification.

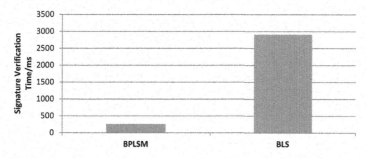

Figure 2.   The average verification efficiency between BPLSM protocol and BLS signature.

5   CONCLUSION

Starting with the multi-party computing scenario, this research offers BPLSM, a general-purpose safe multi-party computing protocol based on blockchain technology for data privacy protection. The protocol uses the Pedersen additive homomorphism to realise the combined transaction commitment under the chain, as well as the verification of the signature on the chain and the guarantee of the promise, the validity of the encrypted value, and the concealing of the address. Simulating the planned protocol experimentally is also done at the same time. In a small-scale multi-party transaction with a fixed number of persons, it has been confirmed that the solution to this privacy calculation can lower the time cost of signature verification when compared to the existing BLS signature. The developed BPLSM protocol is not put into a more realistic multi-party application scenario for analysis because the protocol is just in the experimental simulation stage. Many unpredictable issues exist with off-chain signatures in real-world scenarios, such as signing parties not being online at the same moment, missing signatures, and so on. As a result, the next stage in this research will be to use the BPLSM protocol to apply the protocol to specific circumstances, such as cross-border commerce scenarios, and to integrate the current mainstream zero-knowledge proof algorithm for further optimization.

REFERENCES

Chen F. *et al.* "Blockchain-based Optical Network Slice Rental Approach for IoT," 2020 *IEEE Computing, Communications and IoT Applications (ComComAp)*, 2020, pp. 1–4, doi: 10.1109/ComComAp51192.2020.9398886.
Crowe W. and Oh T. T., "Distributed Unit Security for 5G Base-Stations using Blockchain," *2020 International Conference on Software Security and Assurance (ICSSA)*, 2020, pp. 10–14, doi: 10.1109/ICSSA51305.2020.00010.

Hewa T. M., Kalla A., Nag A., Ylianttila M. E. and Liyanage M., "Blockchain for 5G and IoT: Opportunities and Challenges," *2020 IEEE Eighth International Conference on Communications and Networking (ComNet)*, 2020, pp. 1–8, doi: 10.1109/ComNet47917.2020.9306082.

Hewa T., Braeken A., Ylianttila M. and Liyanage M., "Multi-Access Edge Computing and Blockchain-based Secure Telehealth System Connected with 5G and IoT," GLOBECOM 2020 – 2020 *IEEE Global Communications Conference*, 2020, pp. 1–6, doi: 10.1109/GLOBECOM42002.2020.9348125.

Huang H., Miao W., Min G., Tian J. and Alamri A., "NFV and Blockchain Enabled 5G for Ultra-Reliable and Low-Latency Communications in Industry: Architecture and Performance Evaluation," in *IEEE Transactions on Industrial Informatics*, vol. 17, no. 8, pp. 5595–5604, Aug. 2021, doi: 10.1109/ TII.2020.3036867.

Khujamatov K., Reypnazarov E., Akhmedov N. and Khasanov D., "Blockchain for 5G Healthcare Architecture," *2020 International Conference on Information Science and Communications Technologies (ICISCT)*, 2020, pp. 1–5, doi: 10.1109/ICISCT50599.2020.9351398.

Li H., Gao P., Zhan Y. and Tan M., "Blockchain Technology Empowers Telecom Network Operation," in *China Communications*, vol. 19, no. 1, pp. 274–283, Jan. 2022, doi: 10.23919/JCC.2022.01.020.

Longbing L., Shuaifeng H., Peng J., Jiapeng Y. and Hengji C., "Research on information fusion mechanism algorithm of blockchain nodes based on 5G," *2021 IEEE 3rd International Conference on Civil Aviation Safety and Information Technology (ICCASIT)*, 2021, pp. 674–677, doi: 10.1109/ ICCASIT53235.2021.9633484.

Mirtskhulava L., Iavich M., Razmadze M. and Gulua N., "Securing Medical Data in 5G and 6G via Multichain Blockchain Technology using Post-Quantum Signatures," 2021 *IEEE International Conference on Information and Telecommunication Technologies and Radio Electronics (UkrMiCo)*, 2021, pp. 72–75, doi: 10.1109/UkrMiCo52950.2021.9716595.

Mukesh Soni, Dileep Kumar Singh "Privacy Preserving Authentication and Key Management Protocol for Health Information System", *Data Protection and Privacy in Healthcare: Research and Innovations*, Page-37, CRC Publication,2021.

Ning Z., Chen H., Wang X., Wang S. and Guo L., "Blockchain-Enabled Electrical Fault Inspection and Secure Transmission in 5G Smart Grids," in *IEEE Journal of Selected Topics in Signal Processing*, vol. 16, no. 1, pp. 82–96, Jan. 2022, doi: 10.1109/JSTSP.2021.3120872.

Shi Y., Li G., Xu X., Wu J. and Li J., "PFCC: Predictive Fast Consensus Convergence for Mobile Blockchain over 5G Slicing-enabled IoT," *2021 IEEE Global Communications Conference (GLOBECOM)*, 2021, pp. 1–6, doi: 10.1109/GLOBECOM46510.2021.9685419.

Shobanadevi, A., Tharewal, S., Soni, M. *et al.* Novel Identity Management System using Smart Blockchain Technology. *Int J Syst Assur Eng Manag (2021)*. https://doi.org/10.1007/s13198-021-01494-0

Soni M. and Singh D. K., "Blockchain Implementation for Privacy Preserving and Securing the Healthcare Data," 2021 10th *IEEE International Conference on Communication Systems and Network Technologies (CSNT)*, 2021, pp. 729–734, doi: 10.1109/CSNT51715.2021.9509722.

Weerasinghe N., Hewa T., Liyanage M., Kanhere S. S. and Ylianttila M., "A Novel Blockchain-as-a-Service (BaaS) Platform for Local 5G Operators," in *IEEE Open Journal of the Communications Society*, vol. 2, pp. 575–601, 2021, doi: 10.1109/OJCOMS.2021.3066284.

Wu J, Haider SA, Soni M, Kalra A, Deb N. 2022. Blockchain based Energy Efficient Multi-tasking Optimistic Scenario for Mobile Edge computing. *PeerJ Computer Science* 8:e1118 https://doi.org/10.7717/peerj-cs.1118

Yang H., Zheng H., Zhang J., Wu Y., Lee Y. and Ji Y., "Blockchain-based Trusted Authentication in Cloud Radio OverFiber Network for 5G," *2017 16th International Conference on Optical Communications and Networks (ICOCN)*, 2017, pp. 1–3, doi: 10.1109/ICOCN.2017.8121598.

Yazdinejad A., Parizi R. M., Dehghantanha A. and Choo K.-K.R., "Blockchain-Enabled Authentication Handover With Efficient Privacy Protection in SDN-Based 5G Networks," in *IEEE Transactions on Network Science and Engineering*, vol. 8, no. 2, pp. 1120–1132, 1 April-June 2021, doi: 10.1109/TNSE.2019.2937481.

Yue K. *et al.* "A Survey of Decentralizing Applications via Blockchain: The 5G and Beyond Perspective," in *IEEE Communications Surveys & Tutorials*, vol. 23, no. 4, pp. 2191–2217, Fourthquarter 2021, doi: 10.1109/COMST.2021.3115797.

604

Artificial Intelligence, Blockchain, Computing and Security – Dagur et al. (Eds)
© 2024 The Author(s), ISBN: 978-1-032-49393-0

Parallel Byzantine fault tolerance method for blockchain

Kumar Pradyot Dubey*

Assistant Professor, Department of Computer & Information Science, Jagadguru Ram Bhadracharya Divyanga University, Chitrakoot (U.P.)

C.N. Gnanaprakasam*

Department of Electronics and Instrumentation Engineering St. Joseph's College of Engineering, Chennai, Tamilnadu, India

Ihtiram Raza Khan*

Computer Science Department, Jamia Hamdard, Delhi

Md Shibli Sadik*

Msc Computer Network Administrator and Management, University of Portsmouth, Portsmouth, UK

Liwa H. Al-Farhani*

System Analysis, Control and Information Processing Dep., Academy of Engineering, RUDN University, Moscow, Russia

Samrat Ray*

Senior research scholar, The Institute of Industrial Management, Economics and Trade, Peter The Great Saint Petersburg Polytechnic University, Russia

ABSTRACT: Enterprise-focused blockchain applications that offer safe, anonymous, and immutable transactions are growing as blockchain technology matures. Traditional blockchain architecture has low performance and scalability, making it unsuitable for enterprise-level applications with high concurrency and massive data. This paper studies and proposes a simplified Byzantine fault-tolerant (SBFT) consensus algorithm to improve consensus stage efficiency, a Task parallel smart contract model to maximise multi-core system parallel efficiency, and an improved blockchain system architecture to reflect the lightweight, low coupling nature of blockchain technology. It facilitates corporate application secondary development and the ParaChain blockchain and smart contract technology. Experimental verification compares the parallelization – based ParaChain blockchain's TPS and scalability. PBFT-based blockchain systems have been greatly improved.

Keywords: Blockchain, Immutable Transaction, Multi-Core Systems, Byzantine Fault-Tolerant, Smart Contract

1 INTRODUCTION

As an underlying protocol or technical solution, blockchain can effectively solve the trust problem and realize the free transfer of value (Muneeb *et al.* 2022; Sun *et al.* 2022) and other

*Corresponding Authors: pradyotdubey@gmail.com, gnanacn@gmail.com, erkhan2007@gmail.com, up2029264@myport.ac.uk, liwarussia@gmail.com and samratray@rocketmail.com

DOI: 10.1201/9781003393580-92

fields have broad application prospects. However, performance issues are the key issues that hinder the implementation of blockchain technology, and most public chain architectures sacrifice system performance to ensure data security. Consortium blockchain(Giraldo *et al.* 2020)is a new type of blockchain architecture between the public and private chains and is a blockchain system jointly constructed by multiple institutions. Unlike the public chain, the consortium chain is only open to specific institutions, and the node It is necessary to pass an audit and access to join the chain for consensus, which ensures the security and controllability of nodes, reduces the scale of consensus nodes, and speeds up transactions. At the same time, the alliance chain also solves the "information island" in the private chain", which reduces the degree of privatization of the private chain, reduces the communication cost between enterprises and institutions, and finally achieves mutual benefit and win-win within the alliance.

The alliance chain avoids the performance loss caused by the consensus of the whole network of the public chain system. Further, it improves functions such as permission control, making it more able to meet traditional business needs. Compared with the public chain, the various advantages of the consortium chain make more and more consortium chain systems have been developed and put into use

To solve the above problems, this paper proposes a parallel alliance chain ParaChain. ParaChain implements a blockchain system with low coupling, high throughput and high scalability based on Java language and innovatively combines parallelization technology through a unique consensus. The protocol and intelligent contract model optimizes the data throughput performance while ensuring the standard requirements of the alliance chain. Furthermore, the experimental results show that under the same hardware conditions, the TPS performance of the ParaChain blockchain system based on the parallel method is optimized.

The main contributions of this paper are as follows:

1) Designed the Task intelligent contract model, implemented a checkpoint-based runtime scheduling scheme, supported parallel brilliant contract execution and ensured data correctness;
2) The simplified Byzantine consensus algorithm SBFT is designed, which realizes the simplification and parallelism of the consensus process stage;
3) Use Spring Boot + Netty to develop a Para Chain blockchain system, which realizes lightweight, low coupling and contract extensibility.

2 RELATED WORKS

The consensus protocol is one of the most critical components of blockchain technology. POW consensus protocol (Dwivedi *et al.* 2021) is the theoretical support for Bitcoin (Wang *et al.* 2019). However, based on the CAP principle, consistency, availability, and partitioning in distributed systems, Fault tolerance can only be satisfied with at most two characteristics simultaneously. The POW consensus protocol, while ensuring a high degree of decentralization, also brings problems such as large resource consumption and poor data processing performance. The POS (Wu *et al.* 2022; Zhang *et al.* 2021) proposed by Ethereum and the subsequent Consensus protocols such as DPoS and PBFT (Omar *et al.* 2021; Kushwaha *et al.* 2022) aim to improve the many shortcomings of POW and improve the usability of the blockchain system. This paper proposes the Task smart contract parallel model and the SBFT consensus protocol, aiming at the related problems of blockchain performance optimization. Based on the above research, it implements the ParaChain blockchain system and completes the experiment.

3 PARACHAIN PARALLELIZATION DESIGN

The application of similar technology in ParaChain is a vital innovation point. Similar technology refers to completing two or more tasks simultaneously or within the same time interval, which can be roughly divided into time parallelism, space parallelism and time and space parallelism. Time Parallelism refers to the interleaved use of device resources among multiple processing tasks to improve hardware utilization rate. The most common implementation method is pipeline processing. Spatial parallelism is performed on multiple processors to improve the processing speed of functions, mainly relying on large-scale integration. Circuits. In natural parallel task optimization, both time and space parallelism is often used to improve task processing speed (Dustdar *et al.* 2021; Kemmoe *et al.* 2020; Kushwaha *et al.* 2022).

In the system implementation of ParaChain, the design and implementation of the parallel model is an important part, and it is also the key to improving the TPS performance of the entire blockchain. In addition to network transmission, file I/O and other force majeure conditions limited by hardware performance, ParaChain is In the two crucial, time-consuming functional layers in the blockchain system (Garcia *et al.* 2020; Guo *et al.* 2021; Vangala *et al.* 2021), the consensus layer and the smart contract layer, a system improvement scheme based on the parallel model is designed and implemented, thereby improving the overall data processing performance. For example, in the consensus layer, ParaChain, A simplified Byzantine Fault Tolerance (SBFT) algorithm is created, which reduces the time-consuming of consensus by simplifying and parallelizing the traditional PBFT consensus process. Furthermore, algorithmic process and function implementation support parallel intelligent contract execution(Madhwal *et al.* 2022; Saini *et al.* 2021; Su *et al.* 2022; Wu *et al.* 2022).

4 TASK INNOVATIVE CONTRACT PARALLEL MODEL

ParaChain supports two smart contract models: the Task innovative contract model and the Java smart contract model. The Task innovative contract model is ParaChain's original, intelligent contract system, which connects the operation interface of the data on the chain. The execution of the transaction will trigger the performance of the smart contract. In the calling process, their corresponding virtual machines will execute other intelligent agreements, generate the operation sequence of StateDB, and synchronize to the storage module. Figure 1 is a schematic diagram of this process

Task supports parallel execution of transactions. Since transactions in blocks are ordered, and there may be data dependencies between transactions, mainstream blockchain systems execute trades in serial form. Directly running transactions in parallel can improve The speed of the whole process, but the problem of data consistency on the chain may occur. Figure 2 shows an example of transaction data conflict.

The two transactions, T1 and T2, respectively, call the smart contracts C1 and C2, and the two intelligent agreements are embodied in the form of bytecode. At the bytecode level, both C1 and C2 declare the code fragment using the data A1 on the chain, and C1 reads, Take data, and C2 write data. In the example figure, the use logic of C1 and C2 for data A1 is framed: C1 reads data A1 and assigns it to variable a, and determines the subsequent jump process based on whether the value of variable a is equal to 100; C2 sets A1 to 0.

Assuming that A1 is 100 in the initial state, C1 will execute a jump under serial execution. However, if it is directly parallelized in the form of multiple threads, due to the uncertainty of parallel scheduling, when C2 writes 0 to A1, the operation of writing 0 to A1 is executed first. Then C1 Will read the write result of C2 because a is equal to 0 and is no longer valid; C1 will not jump, resulting in inconsistent execution results of intelligent contracts under serial and parallel.

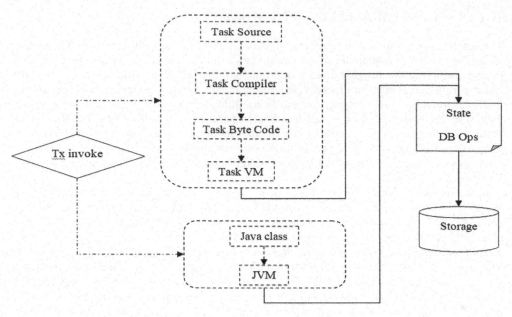

Figure 1.　Contract execution process.

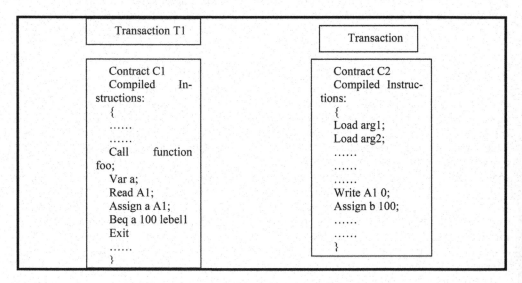

Figure 2.　Example of transaction data conflict.

Based on the orderliness of transactions, parallel execution needs to avoid inconsistencies between results and serial performance. The Task innovative contract model is based on transactional memory and MVCC model, and a fine-grained brilliant contract parallel algorithm-PTR (ParaChainTaskRuntime) algorithm is proposed. On the premise of ensuring data consistency on the chain, it supports the concurrent execution of transactions to improve the data throughput performance of the entire system. The PTR algorithm needs to be supported by a specific TaskRuntime, so ParaChain implements a complete set of runtime mechanisms based on ANTLR tools.

The pseudocode of the Task parallel algorithm is shown in Algorithm 1.

Algorithm 1 Task Parallel Algorithm

Step 1: TQ=getTxQueue(block)/*Get the transaction queue*/
Step 2: forunitinTQ:PTR(unit,wset)/*Parallel execution in batches*/
Step 3: syncDB(wset)/*Sync to the chain*/
Step 4: exit

In Algorithm 1, the unit is an ordered transaction queue with a customizable length, which is used as the input to execute the PTR algorithm once; wset is the final write information of the data on the chain in this process. The PTR algorithm runs the transactions in the unit in parallel, Generates instructions to write data on the chain, and finally synchronizes to the chain.

The pseudocode of the main flow of the PTR algorithm is shown in Algorithm 2.

Algorithm 2 PTR Algorithm – Main Process

Step 1: Init(threads, scheduler, wset)
Step 2: for tx in unit:
a. tbytecode=compile(tx)/*Generate byte code*/
b. threads. add(tbytecode,wset)/*Add thread pool*/
Step 3: scheduler. schedule(threads)/*Parallel scheduling*/
Step 4: wait()/*Wait for all unit transactions to be executed*/
Step 5: exit

The PTR algorithm selects limited parallel scheduling to prevent the detection of shared data access caused by too many similar transactions. The PTR algorithm read request conflict detection pseudocode is shown in Algorithm 3.

Algorithm 3 PTR Algorithm – Read Request Conflict Detection

Step 1: TransactionalRead(txNo,key)
Step 2: if key is in Write Set
a. for item in Write Set:
b. if item. key==key and item. txNo>txNo:
c. Rollback(VM)/*corresponds to virtual machine rollback*/
d. Delete(item)/*Delete this record*/
Step 3: return Latest Value(key)
Step 4: exit

Algorithm 3 describes how the PTR algorithm handles a read request. Among them, txNo is the transaction priority number that initiates the operation, and the key is the key that needs to read the value. The PTR algorithm will maintain a runtime WriteSet for recording parallel runtimes. The write operation of TaskVM to StateDB. When the TaskVM corresponding to the transaction with the serial number of txNo has a read operation, traverse the WriteSet and roll back the TaskVM whose transaction priority is lower than the current transaction to the state where the access data has not been written. Similarly, the conflict detection of the write request will be rolled back to the TaskVM related to the corresponding read operation. Therefore, the PTR algorithm ensures the consistency of the results obtained from the parallel execution of the Task smart contract and the serial execution.

5 EXPERIMENT AND EVALUATION

To test whether the performance of the ParaChain system based on a similar design is improved compared to the traditional blockchain, this paper deploys the ParaChain system on a 4-node cluster to complete the TPS performance comparison experiment. The results are shown in Figure 3 and Table 1. The black part in the figure is based on traditional blockchain, The normal operation mode of the PBFT consensus protocol; the grey region is

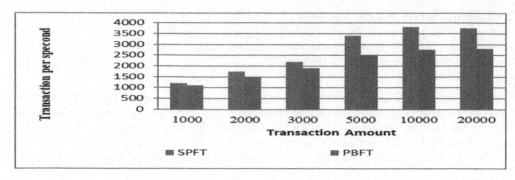

Figure 3.　TPS performance comparison.

Table 1.　TPS performance comparison.

	Transaction per second					
Transaction Amount	1000	2000	3000	5000	10000	20000
SPFT	1200	1750	2200	3400	3800	3750
PBFT	1100	1500	1900	2500	2750	2800

the ParaChain parallel operation mode, and the horizontal axis of the coordinates represents the size of the number of transactions in the block. The vertical axis is the TPS value. It can be seen that with the increase in the block data volume, The acceleration effect obtained by ParaChain using the parallel scheme is more apparent.

Through the analysis of the log records of the experiment in Figure 3 and Table 1, the primary sources of time-consuming in the consensus process can be divided into three parts: data network communication between nodes, internal data calculation of nodes and other work. Figure 4 and Table 2 counts the total time spent in the consensus process.

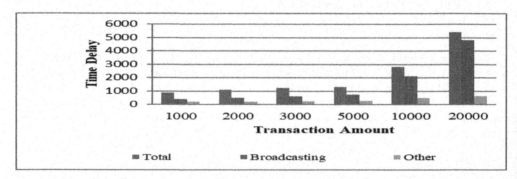

Figure 4.　Time-consuming data trend.

Table 2.　Time-consuming data trend.

	Transaction per second					
Transaction Amount	1000	2000	3000	5000	10000	20000
Total	900	1100	1250	1300	2800	5400
Broadcasting	400	500	600	750	2100	4800
Other	200	220	240	300	500	650

Still, the growth rate is relatively slow, while the total time-consuming and network communication time-consuming expansion. Furthermore, when the number of transactions is significant in the block, the network communication time is dominant the whole time, so the performance optimization obtained by optimizing the network communication delay is more prominent. ParaChain implements SBFT. The parallelization and simplification scheme effectively reduces the delay caused by multiple network transmissions.

6 CONCLUSION

ParaChain, a blockchain solution for enterprise scenarios, provides high-availability blockchain technology services for traditional business scenarios. ParaChain improves and innovates the traditional blockchain system structure and implements the SBFT consensus algorithm. The Task innovative contract model optimizes the system's availability and data throughput performance. ParaChain builds a layered, low-coupling modular structure to facilitate secondary development for different business scenarios; the system supports pluggable consensus protocols and various The innovative contract model provides users with more customization space. ParaChain has been used in production environments such as point systems and financial systems, adding traceability and anti-tampering features to the traditional centralized data management model; In the business scenario, ParaChain has shown good usability and adaptability.

REFERENCES

Dustdar S., Fernández P., García J. M. and Ruiz-Cortés A., "Elastic Smart Contracts in Blockchains," in *IEEE/CAA Journal of Automatica Sinica*, vol. 8, no. 12, pp. 1901–1912, December 2021, doi: 10.1109/JAS.2021.1004222.

Dwivedi V., Norta A., Wulf A., Leiding B., Saxena S. and Udokwu C., "A Formal Specification Smart-Contract Language for Legally Binding Decentralized Autonomous Organizations," in *IEEE Access*, vol. 9, pp. 76069–76082, 2021, doi: 10.1109/ACCESS.2021.3081926.

Garcia P. S. R. and Kleinschmidt J. H., "Sharing Health and Wellness Data with Blockchain and Smart Contracts," in *IEEE Latin America Transactions*, vol. 18, no. 06, pp. 1026–1033, Jun 2020, doi: 10.1109/TLA.2020.9099679.

Giraldo F. D., B. Milton C. and Gamboa C. E., "Electronic Voting Using Blockchain and Smart Contracts: Proof Of Concept," in *IEEE Latin America Transactions*, vol. 18, no. 10, pp. 1743–1751, October 2020, doi: 10.1109/TLA.2020.9387645.

Guo L., Liu Q., Shi K., Gao Y., Luo J. and Chen J., "A Blockchain-Driven Electronic Contract Management System for Commodity Procurement in Electronic Power Industry," in *IEEE Access*, vol. 9, pp. 9473–9480, 2021, doi: 10.1109/ACCESS.2021.3049562.

Kemmoe V. Y., Stone W., Kim J., Kim D. and Son J., "Recent Advances in Smart Contracts: A Technical Overview and State of the Art," in *IEEE Access*, vol. 8, pp. 117782–117801, 2020, doi: 10.1109/ACCESS.2020.3005020.

Kushwaha S. S., Joshi S., Singh D., Kaur M. and Lee H.N, "Ethereum Smart Contract Analysis Tools: A Systematic Review," in *IEEE Access*, vol. 10, pp. 57037–57062, 2022, doi: 10.1109/ACCESS.2022.3169902.

Kushwaha S. S., Joshi S., Singh D., Kaur M. and Lee H. -N., "Systematic Review of Security Vulnerabilities in Ethereum Blockchain Smart Contract," in *IEEE Access*, vol. 10, pp. 6605–6621, 2022, doi: 10.1109/ACCESS.2021.3140091.

Madhwal Y., Borbon-Galvez Y., Etemadi N., Yanovich Y. and Creazza A., "Proof of Delivery Smart Contract for Performance Measurements," in *IEEE Access*, vol. 10, pp. 69147–69159, 2022, doi: 10.1109/ACCESS.2022.3185634.

Muneeb M., Raza Z., Haq I. U. and Shafiq O., "SmartCon: A Blockchain-Based Framework for Smart Contracts and Transaction Management," in *IEEE Access*, vol. 10, pp. 23687–23699, 2022, doi: 10.1109/ACCESS.2021.3135562.

Omar I. A., Jayaraman R., Debe M. S., Salah K., Yaqoob I. and Omar M., "Automating Procurement Contracts in the Healthcare Supply Chain Using Blockchain Smart Contracts," in *IEEE Access*, vol. 9, pp. 37397–37409, 2021, doi: 10.1109/ACCESS.2021.3062471.

Saini A., Zhu Q., Singh N., Xiang Y., Gao L. and Zhang Y., "A Smart-Contract-Based Access Control Framework for Cloud Smart Healthcare System," in *IEEE Internet of Things Journal*, vol. 8, no. 7, pp. 5914–5925, 1 April1, 2021, doi: 10.1109/JIOT.2020.3032997.

Su H., Guo B., Shen Y. and Suo X., "Embedding Smart Contract in Blockchain Transactions to Improve Flexibility for the IoT," in *IEEE Internet of Things Journal*, vol. 9, no. 19, pp. 19073–19085, 1 Oct.1, 2022, doi: 10.1109/JIOT.2022.3163582.

Sun J., Huang S., Zheng C., Wang T., Zong C. and Hui Z., "Mutation Testing for Integer Overflow in Ethereum Smart Contracts," in *Tsinghua Science and Technology*, vol. 27, no. 1, pp. 27–40, Feb. 2022, doi: 10.26599/TST.2020.9010036.

Vangala A., Sutrala A. K., Das A. K. and Jo M., "Smart Contract-Based Blockchain-Envisioned Authentication Scheme for Smart Farming," in *IEEE Internet of Things Journal*, vol. 8, no. 13, pp. 10792–10806, 1 July1, 2021, doi: 10.1109/JIOT.2021.3050676.

Wang S., Ouyang L., Yuan Y., Ni X., Han X. and Wang F. -Y., "Blockchain-Enabled Smart Contracts: Architecture, Applications, and Future Trends," in *IEEE Transactions on Systems, Man, and Cybernetics: Systems*, vol. 49, no. 11, pp. 2266–2277, Nov. 2019, doi: 10.1109/TSMC.2019.2895123.

Wu C., Xiong J., Xiong H., Zhao Y. and Yi W., "A Review on Recent Progress of Smart Contract in Blockchain," in *IEEE Access*, vol. 10, pp. 50839–50863, 2022, doi: 10.1109/ACCESS.2022.3174052.

Wu J, Haider SA, Soni M, Kalra A, Deb N. 2022. Blockchain Based Energy Efficient Multi-tasking Optimistic Scenario for Mobile Edge Computing. PeerJ Computer Science 8:e1118 https://doi.org/10.7717/peerj-cs.1118.

Zhang Y., Yutaka M., Sasabe M. and Kasahara S., "Attribute-Based Access Control for Smart Cities: A Smart-Contract-Driven Framework," in *IEEE Internet of Things Journal*, vol. 8, no. 8, pp. 6372–6384, 15 April15, 2021, doi: 10.1109/JIOT.2020.3033434.

Artificial Intelligence, Blockchain, Computing and Security – Dagur et al. (Eds)
© 2024 The Author(s), ISBN: 978-1-032-49393-0

Fuzzy random proof of work for consensus algorithm in blockchain

Akhilesh Kumar*
Assistant Professor, PG Department of Information Technology, Gaya College, Gaya, Bihar, India

Z.H. Kareen*
Medical Instrumentation Techniques Engineering Department, Al-Mustaqbal University College, Babylon, Iraq

Mustafa Mudhafar*
Department of Anesthesia and Health Care Faculty of Altuff College University, Iraq

Gioia Arnone*
PhD Student, Disaq – Università Degli Studi Di Napoli "Parthenope", Italy

Mekhmonov Sultonali Umaralievich*
Tashkent Institute of Finance, Tashkent, Uzbekistan

Avijit Bhowmick*
Deparment of Computer Science & Engineering, BBIT, Kolkata

ABSTRACT: The consensus method, being one of the most important aspects of blockchain, differs depending on the sector. The commonly used proof-of-work (PoW) consensus technique for public chain application scenarios still has challenges that are difficult to handle, such as security and high computer power. As a result, the PoW method is investigated in terms of enlarging the solution space and improving the adjustment mechanism. A consensus technique based on fuzzy random proof of work (FRMH) is presented. The FRMH algorithm improves the security of the blockchain consensus mechanism by increasing the solution space of the consensus algorithm by incorporating technologies such as a fuzzy transitive closure matrix in fuzzy mathematics. In addition, the FRMH algorithm uses a dual adjustment method to cope with machines with high computational power and thus solves the issue of high computing power being difficult to regulate on the blockchain. Through mathematics, it has been shown that the FRMH algorithm has greatly enhanced solution space and greater processing power control.

Keywords: Blockchain, Consensus Algorithm, Fuzzy Random, Security, Public Chain Application, Computing Power

1 INTRODUCTION

Blockchain which was proposed in 2008, also known as the Internet of value (Cui *et al.* 2020; Zhang *et al.* 2021) is essentially a decentralized distributed database, which is an integrated innovation of various technologies such as cryptography, consensus algorithms,

*Corresponding Authors: getaky123@gmail.com, Zahraa.hashim@mustaqbal-college.edu.iq, almosawy2014@gmail.com, gioia.arnone@studenti.uniparthenope.it, mehmonov_s@tfi.uz and dr.avijit.bhowmick@gmail.com

DOI: 10.1201/9781003393580-93

and peer-to-peer networks. The blockchain realizes the consistency of node data in the blockchain system through the consensus mechanism, and the consensus algorithm is the core element to achieve this goal.

According to different application scenarios, blockchain is divided into public chain, private chain, and alliance chain (Liang *et al.* 2021). The writing authority of the private chain belongs to an organization and institution, and the participating nodes in its network system will be strictly limited. The number of nodes in the alliance chain is relatively fixed, divided into accounting nodes and other nodes. Among them, the accounting nodes are designated by the group in advance, and the accounting nodes jointly decide the generation of each block on the chain. Other nodes can participate in the transaction, but do not ask the accounting process. The consensus algorithm of the alliance chain is divided into Byzantine fault-tolerant consensus algorithm and non-Byzantine fault-tolerant consensus algorithm. The typical consensus algorithm of the former is Practical Byzantine

Tribunal Fault Tolerance PBFT(Lai *et al.* 2019). HotStuffBFT(Jeong *et al.* 2019), etc.; the typical consensus algorithms of the latter include Paxos (Cirstea *et al.* 2018), Raft(He *et al.* 2022), etc. Compared with private chains and consortium chains, public chains are currently the most widely used blockchains.

The public chain system allows all participating nodes to read and write their data, and can freely enter and exit. Its typical consensus mechanisms include Proof of Work (PoW) (Banerjee *et al.* 2021). Proof of Stake (PoS) (David *et al.* 2021) and Proof of Share Authorization (DPoS) (Ray *et al.* 2021) etc. algorithm. Among them, the PoW algorithm, as a relatively mature consensus technology, has been applied to public chains such as Bitcoin and Ethereum. However, there are some urgent problems in practical applications. For example, the emergence of high-speed computers or quantum computers will make it difficult for the encryption algorithm used by PoW to ensure its security Kumutha *et al.* 2021) and it is also difficult to control the computing power. Further improving the security of the blockchain and regulating the computing power of high-speed computers are the hotspots of current blockchain research. This paper aims at the security of the consensus algorithm and the difficulty in controlling the computing power (Han *et al.* 2020; Tan *et al.* 2022) Some researcher are proposing security (Shobanadevi *et al.* 2021; Soni *et al.* 2021) technique using blockchain.

To solve the problem, a new consensus algorithm based on fuzzy mathematics is proposed. This algorithm increases its solution space to solve the security problem by introducing technologies such as fuzzy transitive closure matrix, and also adopts a dual adjustment mechanism to solve the problem of difficult control of computing power.

1.1 *Fuzzy transitive closure matrix*

In 1965, Professor Zadeh, an American cybernetics expert, published the article "Fuzzy Sets" (Whaiduzzaman *et al.* 2021) it laid the foundation for fuzzy mathematics. This pioneering work opens up broad avenues for improving the applicability of mathematics in the soft and hard sciences. The fuzzy transitive closure matrix and other related knowledge used in this paper are defined as follows (Zhaofeng *et al.* 2020).

Define 1 rij $\in$[0,1](i = 1,2, ... ,m; j = 1,2, ... ,n), the matrix R = (rij)m × n is called a fuzzy matrix of order m × n. Definition 2 When m = n and RT = R, R is a fuzzy symmetric matrix. Definition 3. A transitive matrix containing R and contained by any transitive matrix containing R is called the transitive closure of R t(R). The procedure for calculating t(R) uses the multiplication formula:

$$S = Q \circ R = \left(\bigvee_{k=1}^{n} \left(q_{ik} \wedge r_{kj} \right) \right)_{nxn} \tag{1}$$

where "$\circ$" represents a soft algebra calculation operator. Q represents the fuzzy matrix of order n × n, R represents the fuzzy matrix of order n × n, S represents the product of two fuzzy matrices, qik represents the element of the fuzzy matrix Q, rkj represents the element of the fuzzy matrix R, and the two are obtained by formula (1). The multiplication matrix of a matrix, and then through the power algorithm formula

$$t(R) = R^k (k \geq n) \tag{2}$$

Find the fuzzy transitive closure matrix t(R) of R, where:

$$R\hat{\ }k = R\hat{\ }(k-1) \circ R \tag{3}$$

and has the properties:

$$t(R) \circ t(R) = t(R) \tag{4}$$

Theorem 1 The process of finding t(R) goes through n4 times of $\vee$ and $\wedge$ operations. Prove that $2 = R \circ R \overset{\Delta}{=} (s_{ij})_{nxn}$, where $s_{ij} = (\vee_{k=1}^{n}(q_{ik} \wedge r_{kj}))_{nxn} (i = 1, 2, \ldots, n; j = 1, 2, \ldots, n)$.

(1) For fixed $i, j = 1, 2, \ldots, n$: $q_{ik} \wedge r_{kj}$ is n times of $\wedge$ operations;
(2) (2) For i = 1,2, ... ,n, there are $n^2 \wedge$ operations in total;
(3) (3) For k = 1,2, ... ,n, there are n times of $\vee$ operations;
(4) From the above three points, it can be known that for the matrix multiplication operation of formula (1)
(5) R $\circ$ R has a total of n × $n^2 = n^3$ ($\vee, \wedge$) operations.
(6) (4) From formulas (2) and (3), it can be known that the process of finding t(R) has a total of n times
(7) Matrix multiplication operation.
(8) To sum up the above four points, it can be seen that the fuzzy transitive closure matrix of t(R) has n × $n^3 = n^4$
(9) The operation is completed, and the proof is completed.

2 FRMH ALGORITHM DESIGN

With the continuous improvement of the computing speed of the computer and the constant solution space to deal with the high-speed computer, in theory, the encryption algorithm of PoW will eventually be cracked. The new consensus algorithm in this paper introduces the combination of fuzzy transitive closure matrix and hash encryption to achieve secondary encryption, and then obtains a hash value that meets the conditions through random number collision to improve the security of the consensus algorithm. This innovative algorithm is named fuzzy random collision proof-of-work consensus algorithm, also known as FRMH consensus algorithm.

The FRMH consensus algorithm uses the seed and block header hash as input, and then operates the random number to obtain a block that meets the difficulty and passes the verification to complete the entire algorithm. The seed consists of n, p, and H in the previous block, where n is the order of the fuzzy transitive closure matrix, p is the number of 0s in the hash value, and H is the value of the random number in the previous block. The FRMH consensus algorithm mainly includes four stages: construction matrix, operation matrix, job generation, and job verification.

2.1 Constructing the matrix

The technological innovation of the FRMH consensus algorithm is the introduction of fuzzy transitive closure matrix into the consensus algorithm for the first time. In order to obtain a transitive closure matrix that meets the requirements, the first task is to construct a fuzzy symmetric matrix of order n. The main steps are as follows:

(a) Generate fuzzy symmetric matrixA fuzzy symmetric matrix R of order n whose row and column vectors are linearly independent is generated from the seed.

(b) Find the fuzzy transitive closure matrix of RThe fuzzy propagation of R is calculated by formulas (1) and (2) above.Transitive closure matrix t(R) The transitive closure matrix t(R) obtained here is a one-way irreversible process,It combines the characteristics of blockchain hash encryption with fast forward speed and difficult reverse direction.

2.2 Operation matrix

The fuzzy transitive closure matrix t(R) obtained by formula (2) is encrypted by the S encryption algorithm, and a value f with a length of 32 bytes is obtained, where:

$$f = S(t(R)) \tag{5}$$

2.3 Job generation

Use the same S encryption algorithm as the operation matrix stage to encrypt f and the block header, and also get a value with a length of 32 bytes
hf, where:

$$hf = S(f + BlockHead) \tag{6}$$

2.4 Work verification

To verify whether a node has completed its work, it is to compare the work value of the node with the difficulty of the block. Since the S encryption algorithm used in the FRMH consensus algorithm is the SHA256 encryption algorithm, the node work verification is to determine whether the number of 0s in the 256-bit hash value of hf is equal to the p value in the seed. If it is not equal, it means that the work verification is unsuccessful, then assign the nonce value generated after the encryption operation of the work generation stage S to H, and start a new round of calculation process. After the verification is successful, the node will broadcast the block before others broadcast it.

2.5 Solution space

The FRMH consensus algorithm is divided into two processes: solving the fuzzy transitive closure matrix and hashing.

(a) Find the transitive closure matrix
  Formula (1) is the intermediate process of exponentiation to find t(R). Through Theorem 1, it can be known that the calculation times of this process is n4, then the solution space of this process is n4.
(b) Encryption operation
  The encryption operation process uses the SHA256 encryption algorithm to operate the block header and the encrypted value of the fuzzy transitive closure array, and the solution space is 2256. Based on the calculation results of the above two steps, the FRMH consensus algorithm solution is obtained.

The interval is $n^4 \times 2^{256}$.

2.6 Difficulty adjustment

The FRMH consensus algorithm adjusts the difficulty of the block by the order n of the fuzzy transitive closure matrix and the number p of the leading zeros in the hash value. The experimental hardware adopts the Intel Core i7-7700HQ processor with 4 cores and 8 threads, and the hardware platform with 8 GB memory. The computing power simulation curve shown in Figure 3 is obtained by building the gen private chain. The order n of the fuzzy matrix in the figure and the number p of the leading 0 in the hash value are automatically adjusted with the size of the computing power, which can be divided into linear adjustment and geometric adjustment. Mainly divided into the following situations.

(1) When $p < 256$
- Linear adjustment:

$$f(\alpha) = \begin{cases} n = n - 1, p = p, \alpha < 0.8 \\ n = n, p = p, 0.8 \leq \alpha \leq 2 \\ n = n + (k - 1), p = p, k \leq \alpha < k + 1 (k = 2, 3, \ldots, 6) \\ n = n + 6, p = p, 7 \leq \alpha \leq 10 \end{cases} \qquad (7)$$

- Geometry adjustment:

$$f(\alpha) = n = n, p = p + (k - 1), 10^{k-1} < \alpha \leq 10^k (k = 2, 3, \ldots, 251) \qquad (8)$$

(2) When $p = 256$
- Linear adjustment:

$$f(\alpha) = \begin{cases} p = p, n = n + (k - 1)(n = 1, 2, \ldots) \\ 2^{k-1} \leq \alpha < 2^k (k = 1, 2, \ldots) \\ p = p, n = n - 1, 10^{-2} \leq \alpha < 1 \end{cases} \qquad (9)$$

- Geometry adjustment:

$$f(\alpha) = \{ p = p - k(k = 1, 2, \ldots), n = n, \alpha < 10^{-2k}$$

Where α is the difficulty coefficient, calculated by $\alpha = 10 \div (t - t_0)$ (t_0 = previous block time, t = this block time).

2.7 *FRMH algorithm implementation*

(1) According to the four stages of the FRMH consensus algorithm, the algorithm frame diagram is constructed, as shown in Figure 4. According to the two algorithm block diagrams in Figures 1 and 2, the FRMH consensus algorithm increases the variable

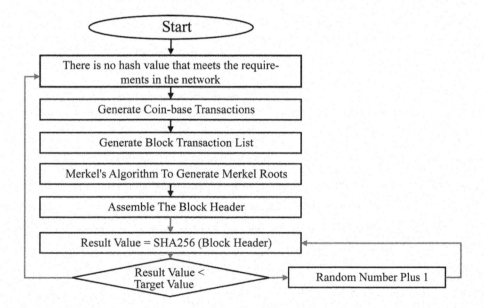

Figure 1.   Bitcoin PoW consensus algorithm working block diagram.

matrix order n compared to the PoW consensus algorithm when assembling the block header; the calculation of the matrix is added during the operation. The difficulty adjustment of the FRMH consensus algorithm is carried out according to a double adjustment mechanism, that is, the values of n (n can be infinite) and p (up to 256 bits) are changed at the same time; while the PoW consensus algorithm is adjusted according to the block time of the latest 2016 blocks Difficulty, i.e. only changing the p-value. However, the variables n introduced by the FRMH consensus algorithm, the change of n reflects the advantages of high computing power regulation.

(2) For the FRMH consensus algorithm solution space problem, when n = 6, the FRMH consensus algorithm solution space is 64 × 2256, the PoW consensus algorithm solution space is 2256, and the RMH consensus algorithm is 1,296 times that of the PoW consensus algorithm, greatly improving security of the blockchain.

3 FRMH ALGORITHM EVALUATION

The FRMH consensus algorithm greatly improves the security performance of the blockchain. From Table 1, it can be seen that the Bitcoin and Ethereum algorithms both use hash encryption algorithms, the solution space is 2256, and the calculation collision takes 248 years (Zhu $et\ al.$ 2021). GeneralIt is almost impossible for an ordinary computer to crack its encryption algorithm, which is why the blockchain has not been cracked so far. But one quantum computer is equivalent to the computing power of 105 ordinary computers (Mukesh Soni $et\ al.$ 2022). For example, using a quantum computer to crack the hash encryption algorithm has no problem at all. The emergence of quantum computers will make the hash encryption algorithm invalid, and the existing blockchain technology will lose its meaning (Yuan $et\ al.$ 2017). The solution space of the double encryption algorithm of the FRMH consensus algorithm is $n^4 \times 2^{256}$, which greatly improves the security of the blockchain system, and with the increase of n, its security performance increases geometrically.

Table 1. Comparison of solution space between FRMH algorithm and mainstream consensus algorithms.

Algorithm	Bitcoin Algorithm	Ethereum Algorithm	Frmh Algorithm
Solation Space	2^{256}	2^{256}	$n^4 x 2^{256}$

The FRMH consensus algorithm can resist the attack of computers with high computing power. It can be seen from Table 2 that both the Bitcoin and Ethereum algorithms adjust the calculation difficulty of the block through the p value, and the adjustment range is 0~256; the FRMH consensus algorithm The double adjustment is achieved by the order n of the fuzzy matrix and the number p of the leading 0s in the hash value, where the value range of p is 0~256, but the value range of n is infinite. No matter how fast a quantum computer operates, there is a limit. In theory, it is always feasible to use the infinite to resist the finite.

Table 2. Comparison of difficulty adjustment between FRMH algorithm and mainstream consensus algorithms.

Algorithm	Bitcoin Algorithm	Ethereum Algorithm	Frmh Algorithm	
Difficulty Adjustment index	p	p	p	n
Difficulty Adjustment range	0~256	0~256	0~256	0~∞

4 CONCLUSION

The consensus mechanism is an important guarantee for the data consistency of blockchain nodes. The existing PoW consensus algorithm has two shortcomings: security and high computing power difficult to control. The FRMH consensus algorithm based on fuzzy mathematics method proposed in this paper solves the security problems brought by high computing power computers and even future quantum computers to the blockchain. The FRMH consensus algorithm incorporates technologies such as fuzzy transitive closure arrays. Through the change of the order of the fuzzy transitive closure arrays, the solution space is increased to n4 × 2256, which greatly improves the security of the blockchain; the algorithm adopts a dual adjustment mechanism (linear adjustment and geometric adjustment), with infinite resistance and limited, it can resist the attacks of high computing power computers and even future quantum computers. In the next step, the FRMH consensus algorithm can be optimized and researched to further improve the block generation time problem on the gen chain, thereby speeding up the transaction speed on the chain, and applying the FRMH consensus algorithm to more public chains for blockchain development. Bring more space.

REFERENCES

Banerjee U. and Chandrakasan A. P., "A Low-Power Elliptic Curve Pairing Crypto-Processor for Secure Embedded Blockchain and Functional Encryption," *2021 IEEE Custom Integrated Circuits Conference (CICC)*, 2021, pp. 1–2, doi: 10.1109/CICC51472.2021.9431552.

Cirstea, Enescu F. M., Bizon N., Stirbu C. and Ionescu V. M., "Blockchain Technology Applied in Health The Study of Blockchain Application in the Health System (II)," *2018 10th International Conference on Electronics, Computers and Artificial Intelligence (ECAI)*, 2018, pp. 1–4, doi: 10.1109/ECAI.2018.8679029.

Cui H., Wan Z., Wei X., Nepal S. and Yi X., "Pay as You Decrypt: Decryption Outsourcing for Functional Encryption Using Blockchain," in *IEEE Transactions on Information Forensics and Security*, vol. 15, pp. 3227–3238, 2020, doi: 10.1109/TIFS.2020.2973864.

David S. and Canessane A., "A Centralized Blockchain-based Data Security System for Electrical Energy against Attacks," *2021 International Conference on Communication, Control and Information Sciences* (ICCISc), 2021, pp. 1–4, doi: 10.1109/ICCISc52257.2021.9484898.

Han D., Chen J., Zhang L., Shen Y., Wang X. and Gao Y., "Access Control of Blockchain based on Dual-policy Attribute-based Encryption," *2020 IEEE 22nd International Conference on High Performance Computing and Communications; IEEE 18th International Conference on Smart City; IEEE 6th International Conference on Data Science and Systems (HPCC/SmartCity/DSS)*, 2020, pp. 1282–1290, doi: 10.1109/HPCC-SmartCity-DSS50907.2020.00200.

He Y. *et al.*, "An Efficient Ciphertext-Policy Attribute-Based Encryption Scheme Supporting Collaborative Decryption With Blockchain," in *IEEE Internet of Things Journal*, vol. 9, no. 4, pp. 2722–2733, 15 Feb.15, 2022, doi: 10.1109/JIOT.2021.3099171.

Jeong Y., Hwang D. and Kim K.-H., "Blockchain-Based Management of Video Surveillance Systems," *2019 International Conference on Information Networking (ICOIN)*, 2019, pp. 465–468, doi: 10.1109/ICOIN.2019.8718126.

Kumutha K. and Jayalakshmi S., "Hyperledger Fabric Blockchain Framework: Efficient Solution for Academic Certificate Decentralized Repository," *2021 Fifth International Conference on I-SMAC (IoT in Social, Mobile, Analytics and Cloud) (I-SMAC)*, 2021, pp. 1584–1590, doi: 10.1109/I-SMAC52330.2021.9640785.

Lai W.-J., Hsueh C.-W. and Wu J.-L., "A Fully Decentralized Time-Lock Encryption System on Blockchain," *2019 IEEE International Conference on Blockchain (Blockchain)*, 2019, pp. 302–307, doi: 10.1109/Blockchain.2019.00047.

Liang W., Zhang D., Lei X., Tang M., Li K.-C. and Zomaya A. Y., "Circuit Copyright Blockchain: Blockchain-Based Homomorphic Encryption for IP Circuit Protection," in *IEEE Transactions on Emerging Topics in Computing*, vol. 9, no. 3, pp. 1410–1420, 1 July-Sept. 2021, doi: 10.1109/TETC.2020.2993032.

Mukesh Soni, Dileep Kumar Singh, Privacy-preserving Secure and Low-cost Medical Data Communication Scheme for Smart Healthcare, *Computer Communications*, Volume 194, 2022, Pages 292–300, ISSN 0140-3664. https://doi.org/10.1016/j.comcom.2022.07.046C.

Ray P. P., Chowhan B., Kumar N. and Almogren A., "BIoTHR: Electronic Health Record Servicing Scheme in IoT-Blockchain Ecosystem," in *IEEE Internet of Things Journal*, vol. 8, no. 13, pp. 10857–10872, 1 July1, 2021, doi: 10.1109/JIOT.2021.3050703.

Shobanadevi, A., Tharewal, S., Soni, M. *et al.* Novel Identity Management System using Smart blockchain Technology. *Int J Syst Assur Eng Manag* (2021). https://doi.org/10.1007/s13198-021-01494-0

Soni M. and Singh D. K., "Blockchain Implementation for Privacy Preserving and Securing the Healthcare data," *2021 10th IEEE International Conference on Communication Systems and Network Technologies (CSNT)*, 2021, pp. 729–734, doi: 10.1109/CSNT51715.2021.9509722.

Tan L., Yu K., Shi N., Yang C., Wei W. and Lu H., "Towards Secure and Privacy-Preserving Data Sharing for COVID-19 Medical Records: A Blockchain-Empowered Approach," in *IEEE Transactions on Network Science and Engineering*, vol. 9, no. 1, pp. 271–281, 1 Jan.-Feb. 2022, doi: 10.1109/TNSE.2021.3101842.

Whaiduzzaman M., Mahi M. J. N., Barros A., Khalil M. I., Fidge C. and Buyya R., "BFIM: Performance Measurement of a Blockchain Based Hierarchical Tree Layered Fog-IoT Microservice Architecture," in *IEEE Access*, vol. 9, pp. 106655–106674, 2021, doi: 10.1109/ACCESS.2021.3100072.

Yuan, M. Xu, X. Si and B. Li, "Blockchain with Accountable CP-ABE: How to Effectively Protect the Electronic Documents," *2017 IEEE 23rd International Conference on Parallel and Distributed Systems (ICPADS)*, 2017, pp. 800–803, doi: 10.1109/ICPADS.2017.00111.

Zhang L. and Ge Y., "Identity Authentication Based on Domestic Commercial Cryptography with Blockchain in the Heterogeneous Alliance Network," 2021 *IEEE International Conference on Consumer Electronics and Computer Engineering (ICCECE)*, 2021, pp. 191–195, doi: 10.1109/ICCECE51280.2021.9342494.

Zhaofeng M., Xiaochang W., Jain D. K., Khan H., Hongmin G. and Zhen W., "A Blockchain-Based Trusted Data Management Scheme in Edge Computing," in *IEEE Transactions on Industrial Informatics*, vol. 16, no. 3, pp. 2013–2021, March 2020, doi: 10.1109/TII.2019.2933482.

Zhu T.-L. and Chen T.-H., "A Patient-Centric Key Management Protocol for Healthcare Information System Based on Blockchain," *2021 IEEE Conference on Dependable and Secure Computing* (DSC), 2021, pp. 1–5, doi: 10.1109/DSC49826.2021.9346259.

Artificial Intelligence, Blockchain, Computing and Security – Dagur et al. (Eds)
© 2024 The Author(s), ISBN: 978-1-032-49393-0

Security model to identify block withholding attack in blockchain

Ismail Keshta*
Computer Science and Information Systems Department, College of Applied Sciences, AlMaarefa University, Riyadh, Saudi Arabia

Faheem Ahmad Reegu* & Adeel Ahmad*
College of Computer Science and Information Technology, Jazan University, Jazan

Archana Saxena*
Assistant Professor, Department of Computer Applications, Invertis University, Bareilly, UP

Radha Raman Chandan*
Associate Professor & Head of Department of Computer Science & Engineering, Shambhunath Institute of Engineering & Technology, College Jhalwa Prayagraj

V. Mahalakshmi*
Assistant Professor, Department of Computer Science, College of Computer Science & Information Technology, Jazan University, Jazan, Saudi Arabia

ABSTRACT: Block interception attack, also known as block withholding attack, is an attack method in the blockchain. The attacker penetrates the target mining pool for passive mining to destroy the target mining pool. This paper briefly introduces the mining mechanism and the working principle of block interception attack, summarizes several models of block interception attack, studies the attack methods and benefits of the existing block interception attack models, and analyses their attack effects. A block interception attack model is constructed to increase the rate of return while increasing the rate of return. The completed model is verified through simulated mining experiments. The typical and made models are compared and analyzed based on the rate of return. Finally, the applicable environments of different attack models are given based on the experimental results.

Keywords: Blockchain, Block Withholding Attack, Block Interception Attack, Security, Target Mining, Passive Mining

1 INTRODUCTION

Blockchain technology originated from Bitcoin and is the underlying core technology of many digital currency systems represented by Bitcoin. It integrates Merkle tree, proof of work (PoW), hash function, digital signature, etc. Therefore, it has the characteristics of decentralization, trustlessness, and data immutability (Sathya *et al.* 2021). Blockchain technology is now widely tried and applied by scholars at home and abroad in financial transactions, Internet of Things, identity authentication and other fields, with broad application prospects (R. G.S. & Dakshayini *et al.* 2020).

---

*Corresponding Authors: imohamed@mcst.edu.sa, farfaheem6211@gmail.com, Akahmad@jazanu.edu.sa, archanabalin@gmail.com, rrcmiet@gmail.com and mahabecs@gmail.com

DOI: 10.1201/9781003393580-94

As an emerging technology, blockchain technology has not been thoroughly studied by scholars on the attack behavior of the blockchain. At the same time, there are many ways of attacking the blockchain, among which the main attacks are selfish mining, Sybil attack, Double-spend attack, block interception attack (block withholding attack), etc. (Naidu *et al.* 2018). Block ithholding (BWH) attack is an attack on virtual currencies such as Bitcoin. By joining the target mining pool, the attacker retains and does not publish the complete block after mining, so as to reduce the income of the target mining pool. Behavior. Block interception attack was first proposed and studied by Rosenfeld in (Su *et al.* 2020). In Rosenfeld's research on block interception attack, it is regarded as a kind of "harming others and not oneself" attack method. Later, in the literature (Wu *et al.* 2021, 2022), new block interception attack models were successively proposed, and it was proved that malicious attackers can increase their own income through block interception attacks. Reference (Latifi *et al.* 2019) creatively proposes a sponsored block interception attack, It is possible for an attacker to hire miners to attack a victim's mining pool and reduce its effective computing power ratio. This tactic can give the attacker's mining pool an advantage by increasing its effective computing power ratio. By hiring miners to attack the victim's mining pool, the attacker can create a situation in which their mining pool has more computing power than the victim's mining pool. This gives the attacker a significant advantage in the mining process, allowing them to mine more blocks and earn more rewards. Literature (Bekman *et al.* 2021; Cheng *et al.* 2020;) proposes a fork after withholding (FAW) attack by combining block withholding attack and selfish mining: the attacker deliberately forks after the block withholding attack to compete for the main chain, so the attacker Not only will you get the additional benefits of block interception attacks, but you will also have the chance to get the increased benefits after becoming the main chain.

From the above analysis, it can be seen that scholars at home and abroad have carried out considerable theoretical research on block interception attacks, and the research focuses are different, but the research on improving the speed of revenue is still insufficient. The speed of gain is an important indicator to measure whether the attacker is profitable, and it is also an important measure of whether the block interception attack is ultimately effective. In this paper, an attack model is designed by analyzing the principle and attack process of block interception attack, which can increase the rate of return while stably increasing the rate of return, and has been verified by experiments that the attack is successful and effective.

2 BLOCK INTERCEPTION ATTACK

2.1 *Symbol description*

Table 1 is the description of the symbols, the ciphers used in this paper and their related denotations are in accordance with Table 1.

2.2 *The principle of block interception attack*

When honest miners locate PPoWs that satisfy the task aim, they submit them directly to the mining pool administrator so that they can be compensated (Dabboussi *et al.* 2021; Yang *et al.* 2019). A malicious miner that launched a block interception attack discovered PPoWs and submitted them to the mining pool administrator, but discovered that the FPoWs had been abandoned rather than submitted. The mining pool administrator has no means of knowing that the malicious miner abandoned the FPoWs while still believing that the miner is struggling to locate FPoWs. Because the mining pool administrator still believes the malicious miner is honest, the malicious miner will continue to reap the mining pool's advantages. Other honest miners in the mining pool lose money as a result of malevolent

Table 1. Symbol description.

Definition	Description
rate of return	The ratio of expected revenue to computing power
Revenue rate	The expected revenue per second of the mining pool
α	Attacker's computing power accounts for the quantity of the total network computing power
β	The proportion of attackers' computing power used for block interception attacks
ρ	The victim's computing power accounts for the proportion of the total network computation power
ρ'	The Hashrate of the Employer Pool
γ	Percentage of additional revenue given to attackers by employer pools
T	Expected time to mine a block, 600 s
t	Actual time when blocks were mined
D	The average number of blocks produced by each hash calculation
h	The number of hashes per second of the entire blockchain network

miners' behaviour since they do not contribute to the discovery of new minerals.(Huang *et al.* 2020; Shi *et al.* 2020).

Rosenfeld's attack model is thought to be more than worth the loss because it will also result in a decline in the income of malevolent miners. Here, it is assumed that there are only two mining pools in the whole network, that the network's combined processing power is one, and that the network's combined revenue is one in order to reach a more logical conclusion. The victim's mining pool contains malicious miners. Malicious miners execute a block interception attack by only sending PPoWs and not FPoWs to the administrator. The victim mining pool's entire computer power is ρ, the honest miners' combined computing power is $1 - \rho$, and the honest mining pool's computing power is $\rho - \alpha$. If everyone mines ethically, each miner will receive an income commensurate with their investment, meaning that the income of the victim mining pool is and the income of the honest mining pool is $1 - \rho$. Because the malicious miner only submits oWs and not FPoWs throughout the mining process, the malicious miner's CPU power will have no practical impact on mining after launching a block interception attack. The true processing power of the blockchain network is thus only $1 - \alpha$ at the moment, and the real computing power of the victim's mining pool is just $\rho - \alpha$. Real computing power indicates that the victim pool's typical income is:

$$R1 = \frac{\rho - \alpha}{1 - \alpha}$$

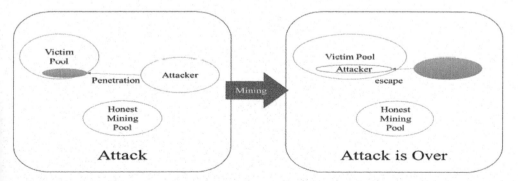

Figure 1. Shows the attack flow of the Rosenfeld attack model.

The malicious miner's revenue is:

$$R = \frac{a}{\rho} \times R_1 = \frac{a(\rho - a)}{\rho(1 - a)}$$

The average return of honest mining pools is:

$$R_2 = \frac{1 - \rho}{1 - a}$$

When all miners mine honestly, their income is proportional to their own computing power, so the income rate of the victim mining pool is:

$$PR_1 = \frac{R_1}{\rho} = \frac{\rho - a}{\rho(1 - a)}$$

Among them, since $\rho < 1$, PR1 < 1, that is, the income of the victim mining pool is reduced. Malicious miner's rate of return:

$$PR = \frac{R}{a} = \frac{\rho - a}{a(1 - a)}$$

Likewise, since $\rho < 1$, PR< 1, i.e., the profit of malicious miners is also reduced. Yield for honest mining pools:

$$P_{R_2} = \frac{R_2}{1 - \rho} = \frac{1}{1 - a}$$

Among them, since $0 < a < 1$, $P_{R_2} > 1$, that is, the revenue of the honest mining pool increasesTaller. The revenue of harmful miners will not rise due to Rosenfeld's block interception attack model, instead the revenue of malicious miners and victim mining pools will fall, resulting in an increase in the revenue of honest mining pools. To that end, the Rosenfeld block interception assault is often used as a metaphor for "harming others and not oneself" in the context of cyberattack".

2.3 *Courtois block interception attack*

The block interception attack model proposed by Courtois and Bahack in (Wu *et al.* 2022) pointed out that malicious miners can improve their own interests by indirectly utilizing the computing power of honest miners. The flow of the Courtois attack model is shown in Figure 2. Suppose there is an attack.

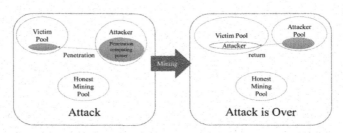

Figure 2.   Courtois block interception attack.

Multiple legitimate mining pools, and one malicious one. One half of the attacker mining pool's computing power is used for legitimate mining operations (labelled as "honest computing power A"), while the other half is used to target other mining pools (labelled as "attacker B"). In this approach, dishonest miners hurt themselves by making it harder for legitimate ones to make money.

2.3.1 *Courtois block interception attack model of dual mining pools*

Assuming that the attacker's mining pool uses the β part of its computing power as infiltrating miners to attack the victim's mining pool, and the remaining $1 - \beta$ part is honestly mining, then the attacker's income will be divided into attack income and honest income. At this time, the real computing power of the blockchain network is $1 - \alpha\beta$, and the real computing power of the victim mining pool is

$$1 - \alpha$$

$$1 - \alpha\beta.$$

The attacker's income is divided into two parts, one part is the income caused by infiltrating the victim's mining pool to launch a block interception attack, and the other part is the income of honest mining.

Attacker's attack benefit:

$$R_1 = \frac{\alpha\beta(1-\alpha)}{(1-\alpha\beta)(1-\alpha+\alpha\beta)}$$

Attacker's honest gain:

$$R_2 = \frac{\alpha(1-\beta)}{1-\alpha\beta}$$

So the total payoff for the attacker is:

$$R = R_1 + R_2 = \frac{\alpha\beta(1-\alpha)}{(1-\alpha\beta)(1-\alpha+\alpha\beta)} + \frac{\alpha(1-\beta)}{1-\alpha\beta}$$

The attacker's rate of return is:

$$P_R = \frac{R}{\alpha} = 1 + \frac{\alpha\beta(1-\alpha)(1-\beta)}{(1-\alpha\beta)(1+\alpha\beta-\alpha)}$$

Obviously available, PR > 1, the attacker's gain increases.

3 SIMULATION VERIFICATION

In order to verify that the block interception attack model proposed in this paper can increase the rate of return while increasing the rate of return, the experimental platform is arranged on Windows10, and the Hyperledger Fabric 1.2 blockchain environment is built through docker for Windows. The experimental results are carried out on Python coding exhibit.

3.1 *Premise assumptions*

The verification experiment of the block interception attack model is carried out under the following preconditions:

Every 2,016 blocks is mined as a stage (about 14 days), the blockchain network will update the difficulty value of the entire network in each new stage, assuming that the attacker launches an attack in a new stage.

It is assumed that after the attacker launches the attack, the total number of blockchain nodes and the total computing power of the entire network remain stable.

Assuming that the network's overall computer power is 1, the victim's mining pool's computing power is 0.1, the attacker's computing power varies from 0.1 to 0.2, and is the proportion of the attacker's computing power that successfully accesses the network's computing power Several ethical mining pools own the power. Each mining pool's mining lucky value is set to 100 percent, meaning that the processing power and money generated by the pool are exactly equal. There are around 10,000 mining simulations total. 3.2.1 Courtois block interception attack revenue analysis In Figure 6, where R is the attacker's output per block of the whole network under the Courtois block interception attack, the abscissa indicates the attacker's computational power and the ordinate is the attacker's rate of return (RRh). When mining honestly, the attacker of a block earns Rh, which represents the amount of money they make. The four curves represent the attacker's earnings when he or she distributes various percentages of the processing power for penetration.

It can be clearly seen from Figure 3 that in the block interception attack model proposed by Courtois, the proportion of the attacker's revenue in the entire network is obviously higher than the proportion of its own computing power in the entire network, but because the attacker needs to invest a lot of computing power. If the victim's mining pool is

Table 2. Data for the rate of return of the attacker in the Courtois model.

β	Value
β = 0.25	97
β = 0.30	94
β = 0.35	93
β = 0.40	90

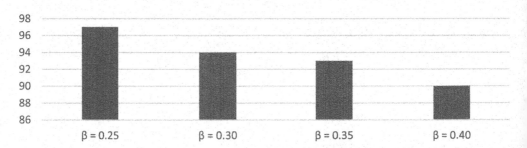

Figure 3.　The rate of return of the attacker in the Courtois model.

Table 3.　Data for rate of gain in the Courtois model.

β	v1	v2	v3	v4
β = 0.25	97	85	78	69
β = 0.30	94	86	76	66
β = 0.35	93	83	75	65
β = 0.40	90	81	72	61

infiltrated by force, the income of the victim's mining pool will be significantly reduced. In the model proposed by Courtois, the proportion of the attacker's revenue is significantly increased, but this strategy ignores the increase in the expected time of mining caused by the reduction of the computing power of the entire network after the attacker launches the block interception attack, thus indirectly It leads to the amount of blocks obtained by the attacker per unit of time, that is, the revenue

The speed limit is lowered. Figure 4 shows that the attacker's rate of gain drops dramatically once they've devoted a lot of computational resources into infiltrating the system. 3.3 A Profitability Study of the Latest Block Interception Attack Model The attacker modified their attack method and decreased the size of the penetration computing power in the created block interception attack model, so it can be seen from Figure 5 that the attacker's return rate is smaller than Courtois The rate of return in the block interception attack model, but at the it also shows that the attacker's rate of gain is higher than that of honest mining and much higher than that of the Courtois model. When β exceeds the threshold, the attacker's revenue rate will be lower than that of honest mining, resulting in a reduction in de facto revenue.

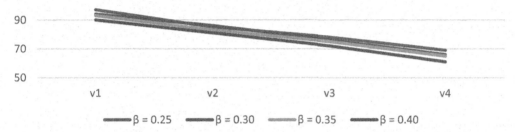

Figure 4.　The attacker's rate of gain in the Courtois model.

Table 4.　Table for rate of return in the new model.

β	v1	v2	v3	v4
$\beta = 0.05$	0.47	0.94	1.03	1.09
$\beta = 0.10$	0.35	0.82	0.91	0.97
$\beta = 0.15$	0.25	0.72	0.81	0.87
$\beta = 0.20$	0.1	0.57	0.66	0.72

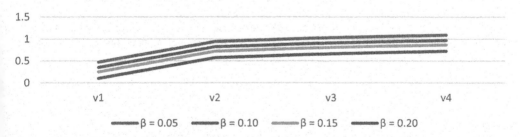

Figure 5.　The attacker's rate of return in the new model.

Since the attacker's attack behaviour decreases the actual computing power of the entire network, the proportion of computing power belonging to the honest mining pool rises, which raises the revenue speed in the designed attack model. Due to the attacker's behaviour of infiltrating computer power, the victim mining pool's own revenue rate has decreased, and it has become the lone victim of the block interception assault in the entire network. The findings of the simulated mining experiment corroborate the predictions made before the trial began: the speed with which the honest mining pool earns money has grown marginally, while the speed with which the victim mining pool earns money has decreased dramatically.

3.2 *Analysis of applicable scenarios*

Compared with the Courtois model, the model constructed in this paper reduces the rate of return while increasing the rate of return. This paper's concept prevents the attacker from spending a big amount of computational power on infiltration, therefore that a modest amount of infiltration power can be used to mine the victim's mine. The income of the pool is maintained within a reasonable range to ensure a high level of anonymity. Due to the increase in the attacker's revenue rate, the mining reward obtained by the attacker will increase in the short term, but at the same time, due to the decrease in the attacker's rate of return, the amount of blocks mined by the attacker is less than the Courtois model. Every time a block is mined, the Bitcoin network will automatically issue a Bitcoin reward to the mining pool. The Courtois model will reduce the amount of Bitcoin revenue in the short term due to the lower revenue rate, but at the same time, due to the attack behavior and its own comparative High yield, Courtois model will increase attacker's Bitcoin holdings in the long run.

There is a ratio, considering the value of Bitcoin itself, a high holding ratio will bring higher benefits. On the contrary, the model constructed in this paper can bring more Bitcoin rewards in the short term through high revenue speed. Therefore, if we consider to increase the income in the short term, the model in this paper will be better than the Courtois model; if we consider to obtain a higher proportion of Bitcoin holdings through long-term attack behavior, the Courtois model will be better than the model in this paper.

4 CONCLUSION

This paper constructs a block interception attack model that can improve the speed of revenue, and re-analyzes the benefits of block interception attacks by introducing time parameters. Simulation experiments are carried out based on the constructed model. The experiments are compared with the yield rate and yield rate of typical Courtois block interception attacks, and it is proved that the attack model constructed in this paper is effective and can indeed improve the yield rate and yield rate at the same time. However, in order to ensure the speed of income, the rate of return of the model in this paper is lower than that of the Courtois attack model. Safe and conservative attack strategy. The model in this paper still cannot solve the Prisoner's Dilemma problem alone, but it can also be combined with FAW attack to solve the Prisoner's Dilemma problem. As an emerging technology, blockchain has broad prospects for development, and its characteristics ensure that it can be applied to many industries. Therefore, blockchain security research will become the focus of future blockchain research. How to effectively defend against block interception attacks while ensuring the characteristics of the blockchain will become the focus of the next research.

REFERENCES

Bekman T., "Use of Blockchain Technology to Fight Trade in Counterfeit Goods," *2021 6th International Conference on Computer Science and Engineering (UBMK)*, 2021, pp. 719–724, doi: 10.1109/ UBMK52708.2021.9558947.

Cheng Y., "Music Information Retrieval Technology: Fusion of Music, Artificial Intelligence and Blockchain," *2020 3rd International Conference on Smart BlockChain (SmartBlock)*, 2020, pp. 143–146, doi: 10.1109/SmartBlock52591.2020.00033.

Dabboussi D., Victor F. and Prinz W., "BCDM - A Decision and Operation Model for Blockchains," 2021 *IEEE International Conference on Blockchain and Cryptocurrency* (ICBC), 2021, pp. 1–3, doi: 10.1109/ ICBC51069.2021.9461146.

G.S R. and Dakshayini M., "Block-chain Implementation of Letter of Credit based Trading System in Supply Chain Domain," *2020 International Conference on Mainstreaming Block Chain Implementation (ICOMBI)*, 2020, pp. 1–5, doi: 10.23919/ICOMBI48604.2020.9203485.

Huang S., "Academic Records Verification Platform Based on Blockchain Technology," *2020 International Conference on Computer Science and Management* Technology (ICCSMT), 2020, pp. 203–206, doi: 10.1109/ICCSMT51754.2020.00048.

Latifi S., Zhang Y. and Cheng L.-C., "Blockchain-Based Real Estate Market: One Method for Applying Blockchain Technology in Commercial Real Estate Market," *2019 IEEE International Conference on Blockchain (Blockchain)*, 2019, pp. 528–535, doi: 10.1109/Blockchain.2019.00002.

Naidu V., Mudliar K., A. Naik and Bhavathankar P., "A Fully Observable Supply Chain Management System Using Block Chain and IOT," *2018 3rd International Conference for Convergence in Technology (I2CT)*, 2018, pp. 1–4, doi: 10.1109/I2CT.2018.8529725.

Sathya D., Nithyaroopa S., Jagadeesan D. and Jacob I. J., "Block-chain Technology for Food Supply Chains," *2021 Third International Conference on Intelligent Communication Technologies and Virtual Mobile Networks (ICICV)*, 2021, pp. 212–219, doi: 10.1109/ICICV50876.2021.9388478.

Shi X., Li G., Chen S. and Wu Z., "Design of Electronic Contract Architecture Based on Blockchain Technology," *2020 IEEE 3rd International Conference of Safe Production and Informatization (IICSPI)*, 2020, pp. 458–461, doi: 10.1109/IICSPI51290.2020.9332420.

Soni M. and Singh D. K., "Blockchain Implementation for Privacy Preserving and Securing the Healthcare Data," *2021 10th IEEE International Conference on Communication Systems and Network Technologies (CSNT)*, 2021, pp. 729–734, doi: 10.1109/CSNT51715.2021.9509722.

Su Z., Wang H., Wang H. and Shi X., "A Financial Data Security Sharing Solution Based on Blockchain Technology and Proxy Re-encryption Technology," *2020 IEEE 3rd International Conference of Safe Production and Informatization (IICSPI)*, 2020, pp. 462–465, doi: 10.1109/IICSPI51290.2020.9332363.

Wu J, Haider SA, Soni M, Kalra A, Deb N. 2022. Blockchain Based Energy Efficient Multi-tasking Optimistic Scenario for Mobile Edge Computing. *PeerJ Computer Science* 8:e1118 https://doi.org/10.7717/ peerj-cs.1118

Wu X., Ai C. and Chen J., "Research on the Development of Computer Network Platform Under Big Data and Blockchain Technology," *2021 IEEE 3rd International Conference on Civil Aviation Safety and Information Technology (ICCASIT)*, 2021, pp. 636–638, doi: 10.1109/ICCASIT53235.2021.9633752.

Yang S., Chen Z., Cui L., Xu M., Ming Z. and Xu K., "CoDAG: An Efficient and Compacted DAG-Based Blockchain Protocol," *2019 IEEE International Conference on Blockchain (Blockchain)*, 2019, pp. 314–318, doi: 10.1109/Blockchain.2019.00049.

Threshold public key-sharing technique in block chain

Sagar Dhanraj Pande*
School of Computer Science and Engineering, VIT-AP University, Amaravati, Andhra Pradesh, India

Gurpreet Singh*
Assistant Professor, Faculty of Computational Science, GNA University, Phagwara

Djabeur Mohamed Seifeddine Zekrifa*
Centre for Quantum Computation and Communication Technology, Sydney, Australia

Shilpa Prashant Kodgire*
Associate Professor, Maharashtra Institute of Technology, Aurangabad, Maharashtra, India

Sunil A. Patel*
Assistant Professor, Computer Engineering Department, L.D. College of Engineering, Ahmedabad

Viet-Thanh Le*
Faculty of Information Technology, Ton Duc Thang University, Ho Chi Minh City, Vietnam

ABSTRACT: A threshold key-sharing technology-based publicly verifiable key-sharing technology is presented to explore blockchain users' private keys' security. After receiving key fragments, participating nodes validate them. When splitting the key, the central splitting node does evil. In the critical recovery stage, the key fragments of the nodes participating in key splicing are publicly verified to prevent them from doing evil. In the critical distribution stage, add IDs to participating nodes to track malicious nodes and update node status in real-time, and a dynamic threshold mechanism is designed. To maintain private key fragment integrity, the owner of the key fragment and the controller node redistribute it to the new participating nodes once the vital fragment node goes offline. Experimental results reveal that this scheme's private key recovery rate is 80% and has threshold properties, traceability, unforgeability, and recoverability.

Keywords: Blockchain, Key-Sharing, Critical Distribution, Publicly Verifiable

1 INTRODUCTION

Blockchain is essentially a decentralized storage system without administrators, where each node owns all data. As a new computing paradigm and Collaboration mode (Lin *et al.* 2018), blockchain is widely used in the global deployment of the Internet of Vehicles (Su *et al.* 2022), Internet of Things (Li *et al.* 2019), financial services (Anjie Peng & Lei Wang 2010; Chowdhury *et al.* 2018), smart grid (Van Der Merwe *et al.* 2007) due to its unique trust establishment mechanism., supply chain (Janbaz *et al.* 2020; Yu *et al.* 2007) and other fields. Blockchain (Soni *et al.* 2022), big data (Su *et al.* 2022), artificial intelligence (Xiong *et al.* 2019), cloud computing and network security are several major directions for the development of the current emerging

*Corresponding Authors: sagarpande30@gmail.com, gurujaswal.phg@gmail.com, d.zekrifa@ieee.org, spkodgire@gmail.com, dp.sapatel@gmail.com and vtle.it22@gmail.com

DOI: 10.1201/9781003393580-95

digital industry. While showing its vigorous vitality, the security problems of its underlying decentralized platform are also increasingly exposed. In 2014, the famous Bitcoin trading platform Mt.Gox claimed to have suffered a malleability attack and lost a total of 850,000 Bitcoins, setting a record high theft record. IOTA (a new micropayment crypto currency newly optimized for the Internet of Things), which is the backbone of the Internet, received an email from the MIT Academic Research Expert Group in 2017, reminding it of Curl-P in its hash algorithm. The existence of loopholes has attracted the attention of the academic community to the security technology of blockchain cryptography. The private key, as the only proof used to identify the user's identity in the blockchain world, cannot be recovered once lost. According to "Vernacular Blockchain" (Zhang *et al.* 2021) According to the information disclosed, in the Bitcoin system, there are many addresses with forgotten private keys, the total value of which is as high as billions of dollars. Therefore, it is urgent to propose a safe and feasible blockchain user private key management scheme solved problem.

2 RELATED RESEARCH

At present, in view of the security management of users' private keys in the blockchain network, the academic community mainly focuses on how to improve the generation of users' private keys, how to store private keys, and the scalability and security in the use of private keys (Cha *et al.* 2020). In the user private key stage, this research (Zheng *et al.* 2021) proposed to use one-way hash chain technology to generate public and private key pairs, and allow the key pair to be self-verified at any time. The single-item hash chain technology increases the difficulty for attackers to steal keys. In the stage of storing the private key, the academic community proposed solutions such as local storage, account custody (Chen *et al.* 2021; Uddin *et al.* 2021; Wu *et al.* 2022), offline storage (Wang *et al.* 2021), cloud storage and encrypted wallet protection. In the stage of using the private key, the academic community proposed threshold-based signatures (Chen *et al.* 2021) and multi-signature (Mukesh Soni *et al.* 2022; Yuan *et al.* 2017) schemes. This esearch is optimized on the basis of the above. After each node votes to elect the master node, the master node splits the private key. When distributing the split private key fragments to each participating node, the identity ID of each participating node is added, so that the identity ID of each participating node can be added according to the identity The ID tracks the participating nodes, and when a new master node is elected in each round of voting, the new master node redistributes the private key fragments to each participating node; after the participating nodes receive the private key fragments, Verification to prevent the master node from doing evil when the private key is split; in the stage of splicing the private key, after the participating nodes 2eryfy the private key fragments held, the verification algorithm broadcasts the verification results in the blockchain network to prevent the participating nodes from splicing The private key stage is malicious, and prevents the master node and the participating nodes from colluding and doing evil. Even if the user accidentally loses the private key, the original private key can be recovered by collecting or splicing key fragments equal to or greater than the threshold. This study verifies both the key segmentation and recovery phases, and it is difficult for an attacker to steal the user's private key by collecting key fragments that exceed the threshold or attacking participating nodes. The owner node of the key fragment and the master node undertake the task of releasing new key fragments together to ensure the dynamic management of the key fragments and the recoverability of the user's private key in the dynamic network. Publicly verifiable threshold encryption in the blockchain the flow chart of key sharing technology is shown in Figure 1.

2.1 *Preliminary knowledge*

Blockchain is a distributed shared general ledger system, which ensures that it has the advantages of tamper-proof, decentralization, openness and transparency, and un-forgeable

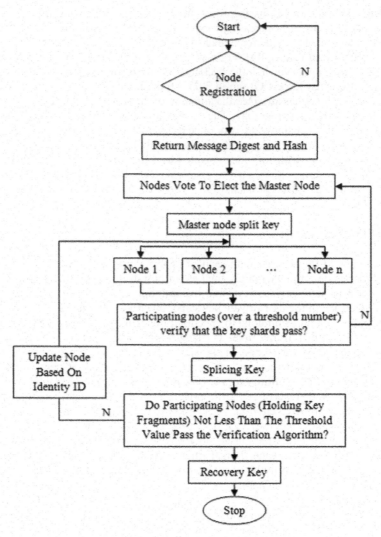

Figure 1. Flow chart of publicly verifiable threshold key sharing technology in blockchain.

block data through cryptography-related technologies. P2P) network protocol and chain structure to realize the distributed storage and decentralization of data. The consensus mechanism is used to constrain each decentralized node in the blockchain network, maintain the order and fairness of the operation of the blockchain network system, so that each an unrelated node can verify and confirm the data in the network, thereby generating trust and reaching a consensus. Cryptography technology is used to ensure the confidentiality, integrity and availability of user keys and transmitted information in the blockchain network. It supports users to generate smart contracts quickly, accurately and securely using automated scripts, which greatly expands the application of blockchain. The security model of blockchain can be abstracted into 3 levels from bottom to top, as shown in Figure 2.

1) Data layer. The data layer uses a variety of cryptographic techniques such as hash functions, encryption algorithms, Merkle trees, key management, etc. to ensure the security of data in the blockchain network.

2) Network and consensus layer. It mainly includes the networking method and consensus mechanism of the blockchain. The blockchain adopts a point-to-point protocol for network transmission. The nodes verify whether the transaction information is reliable and store it in the block. The nodes use the consensus mechanism to make Blockchain consensus.

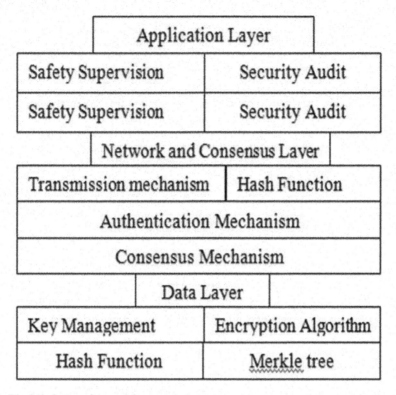

Figure 2.   Blockchain security model.

2.2   Shamir(t, n) threshold key sharing

Shamir (t, n) threshold key sharing technology is a key sharing technology based on Lagrangian interpolation algorithm for how account keys can be distributed to multiple participants credibly and securely. In the scheme, the shared the key s is divided into n parts and distributed to n participants, each participant holds a key, namely the shard s_i, and the shared key s can be recovered as long as at least t shards s_i are collected.

The key distributor randomly selects a polynomial of degree $f(x)$,

$f(x) = a_{t-1}x^{t-1} + a_{t-2}x^{t-2} + \cdots + a_1 x + s$ In that $a_1, \cdots, a_{t-1} \in Z$ and $a_{t-1} \neq 0$.

The key distributor selects a random polynomial that satisfies the condition $f(0) = s$, and then distributes $s_i = f(i)$ to each participant $P_i, i = 1, 2, \cdots, n$.

Any t members of the participants s $P = \{P_1, P_2, \cdots, P_t\}$ are reconstructed using Lagrangian interpolation:

$$f(x) = \sum_{i=1}^{k} s_i \prod_{1 \leq j \leq k, i \neq j} \frac{x - x_j}{x_i - x_j} \qquad (1)$$

In the formula: $f(x)$ is the original key after reconstruction, and k is the threshold value. After construction, the original shared key s can be calculated by $s = f(0)$.

3 PUBLICLY VERIFIABLE THRESHOLD KEY SHARING SCHEME BASED ON BLOCKCHAIN

This research scheme is based on *Shamir*(t, n) threshold key sharing technology and Pedersen's verifiable key scheme. By using publicly verifiable threshold key sharing technology in the blockchain network, the traditional threshold key sharing technology can be solved. The key sharing technology leads to the disclosure of the user's private key due to its own defects, or the problem that the user's private key cannot be recovered due to the offline node holding the key fragments.

3.1 *Initialization*

All nodes in the blockchain network elect the master node by voting. Let p and q be large prime numbers, respectively, where q is the large prime factor of $p - 1$, and is the only G_q order subgroup of the multiplicative cyclic group $\log_g h$. The discrete logarithm is unknown to anyone (except the master node). Assume that the master node elected by voting is the key distributor, D and n child nodes are participating nodes, denoted by $P_1, P_2, \cdots, P_n$, respectively, and the threshold value is k.

3.2 *Key recovery*

One or more sub-nodes participating in the key recovery are executed to prove whether their key fragments are correct. The sub-nodes execute the verification algorithm is $g^{s_{i,t}} = \prod_{j=0}^{k-1} F^{t^j}{}_j$, and the verification algorithm will publish the verification results in the blockchain network after the execution is completed. Only the sub-nodes that pass the verification can use the Lagrangian polynomial interpolation method to perform key splicing, and the blockchain network will track and refresh the node that fails to pass the verification according to the identity ID of the key distribution stage.

3.3 *Key shard append*

When the node holding the key fragment goes offline, the owner node of the key fragment and the master node distribute the key fragment to the unassigned nodes together. The master node summons other participating nodes greater than or equal to the threshold value, after the key fragment is verified, it is restored, and after the original key is restored, the key is re-split, and then distributed to other participating nodes.

4 PROGRAM SECURITY ANALYSIS

4.1 *(t, n) Threshold characteristics*

The (t,n) threshold feature means that the key is divided and distributed to each participating node. The original key t can be recovered as long as it is equal to or greater than the number of correct nodes, and the original key cannot be recovered if it is less than 1. Even if the attacker obtains a key fragment t-1, and t-1 can only construct a system of equations with unknowns:

$$\left. \begin{array}{l} s + F_1(s_1, t_1) + F_2(s_1, t_1)^2 + \cdots + F_{t-1}(s_1, t_1)^{t-1} = F(s_1, t_1), \\ s + F_1(s_2, t_2) + F_2(s_2, t_2)^2 + \cdots + F_{t-1}(s_2, t_2)^{t-1} = F(s_2, t_2), \\ \vdots \\ s + F_1(s_{t-1}, t_{t-1}) + F_2(s_{t-1}, t_{t-1})^2 + \cdots + F_{t-1}(s_{t-1}, t_{t-1})^{t-1} = F(s_{t-1}, t_{t-1}), \end{array} \right\} \quad (2)$$

In the formula: (s_i, t_i) is the key fragment possessed by a certain node. When the number of unknowns is greater than the number of equations, the above equation has no solution, and the specific form cannot be obtained $F(x)$, that is, $F(0)$ the original shared key cannot be obtained. Therefore, this research scheme has the threshold characteristic $s = f(0) = a_0$, and the original shared key can be recovered only when at least one participating node is satisfied.

4.2 Unforgeability and traceability

The unforgeability of each participating node means that no participating node can generate legal key segments in the name of other participating nodes. Assuming that the identity set of participating nodes is $ID = \{ID_1, ID_2 \cdots, ID_t\}$, for participating nodes P_i, its identity is known ID_i.

Attack 1 Attacker P_j impersonates P_i to perform key splicing. Before key splicing, key verification must be performed P_j, and the verification function must be executed $g^{s_i,t} = \prod_{j=0}^{k-1} F_{i,j}^{t^j}$ fail, then P_j the result of the verification failure is published in the block-chain network according to the P_j the identity id traces back to P_j later on P_j node to refresh. Therefore, P_j cannot impersonate P_i.

The security of this research scheme is compared with Scheme 1 (Li et al. 2019), Scheme 2 (M. J. M et al. 2018), and Shamir threshold key sharing scheme, and the analysis results are shown in Table 1. The scheme of this research is resistant to collusion attacks and does not need to be trusted. Nodes participate and allow dynamic addition of participating nodes, which can ensure the security and privacy of the user's private key.

Table 1. Safety comparison between this study protocol and existing typical protocols.

Program	Whether trusted nodes are required to participate in the reconstruction phase	Can resist collusion	Whether to dynamically add participating nodes
Scheme 1	Not Required	Yes	No
Scheme 2	Not Required	Yes	No
Shamir threshold key sharing scheme	Required	No	No
This Paper	Unnecessary	Yes	Yes

5 PROTOCOL EXPERIMENTAL ANALYSIS

The experimental environment of the publicly verifiable key sharing technology and Shamir threshold key sharing technology scheme in the blockchain is as follows: The operating system is Windows10 Home Chinese version 64-bit, and the CPU is Inter(R)Core(TM)i7-10510U CPU@ 1.80 GHz 2.30 GHz, the memory size is 16 GB, implemented using the Java development language.

5.1 Private key recoverability

Private Key recoverability means that when the participating nodes in the blockchain network change dynamically, the user can recover his private key by collecting enough key fragments through the master node. As shown in Figure 3, the comparison of the four schemes is set up. There are 20 participating nodes. In the scheme proposed in this study, all nodes use dynamic allocation to distribute key fragments, and the threshold value is 11; Scheme 1 (Li et al. 2019), Scheme 2 (M. J. M et al. 2018) and Shamir threshold encryption the threshold value of the key scheme is fixed at 10, and nodes join or leave the network randomly in the experiment. In the figure, Ru is the update rate of blockchain nodes, and r is the private key recovery rate. It can be seen that with the node update rate the private key recovery rate of scheme 2 (M. J. M *et al.* 2018) and Shamir threshold key scheme is similar, and the private key recovery rate of scheme 1 1 (Li *et al.* 2019) is relatively higher. When the update rate of nodes reaches 35%, scheme 2 (M. J. M *et al.* 2018) Due to the different weights of key fragments held by nodes, the private key recovery rate is close to 0. The private key recovery rate of the Shamir threshold key scheme is close to 15%, and the private key recovery rate of Scheme 1 1 (Li *et al.* 2019)is slightly different. The solution recommended in this study can effectively deal with the situation of node withdrawal or joining. Through the method of the master node calling participating nodes, the newly joined nodes also have key fragments, which ensure the fragment size and keeps the private key recoverability at a relatively low level high level. Even if the network node update rate reaches 40%, more than 80% of the private keys can still be recovered in the scheme recommended in this study. Therefore, the scheme recommended in this study is more suitable for dynamic blockchain networks and can effectively tolerate carry Exit of key shard nodes and join of new nodes.

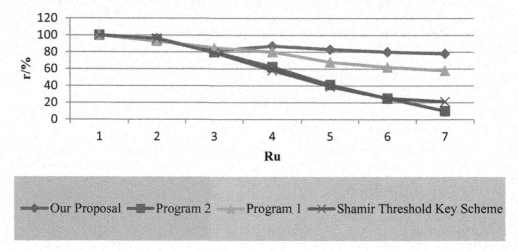

Figure 3.   Private key recovery rate for a single user.

5.2   *Private key recovery time*

When a legitimate user requests to restore the private key, the master node in the blockchain network broadcasts the request to restore the private key to other nodes. The node holding the private key fragment first executes the verification algorithm, and participates in restoring the private key after verification. There are 20 Participating nodes, the threshold value is from 11 to 18, with the threshold value required to recover the key as the abscissa, and the key recovery time as the ordinate. The comparison diagram of the threshold key scheme is shown in Figure 4. In the figure, T is the recovery time. The key fragments of the proposed scheme, scheme 1, scheme 2 and Shamir threshold key scheme are distributed in different nodes in the blockchain network, splicing the private key requires more than a threshold number of key fragments in the entire blockchain network. Even if each key fragment is publicly verified in this research scheme, the key recovery time and The other three schemes are still not much different, and the scheme in this study is ensuring that users the private key will not be leaked while improving the security of the blockchain network.

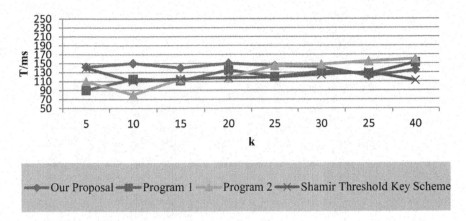

Figure 4.   User private key recovery time.

5.3   *Private key recoverability with different number of users*

The recoverability of the private key when the number of users is different refers to the recoverability rate of the private key when the number of users is different and the number of nodes participating in the recovery of the private key is different in the blockchain network. The private key recoverability rate when the number of user's nu is different. The number of users in the blockchain network increases from 1 to 12, the participating nodes are 15 (threshold value 10) and 20 (threshold value 14), and nu is the blockchain the number of users in the network.

When the number of participating nodes in the blockchain network is 15 and the number of users does not exceed 4, the private key recovery rate is 100%. When the number of user's increases to 12, the private key recovery rate is close to 60%; when the number of participating nodes in the blockchain network is 20 and the number of users does not exceed 3, the private key recovery rate is 100%. When the number of users increases, the private key recovery rate is about 60%. When the number of users increases to 12, the private key recovery rate is 50%. Even if the number of nodes and users continue to increase, the private key of the proposed scheme in this study can be recovered, as shown in Figure 6. The rate is about 50%, so the scheme recommended in this study is suitable for small and medium-sized blockchain networks.

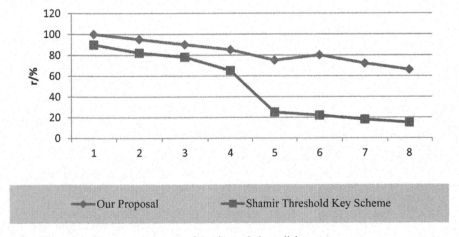

Figure 5.   The private key recovery rate when the node is malicious.

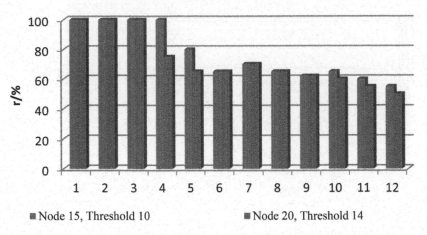

■ Node 15, Threshold 10   ■ Node 20, Threshold 14

Figure 6.   Private key recoverability with different number of users.

6   CONCLUSION

Aiming at the security problem of the loss or leakage of the user's private key in the current blockchain network, a publicly verifiable threshold key sharing technology in the blockchain is designed. The design of the dynamic threshold ensures that even if the node holding the key fragments is offline, it can still be guaranteed. The recoverability of the user's private key. The security analysis shows that the scheme in this study has threshold characteristics, enforceability and traceability, and is suitable for dynamic blockchain networks. In the next research, the key splicing algorithm will be studied, and when the number of users in the blockchain network increases, the recoverability rate of the user's private key will be further improved, making it suitable for large-scale blockchain networks.

REFERENCES

Anjie Peng and Lei Wang, "One Publicly Verifiable Secret Sharing Scheme based on Linear Code," *2010 The 2nd Conference on Environmental Science and Information Application Technology*, 2010, pp. 260–262, doi: 10.1109/ESIAT.2010.5568356.

Cha S., Baek S. and Kim S., "Blockchain Based Sensitive Data Management by Using Key Escrow Encryption System From the Perspective of Supply Chain," *in IEEE Access*, vol. 8, pp. 154269–154280, 2020, doi: 10.1109/ACCESS.2020.3017871.

Chen B., He D., Kumar N., Wang H.and Choo K.-K.R., "A Blockchain-Based Proxy Re-Encryption With Equality Test for Vehicular Communication Systems," *in IEEE Transactions on Network Science and Engineering*, vol. 8, no. 3, pp. 2048–2059, 1 July-Sept. 2021, doi: 10.1109/TNSE.2020.2999551.

Chowdhury M. J. M., Colman A., Kabir M. A., Han J. and Sarda P., "Blockchain as a Notarization Service for Data Sharing with Personal Data Store," 2018 *17th IEEE International Conference On Trust, Security And Privacy In Computing and Communications/ 12th IEEE International Conference On Big Data Science And Engineering (TrustCom/BigDataSE)*, 2018, pp. 1330–1335, doi: 10.1109/TrustCom/BigDataSE.2018.00183.

Janbaz S., Asghari R., Bagherpour B. and Zaghian A., "A Fast Non-interactive Publicly Verifiable Secret Sharing Scheme," *2020 17th International ISC Conference on Information Security and Cryptology (ISCISC)*, 2020, pp. 7–13, doi: 10.1109/ISCISC51277.2020.9261914.

Li G. and Sato H., "A Privacy-Preserving and Fully Decentralized Storage and Sharing System on Blockchain," *2019 IEEE 43rd Annual Computer Software and Applications Conference (COMPSAC)*, 2019, pp. 694–699, doi: 10.1109/COMPSAC.2019.10289.

Lin C., Hu H., Chang C.-C. and Tang S., "A Publicly Verifiable Multi-Secret Sharing Scheme With Outsourcing Secret Reconstruction," *in IEEE Access*, vol. 6, pp. 70666–70673, 2018, doi: 10.1109/ACCESS.2018.2880975.

Mukesh Soni, Dileep Kumar Singh, Privacy-preserving Secure and Low-Cost Medical Data Communication Scheme for Smart Healthcare, *Computer Communications*, Volume 194, 2022, Pages 292–300, ISSN 0140-3664, https://doi.org/10.1016/j.comcom.2022.07.046C.

Soni, M., Singh, D.K. New Directions for Security Attacks, Privacy, and Malware Detection in WBAN. *Evol. Intel.* (2022). https://doi.org/10.1007/s12065-022-00759-2

Su Y., Sun J., Qin J. and Hu J., "Publicly Verifiable Shared Dynamic Electronic Health Record Databases With Functional Commitment Supporting Privacy-Preserving Integrity Auditing," *in IEEE Transactions on Cloud Computing*, vol. 10, no. 3, pp. 2050–2065, 1 July-Sept. 2022, doi: 10.1109/TCC.2020.3002553.

Uddin M. N., Hasnat A. H. M. A., Nasrin S., Alam M. S. and Yousuf M. A., "Secure File Sharing System Using Blockchain, IPFS and PKI Technologies," *2021 5th International Conference on Electrical Information and Communication Technology (EICT)*, 2021, pp. 1–5, doi: 10.1109/EICT54103.2021.9733608.

Van Der Merwe J., Dawoud D. S. and McDonald S., "A Fully Distributed Proactively Secure Threshold-Multisignature Scheme," *in IEEE Transactions on Parallel and Distributed Systems*, vol. 18, no. 4, pp. 562–575, April 2007, doi: 10.1109/TPDS.2007.1005.

Wang Z., Qian L., Chen D. and Sun G., "Sharing of Encrypted Lock Keys in the Blockchain-based Renting House System from Time- and identity-based proxy Reencryption," *in China Communications*, vol. 19, no. 5, pp. 164–177, May 2022, doi: 10.23919/JCC.2021.00.010.

Wu J, Haider SA, Soni M, Kalra A, Deb N. 2022. Blockchain Based Energy Efficient Multi-tasking Optimistic Scenario for Mobile Edge Computing. *PeerJ Computer Science* 8:e1118 https://doi.org/10.7717/peerj-cs.1118

Xiong F., Xiao R., Ren W., Zheng R. and Jiang J., "A Key Protection Scheme Based on Secret Sharing for Blockchain-Based Construction Supply Chain System," *in IEEE Access*, vol. 7, pp. 126773–126786, 2019, doi: 10.1109/ACCESS.2019.2937917.

Yu J., Kong F. and Hao R., "Publicly Verifiable Secret Sharing with Enrollment Ability," *Eighth ACIS International Conference on Software Engineering, Artificial Intelligence, Networking, and Parallel/Distributed Computing (SNPD 2007)*, 2007, pp. 194–199, doi: 10.1109/SNPD.2007.256.

Yuan, M. Xu, X. Si and B. Li, "Blockchain with Accountable CP-ABE: How to Effectively Protect the Electronic Documents," *2017 IEEE 23rd International Conference on Parallel and Distributed Systems (ICPADS)*, 2017, pp. 800–803, doi: 10.1109/ICPADS.2017.00111.

Zhang J., Xin Y., Gao Y., Lei X. and Yang Y., "Secure ABE Scheme for Access Management in Blockchain-Based IoT," *in IEEE Access*, vol. 9, pp. 54840–54849, 2021, doi: 10.1109/ACCESS.2021.3071031.

Zheng W., Wang K.and Wang F.-Y., "GAN-Based Key Secret-Sharing Scheme in Blockchain," *in IEEE Transactions on Cybernetics*, vol. 51, no. 1, pp. 393–404, Jan. 2021, doi: 10.1109/TCYB.2019.2963138.

Artificial Intelligence, Blockchain, Computing and Security – Dagur et al. (Eds)
© 2024 The Author(s), ISBN: 978-1-032-49393-0

5G geological data for seismic inversion data detection based on wide-angle reflection wave technology

M. Thiyagesan*
Department of Electrical and Electronics Engineering, R.M.K. Engineering College, Kavaraipettai, Chennai, Tamil Nadu, India

B. Md. Irfan*
Department of Information Technology, Nalsar University of Law, Hyderabad, Telengana, India

Ramakrishnan Raman*
Symbiosis Institute of Business Management, Pune and Symbiosis International (Deemed University), Pune, Maharashtra

N. Ponnarasi*
Department of Electronics and Communication Engineering, Jkk Nattraja College of Engineering and Technology, Komarapalam, Tamil Nadu, India

P. Ramakrishnan*
Department of Electronics and Communication Engineering, M. Kumarasamy College of Engineering, Karur, Tamil Nadu, India

G.A. Senthil*
Department of Information Technology, Agni College of Technology, Chennai, Tamil Nadu

ABSTRACT: With the development of Geophysics theory, inversion technology has stepped from post stack inversion to pre stack seismic inversion which can reflect abundant reservoir parameter information, and the resolution of conventional deterministic inversion method cannot meet the needs of current exploration and development. Mesozoic and Paleozoic carbonate layers are more difficult to image. For this reason, we detect seismic inversion data of complex geologic structures using wide-angle reflected wave technology. A comparative analysis of real seismic data and inversion simulation data in the South Yellow Sea area verifies that effective and interfering waves can be distinguished with the help of inversion simulation data under complex geological tectonic conditions. This will take into account the factors affecting the imaging accuracy of Mesozoic and Paleozoic marine carbonate layers and lead to the improvement of acquisition parameters in the hard-imaging area of deep target layers in the southern Yellow Sea. To improve the quality of parameters of seismic data.

Keywords: wide-angle reflection wave, seismic survey, inversion data, seismic inversion simulation, interference wave

1   GENERAL INSTRUCTIONS

With the development of inversion technology, geophysical inversion has changed from single attribute, post stack inversion with poor sensitivity to reservoir to pre stack seismic

---

*Corresponding Authors: mtn.eee@rmkec.ac.in, irfan@nalsar.ac.in, raman06@yahoo.com, ponnarasin@gmail.com, ramumkce@gmail.com and senthilga@gmail.com

DOI: 10.1201/9781003393580-96

inversion which can effectively extract various geophysical parameters such as P-wave, S-wave velocity, density and elastic modulus. Through rock physics, pre stack inversion provides effective inversion data for reservoir lithology and fluid identification, so as to master reservoir the distribution, physical properties and petroliferous properties of the layers are of great significance (Liu *et al.* 2019; Pollyea *et al.* 2018)

The seismic data has good lateral continuity, and it is very important to integrate other information to improve the reliability and accuracy of inversion results. Dubrule reduces the uncertainty of inter well estimation by adding seismic data constraints to geostatistical reservoir modeling. Pendrel uses seismic data with reliable transverse continuous information to constrain inversion. Geophysical inversion method is very important for oil and gas exploration and development. At present, deterministic inversion methods have been very mature. These methods take seismic data as the main data and are suitable for the stratum with relatively thick underground medium and slow lateral change. However, due to the limited bandwidth of seismic data, the conventional deterministic inversion lacks high and low frequency information and the resolution is difficult to meet the requirements of high-resolution geophysical exploration. In addition, the model parameters and observation data of geophysical inversion are non-linear. Local optimization methods, such as the steepest descent method and conjugate gradient method, require too much of the initial model, which is easy to make inversion fall into local optimum. Finally, a single local smooth inversion result is output, and the complexity and randomness of underground geological model are not considered (Chang *et al.* 2018; Li & He 2019). This treats model parameters as random variables, and uses random inversion methods to obtain elastic parameters and reservoir parameters to provide technical support for seismic exploration.

2 MATERIAL AND METHODS

2.1 *Problems in earthquake inversion*

The biggest limitation of deterministic inversion is that its inversion results lack high and low frequency information. The missing frequencies are mainly supplemented by logging data. In addition, nonlinear problems are common in geophysical inversion. Seismic records are the signals received by the source pulse after passing through the internal materials and structures of the earth. Converting the signals into the received seismic records is a non-linear system, which determines that the inversion problem and solution should use a nonlinear method. The current linear inversion method weakens the consideration of this problem to a certain extent. When faced with complex geological problems, the reservoir prediction cannot be performed well. Therefore, it is very meaningful to study the nonlinear inversion method. Figure 1 is a schematic diagram of the seismic frequency band range (Ma *et al.* 2020; Yue *et al.* 2018).

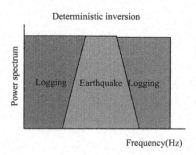

Figure 1. Schematic diagram of deterministic inversion frequency band distribution.

2.2 Analysis of seismic inversion data of complex geological structures based on wide-angle reflected wave technology

The acoustic wave equation inversion simulation technology was used to simulate the geological model of the South Yellow Sea, and the seismic response of the area was obtained. The simulation results were analyzed to study the propagation law and wave field characteristics of seismic waves in underground formations. The difference between the seismic wave field characteristics of the layer area and the area without the high-speed shield layer to analyze the impact of the high-speed shield layer on the collection of effective information. Combining the study of rock geophysical characteristics in the sea area based on the parameters in Table 1, a two-dimensional seismic geological model of the South Yellow Sea in this survey line direction was established. The model is a south-north strike profile of the South Yellow Sea Basin, with a horizontal distance of about 390km, a longitudinal depth of about 12km, and a seawater depth of 50m.From north to south, they are: Qianliyan uplift, northern depression, central uplift, southern depression, and Benansha uplift. The Qianliyan uplift area and the northern depression area in the north of the basin mainly developed Cenozoic and Mesozoic clastic rock strata, and there are also Paleozoic carbonate rock strata in some areas. However, in the central uplift area, southern depression area and Benansha uplift area, Cenozoic and Mesozoic clastic rock formations and Paleozoic carbonate rock formations are relatively developed. There are high-speed shields in the central uplift and the Benansha uplift, which seriously affects the exploration effect in the area.

Table 1. Seismic geological model stratum parameter table.

Stratum	P-wave velocity	Shear wave velocity	Density	Poisson's ratio
Seawater	1500m/s	0m/s	1000kg/m	0.500000
N-Q	2240m/s	1294m/s	2101kg/m	0.249573
E	3100m/s	1788m/s	2215kg/m	0.250747
K	4900m/s	2841m/s	2485kg/m	0.246803
K + J	4500m/s	2605m/s	2425kg/m	0.247993
T	6000m/s	3510m/s	2420kg/m	0.239862
P	3800m/s	2194m/s	2320kg/m	0.249976
O	5000m/s	2900m/s	2500kg/m	0.246534
Z	5500m/s	3205m/s	2575kg/m	0.242916
D	5600m/s	3266m/s	2590kg/m	0.242265
S	5800m/s	3388m/s	2620kg/m	0.241025

For the built geological model, the seismic wave equation inversion simulation is performed according to the observation system parameters shown in Table 2.

Table 2. Observation system parameter table of seismic geological model inversion simulation.

Road spacing	12.5m/25m
Shot spacing	50m/100m
Total length of cable working section	6000m
Number of paths	480
Coverage times	60
sampling rate	4ms
Record length	8s
Minimum offset	200m
Focal sinking depth	6m
Cable sinking depth	10m

3 RESULTS

3.1 *Experimental area selection*

According to the simulation results, the seismic wave propagation law was analyzed and characteristics, compare and analyze the difference of seismic response between the presence and absence of high-speed shielding layer under two geological conditions, and study the adverse effects of high-speed shielding layer on the effective information acquisition of the target stratum below it.

3.2 *Analysis of experimental process*

The seismic exploration work in the South Yellow Sea, a wide-angle seismic study of high-speed shields was carried out on the basis of inversion simulation. According to the principle of wide-angle reflection, wide-angle reflection occurs only when the incident angle is greater than the critical angle. To increase the incident angle of the seismic wave can be achieved by increasing the offset parameter. Under the condition that other parameters remain unchanged, the track spacing is increased from the original 12.5m to 25m, and the wavelet with the main frequency of 10Hz is used to perform acoustic inversion on the central uplift area where the high-speed carbonate formations in the South Yellow Sea are concentrated simulation. Figure 2 is a single shot record obtained by sonic equation inversion of the central uplift area using the observation system parameters of 25m channel spacing and 480 channel reception. The sea surface is set as the absorption surface, and it is more beneficial to remove the free surface multiple waves Identification of effective waves.

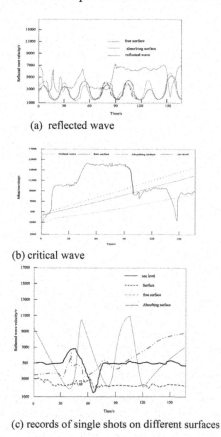

(a) reflected wave

(b) critical wave

(c) records of single shots on different surfaces

Figure 2.   Simulated single shot records of the central uplift area.

4 DISCUSSION

According to the wide-angle reflection principle, by adjusting the inversion simulation parameters and using the large offset observation system, the effective information of the deep formation below the high-speed layer is obtained. Through research and analysis, the following conclusions are drawn:

(A) The South Yellow Sea is an important potential oil and gas area, but due to the complex geological conditions, the degree of exploration is relatively low. There is a middle Paleozoic carbonate high-speed layer in the central uplift area and the Wunansha uplift area of the South Yellow Sea basin.

(B) The existence of the high-speed shielding layer not only makes the the multiple wave is more developed, and it has played a serious role in hindering the propagation of seismic waves to the formation below the high-speed layer and the upward propagation of deep formation information.

(C) The wave impedance difference between deep formations in the South Yellow Sea is small, and a good reflection interface cannot be formed.

(D) Taking advantage of the wide-angle earthquake's advantages in seismic exploration in areas with high-speed shields is conducive to improving the quality of seismic data and improving the exploration degree and accuracy of the South Yellow Sea. In the study of high-speed shielding layer in the South Yellow Sea based on inversion simulation, some further research points were found.

5 CONCLUSION

Seismic applications have gone from structural morphology to reservoir evaluation. Adopting the combination of low-frequency and high-energy multi-layer air gun array is conducive to increasing the depth of seismic wave penetration, improving the energy of seismic wave reaching the deep target stratum, and reducing the seismic resources of seafloor ringing the influence of material quality.The multi-channel and wide-angle seismic method with large array reception is applied to exploration in the shield area of high-speed layer, which can not only avoid all kinds of interference waves such as direct wave and refraction wave in the middle and near offset, improve the signal-to-noise ratio of seismic data, but also receive effective reflection wave with strong energy from the middle and Paleozoic strata in the deep, and improve the reflection wave energy and resolution in the middle and deep strata of the stack section.

REFERENCES

Chang A., Han L.G., Zhang L., Seismic Data Reconstruction Based on an Improved Poisson Disk Sampling, *Science Technology and Engineering*, 2018, 18(15), 28–35.

He Z., Zhou F., Xia X., *et al.*, Interaction Between Oil Price and Investor Sentiment: Nonlinear Causality, Time-Varying Influence, and Asymmetric Effect, *Emerging Markets Finance and Trade*, 2019, 55(12), 2756–2773.

Li Y., Lin S., Lin Z.L., *et al.*, Genesis Classification, Development Mechanism and Sedimentary Model of Deep-Lacustrine Gravity Flow in Fushan Sagof Beibuwan Basin, *Journal of Jilin University (Earth Science Edition)*, 2019, 49(2), 323–345.

Liu X., Greenhalgh S., Zhou B., *et al.*, Frequency-domain Seismic Wave Modelling in Heterogeneous Porous Media Using the Mixed-grid Finite-difference Method, *Geophysical Journal International*, 2019, 216(1), 34–54.

Ma X.G., Zhou J., Cai W.Y., *et al.*, The Combined Application of Reflective Wave Imaging and P-wave Velocity Variation to the Exploration and Development of Fractured Carbonate Reservoirs of North China Oilfield, *Geophysical and Geochemical Exploration*, 2020, 44(2), 271–277.

Pollyea R.M., Mohammadi N., Taylor J.E., *et al.*, Geospatial Analysis of Oklahoma (USA) Earthquakes (2011–2016): Quantifying the Limits of Regional-scale Earthquake Mitigation Measures, 2018, *Geology*, 46(3), 215–218.

Yue Y., Jiang T., Zhou Q., The Application of Variable-frequency Directional Seismic Wave Technology Under Complex Geological Conditions, *Near Surface Geophysics*, 2018, 16(5), 545–556.

Artificial Intelligence, Blockchain, Computing and Security – Dagur et al. (Eds)
© 2024 The Author(s), ISBN: 978-1-032-49393-0

Performance evaluation and comparison of blockchain mechanisms in E-healthcare

Prikshat Kumar Angra & Aseem Khanna
School of Computer Applications, Lovely Professional University, Punjab

Gopal Rana
Department of Mechanical Engineering, Lovely Professional University, Punjab

Manvendra Singh
Department of Government and Public Administration, School of Humanities, Lovely Professional University, Punjab

Pritpal Singh & Ashwani Kumar
Mittal School of Business, Lovely Professional University, Punjab

ABSTRACT: Blockchain has built-in features like distributed ledgers, decentralized storage, identification, security, and the fact that data can't be changed. Blockchain technology has a lot of promise in the healthcare field because it can help bring together different systems, improve the quality of electronic medical data, and make e-healthcare systems more focused on the patient. The latest blockchain study in the healthcare field is looked at and compared to a client-server architecture and other systems. The goal of this study is to show how different mechanisms, such as proof of work (PoW), byzantine fault tolerance (BFT), and practical byzantine fault tolerance (PBFT), work in blockchain-based e-healthcare system research and how the technology could be used. The open-source Docker platform can be used to make, share, and run applications. In this paper, a tool called Hyperledger Caliper is used to measure how well consensus methods work.

Keywords: Blockchain, Consensus mechanism, Docker Platform, PBFT, Proof of work

1 INTRODUCTION

Due to additional regulatory needs to secure patients' medical information, the healthcare sector has specific security and privacy requirements. With cloud storage and the use of mobile health devices, the exchange of records and data is becoming more common in the Internet age, but so is the possibility of hostile assaults and the chance of private information being compromised. The sharing and privacy of this information are issues when patients visit several providers and access to health information via smart devices increases. Authentication, interoperability, data sharing, the transfer of medical information, and concerns for mobile health are the specific requirements that the healthcare sector must meet. Due to these security concerns, our research implements a security framework using blockchain technology that is free from these vulnerabilities. In this paper, the experimental comparison has been discussed with client-server architecture which is also used in E-healthcare systems to secure transactions.

DOI: 10.1201/9781003393580-97

1.1 Blockchain in healthcare

Smart contracts have been getting better for more than forty years. The simplest way to describe blockchain technology is as a decentralized, shared record of where a computer resource came from. Blockchain, also called Distributed Ledger Technology (DLT) (Azaria 2016), makes it easy and impossible to change the history of any computerized resource by using decentralization and cryptographic hashing.

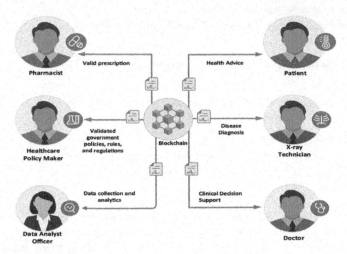

Figure 1. Blockchain structure.

Figure 1 depicts a blockchain-based healthcare system, in the system distinguish peer connected with the blockchain network.

1.2 Precision of the chain

Trades on the blockchain network are supported by an association of thousands of PCs. This wipes out basically all human consideration in the really look at communication, achieving less human screw up and a precise record of information. Whether or not a PC on the association were to submit a computational blunder, the misstep would simply be made to one copy of the blockchain. All together for that slip-up to spread to the rest of the block-chain, it would ought to be made by in any occasion 51% of the association's PCs-a near incomprehensibility for a gigantic and creating association the size of Bitcoin's (Bahga 2013), (Badr 2018).

1.3 Cost reductions

Ordinarily, purchasers pay a bank to really look at a trade, a legitimate authority to sign a document, or a minister to play out a marriage. Blockchain kills the prerequisite for untouchable checks and, with it, their connected costs.

1.4 Decentralization

Blockchain doesn't store any of its information in a central region. In light of everything, the blockchain is reproduced and spread across an association of PCs. Whenever one more square is added to the blockchain, every PC on the association revives its blockchain to reflect the change.

1.5 *Blockchain consensus algorithm*

The blockchain agreement calculation is that it is a methodology by means by which all the companions of a Blockchain network arrive at a typical acknowledgment or agreement about the ongoing condition of the dispersed record. An agreement system empowers the blockchain organization to accomplish dependability and fabricate a degree of trust between various hubs while guaranteeing security in the climate.

2 RELATED WORKS

2.1 *Traditional transaction managementssystem*

In this part, we'll talk about some of the factors you should think about while evaluating different blockchain consensus algorithms. There are several aspects of blockchain technology to think about, such as the transaction rate, scalability, and attacker tolerance model. Latency, throughput, bandwidth, experimental setup assaults, communication model, communication complexity mining, energy usage, consensus category, and finality of the consensus are all important metrics to consider when assessing blockchain consensus algorithms.

2.2 *Blockchain-based transaction management system*

Blockchain is a new technology that lets network users share information without having to trust each other. It does this by using multiple ledgers. Blockchain has been used in many projects to fix problems with traditional methods. Some of these works' (Castro 1999) medical data was saved in the cloud, and the blockchain was used to record the hash of that data to make sure it was correct. Still, a single place where something could go wrong is a problem. Blockchain was used in most of the work (Fan 2018) to store the medical data of the patients. Out of these, (Fernandez 2012), (Gupta 2019), (Guo 2018) suggested ways to protect the security and safety of medical information by encrypting and decrypting data, using digital signatures, transferring data, or making keys. The author demonstrated a smart contract framework for managing access to medical data and provided a blockchain-based healthcare records management system for exchanging medical records. However, these investigations rely on the inefficient and resource-intensive Proof of Work (Pow) agreement method (Kohan 1996), (Liang 2017). Using PBFT (Patel 2018), which consumes less energy than Pow (Rind 1997), the author devised a blockchain-based approach for sharing medical data to address these issues. But (Schoenberg 2000), (Saravanam 2017) claim that patients cannot upload their health records to the blockchain and that the hospital must instead maintain all patient information.

3 PROPOSED METHODOLOGY

The Hyperledger Caliper tool enables such comparisons between blockchain and conventional networks. There are a number of different Hyperledger systems available, including Fabric, Composer, Sawtooth, Iroha, and others. In this research, the Caliper tool was used to measure and evaluate latency, throughput, CPU utilization, memory consumption, disc write/read, network I/O, and other system evaluation metrics.

Several setup factors, including block size, block time, endorsement policy, channel, resource allocation, and ledger database, may be adjusted depending on the evaluation. Here's how the computer systems for modelling are set up:

- Intel core i5 processor (with 3–4 MB cache)
- 8GB RAM

- Minimum 320 GB Hard Disk Space
- Gbit/s network

Hyperledger is an open-source program, can run on any platform like windows, MacOS and Linux. In our study Ubuntu 20.04 and windows 10 operating system used.

3.1 *Basic experiment*

The first experiment is done with different inputs and five rounds of putting 1000 transactions into the ledger network at different rates of 50, 100, 150, 200, 250, and 500 transactions per second. The time it takes to do a transaction show how well the blockchain network works. Figure 2 shows the unusual way the network is set up. It has multiple lines that show how long events took to complete.

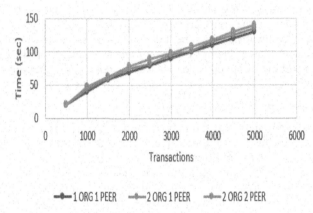

Figure 2.   Time is taken to successfully execute transactions.

Different transactions are shown by the numbers 1org1peer, 2org1peer, and 2org2peer. The results are worked out over five rounds, with 1000 deals at different tps rates in each round. The 1org1peer needs 140 seconds to make 5000 transactions. 2org1peer gets to 3000 transactions in 140 seconds, but 2org2peer only gets to 2000 transactions in the same amount of time. Therefore, it is evident that the time it takes to complete transactions increases as the number of organizations and peers increases.

A mathematical method is used to figure out how long a transaction takes. Assume that TL stands for "transaction latency," which is the amount of time it takes to use the network. CT is the time it takes for the transaction to be confirmed. This time changes based on the network threshold NT. In the blockchain network, as shown in Table 1, an ST is the time a transaction is sent.

Transaction latency $TL = (CT * NT) - ST$.

Table 1.   Basic measurement.

Parameter	Configurations
Transactions	1000 per round
Mode of Transactions	Read/Write and config
Rounds	5
Rate	70 to 250tps
Varied factor	———

Figure 3 depicts the typical lag time encountered during performance testing using the industry standard tool, Caliper. This diagram depicts the delay time in seconds. It reveals the average length of negotiations and the proportion of agreements that are ultimately reached. Compared to 2org 1peer and 2org 2peer, 1org and 1peer have a significantly smaller delay. On the other hand, 2org 1 peer and 2org 2 peer only manage 20 and 10 tps of throughput, respectively. This lengthens lag times and breaks in communication, allowing for enhanced efficiency.

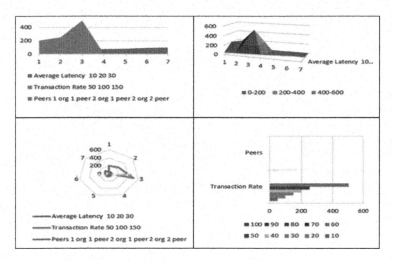

Figure 3.   Average latency.

3.2   *Experiment with varying blocks*

In this experiment, we attempt to improve the network's performance. Results of the Hyperledger Caliper for EHR system vary depending on how the block creation time is measured. The arrangement of the caliper used in the experiment is shown in Table.

Table 2.   Varied block time measurements.

Parameter	Configuration
Transactions	1000 per round
Transactions mode	Write
Rate	50 to 250 tps
Network size	2Org 2 Peer
Varied factor	Block Time
Endorsement policy	2 of: {signed-by: {0,1}}
Round	5

The time it takes for transactions to happen on a two-group, two-peer network The optimization factors were taken into account so that the latency could be as low as possible. The data shows that the network latency dropped by 1.5 times, which helps the EHR system work better. At a rate of 50 transactions per second, the minimum delay is now about 27 seconds. It used to be 52 seconds. The number of 37s for 250 tps is less than the number of 50s. The Hyperledger network settings can be changed to make this system work better.

Five iterations of reading and writing 1,000 transactions each iteration into the ledger's network at 100, 150, 200, 250, and 300 transactions per second.

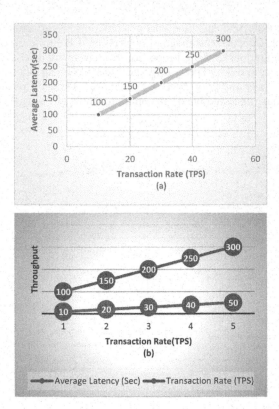

Figure 4. The impact of block time changes on transaction latency (a) and transaction throughput (b).

Figure 4 depicts the transaction rate (b). It demonstrates how network policy influences block time, which improves throughput and performance (how quickly and reliably a transaction is committed). When all of the optimizations were factored in, the average throughput increased from 50 tps to 250 tps, a 1,750% increase. Optimized performance in terms of transaction rate and latency is depicted in the figure. As a result of optimization, the system latency has been reduced by 40%, which is great news for the network's overall performance.

If you alter the policy and the block time by 2 seconds, the average latency lowers from 9 seconds (with the default block time of 250 milliseconds) to 4 seconds. The new configuration reduces the read latency from 13 seconds (the typical block time for a max tps of 250) to 6 seconds.

Table 3. Varied block time measurement with reading mode.

Parameter	Configuration
Rounds	3
Transactions	1000 per round
Transaction mode	Read
Rate	100,200, 250 tps
Network size	2Org 2peer
Varied factor	Block time
Endorsement policy	2 of: {signed-by: {0,1}}

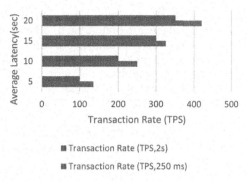

Figure 5.   Optimized performance in terms of transaction rate and latency.

4   COMPARATIVE ANALYSIS

Considerations for designing blockchain consensus algorithms are discussed here. Many factors must be considered, including the blockchain's architecture, transaction speed, scalability, and paradigm for dealing with malicious actors. Important characteristics for evaluating various blockchain consensus algorithms include latency, throughput, bandwidth, trial setup assaults, communication model, communication complexity, mining, energy use, consensus category, and finality of the consensus.

In our research, we measured performance using a Hyperledger caliper. The findings show that the suggested scheme improves blockchain performance using the provided algorithms, and that the latency may be lowered utilizing the proposed approach. This illustrates that Blockchain, like the steam engine and the Internet, may bring about revolutionary change for the betterment of mankind. Efficiency of Financial Dealings Resources used (CPU, Memory, Network IO, etc.) Transaction delay (minimum, maximum, average).

Docker is an open-source platform that streamlines the process of creating, distributing, and running software applications. Docker allows you to isolate your programmes from your server environment, allowing for rapid software deployment. Hyperledger employs local Docker images to partition the network into individual nodes. Factors include the Hyperledger Caliper hosting process and a network configuration file that contains the testing system and network parameters create average memory consumption variation across peers due to the PBFT consensus mechanism. The letters "org" stand for "organization," while "org1" and "org2" represent the first and second organizations, respectively, in a network. Hyperledger's caliper generates in-depth data on a host system's memory, CPU, incoming and outgoing network traffic, and disc writes whenever a PBFT network peer is hosted by that system. After hosting its peers peer0.org1.example.com and peer0.org2.example.com, peer1.org1.example.com saw a decrease in its average memory utilization, from 266.0 MB to 244.0 MB. When peer1 originally hosted, it required 224.1 MB of RAM, but now it just needs 10.1 MB. There is constant use of 205.2 MB and 202.0 MB of RAM and 24.25% and 27.95% of CPU by peer0.org1.example.com and peer0.org2.example.com, respectively. The process of new nodes joining a network and exchanging information with existing ones is known as "in and out traffic."

Permissionless distributed blockchains like Ethereum and Bitcoin are only two examples; in these networks, any node may take part in the consensus process that groups transactions into blocks. Therefore, these systems depend on probabilistic consensus procedures to provide consistent ledgers despite their susceptibility to forks, which occur when two or more ledgers diverge from one another. On the other side, Hyperledger Fabric is operational. An ordered node, sometimes known as a "ordering node," coordinates the ordering of incoming transactions as part of a larger ordering service. Each block approved by the peer is guaranteed to be final and valid due to the deterministic consensus processes used in Fabric's design. Due to their immutability, ledgers are preferable to other permissionless distributed ledger systems. As a service measure, Orderer.example.com had a memory use of 56.9 MB, a CPU utilization of 4.05%, a data throughput of 3.9 MB, and an outbound data transfer of 16.25 MB. To complete the financial operations, a disc write of 5MB is necessary.

Our research used two different consensus techniques at the implementation stage. A Hyperledger fabric with POW and a plugged PBFT. A comparison of the two shows that POW is superior, but it is only applicable to bitcoin deals. Another drawback of POW is the considerable energy it chews up during transaction processing. We found that protecting E-Healthcare data using PBFT on the Hyperledger fabric was the most effective method. The results of a PBFT consensus were tabulated below.

5 CONCLUSION

In this paper, a secure healthcare framework is implemented using blockchain technology. The PBFT consensus mechanism was used and examined for a secure consensus mechanism. Implementation is done using Hyperledger fabric. A blockchain-based transaction management system is proposed for secure transactions in the e-healthcare system. From various experiments, it has been found that blockchain-based healthcare systems are secure, and the performance of blockchain-based systems is very good with the PBFT consensus mechanism as compared to Proof of Work (Pow) consensus.

REFERENCES

Azaria A *et al.* 2016. MedRec: Using Blockchain for Medical Data Access and Permission Management, *2nd Int Conf Open Big Data*
Bahga A *et al.* 2013. A Cloud-based Approach for Interoperable Electronic Health Records (EHRs). *IEEE J Biomed Heal Informatics* 17:894–906.
Badr S *et al.* 2018. Multi-tier Blockchain Framework for IoT-EHRs Systems, *Procedia Comput Sci* 141:159–166
Castro M *et al.* 1999. Practical Byzantine Fault Tolerance. In: *Proceedings of the Third Symposiumon Operating Systems Design and Implementation*, pp 173–186
Fan K *et al.* 2018. MedBlock:Efficient and Secure Medical Data Sharing Via Blockchain, *J Med Syst* 42
Fernández-Cardeñosa G *et al.* 2012. Analysis of Cloud-based Solutions on EHRs Systems in Different Scenarios, *J Med Syst* 36:3777–3782
Gupta R *et al.* 2019. Blockchainbased Telesurgery Framework for Healthcare 4.0. *2019 International Conference on Computer, Information and Telecommunication Systems (CITS)*. 2019.p.1–5
Guo RUI *et al.* 2018. Secure Attribute-Based Signature Scheme with Multiple Authorities for Blockchain in Electronic Health Records Systems, *IEEE Access* 6:11676—11686
Kohane I *et al.* 1996. Building National Electronic Medical Record Systems via The World Wide Web, *J Am Med Informatics Assoc* 3:191–207
Liang X *et al.* 2017. Integrating Blockchain for Data Sharing and Collaboration in Mobile Healthcare Applications, *IEEE 28th Annual International Symposium on Personal, Indoor, and Mobile Radio Communications (PIMRC)*. pp 1–5
Patel V 2018. A Framework for Secure and Decentralized Sharing of Medical Imaging Data via Blockchain Consensus, *Health Informatics J*
Rind DM *et al.* 1997. Maintaining the Confidentiality of Medical Records Shared over the Internet and the World Wide Web, *Am Coll physicians*
Schoenberg R *et al.* 2000. Internet Based Repository of Medical Records that Retains Patient Confidentiality, *BMJ* 321:1199–1203
Saravanan M *et al.* 2017. A Secured Mobile Enabled Assisting Device for Diabetics Monitoring, *IEEE International Conference on Advanced Networks and Telecommunications Systems (ANTS)*.

Blockchain-Aware secure lattice aggregate signature scheme

Motashim Rasool*
Assistant Professor, Computer Science and Engineering, SR Institute of Management and Technology, Lucknow, Uttar Pradesh, India

Arun Khatri*
Professor, JK Business School, Gurgaon

Renato R. Maaliw III*
College of Engineering, Southern Luzon State University, Lucban, Quezon, Philippine

G. Manjula*
Assistant Professor, RMK College of Engineering and Technology, Thiruvallur

M.S. Kishan Varma*
Assistant Professor, Department of Management Studies, VFSTR Deemed to be University

Sohit Agarwal*
Assistant Professor, Department of Computer Engineering and Information Technology, Suresh Gyan Vihar, University, Jaipur

ABSTRACT: In the Forward Secure Ordered Aggregation (FssAgg) signature system, the signer gradually and orderly aggregates the signatures under different keys in different periods into one signature in a hierarchical "onion-like" manner. Among them, the innermost signature is the first signature. In addition, compared with ordinarily ordered aggregate signatures, forward secure requested aggregate signatures are the aggregation of different signatures of the same signer rather than the signatures of additional signers so that the signature verifier can use public key complete verification of all aggregation processes. The forward-safe ordered aggregate signature has the advantages of both forward-secure signature and aggregate signature. It has been widely used in many application scenarios, such as log systems and blockchains. Several existing forward-safe ordered aggregate signatures are based on traditional number theory problems, which will no longer be problematic in the post-quantum era. Therefore, finding a forward-safe requested aggregate signature in the quantum computing environment is imminent. Based on the small integer solution problem on the lattice, a forward-secure lattice-based rated aggregate signature scheme under the standard model is constructed. To achieve high efficiency, the system uses fixed-dimensional lattice-based delegation technology to realize critical updates and achieve forward security; then, the message to be signed and the difficulty problem on the lattice are respectively embedded into the signature using message addition technology and preimage sampling algorithm, making the signature unforgeable under the standard model.

Keywords: Blockchain, Signature Scheme, forward-safe ordered, secure Lattice, aggregation processes

---

*Corresponding Authors: mail2motashim@gmail.com, khatriarun@yahoo.com, rmaaliw@slsu.edu.ph, manjulacse@rmkcet.ac.in, mskishanvarma@gmail.com and sohit.agarwal@gmail.com

DOI: 10.1201/9781003393580-98

1 INTRODUCTION

Logs are a key component of computer information systems, they record "when and by whom what happened" in the system, and have certain forensic value. Therefore, it is particularly important to protect the log security of the system. In order to prevent attackers from tampering with historical logs, traditional audit methods write logs to devices that can only append records, or perform remote recording, but this cannot prevent malicious system administrators from tampering with logs. In order to reduce the threat brought by malicious system administrators, forward security signature technology is introduced into the log system.

The concept of forward secure signature (Quan *et al.* 2022) introduces to reduce the problem of key leakage. In order to achieve forward security, the key generation center (key generation center, KGC) divides time into k discrete time periods, and uses different keys in different periods. Among them, the signature key of period $i+1$ is calculated by the key update algorithm using the signature key of period i as input, where $1 \leq i < k$. Therefore, as long as the key update algorithm satisfies the characteristics of the one-way trapdoor function, even if the key of a given period is leaked, it will not be safe for the signature (the signature here is the signature of the log) of the previous period create a threat. Therefore, the forward security signature can effectively reduce the problem of key exposure, thereby reducing the threat of malicious log tampering by system administrators.

One of the main obstacles to deploying a forward-safe signature scheme in a secure logging system is its overhead. Storing a signature for each message is a heavy workload for a logging system that generates a large number of messages. The introduction of the concept of ordered aggregate signature (Li *et al.* 2022) can effectively solve this problem. In an ordered aggregation signature scheme, the aggregation of the certificate chain is carried out step by step. For example, user U_1 first uses its signature private key SK_1 to calculate and output the signature sgi_1 for message m_1; then, taking sig1 as input, user U_2 uses its signature private key SK_2 to calculate and output the signature sgi_2 for messages m_1 and m_2; Taking the signature sgi_{i-1} of users $U_1, U_2, \cdots, U_{i-1}$ on messages $m_1, m_2, \cdots, m_{i-1}$ as input, U_i uses its signature private key SK_i to calculate and output the signature sgi_{i-1} for messages $m_1, m_2, \cdots, m_i$ Sign sgi_i; until U_k finishes signing. When k users sign in such a step-by-step manner, while retaining the minimum communication overhead of the aggregate signature, it also fully guarantees the non-repudiation of the signature to the message.

In order to take into account the security and efficiency of the log system, the first forward-secure ordered aggregate signature scheme - BLS-FssAgg signature (Zheng *et al.* 2019) came into being. However, since the construction of the BLS-FssAgg signature depends on the bilinear pairing problem, the huge overhead of the bilinear pairing operation will inevitably make the signature verification cost huge. In the following year, Ma proposed two practical FssAgg signature schemes (Zheng *et al.* 2022), namely BM-FssAgg signature and AR-FssAgg signature. In these two FssAgg signature schemes, the multiplication operation on the QRn group is used in the signature construction process, so that the public key, private key and signature size reach a constant level, and the time complexity of secret key update and signature is also constant, which reflects its practicability; and compared with the BLS-FssAgg signature in (Zheng *et al.* 2019), the verification speeds of these two signatures are increased by 16 times and 4 times respectively. In 2018, based on the strong RSA assumption and the discrete logarithm problem on the elliptic curve, Wei Xingjia successively proposed a non-repudiation identity-based aggregate signature scheme with forward security (Xu *et al.* 2022) and an identity-based aggregate signature scheme with forward security properties. Proxy aggregate signature scheme (Zheng *et al.* 2022), and prove that it satisfies forward security and existence unforgeability under the random oracle model. However, the above two aggregated signatures are out of order, so they cannot be applied to the log system. Literature (Qian *et al.* 2021) proposed an efficient forward-secure ordered aggregate signature scheme for secure log systems- FAS signature. In addition to satisfying forward security and unforgeable existence, it is worth mentioning that the size of the public key and private key of the FAS signature is constant, and the key size is the same as that of the forward secure non-aggregate signature BM signature (Quan *et al.* 2022) At the same time, the FAS signature reduces the verification complexity of the BM signature from $O(k^2)$ to $O(lk)$, where l represents the output length of the hash function. Literature (Chan *et al.* 2022) used the Chinese remainder theorem, combined with bilinear pairing technology, to propose an aggregate signature scheme with forward security properties based on elliptic curve cyclic groups. The forward security of the scheme is determined by strong RSA the assumption is guaranteed, and the two-way verifiability and enforceability of the scheme are also provable. However, several existing practical forward-secure ordered aggregate signatures are based on traditional number theory problems. At the same time, research on forward-secure signatures with other special properties (Li *et al.* 2022; Lu *et al.* 2018; Yang *et al.* 2021; Qian *et al.* 2022) is also in full swing. Literature (Yao *et al.* 2019) those classical number theory problems are no longer safe in the quantum computing environment. With the gradual commercialization of quantum computers, the post-quantum era has come to people, and it is imminent to find quantum-immune forward-safe orderly aggregate signatures.

The US National Institute of Standards and Technology has been collecting post-quantum cryptography standards from around the world since 2016. In the competition for the third round of cryptographic standards that just ended in July 2020, there were a total of 7 winning algorithms, of which only lattice public key cryptography schemes accounted for 5 (including 3 encryption algorithms and 2 signature algorithms). In the view of the National Institute of Standards and Technology, these lattice-hard problem-based schemes are the most promising general-purpose algorithms for public-key encryption and digital signature schemes. Therefore, constructing a forward-safe ordered aggregate signature based on the Lattice difficulty problem may open a new situation for ensuring the security of the log system in the post-quantum era.

Based on the small integer solution problem on the lattice, this paper constructs a forward-secure lattice-based ordered aggregation signature scheme under the standard model. In order to achieve high efficiency, the scheme uses fixed-dimensional lattice-based delegation technology to realize key update and achieve forward security; then, the message to be signed and the difficulty problem on the lattice are respectively embedded into the signature by means of message addition technology and preimage sampling algorithm. , making the signature unforgeable under the standard model, and compared with the existing lattice-based ordered aggregation signature scheme (Agyekum *et al.* 2022; Alemany *et al.* 2022; Chen *et al.* 2021; Giraldo *et al.* 2020), the new scheme is still forward secure in the quantum computing environment.

2   RELATED WORK

Definition 1 (Yao *et al.* 2022) (lattice) Let $B = (b_1||b_2|| \cdots ||b_n)$, where $b_1, b_2, \cdots, b_n \in R_m$ are n linearly independent vectors. Then the lattice Λ generated by matrix B is defined as:

$$\Lambda = L(b_1, b_2, \cdots, b_n) = \left\{ \sum_{i=1}^{n} a_i b_i | a_i \in Z \right\}$$

At this time, n and m are the rank and dimension of Λ of the lattice respectively, and B is its basis.

Definition 2 (Yao *et al.* 2022) (q-ary lattice) Define m-dimensional full-rank q-ary lattice as:

$$\Lambda_q^T(A) = \{x \in Z^m | Ax = 0 (\text{mod } q)\}$$

$$\Lambda_q^u(A) = \{x \in Z^m | Ax = u (\text{mod } q)\}$$

Among them, the parameters are prime number q, positive integers n and m, matrix $A \in Z_q^{n \times m}$, and vector $u \in Z_q^n$, and $\Lambda_q^T q(A)$ and $\Lambda_q^u(A)$ are often abbreviated as $\Lambda^T(A)$ and $\Lambda^u(A)$.

Definition 3 (Yao *et al.* 2022) (Gaussian function) Given positive integers n and m, define the Gaussian function with $\sigma \in R^+$ as the parameter and $c \in R^m$ as the center as

$\rho_{\sigma,c}(x) = \exp\left(-\pi\|x - c\|^2/\sigma^2\right)$. Define the discrete Gaussian function with c as the center and σ as the parameter on the n-dimensional lattice Λ as:

$$D_{\sigma,c}(x) = \frac{\rho_{\sigma,c}(x)}{\rho_{\sigma,c}(\Lambda)}$$

Among them, the Gaussian function satisfies $\rho_{\sigma,c}(\Lambda) = \sum_{x \in \Lambda} \rho_{\sigma,c}(x)$. When the center vector $c = 0$, the discrete and Gaussian functions can be abbreviated as $D_{\Lambda,\sigma}(x)$ and $\rho_\sigma(x)$.

Definition 4 (Qu *et al.* 2021) ($D_{m \times m}$ distribution) It is known that q is a prime number, a positive integer $\geq 6nlbq$, and the parameter $\sigma \geq \sqrt{nlbq} \cdot \omega(\sqrt{lbm})$. $D_{m \times m}$ represents the distribution of the invertible matrix $A_i = [a_{i1}\|a_{i2}\| \cdots \|a_{im}]$ on the space $Z_q^{m \times m}$, where $a_{ij} \sim D_{Z^m,\sigma}, j \in [m]$.

Lemma 1 (Welligton dos Santos Abreu *et al.* 2022) (GPV08 Lattice Trapdoor Generation Algorithm) There is a probabilistic polynomial time algorithm TrapGen(1^n) that takes prime number $q \geq 3$ and positive integer $m \geq 6nlbq$ as input, and can output $A \in Z_q^{n \times m}$ and lattice $\Lambda^T(A)$ short basis T of, satisfying the distribution of A is statistically indistinguishable from the uniform distribution on $Z_q^{n \times m}$, and $\|T\| \leq O(nlbq)$, $\|\tilde{T}\| \leq O(nlbq)$. Here, $\tilde{T}$ represents the basis after performing Gram-Schmidt orthogonalization on T.

Lemma 2 (Yao *et al.* 2022)(common sampling algorithm on] lattice) Known positive integer $q \geq 2, m > n$, matrix $A \in Z_q^{n \times m}$, a set of basis T of lattice $\Lambda^T(A)$, Gaussian parameter $\sigma \geq \|\tilde{T}\|\omega(lbm)$, and any vector $c \in R^m$, $u \in Z_q^n$.

1. There is a probabilistic polynomial time algorithm *SampleD*(A, T, σ, c) that can output a vector v whose distribution statistics are close to $D_{\Lambda^T(A),\sigma,c}$.
2. There exists a probabilistic polynomial time algorithm *SamplePre*(A, T, σ, u) that can output a vector v whose distribution statistics are close to $D_{\Lambda^u(A),\sigma}$.

Lemma 3 (Qu *et al.* 2021) (Lattice-based delegation algorithm) Input full-rank matrix $A \in Z_q^{n \times m}$, matrix $R \sim D_{m \times m}$, a set of basis T of lattice $\Lambda^T(A)$, Gaussian parameters satisfy $\sigma > \|T\| \cdot \sigma_R\sqrt{m} \cdot \omega\left(lb^{\frac{3}{2}}m\right)$, and then there is a probabilistic polynomial time algorithm *BasisDel*(A, R, T, σ) that can output a set of basis T' of $\Lambda^T(AR^{-1})$, satisfying $\|T'\| < \sigma/\omega(\sqrt{lbm})$. Where the Gaussian parameter satisfying $\sigma_R = \sqrt{nlbq} \cdot \omega(\sqrt{lbm})$, $D_{m \times m}$ represents the distribution of matrices in $Z^{m \times m}$ that satisfy $\left(D_{\sigma_R}^m\right)^m$ and whose modulus q is invertible. If R is the product of l matrices extracted from $D_{m \times m}$, the parameters satisfy $\sigma > \|\tilde{T}\| \cdot \left(\sigma_R\sqrt{m} \cdot \omega\left(lb^{\frac{1}{2}}m\right)\right)^l \cdot \omega(lbm)$.

Lemma 4 (Liang *et al.* 2021) Input a full-rank matrix $A \in Z_q^{n \times m}$, then there is a probabilistic polynomial time algorithm *SampleRwithBasis*(A) that can output a matrix $R \sim D_{m \times m}$ and a set of basis T of the lattice $\Lambda^T(AR^{-1})$, satisfy $\|T'\| < O(\sqrt{nlbq})$.

Definition 5 (Yao *et al.* 2022)(Small integer solution problem on the lattice, $SIS_{q,m,\beta}$) Given a random matrix $A \in Z_q^{n \times m}$ and a real number $\beta > 0$, the so-called small integer solution (SIS) problem on the lattice That is to find a vector $v \in Z^m$ satisfying $Av = 0 \bmod q$ and $0 < \|v\| \leq \beta$.

Lemma 5 (Xiong *et al.* 2022) (Difficulty Reduction of Small Integer Solving Problems) Knowing any polynomially bounded real number $m, \beta = poly(n)$ and prime number $q \geq \beta \cdot \omega(\sqrt{nlbn})$, solve the $SIS_{q,m,\beta}$ of the average instance the difficulty of the problem is comparable to that of solving the shortest independent vectors problem ($SIVP_\gamma$) under the worst instance on the lattice, where $\gamma = \beta \cdot \tilde{O}(n)$.

3 PROPOSED MODEL

3.1 *Ordered aggregate signature*

This chapter describes the definition of forward-safe ordered aggregate signature and security proof as follows.

Definition 6 (forward secure ordered aggregate signature scheme) A complete forward secure ordered aggregate signature scheme consists of six polynomial time algorithms, which are the system establishment algorithm Setup, the signature key generation algorithm *FssAgg − Keygen*, and the signature private key Extraction algorithm *FssAgg − Extract*, signature key update algorithm *FssAgg − KeyUpdate*, aggregate signature algorithm *FssAgg − Sign* and signature verification algorithm *FssAgg − Verify*.

Setup: Taking the system security parameter n as input, KGC generates and outputs the system public parameter PP.

FssAgg − Keygen: Input PP, and KGC runs the algorithm to output the master/private key pair (msk, mpk).

FssAgg − Extract: Taking user U's identity id as input, the algorithm generates user U's key $sk_{id|0}$.

FssAgg − KeyUpdate : The private key $sk_{id|i-1}$ of the previous time period and the current time period i are input, the algorithm calculates and outputs the signature private key $sk_{id|i}$ of the current period i.

FssAgg − Sign : Input the message m_i to be signed and the previous aggregate signature $sk_{id|i-1}$, the algorithm uses $sk_{id|i}$ to calculate the aggregate signature sigi of the output period i.

FssAgg − Verify : Input aggregate signature sig_i and message set $m_j(j \le i)$, if and only if the aggregate signature sigi is the legal signature of message set $m_j(j \le i)$, the algorithm output is "1", otherwise it outputs "0".

3.2 *Correctness and forward enforceability*

In general, forward-secure ordered aggregate signatures should satisfy the following two conditions: correctness and forward unforgeability.

Definition 7 (Correctness) In the forward-secure ordered aggregation signature scheme above, the correctness of the signature scheme means that *FssAgg − Verify* can output "1" with a probability of approximately 1.

Definition 8(Wang *et al.* 2019) (The existence of forward security cannot be forged) If there is no polynomial time adversary A can win the following game with non-negligible probability, then the *Fss − Agg* signature scheme is said to be forward-secure under the aggregation attack model Existence cannot be forged.

Setup: Input the security parameter n, and the adversary gets the system public parameter PP.
Queries: Adversary A can make the following queries of polynomial order.

1. *FssAgg − ExtractQueries* : Input the identity id of user U, and challenger C returns its key $sk_{id|0}$.
2. *BreakinQueries* : When adversary a makes a Breakin Queries query for period j, challenger C returns the signature private key $sk_{id|j}$ of this period to the adversary.
3. *FssAgg − SignQueries* : Taking the current period i, the message to be signed m_i, the private key $sk_{id|i}$, and the previous signature sk_{i-1} as input, the challenger C returns the aggregated signature sk_i of the current period i to the adversary A.

Forgery : After finishing the above query, adversary A outputs the aggregated signature sig_k for messages $m_1, m_2, \ldots, m_k$. Adversary A wins the above game if and only if the following conditions are met: (1) signature sigk satisfies correctness; (2) signature sig_k is non-trivial, that is, there exists at least one time period $i* \in [k]$ not Perform aggregate signature query on messages $m_1, m_2, \ldots, m_{i^*}$; (3) $1 \leq i^* < j \leq k$.

3.3 Aggregation signature scheme

Based on the difficulty of lattice, this chapter constructs a forward-secure lattice-based ordered aggregation signature scheme under the standard model. Compared with the traditional signature scheme on the lattice, in order to achieve efficient forward security, the scheme uses the fixed-dimensional lattice-based delegation technology to realize the key update; the above hard problem is embedded into the signature, making the signature unforgeable under the standard model.

Let n be a security parameter, prime number $q \geq \beta \cdot \omega(lbn), \beta = poly(n), m \geq 6nlbq, L \leq O(\sqrt{nlbq})$. Then the forward-secure ordered aggregation signature scheme on lattice is described as follows.

Setup: Input the security parameter n of the system, and the KGC generates the system public parameter PP as follows. k is the pre-specified number of time periods and Gaussian parameters $(\sigma_0, \sigma_1, \cdots, \sigma_k)$, Gaussian parameters $(\sigma'_0, \sigma'_1, \cdots, \sigma'_k)$, where $\sigma_i \geq Lm^i \cdot \omega(lb^{i+1}m), \sigma'_i \geq \sigma_i \cdot \omega(\sqrt{lbm}), \eta \geq m \cdot \sqrt{\omega lbm}, t1, t2, l$ is a positive integer. $2(k+1)t_1$ random matrices $R^0_{i,j}, R^1_{i,j} \leftarrow D_{m \times m}(i \in \{0, 1, \cdots, k\}, j \in [t_1]), t_2$ random matrices satisfying $C_i \leftarrow Z^{n \times m}_q(i \in [t_2])$. The three hash functions are $H : \{0,1\}^* \rightarrow \{0,1\}^{t_1}, H' : \{0,1\}^* \rightarrow \{0,1\}^{t_2}, H1 : \{0,1\}^* \rightarrow \{0,1\}^l$ and $G_{f_A} : \{0,1\}^l \rightarrow Z^n_q$, and the trapdoor function $f_A : Z^m_q \rightarrow Z^n_q$ and the encryption function $enc : \{0,1\}^* \rightarrow \{0,1\}^* \times Z^n_q$ and the decryption function $dec : \{0,1\}^* \times Z^n_q \rightarrow \{0,1\}^*$.

FssAgg − KeyGen : Input the public parameter PP, KGC runs the trapdoor generation algorithm $TrapGen(1^n)$ in **Lemma 1** to generate lattice $\Lambda^T(A)$, and its trapdoor base $T \in Z^{m \times m}$, satisfying $A \in Z^{n \times m}_q, \|T\| \leq L$. Finally, KGC saves the matrices A and T as the master public key and master private key, respectively.

FssAgg − Extract : In order to obtain the key of user U (identity is id), KGC first calculates the hash value $H(id|0) = \rho_0 = \rho id|0, R_0 = R^{\rho_0[t_1]}_{0,t_1}, R^{\rho_0[t_1-1]}_{0,t_1-1} \cdots R^{\rho_0[1]}_{0,1}$ Then run the lattice-based delegation algorithm $BasisDel(A, R_0, T, \sigma_0)$ in **Lemma 3** to output the matrix $T_{id|0} = [t_1||t_2||\cdots||t_m]$ as the key $sk_{id|0}$. Finally, KGC sends $T_{id|0}$ to user U.

FssAgg − KeyUpdate : Input the key $T_{id|i-1}$ and the current period i, KGC first calculates $R_i = R^{\rho_i[t_1]}_{i,t_1} R^{\rho_i[t_1-1]}_{i,t_1-1} \cdots R^{\rho_i[1]}_{i,1}, R_{id|i-1} = R_{i-1}R_{i-2} \cdots R_0$ and the matrix $A_{id|i-1} = AR^{-1}_{id|i-1}$. Subsequently, KGC runs the lattice-based delegation algorithm $BasisDel(A_{id|i-1}, R_i, T_{id|i-1}, \sigma_i)$ to generate the signature private key $T_{id|i}$ for the current period.

FssAgg − Sign : Input the current message m_i and the current period i, the signature algorithm works as follows:

(1) If $i = 1$, then
(2) $\Sigma 0 \leftarrow (e, e, e, e, 0^n)$, end.
    Where e represents an empty string or an empty vector, if $i = 2, 3, \cdots, k$, it is performed as follows.
(3) Split Σi - 1 into (f· Aid|i, m · i - 1, α · i - 1, sigi - 1, h'i - 1),
    Where · fAid|i is lattice trap A set of gate functions, satisfying f· Aid|i= fAid|1|fAid| 2|···|fAid|i, sigi - 1 is the aggregate signature of the previous period, αi - 1 is the encrypted signature of sigi - 1, α · i - 1 is the set of encrypted signatures, that is, α · i - 1 = α1|α2|···|αi - 1; hi - 1 is the hash value with a constant length l.
(4) If FssAgg-Verify (Σi - 1) ==($\perp, \perp$), then end.
(5) Calculate (αi,βi) $\leftarrow$ encfAid|i(sigi - 1).
(6) Set $\alpha \cdot i = \alpha \cdot i$ - 1|αi .
(7) Calculate hi = hi - 1$\oplus$H1(f· Aid|i,m · i,α · i - 1, sigi - 1).

658

(8) Calculate gi ← GfAid|i(hi).
(9) Choose a random string ri∈ {0,1}n and calculate the vector

4 RESULT ANALYSIS

Currently, there are schemes (Agyekum *et al.* 2022; Alemany *et al.* 2022; Chen *et al.* 2021; Giraldo *et al.* 2020) for the existing ordered aggregation signatures on the grid, and they are all signature schemes under the random oracle model. This section compares the efficiency of the ordered aggregation signature scheme on the lattice with the previous signature scheme from three aspects: public key, private key, and signature size; see Table 1 for details. Among them, the parameters satisfy L ≤ O(n lb q), M = ω(lb m), m ≥ 5n lb q, k represents the number of periods specified in Section 3.1, I represents the stage of the current classification, and i = 1,2,...,k. It is easy to see from Table 1 that the public key, signature private key, and signature size of the schemes in (Agyekum *et al.* 2022; Alemany *et al.* 2022; Chen *et al.* 2021; Giraldo *et al.* 2020)are all equivalent. However, the forward-secure aggregate signature scheme proposed in this paper adopts the grid-based delegation technology to achieve forward security. Therefore, compared with the literature (Agyekum *et al.* 2022; Alemany *et al.* 2022; Chen *et al.* 2021; Giraldo *et al.* 2020), the private key and signature length will vary with the classification (i.e.) increases with increasing. Therefore, it can be understood that the new scheme sacrifices a slightly larger signature private key size and signature size in exchange for higher security guarantees, thereby adapting to a more stringent security environment. The specific implementation of the signature scheme can be completed in the C++ language with the help of the FLINT library.

Based on the model, numerical simulations are performed to verify the information dissemination performance with blockchain support in the lattice model and compare the results with the information dissemination performance without blockchain support. Assuming

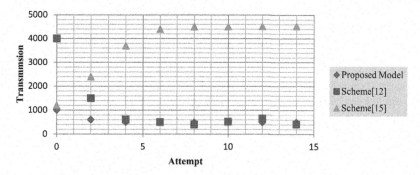

Figure 1.   Attack transmission over the lattice.

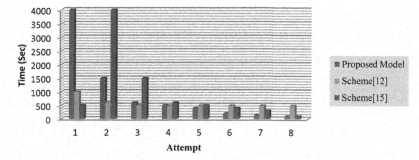

Figure 2.   Attack Time over the lattice.

simulations in a fixed population of T(u) + J(u) + S(u) = 5000, the process of information propagation from the initial stage u = 0 to the end-stage u = 20 is studied. In the two cases of the above information dissemination process, as shown in Figures 1–2, when the contract node exists, the user's dissemination behavior is shown in Figure 2. When the contract node does not exist, as shown in Figure 2 with table, From the analysis of the above figure, it can be seen that the number of communicators in the model without contract nodes exceeds 80% compared with the existence of contract nodes and the non-existence of contract nodes. Conversely, with the addition of contract nodes, the number of information communicators decreases to 20%, indicating that user information dissemination in the blockchain lattice is rationalized. User information in the blockchain environment can better completely curb distorted news than traditional lattice information dissemination, and the dissemination is relatively stable.

5 CONCLUSION

Due to the advantages of both forward security and small storage space, the forward-secure ordered aggregate signature scheme has been widely used in log systems once it was proposed. However, with the gradual commercialization of quantum computers, the existing forward-secure ordered aggregation signature scheme can no longer meet the security requirements. Therefore, finding forward-secure rated aggregation signature schemes with quantum immunity is imperative. It is not difficult to see in the three-round post-quantum cryptography alternatives announced by the National Institute of Standards and Technology that lattice public key cryptography occupies a pivotal position, so compared with hash-based public key cryptography and encoding-based public key In terms of cryptographic systems and multivariate public key cryptosystems, lattice public key cryptography is undoubtedly the best choice for cryptographic standards in the post-quantum era. Based on the lattice-hard problem—SIS problem, this paper proposes a quantum-safe forward-safe ordered aggregation signature scheme under the standard model. The system realizes critical updates using fixed-dimensional lattice-based delegation technology and achieves forward security; then, the message to be signed and the difficulty problem on the lattice are respectively embedded into the signature through the message addition technology and the preimage sampling algorithm so that the signature is in the standard Models are unforgeable. To further compress the block size in the blockchain, the construction of the ordered aggregate signature of the specified verifier will be the focus of the follow-up work.

REFERENCES

Agyekum K. O. -B. O., Xia Q., Sifah E. B., Cobblah C. N. A., Xia H. and Gao J., "A Proxy Re-Encryption Approach to Secure Data Sharing in the Internet of Things Based on Blockchain," in *IEEE Systems Journal*, vol. 16, no. 1, pp. 1685–1696, March 2022, doi: 10.1109/JSYST.2021.3076759.

Alemany P., Vilalta R., Munoz R., Casellas R. and Martinez R., "Evaluation of the Abstraction of Optical Topology Models in Blockchain-based Data Center Interconnection," in *Journal of Optical Communications and Networking*, vol. 14, no. 4, pp. 211–221, April 2022, doi: 10.1364/JOCN.447833.

Chan Y. -C. and Jafar S. A., "Secure GDoF of the Z-Channel With Finite Precision CSIT: How Robust are Structured Codes?," in *IEEE Transactions on Information Theory*, vol. 68, no. 4, pp. 2410–2428, April 2022, doi: 10.1109/TIT.2021.3137151.

Chen R. *et al.*, "BIdM: A Blockchain-Enabled Cross-Domain Identity Management System," in *Journal of Communications and Information Networks*, vol. 6, no. 1, pp. 44–58, March 2021, doi: 10.23919/JCIN.2021.9387704.

Giraldo F. D., B. Milton C. and Gamboa C. E., "Electronic Voting Using Blockchain And Smart Contracts: Proof Of Concept," in *IEEE Latin America Transactions*, vol. 18, no. 10, pp. 1743–1751, October 2020, doi: 10.1109/TLA.2020.9387645.

Li Q., Luo M., Hsu C., Wang L. and He D., "A Quantum Secure and Noninteractive Identity-Based Aggregate Signature Protocol From Lattices," in *IEEE Systems Journal*, vol. 16, no. 3, pp. 4816–4826, Sept. 2022, doi: 10.1109/JSYST.2021.3112555.

Liang W., Zhang D., Lei X., Tang M., Li K.-C. and Zomaya A. Y., "Circuit Copyright Blockchain: Blockchain-Based Homomorphic Encryption for IP Circuit Protection," in *IEEE Transactions on Emerging Topics in Computing*, vol. 9, no. 3, pp. 1410–1420, 1 July-Sept. 2021, doi: 10.1109/TETC.2020.2993032.

Lu X., Yin W., Wen Q., Jin Z. and Li W., "A Lattice-Based Unordered Aggregate Signature Scheme Based on the Intersection Method," in *IEEE Access*, vol. 6, pp. 33986–33994, 2018, doi: 10.1109/ACCESS.2018.2847411.

Qian J., Cao Z., Dong X., Shen J., Liu Z. and Ye Y., "Two Secure and Efficient Lightweight Data Aggregation Schemes for Smart Grid," in *IEEE Transactions on Smart Grid*, vol. 12, no. 3, pp. 2625–2637, May 2021, doi: 10.1109/TSG.2020.3044916.

Qian J., Cao Z., Lu M., Chen X., Shen J. and Liu J., "The Secure Lattice-Based Data Aggregation Scheme in Residential Networks for Smart Grid," in *IEEE Internet of Things Journal*, vol. 9, no. 3, pp. 2153–2164, 1 Feb.1, 2022, doi: 10.1109/JIOT.2021.3090270.

Qu Y., Pokhrel S. R., Garg S., Gao L. and Xiang Y., "A Blockchained Federated Learning Framework for Cognitive Computing in Industry 4.0 Networks," in *IEEE Transactions on Industrial Informatics*, vol. 17, no. 4, pp. 2964–2973, April 2021, doi: 10.1109/TII.2020.3007817.

Quan Y., "Improving Bitcoin's Post-Quantum Transaction Efficiency With a Novel Lattice-Based Aggregate Signature Scheme Based on CRYSTALS-Dilithium and a STARK Protocol," in *IEEE Access*, vol. 10, pp. 132472–132482, 2022, doi: 10.1109/ACCESS.2022.3227394.

Wang S., Ouyang L., Yuan Y., Ni X., Han X. and Wang F.-Y., "Blockchain-Enabled Smart Contracts: Architecture, Applications, and Future Trends," in *IEEE Transactions on Systems, Man, and Cybernetics: Systems*, vol. 49, no. 11, pp. 2266–2277, Nov. 2019, doi: 10.1109/TSMC.2019.2895123.

Welligton dos Santos Abreu A., Coutinho E. F. and Ilane Moreira Bezerra C., "Performance Evaluation of Data Transactions in Blockchain," in *IEEE Latin America Transactions*, vol. 20, no. 3, pp. 409–416, March 2022, doi: 10.1109/TLA.2022.9667139.

Xiong H. *et al.*, "On the Design of Blockchain-Based ECDSA With Fault-Tolerant Batch Verification Protocol for Blockchain-Enabled IoMT," in *IEEE Journal of Biomedical and Health Informatics*, vol. 26, no. 5, pp. 1977–1986, May 2022, doi: 10.1109/JBHI.2021.3112693.

Xu C., Qu Y., Luan T. H., Eklund P. W., Xiang Y. and Gao L., "A Lightweight and Attack-Proof Bidirectional Blockchain Paradigm for Internet of Things," in *IEEE Internet of Things Journal*, vol. 9, no. 6, pp. 4371–4384, 15 March15, 2022, doi: 10.1109/JIOT.2021.3103275.

Yang Y., Chen Y. and Chen F., "A Compressive Integrity Auditing Protocol for Secure Cloud Storage," in *IEEE/ACM Transactions on Networking*, vol. 29, no. 3, pp. 1197–1209, June 2021, doi: 10.1109/TNET.2021.3058130.

Yao S. *et al.*, "Blockchain-Empowered Collaborative Task Offloading for Cloud-Edge-Device Computing," in *IEEE Journal on Selected Areas in Communications*, vol. 40, no. 12, pp. 3485–3500, Dec. 2022, doi: 10.1109/JSAC.2022.3213358.

Yao Y., Zhai Z., Liu J. and Li Z., "Lattice-Based Key-Aggregate (Searchable) Encryption in Cloud Storage," in *IEEE Access*, vol. 7, pp. 164544–164555, 2019, doi: 10.1109/ACCESS.2019.2952163.

Zheng P. *et al.*, "Aeolus: Distributed Execution of Permissioned Blockchain Transactions via State Sharding," in *IEEE Transactions on Industrial Informatics*, vol. 18, no. 12, pp. 9227–9238, Dec. 2022, doi: 10.1109/TII.2022.3164433.

Zheng P., Xu Q., Zheng Z., Zhou Z., Yan Y. and Zhang H., "Meepo: Multiple Execution Environments per Organization in Sharded Consortium Blockchain," in *IEEE Journal on Selected Areas in Communications*, vol. 40, no. 12, pp. 3562–3574, Dec. 2022, doi: 10.1109/JSAC.2022.3213326.

Zheng W., Zheng Z., Chen X., Dai K., Li P. and Chen R., "NutBaaS: A Blockchain-as-a-Service Platform," in *IEEE Access*, vol. 7, pp. 134422–134433, 2019, doi: 10.1109/ACCESS.2019.2941905.

Artificial Intelligence, Blockchain, Computing and Security – Dagur et al. (Eds)
© 2024 The Author(s), ISBN: 978-1-032-49393-0

Developing secure framework using blockchain technology for E-healthcare

Pritpal Singh*, K. Jithin Gangadharan, Ashwani Kumar & Priya Chanda
Mittal School of Business, Lovely Professional University, Punjab

Prikshat Kumar & Aseem Khanna
School of Computer Applications, Lovely Professional University, Punjab

ABSTRACT: Blockchain technology is the most original technology that is growing and being used in many different fields at the fastest rates. In this study, the same goal was reached by using pluggable PBFT (practical byzantine fault tolerance) and the Hyperledger Fabric blockchain framework to make sure that healthcare transfers are safe. This study began by looking at successful blockchain applications and doing basic research to find out why they were needed. With the help of this technology, patients now have full, permanent data and can get to their electronic health records (EHRs) whenever they want. In this study, a secure framework for securing e-healthcare systems with Hyperledger Fabric and the PBFT consensus method was made using blockchain technology. Results of study explains the proposed framework is having better efficiency then existing systems.

Keywords: Blockchain, PBFT Consensus mechanism, Healthcare, Electronic health records

1 INTRODUCTION

In clinical modelling of health care, the amount of data that is made, shared, and used has grown a lot in recent years. Huge amounts of data need to be shared and kept, but it's hard to do so because they are often private and have other rules like privacy and security (Ali 2017). Healthcare is one of the biggest and has a lot of deals happening all the time. Electronic health care is a way to store health information about a person on a computer. The patient-centered records let any authorized user get to the information from any place and at any time. Every year, the internet health system saves $81 billion. Electronic health care improves health and social well-being and cuts down on medical mistakes. An electronic health record (EHR) is a digital copy of a patient's medical information that is kept using computer technology. A computerized health record is computerized information about a patient's health that is made by more than one care delivery organization. The different care delivery organizations can access the patient's full information through the portal, which is part of the EHR. Information about a patient's health, such as prescriptions, lab results, medical background, etc. With EHRs, all of a patient's health information is in one place and can be found when and where it is needed. Suppliers get the information they need when they need it to make a decision. For safe and effective care, it is important to have solid access to complete, long-term health statistics (Ageykum 2018). Due to the fact that blockchain is decentralized and reliable, it has a lot of promise in a number of e-health industries, such as the safe sharing of electronic health records

---

*Corresponding Author: pritpal.16741@lpu.co.in

                                          DOI: 10.1201/9781003393580-99

(EHRs) and the control of data access among many medical institutions (Ananth 2018). Putting blockchain technology to use could offer ways to make healthcare services easier and completely change the healthcare industry. Blockchain is being used more and more in the medical field (Gani 2018). When a treatment patient is mobile, which means they have to go to different medical facilities or hospitals for people with disabilities because they move between MEC nodes, they need autonomous integration and contact with their therapy profile that is safe and works well (Jiang 2018). Many different types of people, such as workers, nurses, insurance companies, and others, have access to healthcare data at the same company. Electronic health records (EHR) have a lot of information that is constantly being added to and changed. So, it is very important to check the EHR. In the e-healthcare system, trillions of deals are done every hour. The people in charge of each transaction are only those who have been verified, and the data from each transaction is saved on a computer. Doctors are the only ones who can see their patients' data, and patients are the only ones who can see their doctors' data. With the information gathered, a consensus-based e-healthcare system based on block chains will be used to provide a safe healthcare option. Blockchain is a big step forward in technology. It makes a lasting record of information that can be sent to all of the squares in a shared network. In healthcare, blockchain technology can be used to fix systemic problems and provide a safety net for patients, pros, hospital staff, and backup plans. Advances is a well-known blockchain development business that makes unique blockchain applications for the healthcare industry. These applications do things like keep track of the patient's own data, report symptoms, and recommend specialists. Blockchain technology is the most original technology that is growing and being used in many different fields at the fastest rate. It is known for being able to apply this technology to a variety of fields, which helps store data in fields that are connected by peer-to-peer networks. In this study, the same goal was reached by using pluggable PBFT and the Hyperledger Fabric blockchain framework to make sure that healthcare transfers are safe (Ozair 2015). This study began by looking at successful blockchain applications and doing basic research to find out why they were needed. It focuses on blockchain apps that are already out there. Proofs of concept in different fields were looked at, and the benefits of the technology in those fields were listed. Since hospitals, not people, are in charge of electronic health records (EHRs), it is hard to get medical advice from more than one hospital. Patients need to take back control of their own medical information and focus on the details of their own care (Ovitskaya 2017). The fast growth of blockchain technology helps with healthcare for the whole population, including patient information and medical records.

With the help of this technology, patients now have full, permanent data and can get to their electronic health records (EHRs) whenever they want. In this study, a secure framework for securing e-healthcare systems with Hyperledger Fabric and the PBFT consensus method was made using blockchain technology. Framework works both ways, from the patient's side to the doctor's side and back again. When a patient registers online, they give their personal information. This information includes all the user data they need to log in. In the e-healthcare system, this information is sent directly to the database along with some of the order actions. When patient registration is accepted, patients can move on in the framework of blockchain which is an open, decentralized, and distributed database that uses cryptography to create data blocks that are often arranged and tied to each other. Blockchain is a new way to think about how computers work together. Its basic design principles and benefits are not centralized, and they can be shared without relying on individual nodes thanks to encryption techniques, time stamping methods, consensus processes, and reward systems. Network technology makes it possible for people to share knowledge, coordinate, and work together on a point-to-point basis (Rachkidi 2017). A blockchain is made up of a list of blocks that are put together in order. The block's information is mostly made up of a transaction counter and data about each transaction. The block header has the Merkle root hash, the value of the parent hash, a time stamp, the level of complexity of the calculation, and a random number (Rahmadika 2018). The maximum number of transactions that may be included in a single block is a function of the size of individual transactions and the block size. Depending on the specifics of the situation, the appropriate data record type may be selected from among those covering Internet of Things data, clearing, asset transactions, and asset issuance.

2 WHAT IS BLOCKCHAIN?

The blockchain, in general, has the following salient characteristics: The first stage is decentralization. Every trade in a conventional trading system must be verified by a trusted central device; therefore, a central server for managing costs and ensuring security is always required. Users of the blockchain network use a consensus mechanism to guarantee data consistency throughout the network. This eliminates the need for the mainframe. Second, persistence Quickly verifying the accuracy of transaction data is feasible, and once a transaction has been put on the blockchain, it can hardly be undone. During the verification phase, the block containing the transaction data will be verified and deleted. Secrecy comes in third. Each user uses their unique user address to connect to the blockchain and conceal their true identity.

The blockchain system runs itself with the help of distributed timestamp servers and peer-to-peer networks. The blockchain's infrastructure and the technology that makes it work include decentralization, openness, liberty, the ability to change information, and anonymity. Figure 1 shows seven-layer analogue computer network model shows that the blockchain system model is made up of the data layer, the network layer, the decision layer, the incentive layer, the contract layer, and the application layer.

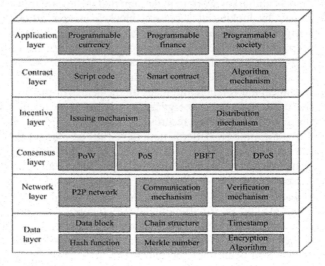

Figure 1.   Multiple layer model.

3 LITERATURE REVIEW

The author of this paper presented a framework plan in which blockchain technology is suggested to be used in medical care. In this plan, important data about clinical exams is split between emergency clinics, clinical centers, and research foundations based on how patients choose to get to them. To protect secret information, our approach uses two different types of chains: a private side chain that stores patient ID data and a public main chain that stores patient healthcare data separated by a temporary ID. The author made the plan and put it to the test using the Hyper Ledger Fabric structure (Rsouli 2019). In this article, the author proposes a method for collecting, exchanging, and collaborating on health information between individuals, medical care providers, and insurance companies by means of a portable medical services system. Modifications may be made so that medical records can be mined for insights. The appropriate information can be found faster when people share their electronic health data, and the systems are built with security and privacy in mind. To better the e-health infrastructure, the author proposes a blockchain-based secure and privacy-saving PHI sharing (BSPP) approach (Shiekh 2012). The PHI is stored on a private blockchain, and the group blockchain keeps track of the encrypted copies of that data. Protected health information (PHI), catchphrases, and character are all public-key encoded with a keyword search for the purposes of ensuring data confidentiality, controlling who has access, conserving resources, and ensuring safe hunting.

In this paper, the author proposes a small blockchain-based medical care stage and looks at how long it takes to run and how much information moves with a customer/worker framework model for updating and looking up health records. This is because there are more records and emergency centers than ever before (Shetty 2018).

4 PROPOSED METHODOLOGY

In the suggested approach, healthcare transactions are done using blockchain technology and a global record. Numerous deals execute on different nodes, as shown in Figure 2. The admin makes a transaction over the network and is directly connected to the doctor through a proper confirmation process. Using the blockchain network, the doctor is linked to the patient's registration and medication. Lab prescription information related to patient prescriptions in the suggested plan, blockchain is used to check all the transactions. Using blockchain, the flow of events from one peer to the next is kept safe. The suggested model would also protect transactions from prescription and lab test prescription data that move between the doctor's and patient's register models. During the "implement" part of our study, we used two ways to reach a group decision. POW (Proof of work) and put PBFT (Practical Byzantine Fault Tolerance) into Hyperledger fabric. In a comparison, it was found that POW is good, but only for deals with cryptocurrencies (Sun 2018). Another problem with POW is that it uses a lot of power to make deals. We found that PBFT was the best way to keep e-healthcare data safe when we used Hyperledger Fabric.

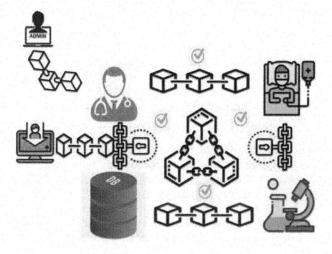

Figure 2. Proposed model.

The Hyperledger Caliper tool is used to measure how well something works. With the proposed algorithm, the proposed scheme improves the performance of the blockchain. Based on the data, the proposed method may also lower the latency. This shows that blockchain, like the steam engine and the Internet, has the power to make big changes for the betterment of humanity.

4 RESULTS

When analyzing and reviewing the hyperledger platform of blockchain technology, many different things are taken into account. The first experiment is done with different inputs and five rounds of putting 1000 transactions into the ledger network at different rates of 50, 100, 150, 200, 250, and 500 transactions per second. The time it takes to do a transaction show how well the blockchain network works. Figure 3 explains the reading of experiment.

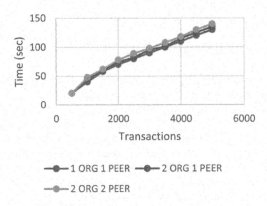

Figure 3.   Time taken to successfully execute transactions.

Each line shows how long it took to complete a transaction140 seconds, but 2org2peer only gets to 2000 transactions in the same amount of time. So, it's clear that as the number of organizations and peers grows, so does the time it takes to finish deals. A scientific method is used to figure out how long a transaction takes. Assume that TL stands for "transaction latency," which is the amount of time it takes to use the network. CT is the completion time for the transaction, which changes based on the network barrier NT.

During the "implement" part of our study, we used two ways to reach a group decision. POW and put PBFT into Hyperledger fabric. In a comparison, it was found that POW is good, but only good for deals with cryptocurrencies. Another problem with POW is that it uses a lot of power to do deals (Wang 2019). We found that PBFT was the best way to keep E-Healthcare data safe when we used Hyperledger fabric. The results of using the PBFT agreement method are shown in Table 1.

The open-source Docker technology can be used to make, share, and run applications. Docker lets you separate your apps from your infrastructure, so you can get software out quickly. Hyperledger used Docker files that were already built in to make separate peers. Table 1 shows that there are five different peers in the experiment. Using the PBFT consensus method, the average amount of memory used by each peer varies depending on the hosting process of Hyperledger Caliper and a network configuration file that includes information about the machine being tested and details about the network link. Org stands for organization, and org1 and org2 are, respectively, organizations 1 and 2 that are linked to the friend of networks.

Table 1. Performance metrics using PBFT.

Type	Name	Memory	Cpu	Traffic In
Docker	Peer 1	267 mb	15.60%	3.1mb
Docke	Peer 1	207 mb	24.25%	5.3 mb
Docke	Peer 1	201.2 mb	27.70%	6.4 mb
Docke	Peer 1	226 mb	15.70%	4.2 mb
Docke	Peer 1	56.6 mb	4.46%	3.7 mb

The optimization of the network is carried out in this experiment. Changing the block formation time measurement in the hyperledger caliper for EHR system produces different results. The arrangement of the caliper used in the experiment is shown in Table 2.

Table 2. Caliper arrangement configuration.

Parameter	Configuration
Transactions	1000 per round
Transaction mode	Write
Rate	50 to 250 tps
Network size	2Org2peer
Varied factor	Block time
Endorsement policy	signed by {0,1}
Round	5

Figure 4 shows how long it takes for a transaction to happen in a 2-org, 2-peer network. So that the delay could be as low as possible, optimization factors were taken into account. When the endorsement policy block creation time is pushed back, the caliper findings default to the block time. The finding shows a 1.5x decrease in network delay, which helps improve the performance of the EHR system. For a transaction rate of 50, the minimum delay is now about 27 seconds, down from 52 seconds. Having 37s for 250 tps is also less than having 50s for 250 tps. This is a system speed that can be reached by changing how the network is set up by default on the hyperledger.

It shows how the policy of the network affects the block time, which leads to better throughput and performance in terms of the time it takes to make a transaction and how often it works. By adding optimization, output went up from 50 to 250 tps by 1.75x. In read transaction mode, the incoming transaction is read after a certain amount of time has passed. In Table 2, you can see how the read transaction is set up. Changes have been made to the network's support policy and block time for reading transactions in order to make the system run better. The blockchain is optimized with different block times for creation, which makes it easier to read or ask about events.

5 CONCLUSION

The semantic model used in this study makes it easier to integrate without making partners do anything. By combining EHR data, it will be easier to make sickness records that will help improve public health. The unified data would need big-data methods for analysis and study. The HER makes it easy to connect data from personal monitors and smart phones so that it can be analyzed in real time. In conclusion, the study work meets its goals of designing and implementing a security framework for e-healthcare using blockchain technology. The

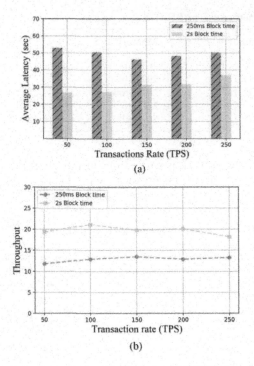

Figure 4. (a) Transaction latencies with varying block time, (b) Transaction throughput with varying block time.

semantic representation makes sure that health data from different sources can be easily joined and shared when needed. Standardized medical language and universal identifiers are used in the study to identify the important information in the semantic EHR. With the EHR system, the patient can take part in how his or her own health data is used. The EHR is easy for health care workers to access, and its main purpose is to help them care for patients in an efficient and ongoing way. After the blockchain security system was put into place in e-healthcare, The blockchain keeps track of every exchange. In the book, the date, time, people involved, and amount of each transaction are written down. Each node in the network has a full copy of the blockchain, and the transactions are confirmed by the Bitcoin miners, who keep the record using cryptography. These rules also make sure that these nodes always and instantly agree on the current state of the record and every transaction in it. If someone tries to change a transaction, the nodes won't be able to agree on what to do, so the transaction won't be added to the blockchain.

REFERENCES

Ali G.H. 2017. How the Blockchain Revolution will Reshape the Consumer Electronics. *Industrial Electronics Magazine*: 19–23. IEEE: Consumer Electronics.

Agyekum, S. 2018. A Blockchain-based Architecture Framework for Secure Sharing of Personal Health Data, in *Proc. IEEE 20th Int. Conf. e-Health Netw. Appl. Services (Healthcom)*:11–18.

Ananth, M. 2018. A Secured Healthcare System Using Private Blockchain Technology, *J. Eng. Technol.*, vol. 6, no. 2. 42–54.

Gani, R. 2018. Decentralized Consensus for Edge-centric Internet of Things: A Review, Taxonomy, and Research Issues, *IEEE Access*, vol. 6, pp. 1513–1524.

Jiang, J *et al.* 2018. BlocHIE 2018: A Blockchain-based Platform for Healthcare Information Exchange, in *Proc. IEEE Int. Conf. Smart Compute.* (SMARTCOMP): 49–56.

Ozair FF *et al.* 2015. Ethical Issues in Electronic Health Records: A General Overview. *Perspect Clin Res.*Apr-Jun;6(2):73–6.

Ovitskaya, Z. *et al.* 2017. Secure and Trustable Electronic Medical Records Sharing Using Blockchain in Proc. *AMIA Annu. Symp*: pp. 650–659

Rachkidi, N. Agoulmine, and N. C. Taher, 2017. Towards Using Blockchain Technology for eHealth Data Access Management, in Proc. *4th Int. Conf. Adv. Biomed. Eng. (ICABME)*, Beirut, Lebanon: 1–4.

Rahmadika and K.-H. Rhee. 2018 "Blockchain Technology for Providing an Architecture model of Decentralized Personal Health Information," *Int. J. Eng. Bus. Manage.* vol. 10, 11–17.

Rasouli J *et al.* 2019. Emerging Blockchain Technology Solutions for Modern Healthcare Infrastructure, *J. Sci. Innov.* Med: 23–32.

Sheikh J, Zuber. 2012. Temporal Blockchain Approach Based Secure Ehealth Framework., *International Journal of Computer Applications*, 174(28).

Shetty, J. Liu, and D. Li, 2017. Integrating Blockchain for Data Sharing and Collaboration in Mobile Healthcare Applications, in *Proc. IEEE PIMRC, Montreal*, QC, Canada, pp. 1–5.

Sun K 2018. MIStore: A Blochain-based Medical Insurance Storage System, *J. Med. Syst.*, vol. 42: 141–147.

Wang b, Yu Zhao a, Samara Kumar Sah Tyagi c, Neeraj Kumar, 2019. Blockchain Data-based Cloud Data Integrity Protection Mechanism., *Future Generation Compute System.* 902–911.

Artificial Intelligence, Blockchain, Computing and Security – Dagur et al. (Eds)
© 2024 The Author(s), ISBN: 978-1-032-49393-0

Blockchain-based trusted dispute resolution service architecture

Ravi Mohan Sharma*

Associate Professor, Department of Computer Science and Applications, Makhanalal Chaturvedi National University of Journalism and Communication, Bhopal, Madhya Pradesh, Indian

V. Rama Krishna*

Associate Professor, CVR College of Engineering, Hyderabad

Tirtha Saikia*

Assistant Professor, Economics, Mittal School of Business, Lovely Professional University

Ashish Suri*

Assistant Professor, School of Electronics and Communication Engineering, Shri Mata Vaishno Devi University

Richard Rivera*

Department of Informatics and Computer Science, Escuela Politécnica Nacional. Quito, Ecuador

Dinesh Mavaluru*

College of Computing and Informatics, Saudi Electronic University, Riyadh, Kingdom of Saudi Arabia

ABSTRACT: In the traditional service architecture (Service-Oriented Architecture, SOA), the Web Service provider registers its service description in the registry for service consumers to discover and call services. Although this architecture can provide open service calls, it lacks a dispute resolution mechanism. The trusted invocation of services between service consumers and providers who do not trust each other cannot be guaranteed. Blockchain technology has significant advantages in decentralization and anti-tampering and can be reasonably applied in traditional SOA to solve the problem of service credibility. Trusted SOA architecture based on blockchain is proposed to realize credible service invocation by combining traditional SOA architecture and blockchain technology. During a credible service invocation, or service consumption, The provider encrypts the parameters and sends them to the target service provider; the service provider receives the encrypted parameters and completes the decryption; the service provider completes the service execution and completes the encryption of the output result; finally, the service provider sends the encrypted result to the service provider Consumers complete the construction and chaining of trusted credentials at the same time. Based on the above, when a service dispute occurs, the ruling smart contract will be triggered to handle the service dispute correctly. The experimental results show that compared with the traditional invocation, the proposed method can correctly handle service disputes between service providers and requesters under the premise of ensuring that the growth rate of most Service trusted call times is not more excellent than 30%. by combining traditional SOA architecture and blockchain technology by combining traditional SOA architecture and blockchain technology.

Keywords: Blockchain, Service-Oriented Architecture, decentralization, anti-tampering, Web Service

*Corresponding Authors: ravi@mcu.ac.in, rama.vishwa@gmail.com, tirthasaikia9@gmail.com, ashish.suri@smvdu.ac.in, richard.rivera01@epn.edu.ec and d.mavaluru@seu.edu.sa

DOI: 10.1201/9781003393580-100

1 INTRODUCTION

With the development of the Internet era, various Webservices' (Chen *et al.* 2016; Viriyasitavat *et al.* 2019) have emerged which facilitate the use of Internet practitioners and ordinary users (Zhang *et al.* 2014). At present, WebService has become a research hotspot that people pay more and more attention to. Service Architecture (Viriyasitavat *et al.* 2019). Keep a loose coupling between service consumers and providers. In traditional SOA, WebService providers register service descriptions in the registration centre for service consumers to discover and call services. However, trust challenges exist in this service provision scheme in an entrusted, distributed scenario (Schall *et al.* 2012). For example, service providers may provide complete or malicious services, and service consumers may refuse to accept correct services because of their interests or even frame services provider. Traditional SOA lacks a dispute resolution mechanism, so the service credibility of the parties who do not trust each other cannot be guaranteed.

Blockchain(Chang *et al.* 2009; Rajan and Hosamani 2008; She *et al.* 2016) solves the problem of data security and credibility by integrating multiple technologies such as cryptography, P2P protocol, and consensus algorithm. For example, the author Data ownership disputes; (Viriyasitavat *et al.* 2019).signed and recorded the event records on the blockchain as evidence. Therefore, when a service dispute occurs, the conflict can be resolved by querying the evidence.

Aiming at the Service's trustworthiness, as mentioned in earlier calls, combined with the unique advantages of the blockchain, this paper proposes a trusted SOA architecture based on the blockchain, which uses the blockchain as an evidence recorder and service registration agent to solve the problems in traditional SOA service call trustworthiness. The specific work of this paper is as follows:

(1) A trusted SOA mechanism based on blockchain is proposed. To ensure the traceability of services, the signatures of service consumers and service providers are uploaded to the chain as trusted credentials during service invocation; In the event of a dispute, the trusted certificate on the chain will be reacquired for verification. At the same time, to ensure the service data's privacy, the Service's input parameters, and results are encrypted while constructing the trusted certificate.
(2) To support the realization of the above Service trusted invocation process, a trusted SOA-oriented supporting framework is proposed, including the service consumer agent and the service provider agent. For a given web service, the service provider agent can convert it into Support trusted SOA service BService. The service consumer proxy shields the differences between services so that service consumers can initially use BService.
(3) From effectiveness and execution performance, the method is evaluated for 13 specific web services. The experimental results show that the way in this paper can realize the conversion from WebService to BService and the trusted invocation of services in about 2s. Furthermore, compared with the traditional Compared with calling methods, the method in this paper can correctly handle service disputes between service providers and consumers in the case of independent malicious behaviors and most compound negative behaviors.

Section 2 of this paper introduces the work related to service computing and blockchain; Section 3 briefly introduces the trusted SOA mechanism based on blockchain; Section 4 presents the supporting framework for trusted SOA, including service provider agent, service consumer agent, and Service trusted invocation mechanism; Section 5 conducts experimental evaluation, discusses the effectiveness and execution performance of the method in this paper; finally summarizes the full text and looks forward to the future.

2 RELATED WORK

2.1 *Service computing and SOA*

Service computing is a computing paradigm that uses services as essential elements to quickly build distributed application systems under SOA (Branca *et al.* 2012). technology, norms, theory, and supporting environment (Branca *et al.* 2012). Service-oriented architecture SOA is the basic logic of service computing, which can solve the problem of "on demand" in enterprise application systems and realize service sharing, reuse, and a new distributed software system architecture for business integration service computing and service-oriented architecture. SOA makes up for the shortcomings of traditional, tightly coupled, and overall software architecture (Slamaa *et al.* 2021).

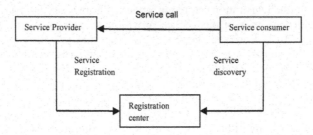

Figure 1.   Traditional SOA architecture.

In this system, WebService(Chen *et al.* 2010). has strong autonomy characteristics, and the service provider, registration centre and service consumer are separated. As shown in Figure 1, traditional SOA includes three core roles of registration centre, service provider, and Service consumer, and service registration three basic operations, service discovery, and service invocation. Among them, service registration means that the service provider registers the Service with the registry; service discovery implies that the service consumer discovers the Service from the registry; service invocation means that the service consumer directly calls the Service to the service provider. In addition, service consumers have the characteristics of "use but not own" between WebServices. Like many other distributed architectures, traditional SOA lacks a unified trust system between service consumers and providers (Chen *et al.* 2010). The trustworthiness of the services of the parties who do not trust each other cannot be guaranteed. Therefore, the traditional SOA is facing trust challenges.

2.2   *Blockchain*

Blockchain (Chen *et al.* 2010). is a trustworthy distributed ledger that keeps track of transactions and links them together using hash values. Through consensus across nodes, this method establishes mutual trust between entrusted entities. Its benefits include decentralization, anti-tampering, and anonymity (Dai *et al.* 2015; Lin *et al.* 2009). POW, POS, POA, and other commonly used consensus procedures are listed below. The blockchain-based dispute resolution system has been extensively researched. The author suggested a blockchain-based service contract management system (Service Contract Management Scheme, SCMS). Before using an additional dispute arbitration protocol to determine whether the service provider has complied with the contract, SCMS first requires the service provider to publish the agreement on the blockchain in advance. After that, the consumer is instructed to submit a certificate of the actual service behavior to the blockchain, suggesting a comparable blockchain-based strategy. All service events are signed and recorded on the blockchain as irrefutable proof in their proposed dispute resolution system. These systems, however, lack the correctness verification of the evidence data on the chain and are overly idealistic because they all rely on users to publish valid proof voluntarily and honestly on the blockchain without using a mandatory mechanism. These biased and flawed approaches can only achieve weak fairness and not produce satisfying outcomes since malicious clients can willfully deny the Service's correctness.

The advancement of blockchain technology has helped smart contracts succeed in recent years. A group of digital promises called "smart contracts" can be activated to carry out

pre-planned actions. As a result, smart contracts can now implement somewhat sophisticated algorithms because they are now a Turing-complete language. Smart contracts have
been used in dispute settlement because of the blockchain's openness and objectivity. Using
smart Contracts as a general conceptual design, the author suggested a replacement for
online dispute resolution. The author ([Nordbotten *et al.* 2009) uses artificial intelligence
contracts to settle arguments. This plan introduces monitoring agents to verify intelligent
contracts periodically for transactions. However, this plan necessitates a lot of information
disclosure from service providers, which can be wasteful and expensive.

Although numerous methods can successfully resolve service disputes, the majority have
glaring flaws, such as being unfairly biased, etc., and their cost for service verification and
dispute resolution in actually distributed scenarios is astronomical. Thus, a more
equitable and effective service supply scheme is needed in a service-oriented architecture.

3 TRUSTED SOA ARCHITECTURE BASED ON BLOCKCHAIN

To support the above Service trusted invocation process, this paper proposes a trusted SOA
architecture based on blockchain, as shown in Figure 2. This method transforms the
WebService service application into a service description model on the blockchain through
service registration, and the Service provider proxy is implemented by changing the service
calling process; querying the service description on the chain is completed through service
discovery, and the service request process is changed to realize service consumer proxy. This
method includes two parts: service provider proxy and service consumer agent.

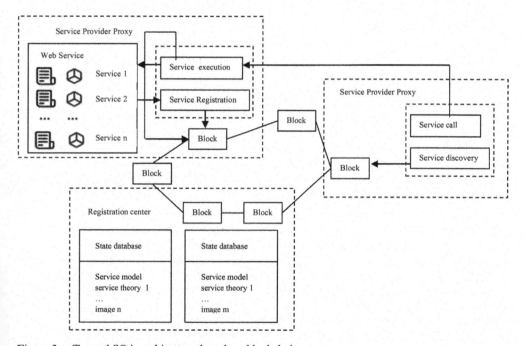

Figure 2. Trusted SOA architecture based on blockchain.

Service provider agent service registration module and service execution module the
service registration module is responsible for mapping the service application of
WebService into a service description model and publishing it to the service registration
agent section. The service execution module is the core of the trusted invocation process.

The module is responsible for receiving the encrypted parameters transmitted by service consumers through Remote Procedure Call (RPC), decrypting them, and completing the application call execution process of WebService service after encrypting the output result of the service call, calling it through RPC Return to the service consumer. At the same time, the service execution module will also complete the construction of trusted credentials during the calling process and publish them to the trusted credential recorder. In addition, the service consumer agent includes a service discovery module and a service invocation module. The service discovery module is Responsible for matching the service discovery request information with the service description model and returning the matched on-chain service description model to the service consumer according to the predefined matching priority. The service call module is responsible for completing the call parameter construction and transmitting it through RPC To the node where the target service is located; the service consumer waits to receive the execution result of the service provider node and decrypts the result.

Based on the service provider agent and service consumer agent, this architecture can realize the trusted invocation of services. When there is a dispute between the two parties, the trusted certificate stored on the chain and the dispute resolution mechanism of innovative contract technology can handle the Service fairly and effectively—service disputes between providers and service consumers.

3.1 Service provider proxy

The service provider agent mainly includes a service registration module and a service execution module.

3.1.1 Service registration

The service registration module is primarily used to realize the automatic registration of the WebService service application to the service description information on the chain. The existing recognized WebService service application usually only contains its description information and lacks the service provider's information required in the trusted call process the public key and the address information of the method responsible for receiving the new service model. This paper redefines the service description model to solve this problem according to the key-value pair storage format at the bottom of the blockchain state database. It clarifies the original WebService service description model. The service registration service is realized through the service registration contract. Among them, the definitions of the WebService description model and the new service description model are as follows.

Definition 1. The WebService service is a quadruple, that is, WS = ⟨WServiceName, WInput, WOutput, WTextDescription ⟩. Among them, WServiceName represents the service name; WInput represents the service input parameter, expressed as {WInput|name=type}, that is, the parameter name and parameter type of key-value pair; WOutput represents the service output result, described as {WOutput|name=type}, that is, the key-value pair of parameter name and parameter type; WTextDescription represents the text description of the Service.

In addition to the basic description involved in the WebService mentioned above service description model, the target service description model in this paper also needs to perform input or output conversion according to the trusted invocation process. In addition, the service provider needs to provide a Uniform Resource Locator (UniformResourceLocator, URL).

Definition 2. The blockchain-based service description model CService is expressed as {CService|CService^k = CService^v} is a key-value pair of CService^k and CServicev. Among them, CServicek is the index key of CService, and CServicev is the value of the Service.

Definition 3. The index key of the blockchain-based service description model CServicek = CServiceNametext⊕CTextDescription, where CServiceName represents the service name,

CTextDescription means the service text description, and the operating rules are shown in formula (1):

$$B \oplus C = \begin{cases} B, & \text{if } C \subseteq B \\ C, & \text{if } B \subseteq C \\ B \cup C, & \text{else} \end{cases} \tag{1}$$

Among them, B and C represent any string; if C is a substring of B, then B⊕C = B; if B is a substring of C, then B⊕C = B; otherwise B⊕C = B∪A.

Definition 4. The value CServicev of the service description model represents a quintuple, that is, CServicev = ⟨CServiceName, CTextDescription, input, output, URI ⟩. Among them, CServiceName represents the service name, CTextDescription represents the service text description, and the input parameter CInput=Qu$_C$⊖WInput; each Service has one or more parameters. The output result COutput = Qr$_C$⊖(Qu$_B$⊖WOutput), and each Service has one parameter. URI indicates the address of the method used to receive the call request of the service consumer and is represented by a uniform resource locator. Among them, Qu$_C$ means the public service key of the provider, Qu$_B$ represents the public key of the service consumer, Qr$_C$represents the private key of the service provider, and the expression B⊖C means that the value of C will be encrypted by A.

3.1.2 Service execution

In the trusted invocation process, the service execution is responsible for receiving the invocation parameter model. According to the encryption and assembly principle of the invocation parameters, define the decryption rules for reverse decryption, locate the service application program according to the service name, and obtain the service application program execution based on the decryption rules. Next, input parameters, complete the execution of the Service and get the result output set, and finally define the Service call trusted certificate model, and publish it to the blockchain as a transaction through the authorized certificate contract. Among them, rule formulation, service call execution output results, and call the Service trusted credentials in the process are defined as follows.

Definition 5. Assume that the mapping relationship C=Map(B), if C=Map(B) and B=Map (A), then the mapping relationship is symmetrical. Let Public_Key represent the public key, and Qu-CA represent the Fabric-CA with general key information Identity certificate. Suppose there is a mapping relationship between Qu-CA and Public_KeyPublic_Key=Map (Qu-CA), then the mapping has a symmetric property, namely:

$$Qu - CA = Map(\text{Public\_Key})$$
$$\text{Public\_Key} = Map(Qu - CA) \tag{2}$$

Suppose Private_Key represents a private key, where Public_Key can be abbreviated as Qu, and Private_Key can be abbreviated as Qr. Then, the data encrypted by the public key can only be successfully decrypted by its corresponding private key and vice versa. Therefore, if the expression QrB (T) represents Service consumer B uses its own private key QrB to encrypt information Y, and the corresponding decryption rule is shown in formula (3):

$$Qu_Y(Q_B(Y)) = \begin{cases} Y, & \text{if } Z = B \\ Y', & \text{if } Z \neq B \wedge Y' \neq Y \end{cases} \tag{3}$$

Equation (3) indicates that the encrypted information Qr$_B$ (Y) can only be decrypted by its corresponding public key Qu$_B$, and the non-corresponding public critical QuZ ≠ B cannot be decrypted; if the expression Qu$_B$ (Y) indicates that the information Y is interpreted by the service consumer's public essential Public_Key encryption, the corresponding decryption rules are shown in formula (4):

$$Qr_Z(Qu_B(Z)) = \begin{cases} Y, & \text{if } Z = B \\ Y', & \text{if } Z \neq A \wedge Y' \neq Y \end{cases} \tag{4}$$

Equation (4) indicates that the encrypted value Qu_B (Y) can only be decrypted by its corresponding private key Qr_B, and the non-corresponding private key $QrZ \neq B$ cannot be decrypted.

Expression Service Execute (serviceName, Input_Info) means that the input parameter Input_Info invokes the Service of executing serviceName after completing the Service. After conduct (serviceName, Input_Info), an output result will be obtained, expressed as Result = $\{S_1, \cdots, S_n\}$. S_i = ⟨name, type, data⟩, name indicates the result name, type indicates the result type, and data suggests the result data.

Definition 6. The credible proof Proof_Call of a credible call process contains a service consumer call request and a service provider execution result, which is expressed as a two-tuple, that is, Proof_Call=⟨Request, Qu_B (Result) ⟩, used to describe from The service consumer starts to call the Service and the service provider receives and completes the whole process of the Service. Among them, Request indicates that the service consumer calls the parameter model (see definition 9 for details); Qu_B (Result) is obtained by encrypting result with the public key Qu_B of the service consumer.

4 RESULT ANALYSIS

This section evaluates the invocation performance of the trusted SOA architecture based on blockchain. The experiment makes multiple invocations to the above 13 service applications, calculates the time cost consumption, and compares it with the actual direct invocation time cost consumption. For comparison, the experimental results are shown in Figure 3 with Table 1. It can be seen from the figure that the time gap between XS3, XS8, and XS13 is relatively large. According to statistics, the growth rate of the additional overhead of all application services is within 30%. In completing a trusted call, the time overhead includes communication overhead and computing overhead, and the proportion of communication overhead is too small to be ignored. Computing overhead refers to the time cost required for request parameter encryption and decryption, and result set encryption and decryption. Since the request parameter Request size is generally No more than 60B, when the size of the

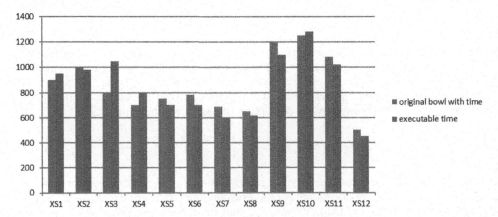

Figure 3.   Comparison of trusted execution time and direct call time.

676

Table 1. Comparison of trusted execution time and direct call time.

Server	Original Bowl with Time	Executable Time
XS1	900	950
XS2	1000	980
XS3	800	1050
XS4	700	800
XS5	750	700
XS6	780	700
XS7	690	600
XS8	650	615
XS9	1200	1100
XS10	1250	1280

result set is much larger than the request parameters, the main calculation overhead will be concentrated on the encryption and decryption of the result set. The relationship between the time result set and time growth is shown in Figure 4 with Table 2. When the size of the result set when is less than 10kB, the growth rate of trusted service execution time is less than 30%, and as the result set grows, the time growth rate increases accordingly.

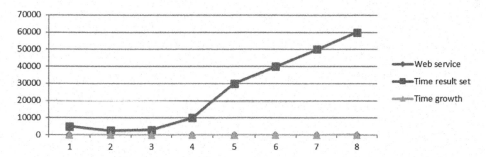

Figure 4. Relationship between service time growth and result set size.

Table 2. Relationship between service time growth and result set size.

Web service	Time result set	Time growth
1	5000	10
2	2500	10
3	3000	10
4	10000	10
5	30000	20
6	40000	30
7	50000	40
8	60000	50

It can be seen from Figure 3 that most of the trusted calls can be completed within 1 second, and when the result set increases, the longest time required reaches about 2 seconds, and this time cost is acceptable.

The fragmentation idea is adopted for the result set to reduce the time consumption in the encryption and decryption of the result set; 2) 13 WebService services are selected in the experiment, which can only be regarded as a small order of magnitude test, and follow-up work can test more and different types of services. To better verify the effectiveness and execution performance of the method in this paper.

5 CONCLUSION

This paper proposes a trusted SOA architecture based on blockchain. Building a unified trust system supports the credibility of service calls between service consumers and service providers. Experimental results show that this method can correctly handle service consumption—the service dispute between the provider and the service provider. The focus of future work mainly includes two aspects: 1) the follow-up work can adopt the fragmentation idea for the result set to reduce the time consumption in the encryption and decryption of the result set; 2) the experiment selected The 13 WebService services can only be regarded as small-scale tests, and the follow-up work can test more and different types of services to verify better the effectiveness and execution performance of the method in this paper.

REFERENCES

Branca G. and Atzori L., "A Survey of SOA Technologies in NGN Network Architectures," in *IEEE Communications Surveys & Tutorials*, vol. 14, no. 3, pp. 644–661, Third Quarter 2012, doi: 10.1109/SURV.2011.051111.00127.

Chang R. N. *et al.*, "Gaining insight into the health of SOA infrastructures," in *IBM Journal of Research and Development*, vol. 53, no. 6, pp. 5:1–5:14, Nov. 2009, doi: 10.1147/JRD.2009.5429033.

Chen H.-M., Kazman R. and Perry O., "From Software Architecture Analysis to Service Engineering: An Empirical Study of Methodology Development for Enterprise SOA Implementation," in *IEEE Transactions on Services Computing*, vol. 3, no. 2, pp. 145–160, April-June 2010, doi: 10.1109/TSC.2010.21.

Chen I.-R., Guo J. and Bao F., "Trust Management for SOA-Based IoT and Its Application to Service Composition," in *IEEE Transactions on Services Computing*, vol. 9, no. 3, pp. 482–495, 1 May-June 2016, doi: 10.1109/TSC.2014.2365797.

Dai W., Vyatkin V., Christensen J. H. and Dubinin V. N., "Bridging Service-Oriented Architecture and IEC 61499 for Flexibility and Interoperability," in *IEEE Transactions on Industrial Informatics*, vol. 11, no. 3, pp. 771–781, June 2015, doi: 10.1109/TII.2015.2423495.

Lin C. *et al.*, "A Reference Architecture for Scientific Workflow Management Systems and the VIEW SOA Solution," in *IEEE Transactions on Services Computing*, vol. 2, no. 1, pp. 79–92, Jan.-March 2009, doi: 10.1109/TSC.2009.4.

Nordbotten N. A., "XML and Web Services Security Standards," in *IEEE Communications Surveys & Tutorials*, vol. 11, no. 3, pp. 4–21, 3rd Quarter 2009, doi: 10.1109/SURV.2009.090302.

Rajan H. and Hosamani M., "Tisa: Toward Trustworthy Services in a Service-Oriented Architecture," in *IEEE Transactions on Services Computing*, vol. 1, no. 4, pp. 201–213, Oct.-Dec. 2008, doi: 10.1109/TSC.2008.18.

Schall D., Skopik F. and Dustdar S., "Expert Discovery and Interactions in Mixed Service-Oriented Systems," in *IEEE Transactions on Services Computing*, vol. 5, no. 2, pp. 233–245, April-June 2012, doi: 10.1109/TSC.2011.2.

She W., Zhu W., Yen I. -L., Bastani F. and Thuraisingham B., "Role-Based Integrated Access Control and Data Provenance for SOA Based Net-Centric Systems," in *IEEE Transactions on Services Computing*, vol. 9, no. 6, pp. 940–953, 1 Nov.-Dec. 2016, doi: 10.1109/TSC.2015.2432795.

Slamaa A. A., El-Ghareeb H. A. and Saleh A. A., "A Roadmap for Migration System-Architecture Decision by Neutrosophic-ANP and Benchmark for Enterprise Resource Planning Systems," in *IEEE Access*, vol. 9, pp. 48583–48604, 2021, doi: 10.1109/ACCESS.2021.3068837.

Viriyasitavat W., Da Xu L., Bi Z. and Sapsomboon A., "New Blockchain-Based Architecture for Service Interoperations in Internet of Things," in *IEEE Transactions on Computational Social Systems*, vol. 6, no. 4, pp. 739–748, Aug. 2019, doi: 10.1109/TCSS.2019.2924442.

Viriyasitavat W., Xu L. D., Bi Z. and Hoonsopon D., "Blockchain Technology for Applications in Internet of Things—Mapping From System Design Perspective," in *IEEE Internet of Things Journal*, vol. 6, no. 5, pp. 8155–8168, Oct. 2019, doi: 10.1109/JIOT.2019.2925825.

Viriyasitavat W., Xu L. D., Bi Z. and Hoonsopon D., "Blockchain Technology for Applications in Internet of Things—Mapping From System Design Perspective," in *IEEE Internet of Things Journal*, vol. 6, no. 5, pp. 8155–8168, Oct. 2019, doi: 10.1109/JIOT.2019.2925825.

Zhang W., Zhang S., Qi F. and Cai M., "Self-Organized P2P Approach to Manufacturing Service Discovery for Cross-Enterprise Collaboration," in *IEEE Transactions on Systems, Man, and Cybernetics: Systems*, vol. 44, no. 3, pp. 263–276, March 2014, doi: 10.1109/TSMCC.2013.2265234.

Deep learning based federated learning scheme for decentralized blockchain

Gowtham Ramkumar*
Department of Commerce, School of Commerce, Finance and Accountancy, Christ University, Bengaluru, India

S. Sivakumar*
Assistant Professor, Department of Electrical and Electronics Engineering, St. Joseph's College of Engineering, OMR, Chennai

Mukesh Soni*
Assistant Professor, Department of CSE, University Centre for Research & Development Chandigarh University, Mohali, Punjab, India

Yasser Muhammed*
College of Technical Engineering, Al-Farahidi University, Baghdad, Iraq

Hayder Mahmood Salman*
Al-Turath Universiy College, Baghdad, Iraq

Arsalan Muhammad Soomar*
PhD Scholar, Faculty of Electrical and Control Engineering, Gdańsk University of Technology, Poland

ABSTRACT: Blockchain has the characteristics of immutability and decentralization, and its combination with federated learning has become a hot topic in the field of artificial intelligence. At present, decentralized, federated learning has the problem of performance degradation caused by non-independent and identical training data distribution. To solve this problem, a calculation method for model similarity is proposed, and then a decentralized, federated learning strategy based on the similarity of the model is designed and tested using five federated learning tasks: CNN model training fashion-mnist dataset, alexnet model training cifar10 dataset, TextRnn model training thusnews dataset, Resnet18 model training SVHN dataset and LSTM model training sentiment140 dataset. The experimental results show that the designed strategy performs decentralized, federated learning under the non-independent and identically distributed data of five tasks, and the accuracy rates are increased by 2.51, 5.16, 17.58, 2.46 and 5.23 percentage points, respectively.

Keywords: Blockchain, Federated Learning, Deep Learning, CNN model, LSTM model

1 INTRODUCTION

Artificial intelligence's algorithm and computing power is developing rapidly, and the lack of data volume has become a bottleneck for further development. In the actual situation, data is scattered

*Corresponding Authors: gowthamphenom@gmail.com, sivashanmugammit@gmail.com, mukesh.research24@gmail.com, yasir.mohammed0086@uoalfarahidi.edu.iq, haider.mahmood@turath.edu.iq and arsalan.muhammad.soomar@pg.edu.pl

DOI: 10.1201/9781003393580-101

across many different organizations. To protect data privacy, all data cannot be shared between organizations to train artificial intelligence models, and federated learning (Cheng *et al.* 2022; Tang *et al.* 2022) came into being. Federated learning allows a large number of edge devices to use private data to calculate their models locally and aggregate the models uploaded by the edge devices in the cloud through an aggregation algorithm to obtain a shared model. Federated learning does not require data sharing between edge devices, which protects data privacy and security to some extent, so it has gradually become a hot research field in machine learning. But there are still challenges in traditional federated education: the central server plays a vital role in aggregating the local models of edge devices. When the primary server is challenged by security, an unstable central server will cause the system to crash. The blockchain has security features such as decentralization and non-tampering, which can solve the security problem of the central server in federated learning, so its combination with federated learning has attracted widespread attention.

Blockchain is a decentralized distributed database. Unlike traditional centralized systems, the blockchain does not require any central server. Due to the existence of the "longest chain rule" and the "Merkel tree" structure, the data it stores cannot be tampered with. These two characteristics make it a reliable and trustworthy system in the distributed environment has also led many research works to use the blockchain as the underlying foundation of federated learning and to meet the needs of model aggregation by designing consensus algorithms or aggregation algorithms on the upper layer of the blockchain. Literature (Cui *et al.* 2022) proposed a blockchain-based federated learning framework. After local training, the local model is uploaded to the blockchain network, and after the miner node consensus, the federated average (Tang *et al.* 2022) is used for aggregation. And rewards can be provided for equipment based on parameters such as the output rate of the equipment. Researchers from NEC Laboratory and Georgia Institute of Technology proposed a blockchain-based free federated learning framework (BAFFLE) (Nguyen *et al.* 2022) they implemented a practical, production-level BAFFLE on the private Ethereum framework and used Large deep neural networks to demonstrate the benefits of BAFFLE. In BAFFLE, each user's contribution to the model can be evaluated to determine the user's reward. Literature (Abdel-Basset *et al.* 2022) proposed a distributed machine learning security framework based on blockchain and a 5G network, using the communication speed of a 5G network to solve the usability problem of federated learning, and at the same time, introduced blockchain fragmentation technology and designed a Set of gradient anomaly detection strategies. Literature (Toyoda *et al.* 2020) proposed the blockchain-based federated learning framework BFLC. The storage method of the model has been formulated in detail, and a new training process and consensus mechanism have been redesigned. The paper discusses the community node management of BFLC, defence against malicious node attacks and storage optimization issues. The effectiveness of BFLC is proved through experiments on federated learning datasets, and the security of BFLC is verified by simulating malicious attacks. In 2021, Literature (Zhang *et al.* 2021) proposed the DAG-FL architecture, using the DAG blockchain architecture to improve the model verification mechanism of decentralized, federated learning, effectively monitor abnormal nodes, and prove its effectiveness through experiments.

The current research on the blockchain-based federated learning framework mainly focuses on reducing system communication and computing costs, improving resource allocation, enhancing data security and reliability, and enhancing the robustness of federated learning. However, with the large-scale use of the framework, tens of thousands of devices and their private data sets are used for model training. Therefore, the problem of non-independent and identical data distribution due to different scenarios and users will also be faced. Federated learning can be divided into three categories: horizontal, vertical, and federated transfer learning. The feature of horizontal federated learning is that the data characteristics of each client are the same, and the users are different. The feature is that the users are the same, but the data features are various, which is an extension of the sample data features. The research on data non-independent and identical distribution of federated learning focuses on the classification of horizontal federated knowledge (Miao *et al.* 2022). The same distribution means that the data trend is stable, does not fluctuate, and the distribution of all data follows the same probability. Independence implies that each sample is considered independent and has no connection with each other. For traditional distributed machine learning, each sub-dataset is divided from the whole data set, and these sub-datasets can represent the overall distribution. But for federated learning, the private data on each device is not randomly collected or generated, which leads to a specific correlation, so it is not independent. Furthermore, the local data volume of each device's private data is different, so it also violates the same distribution. This unbalanced data distribution caused by non-IID data will introduce bias to model training and may lead to performance degradation of federated learning.

There are many solutions to the problem of non-independent and identically distributed data in horizontal federated learning in the framework of centralized federated learning. Literature (Miao *et al.* 2022) proposed a strategy to improve the training accuracy of non-IID data by creating a subset of data shared among all private datasets to close the gap between each model. Literature (Qi *et al.* 2022) proposed a new federated learning mechanism, Hybrid-FL, which designed a heuristic algorithm to allow the central server to collect data from devices with low privacy sensitivity to construct independent and identically distributed data and use the data trains the model and aggregates to the global model. Literature (Ayaz *et al.* 2022) proposed an aggregation algorithm for model selection to improve the aggregated international model degree by identifying and excluding models biased towards local updates. However, these methods and algorithms are all based on the centralized federated learning framework, using the central server to share part of the data or select models for aggregation. The problem of the accuracy drop of the decentralized, federated learning framework for non-IID data still exists.

Therefore, for the non-IID data problem of the decentralized, federated learning framework, this paper, inspired by the literature (Miao *et al.* 2022), proposes a calculation method for model similarity and designs a new decentralized, federated learning model based on the similarity of the model. Strategy. This strategy puts a model trained with a small amount of independent and identically distributed data in a smart contract. It calculates the similarity for each uploaded edge model through the smart contract. The local device can select the model in the blockchain according to the order of model similarity. Models are aggregated using the federated averaging algorithm. Using the test set of smart contracts is constantly updated by the global model.

Replace the model used for comparison. Experiments show that compared with the original model similarity in literature (Miao *et al.* 2022), the improved model similarity in this paper can achieve better results. Comparative experiments verify that this paper's decentralized, federated learning strategy is better than horizontal federated learning. Furthermore, the performance in the non-independent and identically distributed data environment is better than that of traditional decentralized federated learning strategies.

2 FRAMEWORK DESIGN AND ORIGINAL STRATEGY INTRODUCTION OF DECENTRALIZED FEDERATED LEARNING

2.1 *A framework design for decentralized, federated learning*

This section introduces a decentralized, federated learning framework. As shown in Figure 1, the framework mainly comprises an edge learning layer, backend interface layer, smart contract layer and storage layer.

2.1.1 *Edge learning layer*
Edge devices train models with their processors and private data. Each edge device trains its local model and aggregates the global model.

2.1.2 *Backend interface layer*
The backend interface layer processes the model and HTTP data transmitted from the edge device and forwards them to the smart contract layer.

2.1.3 *Smart contract layer*

The intelligent contract layer deploys the intelligent contract nodes of the Hyperledger, an open-source blockchain framework hosted and managed by the Linux Foundation. Special operations can be performed on federated learning at the smart contract layer, such as checking the accuracy, issuing models, etc.

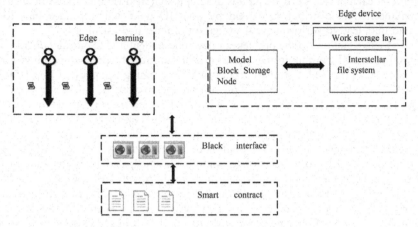

Figure 1. Framework diagram.

2.1.4 *Storage layer*

The storage layer deploys the storage nodes and the interstellar file system of the hyperbook. The hyperedger framework will set up a sub-network for the members of a consortium chain. Only the members of the sub-network can see the chain's transaction data and smart contracts to ensure privacy. The interplanetary file system is a file system that integrates P2P network technology, DHT distributed Hash, BitTorrent transmission technology, self-certification file system SFS and Git version control system. Its sharing and hash storage characteristics are very suitable for federated learning. This kind of model has many parameters and reads and writes a lot of scenes.

2.2 *The process of common decentralized, federated learning strategy*

Most of the research on decentralized, federated learning is based on the BlockFL architecture proposed in (Cui *et al.* 2022), which is improved for horizontal federated learning. The steps of the decentralized, federated learning strategy applied to the BlockFL architecture in the above framework are as follows:

Step 1: The authoritative organization sets the current round number e of federated learning and the initial model M0. Then, the smart contract distributes the initial model M0 to the edge nodes, and the edge nodes use their private data to train the initial model M0 to obtain the local Mn.

Step 2: Each edge node uploads Mn to the smart contract, and the intelligent contract performs workload proof on the model's accuracy based on its own test set and issues an on-chain license. If a round of federated learning time is up or enough edge models are collected, the intelligent contract sends an aggregation signal to the edge nodes.

Step 3: Each edge node reads the edge model queue in a round of federated learning from the blockchain, aggregates the edge model queue using the federated average algorithm, obtains a global model Mg, and stores it in the blockchain.

Step 4: The intelligent contract sends a signal to start the next round of learning to the edge nodes. Finally, the edge nodes start the next game of decentralized, federated learning with aggregated global models.

Compared with the original paper, the above strategy changes miner verification to intelligent contract verification and uses the raft (Zhang *et al.* 2022) algorithm for the consensus of blockchain nodes. However, it still retains its decentralized ideas and mechanisms.

3   DECENTRALIZED FEDERATED LEARNING STRATEGY FOR NON-IID DATA

The previous article explained the process of a common decentralized, federated learning strategy based on a framework design. This chapter introduces an improved method for decentralized, federated learning based on the same framework for horizontal federated learning.

Generally speaking, if the private data of a device has independent and identical distribution characteristics, the parameters of the trained deep learning model should have certain similarities. Suppose the model of a machine is prepared by data that is not independent and identically distributed. In that case, the similarity between its model and the standard model must be smaller than that of the model trained on independent and identically distributed data. In literature [8], to study the impact of non-independent and identically distributed data on the federated average aggregation algorithm, the weight difference of the model is proposed and derived:

$$weight\ divergence = \frac{\| W^{FedAvg} - W^{SCD}\|}{\| W^{SCD}\|} \tag{1}$$

Among them, W^{FedAvg} is the parameter of a layer in the global model obtained after federated learning using the federated average aggregation algorithm, and W^{SCD} is the parameter of a layer in the model received after traditional deep understanding using stochastic gradient descent. Inspired by this, based on formula (1), this paper enlarges the parameter difference between each layer through experimental comparison, sums the weight difference of each layer as the model similarity of the two models, and uses this as an indicator to filter the model:

$$S_{ab} = \sum_{k=1}^{n} \frac{(\| P_{ak} - P_{bk}\|)^2}{\| P_{bk}\|} \tag{2}$$

Among them, n is the number of layers of the model, P_{ak} is the parameter of model a in the kth layer, P_{bk} is the parameter of model b in layer k.

Based on the model similarity, the proposed strategy steps are as follows:

Step 1: The authoritative organization sets the current round number e of federated learning and the initial model M_0 and trains a comparison model M_s with nearly independent and synchronous data collected or purchased to obtain the accuracy acc_0. Then, the smart contract distributes the initial model M_0 to the edge nodes, and the edge nodes use their private data to train the initial model M_0 to obtain the local M_n.

Step 2: Each edge node uploads Mn to the smart contract. After the smart contract receives M_n, it calculates the model similarity with the comparison model for each model: $S_{M_n M_o} = \sum_{k=1}^{n} \frac{(\| P_{ak} - P_{bk}\|)^2}{\| P_{bk}\|}$, sends the model similarity to the edge node, and the edge node compares M_n and $S_{M_n M_o}$ stored in the blockchain. If a round of federated learning time is up or enough edge models are collected, the intelligent contract sends an aggregation signal to the edge nodes.

Step 3: Each edge node reads out the edge model queue in a round of federated learning from the blockchain, sorts the model similarity of the current game of edge models in ascending order, and takes the models corresponding to the first η model similarities to form a new subset Set mi. Finally, the new edge model queue mi is

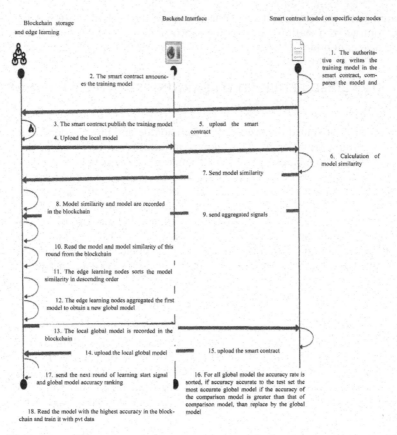

Figure 2.　Flow chart of improvement strategy.

aggregated using the federated average algorithm to obtain a global model M_g, which is stored in the blockchain and sent to the smart contract.

Step 4:　After the intelligent contract accepts the global model sent by the edge node, it uses its test set to obtain the accuracy acc_n of the worldwide model. If $acc_n > acc_0$, take M_g as the comparison model, $acc_0 = acc_n$.

Step 5:　The intelligent contract sends the accuracy ranking of the global model and the start of the next round of learning to the edge nodes. Each edge node uses the global model with the highest accuracy to start the next game of decentralized, federated learning. The above steps are shown in Figure 2.

4　EXPERIMENT AND RESULT ANALYSIS

This chapter will conduct an experimental analysis on the improvement of model similarity and conduct comparative experiments on the performance improvement of decentralized federated learning strategies under non-independent and identical distribution. The following is a discussion from five aspects: experimental environment setting, experimental evaluation index, experimental parameters, and comparative experiment.

4.1 *Experimental environment*

Experimental environment: The GPU server environment has a QuadroP5000GPU (16 GRAM), loaded with the centos7 system environment, and the machine learning function is programmed using the PyTorch framework based on python3.7. The back-end programming uses the gin framework and the Golang language to write the smart contract of the hyper ledger. In terms of the super ledger, the experiment uses the hyperledgefabric1.4 version. It divides the peer nodes into three organizations; each organization has two peer nodes, each organization selects a node to deploy smart contracts, and enables five order nodes to execute raft consensus service. In federated learning, each time a model is trained on a GPU stand-alone machine and uploaded to one of the intelligent poolsabout node.

4.2 *Experimental setup*

A total of 5 training tasks were conducted in this paper to evaluate the strategy's performance. The first training task is to classify the FashionMNIST (Kumar *et al.* 2020, 2021; Qin *et al.* 2022) image dataset using a simple CNN (Kumar *et al.* 2021) model with two 5×5 convolutional layers; the first output channel is 20, the second output channel is 50, and the ReLU function activates the two convolutional layers, and a 2×2 top pooling layer is used between each. Then a linear layer and a softmax layer are used for output. FashionMNIST has a training set of 60,000 samples and 10,000 pieces, which include pictures of 10 categories, such as T-shirts, pants, and coats. The setting method of the independent and identically distributed data group is that the training set data is evenly distributed to each client; each client is randomly assigned 1200 pieces of data. The setting method of the non-IID data group is as follows. First, sort the data labels in the experiment, divide the sorted training set into 1200 groups; each group has 50 pictures, and then put 50 edge nodes each. The edge nodes are divided into four data groups, and the remaining data are randomly allocated to the edge nodes. Using this distribution scheme, each node is assigned a local dataset consisting of two large taxas.

He has some fragmented classifications, and the amount of data in each node is also different. Therefore, in this task, each edge node is trained for ten rounds with a learning rate of 0.005, and the comparison model is obtained by introducing the client assigned the most data for ten games.

The second training task is to classify images on the cifar10 dataset using the Alexnet (Wang *et al.* 2022) model. The cifar10 data set has 50,000 training samples and 10,000 test samples. The difference between the FashionMNIST data collection and the FashionMNIST data collection is that cifar10 is a 3-channel colour picture, and it is a real object in the real world, with significant noise, different object proportions and characteristics. This makes identification more complicated. Similar to the experimental settings of FashionMNIST, the independent and identically distributed data set is distributed to 50 clients evenly; each client is randomly assigned 1000 pieces of data. The location of the non-IID data group first sorts the labels of cifar10 and then divides the sorted training set into 1000 groups; each group has 50 pictures. Set 50 edge nodes, each node is divided into two data groups, and the remaining groups are randomly assigned. This task sets the edge nodes to train for ten rounds with a learning rate of 0.000 1, and the comparison model is obtained by introducing the client assigned the most data for ten games.

The third training task uses the TextRNN (Yazdinejad *et al.* 2022) model to classify news on the THUnews (Wang *et al.* 2022) dataset. The THUnews dataset has 740,000 news documents and is divided into 14 categories: finance, lottery, and real estate. In this task, 50,000 pieces of data are randomly selected as the sum of the training set, 10,000 pieces of data are taken as the test set, and 50 edge nodes are set. The training set data is evenly distributed to 50 clients in independent and identically distributed data sets. In terms of dividing non-IID data, this task first sorts the labels of 10,000 pieces of data, distributes them evenly to the edge nodes, and randomly distributes the remaining data. This task chooses to use the TextRNN model, the edge nodes are trained for five rounds with a learning rate of 0.001, and the comparison model is obtained by introducing the client assigned the most data for five games.

The fourth training task uses the Resnet18 (Wang *et al.* 2022) model to perform image classification on the SVHN (Hua *et al.* 2020) dataset. The SVHN dataset contains more than 200,000 RGB images of house numbers captured by Google Street View cars. Each image has 1 to 3 digits from 0 to 9. This task sets up 50 edge nodes. In independent and identically distributed data sets, the 73 257 training set data are almost evenly distributed to 50 clients; each client is assigned 1 465 or 1 466 pictures. In terms of dividing non-independent and identically distributed data, the experiment sorts the data labels and then divides the sorted training set into

One thousand four hundred sixty-five groups, each group has 50 pictures, and then set 50 edge nodes, each edge node selects four data groups, and the remaining groups are randomly assigned to different nodes. In terms of training parameters, this task sets each edge node to train 1 round with the residual network Resnet18 at a learning rate of 0.1. The comparison model is obtained by introducing one game with the same learning rate on the client assigned the most data.

The fifth training task is sentiment classification on the sentiment140 (Aliyu *et al.* 2020) dataset with the LSTM (Mothukuri *et al.* 2022) model. sentiment140 contains 1 600 000 tweets crawled from Twitter and has three labels: positive, neutral, and hostile. In this task, 30,000 entries are randomly selected as the training set, 10,000 entries are used as the test set, and 50 edge nodes are set. In independent and identically distributed data sets, the training-selected data is evenly distributed to 50 clients; each client has 600 pieces of data. In terms of dividing non-IID data, this task first sorts the labels of 10,000 pieces of data, distributes them evenly to the edge nodes, and randomly distributes the remaining data. This task sets every.

Each edge node uses the traditional LSTM model to train for one round with a learning rate of 0.001, and the comparison model is obtained by introducing the client with the most data with the same learning rate for one game.

4.3 *Evaluation indicators*

The performance evaluation index of model similarity is the variance of the model similarity between the edge model and the standard model in each round of federated learning. The conflict is calculated to observe the degree of dispersion of the similarity between each edge model and the standard model. To judge the performance of the model similarity calculation method. The formula is as follows:

$$\sigma^2 = \frac{\sum\limits_{i=1}^{n} (s_i - \bar{s})^2}{n} \tag{3}$$

Where s_i is the similarity between each edge model and the standard model in one round, $\bar{s}$ is the average similarity between the edge model and the standard model in this round, and n is the number of devices participating in the federated learning task in this round. This paper uses the accuracy rate as the performance index of the decentralized, federated learning strategy in five deep learning tasks. This paper uses acc to represent, calculateas follows:

$$acc = \frac{TP}{TP + FP} \times 100\% \tag{4}$$

Among them, TP is the data correctly labelled by the model, and FP is the data incorrectly marked by the model.

4.4 Model similarity performance test

In the experiment, the original model's performance and the modified model's similarity were tested, respectively. The calculation method of the similarity of the original model and the similarity of the modified model are as follows:

$$S1_{ab} = \sum_{k=1}^{n} \frac{\|P_{a_k} - P_{b_k}\|}{\|P_{b_k}\|} \tag{5}$$

$$S2_{ab} = \sum_{k=1}^{n} \frac{(\|P_{a_k} - P_{b_k}\|)^2}{\|P_{b_k}\|} \tag{6}$$

First, test the variance of the similarity of the original model and the similarity of the modified model under the non-IID data set, and then try the conflict of the similarity of the modified model under the IID data set. Among them, model a is the edge model, and model b is the standard model. The standard model is obtained from each task's training set data

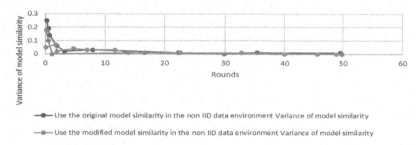

Figure 3. CNN model similarity variance.

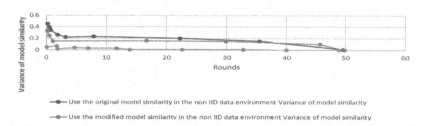

Figure 4. CNN model similarity variance.

and uses the same training parameters as the edge device. In the test, the edge device training parameters in the five studies are the same, and the results are shown in Figures 3–7. Then two different model similarity calculation methods are applied to this paper's decentralized, federated learning strategy. In the case of data set distribution and training models with the same training parameters, five tasks are charged with 50 nodes using different model similarities. The strategy of the calculation method is to carry out 50 rounds of federated learning on the non-IID data set, set the number of model aggregation to $\eta = 25$, and show the change of the accuracy rate of the global model with the highest accuracy in each round of blockchain on the test set. The results are shown in Figures 8~12.

Figures 3 to 5 show that federated learning is performed on the same non-independent and identically distributed data set. In the beginning, due to the difference in each edge data, the

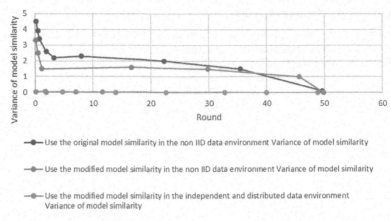

Use the original model similarity in the non IID data environment Variance of model similarity

Use the modified model similarity in the non IID data environment Variance of model similarity

Use the modified model similarity in the independent and distributed data environment Variance of model similarity

Figure 5.　TextRNN model similarity variance.

edge models are very different, and the variance of the corresponding model similarity is also high. As the number of rounds increases, after several model aggregations, the difference between edge models becomes smaller and smaller, and the variance of model similarity tends to be relatively stable. In the comparison experiment between the similarity of the original model and the modified model, the conflict of the similarity of the modified model of the CNN task is 0.0084 higher than the variance of the similarity of the original model on average.

The clash of the similarity of the modified model of the Alexnet task is higher than the variance of the similarity of the original model. Higher by 0.111 5, the conflict of TextRNN task changed model similarity is 1.139 higher on average than the actual model similarity variance 9, the average variance of Resnet18 job modified model similarity is 7 968 higher than the variance of original model similarity, LSTM task modified model similarity The clash of is 0.053 4 higher on average than the variance of the actual model similarity. In different studies, the conflict of the similarity of the modified model is higher than the variance of the similarity of the original model; that is, the separation of the similarity of the modified model is larger than the separation obtained by the similarity of the original model, and this greater separation is more conducive to the strategy. Compare the similarity of the models.

5　CONCLUSION

This paper proposes a decentralized, federated learning strategy under non-IID data. This strategy can enable decentralized, federated learning to overcome the challenge of non-IID data in horizontal federated learning and approach the accuracy of single-machine deep understanding as much as possible. Moreover, it can filter out approximately independent and identically distributed federated learning model sets for aggregation without processing training data, thereby improving federated devices' data quality. The basic idea is to use the improved model similarity at the smart contract to compare the degree of non-independent and identical distribution of the data used for model training. Experiments show that the decentralized network utilizinges this strategy. The state learning performance can be significantly improved on different tasks. The following research direction is the parameter problem in the dynamic selection strategy of the smart contract, and the incentive mechanism based on the model similarity is added to strengthen the role of the blockchain in federated learning.

REFERENCES

Abdel-Basset M., Moustafa N., Hawash H., Razzak I., Sallam K. M. and Elkomy O. M., "Federated Intrusion Detection in Blockchain-Based Smart Transportation Systems," in *IEEE Transactions on*

Intelligent Transportation Systems, vol. 23, no. 3, pp. 2523–2537, March 2022, doi: 10.1109/TITS.2021. 3119968.

Aliyu I., Van Engelenburg S., Mu'Azu M. B., Kim J. and Lim C. G., "Statistical Detection of Adversarial Examples in Blockchain-Based Federated Forest In-Vehicle Network Intrusion Detection Systems," in *IEEE Access*, vol. 10, pp. 109366–109384, 2022, doi: 10.1109/ACCESS.2022.3212412.

Ayaz F., Sheng Z., Tian D. and Guan Y. L., "A Blockchain Based Federated Learning for Message Dissemination in Vehicular Networks," in *IEEE Transactions on Vehicular Technology*, vol. 71, no. 2, pp. 1927–1940, Feb. 2022, doi: 10.1109/TVT.2021.3132226.

Cheng X., Tian W., Shi F., Zhao M., Chen S. and Wang H., "A Blockchain-Empowered Cluster-Based Federated Learning Model for Blade Icing Estimation on IoT-Enabled Wind Turbine," in *IEEE Transactions on Industrial Informatics*, vol. 18, no. 12, pp. 9184–9195, Dec. 2022, doi: 10.1109/TII.2022.3159684.

Cui L., Su X. and Zhou Y., "A Fast Blockchain-Based Federated Learning Framework With Compressed Communications," in *IEEE Journal on Selected Areas in Communications*, vol. 40, no. 12, pp. 3358–3372, Dec. 2022, doi: 10.1109/JSAC.2022.3213345.

Hua G., Zhu L., Wu J., Shen C., Zhou L.and Lin Q., "Blockchain-Based Federated Learning for Intelligent Control in Heavy Haul Railway," in *IEEE Access*, vol. 8, pp. 176830–176839, 2020, doi: 10.1109/ ACCESS.2020.3021253.

Kumar R. *et al.*, "Blockchain-Federated-Learning and Deep Learning Models for COVID-19 Detection Using CT Imaging," in *IEEE Sensors Journal*, vol. 21, no. 14, pp. 16301–16314, 15 July15, 2021, doi: 10.1109/JSEN.2021.3076767.

Miao Y., Liu Z., Li H., Choo K.-K. R. and Deng R. H., "Privacy-Preserving Byzantine-Robust Federated Learning via Blockchain Systems," in *IEEE Transactions on Information Forensics and Security*, vol. 17, pp. 2848–2861, 2022, doi: 10.1109/TIFS.2022.3196274.

Mothukuri V., Parizi R. M., Pouriyeh S., Dehghantanha A. and Choo K.-K. R., "Fabric FL: Blockchain-in-the-Loop Federated Learning for Trusted Decentralized Systems," in *IEEE Systems Journal*, vol. 16, no. 3, pp. 3711–3722, Sept. 2022, doi: 10.1109/JSYST.2021.3124513.

Nguyen D. C., Hosseinalipour S., Love D. J., Pathirana P. N. and Brinton C. G., "Latency Optimization for Blockchain-Empowered Federated Learning in Multi-Server Edge Computing," in *IEEE Journal on Selected Areas in Communications*, vol. 40, no. 12, pp. 3373–3390, Dec. 2022, doi: 10.1109/JSAC.2022.3213344.

Qi J., Lin F., Chen Z., Tang C., Jia R. and Li M., "High-Quality Model Aggregation for Blockchain-Based Federated Learning via Reputation-Motivated Task Participation," in *IEEE Internet of Things Journal*, vol. 9, no. 19, pp. 18378–18391, 1 Oct.1, 2022, doi: 10.1109/JIOT.2022.3160425.

Qin Z., Ye J., Meng J., Lu B.and Wang L., "Privacy-Preserving Blockchain-Based Federated Learning for Marine Internet of Things," in *IEEE Transactions on Computational Social Systems*, vol. 9, no. 1, pp. 159–173, Feb. 2022, doi: 10.1109/TCSS.2021.3100258.

Tang F., Wen C., Luo L., Zhao M. and Kato N., "Blockchain-Based Trusted Traffic Offloading in Space-Air-Ground Integrated Networks (SAGIN): A Federated Reinforcement Learning Approach," in *IEEE Journal on Selected Areas in Communications*, vol. 40, no. 12, pp. 3501–3516, Dec. 2022, doi: 10.1109/JSAC.2022.3213317.

Toyoda K., Zhao J., Zhang A. N. S. and Mathiopoulos P. T., "Blockchain-Enabled Federated Learning With Mechanism Design," in *IEEE Access*, vol. 8, pp. 219744–219756, 2020, doi: 10.1109/ ACCESS.2020.3043037.

Wang P. *et al.*, "Blockchain-Enhanced Federated Learning Market With Social Internet of Things," in *IEEE Journal on Selected Areas in Communications*, vol. 40, no. 12, pp. 3405–3421, Dec. 2022, doi: 10.1109/ JSAC.2022.3213314.

Wang X., Zhao Y., Qiu C., Liu Z., Nie J. and Leung V. C. M., "InFEDge: A Blockchain-Based Incentive Mechanism in Hierarchical Federated Learning for End-Edge-Cloud Communications," in *IEEE Journal on Selected Areas in Communications*, vol. 40, no. 12, pp. 3325–3342, Dec. 2022, doi: 10.1109/JSAC.2022.3213323.

Wang Y., Peng H., Su Z., Luan T. H., Benslimane A. and Wu Y., "A Platform-Free Proof of Federated Learning Consensus Mechanism for Sustainable Blockchains," in *IEEE Journal on Selected Areas in Communications*, vol. 40, no. 12, pp. 3305–3324, Dec. 2022, doi: 10.1109/JSAC.2022.3213347.

Yazdinejad A., Dehghantanha A., Parizi R. M., Hammoudeh M., Karimipour H. and Srivastava G., "Block Hunter: Federated Learning for Cyber Threat Hunting in Blockchain-Based IIoT Networks," in *IEEE Transactions on Industrial Informatics*, vol. 18, no. 11, pp. 8356–8366, Nov. 2022, doi: 10.1109/TII.2022.3168011.

Zhang W. *et al.*, "Blockchain-Based Federated Learning for Device Failure Detection in Industrial IoT," in *IEEE Internet of Things Journal*, vol. 8, no. 7, pp. 5926–5937, 1 April1, 2021, doi: 10.1109/JIOT.2020.3032544.

Zhang Z., Yang T. and Liu Y., "SABlockFL: A Blockchain-based Smart Agent System Architecture and its Application in Federated Learning," in *International Journal of Crowd Science*, vol. 4, no. 2, pp. 133–147, June 2020, doi: 10.1108/IJCS-12-2019-0037.

Artificial Intelligence, Blockchain, Computing and Security – Dagur et al. (Eds)
© 2024 The Author(s), ISBN: 978-1-032-49393-0

Blockchain-aware federated anomaly detection scheme for multivariate data

V. Selvakumar*
Assistant Professor, Department of Maths and Statistics, Bhavan's Vivekananda College of Science, Humanities and Commerce, Hyderabad, Telangana

Renato R. Maaliw III *
College of Engineering, Southern Luzon State University, Lucban, Quezon, Philippines

Ravi Mohan Sharma*
Associate Professor, Department of Computer Science and Applications, Makhanalal Chaturvedi National University of Journalism and Communication, Bhopal, Madhya Pradesh

Rajvardhan Oak*
Department of Computer Science, University of California Davis, USA

Pavitar Parkash Singh*
Dean, Mittal School of Business, Lovely Professional University, Phagwara, Punjab

Ashok Kumar*
Assistant Professor, Department of Computer Science, Banasthali Vidyapith, Rajasthan, India

ABSTRACT: Real-time sensors in airports, power plants, intelligent manufacturing, and healthcare systems make multivariate time-series anomaly detection more critical. Two significant obstacles remain. First, data organisations isolate sensitive data on islands and train high-performance anomaly detection models to preserve privacy and security. Data organisations have statistical heterogeneity. A unified anomaly detection methodology fails with personalised data. Blochchain-aware federated anomaly detection framework (BcFad) for multivariate time series data. BcFad uses the federated learning architecture to aggregate data while respecting privacy and fine-tuning a reasonably personalised model. BcFad improves F1 scores by 6.9% relative to the baseline technique in NASA spacecraft dataset experiments.

Keywords: Anomaly detection, Blockchain, Federated learning, Multivariant data

1 INTRODUCTION

Real-world applications of multivariate time series data have been widespread in numerous fields, such as weather data analysis and forecasting (Tang et al. 2022), healthcare (Qu et al. 2021), finance (Nguyen et al. 2022), etc. Cui et al. 2022; Miao et al. 2022). Anomaly detection is an important problem in multivariate time series analysis, the purpose is to detect sequence data that does not meet the expected behaviour, and it is one of the critical technologies of data mining. As deep learning has shown significant advantages in learning feature representations for complex data (Kalapaaking et al. 2023), anomaly detection using deep learning methods has received increasing

*Corresponding Authors: vskselva79@gmail.com, rmaaliw@slsu.edu.ph, ravi@mcu.ac.in, rvoak@ucdavis.edu, pavitar.19476@lpu.co.in and kuashok@banasthali.in

DOI: 10.1201/9781003393580-102

attention in recent years. For instance, Deep Autoencoder Gaussian Mixture Model (DAGMM) (Quang Hieu et al. 2022; Toyoda et al. 2020; Wang et al. 2022) models the density distribution of multidimensional data using both Deep Autoencoder and Gaussian Mixture Model. In modelling temporal correlation in time series, LSTM has attained a high level of generalisation.

However, anomaly detection in multivariate time-series data still faces challenges. Taking flight data anomaly detection as an example, flight data is a specific multivariate time-series data; effective anomaly detection will improve the safety and reliability of aviation systems and improve maintenance action organization after landing. However, flight data is highly commercially confidential, causing data barriers between general airlines; at the same time, The probability distributions of flight data provided by aircraft of different sorts and tasks are vastly varied, Consequently, there is no universal detection model that can be used in all contexts.

This research introduces FedPAD, a personalised federated learning system for anomaly detection of multivariate time-series data, in response to the aforementioned issues. FedPAD can solve data silos and personalization problems at the same time. Through federated learning (Kumar et al. 2021) and homomorphic encryption (Abdel-Basset et al. 2022), FedPAD aggregates data from different institutions, builds a high-performance deep anomaly detection model in the cloud, and protects private data well. After the cloud model is established, FedPAD uses fine-tuning (fine-tuning) further to train a personalized anomaly detection model for each institution.

2 RELATED WORK

2.1 *Prediction-based multivariate time-series anomaly detection model*

LSTM-NDT (Mothukuri et al. 2022) uses an extended short-term memory network to achieve high-performance time-series data prediction while ensuring the interpretability of the whole system. After the model generates forecast data, residuals are evaluated using a nonparametric, dynamic thresholding method. Specifically, let y_i be the signal value of the i-th time step of the input sequence and y_i' be the signal value of the i-th time step of the output sequence predicted by the model. Then, the prediction error is $e_i = y_i - y_i'$, with multiple time step error to compute the threshold sequence ε:

$$\varepsilon = \mu(e_s) + z\sigma(e_s) \tag{1}$$

Where e_s is the error sequence over multiple time steps, $\mu(\cdot)$ is the mean, $\sigma(\cdot)$ is the standard deviation, and z is the weighting coefficient. The threshold ε_i of each time step is dynamically changed, depending on the maximum value of the previous entire threshold sequence ε; the calculation formula is as follows:

$$\varepsilon_i = argmax(\varepsilon) = \frac{\frac{\Delta\mu(e_s)}{\mu(e_s)} + \frac{\Delta\sigma(e_s)}{\sigma(e_s)}}{|e_a| + |E_{seq}|^2} \tag{2}$$

$$\Delta\mu(e_s) = \mu(e_s) - \mu(\{e_i \in e_s|, e_i < \varepsilon_i\}) \tag{3}$$

$$\Delta\sigma(e_s) = \sigma(e_s) - \sigma(\{e_i \in e_s|, e_i < \varepsilon_i\}) \tag{4}$$

Among them, e_a is the error value of the abnormal sequence, E_{seq} is the error value of the continuous abnormality in the abnormal sequence, and the expression is as follows:

$$e_a = \{e_i \in e_s|, e_i > \varepsilon_i\} \tag{5}$$

$$E_{seq} = \{e_a\} \tag{6}$$

In addition, the pruning method is also used to reduce false positives, and the maximum value e_{max} in all error sequences is arranged in descending order to obtain e_s. Then the sequence is traversed to calculate the drop percentage d^i:

$$d^i = \frac{e_{\text{max}}^{i-1} - e_{\text{max}}^i}{e_{\text{max}}^{i-1}} \tag{7}$$

If di exceeds a minimum percentage p, the associated outlier series remain abnormal; if neither d^i nor all subsequent percentage declines exceed p, these erroneous series are reclassified as usual.

2.2 *Federated learning and personalization*

Google first proposed federated learning in 2016 (Peng et al. 2022). It is built with the intention of facilitating secure data exchange among a large number of participants or computing nodes while also protecting individual privacy and adhering to all applicable laws and regulations. Many fields can benefit from federated learning due to its ability to efficiently address the issue of data islands; for instance, it has demonstrated strong performance and robustness when used to the next prediction problem on mobile devices (Sun et al. 2022). A mountable system for massive federated mobile learning is proposed in the literature (Qi et al. 2022). Using a federated learning framework that was edge-accelerated, the authors of (Wang et al. 2022) were able to attaincompetent recommendation concert while protecting users' anonymity in the POI recommendation task.

The primary motivation for institutions to participate in federated learning is to obtain better models. However, the global model acquired by federated learning may not be suitable for the detection requirements of institutions with sufficient local data to train efficient models. Actors may not be able to improve model performance via federated learning for many tasks, as it has been proven in the literature (Cheng et al. 2022) that the globally shared model is not as accurate as the local model. Further, the global model's validity has been questioned in the published literature (Zhang et al. 2021). Non-IID client data makes it challenging to build a universal model that can be applied to all clients. To overcome this heterogeneity problem, we demonstrate that a standard federated learning approach, called Federated Averaging (FedAvg), can deal with certain types of non-IID data. In the face of severely skewed data distributions, however, FedAvg can lead to significant performance loss. Specifically, on one hand, FedAvg will produce a model with inferior performance compared to the centralised technique (Lu et al. 2021), as non-IID data will cause weight disparities between the federated learning process and the standard centralised training procedure. To contrast, FedAvg can only learn broad aspects from the data and is incapable of learning task-specific details.

To deal with the challenges of statistical heterogeneity and non-IID data, personalized global models become necessary. Most personalization techniques (Ayaz et al. 2022) usually consist of two steps. In the first step, an international model is trained collaboratively. In the second step, the global model is personalized for each client using the client's private data. FedPer was first proposed in the literature (Ayaz et al. 2022), wherein it was argued that the deep learning model should be viewed as the base + personalization layer, with the base layer functioning as a shared layer, and trained using existing federated learning techniques.

In contrast, the personalization layer is trained locally. By first training a global model via conventional federated learning and then sending it to all devices, as described in the aforementioned literature (Lu et al. 2021), it will be possible for each device to create a unique model by fine-tuning the global model with its own local data.Literature. (Zhang et al. 2021) make a trade-off between traditional global and regional models, where each device can learn a local model from its local data without any communication. To achieve personalization, Literature. (Ayaz et al. 2022) did not calculate the average value of model parameters like FedAvg. Still, they figured out how much a client could benefit from another client's model aggregation and obtained the optimal value of each client—weighted Model Portfolio.

3 METHOD OF THIS PAPER

3.1 *Problem definition*

$\{S_1, S_2, \cdots, S_N\}$ represent time series data from N different institutions $\{Q_1, Q_2, \cdots, Q_N\}$, and the data of different institutions have different distributions. Traditional centralized methods use global data S = $S_1 \bigcup S_2 \bigcup \ldots \bigcup S_N$ to train a unified model MALL. This work aims to train a federated anomaly detection model MFED using all available data. No company's data will be shared with another company's during the model training process. To achieve a detection rate that is on par with or better than that of MALL, it is necessary to train the federated anomaly detection model (MFED).

3.2 *Framework overview*

FedPAD aims to achieve high-performance anomaly detection through personalized federated learning while protecting privacy. Figure 1 gives an overview of the framework, taking flight data anomaly detection as an example (it can be extended to other scenarios), assuming that there are three general aviation companies, each with different aircraft types. The framework mainly includes four processes. Initially, train the cloud model on the server using the publicly available dataset. The cloud model is subsequently disseminated to all institutions, with each institution introducing its own model utilising local data on top of the cloud model. Then, return the organization's model to the cloud and update the cloud's model using FedAvg. Repeat the preceding steps until the model converges or the specified training rounds are reached. Last but not least, each institution can utilise cloud models and local data to further train tailored models. Being that there is a significant discrepancy between the global statistics and the institution's local data, a method of fine-tuning is employed to better match the model to the local data. Through homomorphic encryption, all parameter-sharing functions will not leak user information throughout the procedure.

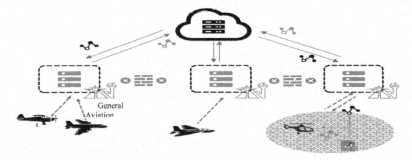

Figure 1. Overview of the FedPAD framework.

3.3 *Federated learning*

In order to solve the problem of data isolation and facilitate the training and distribution of distributed encryption models, FedPAD employs the federated learning paradigm. There are two main parts to this procedure: gathering models from the cloud and teaching models in an institution. FedPAD uses a neural network based on LSTM time series prediction as both the cloud and institutional model. When given access to institutional data, LSTM performs in-depth feature learning. The formulas (7) and (8) below illustrate the educational goals of both the cloud and campus-based models.

$$\underset{w,b}{\text{argmin}} \, L = \sum_{i=1}^{n} l(y_i, f_s(x_i)) \tag{8}$$

$$\underset{w^j, b^j}{\arg\min}\, L = \sum_{i=1}^{n^j} l\left(y_i^j, f_s\left(x_i^j\right)\right) \tag{9}$$

Among these, ω and b stand for all of the factors that need to be learned, which include weights and biases, $l(\cdot, \cdot)$ represents the loss function, j represents the organization number, (x_i, y_i) and (x_i^j, y_i^j) represent the Time-series data instances of j institutions, n and n j represent the size of the data set.

Once each user's model f j has been trained, it is uploaded to the cloud. Get the average model f_S' by aligning user models using the federated averaging approach (Mothukuri et al. 2022) and averaging M user models during each training cycle.

$$f_S'(w, b) = \frac{1}{M} \sum_{m=1}^{M} f_{jm}(w, b) \tag{10}$$

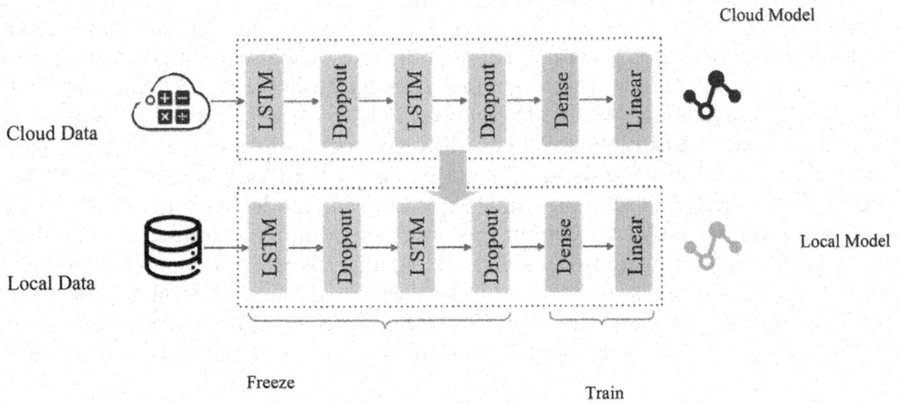

Figure 2. FedPAD fine-tuning process.

3.4 *Personalized learning*

The issue of data silos can be helped along by federated learning, which makes it possible for FedPAD to create anomaly detection models by making use of all institutional data. The statistical diversity of the data is an extra crucial factor that has an effect on performance. Due to the disparity in data distribution between a single institution and global data, the performance of directly employing the cloud model on a specific institution is still subpar. The cloud-based generic model can only learn coarse features from all institutions and cannot learn fine-grained characteristics from the data of specific institutions. Because they emphasise learning standards and lower-level features, (Peng et al. 2022) proved that lower-level features in deep neural networks are highly transferrable. The network's higher layers will learn more specific elements. Consequently, after receiving the cloud model, the institution employs the approach of fine-tuning to actualize the personalised institution model, as depicted in Figure 2. The neural network consists of two LSTM layers, two Dropout layers, one Dense layer, and one Linear layer. We take in multivariate time series data and output forecasted time series data. While altering the parameters of the upper layers, FedPAD keeps the lowest layer settings (LSTM and Dropout) constant (Dense and Linear).

3.5 *Algorithm process*

The model training process of FedPAD is introduced in Algorithm 1. When an institution generates new data, FedPAD can update the institution and cloud models. Therefore, the longer you use FedPAD, the better your model will perform.

Algorithm 1 FedPAD model training process
Input: public dataset D0, datasets $\{Q_1, Q_2, \cdots, Q_j\}$ from different institutions
Output: personalized models $\{f_1, f_2, \cdots, f_j\}$ for different institutions

1. Use the public dataset to train the LSTM model f_S in the cloud
2. FOR round $= 1, 2, \cdots r$ DO
3. Send the cloud model f_S to all institutions
4. Each institution uses local data D_j to train its model f_j and uploads the model parameters (ω, b) to the cloud
5. The cloud uses the federated average algorithm to aggregate all institutional model parameters and update the cloud model
6. END FOR
7. Each institution uses local data for fine-tuning to obtain a personalized model

4 EXPERIMENT AND RESULT ANALYSIS

4.1 *Dataset*

The description of the data set is shown in Table 1.

Table 1. FedPAD model training process.

Data set	Number of channels	Feature dimension	Number of training sets	Number of test sets	Abnormal rate/%
SMAP	56	30	135179	427634	14.45
MSL	25	60	58328	73754	11.12

4.2 *Evaluation indicators*

To make a direct comparison with the benchmark proposed by LSTM-NDT (Abdel-Basset et al. 2022; Mothukuri et al. 2022), the Point-based detection index commonly used in sequence data anomaly detection tasks is used; that is, when the predicted anomaly overlaps with the actual value, it is recorded as a true positive, and the expected anomaly and any true value When there is no intersection between the values, it is recorded as false positive. When there is no intersection between the real deal and any predicted value, it is registered as a false negative. Among them, the calculation of precision rate (Precision), recall rate (Recall) and F1 value are the same as the general detection task:

$$Precision = \frac{TP}{TP + FP} \tag{11}$$

$$Recall = \frac{TP}{TP + FN} \tag{12}$$

$$F1 = \frac{Precision \times Recall}{Precision + Recall} \tag{13}$$

Among them, TP is true positive, FP is false positive, TN is true negative, and FN is false negative.

4.3 *Experimental setup*

The time series data between channels in the NASA spacecraft data set have the same feature dimension. Still, there is a significant difference in feature distribution, that is, statistical heterogeneity, which is in line with the background of this paper. Therefore, each channel's data is regarded as an institution's data; that is, there are 55 institution nodes in the experiment on SMAP, and there are 27 institution nodes in the investigation on MSL for FedPAD model training. The LSTM-NDT method uses each channel data to train a model separately. The same model architecture and parameters as LSTM-NDT are used for a more intuitive comparison, as shown in Table 2.

Table 2.　Model architecture and parameters.

Parameter name	Value
Hidden layer	3
Hidden layer unit	82
Sequence length	255
Cycle	36

4.4 *Result analysis*

FedPAD is compared with LSTM-NDT, while the performance using only federated learning (FED) is recorded.

Table 3.　Telemetry prediction error.

Method	SMAP	MSL	Total
LSTM-NDT	5.7	7.4	6.5
FED	7.6	7.8	7.8
FedPAD	4.5	5.9	5.4

As shown in Table 3, compared with LSTM-NDT, only using federated learning for model training can solve the problem of data islands. Still, due to the statistical heterogeneity of data, the predictive performance of the FED model appears on both datasets. The overall forecast error increased by 1.2 percentage points. At the same time, the general prediction error of FedPAD has dropped by 1.6 percentage points, which is better than the LSTMNDT method. This is because federated learning can indirectly learn data characteristics from multiple institutions and then train a better model, and through fine-tuning, the model becomes better. Be more personalized and more adaptable to the data characteristics of each institution.

In the LSTM-NDT method, the pruning parameter p is vital to control the precision rate and the recall rate. The trade-off between the precision and recall rates is achieved by adjusting the parameter p. In the experiments in this paper, p is used as a control variable to compare the performance of the three methods in anomaly detection in two data sets. As shown in Figures 3 and 4, under different parameters p, the performance of the FED method is the most unstable, and the precision rate and recall rate cannot reach a high level simultaneously. On the other

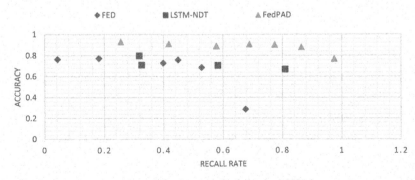

Figure 3.　Performance comparison of the three methods in the MSL dataset.

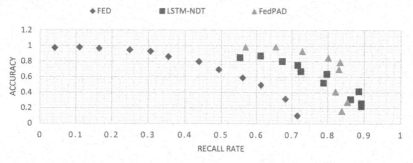

Figure 4. Performance comparison of the three methods in the SMAP dataset.

hand, the Fed-PAD method is slightly lower than LSTM-NDT when the value of p in the SMAP dataset is low. In other cases, the precision and recall are higher than LSTM-NDT, which benefits from the improvement of fine-tuning. Predictive performance of LSTM models.

Tables 4 and 5 show the average performance of the three methods under different parameters, and p is recorded. Compared with LSTM-NDT, FED's F1 scores on the two datasets decreased by 4.3% and 10.9%, respectively, and FedPAD's anomaly detection F1 scores increased by 10.1% on the MSL dataset and 3.6% on the SMAP dataset. the average F1 score increased by 6.9%. This again demonstrates the effectiveness of FedPAD in improving anomaly detection performance. The reason is that during the federated learning and fine-tuning process, the anomaly detection model on each data institution in FedPAD can learn the data characteristics of other institutions, which improves the reasoning performance of the model. Using a unified federated learning model is more prone to the problem of increased false positive rate or reduced model robustness. FedPAD can solve these problems by building a more personalized anomaly detection model for each data institution through fine-tuning.

Table 4. Average performance comparison on the MSL dataset.

Method	Precision	Recall	F1
LSTM-NDT	0.88	0.56	0.62
FED	0.84	0.54	0.65
FedPAD	0.94	0.66	0.77

Table 5. Average performance comparison on the SMAP dataset.

Method	Precision	Recall	F1
LSTM-NDT	0.87	0.86	0.72
FED	0.83	0.64	0.74
FedPAD	0.88	0.82	0.83

5   CONCLUSION

For anomaly detection of multivariate time series data, an anomaly detection framework FedPAD based on personalized federated learning is proposed. Based on the federated learning framework, FedPAD can learn the time series data characteristics of different institutions without disclosing data and privacy and use local data to model fine-tuning at the respective institutions to obtain personalized detection models. Experiments show that by detecting anomalies in NASA spacecraft data, FedPAD improves anomaly detection F1 scores by 6.9% compared to baseline methods.

REFERENCES

Abdel-Basset M., Moustafa N., Hawash H., Razzak I., Sallam K. M. and Elkomy O. M., "Federated Intrusion Detection in Blockchain-Based Smart Transportation Systems," in *IEEE Transactions on Intelligent Transportation Systems*, vol. 23, no. 3, pp. 2523–2537, March 2022, doi: 10.1109/TITS.2021.3119968.

Ayaz F., Sheng Z., Tian D. and Guan Y. L., "A Blockchain Based Federated Learning for Message Dissemination in Vehicular Networks," in *IEEE Transactions on Vehicular Technology*, vol. 71, no. 2, pp. 1927–1940, Feb. 2022, doi: 10.1109/TVT.2021.3132226.

Cheng X., Tian W., Shi F., Zhao M., Chen S. and Wang H., "A Blockchain-Empowered Cluster-Based Federated Learning Model for Blade Icing Estimation on IoT-Enabled Wind Turbine," in *IEEE Transactions on Industrial Informatics*, vol. 18, no. 12, pp. 9184–9195, Dec. 2022, doi: 10.1109/TII.2022.3159684.

Cui L., Su X. and Zhou Y., "A Fast Blockchain-Based Federated Learning Framework With Compressed Communications," in *IEEE Journal on Selected Areas in Communications*, vol. 40, no. 12, pp. 3358–3372, Dec. 2022, doi: 10.1109/JSAC.2022.3213345.

Kalapaaking A. P., Khalil I., Rahman M. S., Atiquzzaman M., Yi X.and Almashor M., "Blockchain-Based Federated Learning With Secure Aggregation in Trusted Execution Environment for Internet-of-Things," in *IEEE Transactions on Industrial Informatics*, vol. 19, no. 2, pp. 1703–1714, Feb. 2023, doi: 10.1109/TII.2022.3170348.

Kumar R. *et al.*, "Blockchain-Federated-Learning and Deep Learning Models for COVID-19 Detection Using CT Imaging," in IEEE Sensors Journal, vol. 21, no. 14, pp. 16301–16314, 15 July15, 2021, doi: 10.1109/JSEN.2021.3076767.

Lu Y., Huang X., Zhang K., Maharjan S. and Zhang Y., "Low-Latency Federated Learning and Blockchain for Edge Association in Digital Twin Empowered 6G Networks," in *IEEE Transactions on Industrial Informatics*, vol. 17, no. 7, pp. 5098–5107, July 2021, doi: 10.1109/TII.2020.3017668.

Miao Y., Liu Z., Li H., Choo K.-K. R. and Deng R. H., "Privacy-Preserving Byzantine-Robust Federated Learning via Blockchain Systems," in *IEEE Transactions on Information Forensics and Security*, vol. 17, pp. 2848–2861, 2022, doi: 10.1109/TIFS.2022.3196274.

Mothukuri V., Parizi R. M., Pouriyeh S., Dehghantanha A. and Choo K.-K. R., "FabricFL: Blockchain-in-the-Loop Federated Learning for Trusted Decentralized Systems," in *IEEE Systems Journal*, vol. 16, no. 3, pp. 3711–3722, Sept. 2022, doi: 10.1109/JSYST.2021.3124513.

Nguyen D. C., Hosseinalipour S., Love D. J., Pathirana P. N. and Brinton C. G., "Latency Optimization for Blockchain-Empowered Federated Learning in Multi-Server Edge Computing," in *IEEE Journal on Selected Areas in Communications*, vol. 40, no. 12, pp. 3373–3390, Dec. 2022, doi: 10.1109/JSAC.2022.3213344.

Peng Z. *et al.*, "VFChain: Enabling Verifiable and Auditable Federated Learning via Blockchain Systems," in *IEEE Transactions on Network Science and Engineering*, vol. 9, no. 1, pp. 173–186, 1 Jan.-Feb. 2022, doi: 10.1109/TNSE.2021.3050781.

Qi J., Lin F., Chen Z., Tang C., Jia R.and Li M., "High-Quality Model Aggregation for Blockchain-Based Federated Learning via Reputation-Motivated Task Participation," in *IEEE Internet of Things Journal*, vol. 9, no. 19, pp. 18378–18391, 1 Oct.1, 2022, doi: 10.1109/JIOT.2022.3160425.

Qu Y., Pokhrel S. R., Garg S., Gao L. and Xiang Y., "A Blockchained Federated Learning Framework for Cognitive Computing in Industry 4.0 Networks," in *IEEE Transactions on Industrial Informatics*, vol. 17, no. 4, pp. 2964–2973, April 2021, doi: 10.1109/TII.2020.3007817.

Quang Hieu N., Tran T. A., Nguyen C. L., Niyato D., Kim D. I. and Elmroth E., "Deep Reinforcement Learning for Resource Management in Blockchain-Enabled Federated Learning Network," in *IEEE Networking Letters*, vol. 4, no. 3, pp. 137–141, Sept. 2022, doi: 10.1109/LNET.2022.3173971.

Sun J., Wu Y., Wang S., Fu Y.and Chang X., "Permissioned Blockchain Frame for Secure Federated Learning," in *IEEE Communications Letters*, vol. 26, no. 1, pp. 13–17, Jan. 2022, doi: 10.1109/LCOMM.2021.3121297.

Tang F., Wen C., Luo L., Zhao M. and Kato N., "Blockchain-Based Trusted Traffic Offloading in Space-Air-Ground Integrated Networks (SAGIN): A Federated Reinforcement Learning Approach," in *IEEE Journal on Selected Areas in Communications*, vol. 40, no. 12, pp. 3501–3516, Dec. 2022, doi: 10.1109/JSAC.2022.3213317.

Toyoda K., Zhao J., Zhang A. N. S. and Mathiopoulos P. T., "Blockchain-Enabled Federated Learning With Mechanism Design," in *IEEE Access*, vol. 8, pp. 219744–219756, 2020, doi: 10.1109/ACCESS.2020.3043037.

Wang P. *et al.*, "Blockchain-Enhanced Federated Learning Market With Social Internet of Things," in *IEEE Journal on Selected Areas in Communications*, vol. 40, no. 12, pp. 3405–3421, Dec. 2022, doi: 10.1109/JSAC.2022.3213314.

Wang Y., Peng H., Su Z., Luan T. H., Benslimane A.and Wu Y., "A Platform-Free Proof of Federated Learning Consensus Mechanism for Sustainable Blockchains," in *IEEE Journal on Selected Areas in Communications*, vol. 40, no. 12, pp. 3305–3324, Dec. 2022, doi: 10.1109/JSAC.2022.3213347.

Zhang W. *et al.*, "Blockchain-Based Federated Learning for Device Failure Detection in Industrial IoT," in *IEEE Internet of Things Journal*, vol. 8, no. 7, pp. 5926–5937, 1 April1, 2021, doi: 10.1109/JIOT.2020.3032544.

Recent advancements and challenges in Artificial Intelligence, machine learning, cyber security and blockchain technologies

Artificial Intelligence, Blockchain, Computing and Security – Dagur et al. (Eds)
© 2024 The Author(s), ISBN: 978-1-032-49393-0

Relative study on machine learning techniques for opinion analysis of social media contents

V. Malik & N. Tyagi

Shobhit Institute of Engineering & Technology (NAAC 'A' Grade Accredited Deemed-to-be University), Meerut, U.P.

ABSTRACT: Social media platforms are used for creating, sharing, and managing information through the network/internet. There are many social networking sites, e.g. Whatsapp, Facebook, Twitter, Instagram, and many more. A variety of social media handlers use these platforms to promote themselves or their work. Technology has made great advancements for identifying opinions, sentiments, likes, dislikes, brand promotions, and many more insights for social media. Also, by mining the text on social media, suicidal, melancholy, farming, or terrorism forecasting can be accomplished. Machine learning has a major impact in this domain. This paper focuses on existing methods and techniques utilized to grab viewpoint from social media, that help in determining decisions/sentiments of the public regarding a specific problem and presents a comprehensive examination of the literature on opinion analysis systems based on machine learning and identifies any holes in the science. Research in the specified domain based on techniques like neural networks, SMO, MaxEnt, and their related algorithms are discussed.

Keywords: Twitter, Sentiments, Opinion, Social media, Analysis, Machine learning

1 INTRODUCTION

Everyone nowadays uses social media. People use these sites for a variety of reasons. They generate and post material on social media based on their career. Celebrities, teachers, doctors, businessmen, politicians, and others advertise themselves or their work through social media networks. As time goes on, people prefer to share their opinions with the entire globe rather than just their family and friends. This has both positive and negative consequences for society. There are certain people who post unpleasant content on social media, which have serious ramifications. On social media, there is religious discrimination material, political content, nation content, and many other types of content. Algorithms of machine learning forecast [25] new output values using historical data as input. Today, machine learning is used by many thriving organizations. Machine learning has become a critical competitive differentiator for many firms. There are various methods of ML. Supervised and unsupervised learning are among those methods. The most effective method for classifying texts is supervised learning. Classifier is used to categorise the text into positive, negative, and neutral statements before extracting the public's sentiment on a given subject. In current scenario, identifying opinion of public is considered as one of the most trendy and important section of ML. According to [31], opinion analysis has been divided down into three levels: document, sentence, and aspect levels. Document based mining is suitable for emotions. While, sentence based mining consider each sentence as independent unit. The goal of aspect level which is the third and last level of opinion mining is to identify the hypothesis. Different classifiers can be used to anticipate how individuals in a specific region will think in a particular language. Texts written in Bangla, Thai, English, and Punjabi are utilised to

DOI: 10.1201/9781003393580-103

categorise the content as neutral, good, or terrible. The use of machine learning algorithms to extract opinions from social media is discussed in this study. It covers following sections.

- A section will provide a summary of the related works by various authors.
- Basic information regarding opinion analysis and general steps followed by different text classifiers.
- A brief discussion about text classification algorithms and a comparative study is done on existing technique.

A conclusion is drawn from the discussed information. Future scope is also highlighted.

2 LITERATURE REVIEW

One of the areas of computer science study that is regarded to be growing the fastest is sentiment analysis. For this many researchers have used various techniques of ML to take opinions of the public on different topics. Below listed works describe machine learning algorithms used to take opinions. Regardless of whether the language is good or negative, Chowdhury *et al.* (2014) [6] employed the MaxEnt (Maximum Entropy) and SVM method to automatically extract emotions from tweets written in Bangla. Then after various modifications and continuous work in this field, Mridula S. Mishra & Ruppal W. Sharma (2019) [15] based on consumer reactions to affected brands' and its main competitor's social media corporate communications, a conceptual model was put forth. Arora, P. & Arora, P. (2019) [5], used Multinomial Naïve Bayes and Support Vector Regression algorithm to classify the tweets related to health for depression and anxiety. People employ hybrid languages, such as English-Hindi or English-Punjabi. Singh *et al.* (2019) [29] extracted English-Punjabi mixed comments related to agriculture. Further, an English-Punjabi code mixed dictionary is created. James Nata Salim *et al.* (2020) [26] used Twitter to gauge public sentiments on whether or not "Smart farming" or "Agriculture 4.0" is practised in Indonesia. Naïve Bayes method is used for this. Sentiment analysis on social media for brands, Bangla text, Persian movies, etc. are used, also, machine learning is used to take opinions of public on very trendy topic COVID-19. Nemes, L., & Kiss, A. (2021) [17] used Natural Language Processing and sentiment classification using Recurrent Neural Networks (RNN) to recognise the public's sentiment on Twitter based on the primary trends (by searching term, the coronavirus topic in articles) and main trends.Nemes *et al.* have used third-party sentiment analyzers like TextBlob and compared its result with the results of RNN. Over 500 Tweets are used as datasets to analyze positive, negative, and neutral emotions from the sentences that contain the word COVID. From their observation, this is clear that in terms of performance and prediction, RNN performs well. Less data were used in the neutral or zero outcomes produced by the RNN model. An analysis of farmer suicide instances led Singh, J., Singh *et al.* (2021) [28] to develop a model that divides Punjabi tokens into four classes with a negative orientation. Following 10-fold cross-validation, the average sentiment prediction accuracy for each of the four classes is 95.45 percent, 93.85 percent, 88.53 percent, and 83.3 percent. Whether it is Facebook, Twitter, Instagram or any other social media platform, the trend of using hashtags(#) is increasing day by day. In context of this hashtag trend Pilař *et al.* (2022) [22], used a software Gephi 0.9.3 for network visualization and analysis. The tweets of food bloggers with hashtags such as #healthy,#vegan, etc. were imported into Gephi and a corpus of hashtags was developed. This helps in identifying the three main clusters of food bloggers namely-healthy lifestyle; fast food and home-made food. Two mini clusters were also identified that involves – breakfast and brunch; food travelling. Rahman *et al.* (2022) [24] used a total of 12,000 UK-related tweets and three independent reviewers thoroughly annotated these. Three alternative ensemble ML models were presented to categorise the evidence gathered from tweets into 3 types: good, terrible, and neutral on the basis of tagged tweets. Based on the study, the voting classifier (VC) came in second with an F1-score of 83.3 percent, while the bagging classifier (BC) came in third with an F1-score of 83.2% and the F1-score of 83.5 percent for stacking classifier (SC) was the highest. Tuli *et al.* (2022) [31] discussed about different levels of opinion mining that involves document level, sentence level and aspect level text mining. Also, concept of privacy preservation is reviewed, including both single privacy preservation and multiple privacy preservation.

3 OPINION ANALYSIS

Simply said, an opinion is a person's belief or way of thinking on a certain subject. When we discuss opinion analysis, we are examining the public's point of view on a certain topic. Social media is growing in popularity in the modern world. Using various machine learning algorithms that are optimized for text classification can help gather opinions from social media. A variety of classifiers are available for processing text documents. A variety of classifiers can be used to categorise text as good, bad, or neutral. We must first take a .csv file while discussing text classification using machine learning. In order to download this file from Twitter, it has provided us with Twitter APIs that enable us to grab public Tweets on a topic.

Here, the fundamental procedures that almost all classifiers adhere to are covered.

1. Pre-processing of data.
2. Fitting the training dataset to the classification algorithm.
3. Making a test result prediction
4. Verify the result's accuracy.
5. Displaying the results of the test set.

4 TEXT CLASSIFIERS

4.1 *Naïve bayes*

The Naive Bayes classifier is predicated on hypotheses and effective with independent features. A method for determining posterior probability is provided by the Bayes theorem P(J|K) from P(J), P(K) and P(K|J). Look at the equation below:

$$P(J/K) = [P(K/J).P(J)]/P(K) \qquad (1)$$

Where,
P(J/K) is the posterior probability,
i.e. probability of J given K. P(J) is the probability of J (hypothesis).
P(K|J) is the likelihood or conditional probability i.e. probability of K given that J hypothesis is true.
P(K) is the marginal probability.
Naïve Bayes classifier is broadly classified into three types-

1) Gaussian Naïve Bayes which is used for continuous variables.
2) Multinomial Naïve Bayes is used for document classification.
3) Bernoulli Naïve Bayes for Boolean variables.

The advantage of Naive Bayes is that it just requires a little quantity of training data to estimate the parameters associated to classification [26] [23]. Word frequency is used as a feature in the Multinomial Naive Bayes classification method used by [23].

4.2 *MaxEnt (Maximum Entropy)*

A machine learning method for classifying text data is called MaxEnt(Maximum Entropy) classifier [6]. Unlike the Naive Bayes classifier, this probabilistic classifier does not start with the same presumption. As compared to Naïve Bayes classifier, MaxEnt requires more training time. This is because there is a need to solve the optimization problem to estimate model's parameters. The format in which MaxEnt collect the training data is- (xi, yi). Where xi represents the spare array which contains contextual information of document and yi represents its class. According to the maximum entropy principle, the MaxEnt algorithm selects the model that has the highest entropy among each of the models that are suitable for our training data. A wide range of text classification issues, including language identification, sentiment analysis, and more, can be handled with the MaxEnt classifier. We frequently employ the Maximum Entropy classifier when we don't know anything about the prior distributions and when making any such assumptions is problematic due to the lack of assumptions it contains.

4.3 *SMO (Sequential Minimal Optimization)*

SentiWordNet and WorldNet [9] are lexicon based dictionaries for sentiment analysis. WorldNet is used to get the sense of words and SentiWordNet is used to get the word polarity. WorldNet is a dictionary that uses a lexicon-based approach. First, lexicons are collected from the document then WorldNet is used to get synonyms, antonyms, adverbs, adjectives of the text, in order to obtain text semantic orientation. SentiWordNet works on the data provided by wordNet. In addition, it gives polarity to each word as positive, negative or neutral and this polarity ranges from 0 to 1. These dictionaries are used by machine learning algorithms/classifiers to analyze or classify the data. There are so many programming optimization problems among which quadratic programming optimization problems are difficult to solve. Also, to handle classification and regression problems SVMs (Support Vector Machines) are required to train properly. So, to train SVMs for quadratic programming problems SMO (Sequential Minimal Optimization) algorithm is used. To train a support vector machine in this, a very large quadratic programming (QP) optimisation problem must be resolved. SMO divides this complicated QP problem into its simplest sub-problems. These tiny QP issues are resolved analytically, avoiding the need for an inner loop that involves a long-drawn-out numerical QP optimization. SMO is capable of handling very large training sets because memory utilisation increases linearly with training set size. Because to the avoidance of matrix computing, The conventional chunking SVM technique ranges within linear and cubic problems, whereas SMO expands across linear and quadratic for a variety of test cases in training set.

4.4 *LSTM (Long Short Term Memory)*

RNN (Recurrent Neural Network) are those networks that contain algorithms that recall its input and with the help of intramural memory. This feature helps RNN to remember things. But, there is a shortcoming of RNN. The demerit is vanishing gradient, this means, it can't remember long term data or dependency. LSTM is the advanced version of RNN as it is capable of thinking about long term dependency. Three components make up LSTM's internal operation. The first part decides whether or not the details from the previous timestamp should be recalled. This cell attempts to learn new information using the input from the second part. In the third section, the cell eventually communicates the updated data from the present timestamp to the future timestamp. The three parts that make up an LSTM cell are known as the gates. The Forget gate is the initial part, followed by the Input gate and the Output gate. The equation for forget gate is,

$$f_t = \sigma(i_t * V_f + H_{t-1} * W_f) \tag{2}$$

Where, i_t: current timestamp input. V_f: weight of the input.
H_{t-1}: previous timestamp's hidden state. W_f: hidden state's weight matrix.
   Then sigmoid function is applied on these variables which will give the value of f_t between 0 and 1. Later, the product of the value of f_t and the state of the previous timestamp cell is calculated. If the value comes out to be 0 that means the network will forget everything and

if the value is 1 that means the network will not forget everything. Similarly, the equation for Input gate is given below,

$$g_t = \sigma(i_t * V_i + H_{t-1} * W_i) \tag{3}$$

where,

I_t: current timestamp input. V_i: weight matrix of input.

H_{t-1}: previous timestamp's hidden state. W_i: hidden state's weight matrix of input.

The value of g at t will range from 0 to 1. Now, new data (N_t) is calculated. The new data that had to be sent to the cell state now depends ona hidden state at timestamp t-1 in the past and input x at timestamp t. Tanh is the activation function in this case. The tanh func tion causes the value of new data to range from -1 to 1. The information is deducted from the cell state if the value of N_t is negative, and added to the cell state at the current timestamp if the value is positive. Then, this new information is added with the forget and input gate as follows,

$$C_t = f_t * C_{t-1} + g_t * N_t \tag{4}$$

And, for output gate the equation is,

$$O_t = \sigma(i_t * V_o + H_{t-1} * W_o) \tag{5}$$

Due to the sigmoid function its value also ranges between 0 and 1. We will now use O_t and tanh of the updated cell state to determine the current hidden state. The updated equa tion is,

$$H_t = O_t * \tanh(C_t) \tag{6}$$

This leads us to the conclusion that the output function and long-term memory (C_t) function both have the ability to reveal concealed states. Also, SoftMax activation function can be applied on hidden state to get the output of current timestamp.

5 COMPARATIVE STUDY OF EXISTING TECHNIQUES

Researchers from many fields employed various classifiers to anticipate public opinions on various subjects through social media. Authors gathered social media statistics and used several categoriza- tion algorithms to analyze them. The following table displays a comparison of the available research.

Table 1. Comparative study of existing work.

Authors	Method Employed	Accuracy(%)
Çoban, Ö., et al. [7]	RNN	91.6%
Khan et al. [13]	SVM;	62%
	Random Forest,	58%
	K-Nearest Neighbors,	55%
	Naive Bayes,	52%
	Neural Network.	50%
Aljuaid et al. [4]	Random Forest	83% for dataset 1
	classification model	67% for dataset 2
Naresh, et al. [16]	SMO(with decision tree)	89.47%
ONAN A. [19]	LSTM inconjunction with GloVe word embedding scheme based representation.	95.80%
Salim et al. [26]	Naive Bayes	97%
Singh et al. [28]	RNN	90.29%

6 CONCLUSION & FUTURE SCOPE

Recent studies show that textual sentiment analysis is one of the rapidly expanding trends. Many researchers believed that supervised learning is best suited for analyzing positive, negative and neutral sentiments of the public. The comparative research mentioned earlier indicates that neural networks give best results for classification of text data. For some global issues, on which the public shares their views on social media a text data classification system can be proposed. Public emotions on COVID-19, agriculture related problems in a country, emotions towards a celebrity, etc. can be analysed by adopting various text classification algorithms. Social media are used globally by humans and different kinds of data are being uploaded by people on it. There are some APIs of different social media platforms which are used to download raw data about any particular issue. This research paper has discussed some machine learning techniques that can be applied on the raw data obtained in .csv file format (whenever we are using python) to take sentiments of people. Here, MaxENT, Naïve Bayes theorem, SMO (Sequential Minimal Optimization), SentiWordNet, WorldNet and LSTM are discussed. So, according to the type of data any machine learning approach can be applied for sentiment analysis. Also, it is concluded that whenever there is data which requires long term information to analyse sentiments, LSTM is well suited for this. As neural networks work well in this regard. Naïve Bayes and MaxEnt works on probabilistic data with some difference in both algorithms. Naïve Bayes requires less training data whereas MaxEnt takes more training time.For future work, the concept of Ensemble Learning can be adopted. Ensemble learning combines several models in order to improve machine learning results. This can help in improving the accuracy of prediction.

REFERENCES

[1] Abayomi-Alli, A., Abayomi-Alli, O., Misra, S., & Fernandez-Sanz, L. 2022. Study of the Yahoo-Yahoo Hash-Tag Tweets Using Sentiment Analysis and Opinion Mining Algorithms. *Information*, 13 (3), 152.

[2] Ahmad, G. I., & Singla, J. 2021. Sentiment Analysis of Code-Mixed Social Media Text (SA-CMSMT) in Indian-Languages. *In 2021 International Conference on Computing Sciences (ICCS)* (pp. 25–33). IEEE.

[3] Al-Hashedi, A., Al-Fuhaidi, B., Mohsen, A. M., Ali, Y., Gamal Al-Kaf, H. A., Al-Sorori, W., & Maqtary, N. 2022. Ensemble Classifiers for Arabic Sentiment Analysis of Social Network (Twitter data) Towards COVID-19-Related Conspiracy Theories. *Applied Computational Intelligence and Soft Computing, 2022.*

[4] Aljuaid, H., Iftikhar, R., Ahmad, S., Asif, M., & Afzal, M. T.,2021. Important Citation Identification Using Sentiment Analysis of In-text Citations. *Telematics and Informatics*, 56, 101492.

[5] Arora, P., & Arora, P. 2019. Mining Twitter Data for Depression Detection. *In 2019 International Conference on Signal Processing and Communication (ICSC)* (pp. 186–189). IEEE.

[6] Chowdhury, S. and Chowdhury, W., 2014. Performing Sentiment Analysis in Bangla microblog posts. In 2014 *International Conference on Informatics, Electronics & Vision (ICIEV)* (pp. 1–6). IEEE.

[7] Çoban, Ö., Özel, S. A., & İnan, A. 2021. *Deep Learning-based Sentiment Analysis of Facebook Data*: The CasPark, S., Strover, S., Choi, J., & Schnell, M.

[8] Cooper, K., Dedehayir, O., Riverola, C., Harrington, S., & Alpert, E. 2022. Exploring Consumer Perceptions of the Value Proposition Embedded in Vegan Food Products using Text Analytics. *Sustainability*, 14(4), 2075.

[9] Hasan, K.A. and Rahman, M., 2014. Sentiment Detection from Bangla Text Using Contex tual Valency Analysis. *In 2014 17th International Conference on Computer and Information Technology (ICCIT)* (pp. 292–295). IEEE.

[10] Heidari, M., James Jr, H., & Uzuner, O., 2021. An Empirical Study of Machine Learning Algorithms for Social Media Bot Detection. In 2021 *IEEE International IOT, Electronics and Mechatronics Conference (IEMTRONICS)* (pp. 1–5). IEEE.

[11] Jain, P. K., Pamula, R., & Srivastava, G., 2021. A Systematic Literature Review on Machine Learning Applications for Consumer Sentiment Analysis Using Online reviews. *Com puter Science Review*, 41, 100413.

[12] Kabir, M., Kabir, M. M. J., Xu, S., & Badhon, B., 2021. An Empirical Research on Sentiment Analysis Using Machine Learning Approaches. *International Journal of Computers and Applications*, 43(10), 1011–1019.

[13] Chaturvedi, R., Kumar, S., Kumar, U., Sharma, T., Chaudhary, Z., & Dagur, A. (2021). Low-cost Iot-enabled Smart Parking System in Crowded Cities. *In Data Intelligence and Cognitive Informatics: Proceedings of ICDICI 2020* (pp. 333–339). Springer Singapore.

[14] Y. Sharma, Shatakshi, Palvika, A. Dagur and R. Chaturvedi, "Automated Bug Reporting System in Web Applications," *2018 2nd International Conference on Trends in Electronics and Informatics (ICOEI)*, Tirunelveli, India, 2018, pp. 1484–1488

[15] Mishra, M. S., & Sharma, R. W., 2019. Brand Crisis-Sentiment Analysis of User-Generated Comments about@ Maggi on Facebook. *Corporate Reputation Review*, 22(2), 48–60.

[16] Naresh, A., & Venkata Krishna, P., 2021. An Efficient Approach for Sentiment Analysis Using Machine Learning Algorithm. *Evolutionary Intelligence*, 14, 725–731.

[17] Nemes, L., & Kiss, A., 2021. Social Media Sentiment Analysis Based on COVID-19. *Journal of Information and Telecommunication*, 5(1), 1–15.

[18] Nimirthi, P., Venkata Krishna, P., Obaidat, M. S., & Saritha, V. 2019. A Framework for Sentiment Analysis Based Recommender System for Agriculture Using Deep Learning Approach. *In Social network forensics, cyber security, and machine learning* (pp. 59–66). Springer, Singapore.

[19] ONAN, A., 2021. Sentiment Analysis on Massive Open Online Course Evaluations: A Text Mining and Deep Learning Approach. *Computer Applications in Engineering Education*, 29(3), 572–589.

[20] Park, S., Strover, S., Choi, J., & Schnell, M. (2021). Mind games: A Temporal Sentiment analysis of the Political Messages of the Internet Research Agency on Facebook and Twitter. *New Media & Society*, 14614448211014355.

[21] Piedrahita-Valdés, H., Piedrahita-Castillo, D., Bermejo-Higuera, J., Guillem-Saiz, P., Bermejo-Higuera, J. R., Guillem-Saiz, J., ... & Machío-Regidor, F., 2021. Vaccine Hesitancy on Social Media: Sentiment Analysis from June 2011 to April 2019. *Vaccines*, 9(1), 28.

[22] Pilař, L., Pilařová, L., Chalupová, M., Kvasničková Stanislavská, L., & Pitrová, J. 2022. Food Bloggers on the Twitter Social Network: Yummy, Healthy, Homemade, and Vegan Food. *Foods*, 11 (18), 2798.

[23] Pinilla, R., & Gaitán-Angulo, M., 2020. *Sentiment Analysis of Facebook Comments Using Various Machine Learning Techniques.*

[24] Rahman, M., & Islam, M. N., 2022. Exploring the Performance of Ensemble Machine Learning Classifiers for Sentiment Analysis of Covid-19 Tweets. In *Sentimental Analysis and Deep Learning* (pp. 383–396). Springer, Singapore.

[25] Rani, S., Gill, N. S., & Gulia, P., 2021. Analyzing Impact of Number of Features on Efficiency of Hybrid Model of Lexicon and Stack Based Ensemble Classifier for Twitter Sentiment Analysis Using WEKA Tool. *Indonesian Journal of Electrical Engineering and Computer Science*, 22(2), 1041–1051.

[26] Salim, J. N., Trisnawarman, D., & Imam, M. C., 2020. Twitter Users Opinion Classification of Smart Farming in Indonesia. In *IOP Conference Series: Materials Science and Engineering* (Vol. 852, No. 1, p. 012165). IOP Publishing.

[27] Shahreen, N., Subhani, M., & Rahman, M. M., 2021. Suicidal Trend Analysis of Twitter Using Machine Learning and Neural Network. *In 2018 international conference on Bangla speech and language processing (ICBSLP)* (pp. 1–5). IEEE.

[28] Singh, J., Singh, G., Singh, R., & Singh, P., 2021. Morphological Evaluation and Sentiment Analysis of Punjabi Text Using Deep Learning Classification. *Journal of King Saud University-Computer and Information Sciences*, 33(5), 508–517.

[29] Singh, M., Goyal, V., & Raj, S., 2019. Sentiment Analysis of English-Punjabi Code Mixed Social Media Content for Agriculture Domain. *In 2019 4th International Conference on Information Systems and Computer Networks (ISCON)* (pp. 352–357). IEEE.

[30] Stephen, J. J., & Prabu, P., 2019. Detecting the Magnitude of Depression in Twitter Users Using Sentiment Analysis. *International Journal of Electrical and Computer Engineering*, 9(4), 3247.

[31] Tuli, P., & Walia, K., 2022. A Review on Opinion Mining Based Iintegrated with Privacy Preservation in Social Media. Available at *SSRN 4019654*.

[32] Mehra, P. S., Jain, K., Chawla, D., Dagur, A., Singh, S., & Sharma, J. (2022). GWO-EFUCA: *Grey Wolf Optimisation and Fuzzy Logic based Unequal Clustering and Routing protocol for sustainable WSN-based Internet of Things.*

707

Artificial Intelligence, Blockchain, Computing and Security – Dagur et al. (Eds)
© 2024 The Author(s), ISBN: 978-1-032-49393-0

Review of permission-based malware detection in Android

Nishant Rawat*
Computer Science & Engineering, Sharda School of Engineering & Technology, Sharda University, Greater Noida, Uttar Pradesh, India

Amrita*
Center for Cyber Security and Cryptology, Computer Science & Engineering, Sharda School of Engineering & Technology, Sharda University, Greater Noida, Uttar Pradesh, India

Avjeet Singh*
Computer Science & Engineering, Sharda School of Engineering & Technology, Sharda University, Greater Noida, Uttar Pradesh, India

ABSTRACT: A market study showed that an average of 70% of smartphone users use an android-based smartphone. The Android operating system draws numerous malware threats as a result of its popularity. The statistic reveals that 97% of malware prey on android due to its lack of security. The malware enters the system and after installing the malware application they ask permission from the user to run. Permissions like Read_Phone, Access_Internet, Write_Log, etc give full control of the victim's phone to the hacker. In recent years more efforts were made to analyze android permissions to detect malware. In this paper, we review some of the already published work in detail which is in the area of malware detection using static analysis.

Keywords: Android, Cyber Security, Malware detection, Machine learning, Permission.

1 INTRODUCTION

In this 21st century, smartphones become the most crucial part of our life. Everyday operations were carried out with the help of smartphones. There are different types of smartphones available based on OS, size, etc. Most of the smartphones which are available in the market majority of them were operated on Android (Bai *et al.* 2020).

A recent survey shows that more than 70% of the smartphone market was captured by android (Smartphone OS market share (Statcounter Global Stats 2022)). This dominance of android happens due to its open-source feature. Any company can put its UI on it and sell it based on their accordance. This gives the flexibility to the companies to control their software. Anyone can upload their application in the play store because it is open-source. Due to this feature of open-source people like cyber-criminals upload virus-infected applications on the play store which give them access to the victim's phone once it is installed in the system. This simplicity of uploading an application into the play store without proper scanning of the application can lead to cybercrimes. Android becomes the favorite target of cyber-criminals due to its open-source feature. Malware is nothing, but a piece that incorporates the legit software code to hide its presence from the antivirus of the system (Durmus *et al.* 2021). Hackers create malicious software to gain access to the victim's systems to steal

---

*Corresponding Authors: nishantrawat575@gmail.com, amritaprasad_y@yahoo.com and avjeet.mnnit.cs@gmail.com

DOI: 10.1201/9781003393580-104

information from them. This type of malicious software can be entered into the system by various means such as the internet, Pendrive, etc. The virus activates itself when some conditions are met or some function is triggered in the system. Before that, it stays hidden in the system and replicates itself in the memory to do as much damage as possible. A recent study, shows that in quarter 2 of 2021, there was new android malware in the market was published every second (Chebyshev 2021). There are many ways to detect malware in the system. The first line of defense in every system against malware is a firewall, but nowadays cyber criminals become so smart and manage to get past the firewall easily. After the firewall antivirus comes into play which stores the signature of every known malware to detect malware as soon as it enters the system and blocks it. There are various ways for malware detection, but the two most commonly used methods are static and dynamic analysis of malware.

In a dynamic analysis of the malware, the application first runs into an isolated environment called Sandbox and its behavior were captured. All the features like abnormal behavior, API calls, runtime behavior, intents, etc were captured and stored in the database to create a signature of that malware. Static is good in detecting only the known malware but performs not as well as dynamic approaches. In this paper, we review papers that detect android malware using permissions. Static approaches were used by most of the authors with a machine learning algorithm. Permissions are the most important part of the application (Kandukuru *et al.* 2019). Due to permissions application can run at its full potential. The permission was granted by the user of that application according to their need. There are various types of permissions in android like android. permission. Internet is used to give access to the internet to the application. Due to the permission malicious apps can access the Internet and download, the malicious code from the hacker's website, and run into the device. Due to this reason, permission becomes the hacker's favorite area to create malware (Saracino *et al.* 2020).

In Android, users can download any application using a third-party store or website which makes Android an easy target for malware. In recent study reveal that 97% of malware was prayed on android due to its lack of security. This malware enters the system and after installing that malware application, they ask permission from the user to run. Permissions like Read_Phone, Access_Internet, Write_Log, etc give full control of the victim's phone to the hacker. In recent years more efforts were made to analyze android permissions to detect malware.

For example, a system called Derbin[25] uses a machine learning algorithm with permission and API calls to statically analyze android malware. The Derbin experiments result shows that using more number of feature yield high accuracy in the detection of android malware, but also include high overhead which makes it slower to detect the malware. The Derby is a lightweight application that uses more than 1 Lakhs of applications to train their model. There are a total of 154 different permission an application can ask and it is well known that no one application can ask for all 154 permissions. Google has listed 24 permission that is most dangerous and frequently used by malware applications. The most common permission used by hackers is Android.Permission.Internet and Android. Permission.WRITE_EXTERNAL_STORAGE. By this combination of permissions, hackers can access the internet from the victim's device and download malware into the device, and store that malware in the phone storage. All permissions were written in the Manifest. xml file of the application. Researchers analyze the permissions listed in the manifest file and classify whether or not it's malware or benign software.

In this paper, we have discussed 25 different android malware detection system which uses permissions and other features for the detection of android malware. The major drawback of these discussed systems is that they use an imbalanced dataset for the training and testing of the model or system which results in a high false positive rate and less accuracy. They use more features for the classification instead of selecting only the significant ones which increases the detection rate and accuracy.

2 LITERATURE REVIEW

In the literature review section, we discuss some of the papers which are already been published in the field of malware detection using static analysis. The proposed system (Almin *et al.* 2015) uses permission for the detection of malware. The system extracts permission from the manifest file and then after extraction, it uses the K-means algorithm for clustering of permission. Each cluster consists of a set of permissions used for each class of malware. The proposed system (Arp *et al.* 2014) is called Drebin. The system uses permissions and API calls as features for the detection of malware. The system classifies the family of the malware as well using a static analysis approach. The system uses over 1 Lakh of applications as a sample for training and testing the system and achieving 94% accuracy in malware detection. The proposed system (Bai *et al.* 2020) is called fast android malware detector (FAMD). The system uses 2 features-Permissions and Dalvik opcode for the detection of the malware. The system uses the N-gram technique for dimensionality reduction and a Fast correlation-based filter (FCBF). The system achieves 97.40% accuracy in malware detection and achieves 97.38% accuracy in classifying the malware family.

The proposed system (Cai *et al.* 2019) is called Joint optimization of weight mapping (JOWMdroid). The system is based on feature weights and joint optimization of weight mapping and classifier parameters for malware detection. They weights each feature according to the information gained of each feature and used it for malware detection. 19. Similarly, the proposed system (Saracino *et al.* 2020) called a Multi-level anomaly detector for android malware (MADAM). The system monitors the app by extracting 5 features at 4 different levels kernel level, application level, user level, and package level. The system detects the misbehavior of the application by analyzing system calls, API calls, and user activity. The system achieves an accuracy of 96.95 in the detection of malware with only 1.4% overhead. Similarly, The system (Sanz *et al.* 2019) is called PUMA (Permission Usage to Detect Malware in Android) which uses permission to detect malware. They use 4031 samples of android applications and apply a classifier like RandomForest, RandomTree, NaiveBayes, etc and find that random-forest gives a high detection accuracy of 86.41%.

Similarly, the proposed system (Durmus *et al.* 2021) uses significant permission from android for the detection of malware. The system at the time of feature selection reduces the feature vector by selecting only the significant permission from the permission list which reduces the training time. The system uses a linear regression method for feature selection. The system uses a balanced dataset of applications of 2000 samples and achieves an accuracy of 96.1% in the detection of malware.

Similarly, the paper (Feizollah *et al.* 2017) shows a system that uses android intents (explicit and implicit) for detection of the malware. The system yields 91% accuracy when only the intents are used for classification and shows 95% accuracy when it uses the intent and permission of the application.

The proposed system (Jannath *et al.* 2018) identifies the repacked application. They use the permission list of the original application and compare it with the permission list of the repacked android application. Random Forest technique got the highest accuracy rate of 88.53% as compared to other classifiers like SVM, KNN, etc. Similarly, the proposed (Talha *et al.* 2015) system called APKauditor uses permissions to identify and classify between benign and malware applications. There are three parts in the system, one is a database used to store the permissions or signature of the database, second is an android application for the end-user, and lastly, a central server that communicates between both the end-user application and the database. All the analyzing is done on the central server and the result is fetched by the end-user application. They use 8762 applications and achieve 88% accuracy in the detection of malware.

The proposed system (Li *et al.* 2018) is based on permission and detects malware by using only the effective permission out of all the permissions in the android manifest file. The system uses only 22 permissions for classification and achieved above 90% accuracy. The proposed system (Lilian *et al.* 2019) uses a meta-ensemble classifier to detect malware. The system uses a random forest classifier to build a meta-ensemble classifier to detect malware. The system took 660 samples of applications and get 97.5% accuracy in the detection of malware. The system uses 200 trees in random forests.

Similarly, the proposed system (Su *et al.* 2019) extracts permission and sensitive functions along with the native permissions and intent priorities for static analysis. It runs the application in a sandbox to capture and record the behavior of the application and store it. The system uses Naïve Bayes, SVM, KNN, logistic regression, etc classifiers for the classification of the malware and achieves 93% accuracy in detection.

The paper (Rovelli *et al.* 2020) introduced a system called PMDS (Permission-based Malware Detection System) which uses the permissions of the application as a behavior marker and based on the marker machine learning classifier is modeled for the detection of the malware. They use 2950 samples of benign and malware samples and achieved 94% accuracy in the detection of new malware with a 3.93% of false positive rate. Similarly, the paper (Qiao *et al.* 2016) proposed a system that decompiled the android application for the extraction of dex byte code and later it is converted into java source code for the extraction of the API calls of the application. They figure out that the manifest file of the android application may seem like an over-approximation because malware usually requests excess permission from the device. They achieve a detection accuracy rate of 94.98% when they use Random Forest, SVM, and ANN.

Similarly, paper (Sato *et al.* 2019) proposed a system that uses the Androidmanifest.xml file and analyses it. It extracts 6 different information items from the manifest file such as permission, intent filter (action), process name, etc. It uses 365 android applications in the database and achieves 90% accuracy. Similarly, a paper (Sandeep 2019) uses an Exploratory Data Analysis approach which uses deep learning techniques for identifying malware or benign application during its installation process in the device. The framework proposed uses permission to replicate the intention of the application and the system gets 94.6% accuracy in malware detection when this framework uses RF as a classifier.

The system proposed (Kandukuru *et al.* 2019) in this paper uses a hybrid model for malware detection. The system uses a decision tree to model the classifier which classifies malware and benign application. The system uses 180 applications (120 malware and 60 benign) and achieves an accuracy of 95% in the detection of malware.

The proposed system (Millar *et al.* 2020) uses deep learning Discriminative Adversarial Network (DAN) which classifies whether an application is obfuscated or not and then detects whether that application is malware or benign application. The system uses three sets of features, 1) Permissions, 2) API calls, and 3) Raw opcodes. The system takes 68.880 samples of applications and achieves 97.8% accuracy in the detection of malware. Similarly, the paper (Suarez-Tangil *et al.* 2017) introduced a system which is called as DroidSeive. The system uses several features like API calls, code structure, permissions, etc for the classification of the malware. The framework detects the malware and later classifies the family of the malware it belongs to. The system uses over 1 lakh of android applications and they achieve 99.44% of accuracy in detection and achieve 99.265 accuracy in the classification of malware with zero false positive rates.

The paper (Sun *et al.* 2021) proposed a tool that needs minimal human interaction and they named the tool WaffleDetector. They use sensitive API calls along with the permissions and get achieved approx 97.14% accuracy. Similarly, the paper (Wang *et al.* 2019) shows a Multilevel Permission Extraction(MPE) approach. This MPE approach extracts and identifies the permissions of the android application and then helps in identifying whether the application is benign or malware based on the permissions interaction. Their approach shows a 97.88% of accuracy in the detection of malware.

The paper (Wu *et al.* 2012) proposed a framework that uses the K-means algorithm for building a classification model and then uses singular value decomposition for clustering, and finally uses the KNN algorithm for the classification of the malware. The system uses permissions along with the intent and API calls and achieves an accuracy of 97.87%. The paper (Ping *et al.* 2021) introduces a system that uses contrasting permission for the detection of malware applications. In this system, a classifier called Enclamald is used for malware detection. The paper proposed a new metric called dism which is used for the classification between malware and benign application. It achieves 94.38% accuracy in the detection of malware.

Similarly, the paper (Zhu *et al.* 2018) proposed a framework that uses system events, sensitive APIs, and permission from the application for the calculation of the application permission rate based on the feature set. Cheng's framework is cost-effective and achieves 88.6% accuracy in identifying the malware.

3 CONCLUSION

In today's world, android smartphones are becoming a very easy target for hackers because of their easy access and lack of security. Hackers use various approaches to take control of the victim's phone through malware applications that can be downloaded from any third-party AppStore. The most common approach malware uses nowadays is via permissions of the application. They request permissions from the user via the malware application and access the device from it. Much research was done in this area to detect malware in android taking the permissions of applications.

In this paper, we discussed and review about 25 different paper which uses static techniques for the detection of malware in android using permissions as their primary feature for classification. We discussed their performances and the techniques they used for the classification of malware. In most of the papers we reviewed and discussed, we found that they are using imbalanced datasets i.e some have more benign applications than malware applications in their dataset and some have an insufficient amount of applications in their dataset. We found that they used all the android permissions available in the manifest file rather than using only those permission combinations which is significant for malware detection and can be optimized by using feature selection.

REFERENCES

Almin, SB.; Chatterjee, M. 2015. A Novel Approach to Detect Android Malware. *ICACTA(2015)* 407–417. Elsevier.

Arp, D.; Spreitzenbarth, M.; Hubner, M.; Gasco, H.; Rieck, K.; 2014. *Drebin: Effective and Explainable Detection of Android Malware in Your Pocket.* Elsevier.

Bai, H.; Xie, N.; Di, X.; Ye, Q. 2020. *FAMD: A Fast Multi-feature Android Malware Detection Framework, Design, and Implementation.* IEEE.

Cai, L.; Li, Y.; Xiong, Z. 2019. *JOWMDroid: Android Malware Detection Based on Feature Weighting with Joint Optimization of Weight-Mapping and Classifier Parameters.* Elsevier.

Chebyshev, V. (2021) IT Threat Evolution in Q2 2021. Mobile Statistics. Available at: https://securelist.com/it-threat-evolution-q2-2021-mobile-statistics/103636/.

Durmus, OS.; Kural, OE.; Akleylek,S.; Akleylek, E. 2021. *A Novel Permission-based Android Malware etection System Using Feature Selection Based on Linear Regression.* Springer-Verlag London Ltd.

Feizollah, A.; Anuar, N.B.; Salleh, R.; Suarez-Tangil, G.; Furnell, S 2017. AndroDialysis: Analysis of Android Intent Effectiveness in Malware Detection. *Comput. Security.*

Jannath. N.O.S.; Bhanu, S.M.S. 2018. S.M.S. Detection of Repackaged Android Applications Based on Apps Permissions. In *Proceedings of the 2018 4th International Conference on Recent Advances in Information Technology (RAIT)*, Dhanbad, India.

Kandukuru, S.; Sharma,RM. 2019. *Android Malicious Application Detection Using Permission Vector and Network Traffic Analysis.* Springer-Verlag Berlin Heidelberg.

Li, J. Sun, L.; Yan, Q.; Li, Z.; Srisa-An, W.; Ye, H. 2018. Significant Permission Identification for Machine-Learning-Based Android Malware Detection. *IEEE Trans. Ind. Inform.*

Lilian, D. 2019. Feature Selection and Ensemble of Classifiers for Android Malware Detection. *IEEE.*

Millar, S.; McLaughlin, N.; Rincon, J. 2020. DANdroid: A Multi-View Discriminative Adversarial Network for Obfuscated Android Malware Detection. *Tsinghua Science and Technology. Mobile Operating System Market Share Worldwide | Statcounter Global Stats 2022.* Available at: https://gs.statcounter.com/os-market-share/mobile/worldwide/.

Ping, X.; Xiaofeng,W. 2021. *Android Malware Detection with Contrasting Permission Patterns*, Elsevier.

Qiao, M.; Sung, A.H.; Liu, Q 2016. Merging Permission and API Features for Android Malware Detection. In *Proceedings of the 2016 5th IIAI international Congress on Advanced Applied Informatics (IIAI-AAI)*, Kumamoto, Japan.

Rovelli, P.; Vigfusson, Y. 2020. *Permission-Based Malware Detection System.* Springer International Publishing Switzerland.

Sandeep, H.R.2019. Static Analysis of Android Malware Detection Using Deep Learning. In *Proceedings of the 2019 International Conference on Intelligent Computing and Control Systems (ICCS)*, Secunderabad, India. pp. 841–845.

Sanz, B.; Santos,I.; Laorden,C.; Pedrero,XU.; Bringas,PG.; Alvarez,G. 2019. *PUMA: Permission Usage to Detect Malware in Android.* Springer Berlin.

Saracino, A.; Sgandurra,D.; Dini, G.; Martinelli, F. 2020. *MADAM: Effective and Efficient Behavior-based Android Malware Detection and Prevention.* Springer Berlin Heidelberg.

Sato, R.; Chiba, D.; Goto, S. 2019. *Detecting Android Malware by Analyzing Manifest Files*, Springer International Publishing Switzerland.

Su, MY.; Fung, KT. 2019. *Detection of Android Malware by Static Analysis on Permissions and Sensitive Functions.* IEEE.

Suarez-Tangil, G.; Dash, S.K.; Ahmadi, M.; Kinder, J.; Giacinto, G.; Cavallaro, L. 2017. Droidsieve: Fast and Accurate Classification of Obfuscated Android Malware. In *Proceedings of the Seventh ACM on Conference on Data and Application Security and Privacy*, Scottsdale.

Sun, Y.; Xie, Y.; Qiu, Z.; Pan, Y.; Weng, J.; Guo, S 2021. Detecting Android Malware Based on Extreme Learning Machine. In Proceedings of the 2021 IEEE 15th Intl Conf on Dependable, *Autonomic and Secure Computing, 15th Intl Conf on Pervasive Intelligence and Computing, 3rd Intl Conf on Big Data Intelligence and Computing and Cyber Science and Technology Congress (DASC/PiCom/DataCom/CyberSciTech)*, Orlando, FL, USA.

Talha, KA.; Alper, I.; Aydin, C. 2015. *APK Auditor: Permission-based Android Malware Detection System.* Elsevier.

Wang, Z.Li, K.Hu, Y.Fukuda, A.Kong, W. 2019. Multilevel Permission Extraction in Android Applications for Malware Detection. *International Conference on Computer, Information and Telecommunication Systems (CITS).*

Wu, D. 2012. Droidmat: Android Malware Detection Through Manifest and API Calls Tracing. In *Proceedings of the 2012 Seventh Asia Joint Conference on Information Security*, Washington, DC, USA.

Zhu, H. 2018. DroidDet: Effective and Robust Detection of Android Malware Using Static Analysis Along with Rotation Forest Model. *Neurocomputing*, 272, 638–646.

Artificial Intelligence, Blockchain, Computing and Security – Dagur et al. (Eds)
© 2024 The Author(s), ISBN: 978-1-032-49393-0

A two-way online speech therapy system

Monika Garg*, Mohini Joshi* & Anchal Choudhary*
Meerut Institute of Engineering and Technology, Meerut, U.P., India

ABSTRACT: Networked accent treatment application can be used as a personalised treatment system & can work to solve various kind of problems. Potential of internet is infinite. In today's time with the expansion of knowledge and technology we can solve any challenge in life with perfection. The purpose behind this research is to help those people who have speech disorders like stuttering. In this we use python libraries and modules to operate different functions like for speaking the text "speak" function and modules like "gtts", "speech recognition", "play sound" and "os" etc. Performing therapy by repeating incorrect word spoken by human until is corrected is main concept of this system. So that human get perfection in their word's pronunciation.

Keywords: Language Therapy, Recommendation System, Artificial Intelligence

1 INTRODUCTION

Speech and Language Disorders: Speaking is the method of conversing particular voice and sound that convey meaning to person who is listening voice. Speech disorder points out situation in which person's capability affectedly speak voice. Human communicate their thoughts, feelings, emotions and ideas by communicating with another human[1,2].

Multiple parts of body including the head, chest, neck, and abdomen used in speaking voice. Human's way of speaking affect by speech disorder to produce voice making them capable to speak so they can interact other humans. Speech disorder are not like text speak disability. Text speaking disorder prevent human from forming right word sounds, while on the other hand text voice disorders make people unable to understand words what is said by the other people[2]. Nowadays there are many families who are rich and have enough money to get resources in order to do the conversations therapy for the child or for adults. But woefully there exist many scenarios in which not all people can afford this type of anthropomorphise medicament that take into account skilled and experienced person who will work alongside a few hours each day with everyone so that they can assure him right feedback given by speech therapy skilled person[4]. As a result, we can use machine oriented mobile device or personal computer, look appropriate. That will not show less engagement of individual skilled expert in operate disordered person medicament to provide daily routine exercise that are important for giving nursing. Process, has various benefits to cure person from isolated community's department by using modern technologies with solution executed by the patient with disorder itself. When we discussed about this to the specialist in speech therapy method, right opinion is finest way it is not good to make this type solutions. The rules will be change according to individual[5]. So, making networked speech therapy application solution is chosen for treatment implementation and monitoring. The application will work as good text acceptance solution is necessary to identify the right pronunciation and make that incorrect word should be changed.

*Corresponding Authors: monika.garg.cs.2019@miet.ac.in, mohini.joshi.cs.2019@miet.ac.in and anchal.choudhary@miet.ac.in

DOI: 10.1201/9781003393580-105

2   LITERATURE REVIEW

On the basis of[4], digital platforms are so much capable that can help people who face problems in speaking. For this[8] organised an interactive workshop based on digital application. These solutions helped human with disabilities by working together with their parents. Aim of this initiative is transferring information to make research better, make good changes in life of human & their families.

Related to educational platforms, [10] researched how problems is treated in many forms online, effect on instructing and schooling by the use of digital platforms. [8] describe how internet can be used as a right platform in acknowledge the problem and providing the solution.

Medicament of speech disorders during childhood is essential. Their solution starts with home speech therapist. But with time it faces many challenges like the challenged person's lack of interest, demotivation, lack of proper skilled speech therapist[11]. Then in digital era like today, many technologies came so we can easily move towards the online solution for this like digital games helping the challenged person in a more interactive way[2]. As children are not part of many research because most of researches are based on adults. Variations are very high and is more dynamics in children's speech(ASR) as compared to adults. Length of signals of particular vowels differ with different human & main reason for this is different speed of speaking. It can be improvised by extending no. of raw data facts[11,14].

Due to excitement, vowel(o & u) can be spoken in similar way because human open mouth wider in this case than usual. Emotions & stress are some factors which can control the way of speaking of vowels. Vowels like o & u connected with opening and closure of mouth, because of this these vowels speaking is quite similar. Speech recognition is more related to speech analysis. To perform better speech signals should be properly coded[7]. There are many automation technologies that can cure text and speech problems and can be used for practice word ability e.g., Online Games. Like some digital games are otsimo, stamurai-stuttering therapy etc. But there are also some limitations with this digital solution as well like: there were lot of obstacles highlighted in research-covers sight of annoyance & modest self-confidence after failures in task, background interruption will cause reduce task performance, and confliction between stages of task. Person who not able to speak properly does not easily read words or sentences due to insufficient details etc. And one more drawback is that these games are one way leading this solution less efficient. This online speech therapy system solve these problems and because of this is two-way system and automated, it increase the accuracy of therapy and give better results[7,8].

It perform speech therapy & give analysis report if spoken word correct and then suggest the accurate therapy for the same. In this way system add new features to previous ones removing drawback and helping human in medicament of speech disorders.

3   PROPOSED METHODOLOGY

General Architecture of proposed system-Working of system starts with register or login into system then chose the exercise as per the human need. If incorrect word spoken by human then system recommend to repeat word until the word is corrected. If word is correctly pronounced then move on to next word and practice more words to achieve perfection.

Algorithm Used-

Step 1. Initialise the system and run the code to practice words.
Step 2. Input first word to practice
Step 3. **While** the user has not yet entered "Exit"
> Start speaking the suggest word as input. /* then system try to recognise human voice using python "gtts" module and provide feedback to human whether the spoken word is right or not*/
> **If** word is incorrectly spoken

Perform exercise on the specific word. //repeat same word until correctly spoken. **else**
Move on to the next word and perform entire process again
Endif
End While

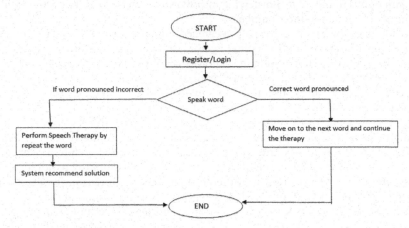

Figure 1. General architecture of model.

4 RESULTS

By taking different no. of words in each set and checking whether system correctly recognise the word and provide correct feedback about the word to user or not and on the basis of this feedback, calculating the precision.

Table 1. Calculating precision.

INPUT	Correctly Pronounced (C)	Incorrectly Pronounced (W)	Precision (P) = [C/(C+W)]*100
		SET 1:	
1.Apple	C		C = 3
2.Cow		W	W = 7
3.Boy	C		P = [3/(3+7)]*100
4.Bad		W	= 30%
5.Right		W	
6.Python		W	
7.Ball		W	
8.Green		W	
9.Ram	C		
10.Cat		W	
		SET 2:	
1.Python			C = 5
2.Car	C		W = 7
3.Cat	C		P = [5/(5+7)]*100
4.Dog		W	= 41.66%
5.Practice		W	
6.Good		W	
7.Continue	C		
8.Hello		W	

(continued)

716

Table 1. Continued

INPUT	Correctly Pronounced (C)	Incorrectly Pronounced (W)	Precision (P) = [C/(C+W)]*100
9.Good to go		W	
10.Direction		W	
11.Class	C		
12.Ball	C		
SET 3:			
1.Boy	C		C = 4
2.Class	C		W = 3
3.Apple		W	P = [4/(4+3)]*100
4.Continue	C		= 57.14%
5.Red		W	
6.Blue	C		
7.Sky		W	
SET 4:			
1.Pyhton	C		C = 8
2.Black	C		W = 2
3.Sand	C		P = [8/(8+2)]*100
4.Sun		W	= 80%
5,Star		W	
6.Ball	C		
7.Blue	C		
8.Cow	C		
9.Apple	C		
10.Cat	C		

Graphical Analysis-Hence we can see in graph when we provide same word in different set its precision increases which means that system helps in improving human way of speaking words. But when provide a new word first system analysed wrong if human speak incorrect word then after repeatedly speaking word again make system understand word correctly by human which at the end affect system efficiency.

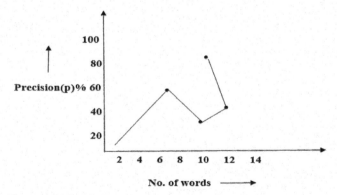

Figure 2. Graphical representation.

5 CONCLUSION

The humanize system is made with the use of compatible manner. This makes us to believe that our life cycle growth can be flexible and give us a better point of view. The use of front-line technology simplifies the architecture of software and development which provide

717

perception of learning continuous rules of method from experienced selected individualize situation. As per gathered result information will be transfer with familiar solutions to expand demand of machine-based speech therapist solution quickly. This research also hold some future scope in the field of digitalisation of solution in medical fields and make easy for the human to treat themselves at easier level as well as on their own also to some extent.

REFERENCES

[1]   Zourmand A. and Nong T. H., "Intelligent Malay Speech Therapy System," *2017 IEEE Conference on Systems, Process and Control (ICSPC)*, 2017, pp. 111–116, doi: 10.1109/SPC.2017.8313031.

[2]   Quisi-Peralta D., Robles-Bykbaev V., López-Nores M., Chaglla-Rodriguez L. and Chiluisa-Castillo D., "Recommendation System of Authorities and Content Based on Twitter for Language Therapy Through Data Mining Techniques," *2018 IEEE Biennial Congress of Argentina (ARGENCON)*, 2018, pp. 1–6, doi: 10.1109/ARGENCON.2018.8646316.

[3]   Georgopoulos V. C., "An Investigation of Audio-visual Speech Recognition as Applied to Multimedia Speech Therapy Applications," *Proceedings IEEE International Conference on Multimedia Computing and Systems*, 1999, pp. 481–486 vol.1, doi: 10.1109/MMCS.1999.779249.

[4]   Yang C.-H., Chang P.-H., Lin K.-L. and Cheng K.-S., "Outcomes Comparison Between Smartphone Based Self-learning and Traditional Speech Therapy for Naming Practice," *2016 International Conference on System Science and Engineering (ICSSE)*, 2016, pp. 1–4, doi: 10.1109/ICSSE.2016.7551624.

[5]   Attawibulkul S., Kaewkamnerdpong B. and Miyanaga Y., "Noisy Speech Training in MFCC-Based Speech Recognition with Noise Suppression Toward Robot Assisted Autism Therapy," *2017 10th Biomedical Engineering International Conference (BMEiCON)*, 2017, pp. 1–5, doi: 10.1109/BMEiCON.2017.8229135.

[6]   Basiron H., Azmi M. A., Abd Latif M. J., Kamaruddin A. I., Zaidi A. I. M. and Badrulzaman W. M. F. W., "Development of Speech Therapy Mobile Application for Speech Disorder Post-Stroke Patients," *2021 IEEE 11th International Conference on System Engineering and Technology (ICSET)*, 2021, pp. 130–133, doi: 10.1109/ICSET53708.2021.9612432.

[7]   Palacios-Alonso D., Meléndez-Morales G., López-Arribas A., Lázaro-Carrascosa C., Gómez-Rodellar A. and Gómez-Vilda P., "MonParLoc: A Speech-Based System for Parkinson's Disease Analysis and Monitoring," in *IEEE Access*, vol. 8, pp. 188243–188255, 2020, doi: 10.1109/ACCESS.2020.3031646..

[8]   Diogo M., Eskenazi M., Magalhães J. and Cavaco S., "Robust Scoring of Voice Exercises in Computer-based Speech Therapy Systems," *2016 24th European Signal Processing Conference (EUSIPCO)*, 2016, pp. 393–397, doi: 10.1109/EUSIPCO.2016.7760277.

[9]   Balaji V and Sadashivappa G., "Speech Disabilities in Adults and The Suitable Speech Recognition Software Tools - A Review," *2015 International Conference on Computing and Network Communications (CoCoNet)*, 2015, pp. 559–564, doi: 10.1109/CoCoNet.2015.7411243.

[10]   Mehra, P. S., Jain, K., Chawla, D., Dagur, A., Singh, S., & Sharma, J. (2022). *GWO-EFUCA: Grey Wolf Optimisation and Fuzzy Logic Based Unequal Clustering and Routing Protocol for Sustainable WSN-based Internet of Things.*

[11]   Mehra, P. S., Mehra, Y. B., Dagur, A., Dwivedi, A. K., Doja, M. N., & Jamshed, A. (2021). COVID-19 Suspected Person Detection and Identification Using Thermal Imaging-based Closed Circuit Television Camera and Tracking Using Drone in Internet of Things. *International Journal of Computer Applications in Technology*, 66(3–4), 340–349.

[12]   Grossinho A., Guimarães I., Magalhães J. and Cavaco S., *"Robust Phoneme Recognition for a Speech Therapy Environment,"* 2016 *IEEE International Conference on Serious Games and Applications for Health (SeGAH)*, 2016, pp. 1–7, doi: 10.1109/SeGAH.2016.7586268.

Artificial Intelligence, Blockchain, Computing and Security – Dagur et al. (Eds)
© 2024 The Author(s), ISBN: 978-1-032-49393-0

Comparative analysis of electronic voting methods based on blockchain technology

Zarif Khudoykulov, Umida Tojiakbarova*, Ikbola Xolimtayeva & Barno Shamsiyeva
Department of Cryptology Tashkent University of Information Technologies, Tashkent, Uzbekistan

ABSTRACT: In recent years, electronic voting has become a very popular and topical topic. Electronic voting technology can speed up ballot counting and provide accessibility for voters with disabilities. Electronic voting can also facilitate electoral fraud, especially given the risks associated with remote voting. Building a secure electronic voting system that offers the fairness and privacy of current voting schemes, while providing the transparency and flexibility offered by electronic systems has been a challenge for a long time. In this work-in-progress paper, we evaluate an application of blockchain as a service to implement distributed electronic voting systems. The paper proposes a novel electronic voting system based on blockchain that addresses some of the limitations in existing systems and evaluates some of the popular blockchain frameworks for the purpose of constructing a blockchain-based e-voting system. In particular, we evaluate the potential of distributed ledger technologies through the description of a case study; namely, the process of an election, and the implementation of a blockchainbased application, which improves the security and decreases the cost of hosting a nationwide election.

Keywords: electronic voting, blockchian, security, confidentiality, hash function, algorithm

1 INTRODUCTION

Today, modern technologies play a major role in the development of all industries. As technologies perform different functions based on their task, they are selected and used based on this feature.

E-voting opportunities are becoming widespread in remote election processes. Choosing blockchain technology in electronic voting ensures data security, limits third-party interference, and ensures data confidentiality and integrity. Blockchain technology can be used not only for electronic voting, but also for collective voting and ensuring anonymity.

2 METHODS

In this article, the blockchain technology, system organization processes, general concepts and the analysis of the scientific works carried out by world scientists are considered in electronic voting. In the article of Laxmi Ashrit *et al.*, the definition of e-voting, types of e-voting and their schemes, applications of e-voting system, disadvantages and advantages of e-voting etc. are given [1].

In the article by W. J. Douglas *et al.*, the problems and benefits of the election for student voting in the state of Iowa, USA, and additional suggestions for software tools are given [2].

Below are the issues raised in the article:

1. Software errors in calculators when counting bluetons;
2. vote falsification;

---

*Corresponding Author: umidatojiakbarova@mail.ru

DOI: 10.1201/9781003393580-106

3. vote on someone's behalf through another ID;
4. weakness, i.e. use of ready-made programming tools that are not suitable for the activity of the enterprise-organization.

A.A. Shmanov's article suggests three different ways to eliminate the possibility of falsifying traditional voting results [3]:

1. to divide voting bulletins into two copies;
2. to ensure randomness, i.e. the random appearance of the voter when receiving bulletins for candidates;
3. There should be special symbols to mark the given candidates in the bulletin [3].

In the article of Umeh MaryBlessing, the function of electronic voting system, authentication process, technologies used for authentication (Smart Card), fingerprint based on biometric parameters are also presented [4].

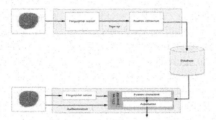

Figure 1.   Implementation of electronic voting system through fingerprint.

Muhammad Adeel Javaid's article presents the main security issues in electronic voting systems, especially the security threats for DRE voting systems. Suggestions for security issues are presented [5].

In the article of Emad Abu-Shanab, 302 students studying at the bachelor's level in the country of Jordan participated in the case related to the election process of the university council, as a result, this convenience was liked by many individuals and created the following opportunity [6]:

1. cost reduction;
2. increase opportunities for participation and voting;
3. speed and accuracy of placing and counting of votes;
4. Accessibility for the disabled.

In the article of Sarah Al-Maaitah, the role of blockchain technology, the types of consensus algorithms, the advantages and disadvantages of the technology, the types of blockchain technology (public, private) are also presented [7].

The proposed system consists of three elements divided into three stages:

1. the initial stage;
2. voting stage;
3. calculation and verification stage.

In Michał Pawlaka's article, among the problems of electronic voting, lack of transparency and auditing, solutions of blockchain technology are mentioned [8].

In the article of Camilo Denis Gonzalez, the shortcomings and problems of the traditional voting system, proposals for increasing the efficiency of the system in electronic voting based on blockchain technology, the issues of ensuring anonymity and reliability in enterprises-organizations are also discussed, and the NFT model is proposed for these cases done [9].

In Baocheng Wang's article, the weaknesses in electronic voting, and to prevent them, it is proposed to use blockchain technology. consists of checking the giving. Homomorphic algorithm from cryptographic algorithm was used in electronic voting [10].

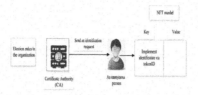

Figure 2.   The process of working the NFT model.

In Marianne Dengo's article, the shortcomings of traditional voting, the architecture of blockchain technology, the SHA-256 algorithm, the Ring signature algorithm used to ensure anonymity, the use of smart cards when blockchain technology is not used in the electronic voting process, and voting based on the Zcash protocol are discussed. mentioned [11].

In Christopher Andrew Collord's article, the working mechanism of the electronic voting process, the RSA algorithm for data exchange and the El-Gamal algorithm for stronger security are used, the election injection attack theory is also presented, the OpenSSL library is used for packet exchange, the voting interface, votes the calculation process is also mentioned, besides the network traffic is analyzed [12].

In the article of Maitha Ali Mohammed Hamad Al Ketbi, the function of blockchain technology, its types, management, monitoring and evaluation of identification and access, keys and certificates, network and vulnerabilities, risks related to the election process and methods of using blockchain types to prevent them are analyzed. done [13].

In Shorouq Alansari's article, the functions of cloud and blockchain technology, management of identification and access control based on blockchain technology, safe data exchange, serve to ensure the security and transparency of information based on the SeTA model [14].

Figure 3.   Authentication process.

Chang-Hyun Roh's article proposes the use of blockchain technology to prevent ballot fraud and theft, the process of blockchain technology operation, and the algorithms used in it are presented [15].

In the article of Kashif Mehboob Khan, the implementation of electronic voting method and improvement problems through blockchain technology, the architecture of the proposed electronic voting system is also mentioned [16].

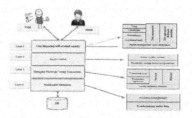

Figure 4.   Proposed electronic voting system architecture

In the article of Yousif Abuidris, the use of blockchain technology in electronic voting, general information about the consensus algorithm used in blockchain technology is presented, and the **PSC-BCHAIN** model is also proposed [17].

In the article of Noha E, El-Sayad, the general concepts of the electronic voting process, the use of facial images for authentication, and the algorithm of voting for electoral candidates are presented [18].

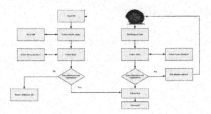

Figure 5.   Online voting system based on face recognition using Gabor filter.

Table 1.   Comparative analysis of articles.

№	The author of the article	Working principle of electronic voting method	The issue of authentication in electronic voting	Blockchain technology working principle	Use of blockchain technology in electronic voting	Advantage	Disadvantage	Algorithm used	Ensuring security	The proposed method, or the price of the offer	
										Cheap	Expensive
1	Laxmi Ashrit	+	-	-	-	Electronic voting has more options than traditional voting	Only types of electronic voting have been analyzed	-	-	-	-
2	Duglas V. Jons,	+	+	-	-	A software product designed for electronic voting, a compilation proposal	Submission of proposals in EV	-	Prevention of corruption within the government	+	-
3	A.A. Shmanov										
4	Umeh Maryblessing	-	+	-	-	Smart card based EV	The value of the proposed bids	RSA	Data security	-	+
5	Muhammad Adeel Javaid	+	+	-	-	Prevention of fraudulent voting in EVB	Lack of voter-approved Audit on DRE systems	MD5/RSA	Prevent threats in EVB	-	+
6	Emad Abu-Shanab	+	+			Accessibility for the disabled	It can be concluded that the number of elements and low strength and the reliability of measurements is a major weakness	Hash function	Data reliability	-	+
7	Sarah Al-Maaitah	-	-	-	+	Electronic voting analysis	-	Not quoted	Dasturiy vositalr xavfsizligi	Not quoted	Not quoted
8	Michal Pawlaka	-	-	-	+	Eliminate problems with paper voting	Whether authentication, privacy or data integrity is well ensured	Not quoted	Electronic voting system security	Not quoted	Not quoted
9	Camilo Denis Gonzalez	-	-	+	+	Based on the NFT Data Model, electronic voting offers low energy consumption	Not quoted	Consensus algorithm	Implementation of voting processes with high standards of auditing and security	Not quoted	Not quoted
10	Baocheng Wang	-	-	-	+	One-time digital signature during the voting process	Too many nodes in system operation	El-Gamal	Ensuring anonymity	-	+
11	Marianne Dengo	+	+	+	+	Electronic voting based on Zcash protocol and smart card	The cost of the proposed system	SHA-256	Ensuring anonymity	-	+
12	Christopher Andrew Collord	+	+	-	-	Network traffic analysis	Not quoted	RSA, El-gamal	Protection against injection attack	-	+
13	Maitha Ali Mohammed Hamad Al Ketbi	-	-	+	-	Application of blockchain technology in various systems	Not quoted	Not quoted	Risk assessment	Not quoted	Not quoted
14	Shorouq Alansari	-	-	+	-	Data exchange based on blockchain	Weakness of data verification mechanism	Hash function	Confidentiality	Not quoted	Not quoted
15	Chang-Hyun Roh	-	-	-	+	Prevent falsification or alteration of ballots	Not using a private blockchain platform	Consensus algorithm	Confidentiality	-	+
16	Kashif Mehboob Khan	-	-	+	-	Creating convenience for voters	Not quoted	Hash function	Data privacy	Not quoted	Not quoted
17	Yousif Abuidris	-	-	-	+	A PSC-blockchain model is proposed	Not quoted	Konsensus algoritmi	Confidentiality	-	+
18	Noha E, El-Sayad	-	+	-	-	Authentication in electronic voting	It is 89.3% using facial authentication method, and increasing its effectiveness requires time-consuming filter calculations.	Hash function	Data integrity	-	+

3 RESULTS

Above, the electronic transmission processes were introduced by various scientists. Based on their work, we can see their status, divided into categories, in the table below.

4 CONCLUSION

In this article, electronic voting systems, their working principles, disadvantages, advantages, used algorithms, authentication issues, as well as general concepts of blockchain technology, its working mechanism, application in industries, various processes of electronic voting based on blockchain technology Scientific works of different scientists were reviewed and analyzed. The result of the analysis shows that a lot of translation and knowledge has been acquired for the ongoing scientific work, which is used as a basis for application in scientific work.

REFERENCES

[1]   Laxmi Ashrit, "How Electronic Voting (e-Voting) Works – Types, Application & Advantage", Electronics & Communication, 2019.

[2]   Duglas V. Jons, *"E-Voting - Prospects and Problems"*, University of Iowa, Iowa City, Iowa April 13, 2000.

[3]   А.А. Шмонов, "Пособы устранения возможности фальсификации результатов тайного голосования", 2009.

[4]   Umeh Maryblessing, *"Design and Implementation of an e-voting System"*, Department of Electronic and Computer Engineering Namdi Azikiwe University Awka, November, 2019.

[5]   Muhammad Adeel Javaid, "Electronic Voting System Security", *SSRN Electronic Journal*, January 2014.

[6]   Emad Abu-Shanab, "E-voting systems: a tool for e-democracy", Management research and practice vol, 2010.

[7]   Sarah Al-Maaitah, "E-Voting System Based on Blockchain Technology: A Survey", *International Conference on Information Technology (ICIT)*, 2021.

[8]   Michał Pawlaka," Towards the Intelligent Agents for Blockchain e-voting System", ScienceDirect Procedia Computer Science 141 (2018).

[9]   Camilo Denis Gonzalez," Electronic Voting System Using an Enterprise Blockchain", *Applied Sciences*, 2022.

[10]   Baocheng Wang, "Large –scale Election Based On Blockchain", Procedia Computer Science 129 (2018) 234–237.

[11]   Marianne Dengo, "Blockchain Voting: A Systematic Literature Review", *University of tartu Institute of Computer Science Computer Science Curriculum*, 2020.

[12]   Christopher Andrew Collord, "Electronic Voting: Methods and Protocols" Computer Sciences Commons, 2013.

[13]   Maitha Ali Mohammed Hamad Al Ketbi, "Establishing a Security Control Framework for Blockchain Technology", *Interdisciplinary Journal of Information, Knowledge and Management*, 2020.

[14]   Shorouq Alansari, *"A Blockchain-based Approach for Secure, Transparent and Accountable Personal Data Sharing"*, Faculty of Engineering, Science and Mathematics School of Electronics and Computer Science, 2020.

[15]   Chang-Hyun Roh, "A Study on Electronic Voting System Using Private Blockchain", *Journal of Information Processing Systems*, 2020.

[16]   Kashif Mehboob Khan,*"Secure Digital Voting System based on Blockchain Technology"*, 2020.

[17]   Yousif Abuidris, *"Secure Large-scale E-voting System Based on Blockchain Contract Using a Hybrid Consensus Model Combined with Sharding"*, 2020.

[18]   Noha E, El-Sayad, "Face Recognition as an Authentication Technique in Electronic Voting", International Journal of Advanced Computer Science and Applications, Vol. 4 No. 6, 2013.

Artificial Intelligence, Blockchain, Computing and Security – Dagur et al. (Eds)
© 2024 The Author(s), ISBN: 978-1-032-49393-0

Insider threat detection of ransomware using AutoML

R. Bhuvaneswari*, Enaganti Karun Kumar, Annadanam Padmasini & K.V. Priyanka Varma
Department of Computer Science and Engineering, Amrita School of Computing, Amrita Vishwa Vidyapeetham, Chennai

ABSTRACT: Insider threats are one of the biggest issues that modern-day organizations and many large-scale companies face. The inside threats are caused by the insiders who are authorized individuals, also may have proper access to sensitive and secret information, and may be aware of the weaknesses in the implemented systems and operational procedures. And, it is confirmed that these insider threats have already shown their great damaging power in securing important information and because of this, organizations require more proactive and modern-day solutions to support existing cybersecurity tools in order to detect and prevent them. Conventional interruption recognition frameworks neglect to be powerful in insider threats because of the absence of a wide range of information for insider ways of behaving. All things considered; a more complex and standard technique is expected to have a more profound comprehension of the ways that insiders speak with the information framework. Ransomware hackers appear to be adopting new methods in response to tightening security. Automated machine learning has been utilised in this work for detection of ransomware and we also included the ensembling method for a few algorithms that are suitable for the given dataset.

Keywords: Ransomware, Machine Learning, AutoML, Ensembling, Data Features.

1 INTRODUCTION

A company's or association's internal operations are where a computer security threat begins. It commonly happens when a former or previous employee, business associate, merchant, or business colleague with genuine client qualifications exploits their admittance to the threat or causes damage to the association's organizations, frameworks, and information. An insider danger might be executed purposefully or unintentionally. Regardless of the aim, the outcome is compromised secrecy, accessibility, as well as uprightness of big business frameworks and information. Contrasted with the outsider attacks the insider attacks, the assaults from insiders are difficult to recognize in light of the fact that the malicious-minded insiders as of now have the approved permissions and ability to get to the inside data frameworks.

A new ransomware assault makes news almost every day. President Biden prioritized cybersecurity as the most important concern for national defense. Ransomware stood out on the news line all through 2021 and keeps on making the news in 2022. There are heard stories of attacks on enormous organizations, and associations, or maybe you as an individual have encountered a ransomware assault on your own gadget. Ransomware is a critical issue and a terrifying possibility to have user's files and information kept on locked and prevents them from accessing those files until some ransom/payment is paid up.

ML finds its application in various fields [1] ranging from security till medical application [2–4]. We will be using Automated ML to find out the ransomware. One of the most compelling reasons to adopt AutoML tools is their ability to greatly increase productivity. The more efficiently you can use your machine learning resources since time equals money, the better. The effectiveness is

---

*Corresponding Author: bhuvanacheran@gmail.com

DOI: 10.1201/9781003393580-107

dependent on choosing the appropriate algorithm for your machine learning model. You can automatically improve your algorithm with AutoML. Without the need for manual effort, it will transfer your data to each training method and choose the optimum design. This implies that algorithm selection may be finished quickly rather than requiring hours.

2 LITERATURE SURVEY

Telecommunications and computer networks are important for the sharing of information. Threats have increased as a result of an increase in valuable information and enabling technological advancements. These risks originate not just from the outsiders but also from within the organization. Such threats pose a significant security risk and are hard to identify. In addition, 27% of the firms that were polled said that internal attacks were the source of the attacks. Organizations should have an insider threat detection system that can identify and stop hostile insiders from spreading threats in order to ward against dangerous insiders. Unfortunately, not enough is known about insider threats.

In context of Opcodes, Zhang *et al.* suggested a technique for ransomware detection[5]. In their method, program Opcodes are obtained by utilizing the IDA Pro Tool. The optimal combination of Opcodes is then separated by the Frequency Inverse Document Frequency (F IDF) approach. At an end, the occurrence of these significant Opcode combinations is taken into account as a characteristic, and the detection model is created using Naive Bayes, Random Forrest, and K-Nearest Neighbor approaches.

Chandola *et al.*'s overview of anomaly detection methods covers a variety of application sectors [6]. The authors provided a discussion for a number of various application areas, such as image processing, fraud detection, medical behavior recognition, detection of intrusion. They divided the currently used approaches into many groups on the basis of distinct principles behind each approach since every area could use a separate set of detecting techniques [7]. Finally, they developed a set of directions for future research, including methods for contextual, collaborative, and distributed anomaly detection. Since it is appropriate for IoT contexts, for his research, they used an "online anomaly detection" type to evaluate the recognition approach.

Machine learning approaches from Jiang *et al.*'s survey[8] may be used in a variety of computer security fields, such as systems for detecting intrusions, application protocols, access control administration, detecting and preventing malware, and so on. They talk about both the insider danger and the larger onslaught that is producing security issues. They reviewed the objective and conceptual system elements in each study's attack detection system analysis, then chose the system's machine learning approach. As a consequence they suggested machine learning techniques centered on human and game theory for configuration. They also proposed a taxonomy of machine-learning utilized in various security areas[9]. The disadvantage of their study is that it does not address internal threats.

The production process of ransomware was determined by analytical model on Android and Windows platforms is depicted and analysis of Windows-based Ransomware are shown in the article. -It also shows how ransomware for Windows has evolved. -Cuckoo Sandbox and the MD5 technique are both utilized for malware investigation[10]. -The usage of RSA and AES for encryption. The primary goal is to identify the ransomware by keeping an eye on unusual file-system registry activity. - For the detection of Windows ransomware, PEID is employed. But there is a limitation in this work, To prevent personal data[11] with making unrepairable, end users must create periodic on- and off-line backup systems of each of their important data, including images. Aside from that, limits of current solutions and potential detection methods are yet investigated. In this research, we'll talk about how to categorize ransomware attacks and how to eliminate them from a company or organization using a fully Automated model.

3 PROPOSED WORK

Automated machine learning also called as AutoML is the technique of automating the challenging, time-consuming, repetitive tasks involved in constructing a machine learning technique. It helps data scientists, researchers, and developers to build ML algorithms featuring great scaling, reliability, and production spread while retaining model quality. When machine learning is automated, it becomes more user-friendly and frequently produces better, more accurate results than manually-coded algorithms, powerful ML model. Traditionally, machine learning, models are created manually, and each stage of the procedure has to be managed independently. Automated Machine Learning automatically identifies and applies the most appropriate ML model for a given task. AutoML does this using two ideas:

Neural network design is automated using neural architecture search. This assists AutoML models in discovering novel architectures for situations that require them.Transfer learning is the process through which previously trained models use fresh data sets to apply their knowledge. Transfer learning enables AutoML to adapt established structures to new situations that need their use.

The proposed architecture diagram for the models is represented in the Figure 1.

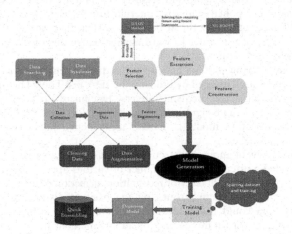

Figure 1.   Proposed threat detection architecture using AutoML.

3.1   Data pre-processing

The first process of the ML workflow is pre-processing data[12]. The workflow may be broken down into three parts: data collecting, removing noise in the data, and data augmentation. To create a new dataset or expand an existing dataset, data collecting is a vital first step. In order to protect model training in the future, noisy data is filtered out throughout the data cleaning procedure. Data augmentation is crucial for boosting model performance and resilience. The three components will be covered in more detail in the subsections that follow.

3.2   Data collection

Working with various dataset types is essential for success in the field of machine learning or to become a great data scientist. Finding a proper dataset for every type of machine learning project, however, is a complex and difficult job[13]. In data collection two types of methods are there to prepare a proper dataset. Those are:

3.3   Data searching

Searching for Web data is an easy technique to compile a dataset because the Internet is an endless source of data. However, employing Web data has several drawbacks. The search results might not initially match the terms precisely. As a result, irrelevant data must be filtered. Moreover, Web data might not be labelled at all or be labelled inconsistently. To address this issue, a learning-based self-labelling approach is frequently applied.

Google Dataset Search, a service for searching dataset repositories on the Web, was recently released. The reason for this is because there are large numbers of data repositories on the Internet that makes difficult-to-search datasets. In order to make their datasets better available, Dataset Search allows dataset providers to characterize their datasets using a variety of information (such as publisher, date of publication, method of data collection, and terms for using the data).

3.4 Data synthesis

Synthetic datasets, as the name implies, are "synthetic" that they have been produced by software processes as opposed to being generated from paperwork of real-world occurrences. Therefore, why should we even think about using fake data for ML models?

Synthetic datasets are highly useful when adequate actual data is impossible to collect. A further key benefit of using synthetic data is the capacity to accurately specify a range of characteristics, such as the range, structure, and level of noise present inside this dataset. It is important to not undervalue the possibility of copyright infringement or privacy problems being eliminated by employing synthetic data.

Furthermore, utilizing generated data sets has a number of significant disadvantages. So first, creating synthetic data is a labor-intensive engineering task, particularly when done by yourself or with a narrow sample. The possibility of adding bias to your data is the other risk. Presently, random data is not enough to train complex or extensive machine learning algorithms.

3.5 Cleaning data

Everything that is inputted must be outputted in ML algorithms. We may anticipate that the outcome will be trash if we input junk into our algorithm. Trash in this context refers to distortion in the statistics.

There will always be noise in the obtained data, but this noise might hurt the model's training. This is common when combining data from numerous sources or receiving data from different clients or organizations. As a result, if necessary, the data cleansing process must be carried out.

But every day, enormous volumes of data are produced in the actual world. In other terms, the question of the challenge of ongoing data cleaning one that merits research, particularly for businesses. In any dataset, outliers are virtues that are uncommon. These may influence your research and because they are so drastically different from previous datasets, shatter your preconceptions. Eliminating these is a decision which depends on the information you want to analyze. In practice, eliminating undesirable outliers will aid in enhancing the performance of the data you're using.

Strange naming practices, typing errors, and erroneous a few instances of systemic issues include capitalization. Any discrepancy will lead in wrongly categorized segments or labels.

When you have both "N/A" and "Not Applicable," it is a nice illustration. Despite the fact that they will both appear in different categories, they should both be assessed as one.

3.6 Data augmentation

To some extent, data augmentation (DA) can also be considered a data collection method. Data augmentation is a collection of ways for increasing the amount of data artificially by producing additional data points from current data[14]. Variations of this include applying deep learning techniques to generate fresh data or modifying small alterations to existing data. By providing more and different examples to trained data, data augmentation can improve the efficiency and outcomes of ML models. Whenever the collection is large and sufficient, a machine learning algorithm works well and is highly accurate.

For ML models, information gathering and categorization may be time-consuming and complex processes. By adopting methods of data augmentation to change data, businesses may save operational costs.

3.7 Future engineering

Feature Engineering is the next buzz word after big data. It is the technique of applying domain behavior of the information to produce features that allow ML model to succeed. The goal of feature engineering is to gather the most characteristics or characteristics possible from raw data for use by modelling techniques and algorithms, also simplifying, accelerating, and improving model correctness in data transformations[15]. There are three sub-topics within this namely selecting, extracting and constructing of features. Usually, feature construction is employed to enlarge initial feature spaces, feature extraction reduces by using specific matching technique, the complexity of characteristics and techniques, and feature selection reduces feature duplication by choosing significant features. The three components will be discussed in further depth in the following subsections.

3.8 *Feature selection*

Features are the inputs which we supply to our ML techniques. In our dataset, each column has a feature. We should guarantee that just the relevant properties are employed while training an optimal model. Several features might cause the model to gather unimportant structures and detect noise-based knowledge acquisition. The method for selecting our information's most key aspects is called feature selection. This is a method for automatically choosing the right traits for our ML approach that relies on the kind of problem it's trying to address. This is accomplished by adding or removing significant features without altering them. This helps reduce the amount of distortion in our dataset as well as the size of our input set of information. Depending on search strategy, a subset of characteristics is chosen and evaluated. Then, to establish whether the subset is legitimate, a validation technique is applied. The preceding stages are continued until the stop requirement is met.

3.9 *Feature extraction*

All of the information we get in real life is substantial. We need a mechanism to analyze this data. It is impossible to analyze them manually. The idea of feature extraction enters the picture at this point. A phase in the dimensionality reduction procedure called feature extraction separates and condenses a large initial collection of raw data into smaller, easier-to-manage groups. Computation is becoming less complicated as a result. The much more significant feature among all of these high volume datasets is that they include a wide variety of various characteristics. That many parameters require more computer resources to analyze and choosing which attributes to combine as features, proposed technique helps to effectively minimize the enormous amount of extracted data is the greatest asset from such large amounts of datasets. The true set of data should be properly & unmistakably described despite yet being easy to utilize.

3.10 *Feature construction*

Feature construction is a technique that identifies lacking data on the connections among characteristics and arguments in the attribute field by assuming rings or generating new features. Assuming there are p features A1, A2, ...,Ap, we may have extra q features after feature building. Ap+1,Ap+2, ...,Ap+q. As a result, no intrinsically new information is added through feature creation because all new created features are specified in terms of original features. The goal of feature creation is to augment the expressive power of the original features. Typically, the new feature set's dimensionality is increased and is larger than the previous feature set. The goal of feature construction is to change originally intended visualization area into a fresh set which can assist in more effectively achieving the goals of data mining, including increased accuracy, clear comprehension, true clusters that reveal hidden patterns, etc. It is difficult to analyze all alternatives individually. Therefore, to increase efficiency even further, some automatic feature construction techniques had been proposed to automate the procedure of looking for and examining the operation combination. It has been demonstrated that these methods produce results that are on par with or better than those produced by human competency.

3.11 *Model generation and selection*

The decision of which model to use is driven by a variety of factors, most obviously accuracy and computability—how accurate the predictions provided by the model are and how computationally demanding they are. The two components of model generation are search space and optimization techniques. The design ideas of neural networks are defined by the search space. Different circumstances call for various search areas. After the search space is established, in order to rapidly find the system model structure with superior efficiency, the architecture optimization (AO) technique explains how to guide the search. [16].

A model's performance must be assessed once it has been created. The most straightforward way to achieve strategy is to evaluate prediction accuracy on the validation data after training the algorithm to combine on the training dataset; nevertheless, this approach is time consuming and asset intensive. Some cutting-edge methods can expedite the procedure but compromise process integrity. So it makes sense to research how to balance an evaluation's efficiency and efficacy.

3.12 *Ensembling model*

It's possible that more than one model won't produce the most precise prediction for a given data set. ML model have limitations, making it challenging to develop a algorithm with high precision. By creating and mixing many models, we may increase precision of entire dataset and model.

ML techniques have issues with variance and/or bias. The difference here between anticipated value and the initial value produced by the model is the bias. Whereas variance is the variability in the model prediction. Because the basic model does not follow data patterns, it makes errors in predicting training and testing data, resulting in a model with high bias and high variance. Therefore, ensemble learning techniques are created to improve the effectiveness of the models and its accuracy. Multiple models are trained using machine learning algorithms in an ensemble, which is a machine learning concept[17]. It mixes the results from each algorithm with classifier that don't function well.to get the final prediction.

3.13 *Dataset*

We gathered a dataset of 138,045 samples, of which 70% are data from files that have been attacked by ransomware and the remaining 30% are ransomware-free. For each file, there are 54 features. We didn't apply any data synthesis process for our dataset.

4 IMPLEMENTATION AND RESULTS

In this paper, we suggested an efficient analytic approach for detecting ransomware. Auto ViML was created to construct High Efficiency Interpretable Models with its least number of required variables. The "V" in Auto ViML stands for Variant since it attempts many models with various features to identify the model that performs the best for one's dataset. The "I" in Auto ViML stands for "interpretable," since it chooses the fewest characteristics required to create a model that is easier to understand. Auto ViML often creates models with 20 to 99 percent less characteristics than a similarly performing model with all features present.

With this methodology, we achieved the best outcomes. In order to detect various encrypted files, we will immediately use We use this approach as our technique and focus on the analysis of better feature extraction through enhancement.

4.1 *Process flow of our model*

4.1.1 *Step 1*
The first step of our model is data analysis of our Ransomware dataset. Reading the dataset using "pandas" library, in our data set there are totally 138,045 samples, considering the 90% percent of dataset for training, in which 66.7% of samples are attacked by ransomware and remaining 33.3% of samples are legitimate. The distribution of our training dataset may be seen in the Figure 2.

4.1.2 *Step 2*
Next, performing the binary classification visualization, if there are any non-binary values, they will be changed to corresponding binary value. At first the data will be shuffled randomly before

training the dataset with a model. Now we have to perform hyper parameter tuning. Hyper-parameters are tuned to provide the best fit after learning model parameters from the data. Due to the time-consuming nature of finding the appropriate hyper-parameters and search techniques like random search and grid search are employed. Due to the size of our dataset, we selected RandomizedSearchCV, which is three times quicker than GridSearchCV[18]. Random The model is trained and scored using variations selected at random from a field of hyper - parameter values created by Search. This enables you to control the number of selected attributes that are tested directly. Timing or insufficient resources will affect how many times the search is iterated.

4.1.3 *Step 3*

If two variables in a dataset are highly correlated, what else should you eliminate and which one should you keep? The choice is not as simple as it appears. To eliminate strongly correlated features, Auto ViML employs the SULOV method[19].

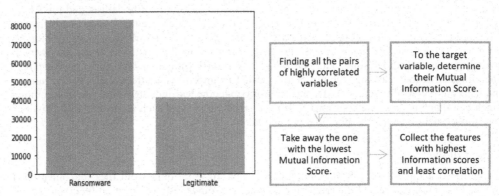

Figure 2.   Dataset analysis.   Figure 3.   Flow of SULOV method.

The process followed in the SULOV Algorithm is shown in below Figure 3. SULOV stands for Searching for Uncorrelated List of Variables. The algorithm operates in the ways listed below.

• The first step is to identify any pairs of strongly correlated variables that above a certain correlation threshold.
• Next, determine their Mutual Information Score with respect to the desired variable. A non-parametric scoring technique is called the Mutual Information Score. It is therefore appropriate for all types of variables and targets.
• The next step is to choose the correlated variable that has the lower Mutual Information Score for each pair of correlated variables.
• The last stage is to compile the results that have the least correlation among them and the greatest information scores.

From the Figure 4, After applying SULOV algorithm we can remove the two highly correlated variables:*['SectionsMeanEntropy', 'ResourcesMeanEntropy']*

4.1.4 *Step 4*

After SULOV has chosen variables with high mutual information scores and low correlation, we utilize XGBoost to determine the best features from the remaining variables.

After performing above mentioned steps to our training dataset, the importance of features is presented in the form of a graph as shown in Figure 5. For prediction, we choose the feature with the highest feature importance, i.e., "ImageBase."

Now with the selected feature, it will find the best threshold value for best F1 score and also gets the best parameters for model like learning rate, number of estimators. For our model the best parameters are: {'*gamma*': 21, '*learning_rate*': 0.08039991715696582, '*max_depth*': 7, '*n_estimators*': 214}. From the below graph the best threshold value for our model is 0.47 for which we got a balanced accuracy of 99.53 percent.

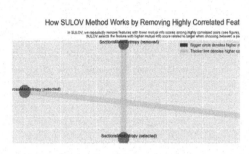

Figure 4. Finding highly correlated features.

Figure 5. Importance of each feature in dataset.

We already understand that a one algorithm might not produce the most accurate prediction. ML models have limitations, making it challenging to develop a model of high precision. By creating and mixing many algorithms, we may dramatically increase accuracy and precision of our model. So result for our ensembling model are as shown in Figure 6.

From the above image, we got the best results for Bagging Classifier in terms of accuracy scores and precision scores. Figure 7 show the classification report of our proposed model.

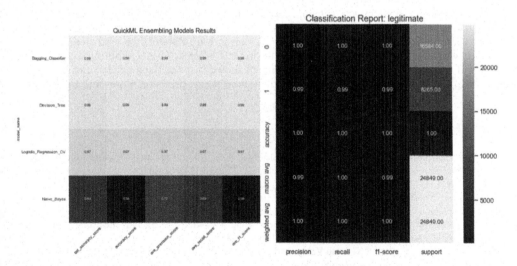

Figure 6: Ensembling model graph

Figure 7. Classification report for target variable "legitimate".

4.2 *Preventive measures to overcome ransomware attacks*

Fortunately, precautions can be taken to guard against this type of infection. The following are few strategies to assist safeguard the business.

- Perform early ransomware analyses: Perform risk evaluations and vulnerability scanning to examine the system vulnerabilities, current degree of cyber resilience, and readiness in terms of equipment, processes, and competencies to fend off attacks.
- Strengthen patch management: Always be on the lookout for flaws. To prevent attacks from employing security vulnerabilities to breach into networks and distribute ransomware, upgrade devices frequently with the appropriate security fixes. Analyze innovations in software and legislation that can enhance patched practises. by auditing them, and if possible, automate them.
- Utilize threat intelligence to supplement your efforts. Being up to date on advanced threat enables you to see attacks early, take swift action, and stop them in their tracks. You may employ actionable insights to restrict incoming traffic at the firewall by using this knowledge to determine the origin of a few of the assaults. Test, test, and test to ensure recovery and Isolate backup data.

5 FUTURE WORK

Cyberthreats are recently emerged as the most important problem affecting organizations' economies, particularly those that operate online. Government and traditional enterprises are among the groups that are being negatively impacted by this financial damage. Not only ransomware, there are so many different types of threats and ways emerging up nowadays. The proposed work can be extended to all kind of malwares and real time attacks.

6 CONCLUSION

Species of ransomware offer serious security risks and critical challenges to cybersecurity, and they might do significant damage to webpages, server farms, application servers. and android application used by a wide range of enterprises and organizations. It is necessary to create an intelligent machine to effectively categorize and identify ransomware, and lower the danger of harmful activities.. In this study, we introduced unique framework based on feature selection and used several machine learning techniques. On a ransomware dataset, we used the framework together with all of the trials, and we analyzed the models' efficiency using a thorough comparison of several techniques. The results of the experiment shown that the Bagging classifier performed better than other classifiers by obtaining the greatest accuracy, precision scores and F-Beta.

REFERENCES

[1] Jin, Wei. (2020). Research on Machine Learning and Its Algorithms and Development. *Journal of Physics: Conference Series*. 1544. 012003. 10.1088/1742-6596/1544/1/012003.
[2] Sarker, I.H. Machine Learning: Algorithms, Real-World Applications and Research Directions. *SN Comput. Sci.* 2, 160 (2021). https://doi.org/10.1007/s42979-021-00592-x
[3] Prabu kanna G, Abinash M.J, S, Suganya E, Sountharrajan S, Bhuvaneswari R, & K. Geetha. (2022). Identification and Diagnosis of Breast Cancer using a Composite Machine Learning Techniques. *Journal of Pharmaceutical Negative Results*, 13(4), 78–85.
[4] Bhuvaneswari, R., and S. Ganesh Vaidyanathan. "Classification and Grading of Diabetic Retinopathy Images Using Mixture of Ensemble Classifiers." *Journal of Intelligent & Fuzzy Systems Preprint* (2021): 1–13.
[5] Hanqi Zhang, Xi Xiao, Francesco Mercaldo, Shiguang Ni, Fabio Martinelli, Arun Kumar Sangaiah, Classification of Ransomware Families with Machine Learning Based on N-gram of Opcodes, *Future Generation Computer Systems*, Volume 90,2019, Pages 211–221, ISSN 0167-739X, https://doi.org/10.1016/j.future.2018.07.052.
[6] Chandola, Varun & Banerjee, Arindam & Kumar, Vipin. (2009). Anomaly Detection: A Survey. *ACM Comput. Surv..* 41. 10.1145/1541880.1541882.

[7] Kumar, A. & Alam, B. (2018). Task Scheduling in Real Time Systems with Energy Harvesting and Energy Minimization. *Journal of Computer Science*, 14(8), 1126–1133.

[8] Kumar, A., & Alam, B. (2014, February). Real Time Scheduling Algorithm for Fault Tolerant and Energy Minimization. In *2014 International Conference on Issues and Challenges in Intelligent Computing Techniques (ICICT)* (pp. 356–360). IEEE.

[9] Chaturvedi, R., Kumar, S., Kumar, U., Sharma, T., Chaudhary, Z., & Dagur, A. (2021). Low-cost IoT-enabled Smart Parking System in Crowded Cities. In *Data Intelligence and Cognitive Informatics: Proceedings of ICDICI 2020* (pp. 333–339). Springer Singapore.

[10] Y. Sharma, Shatakshi, Palvika, A. Dagur and R. Chaturvedi, "Automated Bug Reporting System in Web Applications," *2018 2nd International Conference on Trends in Electronics and Informatics (ICOEI)*, Tirunelveli, India, 2018, pp. 1484–1488

[11] Priyanka Singh, Samir Kumar Borgohain, Lakhan Dev Sharma, Jayendra Kumar, Minimized Feature Overhead Malware Detection Machine Learning Model Employing MRMR-based Ranking, *Concurrency and Computation: Practice and Experience*, 10.1002/cpe.6992, 34, 17, (2022).

[12] Agarwal, Vivek. (2015). Research on Data Preprocessing and Categorization Technique for Smartphone Review Analysis. *International Journal of Computer Applications*. 131. 30–36. 10.5120/ijca2015907309.

[13] Jingang Liu, Chunhe Xia, Haihua Yan, Jie Sun, "A Feasible Chinese Text Data Preprocessing Strategy", *2020 11th IEEE Annual Ubiquitous Computing, Electronics & Mobile Communication Conference (UEMCON)*, pp.0234–0239, 2020.

[14] Sivakumar, Ashwin and Ramalingam Gunasundari. "A Survey on Data Preprocessing Techniques for Bioinformatics and Web Usage Mining." (2017).

[15] Rawat, Tara & Khemchandani, Vineeta. (2019). Feature Engineering (FE) Tools and Techniques for Better Classification Performance. 10.21172/ijiet.82.024.

[16] Mehra, P. S., Mehra, Y. B., Dagur, A., Dwivedi, A. K., Doja, M. N., & Jamshed, A. (2021). COVID-19 Suspected Person Detection and Identification Using Thermal Imaging-based Closed Circuit Television Camera and Tracking Using Drone in Internet of Things. *International Journal of Computer Applications in Technology*, 66(3–4), 340–349.

[17] Kumar, A., & Alam, B. (2016). Real-Time Fault Tolerance Task Scheduling Algorithm with Minimum Energy Consumption. In *Proceedings of the Second International Conference on Computer and Communication Technologies: IC3T 2015, Volume 2* (pp. 441–448). Springer India.

[18] Dagur, A., Malik, N., Tyagi, P., Verma, R., Sharma, R., & Chaturvedi, R. (2021). Energy Enhancement of WSN Using Fuzzy C-means Clustering Algorithm. In *Data Intelligence and Cognitive Informatics: Proceedings of ICDICI 2020* (pp. 315–323). Springer Singapore.

[19] Dagur, A., Kaushik, A., Rastogi, A., Singh, A., Kumar, A., & Chaturvedi, R. (2021). Optimization of Queries in Database of Cloud Computing. In *Data Intelligence and Cognitive Informatics: Proceedings of ICDICI 2020* (pp. 325–332). Springer Singapore.

733

Artificial Intelligence, Blockchain, Computing and Security – Dagur et al. (Eds)
© 2024 The Author(s), ISBN: 978-1-032-49393-0

Multispectral image processing using ML based classification approaches in satellite images

V.V. Satyanarayana Tallapragada
Department of Electronics and Communication Engineering, Mohan Babu University, Tirupati, Andhra Pradesh, India

G. Venkatesan
Department of Civil Engineering, Saveetha Engineering College, Chennai, Tamil Nadu, India

G. Manisha
Department of Computer Science and Engineering, R.M.D Engineering College, Kavaraipettai, Tamil Nadu, India

N. Sivakumar
Department of Computer Science and Engineering, Panimalar Engineering College Chennai City Campus, Chennai, Tamil Nadu, India

Ashok Kumar
Department of Computer Science, Banasthali Vidyapith, Rajasthan, India

J. Karthika
Department of Electrical and Electronics Engineering, Sri Krishna College of Engineering and Technology, Coimbatore Tamil Nadu, India

ABSTRACT: Many academics now choose to use intelligent algorithms over traditional ones when dealing with huge amounts of data, thanks to the introduction of technologies that resemble nature. Among these algorithms, the generative adversarial network (GAN) is one of the most important. The purpose of this research is to investigate whether the GAN can be used to classify multispectral images captured by satellites in the research area. Agriculture is represented by the research area. In addition to researching the GAN's capability to classify multispectral images, an architecture for this domain is being developed. When the GAN was used to classify data, it showed higher accuracy compared to Maximum Likelihood, where the GAN had an overall accuracy of (84.14%), while Maximum Likelihood had an accuracy of (81.66%).

Keywords: Multispectral image processing, Generative adversarial network, machine learning, satellite image, classification

1 INTRODUCTION

The constrained datasets and high expense of data collection are just two ways that satellite image analysis varies from agricultural ground view image analysis. Due of these considerations, the currently available high-resolution satellite datasets are task-specific and only cover a few cities (Mansourifar et al. 2022). As a result, the amount of radiation measured at very specific wavelength ranges for each image is contained in each channel or band in a multispectral image. The primary uses of multispectral image processing include space-based imaging, remote sensing, military target tracking, ballistic missile classification, agricultural land

DOI: 10.1201/9781003393580-108

mine classification, document and painting analysis, farming, and healthcare. Machine learning for classification and regression analysis uses dimensional hyperplane-based learning models for pattern identification.In machine learning, a classification problem is one where a class label is anticipated for a particular example of input data (Bawa *et al.* 2022).

2 RELATED WORKS

(Caraballo-Vega *et al.* 2022) have created the performance of methods that only use cloud spectral data is poor. A multi-regional, multi-sensor deep learning algorithm can be used to identify clouds utilising the extraordinarily high-resolution World View satellite data used in this work. (Ghezzo *et al.* 2022) have used searching for fossiliferous outcrops, this approach concentrates on individual fossil specimens, which may increase the effectiveness and efficiency of fieldwork. (Osman Isa *et al.* 2022) have used Support vector machines (SVMs), multi-layer perceptrons (MLPs), and ensemble learning (EL) classifiers were regarded as the three main categories of machine learning classifiers. Coastlines were calculated within the groups using a variety of formulas and/or classifiers, and their accuracy was evaluated while accounting for different types of coastlines (Celik *et al.* 2022). (Soumia Bengoufa *et al.* 2021) have created RF based on MRS supplied the coastline with the highest level of precision, with 55.5% of the retrieved shoreline being within 1 pixel of the in situ shoreline. (Sana Basheer *et al.* 2022) have used systematically assess the efficacy of various satellite data and classification method combinations in order to determine the most efficient way for LULC classification given the proliferation of geospatial analysis tools, categorization algorithms, and satellite data (Ali *et al.* 2020; Basheer *et al.* 2022)

3 PROPOSED METHODOLOGY

In this section, the proposed generative adversarial network in machine learning(GAN-ML)

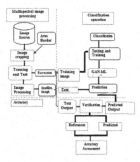

Figure 1. Proposed architecture of GAN- ML classification.

based classification technique has been described.

In Figure 1, the main goal of image processing is to offer output data, such as generative adversarial network ML classifications-based multispectral image processing, and collect input data from bands of crop-clipped images. As a result, the output classes associated with satellite images may be visually recognized.

3.1 *Multispectral image processing*

The GAN-ML method is divided into two components. In order to discover potential satellite image opportunities, the collected images were first run through an image processing pipeline. The classification of satellite image from the background's branchy and vegetated remnants was the aim of the satellite candidate recognition. Due to the complexity of gathering information from images and satellite data, categorising photographs is a difficult task. Due to significant advancements in both natural and artificial intelligence algorithms, it has nevertheless helped in the resolution of classification problems. The goal of the current work is to verify whether using the GAN for land cover classification is practical.

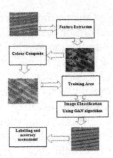

Figure 2.   Steps of multispectral image classification.

3.2   *Feature extraction*

Generally speaking, there is no symbiotic relationship between low-level and high-level traits. Since computer systems can easily extract these elements, the low level character-istics—often separated into two categories: local features and global features—are employed to recover the image's content. The characteristics that have been extracted from the entire image and may be utilised to include all of the image's data are known as global features. However, local features are utilised to extract features from specific visual objects and may be advantageous for uses like handwritten digit and face recognition. Figure 2 shows the feature extraction methods' flowchart.

4   EXPERIMENTAL RESULTS AND DISCUSSION

Kaggle datasets is used for satellite-image-classification. The analysis of satellite images and its numerous applications have made significant advancements in recent years. There is a growing need for the automatic interpretation of satellite images as they are now more

Table 1.   Individual satellite sceneries used for GAN with personal data's spatial resolution and cost.

Satellite	Spatial resolution multispectral(m)	Cost
Ikonos	0.83–4.24	2.67
Quickbird	0.66–3.41	24
Pleiades	0.8–5	6
TerraSar-X	1–5	2.69
WorldView 2	0.47–6	15-64
WorldView 3&4	0.34–1.25	15–64
RapidEye	6	0.98
Sentinel 2	11	free
Landsat 8	32	free

readily available than previously. The benchmark datasets are crucial building blocks for creating and evaluating clever interpretation algorithms in this scenario.

Sensor-dependent spatial resolution is one of this technology's greatest disadvantages (Table 1). Digital aerial cameras, which have a greater radiometric and spatial resolution than regular digital cameras, are the most widely used remote sensing surveying equipment for observations and records of Earth parameters. Data can be categorised both in organised and unstructured formats. In order to classify a set of data, different groups are created. The initial stage in the process is predicting the class of the supplied data points. The terms target, label, and classes are frequently used to describe the classes. A method for roughly anticipating the mapping function from discrete input inputs to output variables is classification predictive modelling. Since classification is a form of supervised learning, the input data are

Table 2. Evaluation of unpaired findings using MIP and non-MIP methods.

Group	Non-MIP	MIP group 2	MIP Group 1
Mean	1.561177	1.986622	2.917181
SD	0.343810	0.672611	0.728281
SEM	0.234589	0.052718	0.637262
N	19	19	19

Table 3. Review of paired accuracy results: MIP vs. Non-MIP methods.

Group	Without GAN-ML	With GANL-ML
Mean	0.3672	0.73672
SD	0.76378	0.27369
SEM	0.37382	0.368829
N	16	16

also made available to the targets. Let's familiarise ourselves with the terminology that is used in machine learning categorization.

SEM, which measures how much the data average is projected to depart from the actual population observations, is the same as SD, which stands for standard deviation, Table 1 provide information on IS related to the evaluation of unpaired findings using MIP and non-MIP approaches. Table 3 reports the results for the 4 groups both with and without GAN-ML (in percent). The mean group displays the best result with just 0.3672 without ML-MIP, demonstrating the classification's representative use of the water/shade endmember.

SEM is for standard error of the mean, which quantifies how much the data average is predicted to depart from the actual population observations, while SD stands for standard

Table 4. Evaluation metrics for the training and testing datasets for the MIP model development.

Parameter	Training	Text
Total predicted satellite image	356	88
Total original satellite image	358	90
Difference	4	5
Percent Error	0.4	3.5
Classification error	4	20

deviation, Table 3 provide information on IS related to the evaluation of unpaired findings using GAN-ML and non GAN-ML approaches.

In computing the total number of satellite images, the MIP model performed well, with just 0.4% error during training and 3.5% error during validation (Table 4).

5 CONCLUSION

The multispectral image processing procedure required for ML classifier based generative adversarial network (GAN) algorithm of satellite images was detailed in this research. Satellite images were produced using an unconventional classification method based on the GAN. It has been shown that using multispectral images is better than using single images; however, there needs to be a method for extracting the characteristics without losing information. As a result, successfully extracted and entered into the classification algorithm based on the GAN and satellite image. It has been found that using algorithms inspired by nature, like the GAN, can help with identifying multispectral images. The analysis using the Kaggle dataset shows that the suggested approach performs better.

REFERENCES

Ahmed Khaleel, Almas, and Joanna Hussein Al-Khalidy. "Multispectral Image Classification Based on the Bat Algorithm." *International Journal of Electrical and Computer Engineering Systems* 13, no. 2 (2022): 119–126.

Al-Ali, Z. M., Abdullah, M. M., Asadalla, N. B., & Gholoum, M. (2020). A Comparative Study of Remote Sensing Classification Methods for Monitoring and Assessing Desert Vegetation Using a UAV-based Multispectral Sensor. *Environmental Monitoring and Assessment*, 192(6). doi:10.1007/s10661-020-08330-1.

Bawa, Arun, Sayantan Samanta, Sushil Kumar Himanshu, Jasdeep Singh, JungJin Kim, Tian Zhang, Anjin Chang *et al.* "A Support Vector Machine and Image Processing based Approach for Counting Open Cotton Bolls and Estimating Lint Yield from UAV Imagery." *Smart Agricultural Technology* (2022): 100140.

Basheer, Sana, Xiuquan Wang, Aitazaz A. Farooque, Rana Ali Nawaz, Kai Liu, Toyin Adekanmbi, and Suqi Liu. "Comparison of Land Use Land Cover Classifiers Using Different Satellite Imagery and Machine Learning Techniques." *Remote Sensing* 14, no. 19 (2022): 4978.

Bengoufa, Soumia, Simona Niculescu, Mustapha Kamel Mihoubi, Rabah Belkessa, Ali Rami, Walid Rabehi, and Katia Abbad. "Machine Learning and Shoreline Monitoring Using Optical Satellite Images: Case Study of the Mostaganem Shoreline, Algeria." *Journal of Applied Remote Sensing* 15, no. 2 (2021): 026509.

Caraballo-Vega, J. A., M. L. Carroll, C. S. R. Neigh, M. Wooten, B. Lee, A. Weis, M. Aronne, W. G. Alemu, and Z. Williams. "Optimizing WorldView-2,-3 Cloud Masking Using Machine Learning Approaches." *Remote Sensing of Environment* 284 (2023): 113332.

Çelik, Osman İsa, and Cem Gazioğlu. "Coast Type Based Accuracy Assessment for Coastline Extraction From Satellite Image with Machine Learning Classifiers." *The Egyptian Journal of Remote Sensing and Space Science* 25, no. 1 (2022): 289–299.

Ghezzo, Elena, Matteo Massironi, and Edward B. Davis. "Multispectral Satellite Imaging Improves Detection of Large Individual Fossils." *Geological Magazine* (2022): 1–10.

Mansourifar, Hadi, Alex Moskowitz, Ben Klingensmith, Dino Mintas, and Steven J. Simske. "GAN-based Satellite Imaging: A Survey on Techniques and Applications." *IEEE Access* (2022).

Prieur, Colin, Antoine Rabatel, Jean-Baptiste Thomas, Ivar Farup, and Jocelyn Chanussot. "Machine Learning Approaches to Automatically Detect Glacier Snow Lines on Multi-Spectral Satellite Images." *Remote Sensing* 14, no. 16 (2022): 3868.

Malicious data detection in IoT using deep learning approach

Srinivas Kolli
Department of Information Technology, Vallurupalli Nageswara Rao Vignana Jyothi Institute of Engineering and Technology, Hyderabad, Telangana, India

Aravindan Srinivasan
Department of Computer Science Engineering, Koneru Lakshmaiah Education Foundation, Vaddeswaram, Andhra Pradesh, India

R. Manikandan
Department of Electronics and Communication Engineering, Panimalar Engineering College, Chennai, Tamil Nadu, India

Shalini Prasad
Department of Electronics and Communication, City Engineering College, Bangalore, Karnataka, India

Ashok Kumar
Department of Computer Science, Banasthali Vidyapith, Rajasthan, India

S. Ramesh
Department of Computational Intelligence, Saveetha School of Engineering, Chennai, Tamil Nadu, India

ABSTRACT: This article compares the efficacy of various DL intrusion detection techniques and identifies the optimal DL methodology for ID in the IoT. In this study, the DL techniques employed were artificial neural networks (ANN), long short-term memory (LSTM), and gated recurrent units (GRUs). The suggested model is assessed using a common dataset for IoT intrusion detection. Next, the experimental findings are scrutinized and contrasted with existing IoT techniques for intrusion detection. When compared to the previously used approaches, the suggested technique appeared to have the highest accuracy (99.9%).

1 INTRODUCTION

The IOT is anticipated to have 60 million gadgets by 2021 due to its extensive use in a variety of new applications, including smart urban, home appliances, automobiles, and smart industrial equipment (Azumah *et al.* 2021). Data privacy, integrity, and availability are greatly in danger due to this increase, which bad actors may take advantage of. Cybersecurity includes securing data and personal information in addition to preventing illegal access to networks and computer systems. The emergence of various innovative applications that rely on connected devices has increased awareness of IoT security in recent times (Thakkar *et al.* 2021). Deep learning-based security systems can operate across devices, underlying operating systems, and data without requiring a network connection to detect threats (Banaamah *et al.* 2022). Choosing an appropriate deep learning technique for IoT can be very beneficial for intrusion detection. Such selection may be carried out by contrasting approaches to ascertain the most accurate one, after which the chosen strategy may be put into practice (Khan *et al.* 2021). In this setting, DL is rapidly becoming a viable and efficient solution to many IoT security issues (Aversano *et al.* 2021).

DOI: 10.1201/9781003393580-109

2 RELATED WORKS

Feed-forward neural networks (FNN) are extensively employed in IoT networks to categorize intruder assaults. This study investigates how efficiently the Self-normalizing Neural Network (SNN) and the FNN categorize intruder threats in an Iot. The FNN surpasses the SNN at detecting intrusions in IoT networks, according to assessment criteria from empirical research (Ibitoye *et al.* 2019). It developed a hybrid DL technique to protect and identify illegal activity (Alkadi *et al.* 2020). Based on how the virus behaved in terms of the sequence of system calls that it made while it was being executed, this model was used to identify the malware. The Ubuntu Strace program is used to collect the system calls of IoT malware (Shobana *et al.* 2020).

3 PROPOSED SYSTEM

The implementation of this research will make use of deep learning technology and a step-by-step methodology to construct an extensive IoT protection mechanism that enhances the accuracy of security threat detection. The preprocessing of the datasets is depicted in the first portion of the picture. Scaling, normalization, and data cleaning were the three sub-steps that made up the pre-processing. After that, the dataset were tagged. The next phase involves classifying data utilizing ANN, LSTM, and GRU. After that, our model was trained, tested, and assessed. Phase 1 to 5 are Bot-IoT, Pre-Processing the Datasets, Feature Selection, Training, Testing, and Assessment, and Classification. In Phase 4, (ANN) in supervised learning, categorization issues are frequently addressed with ANN. Input, output, and numerous hidden layers of neurons make up an ANN. RobustScaler is utilized to scale the training data since it enhances ANN performance. A cutting-edge RNN called the LSTM network is trained using long and short-term dependencies. Modeling time-dependent information is extremely accurate on the DL technique. Three gates make up the LSTM block, which resembles an ANN neuron. Sigmoid functions can be used to express the forget (f), input (i), and output (o) gates. The LSTM and the GRU are very similar recurrent neural networks (RNN), although the GRU has a simpler design. The three gates employed by the LSTM are replaced by the update and reset gates in the GRU. Additionally, it lacks a discrete cell state or memory. Instead, data is transferred through the hidden state. The update gate serves as both an input and a forget gate by deciding which new information should be considered and which should be disregarded. The reset gate is employed to regulate how much prior data should be forgotten.

4 RESULTS AND DISCUSSION

Results of experiments carried out in Collaboratory by Google Research are presented in this section. All of the tests were conducted utilizing Python. Trials use independent training and testing datasets. Test and training sets accounted for around 85%-15% of the Bot-IoT dataset's total data. Each dataset's test set had the same amount of samples as the dataset itself. Our classifiers were created using Kera's library, and Kera's library's backend was TensorFlow. Performance metrics like accuracy, precision, recall, and F1-score are taken into consideration in this work for comparison.

Table 1. Training settings for ANN classifier.

Parameter	Description
Classification	ANN
Layers	(6) Input (8) Hidden and (1) Output
Features of the Input	6
Output	Common (0) Assault (2)
Dataset for training	85% for Training 15% for Testing

Table 1 lists the ANN's training settings. Robust Scaler is utilized to scale the training data since it enhances ANN performance.

Table 2 lists the training settings for LSTM and GRU classifiers. It describes the following parameters.

Table 2. Training settings for LSTM and GRU classifiers.

Parameter	Description
Classification	LSTM Furthermore, GRU
Layers	(6) Input (102) Hidden and (1) Output
Features of the input	6
Output	Common (0) Assault (2)
Dataset for training	85% for Training 15% for Testing

Table 3. Accuracy and False Alarm (FA) test outcomes.

Classification	Accuracy	FA
ANN	0.997	0.993
LSTM	0.999	0.992
GRU	0.996	0.994

The investigation's findings are presented in Table 3, which contrasts the classification methods' accuracy and FA rates. This table shows that when contrasted to ANN and GRU, the LSTM accuracy rate is superior.

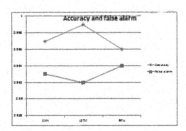

Figure 1. Analyze experimental findings for accuracy and FA.

Figure 1 displays the accuracy and FA rates for ANN, LSTM, and GRU. The accuracy of the LSTM is higher than that of the ANN and GRU, at 99.9%.

Table 4. Outcomes of the investigation in terms of precision, recall, and F1-score.

Classification	Precision	Recall	F1-Score
ANN	0.997	0.998	0.997
LSTM	0.999	1.121	0.997
GRU	0.997	1.121	0.997

Table 4 shows that when compared to ANN and GRU, the LSTM precision rate is superior. Table 4 lists the ANN, LSTM, and GRU's precision, recall, and F1-scores.

Contrast the effectiveness of the suggested methodology to other cutting-edge techniques. Table 5 displays the outcomes. Comparative between simulation outcomes and state-of-the-

741

Table 5. Comparison of the model with other cutting-edge techniques.

Classification	Dataset	Accuracy (%)
ANN	Bot-IoT	99.6%
LSTM	Bot-IoT	99.9%
GRU	Bot-IoT	99.5%
FNN	Bot-IoT	95.2%

art techniques are made for the 3 classifications ANN, LSTM, and GRU. In contrast to the existing cutting-edge methods, The greatest accuracy of the suggested method was 99.9%.

5 CONCLUSION AND FUTURE WORK

This work employs deep learning techniques to detect intrusions into IoT devices, and it is described in this article. The standard dataset Bot-IoT was utilized in this study to detect IoT intrusions. IoT intrusion detection has been carried out utilizing a variety of DL methods, including the ANN, GRU, and LSTM. The Suggested method has been assessed and contrasted with existing methodologies. The findings of the experiments have demonstrated the suggested approach's potential utility for IoT intrusion detection.

REFERENCES

Alkadi, Osama, Nour Moustafa, Benjamin Turnbull, and Kim-Kwang Raymond Choo. "A Deep Blockchain Framework-enabled Collaborative Intrusion Detection for Protecting IoT and Cloud Networks." *IEEE Internet of Things Journal* 8, no. 12 (2020): 9463–9472.

Aversano, Lerina, Mario Luca Bernardi, Marta Cimitile, and Riccardo Pecori. "A Systematic Review on Deep Learning Approaches for IoT security." *Computer Science Review* 40 (2021): 100389.

Azumah, Sylvia Worlali, Nelly Elsayed, Victor Adewopo, Zaghloul Saad Zaghloul, and Chengcheng Li. "A Deep lstm Based approach for Intrusion Detection iot DevicesNetwork in Smart Home." In *2021 IEEE 7th World Forum on Internet of Things (WF-IoT)*, pp. 836–841. IEEE, 2021.

Banaamah, Alaa Mohammed, and Iftikhar Ahmad. "Intrusion Detection in IoT Using Deep Learning." *Sensors* 22, no. 21 (2022): 8417.

Ibitoye, Olakunle, Omair Shafiq, and Ashraf Matrawy. "Analyzing Adversarial Attacks Against Deep Learning for Intrusion Detection in IoT Networks." In *2019 IEEE global communications conference (GLOBECOM)*, pp. 1–6. IEEE, 2019.

Idrissi, Idriss, Mohammed Boukabous, Mostafa Azizi, Omar Moussaoui, and Hakim El Fadili. "Toward a Deep Learning-based Intrusion Detection System for IoT Against Botnet Attacks." *IAES International Journal of Artificial Intelligence* 10, no. 1 (2021): 110.

Li, Yuxi, Yue Zuo, Houbing Song, and Zhihan Lv. "Deep Learning in Security of Internet of Things." *IEEE Internet of Things Journal* (2021).

Khan, Tauseef, Ram Sarkar, and Ayatullah Faruk Mollah. "Deep Learning Approaches to Scene Text Detection: A Comprehensive Review." *Artificial Intelligence Review* 54, no. 5 (2021): 3239–3298.

Shobana, M., and S. Poonkuzhali. "A Novel Approach to Detect IoT Malware by System Calls Using Deep Learning Techniques." In *2020 International Conference on Innovative Trends in Information Technology (ICITIIT)*, pp. 1–5. IEEE, 2020.

Thakkar, Ankit, and Ritika Lohiya. "A Review on Machine Learning and Deep Learning Perspectives of IDS for IoT: Recent Updates, Security Issues, and Challenges." *Archives of Computational Methods in Engineering* 28, no. 4 (2021): 3211–3243.

Detecting cross-site scripting attacks using machine learning: A systematic review

D. Baniya*
Computer Science & Engineering, Sharda School of Engineering & Technology, Sharda University, Greater Noida, Uttar Pradesh, India

Amrita*
Center for Cyber Security and Cryptology, Computer Science & Engineering, Sharda School of Engineering & Technology, Sharda University, Greater Noida, Uttar Pradesh, India

A. Chaudhary*
Computer Science & Engineering, Sharda School of Engineering & Technology, Sharda University, Greater Noida, Uttar Pradesh, India

ABSTRACT: In today s age, many daily tasks are performed through the Internet using various web applications. While using the web application, the information and data are stored in the database of the network which can easily be attacked by the attacker. The attacker used different types of attack patterns to steal the victim s information or any details like their bank information, and business details. Among the different types of attacks over the web application vulnerabilities, the most prominent category of attack is a cross-site scripting (XSS) attack. In the XSS attack, an attacker transmits the script to the victim s web Browser. As the victims are unaware that they were being attacked leading to stealing the valuable information like personal details, bank details, and cookies. Using the cookies of the victim, the attacker can easily access their account without using a username and password. Early, different types of approaches have been done to shoot out the vulnerabilities related to XSS. This paper purports to be a survey on XSS. The primary contribution of this study to the current state-of-the-art XSS is a systematic review of various types of research done to detect XSS vulnerabilities using Machine Learning. This paper also provides the background related to XSS, techniques, difficulties and future guidelines in the field of XSS to provide researchers with significant knowledge for further work.

1 INTRODUCTION

Nowadays the world has become small through the internet connection as we perform our daily tasks through the Internet using the different types of web browsers, and web applications to perform activities like online shopping, online booking tickets, and transaction of money through the bank website using the network. The web application which we are using may result in Cross-Site Scripting (XSS) vulnerabilities leading to stealing our information, which may cause security issues (State of Software Security 2023). So, the security concern has been most important nowadays as we use the internet frequently. In the XSS, the attacker embeds the JavaScript code into the target victim s web application or the website

---

*Corresponding Authors: baniyadurgesh424@gmail.com, amritaprasad_y@yahoo.com and maxashoka3@gmail.com

DOI: 10.1201/9781003393580-110

that the victim visits. Once the victim is under attack, cookies associated with the website become freely accessible to hackers. The hacker can easily access the victim s account without using an ID and Password, leading to a series of security issues (Gupta & Gupta 2017). According to the survey of security made based on the annual global security data in which lots of security events were analysed, all studied web applications exhibit that web attacks and vulnerabilities are becoming more complex to mitigate (IT Digital Media Group 2018). Lots of research have been done to detect the XSS vulnerabilities using Machine Learning (ML) Methods (Guo *et al.* 2015; Marashdih *et al.* 2017). This paper explains the different types of techniques which were employed to detect XSS vulnerabilities. This paper pretends to present a survey of the various work done and techniques trends to detect XSS vulnerabilities using ML. The following is the rest of the paper. The detailed in sight of the related background of the previously mentioned study is presented in the next section. Section three provides a systematic review of related work in the literature. Section 4 discusses the challenges and prospects for the future. Section 5 contains the directions and, finally, the conclusion.

2 RELATED BACKGROUND

2.1 *Cross-site scripting*

The XSS is the type of attack which is commonly seen over the internet. In this, the hacker simply injects the JavaScript code into the user s web browser which leads to executing the malicious JavaScript in the web application. The attacker does not choose the victim he wishes to target instead, he exploits vulnerabilities in a website that the victim visits to perform the work via the network connection. As the victim is unaware of sending the request for the malicious string after that the attacker sends the malicious URL to the victim in form of an email or message and trick the user to visit the URL link provided by the attacker leading to XSS vulnerability.

The XSS attacks and exploitation are categorized into three types—persistent XSS, reflected XSS, and domain object model (DOM) XSS. In Persistent XSS, the attacker merely injects the JavaScript code into the vulnerable web pages which are saved in the web browser and executed in their context. In Reflected XSS, the victim requests the malicious string over the web application and the attacker sends the string back in the response over the website which is simply in the form of the URL link sent through message or email. DOM XSS is the variation of both reflected and persistent XSS.

2.2 *Machine learning*

ML has used a different set of mathematical algorithms which is used to make any set of operations automatic leading to complex work in a simple form. It is also known as Artificial Learning methods, where ML tends to learn the model using different approaches like neural networks, decision trees etc. (Amrita & Kant 2019). In this, different types of data sets are provided for training the model and change according to the input dataset. The learning methods are categorized into following categories—supervised, unsupervised, and reinforcement learning (RL) (Han *et al.* 2011; Medeiros *et al.* 2016).

3 A SYSTEMATIC LITERATURE REVIEW OF RELATED WORK

A lot of research has been done to find the presence of XSS vulnerability in the source code (Guo *et al.* 2015; Marashdih *et al.* 2017). The paper (Guo *et al.* 2015) has presented how an understanding of HTML ideas will lead us to find and protect the programming from the

XSS vulnerability and its detection which is present in the source code of PHP applications. The false detection results are obtained in the ignorance of input data (Zhang *et al.* 2015). The automated tool offers a cost-reliable, high-performance, security-based line considers in the early process of Software Development in its early stage (Zhang *et al.* 2015). Lots of similar works are studied in (Hydara *et al.* 2015), which have been done on XSS and only got the result based on the DOM-XSS with the smaller number of percentages. The analysis in the form of a static and dynamic process approach is created to minimize the risk associated with XSS (Vogt *et al.* 2007). But they lack in focusing tracking unreliable data, and vulnerability and only focus on the user cookie, and potential information leakage.

The optimization technique and dynamic attack vector generation method employing the hidden Markov model are proposed to enhance the detection of XSS vulnerability (Wang *et al.* 2017). The decision tree model is used to generate mutated attack vectors to classify the attack vectors. Whereas, the attack vector is deformed using a code confusion strategy. This method enhanced the detection efficiency and reduced the miss report and response time. The process is created in (Saxena *et al.* 2010), which is utilizing the fuzzy based on the black box method (Sutton *et al.* 2007) implementing enchaining taint. This approach explains how dynamic taint is analysed with automated technology regarding fuzzing and developing the corresponding tool in the same interval of time. A dynamic detection framework utilizing taint tracking to detect DOM XSS on the client side is proposed in (Wang *et al.* 2018). All DOM APIs and JavaScript features are rewritten to spoil the rendering process of browsers. The black box fuzzing was adopted in the Firefox plug-in for DOM-XSS detection. The tool for detection of DOM-XSS is Dominator which is based on taint tracking, by using the spider Monkey JavaScript in modified form in Firefox engine. But it filed to explain vulnerability detection.

The genetic algorithm and static analysis are employed to detect XSS in PHP web applications (Marashdih *et al.* 2017). It removes the inappropriate and infeasible paths from the control-flow graph to find XSS vulnerability. This method outperformed other earlier methods and helps to reduce false positive rates in PHP web applications. The paper (Marashdih & Zaaba 2017) has proposed a study based on approaches for removing XSS vulnerability instead of only detecting of an XSS vulnerability in web Applications. The work is performed on the HTML purifier, PHP commands composed to mitigate XSS vulnerability in its tools. It is used to guard the web application against getting any vulnerability and any sort of threat.

An investigation of the integration of static analysis with other algorithms has been done to detect XSS vulnerability is presented in (Marashdih *et al.* 2019). It has explained XSS in two different ways i.e server and client-side XSS vulnerability. Data mining and ML approaches are presented to use to detect vulnerability from a great number of data sets in PHP web applications. The PHP web applications are the prominent vulnerable applications to XSS. The detection approaches for this are described in (Marashdih *et al.* 2018). The tool is developed in (Vishnu & Jevitha 2014) and is used to predict XSS attacks using the ML Algorithm. The classifiers are performed based on the XSS-URL and XSS-JavaScript.

The comprehensive study and survey on detection methods for XSS vulnerability are presented in (Sarmah *et al.* 2018). This survey presented a detection method based on deployment sites, analysis mechanism, and also the pros and cons of each method. The types of XSS attacks are also described, which are mainly server-side, client-side, and client-server-side detection approaches.

The Blind XSS attacks using ML techniques using SVM for classification of the difference between blind and store XSS is proposed in (Kaur *et al.* 2018). In (Garcia-Alfaro & Navarro-Arribas 2007), how vulnerabilities in web applications can lead to risk for both users and the application itself is presented. Both detection and prevention methods are discussed to protect XSS vulnerability. The issues, research challenges and various methods to conduct XSS attacks are also presented in this research. A test data generator based on a genetic algorithm is designed in (Ahmed & Ali 2016), which generates possible attacks from the XSS attack pattern database. The DOM reflected, and stored-based XSS vulnerability is used. It is utilized to generate the multiple-path for testing for XSS mostly considering only PHP using JavaScript web applications along with ASP.net and JSP platform.

A comparative and calculation-based survey on XSS detection methods proposed by various works published in journals between 2019 & 2020 is reviewed (Stency & Mohanasundaram 2021). The findings of this survey have demonstrated the critical role XSS detection based on the deep learning module plays in generating effective XSS intrusion detection systems. According to the training and testing objectives, the XSS characteristics have been extracted and categorised. This paper will provide guidance and pointers for further research in the area of detection of XSS attack based on the deep learning model. This paper has also included the future work direction calls for the use of powerful AI-based tools and methodologies to defend against XSS attacks. An innovative technique for leveraging RL to modify attack payloads to the target reflected in XSS vulnerability is proposed in (Lee *et al.* 2022). The research provides a RL framework named Link that aims to discover reflected XSS vulnerabilities using states, actions, and reward mechanisms in an automated mode. The experimental findings show that using RL to detect newly discovered XSS vulnerabilities shows encouraging direction for precise and scalable penetration testing of it.

A convolutional neural network employing the NiN model and Modified ResNet has been proposed for the prevention and detection of XSS attacks (Yan *et al.* 2022). The main advancements of the proposed model include pre-processing the URL by the structural semantics and syntax of the XSS attack script enhancing, encoding the ResNet residual module which extracts the features from three different angles, and replacing the fully connected layer with 11 convolution properties. This paper has claimed that the recommended model performs better and converges faster than deep learning detection methods and conventional ML. The model has a detection rate of up to 75% and has an accuracy of 99.23% with a precision value of 99.23% and a recall value of a 98.53% when compared to a baseline. The paper (Lu *et al.* 2022) suggests a fusion verification technique which combines traffic monitoring with XSS payload identification. The outcomes have found a powerful advantage for lowering the false negatives in the case of the uniform sample distribution. The research suggests seven new payload attributes to increase the effectiveness of detection. The outcomes of the experiments demonstrate that the suggested approach has greater accuracy than the single traffic detection model.

A train-tracking method for detection of DOM XSS with high recall and accuracy is suggested in (Melicher *et al.* 2021). The proposed methods check out whether train tracks may be replaced or supplemented by ML classifiers for identifying DOM XSS vulnerabilities. For these 18 billion JavaScript functions were gathered using extensive web scans & employed train tracking to identify over 180,000 routines as possibly susceptible. A deep neural network is trained with this data to investigate JavaScript function & predict its vulnerability to DOM XSS. Data mining is used in (Shar & Tan 2013) for the detection of XSS vulnerability by creating tools based on utilizing the occurrences of the vulnerabilities in the form of a PHP program. This tool is used to detect the set of attributes and features from a program and after that, they used the process based on an algorithm to extract the features. Their research found that false positive rate has been 6% in extracting the XSS vulnerabilities and the 11% in SQL vulnerabilities.

4 CHALLENGES AND FUTURE DIRECTIONS

There are a lot of challenges that occur during the classification of XSS attack patterns using ML. XSS is one of the major important factors when considering web security. The web application we used with XSS vulnerabilities may result in seating our information which may lead to serious security issues. In this review paper, we are trying to find out the XSS attacks using the ML methods done previously. However, the collection of the data set through data mining is a challenging task to train the model.

In the future, the capability of the detection of XSS using ML can be enhanced by using the more advanced algorithm which can be implemented in the model to make it more effective to mitigate the vulnerability present in the web browser. In addition, various types of ML algorithms can be utilized to detect XSS vulnerabilities.

5 CONCLUSION

Cross-Site Scripting (XSS) vulnerability is the major common seen threat and issue which occur in web-based applications. Lots of research is still being carried out to detect and mitigate it. This vulnerability is an effective way to exploit the web applications source code. XSS occurs by injecting harmful JavaScript code into the web applications which causes vulnerabilities for the client and the server side. This research discussed a comprehensive survey of the recent studies on the XSS vulnerability, its detection and mitigation using Machine Learning (ML) by implementing various algorithms. It helps to build a strong base in this field for researchers. The implementation of this study will help what kind of research has been done in analysing and mitigating XSS vulnerability in web browsers based on ML. This research would give insight to the researchers about XSS techniques that can be employed and challenges in XSS for novice researchers. Also, the addition of future directions is presented to researchers to improve detection and also mitigation methods for XSS vulnerability and guide them about the progression in this research field.

REFERENCES

Ahmed, M.A. & Ali, F. 2016. Multiple-path Testing for Cross Site Scripting Using Genetic Algorithms, *Journal of Systems Architecture* 64: 50–62.

Amrita & Kant, S. 2019. Machine Learning and Feature Selection Approach for Anomaly based Intrusion Detection: A Systematic Novice Approach, *International Journal of Innovative Technology and Exploring Engineering (IJITEE)* 8(6S): 434–443.

Garcia-Alfaro, J. & Navarro-Arribas, G. 2007. A Survey on Detection Techniques to Prevent Cross-Site Scripting Attacks on Current Web Applications, *Lecture Notes in Computer Science*: 287–298.

Guo, X., Jin, S. & Zhang, Y. 2015 XSS Vulnerability Detection Using Optimized Attack Vector Repertory, *Cyber-Enabled Distributed Computing and Knowledge Discovery*: 29–36

Gupta, S. & Gupta, B.B. 2017. Cross-Site Scripting (XSS) Attacks and Defense Mechanisms: Classification and State-of-the-art, *International Journal of Systems Assurance Engineering and Management* 8(S1): 512–530.

Han, J., Kamber, M. & Pei, J. 2011. *Data Mining: Concepts and Techniques*, Elsevier.

Hydara, I., Sultan, A.B.M., Zulzalil, H. & Admodisastro, N. 2015. Current State of Research on Cross-site Scripting (XSS) - a Systematic Literature Review, *Information and Software Technology* 58: 170–186.

IT Digital Media Group Discover the New. 2018. Available at: https://discoverthenew.ituser.es/security-and-risk-management/2018/04/el-100-de+lasaplicaciones-web-contienen-vulnerabilidades.

Kaur,G., Malik,Y., Samuel, H. & Jaafar, F. 2018. *Detecting Blind Cross-Site Scripting Attacks Using Machine Learning, International Conference on Signal Processing*: 22–25.

Lee, S., Wi, S. & Son, S. 2022. Link: Black-Box Detection of Cross-Site Scripting Vulnerabilities Using Reinforcement Learning, *Proceedings of the ACM Web Conference*: 745–754.

Lu, J., Wei, Z., Qin, Z., Chang, Y. & Zhang, S. 2022. Resolving Cross-Site Scripting Attacks through Fusion Verification and Machine Learning, *Mathematics* 10(20): 3787.

Marashdih, A. W., Zaaba, Z. F., Suwais, K. & Mohd, N. A. 2019. Web Application Security: An Investigation on Static Analysis with other Algorithms to Detect Cross Site Scripting, *Procedia Computer Science* 161: 1173–1181.

Marashdih, A.W. & Zaaba, Z.F. 2017. Cross Site Scripting: Removing Approaches in Web Application, *Procedia Computer Science* 124: 647–655.

Marashdih, A.W., Zaaba, Z.F. & Omer, H.K. 2017. Web Security: Detection of Cross Site Scripting in PHP Web Application using Genetic Algorithm, *International Journal of Advanced Computer Science and Applications* 8(5): 64–75.

Marashdih, A.W., Zaaba, Z.F. & Suwais, K. 2018. Cross Site Scripting: Investigations in PHP Web Application, *International Conference on Promising Electronic Technologies (ICPET)*: 25–30.

Medeiros, I., Neves, N. & Correia, M. 2016. DEKANT: A Static Analysis Tool That Learns to Detect Web Application Vulnerabilities, *International Symposium on Software Testing and Analysis*: 1–11.

Melicher, W., Fung, C., Bauer, L. & Jia, L. 2021. Towards a Lightweight, Hybrid Approach for Detecting DOM XSS Vulnerabilities with Machine Learning, *The Web Conference*: 2684–2695.

Sarmah, U., Bhattacharyya, D. & Kalita, J. 2018. A survey of detection methods for XSS attacks, *Journal of Network and Computer Applications* 118: 113–143.

Saxena, P., Hanna, S., Poosankam, P. & Song, D. 2010. FLAX: Systematic Discovery of Client-side Validation Vulnerabilities in Rich Web Applications, *Network and Distributed System Security Symposium.*

Shar, L. K. & Tan, H. B. K. 2013. Predicting SQL Injection and Cross Site Scripting Vulner abilities Through Mining Input Sanitization Patterns, *Information and Software Technology* 55(10): 1767–1780.

State of Software Security. 2023. Available at: https://www.veracode.com/state-of-software-security-report.

Stency, V.S. & Mohanasundaram, N. 2021. A Study on XSS Attacks: Intelligent Detection Methods, *Journal of Physics: Conference Series* 1767.

Sutton, M., Greene, A. & Amini, P., 2007. *Fuzzing: Brute Force Vulnerability Discovery.* Pearson Education.

Vishnu, B.A. & Jevitha, K.P. 2014. Prediction of Cross-Site Scripting Attack Using Machine Learning Algorithms, *Proceedings of the International Conference on Interdisciplinary Advances in Applied Computing* 55: 1–5.

Vogt, P., Nentwich, F., Jovanovic, N., Kirda, E., Krugel, C. & Vigna, G. 2007. Cross Site Scripting Prevention with Dynamic Data Tainting and Static Snalysis, *Network and Distributed System Security Symposium, NDSS* : San Diego, California, USA.

Wang, D., Gu, M. & Zhao, W. 2017. Cross-site Script Vulnerability Penetration Testing Technology, *Journal of Harbin Engineering University* 38(11): 1769–1774.

Wang, R., Xu, G., Zeng, X., Li, X. & Feng, Z. 2018. TT-XSS: A Novel Taint Tracking Based Dynamic Detection Framework for DOM Cross-Site Scripting, *Journal of Parallel and Distributed Computing* 118: 100–106.

Yan, H., Feng, L., Yu, Y., Liao, W., Feng, L., Zhang, J., Liu, D., Zou, Y., Liu, C., Qu, L. & Zhang, X. 2022. Cross-site Scripting Attack Detection Based on a Modified Convolution Neural Network, *Frontiers in Computational Neuroscience* 16: 1–13.

Zhang, N., Wu, B. & Bao, X. 2015. Automatic Generation of Test Cases Based On Multi-population Genetic Algorithm, *International Journal of Multimedia and Ubiquitous Engineering* 10(6): 113–122.

Artificial Intelligence, Blockchain, Computing and Security – Dagur et al. (Eds)
© 2024 The Author(s), ISBN: 978-1-032-49393-0

A speech emotion recognition system using machine learning

Reshma Kanse, Supriya Ajagekar, Trupti Patil, Harish Motekar, Vinod Rathod,
Rahul Papalkar & Shabir Ali
Bharti Vidyapeeth Deemed University Pune, DET Navi Mumbai, India

ABSTRACT: As we know speech is very direct way of expression our emotions. When it comes to human to human talks we can easily recognize each other emotions, feelings and views but when it comes to human to machine communication, it's difficult for system or machine to get exact emotions of the person. So here is when SERtion comes into picture. The SERtion is a machine learning project built on various classifiers and feature extraction techniques. SERtion aims to create a machine learning model that will recognize the exact emotions hidden in the audio signals. It follows various processing steps to accomplish the goal of creating a successful model. SERtion can be used in call centers where machine can predict the emotions of the customer that calls and gives feedback of their service or those who enquiry and even fire complaints for their service.

Keywords: Speech, Emotion, Recognition, SER, Machine Learning.

1   INTRODUCTION

SERtion is a speech recognition model in which a speech's emotion is detected. This model uses libraries like librosa, soundfile and sklearn. These libraries help to extract the features of the speech. These libraries are used to create a MLPClassifier which will recognize emotions from the sound. Some training is required for the system to work accurately.This model has a great scope in many fields, like chatbot, robotics. In all the fields where human interaction is required this model can be used coupled with different AI algorithms and can detect the emotions of human speech. This project has great potential with chatbot systems. If a model detects the emotion of a speech then it can increase the interaction with the person using it. In real life people can easily judge someone's emotions by their voice, theri tone, the pauses. But for a computer it is difficult because it lacks human intelligence. Sometimes the person can mean different things but the words that he/she uses may be different. So in such a case this model helps to detect the emotion of the human speech which helps to understand the meaning between the lines. It will predict if the speech is joyful or angry, or happy or sad. The visualization will be done for better understanding.

In SERtion, we are going to create a model which takes human audio as input and visualizes it according to the emotion. The speech which is collected as input will then be divided into different parts, all the feature will be extracted from it using ML algorithm and then data augmentation will take place. Firstly the system will train it self for a good amount of time. Once ample training is done, then the system will be ready to perform the prediction of the emotions. There will be total 8 emotions: Surprise, Neutral, Disgust, Fear, Sad, Calm, Happy, Angry.Once the system is ready, if any input audio is given, then the system will predict the emotion of the speech from above 8 types. Speech signals from a call center application were used in a proposal for domain-specific emotion identification years ago. This study's major objective is to identify both positive and negative emotions, such as rage and happiness. Acoustic, lexical, and discourse information, among other forms, are utilised to identify emotions. In addition, information-theoretical emotional salience tasks are offered in order to collect information on emotional information at the language level. Both linear discriminant classifiers and k-NNs are capable of using several types of features. A mix of auditory and linguistic data yields the greatest results, according to experimental findings. The findings indicate that using three sources of data as opposed to only one increases classification accuracy for both men and women by 40.7percent and 36.4%, respectively. Improvement range in accuracy when compared to previous work.

2 LITERATURE SURVEY

This paper gives us the brief idea about the Novel Emotion Recognition Algorithm that combines speaker gender information. This is basically a study of how we can recognize the emotions of both the audio types i.e Male version and even a Female Version. Algorithm talks about various characteristics and features that defers in the voice of male and female [1].This paper gives an overview of the various classifier method that helps to recognize emotions and also about the various types of features that can be extracted from the audio or the voice input. This paper also talks about how feature extraction plays an important role in model creation and how the output totally relays on the features that we extract[2].This paper has overview of feature extraction using CNN. It briefs us about how Deep learning is used in recognition of the emotion. It has an overview of different types of Neural Network[3]. This essay discusses the categories of speech features. The language used, the speaker's facial expressions, and auditory features are the three main types of aspects that can be found in speech (sound properties like pitch, tone, jitter, etc.). Multiple groups of speech traits are discussed in the study. Lexical features (the vocabulary used), visual data (the speaker's face expressions), and auditory characteristics (sound characteristics like pitch, tone, jitter, etc.) are the three main kinds of features discovered in speech[4]. a a convolutional online speech recognition system that outperforms two trustworthy benchmarks in throughput, WER, & latency while showing the trade-offs in improving one metric over another and how measures may be adjusted to the application's requirements. To further boost performance, future research will include 8-bit fixed precision quantized as well as other techniques like weight sparsity and model depth reductions on demand. The aforementioned methods might also shorten inference time while reducing model size. Also, models were delivered to devices; however, we reserve a detailed analysis with comparative benchmarks for future work. [5]. According to study, modern IDS can now accurately forecast malware attacks on virtual machines. Initially, ePIA used Shannon entropy to filter out low-risk malicious packets and knew the characteristics of the packet flow. The feature selection technique is then used to choose the best features from the filtered packets. The IBES technique is used to extract the beast traits; BES is enhanced by the OBL approach and the best fitness function in this case.[6]. This essay analyses speech to identify the emotions spoken. The feature vector is made up of audio signal elements that represent unique characteristics of the speaker, such as tone, pitch, and energy. This information is crucial for setting up the classifier model to accurately identify a given emotion. Mel-frequency cepstral coefficients (MFCC), which were taken from the speech signals, were then used to create the dataset for the study. The classifier model is then given the feature vector that was extracted from the training dataset. The extraction method will be applied to the test dataset before the classifier makes a decision regarding the hidden emotion in the test audio. Four distinct datasets (SAVEE, RAVDESS, TESS, and CREMA-D) have been prepared as part of the dataset preparation[8]. two methods for recognizing emotions in human speech. With an overall accuracy of 88.7%, the first method, which uses cross-correlation between audio samples for emotion recognition, is highly effective. It does a great job of identifying the happy emotion (100% accuracy) and the angry emotion (86% accuracy). With the Cubic SVM Classifier, the second approach, which uses feature selection of six characteristics, achieves a substantially higher accuracy of 91.3%. In the current work, three emotions—angry, pleased, and neutral—are recognized. The study lays the door for the quick creation of a real-time system that can accurately identify the emotions expressed in human speech.[9]. An algorithm based on DWT coefficients managed to recognize seven emotions with an overall accuracy of 85.71%. When compared to the other emotions, neutral and disgust have slightly lower recognition accuracy. Because the entropy levels of disgust and boredom are nearly equivalent, boredom is sometimes mistaken for disgust. The findings show a significant improvement over the relevant previous research, both in terms of the quantity of emotions taken into account and the level of accuracy attained. Additionally, this study can be advanced utilizing a hybrid classifier and tested using genuine speech database emotion samples. Additionally, the misclassification of neutral and disgust emotions can be decreased, improving overall accuracy[10]. The work can be expanded to include the robot to help it understand the matching human's mood, which will assist it have a conversations, as well as it can be connected with other music apps to offer music to its users according to their emotions[11]. Some SER components that could be considered in future works are the utilisation of stronger data bases, studies on under-resourced various languages, real-time acknowledgment, and the addition of new types of emotions to fulfil both dimensions of valence and arousal[12].

3 PROPOSED SYSTEM

3.1 *Problem statement*

SERtion is an Machine Learning based project that deals with Recognizing Speech Emotion by developing an machine learning model. Creating a Machine Learning Model, doing Exploratory Data Analysis and Then predicting the Emotions by using suitable Classifier. It's quite simple: *Dataset as Input, EDA, Feature Extraction, Training Model,* Predicting Output.

SERtion is the process of attempting to distinguish emotional states and human emotion from speech. This makes use of the frequent use of tone and pitch in the voice to convey underlying emotion. This phenomena is used by animals like horse and dogs to interpret human emotion. In this mini-project, we'll build an MLPClassifier model utilising a variety of libraries, including librosa, soundfile, and sklearn. In audio files, this technology will be able to detect emotions. This dataset will be split into testing and training sets when the data has been loaded and its characteristics have been extracted. Once an MLPClassifier has been initialised, the model will next be trained. Next, we'll assess the model's precision.

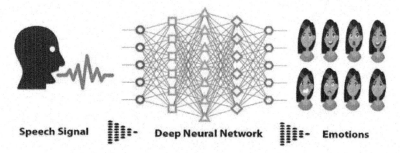

Figure 1. Proposed Model.

3.2 *Objectives*

1. To Create a Machine Learning Model
2. Reading Dataset that has both Male/Female actor's voice

3. Performing Exploratory Data Analysis
4. Performing Data Visualization
5. Performing Data Augmentation
6. Extracting Feature
7. Training Model
8. Predicting Output

3.3 *Project design*

3.3.1 *System block diagram*

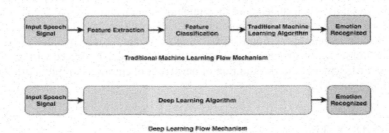

Figure 2.   Process Involve in Emotion Recognition.

3.3.2 *Flow diagram*

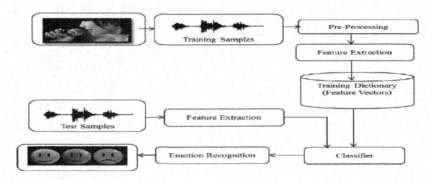

Figure 3.   Step Involved to Recognize Emotion.

3.3.3 *Algorithm*

Step 1: Start
Step 2: Collection of Dataset
Step 3: Performing Data Pre-processing on input data
Step 4: Performing Exploratory Data Analysis
Step 5: PerformingDataVisualization
Step 6: PerformingDataAugment tion
Step 7: Extraction of Features
Step 8: Splitting of Dataset into Train and Test
Step 9: Creating and Training Model.
Step 10: Prediction and Accuracy Analsis.
Step 11: Stop

4 IMPLEMENTATION

4.1 *Module and description*

The proposed system involves five important Phases:

Input Dataset:-The RAVDESS is a multimodal database of validated emotional speech and music. 24 professional actors, representing a gender-balanced database, vocalise lexically-matched statements with a neutral N-American accent. A variety of emotions, such as those linked to calmness, joy, sorrow, rage, fear, surprise, and disgust, can be found in both speech and music. Each expression has a neutral expression as well as two levels of emotional intensity. All conditions are offered in voice-only, voice-and-face, and voice-only formats. Each of the 7356 recordings was given a score out of 10 for emotional authenticity, intensity, and validity. 247 persons who typified untrained research volunteers from North America provided ratings.

EDA:- EDA refers to Exploratory Data Analysis. In this phase Data is analyze using various techniques to obtain some useful relationships between various aspects to Dataset. To that end, plotting is done. Data analysis is a methodology for assessing sets of data to highlight their main characteristics using additional information and statistical graphic visualization approaches. EDA is mostly employed to determine what the data can reveal to us without first engaging in formal modeling or testing hypothesis. It is possible to use or not use a statistical model.

Data Augmentations:- Dataset is the process of adding partial shading condition to the basic training set to generate new artificial data samples. Sound syntactic information can be created using a variety of approaches, including sound injection, time shifting, modifying the pitch and tempo, and more. Our aim is to strengthen the generalizability of our model by making it insensitive to those perturbations. Processes known like "data augmentation" within data analysis are utilized to expand the amount of data by collecting greatly altered copy of just either information or wholly new synthetic data originates from that is generated from existing data. It serves as a regularize during the training of a machine-learning model and aids in minimizing over fitting. It is closely related to sampling in data analysis.

Feature Extraction: - Feature Extraction refers to mine some useful characteristic of each voice based on which we can categorize it into various emotional states. It involves various feature extraction such as Chroma, MFCC, Mel. By generating new characteristics from the current ones in a dataset, feature extraction tries to reduce the total amount of characteristics in the dataset. Therefore, this new, more condensed set of features should be able to capture the bulk of the data in the entire collection of characteristics.

Training Model:-

ML algorithms are trained using training models, which are datasets. It includes collections of pertinent input data that influence the outcome as well as representative output data. Models for machine learning come in a wide variety, but the most common ones are supervised and unsupervised models. Learning, actual building of the models and the training will take place in this phase. Training and testing datasets will be created from the original dataset. For model training, Librosa Library and MLP Classifiers will be used.

5 RESULTS & ANALYSIS

5.1 *EDA and visualization*

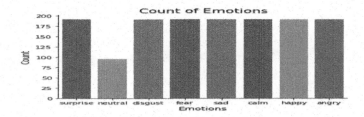

6 CONCLUSION

By employing a variety of Python modules and an MLP Classifier to categorise distinct emotions, the SERtion ensures that the emotions of the audio signals is recognized. Five different kinds of features, including ZCR, MFCC, Chroma shift, Root Mean Squared, and MelSpectogram, were thus retrieved. Processing and proper formatting of the data were done at the beginning. Then comes EDA(Exploratory Data Analysis) that was done to get the proper idea about co-relations between the data. Then data Augmentation to manipulate the audio, so that our training data is not over fitted. So this how SERtion will ensure a perfect accuracy of the Model.

Further the Project can be extended by extracting more features and deploying the model and turning it into software. Further other classifier can be used to get more accurate prediction.

REFERENCES

[1] Mr. Ting-Wei "End-to-End Ser with Gender Information" - August 2020j. Clerk Maxwell, *A Treatise on Electricity and Magnetism*, 3rd ed., vol. 2. Oxford: Clarendon, 1892, pp.68–73.

[2] Mr. Mohamed Mbarki, Mr. Catherine Cleder *"Automatic Ser Using Machine Learning"*- March 2019

[3] Mr. Suresh Dara, Ms. Priyanka Tumma *"Feature Extraction Using Deep Learning"*- 2018

[4] Mr. Mohit Wadhwa, Mr. Anurag Gupta *"Speech Based Ser Using Machine Learning"* - july 2020

[5] Vineel Pratap, Qiantong Xu, Jacob Kahn, Gilad Avidov, Tatiana Likhomanenko, Awni Hannun, Vitaliy Liptchinsky, Gabriel Synnaeve, Ronan Collobert *"Scaling Up Online Speech Recognition Using Convnets"*, January 10, 2020

[6] Jian Wang, Zhiyan Han, *"Research Onspeech Emotion Recognition Technology Based on Deep and Shallow Neural Network"*, Proceedings of the 38th Chinese Control Conference July 27–30, 2019, Guangzhou, China.

[7] Taiba Majid Wani, Teddy Surya Gunawan, Syed Asif Ahmad Qadri, Mira Kartiwi,Eliathamby Ambikairajah "A Comprehensive Review of Speech Emotion Recognition Systems", *IEEE Access*

[8] Raghav Mittal, Satya Vart, Prayag Shokeen, Manoj Kumar."Speech Emotion Recognition", *2022, 2nd International Conference on Intelligent Technologies (conit)*.

[9] Joyjit Chatterjee, Vajja Mukesh, Hui-huang Hsu, Garima Vyas, Zhen Liu "Speech Emotion Recognition Using Cross Correlation and Acoustic Features", *2018 IEEE 16th int. Conf. on Dependable, Autonomic & Secure Comp., 16th Int. Conf. on Pervasive Intelligence & Comp., 4th Int. Conf. on Big Data Intelligence & Comp., and 3rd Cyber Cci. & Tech. Cong,.*

[10] Lalitha S., Anoop Mudupu, Bala Visali Nandyala, Renuka Munagala "Speech Emotion Recognition Using dwt", *2018 IEEE 16th Iintl Conf on Dependable, Autonomic and Secure Computing, 16th Intl Conf on Pervasive Intelligence and Computing, 4th Intl Conf on Big Data Intelligence and Computing and Cyber Science and TechnologyCongress(dascl/picom/datacom/cyberscitech)*

[11] Apoorv Singh, Kshitij Kumar Srivastava, Harini Murugan "Speech Emotion Recognition Using Convolutional Neural Network (cnn)" June 2020 *International Journal of Psychosocial Rehabilitation* 24(8)

[12] Mustafa, M.B., Yusoof, M.A.M., Don, Z.M. "Speech Emotion Recognition Research: An Analysis of Research Focus", *International Journal of Speech Technology* Volume 21, Pages137–156 (2018)

Artificial Intelligence, Blockchain, Computing and Security – Dagur et al. (Eds)
© 2024 The Author(s), ISBN: 978-1-032-49393-0

A predictive approach of property price prediction using regression models

Saad Khan, Shikha Singh, Bramha Hazela & Garima Srivastava
Amity School of Engineering and Technology, Amity University (Uttar Pradesh), Lucknow Campus

ABSTRACT: Machine learning has played a crucial role in recent decades in the practice of medicine, in everyday speech commands and promotional messages. Machine learning is crucial to the operations of many of the leading companies of today, like Uber, Google and Facebook. Machine learning has become a major competitive differentiator for many firms. The property market is one of the most competitive and price-focused in existence today. People are attempting to purchase a new home by considering their budgets and market strategy. The primary drawback of the existing method is that it determines the price of a house without the essential foresight into potential future market movements, which leads to an increase in price. This paper's primary goal is to accurately anticipate the price of a house without incurring any losses. For the purpose of estimating home prices and attempting to provide clients with effective house pricing that respects both their goals and their budget, numerous elements must be taken into account. Therefore, the paper focuses on a model to estimate housing costs by utilising algorithms such as Multiple Linear Regression, Lasso Regression and Polynomial Regression. This strategy will enable consumers to fund bequests without turning to a broker. The findings of this study declare Polynomial Regression model as the best fit as it has the highest degree of accuracy.

Keywords: Multiple Linear Regression, Lasso Regression, Polynomial Regression, RMSE, RSS, R2.

1 INTRODUCTION

Buying a house has always been a dream of every common man. With all his hard-earned money one goes into market to purchase one for him. A lot of web sites are available online which claims that they offer a variety of homes that suits once demands. There are various parameters on which a genuine buyer is confined to, like its location, how far is it from a hospital, a school, a cinema hall, shopping complex, railway station, bus stand or an airport. How connected that area is with local convenience etc. These are some points which every buyer has it in his mind before he goes for purchasing a house. Second is commercial property buyers. Their viewpoint of buying captures certain different issues. Real estate market has been at its boom. Investment in real estate is also considered for gaining monetary benefits. An area of artificial intelligence (AI) called machine learning employs tools and formulas to extract knowledge from data. Today, a variety of machine learning techniques are employed to address difficulties that arise in the actual world. Hundreds of homes are sold each day. In this paper, a machine learning algorithm is proposed to predict the cost of a house utilising data about the home (its size, the year it was built in, etc.). Zulkifley *et al.* (2020) The real estate market aids a great deal to a country's economy. The house purchase leads to the buying of utensils, furnitures, building houses provides employment to builders and construction labourers more over it also boost the sale of raw materials such as bricks,

DOI: 10.1201/9781003393580-112

cement, plaster etc. Thus it clear adds a great deal to other different industries as well Various Machine Learning techniques, including Lasso Linear Regression, Multiple Linear Regression, and Polynomial Regression, are used in this paper to forecast the house's price. The goal of this effort is to accurately recommend and anticipate the house price based on user parameters by employing various machine learning techniques, as the field desperately requires automation and expertise.

The paper is set up as follows. The Introduction is Section 1, and the Literature Review, which summarises and provides the results of earlier work that served as the background research for this paper, is Section 2. The Methodology used for this project and a flowchart of the suggested procedure, the Machine Learning Models used in this project are described in Section 3; the Model Evaluation component of each model, premised on several Evaluation Metrics and Result with discussion, is highlighted in Section 4, followed by Conclusion in section 5.

2 LITERATURE REVIEW

All the papers enlisted in the Table 1 were thoroughly studied, referred to and analysed during the course of this work. Every paper had its own set of evaluation metrics, techniques, methodology, implementation, and result.

Table 1. Various papers on house price prediction using different ML algorithm.

Authors	Methodology Used	Result/ Accuracy
Zhou et al. (2021)	Polynomial Regression	R2 score Polynomial Regression-0.88
Wang et al. (2018)	Random Forest, Linear Regression	R2 Random Forest-0.6391 Linear Regression-0.344 RSME Random Forest- 367.054 Linear Regression- 407.904
Varma et al. (2018)	Random Forest, Ridge Regression, Lasso Regression	Root Mean Square Error: Simple Linear Regression-4.701 Ridge Regression-1.956 Lasso Regression-2.833
Truong et al. (2020)	Random Forest, XGBoost, Hybrid Regression, Stack Generalized Regression	Root Mean Square Error: Random Forest-0.16568 XGBoost-0.16603 Hybrid Regression-0.16372 Stack Generalized Regression-0.1635
Gao et al. (2022)	XG Boost, SVR, MLR, and ANN	Random Forest error: Root Mean Square: 0.156 Regression on the Ridge: 0.102 Lasso Regression-0.116

3 METHODOLOGY AND IMPLEMENTATION

The aim of this research is to gain a better understanding of machine learning regression methods by analysing how accurately they can forecast house prices. Multiple linear regression, Lasso, polynomial regression, and elastic net regression will be utilized to assess the precision of the forecasts. In this study, multiple linear regression, Lasso, polynomial regression, and elastic net regression are used to assess the accuracy of house price predictions. Consequently, the goal of this research is to comprehend machine learning regression methodologies in greater detail. Mahale & Aditi *et al.* (2022), In order to improve performance, the given datasets need also be processed. This is accomplished by determining the necessary features and using one of the sampling techniques to remove the undesirable variables because each house has certain traits that are used to estimate its price. These attributes might or might not be common to all homes, therefore they have a different impact on house pricing, which leads to misleading results. The research provides answers to the following questions: Which machine learning approach probably works better and produces the most accurate results for predicting property prices? Then why? Figure 1 shows the basic flow diagram of the methodology adopted for proposed model.

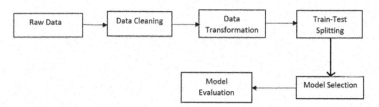

Figure 1.   Proposed model flow diagram.

3.1   *Multiple linear regression*

To calculate the association among two or more independent variables and one dependent variable, utilise multiple linear regression. In Jha *et al.* (2020) and Thamarai *et al.* (2020) demonstrated a statistical method known as multiple linear regression (MLR), usually referred to as multiple regression, employs several explanatory factors to forecast the result of a response variable. OLS (linear) regression with only one explanatory variable is expanded in multiple regression.
   Training Metrics of the Multiple Linear Regression Model:
   R2: 0.6789097089550895, RSS: 466429810296572.75, MSE: 1094905657973.1754,
   RMSE: 1046377.3974877207
   Testing Metrics of the Multiple Linear Regression Model:
   R2: 0.6866794976385522, RSS: 116042808105904.73, MSE: 1084512225288.8295
   RMSE: 1041399.1671250891

3.2   *Lasso regression*

The linear regression process known as "lasso regression" employs shrinkage. Madhuri *et al.* (2019) Shrinkage is the term used when data values move closer to a middle value, such as the mean. The lasso technique encourages simple, sparse models (that is models with lesser attributes and features). Satish *et al.* (2019) In order to create a regression model that is simpler to understand, some of the s are reduced to exactly zero. The L1 penalty's severity is controlled by the tuning parameter. Basically, represents the shrinkage: No arguments are deleted when $\lambda = 0$. The estimate matches the one obtained using linear regression. As rises, an increasing number of coefficients are zeroed out and removed (theoretically, when $\lambda =$, all coefficients are removed). Bias increases as λ rises. Variance increases as λ falls. Training Metrics of the Lasso Regression Model:
   R2: 0.6789097088318172, RSS: 466429810475643.5, MSE: 1094905658393.5294, RMSE:
1046377.3976885823
   Testing Metrics of the Lasso Regression Model:
   R2: 0.6866804541048077, RSS: 116042453864706.62, MSE: 1084508914623.4265
   RMSE: 1041397.5775962927

3.3 *Polynomial regression*

An nth degree polynomial is how the "Polynomial Regression" regression method depicts the relationship between a dependent variable (y) and an explanatory variable (x). In machine learning, it is sometimes referred to as the Multiple Linear Regression Special Case. It is a linear model that has undergone many modifications to increase accuracy. The polynomial regression training set has a non-linear structure. It makes use of a linear regression model to fit the complicated and nonlinear functions and datasets. Zhou *et al.* (2021) said that this leads to the statement that "In polynomial regression, the principal components are transformed into polynomial features of the requisite degree (2, 3,..,n) and then modelled using a linear model." Phan & The Danh (2018) On the contrary hand, a curve that fits the polynomial model is appropriate to cover the majority of the data points. Therefore, we should utilise the Polynomial Regression model rather than the Simple Linear Regression if the datasets are organised in a non-linear way.

Training Metrics of the Polynomial Regression Model:

R2: -6.504867165254966e+17, RSS: 2.4091722262761236e+32, MSE: 2.2515628282954427e+30, RMSE: 1500520852336095.5

Testing Metrics of the Polynomial Regression Model:

R2: 0.6866804541048077, RSS: 116042453864706.62, MSE: 1084508914623.4265, RMSE: 1041397.5775962927

4 RESULT AND DISCUSSION

The R2 is also called the coefficient of Determination. In any regression model, the R2 plays a major role in terms of the determining the errors in as well as the accuracy of the model. The R2 metric elaborates the variance that the independent variables have in the model. It is also of major use in calculating the model's accuracy. Table 2 reflects the comparative evaluation metrics for different regression models. Figure 2 shows the RMSE training and test score of Multiple Linear Regression, Lasso Regression and Polynomial Regression respectively.

Table 2. Comparison among all models based on various evaluation metrics.

Model	Train R2	Test R2	Train RSS	Test RSS	Train MSE	Test MSE	Train RMSE	Test RMSE
MLR	0.687910	0.668695	4.664298	1.1604	1.0949	1.0284	1.0463	1.0413
LR	0.6549	0.66416	4.246	1.1623	1.0846	1.069	1.0458	1.0628
PR	0.9953	5.34715	6.7324	2.093	1.593	1.135	1.2951	1.3796

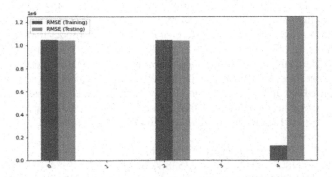

Figure 2. Bar graph showing the RMSE training and test score of MLR, LR and polynomial regression.

Table 3. Accuracy and RMSE of all models.

S. No.	Algorithm	Accuracy	RMSE
1.	Linear Regression	67.98	0.1304
2.	Lasso Regression	67.89	0.1304
3.	Polynomial Regression	99.53	0.1061

Table 3 shows, accuracy and RMSE values for three of the regression models. Better the model, lower the RMSE! Additionally, the model should closely match the training and testing results. It can be claimed that polynomial regression significantly overfit the current problem for this issue. The MLR Model's surprising simplicity produced the best outcomes.

5 CONCLUSION

We were able to gain some understanding of the feature-set by graphing the distribution of the data and their relationships. The features had a high degree of multicollinearity, thus in the feature extraction process, we used the VIF Technique to narrow down the suitable features. Testing numerous algorithms using the default hyperparameters helped us discover how different models performed on this particular dataset. While many regression algorithms are safe to utilise because their scores were very close and they are more generalizable, polynomial regression (order-2) was indeed the overfitting method. In light of our investigation, we decided to incorporate the algorithms (Linear Regression, Lasso regression, and Polynomial Regression) with RMSE values less than 0.106 and set the threshold value of RMSE at 0.106. The precision is definitely improved by this. The future development of these places may benefit from the information in this paper.

REFERENCES

Gao, Guangliang, *et al.* "Location-centered House Price Prediction: A Multi-task Learning Approach." *ACM Transactions on Intelligent Systems and Technology (TIST)* 13.2 (2022): 1–25.
Ho, Winky KO, Bo-Sin Tang, and Siu Wai Wong. "Predicting Property Prices with Machine Learning Algorithms." *Journal of Property Research* 38.1 (2021): 48–70.
Jha, Shashi Bhushan, *et al.* "Housing Market Prediction Problem using Different Machine Learning Algorithms: A Case Study." *arXiv preprint arXiv*:2006.10092 (2020).
Madhuri C. R., Anuradha G. and Pujitha M. V., "House Price Prediction Using Regression Techniques: A Comparative Study," *2019 International Conference on Smart Structures and Systems (ICSSS)*, Chennai, India, 2019, pp. 1–5, doi: 10.1109/ICSSS.2019.8882834.
Mahale, Aditi, Vidya Bhistannavar, Nikita Chauhan, Vedang Matey, and AjitKumar Shitole. "*Housing Price Prediction Using Supervised Learning.*"
Phan, The Danh. "Housing Price Prediction Using Machine Learning Algorithms: The Case of Melbourne City, Australia." *2018 International Conference on Machine Learning and Data Engineering (iCMLDE)*. IEEE, 2018.
Satish, G. Naga, *et al.* "House Price Prediction Using MachineLearning." *Journal of Innovative Technology and Exploring Engineering* 8.9 (2019): 717–722.
Thamarai, M., and S. P. Malarvizhi. "House Price Prediction Modeling Using Machine Learning." *International Journal of Information Engineering & Electronic Business* 12.2 (2020).
Truong, Q., Nguyen, M., Dang, H. and Mei, B., 2020. Housing Price Prediction via Improved Machine Learning *Techniques*. *Procedia Computer Science*, 174, pp.433–442.
Varma, A., Sarma, A., Doshi, S., & Nair, R. (2018, April). House Price Prediction Using Machine Learning and Neural Networks. In *2018 Second International Conference on Inventive Communication and Computational Technologies (ICICCT)* (pp. 1936–1939). IEEE.
Wang, Changchun, and Hui Wu. "A New Machine Learning Approach to House Price Estimation." *New Trends in Mathematical Sciences* 6.4 (2018): 165–171.
Zhou, Chenhao. "House Price Prediction Using Polynomial Regression with Particle Swarm Optimization." *Journal of Physics: Conference Series*. Vol. 1802. No. 3. IOP Publishing, 2021.
Zulkifley, Nor Hamizah, *et al.* "House Price Prediction using a Machine Learning Model: A Survey of Literature." *International Journal of Modern Education & Computer Science* 12.6 (2020).

Artificial Intelligence, Blockchain, Computing and Security – Dagur et al. (Eds)
© 2024 The Author(s), ISBN: 978-1-032-49393-0

Biological immune system based risk mitigation monitoring system: An analogy

Nida Hasib
Amity Institute of Information Technology, Amity University Uttar Pradesh, Lucknow

Syed Wajahat Abbas Rizvi
Amity School of Engineering and Technology, Amity University Uttar Pradesh, Lucknow

Vinodani Katiyar
Dr. Shakuntala Misra National Rehabilitation University, Lucknow

ABSTRACT: Risk mitigation, monitoring, and management is a detailed and ongoing procedure that seeks to move a project that may fail to a safer shore. Researchers developed the theory and mechanisms of the Artificial Immunity based system for successful risk free software development using the concepts and theories of the Biological Immune System. The authors discovered that there is a deficiency of effective risk mitigation monitoring in software project development compared to the recommended alternatives. The results of this study, which looked at immune system activity and existing risk mitigation approaches, are crucial in assisting firms with enhancing their risk mitigation processes. The limitations of previous research as well as potential future possibilities in these areas are highlighted. As a result, the information as the findings of this work could be published and further, in-depth research might be planned.

Keywords: Risk mitigation monitoring system, RMMS, Biological Immune System, BIS, Artificial immune system, AIS

1 INTRODUCTION

Risk arises due to interaction between the system and environmental conditions. Before any action is made, the system is created in such a way that the risk can either not exist or, if it does, will not lead to an accident. Mitigation of risk during the project development procedure will be very useful for project managers in an organization to produce the best quality software product on time and within budget. Mitigation and elimination will boost project success rates, give a timeline estimate, and enhance the quality of the final result. Various risk issues can arise when developing software in businesses or organizations. In the Artificial Immune system, we can use a variety of computational tools to assess and reduce numerous hazards that could obstruct an organization's development project. Lowering, removing, or transferring risks, will undoubtedly assist managers of that organization in running the system of project development smoothly. Numerous approaches have been developed for risk management, although some of them include failures or are incomplete (Masso *et al.* 2020); (Hilali *et al.* 2020); (Henri Jason & Evans 2020); (Felderer & Michael 2017); (Husin Wan Suzila *et. al* 2019); (Firdose *et al.* 2018); (Banu & Roja 2013). In order to address these issues, research is looking for a connection between the immune system and risk mitigation monitoring. This will deliver a new approach with a new description of the Risk mitigation process and project development process in organizations and enterprises (Al-Enzi *et al.* 2010); (Floreano *et al.* 2008); (Aickelin *et al.* 2013). Risk is the likelihood that

DOI: 10.1201/9781003393580-113

a selected course of action will result in a loss (an unfavourable result). The idea suggests that there is (or was) a decision that affected how things turned out. The word "risks" alone can refer to potential losses. Almost all human endeavours include some risk, although some are significantly more risky than others. Risk is also an unknown situation in which certain potential outcomes might result in a loss, catastrophe, or other unfavorable events. Measurement of risk is a set of possibilities each with quantified probabilities and quantified losses (Rich, Elaine *et al.*, Artificial Intelligence); (Sommerville, Software Engineering); (Sarkheyli *et al.* 2010). One element that can determine a project's success or failure is risk. The way we manage the risk has an impact on the project's budget and timeline as well. Our software sector can benefit much if the risk is anticipated, managed, and handled timely and in a structured manner. We found that BIS matches Boehm's risk management approach. When explored, a field of AIS that may be used to abstract the structure and operation of BIS suggested that BIS and risk mitigation monitoring were clearly analogous (Sarkheyli *et al.* 2010); (Sarkheyli *et al.* 2011).

On the basis of the literature review, it has been seen that AIS provides efficient solutions to many complex problems. Reduced risk of infection as a result of pathogen attacks is the main goal of the countermeasures used against pathogens. Physical barriers within the human body can also make it less likely for an organism to enter, decreasing its chances of surviving and procreating. infections are able to quickly adapt strategies to neutralize the defences put in place by the host to keep infections out of its body. The immune system is made up of this group of defences. The immune system offers a defense against the potential for each individual host cell to malfunction and fail (Hasib *et al.* 2023); (Barkley, Bruce, Project risk management).

2 ADAPTIVE IMMUNE SYSTEM

Once this mechanism is activated, only certain infections result in the production of antibodies. These antibodies are a sign of the type of infection to which the body has been exposed. If the same antigen stimulation occurs repeatedly, these antibodies can handle the infection and stop the development of disease in the body. Cells of the Adaptive Immune System exhibit a kind of immune memory and tend to improve their attack on specific antigens at every encounter. Lymphocytes perform the job of both recognition and elimination of the antigens. They are activated by interactions with antigenic material that result in the generation of immune memory. This happens after vaccination or exposure to a disease. In this particular system antigen presenting cells identify and detect an antigen from which two types of selections are made. When the cells exhibit high enough binding to MHC protein and self peptide, positive selection(non-self) occurred. When the cell doesn't have binding capacity to MHC protein and peptide, but binds to self-cells, the negative selection (self) is said to have occurred and results in apoptosis of most of those cells in the early stages of development. After positive selection(non-self) is made T-cell is generated, and it is generated in thymus. If the presentation yields recognition, the T-cell becomes activated and its lysis occurs. These cells then secrete lymphokines, that indicate the signal and stimulate several other components of the immune system. Lymphokines mix in the blood and attach to the B-cell which constitutes of Peptide receptor. B-cell formation starts from the fetal liver and continues in bone marrow throughout life. B-cell recognizes the signal of antigen given by lymphokines. That antigenic substance can bind to the B-cell peptide receptor, which triggers positive selection (non-self) and causes the B-cell to become activated and create memory cells and plasma cells. Memory cell stores the antigen in their memory whereas plasma cell creates antibodies. This whole procedure signifies the primary immune response. The memory cell is able to identify the same infection if it attacks again. As a result, plasma cells start making antibodies. This is referred to as a secondary immunological response. This is significantly more potent than the body's natural immunological reaction (Figure 1). If the vaccine is given to the person, to prevent the future attack of infections (antigen), it will

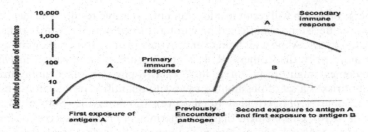

Figure 1.   Changes in antibody levels following exposure to antigens A and B.

activate T-cells & B-cell which in turn produce the memory cell and plasma cells. If the same or similar antigen attacks again then it can be protected by antibodies produced by memory cells and mutated or proliferated detectors by plasma cells to overcome changes in the antigen. Certain diseases and antigens cannot be vaccinated against and are therefore exceedingly dangerous. These antigens rapidly alter. As a result, the antibodies made by plasma cells have a decreasing affinity for that antigen over time (Figure 2).

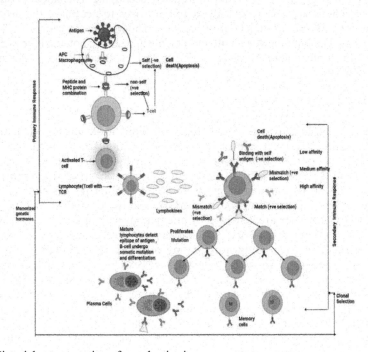

Figure 2.   Pictorial representation of an adaptive immune system.

3   BACKGROUND

There are several areas where the usage of AIS, or intelligent approaches, is employed to solve issues. They draw inspiration from the methods, principles, and models of the biological immune system. Theories of AIS serve as the foundation for a wide variety of techniques and algorithms, which in turn aid in enhancing the effectiveness of several technical procedures and processes.

Our analysis of AIS techniques and algorithms revealed several characteristics.

1. Adaptive: Capable of learning over time and forming adaptive behaviour.

2. Robust: able to process in the presence of ambiguous and imperfect data.
3. Resiliency: the immune system created with a bottom-up approach of Agent based paradigm, to realize a dynamic, heterogeneous, and distributed environment (Sycara *et al.* 1996).
4. Self-tolerance: The ability to control the immune system's self-reactive reaction to a particular antigen.
5. Lightweight: capable of processing with minimal computing complexity.
6. Distributed: able to process decentralized rather than centralized.

Inspiration from the remarkable properties expressed by the natural immune system led to the conception and design of Artificial Immune Systems exhibiting similar functionalities (Rich, Elaine *et al.*, Artificial Intelligence).

1. Recognition
   Numerous diverse patterns can be recognized, identified, and dealt with by the immune system.
2. Feature Extraction
   Before being delivered to other immune cells, such as lymphocytes, antigen-presenting cells (APC) in the immune system can extract antigen characteristics employing an antigen, which is a group of molecules that cause sickness, to filter out molecular noise.
3. Diversity
   The creation and preservation of immune system diversity involve two fundamental processes. In the first, gene segments from gene libraries are recombined to create receptor molecules. Somatic hypermutation is the second mechanism, which helps with immune system diversity. In response to invasive antigens, immune cells divide.
4. Self-regulation

The dynamics of immune systems make it so that local interactions, rather than a single point of control, determine the population. The immune system returns to its regular stable state after successfully combating a disease until it is required to respond to another antigen. This kind of self-control mechanism is specifically explained by the immunity network hypothesis.

Improved comprehension and research have led to the development of immune system ideas under a new area of artificial intelligence called the Artificial Immune System (AIS). The biological immune systems have a number of appealing qualities, including the capacity to remember, categorize, and reduce the impact of invading particles. The fundamental concept of the biological immune system is closely modeled in AIS. The creation of the Artificial Immune System makes use of numerous immune system components. The detector combines the characteristics of antibodies, T cells, and B cells. On the surface of lymphocytes are hundreds of thousands of detection symbols. These sensors attach to spots (called epitopes) on viruses. Since binding is based on chemical structure and charge, receptors are predicted to bind to a limited number of epitopes of the same type. The likelihood of a link forming increases with the receptor and epitope's affinities. (Timmis *et al* 2004); (Hofmeyr *et al* 2000). Memory cells and plasma cells are produced by the biological immune system's processes of positive and negative selection. Clonal selection then multiplies clones and differentiates them into distributed detectors based on modifications made to the antigen's nature. Clonal selection benefits from a scattered population of detectors and is adaptive in nature. An artificial immune system is built on the theories of immunological networks, clonal selection, and affinity maturation. In an AIS, the process of altering the network of antibodies to train the antigen patterns using the clonal selection theory involves the construction of a network of memory cells that can detect the presence of data clusters. To determine the affinity between the training pattern and the antibody, the Euclidean distance is calculated through training. The antibody with the highest affinities is chosen for cloning. To enhance antigen recognition, the mutation phase is added to the cloning process. The memory set is chosen from the modified clone of the antibody with the highest affinity.

763

Global searches benefit from an AIS. It also does a great job of postponing the introduction of regionally advantageous alternatives (Kuby, Immunology); (Aickelin & Dasgupta 2013); (Silva Costa & Dasgupta D 2015).

4 THE ANALOGY BETWEEN THE ADAPTIVE IMMUNE SYSTEM AND RISK MONITORING SYSTEM

The abstract characteristics of the immune system's cells are comparable to those of a risk monitoring environment. We have noticed that risk management monitoring in software development mimics the biological immune system. Prior to the software development process, primary precautions are taken and system analysis is carried out to generate solutions for risk reduction and elimination. The preterm body is vaccinated against a variety of diseases so that it can create antibodies to combat those antigens. If the same or similar antigens assault during the life cycle, the organism responds by using remembered or modified antibodies to combat the antigens. Antigen effects on organisms are diminished or completely abolished during the course of their life cycles as a result of the antigens' decreased affinity for antibodies as they change form. Antigen in the biological immune system equates to risk in risk mitigation monitoring. Negative and positive selections are made as the antigen is recognized by the antigen-presenting cell and are based on the analysis and ranking of risks once they have been discovered. Low ranked risks are negatively selected and are removed naturally (no need of any specific system to remove it, as it tries to combine with self cells. Positive selections in lead to the generation of T-cells which in turn leads to the activation of T-cell & B-cell and in turn production of antibodies, which leads to controlling decisions (solutions). Activated B-cell produces antibodies for antigens, those that are perfect or loose matching undergo cloning and mutation procedures for those, if the same or similar antigens attack in the future again then it is protected from antibodies that are cloned or mutated according to the removal of similar antigens. These procedures are done under memory cells and plasma cells, accordingly during the life cycle the same antibodies or mutated antibodies are used for similar antigens and day by day cause of infections is reduced or eradicated for a large number of infections during the lifecycle. Similarly, this procedure is very helpful in minimizing or eradicating risks that will arise during software development. It is applied to the risk mitigation monitoring process to reduce risks and apply connective action where necessary. During project development, secondary measures are taken into account, if risks occur during the software development process and it resembles the documented risks with distributed solutions then on the basis of the best possible solution selection criteria, the solution is applied. Those risks whose resemblance is very low or negligible are moved through the primary measures. Clones of the best possible solutions with similar functionalities are produced for other similar risks that can occur in the future.

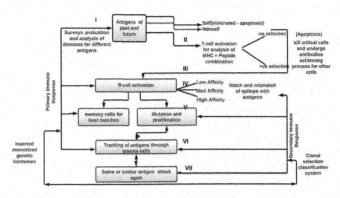

Figure 3.   Adaptive immune system.

If the antigens whose antibodies are created beforehand, attack the body again during the life cycle, then the corresponding infection is removed through the selected antibodies. For new antigens, it needs a dose of primary immune response and it also refers to other cloned and mutated antibodies. Antigens having less affinity with the antibodies needs more dose of Primary immune response (Figure 3).

There are some risk factors, whose nature changes continuously, thus there can't be any fixed possible solution. There is no way out to develop the measures for this risk. As if measures are taken for a possible solution, these risk changes its effect on the future of the project. On the basis of this analogy with the immune system, we found out that by using the concept of an increase in the population of detectors, managers can able to solve this problem (Sarkheyli & Ithnin, A. Binti 2010; Sommerville. Software Engineering.: Pearson education). Best detectors are cloned into a number of similar detectors and mutated and differentiated detectors according to changes in the nature of antigens, as described in (Figures 3–4).

5 CONCLUSION AND FUTURE SCOPE

The Biological Immune System is a highly evolved system that protects our bodies against the illnesses to which they are occasionally exposed. This motivates us to create a comparable risk mitigating strategy while we are developing software. We drew an analogy to show how this might result in a functional system. As seen in (Figures 3–4) there is step-to-step matching between BIS and RMMS. As a result, we can draw the conclusion that a BIS based, effective framework for risk mitigation is feasible. The BIS approach can be used to produce an AIS based risk mitigation model that can be integrated into the software development process and useful at different phases of risk reduction during the development process. In this study, the main emphasis is on developing and offering an efficient reaction to risk in order to prevent it rather than treat it. There has been a lot of effort done to assess the risks, but little of it is related to the creation of effective remedies to the hazards. We have also considered risk mitigation strategies while taking into account each risk seriousness, the likelihood of occurrence, and potential effects at any or all stages of the software development life cycle. For each type of risk, a corrective solution is provided as part of this study contribute to risk avoidance. Many risk factors have been identified as the root of a sizable fraction of software project failures. These risk variables have previously undergone thorough identification and evaluation. But even so, these risks develop into issues that ultimately lead to project failure. For any type of risk, we have solutions and plans for risk mitigation. It is reasonable to assume that each risk likelihood and impact factor will have an impact on the mitigation and contingency strategy. It is crucial to understand how many phases of the software development life cycle those plans should cover. It may become more frequent in one phase and inevitably pick up speed in the other, making risk difficult to completely remove. This study

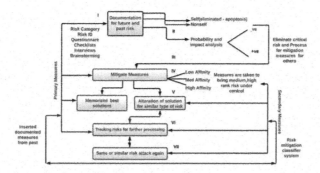

Figure 4. Proposed Risk Mitigation Monitoring System (RMMS).

might provide the foundation for further developing risk mitigation monitoring strategies. Despite the fact that there are methodologies for risk management that are currently known, they frequently fail due to factors such as lack of a specific function to recover from or avoid risks or attacks against organizations, poor executive support, high implementation costs, tardy responses, insufficient accountability, inability to qualitatively measure the control environment, infrequent assessment, and inaccurate data. As a comprehensive system, the immune system, for example, may serve as the finest model for replicating it with risk mitigation monitoring in companies for software development. Research is being done to enhance the current approaches by making them intelligent like the natural management systems as a result of these problems and the significance of risk management.

In this study, we provide a new method for a risk mitigation monitoring system that uses the algorithms and features of AIS to replicate the biological immune system, one of the most comprehensive systems in nature. Despite the fact that there are numerous approaches for controlling risk, some of them have issues. Therefore, the study aims to identify them through comparison and activity research. As a result, risk mitigation monitoring is the new definition of risk management that applies to this study. However, a strategy for further research is to try to develop a procedure using a synthetic immune system and test the suggested Risk mitigation monitoring procedure against specific threats in a real world environment.

REFERENCES

Aldhaheri, Sahar & Alghazzawi, Daniyal. 2020. Artificial Immune Systems Approaches to Secure the Internet of Things: A Systematic Review of the Literature and Recommendations for Future Research. *Journal of Network and Computer Applications*, DOI: 10.1016/j.jnca.2020.102537: 1084–8045.

Aickelin, U. & Dasgupta, D. 2013. Artificial Immune Systems. *Introductory Tutorials in Optimization and Decision Support Techniques*, DOI: 10.1007/978-1-4614-6940-7_7

Al-Enezi, J.R. & Abbod, M.F. 2010. Artificial Immune Systems-Models, Algorithms and Applications. *International Journal of Research and Reviews in Applied Sciences*, 118–131.

Banu, Roja & MS. 2013. *An Analysis of Development of rmmp by Focusing Risk Evaluation and Continuous Mitigation Plan for Unlike Risks*. http://hdl.handle.net/10603/133625

Barkley, Bruce T. 2004. *Project Risk Management*.: TMH companies.

Brosseau, Jim. *Software Teamwork Taking Ownership for Success*.: Pearson education.

Brownlee, Jason. 2005. *Artificial Immune Recognition System (AIRS) a Review and Analysis*. Technical Report No. 1-02:1–44.

Caldas, B. J. B., Oliveira, F. R. S. & Neto, F. B. d. L. 2008. Improving Support of Appropriate Executive Decisions by Combining Artificial Immune Systems and Fuzzy Logic. *10th Brazilian Symposium on Neural Networks*, Salvador, Brazil, doi: 10.1109/SBRN.2008.13.: 99–104.

Cunha, Jose C.S. & Demirdal, B. 2005. Use of Quantitative Risk Analysis for Uncertainty Quantification on Drilling Operations-Review and Lessons Learned. *SPE Latin American and Caribbean Petroleum Engineering Conference*, Rio de Janeiro, Brazil. doi: https://doi.org/10.2118/94980-MS

Felderer, Michael & Auer, Florian .2017. Risk Management During Software Development: Results of a Survey in Software Houses from Germany, Austria and Switzerland. *International Workshop on Risk Assessment and Risk-driven Testing*, DOI: 10.1007/978-3-319-57858-3_11: 143–155.

Firdose, Salma & Rao, L. Manjunath 2018. PORM: Predictive Optimization of Risk Management to Control Uncertainty Problems in Software Engineering. *International Journal of Electrical and Computer Engineering* Vol. 8, ISSN: 2088-8708, DOI: 10.11591/ijece. v8i6.: 4735–4744

Floreano, Dario & Mattiussi, Claudio. 2008. Bio-Inspired Artificial Intelligence Theories, Methods, And Technologies. *Intelligent Robotics and Autonomous Agents series*, ISBN:9780262062718, The MIT Press: 674

Garrett, Simon M. 2005. How Do We Evaluate Artificial Immune Systems? *Evolutionary Computation*, DOI: 10.1162/1063656054088512, 145–177.

Gong, Tao. 2008. Artificial Immune System Based on Normal Model and Immune Learning. *IEEE International Conference on Systems, Man and Cybernetics*, ISSN:1062-922X, DOI:10.1109/ICSMC.2008.4811468.: 1320–1325

Gupta, Shikha, & Saini, A. K. 2018. *An Artificial Intelligence-Based Framework for Risk Management of IT Systems*. Teerthanker Mahaveer University- A Theses, Http://Hdl.Handle.Net/10603/226798.

Hasib, N., Rizvi, S.W.A. & Katiyar, V. 2023. Artificial Immune System: A Systematic Literature Review. *Journal of Theoretical and Applied Information Technology, Publisher Little Lion Scientific*, ISSN: 1992–8645, 101(4): 1469–1486, Retrieved from www.scopus.com

Henri, Evans Jason 2020. *A Review of Risk Management in Different Software Development Methodologies.* https://www.researchgate.net/publication/343584508: 1–10

Hilali, Raqiya Ahmed Al & Sudevan, Smiju. 2020. *Software Project Risk Management Practice in Oman.* DOI: 10.22161/eec.563: 32–47.

Hofmeyr, Steven A. & Forrest, Stephanie A. 2000. Architecture for an Artificial Immune System", *Journal of Evolutionary Computation*, DOI: 10.1162/106365600568257: 443–473.

Hosseinpour, Farhoud & Abu Bakar, Kamalrulnizam. 2010. Survey on Artificial Immune System as a Bio-Inspired Technique for Anomaly Based Intrusion Detection System. *IEEE International Conference on Intelligent Networking and Collaborative Systems*, ISBN: 978-0-7695-4278-2/10, DOI 10.1109/INCOS.2010.40.

Husin, Wan Suzila Wan & Yahya, Yazriwati. 2019.Risk Management Framework for Distributed Software Team: A Case Study of the Telecommunication Company. *Fifth information systems international conference*, https://doi.org/10.1016/j.procs.2019.11.113: (169) 178–-186

Kuby. Immunology: W.H. Freeman.

Liu, Caiming & Guo, Minhua. 2010. Artificial Immunity-based Model for Information System Security Risk Evaluation. *International Conference on E-Health Networking Digital Ecosystems and Technologies*, Shenzhen, doi: 10.1109/EDT.2010.5496552.: 39–42

Masso, Jhon & Pino J., Francisco. 2020. Risk Management in the Software Life Cycle: A Systematic literature review. *Computer Standards & Interfaces* 71, ISSN 0920-5489, https://doi.org/10.1016/j.csi.2020.103431.

Negi, Pallav. 2006. *Artificial Immune System Based Urban Traffic Control.* A Theses, Texas A&M University

Novotny, Alma, Biochemistry and Cell Biology Lecturer Department of Biosciences, Ph.D. Rice University, "*Fundamentals of Immunology: Innate Immunity and B-Cell Function*", Coursera.Org/Verif Y/8PVNX66M8R8J, A Course Authorized by Rice University and offered through Coursera.

Rich, Elaine & Knight, Kevin. *Artificial Intelligence.*: McGraw Hill.

Roy, Bibhash & Dasgupta Ranjan. 2015. A Study on Risk Management Strategies and Mapping with SDLC. *2nd International Doctoral Symposium on Applied Computation and Security Systems*, DOI:10.1007/978-81-322-2653-6_9: 1–12

Sarkheyli & Ithnin, A. Binti. 2010. Improving the Current Risk Analysis Techniques by Study of their Process and Using the Human Body's Immune System. *5th International Symposium on Telecommunications*, ISBN:978-1-4244-8183-5: 651–656.

Sarkheyli, Azadeh & Ithnin, Binti. 2011. Study of Immune System of Human Body and Its Relationship with Risk Management in Organizations. *5th International Symposium of Advances on Science and Technology*, SASTech

Shahzad, B., & Safvi, S.A. (2008). Effective Risk Mitigation: A User Perspective. Corpus ID: 112134333

Shuqin, C. & Ge, W. 2008. Research on the Application of Immune Network Theory in Risk Assessment. *International Colloquium on Computing, Communication, Control, and Management*, Guangzhou, China, doi: 10.1109/CCCM.2008.324.: 230–-235

Silva, G. Costa & Dasgupta, D. 2015. A Survey of Recent Works in Artificial Immune Systems. *Handbook on Computational Intelligence*. World Scientific, DOI: 10.1142/9789814675017_0015: 547–586.

Sommerville. *Software Engineering.*: Pearson education.

Timmis, J. & Knight, T. 2004. An Overview of Artificial Immune Systems. *Computation in Cells and Tissues*, Chapter of Natural Computing Series, DOI: 10.1007/978-3-662-06369-9_4: 51–91

Trivedi, Ankita & Shrivastava, Amit 2018. Survey Analysis on Immunological Approach to Intrusion Detection. *IEEE International Conference on Advanced Computation and Telecommunication*, ISBN:978-1-5386-5367-8, DOI: 10.1109/ICACAT.2018.8933710

Vasilyev, Vladimir & Shamsutdinov, Rinat. 2019. Distributed Intelligent System of Network Traffic Anomaly Detection Based on Artificial Immune System. *7th Scientific Conference on Information Technologies for Intelligent Decision-Making Support*, ISSN 1951-6851https://doi.org/10.2991/itids-19.2019.7, ISBN 978-94-6252-728-7: 40–45.

Watkins, Andrew & Timmis, Jon 2004. Artificial Immune Recognition System (AIRS): An Immune Inspired Supervised Learning Algorithm. *Genetic Programming and Evolvable Machines*, 5, DOI 10.1023/B:GENP.0000030197.83685.94: 291–317.

Yang, Bo & Yang, Meifang 2021. Data-Driven Network Layer Security Detection Model and Simulation for The Internet of Things Based on An Artificial Immune System. *Neural Computing and Applications* 33, DOI: 10.1007/s00521-020-05049-5: 655–666.

Artificial Intelligence, Blockchain, Computing and Security – Dagur et al. (Eds)
© 2024 The Author(s), ISBN: 978-1-032-49393-0

A review on malicious link detection techniques

Ashim Chaudhary, K.C. Krishna, Md Shadik & Dharm Raj
School of Engineering & Technology, Sharda University, Greater Noida

ABSTRACT: Recent advancement in technology has led to more and more use of the internet and its services. Many people use the internet for their day-to-day tasks like buying things from the internet, chatting with friends through social media, sending mail, and it is also used by many businesses and corporations. We all know that the Internet has made our connectivity more efficient and we are totally dependent on it. Though the internet has many advantages, one of its disadvantages can be taken as the privacy issue and theft of data via unauthorized access. There are many ways through which hackers or eavesdrops try to get people's personal data and one of the ways is by sending or embedding malicious links or phishing links via mail attachments or embedding it on the website. So different ways have been developed to detect such malicious links. Their design includes features for a variety of harmful detection techniques, including blacklist approach, whitelist approach, visual-similarity based, content-based and mostly URL-based approach. Each has unique benefits and disadvantages. In this overview study, we mainly focus on machine learning-based strategies for detection of malicious URLs. In this study, we explore the various detection processes and approaches used by URL-based features in order to comprehend their structure. The performance based on various dataset's combinations of URL characteristics is then examined. In order to encourage the development of improved URL-based phishing detection systems, we wrap up our research.

Keywords: URL Detection, Machine Learning, Security. Phishing Detection

1 INTRODUCTION

Spreading malware and getting hacked with a simple click on the malicious URL is a growing concern for people using the internet. Phishing attacks, which use malicious websites linked to email, SMS, or other forms of communication to trick people, are based on social engineering and malware [1]. The most common was the Phishing prevalent sort of cyber attack in 2021, and the number of instances nearly quadrupled from 241,342 in 2020 to 323,972 in 2021, according to the FBI Crime Report 2021. [2]. In phishing, attackers create fake versions of well-known websites and lure users there, where they deceive them into divulging personal information such as giving their username and password, banking information and credit card numbers, and other credentials [3].

Attackers try to obtain a lot of data and/or money by targeting a broad spectrum of users. As per Kaspersky's analysis, how much an attack costs on average in 2019 varies depending on the attack's scope from 108K to $ 1.4 billion. A further spent amount is about 124 billion dollars on products and services connected to international security every year [4].

As per the APWG, there were 1,025,968 phishing attacks alone in the inital quarter of 2022. The second quarter of 2022 beat the previous record for the most phishing attacks in a single quarter in APWG history with a total of 1,097,811 attacks. Since the APWG started keeping track of them in early 2020, phishing attacks have grown. Between 68,000 to 94,000 assaults occurred per month at that time. [5].

DOI: 10.1201/9781003393580-114

Phishing Attacks, 3Q2021-Q2022

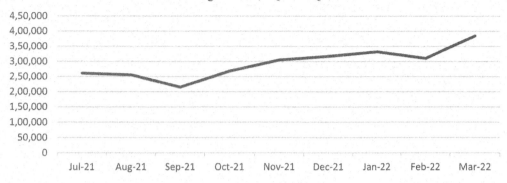

Figure 1. Phishing attacks from 3rd quarter 2021 to quarter 2022 [5].

So, detecting and blocking phishing websites is always an important area of research.

In this paper, we have presented various methods and techniques that can be used for the detection of malicious URL. We have also tried to present different algorithms that can be used for training the model and extract the feature from the raw data sets.

The structure of this paper has been divided into the following sections: An overview of the literature survey on phishing and its different detection methods is presented in Section 2. The anatomy of URL-based phishing is covered in Section 3, along with a study of the various features, nature of datasets, techniques, and assessment metrics. Section 4 offers different algorithms that can be used for the training of the model, Section 5 contains the summary and our insights. The conclusion and a list of references are included in Section 6.

2 LITERATURE REVIEW

The technology for identifying and stopping phishing attacks is continually changing, and they pose a severe threat. We have studied and tried to present different methods and techniques for detecting malicious URLs. Some of the techniques include blacklist, whitelist, content-based, URL-based approaches etc.

A brief summary of relevant research to detect malicious websites is provided in this section.

2.1 *Blacklist-based approach*

The blacklist strategy entails locating harmful websites, reporting the URLs, and logging them into a different database with pertinent information about the bad URL. By keeping a list of websites that have been blacklisted, such as Google Safe Browsing, web browsers employ this process to defend themselves against phishing attacks. It's a very old technique for detecting the malicious URL and much research has been done in the past. One of the papers by Fukushima *et al.* [6] presented a proactive blacklisting-based malicious URL detection where the reputation of the registrars and IP address block used by the attackers was analyzed. Additionally, PhishNet [7] foresees phishing attempts using a blacklisting strategy.

To identify new phishing websites, it uses five different algorithms: top-level domain , brand name, IP address, query string and directory structure. Over large datasets, it obtains a rate as 95% true positive and a 3% false positive, while being unable to detect zero-hour phishing sites.

Limitation to the blacklisting-based approach is also apparent, there is no enhanced generalization ability, making it unable to discover newly developing malicious domains.

The blacklist technique is simple to use, but storage of data and updating will be difficult because a significant number of malicious websites are added every day.

2.2 Whitelist-based approach

A white-list technique lists all trustworthy websites; any website that isn't on the list is deemed illegitimate and warns the user. Azeez et al. [8] in their paper has presented an automated whitelist approach to detect phishing URLs. In their model, prior to making a choice, the domain name and the items on the whitelist are compared, and then the IP address is matched. The similarities of the known trustworthy site are next determined by examining the actual link and the visual connection, and then further analyze them. Likewise, Mityukov et al. [9] have created a way to tackle phishing that consists of three primary modules: one that gets the URL of the web page viewed, one that filters based on a whitelist, and one that searches the login form.

Due to its size and rapidly rising traffic, creating a global whitelist of the whole World Wide Web is impractical.

2.3 Content-based approach

Mahmood Moghimi et al. [10] proposed an approach that consists of two different feature sets, each of which has four features. The first feature set assesses the web page's resource identification, and the second feature set determines the components' access protocols. Unlike third-party services like search engines, blacklists, and whitelists, these capabilities are unique. These two feature sets will be derived from the page document object model and the contents of web pages (DOM).Zuraiq et al. [11] also proposed a model based on content-based phishing URL detection. In the proposed model a weighted TF-IDF technique is then used to the text that was retrieved from the meta tag, title tag, and body tag after the system has extracted the identity words from the webpage and title tag, the domain name, and meta tag.

2.4 Visual similarity-based approach

By contrasting the purportedly dangerous site with the existing collection of resources, including logos, screenshots, favicons, and styles, visual similarity-based approaches identify phishing sites. Using a visual resemblance strategy, Rao and Pais [12] devised a method to identify phishing. As a first-level filter, they used a simple visual similarity-based blacklist technique to find almost identical phishing sites. To overcome the drawbacks of conventional blacklists, they have incorporated similarity-based characteristics (scripts, tag attribute values, anchor links, directories, pictures, styles, and filenames). Likewise, Sahar Abdelnabi et al. [13] proposed VisualPhishNet, the first similarity-based detection methodology to employ deep learning (specifically, convolutional triplet neural networks) to generate a stable visual resemblance among any two websites that are the same webpage screenshots, as opposed to depending on one-to-one matching.

2.5 URL-based approach

A model based on hyperlink analysis was developed by Ankit Kumar Jain et al. [14]. The method covered in this study uses links data taken directly from the source code of websites to distinguish between trustworthy and counterfeit websites. In order to assess whether the supplied website is a phishing scam, the recommended approach looks at the hyperlinks that were extracted from the page source. Total hyperlinks, internal hyperlinks, no hyperlink, internal error, external error, external hyperlinks, null hyperlink, internal redirect, login form link, external redirect, CSS external/internal favicon, and external/internal are the 12 categories into which we have broken down hyperlink characteristics. Harshal Tupsamudre *et al.*'s research [15] examined a variety of feature extraction methods which are basically based on word segmentation as well as n-grams.

2.6 *Other approaches*

Researchers employ a range of various methods for phishing detection. Different approaches include Deep Vision based approach [16], rule-based tree model [17], meta-heuristic-based approach [18], DNS based [19]. Many other researches are going on to make the phishing detection more robust.

3 ANATOMY OF PHISHING URL

We are using a URL-based approach to detect the phishing links. To better understand how URL-based detection is done, it is important to know the architecture of a phishing URL. In order to conduct URL-based phishing attacks, critical words or characters that:

  (i)   Mimic similar but misspelled terms,
 (ii)   Include redirecting special characters.
(iii)   Use URL shorteners.
(iv)   Use delicate, trustworthy terms.
 (v)   Insert a malicious file inside the URL, etc.

Figure 1 demonstrates URL phishing in action. Users give credential information to attackers when phishers pose as trustworthy websites because they are unaware that the website is a phony.

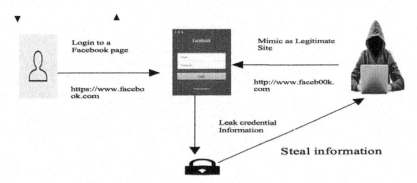

Figure 2.   Overview of phishing links.

3.1 *Types of URL*

Five main methods of URL obfuscation are used, according to PhishStorm [20]. The following list of instances and methods for obfuscating URLs with keywords in low-level domains, queries, and paths is provided in Table 1:

Type 1: Using different websites to obfuscate
Type 2: Using terms to obfuscate
Type 3: Typing-related domains
Type 4: IP address obfuscation
Type 5: Making use of URL shorteners to obfuscate

Table 1. Types of URL obfuscation.

Type	Sample
Type 1	https://sch00l497.ru/2022/www.pa-ypal. com/293745274276105805
Type 2	http://4uadr0deofertas.com.br0/www1. pay-pal/encrypted/ssl2018
Type 3	http://cgei-3.paypalsecure.de/info2/vedrikerdit.html
Type 4	http://89.72.140.98/java-seva/https://p ay-pal.com/uk/onepage-paypal.html
Type 5	http://g00.gl/HQpx5g

3.2 URL-based features

Researchers use a variety of features in the phishing detecting depending on their detection methods. In Table 2, we list the more popular characteristics in URL-based detection.

Table 2. Common features of URLs.

No.	Feature Name	Description
1	having_ip_address	This feature will determine whether or not the URL has an IP address.
2	abnormal_url	From the WHOIS database, this feature can be obtained. Identity is often included in a respectable website's URL.
3	google_index	This feature determines whether or not the URL has been indexed by Google Search Console.
4	Count dots	It counts for the occurrence of dots in the URL.
5	Count-www	If the URL has no www or more than one www, this feature aids in the detection of fraudulent websites.
6	count@	Counts "@" characters present in the URL.
7	Count_dir	Counts for multiple directories. Multiple directories in the URL are typically a sign of suspect websites.
8	Count_embed_domain	The quantity of embedded domains can aid in the identification of malicious URLs. You can do it by looking for the character "//" in the URL.
9	Suspicious words occurred in URL	Occurrence of these frequently used suspicious words in the URL has been identified as a binary variable, i.e., whether such words are present in the URL or not.
10	Short_url	The purpose of this feature is to show whether a URL has been shortened using a service, such as bit.ly, goo.gl, go2l.ink, etc.
11	Count_https	The HTTPS protocol's inclusion or exclusion in the URL is a crucial feature.
12	Count_http	Phishing or malicious websites frequently have many HTTPs in their URLs, whereas trustworthy websites typically have just one HTTP.
13	Count%	The number of "%" present in the URL are counted here.
14	Count?	The (?) symbol in the URL denotes the query string, which includes the data to be sent to the server. A URL with more ?s is unquestionably suspicious.
15	Count-	In order to make the URL appear legitimate, phishers and cybercriminals typically append dashes (-) to the brand name's prefix or suffix. An illustration. www.flipkart-india.com.
16	Count=	The equals sign (=) in a URL denotes that variables are being sent from one form page to another.Since anyone may alter the values in a URL to change the page, it is regarded as being riskier.
17	url_length	Safe URLs often have a short length. Therefore, the URL length is measured using this feature.
18	hostname_length	The hostname's length is a crucial element in identifying fraudulent URLs.
19	First directory length	This function helps determine how long the first directory of the URL should be.
20	Length of top-level domains	When identifying malicious URLs, the length of the TLD is also crucial. Since . com is the most common extension for URLs. TLDs between 2 and 3 typically denote secure URLs.
21	Count_digits	Counting the amount of digits in a URL is a key characteristic for identifying phishing URLs because safe URLs often do not include digits.
22	Count_letters	When recognising phishing URLs, the URL's letter count is also a key factor.

3.3 URL-based detection schemes

In prior research following two approaches have received the majority of attention:

- Engineering-based features (feature selection + feature extraction + classification)
- Algorithms-based (feature extraction + classification) detection

In the parts that follow, we talk about these plans.

3.4 Evaluation matrices

Here, we suppose that P stands for phishing and L stands for genuine websites and that N indicates the ratio of legitimate to phishing websites.

True Positive rate (TPR) measures how many phishing attempts were really recognized properly ($N_{P \to P}$) to all total phishing attempts ($N_{P \to P} + N_{P \to L}$). For more detail, see Equation (1).

The False Positive Rate (FPR) is the proportion of valid websites misidentified as phishing attacks ($N_{L \to P}$) to the total number of legitimate websites ($N_{L \to L} + N_{L \to P}$). For detail, see Equation (2).

The ratio of correctly categorized legitimate sites ($N_{L \to L}$) to all legitimate sites currently in existence ($N_{L \to L} + N_{L \to P}$) is known as the true negative rate (TNR). For more detail, see Equation (3).

False Negative Rate (FNR): The proportion of phishing attempts that are mistakenly identified as legitimate attacks ($N_{P \to L}$) to all phishing attacks ($N_{P \to P} + N_{P \to L}$). For more information, see Equation (4).

Precision (P) quantifies the ratio of successfully classified phishing assaults ($N_{P \to P}$) to all classified phishing attacks($N_{L \to P} + N_{P \to P}$). look at equation (5).

Comparable to the TP rate is Recall (R). Look at Equation (6).

Accuracy (ACC) is the percentage of accurately categorized phishing and legitimate websites ($N_{L \to L} + N_{P \to P}$) to total websites ($N_{L \to L} + N_{L \to P} + N_{P \to P} + N_{P \to L}$). To learn more, refer to Equation (7).

$$TPR = \frac{N_{P \to P}}{N_{P \to P} + N_{P \to L}} \tag{1}$$

$$FPR = \frac{N_{L \to P}}{N_{L \to L} + N_{L \to P}} \tag{2}$$

$$TNR = \frac{N_{L \to L}}{N_{L \to L} + N_{L \to P}} \tag{3}$$

$$FNR = \frac{N_{P \to L}}{N_{P \to P} + N_{P \to L}} \tag{4}$$

$$P = \frac{N_{P \to P}}{N_{L \to P} + N_{P \to P}} \tag{5}$$

$$R = TP \tag{6}$$

$$ACC = \frac{N_{L \to L} + N_{P \to P}}{N_{L \to L} + N_{L \to P} + N_{P \to P} + N_{P \to L}} \tag{7}$$

3.5 Datasets nature

Researchers use well-known websites to gather data, like OpenPhish and PhishTank for phishing and Alexa and DMOZ for reputable websites. Table 3 displays some of the popular websites.

Table 3. Sources of data.

Type of URL	Source of Data
Legitimate	DMOZ, digg58.com, gateway of the payment, Alexa, leading website for banking
Phishing	OpenPhish, PhishTank, VirusTotal, Malware Domains, MalewareDomainList, jwSpamSpy

Since the quantity of phishing sites and legal websites cannot be compared, we find that the majority of study relies on unbalanced data. To counteract dataset bias, numerous research employs balanced datasets.

4 ALGORITHMS USED FOR URL-BASED DETECTION

Here, we outline the machine learning techniques of phishing detection literature which are frequently used.

4.1 Naïve Bayes

The Naive Bayes (NB) classifier is a simple yet powerful classifier. P (x|y) in an NB classifier represents the feature vector's conditional probability given its label. x stands for the feature vectors, and y is a label designating a legitimate or phishing website (phishing denotes x = 0 and legal denotes y = 0). Considering that it can be challenging to discern between a trustworthy and phishing website, the following is the posterior probability that x belongs to y = 1:

$$P(y = 1|x) = \frac{P(x|y = 1)}{P(x|y = 1) + (x|y = 0)} \tag{8}$$

4.2 Support Vector Machine

A common classification and regression approach in machine learning is the Support Vector Machine (SVM). The best-separating hyperplane between two labels is found using SVM. It is represented by the nonnegative coefficients 3 and the kernel function K(x, x'), where they assesses the similarity of two different feature vectors. Which training cases are close to the decision border may be seen using SVM. Data is categorized based on how close it is to the decision boundary.

$$h(x) = \sum_{1}^{n} \propto i(2yi - 1)K(xi, x) \tag{9}$$

4.3 Random Forest

With the use of bagging, a Random Forest is constructed utilizing random attribute selection. Divide and conquer tactics (ensemble mechanism) are used by Random Forests to boost performance. In a random forest, several random subsets of trees are combined using the method. Averaging or weighing the average of the various results is used to get the final result. The strength of each individual classifier as well as the reliance between them affect accuracy, and they help to solve the overfitting issue with decision trees.

4.4 Convolutional Neural Network (CNN)

Convolutional Neural Networks (CNN) are a particular class of Neural Networks with Deep Learning used to research image processing. Compared to other image classification

Table 4. Commonly used algorithms.

No.	Algorithms	References
i	Logistic Regression	[21], [22], [23]
ii	Random Forest	[24], [25], [26]
iii	Support Vector Machine	[28], [29], [30]
iv	Naïve Bayes	[31], [32]
v	LSTM	[33], [34], [35]
vi	Neural Network	[36], [37], [38]
vii	k-means	[39], [40]
viii	Decision Tree	[27], [29]

techniques, CNN requires very little preprocessing. It differs from previous classes researchers' predetermined criteria for traditional phishing detection in that it learns features 4 on its own, which is a significant advantage in feature engineering. It is used to detect phishing by operating on character-level embeddings because it is primarily intended for picture classification. Convolution, pooling, and fully linked networks with non-linear activation functions are all present in CNN networks. Several algorithms used often in the field of phishing detection are listed in Table 4.

5 SUMMARY

Two views from the present detection systems were observed in our survey: the dataset perspective and the feature perspective.

Researchers typically examine the detection approach on imbalanced data, where the majority of sites are trustworthy, from a dataset perspective. As a result, the majority class becomes biased. In other words, the result is biassed even though there are many false positives in it. A workable solution to this issue is data oversampling on minorities, which strikes a compromise between data size and accurate automated minority-class data.

A variety of Subdomain count and URL length are two URL-based features that may also be skewed from a feature standpoint because of how heavily they rely on the dataset. To put it another way, many academics consider Alexa.com as a reliable dataset that only includes index pages from highly ranked websites. The whole URLs of the phishing webpages, on the other hand, are listed in datasets from the websites PhishTank.com and OpenPhish.com, where the phishers make use of highly rated free hosting services according to Alexa. As a result, legitimate Alexa.com websites won't have any subdomains, whereas fraudulent websites have. Additionally, other than the domain name, phishers have total control over how a URL is put together. It is simple to change features like URL length. Researchers have recently concentrated on domain name-based attributes rather than the entire URL in order to extract features of the domain name and the current page content.

6 CONCLUSION

This study presented the results of an extensive analysis of different URL-based phishing detection techniques. We put a lot of effort into comprehensive URL-based detection in terms of attributes, whereas earlier survey publications mainly focused on overall phishing detection strategies. We started out by reading up on common phishing detection methods in the literature. Second, we discussed URL-based phishing's structure as well as frequently used features and algorithms. Third, common data sources were mentioned, and for a more thorough study, comparative evaluation results and matrices were displayed. Finally, we offered our suggestions for future phishing detection technologies that will be even more effective.

REFERENCES

[1] Singh C. and Meenu, "Phishing Website Detection Based on Machine Learning: A Survey," *2020 6th International Conference on Advanced Computing and Communication Systems (ICACCS)*, 2020, pp. 398–404, doi: 10.1109/ICACCS48705.2020.9074400.

[2] *FBI Internet Crime Report 2021*, https://www.ic3.gov/Media/PDF/AnnualReport/2021_IC3Report.pdf

[3] Arathi Krishna V, Anusree A, Blessy Jose, Karthika Anilkumar, Ojus Thomas Lee, 2021, Phishing Detection Using Machine Learning Based URL Analysis: A Survey, *International Journal Of Engineering Research & Technology (IJERT) Ncreis – 2021* (Volume 09–Issue 13)

[4] Loftus R., "What Cybersecurity Trends Should You Look Out for in 2020?" *Daily English Global blogkasperskycom. [Online]*. Available: https://www.kaspersky.com/blog/secure-futures-magazine/2020-cybersecurity-predictions/32068/.

[5] Anti-Phishing Working Group, "*Phishing Activity Trends Report,*" 2022.

[6] Fukushima Y., Hori Y. and Sakurai K., "Proactive Blacklisting for Malicious Web Sites by Reputation Evaluation Based on Domain and IP Address Registration," *2011IEEE 10th International Conference on Trust, Security and Privacy in Computing and Communications*, 2011.

[7] Prakash P., Kumar M., Kompella R. R., and Gupta M., "Phishnet: Predictive Blacklisting to Detect Phishing Attacks," *Proc. IEEE INFOCOM*, 2010, pp.1–5, 2010.

[8] Azeez, Nureni Ayofe, *et al.* "Adopting Automated Whitelist Approach forDetecting Phishing Attacks." *Computers & Security* 108 (2021): 102328.

[9] Mityukov, E. A., *et al.* "Phishing Detection Model Using the Hybrid Approach to Data Protection in Industrial Control System." *IOP Conference Series: Materials Science and Engineering*. Vol. 537. No. 5. IOP Publishing, 2019.

[10] Mahmood Moghimi and Ali Yazdian Varjani, "New Rule-based Phishing Detection Method," *Expert Systems With Applications*, vol. 53, pp. 231–242, 2016.

[11] Zuraiq, AlMaha Abu, and Mouhammd Alkasassbeh. "Phishing Detection Approaches." *2019 2nd International Conference on new Trends in Computing Sciences (ICTCS)*. IEEE, 2019.

[12] Rao, R.S., Pais, A.R. Two Level Filtering mechanism to Detect Phishing Sites Using Lightweight Visual Similarity Approach. *J Ambient Intell Human Comput* 11, 3853–3872 (2020). https://doi.org/10.1007/s12652-019-01637-z

[13] Abdelnabi, Sahar, Katharina Krombholz, and Mario Fritz. "VisualPhishNet: Zero-day Phishing Website Detection by Visual Similarity." *Proceedings of the 2020 ACM SIGSAC conference on computer and communications security*. 2020.

[14] Jain, Ankit Kumar, and Brij B. Gupta. "A Machine Learning Based Approach for Phishing Detection Using Hyperlinks Information." *Journal of Ambient Intelligence and Humanized Computing* 10.5 (2019): 2015–2028.

[15] Tupsamudre, Harshal, Ajeet Kumar Singh, and Sachin Lodha. "Everything is in the Name–a URL Based Approach for Phishing Detection." *International symposium on cyber security cryptography and machine learning*. Springer, Cham, 2019.

[16] Liu, Ruofan, *et al.* "Inferring Phishing Intention via Webpage Appearance and Dynamics: A Deep Vision Based Approach." *30th {USENIX} Security Symposium ({USENIX} Security 21)*. 2022.

[17] da Silva, Carlo Marcelo Revoredo, *et al.* "Piracema. IO: A Rules-based Tree Model for Phishing Prediction." *Expert Systems with Applications* 191 (2022): 116239.

[18] Samarthrao, Kadam Vikas, and Vandana M. Rohokale. "A Hybrid Meta-heuristic-based Multi-objective Feature Selection with Adaptive Capsule Network for Automated Email Spam Detection." *International Journal of Intelligent Robotics and Applications* (2022): 1–25.

[19] Fernandez, Simon, Maciej Korczyński, and Andrzej Duda. "Early Detection of Spam Domains with Passive DNS and SPF." *International Conference on Passive and Active Network Measurement*. Springer, Cham, 2022.

[20] Marchal, Samuel, *et al.* "PhishStorm: Detecting Phishing with Streaming Analytics." *IEEE Transactions on Network and Service Management* 11.4 (2014): 458–471.

[21] Chiramdasu, Rupa, *et al.* "Malicious URL Detection Using Logistic Regression." *2021 IEEE International Conference on Omni-Layer Intelligent Systems (COINS)*. IEEE, 2021.

[22] Vanitha, N., and V. Vinodhini. "Malicious-URL Detection Using Logistic Regression Technique." *International Journal of Engineering and Management Research (IJEMR)* 9.6 (2019): 108–113.

[23] Feroz, Mohammed Nazim, and Susan Mengel. "Examination of Data, Rule Generation and Detection of Phishing URLs Using Online Logistic Regression." *2014 IEEE International Conference on Big Data (Big Data).* IEEE, 2014.

[24] Patgiri, Ripon, *et al.* "Empirical Study on malicious URL Detection Using Machine Learning." *International Conference on Distributed Computing and Internet Technology.* Springer, Cham, 2019.

[25] Catak, Ferhat Ozgur, Kevser Sahinbas, and Volkan Dörtkardeş. "Malicious URL Detection Using Machine Learning." *Artificial intelligence paradigms for smart cyber-physical systems.* IGI Global, 2021. 160–180.

[26] Bazzaz Abkenar, Sepideh, *et al.* "A Hybrid Classification Method for Twitter Spam Detection Based on Differential Evolution and Random Forest." *Concurrency and Computation: Practice and Experience* 33.21 (2021): e6381.

[27] He, Shen, *et al.* "An Effective Cost-sensitive XGBoost Method for Malicious URLs Detection in Imbalanced Dataset." *IEEE Access* 9 (2021): 93089–93096.

[28] Ma, Qian, *et al.* "A Novel Model for Anomaly Detection in Network Traffic Based on Kernel Support Vector Machine." *Computers & Security* 104 (2021): 102215.

[29] Butnaru, Andrei, Alexios Mylonas, and Nikolaos Pitropakis. "Towards Lightweight URL-based Phishing Detection." *Future Internet* 13.6 (2021): 154.

[30] Odeh, Ammar, Ismail Keshta, and Eman Abdelfattah. "Machine Learning Techniques for Detection of Website Phishing: A Review for Promises and Challenges." *2021 IEEE 11th Annual Computing and Communication Workshop and Conference (CCWC).* IEEE, 2021.

[31] Salloum, Said, *et al.* "Phishing Website Detection from URLs Using Classical Machine Learning ANN Model." *International Conference on Security and Privacy in Communication Systems.* Springer, Cham, 2021.

[32] Itoo, Fayaz, and Satwinder Singh. "Comparison and Analysis of Logistic Regression, Naïve Bayes and KNN Machine Learning Algorithms for Credit Card Fraud Detection." *International Journal of Information Technology* 13.4 (2021): 1503–1511.

[33] Ozcan, Alper, *et al.* "A Hybrid DNN–LSTM Model for Detecting Phishing URLs." *Neural Computing and Applications* (2021): 1–17.

[34] Afzal, Sara, *et al.* "URL Deep Detect: A Deep Learning Approach for Detecting malicious URLs Using Semantic Vector Models." *Journal of Network and Systems Management* 29.3 (2021): 1–27.

[35] Srinivasan, Sriram, *et al.* "DURLD: Malicious URL Detection Using Deep Learning-based Character Level Representations." *Malware Analysis Using Artificial Intelligence and Deep Learning.* Springer, Cham, 2021. 535–554.

[36] Yuan, Jianting, *et al.* "Malicious URL Detection Based on a Parallel Neural Joint Model." *IEEE Access* 9 (2021): 9464–9472.

[37] Tekerek, Adem. "A Novel Architecture for Web-based Attack Detection Using Convolutional Neural Network." *Computers & Security* 100 (2021): 102096.

[38] Yang, Xiaocui, *et al.* "Multimodal Sentiment Detection Based on Multi-channel Graph Neural Networks." *Proceedings of the 59th Annual Meeting of the Association for Computational Linguistics and the 11th International Joint Conference on Natural Language Processing (Volume 1: Long Papers).* 2021.

[39] Gupta, Brij B., *et al.* "A Novel Approach for Phishing URLs Detection Using lexical Based Machine learning in a Real-time Environment." *Computer Communications* 175 (2021): 47–57.

[40] Indrasiri, Pubudu L., Malka N. Halgamuge, and Azeem Mohammad. "Robust Ensemble Machine Learning Model for Filtering Phishing URLs: Expandable Random Gradient Stacked Voting Classifier (ERG-SVC)." *IEEE Access* 9 (2021): 150142–150161.

Application of state-of-the-art blockchain and AI research in healthcare, supply chain, e-governance etc.

Artificial Intelligence, Blockchain, Computing and Security – Dagur et al. (Eds)
© 2024 The Author(s), ISBN: 978-1-032-49393-0

V2E: Blockchain based E-voting system

Bipin Kumar Rai*
ABES Institute of Technology, Ghaziabad, India
ORCID ID: 0000-0002-9834-8093

Mukul Kumar Sahu & Viraaj Akulwar
ABES Institute of Technology, Ghaziabad, India

ABSTRACT: The Information Age, embraced by the 21st century, has seen the rapid adoption of new technologies, including in the election process. However, the security and transparency of both voters and their votes remain a concern with the centralized traditional system. The database can be tampered with, vote rigging, hacking, and booth capturing are possible. Blockchain technology provides a solution based on a decentralized and distributed system, where the database is controlled and owned by multiple users. E-voting is a quick, inexpensive, and effective way to cast a vote. Our proposed V2E system uses a Peer-to-peer network that enhances accountability and security. We will employ distributed ledger technologies to prevent vote faking and design a user credential model based on cryptographically secure Hash Functions to provide authenticity and non-repudiation. Since votes will be updated synchronously to the ledger, vote rigging, hacking, or system destruction will be nearly impossible.

1   INTRODUCTION

A current approach to increase the propulsion of systems employed in several areas is blockchain. Blockchain is decentralized, unchangeable database that permits for the tracking of network assets and the recording of transactions. Practically everything of value may be tracked, recorded, and exchanged on a blockchain network, minimizing costs, and reducing risk for all parties. It's a certain form of distributed ledger (the distributed character of blockchain, which helps to address the issue of information sharing without being compromised [1]) that includes a growing array of records, identified as blocks which are securely connected to each other via cryptographic algorithms. Pseudonymization, encryption, role-based access control models, and other mechanisms for data security and access controls have all been developed to secure sensitive data. Blockchain was firstly used for keeping track of bitcoin transactions. On the contrary hand, innovative uses and applications have emerged lately, and healthcare is also one of them. Electronic Health Record systems available today fall short of giving patients secure, traceable ownership of their medical data. PcBEHR (patient-controlled blockchain enabled electronic health records) is an idea put forth by the authors to give patient a safe management of their data which is decentralised, immutable, transparent, traceable, and reliable [2] [3]. Gradually blockchain based e-voting has become an increasingly relevant alternative for overcoming some of the issues involved with e-voting. Governments are encouraged to use intelligent sustainable voting methods and the incorporation of sustainability data into voting systems thanks to blockchain technology. Election integrity is crucial for nations with democratic governments or other consensual electoral

---

*Corresponding Author: bipinkrai@gmail.com

DOI: 10.1201/9781003393580-115

systems, but it is also important for boosting voter confidence and accountability. Political voting procedures are crucial in this regard. From the government point of view, electronic-voting systems might boost voter turnout, confidence, and enthusiasm for the electoral process. Voting electronically either assists or handles the casting and tallying of ballots. It can take place on an Electronic Voting Machine, or a computer with an Internet connection, relying on implementation. The vast majority of these requirements includes security and protection, Accuracy, Integrity, privacy protection, Auditability, easy accessibility, and Cost-Effectiveness, must be met by a reliable electronic voting system. If a qualified voter uses an electronic voting machine (EVM) at a polling location, and since the EVM is a circuitry and there is no way to track back votes, it is possible that the voter will never know whether their vote reached the candidate they intended to support or was misdirected into another candidate's tally. However, if blockchain technology is employed, it will store everything as a transaction and provide a receipt for a voter's vote (as a Transaction ID) that may be utilized to confirm if the vote was securely added. If, however, a digitally electing process (web or dapp) is introduced to digitize the electing system and on a single administrative server or workstation, all critical information is stored, a hacker or snooper might alter the candidate's vote total from 1 to 28. People might not be certain that malware has been installed, that votes have been stolen or invalidated, or that the primary server

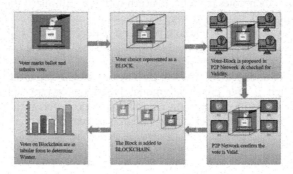

Figure 1.   Shows the stages of E-voting system.

has been targeted by a hacker. A system will be safeguarded by a special property called as immutability if it is connected to blockchain to prevent this.

2   RESEARCH OBJECTIVE

Objective of our proposed work are as follow:

- To build electronic voting system based on blockchain that ensures security and transparency, along with building trust among the voters in election process.
- To build secure and trusted e-voting system that ensures data integrity, voter's anonymity, uniqueness, verifiable and reliability.

3   RELATED WORK

Not only is electoral integrity necessary for democracies to function, but it also influences public opinion, voter trust, and accountability[4]. In a research paper, author Tas *et al.*, explained how a decentralized e-voting system could use a blockchain as a viable method. Additionally, the voting records stored in these proposed systems are open to all voters and impartial observers, but the reality is that electronic voting methods bring weaknesses in the communications infrastructure, software, and hardware. Author Anasune *et al.*, simply provides security and anonymity using homomorphic encryption. The exponential ElGamal cryptosystem is used in the voting system based on this encryption to encrypt each cast ballot's content. By using this cryptosystem's additive homomorphism property, it is possible to directly tally encrypted votes without having to first decrypt them [5]. Author Abuidris et al., compares the privacy and confidentiality requirements of the current blockchain-based electronic voting system. To establish voter anonymity, lower polling costs, ensure poll's integrity and end-to-end verifiability by handling privacy and security concerns in accordance with needs [6] but implementing voting via blockchain procedures may bring undetected security issues and flaws. Al- Maaitah et al., describes about the preferred way to tackle trust issues in electoral system. From the past to the present, there has been a growing concern about how the government is perceived, as well as how third parties that can influence elections in democratic societies, are interfering with those countries' procedures [7]. Infringing on this right to vote and choose the best candidate to represent them will always be a major problem, but the issue is that the growing concern about privacy and security measures may act as a roadblock to the development of practical block-chain applications. Author also explained that Blockchain technology is the most preferred technique to tackle trust issues in the electoral system, but the research demonstrates that the majority of earlier iterations of Ethereum did not build their consensus on the application, roles, and shareholders. Author Nir Kshetri, Jeffrey Voas, address the fact that blockchains produce cryptographically secure voting records, which makes them resistant to voter tampering. Votes are accurately, thoroughly, securely, and transparently recorded. Therefore, votes cannot be changed or manipulated. E-voting with block-chain support may expand voter access while reducing voter fraud [8]. BVE can overcome all traditional problem, but its complexity might hinder the public to accept it. Eligible voters cast vote using a computer or smartphone anonymously. It uses an encrypted key and tamper-proof individual IDs. To prevent voter fraud, blockchains create encrypted records of votes. Votes are accurately, thoroughly, securely, and openly recorded. Consequently, nobody can alter votes. Author Rifa Hanifatunnisa Budi Rahardjo, says that the use of hash values and digital signatures increases the security and dependability of the system used to record the results of voting at connected polling places. By using blockchain technology to distribute data on electronic voting machines, one of the primary sources of database manipulation can be lessened. Author provided a method based on a pre-determined turn on the system for each node in the built-in blockchain, unlike Bitcoin that uses Proof of Work. Electronic voting applications are supported by blockchain. A blockchain transaction can be made from each voter's vote in order to track voice counting [9]. Due to the free blockchain auditing and testing platform, the vote added can be confirmed that no data has been added without authorization, deleted, or manipulated in any way. As a result, everyone may approve the final computation. Yi Liu, Qi Wang, developed a workable and widespread e-voting protocol without the aid of a trusted third party (TTP). Everyone interested in the election, including observers, can confirm the entire election's process and result by viewing the voting process (recorded on blockchain) [10] unless implemented with access restriction utilizing permissioned blockchain, intentional transparency of blockchain may seem challenging to satisfy coercion-resistance (Voters shouldn't be allowed to show how they cast their ballots.). Hardwick et al., meets the basic requirements for electronic voting while also offering some decentralising level and giving voters as much power over the procedure as possible was con-sidered practical. A number of functional as well as security criteria, including availability, auditability, secrecy/privacy, accuracy, and auditability, as well as system and data integrity, are listed for a reliable e-voting system [11]. Author Koc et al., concentrating solely on smaller-scale implementation tasks and developing solutions. Ethereum and smart contracts, one of the most significant developments since the blockchain itself, contributed to changing the relatively lim-ited impression of blockchain as a cryptocurrency (coin) and expanding it to become a broader source of solutions for many current Internet-related issues. This may also make it possible for blockchain to be used globally [12]. Hsiao et al., focuses on using homomorphic encryption, secret sharing schemes, to create a dapp on e-voting without the need for 3rd party to be trusted. It increases voter trust and decreases the expenditure of election resources [13]. Author Garg

783

et al. concentrates on several issues such forced voting, duplicate voting, machine capture, non-authentic voter voting, and more [14]. The process is decentralised and do not depend on faith, according to author Ahmed Ben Ayed. Any qualified participant will be allowed to cast a vote via any device with an Internet connection [15].

4 PROPOSED WORK

We suggest a decentralized blockchain-based electronic voting system named "V2E: Electronic Voting System on Blockchain". Given that the database as a whole will be owned and governed by numerous individuals under a decentralised system, blockchain technology may be the all-encompassing solution. Voters can cast and have their ballots counted electronically (thus the term "e-voting"). It is a quick, inexpensive, and effective way to give a vote. It will be a time and money efficient method of performing a voting process. In an effort to upgrade transparency and security of e-voting, we will use a P2P network. To prevent vote fraud, we will first create a global record of transactions approach for synchronized vote records (DLT). Secondly, we will design an individual credential framework based on cryptographically secure

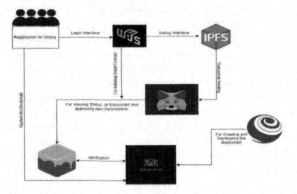

Figure 2.   System architecture diagram of V2E: Electronic voting system on blockchain.

hash functions to achieve authenticity and not being repudiated. A e-voting structure based on blockchain on the Peer-to-Peer network is suggested for the fundamental needs of the electronic voting system by merging the aforementioned designs.

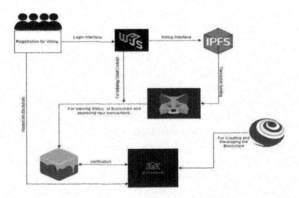

Figure 3.   Sequence diagram of V2E: Electronic voting system on blockchain.

4.1 System architecture

Above diagram shows the architecture of V2E where MetaMask, Ganache, IPFS, Truffle etc are used for E-voting.

4.2 Sequence diagram

Above diagram shows the sequence of communication in between the individual Voter, Candidate, System (IPFS) and Organisation or Admin.

5 CONCLUSION

This work suggests V2E: Electronic Voting System on Blockchain which enables e-voting through smart contracts and a number of other, Blockchain characteristics to guarantee security, transparency, correctness, integrity, privacy, auditability, and anonymity.

REFERENCES

[1] Rai, B. K. BBTCD: Blockchain Based Traceability of Counterfeited Drugs. *Health Serv Outcomes Res Methodol*, 2022. https://doi.org/10.1007/s10742-022-00292-w.

[2] Rai, B. K.; Srivastava, A. K. *Patient Controlled Pseudonym-Based Mechanism Suitable for Privacy and Security of Electronic Health Record*; 2017; Vol. 07.

[3] Rai, B. K.; Fatima, S.; Satyarth, K. Patient-Centric Multichain Healthcare Record. *International Journal of E-Health and Medical Communications*, 2022, *13* (4), 1–14. https://doi.org/10.4018/IJEHMC.309439.

[4] Taş, R.; Tanrıöver, Ö. Ö. A Systematic Review of Challenges and Opportunities of Blockchain for E-Voting. *Symmetry. MDPI AG August* 1, 2020, pp 1–24. https://doi.org/10.3390/sym12081328.

[5] Anasune, V.; Choudhari, P.; Kelapure, M.; Shirke, P.; Halgaonkar, P. *Literature Survey-Online Voting: Voting System Using Blockchain*.

[6] Abuidris, Y.; Kumar, R.; Wenyong, W. A Survey of Blockchain Based on E-Voting Systems. In *ACM International Conference Proceeding Series*; Association for Computing Machinery, 2019; pp 99–104. https://doi.org/10.1145/3376044.3376060.

[7] Al-Maaitah, S.; Qatawneh, M.; Quzmar, A. E-Voting System Based on Blockchain Technology: A Survey. In *2021 International Conference on Information Technology, ICIT 2021 - Proceedings*; Institute of Electrical and Electronics Engineers Inc., 2021; pp 200–205. https://doi.org/10.1109/ICIT52682.2021.9491734.

[8] Kshetri, N.; Voas, J. Blockchain-Enabled E-Voting. *IEEE Softw*, 2018, *35* (4), 95–99. https://doi.org/10.1109/MS.2018.2801546.

[9] Hanifatunnisa, R.; Rahardjo, B. *Blockchain Based E-Voting Recording System Design*.

[10] Liu, Y.; Wang, Q. *An E-Voting Protocol Based on Blockchain*.

[11] Hardwick, F. S.; Gioulis, A.; Akram, R. N.; Markantonakis, K. *E-Voting with Blockchain: An E-Voting Protocol with Decentralisation and Voter Privacy*. 2018.

[12] Koç, A. K.; Yavuz, E.; Çabuk, U. C.; Dalkiliç, G. Towards Secure E-Voting Using Ethereum Blockchain. In *6th International Symposium on Digital Forensic and Security, ISDFS 2018 - Proceeding*; Institute of Electrical and Electronics Engineers Inc., 2018; Vol. 2018-January, pp 1–6. https://doi.org/10.1109/ISDFS.2018.8355340.

[13] Hsiao, J. H.; Tso, R.; Chen, C. M.; Wu, M. E. Decentralized E-Voting Systems Based on the Blockchain Technology. In *Lecture Notes in Electrical Engineering*; Springer Verlag, 2018; Vol. 474, pp 305–309. https://doi.org/10.1007/978-981-10-7605-3_50.

[14] Garg, S.; Kumar, D.; Kumar Manna, S.; Chauhan, S.; Astya, R. *Blockchain Based Decentralized Voting System*; Vol. 8.

[15] Ben Ayed, A. A Conceptual Secure Blockchain Based Electronic Voting System. *International Journal of Network Security & Its Applications*, 2017, 9 (3), 01–09. https://doi.org/10.5121/ijnsa.2017.9301.

Artificial Intelligence, Blockchain, Computing and Security – Dagur et al. (Eds)
© 2024 The Author(s), ISBN: 978-1-032-49393-0

Blockchain based supply chain management system

Bipin Kumar Rai, Dhananjay Singh & Nitin Sharma
ABES Institute of Technology, Ghaziabad, India

ABSTRACT: The authenticity of a product is crucial for customers to justify their purchase, but counterfeit and rising prices create problems for the coffee industry. The lack of transparency regarding the complete production of coffee and labeling issues further add to the issue. To address this problem, we are proposing a blockchain-based decentralized system called RCSC (Real Coffee Supply Chain). RCSC will enable customers to track and trace the entire coffee supply chain, from the farmer to the cup. It will display information such as the location of origin, quality assessment methods, storage, and roasting locations, as well as manufacturing and expiry dates. Additionally, the system will provide details through barcode/QR scans and include certificates of authenticity and uniqueness. By implementing this system, companies can ensure transparency, prevent fraudulent activity, and provide customers with the confidence that they are purchasing authentic coffee products.

Keywords: Supply chain, Blockchain, SCMs, Transparency

1 INTRODUCTION

One of the most popular drinks in the world and a major source of caffeine for many students and employees, coffee continues to play a vital role in society's daily routine. According to Business Insiders, coffee is the second-most valuable commodity in the world, with a global market worth more than $100 billion. Since 500 billion cups of coffee are typically consumed annually around the world, the industry is currently valued at 20 billion dollars and is only anticipated to increase. Starbucks, Panera Bread, McCafé, Dunkin' Brands, Tim Hortons, Dutch Bros Coffee, etc. are a few well-known companies. The top 50 businesses in the coffee industry account for 70% of sales, while the bottom of the market is severely fragmented. India is one of the world's major producers of coffee, with 114 Cr people drinking it every day. Despite its importance, there are still a number of difficulties in the supply chain, such as location, social, economic, market crises, environmental, time and cost, traceability, transparency, etc. A roaster may buy coffee and have access to all the details about it, from manufacturing to delivery, when employing Blockchain technology for coffee commerce. Blockchain technology can be used to track the origin of coffee, its price, purchasers, and the amount of time it took for the coffee to get from the farm to the cup. It can also facilitate international transfers of raw materials and provide transparency for the final consumers. Blockchain technology can improve the efficiency and transparency of the modern coffee supply chain by creating blockchain-based distributed ledgers that trace every batch of coffee beans' travel and identify their precise origin. Blockchain is being used by a number of coffee firms to support their inventory optimization software to find the optimum places to buy, cook, and store each cut of pork before it is transported to a store. Businesses can more easily communicate information and data with suppliers, manufacturers, and vendors. Blockchain transparency prevents stock from building up in the supply chain, which prevents disputes and holdups. As a result of real-time tracking, there is a reduced chance of product misplacements. By using IPFS (Interplanetary File System), there is a way to completely

DOI: 10.1201/9781003393580-116

decentralize the system and get beyond the high cost and scalability issues with blockchain-based apps[1][2]. Real-time product tracking encouraged by blockchain reduces the total cost of moving items along the coffee supply chain. It also offers scalability, allowing for remote access to any significant database from different locations. Middlemen and intermediaries must be eliminated from the supply chain to limit the possibility of fraud and product duplication [3] while also making financial savings. We have introduced RCSC (Real Coffee Supply Chain) which use the blockchain technology to reduce the sustainability and provide the transparency to the customers. A supply chain management system (RCSC) based on blockchain technology can be developed for coffee to enable users to track and trace the entire journey of coffee from its origin on the farm to the final product in the cup. This system will provide comprehensive information about the coffee, such as its place of cultivation, refinement process, roasting details, manufacturing date, and expiry date. Users can view the information by scanning a QR or barcode.

In Figure 1, The supply chain of coffee beans is a complex and multi-stage process that involves several stages, such as planting, harvesting, hulling, drying, packaging, bulking, mixing, roasting, exporting, and distribution to merchants. In addition, immediate steps are also included in the process.

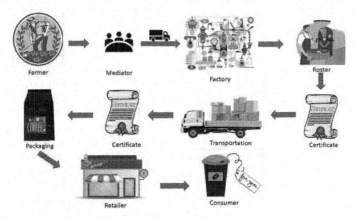

Figure 1.   Traditional- coffee supply chain.

2   RESEARCH OBJECTIVE

Objectives of our proposed work are as follows:

1. Develop a blockchain-based supply chain management system (RCSC) for coffee to track and trace its journey from the farm to the cup, providing detailed information on culti-vation, refining, roasting, and relevant dates.
2. With a QR/Barcode scan, the user will be able to see all the above-mentioned details along with the quality certificates issued by the different entities of the chain.

3   RELATED WORK

The notion of supply chain management has been of significant importance for a long time, dating back to the early 20th century, particularly with the advent of the assembly line[4]. This paper provides the information about the Hyperledger how it works. Regarding the coffee Additionally, the study focused on the coffee business globally. By the help of [5] this provide that coffee is made by cultivating the beans, harvesting them, hulling them, drying them, packaging them, bulking them up, mixing them, and then roasting them. [6] How the coffee SC functions from start to finish is explained in this article. With the help of this [7] web site article the global status of coffee is been visualized. How country grow there turnover in coffee industry. In our research paper it usually based on fake coffee product which is been sell large scale by using the famous company brand name and sell the bad quality to the customers [8]. There are many challenges are faced among the production [9]. With the help of [9] they can be restored. Trado industry explain [10] how the transparency can be reduced by fintech for sustainability. The article [11] show the working of SC management in the industry and sustainability of SC. Burundi [12] coffee proposed the challenges the face in their industry as they are the top seller of the coffee in the market. The Brazilian coffee [13] industry is a key export for both the Brazilian economy and the global market, and a bibliometric assessment and survey revealed the drivers and barriers of SSCs. The utilization of blockchain technology [14] is anticipated to address the challenge of allocating transaction resources among involved parties in the supply chain of fresh fruits. Blockchain technology can help improve sustainability in agro-food commodity supply-chains by providing traceability and transparency.

4 PROPOSED WORK

We advise using open-source blockchain-based SCMs. The suggested method uses a A supply chain management system (SCM) for coffee, based on blockchain technology, can be established to track and trace the entire journey of coffee from a farmer to a cup, ensuring transparency in every stage of the supply chain. This transparency will serve as proof of the superior quality of the coffee. It will display the location of the coffee's origin, the method used to assess its quality, the location of the coffee's storage and roasting, the location of the packaging's production, and the precise dates of its creation and expiration. Additionally, it aids other production firms in increasing demand for their products by demonstrating how well- made goods they produce. Mutual obligations between supply chain components are necessary to maintain the sustainability of corporate operations, particularly when deciding on the quality standards and quantity of products. Moreover, this proposed system will show the details with a Barcode/QR scan and the certificates of originality will also be provided in this system.

4.1 Blockchain based system

In Figure 2, shows the Blockchain Based System RCSC where the Mediator has been removed and the transparency with the authenticated Certificates are provided.

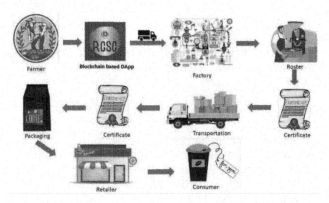

Figure 2. Blockchain based system.

4.2 Sequence diagram

In Figure 3., it shows the sequence through which communication will be taking place in between various entities involved in Blockchain based Coffee SCMs.

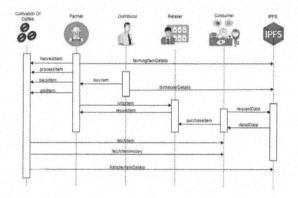

Figure 3.   Sequence diagram.

4.3 QR scanner diagram

Figure 4 shows how the complete system will work and how the user can access all the details and information with a quick scan.

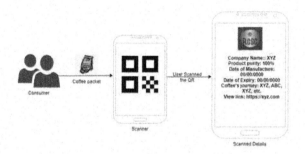

Figure 4.   QR scanner diagram for our supply chain.

5   CONCLUSION

In this paper, we proposed a Blockchain based Coffee SCMs that implements the Supply Chain through Hyperledger Fabric, Smart contracts and various other Blockchain capabilities to ensure a better Supply chain with high transparency.

REFERENCES

[1]   Rai, B. kumar; Fatima, S.; Satyarth, K. Patient-Centric Multichain Healthcare Record. *International Journal of E-Health and Medical Communications*, 2022, 13 (4). https://doi.org/10.4018/ IJEHMC.309439.

[2]   Rai, B. K. PcBEHR: Patient-Controlled Blockchain Enabled Electronic Health Records for Healthcare 4.0. *Health Serv Outcomes Res Methodol*, 2022. https://doi.org/10.1007/s10742-022-00279-7.

[3]   Rai, B. K. BBTCD: Blockchain Based Traceability of Counterfeited Drugs. *Health Serv Outcomes Res Methodol*, 2022, No. Springer Link. https://doi.org/10.1007/s10742-022-00292-w.

[4]   Ravi, D.; Ramachandran, S.; Vignesh, R.; Falmari, V. R.; Brindha, M. Privacy Preserving Transparent Supply Chain Management through Hyperledger Fabric. *Blockchain: Research and Applications*, 2022, 3 (2), 100072. https://doi.org/10.1016/J.BCRA.2022.100072.

[5]   Melanie. *The Remarkable Supply Chain of the Coffee Bean* https://www.unleashedsoftware.com/blog/remarkable-supply-chain-coffee-bean#:~:text=The%20supply%20chain%20of%20coffee,stores%2C%20cafes%20and%20specialty%20shops. (accessed Dec 31, 2022).

[6]   Caitlyn Hutson. *From Bean to Cup: How the Coffee Supply Chain Works.*

[7]   Vayola Jocelyn. *Coffee - Worldwide* https://www.statista.com/outlook/cmo/hot-drinks/coffee/worldwide (accessed Dec 31, 2022).

[8]   FPJ Web Desk. *Fake Coffee to be Filtered Out with Acviss' Blockchain-based Tech* https://www.freepressjournal.in/business/fake-coffee-to-be-filtered-out-with-acviss-blockchain-based-tech (accessed Dec 24, 2022).

[9]   Perfect Daily Grind. *What are the Biggest Coffee Production Challenges Facing Producers Today?* https://perfectdailygrind.com/2017/05/what-are-the-main-challenges-faced-by-coffee-producers/ (accessed Dec 24, 2022).

[10]   Verhagen, T.; Freshfields, J. R. Trado: New Technologies to Fund Fairer, More *Transparent Supply Chains Fintech for Sustainability Taskforce View Project.* 2019. https://doi.org/10.13140/RG.2.2.29720.34568.

[11]   Sukati, I.; Hamid, A. B.; Baharun, R.; Yusoff, R. M. The Study of Supply Chain Management Strategy and Practices on Supply Chain Performance. *Procedia Soc Behav Sci*, 2012, 40, 225–233. https://doi.org/10.1016/j.sbspro.2012.03.185.

[12]   Bamber, P.; Gereffi, G. Burundi in the Coffee Global Value Chain: Skills for Private Sector Development Decent Work and Global Supply Chains View *Project Industries without Smokestacks View Project.* 2014. https://doi.org/10.13140/RG.2.1.1097.4808.

[13]   Guimarães, Y. M.; Eustachio, J. H. P. P.; Filho, W. L.; Martinez, L. F.; do Valle, M. R.; Caldana, A. C. F. Drivers and Barriers in Sustainable Supply Chains: The Case of the Brazilian Coffee Industry. *Sustain Prod Consum*, 2022, 34, 42–54. https://doi.org/10.1016/J.SPC.2022.08.031.

[14]   Zhang, Y.; Chen, L.; Battino, M.; Farag, M. A.; Xiao, J.; Simal-Gandara, J.; Gao, H.; Jiang, W. Blockchain: An Emerging Novel Technology to Upgrade the Current Fresh Fruit Supply Chain. *Trends Food Sci Technol*, 2022, 124, 1–12. https://doi.org/10.1016/J.TIFS.2022.03.030.

Artificial Intelligence, Blockchain, Computing and Security – Dagur et al. (Eds)
© *2024 The Author(s), ISBN: 978-1-032-49393-0*

Sign language detection using computer vision

Shivani Sharma, Bipin Kumar Rai, Manak Rawal & Kaustubh Ranjan
ABES Institute of Technology, Uttar Pradesh, India

ABSTRACT: Sign Language is a critical tool to aid the hearing-impaired population and allow them to meet the ends with the normal people. For the purpose of translating sign language, the camera is the primary element used in Sign Language Recognition (SLR). Mostly existing SLR through image processing uses high quality cameras whereas to reach larger audience, we need a better approachability, hence this paper proposes using of normal cameras like of smartphones and webcams. This paper depicts the significance of encompassing intelligent solution into the SLR systems and meets the requirement of a SLR web portal that is universally and easily available for every needful person. Altogether, it is anticipated that this study will promote the production of intelligent-based SLR, the accumulation of knowledge, and will give readers, researchers and practitioners a roadmap for future direction.

Keywords: Computer Vision, Machine Learning, CNN, Sign Language, Gesture Detection

1 INTRODUCTION

Sign language is a kind of conversing dialect that conveys meaning visually using facial expressions, hand gestures, and body language. For those who have trouble hearing or speaking, sign language is quite helpful. The gesture translation into the alphabets or words of formal spoken languages that already exist is called sign language recognition (SLR). Therefore, utilizing an algorithm or model to translate sign language into words can help close the communication gap between persons who have hearing or speech impairments and the rest of society.

Researchers working in the fields of computer vision and machine learning are now focusing intensively on image-based hand gesture awareness. Many researchers are working in this field since it is a natural method of human connection, having the aim of creating the interaction of human and computer, simple and natural by cutting off the need of additional gadgets. Therefore, the primary intent of research in gesture identification is to develop systems that are able to recognize and utilize particular human gestures. For example, to deliver information. For the same, interfaces that are vision based, need quick, extremely authentic hand detection as well as real-time sign identification for this.

Sign Language is a popular & effective way to facilitate the deaf and dumb community, but the primary limitation of lack of sign language interpreters results in non-availability and non-reachability of the interpreters. The expected achievement is to contribute towards research and development towards solutions for deaf and dumb community. The aim of this paper is to model a Sign Language Recognition System accessible via a web portal, which takes video input and predicts hand actions into signs, achieving higher accuracy than existing solutions.

Within this frame of reference, we have sign language recognition (SLR), the mode of communication utilised by the deaf and the silent. Hand gestures constitute a potent human communication method that has many possible uses.[1]

DOI: 10.1201/9781003393580-117

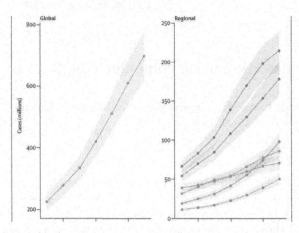

Figure 1. Hearing impairment ubiquity of 35 dB or more, 1990-2019 by WHO, including predictions to 2050.95%. UI is rendered as shading. UI = uncertainty interval.

Figure 1 depicts the estimate ubiquity account in the interval of 1990 and 2019, resulting in rise of the rate by 27·8% (95% UI 26·6–29·0), from 15·9% (15·3–16·6) in 1990 to 20·3% (19·5–21·1) in 2019. The prevalence rate of age-standardized hearing impairment stays steady globally, on comparison. In conclusion, the noted pattern signifies that growth of prevalent instances, when there is rise in cases of hearing loss but stability in age standardized rates, are causes of ageing and growth in population.

2 METHODOLOGY

The architectural diagram in Figure 2 describes working infrastructure of the plan. It depicts a user logging in to the web portal provided for the user interaction and after accepting requests for required permissions, it briefly introduces to the functioning of camera as the primary means of interplay between detection system and user. Fleeting operations including pre-processing, key points extraction and database accessing takes place. Data from database is used for training and testing, where testing data is also the feeded input data that is converted from gestures to signs and finally the user experience is received as feedback.

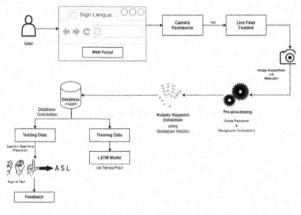

Figure 2. Architectural diagram for sign language detection.

3 LITERATURE SURVEY

The procedures followed in our work focuses on use of different methods mentioned below. Following observations were made after having a detailed study of previously acquired results in image processing.

3.1 *Image acquisition*

[2,3]Camera/Webcam eliminates the requirement of equipping other devices and is easy to use, making it comparatively the best image acquisition method. [4]Data glove is a feasible and easy feature extraction device as it prevents influence from outside environment. They are highly expensive with low convenience and fails to maintain originality of interaction. Kinetic is another image acquisition device which has high usage in various applications involving human computer interaction with limited range depth detection. Leap motion controller has elevated speed processing with high accuracy in recognition, that facilitates in detecting hand and fingers.

3.2 *Image enhancement*

[5]Using Histogram Equalization (HE) for images in grayscale gets best result and it has simple implementation. The downside is that image brightness gets modified due to which featured and noise become hard to distinguish. [6]Adaptive Histogram Equalization (AHE) has better performance than histogram equalization and is best suited to amplify edges and local contrast of image. [7]Original brightness is retained in Contrast Limited Adaptive Histogram Equalization, and reduced noise can be noticed when comparing to HE and AHE. [8]Logarithmic Transformation is useful when high intensity pixel values are to be reduced into lower intensities pixel values.

3.3 *Image filtering*

[9]Mean filter implementation is easy, but shows significant impact in incorrect pixel value representation. [10]Median filter counters problem of mean filter by retaining image sharpness and thin edges. [11]Gaussian type of noise is best removed using Gaussian filter.[12] Adaptive filer is better at preserving high frequency parts like edges than linear filter. [13] Wiener filter is a well-liked picture repair filter. Noise has no effect on it, so it is suitable for utilizing the image's statistical features.

3.4 *Image segmentation*

[14]Thresholding method as segmentation technique is an efficient and simple strategy. It can be used without any prior knowledge and requires less computation power. [15]Edge based method is suitable for pictures with better object contrast. [16]Region based method is more useful and less sensitive to noise when the similarity criteria are simple to define. [17] Because of the usage of the fuzzy partial membership, Clustering method is more applicable to real-world problems. [18]Artificial Neural-Network based method can function without a complicated program, and it is less noisy in nature.

4 WORKING STEPS

MP Holistic- Media Pipe Holistic combines distinct model of hands, poses and face.
Keypoints- Landmarks for object detection.
LSTM- Long Short-Term Memory Model processes sequence of data like video input

i. Install and import dependencies.
ii. Detect keypoints and extract their values using MP Holistic.
iii. Collect keypoint values in folders.
iv. Create labels and features.
v. Build and Train LSTM Neural Network.
vi. Initiate predictions and save weights.
vii. Evaluate and test in real time.

5 RESULTS

Table 1.   Sensitivity (Se) results of the ASL letters.

Letters	LSTM Se(%)
A	92.35
B	87.89
C	100.00
D	89.87
E	87.87
F	100.00
G	98.97
H	56.76
I	100.0
J	79.09
K	86.65
L	100.00
M	97.23
N	97.67
O	97.72
P	88.23
Q	78.93
R	99.78
S	68.38
T	86.67
U	56.78
V	100.0
W	100.0
X	81.80
Y	100.0
Z	87.98

6 CONCLUSION

Sign Language Recognition is a subject of in-progress research which still has no extensive deployed system. This paper provides computer vision-based solution with better accuracy and accords to the associated knowledge domain. Literature Survey of this paper focuses on use of distinct methods for the solutions of gesture recognition and provides various perspectives on use of dissimilar methods and techniques.

Future research will include relative analysis of this study with other existing and newly gleaned solutions incorporated with ever-evolving technologies.

REFERENCES

[1] *Articles*; 2021.

[2] Adeyanju, I. A.; Bello, O. O.; Adegboye, M. A. Machine Learning Methods for Sign Language Recognition: A Critical Review and Analysis. *Intelligent Systems with Applications*, 2021, 12, 56. https://doi.org/10.1016/j.iswa.2021.20.

[3] Kesarwani, A.; Maheshwari, S.; Sharma, S.; Rai, B. K. Hand Talk: Intelligent Gesture Based Communication Recognition & Object Identification for Deaf and Dumb. In *AIP Conference Proceedings*; American Institute of Physics Inc., 2022; Vol. 2424. https://doi.org/10.1063/5.0076796.

[4] Ravindran, U.; Alam, T.; Vashishtha, R.; Rai, A.; Sharma, S. *An Iot Based Smart Glove*; 2019.

[5] Verma, N.; of Engineering, M. D.-J.; (IRJET), T.; undefined 2017. *Contrast Enhancement Techniques: A Brief and Concise Review*. academia.edu, 2017.

[6] Kamal, S. M.; Chen, Y.; Li, S.; Shi, X.; Zheng, J. Technical Approaches to Chinese Sign Language Processing: A Review. *IEEE Access*, 2019, 7, 96926–96935. https://doi.org/10.1109/ACCESS.2019.2929174.

[7] Suharjito; Anderson, R.; Wiryana, F.; Ariesta, M. C.; Kusuma, G. P. Sign Language Recognition Application Systems for Deaf-Mute People: A Review Based on Input-Process-Output. *In Procedia Computer Science*; Elsevier B.V., 2017; Vol. 116, pp 441–448. https://doi.org/10.1016/j.procs.2017.10.028.

[8] Chourasiya, A.; Khare, N. A Comprehensive Review of Image Enhancement Techniques. *International Journal of Innovative Research and Growth*, 2019, 8 (6). https://doi.org/10.26671/ijirg.2019.6.8.101.

[9] Kasmin, F. A Comparative Analysis of Filters towards Sign Language Recognition. *International Journal of Advanced Trends in Computer Science and Engineering*, 2020, 9 (4), 4772–4782. https://doi.org/10.30534/ijatcse/2020/84942020.

[10] Dhanushree, M.; Priyadharsini, R.; Sharmila, T. S. Acoustic Image Denoising Using Various Spatial Filtering Techniques. *International Journal of Information Technology 2019 11:4*, 2019, 11 (4), 659–665. https://doi.org/10.1007/S41870-018-0272-3.

[11] Basu, M. Gaussian-Based Edge-Detection Methods - A Survey. *IEEE Transactions on Systems, Man and Cybernetics Part C: Applications and Reviews*, 2002, 32 (3), 252–260. https://doi.org/10.1109/TSMCC.2002.804448.

[12] Kaluri, R.; Pradeep Reddy, C. A Framework for Sign Gesture Recognition Using Improved Genetic Algorithm and Adaptive Filter. *Cogent Eng*, 2016, 3 (1), 1251730. https://doi.org/10.1080/23311916.2016.1251730.

[13] Maru, M.; Scholar, P. G.; Parikh, M. C. *Image Restoration Techniques: A Survey*; 2017; Vol. 160.

[14] Xu, W.; Li, Q.; Feng, H. J.; Xu, Z. H.; Chen, Y. T. A Novel Star Image Thresholding Method for Effective Segmentation and Centroid Statistics. *Optik (Stuttg)*, 2013, 124 (20), 4673–4677. https://doi.org/10.1016/J.IJLEO.2013.01.067.

[15] Rashmi; Kumar, M.; Saxena, R. Algorithm and Technique on Various Edge Detection: A Survey. *Signal Image Process*, 2013, 4 (3), 65–75. https://doi.org/10.5121/sipij.2013.4306.

[16] Garcia-Lamont, F.; Cervantes, J.; López, A.; Rodriguez, L. Segmentation of Images by Color Features: A Survey. *Neurocomputing*, 2018, 292, 1–27. https://doi.org/10.1016/J.NEUCOM.2018.01.091.

[17] Cebeci, Z.; Yildiz, F. Comparison of K-Means and Fuzzy C-Means Algorithms on Different Cluster Structures. *Journal of Agricultural Informatics*, 2015, 6 (3). https://doi.org/10.17700/JAI.2015.6.3.196.

[18] Khan, W. Image Segmentation Techniques: A Survey. *Journal of Image and Graphics*, 2014, 166–170. https://doi.org/10.12720/JOIG.1.4.166-170.

Artificial Intelligence, Blockchain, Computing and Security – Dagur et al. (Eds)
© 2024 The Author(s), ISBN: 978-1-032-49393-0

An overview of thalassemia: A review work

Ruqqaiya Begum* & G. Suryanarayana*
Department of Computer Science & Engineering, Vardhaman College of Engineering Hyderabad, Telangana, India

B.V. Saketha Rama*
Department of Computer Science Engineering, Indian Institute of Information Technology, Design and Manufacturing, Kancheepuram, Chennai, Tamil Nadu, India

N. Swapna*
Department of Computer Science & Engineering, Vijay Rural Engineering College, Nizamabad, Telangana, India

ABSTRACT: Thalassemia is a Disease that passes from parents to children through genes. It is a red blood cell disorder caused when the body doesn't make enough of a protein in the blood called hemoglobin. Three classifications of Thalassemia are Alpha(α), Beta(β), Delta-Beta($\delta\beta$) And Beta thalassemia is the most effective disease compared to other types. Out of 5% of patients 3% comes in Beta Thalassemia. It is first identified in the early 19's with the symptoms like poor growth, severe anemia, and huge abdominal organs. It is a disease of which most people are deceased due to unawareness, and not getting proper Medication, especially in rural areas as they cannot find symptoms because other disease symptoms differ from thalassemia. Out of 10000 newborns, approximately an average of 4.4 will be affected by thalassemia. Out of 32 districts In India, 10 to 12 thousand children were born with this disease. To date, no research work has shown the exact death rates of thalassemia worldwide. Blood transfusion is the major method used for the prevention of thalassemia-affected patients. Apart from that Bone marrow is another prevention method used for Beta Carriers. In this work, we gathered detailed Information about thalassemia and its types so that people will get aware of this disease. This work provides details of how this disease is getting inherited from parents to children, the depth of the characteristics of every subtype, and the statistics about causes and death rates from this disease.

Keywords: Thalassemia, Blood transfusion, Alpha(α), Beta(β), Delta-Beta($\delta\beta$) Thalassemia.

1 INTRODUCTION

The name "Thalassemia" is a mixture of the Greek words "Thalassa" (which means "sea") and "Hema" (which means "blood") (Kallenbacha 2015). Body creates less hemoglobin than is expected due to a hereditary blood disorder called thalassemia. Due to the presence of hemoglobin, red blood cells may carry oxygen. You could feel worn out due to anemia caused by thalassemia. This disease was found in early nineteen Hemoglobin contains four genes that are two Alpha genes and two Beta genes. And any Damage to these four genes leads to anemia. Alpha thalassemia can be caused if there is a disorder in alpha genes and

---

*Corresponding Authors: ruqqaiya1224@gmail.com, surya.aits@gmail.com, sakethram9999@gmail.com and swapnanarasalas@gmail.com

DOI: 10.1201/9781003393580-118

Beta thalassemia is caused when there is a disorder in Beta genes. Apart from these two types, there is one more type of thalassemia which is δβ that occurs due to chromosome 11 deletion of the Delta and Beta genes and a very rare disorder. Unlike Alpha and Beta Thalassemia Delta-Beta shows very mild symptoms. Some symptoms are weakness, yellowish complexion, facial abnormalities, slow development, fatigue, swelling of the abdomen, dark urine, etc. The picture shows the differences between Normal Red blood cells and Thalassemia cells (www.omniahealth.com). The Mediterranean region has greater incidence of the hereditary illness thalassemia than other regions of the world and also effects in other regions like Southeast Asia, West Africa and Indian subcontinent (Alain J. Marengo-Rowe 2007). In India, out of 10000 newborns approximately 4.4 are affected with this disease. Apart from this, there are about 128 million pregnancies, and of them, 55,875 give birth to children with significant thalassemia in South, East, and Southeast Asia is the birthplace of at least 40,000 of these impaired newborns. Over 20,000 thalassemia babies are anticipated to be born each year in China. This is one of the most prevalent genetic red blood cell illnesses in China, with an estimated 47.48 million carriers. About 3% of carriers have alpha thalassemia and 5% have beta-thalassemia or HbE (Kent Corley & Spring 2015). Many afflicted youngsters pass away in their early years. The condition's kind and severity will determine the indications and symptoms of the Disease. Figure 1 shows Normal RBCs and Thalassemia cells.

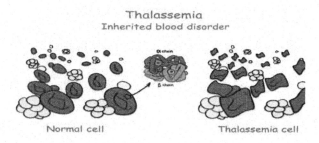

Figure 1.   Thalassemia effected cells.

1.1   *Thalassemia root causes*

Thalassemia is brought on by a mistake or mutation in a gene that makes haemoglobin. Your parents gave you this genetic abnormality. If only one parent is a carrier, thalassemia minor, a condition, may be experienced. You will still be a carrier if this occurs even though you most likely will not exhibit any symptoms. Thalassemia minor individuals can experience mild symptoms. If both of your parents are thalassemia carriers, there is a higher chance that you may inherit a more severe form of the illness. (Shilpa Amin 2019).

2   THALASSEMIA TYPES

2.1   *Alpha thalassemia*

Alpha thalassemia, which is inherited within families, is a kind of thalassemia. It is a blood condition that prevents the body from producing healthy red blood cells and hemoglobin. All bodily cells are supplied with oxygen by hemoglobin, an iron-rich protein present in red blood cells. The alpha globins and beta globins subunits of this protein are both present (Cornelis L Harteveld & Douglas R Higgs. 2010).

The quantity of gene mutations inherited from parents determines the level of thalassemia if an individual in the family has alpha-thalassemia. A more severe form of this thalassemia will arise from further gene mutations. **One mutated gene** has no thalassemia symptoms or indications. However, conditions can be transmitted to Offspring by its carriers. The indications and symptoms of two mutant genes are not severe. The alpha thalassemia trait is the name given to this disorder. The indications and symptoms of three mutant genes range from mild to severe. Extremely uncommon mutations in four genes commonly cause stillbirth. Most infants with this illness die soon after delivery or need transfusion treatment for the rest of their lives. Rarely, stem cell transplants and transfusions can be used to treat a kid who was born with this condition. (www.mayoclinic.org/diseases-conditions/thalassemia). Figure 2 represents types of Thalassemia and Figure 3 shows Alpha Thalassemia (www. osmosis.org/learn/Alpha-thalassemia). Figure 4 shows how Alpha Thalassemia inherits in Children from parents (www.nhlbi.nih.gov/health/thalassemia). The gene that determines how much alpha globin should be produced has a mutation that leads to alpha thalassemia. Two beta globin's and two alpha globin's combine to form hemoglobin.

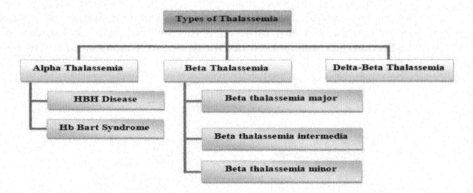

Figure 2.   Types of thalassemia.

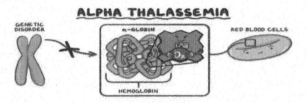

Figure 3.   Alpha thalassemia.

The gene mutation that causes alpha thalassemia causes the body to produce less alpha globin than beta globin. Anaemia and other medical issues from alpha thalassemia are brought on by the imbalance in alpha and beta globin.

The instructions (or genes) needed to produce alpha and beta globin are passed down from one generation to the next. The hemoglobin found inside red blood cells is created by the combination of alpha and beta globin's. Every kid receives two genes from each parent to make the four genes that make up alpha globin (Saleem *et al.* 2021). A alteration or mutation in the alpha globin gene causes less alpha globin to be produced than is expected in a person with alpha thalassemia. Alpha and beta levels of hemoglobin become unbalanced as a result of the decline in alpha-globin. Anemia and other alpha thalassemia-related health issues are brought on by this imbalance.

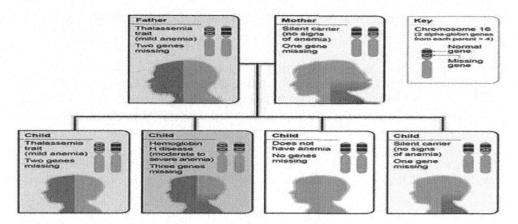

Figure 4. Alpha thalassemia inherits in children's.

2.1.1 *Alpha thalassemia types*
Carriers of alpha thalassemia do not exhibit any anemia-related symptoms or indications. They do not require any medical attention. Alpha thalassemia minor, commonly known as the alpha thalassemia trait, causes mild to moderate anaemia, however most people don't need any form of medical care.

2.1.1.1 *HBH disease*
Hemoglobin H can also be seen on newborn screens by medical professionals. With this diagnosis, a hematologist will regularly monitor the health of the kid. Additionally, some people are discovered later in life while receiving treatment for anemia.

2.1.1.2 *Hb bart syndrome*
On a newborn ultrasound, hydrops fetalis can be identified by its distinctive characteristics rather than by a precise diagnosis. A doctor will do a work-up to determine the cause of a fetus developing fluid buildup (also known as hydrops). Four alpha-globin genes are absent in Hb Bart syndrome (Cornelis L Harteveld & Douglas R Higgs 2010).Table 1 gives brief information about Different Alpha Thalassemia Syndrome and the number of genes affected by that syndrome, Clinical features like anemia and death rates for each syndrome and also describes Hemoglobin patterns for Alpha thalassemia.

2.1.2 *Treatment for alpha thalassemia*
People who are silent carriers (minima) or have mild cases of alpha thalassemia do not require therapy. Although moderate anemia will persist throughout one's life if they have

Table 1. Alpha thalassemia syndromes.

Syndrome	No of α-genes affected	clinical traits	Pattern of he-moglobin
Silent carrier	1	Minimal anemia	1 to 2 per HB
Thalassemia trait	2	Hypochromic microcytic, Mild anemia.	5 to 10 per HB
HbH illness	3	Hypochromic microcytic, Moderate anemia, RBC inclusion bodies.	10 to 30 per HBH
Foetal hy-drops	4	Severe anemia, Death at birth or in utero	97 per HB 3 per HBH

alpha thalassemia minor. Blood transfusions or chelation treatment may be necessary for those with more moderate to severe instances.

2.1.2.1 Transfusions

The mild anemia that HbH illness patients often have is well-tolerated. However, due to rapid red blood cell breakdown during diseases with a fever, transfusions are occasionally required. Adults could need transfusions more often. Hemoglobin H-Constant Spring illness, a more severe type of HbH sickness, can cause severe anemia and necessitate repeated blood transfusions throughout a person's lifetime.

2.1.2.2 Iron chelation therapy

With HbH illness, iron excess is possible. Even without blood transfusions, this increased iron absorption in the small intestine. chelators help the body get rid of excess iron. (Bender *et al.* 2020).

2.2 Beta thalassemia

A blood disorder known as beta-thalassemia causes the body to create less hemoglobin. A paucity of mature red blood cells and a deficiency in oxygen in the body are caused by low hemoglobin levels. People with beta-thalassemia who are anemic may have paleness, weakness, fatigue, and other serious issues. Since thalassemia is occurs by mutations in either the alpha or beta-globin gene, they are categorized as quantitative hemoglobin diseases. These modifications induce the formation of little to no globin. Anomalies in hemoglobin subunit beta (HBB) are the cause of beta-thalassemia. Frequent blood transfusions can lead to significant iron overload disorders in certain people (Vichinsky 2009). When two beta hemoglobin genes are damaged, most newborns are healthy at birth, but within few years, symptoms will get start. Two faulty genes can also result in thalassemia intermedia, a less severe kind. Figure 5 shows the genetic disorder of Beta Thalassemia (Kohli-Kumar 2001). Figure 6 explain how Beta Thalassemia inherit in Child from their parents (www.nhlbi.nih.gov/health/thalassemia). Beta thalassemia is formed with the absence of Beta genes. Beta has 2 genes one from each parent. When two carrier parents have thalassemia minor, the kid will acquire thalassemia major. There is a 25% probability that each offspring of two carriers will have beta thalassemia major. When both parents are carriers, the kid will inherit thalassemia minor. Beta thalassemia minor has a 50% probability of developing in each kid of two carriers. (www.cedars-sinai.org/health-library).

2.2.1 Complications of beta thalassemia

Complications of this type may vary from Moderate to Severe. In moderate conditions people may get affected by the:

- **Iron Overload**: Beta thalassemia and frequent blood transfusions both increase the body's iron levels. the liver, endocrine system, and Heart which contain hormone-producing glands that regulate physiological functioning can be affected by excess iron.

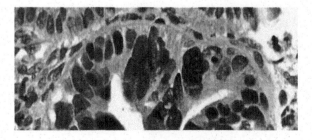

Figure 5.   Beta thalassemia.

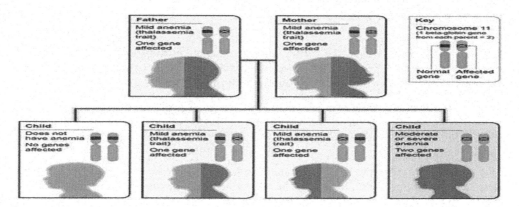

Figure 6.   Beta thalassemia inherits in children's.

- **Infection**: Patients with thalassemia are more susceptible to infections. If your spleen has been removed, this is especially true.

In Severe conditions people may suffer from:

- **Bone anomalies**: The marrow in the bones may enlarge as a result of thalassemia. Particularly in the face and skull, this might result in aberrant bone structure. Bone fracture risk is also increased by bone marrow expansion, which thins and makes bones brittle.
- **Enlarged spleen:** The spleen serves as the body's filter for unwanted materials like old or broken blood cells and to fight infection. One of the frequent side effects of thalassemia is red blood cell apoptosis. An enlarged spleen has the potential to reduce the half-life of transfused red blood cells and to exacerbate anemia. If the spleen becomes too large, the doctor could suggest removing it surgically.
- **Slow growth rates:** Children with anemia may have poor development and delayed puberty.
- **Heart problems:** Congestive heart failure and abnormal cardiac rhythms may accompany severe thalassemia. (www.nhlbi.nih.gov/health/thalassemia)

2.2.2 *Types of beta thalassemia*

- **Beta thalassemia major:** One of the worst types. Two Beta-globin genes are either absent or broken in this condition. Those with Beta thalassemia major needs regular blood transfusions, these illnesses are referred to as "transfusion-dependent thalassemia."
- **Beta thalassemia intermedia:** may result in mild to moderate signs of anemia. Two beta-globin genes must also be absent or damaged. With beta thalassemia intermedia, you probably won't require lifelong blood transfusions.
- **Beta thalassemia minor (beta thalassemia trait):** usually results in mild anemic symptoms. One beta-globin gene is either missing or damaged in this condition. The symptoms of beta thalassemia minor might be completely absent in certain persons (Sadiq *et al.* 2021)

The Table 2 Gives brief information about Different Beta Thalassemia Syndrome and the number of genes affected by that syndrome, Clinical features like anemia, Death rates for each syndrome, and also describes Hemoglobin patterns for Beta-thalassemia.

Table 2.   Different Beta thalassemia syndrome.

Syndrome	No Beta-genes affected	Clinical features	Hemoglobin pattern
Major	2 Broken	Severe anemia	10 to 30 per HBH
Intermediate	2 Damaged	Moderate anemia	5 to 10 per HB
Minor or silent carriers	1 Broken	Mild anemia.	1 to 2 per HBH

2.2.3 Treatments for beta thalassemia

2.2.3.1 Blood transfusions
With Beta thalassemia major, the carrier needs regular blood transfusions. they get blood during the surgery from a donor. Red blood cells required to deliver oxygen to human tissues are supplied by the blood that is drawn during a transfusion.

2.2.3.2 Iron chelation therapy
Iron is essential for the hemoglobin protein to carry oxygen. However, too much iron could be harmful. Chelation therapy for iron can help prevent iron excess.

2.2.3.3 Supplemental folic acid
Folic acid may improve the ability to manufacture red blood cells in the patient. The physician could recommend supplements if the patient has a minor case of beta thalassemia. In addition to receiving regular blood transfusions, folic acid must be taken if the illness is more severe.

2.2.3.4 Luspatercept
The body may create more red blood cells if you receive an injection of luspatercept every three weeks if you have severe thalassemia. Lupatercept helps beta thalassemia patients who are anemic get blood transfusions.

2.2.3.5 Bone marrow and stem cell transplant
Stem cells from bone marrow from a donor may be given to the carrier. Stem cells in the bone marrow ultimately mature into red blood cells. The bone marrow stem cells from a healthy donor can be used to treat beta thalassemia. The search for a suitable donor may, regrettably, be challenging. A high-risk procedure, this form of transplant is also considered.

The Table 3 describes the characteristics of Alpha and Beta Thalassemia that differentiate this disease from each other.

2.3 Delta-beta thalassemia

Delta-Beta is an uncommon variant of thalassemia characterized by increased amounts of hemoglobin subunit gamma and decreased synthesis of hemoglobin subunits beta and delta. This condition is autosomal recessive. The red blood cells of someone with delta-beta thalassemia are typically tiny and asymptomatic, however, microcytosis can happen. $\delta\beta$-thalassemia shows the characteristics of thalassemia minor with mild Anemia (Mansoori et al. 2016). Figure 7 shows $\delta\beta$ thalassemia cells (thalassemia-awareness-campaign.blogspot.com).

Table 3. Characteristics to differentiate alpha, beta, thalassemia.

Characteristics	Alpha	Beta
Definition	Reduced formation of Alpha genes	Reduced formation of Beta genes
Symptoms	Small red blood cells, mild anemia, paleness, fatigue, jaundice, enlarged spleen	Severe Anemia, paleness, fatigue, jaundice, enlarged spleen
Diagnosis	Smaller than usual red blood cells	High fetal hemoglobin, hemoglobin A2, or normal overall hemoglobin.
Causes	Genetic mutations of Alpha genes on chromosomes 16.	Genetic mutations of Alpha genes on chromosomes 11.
Risks	Homozygous condition hydrops fetalis occurs and the fetus dies in utero.	Homozygous condition fetus survives in the uterus because there is hemoglobin, but the child will have severe complications.

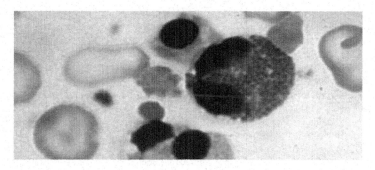

Figure 7.   Delta-beta thalassemia.

This condition is characterized by both Delta and Beta globin chain formation, which is often caused by deletions of Delta and Beta structural genes. Numerous ethnic groups, including German, Sicilian, Turkish, Japanese, and Spanish people, have been documented to carry the delta beta thalassemia mutation (Velasco-Rodriguez D 2014). Recently, non-deletional $\delta\beta$ thalassemia is been found [17]. In contrast to traditional Beta-thalassemia, both heterozygotes and uncommon homozygote individuals have modest clinical manifestations of Delta-Beta thalassemia. However, the combination of the hemoglobin measurement (electro-phoresis or HPLC) and the thalassemia red cell indices, that homozygotes lack HbA and HbA2 whereas heterozygotes have high HbF with normal HbA2 levels, helps to corroborate the diagnosis. The definitive diagnosis for this uncommon disorder's diagnosis is mutation analysis (Carrocini *et al.* 2011). Since both of these illnesses have 100% HbF, homozygous hereditary persistence of fetal hemoglobin (HPFH) is the primary differential diagnosis for homozygous thalassemia (Pirastu *et al.* 1984). It is more likely that someone has Delta and Beta thalassemia than HPFH based on the study of moderate anemia with hemolytic characteristics like indirect hyperbilirubinemia, decreased haptoglobin, reticulocytosis, etc. Studies on both parents' families will reveal thalassemic red blood cell indicators, with or without anaemia, and a 30% rise in HbF among heterozygotes for both beta and delta thalassemia. (Cao *et al.* 1982). Other diagnostic options for 100% HbF include homozygous Beta thalassemia variations, Delta and Beta thalassemia heterozygosity, and classical Beta thalassemia (Cao *et al.* 1982).

3   CONCLUSION

The purpose of this article is to give a detailed description of Thalassemia disease and how it gets inherited in the child from their parents, the depth of the characteristics of every subtype, and the statistics about the causes and prevention methods of this disease. It is caused due to disorder in red blood cells when the body doesn't make enough of a protein in the blood called hemoglobin (Thalassemia). Usually, they are three types Alpha, Beta, and Delta-Beta among three more Patients are affected by Beta thalassemia with Severe Anemia, paleness, fatigue, jaundice, and enlarged spleen. From 32 Districts of India approximately 10 to 12 thousand children have been affected by this disease and among these only fifty percent will not survive till the age of twenty due to poverty and lack of treatment. From this work, people will get an awareness of thalasse-mia, its causes, and different treatments that have been adapted for the prevention of this disease.

REFERENCES

Alain J. Marengo-Rowe (2007) The Thalassemias and Related Disorders, *Baylor University Medical Center Proceedings*, 20:1, 27–31, DOI: 10.1080/08998280.2007.11928230

Bender M, Yusuf C, Davis T, *et al.*. 2020; Newborn Screening Practices and Alpha-thalassemia Detection—United States, 2016. *MMWR Morb Mortal Wkly Rep.* 69:1269–1272.

Cao A, Melis MA, Galanello R, Angius A, Furbetta M, Giordano P, *et al.* 1982. Delta beta (F)-Thalassaemia in Sardinia. *J Med Genet.* ;19:184–92.

Cause of Elevated Fetal Hemoglobin. *Iran J Ped Hematol Oncol.*;3(1):222–7. Epub 2013 Jan 22. PMID: 24575268; PMCID: PMC3915439.

Cornelis L Harteveld and Douglas R Higgs. 2010, α-thalassaemia, *Harteveld and Higgs Orphanet Journal of Rare Diseases.* 2010, 5:13.

G.C.S. Carrocini, L.S. Ondei, P.J.A. Zamaro, C.R. Bonini-Domingos (2011). Evaluation of HPFH and δβ-thalassemia Mutations in a Brazilian Group with High Hb F Levels. *Genet. Mol. Res.* 10(4): 3213–3219. https://doi.org/10.4238/2011.December.21.3

https://thalassemia-awareness-campaign.blogspot.com/2016/05/delta-thalassemia-is-reason-of.html.

https://www.cedars-sinai.org/health-library/diseases-and-conditions-pediatrics/b/beta-thalassemia-in-children.html.

https://www.nhlbi.nih.gov/health/thalassemia/causes.

https://www.omniahealth.com/product/trupcr%C2%AEbeta-thalassemia-kit.

https://www.osmosis.org/learn/Alpha-thalassemia.

Kallenbacha, T 2015. Anaesthesia for a Patient with Beta Thalassaemia Major Southern Afr. *J. Anaesthesia Analgesia*, vol. 21, no. 5, pp. 21–24.

Kent Corley, Spring 2015. Blood Donors–Partners for Life. Perspectives Newsletter of The Northern California Thalassemia Center.

Kohli-Kumar M. 2001. Screening for Anemia in Children: AAP Recommendations-a critique. *Pediatrics.* 108 (3):e56. doi: 10.1542/peds.108.3.e56.

Mansoori H, Asad S, Rashid A, Karim F. 2016 Sep.Delta Beta Thalassemia: A Rare Hemoglobin Variant. *Blood Res.* 51(3):213–214. doi: 10.5045/br.2016.51.3.213. Epub 2016 Sep 23. PMID: 27722137; PMCID: PMC5054258.

Pirastu M, Kan YW, Galanello R, Cao A. 1984 Mar 2.Multiple mutations produce delta beta 0 thalassemia in Sardinia. *Science.* 223(4639):929–30. doi: 10.1126/science.6198720. PMID: 6198720.

Reviewed. https://www.healthline.com/health/thalassemia.

Sadiq.S *et al.*, 2021. Classification of β-Thalassemia Carriers From Red Blood Cell Indices Using Ensemble Classifier. in *IEEE Access*, vol. 9, pp. 45528–45538, doi: 10.1109/ACCESS.2021.3066782.

Saleem N, Anwar A, Shahid N, *et al.* August 31, 2021. Perception of Parents of Thalassemic Child to Thalassemia in Pakistan. *Cureus* 13(8): e17615. DOI 10.7759/cureus.17615.

Shilpa Amin, M.D. Nov 14, 2019. Everything You Need to Know About Thalassemia. Medically

Velasco-Rodríguez D. 2014 Oct. δβ-Thalassemia Trait: How Can We Discriminate it from β-Thalassemia Trait and Iron Deficiency Anemia? *Am J Clin Pathol.* ;142(4):567–73. doi: 10.1309/AJCPPBQ8UB1WHXTS. PMID: 25239426.

Verma S, Bhargava M, Mittal S, Gupta R. 2013. Homozygous delta-beta Thalassemia in a Child: A Rare

Vichinsky EP. 2009. Alpha Thalassemia Major–New Mutations, Intrauterine Management, and Outcomes. *Hematology Am Soc Hematol Educ Program*:35–41. doi: 10.1182/asheducation-.1.35. PMID: 20008180.

www.mayoclinic.org/diseases-conditions/thalassemia/symptoms-causes/syc-20354995.

Cloud computing architecture and adoption for agile system and devOps

Artificial Intelligence, Blockchain, Computing and Security – Dagur et al. (Eds)
© *2024 The Author(s), ISBN: 978-1-032-49393-0*

Smart face recognition attendance system using AWS

Nidhi Sharma*, Samarth Gaur* & Preksha Pratap*
Meerut Institute of Engineering & Technology, Meerut, U.P., India

ABSTRACT: Taking attendance is essential process to perform during lecture or to attend any event, but it is not an easy task to mark attendance of each one without any kinds of error. Still attendees are unaware whether their attendance get marked or not. This research paper presents a model that helps to build a system` that is used to marked attendance using face recognition technique and send email to registered person for confirmation purpose. This project is entirely based on cloud services provided by amazon web services

Keywords: Face recognition technique, cloud services, amazon web services, attendance system, email

1   INTRODUCTION

Attendance is a process that perform regularly in classes during lectures. It is a complex and coordinative task to mark attendance of each individual one by one and put the entire record for a long time in hardcopy format. Manual process is not only time-consuming process but also required a proper management.

Due to overcome such problems bio-metrics attendance system has been introduce in which system extract key points from palm and mark attendance by using fingerprint of the students, but after covid pandemic people usually avoid to use these kinds of systems.

After that face recognition-based attendance system has been come into picture, which marks attendance by matching facial features. It avoids to touch the device to mark attendance or call out the name of individuals one by one. There are various systems exist and used in real world with various technologies. System builds in this research paper build on cloud platform, developed by using different-different services provided by AWS. After marking the attendance system sends mail to registered mail id for confirmation.

2   LITERATURE REVIEW

R.Kodali and R.V.Hemadri [a],in this research paper model is divided into two parts – face detection and face identification. Multi-task Cascaded Convolutional Neural Networks algorithm is applied for face extraction and EfficientNet is used for face identification.

An attendance system through advance algorithms of face recognition was proposed by P.Pattnaik and K.K.M. [b]. In this model, there are primarily two key phases: face detection and face recognition.

Using a real-time system, S.Sawhney, K.K., S.J., S.N.S., and R.G.[c] suggested a system for recording classroom attendance, Two cameras were employed in this system, one for the classroom's exterior and the other for the interior.

---
*Corresponding Authors: nidhi.sharma.csit.2019@miet.ac.in, Samarth.gaur.csit.2019@miet.ac.in and Preksha.pratap@miet.ac.in

DOI: 10.1201/9781003393580-119

A.Mittal, F.S.K., P.K. and T.C. [d] proposed a system to mark attendance by using Viola Jones Object Detection algorithm which required complete frontal image of faces.

Dr.M.G. and Dr.D.P. [e] proposed a system in which the system first create a dataset of 100 images for a person during registration, mark attendance by face recognition and update attendance database. It uses shape prediction method, Dlib and OpenCV to find facial features.

H.Zhang, X.F., H.L., P.G., S.K. and C.Z.[f] proposed a model in which they use baidu cloud for training the image during registration process by json. In face recognition process you will get result in json format by processs data store on baidu cloud.

V.Yadav, G.P.B.[g] developed a model to mark attendance using raspberry pi 3 dev, GT511C3 optical fingerprint sensor, a LCD panel to display message and a keypad consisting 16 keys 0-9, A-D,*,# and 8 wires.

In their approach, P. Patil and Prof.Dr.S.S.[h] use the CNN algorithm for face evaluation and the Haar cascading technique for extraction of face points.

K.Preethi, s.v.[i] developed the model to mark attendance using local binary pattern algorithm. System consists various phase to mark attendance and create a dataset of 200 images.

S.Kakarla, P.G., M.S.R., C.S.C.S.and T.H.S.[j] developed model using CNN algorithm and OpenCV.It follow few steps as data collection, data augumentation, data training data testing and final result.

B.T. Chinimilli, A.T, A.K., V.R., K., J.V.M. [k] develop a model for attendance that used haar cascade algorithm face detection purpose and LBPH for face recognition purpose.

N.K.Jayant, S.B.[l] use voila jones and gabor ternary pattern in their model, it captures image and match with the image exist in the database.

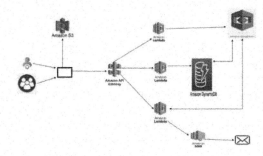

Figure 1   System workflow.

3   PROPOSED SYSTEM

This research paper discusses about the system that helps to take attendance using face recognition, a well-known concept of machine learning. This project is developed by using various services provided by aws and deployed on cloud.

Model consist of two phase first registration phase and second attendance phase.

3.1   *Registration phase*

In first phase put all required details by using webpage developed by using html, css, java-script. All details enter by end user will get store in s3 bucket. API gateway hides upload lambda. When upload lambda get event which have images, user details. Images get store in collection id in Rekognition. Rekognition returns unique face id, which get store along other details in DynamoDB.

808

Figure 2.   Registration webpage.

Figure 3.   Attendance webpage.

3.2   Attendance phase

In second phase, to mark attendance we use web camera to capture image and upload this image in s3 bucket. At that time search event gets invoke by search lambda function using API gateway. Now search lambda gets image knows s3 bucket location and upload that into recognition and does a search to find a valid match. If match is found then it gets face id from Rekognition. Using that face id to look for the details in database and if get a valid match. It means that face id existed previously and then collect other user details to generate SNS notification and send mail to registered mail id.

3.3   AWS services to be used

3.3.1   AWS s3
Aws S3 Bucket is a service that may be used for storage at a large scale of various category of data and is flexible to access and get back at any time. It functions as a database. In this project will store student profile photos and images for facial recognition processing.

3.3.2   AWS api gateway
API Gateway, a service offered by Amazon is a fully arranged and automated service, in general which is used by developers for various purpose as development, management, analysis, and safety APIs at any level.

3.3.3   AWS codepipeline
A continuous delivery service called CodePipeline makes it feasible to describe, virtualize and autonomous all steps involved in delivering software. CreatePipeline, used for creates a pipeline. DeletePipeline, used for deletes pipeline. GetPipeline, provide all information about pipeline structure and pipeline metadata.

3.3.4   AWS cloudformation
The CloudFormation service aids in development by giving offering a mechanism to gather necessary AWS services and resources from other service providers, supply and ordered them in a systematic and predictable manner.

3.3.5   AWS DynamoDB
AWS Database services, often known as DynamoDB, is a facility used to produce and keep data in manageable form for storage purposes. It is an entirely managed SQL service. It is simple to develop and re access recorded data with this service.

809

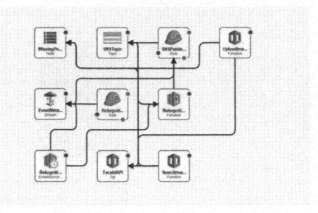

Figure 4.   Cloud formation template of system.

3.3.6   *AWS Rekognition*
AWS Rekognition, which is employed to develop the model using information such as images or even some videos by providing which aid in rekognition the items like individuals, items, creatures or even other, faces, masked things etc.

3.3.7   *AWS sns*
The automated feature Amazon Simple Notification Service notifications from one to another point of communication release alert using various application servers of different kinds, including Web services, Amazon SQS, Amazon Lambda, Amazon Kinesis Data Firehose, mobile messaging alerts, and mobile messages

3.3.8   *AWS lambda*
AWS Lambda is a serverless computing solution that makes it possible to process code in response to action and process it timely configure every action using processing power. Code may execute in Lambda's cloud hosting interface without the necessity to monitor server.

3.3.9   *AWS SAM*
Serverless Application Model, sometimes known as AWS SAM, is a fully accessible platform for building cloud-based services. It offers language for expressing functions, APIs, databases, and translations of information sources.

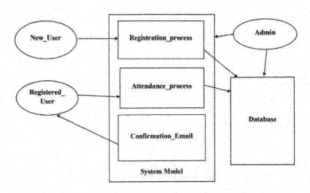

Figure 5   Overview of system design.

Model proposed in this paper is completely based in CICD approach that make easy for developer to make any kind of changes in the system that does not effect on the performance of the existing module. For this purpose, Aws DevOps services has been used as known by code pipeline (use to connect various development phase to each other), code build (used to develop the code), code deploy (end phase use for code deployment purpose), clouds formation (consist template to develop the structure of project). Web application is developed with the help of html, css and javascript that is used for registration purpose, it collects all required data of an end user for correctly execution of the model.

4 RESULTS

By using all the concepts and services describe in this research paper, we make it possible to build a system. Student will get email for successfully marked attendance as notification only if he has registered all details in registration process and cloud database get updated in attendance process after matching captured face image with already saved facial feature detail of registered face image.

5 CONCLUSION AND FUTURE SCOPE

This system model can be scale up and down as per need easily. It can be used for school, colleges, industries, companies, various organizations, events to collect personal details and mark presence. It is able to verify that database get updated by sending mail. This system can be enhanced by adding more functionality without effecting on existing modules.

Project develop in this paper is based on advance technologies as face recognition, cloud computing that have a lot to make it more efficient for its end user. In this project we can add a module that might be helpful to recognize identical twins, or a module that is able to identify facial expression of the individual.

REFERENCES

[1]   Kishore, K. And R.V.H. 2021 "Attendance Management System", 'International Conference on Computer Communication and Informatics" Coimbatore.
[2]   Pattnaik, and K.K.M., 2020 "AI-Based Techniques for Real-Time Face Recognition-based Attendance System- A comparative Study" "Fourth International Conference on Electronics, Communication and Aerospace Technology".
[3]   Sawhney, K.K., S.J., S.N.S. and R.G. 2019 "Real-Time Smart Attendance System using Face Recognition Techniques.
[4]   Mittal, F.S.K., P.K. and T.C. 2017 "Cloud Based Intelligent Attendance System through Video Streaming" "International Conference on Smart Technology for Smart Nation".
[5]   Gopila and Dr. D.P. 2020 Machine Learning Classifier Model for Attendance Management System.
[6]   Zhang, X., F., H.L., P.G., S.K. and C.Z. 2019 Cloud-Based Class Attendance Record System.
[7]   Yadav, G.P.B. 2019 Cloud Based Smart Attendance System for Educational.
[8]   Patil, Dr.S.S. 2020 Comparative Analysis of Facial Recognition Models using Video for Real Time Attendance Monitoring System.
[9]   Preethi, S.V. 2021 "Automated Smart Attendance System Using Face Recognition" "Fifth International Conference on Intelligent Computing and Control System".
[10] Kakarla, P.G., M.S.R., C.S.C.S. and T. H.S. 2020 Smart Attendance Management System Based on Face Recognition Using CNN.
[11] Chinimilli, A.T., A.K., V.R.K., J.V.M. 2022 Face Recognition based Attendance System using Haar Cascade and Local Binary Pattern Histogram Algorithm.
[12] Jayant, S.B., 2016. Attendance Management System Using Hybrid Face Recognition Techniques.

A review on identification of fake news by using machine learning

Naeema Ahmed* & Mukesh Rawat*
Meerut Institute of Engineering & Technology, Meerut

ABSTRACT: Mostly person use internet to connect with people, for entertainment, to connect with the world with the help of news etc. In this way, it is necessary that a person can get accurate news rather than false news. With the help of internet publishers can publish fake and misleading news to become popular or just for the clicks. The fake news can harm people's reputation, damage politicians' image. So, the main problem is to detect fake news. In this survey, existing ML strategies like Decision Tree, Naive Bayes, SVM, Passive Aggressive, LR, KNN, LSTM, Bert are reviewed. According to this paper the highest accuracy predicts with the help of Bert and the lowest accuracy predicts with the help of LR. As a result, this analysis contrasts the various methods for developing models now in use and discusses potential future advancements that could be made by combining several machine learning techniques.

1   INTRODUCTION

Social media is the best platform for interacting with our loved ones, friends, and coworkers. On the internet, the available news can be either fake or real. Smitha (2020) Mostly, the fake news can be used to get more clicks and visits on the sites. The fake news can also be used to destroy someone's reputation, change the thinking of human beings or property. On the internet, fake information is purposefully created, either on purpose or accidentally. Whether it is fraud or not, the development and usage of information online has improved over time.

We frequently hear and read about mob lynching that end in a person's death; fraud news detection goals to identify the reports as false and put a stop to such actions, shielding society from these senseless acts of violence. The primary motive of those websites which provide incorrect news to influence public views on particular issues specially on politics. Fraud news lay out faster than the original news. Therefore the global problem and challenge is fake news. To identify between fraud and real news, it is necessary to build a model. In this study, we represent the survey of various machine learning methods to identify fraud news.

Here is an example of fake news based on Corona Virus. With multiple instances of fake news throughout the Covid-19 outbreak, the internet served as a breeding ground for incorrect information. The statement that 5G technology was connected to the growth of the virus was an excellent example of fraud news on social media. The theory goes like 5G technology weakened the immune system while the virus spreaded by radio waves, hence the two technologies were incompatible. These assertions were false and were frequently refuted by reliable sources, yet they were widely disseminated.

2   LITERATURE REVIEW

Shikun Lyu *et al.* (2020) propose that with the help of decision trees, support vector machines (SVMs), doc2vec, Fake News Tracker, and other tools, author proposes to examine the performance of numerous machine learning techniques in this project. According to their first

---

*Corresponding Authors: naeemaahmed002@gmail.com and Mukesh.rawat@miet.ac.in

DOI: 10.1201/9781003393580-120

findings, decision trees and the SVM can distinguish bogus news with a satisfactory accuracy of 94.8%. The decision trees approach typically yields better results than SVM. The dataset that they used can be obtained from JSON documents and it contains four features.

Della Vedova et al. (2018) presents the first unique machine learning (ML) fake news detection technique, which beats other systems in literature by improving the mode up to 79%. Second, they tried their methodology within a Facebook Chabot and tested it with a real-world app, achieving a accuracy of 82% in detection of fraud news. First of all, dataset was collected which was used in the test, then contact-based approach was applied and suggested to combine it with the social-based strategy was proposed previously in literature. The goal was to label a news item as real or fake. The dataset contains 15,000 posts with 2,000,000 likes approximately from 9M persons.

For the categorization of fraud news, Ratkiewitz et al. (2011) have suggested a variety of ML and DL based approaches. In order to identify the propagation of political disin-formation, established an ML approach that integrates topological, crowd-sourced and content-based elements retrieved from information diffusion networks on Twitter. Despite having a 96% accuracy rate, this method only considers one social media platform, whereas our framework considers public political communication on a variety of platforms.

For the purpose of identifying false news, Hannah Rashkin et al. (2017) suggested a model based on the linguistic traits of unreliable material. By using Wiktionary, the model exam-ines the tone of the news with that of hoaxes and satire, and propaganda in order to extract linguistic traits. Various models of machine learning like Naive Bayes, Maximum Entropy (MaxEnt) and LSTM were trained using the data. MaxEnt had a classification accuracy of 22% for the six classes using samples from PolitiFact, while LSTM and Naive Bayes models had a classification accuracy of 56% for binary classification.

By contrasting 2 different feature extraction techniques and 6 different classification strategies, Hadeer Ahmed et al. (2017) present a fake news detection model that uses ML methods and n-gram. The results of the tests indicate that the so-called features extraction method obtained the best results (TF-IDF). They employed the 92% accurate LSVM classifier. The LSVM used in this model is restricted to handling only the situation where two classes are linearly separated.

A new dataset for automatic detection of fraud news, LIAR, was introduced by William Yang Wang et al. (2017). This corpus can also be utilised for political NLP research, topic modelling, rumour detection, argument mining, and position classification. The majority of works in this field have utilised this standard. It is known that past studies have included data from a variety of fields, whereas this one focuses solely on political information.

Rajeswari et al. (2022) represents a model to detect fake news with the help of machine learning and determines that which news is fake or real using LOGISTIC REGRESSION ALGORITHM. The model is working well and provides an accuracy up to 97.21 %.

In this research, Sachin Kumar et al. gathers 1357 news instances from different people via Twitter and news sources like PolitiFact, and then creates multiple datasets for the true and fake news stories. CNN and LSTM, and attention mechanisms are just a few of the cutting-edge techniques compared in their study. The bidirectional LSTM and CNN grouped net-work with attention mechanism, according to our findings, had the maximum accuracy of 89%, while Ko et al. attempted to identify bogus news and had a detection rate of 85%.

According to the Yang, Shuo, et al. (2019), a Pew Research Center USA survey indicates that social media accounted for about 70% of individuals' news consumption. The news that Donald Trump will be the next president has resulted in a rise of 9.6 lacks Facebook users. Language or visual qualities both have a part to play in their paper. We now turn our attention to network features, which cover both co-occurrence and diffusion networks. Therefore, the accuracy rate for the authors is roughly 83 percent.

3 FAKE NEWS DETECTION TECHNIQUES

In our daily lives, fake news has a significant impact. The effect of false news on our daily lives is considerable.

For humans, identifying fake news is a crucial step. In order to identify fake news, eight different ML algorithms have been deployed.

3.1 Decision tree

Both regression and classification problems can be done by "Decision Tree. Still, it is mainly frequently use to solve classification problems. The aim of using Decision Tree is to create a ML model which is be used to predict the output with the help of simple rules that are inferred from training data.

3.2 Naive bayes

A supervised algorithm based on Bayes theorem which is used for classification problems is Naive Bayes. Naive Bayes Classifier is the simplest and most efficient classification model in today's world. It helps in the development of efficient ML models that can make accurate predictions.
Bayes' theorem:

$$Prob(k|l) = \frac{Prob(l|k) * Prob(k)}{Prob(l)} \tag{1}$$

where,
Prob(k|l) = Posterior Probability, Prob(k) = Prior Probability
Prob(l) = Marginal Probability, Prob(l|k) = Likelihood Probability

3.3 SVM

Murari Choudhary (2021) SVM is typically used for classification, and it is a powerful and flexible supervised method [13]. SVM is a useful algorithm for determining the binary class from the input data. SVM is a supervised ML technique used for classification and regression both. The term "Support Vector Machine" refers to a separating hyperplane and a set of specific data points known as "Support Vectors" in support vector machines.

3.4 Passive aggressive

Uma Sharma et al. (2021). Among beginners and intermediate ML algorithms, the Passive-Aggressive algorithms are those which are not used widely and it's a complete family of ML algorithms. For some purposes, they can be quite helpful and effective. Passive Aggressive can be obtained from two words i.e., Passive and Aggressive. If the classification event is incorrect, such an algorithm remains passive but it becomes aggressive, updating and adjusting. Unlike most of the algorithms, it does not converge. The aim of this algorithm is to make the update that removes the loss without making any change in weight vector's norm.

3.5 Logistics regression

Logistics Regression can provides results either in discrete or categorical value. The result obtained from Logistics Regression can be either 0 or 1, True or False, Yes or No. It cannot provides the exact value of 0 and 1 but it gives probability value that comes under 0 and 1. Linear Regression can be used to solve regression problems, whereas Logistics Regression can be used to solve classification problems.

3.6 K-Nearest Neighbor(KNN)

A supervised learning algorithm used for regression and classification is KNN. It operates by identifying the K nearest data points, and then assigning the test data point a class label based on the majority class of those K nearest data points. It is a straightforward algorithm that anyone may use to solve problems in the real world.

3.7 LSTM

It is a type of recurrent neural network in which the output from the preceding step of an RNN is provided as input into next phase. It raise the problem of long-term RNN dependency, where the RNN is not capable to predict words stored in long-term memory but can make more accurate predictions using more current data. As the gap length grows, RNN cannot function efficiently. The information can often be kept in LSTM for a long time. On the basis of time-series data, it is utilized for predicting, forecasting, and categorization.

3.8 Bert

BERT's architecture is based on the deep learning model Transformers. Each output element in Transformers is connected to each input element, and weightings between them are constantly decided depending on this relationship. In the past, language models could only concurrently analyse input text if it was read sequentially from right to left or from left to right. Because BERT can read simultaneously in both directions, it is unique. Bidirectionality is the term for this capacity, which was made possible by the appearance of Transformers.

3.8.1 System Architecture

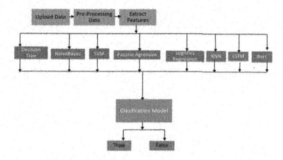

Figure 1.   Working model of machine learning algorithm.

4   COMPARATIVE ANALYSIS OF EXISTING ALGORITHM

Table I shows the analysis of Machine Learning Algorithms. In 2020 authors Shikun Lyu & Dan Chia-Tien Lo achieved highest accuracy up to 95% by Decision Tree; the other researchers had not performed better than Shikun Lyu & Dan Chia-Tien Lo. By Naive Bayes the authors achieved 74% accuracy rate. The authors of SVM technique gives the accuracy up to 93.6% and with the help of Passive Aggressive technique the author produced a result of 90%. All techniques which had been discussed in this table used feature extraction. With the

Table I.   Comparison of result of existing algorithms.

Author	Algorithm	Accuracy
Shikun Lyu, Dan Chia-Tien Lo (2020)	Decision Tree	95%
Mykhailo Granik, Volodymyr Mesyura (2017)	Naive Bayes	74%
Anjali Jain, Avinash Shakya, Harsh Khatter, Amit Kumar Gupta (2019)	SVM	93.6%
Jayashree M Kudari, Varsha V, Monica BG, Archana R (2020)	Passive Aggressive	90%
Fathima Nada, Bariya Firdous Khan, Aroofa Maryam, Nooruz-Zuha, Zameer Ahmed (2019)	Logistics Regression	72%
Ankit Kesarwani, Sudakar Singh Chauhan, Anil Ramachandran (2020)	K-Nearest Neighbor (KNN)	79%
Tejaswini Yesugade, Shrikant Kokate, Sarjana Patil, Ritik Varma, Sejal Pawa5 (2021)	LSTM	91%
Rohit Kumar Kaliyar, Anurag Goswami, Pratik Narang (2021)	Bert	98.90%

help of feature extraction and certain trained model researchers try to increase performance rate to detect fake news. Which algorithms perform better than others is shown in this table. Decision Tree is the best algorithm out of all of them and produces the best results.

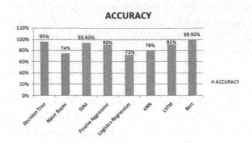

Figure 2. Accuracy according to different algorithms.

5 CONCLUSION

ML-based categorization algorithms play a very crucial part in this process. The process of identifying false news on internet is sophisticated and challenging due to the numerous varied political, social, economic, and other related variables. This study is completely based on the previous existing work based on fraud news detection. In this paper, we discussed many ML classification-based algorithms such as Decision tree, Naive Bayes, SVM, Passive Aggressive, Logistics Regression, KNN, Bert, LSTM. By using classification algorithm, the best accuracy obtained from BERT i.e. 98.90%.

REFERENCES

Della Vedova M. L., Tacchini E., Moret S., Ballarin G., DiPierro M. and de Alfaro L. 2018 "Automatic Online Fake News Detection Combining Content and Social Signals," In *2018, 22nd Conference of Open Innovations Association (FRUCT)*, Jyvaskyla, pp. 272–279.

Hadeer Ahmed, Issa Traore, and Sherif Saad. 2017. "Detection of Online Fake News Using n-gram Analysis and Machine Learning Techniques." In *2017, International Conference on Intelligent, Secure, and Dependable Systems in Distributed and Cloud Environments*, pages 127–138. Springer.

Hannah Rashkin, Eunsol Choi, Jin Yea Jang, Svitlana Volkova, and Yejin Choi. 2017. Truth of Varying Shades: "Analyzing Language in Fake News and Political Fact-checking." In *2017, Proceedings of the Conference on Empirical Methods in Natural Language Processing*, pages 2931–2937.

Jacob Ratkiewicz, Michael D Conover, Mark Meiss, Bruno Gonçalves, Alessandro Flammini, and Filippo Menczer Menczer. 2011."Detecting and Tracking Political Abuse in Social Media." In *2011, Proceedings of the International AAAI Conference on Weblogs and Social Media.*

Kumar, A. & Alam, B. (2018). Task Scheduling in Real Time Systems with Energy Harvesting and Energy Minimization. *Journal of Computer Science*, 14(8), 1126–1133

Mehra, P. S., Jain, K., Chawla, D., Dagur, A., Singh, S., & Sharma, J. (2022). *GWO-EFUCA: Grey Wolf Optimisation and Fuzzy Logic based Unequal Clustering and Routing Protocol for sustainable WSN-based Internet of Things.*

Sachin Kumar, Rohan Asthana, Shashwat Upadhyay, Nidhi Upreti Mohammad Akbar1. "Fake News Detection Using Deep Learning Models: A Novel Approach". https//DOI: 10.1002/ett.3767,.

Sharma Y., Shatakshi, Palvika, Dagur A. and Chaturvedi R., "Automated Bug Reporting System in Web Applications," *2018 2nd International Conference on Trends in Electronics and Informatics (ICOEI)*, Tirunelveli, India, 2018, pp. 1484–1488.

Shikun Lyu, Dan Chia-Tien Lo. 2020. " Fake News Detection by Decision Tree." In *2020, IEEE SoutheastCon.*

Smitha. N, Bharath R. 2020."Performance Comparison of Machine Learning Classifiers for Fake News Detection". In *2020, IEEE (ICIRCA-2020)* Part Number: CFP20N67-ART; ISBN: 978-1-7281-5374-2, pages 696–700.

William Yang Wang. 2017. " Liar, Liar Pants on Fire": A New Benchmark Dataset for Fake News Detection. In 2017, *arXiv* preprint arXiv:1705.00648.

Yang, Shuo, *et al.* 2019."Unsupervised Fake News Detection on Social Media: A Generative Approach." In *2019, Proceedings of the AAAI Conference on Artificial Intelligence*. Vol. 33.

Artificial Intelligence, Blockchain, Computing and Security – Dagur et al. (Eds)
© 2024 The Author(s), ISBN: 978-1-032-49393-0

Optimal resource allocation in cloud: Introduction to hybrid optimization algorithm

Shubham Singh & Pawan Singh
Amity School of Engineering and Technology Lucknow, Amity University Uttar Pradesh, India

Sudeep Tanwar
Institute of Technology, Nirma University, Ahmedabad, Gujarat, India

ABSTRACT: Cloud resource requirements, particularly certain emergent and unclear resource demands, are rising speedily with the advancement of Cloud Computing (CC), and big data. The conventional cloud resource allocation schemes could not handle the emergency mode and it does not guarantee the optimization and timeliness of resource allotment. This paper intends to execute a novel optimization aided resource allocation model. Initially, the workload clustering is done by means of an optimal k-means clustering, where centroid is tuned using Blue Monkey Updated Pelican Optimization (BMU-PO). The tasks are clustered depending upon execution time and QoS (trust). In the end, optimal resource allocation is done using BMU-PO algorithm by considering CPU usage, PUE, and execution time.

Keywords: Cloud Computing, Execution time, k-means clustering, QoS, BMU-PO Algorithm

NOMENCLATURE

Abbreviation	Description
API	Application Programming Interface
BMO	Blue Monkey Optimization
BMU-PO	Blue Monkey Updated Pelican Optimization
CC	Cloud Computing
CSP	Cloud Service Providers
EE	Energy Efficiency
ETC	Expected Time To Complete
MOSOS	Multi-Objectives Symbiotic Organism Search Algorithm
MI	Million Instruction
VM	Virtual Machines
OBSAs	Options Based Sequential Auctions
PM	Physical Machines
POA	Pelican Optimization Algorithm
PUE	Power Usage Effectiveness
MS	Make Span
SLA	Service Level Agreement
RA	Resource Allocation
RAS	Resource Allocation Strategy
QOS	Quality Of Service

DOI: 10.1201/9781003393580-121

817

1 INTRODUCTION

As a new paradigm for computing, CC seeks to give the end users a dependable, personalized, and QoS assured compute dynamic environment (Belgacem *et al.* 2020; Hosseinalipour & Dai 2019; Xu & Palanisamy 2018). CC is a combination of distributed processing, parallel processing, and grid computing. The main principle of CC is that user data is kept online, rather than locally (Gill *et al.* 2020; Joseph & Chandrasekaran 2020; Gong *et al.* 2019; Mireslami *et al.* 2019; Zhang *et al.* 2019).

RA in CC refers to the process of online resource allocation to the required cloud applications (Mergenci & Korpeoglu 2019; Shrimali & Patel 2020; Thein *et al.* 2020). The services will degrade if resource allocation is not handled carefully. The aim of RAS is to integrate cloud provider operations for utilizing and assigning limited resources within the limitations of the cloud environment in order to satisfy the requirements of the cloud platform (Abohamama & Hamouda 2020; Feng *et al.* 2020; Goswami *et al.* 2019; Hejja & Hesselbach 2019; Santos *et al.* 2019).

Cloud users estimates the resource demands to finish a task before the approximated time might lead to an over provisioning of resources. The distribution of resources by resource providers may result in under-provisioning of resources (Asghari *et al.* 2020; Lee *et al.* 2019; Soltanshahi *et al.* 2019). The application need and SLA are two important inputs to RAS from the perspective of the cloud user. Although the cloud offers trustworthy resources, it also presents a significant challenge in dynamically assigning and managing resources across applications (Mao *et al.* 2018).

The contribution of the research work is depicted below:

- In this research, workload clustering is performed with the aid of optimal k-means clustering.
- Introduces a new Blue Monkey Updated Pelican Optimization (BMU-PO) algorithm for optimal centroid selection and optimal resource allocation.
- The tasks are clustered depending upon execution time and QoS. The resource allocation is performed by considering CPU usage, PUE, and execution time.

Organization of the paper-Section 2explains the review on allocation of resources. Section 3provides an explanation of the proposed resource allocation work. The result and conclusion are in Sections4 and 5.

2 LITERATURE REVIEW

For integrated geo-distributed cloud, Xu & Palanisamy (2018) suggested a novel contracts based resource sharing paradigm that enabled resource sharing contracts amongst CSPs. CSPs used contract duration and job scheduling based on the established contracts.

In Hosseinalipour & Dai (2019) used a two-stage auction approach to study the relationships between CSPs and CPs. The OBSAs was used to build the cloud resource allocation paradigm for interactions between clients and cloud administrators.

In Belgacem *et al.* (2020) presented a dynamic allocation of resources for the CC platform. The MOSOS was suggested as a meta-heuristic method for resource allocation.

Gong *et al.* (2019) presented a resource allocation strategy with adaptive control that responded to changing request demands and resource needs.

In Mireslami *et al.* (2019) presented hybrid strategy for allocating cloud resources based on changing user needs. The allocation and dynamic provision stages made up the two phases of this algorithmic strategy. Thus, by resolving the optimization issues, QoS requirements were met and the overall deployment cost was minimized.

Some limitations of the existing work are listed as follows. In Xu & Palanisamy (2018), the auctioneer indicates a single point of failure. The two-stage auction approach (Hosseinalipour & Dai 2019) requires an optimization model to improve the performance. The MOSOS algorithm (Belgacem *et al.* 2020) was suffered from slack convergence rate.

Moreover, the traditional cloud resource allocation schemes could not handle the emergency mode, and they do not guarantee resource optimization and timeliness. Therefore, a hybrid optimization based resource allocation is introduced in this study.

3 PROPOSED RESOURCE ALLOCATION APPROACH IN CLOUD

In this research, an optimization based resource allocation approach is established. The major intent of resolving the resource scheduling issue in cloud environment is to allocate the resources. Then BMU-PO is deployed for the optimum resource allocation with3diverse objectives: (a) CPU utilization (b) PUE and (c) execution time. Figure 1. shows the general architecture of developed work.

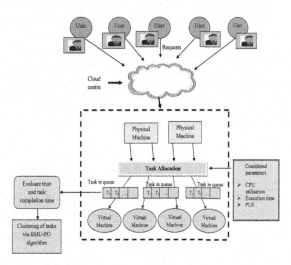

Figure 1. Overall design of resource allocation in cloud.

3.1 Setup of cloud

In this research, cloud sim 3.0 is utilized for implementation. Assume a CC surroundings C that comprises of PM count denoted by M_{PM} that consists of VM count denoted by M_{VM}: $C = \{PM_1,, ..PM_i., PM_{M_{PM}}\}$ $PM_i = \{VM_1,, ..VM_j., VM_{M_{VM}}\}$. The symbol VM_j implies VM, which is assigned by PM for carrying out the particular task. VM_j is described depending upon its unique identifier $VMID_j$ and processing $MIPS_j$. The submitted task set by user is implied as TSK. The task count requested by end user is M_{task}. $task = T_1, T_2, T_3, ...T_k., T_{M_{task}}$. The task contain priority, unique identifier, task length, anticipated time to finish, and QoS ($Pri_{task_k}, TID_{task_k}, len_{task_k}, etc_{task_k}, QoS_{task_k}$) correspondingly. Allotment of T_k onto VM_j has a note worthy effect on performance of the system. The executing time of task on VM_j is $etc_{kj} = \frac{len_j}{MIPS_j}$. In addition, etc is specified in matrix form is known as ETC matrix that is specified as per Eq. (1).

$$etc_{M_{TSk JK}, N_{VM}} = \begin{cases} etc_{1,1} & etc_{1,2} & \cdots & etc_{1,j} & \cdots & etc_{1,M_{VM}} \\ etc_{2,1} & etc_{2,2} & \cdots & etc_{2,j} & \cdots & etc_{2,M_{VM}} \\ etc_{3,1} & etc_{3,2} & \cdots & etc_{3,j} & \cdots & etc_{3,M_{VM}} \\ \vdots & \vdots & \vdots & \vdots & \cdots & \vdots \\ etc_{M_{TSK},1} & etc_{M_{TSK},2} & \cdots & etc_{M_{TSK},j} & \cdots & etc_{M_{TSK},j} \end{cases}$$

(1)

A particular PM is assumed with the count of VM M_{VM}. Prior to assigning the tasks to VM, the task set is clustered depending upon its task completing time and QoS. This is achieved with improved K-means scheme. The pre-determined value K = 2 (cluster count). The VM's is united with the cluster depending upon their task completing time and QoS.

3.2 Improved K-means clustering

Clustering is a technique for breaking up a huge collection of data into a number of smaller groupings. An established clustering method called K-means clustering separates a set of data into k groups.

Let *task* be tasks set and $H_N = \{H_1, H_2, ...H_n\}$ be centroid set.

1) Choose H_N centroid of cluster in an optimal manner by means of BMU-PO algorithm.
2) With Eq. (2), calculatetask completing time and QoSof every task related to centre point that is implied by Di.

$$Di = \sum_{k=1}^{M_{task}} \sum_{n=1}^{N} (\|T_k - H_N\|)^2 \tag{2}$$

3) The tasks approximately nearer to centre point (tasks) with identicaltask completing time and QoSare chosen as the cluster.
4) Re-compute the novel cluster centre as per Eq. (3).

$$H_N = \left(\frac{1}{k}\right) \sum_{T_k \in H_N} T_k \tag{3}$$

5) Re-computetask completing time and QoS among every novel task and cluster centre.
6) If tasks were not reassigned, end the function, otherwise, follow from step 3.

When the clusters are created, the resource assignmentoccursdependingupon the described3-fold objectives.

3.3 Fold objectives for optimal allocation of resources

The objective is shown in Eq. (4). The resources are assigned via BMU-PO algorithm.

$$obj = \min\{CPU + PUE + ET\} \tag{4}$$

3.4 Proposed BMU-PO

As said above, the optimal centroid selection and optimal resource allocation are carried out using new BMU-PO algorithm. The existing POA (Trojovský & Dehghani 2022) model results in lots of advantages. Nevertheless, POA does not reach high optimal solution. For that, BMO(Mahmood & Al-Khateeb 2019) is mingled with POA to form BMU-PO. Combining hybrid algorithms results in more precise results (Beno *et al.* 2014; Devagnanam & Elango 2020; Talasilaand Narasingarao 2022; Thomas & Rangachar 2018).

The POA model depends upon the pelican population, where every individual offers a feasible solution. The initialized members in POA are modeled in Eq. (5)(Mahmood & Al-Khateeb 2019).

$$\Re_{i,j} = q_j + ran(u_j - l_j), \ i = 1, 2, ...H, j = 1, 2...h \tag{5}$$

In Eq. (7), $\Re_{i,j}$ implies j^{th} variable for i^{th} solution, B implies member count, h implies problem variable count, *ran* implies arbitrary integer, u_j and l_j implies upper and lower limit.

3.4.1 *Proposed exploration*

The creation of random prey location enhances POA capacity to solve the problems precisely. Eq. (6) shows the movement of pelican to target prey(Mahmood & Al-Khateeb 2019).

$$\mathfrak{R}_{new2} = \mathfrak{R}_{i,j}^{p_1} = \begin{cases} \mathfrak{R}_{i,j} + ran\left(p_j - I.\mathfrak{R}_{i,j}\right), \ Q_p < Q_i; \\ \mathfrak{R}_{i,j} + ran\left(\mathfrak{R}_{i,j} - p_j\right), \ else \end{cases} \tag{6}$$

As per BMU-PO, if $Q_p > Q_i$, the update occurs by combining Eq. (7) of BMO with POA.

$$\mathfrak{R}_{i+1} = \mathfrak{R}_i + rate_{i+1} * rand \tag{7}$$

$$rand = \frac{\mathfrak{R}_{i+1} - \mathfrak{R}_i}{rate_{i+1}} \tag{8}$$

If the condition $Q_p > Q_i$ is not satisfied, then substitute *rand* value and it is shown in Eq. (8).

$$\mathfrak{R}_{new\,2} = \mathfrak{R}_{i,j}^{p_1} = \begin{cases} \mathfrak{R}_{i,j} + ran\left(p_j - I.\mathfrak{R}_{i,j}\right) * BM, \ Q_p < Q_i; \\ \mathfrak{R}_{i,j} + \left(\dfrac{\mathfrak{R}_{ij+1} - \mathfrak{R}_{ij}}{rate_{i+1}}\right) * \left(\mathfrak{R}_{i,j} - p_j\right), \ else \end{cases} \tag{9}$$

In Eq. (9), I lies among 1 or 2, $\mathfrak{R}_{i,j}^{p_1}$ denotes novel position of ith pelican, Q_p denotes objective and p_j denotes prey position. A new location is permitted only if the objective is enhanced in that position. This updating avoids the approach from moving to non-optimal regions. This is modeled in Eq. (10)(Mahmood & Al-Khateeb 2019).

$$\mathfrak{R}_i = \mathfrak{R} \begin{cases} \mathfrak{R}_i^{p_1}, \ Q_i^{p_1} < Q_i; \\ Z_i, \ else \end{cases} \tag{10}$$

In Eq. (10), $\mathfrak{R}_i^{p_1}$ denotes new position and $Q_i^{p_1}$ denotes objective at phase 1.

3.4.2 *Proposed exploitation*

The hunting behavior of pelican's in Eq. (11), in which, $\mathfrak{R}_{i,j}^{p_2}$ denotes position at 2nd phase, $R = 0.2$, $R\left(1 - \frac{t}{t_{max}}\right)$ denotes neighbor radius of $\mathfrak{R}_{i,j}$, t and t_{max} denotes iteration, and maximal iteration(Mahmood & Al-Khateeb 2019).

$$\mathfrak{R}_{i,j}^{p_2} = \mathfrak{R}_{i,j} + R\left(1 - \frac{t}{t_{max}}\right).(2.ran - 1)\mathfrak{R}_{i,j} \tag{11}$$

As per BMU-PO, levy flight implied by *LF* is combined with Eq. (11) and modeled as shown in Eq. (12). Here, R_{w_i} is updated by the child monkey weight update of BMO and R_{w_i} lies among 4 and 6. Moreover, average crossover in Eq. (13) is performed to ensure the better allocation of resources.

$$\mathfrak{R}_{i,j}^{p_2} = \mathfrak{R}_{i,j} + R_{w_i}\left(1 - \frac{t}{t_{max}}\right).(2.ran - 1)\mathfrak{R}_{i,j} + LF \tag{12}$$

$$AC = \frac{x_i + y_i}{2} \tag{13}$$

Eq. (14) denotes new pelican position, wherein, $Q_i^{p_2}$ and $\mathfrak{R}_i^{p_2}$ denotes objective and novel position.

$$\mathfrak{R}_i = \begin{cases} \mathfrak{R}_i^{p_2}, \ Q_i^{p_2} < Q_i; \\ \mathfrak{R}_i, \ else \end{cases} \tag{14}$$

4 RESULT AND DISCUSSION

4.1 *Simulation procedure*

The BMU-PO work was implemented in PYTHON. The performance of the BMU-PO was validated over the clustering algorithms (FCM and K-means) and optimization methods (AOA, BES, COOT, BMO and POA) in terms of energy consumption, resource utilization, makespan and execution time. Table 1 shows the system configuration

Table 1. System configuration.

Name	Specifications
Processor name	AMD Ryzen 5 3450U with Radeon Vega Mobile Gfx 2.10GHz
Installed RAM	16.0 GB
Edition	Windows 11 home single language
Version	21H2
OS build	22000.1098

4.2 *Assessment on energy consumption*

Figure 2. represents the energy consumption analysis on the BMU-PO over the FCM and K-means methods. Simultaneously, it is examined by adjusting the nodes from 100 to 200 for both the 20 and 40 VM. In accordance with the Figure 2, the BMU-PO consumed minimal energy than the FCM and K-means clustering algorithm for resource allocation in cloud. For instance, considering the Figure 2(a), in the 175th task the BMU-PO consumed the energy of $5.35 \times e^{-01}$, whilst the FCM is$7.65 \times e^{-01}$ and K-means clustering is $8.66 \times e^{-01}$.When the VM is set to as 40, the BMU-PO consumed very low amount of energy than others, thereby making it appropriate to allocate the resource in the cloud.

Therefore, the BMU-PO exhibits that it has a consumed lesser energy and can allocate the resource more effectively.

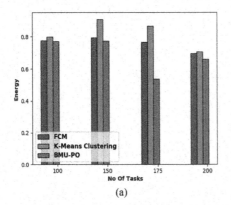

(a)

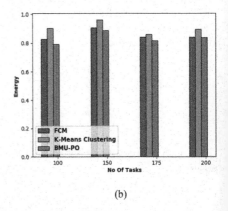

(b)

Figure 2. Analysis on energy consumption of the BMU-PO, FCM and K-means clustering a) number of VM = 20 b) number of VM = 40.

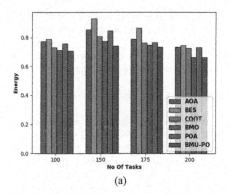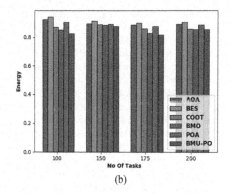

Figure 3.   Analysis on energy consumption of the BMU-PO and the traditional algorithms a) number of VM = 20 b) number of VM = 40.

5   CONCLUSION

This paper executed a novel optimization aided resource allocation model. Initially, the workload clustering was done by means of an optimal k-means clustering, where centroid was tuned using BMU-PO. The tasks were clustered depending upon execution time and QoS. In the end, optimal resource allocation was done using BMU-PO algorithm under the consideration of CPU usage, PUE, and execution time. For the VM 20, the BMU-PO obtained the makespan in the 200th task WAS $2.31 \times e^{-01}$, which WAS superior to AOA = $2.86 \times e^{-01}$, BES = $3.17 \times e^{-01}$, COOT = $2.66 \times e^{-01}$, BMO = $2.57 \times e^{-01}$ and POA = $2.85 \times e^{-01}$, respectively. Overall, the BMU-PO achieved minimum makespan while compared to the standard approaches and has the prospect to allocate the resource in an optimal manner.

REFERENCES

Abohamama, A.S. &Hamouda, E., 2020. A Hybrid Energy–aware Virtual Machine Placement Algorithm for Cloud Environments. *Expert Systems with Applications, 150*, p.113306.
Asghari, A., Sohrabi, M.K. &Yaghmaee, F., 2020. A Cloud Resource Management Framework for Multiple Online Scientific Workflows Using Cooperative Reinforcement Learning Agents. *Computer Networks, 179*, p.107340.
Belgacem, A., Beghdad-Bey, K. &Nacer, H., 2020. Dynamic Resource Allocation Method Based on Symbiotic Organism Search algorithm in cloud computing. *IEEE Transactions on Cloud Computing.*
Beno, M.M., I. R, V., S. M, S. & Rajakumar, B.R., 2014. ThresholdPrediction for Segmenting Tumour from Brain MRI scans. *International Journal of Imaging Systems and Technology, 24*(2), pp.129–137.
Devagnanam, J. &Elango, N.M., 2020. Optimal Resource Allocation of Cluster Using Hybrid Grey Wolf and Cuckoo Search Algorithm in Cloud Computing. *Journal of Networking and Communication Systems, 3*(1), pp.31–40.
Feng, H., Guo, S., Zhu, A., Wang, Q. & Liu, D., 2020. Energy-efficient User Selection and Resource Allocation in Mobile Edge Computing. *Ad Hoc Networks, 107*, p.102202.
Gill, S.S., Tuli, S., Toosi, A.N., Cuadrado, F., Garraghan, P., Bahsoon, R., Lutfiyya, H., Sakellariou, R., Rana, O., Dustdar, S. &Buyya, R., 2020. ThermoSim: Deep Learning Based Framework for Modeling and Simulation of Thermal-aware Resource Management for Cloud Computing Environments. *Journal of Systems and Software, 166*, p.110596.
Gong, S., Yin, B., Zheng, Z. &Cai, K.Y., 2019. Adaptive Multivariable Control for Multiple Resource Allocation of Service-based Systems in Cloud Computing. *IEEE Access, 7*, pp.13817–13831.

Goswami, B., Sarkar, J., Saha, S., Kar, S. &Sarkar, P., 2019. ALVEC: Auto-scaling by Lotka Volterra Elastic Cloud: A QoS Aware Non Linear Dynamical Allocation Model. *Simulation Modelling Practice and Theory, 93*, pp.262–292.

Hejja, K. &Hesselbach, X., 2019. Evaluating Impacts of Traffic Migration and Virtual Network Functions Consolidation on Power Aware Resource Allocation Algorithms. *Future Generation Computer Systems, 101*, pp.83–98.

Hosseinalipour, S. & Dai, H., 2019. A Two-stage Auction Mechanism for Cloud Resource Allocation. *IEEE Transactions on Cloud Computing, 9*(3), pp.881–895.

Joseph, C.T. &Chandrasekaran, K., 2020. IntMA: Dynamic Interaction-aware Resource Allocation for Containerized Microservices in Cloud Environments. *Journal of Systems Architecture, 111*, p.101785.

Lee, J.W., Jang, G., Jung, H., Lee, J.G. & Lee, U., 2019. Maximizing MapReduce Job Speed and Reliability in the Mobile Cloud by Optimizing Task Allocation. *Pervasive and Mobile Computing, 60*, p.101082.

Mahmood, M. & Al-Khateeb, B., 2019. The Blue Monkey: A New Nature Inspired Metaheuristic Optimization Algorithm. *Periodicals of Engineering and Natural Sciences (PEN), 7*(3), pp.1054–1066.

Mao, L., Li, Y., Peng, G., Xu, X. & Lin, W., 2018. A Multi-resource Task Scheduling Algorithm for Energy-performance Trade-offs in Green Clouds. *Sustainable Computing: Informatics and Systems, 19*, pp.233–241.

Mergenci, C. & Korpeoglu, I., 2019. Generic Resource Allocation Metrics and Methods for Heterogeneous Cloud Infrastructures. *Journal of Network and Computer Applications, 146*, p.102413.

Mireslami, S., Rakai, L., Wang, M. & Far, B.H., 2019. Dynamic Cloud Resource Allocation Considering Demand Uncertainty. *IEEE Transactions on Cloud Computing, 9*(3), pp.981–994.

Santos, I.L., Pirmez, L., Delicato, F.C., Oliveira, G.M., Farias, C.M., Khan, S.U. & Zomaya, A.Y., 2019. Zeus: A Resource Allocation Algorithm for the Cloud of Sensors. *Future Generation Computer Systems, 92*, pp.564–581.

Shrimali, B., & Patel, H. 2020. Multi-objective Optimization Oriented Policy for Performance and Energy Efficient Resource Allocation in Cloud Environment. *Journal of King Saud University-Computer and Information Sciences, 32*(7), 860–869.

Soltanshahi, M., Asemi, R. & Shafiei, N., 2019. Energy-aware Virtual Machines Allocation by Krill Herd Algorithm in Cloud Data Centers. *Heliyon, 5*(7), p.e02066.

Talasila, V. & Narasingarao, M.R., 2022. Optimized GAN for Text-to-Image Synthesis: Hybrid Whale Optimization Algorithm and Dragonfly Algorithm. *Advances in Engineering Software, 173*, p.103222.

Thein, T., Myo, M.M., Parvin, S. & Gawanmeh, A., 2020. Reinforcement Learning Based Methodology for Energy-efficient Resource Allocation in Cloud Data Centers. *Journal of King Saud University-Computer and Information Sciences, 32*(10), pp.1127–1139.

Thomas, R. & Rangachar, M.J.S., 2018. Hybrid Optimization Based DBN for Face Recognition Using Low-resolution Images. *Multimedia Research, 1*(1), pp.33–43.

Trojovský, P. & Dehghani, M., 2022. Pelican Optimization Algorithm: A Novel Nature-inspired Algorithm for Engineering Applications. *Sensors, 22*(3), p.855.

Xu, J. & Palanisamy, B., 2018. Optimized Contract-based Model for Resource Allocation in Federated Geo-distributed Clouds. *IEEE Transactions on Services Computing, 14*(2), pp.530–543.

Zhang, X., Wu, T., Chen, M., Wei, T., Zhou, J., Hu, S. & Buyya, R., 2019. Energy-aware Virtual Machine Allocation for Cloud with Resource Reservation. *Journal of Systems and Software, 147*, pp.147–161.

Artificial Intelligence, Blockchain, Computing and Security – Dagur et al. (Eds)
© 2024 The Author(s), ISBN: 978-1-032-49393-0

Fake news detection on social-media: A 360 degree survey view

Vivek Kumar* & Satveer*
Quantum University, Roorkee, Uttarakhand

Waseem Ahmad*
M.I.E.T, Meerut

Satyaveer Singh*
M.M.E.C, Maharishi Markandeshwar (Deemed to be University)

ABSTRACT: Identifying fake news spread on social media has been a hot topic in academia for years, especially now that both social media and professional journalism are at their highest points. Experts worldwide are trying to figure out the essential parts of the fake news detection study topic, which is very hard. This study aims to shed light on the unique characteristics of news stories in the contemporary diaspora, as well as the many kinds of content in news stories and their effects on their audiences. Next, we show some well-known fake news datasets and talk about the current ways to spot fake news, most of which are based on text analysis. In the last section of the work, we outline four major unanswered questions that might direct future study.

Keywords: Phishing. Random forest, NLP, Word Vector.

1 INTRODUCTION

Researchers globally study identifying fake news and its effects on people. Fake news is any information that is untrue and meant to mislead. (Parikh *et al.* 2018).

Social media surpasses traditional news sources but is less reliable due to modification for various motives. It leads to fast spread of fake news in countries like UK, US, Russia, Romania, and Macedonia. (Shu *et al.* 2019).

Social media news, opinions and reviews impact users' choices and can spread false information with negative consequences for individuals, organizations, and society. Thus, researchers prioritize identifying fake news. (Chen *et al.* 2015).

Automatic false- news identification puts a strain on the existing content-based analysis techniques. The main reason is that understanding information requires knowledge of the social context, a shared cause, or political concern. The most cutting-edge natural language processing algorithms cannot yet be learned (Ali *et al.* 2022). Social media are essential for economic, political, and social changes, like the formation of pro-democracy movements (like in Egypt in 2011 and 2013 or Ukraine in 2013–2014), but they also make it easier for false information to spread than with traditional news sources. Assume incorrect information spreads via the internet and social media. In that case, it could affect how well the financial markets work, how quickly people respond in emergencies, how often terrorist acts happen, and other things (Parikh *et al.* 2018). The main goal of this kind of false information is to hurt politically or financially essential people, groups, or even whole countries. By making a catchy, sometimes made-up, headline, you may get people to read it, making it a popular topic. Misleading information may also be used in information warfare between nations, businesses, and organizations (Chen *et al.* 2015).

---
*Corresponding Authors: vivekkumarknit@gmail.com, drsatveer91@gmail.com, waseemahmad.ahmad@gmail.com and satyaveer.sangwan@gmail.com

DOI: 10.1201/9781003393580-122

The term "fake news" became widespread during the 2016 US election. It encompasses false news, hoaxes, propaganda, rumors, and junk news, all of which refer to misleading information. It is important to agree on a definition and categorization of these terms. (Bhogade *et al.* 2021).

Researchers search for automated, accurate, and reliable methods to detect fake news on social media. Fake news detection involves determining the truthfulness of news using machine learning algorithms to help filter and identify false reports. (Zubiaga *et al.* 2016).

2 PLATFORMS FOR NEWS MEDIA

Two-thirds of American news consumption comes from social media, an increase of 5% from 2016. Popular platforms serve as primary news sources for most of the audience and are categorized based on their source of news content. (Mishra *et al.* 2022; Parikh *et al.* 2018).

1) Independent website: Each news website has a unique URL that can be used to share content on social media. (Mishra *et al.* 2022; Parikh *et al.* 2018).
2) Social media: Sharing is the most common method of spreading information on social media, with over 70% of users relying on it as their primary source of daily news. (Mishra *et al.* 2022; Parikh *et al.* 2018).
3) Emails: Consumers could also receive news through emails, but verifying the reliability of news emails can be challenging (Parikh *et al.* 2018).
4) Broadcast networks (PodCast): A few consumers still use podcasts as audio multimedia to get news (Mishra *et al.* 2022; Parikh *et al.* 2018).
5) Radio service: It can be challenging to verify the authenticity of the audio from radio talk programs, which are familiar news sources (Mishra *et al.* 2022; Parikh *et al.* 2018).

3 RELATED TERMS AND DEBUNKING TOOLS FOR FALSE NEWS

The Cambridge Dictionary says false news is "made and spread online to manipulate public opinion and seem genuine." "Fake news" has been around since the birth of writing. Social media has changed how news is conveyed in the past ten years compared to the past. On social media, trolling and computer- generated falsehoods thrive. People sometimes use the terms satire, "yellow journalism," propaganda, hoax, disinformation, and rumor to mean false news (Hangloo *et al.* 2021).

- Propaganda: News stories generated and spread by a political organization to influence political opinion are referred to as "propaganda" (Hangloo *et al.* 2021).
- Misinformation: It is purposely false information that is spread whether intentionally or accidentally, ignoring the genuine objective (Hangloo *et al.* 2021).
- Disinformation: It describes the dissemination of incorrect information to influence facts and deceive the intended audience (Hangloo *et al.* 2021).
- Rumors and hoaxes: The terms "rumors" and "hoaxes" are interchangeable when referring to the intentional creation or falsification of facts. They portray unsubstantiated and false information as having been verified by established news organizations (Hangloo *et al.* 2021).
- Satire and parody: often imitate primary news sources and utilize humor to provide news updates (Hangloo *et al.* 2021).
- Clickbait: Sensational headlines frequently use clickbait to grab readers' attention, get them to click, and send them to another website. More significant ad clicks translate to more revenue (Hangloo *et al.* 2021).

Regular Internet users need help discerning between actual and false news information since propaganda, hoaxes, and satire are increasingly used alongside reliable news (Hangloo *et al.* 2021).

4 TYPE OF DATA IN NEWS

In this section, there are three primary ways in which people receive news, and we will cover the types of data used to create news articles (Rubin *et al.* 2015).

- Text: Text Linguistics is a subfield of linguistics that studies text as a means of communication. It analyzes the content, tone, syntax, and pragmatics of texts to allow for discourse analysis. (Parikh *et al.* 2018).
- Multimedia: Multimedia combines various media formats like graphics, music, video, and photographs to create visually appealing content that immediately captures the audience's attention. (Li *et al.* 2013).
- Embedded Content: Hyperlinks allow authors to connect to multiple sources and build readers' confidence by verifying the news story's concept. Since the rise of social media, authors often include screenshots of relevant posts in their works (Parikh *et al.* 2018).

5 METHODS FOR DETECTING FAKE NEWS

The growth in global use and acceptance of social media platforms has made it easier for false information to spread. The massive and diverse flow of information on these platforms spreads quickly and significantly impacts the entire community (Guimarães *et al.* 2021). This information comprises both accurate and false information. Many academics and IT professionals discovered fake news on the internet. Due to large amounts of user generated data, typical automated rumor detection systems now incorporate deeper level characteristics. This is possible using user data. This section provides cutting-edge research on recognizing false news based on the content and social context of the news story (Zhou *et al.* 2019).

5.1 *Content-based*

The content based fake news detection strategy uses the post's content. It seeks to identify false news by looking at the text, the photos, or any combination of these aspects. Researchers often rely on latent or hand-crafted content features to detect false news automatically (Alonso *et al.* 2021; Azad *et al.* 2021).

1) Knowledge: Fact-checking uses external data sources to verify claims for accuracy, using both manual methods (expert or public) and automated methods (AI)(Ali *et al.* 2021).
   - Manual facts checking: Manual fact-checking has two types: expert-based and crowdsourced. Expert-based uses trained specialists, but is slow and inefficient for large amounts of data like on social media. Fact-checking websites like Snopes and PolitiFact are reliable but slow. (Ali *et al.* 2021). Crowdsourcing uses collective knowledge to check news reports, but can have reliability issues from management challenges, bias, and inconsistent annotations. Platforms like Fiskkit rate news article trustworthiness (Apuke *et al.* 2021; Khan *et al.* 2022).
   - Automatic Fact-Checking: Automated fact-checking uses NLP, data mining, ML, etc., to quickly check social media data. The process involves obtaining data and building a knowledge base, then comparing news to the knowledge base for accuracy. It relies on open sources and databases, but fake news databases can be messy and hard to identify. (Conroy *et al.* 2015).
2) Visual-Based: People who make fake news believe that pictures make an article more trustworthy (Guimarães *et al.* 2021), so they use controversial images to attract and mislead readers. (Orabi *et al.* 2020) use statistical modeling to extract visual and statistical elements. Unigrams and bigrams are retrieved from story words. TFIDF (Term Frequency Inverse Document Frequency) values are used for information retrieval, indicating a word's importance in a text through a numerical statistic.
3) Linguistic Features based Methods: Language techniques extract significant linguistic features from false news, as mentioned below.

- Ngrams: Unigrams and bigrams are extracted from text and stored as TFIDF values to indicate a word's importance in information retrieval. TFIDF is a numerical statistic (Ahmed *et al.* 2017).
- Punctuation: Punctuation helps false news detection algorithms distinguish between misleading and true content. Punctuation feature gathers 11 forms of punctuation via detection (Parikh *et al.* 2018).
- Psycholinguistic features: Some suggest using the LIWC lexicon to obtain psycho-linguistic aspects and assess the language's tone, statistics, and part-of-speech categorization (Parikh *et al.* 2018).
- Readability: Content features include extracting letters, words, syllables, categories, and paragraphs to conduct readability measurements such as Flesch Reading Ease, Flesch-Kincaid, ARI, and Gunning Fog (Parikh *et al.* 2018).
- Syntax: This method retrieves CFG (context-free grammar) characteristics. These features rely significantly on lexicalized steps and their parents and grandparents. This set's functions are TFIDF-encoded for data retrieval (Parikh *et al.* 2018).

4) Style Based: Style based fake news identification is comparable to knowledge-based fake media identification. This technique evaluates the writer's intent to mislead, not the news's authenticity (Orabi *et al.* 2020). Most phony news sources want to influence large audiences by spreading correct and deceptive information (Figueira *et al.* 2017).

5.2 *Social context based*

User profiles, postings and responses, and network architecture are essential characteristics of the social environment (Hoy *et al.* 2017). It shows how news circulates over time and helps determine veracity and political stance. Recent research has explored context-based fake news detection algorithms (Bhogade *et al.* 2021).

1) Network Based: Network-based fake news identification evaluates relationships between connections, tweets, and comments to identify fake news and understand how false information spreads on social media through networks of users with similar interests (Oshikawa *et al.* 2018)
2) Temporal Based: According to research, online news stories change when new information is added, or the original assertion is altered. This is particularly true when rumors arise after the initial news story (Zhang *et al.* 2020).
3) Reliability based: According to several sources, a claim's newsworthiness and trustworthiness are signs of a spreader's credibility. (Pilkevych *et al.* 2021) use the idea of credibility to figure out who is spreading false information. (Preston *et al.* 2021) worry about the validity of a specific claim. (Bhavani *et al.* 2021) suggest analyzing a tweet's reliability. This prevents erroneous or dangerous information from spreading. TweetCred examines a tweet's trustworthiness in real-time online.

6 CONCLUSION

Several researchers have worked to automatically recognize fake news and create realistic benchmark datasets of fake and authentic social media content. As social media grows, more people will use it to get their news instead of traditional media. According to the study, erroneous information distributed through social media hurts people and society. In this study, we studied false news by conducting a literature review in two phases: characterization and detection. This project is about finding fake news using new methods at the news, user, and social levels.

REFERENCES

Ahmed, H., Traore, I. and Saad, S., 2017, October. Detection of Online Fake News Using n-gram Analysis and Machine Learning Techniques. In *International Conference on Intelligent, Secure, and Dependable Systems in Distributed and Cloud Environments* (pp. 127–138). Springer, Cham.

Ali, I., Ayub, M.N.B., Shivakumara, P. and Noor, N.F.B.M., 2022. Fake News Detection Techniques on Social Media: A Survey. *Wireless Communications and Mobile Computing*, 2022.

Alonso, M.A., Vilares, D., Gómez-Rodríguez, C. and Vilares, J., 2021. Sentiment Analysis for Fake News Detection. *Electronics*, 10(11), p.1348.

Apuke, O.D. and Omar, B., 2021. Fake news and COVID-19: Modelling the Predictors of Fake News Sharing Among Social Media Users. *Telematics and Informatics*, 56, p.101475.

Azad, R., Mohammed, B., Mahmud, R., Zrar, L. and Sdiqa, S., 2021. Fake News Detection in Low-resourced Languages "Kurdish Language" Using Machine Learning Algorithms. *Turkish Journal of Computer and Mathematics Education (TURCOMAT)*, 12(6), pp.4219–4225.

Bhavani, A. and Kumar, B.S., 2021, April. A Review of State Art of Text Classification Algorithms. In *2021 5th International Conference on Computing Methodologies and Communication (ICCMC)* (pp. 1484–1490). IEEE.

Bhogade, M., Deore, B., Sharma, A., Sonawane, O. and Singh, M., 2021. A Research Paper on Fake news detection. *InternationalJournal*, 6(6).

Chen, Y., Conroy, N. K. and Rubin, V. L. (2015) "News in an Online World: The Need for an 'Automatic Crap Detector': News in an Online World: The Need for an 'Automatic Crap Detector,'" *Proceedings of the Association for Information Science and Technology*, 52(1), pp. 1–4. doi: 10.1002/pra2.2015.145052010081.

Conroy, N.K., Rubin, V.L. and Chen, Y., 2015. Automatic Deception Detection: Methods for Finding Fake News. *Proceedings of the Association for Information Science and Technology*, 52(1), pp.1–4.

Figueira, Á. and Oliveira, L., 2017. The Current State of Fake News: Challenges and Opportunities. *Procedia Computer Science*, 121, pp.817–825.

Guimarães, N., Figueira, Á. and Torgo, L., 2021. An Organized Review of Key Factors For Fake News Detection. *arXiv preprint arXiv:2102.13433*.

Guimarães, N., Figueira, Á. and Torgo, L., 2021. Can Fake News Detection Models Maintain the Performance Through Time? A Longitudinal Evaluation of Twitter Publications. *Mathematics*, 9(22), p.2988.

Hangloo, S. and Arora, B., 2021. Fake News Detection Tools and Methods–A Review. *arXiv preprint arXiv:2112.11185*.

Hoy, N. and Koulouri, T., 2021. A Systematic Review on the Detection of Fake News Articles. *arXiv preprint arXiv:2110.11240*.

Khan, S., Hakak, S., Deepa, N., Prabadevi, B., Dev, K. and Trelova, S., 2022. Detecting COVID-19-related Fake News Using Feature Extraction. *Frontiers in Public Health*, p.1967.

Li, Y., 2013. Image Copy-move Forgery Detection Based on Polar Cosine Transform and Approximate Nearest Neighbor Searching. *Forensic Science International*, 224(1–3), pp.59–67.

Mishra, S., Shukla, P. and Agarwal, R., 2022. Analyzing Machine Learning Enabled Fake News Detection Techniques for Diversified Datasets. *Wireless Communications and Mobile Computing*, 2022.

Orabi, M., Mouheb, D., Al Aghbari, Z. and Kamel, I., 2020. Detection of Bots in Social Media: A Systematic Review. *Information Processing & Management*, 57(4), p.102250.

Oshikawa, R., Qian, J. and Wang, W.Y., 2018. A Survey on Natural Language Processing for Fake News Detection. *arXiv preprint arXiv:1811.00770*.

Parikh, S. B. and Atrey, P. K. (2018) "Media-rich Fake News Detection: A Survey," in 2018 *IEEE Conference on Multimedia Information Processing and Retrieval (MIPR)*. IEEE.

Pilkevych, I., Fedorchuk, D., Naumchak, O. and Romanchuk, M., 2021, September. Fake News Detection in the Framework of Decision-Making System through Graph Neural Network. In *2021 IEEE 4th International Conference on Advanced Information and Communication Technologies (AICT)* (pp. 153–157). IEEE.

Preston, S., Anderson, A., Robertson, D.J., Shephard, M.P. and Huhe, N., 2021. Detecting Fake News on Facebook: The Role of Emotional Intelligence. *PLoS One*, 16(3), p.e0246757.

Rubin, V.L., Chen, Y. and Conroy, N.K., 2015. Deception Detection for News: Three Types of Fakes. *Proceedings of the Association for Information Science and Technology*, 52(1), pp.1–4.

Shu, K., Bernard, H. R. and Liu, H. (2019) "Studying Fake News via Network Analysis: Detection and Mitigation," in *Lecture Notes in Social Networks*. Cham: Springer International Publishing, pp. 43–65.

Zhang, X. and Ghorbani, A.A., 2020. An Overview of Online Fake News: Characterization, Detection, and Discussion. *Information Processing & Management*, 57(2), p.102025.

Zhou, X., Jain, A., Phoha, V.V. and Zafarani, R., 2019. Fake News Early Detection: An Interdisciplinary Study. *arXiv preprint arXiv:1904.11679*.

Zubiaga, A., Liakata, M., Procter, R., Wong Sak Hoi, G. and Tolmie, P., 2016. Analysing How People Orient to and Spread Rumours in Social Media by Looking at Conversational tThreads. *PLoS One*, 11(3), p.e0150989.

Cloud computing in education

S. Singh
Central University of Jharkhand, Ranchi, India

A. Singh
BBS College of Engineering and Technology, Prayagraj, India

A. Singh
GLA University, Mathura, India

ABSTRACT: India has grown tremendously in Information Technology (IT) and IT-enabled services over the decade. Cloud technology is also witnessing rapid growth in India. This paper discusses the importance and use of cloud computing technology in education and explores the trend and growth of cloud computing in the higher education market. The move to cloud computing has provided educational institutions with several benefits, including improved student-institution collaboration, innovation and learning, educational outcomes, efficient supervision, information sharing, and student counselling brought additional benefits. Now leading institutions are trying to use cloud technology to gain a competitive advantage.

1 INTRODUCTION

India is among the most significant offshore destination for IT companies worldwide. The emerging technology presents various opportunities for leading IT companies in India. According to India Brand Equity Foundation (IBEF), Indian IT and IT-enabled services will grow to US$ 19.93 billion by 2025. The report further says India will spend US$ 144 billion in 2023 (https://www.ibef.org/industry/information-technology-india). The contribution of the IT sector to the Indian GDP has occupied a prominent space. In FY 22, it was 7.4% of India's GDP, expected to increase by 10% by 2025. India's digital ecosystem is also expected to see phenomenal progress as Google announced its plans to invest US$ 1 billion in Bharti Airtel Ltd – an Indian mobile company (Information Technology India, TOP IT companies in India – IBEF).

Cloud technology is also witnessing rapid growth in India. The NASSCOM Cloud: Next Wave of Growth in India 2019 Report says that India's cloud market could see a three-fold increase to US$ 7.1 billion by 2022 due to the increasing acceptance of artificial intelligence, big data analytics, and the Internet of things (NASSCOM 2019).

The importance of technology in education came to the forefront during the Covid 19 outbreak. The priority was more dearly felt in India when many educational institutions, with little or no technology integration, struggled to impart education to their pupils. This was when suddenly, the importance of cloud computing was recognized by one and all related to the education sector. This pandemic has shown that education delivery mode differs from brick-and-mortar arrangements.

The paper discusses the significance of cloud computing technology in education. The paper, in sequence, describes the prevailing situation in Indian Education, then discusses the overview of cloud computing and further moves on towards the implementation possibilities, issues and challenges of cloud computing in education. Towards the end of the paper, the global and Indian cloud computing market overview is discussed.

DOI: 10.1201/9781003393580-123

2 INDIAN EDUCATION SYSTEM SCENARIO

The Indian Education System has long been confronted with its share of issues and challenges. Better education can only be provided when the problems and challenges are resolved. Over the years, India has progressed and improved considerably, but many problems persist. India has witnessed a gradual decline in children's learning outcomes over the last decade despite an increased expenditure on education. Spending on education increased from INR 3600 Billion to INR 4600 billion over this period (2006–2016) (Vora 2021). Problems of enrolment at the primary level, the dropout rate in secondary education, poor infrastructure, lack of quality and the appropriate number of teachers, and crunch of financial resources are the issues that have been posing challenges to the education system in India (Shah & Agrawal 2008).

According to the All India Survey of Higher Education (AISHE 2019–20), there are 1043 Universities, 42343 Colleges and 11779 independent institutions. Of all universities, 396 are private universities, and 422 are in rural areas. Total enrollment in higher education is approximately 38.5 million, with women accounting for approximately 49% of all students. The Gross Enrollment Rate (GER) in Indian higher education is around 27.1. This is a reflection of the current state of Indian education. Many researchers have observed unequal access to higher education (Bluntzer 2008; Peterson & Altbach 2007; Usher & Cervenan 2005).

Indian policymakers are trying to expand online and distance education to the extent possible to increase GER. The government intends to use technology and cloud computing facilities, and infrastructure can be gauged with the announcement of establishing a Digital University in its budget 2022–23 (https://pib.gov.in/PressReleasePage.aspx?PRID= 1814135). Policymakers are shifting towards using technology to the maximum to provide quality and personalized learning experiences at the doorstep.

3 CLOUD COMPUTING IN EDUCATION – A TECHNICAL OVERVIEW

Cloud computing allows users to access resources, applications, databases, email, file services, and other services stored on another server. Users can view the information with the help of an internet connection and electronic devices. Thus, cloud computing shares resources, programs, storage, and information over networks. The important part is that no license is needed; users can access these services by paying charges per their usage.

Cloud computing has two parts: front end and back end. The front end provides the user interface, and the back end includes the cloud environment comprising the platform, the applications, and the infrastructure. The front-end interface helps users access the cloud, where resources, applications, and data are stored. The back-end interface is the backbone of cloud computing technology and is responsible for information security. The back end houses databases, computers and services. Cloud computing makes things fast and super-performing as it distributes many works across multiple resources. Cloud computing also helps reduce the user's software and hardware resources as applications, services and databases are stored in the cloud. Users can call these resources as per their requirements.

There are three main models of cloud computing service. They are differentiated based on degrees of control and autonomy. These are infrastructure as a service (IaaS), platform as a service (PaaS), and software as a service (SaaS).

4 CLOUD COMPUTING IN EDUCATION

Cloud computing technology brings the education system into virtual mode. Cloud computing provides educational institutions an online platform through various applications and subscription-based models. Teachers, administrators and students can use information technologies and technology-enabled services in their education-related activities. Each component/ stakeholder of the education system can access or perform a task related to his role and responsibilities on a single platform. A teacher can upload learning materials and videos, a student can access these materials and homework, and administrators can easily collaborate and monitor the system's progress and functioning (The Cloud: a smart move for higher education, 2018).

Increased functionality of mobile communication devices and technological development have equipped students, teachers and academic institutions to adopt cloud computing. Smart cellphones, tablets, and laptops with fast internet accessibility have made cloud services more popular. Figure 1 presents the primary uses of cloud computing in education.

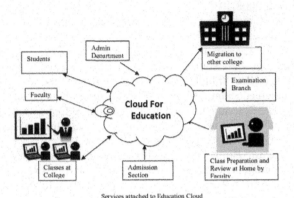

Figure 1.   Use of cloud computing in education.
Source: http://www.ijircce.com/upload/2014/february/21_Role.pdf.

For advanced education, cloud computing benefits administrators, educators, and students (Olanrewaju *et al.* 2015). Higher education institutions dynamically switch to cloud computing to lower costs and exploit the most recent innovation (Burkhardt *et al.* 1995; Writers 2022).

Cloud computing technology is expected to give impetus to schools and colleges. Educational institutions can access and implement technology with little cost and complexities in maintaining hardware, software and IT staff. Standalone/ private installation of the software would incur large technical and financial requirements, which becomes a significant challenge. Educational institutions can bypass these challenges by using cloud computing. Cloud computing adoption in learning management systems is also helping learner – tutor relationship. Students can easily access course materials and submit assignments etc.

5 CLOUD COMPUTING IN EDUCATION – GLOAL & INDIAN MARKET OVERVIEW

Astute Analytica's report estimates global cloud computing revenue in higher education and predicts a remarkable growth from US$ 2,693.5 million in 2021 to US$ 15,180.1 million by 2030, a probable CAGR growth of 22% during the period (2022–30). While charting the segment-wise growth, the report says that technical schools, private institutions and administration applications will see the maximum share of the growth in usage of cloud computing in coming years worldwide. By being technically savvy, technical institutions will lead cloud computing usage. Private institutions will lead the adoption of cloud computing facilities to minimise cost and their abilities to increase funding. A trend of e-learning is taking place; the report suggests that unified communication is expected to see the highest growth rate. It is also estimated that the hybrid cloud segment would be the most chosen deployment model (https://www.astuteanalytica.com/industry-report/cloud-computing-higher-education-market). Reports also suggest that India's e-learning market will soon become the second largest market after the US and is believed to reach US$ 1.96 Billion by 2021. Below is the region-wise cloud computing market as per this report.

5.1 *Cloud computing in higher education – market country wise insights*

5.1.1 *North America region*
The United States holds the region's largest share of revenue in 2021 and is likely to maintain its supremacy during the forecasted period. Among the type of institutions, technical institutions hold the largest market share.

5.1.2 *Europe region*
Region of Western European is likely to register the region's highest compound annual growth rate (CAGR). Germany had the most prominent European market share in 2021 due to its focus on innovation from R&D and technology implementation.

5.1.3 *Asia Pacific region*
India is the region with the largest shareholder in the Asia-Pacific Higher Education Cloud Computing Market in 2021. It is expected to remain at the forefront in the future, too, as education systems are leaning towards cloud services. It is anticipated to record the highest CAGR during the period.

5.1.4 *South America region*
Brazil is anticipated to grow at the region's highest compound annual growth rate during the coming periods. Moreover, based on ownership structure, the private institutions' segment will account for the largest share of the South American higher education cloud computing market in 2021 as private institutions increasingly allocate funds to introducing cloud computing services.

5.1.5 *Middle East region*
Egypt will be the region with the maximum shareholder share in 2021, and the UAE is likely to grow at the highest CAGR during the forecast period. On the other hand, unified communications are anticipated to register the highest CAGR, owing to the rising trend of e-learning during the forecast period.

5.1.6 *Africa region*
South Africa is the region with the highest share in 2021. Due to the security advantages of private cloud deployment, private cloud is expected to witness the highest CAGR during the forecasted period.

6 CONCLUSION

Post Covid 19 brought out the importance of technology in the education system. Technology has been integrated into the education system over the years, but the pandemic provided a much-needed realization. The teaching and learning process is no longer confined to brick-and-mortar arrangements. Textbooks and classrooms are becoming redundant, and computers, laptops and mobile are becoming an essential part of the education system. The necessity of fixed time and timetable is also diminishing as tutors and learners are always connected on and off campus. Technology is disrupting the education system significantly, and those who adapt to this change can gain a lot. Technology provides career-relevant skills offered by a university far away over the Internet (Chibaro 1970; Stone 2022).

Cloud computing has added wings to innovation in the education field. Now, teachers can use cloud computing to create new and creative classrooms. The cloud optimizes innovative teaching models such as blended and flipped courses. Both features allow tutors to interact more with their children at school while accessing classes and assignments from home via the cloud. The cloud can help create a completely modern, state-of-the-art classroom. Cloud computing is also enabling students to become owners of their learning journey. It helps them in personalized learning and making them empowered. Access to technology and innovation beyond the confines of the classroom enables students to become problem solvers, thinkers, collaborators and creators. Students develop a lifelong love of learning by incorporating technology into the classroom. The move to cloud computing has provided educational institutions with several benefits, including improved student-institution collaboration, innovation and learning, educational outcomes, efficient supervision, information sharing, and student counselling brought additional benefits. Now leading institutions are trying to use cloud technology to gain a competitive advantage over their competitors (Britto 2012).

REFERENCES

Bluntzer, A. L. (2008). *Left Behind: An analysis of Higher Education Accessibility in the Eveloping World and Comparisons with Developed Countries.* ProQuest Dissertations and Theses: The Humanities and Social Sciences Collection.

Britto, M. (2012). *Cloud Computing in Higher Education. Library Student Journal.* Retrieved January 15, 2023, from https://www.librarystudentjournal.org/index-php/lsj/article/view/289/321/

Burkhardt, G., Petri, M., & Roody, D. S. (1995). The Kite: An Organizational Framework for Educational Development in Schools. *Theory Into Practice, 34*(4), 272–278. https://doi.org/10.1080/00405849509543691

Chibaro, N. (1970, January 1). *Adoption of Cloud Pedagogy by Higher Learning Institutions in Southern Africa.* Semantic Scholar. Retrieved January 15, 2023, from https://www.semanticscholar.org/paper/Adoption-of-cloud-pedagogy-by-higher-learning-in-Chibaro/8621887f6bce534b8bd7979f29ec8e38797241d3

Cloud Computing in Higher Education Market – Industry Analysis and Forecast. (n.d.). *Cloud Computing in Higher Education Market – Industry Analysis and Forecast.* Retrieved January 15, 2023, from https://www.astute-analytica.com/request-sample/cloud-computing-higher-education-market

India Brand Equity Foundation. (n.d.). *Information Technology India, TOP IT companies in India – IBEF.* Retrieved January 15, 2023, from https://www.ibef.org/industry/information-technology-india#:~:text=The%20Indian%20IT%20%26%20business%20services380%20billion%20to%20India's%20GDP.

Ministry of Education, Government of India (2021). *All India Survey on Higher Education 2019–20.* Retrieved June 26, 2021. (https://www.education.gov.in/sites/upload_files/mhrd/files/statistics-new/aishe_eng.pdf)

NASSCOM. (n.d.). *NASSCOM Cloud: Next Wave of Growth in India 2019.* Retrieved January 15, 2023, from https://nasscom.in/index.php/knowledge-center/publications/nasscom-cloud-next-wave-growth-india-2019

Olanrewaju, R. F., Mir, R. N., Islam Khan, B. U., Baba, A. M., & Gannie, M. R. (2015). Raed: Response Analysis of Educational Data for Leveraging Knowledge Dissemination System. *2015 IEEE Conference on e-Learning, e-Management and e-Services (IC3e).* https://doi.org/10.1109/ic3e.2015.7445577

Peterson, P. M. G., & Altbach, P. G. (2007). *Higher Education in the New Century: Global Challenges and Innovative Ideas.* Sense Publishers.

Press Information Bureau. (n.d.). *Digital University.* Retrieved January 15, 2023, from https://pib.gov.in/PressReleasePage.aspx?PRID=1814135

Shah, H. & Agrawal, S. (2008), *Educating India – Education Sector Report, Angel Broking,*

Stone, W. (2022, October 17). *8 Surprising Ways Cloud Computing is Changing Education.* Cloud Academy. Retrieved January 15, 2023, from https://cloudacademy.com/blog/surprising-ways-cloud-computing-is-changing-education/

The cloud: A Smart Move for Higher Education – Ellucian. (2018). Retrieved January 15, 2023, from https://www.ellucian.com/assets/en/white-paper/whitepaper-cloud-smart-move-higher-education.pdf

Usher, A., & Cervenan, A. (2005). *Global Higher Education Rankings Affordability and Accessibility in Comparative Perspective, 2005.* Distributed by ERIC Clearinghouse.

Vora, R. (2021, July 7). *The State of Education in India Pre-COVID-19 I IDR.* India Development Review. Retrieved January 15, 2023, from https://idronline.org/state-of-school-education-india-pre-covid/

Writers, W. S. (2022, November 10). *Higher Education: A Look Ahead to 2026.* Workday Blog. Retrieved January 15, 2023, from https://blog.workday.com/en-us/2022/higher-education-look-ahead-2026.html

Artificial Intelligence, Blockchain, Computing and Security – Dagur et al. (Eds)

Cloud economics and its influence on business

F. Nadeem & A. Singh

BBS College of Engineering and Technology, Prayagraj, Uttarpradesh, India

ABSTRACT: Modern-day business revolves around data storage and data management. Data storage, transfer, security, and analysis were among the most critical business needs. Private servers and electronic data processing (EDP) sections with a skilled and equipped team of experts are an important part of the corporate layout. With the advent of cloud computing, everything has changed, and today according to the Forbes report, only 18% of businesses are left from cloud coverage. This research attempts to understand the need for the cloud, the cost involved in cloud computing, and the cost structure change. An attempt has also been made to understand the influence of the cloud on business efficiencies. This research has also tried to find out the transition cost of the cloud in light of return on investments.

1 INTRODUCTION

With the advent of cloud computing, blockchain and machine learning, the business management landscape has changed forever. The pandemic has acted as a catalyst in making these terminologies popular as a part of the business management process. The cloud is a network of servers accessible from anywhere in the world. With cloud computing, users and companies no longer need to manage physical servers or run software applications on their machines. Files and applications can be fetched from anywhere on any device, as the data is stored in the data centre instead of local servers (https://www.cloudflare.com/learning/cloud/what-is-the-cloud/).

A Facebook or Instagram user can log in to their account on any phone or computer from anywhere without changing their stored photos, videos and conversations history. Similarly, Microsoft Office 365 or Gmail are examples of cloud. Email and Dropbox or Google Drive are also examples of cloud storage. The presence of social platforms (Facebook, Instagram, Dropbox) for over a decade indicates that cloud technology is familiar and has been used seamlessly for quite some time. The only reason it took so long to gather acceptance among the business fraternity was the security of the data, software, usage cost of the cloud and management cost.

Cloud computing technology is also gaining momentum as it provides cost-saving benefits to organizations. Many organizations are moving to adopt technology to reduce their capital expenditure. The adoption of cloud computing is set to increase the overall profitability of the companies in the long term. This paper attempts to study cloud computing technology from an economic point of view.

2 REVIEW OF LITERATURE

Cloud technology has motivated customers to adopt cloud services as it reduces cost and minimizes management effort (Geyskens *et al.* 2006; Hsu *et al.* 2014). Babu and P S (2013) proposed that government should encourage the acceptance of cloud technology in Small and Medium Enterprises (SMEs) by providing subsidies as the cloud will help create hundreds of SMEs and thousands of jobs. Among the many benefits of the cloud, scalability and

DOI: 10.1201/9781003393580-124

agility are the important reasons for cloud adoption (Weinhardt *et al.* 2009). Wood & Buckley (2015) also presented the same line of thought. They concluded that agility and scalability help the transfer of data anywhere with ease.

Cloud services are cost-effective compared to on-site services. Competition and economies of scale have helped lower the cost of these services more than expected (Foster *et al.* 2008; Kashef & Altmann 2011; Pal & Hui 2012). Alford & Morton (2009) elaborated on cost-effectiveness, finding that governments can achieve significant cost savings by moving information and communication technology (ICT) to the cloud. They even claim that the cost savings due to cloud adoption will be as much as two-thirds over a decade. Etro (2011) proposed a pay-as-you-go model for cloud computing to save costs. The cloud allows businesses to convert some fixed costs into marginal product costs. Examining the economic impact of the cloud on the European economy, McWilliams (2012) found that the cloud contributed EUR 763 billion to GDP and approximately 2.4 million jobs in five countries. A Deloitte (2017) report also reported similar findings. Deloitte (2018) also stated an average return of $2.5 for every $1 invested in cloud services.

The literature also highlights cloud technology's hidden, indirect, and allied costs. Literature has cautioned the end user before adopting the technology. McCafferty (2015) forwarded a few factors like ongoing maintenance costs, cloud implementation problems, and vendor transparency playing a role in unpredicted costs. In the study, Araujo *et al.* (2018) noted that choosing a cloud service can be difficult and expensive. Bildosola *et al.* (2015) also pointed out that customers generally need more clarity and guidelines regarding their requirements. Such ignorance and assumptions lead to unfulfilled profits (Ellram 1993; Ramchand *et al.* 2018).

3 OBJECTIVES OF THE STUDY.

To investigate the relationship of cloud with economics.
To understand CAPEX and OPEX in the cloud context
To find out the cost components of cloud management.
To understand the role of the cloud in improving business efficiencies.
To find out the Returns on Investment (ROI) in cloud management.

4 RESEARCH METHODOLOGY

It is an exploratory analysis of the studies undertaken by various cloud service providers, research organizations, international research organizations, independent research agencies, research blog writers, consultancy firms and research scholars. This study is based on available secondary data on cloud computing, capital expenditures, return on investment and cloud marketing. The conclusion is drawn based on past studies and prevailing conditions in the cloud computing arena.

5 CLOUD ECONOMICS

It is all about migrations from a privately owned and managed server system to a publicly available one. These servers are available for rent on a "pay as you go" model, meaning that users need to pay only for the cloud service they use. Cloud economics is about the cost-benefit analysis of this transition. It helps managers to analyze the cost of management, complaints, security, applications, software and infrastructure. It helps determine whether migration to the cloud will be a good idea. Cloud economics studies the costs and benefits of cloud computing and the financial standards that support them (What is cloud economics? Vmwareglossary 2023). It helps small businesses decide whether to migrate to the cloud to outsource their infrastructure-related needs. Cloud economics also tries to find the ROI of migration to the cloud from the old method and the total cost of ownership (TCO) of a cloud solution against the traditional on-premises solution.

5.1 Cloud's relationship with economics

Organizations today require internal data security, backup, and networking, apart from ready data analysis for decision-making. Cloud provides innovative solutions for managing server-related issues, data storage and management beyond geographical boundaries. On the other hand, cloud economics refers to the cost expected to be incurred while migrating to the cloud and the operational benefits that the cloud provides in terms of efficiencies of operations, analytics, cost, inventory, forecasting, etc. Commercial use of the cloud relates directly to opportunity cost and financial efficiencies. Cloud economics is not only about the actual monetary costs but also about the opportunity costs of the cloud and the intricacies of managing costs in a highly dynamic and complex environment (CloudZero, n. d.). Thus, there is an important relationship between cloud computing and economics. As economics favours cloud computing, the fundamental nature of these services must ensure the highest level of end-user satisfaction. It is estimated to reach $448.34 billion by 2026, at a CAGR of 28.89% over the upcoming years (How to calculate the cost of cloud migration: Consider these factors! 2022). Although three prominent cloud service providers, Amazon Web Services (AWS), Microsoft Azure, and Oracle Cloud, claim they have the highest level of data security, users still feel apprehensive. Data security and cloud cost management will be the key driver for the future economic progress of cloud computing.

5.2 CAPEX and OPEX

Capital expenditure (CAPEX) costs in business is a broader term for all kinds of expenditures on property, plant, equipment, maintenance and acquiring assets that can help generate business beyond the current year. This initial cost gets reduced with time as the physical assets attract depreciation. The business also needs data management and information transfer mechanism to run profitable commercial operations. The need varied from industry to industry, but from the time computers became a part of business, EDP sections and server rooms became a part of corporate layouts. Such facilities required heavy investment in CAPEX related to computers, software, licenses, servers, hardware, and a dedicated operating team. These costs were paid upfront and adjusted through depreciation. It is argued that these privately owned server systems have advantages like data security, the safety of business sensitive information etc. Big business houses still prefer these private servers despite huge CAPEX costs and operational issues like equipment failure, electricity fluctuation, configuration errors, limited automation etc. Cloud computing now handles these issues, but operating expenses (OPEX) have risen.

The incredible efficiencies and cost-saving abilities of cloud computing have made it preferred in medium and small-size organizations. Those small organizations that are unable to make big capital expenditures are now using cloud computing to do business because of low OPEX. OPEX are the operating costs, which are the expenses to run the day-to-day business. (Vanderweide 2019). OPEX of cloud computing is low due to the following reasons:

1. Cloud provides on-demand services, which are on a plug-and-play basis.
2. Buy what one needs.
3. No maintenance issues and costs.
4. Total cost of ownership (TCO) incurred on cloud migration, architecting, planning, and operations are very transparent.

5.3 Cost components and cloud management

Most companies already have a server system, networking arrangement, software, and data security. The cost components of cloud computing will depend on the objective of migration and cloud deployment model (single cloud, multi or hybrid cloud) as they play a significant role in decisions related to migration strategies. Cloud cost optimization helps the organization manage and understand the cost implication of cloud technology cost-effectively, along with efficiently utilizing this technology. With private, public and hybrid cloud services coming into the picture, the infrastructure complexity has increased, and cost calculations have become complicated as the customer may choose from several cloud migration strategies (retain, rehosting, replatforming, retire, refractor, repurchasing) the cost will also vary (https://www.scnsoft.com/cloud/migration).

5.4 Cloud computing and return on investment

The financial impact cloud computing has on an organization that migrates to the cloud from traditional servers is measured through return on investment. A positive ROI is a success indicator for most businesses, indicating increased margins and net profits. If the financial benefit outweighs the original investment, the result is a positive ROI (https://www. vmware.com/topics/glossary/content/cloud-roi.html). The ROI in an organization's context includes quantitative and qualitative elements. The ROI is impacted by several quantitative factors, like decreased capital expenditure, increased monthly costs, cost reduction due to cloud migration, revenue improvement due to cloud operations, etc. The ROI is also affected by qualitative factors like product improvement and enhanced customer satisfaction resulting in increased goodwill, corporate value, brand value etc. Calculating cloud ROI provides an opportunity to understand the impact of cloud transition on an organization's bottom line. It also helps understand the substantiality of the payback if investments are made in cloud computing today. Migration is challenging but becomes easy once it gets translated into cost savings, improved scalability, and operational excellence.

6 CONCUSION

With the worries of data theft and myths of cloud computing fading with time, cloud computing is a new buzzword among the business fraternity. The acceptance of the cloud as an alternative to traditional server systems has improved business ROI and added to business efficiencies. It has not only brought the CAPEX costs down but also improved the overall financial efficiencies of the organization. The cloud is ready to play a significant role in all data-sensitive business activities like marketing, sales, and reporting.

REFERENCES

Alford, T. and G. Morton. 2009. "*The Economics of Cloud Computing Addressing the Benefits of Infrastructure in the Cloud.*" Report. Booze Allen Hamilton. Available at: http://www.federalnewsradio.com/wpcontent/uploads/pdfs/EconomicsofCloudComputingSECURITYOct2009.pdf

Araujo, J., Maciel, P., Andrade, E., Callou, G., Alves, V., & Cunha, P. (2018). Decision Making in Cloud Environments: An Approach Based on Multiple-criteria Decision Analysis and Stochastic Models. *Journal of Cloud Computing*, 7(1). https://doi.org/10.1186/s13677-018-0106-7

Babu, B., & P S, C. (2013). Impact of Cloud Computing on Business Creation, Employment and Output with Special Reference to Small and Medium Enterprises in India. *Management Today*, 3(4). https://doi.org/10.11127/gmt.2013.12.06

Bildosola, I., Río-Belver, R., Cilleruelo, E., & Garechana, G. (2015). Design and Implementation of a Cloud Computing Adoption Decision Tool: Generating a Cloud Road. *PLoS One*, 10(7). https://doi.org/10.1371/journal.pone.0134563

cloudflare (n.d.). *What is the Cloud?* | *Cloud Definition* | Retrieved January 29, 2023, from https://www.cloudflare.com/learning/cloud/what-is-the-cloud/

CloudZero. (n.d.). *What is Cloud Economics? An Intro Guide to Measuring Costs.* CloudZero. Retrieved January 29, 2023, from https://www.cloudzero.com/blog/cloud-economics

Deloitte. 2017. *"Measuring the Economic Impact of Cloud Computing in Europe."* European Commission. Available at: https://ec.europa.eu/digital-single-market/en/news/measuring-economic-impact-cloudcomputing-europe

Deloitte. 2018. *"Economic and Social Impacts of Google Cloud."* September 2018. Available at: https://www2.deloitte.com/content/dam/Deloitte/es/Documents/tecnologia/Deloitte_ES_tecnologia_economic-and-social-impacts-of-google-cloud.pdf

Ellram, L. (1993). Total Cost of Ownership: Elements and Implementation. *International Journal of Purchasing and Materials Management, 29*(3), 2–11. https://doi.org/10.1111/j.1745-493x.1993.tb00013.x

Etro, F. 2011 "The Economics of Cloud Computing." International Think-Tank on Innovation and Competition. *Prepared for the Annual Conference on European Antitrust Law 2011. The Future of European Competition Law in High-tech Industries (Brussels, March 3–4, 2011).* Available at: http://citeseerx.ist.psu.edu/viewdoc/download;jsessionid=765AC87A1493E7F9E0DB4F4CE459B820?doi=10.1.1.656.7166&rep=rep1&type=pdf

Foster I., Zhao Y., Raicu I. and Lu S., "Cloud Computing and Grid Computing 360-Degree Compared," 2008 Grid Computing Environments Workshop, Austin, TX, USA, 2008, pp. 1–10, doi: 10.1109/GCE.2008.4738445

Geyskens, I., Steenkamp, J.-B. E., & Kumar, N. (2006). Make, Buy, or Ally: A Transaction Cost Theory Meta-analysis. *Academy of Management Journal, 49*(3), 519–543. https://doi.org/10.5465/amj.2006.21794670

Hsu, P.-F., Ray, S., & Li-Hsieh, Y.-Y. (2014). Examining Cloud Computing Adoption Intention, Pricing Mechanism, and Deployment Model. *International Journal of Information Management, 34*(4), 474–488. https://doi.org/10.1016/j.ijinfomgt.2014.04.006

Kashef MM, Altmann J. (2011). A Cost Model for Hybrid Clouds. In *International Workshop on Grid Economics and Business Models* (pp. 46–60). Springer, Berlin, Heidelberg

McCafferty D. (2015). *How Unexpected Costs Create a 'Cloud Hangover'.* CIO Insight, URL: https://www.cioinsight.com/it-strategy/cloud-virtualization/slideshows/how-unexpected-costs-create-a-cloud-hangover.html.

McWilliams, Douglas. 2012. *"The Cloud Dividend: Part 1. The Economic Benefits of Cloud Computing to business and the wider EMEA Economy, France, Germany, Italy, Spain and the UK."* Report. Centre for Economics and Business Research. Available at: https://www.cebr.com/wp-content/uploads/2013/03/2010-12-10-Economic-impact-of-cloudcomputing_Cebr-report_final-final_clean.pdf

Pal R & Hui P. (2012). Economic Models for Cloud Service Markets. In *International Conference on Distributed Computing and Networking* (pp. 382–396). Springer, Berlin, Heidelberg

Platforms, Costs. (n.d.). *Cloud Migration Step-By-Step Process.* Guide: Steps, Retrieved January 29, 2023, from https://www.scnsoft.com/cloud/migration

Ramchand K, Chhetri MB, Kowalczyk R (2018) Towards a Comprehensive Cloud Decision Framework with Financial Viability Assessment. In: *Proceedings of the 22nd Pacific Asia Conference on Information Systems (PACIS 2018)*, P(32)

Stefanini. (2022, August 25). *How to Calculate the Cost of Cloud Migration: Consider These Factors!.* Retrieved January 29, 2023, from https://stefanini.com/en/insights/articles/how-to-calculate-cloud-migration-cost-factors-to-consider

Vanderweide, D. (2019, August 29). *OPEX vs. CAPEX: The Real Cloud Computing Cost Advantage.* 10th Magnitude. Retrieved January 29, 2023, from https://www.10thmagnitude.com/opex-vs-capex-the-real-cloud-computing-cost-advantage/

VMware. (2023, January 25). *What is Cloud Economics?: Vmware Glossary.* Retrieved January 29, 2023, from https://www.vmware.com/topics/glossary/content/cloud-economics.html#:~:text=Cloud%20economics%20is%20the%20study,or%20switching%20current%20cloud%20providers%3F

VMware. (2023, January 25). *What is Cloud ROI?: Vmware Glossary.* Retrieved January 29, 2023, from https://www.vmware.com/topics/glossary/content/cloud-roi.html

Weinhardt, C., Anandasivam, A., Blau, B., Borissov, N., Meinl, T., Michalk, W., & Stößer, J. (2009). Cloud computing – A Classification, Business Models, and Research Directions. *Business & Information Systems Engineering, 1*(5), 391–399. https://doi.org/10.1007/s12599-009-0071-2

Wood, K., & Buckley, K. (2015). Reality vs Hype – Does Cloud Computing Meet the Expectations of Smes? *Proceedings of the 5th International Conference on Cloud Computing and Services Science.* https://doi.org/10.5220/0005472401720177

839

Artificial Intelligence, Blockchain, Computing and Security – Dagur et al. (Eds)
© 2024 The Author(s), ISBN: 978-1-032-49393-0

Live virtual machine migration towards energy optimization in cloud datacenters

Rohit Vashisht, Gagan Thakral & Rahul Kumar Sharma
KIET Group of Institutions, Delhi-NCR, Ghaziabad

ABSTRACT: Cloud computing is a fast emerging utility- oriented paradigm providing services to a large number of users across World Wide Web based on spend-as-per-your-utilization model. To offer its users computational services, data centers eat up a lot of energy. Such excessive consumption of power by those virtualized data centers has increased the operating cost of cloud environment for both cloud service provider as well as for intended user. Normally, a typical data center wipes out energy which generally equals to energy consumption of 25000 houses. Along with increasing cost, it also degrades the environment badly by emitting large amount of carbon dioxide gas and thus becoming one of the major causes for environmental issues like global warming and green house effects. In this research paper, the proposed technique not only trying to meet energy efficiency requirement but will also prevent the breaching of Service Level Agreement (SLA) and thus provide cloud users with Quality of Service (QOS).

Keywords: Cloud computing, CloudSim, Dynamic consolidation, Migration, Virtualization

1 INTRODUCTION

Cloud computing is becoming a popular dispersed computing model across World Wide Web and one can get rights to access the applications as utilities by using this model. Cloud provides computing resources to the users over the internet so users do not required buying these resources but they have to pay by following the principle "Pay-as-you-go" (Fu *et al.* 2011). One of the most significant components in the modern cloud environment is energy efficiency that is used to estimate the operational cost requirement and capital investment. Cloud Computing is not only providing these estimates but also indicating the performance and carbon residues in the industry. So, there is a need of the energy efficient technologies for communication networks. In accordance with optimization of energy utilization of a data center, one has to investigate its energy consumption patterns deeply. As, good amount of energy is misused in idle systems: in ordinary usage scenario, server utilization is under 30%, but idle servers still devour 60% of their apex power (Yadav *et al.* 2014). Cloud datacenters are the backbone of today's Information and Communication Technology (ICT) infrastructure, so there is a dire requisite to upgrade their energy efficiency requirement.

2 PROPOSED WORK

The methodical process of moving a Virtual Machine (VM) from a single machine to a different one is called migration which is done under occurrence of either under loaded host or overloaded host for the sake of energy optimization and efficient resource utilization at a

DOI: 10.1201/9781003393580-125

data center. Furthermore, a live migration is the transfer of a VM from one physical node to another without pausing cloud user activity. By doing this, resource usage can be raised as well as operational cost can be reduced in a cloud environment. Following policies are acted in the stated order to do the migration in energy efficient manner.

2.1 *VM allocation policy*

Here, the proposed technique is considering dual threshold limits- Upper threshold limit and Lower threshold limit, between which one has to limit the CPU utilization of all hosts running at a data center to achieve least energy consumption. When a host's utilization is higher than the upper threshold limit, it is said to be over-loaded, and when it is lower than the lower threshold limit, it is said to be under-loaded. In both cases, migration should take place to make utilization rate between upper and lower threshold limits. However, static thresholds do not provide better results for systems with continuously changing and unknown workloads (Beloglazov *et al.* 2010). Through this paper, a dynamic technique is proposed to determine the threshold values using present slot of running VMs and by doing analysis of service usage patterns of past events for each VM.

The following terms and equations are used to decide dynamic T_{upper} and T_{lower} values for a particular physical host:-

UTIL = Total Requested MIPS by that particular VM / Total Requested MIPS by all VMs billed on a particular host
Besides using only CPU utilization to determine threshold values, the proposed technique is also using three more parameters that are allocated RAM, bandwidth and memory storage for both virtual machine and host to determine the threshold values as stated below,
TBW = sum of current bandwidth required by all VMs billed on a particular host
TRAM = sum of current RAM required by all VMs billed on a particular host
TUTIL = sum of all individual VM utilization (ΣUTIL)
TSTOR = sum of allocated memory storage to all VMs billed on a particular host.

2.1.1 *Determination of T_{UPPER}*
When a host's CPU usage surpasses the upper threshold limit, it is considered to be over-loaded. To lower the CPU utilization below the upper threshold figure, certain virtual machines must be moved. Here, the proposed study calculated T_{upper} value for each host separately as follows:-

$$\text{Total} = \text{TUTIL} + (\text{TBW}/\text{TBW(host)}) + (\text{TRAM}/\text{TRAM(host)} + (\text{TSTOR}/\text{TSTOR(host)})$$

To calculate the dynamic upper threshold, proposed technique is using robust statistics rather than classical statistics techniques because the former technique is not affected by small departures from model inputs and assumptions.

$$T_{upper} = 1 - s * \text{Total} \tag{1}$$

where, the variable Total gives the aggregation of four parameters that have been considered and the variable s defines a security framework and also explains how likely the system stabilizes VMs in a determined way (Sinha *et al.* 2015). The lower value of s, lesser will be the energy utilization, but at the cost of higher SLA's level breaching caused by consolidation (Beloglazov *et al.*2012). A higher value of s will provide outcomes similar to those of a static threshold policy in terms of excessive usage of energy (Adhikari *et al.* 2013). Here, this paper has taken s = 0.5 for evaluation of the experimental results to achieve equilibrium between the parameters energy usage and SLA breaching for a given workload.

2.1.2 Determination of T_{LOWER}

A physical node (host) is termed as underutilized when its CPU utilization remains below lower threshold limit. To overcome this problem, one can adopt two major strategies:-

Firstly, all VMs should be moved to other physical node and underutilized machine should be moved to inactive (sleeping) mode to remove idle power usage issue.

Secondly, few VMs from other host can be moved on an underutilized host until its CPU utilization rate exceeds the lower threshold limit.

From the studies (Beloglazov *et al.* 2010), it was observed that if the CPU usage of a particular host is greater than or equal to 30%, lower threshold (T_{lower}) is always set as 0.3. Hence, lower threshold limit can be decided as:-

$$If TUTIL >= 30\%, then T_{lower} = 0.3 \tag{2}$$

This paper has adopted second strategy to overcome the issue of underutilized host after determining the lower threshold dynamically.

2.2 VM selection policy

After using (1) and (2) to calculate lower and upper threshold value respectively, the analysis has found out the over utilized host if any. Now an issue arises:-

"Which VM should we migrate from an over utilized host to make its utilization rate below upper threshold?"

This problem of VM selection can be resolved by using VM Selection Policy. Here, this paper uses four popular VM selection policies – Minimization of Migration Time [MMT], Minimum Utilization [MU], Maximum Correlation [MC], Random Selection [RS] (Beloglazov *et al.* 2012).

2.3 VM placement policy

After selecting the VM for migration, the next issue is to determine the host where the selected VM can be hosted in a power efficient manner. The hosting of VM is considered as a bin packing type of problem (Beloglazov *et al.* 2012). To overcome this problem of live migration of VMs, one can use VM Placement Policies. There are various VM Placement policies available (Chowdhury *et al.* 2015).

In this paper, the authors have employed Best Fit Decreasing (BFD) algorithm (Beloglazov *et al.* 2010). This algorithm place iteratively selected VM on each host one by one and finally place the selected VM on that physical node showing minimum increment in the energy utilization after the placement of VM. In other words, this algorithm finds the most power efficient host among all available hosts.

3 VARIOUS PERFORMANCE METRICS

3.1 Energy consumption

The first parameter which is used to estimate the performance of VM movement is aggregated energy utilization by all physical nodes of the data centers under given application workload (Beloglazov *et al.* 2012).

3.2 SLA violation

The second metric used to measure the performance is SLA Violation which can occurred when a particular VM cannot provide the required no. of requested MIPS by currently available set of resources. The aim is to minimize SLA violation along with focus on energy management (Beloglazov *et al.* 2012). SLA Violation metric is product of another two parameters described below:-

$$SLAV = SLATAH.PDM$$

where, SLATAH is SLA violation time per active host and PDM is Performance Degradation due to migrations (Beloglazov *et al.* 2012).

3.3 VM migrations

The third evaluation parameter is no. of VM migrations proposed by VM consolidator under detection of either underutilized or over utilized host at a data center (Beloglazov *et al.* 2012).

4 EXPERIMENTAL RESULTS

The proposed work has compared each combination of VM selection policy along with Static Double threshold policy (SDT) (Adhikari *et al.* 2013), Dynamic Threshold Policy (Dthr1) proposed by (Beloglazov *et al.* 2010), Dynamic Threshold Policy(Dthr2) proposed in this paper on CloudSim Toolkit (Calheiros *et al.* 2011). To evaluate these algorithms, a cloud environment is considered with just one datacenter. In addition to this, it has 5 hosts which are further running 5 VMs billed on these 5 hosts.

Here, this paper has taken static upper threshold and static lower threshold value 0.8 and 0.3 respectively to evaluate results for Static Threshold policy (SDT) (Adhikari *et al.* 2013).

(Beloglazov *et al.* 2010) used only one parameter i.e. CPU utilization rate of various VMs and host to determine dynamic upper and lower threshold for unpredictable workload in their dynamic threshold policy (Dthr1).

Rather than using only one parameter i.e. CPU utilization of various VM and various hosts to decide T_{upper} and T_{lower} dynamically, the proposed technique (Dthr2) as stated above also uses three more parameters that are Bandwidth, RAM and memory storage of a VM.

Table 1. Comparison of energy consumption in KWH.

Policy used	SDT	Dthr1	Dthr2
MMT	5.18	6.14	5.95
MU	5.83	6.74	6.52
MC	5.59	6.42	6.18
RS	5.41	6.34	6.16

Table 1 and Figure 1 results show the energy consumption in KWH for various VM selection policies along with three VM allocation policies that are static double threshold policy (SDT), Dynamic threshold policy (Dthr1) proposed by (Beloglazov *et al.* 2010) and the proposed dynamic threshold policy (Dthr2). Dthr2 consumes less energy in comparison to existing technique Dthr1. Figure 1 shows that proposed technique Dthr2 gives comparable results to other two existing techniques in terms of energy usage.

Table 2 results show the VM migrations required for every combination of VM allocation policies and VM selection policies to make CPU utilization of under loaded hosts and overloaded hosts in between upper and lower threshold limits corresponding to determined threshold values. The results show that the proposed technique Dthr2 gives less no. of VM migration in comparison to Dthr1 that helps in load balancing, to achieve fault tolerance, for continuation of low level system conservation activities & to minimize energy usage (Kaur *et al.* 2015) (Kaur *et al.* 2017).

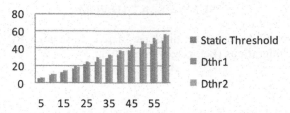

Figure 1.   Energy consumption in KWH v/s number of hosts for MMT.

Table 2.   Comparison of number of VMs migration.

Policy used	SDT	Dthr1	Dthr2
MMT	277	290	261
MU	431	428	396
MC	354	320	318
RS	305	312	310

Table 3 results show how much SLA violation (%) occurred in each combination of VM allocation and VM selection policies. Figure 2 depicts that the proposed logic (Dthr2) minimizes both energy usage and count of VM migrations with slightly compromising the parameter SLA breaching in comparison to Dthr1 policy (Beloglazov *et al.* 2010), whereas

Table 3.   Comparison of SLA violation (%).

Policy used	SDT	Dthr1	Dthr2
MMT	2.75	1.20	1.90
MU	3.15	1.73	1.92
MC	3.08	1.26	1.49
RS	3.04	1.20	1.90

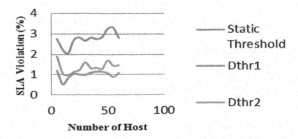

Figure 2.   SLA violation (%) v/s number of hosts for MMT.

Dthr2 offers much lesser SLA violation than static threshold policy which leads to better resource utilization and customer satisfaction.

5 CONCLUSION AND FUTURE DIRECTIONS

The dynamic consolidation of VM and maximization of CPU utilization of an underutilized host by placing additional VMs on it maximizes the energy efficiency of the resources is being studied in this paper. The authors have proposed dynamic consolidation of VMs which predict the upper and lower threshold dynamically using parameters CPU utilization rate, RAM, bandwidth and memory storage for dynamic and unpredictable workload in a cloud environment. The experimental results depict that the presented technique minimizes energy usage and no. of VM migrations slightly compromising SLA violation parameter. By setting the safety parameter (s), one can get tradeoff between energy consumption and SLA infringement. The proposed work focuses to lessen energy usage which contributes towards emerging Green cloud computing technology.

For the future work, one can try to introduce an optimization policy to minimize the SLA violation and operational cost requirement too for a data center. Secondly, the introduction of a novel VM selection policy rather than simulating proposed technique (Dthr2) with existing VM selection techniques (MMT, MU, MC and RS) for betterment of experimental results can also be a robust future direction in the same domain.

REFERENCES

Adhikari J. and Prof. Patil S., "Double Threshold Energy Aware Load Balancing in Cloud Computing", *Fourth International Conference on Computing, Communications and Networking Technologies (ICCCNT)*, pp. 1–6, July 2013.

Beloglazov A. and Buyya R., *"Adaptive Threshold-based Approach for Energy-efficient Consolidation of Virtual machines in Cloud Data Centers"*, CLOUDS Lab, Dept. of Computer Science and Software Engineering, The University of Melbourne, Australia,November 2010.

Beloglazov A. and Buyya R., "Energy Efficient Allocation of Virtual Machines in Cloud Data Centers", *10th IEEE/ACM Intl. Symp, on Cluster, Cloud and Grid Computing*, pp. 577–578, 2010.

Beloglazov A. and Buyya R., "Energy-aware Resource Allocation Heuristics for Efficient Management of Data Centers for Cloud Computing", *Journal of Cloud Computing*, vol. 28, pp 755–768, 2012.

Beloglazov A. and Buyya R., "Optimal Online Deterministic Algorithms and Adaptive Heuristics for Energy and Performance Efficient Dynamic Consolidation of Virtual Machines in Cloud Data Centers", *Concurrency and Computation: Practice & Experience*, vol. 24, no. 13, pp. 1397–1420, September 2012.

Calheiros R. N., Ranjan R., Beloglazov A., De Rose C.A.F. and Buyya R., "CloudSim: A Toolkit for Modeling and Simulation of Cloud Computing Environments and Evaluation of Resource Provisioning Algorithms", *Software- Practice and Experience*, pp. 23–50, 2011.

Chowdhury M. R., Mahmud M. R. and Rahman R. M., "Implementation and Performance Analysis of Various VM Placement Strategies in CloudSim", *Journal of Cloud Computing*, vol. 4, no. 1, pp.1–21, November 2015.

Hu F. *et al.*, "A Review on Cloud Computing: Design Challenges in Architecture and Security", *Journal of Computing and Information Technology*, vol.1, pp 25–55, CIT 19, 2011.

Kaur A., Diwakar A. and Vashisht R., "Alternatives to VM Consolidation Techniques for Energy Aware Cloud Computing," *2017 International Conference on Advances in Computing, Communications and Informatics (ICACCI)*, Udupi, India, 2017, pp. 2005–2009, doi: 10.1109/ICACCI.2017.8126139.

Kaur P. and Rani A., "Virtual Machine Migration in Cloud Computing", *International Journal of Grid Distribution Computing*, vol. 8, no. 5,pp. 337–342, 2015.

Sinha R., Purohit N. and Diwanji H., "Power Aware Live Migration for Data Centers in Cloud Using Dynamic Threshold", *International Journal of Computer Technology and Applications*, vol. 2, no. 6, pp. 2041–2046, 2015.

Yadav V., Malik P., Kumar A. and Sahoo G., "Energy Efficient Data Center in Cloud Computing", *International Journal of Computer Applications*, vol. 106, no. 7, pp. 23–28, November 2014.

Artificial Intelligence, Blockchain, Computing and Security – Dagur et al. (Eds)
© 2024 The Author(s), ISBN: 978-1-032-49393-0

Flexible-responsive data replication methodology for optimal performance in cloud computing

Snehal Kolte & Madhavi Ajay Pradhan
Computer Engineering, AISSMS COE, SPPU, Pune, India

ABSTRACT: Cloud computing has gained significant attention from researchers due to its ability to offer reliability, fast data access and intense data availability through the use of replicas. However, managing replicas in large, dynamic, and virtualized data centers can be complex. In order to minimize response time and deliver worthwhile availability to achieve load balancing in cloud, a placement strategy has been newly put forward. This strategy takes into account two key factors: average response time, and storage capacity utilization. It is important to use replication wisely, as storage space at each site is limited and only important replicas should be kept it plays a vital role in the education field and in any business environment. Additionally, a new replica replacement approach has been developed based on the amount of copies, count replica accessed, file availability, based on replica last accessed, and the size of the replica. The effectiveness of this approach has been evaluated using the CloudSim simulator, and it has been shown effective in average response time, and storage capacity utilization.

Keywords: Data Replication, scheduling algorithms, Round Robin, Replication Management, CloudSim.

1   INTRODUCTION

Cloud is a widespread paradigm in Industry and academia [1–3], and data centers [4] are a fundamental component of the cloud. These centers consist of many storage elements and commodity computing, which can reach in a scale [5]. In cloud, Data that is kept on several nodes inside a single data center or across multiple data centers is stored on numerous nodes. This makes a user to the access data without having to create their own infrastructure, reducing costs. Many major IT establishments have started own private cloud storage platforms on-premises, like AmazonS3 [6] and Google's GFS [7]. It's very much vital to enlarge data consistency in cloud systems, which is why both the GFS(Google file system) and HDFS(Hadoop distributed file system) use the replications to increase accessibility. If replicas of the data file are kept on diverse nodes, the Data remains available even if one node fails. As the replicas go on increasing, it affects the most of esteemed organization to increase the cost of maintaining it. Incorporating the applicable criteria that follow.

In large scale heterogeneous storage systems, replica placement becomes a complex task due to the varying capabilities of each data node. Using a random replica placement strategy can lead to an uneven distribution of workload, resulting in access skew and load imbalance across the cloud storage system. This can negatively impact parallelism and performance. To address these issues, we propose an adaptive replica placement strategy that aims to distribute workloads effectively by storing replicas on the best site. This strategy helps to balance the load in the system and improve overall performance.

DOI: 10.1201/9781003393580-126

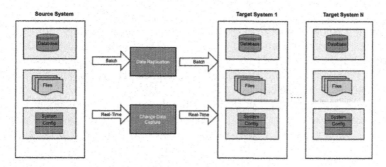

Figure 1. Data replication model.

Valid to optimize performance of storage system, A successful replica replacement strategy must be in place. To make the most use of the resources available, it is necessary to manage which copies should be removed due to each node's limited storage space. To address this issue, the proposed replica replacement algorithm that aims to reduce replicas that are minimum possible to be needed in the future. This approach helps to ensure general operation of the approach.

To measure performance of the new study, started using Cloud Sim [8] simulator. This toolkit supports the modelling of a Clouds systems component like VMs, data centers, RP policy, and allows for the implementation of common resource provisioning policies with minimal effort. We have sub classes of the Cloud Sim toolkit and used to the simulate this new study algorithm, and the results of this new study simulation measure show that this new study approach is much better when study equated to existing algorithm's. This new study algorithm demonstrates significant improvements to the response time and the effective network utilization.

This study is the briefed as follows:

- Measured models to describe the studied objectives including response time and storage capacity utilization. These models take into account the capacity of each data node in order to accurately capture the relevant factors.
- An approach for placing replicas mid data nodes with respect to the study objectives described above. This allows for more effective distribution of workload and helps to optimize the performance of the system. Here proposes a fresh the replica replacement methods that consider following important parameters: based on number of the replica, count of the copy accessed and files availability, based on the replica last accessed. This proposed FDRS (Flexible/Responsive Data Replica Selection) algorithm improves the progressive locality property, makes decisions based on factors like the availability of the data file to direct which replicas should be replaced (also known as the victim file). This approach helps to optimize the functioning of the system.
- Conducted an assessment of the FDRS strategy and the results of our simulations show that the FDRS strategy leads to the lowest relative delay time for job completion. Additionally, our analysis demonstrates that the FDRS strategy performs better and is more efficient in terms of data access on the cloud.

2 RELATED WORK

Numerous research determinations that have been concentration on task to meet scheduling to show ideas, such as time response along with rate that to be reduced [10]. Shortest path find is the best way to manage the data on the cloud, if the data is located near nodes it would be easy to access those data without any interruption. Response time is decreased if data is reside in a nearby location [11]. T. Yuan *et al.* suggested fully along with partially replicated data. The determination in the large networks distributes the data which are copied as partial and absolute [12].

Researchers Proposed Combining the data and replica placement which provides a single point single optimization problem.

The framework which is UnifyDR allocates the data to the various nodes which helps to minimize the storage cost and traffic [13]. In Data replication which improves enactment and consistency by creating several copies of the same file or data which is very effective In tradition networks, it may be wired or wireless data replication is used to improve the performance. End users can reduce the overburden of the load and can retrieve the data [14]. Data replication gives the metaheuristics approach reviews which have high availability, minimum response time, low bandwidth consumption, and reliability [15].

As several replication strategies which have been projected to come across numerous aims [16]. such as Considering balancing of load. [17] response time that to be reduces and transfer amount of data [18]. The several work has been proved to combining both and replication and scheduling strategies which is extremely beneficial to gain performance effectively it aims to enhance amount of data transfers along with task's response time that is reducing [19,20].

3   FLEXIBLE-RESPONSIVE DATA REPLICATION METHODOLOGY

Here, proposing the data replication strategy for cloud storage that is optimized for multiple objectives. This approach considers provide average response time, and storage capacity utilization. The goal is to understand how the layout of replicas impacts these performance metrics.

Attempting to optimize multiple objectives related to average response time, and storage capacity utilization to identify the tradeoffs between these objectives. Because we are considering multiple objectives, Using multiobjectives optimization to determine the optimal replications facto which layout for each data a file. Which tallows to find resolutions that approach optimal values for these new vigorous parameters. The proposed strategy is flexible and can be customized to the user's specific needs by allowing to assign higher weights to the performance metrics that are most important to them.

This study calls it as the Flexible Responsive data replication strategy (FDRS), which also has an efficient method for placing replicas but also involves a strategy for selecting the locations for replicas.

This strategy is divided into three parts:

3.1   *Replica creation and management*

To ensure fault tolerance, high availability plus efficiency, it is important to adjust the locations for new replicas based on the current conditions in the cloud. Placing replicas randomly can lead to uneven access patterns, where some sites are heavily used while others are idle. This can cause imbalanced workloads and poor performance in the cloud storage system. However, it can be challenging to effectively place replicas in highly scalable, ultralarge, and virtualized data centers. One way to improve response time is to generate new replicas and store them on sites with lower workloads. Selecting an appropriate location for placing a new copy of data in a cloud storage system can be a complex task that may involve solving an NP-hard problem. An effective replica strategy can help to ensure that the copy is created in a timely manner, while minimizing overhead combined with minimizing need for copies migration in the future. A good replica strategy can help to optimize the performance and efficiency of the cloud storage system, enabling it to handle the demands of a wide range of workloads and applications. Overall, the goal is to choose a location for the new copy that maximizes the benefits and minimizes the costs and risks. While creating a new replica and the placement of the newly created replica, consider a few parameters.

3.1.1 Response time
To minimize latency in a storage system, it is important to maximize the bandwidth capacity of the data nodes where the files are stored. This will reduce latency during read operations. In this study, we are focused on minimizing read latency specifically. Each file can have more than one replica. When latest replication is created, FDRS determine to select a site with a lighter load to store the replica. This makes our strategy adaptable to the specific needs of the user.

3.1.2 Storage capacity utilization
To minimize the waiting time for frequently accessed files, it is advisable to locate them on data nodes with low storage usage.

3.2 Storage management using aging of files

If there is sufficient storage space at a chosen location, the a replica will be stored there. If there is not enough space available, one or more files will need to be aged out with the help of the following criteria:

- based on the size of the replica
- count of the replica accessed
- files availability
- based on the replica last accessed.

The likelihood of a replica being requested again can be determined with the number of time's it has been accessed and the last time it was requested. It is more worthwhile to replace larger files because doing so will result in fewer replica replacements being needed.

3.3 Identify and track changes to replicate

Once a new replica is available and done with the initial sync then it doesn't make sense to sync the whole data again to other nodes. It is important to identify and this tracks a data changes in a source primary system. Along with the full replication, CDC(Change Data Capture) is the only way to make sure all the replicas are synced across hybrid environments.
Identifying a correct change to replicate on other nodes to minimize the network latency, storage usage. CDC(Change data Capture) is the only way to make sure, all other nodes are update with only change and not all the files. There are several ways to identify only changes by means of systems logs, transaction logs or using events to identify the exact changes.
The proposed algorithm of FDRS as below:

Input: Initial data set (D), Number of nodes or locations (N), Data access patterns (A)
Output: Optimally replicated data set (D') and response to each requests
Step 1: Instantiate replication
   Replicate the data set (D) across the specified number of nodes or locations (N).
Step 2: Data access monitoring
   Monitor the access patterns (A) of the replicated data set.
Step 3: Replica Creation
   For each missing file on the new Site. Locate the site with the less storage for storing a replica of file.
Step 4: Replica Management
   Based on the data access patterns, adapt the replication scheme by moving copies of the data to the nodes or locations that are receiving the most requests.

Step 5: Continuous monitoring

Continuously monitor the access patterns and adapt the replication scheme as necessary to ensure optimal performance and availability.

Step 6: Optimal Replication

Continuously monitor the numeral replicas, their location and network latency between replica and client, to make sure optimal replication is maintained. Return the optimally replicated data set (D').

4 EXPERIMENTS AND COMPARISON

We will be discussing the experiment and performance or result of the algorithm in this section:

4.1 CloudSim

CloudSim offers an adaptable imitation policy that can be used to analysis of the operation of different cloud computing architectures and configurations. It can be used to analyze the trade-offs between different design choices, such as the number and types of servers, the allocation of VMs to servers, and the provisioning of storage and networking resources.

To use CloudSim, users must first define the simulation scenario, including the characteristics of datacenters, servers, VMs, and other resources. They can then specify the workloads that will be work on the simulators, such as the number and type of VMs that be created, and the workloads that will be run on these VMs. Once the simulation scenario has been defined, users can run the simulation and analyze the results.

CloudSim can be used in a variety of contexts, including research and development, education, and system design. It is an open source toolkit that is available for free download and can be extended and customized by users to meet their specific needs. Developer can do:

- Create a variety of a workloads requests distributions and applications configurations.
- Model different cloud scenarios on custom settings.
- Execute pattern approaches for provision the applications in clouds.

4.2 CloudSim configurations

Users provide basic information about the virtual machine, including it's the processing speed, RAM size, the image size, the bandwidth, and number of CPUs. There are also properties of the cloudlet, such as it's the length, the output size and the file size. CloudSim uses this information to set up the datacenter's, brokers, and virtual machines. To run the simulation, the users code calls the beginSimulation() method. This method then calls several other methods in sequence, including begin(), startTimer(), and halt(). These methods process the entered cloudlets using a First Come, First Serve (FCFS) scheduling policy.

For this simulation, configured 16 datacenters with a geographically distributed topology as shown in Figure 2. These datacenters are signified by 100 virtual machines, each of which has 32MB of RAM, 512 MBPS processing power, 512MB of bandwidth, and a single processor. The cloud storage environment contains thousands of different data files with sizes ranging from 0.01 to 1 GB. Made sure that the service providers processing requirements were larger than 10 MB by distributing 10,000 jobs to them. The task file sizes and output file sizes were likewise assigned using a random differential function, and each distinct task has a randomly assigned requirement for one or two data files. The VMSchedulerTimeShared policy is used by the host to distribute query requests to each virtual machine. Each data file initially has one replica that is distributed at random.

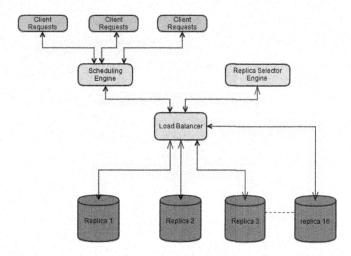

Figure 2.   Data center diagram.

5   RESULTS AND ANALYSIS

As discussed previously, the Criteria used for evaluation used are response time and storage capacity utilization

5.1   *Average response time*

The response time considerably increases as number of tasks grows, especially when that number approaches 10,000 in Figure 3. The round robin and Random schedule algorithm shown increase in response time while FDRS algorithm shows less response time as part of this The files which are most likely accessed are stored on nodes.

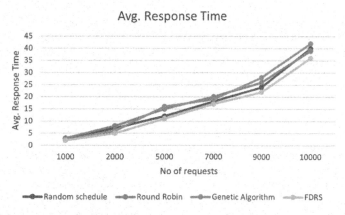

Figure 3.   Average response time comparison.

Measured the average response time for various tasks sizes ranging from 1000 to 10000 tasks and for each dataset the FDRS shown comparably less average response time (seconds) than other compared algorithms.

The proposed replication strategy, FDRS, reduces data movement by strategically placing datasets with dependencies in flexible locations. This results in an improvement in response time and maintains a stable level of response time over only a brief time. This is because FDRS increases the number of the local accesses by locating clones in the ideal location and minimizing pointless replication. As a result, all numbers of replications and the remote accesses is reduced, leading to an increase in the hit ratio.

5.2 *Storage resources*

To assess how storage resources are being used at locations, it is possible to monitor the proportion of storage that was used during the simulation.

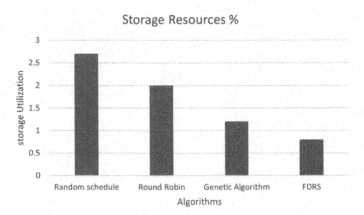

Figure 4.   Storage resource utilization comparison.

This might be helpful in examining a strategy from many angles: on the one hand, the goal can be to increase storage use if the cost of storage is fixed; on the other hand, the goal might be to limit storage usage if the cost of storage is proportional to the amount being utilized.

By dynamically creating copies in advance with FDRS method, files can be kept in a single location rather than being spread across several locations. Which can reduce storage usage. This suggests that FDRS produces fewer replicas, leading to better performance compared to other algorithms in Figure 4.

The study showed that FDRS was using less storage utilization than the other algorithms compared in various experiments. The average storage usage numbers shows the same for each different algorithm.

6   CONCLUSION

In conclusion, replication is an important aspect of cloud computing, because it enhances the performance, dependability, and accessibility of data and applications. There are various replication strategies that can be used, each with its own trade-offs and benefits. In this study, we have evaluated Random schedule Round Robin, genetic algorithms and FDRS. Average response time, and storage capacity utilization are strong points for FDRS. But it's crucial to remember that the appropriate replication strategy dependent upon specific needs and requirements within cloud computing system. Factors for instance data access patterns, network bandwidth, and storage replication capacity should be considered when selecting a replication strategy.

In this experiment, there are several potential directions for future studies using CDC (change data capture) on FDRS to identify and track the changes to replicate. This could involve optimizing the algorithms used to detect and replicate data changes, or finding ways to minimize the impact of CDC on system performance. Another area of focus could be on developing strategies for efficiently replicating large data sets using CDC on FDRS. This could involve exploring technique such as data compression, or developing ways to divide large data sets into smaller chunks for more efficient replication.

REFERENCES

[1] Aa Fu X, Zhou C. Virtual Machine Selection and Placement for Dynamic Consolidation in Cloud Computing Environment. *Frontiers of Computer Science*, 2015, 9(2): 322–330

[2] A Mi H B, Wang H M, Zhou Y F, Rung Tsong Lyu M, Cai H, Yin G. An Online Service Oriented Performance Profiling Tool for Cloud Computing Systems. *Frontiers of Computer Science*, 2013, 7(3): 431–445

[3] Chen T, Bahsoon R, Tawil A R. Scalable Service Oriented Replication with Flexible Consistency Guarantee in the Cloud. *Information Sciences*, 2014, 264: 349–370

[4] Wu H, Zhang W B, Zhang J H, Wei J, Huang T. A Benefit Aware Ondemand Provisioning Approach for Multi-tier Applications in Cloud Computing. *Frontiers of Computer Science*, 2013, 7(4): 459–474

[5] Al-Fares M, Loukissas A, Vahdat A. A Scalable, Commodity Data Center Network Architecture. *Computer Communication Review*, 2008, 38:63–74

[6] Amazon-S3. *Amazon Simple Storage Service (Amazon s3)*. http://www.amazon.com/s, 2009 Ghemawat S, Gobioff H, Leung S. The Google file system. In: *Proceedings of the 19th ACM Symposium on Operating Systems Principles*. 2003

[7] Calheiros R N, Ranjan R, Beloglazov A, Rose C, Buyya R. CloudSim: A Toolkit for Modeling and Simulation of Cloud Computing Environments and Evaluation of Resource Provisioning Algorithms. *Software: Practice and Experience*, 2011, 41(1): 23–50

[8] Qiu L L, Padmanabhan V N, Voelker G M. On the Placement of Web Server Replicas. In: *Proceedings of the 20th Annual Joint Conference of the IEEE Computer and Communications Societies*. 2001, 1587–1596

[9] Ramezani, F., Lu, J., Taheri, J., Hussain, F.K., 2015. Evolutionary Algorithm Based Multi Objective Task Scheduling Optimization Model in Cloud Environments. *World Wide Web* 18, 1737–1757

[10] Mansouri Y., Toosi A. N., and Buyya R., "Cost Optimization for Dynamic Replication and Migration of Data in Cloud Data Centers," *IEEE Trans. Cloud Comput.*, vol. 7, no. 3, pp.705–718, Sep. 2019.

[11] Hsu T.Y., Kshemkalyani A. D., and Shen M., "Causal Consistency Algorithms for Partially Replicated and Fully Replicated Systems," *Future Gener. Comput. Syst.*, vol. 86, pp. 1118–1133, Sep. 2018 Atrey A, Van Seghbroeck G, Mora H, Volckaert B, De Turck F. UnifyDR: A Generic Framework for Unifying Data and Replica Placement. *IEEE Access* 2020;8(1):216894–910

[12] Mehra, P. S., Mehra, Y. B., Dagur, A., Dwivedi, A. K., Doja, M. N., & Jamshed, A. (2021). COVID-19 Suspected Person Detection and Identification Using Thermal Imaging-Based Closed Circuit Television Camera and Tracking Using Drone in Internet of Things. *International Journal of Computer Applications in Technology*, 66(3–4), 340–349.

[13] Kumar, A., & Alam, B. (2016). Real-Time Fault Tolerance Task Scheduling Algorithm with Minimum Energy Consumption. In *Proceedings of the Second International Conference on Computer and Communication Technologies: IC3T 2015*, Volume 2 (pp. 441–448). Springer India.

[14] Dagur, A., Malik, N., Tyagi, P., Verma, R., Sharma, R., & Chaturvedi, R. (2021). Energy Enhancement of WSN Using Fuzzy C-means Clustering Algorithm. In *Data Intelligence and Cognitive Informatics: Proceedings of ICDICI 2020* (pp. 315–323). Springer Singapore.

[15] Dagur, A., Kaushik, A., Rastogi, A., Singh, A., Kumar, A., & Chaturvedi, R. (2021). Optimization of Queries in Database of Cloud Computing. In *Data Intelligence and Cognitive Informatics: Proceedings of ICDICI 2020* (pp. 325–332). Springer Singapore.

[16] Kushwaha, A., Amjad, M., & Kumar, A. (2019). Dynamic Load Balancing ant Colony Optimization (DLBACO) Algorithm for Task Scheduling in Cloud Environment. *Int J Innov Technol Explor Eng*, 8 (12), 939–946.

[17] Mokadem, R., Hameurlain, A., 2020. A Data Replication Strategy with Tenant Performance and Providereconomic Profit Guarantees in Cloud Data Centers. *Journal of Systems and Software* 159, 110447.

[18] George S, Edwin EB. A Review on Data Replication Strategy in Cloud Computing. *IEEE Int Conf Comput Intell Comput Res ICCIC* 2018;1–4

Trends in cloud computing and bigdata analytics

Artificial Intelligence, Blockchain, Computing and Security – Dagur et al. (Eds)
© 2024 The Author(s), ISBN: 978-1-032-49393-0

An innovative technique for management of personal data using intelligence

A. Sardana & A. Moral
Apex Institute of Technology (CSE), Chandigarh University, Gharuan, Mohali, Punjab, India

S. Gupta
Chitkara University Institute of Engineering and Technology, Chitkara University, Punjab, India

V. Kiran
Ch Devi Lal State Institute of Engineering and Technology, Sirsa, Haryana, India

ABSTRACT: Cloud computing provides faster and better services to organizations of both type's public and private at low cost. Despite many benefits, challenges in this domain are security and privacy of data on cloud. The content analysis revealed that cloud security and privacy inherit most of its challenges from cyber security; however, it also introduces several new challenges around trust, legal ethics, and data interoperability. This paper talks about the security of the data on the cloud and privacy which get in the way of making it ideal to use. We have proposed a new cloud computing environment based on double locking principle on a narrower level in an education sector. A case study is disclosed in this paper to further support the research.

1 INTRODUCTION

Cloud Computing [6] has emerged as a subject for study and used wisely and effectively in these recent years. Cloud Computing is a service-oriented platform that provides users numerous ways to store data, share data and back up data. For some users, cloud is a hub (storage house) of computing resources and for others cloud is a path to access software and data. This advantage itself poses a risk to data. During usage of the various clouds, the data can be personal like medical history, bank details. In order to avoid risks to the data, it is must to protect it. Further, load balancing [1, 4, and 2] can be adopted to suppress the load among various virtual machines in a cloud environment.

In such cases, a high-security lock is important on data to keep it safe, by applying passwords to records of the users that want to have their data on cloud and making it private even on the public cloud. This paper discloses data security techniques by using passwords for protecting and securing data in the cloud. It discloses the emerging threats to the data in the cloud and the ways by which it can be secured in a better way.

The General Data Protection Regulation (GDPR) [8, 9, and 14] is the difficult privacy [13] and security law in the world for implementation. Though it was written and passed by the European Union (EU), it imposes boundaries onto organizations anywhere in the world, so long as organizations target or collect data related to people in the EU. The regulation is effective from May 25, 2018. The GDPR will levy harsh fines against those who violate its privacy and security standards, with penalties reaching into the tens of millions of Euros. If this law is so rigid there might be several reasons behind it. Personal data [12] as the name suggest it is personal. Nobody can claim on anybody's personal data. Personal data is

DOI: 10.1201/9781003393580-127

property of that particular person whose data is that. Law is one thing that brings fear factor. Other factor is implementing security [3, 5, and 11] policies through software/application/logical methods. These security policies are must in Internet of things (IOT) [10] environment and for handling unstructured data like big data [7]. First factor is to be followed when personal data is utilized for creating some insight, report, survey analysis, and consulting data. If this kind of data is to be used then taking permission is a must. Personal data can be related to a person or any organization. Tracking whether any personal data of any person is used somewhere or not is a toughest task as compared to tracking whether any personal data of any organization is used by anybody or not. Personal data of any person may be diseases of that particular person, bank account details, mobile number, web cookies information etc. On the other side, personal data of any organization may be financial data of current year, financial data of previous years, stock details, military data, etc. Organizational data is available on company website for public but it cannot be used in any consultancy report by third party for providing that data to someone else.

In this paper, an analysis is presented for education sector for securing the personal data of faculty, students, or staff by utilizing various techniques. In this paper we have categorized our work into different sections. First section is brief introduction of personal data. Next section discloses literature survey in this field. Third section provides glimpse of proposed work for dealing problems of education sector by considering case study. Additionally this section discloses implementation of the proposed work using some existing applications. The last section highlights the conclusion of the paper.

2 LITERATURE SURVEY

Shared sheets are a mechanism by which different remote users can enter data in the shared sheets. Their timings may be same or different. In the existing sheets, security is ensured but privacy [13] is still a concern as per GDPR requirements. Specifically, a creator of the sheet can invite contributors to edit the sheet. Only those specific contributors can participate and enter data in that particular sheet. That is also possible when they will login to their specific mail account. After successful authentication only, they can enter or edit the shared sheet. Here, problem arises when protection is needed from each contributor so that personal data such as banking data, military data, and marks related data is not leaked by any means. A survey was conducted among students to identify a count who faced this problem. From that survey, it is analyzed that out of 476 students, 398 students want to protect their personal data. Leave behind this survey analysis, European law itself demand protection of personal data.

3 PROPOSED WORK

In this paper we are proposing methodology which is based on password only but it is beneficial for persons for protecting their personal data from third party without their permission. Specifically, a case study is discussed related to a problem in the education sector. Earlier was the time when offline work was in limelight but since March 2019 (COVID) online work or hybrid work is the need of the time. In education sector, coping up with these two modes was a problem in beginning for students and faculty both but gradually everybody coped up by learning from their mistakes. In this case study we are going to discuss one problem with proposed solution.

In this case study, contributors are faculty/staff member and students. Contributor 1 creates a shared sheet having different sheets academic details and personal details in the sheet named student data. In the academic sheet, marks are to be inserted in the specific columns for different semesters. Here, issue was nobody wants that his or her academic details should be visible to other students except Contributor 1 and contributor 2 himself. As academic details including marks are personal data of any student. Some students are brilliant one, some are average students, and some are below average students. In short, only contributor 1 should be able to access contributor 2 data and no other contributor 2 should be able to access contributor 2 data.

This is good as per GDPR policy as well as contributor 2 point of view. The contributor 1 has all the permissions by default in shared sheet like read, write, delete and edit. On the other side, contributor 2 has permissions which are assigned by contributor 1. Contributor 2 permissions are at least one of the following: write, read, delete, append. Based on these permissions contributor 2 is allowed to share his or her data with contributor 1.

Initial requirement for same is lock is to be for every contributor 2 by default. Specifically, if publication is on cloud then public cloud is more dangerous as there are chances of leakage of data like academic data. To overcome this problem, lock is to be opened by the contributor 1 only. This is one scenario. Other scenario is how lock can be implemented on only one contributor 2 from a list of contributors 2. We have proposed here methodology for protection. Name of the methodology is double locking scheme. This methodology is defined below with an example. In Figure 1 below, test data has been taken for the purpose of implementing proposed work. In this snapshot, data pertains to personal data of different users namely A to O. This presentation of data is open to everybody so, anyone can access anyone's personal data. To get rid of this problem we have proposed two way techniques.

Sr. No.	Roll no.	First name	Last Name	Gender	DOB	Domicile	Category	Blood Group	12th Marks	10th Marks	Scholar (Day Scholar / Hosteler)	Mess (Yes,
1	10	A	P	Male	14/03/2003	Haryana	General	A+	83.60%	87.80%	Day Scholar	-
2	20	B		Female	15/02/2003	Punjab	General	B+	92.20%	81.20%	HOSTELER	Yes
3	30	C	Q	male	10/01/2004	haryana	general	o+			hosteler	yes
4	40	D	R	Male	10/04/2003	Haryana	General	O+	79	77.4	hosteler	Yes
5	50	E		Male	25/08/2003	haryana	General	o+	80%	80%		
6	60	F	S	Male	08/05/2003	Chandigar	General	A+ve	93%	92.50%	Day Scholar	
7	70	G	T	Male	29/05/2004	Haryana	BC	O+	77%	73%	Day Scholar	
8	80	H	U	Male	08/05/2003	Chandigar	General	A+	80.2	82.8	Day Scholar	
9	90	I	V	Female	29/07/2003	Punjab	General	A+	80	97%	Hosteller	Yes
10	100	J	W	Female	13/09/2003	Himachal	General	A+	86.20%	85.20%	Hosteller	Yes
11	110	K	X	Female	26/08/2003	Haryana	General	Ab+ve	87.20%	84%	Hosteller	Yes
12	120	L	Y	male	02/07/2003	Punjab	General	B+	78.20%	61.20%	Day Scholar	
13	130	M	Z	Female	11/07/2003	Punjab	General	B+	71%	71%	Day Scholar	
14	140	N	A	male	07/01/2003	Haryana	general	A+	78.60%	77%	Hosteler	Yes
15	150	O	C	female	01/08/2003	rajasthan	general	ab-	78%	75%	hosteler	yes

Figure 1.   Snapshot 1.

Double Locking scheme is that scheme which protects personal data from leakage. In this methodology, first lock is when contributor 2 gets to see any mail from contributor 1. At that time, every row is locked (refer Figure 2 below). This is the first lock. Whenever contributor 2 will try to open the shared application like excel, lock will be open up by contributor 1. Contributor 1 will provide appropriate permission to the contributor 2 as soon as

Sr. No.	Roll no.	First name	Last Name	Gender	DOB	Domicile	Category	Blood Group	12th Marks	10th Marks	Scholar (Day Scholar / Hosteler)	Mess (Yes	First Lock
													LOCK 1
													LOCK 1
													LOCK 1
													LOCK 1
													LOCK 1
													.
													.
													.
													.
													.
													LOCK 1

Figure 2.   Snapshot 2.

contributor 2 will try to write anything in any cell of any row. Figure 3 below show open lock for particular contributor 2 by contributor 1.

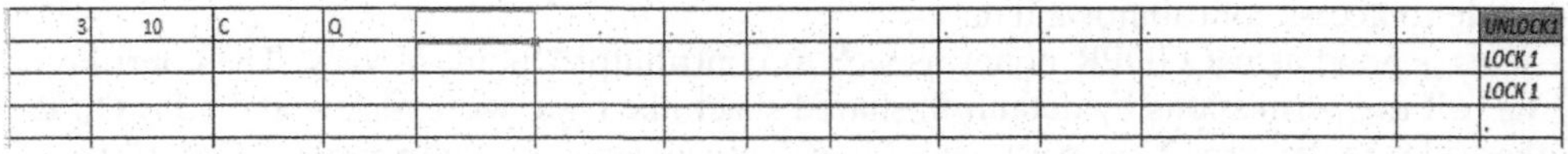

3	10	c		q														UNLOCK1
																		LOCK 1
																		LOCK 1

Figure 3. Snapshot 3.

In this scheme, second lock is that lock which is exclusively for that particular user who is entering the data in the shared sheet. This second lock is also set by the contributor 1 when contributor 2 is doing entry in any particular row. In this scheme, although there might be many contributors 2 who are simultaneously working in the shared sheet but these contributors 2 cannot see entry done by other contributor 2 so as to protect personal data of each contributor 2 from each other. Here, one issue comes many contributor 2 can enter data in same row of shared sheet. To resolve this issue, contributor 1 can hide those rows from the sheet which have first lock for contributor 2. By chance, if two contributors 2 are trying to access same row, then it is the responsibility of contributor 1 to provide access to them by way of notification in which row number is mentioned by the contributor 1 so that conflict between two contributors 2 can be broken. This way scheduling of rows is planned for contributors 2. There might be so many methods for scheduling like first come first serve or random basis using randomized function.

This double locking scheme solves two objectives. First, protection of personal data of various collaborators under the leadership of creator of shared sheet. Second, simultaneous storage of personal data with lock principle. Lock on a particular row can be a password or a biometric. If password is a way of protection then there are chances of password leakage. To overcome that the biometric is the best solution for protecting personal data from each other collaborators 2.

In this manner, protection can be enforced in a closed group of collaborators, whereby these collaborators are chosen by the creator and also, permissions to the collaborators are assigned by the creator. Shared excel sheet is a case study where double locking can be implemented for better protection assuming internet is of stable speed. Further, collectively, data can be stored on a cloud so that load can be shared among various virtual machines to handle unstructured/structured data. Although unstructured data is more in number still stress is put on structured data as unstructured data cannot exist without managing structured data. Personal data of related domain such as student data, banking data, purchase data, manufacturing data, energy usage data is utmost important to the connected persons. To protect this structured data, proposed scheme of double locking is of great use. Further as a modification structured data can be stored on cloud if millions of users are entering data in the shared sheet.

Personal data in education sector may also be related with energy usage per day by student or faculty so as to have clear understanding of energy consumption in a university of college. The energy consumption based upon individual data is personal and need to be secure. Like previous example, protection of energy data is possible using double locking scheme. Some of the reflection of energy data for June 14, 2022 is shown in Table 1.

After collecting data on the shared platform, there is a need to protect data and analyze that protected data. The analyzed data is shown in Figures 4 and 5 for column names S. No. and energy consumption per day and S. No. and previous month energy consumption by energy users in the campus.

Cognitive intelligence engineering is used to further provide the insights to the university professionals so as to have clear reflection of metric to display total energy consumption by

Table 1. Energy usage data.

S. No.	Name	Department	Energy Consumption/Day	Previous Month
1	Anjali (Faculty)	CSE	4 Hrs	100 Hrs
2	Arjun (Faculty)	ECE	5 Hrs	112 Hrs
3	Himnish (Staff)	CSE	7 Hrs	98 Hrs
4	Gaurav (Student)	ME	6 Hrs	87 Hrs

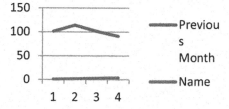

Figure 4. Analysis 1.

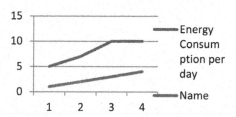

Figure 5. Analysis 2.

each category (Faculty, student, and staff) in a month. Further, insight may reflect in which category energy is wasting due to factors like idle time, fully charged laptop identification. One of the insights can be effective utilization which is shown in Figure 6 below.

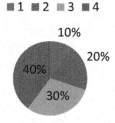

Figure 6. Insight 1-effective usage.

4 CONCLUSION

In this paper, proposed work is for the purpose of protecting personal data of users from each other. Specifically, the proposed technology is meant for saving space on local machines and saving on cloud is discussed. Application area of the proposed technology is when personal data is to be entered in an open source platform for the creator of the file. Other area can be banking, airlines, banking, sales, human resources, energy, and manu-facturing etc. These areas are specified for the purpose of managing structured data on unstructured platform.

REFERENCES

[1] Alankar Bhavya, Gaurav Sharma, Harleen Kaur, Raul Valverde, and Victor Chang 2020. "Experimental Setup for Investigating the Efficient Load Balancing Algorithms on Virtual Cloud." *Sensors* 20, no. 24: 7342.

[2]   Sambit Kumar Mishra, Bibhudatta Sahoo, and Priti ParamitaParida 2020. Load Balancing in Cloud Computing: A Big Picture. *Journal of King Saud University Computer and Information Sciences*, 32 (2):149–158.

[3]   Kumar, A., & Alam, B. (2015, February). Improved EDF Algorithm for Fault Tolerance with Energy Minimization. In *2015 IEEE International Conference on Computational Intelligence & Communication Technology* (pp. 370–374). IEEE.

[4]   Alam, B., & Kumar, A. (2014, March). A Real Time Scheduling Algorithm for Tolerating Single Transient Fault. In *2014 International Conference on Information Systems and Computer Networks (ISCON)* (pp. 11–14). IEEE.

[5]   Rishu Chhabra, Seema Verma, and C. Rama Krishna 2017. A Survey on Driver Behavior Detection Techniques for Intelligent Transportation Systems. *7th International Conference on Cloud Computing, Data Science Engineering-Confluence*, pages 36–41.

[6]   Vivek Manglani, Abhilasha Jain, and Vivek Prasad 2017. Task Scheduling in Cloud Computing. *International Journal of Advanced Research in Computer Science*, 8(3).

[7]   M. Cohen 2018, Big Data and Service Operations, *Production and Operations Management* 27(9) 1709–1723. DOI: https://doi.org/10.1111/poms.12832

[8]   M. Porcedda 2018, Patching the Patchwork: Appraising the EU Regulatory Framework on Cyber Security Breaches, *Computer Law & Security Review* 34(5) 1077–1098. DOI: https://doi.org/10.1016/j.clsr.2018.04.009

[9]   M. Phillips 2018, International Data-sharing Norms: from the OECD to the General Data Protection Regulation (GDPR), *Human Genetics* 137(8) 575–582. DOI: https://doi.org/10.1007/s00439-018-1919-7

[10] Z. Almusaylim, N. Zaman 2019, A Review on Smart Home Present State and Challenges: Linked to Context-awareness Internet of Things (IoT), *Wireless Networks* 25(6) 3193–3204. DOI: https://doi.org/10.1007/s11276-018-1712-5

[11] M. Ogonji, G. Okeyo, J. Wafula 2020, A Survey on Privacy and Security of Internet of Things. *Computer Science Review* 38 100312. DOI: https://doi.org/10.1016/j.cosrev.2020.100312

[12] D. Setiawati, H. Hakim, F. Yoga 2020, Optimizing Personal Data Protection in Indonesia: Lesson Learned from China, South Korea, and Singapore, *Indonesian Comparative Law Review* 2(2) 19–109. DOI: http://dx.doi.org/10.18196/iclr.2219

[13] Quach, S., Thaichon, P., Martin, K.D. *et al*, 2022. Digital Technologies: Tensions in Privacy and Data. *J. of the Acad. Mark. Sci.* 50, 1299–1323. https://doi.org/10.1007/s11747-022-00845-y

[14] Perumal, Vanessa 2022 "The Future of U.S. Data Privacy: Lessons from the GDPR and State Legislation," *Notre Dame Journal of International & Comparative Law*: Vol. 12 : Iss. 1 , Article 7

Artificial Intelligence, Blockchain, Computing and Security – Dagur et al. (Eds)
© 2024 The Author(s), ISBN: 978-1-032-49393-0

Crypt Cloud+: Cloud storage access control: Expressive and secure

Rajesh Bojapelli, Manikanta, Srinivasa Rao, Chinna Rao & R. Jagadeeswari
Bharath Institute of Science and Technology, Chennai, India

ABSTRACT: Secure cloud storage, a forthcoming cloud solution, is designed to protect the privacy of outsourced data while also granting cloud users with information that is not physically theirs flexible control over access. Policy-based encryption in text with attributes, or CP-ABE, is one of the best methods for protecting the service guarantee. However, utilizing CP-ABE could result in an inescapable security violation known as access credentials misuse (i.e., decryption rights), as it includes an intrinsic "all-or-nothing" decryption functionality. In this, we examine the two most typical scenarios in which access credentials have been misused: one on the side of the semi-trusted authority, and the other on the side of the cloud user. We propose Crypt Cloud, the first revocable cloud storage employing CP-ABE solution with accountable authority and white-box auditing, as a solution to the issue. To demonstrate the value of our solution, we additionally do a security study and conduct further testing.

1 INTRODUCTION

Cloud computing may unintentionally compromise users' privacy as well as the preservation of outsourced data privacy. It is particularly challenging to guarantee that the data that was outsourced to the cloud can only be accessed by authorized personnel, wherever they want and whenever they want [1]. Data can be encrypted before being sent to the cloud, but this is a cumbersome workaround. However, there is a limit to the volume of data that can be transmitted and processed through the system. This is because data owners must first download encrypted data from the cloud before exchanging it. They must re-encrypt the data after downloading it (assuming they don't already have a local copy) [2]. Exact access controls on encrypted data are essential in cloud computing. Data secrecy can be enabled and guaranteed via Cypher Text Attribute-Based Policy-Based Encryption (CPABE). As an illustration, in Organisations Users of CP-ABE-based cloud storage systems, like the University of Texas in San Antonio, might be both individuals and institutions. (Such as university employees, pupils, and visiting academics) can first set up access control based on the calibre of a potential cloud user [3]. Following that, cloud users are provided with approved Access credentials depending on the appropriate characteristic sets (such as student, professor, or visitor). (Decryption keys), which can be used to access data that has been outsourced [4]. It provides robust protection for cloud data. One-to-many encryption and exact access control are also made possible.

2 METHODOLOGIES

2.1 *System design*

Data Owners (DOs) encrypt data prior to sending it (encrypted) to a public cloud. according to the relevant access laws (PC). The encrypted data from Dos is kept up to date by PC,

DOI: 10.1201/9781003393580-128

which also responds to user requests for data access (DUs). Authorised DUs have access to the outsourced data and can download and decrypt it, for example [5]. A semi-trusted authority (AT) generates system settings and provides DUs with access credentials, also known as decryption keys. Other organisations have faith in the auditor (AU), who supervises the audit and revoke procedures and gives the YDOs and DUs the results of the audit and trace [6].

2.2 Organization profile creation & key generation

A simple registration method for users is available on the web end. Users provide their own confidential information for this process [7]. After that, the data is stored in the server's database. Now, the Accountable STA (semi-trusted Authority) generates decryption keys for the users depending on their Attributes Set (e.g., name, mail-id, contact number, etc.). The User has the authority to view Organisation data once they have obtained the decryption keys from the Accountable STA [8].

2.3 Data owners file upload

Data owners create accounts in this module's public cloud and upload their data there [9]. In advance of submitting their files to a public cloud For the RSA encryption algorithm, data owners will create a secret key and a public key, and use these to encrypt their data as shown in Figure 1. It also generates a unique file access authorisation key that staff members can use to access corporate information.

2.4 File permission & policy file creation

Different file authorization keys will be generated by different data owners for their files, and users inside the organization will receive those keys to access their files. Additionally, it creates policy files to control who has access to their data. The key for read, write, download, and delete files will each have an own policy file [10].

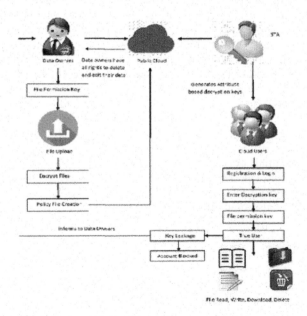

Figure 1.   PANet architecture.

2.5 Tracing who is guilty

Authorised DUs have access to the outsourced data and can read, write, download, delete, and decrypt it [11]. Depending on their experience and position, the organization's staff members are given file permission keys in this situation. The only individuals with access to the files are senior staff (read, w Users within the business will be granted access to unique file authorization keys that various data owners have generated for their files. It also creates policy files to restrict who has access to their data. The read, write, download, and delete file-related keys each have their own policy file [10]. remove, download, and edit). Only viewing the files is allowed for new students. Reading and writing are permitted for some Workers. However, certain employees have the authority to remove data [12]. If any Senior Employee provides their Junior Employees with access to their Secret Permission Keys or discloses them to others, the Data Owners' Data will be requested to be retrieved or erased [13].

To ensure that the user has complete access credentials to the data, the system will generate an attribute set for their role upon key entry [14]. If the characteristics set does not match the data owners' policy files, they will be held responsible. Asking the less experienced employees will allow us to learn who gave them the key [15].

3 RESULTS AND DISCUSSION

3.1 System analysis

3.1.1 Existing system
The CP-ABE might assist us in preventing foreign attackers from getting past our defences. In the existing framework. However, how could we unambiguously demonstrate that a company insider is responsible for the "crimes" of disseminating access to decryption keys and exchanging user data in plain-text for illicit financial benefit [16]? Could you also revoke the stolen access credentials for us? In addition to the questions discussed above, we also have one that relates to key generation authority [17]. The access credentials for a cloud user (i.e., the decryption key) are often released by a semi-reliable authority and may change based on the properties the person possesses.

3.1.2 Proposed system
In order to solve the issue of credential leakage in cloud storage systems built on CP-ABE, this study created Crypt Cloud+, a white-box traceable, auditable, accountable authority, and revocable cloud [18]. This is the first cloud storage system. System based on CP-ABE that simultaneously offers auditing, responsible authority, white-box traceability, and effective revocation. In particular, Crypt Cloud+ gives us the ability to track down and stop rogue cloud users (who leak credentials).

3.2 Project scope

To solve the issue of credential leakage in CP-ABE-based cloud storage systems, this study developed Crypt Cloud+. White-box audits and traceability are made possible by Crypt Cloud+, a responsible, authoritative, and transient cloud [19]. First to provide successful revocation, auditing, responsible authority, white-box traceability, and responsible authority all at once is the aforementioned CP-ABE-based cloud storage solution. We are able to locate and stop malicious cloud users (leakers of credentials) in particular thanks to Crypt Cloud+. Our method continues to work when a source of user credentials is spread [20].

3.3 Product perspective

Owners of the data will keep it in a combination of the public cloud, encryption, and a particular set of access control mechanisms. Data can be downloaded and deleted at any time by the owners of cloud computing services. They will give a public cloud some attribute set before transferring their data there [21]. Enter that particular attribute set when doing additional operations on the data owned by the data owner. if any authorized cloud user want access to their data. To access the data of the data owner, a cloud user wants to register. Their cloud computing company information [22]. In order to access the owner's data, users must supply their information as well as their designation as attributes based on that information; a semi-trusted authority generates the decryption keys. The user can perform a variety of operations on cloud data. The user must specify read-related qualities if he wants to read cloud data and write-related qualities if he wants to write cloud data [23]. Each and every action a user takes within an organisation confirms their unique attribute set. These characteristics would be supplied to the authorised users in a cloud organisation by the administrators. The cloud-based policy files will retain these properties.

3.4 System features

1) The ability to track out harmful cloud users – It is possible to track down and identify users who reveal their login credentials.
2) Accountable power – It is possible to identify a semi-trusted authority that generates and distributes access credentials to unauthorized users without the required authorization. This enables additional what needs to be done (e.g. criminal investigation or civil litigation for damages and breach of contract).
3) Accounting – If a cloud user is (suspected to be) guilty of disclosing his or her login credentials, the auditor can find out.
4) Tracing – requires "almost" no storage. In order to track down hostile we employ a Paillier-like encryption for cloud users. Realistically, we don't need to maintain a table of user identities to track them (unlike the approach used in).
5) Revocation of malicious cloud users – It is possible to cancel access privileges for people who have been identified as "compromised" after being tracked down. To effectively revoke the "traitor(s)," we develop two techniques. While the ATIRCP-ABE gives an implicit revocation mechanism, the ATER-CP-ABE offers an expressly revocation method where a revocation list is explicitly provided into the encryption algorithm Encrypt. Is required to perform periodic key update operations. Owners of the data will keep it encrypted and with a certain set of access control measures on a public cloud. They'll give their data a certain set of attributes. Before uploading it to a public cloud. Enter that specific attribute set to carry out additional operations on the data of the data owner if any permitted cloud user wants to access their info. A cloud user wants to register their details with the cloud organization in order to access the data of the data owner. Users wish to provide their information together with their designation as attributes. To gain access to the owner's data, a Semi-Trusted Authority generates decryption keys based on the user details. Numerous procedures can be carried out on cloud data by the user.

4 CONCLUSIONS

This study developed Crypt Cloud+, and to fix the CP-ABE-based system's issue with credential leakage cloud storage systems, we propose White-box traceability and auditing are provided by responsible authority and revocable cloud. The initial is this CP-ABE-based cloud storage system that provide auditing, responsible authority, white-box traceability, and successful revocation all at once. Particularly, Crypt Cloud+ enables us to recognize and thwart malicious cloud users (leaking credentials). Our approach works just as well when a semi-trusted authority redistributes user credentials. We draw attention to the possibility that a better idea than white-box traceability is black-box traceability might be required for Crypt Cloud. Our forthcoming initiatives include one that will examine black-box audits and traceability.

REFERENCES

[1] Mazhar Ali, Revathi Dhamotharan, Eraj Khan, Samee U. Khan, Athanasios V. Vasilakos, Keqin Li, and Albert Y. Zomaya. Sedasc: Secure Data Sharing in Clouds. *IEEE Systems Journal*, 11(2):395–404, 2017.

[2] Mazhar Ali, Samee U. Khan, and Athanasios V. Vasilakos. Security in Cloud Computing: Opportunities and Challenges. *Inf. Sci.*, 305:357–383, 2015.

[3] Michael Armbrust, Armando Fox, Rean Griffith, Anthony D Joseph, Randy Katz, Andy Konwinski, Gunho Lee, David Patterson, Ariel Rabkin, Ion Stoica, *et al.* A View of Cloud Computing. *Communications of the ACM*, 53(4):50–58, 2010.

[4] Nuttapong Attrapadung and Hideki Imai. Attribute-based Encryption Supporting Direct/indirect Revocation Modes. In *Cryptography and Coding*, pages 278–300. Springer, 2009.

[5] Amos Beimel. *Secure Schemes for Secret Sharing and Key Distribution*. PhD thesis, PhD thesis, Israel Institute of Technology, Technion, Haifa, Israel, 1996.

[6] Mihir Bellare and Oded Goldreich. On Defining Proofs of Knowledge. In *Advances in Cryptology-CRYPTO'92*, pages 390–420. Springer, 1993.

[7] Dan Boneh and Xavier Boyen. Short Signatures Without Random Oracles. In *Eurocrypt – 2004*, pages 56–73, 2004.

[8] Hongming Cai, Boyi Xu, Lihong Jiang, and Athanasios V. Vasilakos. Iot-based Big Data Storage Systems in Cloud Computing: Perspectives and Challenges. *IEEE Internet of Things Journal*, 4(1):75–87, 2017.

[9] Jie Chen, Romain Gay, and Hoeteck Wee. Improved Dual System ABE in Prime-order Groups Via Predicate Encodings. In *Advances in Cryptology* – EUROCRYPT 2015, pages 595–624, 2015.

[10] Angelo De Caro and Vincenzo Iovino. Jpbc: Java Pairing Based Cryptography. In *ISCC 2011*, pages 850–855. IEEE, 2011.

[11] Hua Deng, Qianhong Wu, Bo Qin, Jian Mao, Xiao Liu, Lei Zhang, and Wenchang Shi. Who is Touching My Cloud. In *Computer Security-ESORICS 2014*, pages 362–379. Springer, 2014.

[12] Zhangjie Fu, Fengxiao Huang, Xingming Sun, Athanasios Vasilakos, and Ching-Nung Yang. Enabling Semantic Search Based on Conceptual Graphs Over Encrypted Outsourced Data. *IEEE Transactions on Services Computing*, 2016.

[13] Vipul Goyal. Reducing Trust in the PKG in Identity Based Cryptosystems. In *Advances in Cryptology-CRYPTO 2007*, pages 430–447. Springer, 2007.

[14] Vipul Goyal, Steve Lu, Amit Sahai, and Brent Waters. Black-box Accountable Authority Identity-based Encryption. In *Proceedings of the 15th ACM Conference on Computer and Communications Security*, pages 427–436. ACM, 2008.

[15] Vipul Goyal, Omkant Pandey, Amit Sahai, and Brent Waters. Attribute-based Encryption for Fine-grained Access Control of Encrypted Data. In *Proceedings of the 13th ACM Conference on Computer and Communications Security*, pages 89–98. ACM, 2006.

[16] Qi Jing, Athanasios V. Vasilakos, Jiafu Wan, Jingwei Lu, and Dechao Qiu. Security of the Internet of Things: Perspectives and Challenges. *Wireless Networks*, 20(8):2481–2501, 2014.

[17] Allison Lewko. Tools for Simulating Features of Composite Order Bilinear Groups in the Prime Order Setting. In *Advances in Cryptology–EUROCRYPT 2012*, pages 318–335. Springer, 2012.

[18] Allison Lewko, Tatsuaki Okamoto, Amit Sahai, Katsuyuki Takashima, and Brent Waters. Fully Secure Functional Encryption: Attribute-based Encryption and (Hierarchical) Inner Product Encryption. In *Advances in Cryptology–EUROCRYPT 2010*, pages 62–91. Springer, 2010.

[19] Allison Lewko and Brent Waters. New Proof Methods for Attribute-based Encryption: Achieving Full Security Through Selective Techniques. In *Advances in Cryptology–CRYPTO 2012*, pages 180–198. Springer, 2012.

[20] Jiguo Li, Xiaonan Lin, Yichen Zhang, and Jinguang Han. KSFOABE: Outsourced Attribute-Based Encryption with Keyword Search Function for Cloud Dtorage. *IEEE Trans. Services Computing*, 10 (5):715–725, 2017.

[21] Sood, K., Dhanaraj, R. K., Balusamy, B., & Kadry, S. (Eds.) (2022). *Blockchain Technology in Corporate Governance* Wiley. https://doi.org/10.1002/9781119865247

[22] Janarthanan, S., Ganesh Kumar, T., Janakiraman, S., Dhanaraj, R. K., & Shah, M. A. (2022). An Efficient Multispectral Image Classification and Optimization Using Remote Sensing Data. In S. Bhattacharya (Ed.), *Journal of Sensors* (Vol. 2022, pp. 1–11). Hindawi Limited. https://doi.org/10.1155/2022/2004716

[23] Rajendran, S., Sabharwal, M., Ghinea, G., Dhanaraj, R. K., & Balusamy, B. (2022). *IoT and Big Data Analytics for Smart Cities*. Chapman and Hall/CRC. https://doi.org/10.1201/9781003217404

Artificial Intelligence, Blockchain, Computing and Security – Dagur et al. (Eds)
© 2024 The Author(s), ISBN: 978-1-032-49393-0

Application of MCDM methods in cloud computing: A literature review

A. Kumar
KIET Group of Institutions, Delhi-NCR, Ghaziabad, India

A.K. Singh
Adani University, Ahmedabad, India

A. Garg
University of Engineering & Technology (UETR), Roorkee, India

ABSTRACT: Cloud computing has become one of the most prominent technologies due to recent developments in information technology. Because there are so many cloud service providers (CSPs), it might be difficult for users to choose a CSP that suits all of their needs. This study's aim is to review the application of multi-criteria decision-making (MCDM) for selecting the best CSP. The contribution of this study is many-fold. First, this study reviewed various conferences and journal papers related to the application of MCDM in the area of cloud computing between 2012–2023. Second, a brief description of MCDM methods used in cloud computing is provided, along with a description of other popular MCDM methods that can also be used for CSP selection. Third, this study also highlights future research guidelines for researchers who want to do research to address the problem of choosing the best service provider in cloud computing. The review done in this study will aid the users in choosing cloud service providers. Moreover, this review will help research scholars who want to work to address decision-making problems in cloud computing.

1 INTRODUCTION

Cloud computing has shown to be an effective method of delivering information technology services over the past few years. It offers it's consumers flexible and extensive IT resources via the Internet. The NIST (National Institute of Standards and Technology) describes it as a model that enables simple, quick, and low-effort use of on-demand services from a computing resources' shared pool. Additionally, it offers a platform for the deployment and creation of applications. Various CSPs such as windows, Microsoft, and amazon offer users many services through cloud infrastructure.

Users have a wide range of alternatives for choosing the appropriate cloud service that satisfies their quality-of-service needs because many CSPs provide comparable services. The difficult part is deciding which CSP will best serve users' needs and goals. MCDM approaches are employed to identify the optimal alternative by weighing several criteria and options before making a final choice. In previous studies (Kumar 2016) MCDM has been used for making decisions in various fields. Similarly, MCDM can play a significant role in selecting the best CSPs considering various evaluation criteria. MCDM is a tool to deal with decision-making problems in a multi-criteria environment.

This study reviewed research papers published between 2012 and 2023 on the application of MCDM methods in choosing the most suitable cloud service provider. The contribution of this study is summarized as follows:

DOI: 10.1201/9781003393580-129

- First, this study formulates the two research questions (RQ1 and RQ2). RQ1: what are the most common MCDM methods used in cloud computing? RQ2: What are the different application areas in the cloud computing environment for which MCDM has been applied to choose the best service provider?
- Second, to answer both research questions, this study reviewed various research papers on the use of MCDM methods for selecting the best CSP for multiple applications.
- Third, A brief description of the MCDM methods used in previous studies in the area of cloud computing and other popular MCDM methods that can also be used in decision-making in cloud computing.
- Fourth, this study also presents some future guidelines for using MCDM methods to solve decision-making problems in cloud computing.

The organization of the rest study in this paper is as follows: section 2 describes the methodology, a review study is presented in section 3, section 4 describes the MCDM methods, section 5 highlights future research directions, and finally, section 6 concludes the paper.

2 METHODOLOGY

This section discusses the methodology adopted for reviewing the papers related to the applications of MCDM methods in cloud computing. Different databases, including google scholar, web of science, and Scopus, were searched for the research articles. The keywords such as MCDM + cloud computing, MCDM + cloud service provider, and MCDM + evaluation + cloud computing were used to search the related articles. The review was limited to the last decade, from 2012 to 2023. Next, the work done in the related papers is summarized into a tabular form having the columns- author name, publication year, MCDM method used, criteria, and the application area. Next, MCDM methods are explained with their advantages and disadvantages. Next, future research guidelines for the MCDM's use in cloud computing are presented. The above-stated methodology is indicated in Figure 1.

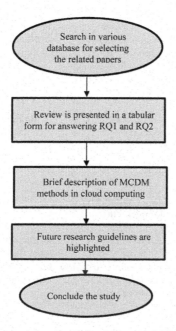

Figure 1. An overview of the methodology used for review on MCDM methods in cloud computing.

3 REVIEW OF MCDM METHODS IN CLOUD COMPUTING

This study reviewed the research papers on the application of MCDM methods in cloud computing from 2012 to 2023. Table 1 summarizes the review listing the authors, MCDM methods used, evaluation criteria, and application areas. From Table 1, it can be observed that AHP, TOPSIS, and VIKOR are the commonly used MCDM methods in cloud computing (Answer of RQ1). In Table 1, the column application area provides the answer to RQ2.

Table 1. A summarized review study on the application of MCDM in cloud computing.

Authors	MCDM Methods	Evaluation Criteria	Application Area
(Huang et al. 2012)	DEMATEL ANP	1. Organizational change 2. Scaling quickly 3. Scalable storage 4. Service migration 5. Continuity of service 6. Security and Privacy	Evaluation of telecommunication service quality.
(Tomar et al. 2012)	AHP TOPSIS	1. Cost 2. Throughput 3. Reliability 4. Availability 5. Response time	Selection of high CPU configured cloud service.
(Lee et al. 2014)	AHP TOPSIS	1. Storage size 2. Storage usage 3. RAM size 4. Clock speed	Reallocation of virtual machine.
(Lee and Seo 2016)	Fuzzy AHP Fuzzy TOPSIS	1. Usability 2. Security 3. Cost 4. Assurance 5. Agility 6. Accountability	Selection of cloud products.
(Wibowo et al. 2016)	TOPSIS	1. Adaptability 2. Scalability 3. Usability 4. Service Accessibility 5. Performance 6. Security	Evaluating cloud service for e-learning content.
(Li et al. 2018)	ANP TOPSIS	1. Coordination capability 2. Resources input capability 3. Product output capacity 4. R & D capability 5. Management capability	Evaluation of four manufacturing services.
(Jatoth 2019)	AHP Grey TOPSIS	1. Performance of memory 2. Consistency of I/O operations 3. Disc storage	Evaluation of seven cloud service providers. Rackspace, HP, Google, City Cloud, CenturyLink Azure, Amazon

(continued)

Table 1. Continued

Authors	MCDM Methods	Evaluation Criteria	Application Area
		4. Processing performance 5. Memory 6. Virtual core 7. Cost	
(Youssef 2020)	BWM TOPSIS	1. Reliability 2. Device response time 3. Maintainability 4. Cost 5. Security management 6. Interoperability 7. Usability 8. Sustainability Scalability	Ranking of eight cloud service providers. Linode, Joynet, Rackspace, Azure GoGrid, Google Amazon
(Naveed et al. 2021)	AHP Fuzzy AHP	1. Environmental impact 2. Innovational ideas 3. Technological advancement 4. Cloud ERP essentials 5. Organizational behavior	A composite ranking of five-dimensional criteria for resource planning of cloud enterprise.
(Saha et al. 2021)	ANP VIKOR	1. Cost 2. Storage 3. Serviceability 4. Stability 5. Availability 6. Performance 7. Security 8. Agility 9. Accountability	Evaluation of three cloud service providers. Rackspace Windows Azure Amazon EC2
(Zhang & Chen 2022)	TOPSIS	1. Social adaptability 2. Economical adaptability 3. Environmental adaptability 4. Technical adaptability	Evaluation of five tunneling methods. Soft rack TBM Shield TBM Mine tunneling method New Austrian tunneling method Hard rack TBM
(Kamanga et al. 2023)	MCDM-based three-variant algorithm.	1. Makespan 2. Cost	Scheduling of three real workflows-Epigenomics, Inspiral, and Montage.

4 BRIEF DESCRIPTION OF MCDM METHODS

MCDM is a method that helps in various decision-making problems in a multi-criteria environment. Here multi-criteria environment means the presence of more than one performance measure for selecting the alternative (Best) among the different available alternatives to solve a particular problem. A variety of MCDM methods have been proposed in previous studies. Every MCDM method has advantages and some disadvantages. Table 2 describes MCDM methods used in cloud computing along with other popular MCDM methods with their advantages and disadvantages.

Table 2.   Brief description of MCDM methods.

MCDM method	Brief description	Advantages	Disadvantages
AHP	Works for the pairwise comparison of criteria.	Simple model, consistency check	Use probability measures, not suitable for complex problems.
ANP	Use network structure for finding weights of criteria.	Suitable for complex problems, consistency check.	More reliable on opinion of experts, not suitable for large number of criteria.
TOPSIS	Works on the distance of alternatives from the ideal solutions.	Widely used method, use full information.	Absence of consistency check.
BWM	Works on the concept of best vs all and worst vs all criteria.	Easy to apply, require less comparison data.	Limited comparison points scale (9 point).
DEMATEL	Works on the concept of influence matrix.	Effective analysis of the mutual interaction between criteria.	Consider independent relationship among between alternatives.
MAUT	Use the concept of utility theory for the ranking of alternatives.	It takes into consideration any variations in the criteria.	The results of the decision criteria are somewhat uncertain.
PROMETHEE	Works on the concept of a preference function.	Useful for complex problems and for a large number of criteria.	Dependent on experts to assign the weight for criteria.
VIKOR	Works on the concept of minimum individual regret and maximum group utility. Provides a compromising solution based on VIKOR index value Q.	Widely used MCDM method in recent studies. Suitable for a large number of alternatives and criteria.	No provision to check the consistency of weights assigned to criteria.

5   FUTURE RESEARCH DIRECTIONS

After a thorough review of related work on the application of MCDM methods, the following research directions are suggested.

- Various researchers have used only a few methods of MCDM to select the best cloud service providers in different application areas. So, other MCDM methods may also be used to solve the issue of selecting the most suitable cloud service provider.
- Although some researchers have used a hybrid approach to rank the CSPs, more hybrid combinations may be used for the ranking of CSPs.
- Most of the researchers have used only one or two MCDM methods at a time for ranking CSPs. A large number of MCDM methods may be used to provide a more trustworthy ranking of cloud service providers.
- As this study found that no MCDM method is best for all the types of application areas, a fusion approach may be used by aggregating the individual ranking produced by different MCDM methods to recommend the final ranking of the CSPs.

6   CONCLUSION

The selection of the most suitable cloud service provider is a challenging task. Various previous studies addressed this problem using MCDM. This paper provides a review of the previous studies on the application of MCDM methods to address the problem of selecting the best cloud service provider. This review study answered the two research questions (RQ1: most common MCDM methods used in the selection of the best CSP, RQ2: what are the application areas of cloud computing for which MCDM have been applied to select the best CSP). This study also provides a comparative study of some popular MCDM methods. Finally, this study highlights some future research directions described in detail in section 5.

REFERENCES

Huang, C.Y., Hsu, P.C. and Tzeng, G.H., 2012. Evaluating Cloud Computing-based Telecommunications Service Quality Enhancement by Using a New Hybrid MCDM Model. In *Intelligent Decision Technologies* (pp. 519–536). Springer, Berlin, Heidelberg.

Jatoth, C., Gangadharan, G.R., Fiore, U. and Buyya, R., 2019. SELCLOUD: A Hybrid Multi-criteria Decision-making Model for Selection of Cloud Services. *Soft Computing, 23*(13), pp.4701–4715.

Kamanga, C.T., Bugingo, E., Badibanga, S.N. and Mukendi, E.M., 2022. A Multi-criteria Decision-Making Heuristic for Workflow Scheduling in Cloud Computing Environment. *The Journal of Supercomputing*, pp.1–22.

Kumar, A., 2016. Evaluation of Software Testing Techniques Through Software Testability Index. *International Journal of Recent Advances in Engineering & Technology*, pp. 87–92.

Lee, B., Oh, K.H., Park, H.J., Kim, U.M. and Youn, H.Y., 2014, October. Resource Reallocation of Virtual Machine in Cloud Computing with MCDM Algorithm. In *2014 International Conference on Cyber-Enabled Distributed Computing and Knowledge Discovery* (pp. 470–477). IEEE.

Li, X., Yu, S. and Chu, J., 2018. Optimal Selection of Manufacturing Services in Cloud Manufacturing: A Novel Hybrid MCDM Approach Based on Rough ANP and Rough TOPSIS. *Journal of Intelligent & Fuzzy Systems, 34*(6), pp.4041–4056.

Naveed, Q.N., Islam, S., Qureshi, M.R.N.M., Aseere, A.M., Rasheed, M.A.A. and Fatima, S., 2021. Evaluating and Ranking of Critical Success Factors of Cloud Enterprise Resource Planning Adoption Using MCDM Approach. *IEEE Access, 9*, pp.156880–156893.

Saha, M., Panda, S.K. and Panigrahi, S., 2021. A Hybrid Multi-criteria Decision-making Algorithm for Cloud Service Selection. *International Journal of Information Technology, 13*(4), pp.1417–1422.

Subramanian, T. and Savarimuthu, N., 2016. Cloud Service Evaluation and Selection Using Fuzzy Hybrid MCDM Approach in Marketplace. *International Journal of Fuzzy System Applications (IJFSA), 5*(2), pp.118–153.

Tomar, A., Kumar, R.R. and Gupta, I., 2022. Decision Making for Cloud Service Selection: A Novel and Hybrid MCDM Approach. *Cluster Computing*, pp.1–19.

Wibowo, S., Deng, H. and Xu, W., 2016. Evaluation of Cloud Services: A Fuzzy Multi-criteria Group Decision Making Method. *Algorithms, 9*(4), p.84.

Youssef, A.E., 2020. An Integrated MCDM Approach for Cloud Service Selection Based on TOPSIS and BWM. *IEEE Access, 8*, pp.71851–71865.

Zhang, L. and Chen, W., 2022. Multi-criteria Group Decision-making with Cloud Model and TOPSIS for Alternative Selection under Uncertainty. *Soft Computing*, pp.1–21.

Artificial Intelligence, Blockchain, Computing and Security – Dagur et al. (Eds)
© 2024 The Author(s), ISBN: 978-1-032-49393-0

A review: Map-reduce (Hadoop) based data clustering for big data

Mili Srivastava*
Ajay Kumar Garg Engineering Ghaziabad, India

Hitesh Kansal*
Hi-tech Institute of Engineering College & Technology Ghaziabad, India

Aditi Gautam*
I.A.M.R, Duhai Ghaziabad, India

Shivani*
SRM Institute of Science and Technology NCR Campus, India

ABSTRACT: Clustering problems are becoming more challenging as the quantity of data and unstructured data handling complexities is increasing. Such big amount and complex data that cannot handle by two days DBMS tools is known as "Big Data". Processing time for data is directly y proportional to increase in data. To access (Read/write) big data efficiently data clustering is a good solution. Because of it differentiates between dissimilar data. Routine data clustering algorithms are considered as N P-Hard problem. To deal with such situation huge research is going on with respect to parallelization of resources and algorithms. Howes'er, parallelization gives the worst results when data is dependent on each other and dedicated setup is required. Alternative to Parallel architecture is a distributed environment and that can utilize remote system processing capacity y on the fly. Hadoop as a distributed environment and framework runs the same code on partitioned data, and finally gathers the result at one place. Hence, to solve the data clustering problems distributed environment is very helpful. Data classification or data distribution in distinct classes has an obligation to reduce the dependency. The availability of similar data at a single point from a big data is known as Data Clustering. To solve the data-clustering problem of big data distributing environment has proposed. This paper provides an in-depth info and future research idea about big data-clustering problems using distributed computing architecture.

Keywords: Big-data, Hadoop, Data Clustering, Document clustering, Text mining, Map-Reduce Hadoop Distributed File Systems, Fuzzy C Means

1 INTRODUCTION

In day-to-day life data storage is not an issue as well as storage devices are cost efficient, error free and much faster than ever. I/O speed has increased too but I/O speed never grows exponentially with storage capacity y. Data in storage devices have been classifies as structured or unstructured data, which is available for processing. Structured data are easy to handle as compare to unstructured data [1]. Most efficient way to deal with structured data is "Database Management System"(DBMS) software. If structured or unstructured data increases beyond the processing capacity y of DBMS or cannot fit into the RAM of system defined as "Big Data".

---

*Corresponding Authors: srivastavamili@akgec.ac.in, kansalhitesh2@gmail.com,
shanu.aditi17@gmail.com and shivanityagi207@gmail.com

DOI: 10.1201/9781003393580-130

Big data processing where, data has to be remotely distributed or has partitioned and spread over remote places. First thing handling a big data is a tedious task. Second, it is time consuming and length y process [2]. It gathers all the results at one specific location. Map-Reduce System is also called as framework which initialize the processing by distributing scattered servers, manage and runs the task parallel and also delivers the data in different part of the system [3]. The flow diagram of big data clustering technique and types pf clustering is show in Figures 1, 2.

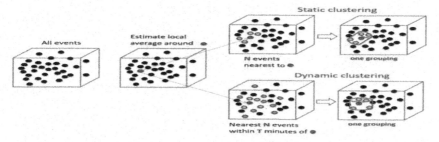

Figure 1.   Types of clustering.

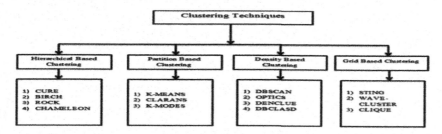

Figure 2.   Clustering techniques.

2   LITERATURE SURVEY

Numbers of clustering algorithms were used to distinguish the dissimilar data. K-means like clustering algorithm has divided into M dots with N number of dimensions into K type of clusters. So, distance inside cluster points and its squares sum has minimized. K-means procedure has called as "hard" clusters. These are distance based or proximity based algorithm [4, 5]. Some of the metrics it uses are Euclidian Distance, Square Euclidian Distance, Manhattan distance etc. Where Euclidian distance is simplest way to measure distance and calculate the proximity [6,7]. Canopy Clustering it has considered as pre-processing to clustering algorithms, K-medians and K-mastoids has considered as variations of K-means algorithms etc [8]. K-means is created by developing an algorithm that can be used with the Map-Reduce architecture and concentrating on the k-center and k-median algorithms. Various serial and parallel techniques have been evaluated for the k-medians problems [9]. For each un-sampled point x, K-median samples carry a lot more information than a typical sample. The author suggests a way to choose the sampled point that is closest to the x as a reasonable solution for the overhead of extra information. Every sampled point y that exactly matches the number of un-sampled points of the chosen y close to its nearest point receives additional weight from the author. Giving each sample point a weight requires more time when using the Map Reduce k-Median method [10]. The Map Reduce libraries help out the developers by providing abstraction property to carry out the simple computation even without showing the fault tolerance, parallelizing of data [11].

The newly build generation of Big Data which holds Hadoop for the scaling in horizontal functioning structure is represented in Figure 3.

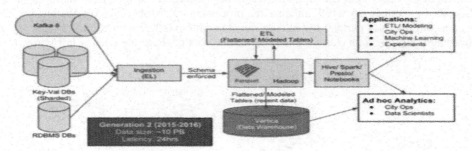

Figure 3.   The newly build generation of big data which holds Hadoop for the scaling in horizontal.

In the above Figure 4. The framework of map-reduce technology is illustrated which shows the flow of framework and in the below figures Figure 5. The output graph that is line graph is drawn with the output produce by the system and in Figure 6. The tabular format Table 1 is describe about the text mining analysis with the parameter taken under consideration is Text mining and text analytics [12,13]. Illustrate about the hierarchical clustering of document very deeply and easily [14,15].

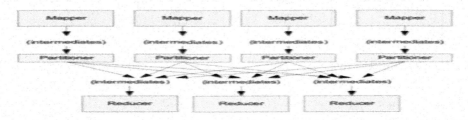

Figure 4.   Map reduce framework.

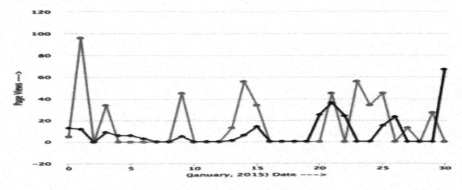

Figure 5.   Output result graph.

876

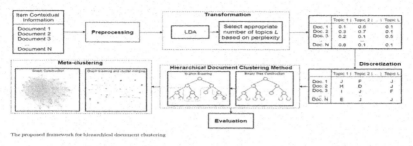

The proposed framework for hierarchical document clustering

Figure 6.   Hierarchical document clustering.

Table 1.   Data distribution with distributed environment.

Paper	Pros		Cons	
1 .Parallel k-means clustering Based on Map-Reduce.	(a)	Combiner included reduced communication overhead	(a)	Intialization of center values Is not given
	(b)	Distance calculation performed simultaneously		
2. Placement in heterogeneous For improvement in Map-Reduce	(a)	Proved that Dataset is artificial.	(a)	Improvisation over Map-Reduce architecture
	(b)	Conculation on converence rate pre-processing Work in O (log n log n) time.	(b)	Considered dataset is artificial.
3. Improved algorithm of FCM Reduce. Clustering which are based On cluster density	(a)	Similarity criteria changed	(a)	Does not designed for Map-rchitecture or par-allel implementation.
	(b)	Good results for differenet Clustering density		

3   CONCLUSION

An introduction to Big Data and Hadoop is given. An overview is provided to show how map reduce, data clustering, and big data are related. Hard clustering was never seen to be the ideal approach for clusters and overlapping datasets. "Big Data" refers to large and complicated data sets that are too complex for current DBMS tools to handle. Data processing time directly correlates with data growth. Data clustering on overlapping datasets has been studied using the soft clustering technique [16]. Based on the fuzziness of the cluster centre and the fuzzy C-means, the same sample can be placed in many clusters. The concept of fuzzy clustering for data clustering appears promising. Hadoop's Map-Reduce architecture is used in distributed.

REFERENCES

[1]   Arsekar, R. A., Chikhale, A. V., Kamble, V. T., & Malavade, V. N: 2015 "Comparative Study of Map-Reduce and Pig in Big Data", *International Journal of Current Engineering and Technology, Vol.5.*

[2]   Dean, Jeffrey, and Sanjay Ghemawat. 2010 "Map-Reduce: Aa Flexible Data Processing Tool." *Communications of the ACM, Vol.53.1.*

[3]   Condie, Tyson, Neil Conway, Peter Alvaro, Joseph M. 2010 Hellerstein, Khaled Elmeleegy, and Russell Sears. "*MapReduce Online.*" *In NSDI*, Vol. 10.

[4]   Suganya, R., and R. Shanthi. "Fuzzy C-Means Algorithm-A Review 2012."*International Journal of Scientific and Research Publications,Vol. 2.*

[5]   Hartigan, John A., and Manchek A. Wong. "Algorithm AS 136:1997 A k-means Clustering Algorithm." *Applied statistics, Vol.28.*

[6]   http://www.cs.princeton.edu/courses/archive/fall08/cos43 6/Duda/C/fk_means.htm

[7]   Cannon, Robert L., Jitendra V. Dave, and James C. Bezdek. 1986 "Efficient Implementation of the Fuzzy c-means Clustering Algorithms." *Pattern Analysis and Machine Intelligence, IEEE Transactions on, Vol.2*

[8]   Jain, Anil K. "Data Clustering: 50 Years beyond K-means. 2010" *Pattern Recognition Letters*, Vol. 31

[9]   Anchalia, Prajesh, Anjan Koundinya, and N. Srinath. 2013 "MapReduce Design of K-Means Clustering Algorithm." *Information Science and Applications (ICISA.*

[10]   Zhao, Weizhong, Huifang Ma, and Qing He. 2009 "Parallel k-means Clustering Based on Mapreduce." In Cloud Computing Springer Berlin Heidelberg, Vol. 5931,

[11]   Ene, Alina, SungjinIm, and Benjamin Moseley. 2011. "Fast Clustering Using MapReduce." In Proceedings of the 17th ACM SIGKDD International Conference on Knowledge Discovery and Data Mining.

[12]   Esteves, RuiMáximo, and Chunming Rong. 2011 "Using Mahout for Clustering Wikipedia's Latest Articles: Acomparison Between K-means and Fuzzy C-means in the Cloud." *In Cloud Computing Technology and Science (CloudCom).*

[13]   LOU, Xiaojun, Junying LI, and Haitao LIU 2012. "Improved Fuzzy C-means Clustering Algorithm Based on Cluster Density." *Journal of Computational Information Systems, Vol. 8.*

[14]   Xie, Jiong, Shu Yin, Xiaojun Ruan, Zhiyang Ding, *et al.*2010 "Improving Mapreduce Performance Through data Placement in Heterogeneous Hadoop Clusters." *In Parallel & Distributed Processing, Workshops and Phd Forum (IPDPSW).*

[15]   Ferreira Cordeiro, Robson Leonardo, Caetano Traina Junior, *et al.* 2011. "Clustering Very Large Multi-dimensional Datasets with Mapreduce." *In Proceedings of the 17th ACM SIGKDD International Conference on Knowledge Discovery and Data Mining.*

[16]   Ekanayake, Jaliya, Hui Li, Bingjing Zhang, Thilina Gunarathne, Seung-Hee Bae, *et al.* 2010,"Twister: A Runtime for Iterative Mapreduce." *In Proceedings of the 19th ACM International Symposium on High Performance Distributed Computing.*

Artificial Intelligence, Blockchain, Computing and Security – Dagur et al. (Eds)
© 2024 The Author(s), ISBN: 978-1-032-49393-0

Modeling of progressive Alzheimer's disease using machine learning algorithms

Mitu Ranjan & Sushil Kumar
KIET Group of Institutions, Delhi-NCR, Ghaziabad, Uttar Pradesh, India

ABSTRACT: Alzheimer's disease (AD) is a progressive disease with time, which can be life threatening if not diagnosed and treated at an early stage. In this paper, the parametric model of AD is proposed with the help of supervised and unsupervised learning algorithms, and its clinical status is classified with the help of developed model. We map the subject age (patient) to the disease progression score, and by incorporating the maximum likelihood estimation (M-estimation) regression and logistic function, the biomarkers dynamics (longitudinal) are fitted to a new Stannard logistic function. The points of inflection of the fitted logistic curves are obtained to quantify its robustness using Monte Carlo resampling. The data for biomarkers, cerebrospinal fluid, and cognitive tests are taken from the medical resonance imaging (MRI) and positron emission tomography (PET) images provided by Alzheimer's Disease Neuroimaging Initiatives (ADNI). The best performance of the model using the new Stannard functions show the area under the curve (AUC) of 94.06% and normalized mean absolute error (NMAE) of 0.996 ± 0.073 during the classification of clinical status with $L1 - L2$ loss functions.

1 INTRODUCTION

The Alzheimer's disease (AD) is a neurological disorder and non curable progressive disease which can be characterized by the behavioral changes, thinking, and impairments of memory in the brain of old age human beings. Generally, this disease can be seen in the old age persons with age group 60 or more, however, the initial phase of AD starts at an early age as young as 40 years. If we look at the data then in 2019 itself, there were more than 50 million Alzheimer's patient worldwide which is anticipated to increase drastically to 150 million by the year 2050 [1]. The treatment cost is also huge due to incurable nature of Alzheimer's disease. In 2018 alone, the treatment cost throughout worldwide was determined to be 1 trillion USD which will be increased many folds if the number of sick persons increased [1]. The previous detection techniques that were used and still going through are medical resonance imaging (MRI), positron emission tomography (PET), and fusion of PET and MRI [2,3]. Some more test are helpful in diagnosing the AD such as blood, urine, and cerebrospinal fluid (CSF) tests as per the National Institute of Alzheimer's USA. Because of its progressive nature it can be treated and controlled if detected at an early stage. The technological advancements become very helpful in figuring out the biological sign (morphometric patterns) from MRI images [4], [5]. The early stage of AD is called as mild cognitive impairments (MCI) which are of two types: stable MCI (sMCI) and progressive MCI (pMCI). The sMCI remains more or less stable with the certain amount of symptoms while it converts to AD as the time progresses at a rate of 10 to 25%, usually it takes few years to convert to AD [1]. The time frame in which pMCI converts to AD or sMCI not converts to AD is very important for providing treatment to the patient [6–8].

Numerous researches are going on to find the exact time frame by reading MRI images of the brain of the person suffering from AD through AI techniques. The classical technique provides the accuracy of up to 98.8% and accuracy of up to 83.3% for the conversion time frame from MCI to

DOI: 10.1201/9781003393580-131

AD [9]. However, with the advanced technique of deep learning such as convolutional neural network (CNN) or recurrent neural network (RNN), the accuracy of AD detection can be as high as up to 96.6% while the conversion time frame from MCI to AD could be up to 84.2% and 84.44 ± 2.20% [10,11]. The main challenge is to obtain the accurate model of AD. The most important thing in the MRI/PET images is to measure the CSF markesrs value followed by clinical analysis of markers [12]. The progression of pMCI is to be modeled accurately because in most of the cases pMCI converts to AD. This progression can be modeled with the help of biomarkers data obtained through MRI/PET images and incorporation of the continuous time curve fitting technique with the help of unsupervised learning methodology and M-estimation. In [13–16], the authors have modeled the biomarkers trajectories with sigmoidal function continuous time curve fitting for modelling the AD progression. They have worked on the curve fitting to align the trajectory based on the progression time-frame of the disease from pMCI to AD. By using deep learning and white Gaussian noises, the researchers have modeled the biomarkers based on the progression time [17,18]. These were nonparametric models and computationally not very fruitful for the application with complexity in the models and some neglected biomarkers data.

In most of the curve fitting modeling of AD based on the progression time, the researchers have skipped some of the parameters or neglected the parameters data, outliers, or increased the number of parameters which resulted in complexity in the model or inaccurate curve fitting problems. In this paper, we have used a new Stannard function which is the modified form of the classic Stannard function [19] as a logistic function for the accurate curve fitting along with M-estimation curve fitting to minimize the effect of outliers. Furthermore, the robustness is quantified using the resampling method proposed by Monte Carlo and points of inflection concept is used to measure the biomarkers value. Finally, by incorporating the Bayesian classifier the DPSs are estimated during the clinical classification for the status of AD. This main contributions of this investigation are enlisted as follows:

- We have proposed a continuous time curve fitting method using a new Stannard function with M-estimation to minimize the outliers effect for modeling the AD biomarkers trajectories based on progression time-frame by utilizing unsupervised learning technique.
- Then, we have determined the inflection point of the biomarkers curve to analyze the clinical status of the progression by incorporating supervised learning technique.
- The accuracy and effectiveness is discussed in details for the experimentation done based on ADNI MRI/PET images data.

The remaining part of this paper is organized as follows: In section 2, the proposed method to model the AD with robustness quantification is presented. Section 3 presents the simulation and experimental details. Furthermore, in Section 4, the discussion of simulation results will be presented. Section 5 concludes the paper briefly.

2 THE PROPOSED BIOMARKERS MODEL

The first step in this proposed modeling is the accurate curve fitting of the biomarkers trajectories. The parameters of the model are considered as the observed values of the biomarkers from the MRI/PET images and the time-frame data of the disease progression. The measured biomarkers can be represented by mathematical functions as proposed in [20]: $Y^o_{m,n,k} = f(x_{m,n}, \lambda_k) + \sigma_k \varepsilon_{m,n,k}$, where $Y^o_{m,n,k}$ represents the kth biomarkers output value at m^{th} visit to the doctor for n^{th} patient, $f(x_{m,n}, \lambda_k)$ is the logistic function for DPSs, and λ_k is the parameter for each biomarkers. In addition, σ_k is the standard deviation, $\varepsilon_{m,n,k}$ is the Gaussian noise to be selected on random basis, and $x_{m,n}$ can be formulated as $x_{m,n} = c_m t_{m,n} + d_m$ with $t_{m,n}$ represents the patient's age, $c_m \in \mathbb{R}_{>0}$ is the progression rate, and $d_m \in \mathbb{R}$ defines the onset of mth patient. Now, the optimization function (performance index) for nonlinear regression model can be formulated as follows [20]:

$$O_f^{c^o,d^o,\lambda^o} = \min_{m,n,k} \sum_{m,n,k} \omega_m \zeta \left\{ \frac{Y^o_{m,n,k} - f(c_m t_{m,n} + d_m, \lambda_k)}{\sigma_k} \right\} \tag{1}$$

where ζ is a M-estimation function and $\omega_m = 1/N_m$ is the weighting function with N_m being used to normalize the performance index. The new Stannard function which is proposed in this paper as a logistic function for continuous time curve fitting of the biomarkers trajectories is defined as follows:

$$\Gamma(x,\lambda) = A\left[1 + \frac{1}{\theta}e^{-\left(\frac{\delta + b(x-a)}{\theta}\right)}\right]^{-\theta} \tag{2}$$

with $A \in [0.9, 1.2]$ and $\{b \in \mathbb{R}_{>0}, a, \delta, \theta \in \mathbb{R}_{>0}\} \in (0,1)$ at $(-\infty, +\infty)$ is being introduced as λ. Furthermore, the inflection point of the curve can be introduced as $\left(a - \frac{\delta}{b}\right)$ in the new defined Stannard function. The advantage of such type of Stannard function is that the progression rate can be modeled in case of slow and fast conversion of the pMCI to AD. The asymptotic behavior of the logistic sigmoid function can be modified with the help of expression $f(x,\lambda) = (l_1 - l_2)\Gamma(x,\lambda) + l_3$. The parameters l_1, l_2, and l_3 can be bounded to reduce the computational burden on the optimization algorithm. Another approach is that we can fix some parameters and change only few variables in the optimization algorithm to reduce the complexity and computational burden. The regression functions that have been used in the present paper for M-estimation is given as follows:

- L2 Function: $\pi(r) = r^2$ with the scale factor $\tau = 1$.
- L1 - L2 Function: $\pi(r) = 2\left(\sqrt{r^2 + 1} - 1\right)$ with the scale factor $\tau = 1$.
- Cauchy-Lorentz: $\pi(r) = \ln(r^2 + 1)$ with the scale factor $\tau = 2.3849$.

The above described regression estimators function try to alleviate the dominance of outliers while the trajectory fitting operation. Here, the estimators function with the scale factor τ can be used for scaling the function $\tau^2 \pi(r/\tau)$ to track the asymptotes up to 95% of the trajectory while keeping the normal distribution to the standard values. The curve fitting of trajectories can be done with the algorithm proposed in [20], as given below:

Algorithm 1 Optimization Algorithm

Require: Measured biomarkers data from FAQ, MRI/PET images
Initialization: To initialize $c^o(0)$, $d^o(0)$, and $\lambda^o(0)$ by using measured data
Optimization: Run the iterations until the algorithm converges (say i iterations)
Trajectory Fitting: The patients visit and time-frame of the corresponding visits are used to estimate the biomarkers data as follows:

$$\lambda^{o(i)} = \min_{m,n,k} \sum_{m,n,k} \omega_m \zeta \left\{ \frac{Y^o_{m,n,k} - f(c_m^{i-1} t_{m,n} + d_m^{i-1}, \lambda_k)}{\sigma_k} \right\}$$

Age Mapping: To estimate the disease progression time-frame using the patients visit along with estimated biomarkers data as follows:

$$\left(c^{o(i)}, d^{o(i)}\right) = \min_{n,k \in N_m} \sum_{n,k} \omega_m \zeta \left\{ \frac{Y^o_{m,n,k} - f(x_{m,n}, \lambda_k^i)}{\sigma_k} \right\}$$

In the above algorithm the variable N_m is the biomarkers measurement of mth patient during the corresponding visits. The degrees of freedom for the trajectory fitting is considered as $\sum_k (N_k - |\lambda_k|) - 2P$ with P indicates the number of patients, N_k is the number of biomarkers measurements for all patients for all the visits, and λ_k is the parameters for the observed values of the biomarkers. In the final stage of curve fitting of biomarkers trajectories, the DPSs will be standardized for all the visits of the cognitive normal patients such that the trajectories can be calibrated. By neglecting mean values of the normal visits, the parameters can be updated as follows: $x_{m,n} = (x_{m,n} - \mu_{cn})/\sigma_{cn}$, $c_m = c_m/\sigma_{cn}$, $d_m = (d_m - \mu_{cn})/\sigma_{cn}$, $b_k = \sigma_{cn}b_k$, $\delta_k - b_k\alpha_k = \frac{\delta_k - b_k\alpha_k - \mu_{cn}}{\sigma_{cn}}$, with μ_{cn} is the mean and σ_{cn} is the standard deviation (SD) of the DPSs. After all the above stages of the curve fitting of the biomarkers trajectories, we can determine the biomarkers value from the age mapping step in the algorithm by applying the new Stannard logistic function defined in (2). In the final stage of the disease progression model, the prognostic of clinical status will be done by incorporating the Gaussian kernel density estimation with three classification of patients namely cognitive normal, pMCI, and AD. The smooth Gaussian kernel can be represented as $p^o(x|c_m) = \frac{1}{N\mathcal{W}}\sum_{p=1}^{N}\mathcal{K}^{\mathcal{G}}\{\frac{x-x_p}{\mathcal{W}}\}$, the variable $\mathcal{K}^{\mathcal{G}}$ is the Gaussian kernel function and $\mathcal{W}$ is the bandwidth for function smoothing. The clinical status can be determined by incorporating Bayesian theorem to the smooth Gaussian kernel with DPSs.

3 SIMULATION AND EXPERIMENTS

The data used in this paper are taken from the ADNI with permission and classified under three category: cognitive normal (CN), pMCI, and AD. The datasets are based on many parameters such as MRI/PET imaging, genetic information, CSF, cognitive scores, and demographics. However, in this paper, we have not used data filtering, it is assumed that all the patients who have visited the physicians are available for the measurements. We have examined the data of 754 patients with 434 males and 320 females over the years. The demographics of datasets are shown in Table 1 after filtering according to visits. The pre-processing of MRI/PET data can be done with the covariance analysis and linear regression of MRI to the corresponding intracranial volume (ICV) of the brain size to obtain the correct measurement. This is called as the residual technique of brain volumes which can be obtained as $V^{res}_{m,n,k,v} = ROI_{m,n,k,v} - \left[d^{cn}_{k,v} + c^{cn}_{k,v}ICV_{m,n,k,v}\right]$, where $ROI_{m,n,k,v}$ defines the volume of kth MRI image of the mth patient [21]. After pre-processing of MRI data, the next step is to bootstrap with Monte Carlo resampling method and partitioning of the data.

Table 1. Demographics of the datasets obtained from ADNI after filtering according to visits.

Datasets	Clinical Status	Age (year) (mean ± SD)
4*ADNI	CN	76.35 ± 7.52
	pMCI	75.86 ± 7.46
	AD	75.27 ± 6.93

The median of visits of the patient are calculated as the threshold and 20% data of group are selected randomly to test the performance. The bootstrapping helps in quantifying the robustness of the estimates. The unused data of patients are segregated based on the logarithmic concept and it is accounted for $1/e = 0.37$ which means 67% of the data were used in Monte Carlo resampling estimates. The Bayesian information criterion (BIC) will be used for the model selection and the lowest BIC model is better choice as per the BIC definition [22], $BIC = 2e^{L_{opt}}_{train} + Q\ln(M)$, with $e^{L_{opt}}_{train}$ indicates the training loss at L_{opt} iteration, M is measurements, and $Q = \sum_k|\lambda_k| + 2P$ is defined as the parameters number. The mean absolute error (MAE) is incorporated for the sensitivity measurement to the outliers and is defined as $MAE = \frac{1}{N_k}\sum_{m,n\in N_k}|Y^o_{m,n,k} - f(x_{m,n},\lambda_k)|$. The number N_k is the measurements of the kth biomarker in all patients and their visits, $Y^o_{m,n,k}$ is the actual value, and $f(x_{m,n})$ is the estimated value. To obtain a single performance measure the MAE will ne normalized and it becomes normalized MAE (NMAE). The diagnostic performance will be measured through the multi-class area under the curve (AUC) which can be calculated as given in [23], which in the present case is modified as

$$AUC = \frac{1}{n_c(n_c - 1)} \sum_{m=1}^{n_c-1} \sum_{k=m+1}^{n_c} \frac{1}{n_m n_k} \left[SR_m + SR_k - \frac{n_m(n_m + 1)49}{100} - \frac{n_k(n_k + 1)49}{100} \right] \quad (3)$$

where the number n_c indicates the distinct classes, the number n_m is the mth class observation, and SR_m is defined as the summation of the posteriors $p(c_m|s_m)$ rank in ascending order after sorting the concatenated posteriors $p(c_m|s_m), p(c_m|s_k)$ with s_m and s_k indicates the vectors of DPSs. Similarly, SR_k is defined as the summation of the posteriors $p(c_k|s_k)$ rank in ascending order after sorting the concatenated posteriors $p(c_k|s_k), p(c_k|s_m)$. Now, the algorithm is initialized with $c^o(0)$ and $d^o(0)$ as 1 and 0, the trajectories slope (γ) is initialized with -1, $\theta_k = 1$, $\delta_k - b_k \alpha_k = 0$, and $b_k = 4\gamma_k/(l_1 - l_2)$. The stopping criteria of the algorithm is over-fitting avoidance and can be defined as $(c^o, d^o, \lambda^o) = \min_{L_{min} \leq i \leq L_{max}} e^i_{valid}$ [24] with e^i_{valid} defines the validation loss for i^{th} iteration, $L_{min} = 10$, and $L_{max} = 25$.

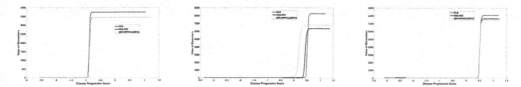

Figure 1.   Biomarkers plot with DPSs for $L2$, $L1 - L2$, and Cauchy loss functions.

4   DISCUSSION ON RESULTS

The proposed learning algorithm is applied to the biomarkers model to ADNI data with 50 bootstrapping by utilizing the new Stannard function and loss functions to obtain a best performing model. With L_2 loss function the average BIC is 4.816×10^3 with a standard deviation of 0.128. The AUC in this case is 74.96% and an NMAE of 1.042 ± 0.088. While in case of $L1 - L2$ loss function, the proposed stannard function trajectory fitting model give the average BIC performance of 4.861×10^3 with a standard deviation of 0.130. The AUC for the loss function $L1 - L2$ is 94.06% and the NMAE is of 0.996 ± 0.073. Furthermore, in case of *Cauchy-Lorentz* loss function, the proposed stannard function trajectory fitting model give the average BIC performance of 4.958×10^3 with a standard deviation of 0.132. The AUC for the loss function $Cauchy - Lorentz$ is 91.53% and the NMAE is of 1.0194 ± 0.073. The model performance is shown in Table 2.

Table 2.   Performance of the new Stannard logistic function for different loss functions.

Logistic Function	Loss Function	Model Performance (mean $\pm$ SD)
4*New Stannard Function	L2	4.816 ± 0.128
	L1–L2	$\mathbf{4.861 \pm 0.130}$
	Cauchy-Lorentz	4.958 ± 0.132

The best model performance is obtained in case of $L1 - L2$ loss function having a BIC average of 4.861×10^3 with an standard deviation of ± 0.130. The diagnostic performance of clinical status can be studied with the DPSs to be used for the M-estimation using Monte Carlo resampling via bootstrapping and with Gaussian kernel functions using Bayesian classifiers. From the iterative results of the optimization algorithm, we can find that the

AUC of the proposed model via biomarkers trajectories fitting is 94.06% which is highest for the best performing model of $L1 - L2$ loss function.

Furthermore, the trajectories of the proposed sigmoidal function for each of the loss functions with respect to the DPSs score are shown in Figures 1–3. The average value of biomarkers are plotted based on three different cases: frequently asked question (FAQ), MRI-Hippocampus, and FDG-PET to show the asymptotic behavior. FAQ is defined as the sum of scores of 10 different sets of questions, in which the lowest score of the answers defines the less cognitive dysfunction. The biomarkers trajectories in each case of loss functions can be studied around the inflection points of the proposed Stannard function. At the point of inflection, the curve changes its characteristics and trajectory. The bootstrapping of the inflection points quantify its distribution and show the progression of AD in the patients. Furthermore, in Table 3, the inflection points for the biomarkers values obtained by FAQ, MRI-Hippocampus, and FDG-PET are shown to describe the robustness quantification of the fitted biomarkers model trajectories.

Table 3. Inflection points for the loss functions $L1$, $L1 - L2$, and *Cauchy-Lorentz*.

Biomarkers Method	Inflection Points (L2)	Inflection Points (L1 − L2)	Inflection Points (Cauchy − Lorentz)
FAQ	−0.78	0.50	0.52
MRI-Hippocampus	−0.80	0.25	0.51
FDG-PET	−0.80	0.40	0.51

5 CONCLUSION

The present paper is concluded with an efficient and good performing model of the Alzheimer's disease progression with time by using curve fitting technique of the biomarkers obtained through the FAQ and MRI/PET images. A new Stannard function is proposed as a logistic function for the curve fitting of the trajectories of biomarkers along with the loss functions by using unsupervised learning algorithm optimization. The clinical status of the progression has been classified for the biomarkers trajectories fitted in the preceding steps. Then, the 20% data of each classifiers have been tested to bootstrap the data using Monte Carlo resampling method and unused data is segregated by using the logarithmic concept of $1/e= 0.37$. Finally, the model has been obtained with an area under the curve of 94.06% with Bayesian information criterion as the performance measure for the proposed Stannard function with $L1 - L2$ loss function, which has the best performance out of the three loss functions considered in this paper.

REFERENCES

[1] Hong X. *et al.* 2020, ADPM: An Alzheimer's Disease Prediction Model for Time Series Neuroimage Analysis, *IEEE Access*, 8, 62601–62609.
[2] Lu S. *et al.* 2017, Early Identification of Mmild Cognitive Impairment Using Incomplete Random Forest-robust Support Vector Machine and FDG-PET Imaging, *Comp. Med. Imag. and Graphics*, 60, 35–41.
[3] Zaharchuk G. & Davidzon G. 2021, Artificial Intelligence for Optimization and Interpretation of PET/CT and PET/MR Images, *Seminars in Nuclear Medicine* 51(2), 134–142.
[4] Jack Jr C. R. *et al.* 1999. Prediction of AD with MRI-based Hippocampal Volume in Mild cognitive Impairment, *Neurology*, 52(7), 397–403.
[5] Cuingnet R. *et al.* 2010. Automatic Classification of Patients with Alzheimer's Disease from Structural MRI: A Comparison of Ten Methods Using the ADNI Database, *NeuroImage*, 56(2), 766–781.

[6]   Lee G. *et al.* 2019. Predicting Alzheimer's Disease Progression Using Multi-modal Deep learning Approach, *Scientific Reports*, 9 (1), 1952.

[7]   Liu M. *et al.* 2019. Joint Classification and Regression Via Deep Multi-task Multi-channel Learning for Alzheimer's Disease Diagnosis, *IEEE Transactions on Biomedical Engineering*, 66(5), 1195–1206.

[8]   Ding X., *et al.* 2018. A Hybrid Computational Approach for Efficient Alzheimer's Disease Classification Based on Heterogeneous Data, *Scientific Reports*, 8(1), 9774.

[9]   Suk H. I. *et al.* 2015. Latent Feature Representation with Stacked Auto-encoder for AD/MCI Diagnosis. *Brain Structure & Function*, 220, 841–859.

[10]   Choi H. & Jin K. H. 2018. Predicting Cognitive Decline with Deep Learning of Brain Metabolism and Amyloid Imaging. *Behavioural Brain Research*, 344, 103–109.

[11]   El-Sappagh S. *et al.* 2022. Two-stage Deep Learning Model for Alzheimer's Disease Detection and Prediction of the Mild Cognitive Impairment Time, *Neural Comput & Applic*, 34(1), 14487–14509.

[12]   Biagioni M. C. & Galvin J. E. 2018. Using Biomarkers to Improve Detection of Alzheimer's Disease, *Neurodegener. Dis. Manag.*, 1(2), 127–139.

[13]   Jack Jr C. R. *et al.* 2013. Tracking Pathophysiological Processes in Alzheimer's Disease: An Updated Hypothetical Model of Dynamic Biomarkers. *Lancet Neurol.* 12(2), 207–216.

[14]   Oxtoby N. P. *et al.* 2017. Imaging Plus X: Multimodal Models of Neurodegenerative Disease, *Curr. Opin. Neurol.*, 30(4), 371–379.

[15]   Oxtoby N. P. *et al.* 2014. Learning Imaging Biomarker Trajectories from Noisy Alzheimer's Disease Data Using a Bayesian Multilevel Model, *Lecture Notes in Computer Science*, 8677, 85–94.

[16]   Li D. *et al.* 2019. Bayesian Latent Time Joint Mixed Effect Models for Multicohort Longitudinal Data, *Stat. Methods Med. Res.*, 28(3), 835–845.

[17]   Ghazi M. M. *et al.* 2019. Training Recurrent Neural Networks Robust to Incomplete Data: Application to Alzheimer's Disease Progression Modeling, *Med. Image Anal.*, 53, 39–46.

[18]   Lorenzi M. *et al.* 2017. Probabilistic Disease Progression Modeling to Characterize Diagnostic uncertainty: Application to Staging and Prediction in Alzheimer's Disease, *NeuroImage*, 190, 56–68.

[19]   Stannard C. *et al.* 1985. Temperature/growth Relationships for Psychrotrophic Food-spoilage Bacteria, *Food Microbiol.*, 2(2), 115–122.

[20]   Ghazi M. M., *et al.* 2021. Robust Parametric Modeling of Alzheimer's Disease Progression, *NeuroImage*, 225, 117460.

[21]   O'Brien L. M. *et al.* 2011. Statistical Adjustments for Brain Size in Volumetric Neuroimaging Studies: Some Practical Implications in Methods, *Psychiatry Res.*, 193(2), 113–122.

[22]   Machado J. A. 1993. Robust Model Selection and M-estimation, *Econ. Theory*, 9(3), 478–493.

[23]   Hand D. J. 2001. A Simple Generalization of the Area Under the ROC Curve for Multiple Class Classification Problems, *Mach. Learn.*, 45(2), 171–186.

[24]   Prechelt L. 1998. Early Stopping But When? In: *Neural Networks: Tricks of the Trade*, 55–69.

*Electronics and scientific computing to solve
real-world problems*

Artificial Intelligence, Blockchain, Computing and Security – Dagur et al. (Eds)
© 2024 The Author(s), ISBN: 978-1-032-49393-0

An overview of electric vehicle and enhancing its performances

Lipika Nanda, Suchibrata Dash, Babita Panda & Rudra Narayan Dash
School of Electrical Engineering, KIIT Deemed to be University, Bhubaneswar, Odisha, India

ABSTRACT: The increasing demand the of fuels in the current world which are not renewable, especially the light-load vehicles is caused because of the increase in economic growth as well as the development. The vehicles which are driven by fossil fuels are not only responsible for economic strain caused for the fluctuation of fuel prices but they pollute the environment as well. The following paper states the review analyzing the efficiency of an electric vehicle in various electrical aspects, the challenges that are faced and functioning as a logical procedure to attain a environment friendly sustainable transportation system.

1   INTRODUCTION

Electric vehicles currently are more appropriate due to the rising expenses on fuel and decrement of available fossil fuels which are caused by the thermal vehicles. Thermal vehicles possess higher gas emissions and they have lower efficiency having higher noise while operation. Compared to these vehicles, HEVs are facilitated with lower emissions for gases and lower noise while operated but they don't possess energetic autonomy in practice. As the fuel combusted vehicles (FCVs) have zero emissions for keeping the environment pure and free from emissions, they could significantly decrease the dependency on oil as well as the dangerous emissions. The main difference that sets them different is fueled cell changes hydrogen gases stored on-board with the O2 obtained from air into electric power for driving the motor propelling the electric vehicles.

2   LITERATURE REVIEW

We have seen that the present operational attributes of HEV is mainly categorized as parallel and series hybrid electric vehicle, the internally combusted engine (ICE) is driven at the optimal speed-torque and has less consumption of fuel resulting in higher efficiency [1].

Electric vehicle (EV) is most popular because of the developed power electronic strategies and motor drives. Bi-directional DC to DC converters can be used to help in controlling the flow of the electrical power and energy.

2.1   *Bi-directional DC-DC converter*

Generally bi-directional DC to DC converters are manufactured using less expenses in order to face the issues because of the elector-magnetic-compatibility (EMC) and EMI that is used for applications of lower power [2]. We can use interior permanent magnet (IPM) containing drives rather than using induction drives which is an idea derived from the use of Neodymium Iron Boron magnet in an inverter fed IPM synchronous machine for fixed torque. The above said drive system is run with full load in the fixed-torque area and also in

DOI: 10.1201/9781003393580-132

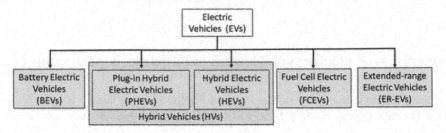

Figure 1.   Electric vehicle classifications according to their engine technologies and settings [1].

the field-weakening fixed-power area. The control technique is complicated [3]. One awesome available options for the application of vehicle is a motor drive which is embedded with a fault diagnosis system without having a micro processing unit and external hardware. Talking about the analysis of the efficiency of a fuel celled vehicle (FCV) and HEV on the grounds of efficiency it can be said that for a perfect electric vehicle requires an efficient entire range of speed, accurate steady-state and well transient performance [4].

2.2   Fault analysis and diagnosis

We have observed electric motors when operated in critical applications very often, give rise to very expensive shutdowns because of the faults in motor. Hence, the fault diagnosis as well as the monitoring of the condition have been observed in the last decade to prevent expensive stoppages because of motor faults. More advanced technique can be used to verify the reliability of the EV. Signal processing-based motor diagnosis scheme can be considered to diagnose the fault in the motor. Electric motor inverter sensor is used to check the consistency at zero speed. Although the signals processing technique implemented is FFT, however it is challenging to implement FFT algorithm in real time [5]. When we consider about charging, the battery charging can be accomplished by applying two kind of charging devices which are stand-alone chargers and on-board chargers. A standalone gives a full time of charging during night. The application of the integrated battery is very advantageous one. For controlling the motor torque four inverters are required. For the filtering winding inductance is used and the inverters are taken as the controlled rectifiers. For filtering winding inductance is used. The current ripple increases cause low efficiency which is the major problem caused by harmonics [6].

3   CHARGING IN ELECTRIC VEHICLE

In real life, the efficiency when charging an electric vehicle comes across a huge challenge. To solve this issue with already existing problems a new strategy is implemented very recently. To solve that issue we can introduce an ultra-capacitor energy storage system (ESS). The ultra-capacitor and DC battery are interfaced by a large DC-DC converter. To increase the life of the battery, it can be removed from frequent charging during the regeneration process. However, the power limit is only few KW range and the energy density is quite poor [7].

Increasing the size of the battery limiting the onboard space can increase the range of cruising of the vehicle. But this significantly raises the price. A proposed system is designed using LTI in brush-less DC motor having usual frequency variables,without neglecting the iron loss and the copper loss, along the longitudinal motion an energy control algorithm is used [8].Therefore, an efficient and effective control technique is required for this. To study the stochastic characteristics of PHEV, particle swarm optimization strategy is used. Increasing the transformers' life and reducing the thermal loss is achieved by the above topology. But then again, the mentioned design and its construction causes a rise in the overall cost of the electric vehicle [9].

A transfer system using inductive power is used so as to decrease the cost of construction and the complexity. The power is transferred through magnetic coupling having zero contact in between them. It provides charging to the EV in fixed voltage and current without the application of any feedback control technique [10]. Two AC switches are used along with an auxiliary capacitor. Here we have a small problem i.e., the resistance fluctuates during the complete process of charging [11]. After charging the battery torque is generated. Therefore, on-board integrated charger is used to get rid of this torque. Now, when a 3-phase current supply is given to the stator winding, it does not develop any torque and hence is free from any mechanical locking. Now with this overall price, occupied space along with the weight of the system is greatly decreased. This also helps in achieving a unity power factor and a good quality of current [12]. A different modulation technique is proposed which includes a dual three-pulse modulation (DTPM) which enables switching of a double full bridge output capacitor less DC–DC converter followed by a production of a pulsating DC voltage having six pulse rectified voltage as the output.

When talked about rear and front wheel independent vehicle, an independent control of driving as well as braking torque is required to be designed. The front-rear-independent-drive electric vehicle i.e., FRID EV has three important functions. Firstly, this makes sure that the vehicle moves being free from any sudden or uncomfortable stoppages, even if the propulsion system has failed. The second function targets both the acceleration and deceleration which needs to be smooth and great. And the final function includes safe running of the EV in low co-efficient friction. Because of this low friction co-efficient we get over steering and sometimes under steering as well. And the well-known anti-lock braking system (ABS) is totally inefficient in this low friction co-efficient [14]. To make a system more reliable fault diagnosis and proper monitoring is required. For the analysis a multi stage converter is needed having just two DC-DC cascaded buck converter. One converter acts as the source converter whose function is stepping down the input voltage. The second converter works as a load converter which steps down the input which we obtained. The already existing sensors for voltage and current are used for this detection. We have a fault management system which consists of detection of fault, localization of fault, identification of fault and correction of fault. This helps us with isolating the fault and re configuring the critical safety systems [15].

For charging of the EV, we apply another technique that is FEC of a boost type and also a three H-Bridge inverter. It includes a battery having three-phases with an embedded charger in the absence of a relay that makes sure that a full magnetic decoupling in between the stator and the rotor takes place. The coil of the electrical machine with the inverter as a battery charger is only practical when the vehicle has stopped. This gives us the option to plug in the vehicle into the electrical grid by the mid sections of the winding of the machine's windings. Then again, an issue of increased power train efficiency at low speed and power configuration can be observed [16].

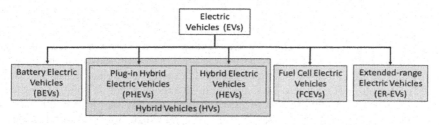

Figure 2.   Caption of a typical figure.

4   BIDIRECTIONAL Z-SOURCE NINE SWITCH INVERTER

We introduce a Z-source nine switch inverter which is bi-directional to save more power in the EV. It includes two AC bidirectional terminals and one DC terminal of the same kind which are substituted by nine switch inverters (NSI). Now we can easily perform the four power handling modes. The battery saves the additional power used by the system. However, BZS-NSI increases the rated values of the switches which result in reduction of the efficiency. [17] We use a developed quasi-Z-source inverter technique to increase the efficiency of the DC-AC inverter. This technique does not require an extra filter on the front-end segment. The bus DC voltage can be stepped up by the application of additional shoot. Here a flat-based control is introduced as well which provides constant frequency giving a high dynamic performance and makes the system robust. There is a reduction of size and the weight of the system. [18]

For a good reliability and life of the system requires the dual traction inverters consisting several numbers of sensors. If the phase current reconstruction strategy is used, the problem can be solved by the use of a sensing device for a single DC-link current. The crucial benefit this provides is: it can also operate even when the phase current sensors are in faulty condition. With the number of sensing devices decreased the overall expenses and the system's volume is also reduced. While the modulation index is improved with the utilization of DC-link voltage [19].

5   MARKET POSITION OF ELECTRIC VEHICLE

Despite of the purchase price is higher compared to other vehicle, the electric vehicle sales volume increases significantly especially from last 10 years. Many countries discourage using the fossil fuel dependent vehicles thus stimulating the hybrid electric vehicle's application. Table 1 explains the market shares of the complete new cars sales during 2013 to 2020. It can observed that Norway has more values in market shares as around more than half of the vehicles sold in the year 2020 are electric vehicles [19].

This picture depicts China and USA have most electric vehicle sales in 2020 [22].

Country	2013	2014	2015	2016	2017	2018	2019	2020
Norway	6.10%	13.84%	22.39%	27.40%	29.00%	39.20%	49.10%	55.90%
Iceland	0.94%	2.71%	3.98%	6.28%	8.70%	19.00%	22.60%	45.00%
Sweden	0.71%	1.53%	2.52%	3.20%	3.40%	6.30%	11.40%	32.20%
The Netherlands	5.55%	3.87%	9.74%	6.70%	2.60%	5.40%	14.90%	24.60%
China	0.08%	0.23%	0.84%	1.31%	2.10%	4.20%	4.90%	5.40%
Canada	0.18%	0.28%	0.35%	0.58%	0.92%	2.16%	3.00%	3.30%
France	0.83%	0.70%	1.19%	1.45%	1.98%	2.11%	2.80%	11.20%
Denmark	0.29%	0.88%	2.29%	0.63%	0.40%	2.00%	4.20%	16.40%
USA	0.62%	0.75%	0.66%	0.90%	1.16%	1.93%	2.00%	1.90%
United Kingdom	0.16%	0.59%	1.07%	1.25%	1.40%	1.90%	22.60%	45.00%
Japan	0.91%	1.06%	0.68%	0.59%	1.10%	1.00%	0.90%	0.77%

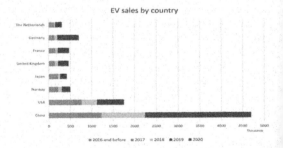

Table 1.   Market share of new sales of electric vehicles.

Figure 2. Electric vehicle (EV) sales in various countries

6   CONCLUSION

This paper discusses various types of electric vehicles and their performance enhancement for various applications. The objective is to look for a newer approach for the pre-existing topologies and also to develop and implement a less expensive, smaller in size, fast charging EV. A novel topology [17] having LCC on the inverter side which is coupled to the CLC-LCL topology can be a suitable area of research in future. By the help of this CLC-LCL strategy a constant current and voltage can be achieved which can be the preferable future work. IPT charging technique can be adopted to get a better efficiency and decrease the interference caused by the magnetic field. The proposed technique may be a very good and highly efficient available solution for charging several electric vehicles with a single inverter in comparatively less expense to face the rise demand electric vehicle and its market.

REFERENCES

[1] Ali Emadi, Kaushik Rajashekara, Sheldon S. Williamson and Srdjan M. Lukic, "Topological Overview of Hybrid Electric and Fuel Cell Vehicular Power System Architectures and Configurations," *IEEE Transactions on Industrial Electronics*, Vol. 54, No. 3, May.2005.

[2] Kumar, A. and Alam, B. (2018). Task Scheduling in Real Time Systems with Energy Harvesting and Energy Minimization *Journal of Computer Science*, Vol. 14, (8), 1126–1133.

[3] Ali Emadi, Sheldon S. Williamson and Alireza Khaligh, Power Electronics Intensive Solution for Advanced Electric, Hybrid Electric and Fuel Cell Vehicular Power Systems," *Power Electronics, IEEE Transactions on*, vol. 21, no. 3, May 2008.

[4] Bimal K Bose and Paul M Szczesny, "A Microcomputer Based Control and Simulation of an Advanced IPM Synchronous Machine Drive System for Electric Vehicle Propulsion," *IEEE Transactions on Industrial Electronics*, Vol. 35, No. 4, November 1988

[5] Sheldon S Williamson, Srdjan M Lukic and Ali Emadi, "Comprehensive Drive Train Efficiency Analysis of Hybrid Electric and Fuel Cell Vehicles Based on Motor – Controller Efficiency Modelling," *Power Electronics, IEEE Transactions*, vol. 21, no. 3 may 2006.

[6] Bilal Akin, Salih Baris Ozturk, Hamid A Toliyat and Mark Rayner, "DSP Based Sensor Less Electric Motor Fault Diagnosis Tools for Electric and Hybrid EV Power Train Applications," *Vehicular Technology, IEEE Transactions*, vol. 58, no. July 2009.

[7] Abdallah ani, Mamadou Bailo Camaro and Baryima Dakyo, "Energy Management Based on Frequency Approach for Hybrid EV Application: Fuel cell/lithium-battery and Ultra-capacitor" Vehicular Technology, *IEEE Transactions*, vol. 61, no. 8, October 2012.

[8] Alireza Khaligh and Serkan Dusmez, "Comprehensive Topological Analysis of Conductive and Inductive Solutions for Plug-in EV," *Vehicular Tech., IEEE Trans*, Vol.61, No.8 Oct. 2012.

[9] Chang-Yeol Oh, Dong-Hee Kim, Dong-Gyun Woo, Won-Yong Sung, Yun-Sung Kim and Byoung-Kuk Lee, "A Highly Efficient Non-isolated Single-stage on Board Battery Charger of EV,". *IEEE Transactions On Power Electronics*, Vol. 28, No. 12, December 2013.

[10] Dong-Gyun Woo, Dong-Myoung Joo, and Byoung-Kuk Lee, "On the Feasibility of Integrated Battery Charger Utilizing Traction Motor and Inverter in Plug-in Hybrid Electric Vehicle,". *Power Electronics, IEEE Transactions*, vol. 30, No.12 Dec 2015

[11] Aree Wangsupphaphol, NikRumziNikIdris, Skudai and Johor, "Acceleration-based Design of EV Auxiliary Energy Source "A& E System Magazines, IEEE Transactions on, No. 10.1109/MAES.2016.140011

[12] Jian Cao, and Ali Emadi "A New Battery/Ultra-capacitor Hybrid Energy Storage System for Electric, Hybrid and Plug-in-hybrid Electric Vehicle" *Power Electronics, IEEE Transactions*, vol. 27, 1 January2017

[13] Jorge Moreno, Micah E. Ortúzar and Juan W. Dixon, "Energy Management System for a Hybrid EV, Using Ultra Capacitor and Neural Network" *IEEE Transactions on Industrial Electronics*, Vol.53, No. 2 April 2006.

[14] Erik Schaltz, Alireza Khaligh and Peter Omand Rasmussen, "Influence of battery/ultra-capacitor energy-system sizing on battery lifetime in a fuel cell HEV," *IEEE Transactions on Vehicular Technology*, vol. 58, no. 8, october 2009.

[15] Farshid Naseri, Ebrahim Farjah and Teymoor Ghanbari, "An Efficient Regenerative Braking System Based on Battery/Super Capacitor for EV, Hybrid and Plug-in-HEV with BLDC Motor," *Vehicular Tech., IEEE Trans*, Vol. 66, No. 5, May. 2017.

[16] Micah Ortúzar, Jorge Moreno and Juan Dixon, "Ultra-capacitor-based Auxiliary Energy System for an EV Implementation and Evaluation," *Ieee Transactions On Industrial Electronics*, vol. 54, no. 4, august 2007.

[17] Yafei Wang, Hiroshi Fujimoto and Shinji Hara, "Torque Distributed-based Extension Control System for Longitudinal Motion of EV by LTI Modelling with Generalised Frequency Variable "*Mechatronics, IEEE/ASME Transactions*, vol. 21, no.1, Feb. 2016.

[18] Dagur, A., Malik, N., Tyagi, P., Verma, R., Sharma, R., & Chaturvedi, R. (2021). Energy Enhancement of WSN using Fuzzy C-means Clustering Algorithm. In *Data Intelligence and Cognitive Informatics: Proceedings of ICDICI 2020* (pp. 315–323). Springer Singapore.

[19] Dagur, A., Kaushik, A., Rastogi, A., Singh, A., Kumar, A., & Chaturvedi, R. (2021). Optimization of Queries in Database of Cloud Computing. In *Data Intelligence and Cognitive Informatics: Proceedings of ICDICI 2020* (pp. 325–332). Springer Singapore.

Artificial Intelligence, Blockchain, Computing and Security – Dagur et al. (Eds)
© 2024 The Author(s), ISBN: 978-1-032-49393-0

Prophet-based energy forecasting of large-scaled solar photovoltaic plant

Akash Tripathi* & Brijesh Singh*
Department of Electrical & Electronics Engineering, KIET Group of Institutions, Ghaziabad, India

Jitendra Kumar Seth*
Department of Information Technology, KIET Group of Institutions, Ghaziabad, India

ABSTRACT: Solar power plants harness photonic energy from the Sun, which is plentiful, accessible, sporadic, and cheap. Due to the increase in installed utility-scale PV plants and the demand for power generation, the solar power generation industry has begun to pay attention to the solar power forecast. Forecasting ensures future energy consumption, economic growth and environmental protection. It also plays an important role in energy demand and supply management. Solar energy forecasting helps private companies and government agencies ensure the proper allocation of available energy resources and assists in decision-making for setting up the energy supply and management infrastructure to meet future demands. Typically, industry and government agencies estimate energy consumption a day in advance. Various types of soft-computing methods are used for this. In this work, using the available statistical data, the forecasting model "fbProphet" is employed to forecast the energy consumption for the next two days. R2 score, MAE score (Mean Absolute Error), and RMSE score (Root Mean Squared Error) criteria have been used to assess the performance of the suggested model. In the proposed work, the prediction for the energy requirements for the next two days in a PV plant in India is simulated based on the available data. The simulation results demonstrate the effectiveness and accuracy of the "fbProphet" forecasting model.

1 INTRODUCTION

The Sun's radiation and available land determine each country's solar potential. This type of power plant is seen as a renewable option because it harnesses the sun's energy, which is a clean, renewable, plentiful and economical source. Sunlight is converted into energy using solar photovoltaics. Materials used in semiconductors are used to make solar photovoltaic cells (for example, silicon). Sunlight causes the atoms of the semiconductor material in the photovoltaic cell to lose their electrons. The free electrons then move through the material to create a direct current (DC) electric current. In short, the light removes electrons from atoms to generate an electric current. The term "photo effect" refers to this phenomenon in physics. Using a special device known as an inverter, the produced direct current (DC) is converted to an alternating current (AC) [1–4]. Although the daily cycle of PV generation is fairly predictable, the inability to accurately predict climatic conditions such as cloud cover will always be an intermittent and non-dispatched force. As a result, the research community has recently focused on creating sophisticated photo voltaic (PV) forecasting models and tools, which can be broadly classified into two primary categories. The first category includes pure physical models, which convert the output of numerical weather prediction (NWP) models

---

*Corresponding Authors: akatripathi11@gmail.com, singhb1981@gmail.com and drjkseth@gmail.com

DOI: 10.1201/9781003393580-133

into PV power generation by carrying out the necessary post-processing computations and often use model output statistics (MOS) to adjust the predictions. use [5–12].

There is little literature on using SARIMAX (Seasonal Autoregressive Integrated Moving Average Model with Exogenous Components) in PV power generation forecasting. Therefore, in the proposed work, an alternative forecasting algorithm, "FacebookProphet", is chosen to observe the algorithm's effectiveness. The performance is evaluated two days before forecasting models [13–16]. This article discusses a useful technique for forecasting the power output of grid-connected PV systems. The FebookProphet model was made available by Facebook as an open-source library and is based on a decomposable (trend + seasonal + holiday) model. This encourages the inclusion of ad hoc seasonal and holiday effects and allows accurate time-series predictions with simple, categorical parameters. According to the parameters R2 score, MAE score (Mean Absolute Error), and RMSE score (Root Mean Squared Error) [17–22], the performance of the suggested model is assessed. The observation shows the effectiveness of energy forecasting for the next two days on large-scale solar plants.

2 PREDICTION MODEL FOR ENERGY CONSUMPTION USING PROPHET

"Facebook Prophet" is an open-source library meant to predict univariate time-series datasets. Univariate is a term most commonly used to describe a data type consisting of observations on only one attribute or characteristic. It is easy to use and designed to automatically find a good set of hyperparameters for the model to make efficient predictions for data with default trend and seasonal structure. The Prophet forecasting model uses a discretized time series model with three main model components: trend, seasonality and holidays. They are combined in the following equation

$$y(t) = g(t) + s(t) + h(t) + \in t \tag{1}$$

where the term "g(t)" refers to a piecewise linear or logistic growth curve used to model non-periodic changes in time series. Holiday impacts (user-provided) with erratic schedules are defined as h(t), where s(t) de-notes periodic changes (e.g. weekly/yearly seasonality). The error term, or t, accounts for anomalous changes that the model cannot. Prophet aims to fit many linear and nonlinear functions as components using time as a regressor. For Prophet to be used for forecasting, a Prophet () object must be defined and configured; this object accepts arguments for configuring the required model type, such as growth type and seasonality. The essential characteristics of the Prophet are next examined.

2.1 Trend parameters

When modelling a trend, a piecewise linear curve is fitted to the non-periodic portion of the trend or time series. The spikes and missing data caused by the linear fitting exercise affect it the least. Growth, changepoints, n_changepoints, and changepoint_prior_scale are described in Table 1 under trend parameters [22].

Table 1. Trend parameters of Prophet.

Parameter	Description
growth	For a linear or logistic trend, use the terms "linear" or "logistic"
changepoints	Dates in a list where potential changepoints should be included (automatic if not specified)
n_changepoints	It may specify the number of changepoints that will be automatically included if changepoints are not specified
changepoint_prior_scale	Changeable automated changepoint flexibility parameter

2.2 Seasonality & holiday parameters

Prophet uses the Fourier series as a flexible model to fit and predict the impacts of seasonality. The following function roughly represents seasonal impacts s(t)

$$s(t) = \sum_{n=1}^{N} \left(a_n \cos \left(\frac{2\pi nt}{P} \right) + b_n \sin \left(\frac{2\pi nt}{P} \right) \right) \tag{2}$$

In order to represent seasonality, parameters [a1, b1, ..., aN, bN] must be determined for a given N. The period (P) is 365.25 for yearly data and 7 for weekly data. A time series experiences predictable shocks related to holidays and events. For instance, a different day is chosen for the Indian festival of Holi, during which a large portion of the population purchases many new goods. Such Seasonality and Holiday Parameters are shown in Table 2 and a thorough discussion.

Table 2.   Seasonality with holiday parameters and description.

Parameter	Description
yearly_seasonality	The fitness value of yearly seasonality
weekly_seasonality	The fitness value of weekly seasonality
daily_seasonality	The fitness value of daily seasonality
holidays	Name and date of feed data frame containing holiday
seasonality_prior_scale	Strength of the seasonality model parameter with changes
holiday_prior_scale	Strength of the holiday model parameter with changes

The Prophet library also offers capabilities for automatically evaluating models and visualizing findings. It is undoubtedly a wise choice for making prompt, precise projections. Anyone with a sufficient subject understanding of forecasting models can adjust its intuitive parameters.

3   EXPERIMENT AND RESULTS

The chosen forecasting models were intentionally applied for forecasting PV power generation under a day-ahead PV plant near Gandikota, Andhra Pradesh, India. Its nominal capacity is 2 KMW. For the Experiment, Solar Power Generation Data and Weather Sensor Data were recorded at 15-minute intervals over 34 days (From 15 May 2020 to 17 June 2020) from a solar power plant in India. This Solar Plant is near Gandikota, Andhra Pradesh, India. The statistics for power generation are obtained at the inverter level; each inverter has many solar panel lines attached. The single array of sensors at the facility, which is arranged optimally, is where the sensor data is acquired. Two pairs of datasets are created for trials:

- DC power, ac power, daily yield, and total yield are the variables in the power-generating dataset.
- The variables in Weather Sensor Dataset are ambient temperature, module temperature, and irradiation.

A persistence model is constructed to assess the efficacy of the suggested pre-casting algorithm. This study defines persistence as the PV generation at each 15-min interval equal to the total PV generation earlier that day. The Facebook Prophet model has been used to analyze the polynomial sequence of the model for seasonal and non-seasonal components. The performance has been evaluated using RMSE. In this work, the forecast for the next day is made for two specific times of the year, May and June. The simulation for forecasting started at 0:00 hrs on 15th May and is simulated for 24 hours at 15-minute intervals.

3.1 Prediction using fbProphet

One of the benefits of Prophet is that it can be considerably faster than the ARIMA forecasting process, especially when dealing with a long time series; nevertheless, we must keep in mind that this method is still in progress and has not been confirmed. Figure 1 depicts the variation in the daily output of Solar Power Plant 1 from 17 May 2020 to 17 June 2020. Figure 2 displays the replay results for the 15-minute interval forecast (18/06/2020 – 19/06/2020) and compares the performance of Mod-L to the original Prophet forecast data. Search results were obtained using the predict() function of Facebook Prophet and a data frame with several columns. The forecast uncertainty intervals in Table 3 consist of the forecast date-time ('ds'), the estimated or anticipated value ('th'), and the lower and upper bounds on the predicted value ('lower' and 'upper'). These three columns are crucial for forecast comprehension.

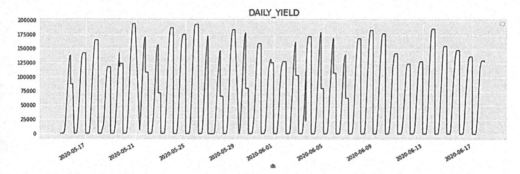

Figure 1.   Daily yield.

3.2 Test results and discussions

It is vital to analyze the accuracy of forecasts to determine which model is superior. Therefore, residual size may not adequately represent the number of underlying forecast errors. Considering only how well a model predicts new data is insufficient for determining its accuracy. Any approach cannot be utilized unless tailored to the appropriate model. This is the most important condition for evaluating the forecast's accuracy. Selecting the appropriate model to divide the supplied data into training and test segments. It is known that training data is used to estimate the parameters of any forecasting technique, while test data is used to evaluate the accuracy of the forecast. Since test data are not used to create predictions, they should demonstrate how well the model predicts new data. The difference between actual and expected numbers is referred to as the "error" in the forecast. Rather than a mistake, "error" here refers to an unanticipated element of an observation. This equation can be expressed as $eT + h = yT + h\ T + h|T$. The training data is represented by $y1, \ldots, yT$, while the test data is represented by $yT + 1, yT + 2, \ldots$ We can assess forecast accuracy by summing forecast errors in several different ways. By comparing the statistical criteria R2 score (Coefficient of Determination), MAE score (Mean Absolute Error), and RMSE score (Root Mean Square Error), the best model was determined (Root Mean Squared Error). As seen in Table 4, it is readily apparent that the suggested FbProphet prediction produces superior results. -Predict/Edit with open-source code [22].

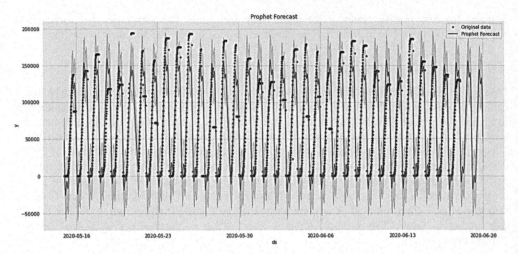

Figure 2. Prophet forecast.

Table 3. Seasonality and holiday parameters and description.

Forecast date-time (d.s.)		Forecasted or predicted value	Lower and upper bounds on the predicted value	
			LOWER	UPPER
2020-05-15	00:00:00	47997.735870	14689.351070	83166.924283
2020-05-15	00:15:00	35600.943422	2825.758906	68950.185576
2020-05-15	00:30:00	23659.660718	−8438.283130	57763.496562
2020-05-15	00:45:00	12513.424255	−18633.258224	44829.977858
2020-05-15	01:00:00	2462.536479	−31847.507334	33889.962331

Table 4. Prophet based forecasting.

Model	MAE Score	RMSE Score	R2 Score
FbProphet	13359.251989	18427.215350	0.895611

4 CONCLUSION

This study evaluated the predictive model FbProphet for PV generation for efficacy. FbProphet's implementation is simple and sophisticated. With a little work and time-series analysis subject matter expertise, this model produces accurate forecast results even when using default arguments or parameters. The FbProphet-based algorithm offers next-day electricity markets, including solar power plants, accelerated convergence and worldwide solutions.

REFERENCES

[1] Sukhatme S. P., Book *"Solar Energy"*, Tata McGraw Hill Publication.
[2] Udi Helman 2014, *Economic and Reliability Benefits of Large-Scale Solar Plants in Renewable Energy Integration*

[3]   https://medium.com/@solar.dao/how-pv-solar-plants-work-a-beginners-guide-79f085b8ee88

[4]   IEA-PVPS 2015, *Snapshot of Global PV Markets*," Report IEA PVPS T1-26. Available: http://www.ieapvps. org/fileadmin/dam/public/report/technical/PVPS_report_A_Snapshot_of_Global_PV_-_1992-2014.pdf

[5]   Diagne HM, David M., Lauret P., and Bolan J. 2012 Solar Irradiation Forecasting: State-of-the-art and Proposition for Future Developments for Small-scale insular Grids, *Proc. of WREF 2012*, Denver, Colorado.

[6]   Kardakos E. G, Alexiadis M. C., Vagropoulos S. I., Simoglou C. K., Biskas P. N., Bakirtzis A. G., 2013 Application of Time Series and Artificial Neural Network Models in Short-term Forecasting of PV Power Generation, *Power Engineering Conference (UPEC), 2013 48th International Universities', Dublin.*

[7]   Rob J Hyndman 2013 *Forecasting: Principles and Practice, George Athanasopoulos.*

[8]   N. Liu, V. Babushkin and A. Afshin 2014 Short-Term Forecasting of Temperature Driven Electricity Load Using Time Series and Neural Network Model, *Journal of Clean Energy Technologies*, vol. 2, no.4.

[9]   https://otexts.com/fpp2/accuracy.html

[10] www.analyticsvidhya.com/blog/2018/05/generate-accurate-forecasts-facebook-prophet-python

[11] https://machinelearningmastery.com/sarima-for-time-series-forecasting-in-python/

[12] Rob J. Hyndman 2010 *The ARIMAX Model Muddle.* Available: http://robjhyndman.com/hyndsight/ arimax/

[13] Alam, B., & Kumar, A. (2014, March). A Real Time Scheduling Algorithm for Tolerating Single Transient Fault. In *2014 International Conference on Information Systems and Computer Networks (ISCON)* (pp. 11–14). IEEE.

[14] Palvika, Shatakshi, Sharma, Y., Dagur, A., & Chaturvedi, R. (2019). Automated bug Reporting System with Keyword-driven Framework. In *Soft Computing and Signal Processing: Proceedings of ICSCSP 2018*, Volume 2 (pp. 271–277). Springer Singapore.

[15] Kumar, A., & Alam, B. (2019). Energy Harvesting Earliest Deadline First Scheduling Algorithm for Increasing Lifetime of Real Time Systems. *International Journal of Electrical and Computer Engineering*, 9(1), 539.

[16] Kumar, A. & Alam, B. (2018). Task Scheduling in Real Time Systems with Energy Harvesting and Energy Minimization. *Journal of Computer Science*, 14(8), 1126–1133.

[17] Kumar, A., & Alam, B. (2014, February). Real Time Scheduling Algorithm for Fault Tolerant and Energy Minimization. In *2014 International Conference on Issues and Challenges in Intelligent Computing Techniques (ICICT)* (pp. 356–360). IEEE.

[18] Kelly R., Modarres R. and Ali Sarhadi 2014 *Snow Water Wquivalent Timeseries Forecasting in Ontario, Canada, in Link to Large Atmospheric Circulations Hydrological Processes*, vol. 28, no. 16, pp. 4640–4653.

[19] Reikard G. 2009 Predicting Solar Radiation at High Resolutions: A Comparison of Time Series Forecasts *Solar Energy*, vol. 83, issue. 3, pp. 342–349.

[20] Mellit A., Benghanem M., Arab A. H., and Guessoum A. 2005 A Simplified Model for Generating Sequences of Global Solar Radiation Data for Isolated Sites: Using Artificial Neural Network and a Library of Markov Transition Matrices Approach *Solar Energy*, vol. 79, no. 5, pp. 469–482.

[21] Kaggle: https://www.kaggle.com/anikannal/solar-power-generation-data

[22] Subashini A., Sandhiya K, Saranya S, and Harsha U 2019 Forecasting Website Traffic Using Prophet Time Series Model", *International Research Journal of Multidisciplinary Technovation.*

Artificial Intelligence, Blockchain, Computing and Security – Dagur et al. (Eds)
© *2024 The Author(s), ISBN: 978-1-032-49393-0*

Skin cancer detection by using squeeze and excitation method

Shaurya Pandey & Rishabh Dhenkawat
National Institute of Technology, Hamirpur

Shekhar Yadav
Madan Mohan Malaviya University of Technology, Gorakhpur

Nagendra Pratap Singh
Dr. B.R. Ambedkar National Institute of Technology, Jalandhar

ABSTRACT: Skin cancer is a serious medical problem. So, finding a clear clinical diagnosis is the top priority for doctors. To identify and categorize skin cancer, some processes are now being developed in the field of image processing with the aid of algorithms and systems. Skin cancer signs can be easily, affordably, and quickly diagnosed thanks to computer-based technology. Numerous non-invasive techniques have been suggested to look into the signs of skin cancer, whether they are caused by melanoma or not. Picture capture, pre-processing, segmentation of the acquired pre-processed image, extraction of the necessary features, and classification of the illness of skin cancer comprise the general procedures used in skin cancer detection.

1 INTRODUCTION

The need for early detection and the increased incidence of skin cancer make it imperative to create a reliable system for automatically classifying skin cancer. The skin is the largest organ in the human body, and since it is tasked with guarding other bodily systems, it is more susceptible to illness. With around 300,000 new cases identified globally in 2018, melanoma was the most common kind of cancer in both men and women. Melanoma was the most common form of cancer. Basal cell carcinoma (BCC) and squamous cell carcinoma (SCC), two additional prominent skin cancer diseases, both had a very high incidence, with over 1 million occurrences in 2018. Melanoma was the only significant skin cancer illness that had a reasonably high prevalence. According to reports, the United States diagnoses more cases of skin cancer each year. than all other malignancies combined. Fortunately, there is a much better chance of fewer errors detected in the findings produced by the approaches recovery if the condition is caught early. When cancer does not metastasize, melanoma has a survival rate of 99 percent after five years, but after that, the likelihood lowers to 20 organs. However, due to the fact that the early indicators of skin cancer are not always easy to spot, precise Diagnostic outcomes sometimes require the expertise of a dermatologist. Unskilled practitioners can make more accurate diagnosis with the use of a critical instrument called an automated diagnosis system. Beyond that, it is extremely subjective and infrequently generalizable to diagnose. skin cancer with the naked eye detection of skin cancer On the dataset used in this study, we will apply a well-known squeeze and excitation block model.

DOI: 10.1201/9781003393580-134

The main contributions of this paper are as follows:

1. We carefully analysed the research on the identification of cancer disorders.
2. Fewer errors are detected in the findings produced by the approaches.

2 RELATED WORKS

Deep learning has become the new benchmark because it is obvious that it handles the lion's share of the job. Deep learning can automatically extract features; however, handmade features need to be manually extracted, which is the fundamental cause of this. As a result, we give a summary of the associations that the idea of deep learning contains in this section. According to (Ahn *et al.* 2017) it was possible to detect a new background and more precisely distinguish the lesion from the surrounding areas by leveraging reconstruction errors brought on by a sparse representation model. A Bayesian model was also developed in this work to more accurately describe the lesion's boundaries and size. On two publicly available datasets that each contained 1100 dermoscopic images, the proposed approach was assessed. Additionally, it was contrasted with alternative approaches. In comparison to other approaches, the findings of the current scheme exhibited more accuracy and robustness in segmenting lesions. This work also held a discussion on a universal expansion of new architecture as a saliency optimization approach for segmentation of lesions.

A mobile healthcare strategy for melanoma detection using mobile picture analysis was proposed by (Do *et al.* 2018) after two years. The key components of the proposed model included an efficient categorical segmentation method suitable for a platform with constrained resources, a novel feature set that successfully captured colour difference and bound anomaly from an image taken with a smartphone, and a new tool for selecting a small group of the most discriminatory set of attributes. The outcomes showed that the suggested the approach was correctly classified and segmented 184 camera images of skin lesions. (Shahi *et al.* 2018) used SVM, KNN, decision trees, and ensemble classifiers, among other techniques, were used to categorise the recovered characteristics of the lesions in the same year. However, when using other classifiers, the accuracy was lower. Here, they employed an SVM classifier to achieve an accuracy of 100.0. Additionally, (Díaz *et al.* 2019) proposed an embedded hardware-based commodity device. This system applied an automated technique for the first screen for skin cancer with the use of dynamic thermal imaging. image segmentation in the visible range and in the actual range were both utilised in the technique. Using multimodal registration, The temperature recovery curve (TRC) of problematic skin lesions is generated. TRC stands for"Temperature Recovery Curve. Two approaches were used by the programme to evaluate the lesion. lethality using the derived TRCs. In order to create a model of the device and get a visual image Of course, a Raspberry Pi 3 Model B+ board was employed in this project. Based on a sample size of 116 willing participants, the statistical detection algorithmic technique produced results. with a specificity and sensitivity of 95 and 98, respectively. In order to increase the accuracy of the results, In various skin (semantic segmentation) lesion datasets made up of noisy expert observations, (Goyal *et al.* 2019) designed fully automated ensemble deep learning schemes using two efficient segmentation approaches, namely DeepLab v3+ and Mask R-CNN (instance segmentation). The productivity of the merged network was examined using two datasets, the PH2 and the ISIC.2017 test set. The findings confirmed that the developed procedures were capable of segmenting. skin lesions with a sensitivity of up to 89.93, and a specificity of around 97.94. An efficient and economical near-field investigation was suggested by (Mansutti *et al.* 2020) in order to diagnose skin cancer in its early stages. Additionally, by utilising substrate-integrated waveguide (SIW) technology, our work ensured an uncomplicated and affordable fabrication operation. This property was necessary to verify the usefulness and efficacy of the proposed framework as a technique for quickly scanning a variety of dubious skin locations.

The main goal of the examine design was to operate at roughly 40 GHz. to detect small skin cancer tumours while preventing the fields from interacting with the tissues. of basic organisms The probe might be effective on all skin types and body parts. The mechanical gadget used for the demonstration has a 0.20 mm lateral sensitivity and a 0.55 mm detecting depth.

The very next year, (Song *et al.* 2020) proposed a multi-task E2E (End-to-End) DL (Deep Learning) model to autonomously analyse the skin lesion. The skin lesion may be detected, classified, and segmented using this approach. When the suggested model was trained, a three-fold joint training technique was used to make sure that the feature learning was successful. The results on the melanoma classification challenge dataset from ISBI 2016 and the melanoma segmentation challenge dataset from ISIC 2017 showed that the suggested model outperformed the traditional methods. Additionally, this model was used as the best computer-aided method for melanoma diagnosis. Additionally, a new model for segmenting and categorising skin lesions for the automatic diagnosis of skin cancer was provided by (Adegun & Viriri 2020) The initial phase made use of an encoder-decoder FCN (Fully Convolutional Network), where the encoder stage learned the coarse form to understand the traits of complicated and irregular skin lesions and the decoder stage learned the specifics of the lesion borders. A brand new dense net architecture built on FCN, was unveiled in the next phase. The concatenation approach, a transition layer, and dense blocks were used to combine and couple the blocks in this architecture. In order to evaluate the provided framework, more than 10,000 freely available HAM10000 datasets were employed, yielding accuracy, recall, and AUC scores of 98, 98.5, and 99, respectively. A 77 GHz low-energy multitone continuous-wave radar system for skin cancer detection in biomedical imaging was presented by (Arab Salmanabadi *et al.* 2020) in the same year once more. An cheap MHMIC (Miniature Hybrid Microwave Integrated Circuit) was required for the development of the sensor. The baseband voltage at each power detector's output was amplified by this approach using two cascade amplifier circuits. The suggested radar design has the capability to identify the dielectric characteristics of tissues to boost its adaptability for precisely detecting melanoma, approximately tens of microns in size. Additionally, (Ashraf *et al.* 2020) showed a ROI (region of interest). inspired model that can recognise and separate melanoma from nevus cancer using a transfer learning technique. This model used an enhanced k-mean technique to extract the necessary image regions. As only photos containing melanoma cells were used for the model training, These ROI-centric approaches assisted in discovering discriminating features. In addition, a transfer learning model built on ConvNet and data augmentation for the targeted image regions were used in this study. The findings showed that the new system successfully detected skin cancer with an accuracy of up to 97.9 for DermIS and 97.4 for DermQuest. A fine-grained classification rule based on a mild skin cancer detection architecture with feature differences was presented by (Wei *et al.* 2020) in Two Lesion Classification Network universal feature extraction units made up the new framework. This study's recognition method had the ability to eliminate more stigmatising lesion features. In order to achieve edge-to-edge high precision lesion region segmentation without involved image pre-processing operations, a light semantic segmentation specimen of the lesion region of the dermos copy picture was generated. The results of the analysis done on ISBI 2016 showed that the framework under investigation is better than others at spotting skin cancer. In order to successfully detect the malignant melanoma, (Kwasigroch *et al.* 2020) concentrated on building an architecture search algorithm. The hill-climbing search technique with network morphism operations was used to explore the search space and find an acceptable network structure. Through the previously trained networks, these procedures helped to maximise the network size. By reusing this knowledge, the computational cost was reduced. The experiment's findings demonstrated how the suggested strategy may lower the computing cost. A method to identify and segment melanoma was suggested by (Albahli *et al.* 2020) From an accuracy standpoint, this method produced significant improvements. Sharpened image areas were produced in the first stage, which

902

comprised eliminating artefacts from dermoscopic images using morphological methods. This work used a correctly calibrated YOLOv4 object detector for melanoma diagnosis by differentiating between highly correlated diseased and unaffected areas in order to find the impacted area. Two datasets, ISIC 2018 and ISIC 2016, were used in this study to evaluate the proposed system. The YOLOv4 detector had the ability to identify numerous skin conditions in the same patient as well as numerous conditions in various people. The problem of fusing images and metadata attributes to categorise skin cancer with deep learning was tackled by (Pacheco & Krohling 2021) This work presented the Metadata Processing Block, also known as MetaBlock, which is a revolutionary method that makes use of metadata to assist in the classification of data This method does this by enhancing the most important features that were extracted from the images while the data was being classified. Two different combination strategies, one of which made use of feature concatenation and the MetaNet, were compared to the suggested strategy. Results from two separate datasets on skin lesions show that our method works better than earlier combinations approaches in six out of ten cases and improves categorization for each model tested. In addition, (Thurnhofer-Hemsi *et al.* 2021) introduced a method known as test-time regularly spaced shifting and an ensemble of augmented convolutional neural networks for the categorization of skin lesions. Both of these were done in order to study the classification of skin lesions. The shifting approach produces numerous copies of the test input image by shifting the relocation vector using a regular lattice in the potential shift zone. These altered versions of the training photos will then be distributed to each ensemble member. The end result is created by combining the output from each classifier. The experiment findings show a significant improvement in accuracy and F-score over the well-known HAM10000 dataset. It has been demonstrated that our approach, which combines evenly spaced test-time shifting with clothes, generates results that are superior to those obtained using each of the two strategies separately. An innovative single-model-based strategy for diagnosing Skin lesions on the basis of sparse and unbalanced datasets were also developed by (Yao *et al.* 2022) In order to demonstrate that models of intermediate complexity outperform the more sophisticated models, a range of DCNNs were first trained on a variety of small and unbalanced datasets. Second, regularisation methods like DropOut and DropBlock were employed to reduce overfitting in order to address the shorter dataset's sample under representation concerns. Additionally suggested is a modified RandAugment augmentation strategy. An end-to-end cumulative learning approach (also known as CLS) was developed in order to overcome the problem of uneven sample size and classification complexity, as well as to decrease the impact of aberrant data on training. The application of this method to mobile devices for automated screening of skin lesions and many other malignancies in low-resource conditions had the potential to be significant. This was due to the great classifier performance it offered at a low cost in terms of both system resources and the amount of time required for inference.

3 METHODOLOGY

After a thorough analysis of the human skin cancer illness identification, the study will get under way. We'll use a dataset of malignant and non-cancerous photos from Kaggle to train the CNN model (SENet) On the basis of experiments, the data collected will be assessed, and the performance will be decided. The following is a list of the essential steps that need to be taken.

3.1 *Dataset distribution*

9,605 photographs make up the dataset gathered through Kaggle, which is further divided into 4,605 images of malignancy and 5,000 images of benignity Table 1 shows how these subdivisions are once again divided into two datasets that are utilised in the ratio 80:20 to both train and Test the models.

Dataset (9605 images)			
Malignant images(4605)		Benign images(5000)	
Train images (3684)	Test images (921)	Train images (4000)	Test images (1000)

Figure 1.   Dataset distribution.

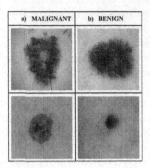

Figure 2.   Images of a) malignant cancer b) benign cancer.

3.2   *Convolutional Neural Network (CNN)*

Convolutional neural networks are one of the deep learning methods for image processing that are most frequently employed (CNNs). CNNs were modelled after the animal visual cortex. They are among the first deeply successful systems that are excellent at both video and image processing. With the aid of modern GPU-accelerated computing approaches, decision support systems for Picture classification and object recognition have effectively implemented CNNs. According to in the current study, deep networks are effective methods for interpreting medical images. They therefore offer a great deal of potential for the classification of melanoma. Particularly effective for this purpose have been CNN ensemble methods.

3.2.1   *Convolution layer*
The foundation of CNN is the convolutional operation. It manages the total computational complexity of the network. In this stage, dual matrices are merged to create a dot product, which is the kernel—a collection of trainable parameters in the perceptron's restricted region. The kernel is larger than an image yet takes up less space.

3.2.2   *Pooling layer*
The pooling layer is responsible for calculating overall statistics from the outputs of neighbouring nodes in order to substitute the output of the network in certain regions. The

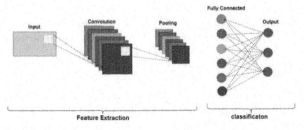

Figure 3.   The architecture of deep CNN model.

amount of calculation and weights that are necessary can be cut down significantly if the size of the picture that is being represented by decreased. The pooling approach is implemented in its entirety by each individual. slice that makes up the representation.

3.2.3 *Fully connected layer*
Every neuron in this layer is closely connected to every other neuron in every layer before and after it, much like a traditional CNN. The bias effect and standard matrix multiplication can be used to calculate it as a result. The FC layer transfers the expression to the distance between the input and output.

3.3 *SENet*

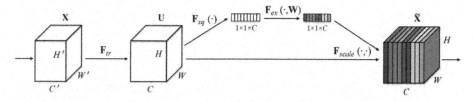

Figure 4.   The architecture of SENet.

A network's representational power can be increased by using the squeeze-and-Excitation Block is an architectural component that enables dynamic channel-wise feature recalibration.

A convolutional block is used as the block's input. The process of applying average pooling"compresses" the data from each channel into a single numerical number.

Because of an additional dense layer that is followed by a sigmoid, each channel possesses a gating function that is completely smooth.

The"Excitation" side network is used to weight each feature map of the convolutional block.

3.4 *SE Block*

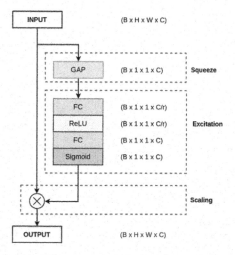

Figure 5.   The architecture of SE block.

In conventional CNNs, the pooling method known as global average pooling is meant to fill the role of completely connected layers.

The required nonlinearity is added by a fully linked layer, followed by a ReLU function. It also has a certain proportional reduction in the complexity of the output channel.

Following the application of a sigmoid activation, a smooth gating function is provided for each channel by a second layer that is totally connected.

The "Scale" layer multiplies the output values of the sigmoids with each corresponding channel of the original layer.

4 RESULTS

At different epochs, various experiments are conducted. The outcomes that were attained in Diverse contexts are listed in the sections that follow.

Epoch	10	15	20	25
Accuray	82.95	84.93	89.56	88.49

We were able to achieve maximum accuracy. during the experimentation at 20 epochs. The graphs representing the results are shown below.

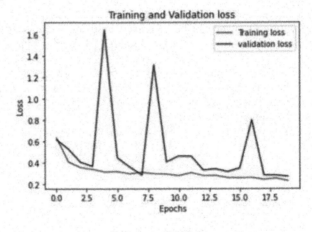

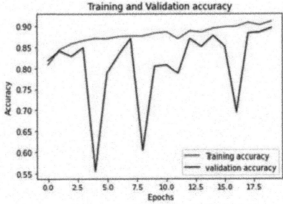

Figure 6.   Graphs of SENet based on loss and accuracy respectively.

REFERENCES

Adegun, Adekanmi, and Serestina Viriri. 2020. "FCN-Based DenseNet Framework for Automated Detection and Classification of Skin Lesions in Dermoscopy Images." *IEEE Access* PP (August) 1–1. https://doi.org/10.1109/ACCESS.2020.3016651.

Ahn, Euijoon, Jinman Kim, Lei Bi, Ashnil Kumar, Changyang Li, Michael Fulham, and David Dagan Feng Feng. 2017. "Saliency-Based Lesion Segmentation Via Background Detection in Dermoscopic Images." *IEEE Journal of Biomedical and Health Informatics* 21 (November) 1685–1693. https://doi.org/10.1109/JBHI.2017.2653179.

Albahli, Saleh, Nudrat Nida, Aun Irtaza, Muhammad Haroon Yousaf, Muhammad Tariq, and Muhammad Mahmood. 2020. "*Melanoma Lesion Detection and Segmentation Using YOLOv4-DarkNet and Active Contour*" (November) https://doi.org/10.1109/ACCESS.2020.3035345.

Arab Salmanabadi, Homa, Lydia Chioukh, Mansoor Dashti Ardakani, Steven Dufour, and S. Tatu. 2020. "Early-Stage Detection of Melanoma Skin Cancer Using Contactless Millimeter-Wave Sensors." *IEEE Sensors Journal* 20 (July) 7310–7317. https://doi.org/10.1109/JSEN.2020.2969414.

Ashraf, Rehan, Sitara Afzal, Attiq Rehman, Sarah Gul, Junaid Baber, Maheen Bakhtyar, Irfan Mehmood, Oh-Young Song, and Muazzam Maqsood. 2020. "Region-of-Interest Based Transfer Learning Assisted Framework for Skin Cancer Detection." *IEEE Access* PP (August) 1–1. https://doi.org/10.1109/ACCESS.2020.3014701.

Díaz, Silvana, Thomas Krohmer, Alvaro Moreira, Sebastián Godoy, and Miguel Figueroa. 2019. "An Instrument for Accurate and Non-Invasive Screening of Skin Cancer Based on Multimodal Imaging." *IEEE Access* PP (November) 1–1. https://doi.org/10.1109/ACCESS.2019.2956898.

Do, Thanh-Toan, Tuan Hoang, Victor Pomponiu, Yiren Zhou, Zhao Chen, Ngai-Man Cheung, Dawn Koh, Aaron Tan, and Suat-Hoon Tan. 2018. "Accessible Melanoma Detection Using Smartphones and Mobile Image Analysis." *IEEE Transactions on Multimedia* 20 (10) 2849–2864. https://doi.org/10.1109/TMM.2018.2814346.

Goyal, Manu, Amanda Oakley, Priyanka Bansal, Darren Dancey, and Moi Hoon Yap. 2019. "Skin Lesion Segmentation in Dermoscopic Images With Ensemble Deep Learning Methods." *IEEE Access* PP (December) 1–1. https://doi.org/10.1109/ACCESS.2019.2960504.

Kwasigroch, Arkadiusz, Michał Grochowski, and Agnieszka Mikołajczyk. 2020. "Neural Architecture Search for Skin Lesion Classification." *IEEE Access* PP (January) 1–1. https://doi.org/10.1109/ACCESS.2020.2964424.

Mansutti, Giulia, Ahmed Mobashsher, Konstanty Bialkowski, Beadaa Mohammed, and Amin Abbosh. 2020. "Millimeter-Wave Substrate Integrated Waveguide Probe for Skin Cancer Detection." *IEEE Transactions on Biomedical Engineering* 67 (September) 2462–2472. https://doi.org/10.1109/TBME.2019.2963104.

Pacheco, Andre, and Renato Krohling. 2021. "An Attention-Based Mechanism to Combine Images and Metadata in Deep Learning Models Applied to Skin Cancer Classification." *IEEE Journal of Biomedical and Health Informatics* PP (February) 1–1. https://doi.org/10.1109/JBHI.2021.3062002.

Shahi, Preeti, Shekhar Yadav, Navdeep Singh, and Nagendra Pratap Singh. 2018. "Melanoma Skin Cancer Detection Using Various Classifiers." In *2018 5th IEEE Uttar Pradesh Section International Conference on Electrical, Electronics and Computer Engineering (UPCON)* 1–5. https://doi.org/10.1109/UPCON.2018.8597093.

Song, Lei, Jianzhe Lin, Z. Jane Wang, and Haoqian Wang. 2020. "An End-to-End Multi-Task Deep Learning Frame-work for Skin Lesion Analysis." *IEEE Journal of Biomedical and Health Informatics* 24:2912–2921.

Thurnhofer-Hemsi, Karl, Ezequiel López-Rubio, Enrique Domínguez, and David Elizondo. 2021. "Skin Lesion Classification by Ensembles of Deep Convolutional Networks and Regularly Spaced Shifting." *IEEE Access* PP (August) 1–1. https://doi.org/10.1109/ACCESS.2021.3103410.

Wei, Lisheng, Kun Ding, and Huosheng Hu. 2020. "Automatic Skin Cancer Detection in Dermoscopy Images Based on Ensemble Lightweight Deep Learning Network." *IEEE Access* 8:99633–99647.

Yao, Peng, Shuwei Shen, Mengjuan Xu, Peng Liu, Fan Zhang, Jinyu Xing, Pengfei Shao, Benjamin Kaffenberger, and Ronald X. Xu. 2022. "Single Model Deep Learning on Imbalanced Small Datasets for Skin Lesion Classification." *IEEE Transactions on Medical Imaging* 41 (5) 1242–1254. https://doi.org/10.1109/TMI.2021.3136682.

Artificial Intelligence, Blockchain, Computing and Security – Dagur et al. (Eds)
© 2024 The Author(s), ISBN: 978-1-032-49393-0

Augmentation of medical image dataset using GAN

Harsh Sheth & Samridhi Singh
Department of Computer Science and Engineering, National Institute of Technology, Hamirpur, H.P., India

Nagendra Pratap Singh
Dr. B.R. Ambedkar National Institute of Technology Jalandhar

Priyanka Rathee
Department of Computer Science and Engineering, National Institute of Technology, Hamirpur, H.P., India

ABSTRACT: The medical image dataset is generally very scarce due to laws that do not allow the sharing of patients' data without consent. The cost of annotating and gathering medical data is also very high and time-consuming. Because of the above-mentioned reasons, datasets in medical domains are generally not balanced. There is no guarantee that a deep neural network model trained These types of datasets will correctly classify the medical condition and deliver the desired results. It may also overfit the majority-class samples. To solve this problem, data augmentation is often performed on data to increase dataset sizes using techniques of position augmentation. like cropping, scaling, rotation, flipping, translation, and augmentation of colour techniques like contrast, saturation, and hue These augmentation steps may cause further overfitting in certain regions. with insufficient data, such as medical image data, and are not guaranteed to be advantageous in such domains. We experimented and tried to generate artificial images from the retinal image dataset. using generative adversarial networks that generate new instances with features similar to the original data. The model evaluation produced a FID score of 3.29.

1 INTRODUCTION

Privacy concerns limit the quantity of medical imaging data sets, and annotating medical pictures is costly and time-consuming, which frequently results in modest volumes of labelled medical imaging data, which cannot be efficiently used in ML/DL-based image processing tasks.

In general, deep learning requires a vast amount of data because it is applied on unsupervised data, i.e., it is the job of the network to learn about the features of the dataset. Therefore, we need vast amount of data so that our model can understand the unique features present in the images. To mitigate this snag, scientists often use data augmentation. It is a technique to expand the size of the dataset. Our goal in this study is to implement a working model using Deep Convolutional Generative Adverserial Networks to enrich the original dataset with retinal images. In 2014, Ian Goodfellow and his colleagues introduced GANs (Goodfellow *et al.* 2014) Figure 1 gives us a high-level description of the architecture of GAN. The primary components of a GAN are:

- Latent tensors
- Generator
- Discriminator

DOI: 10.1201/9781003393580-135

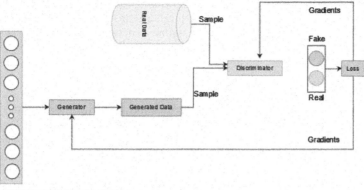

Z("n" dimensional RandomNoise)

Figure 1.   Architecture of generative adverserial network.

- Loss function
- Dataset containing real images.

As depicted in Figure 1, GANs consist of two neural networks that compete against each other. The job of the generator is to construct samples (data) from random noise, while the job of the discriminator is to distinguish those samples between genuine and fake data.

2   RELATED WORKS

GANs have emerged as a promising technology for researchers all over the globe, who often face a shortage of data. GANs can be used to augment datasets artificially and are widely used in medical imaging, where data is scarce. Here we present some of the notable works in the field of medical image synthesis.

In 2014, (Goodfellow *et al.* 2014) proposed a new framework for approximating models that are generative in nature via an adverserial process in which they train two neural networks simultaneously. This framework later came to be known as GAN (Generative Adverserial Network). (Costa *et al.* 2018) proposed a model capable of synthesising novel vascular networks and associated pictures of the eye's fundus. Their approach discovers the fundamental structure of the multitude of potential retinal pictures by analysing instances of vascular networks and eye fundus images. (Chuquicusma *et al.* 2017) in their study, they created realistic lung nodule samples for the first time using a DC-GAN design. Radiologists were fooled by the manufactured samples in their trials. (Costa *et al.* 2017) Applying existing generative adversarial models, the visual and quantitative findings indicate that GAN can learn to synthesise new retinal images if retinal artery trees and matching retinal images are given in a dataset. In addition, the created pictures have a resolution of 512 by 512 pixels, which is larger than images typically generated for ordinary computer vision tasks. (Kora & Ravula 2020) used DCGAN to generate chest X-ray images and got a phenomenal FID score of 1.289. (Baur *et al.* 2018) compared the DCGAN, the LAPGAN, and the DDGAN and showed that even with a limited training dataset, the LAPGAN and the DDGAN can produce varied and realistic samples that resemble the distribution of the training dataset. (Borji 2018) in his paper, provides and lists the strengths and limitations of all qualitative and quantitative measures available for GAN. Jelmer M. Wolterink *et al.* (Wolterink *et al.* 2018) created synthetic blood artery shapes and introduced a generative approach in which the latent space dimension is kept minimal and the generative model is trained on this latent space that generates the shape of whole coronary arteries. (Karras *et al.* 2020) proposed a unique approach to augmenting training data for GAN such that the augmentation does not

909

leak to the generated samples. The method works as well for training a GAN from the beginning as it does for an existing GAN on another dataset that is to be fine-tuned, and it does not need any modifications to loss functions or network structures. Their study has achieved excellent results with datasets of the order of a few thousand images.

3   MATERIALS AND METHODS

The experiment utilised retinal imaging data collected using (Costa *et al.* 2017) The dataset consists of 400 raw images of the retina in ppm (Portable Pixmap Format). As the number of images is few, which is insufficient for training GAN. We suggest augmenting the images manually using any augmentation library (such as Augmentor). We first convert the images to png (Portable Network Graphics), which is more widely recognised and supported. After converting images to PNG, we create an augmentation pipeline and perform operations such as rotation, perspective skewing, shearing, mirroring, cropping, etc. We can increase our dataset from 400 images to 4000 images this way.

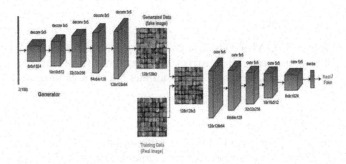

Figure 2.   Deep convolutional generative adverserial network.

Due to GPU memory limits, the photos were shrunk to 64 × 64 pixels. The images were then normalised so that the pixel values lie in the close interval of [1, 1], which helps in the training of the discriminator network.

3.1   *Discriminator network*

The discriminator attempts to categorise an image as"genuine" or"manufactured" based on its input. Here we are using a convolutional neural network that gives a single number for every image. We will employ convolutional neural networks (CNN) that provide a single number for each image. The output feature map's size can be gradually lowered using a stride of value 2. We are using LeakyRLU activation for the discriminator. Leaky ReLU permits a modest gradient signal for negative values to pass through. Consequently, it strengthens the gradients from the discriminator into the generator. The output of the discriminator is in the close interval [0, 1] which is the probability of the image being genuine, i.e., it belongs to the real dataset.

3.2   *Generator network*

The generator is fed with an input vector of random values, also known as a latent tensor. It serves as a starting point for creating a picture. We can control different values of latent tensors to change the features of the generated image. The latent tensor has a dimension of 128x1x1. The generator transforms the dimension of the latent tensor from 128x1x1 to a 3x28x28 image tensor.

For the output layer of the generator, we employ a function that uses the tangent trignometric ratio, also known as the TanH.

4 GAN OBJECTIVE FUNCTION

In GAN, latent space is a hidden representation of a point. Here, we denote the latent space vector with ψ which is taken from the standard normal curve. The generator function that matches ψ to data space is represented as $G\,\psi$, the image data is represented with χ. Let $D\,\chi$ be the function that will give us the prediction whether the χ belongs to real data or fake data.

The aim of our generator function is to learn about the training data distribution so as to generate fake images from the learned distribution.

GANs learn by training a discriminator and a generator at the same time. From the perspective of game theory, it is a min-max game between the two neural networks, where the discriminator tries to select the strategy that will diminish the worst-case loss. Therefore, it tries to maximise the loss function given by $\log(D\langle\chi\rangle)$, on the other hand, the generator tries to minimise the loss from the function, which is the probability that D will predict that its outputs are fake. $\log(1-D\langle G\langle\psi\rangle\rangle)$

$$\text{minmax}\,V_{GAN}(D, G) = E_{\chi\sim pdata(\chi)}[\log D\langle\chi\rangle] + E_{\psi\sim p\psi\,(\psi)}[\log(1 - D\langle G\langle\psi\rangle\rangle)] \qquad (1)$$

This min-max game converges when $p_g = p_{data}$ in theory. This enables the discriminator to make a random estimate as to whether the input photos are genuine (training data) or forgeries (generated data). However, research into the concept of convergence in GANs is still underway, and model training is not always up to this standard.

5 EVALUATION

Until neural networks converge, loss functions are typically employed to train deep learning models. The evaluation of GAN is different from that of other ML/DL models. In GANs, there is nothing like a loss function for training GAN generator models.

There are two approaches that are widely used to measure the quality of generated images.

- Measures that are quantitative in nature, such as the Inception Score (IS), the Fre´chet Inception of Distance, etc.
- Qualitative measures like mode collapse

In this paper, we have used the FID evaluation measure to evaluate our model.

6 RESULTS

Using the hardware that came with the Google Collab free-tier subscription, the proposed working model underwent 200 epochs of training. The model was able to generate images

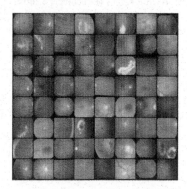

Figure 3.   Images from real dataset vs images generated using GAN.

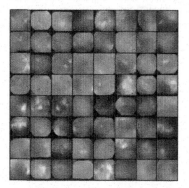

Figure 4.   Loss and accuracy during training of GAN.

that could easily fool average human eyes. The below figure shows two grids consisting of real images from the dataset and phoney images generated by our model.

7 CONCLUSIONS

GAN has proven to be a valuable asset for data, machine learning, and deep learning enthusiasts all over the world. It has shown great potential in the image processing field. In the field of medical imaging, where access to medical imaging data is restricted and labelled data is expensive to collect, GAN has appeared as a ray of hope.

The current aim of the research is to pick a medical image dataset, such as the STARE dataset of retinal images, and augment it artificially using the procedure mentioned in the paper. We have shown that it is possible to apply GANs to datasets with few images (the STARE dataset contains less than five hundred images). During this process, we have achieved a FID of 3.29.

REFERENCES

Baur, Christoph, Shadi Albarqouni, and Nassir Navab. 2018. "*MelanoGANs: High Resolution Skin Lesion Synthesis with GANs*" (April)

Borji, Ali. 2018. "Pros and Cons of GAN Evaluation Measures." *Computer Vision and Image Understanding* 179 (February) https://doi.org/10.1016/j.cviu.2018.10.009.

Chuquicusma, Maria J. M., Sarfaraz Hussein, Jeremy Burt, and Ulas Bagci. 2017. "How to Fool Radiologists with Generative Adversarial Networks? A Visual Turing Test for Lung Cancer Diagnosis." *CoRR* abs/1710.09762. arXiv: 1710.09762. http://arxiv.org/abs/1710.09762.

Costa, Pedro, Adrian Galdran, Maria Ines Meyer, Meindert Niemeijer, Michael Abràmoff, Ana Maria Mendonça, and Aurélio Campilho. 2018. "End-to-End Adversarial Retinal Image Synthesis." *IEEE Transactions on Medical Imaging* 37 (3) 781–791. https://doi.org/10.1109/TMI.2017.2759102.

Costa, Pedro, Adrian Galdran, Maria Meyer, Michael Abràmoff, Meindert Niemeijer, Ana Mendonça, and Aurélio Campilho. 2017. "Towards Adversarial Retinal Image Synthesis." *IEEE Transactions on Medical Imaging* (January)

Goodfellow, Ian, Jean Pouget-Abadie, Mehdi Mirza, Bing Xu, David Warde-Farley, Sherjil Ozair, Aaron Courville, and Y. Bengio. 2014. "Generative Adversarial Networks." *Advances in Neural Information Processing Systems* 3 (June) https://doi.org/10.1145/3422622.

Karras, Tero, Miika Aittala, Janne Hellsten, Samuli Laine, Jaakko Lehtinen, and Timo Aila. 2020. "Training Generative Adversarial Networks with Limited Data." *CoRR* abs/2006.06676. arXiv: 2006.06676. https://arxiv.org/abs/2006.06676.

Kora, Sagar, and Sridhar Ravula. 2020. "Evaluation of Deep Convolutional Generative Adversarial Networks for Data Augmentation of Chest X-ray Images." *Future Internet* 13 (December) 8. https://doi.org/10.3390/fi13010008.

Wolterink, Jelmer M., Tim Leiner, and Ivana Isgum. 2018. *Blood Vessel Geometry Synthesis Using Generative Adversarial Networks.* arXiv: 1804.04381.

Artificial Intelligence, Blockchain, Computing and Security – Dagur et al. (Eds)
© 2024 The Author(s), ISBN: 978-1-032-49393-0

A review on tunable UWB antenna with multi-band notching techniques

Amit Madhukar Patil* & Om Prakash Sharma*
Suresh Gyan Vihar University, Jaipur, Rajasthan, India

ABSTRACT: The Ultra-wideband technology has wide range of applications in huge data transmission across a wide bandwidth, specially greater than 500 MHz and short-range communications due to its low energy levels. Federal Communication Commission has allocated 3.1 GHz - 10.6 GHz frequency band for commercial ultra-wideband communication systems. Different structures are used to get the triple notch characteristics and multiband operation that are needed. Microstrip patch antennas are now used in mobile phones, defence equipment, wireless wearable devices, and other places. The paper presents a review of different UWB antennas with triple-band notch using different notching techniques and its application over the traditional antennas.

Keywords: Ultra-Wideband antenna design, Electromagnetic bandgap (EBG), Band notch techniques, WLAN, 5G

1 INTRODUCTION

In most communication systems, the antennas are the primary energy source. To design antenna with compact dimensions, proper radiations, excellent impedance matching and maintaining low costs is a difficult task. Because of their ability to be easily integrated with other electronic components, as well as their lightweight and flat designs, microstrip patch antennas are excellent option for use in ultra-wideband (UWB) applications [2]. There is already interference in the (5.15 – 6.84) GHz band from WLANs (IEEE802.11a, HIPERLAN/2), X-band satellite transmission (7.94 – 8.49) GHz and 5G sub-6GHz (3.4 – 3.9) GHz while building UWB antennas [3]. The most effective and cost-effective solution to this issue is to reject the interference at the UWB system.

An electromagnetic band-gap (EBG) structure creates a stopband by forming a fine and periodic pattern of small metal patches on dielectric substrates to block electromagnetic waves of specific frequency bands. As a result, the notching in the UWB antenna increases their dependability. Resonant cells in microstrip lines can be inserted into the microstrip lines, and split rings can be used for band rejection. Parasitic elements coupled to the radiator were used by some researchers to achieve band-notched capabilities [4]. Wide interference bands cannot be rejected by typical notch methods which has a bandwidth of 1 GHz. Designers of UWB antennas with great rejection features are paying greater attention. In order to successfully eliminate the interference that is caused by undesirable bands in practice, the utilization of a rectangular notch that possesses a high selectivity is essential [5]. Some of the rectangular antennas have designed for high selectivity in rejecting unwanted bands. At frequencies (5.2–5.8) GHz, typical notch bands were achieved without filtering bandwidth between the two bands to reject WLAN [6]. A rectangular notch band antenna

---

*Corresponding Authors: amit.patil@bvucoep.edu.in and om.sharma@mygyanvihar.com

DOI: 10.1201/9781003393580-136

913

obtained by introducing EBG structures in the WLAN band. Notching characteristics at the satellite downlink bands and WLAN achieved by using stepped slots as its radiator. It is important to designed more than one band rejecting antennas in the UWB spectrum. In order to avoid interference, triple rectangular notch band micro strip patch antenna with two EBG and one SRR, antenna have achieved triple rectangular notch bands [7].

2 THE NEED FOR OPERATIONAL BANDS UWB ANTENNA WITH TRIPLE-BAND NOTCH

The microstrip antennas have a low profile, are simple to fabricate, and are cheaper. However, these microstrip antennas are susceptible to the drawbacks associated with having a limited bandwidth, and in order to attain a larger bandwidth ratio, specific designs are required. Several researchers have developed numerous types of antennas over the course of years in order to obtain a greater bandwidth ratio [8–10]. These ultra-wideband antennas can be designed using a variety of methods. UWB antennas for wearable applications, Fractal UWB antennas, Planar wideband antennas and multiband UWB antennas with band notch capabilities, metamaterial-inspired UWB antennas have all been developed using a variety of different methods. [11–14].

In order to take advantage of the benefits that come with a smaller size, numerous fractal UWB antennas are now under development for wireless components [15–18]. To design these types of UWB antenna was one of the primary challenges that arise, having interference difficulties with other adjacent systems that make use of the frequency of operation that falls inside the coverage area of the UWB antennas [19–26]. In a manner analogous to that of UWB antennas that are used for band rejection of particular bands have been constructed. The development of these antennas with notch characteristics is possible through the utilization of slots, frequency selective surfaces, meander lines, resonator structures, electromagnetic bandgap structures and other such methods [3,21,27–29]. These antennas have the capability of rejecting either a single band or many bands. The efficiency of the procedures that were implemented plays a direct role in determining the accuracy of the bands rejection.

3 RESEARCH APPROACH

The main purpose of our study and review is to summarize the triple-band notch with different operational bands UWB antenna using different notching techniques. In the present review, the analysis is done based upon the following categories.

1. The impedance bandwidth of the antenna
2. High gain compact UWB antenna.
3. Continuously tunable, band-notch UWB antenna using notching techniques.

4 REVIEW AND DISCUSSION

We have undergone several research articles based on the operational bands and notching techniques, Some of them in brief are discussed previously. Now, we are going to carry forward certain terminologies described by different researchers.

4.1 *Notching techniques used for antennas*

The most commonly used notching techniques are as follows:

(a) *EBG*: EBG structures block electromagnetic waves by producing a thin, periodic pattern of small metal patches over dielectric surfaces [7,10,12,23,24,26,29].

(b) *Meander Slot*: An expansion of the basic folding antenna, a meander antenna has frequencies that are substantially lower than the resonances of a single element antenna of the same length. When compared to more traditional antennas that are half or quarter of a wavelength in length, the radiation efficiency of a meander line antenna is quite high [3,10].

(c) *SRR*: A split-ring resonator, also known as an SRR, is a structure that is generated intentionally and is typical of metamaterials. It is important to achieve the appropriate magnetic response in a wide variety of metamaterials operating at frequencies up to 200 terahertz [7,11,19,20,27,29].

(d) *Rectangular slot*: Microstrip patch antennas are rectangular microstrip antennas made up of a rectangular patch. A typical portion of the dielectric substrate and a patch of any planar or non-planar geometry is applied, with a ground plane on the other [8].

(e) *U slot*: U-shaped slot patch antennas are widely recognized for their wide impedance bandwidths. Micro strip patch antenna for WiMAX/WLAN applications has been designed, optimized, and simulated in this paper [10,29].

(f) *Parasitic Strips*: The antenna is capable of successfully creating a dual notched band for both lower and upper WLAN by attaching two parasitic strips to the radiator [21].

(g) *Stepped Slot*: Stepped slot can be used to achieve ultra-wideband operation at its radiator. Multiple resonant modes can be generated by arranging the stepped slot to the appropriate size and feeding it at the appropriate position. This allows the modes to transition smoothly from one to another, resulting in a bandwidth that is significantly wider [28].

(h) *Etched Slot*: Etching slots of varying shapes into the radiator or the ground is typically one of the methods used to realize a notched band. Other approaches include the addition of folded parasitic strips in close proximity to the radiator. The stepped slot is close to a short-ended half-wavelength split-ring slot allowed for the creation of a notched band that was able to cover the WLAN [15].

In the review paper, we have studied the several notching techniques. The researchers have used following techniques for notching, and they are categorized with the help of pie chart.

From Figure 1. It is observed that different notching techniques used by the researchers for rejecting multiple bands. Most of the researchers have preferred Electromagnetic Band Gap structure, Meander slot and Slip Ring Resonators for notching.

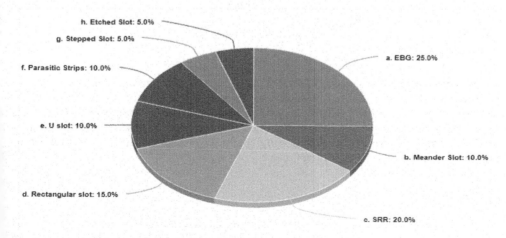

Figure 1.   Notching techniques used for antenna.

4.2 *The performance comparison of the literature*

This review paper presents a UWB antenna with two notch bands that can be switched between while tuning the central frequency. Here the band-notched UWB antennas combine switchable and adjustable characteristics. The majority of the researchers have used EBG structure for notching the frequencies. This information is useful for categorization and designing the new antenna. After studying different notching techniques such as Electromagnetic band gap (EBG) structure, Slip ring resonator and Meander slots, we observed that there is reason to increase notching and tuning accuracy, since the existing research approach needs more rigorous investigation. For achievement of compactness of antenna, better gain, proper tuning and quad band, we need a mechanism to improve accuracy.

Table 1. Lists the various types of notching techniques utilized by various researchers.

Table 1.   The performance comparison chart.

Reference Antennas	Notching Techniques	Antenna Size (mm)	Notch Bands	Notch Bands (GHz)
[03]	Meander Slot	40 × 38 × 1.6	3	3.29 – 4.83, 5.15 – 6.84, 7.94 – 8.49
[07]	SRR and EBG	20 × 26 × 1.52	3	3.4–3.9, 5.15–5.825, 7.25–7.75
[08]	Rectangular slots	26 × 30 × 1.6	2	5.18–5.82, 7.25–8.39
[10]	Meander line EBG and U slots	34.9 × 31.3 × 1.6	4	2.53–3.15, 3.23–3.68, 3.92–4.30, 5.49–6.19
[11]	Slots and SRR	54 × 47 × 1	3	2.23-2.45, 3.26–3.48, 5.54–5.88
[12]	EBG$_s$	24 × 24 × 1.6	3	3.3–4.0, 5.1–5.8, 7.2–7.8
[19]	SRR	40 × 30 × 0.78	4	3.39–3.82, 50.13–5.40, 5.71–5.91, 7.5–8.61
[20]	Slots and SRR	41.5× 32 × 1.6	4	3.3–3.7, 5.15–5.35, 5.725–5.825, 7.25–7.75
[21]	Parasitic Strips	30 ×22 × 3	3	3.26–3.71, 5.15–5.37, 5.78–5.95
[26]	EBG$_s$	50 × 42 × 1.6	3	3.3–3.8, 5.15–5.82, 7.1–7.9
[27]	SRR on the feed line	50 × 50 × 1.57	2	5.15–5.82, 6.2–6.9
[28]	Stepped Slot	28 × 18 × 0.8	2	5.1–6.0, 7.83–8.47
[29]	EBG, U slot, SRR	25 × 25 × 1.6	2	5.94–7.5, 8.02–10.46

5   CONCLUSION

In this paper, different types of UWB antennas with band notch techniques have been studied and described to avoid narrowband interference. UWB antennas with a notch band have been achieved by EBG structure, Slip ring resonator and Meander slots. Most of the researchers have used EBG structure to reduce electromagnetic noise in electronic equipment's. A band-notched UWB antenna simultaneously switched between two notch bands and continuously tunes the operational bands.

REFERENCES

[1]   Gayatri T.; Srinivasu G.; Chaitanya D.M.K.; Sharma V.K., "Design and Analysis of a Compact Wrench Shaped UWB Antenna for Spectrum Sensing in 3.1GHz to 10.6GHz", *IEEE International RF and Microwave Conference* (RFM), DOI: 10.1109/RFM50841.2020.9344798,2020.

[2] Hasan Mahfuz M. M.; Md Mohiuddin Soliman; Md Rafiqul Islam, "Design of UWB Microstrip Patch Antenna with Variable Band Notched Characteristic for Wi-MAX Application", *IEEE Student Conference on Research and Development (SCOReD)*, DOI: 10.1109/SCOReD50371.2020. 9250947, 2020.

[3] Rani R. Kodali1, Polepalli Siddaiah2, and Mahendra N. G. Prasad3, "Design of Quad Band Operational UWB Antenna with Triple Notch Bands Using Meander Line Slot", *Progress In Electromagnetics Research M*, Vol. 109, 63–73,2022.

[4] Zhijian Chen; Weisi Zhou; Jingsong Hong, "A Miniaturized MIMO Antenna With Triple Band-Notched Characteristics for UWB Applications", *IEEE Access* (Volume: 9), DOI: 10.1109/ACCESS.2021.3074511,2021.

[5] Srinivasu G.; Gayatri T.; Chaitanya D.M.K.; Sharma V.K., "Performance Analysis of a Compact High Gain Antenna for RF Energy Harvesting in 1.71GHz to 12GHz", *IEEE International RF and Microwave Conference (RFM)*, DOI: 10.1109/RFM50841.2020.9344793,2020.

[6] Issa Elfergani1, Jonathan Rodriguez1,2, Ifiok Otung2, Widad Mshwat3, Read Abd-Alhameed3, "Slotted Printed Monopole UWB Antennas with Tunable Rejection Bands for WLAN/WiMAX and X-Band Coexistence", Radio engineering doi.org/10.13164/re.2018.0694,2018.

[7] Anees Abbas; Niamat Hussain; Jaemin Lee; Seong Gyoon, "Triple Rectangular Notch UWB Antenna Using EBG and SRR", *IEEE Access* (Volume: 9) DOI: 10.1109/ACCESS.2020. 3047401,2020.

[8] Yadav, A., D. Sethi, R. K. Khanna, "Slot loaded UWB antenna: Dual band notched characteristics," *AEU — International Journal of Electronics and Communications*, Vol. 70, No.3, 331–335, March 2016.

[9] Deqiang Yang, Huiling Zeng, Sihao Liu*, and Jin Pan, "A Vivaldi Antenna with Switchable and Tunable Band-Notch Characteristic", *Progress In Electromagnetics Research C*, Vol. 68, 75–83, 2016.

[10] Modak, S., T. Khan, and R. H. Laskar, "Loaded UWB Monopole Antenna for Quad Band-Notched Characteristics," *IETE Technical Review*, 1–9, 2021, doi:10.1080/02564602.2021.1878942.

[11] Iqbal A., Smida A., Mallat N., Islam M., and Kim S., "A Compact UWB Antenna with Independently Controllable Notch Bands," *Sensors*, vol. 19, no. 6, p. 1411, Mar. 2019.

[12] Trimukhe M. A. and Hogade B. G., "Compact Ultra-wideband Antenna with Triple Band Notch Characteristics using EBG structures," *Prog. Electromagn. Res. C*, vol. 8, no. 7, pp. 1069–1075, Nov 2016.

[13] Xu F., Tang L., Chen X., and Wang X.-A., "Active Uwb Printed Antenna With Tunable And Switchable Band-notched Functions", *Progress In Electromagnetics Research Letters*, Vol. 30, 21–28, 2012.

[14] Durbin J. L. and Saed M., "Tunable Filtenna Using Varactor Tuned Rings Fed With An Ultra Wideband Antenna", *Progress In Electromagnetics Research Letters*, Vol. 29, 43–50, 2012.

[15] Chu Q.-X., Mao C.-X., and Zhu H., "A Compact Notched Band UWB Slot Antenna with Sharp Selectivity and Controllable Bandwidth," *IEEE Trans. Antennas Propag.*, vol. 61, no. 8, pp. 3961–3966, Aug. 2013.

[16] Zhou Z.L., Zhang H.B., Wang L. and Li D.M., "A Novel Compact UWB Antenna with Dual Independent Tunable Band-notches", *IEEE*, DOI: 10.1109/ICMMT.2018.8563722,2018.

[17] Shyam Goyner;Abhishek Patel;Manoj Singh Parihar, "*Design of Monopole UWB Antenna with Controllable Band-Notch Function*", DOI: 10.1109/INDICON47234.2019.9029032,2019.

[18] Ameya A Kadam;Amit A Deshmukh;Kamla Prasan Ray;S B *Deshmukh*. "*Dual Band-Notched UWB Antenna with L-Shaped Slots and Triangular EBG Structures*", DOI: 10.1109/InCAP47789.2019. 9134452,2019.

[19] Jeong M. J., Hussain N., Bong H., Park J. W., Shin K. S., Lee S. W., Rhee S. Y., and Kim N., "Ultrawideband Microstrip Patch Antenna with Quadruple Band Notch Characteristic Using Negative Permittivity Unit *Cells*," *Microw. Opt. Technol. Lett.*, vol. 62, no. 2, pp. 816–824, Feb. 2020.

[20] Chakraborty, M., S. Pal, and N. Chattoraj, "Quad Notch UWB Antenna Using Combination of Slots and Split-Ring Resonator," *International Journal of RF and Microwave Computer-Aided Engineering*, 2019, doi:10.1002/mmce.22086.

[21] Islam M. T., Azim R., and Mobashsher A. T., "Triple Band-Notched Planar UWB Antenna Using Parasitic Strips," *Prog. Electromagn. Res.*, vol. 129, pp. 161–179, 2012.

[22] Ali Karami Horestani; Zahra Shaterian; Jordi Naqui, "*Reconfigurable and Tunable S-Shaped Split-Ring Resonators and Application in Band-Notched UWB Antennas*", DOI: 10.1109/TAP.2016.2585183,2016.

[23] Peng L., Wen B.-J., Li X.-F., Jiang X., and Li S.-M., "CPW Fed UWB antenna by EBGs With Wide Rectangular Notched-Band," *IEEE Access*, vol. 4, pp. 9545–9552, 2016.

917

[24] Abbas A., Hussain N., Jeong M.-J., Park J., Shin K. S., Kim T., and Kim N., "A Rectangular Notch-Band UWB Antenna with Controllable Notched Bandwidth and Centre Frequency," *Sensors*, vol. 20, no. 3, p. 777, Jan. 2020.

[25] Deqiang Yang; Mi Wan; Sihao Liu, *"A Monopole Antenna with Switchable and Tunable Band Notch for UWB Applications"*, DOI: 10.1109/ICMMT52847.2021.9618473,2021.

[26] Jaglan N., Kanaujia B. K., Gupta S. D., and Srivastava S., "Triple Band Notched UWB Antenna Design Using Electromagnetic Bandgap Structures," *Prog. Electromagn. Res. C.*, vol. 66, pp. 139–147, 2016.

[27] Siddiqui J. Y., Saha C., and Antar Y. M. M., "Compact Dual-SRR-loaded UWB Monopole Antenna with Dual Frequency and Wideband Notch Characteristics," IEEE Antennas Wireless Propag. *Lett.*, vol. 14, pp. 100–103, Sep. 2015.

[28] Li W.-A., Tu Z.-H., Chu Q.-X., and Wu X.-H., "Differential Stepped- Slot UWB Antenna with Common-mode Suppression and Dual Sharp- Selectivity Notched Bands," *IEEE Antennas Wireless Propag. Lett.*, vol. 15, pp. 1120–1123, Oct. 2016.

[29] Kumar, G., Singh D., and Kumar R., "A Planar CPW Fed UWB Antenna With Dual Rectangular Notch Band Characteristics Incorporating U-slot, SRRs, and EBGs," *International Journal of RF and Microwave Computer-Aided Engineering*, Vol. 31, No.7, doi:10.1002/mmce.22676,2021.

Artificial Intelligence, Blockchain, Computing and Security – Dagur et al. (Eds)
© 2024 The Author(s), ISBN: 978-1-032-49393-0

System for predicting soil moisture using Arduino-UNO

A. Kapahi, V. Kapahi, D. Gupta, H. Verma & M. Singh
Manav Rachna International Institute of Research & Studies, Faridabad, Haryana, India

ABSTRACT: Agriculture has a very significant impact on India's economy and is a major source of income for the majority of the population. This paper proposed a model to monitor soil moisture levels and irrigating the soil as needed using an Arduino-UNO smart irrigation system which results in an effective system for optimizing production while using the least amount of water, hence conserving scarce water resources. The proposed model is cost-effective, portable, and user-friendly, and it has the potential to save water, time, and human effort.

Keywords: Arduino-UNO, smart irrigation system, soil moisture sensor, microcontroller

1   INTRODUCTION

Agriculture uses 85% of the world's water resources, while this ratio is unlikely to alter given the rate of population expansion and the associated high demand for food. Some nations, such as Bangladesh, China, and India, are recognized primarily for agriculture, and it may play a significant role in economic growth and development.

With recent technical breakthroughs, boosting yearly crop production output in an agricultural-based economy has become crucial. The purpose of this project is to create a prototype which uses the moisture content detector & sensor to find the water level & moisture content in any agricultural / non- agriculture field.

This paper offers and illustrates a cost-effective and easy-to-use controlled irrigation system based on Arduino. To solve environmental difficulties such as humidity, temperature, and agricultural water supply amount, the proposed device incorporates sensors like ultrasonic sensor and moisture in the soil detectors. With minimum human involvement, an automated system can maintain an appropriate flow of water towards the surface. This method can also save you money, energy, and time while conserving water. It will also make it easier for the elderly and disabled to carry out their duties.

2   REQUIREMENT SPECIFICATIONS

2.1   *Software requirements*

In the prototype model, the Arduino IDE (Integrated Development Environment) software is utilized to code in the C++ programming language which is compatible with many operating systems. This program is intended to run on Windows. As IDE is Integrated Development Environment so it will provide the complete environment which includes menu, code editor, toolbar, text terminal etc.. It is a simple programming language used mostly for program generation and compilation.

2.2   *Hardware requirements*

The hardware requirements are as follows:

DOI: 10.1201/9781003393580-137

2.2.1 Arduino-uno board

The Arduino Uno board has several input /output pins −14 pins for digital and 6 pins for analog that may be used to connect to several expansion boards (shields) as well as other circuits. Arduino-Uno board can be programmed using its IDE with the help of type B USB as described in Dogra *et al.* (2017).

2.2.2 Soil moisture sensor

It is a device which is used to monitors the soil's moisture content. If device detects that moisture is less then module output will become high & if moisture is high then the module output becomes low and then it will notify the users when to water their crops and analyzes moisture levels in the soil

2.2.3 Temperature and humidity sensors (DHT11 sensor)

DHT11 is a Humidity and Temperature Sensor that produces calibrated digital output (Figure 3). It can instantly provide results and interface with any microcontroller, including Arduino and Raspberry Pi. Having great long-term reliability and high dependability, the DHT11 sensor is a low-cost option.

2.2.4 Relay module

Relays may be operated with low voltages, such as the 5V provided by the Arduino pins, and are switches that can be turned on or off electronically to allow or block current flow. The concept of a relay is electromagnetic induction, and when an electromagnet is charged with energy, it creates a magnetic field surrounding

2.2.5 Mini solar panel

Mini solar panels are similar in working as bigger solar panels but due to the size they are cost effective & easy to use. They gather heat energy from the sun & converted it into electricity which can be utilized for the working of the designed prototype.

2.2.6 Adapter

The adapter lowers the higher input voltage to produce the output voltage. It has a 120V to 12V converter that works well for tiny electronic gadgets.

2.2.7 Breadboard

Breadboard is a solderless construction base for electronic circuits and wiring. In accordance with the structure of an electrical circuit, it serves as a cluster. Since breadboard connections are temporary, it is simple to start over or replace an item if a mistake is made.

2.2.8 Pump

A mechanical device called a pump converts electrical energy into hydraulic energy to transfer fluids. The experimental plots are supplied with water with the help of miniature submersible pump that is driven by very less DC voltage between 3–6V.

3 LITERATURE SURVEY

Islam *et al.* (2022) plans to develop an automated irrigation system that detects the moisture level of the ground and switches the pumping motor ON and OFF. The advantage of using these systems is that they reduce human meddling while ensuring proper watering.

Praveena (2022) addresses the use of IoT to construct a smart irrigation system. This technology is driven by an on/off table that runs on electricity.

Hasanudin *et al.* (2022) intends to develop a "Moisturizer Sprinkler" that will water plants automatically using a timer. A soil moisture sensor is used to monitor soil temperature, and the data from the sensor is analyzed and sent to a temperature sensor. This project helped us in figuring out how we can use mini solar panel in Arduino projects to maximize compatibility.

Singh et al. (2021) include designing and building a low-cost factor for a water irrigation system based on an embedded platform. This project's main purpose is to create an autonomous irrigation system that will help the farmer to save his time, money, and energy.

Al Mamun et al. (2021) created an automated model employing sensors and microcontroller for efficient irrigation water supply. This paper proposed a model with lesser human interaction to save water & time. However this model could not run at places where it is difficult to provide electricity supply, which was resolved in our prototype with use of a mini solar panel.

Athani et al. (2017, February) aims to outline a manageable, simple-to-implement method for detecting and specifying moisture levels which was managed to achieve peak growth of the plants and parallely rise the available irrigation resources. Although it was not tested in some regions, hence to avoid any future problem, we have designed our prototype in such a way that it can work in any region.

Kumar et al. (2016, January) monitored the soil moisture with the help of low-cost self-designed sensor and Arduino UNO. They offer a method for manufacturing soil moisture sensors that evaluate soil moisture content and provide information about the necessary water supply for optimal farming.

Kumar et al. (2019, March) discussed an efficient irrigation system composed of Arduino Nano and can be adjusted for any crop. There have been multiple projects for automatic irrigation using Arduino, however those are targeting specific crops but the since they require varied moisture levels for growth. This paper uses the model for different crops.

Taneja & Bhatia (2017, June) concluded that the typical way of watering plants wastes more water than is necessary for adequate plant yield, resulting in water waste. On the other hand, by implementing their Arduino project to irrigate plants, the correct amount of water is delivered by estimating the moisture level of the plant soil, resulting in no water waste.

Kesarwani et al. (2019, February) develop a cost efficient model for monitoring the soil moisture which will check the status of moisture level and intimate the same to the user through the Blynk mobile application. As this system requires stable internet connectivity hence it could not assist the rural farmers.

4 PROPOSED SYSTEM

The suggested prototype will be deployed to analyze soil moisture levels and water the land using an Arduino-UNO smart irrigation system as described in Singh et al. 2021, with some more improvisation. This project will have two functional components. They are the moisture and temperature sensor modules, and the pump's motor driver as shown in Figure 1 below. The sensors will measure soil moisture and water level in the surrounding region. The sensor readings and data will be sent to the microcontroller unit, that will begin the irrigation process via the pump based on the requirements and will stop accordingly when the requirement has been met as described in Figure 2.

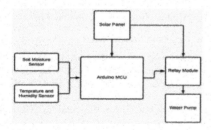

Figure 1.   Proposed system.

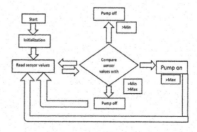

Figure 2.   Program flow.

5 DESIGN & IMPLEMENTATION

An autonomous irrigation model based on sensor technologies is created to decrease water loss and enhance irrigation efficiency shown below in Figure 4. The prototype has enhanced the efficiency of irrigation as result agriculture has become more sustainable & competitive in nature. In the prototype, existence of water is detected by moisture sensor and the present status of water level is transmitted to the microcontroller. Then the status of water level is compared with the threshold value, if it is less than the threshold then signals are transmitted from microcontroller to the pump. Then pump start pumping the water into the field to maintain the desired threshold value. When the ultrasonic sensor identified the desired moisture level, then signals are again sent to the microcontroller and it transmit the signal to turn-off the pumping of the water through the water pump. As this prototype is automatic which in turn reduces the manual intervention and hence reduced the labour cost. The components of the system whether it is the mini solar panel or the microcontroller itself are dependable and low-cost, enabling it to be simply accessible. It also possesses self-intelligence, allowing it to achieve the goals. The circuit diagram of the prototype is elaborated in Figure 3.

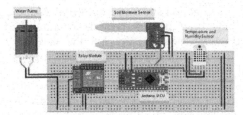

Figure 3.   Circuit diagram.   Figure 4.   Prototype model.

6 CONCLUSION & FUTURE SCOPE

The goal of this project is to reduce time, money, and water use by offering a smart control irrigation system powered by environmentally friendly solar power. The proposed system can be extended in the future by adding the feature where we will be able to fetch live data on a mobile app using 5g connectivity and precise watering using machine learning models to meet the needs of different types of plants. Another idea that can be added to this project is a time and location-based automatic sun tracking system to ensure uninterrupted power supply to the system along with a high-capacity electricity storage unit to ensure operations during night time.

REFERENCES

Al Mamun, M. R., Soeb, M. J. A., Mia, M. S., Rabbi, M. R. I., & Azmir, M. N. (2021). Design and Development of an Automatic Prototype Smart Irrigation Model. *Aust. J. Eng. Innov. Technol*, 3(6), 119–127.

Athani, S., Tejeshwar, C. H., Patil, M. M., Patil, P., & Kulkarni, R. (2017, February). Soil Moisture Monitoring Using IoT Enabled Arduino Sensors with Neural Networks for Improving Soil Management for Farmers and Predict Seasonal Rainfall for Planning Future Harvest in North Karnataka—India. In *2017 International Conference on I-SMAC (IoT in Social, Mobile, Analytics and Cloud) (I-SMAC)* (pp. 43–48). IEEE.

Dogra, M., Singh, Y., Bhateja, N., & Singh, M. (2017). Mobile Application for Vehicular Auto Locking System. *International Journal of Advanced Research in Computer Science*, 8(7).

Hasanudin, Zulkefli & Sulaiman (2022). *Moisturizer Sprinkle*.

Islam M., Islam R. & Sarkar (2022). Structure of Soil Moisture Sensing Electronic Irrigation System. *Asian Journal For Convergence In Technology (AJCT) ISSN- 2350-1146*, 8(1), 12–16.

Kesarwani, S., Mishra, D., Srivastva, A., & Agrawal, K. K. (2019, February). Design and Development of Automatic Soil Moisture Monitoring with Irrigation System for Sustainable Growth in Agriculture. In *Proceedings of International Conference on Sustainable Computing in Science, Technology and Management (SUSCOM), Amity University Rajasthan, Jaipur-India.*

Kumar, M. S., Chandra, T. R., Kumar, D. P., & Manikandan, M. S. (2016, January). Monitoring Moisture of Soil Using Low Cost Homemade Soil Moisture Sensor and Arduino UNO. In *2016 3rd International Conference on Advanced Computing and Communication Systems (ICACCS)* (Vol. 1, pp. 1–4). IEEE.

Kumar, N. K., Vigneswari, D., & Rogith, C. (2019, March). An Effective Moisture Control based Modern Irrigation System (MIS) with Arduino Nano. In *2019 5th International Conference on Advanced Computing & Communication Systems (ICACCS)* (pp. 70–72). IEEE.

Malik, A., Magar, A. T., Verma, H., Singh, M., & Sagar, P. (2019). A Detailed Study of an Internet of Things (IoT). *Int. J. Sci. Technol. Res*, 8(12), 2989–2994.

Praveena (2022) *An Implementation of Agricultural Sensors Based Artificial Intelligent Irrigation System (AIIS) Using Dedicated Machine Learning Algorithm.*

Singh, G.S., R.S. & S.I. (2021). *Automatic Irrigation System using Arduino Microcontroller.*

Taneja, K., & Bhatia, S. (2017, June). Automatic Irrigation System Using Arduino UNO. In *2017 International Conference on Intelligent Computing and Control Systems (ICICCS)* (pp. 132–135). IEEE.

Artificial Intelligence, Blockchain, Computing and Security – Dagur et al. (Eds)
© 2024 The Author(s), ISBN: 978-1-032-49393-0

Tank water flow automation

Aniketh Santhan*, Aaditya Kumar Tomar*, Vikalp Arora* & Dharm Raj*
School of Engineering and Technology, Sharda University, Greater Noida, Uttar Pradesh

ABSTRACT: Water is the most valuable thing on earth that life sustains on this planet. We use water in almost every home activity for drinking, washing, and many more. With the increase in population and industries, the availability of clean water has become a problem. Therefore it is crucial to find an efficient solution for water monitoring and monitoring system. In this paper, we have mentioned our research based on an autonomous water tank filling system using IoT with the help of embedded sensors to monitor the tank status with keeping in mind saving water as well as electricity.

Keywords: Monitoring, water saving, IoT, Sensors, Cloud

1 INTRODUCTION

Water is the most crucial natural resource for life to sustain on earth, On the surface of the earth, it exists in three different states: solid, liquid, and gas. Ice is one type of it that exists in solid form. Seawater makes up the majority of it in liquid form, and the remaining fraction can be found in lakes, rivers, and undersea. In gaseous form, it can be found in the atmosphere. Water is essential for life on Earth. In plants, it is used in photosynthesis. In humans, it is used in cooking, washing, drinking, bathing, and other activities. Mars, Venus, and Mercury are the only planets in our solar system without water, which is why life cannot exist on them. The earth is the only planet in our solar system with water in its environment. as a needed for us to recognise the significance of water. Fresh water is becoming polluted as a result of increased global warming, water pollution, and numerous other issues, and our supply is getting smaller. Since there are too many people, the water is being wasted..The clean water ratio is just depleting day by day, it is just that we innovate new technologies for water conservation to fulfill the need for clean water for upcoming generations and backward areas. In our daily life, we have seen many areas where water is spilled or wasted unnecessarily. Factories, industries, and home utilities are the main areas where we see water wastage more often. We have seen factories lay out their bi-product to open water bodies like lakes, rivers, etc. This technique is causing a lot of water pollution, whereas in societies we are often seeing people using a large quantity of fresh water for their daily needs, like washing their vehicles, bathing, and toothbrushing while the tap is running which eventually makes their private tanks empty even faster. For this problem with the help of IoT, we want to contribute to society to save water for the upcoming generation and save the earth. Modern technology has contributed a lot to water conservation techniques, in many cities and villages there is a water supply shortage problem and in some, people are taking the water supply for granted and wasting it. We can still manage to save the planet by saving water for the generation and upcoming ones, only possible when we start from the ground

*Corresponding Authors: anikethsanthan999@gmail.com, aadityatomar1000@gmail.com, aadityatomar1000@gmail.com and dharmraj4u@gmail.com

DOI: 10.1201/9781003393580-138

roots in our daily life in societies or neighboring homes we see private submersibles are used for water extraction for personal use and in that hurry people forgot to switch off the motor which eventually causes extreme water loss.Water depicts the world's natural beauty in its numerous forms. We shall create new ideas to save water and its resources so that life can continue on this planet with the aid of contemporary technology and the Internet of Things (IoT). Water also sculpts the beauty of nature.

The internet of things (IoT)has connected the world from anything to anywhere so for this problem automation is the key which can save both waters as well as electricity. The use of automation in homes, factories, buildings, etc. will allow us to conserve water in a drastic way that we were wasting unknowingly. We are going to bring a drastic change in the automation sector where we will use IoT to solve these kinds of problems where we are facing the tank overflowing problem. Our main motivation for doing this project is to save water and the earth for our upcoming future generations. No life can sustain on earth without water. Our product will be contributing to saving water in a decent manner where it will be providing home automation to the common man at a reasonable price as well as contributing to saving water. Our product will be installed in different areas like houses, factories, industries, and wherever there is a submersible that pumps water to a private tank because it will not only provide automation but it will also reduce the man work that will be required for the switching on and switching off of the motor in the right time, where there is a huge chance of error of not switching off the motor in the right time and eventually losing liters of water and electricity. Our product will also indirectly save electricity through its smart end technology which we will see next in the paper description.

The rest of the paper contains the following details, the related previous work on the automation field in this area, the architecture of the proposed model, results, and discussion, and finally the future directions of this article.

2 LITERATURE REVIEW

The internet of things(IoT)has connected the world from anything to anywhere. Nowadays automation has become a very handy thing that eases the workload of human beings, automation is being used for most of the things that we are using in our daily life like remote starting our car, security checks of our home via digital cameras, anti-theft alarms, etc.

Autonomous water filling system by S.Nalini Durga, M.Ramakrishna, and G. Dayanandam created a model using Arduino Uno and IoT, with the help of a water sensor, potentiometer, solenoid valve, and GSM shield. The system works in the manner that the main flow of the water is detected by the water sensor and it intimates the same to Arduino then the availability of electricity is checked with the help of a potentiometer. After the validation, the system checks for the water level using an Ultrasonic sensor if the water level is not full up to the mark it will trigger the motor once the level is reached the motor is been shut off. This is a great invention that will solve the problem of water overflowing and saves electricity as well. But this product is a high-budget product that many cannot afford, we need a low-cost product with the same qualities or better.

The second paper is from the Engineering physics international Conference in this model they used an Arduino Uno microcontroller, HC-Sr04, a Relay, a water pump, some connector cables male and female, a plastic box for fixing the architecture, an LCD, a potentiometer, a printed circuit board (PCB) this system is suitable for to be used in our homes, factories, and industries to decrease the value of energy consumption and water overflowing. The third paper is Automated Water Management System (WMS) by Rakib Ahmed and Mahfida Amjad. This article uses the WMS framework to monitor whether the water tank level is high or low. if any client needs to change the water temperature then he/she can likewise make it happen. Smart Water Flow Control and Monitoring System, by Janhavi Sawanth V, Lourd Mary J, Madduleti Vidya, and Mounika D, is the fourth paper.

Equipment with a naturally aspirated engine (Pipe and ball). In order to organically control the water, this paper used natural equipment. David Muscarine and Robert B. Donner's automated multiport flow-through water pumping and sampling.This project uses an AMPS SYSTEM. The AMPS is convenient, smaller, flexible, straightforward, and dynamic, permitting it to be utilized in numerous settings furthermore, effortlessly custom fitted to address specialist issues. Automation of Residential Water Flowmeter by Aditya Kumar Thakur, Shiwanshu Shekhar, Akash Priyadarshee, Deepak Kumar, Shashi Kumar, Nitish Shrivastava, Aniket Sinha. This paper has several advantages such as user-friendly, cost-effectiveness, etc. and hence it can be used as an alternative for saving water.5). Urban water supply automation – today and tomorrow by Gustaf Olsson the paper reflects encounters most memorable endeavor to apply programmed control in water frameworks was back in 1973. Automation of submersible pumps and designs by M. Babu Prasad, Dr. M. Sudha Siphon framework application. This paper proposes and recommends a plan answer for sub siphon framework application particularly at the point when they are utilized for series and equal blend for seepage, water system, and salvage.The Automatic Control Model of the Water Filling System with Allen Bradley Micrologix 1400 PLC is the fifth research paper we have chosen for our study. The system's formulation entails understanding how to enter and execute ladder logic in a PLC. A water filling automation system was designed using Allen Bradley Micrologic 1400 type 1766-L32BXB PLC.

Samuel C. Olisa, Christopher N. Aseigbu, Juliet E. Olisa, Bonaventure O. Ekengwu, Abdulhakim A. Shittu, and Martin C. Eze's "smart two-tank quality and level detection system via IoT" is the sixth research study we have chosen for our study. In many cultures and colonies, the two-tank water system—which uses two overhead tanks to maintain the water flow of the residences in that area—is a typical practise. The water flow in these is handled by electronic pumps in pipeline networks, but there is a disadvantage in that we have little control over the pumping system and the water quality is unknown to us in the current system. In this work, control is provided using an Android mobile application, Ultrasonic pulse-echo technique was used in this system. Three-level conditions (LC_1, LC_2, LC_3), and water quality check conditions (QC_1, and QC_2)were devised and used in an intelligent control algorithm system.

The seventh paper we selected for our research is Bernadette Coelho, Ph.D.'s "Efficiency achievements in water supply systems- A review." The author of this paper has discussed the WSS (Water Supply System), how to enhance the "Water Filling System," how to create an effective leakage-proof system, how to estimate water demand in various locations, how to optimise the pump system, and real-time operation. They have utilised renewable energy sources to ensure that all of these services operate effectively. Hydraulic simulation: These models, which reproduce the nonlinear dynamics of a network depicted at particular periods, are computerised representations of systems that play an essential role in management and operational control.GIS integration enables the collection, administration, analysis, and presentation of information with a geographic context. Pumping apparatus Energy expenses for WSS pumping are typically Although pumps can be adjusted accordingly, in most situations they are only turned on when the identical reservoirs are at their maximum permissible level. The accompanying expenses would be greatly decreased if the pumps were run in accordance with the day-to-day variations in the energy tariff as well as the patterns of water usage. They have done excellent studies to promote water conservation measures.

3   METHODOLOGY

In this research, we tried to solve the problem of water conservation using automation by IoT. In this project, we used various materials listed in customized Printed Circuit board (PCB)which Contains (555 timer IC, BC547(NPN) transistor, 1N4007 Diode, LED (1.5),

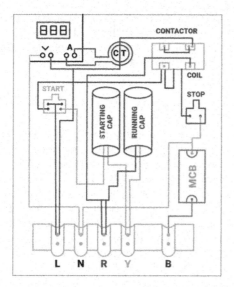

Figure 1.   Structure of the model.

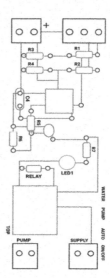

Figure 2.   PCB design.

1k resistor, 22k resistor, 180k resistor, 1M resistor, 100 nF(104)Capacitor, 12v SPDT, and Connectors) which is the main central processing unit of the system.

With the help of connectors and male and female cables, we have done the necessary connections to the sensors that have to be connected to the overhead tank, the stop switch, and for the submersible. The system works properly after all these connections are done systematically in the scheduled test whenever the water touches the sensor it intimates the PCB and stops the submersible immediately every time which makes our product a huge success. The system works in an order in which first we have to switch on the submersible which will command the water to flow towards the tank and the tank will get filling, once the water level reaches the marked area where the sensor is placed the sensor will directly intimate the system to shut down that will make the submersible to stop pumping water and the submersible turns off. The circuit diagrams show's how the connectors and different components of the system join with each other and make the system work. This system will shutdown the submersible once the water level reaches the sensor it will intimate the PCB to shutdown the motor immediately. Our system is a huge success as it does the same thing without fail everytime the circuit completes and the system always shuts down to stop the motor.

4   RESULTS AND DISCUSSION

Our proposed system produced the best outcome as it was tested on various overhead tanks and each time the results were positive.It gave us the liberty to be tension free about the submersible shutting down. Various similar products in the market offer similar kinds of technology but with less efficiency, there is a floating ball technique that is often used in Aquagards and overhead tanks also, but the drawback of it is that it's not efficient always and the chances of the system to fail are higher as the ball may be stuck somewhere and the functionality of the system effects. but on the other hand, our system works on sensors that whenever the water level touches the sensor initiate the PCB to stop the motor without fail, the system also has a low price range that will attract. common public to install this system in their homes and make their home a more automated space where they will be eventually contributing to saving planet earth. Our system limits in the areas where every time the water level goes down we have to manually switch on the motor and make it work.

5 CONCLUSION

Our intention in building this project was to use modern-world technology to solve modern problems. we wanted to build a system that will eventually help us to make our home water supply waste minimal and parallel electricity saving, we came up with a smart "Tank Waterflow Automation System", which solved the problem eventually. Our system enables the user to be free after they have switched on the system for once, Our system took care of the system shut down after the tank is full automatically which eventually reduced the workload of the user, saved water, and saved electricity. Our system is a low-budget product that every common man can install in their home to automate their house by these functions and have a futuristic great feeling. We are working more toward this product and we will be focusing on making the product more advanced where it will be fully automatic by keeping in mind the low price range for the common people.

REFERENCES

Ahemed, R., & Amjad, M. (8 May 2019). Automated Water Managemenrt System (WMS). *I.J. Education and Management Engineerin.* doi:10.5815/ijeme.2019.03.03

Bitella, G., Rossi, R., Bochicchio, R., Perniola, M., & Amato, M. (21 october 2014). *A Novel Low-cost Open-hardware Platform for Monitoring Soil Water Content and Multiple soil-air-vegetation Parameters.* National Library of Medicine. doi:10.3390/s141019639

D'Ausilio, A. (25 october 2011). *Arduino: A low-cost Multipurpose Lab Equipment.* Psychonomic Society, Inc. Retrieved from https://link.springer.com/content/pdf/10.3758/s13428-011-0163-z.pdf

Durga, N. S., Ramakrishna, M., & Dayanandam, G. (9 september 2018). Autonomous Water tank Filling System using IoT. *International Journal of Computer Sciences and Engineering.* Retrieved from https://www.researchgate.net/publication/328960063_Autonomous_Water_tank_Filling_System_using_IoT

Glass, R. L., Vessey, I., & Ramesh, V. (16 April 2002). *Research in Software Engineering: An Analysis of the Literature.* Retrieved from https://www.sciencedirect.com/science/article/abs/pii/S0950584902000496

Guha, A., Ganveer, A., Kumari, M., & Rajput, A. S. (6 June 2020). Automatic Bottle Filling Machine. *International Research Journal of Engineering and Technology (IRJET).* Retrieved from https://www.irjet.net/archives/V7/i6/IRJET-V7I6184.pdf

Gupta, A., Kumawat, S., & Garg, S. (2016). Automatic Plant Watering System. *Imperial Journal of Interdisciplinary Research.* Retrieved from https://www.researchgate.net/publication/321307351_Automatic_Plant_Watering_System

Nallani, S., & Hency, V. B. (2015). Low Power Cost Effective Automatic Irrigation System. *Indian Journal of Science and Technology,* 1–6. doi: 10.17485/ijst/2015/v8i23/79973

Prima, E. C., Munifaha, S. S., Salam, R., Muhammad, H. A., & Suryani, A. T. (19 April 2017). *Automatic Water Tank Filling System Controlled Using ArduinoTM Based Sensor for Home Application.* Retrieved from https://www.sciencedirect.com/science/article/pii/S1877705817311761

Samuel C. Olisa a b, C. N., Olisa, S. C., Asiegbu, C. N., Olisa, J. E., Ekengwu, B. o., Abdulhakim, & Eze, M. C. (24 July 2021). *Smart Two-tank water Quality and Level Detection System Via IoT.* Retrieved from https://www.sciencedirect.com/science/article/pii/S2405844021017540

Thakur, A. K., shekhar, s., Priyadarshee, A., Kumar, D., Kumar, S., Shrivastava, N., & Sinha, A. (2014). Automation of Residential Water Flowmeter. *JETIR.* Retrieved from https://www.jetir.org/papers/JETIR2108296.pdf

V, J. S., J, L. M., Maduletti, V., & V, M. D. (24/04/2018). Smart Water Flow Control and Monitoring System. *IJERT.* doi:10.17577/IJERTCONV6IS13071

Artificial Intelligence, Blockchain, Computing and Security – Dagur et al. (Eds)
© 2024 The Author(s), ISBN: 978-1-032-49393-0

Recognition techniques of medicinal plants: A review

Nidhi Tiwari
Research Scholar, Department of Computer Science and Information Systems, SRMU-Barabanki, India

Bineet Kumar Gupta
Associate Professor, Department of Computer Science and Information Systems, SRMU-Barabanki, India

Rajat Sharma
Assistant Professor, Department of Computer Science and Information Systems, SRMU-Barabanki, India

ABSTRACT: Pharmaceutical companies are gaining interest in medicinal plants due to their lower costs and fewer side effects as compared to modern drugs. These facts have led to many researchers expressing an interest in the study of automatic medicinal plant recognition. As a result of the development of strong classifiers that can categorize medicinal plants reliably in real time, several avenues for progress are available. This study focuses on the effectiveness and reliability of various machine learning techniques used in recent years for classifying plants using leaf pictures. In the review, several machine learning classifiers are discussed for detecting leaves and extracting significant leaf features. In this study, we investigate how these classifiers perform when classifying leaf ismages according to standard features such as shape, veins, texture, and a combination of these factors. A discussion of some of the most important ongoing studies and potential for improvement in this area wraps up our discussion of the leaf databases that is publicly available for pre-programmed plant recognition.

1 INTRODUCTION

As synthetic drugs are limited in their ability to control and cure chronic diseases, many first world countries have begun turning to traditional medicine for medical treatment (WHO 1999). Pharmaceutical industry uses traditional medicines extensively, according to (Karami *et al.* 2017), with a quarter of all drugs prescribed worldwide derived from medicinal plants. In comparison to synthetic drugs, medicinal plants offer a significantly lower rate of adverse reactions and are substantially cheaper (Lulekal *et al.* 2008). Bioactive compounds derived from medicinal plants like phenolics, carotenoids, anthocyanins, and tocopherols also have anti-inflammatory, antioxidant, antibacterial, and antihepatic properties (Altemimi *et al.* 2017). The process of recognizing medicinal plants manually is difficult and time-consuming, alike to other plant recognition, because expert opinions are available (Singh *et al.* 2017; Sladojevic *et al.*2016; Wäldchen *et al.* 2018). Numerous automatic leaf or plant recognition systems were developed as a result of these problems, most of which used machine learning techniques.

As a result of lack of knowledge, humans are unable to identify medicinal plants, which can save a life if used correctly. People can recognize these medicinal plants with the aid of an automated system and take the necessary steps to save the patients' lives. This makes the need for an automated system extremely vital for the betterment of the planet. Agricultural production will also be boosted.

More than 2 million of individual have contracted with the novel Coronavirus disease (COVID-19), resulting in more than 100 thousand deaths. Diabetes, hypertension, kidney disorders, as well as other infectious and non-infectious diseases, are considered to be at higher risk of contracting this virus. As a result, enhancing immunity (natural body system) may contribute to maintaining optimal health and preventing multiple pathogenic conditions.

Photosynthesis is the process by which these organisms make their food on mother earth. The food and oxygen we consume are entirely reliant on plants. Humans have used different methods to learn about plants and their properties throughout history. Information Technology, which is the science of plants, should be integrated with Botany through an automated plant identification system that enables people without deep or in-depth knowledge or training to be able to identify plants and their properties. An automated system will assist the user in identifying the species of plants by taking images of the plants and using the images to identify the species.

Our country lacks the knowledge necessary to properly cultivate medicinal crops. Farmers will be able to grow their business as a result of it. According to Salman *et al.* 2017 the proposed strategy utilized shrewd edge locator and Support Vector Machine as the classifier to separate 15 highlights of leaf pictures. In addition to the convex area, the filled area, circumferences (P), eccentricities (E), firmness, circumference-to-diameter ratios, orientation, narrow factors, boundaries, and Euler numbers were extracted.

1.1 Datasets

Research on the identification of medicinal plants used datasets created by the researchers. A number of medicinal plants from various countries have been studied, including Andrographispaniculata, Morindacitrophilia, Persicaria minor, Micromellumminutum, and Chromolaenaodorata, however very few studies have been conducted on their automatic identification. As a result of the different number of images used in the existing studies, the accuracy obtained is not comparable.

Table 1. Dataset name and their feature.

Sr.No.	Name of Database	Feature of Database
1	Image-CLEF11/ Image-CLEF12/ Image-CLEF13, Plant-CLEF14, PlantCLEF15, Plant-CLEF16, Plant-CLEF17	There are a variety of lighting conditions, backgrounds, viewpoints, and obstructions represented in the dataset that was compiled through crowd sourcing (Sanderson & Clough 2019). Among the databases, ImageCLEF11 has 5436 images from 71 species; ImageCLEF12 has 11,572 images from 126 species; ImageCLEF13 has 26,077 images from 250 species. PlantCLEF14 contains 60,961 images from 500 species, PlantCLEF15 includes 113,205 images from 1000 species, PlantCLEF16 includes 121,205 images from 1000 species, and PlantCLEF17 includes 256,287 images from 10,000 species.
2	Flavia	There are 1907 images in this dataset representing 33 plant species, which contain no noise, are aligned well on different backgrounds, and have small or no color and luminance differences (Wu *et al.* 2007).
3	Swedish Leaf Dataset	A few of the images are skewed and some are clean images with similar leaf features to Flavia data. A total of 75 images are included in each tree class in this dataset (Soderkvist 2001).
4	ICL	There are images of leaves that were grown exclusively in China in the dataset. A total of 16,851 images cover 220 plant species (Wang *et al.* 2017).

By combining a global layer pooling layer and an establishment structure, Hang *et al.* propose an improved Conventional CNN method to identify leaf diseases. Through the combined installation structure, 91.7% of leaf disease detection parameters were reduced and the performance of the model was improved.

Kherkhah and Aaghari found that KNN classifiers and PCAs performed better than Pattern Net neural networks and SVMs for GIST feature vectors Pankaja *et al.* propose that hierarchical centroids, discrete wavelet transforms, and gray level co-occurrence matrixes used to improve leaf recognition accuracy. Based on the Flavia dataset, 300 leaf tests were made for 30 unique classes, with an accuracy of 96.7%. After preprocessing, highlight extraction, and characterization of leaf inspection, this result was collected using dim level co-event frameworks and progressive centroid-based methodologies.

A smart edge locator and Support Vector Machines were used as classifiers in Salman *et al.* 2017 approach to segregate 15 highlights of leaf pictures using their approach. They extracted multiple characteristics: convexity, filledness, circumference ratio between diameters, circumference (P), eccentricity (E), firmness, orientation, narrow factor, boundary, euler number, diameter, circular, rectangular, circumference ratio between length and width, complexity, and compactness. In the Flavia dataset, 2220 images were examined from 22 species of plants with a precision of between 85 and 87%.

It has been found that ML and DL techniques used to detect, recognize, and manage weeds. DL techniques were applied to various agricultural problems, including weed detection, in a study published in 2018, Kamilaris & Prenafeta-Boldii (2018). Studies have shown that DL-techniques perform better than conventional image processing techniques.

Developing an autonomous system for mechanical weed control, Merfield (2016) explained ten components and possible obstacles. The breakthrough in Deep Learning have now made it possible to address the problems raised. The studies of Amend *et al.* (2019) have illustrated that plant classification based on Deep Learning can be used in fertilization, irrigation, and phenotyping systems as well. According to their study, "Deepfield Robotics" systems could decrease agricultural and horticultural labour costs.

According to Wang *et al.* (2019), weed detection techniques have the greatest difficulty distinguishing between weeds and crop species. Their research focused on weed detection techniques using machine vision and image processing. Brown & Noble (2005) made a similar observation. They reviewed techniques for mapping and detecting weeds using remote sensing. Both spectral and spatial features can be used to identify weeds in crops, but both have limitations. Both features are preferable, according to their study.

MRLBP-TOP is used in to present dynamic temporal facial dynamics. By learning more compact intermediate representations of MRLBP-TOP features on sub-sequences, the Dirichlet Process Fisher Vector (DPFV) can make a feature vector representation. As a result of which a discriminative representation is created with statistical aggregation approaches for each video sample.

The preceding frameworks rely on handcrafted features. They outperform most state-of-the-art methods with lower error rates, and they performed experiments on the AVEC 2013 and AVEC 2014 depression databases.

The concise overview of deep learning methods employed in video-based depression recognition. Zhu *et al.*(2021) developed a convolutional neural network (CNN) architecture that combines facial appearance and dynamics to evaluate the gravity of depression, achieving superior performance compared to other graphical techniques. Another noteworthy advancement involves the utilization of hyperspectral imaging in an AI system designed to estimate the degree of depression, which effectively integrates complementary patterns. In a separate study by Karimi *et al.* (2006), support vector machines (SVM) were successfully employed to detect weeds in corn fields. Similarly, Wendel and Underwood *et al.* (2016) employed SVM and linear discriminant analysis (LDA) to classify plants, presenting an innovative self-supervised discrimination approach. Their methodology involved segregating vegetation, preprocessing spectral data, and applying Principal Component Analysis (PCA) for feature extraction prior to model training. Additionally, Ishak *et al.* (2007) employed shape features and feature vectors to differentiate narrow-leafed weeds from broad-leafed ones.

In conventional machine learning techniques, a feature extractor must be constructed from raw data with substantial domain expertise. A representation-learning approach is used in the DAI approach for detecting classifications or objects from unprocess data using a machine learning method (LeCun *et al.* 2015). Patternson & Gibson (2017) report that machines can be taught to classify images, texts, and sounds directly. In feature learning, features are extracted automatically based on the task. As a hierarchical architecture of learning, deep learning has features that are unique to each level. The features utilized in this study consist of both hand-crafted and deep-learned representations. Visual cues are extracted through deep learning techniques, capturing discriminative information relevant to depression. Audio cues are captured using spectral Low-Level Descriptors (LLDs) and Mel-frequency Coefficients (MFCCs). Temporal movements across various feature spaces are described using Feature Dynamic History Histograms (FDHH). The authors introduce DepressNet, a deep regression network that predicts depression severity based on single images. To assess the level of depression, depression activation maps (DAM) are employed to highlight salient regions in facial images. Furthermore, a multi-region DepressNet is developed to model distinct patterns across different regions, enhancing overall performance. Large scale investigations conducted on AVEC2013 and AVEC2014 databases demonstrates that the proposed approach surpasses most state-of-the-art visual-based depression recognition methods, demonstrating improved performance through the extraction of local-global features from convolutional 3D networks. For depression detection, the proposed network incorporates spatiotemporal patterns. Notably, combining local and global features in C3D networks yields promising results.

There is a two-stream spatio-temporal framework for depression recognition described in. To generate temporal fragment features, the authors also propose TMP method. In experiments on the AVEC2013 and AVEC2014 databases, the proposed framework performed comparably well at identifying depressions. Currently, deep-learned features are trained on large databases such as CASIA WebFace Database and then fine-tuned on AVEC2014 and AVEC2013 databases. In contrast the authors train the deepmodels from scratch to detect depression. Despite the fact that these approaches have something in common, they cannot be considered as an end-to-end scheme for depression recognition. It is more pertinent to note that there is no guideline as to which network structure is appropriate for modeling depression representations (i.e., the depth of the network).

3 VARIOUS IMAGING TECHNIQUES USED

3.1 *Magnetic resonance imaging*

Magnetic resonance imaging (MRI) devices, also known as NMRs or nuclear magnetic resonance scanners, are commonly recognized for their robust magnets. These magnets efficiently polarize and excite the targeted protons found in water molecules within the tissue, enabling the generation of distinct body images through spatial signal encoding. The procedure begins with the generation of a pulse, which is subsequently directed towards the area under examination. Utilizing radio frequency (RF) pulses, MRI machines selectively interact with oxygen, causing its spin to change direction upon absorbing the transmitted energy. This transformative process in MRI is referred to as resonance.

3.2 Photo acoustic imaging

It is a biomedical imaging technique based on the photo acoustic effect that has been developed from hybrid biomedical imaging. In deep imaging of diffusive and other regimes, it merges optical absorption contrast with ultrasonic spatial resolution. Studies have shown that photo acoustic imaging can be used to diagnose tumors, map blood oxygen levels, examine brain activity, and detect disease.

3.3 Tomography

Tomography is an imaging technique that allows the visualization of a single plane or object. It encompasses various types of tomography, such as linear, polytomography, zonagraphy, computed type (also known as computed axial tomography), and positron emission type. These techniques enable the detailed examination and imaging of specific structures or regions of interest.

3.4 Thermography

In the case of breast imaging, it is one of the most commonly used applications. It is most common to use one of three approaches, telethermography, dynamic angio-thermography, or contact thermography. The principles derived from metabolic activity are used in the imaging thermographic digital methods. A higher value is also detected by studying the vascular circulation around a breast cancer developing.

According to this paper (Anne-KatrinMahlein *et al.* 2012), non invasive sensors can be used to detect plant diseases. There is a comparison and study of different types of sensors, including thermography, chlorophyll fluorescence, and hyperspectral sensors. Due to the large amount of data collected by hyperspectral systems, different approaches are necessary to obtain results. The temperature is the most important parameter in thermography. Yet, these technologies have not yet reached their full potential. It is also challenging to interpret sensor data.

Researchers Fang & Ramasamy present various methods for describing plant disease classifications in the year 2015. A direct method for detecting a particle includes polymerase chain reactions (PCR), immunofluorescence (IF), fluorescent in situ hybridization (FISH), flow cytometry (FCM) and enzyme-linked immunosorbent assays (ELISA). There are several types of indirect methods, including fluorescence imaging as well as hyperspectral imaging. Although the direct method is widely available, it requires expert technicians to operate, analyze data and is time-consuming. In addition, they are not much suited to different types of testing. With the advent of nanotechnology, there have been vast advances in sensitive biosensors whose specification can be further improved by using enzymes, DNA, and antibiotics as the detection element. Indirect methods can be used to detect diseases on the field, but lack specific information about different disease types.

3.5 Spectroscopic and imaging technologies

Using imaging technologies based on spectroscopy and profiling-based techniques, the paper compares methods for assessing leaf health and disease. One advantage of employing these technologies is their accuracy in detecting plant diseases. However, there are challenges associated with these techniques, such as the need to optimize solutions for specific plant diseases and the automation of techniques for continuous disease monitoring (Sankaran *et al.* 2010).

Peanuts are important agricultural products because of their high oil content. A method for early detection of plant diseases is presented in this paper (EwisOmran *et al.* 2016). This study examines how fungal diseases affect peanut plant leaves as leaf spots. A technique

933

called in situ spectroscopy was used to identify the early and late leaf indices. Moreover, thermal and spectral calculations were applied to make the distinction between healthy and infected plant leaves. Infected leaves show later chlorophyll decreases as stress indicators.

It is described in this paper (Federico Martinelli *et al.* 2015) how nucleic acids and proteins can be used to identify plant diseases. In this paper, we discuss lateral flow devices and mobility spectrometers that are used to detect early infections on fluid directly. Through the use of remote sensing technologies in combination with spectroscopy-based methods, high spatialization and high spatial accuracy are achieved, which helps in identifying infections in plants early.

3.6 *Multispectral imaging*

The use of multispectral imaging techniques is used to capture images in different wavebands, whether it is in the form of visible images of fruits and crops, or in near-infrared wavebands, to capture all kinds of images. Machine learning algorithms and classification algorithms are used to convert multispectral images into meaningful data for plant diseases detection.

3.7 *Hyperspectral imaging*

In hyperspectral imaging, various spectral information is gathered simultaneously using both traditional imaging techniques and spectroscopy. In this technique, pixel spectrums contributing to an image are determined. Similar to the traditional computer vision system, a hyperspectral imaging device consists of a dispersion of wavelengths and a subsequent transportation phase.

4 CONCLUSION

Human error is common when identifying medicinal plants manually. To solve these issues, an automated plant identification system may be developed, but this requires an extensive database, an in-depth understanding of plant morphology, and coding knowledge. Autonomous plant identification systems are currently mainly tested on pre-existing datasets developed in controlled environments. Thus, more research is needed into photos with complex backgrounds and diverse lighting situations. Furthermore, the dataset should be fairly large so that training can be more effective. As a result, the established identification system would be more accurate. An accurate classifier would greatly benefit from identifying variables such as form, vein, colour, and texture, according to the review.

REFERENCES

Alexender Wendel and Jamses Underwood, Self-supervised Weed Detection in Vegetable Crops Using Ground Based Hyperspectral Imaging, 10.1109/ICRA.2016.7487717 *Deep Learning: A Practitioner's Approach*, J Patterson, A Gibson book

Altemimi A, Lakhssassi N, Baharlouei A, Watson DG, Phytochemicals: Extraction, Isolation, and Identification of Bioactive Compounds from Plant Extracts. https://doi.org/10.3390/plants6040042

Anne-Katrin Mahlein, Erich-Christian Oerke, Ulrike Steiner &Heinz-Wilhelm Dehne , Recent Advances in Sensing Plant Diseases for Precision Crop Protection European Journal of Plant Pathology volume 133, pages197–209 (2012).

ESE Omran, Early Sensing of Peanut Leaf Spot Using Spectroscopy and Thermal Imaging, https://doi.org/10.1080/03650340.2016.1247952

Kamilaris, A., Prenafeta-Boldii, F.X., 2018. Deep Learning in Agriculture: A Survey. Comput. Electron. Agric. 147, 70–90.

Karimi Y, Prasher SO, Patel RM, Kim SH - Computers and Electronics, Application of Support Vector Machine Technology for Weed and Nitrogen Stress Detection in Corn, https://doi.org/10.1016/j.compag.2005.12.001

Kherkhah, F.M.; Asghari, H. Plant Leaf Classification Using GIST Texture Features. IET Comput. Vis. 2019, 13,

Kumar, A. & Alam, B. (2018). Task Scheduling in Real Time Systems with Energy Harvesting and Energy Minimization. Journal of Computer Science, 14(8), 1126–1133.

Kumar, A., & Alam, B. (2019). Energy Harvesting Earliest Deadline First Scheduling Algorithm for Increasing Lifetime of Real Time Systems. International Journal of Electrical and Computer Engineering, 9(1), 539.

LeCun Y, Bengio Y, Hinton G, Deep Learning, Nature volume 521, pages436–444(2015).

Lulekal E, Kelbessa E, Bekele T, Yineger H (2008) An Ethnobotanical Study of Medicinal Plants in Mana Angetu District, Southeastern Ethiopia. J EthnobiolEthnomed. . https://doi.org/10.1186/1746-4269-4-10

Martinelli F, Scalenghe R, Davino S, Panno S, Advanced Methods of Plant Disease Detection. A Review, Agronomy for Sustainable Development volume 35, pages1–25 (2015)

Palvika , Shatakshi, Sharma, Y., Dagur, A., & Chaturvedi, R. (2019). Automated Bug Reporting System with Keyword-driven Framework. In *Soft Computing and Signal Processing*: Proceedings of ICSCSP 2018, Volume 2 (pp. 271–277). Springer Singapore.

Salman A, Semwal A, Bhatt U, Leaf Classification and Identification Using Canny Edge Detector and SVM Classifier, 10.1109/ICISC.2017.8068597

Sankaran S, Mishra A, Ehsani R, Davis C, A Review of Advanced Techniques for Detecting Plant Diseases, https://doi.org/10.1016/j.compag.2010.02.007

Sladojevic S, Arsenovic M, Anderla A, Culibrk D, Stefanovic D (2016) Deep Neural Networks Based Recognition of Plant Diseases by Leaf Image Classification. ComputIntellNeurosci 2016:1–11. https://doi.org/10.1155/2016/3289801

Wäldchen J, Mäder P, Machine Learning for Image Based Species Identification, Methods in Ecology and Evolution, 2018 - Wiley Online Library, https://doi.org/10.1111/2041-210X.13075

Zhu, H., Xie, C., Fei, Y., & Tao, H. (2021). Attention Mechanisms in CNN-based Single Image Super-resolution: A Brief Review and a New Perspective. Electronics, 10(10), 1187

Security and privacy in the cloud computing

Artificial Intelligence, Blockchain, Computing and Security – Dagur et al. (Eds)
© 2024 The Author(s), ISBN: 978-1-032-49393-0

Secure data storage using erasure-data in cloud environment

K. Bala*, Balakrishna Reddy Mule*, Rishi Raj Kumar*, Srinivasulu Gude* &
Ranga Uday Sudheer Gaddam*
Bharath Institute of Higher Education and Research, Chennai, India

ABSTRACT: As of right now, the present storage model stores user data on cloud servers. So, users forfeit their rights. You are at risk of a privacy violation and don't have control over the data. Since it cannot be completely evaluated from within a cloud server, encryption technology is typically the foundation of conventional privacy protection techniques, although it is not particularly successful in fending off attacks from within. The purpose of this paper is to propose a three-layer storage framework for solving the problem of cloud storage, which is based on fog computing. Frameworks such as the one proposed in this paper can be used to fully utilize cloud storage while protecting the privacy of users. Furthermore, Hash-Solomon code algorithm allows for the division of data into segments. A small amount of data can then be put on local machines and fog servers to ensure privacy. The viability of our plan has been confirmed through theoretical safety analysis and experimental evaluation, which is a significant improvement over the current cloud storage plan. Data information is lost if one data component is absent. As we secure the scheme's security and effectiveness will be demonstrated. The distribution percentage stored in clouds, fog, and local computers can also be computed using this computational intelligence-based approach. Our strategy, which is a considerable advance over the present cloud storage strategy, has been shown to be feasible through theoretical safety analysis and experimental evaluation. Data information is lost if one data component is absent. Additionally, computational intelligence used in this approach allows for the determination of the distribution fraction saved.

1 INTRODUCTION:

Cloud computing is a term used in computer science to describe a sort of outsourcing of computer services, much like how energy is outsourced [1]. It is easy for users to utilize. They don't have to be concerned about how or from where the electricity is created. They pay for their consumption each month. A similar concept underlies cloud computing: By distributing the user's access to storage, processing power, or specially created development environments, each server's ability to protect user privacy can be ensured [2]. This allows users to take advantage of these resources without having to consider how they work [3]. Internet-based systems make up the majority of cloud computing [4]. Computer network diagrams show the internet as a cloud, which is a metaphor for the internet and, as such, is an abstraction that hides the complex internet infrastructure [5]. It is a method of computing in which associated tasks are made "as a service," allowing users to access technology-enabled services via the Internet (or "in the cloud") without being aware of or

---

*Corresponding Authors: bala.dharshinipb@gmail.com, balakrishna.m54@gmail.com, rishirajkumar163@gmail.com, srinivaschowdary2002@gmail.com and udaysudheer2001@gmail.com

DOI: 10.1201/9781003393580-140

in control of the technologies supporting these servers. Massive data structures and expansive cloud systems are examples of fog computing, which alludes to the growing difficulties in objectively acquiring information [6]. This results in low-quality content being created. Numerous effects of fog computing may affect big data and cloud systems. The creation of metrics that seek to boost accuracy has addressed the issue of inaccurate content distribution, although this issue can be viewed from one consistent angle [7]. A control plane and a data plane make up for networking. For instance, fog computing on the data plane enables computing services to be located at the network's edge rather than on servers in a data center. The creation of metrics that seek to boost accuracy has addressed the issue of inaccurate content distribution, although this issue can be viewed from one consistent angle. Fog computing, in contrast to cloud computing, emphasizes proximity while being able to be applied in AAL settings. We suggest utilizing a fog computing-based TLS architecture to safeguard user privacy [8]. The TSL architecture can successfully preserve users' privacy while granting them some level of administrative power. As was already stated, it is challenging to fend off the internal assault [9]. Conventional methods can prevent external attacks, but they are ineffective when CSP is having issues. Instead of employing conventional methods, our concept uses encoding technology to split user input into three portions of various sizes [10]. They will all lack some crucial information needed for concealment. The user's local workstation will store the three data components [11]. As well as the fog computing concept. Furthermore, cloud storage has a lot of security problems. Because users of cloud storage have no real control over how their data is physically stored, there is a separation of data ownership and management for these users [12]. The proposal is workable, according to a theoretical safety analysis. The privacy of data on each server can be protected by carefully distributing the ratio of data blocks kept on different servers. Theoretically, the encoding matrix cannot be broken. Additionally, it is possible to use hash transformation to preserve the partial information [13]. The experiment test revealed that this method of encoding and decoding could be accomplished without impairing the efficiency of cloud storage.

2 METHODOLOGIES

2.1 Login module

The user is presented with this activity as soon as they access the website. For the purpose of logging onto the website, the user must supply a valid phone number as well as the password they choose while enrolling [14]. The user can access the website successfully if their information matches the data in the database table; otherwise, a message stating that the login attempt was failed is displayed, and the user must enter the correct information once more [15]. For new user registration, a link to the register activity is also made available.

2.2 Registration module

An unfamiliar user must register before they may log in if they wish to access the website [16]. If you select the register option during the login process, the register activity will start.

2.3 Storage module

After the data has been uploaded to the cloud, the owner has no more control over it. Three separate layers of encryption are applied to the original data in this module [17]. Prior to being stored in the Cloud, the data in each layer can be encrypted using various cryptographic algorithms and an encryption key.

2.4 *Recovery module*

This module allows users to retrieve their files from three separate storage servers: local machines, fog servers, and cloud servers [18]. To divide data into separate pieces, we are utilizing the Hash-Solomon code technique in this case as Shown in figure given below. If one data part is missing, we lose the data information.

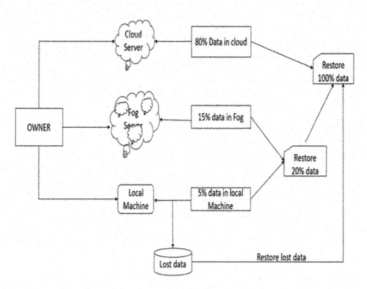

Figure 1.   Computer architecture.

3   ALGORITHMS

3.1 *Bucket*

With the use of ACLs (Access Control Lists), you can limit who and how much of your data is accessible. Data information is lost if one data component is absent. The proposed framework employs algorithms based on the bucket notion.

3.2 *BCH code algorithm*

A sizable family of potent random error-correcting cyclic codes called the Bose, Chaudhuri, and Hocquenghem (BCH) codes comprises these codes. This class of codes is an impressive generalization of the multiple-error correcting Hamming code.In this lecture note, we solely take binary BCH codes into account. In the following lecture note, non-binary BCH codes like Reed-Solomon codes will be covered.

4   RESULTS AND DISCUSSION

4.1 *System analysis*

4.1.1 *Existing system*

- The technology of cloud computing has evolved in recent years.
- Unstructured data has grown quickly, generating more interest in and driving innovation in cloud storage.

- Computer technology has developed rapidly.
- The aforementioned activities employ as a result of the efforts of many people, hash-cloud computing has gradually become more advanced.
- If consumers are put in danger of losing their privacy and control over their data, the present storage model stores all of their data entirely in cloud servers.
- The basis of privacy protection techniques is typically encryption technology.
- These techniques are unable to successfully fend off attacks coming from a cloud server's inside.

4.1.2 *Disadvantages*

Changes in the understanding of risk from extending the datacenter into the cloud. Low latency and location awareness.

4.2 *Proposed system*

- The framework can use the entire cloud storage capacity while preserving data privacy.
- We lose the data information if one data component is absent.
- Algorithms based on the bucket notion are used in the suggested framework.
- Our system employs a bucket idea to cut down on data loss and processing times.
- The Bose-Chaudhuri-Hocquenghem (BCH) code algorithm is what we are employing. Highly flexible.
- BCH code has low levels of redundancy and is employed in many communications applications.

4.3 *Inputs and outputs*

The input design serves as the interface between the user and the information system. It involves developing guidelines and practices for data preparation, which requires taking the necessary actions to transform transaction data into a format that can be processed. Users can either directly enter the data into the system by typing it in or by glancing at the computer to read it from a written or printed document. The input design facilitates user contact with the information system. It entails formulating policies and procedures for data preparation, which calls for carrying out the required procedures to convert transaction data into a format that can be processed. The input design facilitates user contact with the information system. It entails formulating policies and procedures for data preparation, which calls for carrying out the required procedures to convert transaction data into a format that can be processed. What order or coding scheme should the data follow? The conversation serves as a direction for operating staff feedback. Procedures for creating input validations and what to do in case of errors.

An output is considered high-quality if it satisfies the user's needs and provides the data in an understandable manner. Any system has outputs that let users and other systems know what processing results were received. In output design, it is decided how the data as a direct source of information for the user, it is the most significant. The system's ability to assist users in making decisions is improved through efficient and intelligent output design.

4.4 *Data flow diagram*

A two-dimensional diagram known as a data flow diagram (DFD) shows how data is processed and transferred in a system. Each data source is shown graphically along with how it interacts with other sources of data to produce a common outcome.

To create a data flow diagram, one must first:

- Identify external inputs and outputs;
- Ascertain the relationships between the inputs and outputs; and
- Illustrate with drawings how these connections relate to one another and what results.

4.4.1 *Role of DFD*

- Both programmers and non-programmers can understand this documentation support. DFD solely considers what processes accomplish, not how they are carried out.
- A physical DFD assumes who processes the data and where it flows.
- By studying the data that enter the process and observing how they are altered when they leave, it enables analysts to isolate areas of interest in the business and examine them.

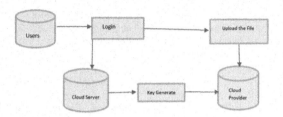

Figure 2.   Data flow diagram.

5   CONCLUSIONS

Many advantages result from the growth of cloud computing. Cloud storage is a piece of technology that facilitates consumers' ability to enhance their storage capacity. But there are also several security issues with cloud storage. Users that store their data on the cloud typically perceive a separation of ownership and management of their data because they do not actually have control over how their data is physically stored. This is where cloud computing has drawn tremendous attention. We develop a BCH Code approach and provide theoretical safety analysis that has shown the idea's viability. By judiciously dispersing the amount of data blocks stored across different servers, the privacy of data on each server can be ensured. Decoding the encoding matrix is theoretically not conceivable. Furthermore, it is possible to use this method of encoding and decoding may be carried out successfully without impairing the efficiency of cloud storage. Additionally, we develop a sensible, all-inclusive efficiency strategy to achieve optimum efficiency. Additionally, we find that coding is improved by the Cauchy matrix.

REFERENCES

[1] Mell P.and Grance T., "The NIST Definition of Cloud Computing," *Nat.Inst. Stand. Technol.*, vol. 53, no. 6, pp. 50–50, 2009.

[2] Dinh H. T., Lee C., Niyato D., and Wang P., "A Survey of Mobile Cloud Computing: Architecture, Applications, and Approaches," *Wireless Commun. Mobile Comput*, vol. 13, no. 18, pp. 1587–1611, 2013.

[3] Chase J., Kaewpuang R., Yonggang W., and Niyato D., "Joint Virtual Machine and Bandwidth Allocation in Software Defined Network (SDN) and Cloud Computing Environments," in *Proc. IEEE Int. Conf. Commun.*, 2014, pp. 2969–2974.

[4] Li H., Sun W., Li F., and Wang B., "Secure and Privacy-preserving Data Storage Service in Public Cloud," *J. Comput. Res. Develop.*, vol. 51, no. 7, pp. 1397–1409, 2014.

[5] Li Y., Wang T., Wang G., Liang J., and Chen H., "Efficient Data Collection in Sensor-cloud System with Multiple Mobile Sinks," in *Proc. Adv. Serv.Comput, 10th Asia-Pac. Serv. Comput. Conf.*, 2016, pp. 130–143.

[6] Xiao L., Li Q., and Liu J., "Survey on Secure Cloud Storage," *J. Data Acquis. Process.* vol. 31, no. 3, pp. 464–472, 2016.

[7]   McEliece R. J. and Sarwate D. V., "On Sharing Secrets and Reed-Solomon Codes," *Commun. ACM*, vol. 24, no. 9, pp. 583–584, 1981.

[8]   Plank J. S., "T1: Erasure Codes for Storage Applications," in *Proc. 4th USENIX Conf. File Storage Technol.*, 2005, pp. 1–74.

[9]   Kulkarni R., Forster A., and Venayagamoorthy G., "Computational Intelligence in Wireless Sensor Networks: A Survey," *IEEE Commun. Surv. Tuts*, vol. 13, no. 1, pp. 68–96, First Quarter 2011.

[10]   Xia Z., Wang X., Zhang L., Qin Z., Sun X., and Ren K., "A Privacy Preserving and Copy-deterrence Content-based Image Retrieval Scheme in Cloud Computing," *IEEE Trans. Inf. Forensics Security*, vol. 11, no. 11, pp. 2594–2608, Nov. 2016.

[11]   Shen J., Liu D., Shen J., Liu Q., and Sun X., "A Secure Cloud-assisted Urban Data Sharing Framework for Ubiquitous-cities," *Pervasive Mobile Comput*, vol. 41, pp. 219–230, 2017.

[12]   Fu Z., Huang F., Ren K., Weng J., and Wang C., "Privacy-preserving Smart Semantic Search Based on Conceptual Graphs Over Encrypted Outsourced Data," *IEEE Trans. Inf. Forensics Security*, vol. 12, no. 8, pp. 1874–1884, Aug. 2017.

[13]   Hou J., Piao C., and Fan T., "Privacy Preservation Cloud Storage Architecture Research," *J. Hebei Acad. Sci.*, vol. 30, no. 2, pp. 45–48, 2013.

[14]   Ali, Md. A., Balamurugan, B., Dhanaraj, R. K., & Sharma, V. (2022). IoT and Block chain based Smart Agriculture Monitoring and Intelligence Security System. In *2022 3rd International Conference on Computation, Automation and Knowledge Management (ICCAKM)*. 2022 3rd International Conference on Computation, Automation and Knowledge Management (ICCAKM). IEEE. https://doi.org/10.1109/iccakm54721.2022.9990243.

[15]   Rajesh Kumar Dhanaraj, S. K., Seifedine Kadry, B.-G. K., & Byeong-Gwon Kang, Y. N. (2022). Probit Cryptographic Block chain for Secure Data Transmission in Intelligent Transportation Systems. In 網際網路技術學刊 (Vol. 23, Issue 6, pp. 1303–1313). Angle Publishing Co., Ltd. https://doi.org/10.53106/160792642022112306013.

[16]   Chandraprabha, M., & Kumar Dhanaraj, R. (2022). An Empirical View of Genetic Machine Learning Based on Evolutionary Learning Computations. In *Machine Learning Methods for Engineering Application Development* (pp. 59–75). Bentham Science Publishers. https://doi.org/10.2174/9789815079180122010008.

[17]   Prasanth, A., D, L., Dhanaraj, R. K., P C, S., & Balusamy, B. (2022). *Cognitive Computing for Internet of Medical Things*. Chapman and Hall/CRC. https://doi.org/10.1201/9781003256243.

[18]   Roobini, S., Kavitha, M., Sujaritha, M., & Rajesh Kumar, D. (2022). Cyber-Security Threats to IoMT-Enabled Healthcare Systems. In *Cognitive Computing for Internet of Medical Things* (pp. 105–130). Chapman and Hall/CRC. https://doi.org/10.1201/9781003256243-6.

944

Artificial Intelligence, Blockchain, Computing and Security – Dagur et al. (Eds)
© 2024 The Author(s), ISBN: 978-1-032-49393-0

Security aspects in E-voting system using cloud computing

Shreyas Agrawal*
M.E. CSE(CC), Chandigarh University, Gharuan, Mohali, Punjab, India

Mohammad Junedul Haque*
AIT-CSE, Chandigarh University, Gharuan, Mohali, Punjab, India

ABSTRACT: In recent years, academics have paid a lot of attention to the implementation of the electronic voting (e-voting) system. In most democratic nations (developed and developing), political party leaders and stakeholders are interested in e-voting. Researchers and technologists have examined technical problems with the e-voting system that may facilitate its seamless adoption, which has prompted its full acceptance in many nations. The difficulty, however, is in securing and maintaining a reliable electronic voting system free from security flaws, particularly from hacking and hijacking. To fully win the public's trust, approval, and adoption of electronic voting systems, the problem still remains unsolved and demands for innovative designs into high level security infrastructure. The suggested approach offers greater voter identification security. All voter security passwords have been verified with the E-voting Commission of India's primary database, and following voter authentication, the voter will be able to cast a ballot. Voters in this model are allowed to cast ballots from any Indian constituency. The major goal of this suggested strategy is to offer security at various levels. Vote counting will be automated under this methodology. This technique saves a tonne of time, and the Indian E-voting Commissioner may quickly announce the outcome.

Keywords: Cloud Computing, Security, Attacks, Cyber Attacks, Firewall, Cloud Security, E-Voting

1 INTRODUCTION

1.1 *Utilising the cloud*

Cloud computing is simply the transmission of a range of services over the internet, sometimes known as "cloud computing." It involves storing and obtaining data from remote servers as opposed to storage devices at home and private servers. With the cloud, businesses needed to purchase and operate their own computing equipment to meet their needs. It was necessary to buy enough server capacity to manage the highest traffic demand while reducing the possibility of interruptions and downtime. As a consequence, the storage capacity of the server is typically neglected to a significant extent. Companies can lessen the demand on pricey desktop computers, technicians, and additional technological assets by utilising cloud service providers today. [1]

1.2 *Different cloud provider models*

These are often known as the "cloud computer stack" because of the way they are built on top of each other. When you comprehend what these are and how things vary from one another, achieving your company's goals becomes easier.

Cloud computing is being embraced by an increasing number of users, including individuals and corporate organisations. According to a Right Scale1 survey, the typical user runs at least four cloud-based applications and is always looking at four more.

---

*Corresponding Authors: shreyas222ag@gmail.com and mohammad.e12447@cumail.in

DOI: 10.1201/9781003393580-141

Figure 1.   Cloud services.

Figure 2.   Different cloud services models.

1.3 *Cloud safety*

Cloud assurance, also known as internet of things security, is an array of safety measures that safeguards the equipment, applications, and information kept in the cloud. These procedures make it possible to govern asset and knowledge availability, as well as to secure the privacy of data. They also encourage the compliance with regulatory data. Cloud systems use security measures to guard against viruses, attackers, denial of service issues, and unauthorised user access and use.

In implementing virtualization instead of locally equipment, the business can position themselves for future. Additional applications are available to you now because of the availability of the internet. Additionally, it promotes greater collaboration, increases data accessibility, and simplifies content administration. Due to concerns about security, certain customers might be hesitant to migrate their data into the cloud. Your concerns can be allayed and your private data can be secured as effectively as is possible without the help of a reliable cloud computing provider.

1.4 *Cloud layer types*

There are three basic types of settings available when looking for security via the cloud. Understanding how these settings contrast with one another is essential given that each has significant security advantages and disadvantages.There are three basic types of settings available when looking for security via the cloud. Public clouds, cloud computing for individuals, and mixed clouds are some of the best possibilities. Understanding how these settings differ from each other is essential given that each has significant security advantages and disadvantages. An important security system rule to re- member is that users shouldn't have to discover workarounds to complete tasks because your security measures are too restrictive. Users frequently find methods to get past security constraints that make a cloud computing service challenging to utilise. According to experts, humans are frequently the weakest link in any security system, and these workarounds make the system insecure.

2 IMPORTANT ELEMENTS TO CONSIDER FOR CLOUD SECURITY

Most frequently, people are curious about the security features that cloud computing offers. Without the essential technologies that make it more secure, cloud security is worthless. These technologies include-

2.1 *Encryption*

It is a method that restricts data comprehension to the appropriate authorities. Its main objective is to provide the data a format that is challenging to grasp. Encryption is the name given to this technique or procedure. Encryption is a key component of cloud computing technology for data security. Unencrypted data might pose significant risks to a business. There are unanticipated issues of data security in cloud computing because of encryption.

2.2 Firewall

Security aspects in cloud computing go beyond basic encryption. A firewall is a very secure method of data protection because it adds an additional layer of security. It guarantees that any dangerous assaults are halted. Such damaging attacks on web traffic happen often. Cloud firewalls are hosted through the cloud, as opposed to being housed on-premises like traditional firewalls that were ineffectual.

2.3 Security policies

Data security issues in cloud computing have no boundaries. Across the board, security guidelines are used for the whole cloud architecture. For better cloud security, security settings must be configured with strict security regulations. When a company doesn't carefully consider its security procedures, data breaches happen.

2.4 Backup plans

To ensure that no single bit of information is at risk, backup processes are also necessary for data security. Data ought to be kept up on-premises or in a different cloud to prevent any form of data loss. Constantly have a backup strategy ready in case there is a data loss. Cloud computing has produced multiple clouds and hybrid cloud platforms to bolster data security [6].

3 CLOUD CYBER ATTACK

Any cyberattack which targets remote service providers who use their own cloud infrastructure to offer web hosting, computing, or storage solutions is referred to as a cloud cyberattack [7].
    Data security also necessitates backup plans in order to guarantee that not a single piece of data is in danger. In order to prevent any type of data loss, data should be backed up either locally or on another cloud. To prevent losses in the event of a data loss, always have a backup plan in place. To boost data security, multi-cloud and hybrid cloud infrastructure have been created.

4 THE CAUSES OF CLOUD COMPUTING CYBERATTACKS

4.1 Misconfiguration

CSPs offer a variety of service tiers based on the amount of control an organisation needs over their cloud installation. Organisations must build these installations in line with their requirements to provide improved cybersecurity. Sadly, this leads to deployment problems because the vast majority of businesses lack the requisite cloud security posture to guarantee the security of these offerings. According to an IBM research, 86 percent of data breaches are the result of improperly setup servers. One can modify the particular installation you're utilising to meet your safety needs via the safety solutions offered by cloud service providers.

4.2 Compromised user databases

Weak password policies are one of the biggest causes of user accounts being compromised. Because users either create weak passwords, recycle old ones, or don't routinely change them, many cloud service users don't have strong password security.

4.3 API vulnerability

Any holes in the connection and configuration of such APIs will give hackers a backdoor to use. When developing and configuring APIs, according to the documentation might help you prevent security mistakes. In order to identify any flaws, organisations must carefully monitor way the APIs are utilised. [8]

947

5 E-VOTING

In information security research in India, the E-voting system employing the cloud technology is a novel and more secure approach of the highly secure voting system. The electronic voting method offers a forum for voters to select their preferred candidate for government and make that choice known. The public's confidence in the electronic voting procedure is crucial. The cloud-based electronic voting system offers complete media coverage of the election process, which is useful for the general public and the election commission if something is going wrong. Voters will feel a great deal of comfort and faith thanks to this concept.

E-voting processes can only be made trustworthy if they adhere to strict security standards for secrecy, confidentiality, and honesty. Secrecy and confidentiality imply that all voters cast thoughtful, private ballots for their preferred electorates and that all voters are aware of their preferred elector. Integrity entails verification; voter authentication and election results must be handled correctly. Secret electronic voting includes confidentiality. This sample procedure urges the voter to cast their vote in a secure and composed manner. [9]

There are primarily two categories of electronic voting:

- Electronic voting which is physically supervised by election officials from a government or non-government organisation.
- Remote electronic voting is done at a distance under the the voter's complete supervision and is invisible to government personnel.

Having computer assistance accurate voting records are made possible by electronic voting, which can also speed up ballot counting, lower election costs, and increase accessibility for voters with disabilities. India began using electronic voting machines (EVMs) in 2004 for its parliamentary elections. The Electronic Voting Machines (EVM) include two basic parts. The voter uses the Ballot unit, which is connected to another unit by a cable, to cast their ballot. To allow voters to cast their ballots, within the voting space, the casting vote unit is kept. The second part is the Command Unit, which the polling officer uses. Each candidate is represented by a button in blue on the voting unit, and voters can choose which button they want to click to support that candidate.

6 PROJECT DESCRIPTION

The development of an online fingerprint voting system, which is now needed, is the paper's main goal. The online system must meet the fundamental requirements, such as having reliable and secure software. This system is automated. This online voting mechanism is incredibly user-friendly and effective. Because each user has a unique fingerprint pattern and because the database will include all the information and fingerprint patterns of those who are qualified to vote, the system will be secure because each user can only cast one vote. Large databases should be supported by this system. Across all these features it also has some drawbacks like software issues, internet problems etc. The project seeks to increase voting equipment security in order to counter fraudulent votes. We introduce Aadhar card registration to replace manual registration. The biometric information is compared to the data on the Aadhar card to access the electronic voting machine. We develop a safe cloud storage system that receives the data from every voting booth in order to save time and labour. Voting, security, and mobile applications are the work's main pillars.

Voting is recorded using the voting machine. Initially mechanical, the voting apparatus has subsequently been transformed into a voting system that is electronic. During voting, the voter registration procedure is manually verified and is fully manual. Voter registration and verification by hand requires more time and labour. The voters profit from the introduction of automated voting operations. Voters are permitted to cast their votes using a real-time hardware setup that is organised in a voting booth during the vote casting phase. When a voter's biometric matches the information on their Aadhar card, access to the electronic voting machine is granted. India's Unique Identification Authority (UIDAI) is a facility for gathering data where the Aadhar holder's information is kept on file. After the voter has been approved and validated, that database is able to be used for comparing and validating the voter's biometric. They can use the computerised polling machine to cast their ballots. The ballots are translated into numbers, encrypted, and then tabulated. Votes are the encrypted data that is transmitted to the cloud. With a unique ID assigned to each party, The vote security is ensured by the Double Encrypt-Decrypt method. The data is encrypted with the usage of random key creation. When a key is powerful, the cryptography also becomes powerful, making it tough to crack. Key generation is the process of creating keys in cryptography. In cryptography, the process that produces keys is known as key generation. The concept behind random key creation is to generate the encryption key randomly each time that encryption is performed. In this round, in order to cast a vote, a voter must show their Aadhar cards. After scanning the card's QR code as the first stage of authentication, the ballot registrant must verify their fingerprint. In the polling booth, when a voter's fingerprints are checked against to a database. He or she may proceed to the voting phase of the procedure at this time. After every vote is cast, the voter database gets revised with the results. The entire network of voting booths displays the vote total in real time. For voters who complete their ballots, we maintain a separate database. The vote list is used to extract the voter data, which is then added to the completeness list. Our main objective in this case is to reduce the amount of votes cast fraudulently.

The security phase, which concludes the process, focuses primarily on data transfer and cloud storage security. The information that is conveyed is a code for the involved parties. An encryption method is used to encrypt the code and save it on the cloud. The encryption algorithm is constructed in such a way as to have a minimal and effective time complexity. Each party receives a special identification number that is kept secret. With the aid of randomly generated keys, the unique ID is used in the encryption. To make encryption and decryption more difficult, the key generation is random.

7   PROJECT DESCRIPTION

Based on client-server architectures, the systems that use biometric techniques can be managed by a system administrator. The system integrates fingerprint control using biometric fingerprint devices with built-in libraries. The Election Commission, which maintains a database regarding all voters, is the server side. Information about candidates that has been verified by the National Database and Registration Authority (NADRA). In order to reduce network traffic and speed up the election process, a local database server is kept on each client workstation. This server holds voter information. Microsoft's Fingerprint Scanner SDK is integrated with the system we designed using Visual Basic.NET, which accepts a fingerprint scan as input and identifies the user by performing authentication process. Once the prints have been obtained, the voter's record is opened from the database to verify the voter's details, such as whether or not the voter is registered and whether or not the person has previously cast a ballot. Working principle of the system during elector's voting procedure is as follows:

- A fingerprint scanner is used to scan the voter's finger.
- The voter database is checked for the voter's fingerprint.
- Voting is not permitted if the voter's fingerprint is not stored in the voter database.

  Otherwise, the elector's information on the screen is checked:

- After verifying, the voter clicks the "Vote" button. The algorithm then determines whether the voter cast a ballot for the election at issue earlier. In that case, the system issues a warning. If not, the voting screen is displayed by the system.
- An elector votes for any party.
- Voting is complete if the voter is confident in their choice.

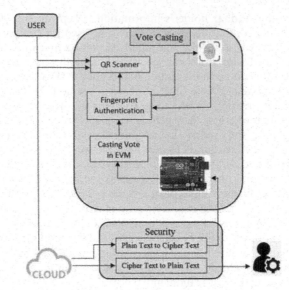

Figure 3.   Block diagram.

REFERENCES

[1] Ashwini Sheth, Sachin Bhosale, Harshad Kadam, *"Research Paper On Cloud Computing"*, ISSN 2231–2137

[2] https://blog.knoldus.com/know-about-cloud-computing-architecture/

[3] https://www.researchgate.net/figure/Cloud-computing-services-IaaS-PaaS-and-SaaSfig1342492294/

[4] https://www.box.com/en-in/resources/what-is-cloud-security

[5] https://www.triskelelabs.com/blog/cloud-cyber-attacks-the-latest-cloud-computing-security-issues:: text=Any 20cyber 20attack 20that 20 targets,SaaS2C20IaaSC20and 20PaaS

[6] https://www.triskelelabs.com/blog/cloud-cyber-attacks-the-latest-cloud-computing-security-issues.

[7] Pankaj kumar Malviya *"E-Voting System Using Cloud In Indian Scenario"* Volume-3, Issue-3, 171–175, ISSN: 2250–3676

[8] Vivek S K, *et al.*, "E-Voting System using Hyperledger Sawtooth", 2020

[9] Jehovah Jireh Arputhamoni S., Gnana Aravanan A., *"Online Smart Voting System Using Biometrics Based Facial and Fingerprint Detection on Image Processing and CNN"*, 2021

[10] Sudhakar M, Divya Soundarya Sai B., "Biometric System Based Electronic Voting Machine Using Arm9 Microcontroller" *IOSR Journal of Electronics and Communication Engineering (IOSR-JECE)* eISSN: 2278-2834,p- ISSN: 2278-8735.Volume 10, Issue No 1, Verion II (January - February 2015), PP 57–65

[11] Amit Taneja R.K. Shukla "Comparative Study of RSA with Optimized RSA to *Enhance Securiy*", pp 975–996

[12] Taneja A, Shukla R K and Shukla R S, *"Improvisation of RSA Algorithm in Respect to Time and Security with the Proposed (AEA) Algorithm"* 1998 (2021) 012036 doi:10.1088/1742-6596/1998/1/012036

[13] Nilam Kate, Katti J.V., "Security of Remote Voting System based on Visual Cryptography and SHA" 2016

[14] Gandhi Usha Devi, K. Anusha, G. V. Rajyalakshmi "An Enhanced e-Voting System in Cloud Using Fingerprint Authentication" 2014

[15] Gurpreet Singh Matharu, Anju Mishra, Pallavi Chhikara "CIEVS: A Cloud-based Framework to Modernize the Indian Election Voting System"

[16] Deepika Verma, Er. Karan Mahajan "To Enhance Data Security in Cloud Computing using Combination of Encryption Algorithms", *International Journal of Advances in Science and Technology*, ISSN 2348-5426, Vol 2, Issue 4, Dec 2014

[17] Dandekar, Girish Umratkar D.B., Pranali Manwar, Sneha Sute, Zain Ansar, Hamza Khan, Baseer Ansari, Bhuvanesh Joshi *"Online Voting System Using Cloud Computing"*

Artificial Intelligence, Blockchain, Computing and Security – Dagur et al. (Eds)
© 2024 The Author(s), ISBN: 978-1-032-49393-0

Enhanced-honey bee based load balancing algorithm for cloud environment

Saurabh Singhal
Department of Computer Engineering and Applications, GLA University, Mathura, Uttar Pradesh, India

Shabir Ali
Bharati Vidyapeeth (D.U.), Department of Engineering and Technology, Navi Mumbai, Maharastra, India

Dhirendra Kumar Shukla & Arvind Dagur
Department of Computer Science and Engineering, Galgotias University, Greater Noida, Uttar Pradesh, India

Rahul Papalkar, Vinod Rathod & Mohan Awasthy
Bharati Vidyapeeth (D.U.), Department of Engineering and Technology, Navi Mumbai, Maharastra, India

ABSTRACT: Cloud computing is an emerging paradigm that brings services to consumers when needed at a lower cost. Adopting cloud computing causes major problems regarding cost-effective load balancing. Effective load-balancing solutions in data centers prevent underloaded and overcrowded conditions. When certain services are full of tasks, to ensure Quality of Service (QoS) some of these tasks are transferred to underloaded resources in the same data center. This paper proposes a load-balancing algorithm using an improved honey bee colony algorithm. To balance the load between resources, the algorithm employs the foraging behavior of honey bees. Jobs in overloaded resources are referred to as beehives, while underloaded resources serve as food sources. The approach decreases both the makespan and response time. The experiment was carried out on the CloudSim simulator, and the result shows that makespan was reduced by 11% and response time by 13% when compared with other static and dynamic algorithms.

Keywords: Cloud Computing, Honey Bee Optimization, Enhanced-Honey Bee Optimization Load balancing, makespan, response time

1 INTRODUCTION

Cloud computing is a computing paradigm based on the Internet. All data and applications are housed in data centers intricately linked by hundreds of computers. A data center is a cluster of servers that hosts various applications and provides storage. Service providers employ a pay-as-you-go approach for resource utilization. Customers can utilize resources like computing, software, storage, and others through the Internet by paying only for the time they use the resource. End-users, data centers, and servers are other key components of a cloud environment besides the Internet. End-users must connect to the data center to subscribe to various applications. A data center is usually far from the clients. Servers in a cloud environment are distributed and available everywhere on the Internet and host various applications. Cloud computing makes it possible to share processing and data resources. This may be due to the availability of a host application, which eliminates the requirement for the user to purchase resources. It is also helpful for remote users, who can use resources across the globe [1]. As users increase in cloud computing, so does the

DOI: 10.1201/9781003393580-142

951

need for shared resources. To make matters worse, job planning requires a balance between these resources. Effective load balancing on all resources in the cloud environment is essential to maintaining QoS. Ensuring that QoS is vital for client satisfaction in cloud computing.

Traditional load-balancing techniques have several issues in the cloud environment due to the instability of demand. To solve these issues, various metaheuristic algorithms like Ant Colony Optimization (ACO) and Artificial Bee Colony (ABC) have been used to balance load among resources [2]. These algorithms are based on the local exchange of numerous agents to accomplish a specific objective.

This study aims to introduce a load-balancing system that efficiently distributes flexible work to all cloud hosts, improving resource utilization and performance time. Jobs are distributed among all accessible VMs. To ensure fairness, the proposal allocates tasks to a less-loaded VM. Also, the proposal stops the execution of workloads on the VM if the processing time difference exceeds the average processing time. Consequently, both the overall response time and the makespan time are lowered. An Enhanced Load Balancing Honey Bee Based Algorithm (ELBHB) is proposed in this study. The work of honey bees' search for natural food balances the load. The shared function informs other functions in the VM government in the same way that bees identify food intake and update other bees in the hive with their waggle dance [3]. The Bee Colony method is a swarm intelligence program [4] that solves numerical problems for improving performance by mimicking the diet of bee colonies. It imitates the action of bees in discovering food. The algorithm comprises scout bees, food-selling bees, and food-producing bees. Scout bees forage for food in bee colonies. After finding a food source, it returns to the hive and performs a beautiful dance. Some bees in the hive use a moving dance to gather information about food availability and the distance from the colony. The food bees begin collecting them after the scout bees reach the beehive. Bees choose food-source locations at random. CloudSim [5] has been used to model the proposed ELBHB algorithm. The proposed method is compared to some well-known metaheuristic algorithms. Experimental findings revealed that ELBHB reduces response time for different tasks.

2   LITERATURE REVIEW

In the cloud environment, the number of jobs submitted is constantly changing. Thus, there is a requirement for dynamic algorithms like metaheuristic algorithms, which can balance load efficiently. This section presents load-balancing algorithms given in the past by various researchers. Honeybees are the inspiration for a proposed load-balancing technology called HBB-LB [6], which distributes the load across multiple virtual machines. An algorithm for load balancing based on the optimization of ant colonies taking comparable jobs was proposed [7]. An approach based on Ant Colony Optimization (ACO) for load balancing was developed by [8]. In [9], the Genetic Algorithm (GA) was used to devise a natural selection load-balancing approach.

The authors in [10] presented a proposal for an agent-based load balancer that could be used in scenarios with more than one cloud. In [11], the authors developed an innovative strategy for dynamically balancing load in the cloud. The authors in [12] developed a novel load-balancing algorithm predicated on a method for estimating when service would come to an end. A novel technique for VM load balancing was given by [13], guaranteeing that requests are distributed uniformly across VMs even during peak times. Application scheduling in mobile cloud computing can benefit from the MAX-MIN ant system developed by [14]. The authors in [15] used machine learning to balance the load in a cloud environment. The authors used the adaptation of Bat optimization [16], PSO [17], and ABC [18] for balancing load in the cloud environment.

3   PROPOSED WORK

The proposal employs honey bee foraging behavior to provide efficient load balancing across resources and transfer jobs to underutilized resources. To balance the load in the cloud environment, the foraging behavior of honeybees is implemented as a honey bee algorithm. The tasks are modeled like honeybees in the proposal. When the honeybees start food foraging, the jobs are assigned to resources for execution. Because the processing capacity of each resource varies, some resources may be overloaded while others may be underloaded. In these conditions, an efficient load-balancing method is required to improve performance. When a resource is overloaded with numerous jobs, some jobs must be moved and assigned to an underloaded resource. In this scenario, the job that is to be transferred is determined by priority. The proposal selects tasks with the lowest priority as candidates for relocation. This plan is a lot like the last one: the bees will move away from the food source when the amount of honey in the nectar gets lower. When honeybees return to the hive from a trip to find food, they tell their fellow bees where they found it. When a potential forager bee starts working, she is a naive worker who is out of work, which means she doesn't know where to find food. She can become a scout by seeking a source and acting on what she sees. The waggle dance isn't the first sign that it's time to fly out and start looking for food. Instead, it's some unknown internal motivation or external signal. On the other hand, a bee might join the colony after seeing a waggle dance, which makes them start looking for food, which leads them to join the colony.

A recruit's memory has information about where they think they are, but a scout's memory doesn't. The main difference between a recruit and a scout is: When a bee finds out about any source, it immediately starts using it and remembers its basics. The suggested ELBHB is inspired by honey bee foraging behavior. The assigned job tells the other tasks about the state of the VM in the same way that bees tell the other bees in their hive about a lot of food by waving their wings. The operation refreshes the status of VM availability and VM load.

The following section mentions the QoS parameter for measuring the proposed algorithm's performance.

3.1 Response time

It is in addition to submission time and a job waiting time in the queue. It is the response obtained by the submitted job for a specified input [18]. This can be represented as:

$$RT_T = T_i(ST_T + WT_T) \tag{1}$$

where, RT_T is the response time of task$_i$, ST_T is the submission time of task$_i$, and WT_T is the waiting time of task$_i$.

3.2 Problem algorithm

The problem can be represented as a graph, $G = (N, E)$ where N represents the environment's virtual machine, and E denotes jobs' mapping to VMs. Initially, all the artificial ants are present in the virtual machine. At the iteration, honey bees(jobs) move from one food source (virtual machine) to another, building the load-balancing solution until they have visited the entire environment. The process flow of the algorithm is shown in Figure 1.

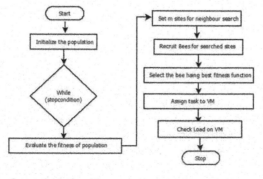

Figure 1.   Process flow of proposed algorithm.

4 RESULT ANALYSIS

The proposed ELBHB algorithm considers the minimization of response time as the objective function. This algorithm was motivated by algorithms that were influenced by nature. The program uses the universal honey bee concept and the fitness function calculation to strike a healthy balance between the demands placed on the cloud infrastructure. The universal honey bees with the highest fitness value are discovered during this iteration.

4.1 Experimental setup

The solution was put into practice on the simulator CloudSim version 3.0.3. A simulation of the cloud computing environment can be created using the CloudSim framework. To simulate our load balancing method, we constructed a virtual computer, data centers, and cloudlets to stand in for individual jobs using CloudSim.

4.2 Experiment results

In this section, the results of the proposal are compared to both static algorithms like RR [3] and dynamic algorithms like Ant Colony Optimization (ACO) and Artificial Bee Colony (ABC). The total number of tasks that need to be executed and the number of instructions that can be run greatly affect how well the system works. Then, to show how the suggested algorithm works, the number of jobs by default is raised to a very high number. It's the same as 1000, but these activities could have 2,000 to 20,000 instructions. Because of this, each VM in the system has a different amount of work to do.

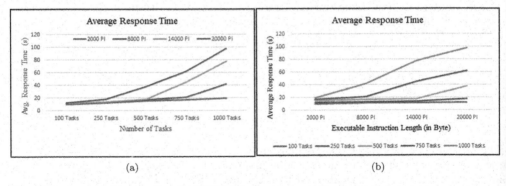

Figure 2.   Average response time for ELBHB for processing instructions

Figure 2a presents the response time of ELBHB while the practical command length of each task varies from 1000 to 25000 Processing Instructions (PI). The response time reflects the time between sending a request and generating the first response. When PI is 2000, the average response time is about 4 seconds, but this number increases slightly to reach 20 seconds when PI is equal to 8000. However, it increases to 100 seconds when the instruction size reaches 20000 PI. This is due to increased system load, the number of tasks, and the processing of instructions. Then, response time increases.

Figure 2b presents the ELBHB response time ratio compared to the PI per user while the tasks vary from 100 to 1000. The response time remains the same when tasks are at a minimum; however, the response time increases significantly with an increase in tasks. When jobs are equal to 100, the average response time in ELBHB increases linearly with an increase in PI. However, an increase in PI increases the response time to 1000 requests. This is because as the duration of the processing instructions grows, the system load increases. Thus, while assigning tasks, the availability and load of each VM become critical.

5 CONCLUSION AND FUTURE WORK

In this paper, an algorithm to measure the load on a cloud computing site based on the forgiving behavior of the honey bee is proposed. The proposed ELBHB reduces both the time it takes for people to respond and the time it takes for data centers to process the work, considering the different availability and load of each VM. The proposal restricts the distribution of requests for a VM when the average processing time for VMs is greater than or equal to the predefined limit. The experimental results were carried out on the ClouSim simulator, showing that the proposed algorithm minimizes the response time. For future work, the migration of tasks can be considered, as it does not work well in the proposed algorithm.

REFERENCES

[1] Endo P. T., Rodrigues M., Gonçalves G. E., Kelner J., Sadok D. H., and Curescu C., "High Availability in Clouds: Systematic Review and Research Challenges," *Journal of Cloud Computing: Advances, Systems and Applications*, pp. 5–16, 2016.

[2] Gamal M., Rizk R., Mahdi H., and Elhady B., "Bio-inspired Load Balancing Algorithm in Cloud Computing," in *Proc. of The International Conference on Advanced Intelligent systems and Informatics (AISI)*, Cairo, Egypt, pp. 579–589, September 2017.

[3] de Vries H., and Biesmeijer J. C., "Modelling Collective Foraging by Means of Individual Behaviour Rules in Honey-bees," *Behavioral Ecology and Sociobiology*, Springer-Verlag, vol. 44, no.2, pp. 109–124, 1998

[4] Kumar R. & Chaturvedi A., Improved Cuckoo Search with Artificial Bee Colony for Efficient Load Balancing in Cloud Computing Environment. In *Smart Innovations in Communication and Computational Sciences* pp. 123–131, 2021.

[5] Calheiros R. N., Ranjan R., Beloglazov A., Rose C. A. F. D., and Buyya R., "CloudSim: A Toolkit for Modeling and Simulation of Cloud Computing Environments and Evaluation of Resource Provisioning Algorithms," *Software: Practice and Experience*, vol. 41, no. 1, pp. 23–50, January 2011.

[6] D.B. LD, Krishna P. V., Honey Bee Behavior Inspired Load Balancing of Tasks in Cloud Computing Environments, *Applied Soft Computing*, 13(5), pp. 2292–2303, 2013

[7] Li K., Xu G., Zhao G., Dong Y., Wang D., Cloud Task Scheduling Based on Load Balancing Ant Colony Optimization, In: *IEEE Sixth Annual ChinaGrid Conference*, pp. 3–9, 2011.

[8] Nishant K., Sharma P., Krishna V., Gupta C., Singh K. P., Nitin N., Rastogi R., Load Balancing of Nodes in Cloud Using Ant Colony Optimization. In: *Proceedings of the 14th International Conference on Modelling and Simulation*, pp. 3–8, 2012.

[9] Dasgupta K., Mandal B., Dutta P., Mandal J. K., Dam S., A Genetic Algorithm (GA) Based Load Balancing Strategy for Cloud Computing, *Procedia Technology*, 10, pp. 340–347, 2013.

[10] Tasquier L., Agent Based Load-balancer for Multi-cloud Environments. *Columbia International Publication Journal of Cloud Computing Research*, 1(1), 35–49, 2015.

[11] Komarasamy D., Muthuswamy V., A Novel Approach for Dynamic Load Balancing with Effective Bin Packing and VM Reconfiguration in Cloud. *Indian J. Sci. Technol.* 9 (11), pp. 1–6, 2016.

[12] Chien N. K., Son N. H., Loc H. D., Load-balancing Algorithm Based on Estimating Finish Time of Services in Cloud Computing, *International Conference on Advanced Commutation Technology (ICACT)*, pp. 228–233, 2016.

[13] Kulkarni A. K., Annappa B., Load Balancing Strategy for Optimal Peak Hour Performance in Cloud Datacenters. In 2015 *IEEE International Conference on Signal Processing, Informatics, Communication and Energy Systems (SPICES)*, pp. 1–5, 2015.

[14] Wei X., Fan J., Wang T., Wang Q., Efficient Application Scheduling in Mobile Cloud Computing Based on MAX–MIN Ant System. *Soft Computing*, 20(7), pp. 2611–2625, 2016.

[15] Agarwal R., R. & Sharma D. K., Machine Learning & Deep Learning Based Load Balancing Algorithms techniques in Cloud Computing, In 2021 *International Conference on Innovative Practices in Technology and Management (ICIPTM)*, pp. 249–254, 2021.

[16] Kumar R., Bhardwaj D., Joshi R., Adaptive Bat Optimization Algorithm for Efficient Load Balancing in Cloud Computing Environment. In *Advances in Computational Intelligence and Communication Technology* pp. 357–369, 2022.

[17] Singhal S., Sharma A., Load Balancing Algorithm in Cloud Computing Using Mutation Based PSO Algorithm. In *International Conference on Advances in Computing and Data Sciences* pp. 224–233, 2020

[18] Singhal S., Mangal D., Mutative ABC Based Load Balancing in Cloud Environment. In *International Conference on Futuristic Trends in Networks and Computing Technologies*, pp. 546–555, 20220.

Artificial Intelligence, Blockchain, Computing and Security – Dagur et al. (Eds)
© 2024 The Author(s), ISBN: 978-1-032-49393-0

A detail study on feature extraction technique for content based image retrieval for secure cloud computing

J. Sheeba Selvapattu*
PhD Scholar, JAIN Deemed to be University, India

Suchithra R. Nair*
Head of CS&IT, JAIN Deemed to be University, India

ABSTRACT: Due to the accessibility of internet technology and the accessibility Organizations and individuals are motivated to outsource picture storage and computing to the cloud by the explosive proliferation of digital images. However, uploading the image data will increase the chance of privacy breach. The proposed effective picture retrieval search techniques that satisfy user needs. That can be managed and retrieved with the help of Content-Based Image Retrieval (CBIR). This paper focus on various feature extraction technique for the picture retrieval. This paper discuss about the global features for the image retrieval. The numerous edge detection strategies to effectively extract the edges and compare them are discussed in this study. Robert's, Prewitt, Sobel, and Canny Edge Detection are the techniques. The software Mathlab is used to display the output of the photographs.

1 INTRODUCTION

The cloud computing platforms enable processing and storing big data are becoming more important. Visual data today makes up one of the largest components of global Internet traffic, both for business and personal. Educational, industrial, medical, social, and other life institutions have produced vast image collections. The numbers of images, graphics, and photos are made and exchanged every day are increasing. It becomes necessary to store a lot of data in the internet-based environment. However, the rapid expansion of image data can soon delete local storage on devices, which encourages consumers to upload the photographs to cloud services. The image owners can conveniently retrieve the desired images from the cloud through the Internet in this fashion, eliminating the need for them to maintain the image collection locally. However, the owners of the photographs lose ownership of them once they are transferred to the cloud. The primary issue with outsourcing is image privacy. First, hackers are constantly interested in attacking cloud servers. A well-known example is the 2014 iCloud breach, which resulted in the exposure of approximately 500 celebrity private photos. The cloud server cannot be completely trusted because they might also be interested in the images' contents. Researchers have proposed privacy-preserving CBIR methods, which can secure the image content and facilitate the search for related photos at the same time. The feature-encryption-based and image-encryption-based schemes can be used to categorise these methods. The picture owners first extract the visual characteristic from the photographs in the feature-encryption based approach. Therefore, the photos must be secured before being stored on the cloud. The primary attributes utilised for general-purpose image retrieval are colour, texture, shape, spatial etc. This paper compares various feature extraction methods in-depth on a image groups. The rest of this paper is structured as

*Corresponding Authors: sheebaselvapattu@gmail.com and r.suchithra@jainuniversity.ac.in

DOI: 10.1201/9781003393580-143

follows. The literature review is briefly described in Section 2. Techniques for feature extraction have been covered in section 3. Focus on the proposed scheme of the preserved framework in Section 4. Image segmentation and edge detection technique is discussed in Section 5. Finally, the conclusion is summarized in Section 6.

2 REVIEW OF LITERATURE

Sajjad [1] proposed CBIR system for texture rotation and colour change. The suggested solution uses a feature vector created by concatenating colour and texture features. To extract colour information, images are converted to HSV colour space and normalized using a colour histogram. the author used Hue and Saturation channels to resistant the changes in illumination. To extract rotation-invariant texture characteristics, one uses rotated local binary patterns (RLBP). Experiments are conducted on the Zurich Building (ZB) to determine the performance. Pavithra and Sharmila [2] proposed a multistage CBIR technique. In initial stage the data is created. In the second stage, texture and shape (edge) properties from the photos are extracted. the texture information was retrieved using LBP, edge information was extracted using the Canny edge detector. Manhattan distance was used to gauge the search's size.The proposed multistage approach has increase performance by reducing running time and boosting precision, however the amount of running time needed is dependent on the dataset's number of images. The proposed system can be used to search datasets of various sizes and types provided some machine learning techniques have been included into it.Pavithra and Sharmila [3] proposed a color-based picture retrieval algorithm. The author used four image datasets were employed in the trials to evaluate the proposed dominant colour descriptor, the proposed algorithm performed better. However, there is still a lack in the algorithm, the color is applied to the all the images in different classes, The suggested method must be used with other feature extraction techniques because the semantic gap still remains because the same colour information may be applied to photos in many semantic classes (shape, texture, and spatial information). Ashraf [4] proposed a CBIR with the colour and edge features. Author extracted colours using a colour histogram. For edge extraction, the proposed YCbCr colour space was used. The authors used Haar wavelet, which is faster in terms of calculations, to speed up the discrete wavelet computation in order to gain better enhancement. The results obtained of mean precision and recall, which used the Manhattan distance as its similarity assessment, showed the effectiveness of the method. But there are downsides as well. It also has a problem with a lack of spatial data, and it is unknown how effective its computational method is.Nazir [5] proposed a content-based picture retrieval techniques, the methodology used are based on colour and texture feature extraction. The local distribution of the image's edges are included in the produced feature vector for texture using discrete wavelet transform (DWT) and EHD.Tadi Bani [6] describe CBIR system that relies on the extraction of colour characteristics in the spatial domain as well as global and local texture data in the frequency and spatial domains. The images are first filtered with a Gaussian filter to reduce the effects of noise. A quantized colour histogram in RGB colour space is used to extract colour features. The Gabor filter is used to extract local texture features, which enhances retrieval efficiency. The proposed method demonstrated high precision values when evaluated against existing state-of-the-art methods using the Simplicity dataset. Although it had a long run time due to the utilisation of various features, it was also said to be low sensitive to noise and invariant to rotation. Rana [7] presented an approach for CBIR based on the combination of parametric and nonparametric features (texture) (color and shape). The extraction of parametric features was done using colour moments and moment invariants, and the extraction of nonparametric features was done using ranklet transformation.The length of the created feature vector, which is 247, increases running time and is viewed as a significant flaw in the proposed technique. The method was tested against five datasets. Alsmadi [8] proposed a method for retrieving content-based images that benefits from combining colour, shape, and texture. The Canny edge histogram and DWT transform of the YCbCr colour space were used to extract colour features, while GLCM was utilised to extract texture data. The RGB colour space's clever edge method was used to extract shape details. The recommended approach combines simulated annealing (SA) and a genetic algorithm (GA), which raises the fitness number and enhances the quality of the solutions. The suggested CBIR system outperforms current state-of-the-art systems in terms of average precision and recall (0.9015 and 0.1803, respectively).

3 CBIR FRAMEWORK

The parties engaged in the suggested privacy-preserving content-based picture retrieval technique, are the cloud, the data owner. Image encryption (i.e., cloud storage, upload, and outsourcing) and data user outsourcing are the two main components of our strategy (i.e., searching and indexing). With the help of this framework frm Figure 1, we can outsource photos from CBIR depending on elements like colour, texture, size, and form while still protecting the privacy of both the data owner and the data user. The most crucial stage, feature extraction, which transforms a visual concept into a numerical representation. Low-level features, such as colour, shape, texture, and spatial information, as well as local descriptors are examples of extracted features. After feature extraction. The final step is to compare the retrieved features from the query image to every other image in the collection in order to find the images that are most pertinent.

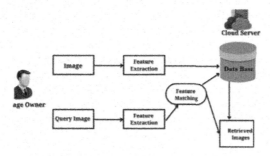

Figure 1.   Proposed cloud based CBIR framework.

4 FEATURE EXTRACTION TECHNIQUE

The feature extraction and selection that represent the semantic information of images is the main purpose of CBIR. These features can be further divided into two groups: local features, which are typically obtained by segmenting the image or by computing some significant points like corners, blobs, and edges, and global features, which describe the entire image and include attributes like colour, texture, shape, and spatial information. Scale, translation, and rotation modifications have no impact on local features.

4.1   Low-level image features

The proposed CBIR systems are built on the extraction of low-level picture features. Image features can be retrieved from either the complete image or from specific parts to perform CBIR. The majority of modern CBIR systems are region-based since it has been discovered that users are typically more interested in particular sections than the full image. Global feature-based retrieval is far less complicated. The segmented sections can then be used to extract low-level information from, such colour, texture, shape, or spatial placement.

4.2 Global features

The most often employed features in image retrieval tasks include colour, texture, shape, and spatial data. The classification of global characteristics and a few feature extraction techniques are shown in Figure 2.

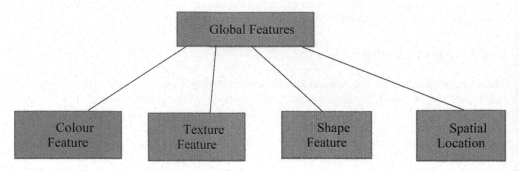

Figure 2.   Classification of global features.

4.2.1 Colour features
One of the most often utilised features in image retrieval is the colour feature. A particular colour space is used to designate colours. There are many different colour spaces, and they are frequently used for various applications. Color spaces including RGB, LAB, LUV, HSV (HSL), YCrCb, and the hue-min-max-difference (HMMD) are frequently utilised in RBIR because they have been demonstrated to be more closely related to human perception. The color-covariance matrix, colour histogram, colour moments, and colour coherence vector are often used colour characteristics or descriptors in RBIR systems. Applying an efficient filter to remove the colour noise can considerably increase retrieval accuracy because colour images are frequently damaged with noise as a result of capturing devices or sensors [11]. Particularly when the retrieval findings are required for human interpretation, the pre-process can be crucial.

4.2.2 Texture feature
The characteristics of the texture it describes the substance of many real-world images, such as fruit skin, clouds, trees, bricks, and cloth, texture contributes significant information in image categorization. Therefore, texture plays a key role in developing high-level semantics for the purposes of picture retrieval. The features obtained using Gabor filtering [9] or wavelet transform [10]. The wavelet and Gabor features are frequently utilised for image retrieval and have been found to closely match the findings of studies on human vision.

4.2.3 Shape feature
One of the basic characteristics used to identify items is shape. The shape extraction can be done using either a region or a border. While the boundary-based technique performs extraction on the region's edge, the region-based approach performs extraction over the entire region. To extract shape information, a variety of techniques are utilised, including the Fourier descriptor,moment invariants. In many specialised photos, such as images of man-made items, shape aspects have proven to be helpful. However, due of the inaccurate segmentation of colour images utilised in most research, shape features are more challenging to apply than colour and texture.

4.2.4 Spatial feature
Spatial position is helpful in classifying regions. For instance, the colours and textures of "sky" and "sea" may be identical, but their spatial placements may differ, with sky typically appearing at the top of an image and sea at the bottom.

The lack of geographical information affects many of the strategies for extracting low-level features. In a two-dimensional image, a spatial feature is essentially connected to where the objects are situated. For instance, two distinct sections in the same image with two distinct spatial contents could have an equal histogram. Computational complexity typically has a negative impact on spatial information. One of the most effective ways to capture an image's spatial characteristics is by spatial pyramid matching.

5 IMAGE SEGMENTATION

Image segmentation is a technique that divides a digital image into several subgroups known as Image segments. By making the original image less complex, this technique makes it easier to handle or analyse the image in the Figure 3. Giving labels to pixels is the process of segmentation, to put it simply.

Figure 3. Orginal image.

5.1 *Edge based segmentation*

To find edges in a picture, edge-based segmentation employs a variety of edge detection operators. These borders highlight regions in a picture where the texture, colour, or grey levels change. As we move from one place to another, the level of grey could change. [12] So long as we can discover that discontinuity, we can locate the edge. Typically, edges are connected to "Magnitude" and "Direction." Some edge detectors offer both directions and magnitude. The number of edge detectors, including the Sobel edge operator, the canny edge detector, the Prewitt edge operator, the Robert's edge operator, etc.

5.1.1 *Robert technique*
Lawrence Roberts is the author of the Robert cross method. This approach was developed to approximate the gradient of an image using separate differentiation, which is achieved by computing the sum of the squares of the differences between diagonally adjacent pixels. One of its key drawbacks is that it is sensitive to noise. Real edges do not offer good results unless they are really sharp. Robert Cross Method picture segmentation is shown in Figure 4 along with other image filtering approaches.

Figure 4. Robert technique.

5.1.2 *Sobel technique*

The Sobel method works by vertically and horizontally filtering the image using a small, continuous, integer-valued filter. It suppresses noise better. The sole difference between the Sobel method and Prewitt's is that the mask coefficient can be changed. The Prewitt approach takes slightly longer to compute. Images are segmented using the Sobel method methodology in Figure 5 together with other image filtering methods.

Figure 5.   Sobel technique.

5.1.3 *Prewitt edge detection technique*

The Prewitt approach relies on the image being convolved in both the vertical and horizontal directions with a tiny, discrete, integer-valued filter. Compared to the Robert approach, it yields more accurate results. It is a less expensive and quicker method for edge detection in terms of computing time. Edge chooses the value on its own the image in Figure 6 is segmented.

Figure 6.   Prewitt edge detection technique.

5.1.4 *Canny technique*

The canny technique involves a Gaussian filter to smooth the image, then computing the edge strength by measuring the intensity of the image's gradients. Next, non-maximum threshold is applied to the gradient magnitude. Figure 7 displays Canny method-based picture segmentation along with several image filtering approaches.

Figure 7.   Canny technique.

6 CONCLUSION

In order to safeguard and improve the security of the image data on the cloud platform, this study has presented the Cloud based CBIR Framework as a new technique. As a result, it is impossible to trust a third-party resource because it could be compromised by any hacker on the same network. Aside from that, the paper focus on Various edge detection methods, including Robert, Prewitt, Sobel, and Canny, that are employed in picture segmentation were discussed in this work. A powerful reaction to diagonal edges is the Sobel. The Prewitt is Noise Sensitive. When compared to other edge detection algorithms, the Canny Edge provides less noise sensitivity.

REFERENCES

[1] Sajjad, M., Ullah, A., Ahmad, J., Abbas, N., Rho, S., & Baik, S. W. (2018, February). Integrating Salient Colors with Rotational Invariant Texture Features for Image Representation in Retrieval Systems. *Multimedia Tools and Applications*, 77(4), 4769–4789.

[2] Pavithra, L. K., & Sharmila, T. S. (2018, August). An Efficient Framework for Image Retrieval Using Color, Texture and Edge Features. *Computers & Electrical Engineering*, 70, 580–593.

[3] Pavithra, L. K., & Sree Sharmila, T. (2019, December). An Efficient Seed Points Selection Approach in Dominant Color Descriptors (DCD). *Cluster Computing*, 22(4), 1225–1240.

[4] Ashraf, R., Ahmedm, M., Jabbar, S., Khalid, S., Ahmad, A., Din, S. & Jeon, G. (2018, March). Content Based Image Retrieval by Using Color Descriptor and Discrete Wavelet Transform. *Journal of Medical Systems*, 42(3), 44.

[5] Nazir, A., Ashraf, R., Hamdani, T., & Ali, N., "Content Based Image Retrieval System by Using HSV Color Histogram, Discrete Wavelet Transform and Edge Histogram Descriptor," in 2018 *International Conference on Computing, Mathematics and Engineering Technologies* (iCoMET), Sukkur, Pakistan, 2018, January, pp. 1–6.

[6] Tadi Bani, N., & Fekri-Ershad, S. (2019, August). Contentbased Image Retrieval Based on Combination of Texture and Colour Information Extracted in Spatial and Frequency Domains. *Electronic Library*, 37(4), 650–666.

[7] Rana, S. P., Dey, M., & Siarry, P. (2019, January). Boosting Content Based Image Retrieval Performance Through Integration of Parametric & Nonparametric Approaches. *Journal of Visual Communication and Image Representation*, 58(3), 205–219.

[8] Alsmadi, M. K. (2020, April). Content-based Image Retrieval Using Color, Shape and Texture Descriptors and Features. *Arabian Journal for Science and Engineering*, 45(4), 3317–3330.

[9] Mehrotra, R., Namuduri, K. R., & Ranganathan, N. (1992). Gabor Filter-based Edge Detection. *Pattern Recognition*, 25(12), 1479–1494.

[10] Kokare, M., Biswas, P. K., & Chatterji, B. N. (2005). Texture Image Retrieval Using New Rotated Complex Wavelet Filters. *IEEE Transactions on Systems, Man, and Cybernetics, Part B (Cybernetics)*, 35(6), 1168–1178.

[11] Vyas, A. M., Talati, B., & Naik, S. (2013). Colour Feature Extraction Techniques of Fruits: S Survey. *International Journal of Computer Applications*, 83(15).

[12] Liu, D., Zhou, S., Shen, R., & Luo, X. (2023). Color Image Edge Detection Method Based on the Improved Whale Optimization Algorithm. *IEEE Access*.

Artificial Intelligence, Blockchain, Computing and Security – Dagur et al. (Eds)
© 2024 The Author(s), ISBN: 978-1-032-49393-0

Secure data storage based on efficient auditing scheme

R. Selvaganesh, K. Akash Sriram, K. Venkatesh & K. Sai Teja
Bharath Institute of Science and Technology, Chennai, India

ABSTRACT: Cloud service companies now hold enormous volumes of data. Cryptography is frequently used by third-party auditors (TPAs) to validate this data. Techniques for Cloud Data Auditing with a Focus on Security and Privacy. It seeks to offer an on-demand resource. It reduces the hassle of having to obtain info online. Cloud storage enables the safe storing of data online. A group of computers that are used to store information and operate applications in a cloud platform are connected to the cloud. We can access any file or document from any user from anywhere in the globe thanks to cloud computing. The primary benefits of using the cloud are cost savings, high scalability, and vast storage capacity. But security in cloud computing is a big problem. It is necessary to create an efficient public auditing protocol that gets around the current audit plan's drawbacks. The suggested solution is designed to allow TPA too periodically or on-demand verifies the accuracy of cloud data without having to download the complete database or place an additional online burden on cloud servers and users. This guarantees that no data content is revealed to TPA during the audit. It upholds the accuracy, reliability, and secrecy of the data being stored.

1 INTRODUCTION

The cloud allows users to save time and money. The term "cloud computing" refers to a type of Internet-based computing where numerous services are provided to the computers and other devices of an organisation over the Internet [1]. In this usage, the term "cloud" refers to the Internet as a whole. Before people and companies may store data and install apps in the cloud, there are still several issues that need to be fixed, despite the fact that cloud computing holds immense potential for Information Technology (IT) applications. Data security is one of the major barriers to deployment, followed by worries about compliance, privacy, trust, and legal issues. Because cloud computing is essential and contains a lot of complicated data, maintaining the security and integrity of data kept there is one of the key objectives. To make the cloud environment trustworthy and encourage widespread adoption by users and businesses, security issues raised by users must first be addressed [2]. Privacy, protection, availability, location, and secure transfer of data are the main concerns in cloud data security. Threats, data loss, service interruption, hostile assaults from outside, among the security issues with the cloud are multi-tenancy issues [3]. Maintaining the veracity of information that has been stored is referred to as data integrity in cloud system [4]. Unauthorized users shouldn't be able to alter or lose the data. It is common knowledge that cloud computing services uphold data accuracy and integrity [5]. Since users put their private or personal data in the cloud, data confidentiality is a crucial issue from their perspective. Data confidentiality is ensured via authentication and access control techniques [6]. By enhancing cloud dependability and trustworthiness in cloud computing, the data confidentiality issue could be resolved. Therefore, from the viewpoint of the user, security,

DOI: 10.1201/9781003393580-144

integrity, privacy, and confidentiality of the data stored in the cloud should be taken into account [7,8]. New procedures or approaches should be created and used to fulfill all of these needs. An efficient public auditing protocol must be created in order to get around the shortcomings of the current auditing system [9]. The suggested system was created to allow TPA too periodically or on-demand test the accuracy of cloud data without having to download the complete database or place an additional online burden on cloud users or cloud servers [10]. It guarantees that during the auditing process, no data content will be disclosed to TPA. It preserves the secrecy, integrity, and accuracy of stored data [11].

2 METHODOLOGIES

2.1 Secure dynamic auditing

The owners of the data will dynamically update it in cloud storage systems. The auditing protocol should be created as an auditing service to accommodate both static archive data and dynamic data. However, the auditing techniques might become vulnerable due to the dynamic processes. The server may specifically carry out the following two attacks: First, Replay Attack. The server can use an outdated version of the owner's data instead of correctly updating it in order to pass audits. 2"Forge Attack" The server might have access to sufficient information from the dynamic processes to change the data tag when the data owner upgrades the data to the most recent version. Any data may be used by the server if it could counterfeit the data tag. It would be considered fabricated. Data tag to satisfy audits.

2.2 Batch auditing for multiowner

A key element of cloud computing, data storage auditing enables data owners to maintain the accuracy of their data on cloud servers. A lot of requests from different data owners would be made to conduct audits on the data stored on cloud servers; in order to optimize efficiency, it would be best to aggregate the requests the auditor must handle. The auditor was previously unable to combine the tags made from the data by the different owners because the data tags were dissimilar. To secure the data integrity of the data owners, auditing challenges must be sent to every cloud that stores the data in order to verify the proofs from each of them. In order to save money by conducting each of these processes separately for each request, combining all of the responses and performing batch verification is advised.

2.3 Experimental results and analysis

The evidence and challenge shown in Figure 1 below are included in the cost of communication between the auditor and the server. It would be interesting to perform a batch audit the sign t stands for the total number of data blocks in contention from all owners across all cloud servers, which is equal. Sectors are created by further dividing the data blocks.

2.4 Computer architecture

System design is that the most challenging and innovative step within the life cycle of the device. For any engineered product or device design n is that the initiative within the development process. Input Design assumes a critical job inside the product improvement life cycle and needs close consideration from engineers. Programming execution is required fundamentally to make sense of a beneficial method of correspondence inside the affiliation essentially between the endeavor head and his associates, that is, the administrator and during this way the buyers. System architecture is often considered to possess two main

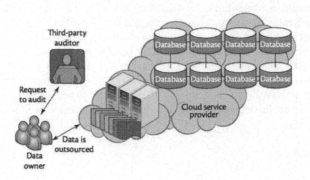

Figure 1. Computer architecture.

segments: framework investigation and framework structure. Framework configuration is that the structure procedure for a substitution business framework or one which can supplant or supplement a current framework.

3 RESULTS AND DISCUSSION

3.1 *Existing system*

Due to the high cost of I/O and network transmission, downloading all the data in order to verify its integrity isn't a workable solution. To the best of our knowledge, our strategy offers scalable and effective public auditing for cloud computing for the first time. We support a user's capacity to audit their outsourced data in the following ways: of the outsourced data and the user's constrained resource capability. Additionally, to avoid requiring users to execute an excessive amount of work in order to consume the data, the cost of accessing cloud storage should be as low as possible (in additional to retrieving the data). Users in particular might not want to go through the challenging process of determining the reliability of the data. Multiple users may potentially access the same cloud storage, for example in an office scenario. The same cloud storage may also be accessed by several users, for instance in an office setting. To make management easier, Cloud should only accept verification requests from one particular party. The Main Advantages are Data loss or leakage, Insecure interfaces and APIs Malicious insiders, Shared technology problems, Abuse and nefarious use of cloud computing.

3.2 *Proposed system*

The following are the three elements of the suggested system: We believe that our approach provides scalable and efficient public auditing for cloud computing for the first time. We support a user's capacity to audit their outsourced data in the following ways: Cloud without being aware of the data's content by motivating the public auditing system of data storage security in cloud computing. Our system specifically enables batch auditing, enabling the TPA to carry out multiple delegated auditing duties from various users at once. The main Advantages are the proposed architecture is platform independent and extremely decentralized in that it does not call for a separate authentication or storage system to be in existence. It also has a novel automatic and enforced logging method in the cloud. After delivering the protected data to the recipient, provide some level of usage control. The outcomes show how effective, scalable, and detailed our strategy.

966

3.3 Inputs and outputs

- The input design establishes a connection between the user and the information system. In order to transform transaction data into a format that can be processed, it includes developing requirements and methods for data preparation. To do this, users can input the information input technique was developed with the goal of lowering the quantity of input required, as well as errors, delays, and unnecessary steps, while maintaining a clear workflow. The entry of the data protects its privacy, usability, and security.

- Any system has outputs that communicate the processing outcomes relating to users and other systems how the data will be displaced for both the hard copy output and the immediate demand is decided during output design. As a direct source of information for the user, it is the most significant. The system's ability to assist users in making decisions is improved through efficient and intelligent output design. Signal significant events, opportunities, challenges, or warnings. Provide information about past activity, current condition, or future estimates. The system's ability to assist users in making decisions is improved through efficient and intelligent output design.

4 CONCLUSIONS

In this research, we suggested a dynamic auditing technique that is both effective and fundamentally secure. Instead of employing the mask technique, it preserves the privacy of the data against the auditor by fusing the bilinear paring's linearity feature with the cryptography technique. Thus, no additional organizer is needed for our multi-cloud batch auditing methodology. Additionally, batch auditing for many owners is supported via our batch auditing protocol. Additionally, by shifting the auditor's processing burdens are transferred to the server, our auditing method results in decreased auditor computation expenses as well as lower communication costs. This has a significant impact on auditing performance and is compatible with massive cloud storage systems.

REFERENCES

[1]   Akinyele J.A, Garman C, Miers I, Pagano M. W, Rushanan M, Green M, and Rubin A. D, "Charm: A Framework for Rapidly Prototyping Cryptosystems," *J. Cryptogr. Eng.*, vol. 3, no. 2, pp. 111–128, Jun. 2013.

[2]   Agrawal. S and Boneh D, "Homomorphism MACs: MAC-based Integrity for Network Coding," in *Applied Cryptography and Network Security, (Lecture Notes in Computer Science)*, vol. 5536. Berlin, Germany: Springer, 2009, pp. 292–305.

[3]   Ateniese G, Burns R, Curtmola R, Joseph Herring, Kissner L, Peterson Z, and Song D, "Provable Data Possession at Untrusted Stores," in *Proc. 14th ACM Conf. Comput. Commun. Secur., Alexandria, VA, USA*, 2007, pp. 598–609.

[4]   Ateniese G, Kamara S, and Katz J, "Proofs of Storage From Homomorphic Identification Protocols," in *Advances in Cryptology*. Berlin, Germany: Springer, 2009, pp. 319–333.

[5]   Barsoum A. F and Hasan M. A, "Provable Multicopy Dynamic Data Possession in Cloud Computing Systems," *IEEE Trans. Inf. Forensics Security*, vol. 10, no. 3, pp. 485–497, Mar. 2015.

[6]   Bonomi F, Milito R, Zhu J, and Addepalli, "Fog Computing and Its Role in the Internet of Things," in *Proc. 1st Ed. MCC Workshop Mobile Cloud Comput., New York, NY, USA*, 2012, pp. 13–16.

[7]   Bowers, A. Juels, and Oprea A, "Proofs of Irretrievability: Theory and Implementation," in *Proc. ACM Workshop Cloud Comput. Secur*, 2009, pp. 43–54.

[8]   Chang. J, Ji Y., Xu M., and Xue R., "General Transformations from Single-generation to Multi-generation for Homomorphic Message Authentication Schemes in Network Coding," *Future Gener. Comput. Syst.*, vol. 91, pp. 416–425, Feb. 2019.

[9] Chang. J, Wang. H, Wang. F, Zhang. A, and Ji. Y, "RKA Security for Identity-based Signature Scheme," *IEEE Access*, vol. 8, pp. 17833–17841, 2020.

[10] Chen F., Xiang T., Yang Y and Chow S. S. M., "Secure Cloud Storage Meets with Secure Network Coding," *IEEE INFOCOM 2014 – IEEE Conference on Computer Communications, Toronto, ON, Canada*, 2014, pp. 673–81, doi: 10.1109/INFOCOM.2014.6847993.

[11] Dhanaraj, R. K., Ramakrishnan, V., Poongodi, M., Krishnasamy, L., Hamdi, M., Kotecha, K., & Vijayakumar, V. (2021). Random Forest Bagging and X-Means Clustered Antipattern Detection from SQL Query Log for Accessing Secure Mobile Data. In D. K. Jain (Ed.), *Wireless Communications and Mobile Computing* (Vol. 2021, pp. 1–9). Hindawi Limited. https://doi.org/10.1155/2021/2730246.

[12] Zhang X. and Si W., "Efficient Auditing Scheme for Secure Data Storage in Fog-to-Cloud Computing," in *IEEE Access*, vol. 9, pp. 37951–37960, 2021, doi: 10.1109/ACCESS.2020.2971630.

[13] Prakash, A.A., Kumar, K.S., "Cloud Serverless Security and Services: A Survey", in *Lecture Notes in Mechanical Engineering*, 2022, pp. 453–462.

Artificial Intelligence, Blockchain, Computing and Security – Dagur et al. (Eds)
© 2024 The Author(s), ISBN: 978-1-032-49393-0

Decision trees to detect malware in cloud computing environment

Poovidha Ayyappa*, Govindu Kiran Kumar Reddy, Katamreddy Siva Satish,
Prakash Rachakonda & Pujari Manjunatha
Bharath Institute of Science and Technology, Chennai, India

ABSTRACT: The internet has become a source of many security threats with the development of wireless communication. Malware detection system (MDS) finds attacks on a system and is able to detect intruders. The MDS was previously applied with various machine learning (ML) techniques to improve its ability to detect intruders and increase its accuracy with the detection of intruders. This paper's goal is to suggest a methodology for creating effective MDS utilizing principal component analysis (DECISION TREE ANALYSIS) and random forest classification. The dataset's dimensions will be lowered, and categorizing the data will aided by random forest. According to the results obtained, the accuracy of this approach is superior to SVM, Nave Bayes, and Decision Trees. This technique results in 3.24 minute performance times, a 96.78% accuracy rate, and a 0.21% error rate.

1 INTRODUCTION

An effort to break into or abuse a computing system is called malware. Malware is any action that compromises the accessibility, privacy, and integrity of data or computer resources [1]. The attacker enters using the system architecture's vulnerabilities to get around the authentication or authorization process [2]. Network security is growing more and more crucial than ever before as resulting from the enormous expansion of network-based services and secure network data. Utilizing a network malware detection system (NMDS), which tracks numerous network events to identify assaults, is one way to deal with this. Therefore, it is increasingly crucial that such systems be more precise in spotting assaults, rapid to train, and provide the fewest false positives feasible [3]. A network is protected by a malware detection system (MDS), which finds malicious anomalies. As a result, MDS are now an essential part of computer networks [4]. Responsiveness and Effectiveness are two prerequisites for MDS as a result of the remarkable expansion of network-based services and secure network data. Additionally, MDS should be able to identify between assaults that are internal to the organization (originating from own employees, customers, or any other) and those that are external (attacks posted by hackers). Malware Detection Systems (MDS) of two popular varieties are host-based and network-based (HMDS) [5]. Network-based MDS makes an effort to spot illegal, unauthorized, and abnormal activities based purely on network data. In contrast, HMDS makes an effort to locate unauthorized, illegal, and uncharacteristic behavior on a certain gadget [6]. In this article by providing a novel method for network-based malware detection systems. One of the two detection methods signature-based or anomaly-based is employed by the majority of malware detection systems [7]. Any divergence from "normal" behavior is identified and detected by anomaly-based MDS. It gains knowledge from the regular data that is gathered when there are no unusual behaviors [8]. In contrast, an MDS based on signatures analyses network traffic and checks for pre-configured and pre-planned attack patterns [9].

---

*Corresponding Author: poovithabharathuniv@gmail.com

DOI: 10.1201/9781003393580-145

2 METHODOLOGIES

2.1 *Data collection*

The actual process of developing a machine learning model and gathering data officially begins. The amount and quality of data we can gather at this phase will determine how effectively the model works [10]. Web scraping, hand-input, and other techniques are used for data collection.

2.2 *Dataset*

42 columns make up the dataset, which contains 125974 distinct pieces of data [11].

2.3 *Data preparation*

The data will be transformed by eliminating any missing data and some columns. The column names that we want to keep or retain will first be listed [14]. After that, we drop or eliminate all columns save for the ones we wish to keep. The last step is to remove the rows with missing values from the data set, or we erase them and divided into sets for training and evaluation [15].

2.4 *Proposed algorithm*

2.4.1 *Principal component analysis*

Reduce the amount of variables in machine learning by using the unsupervised learning technique known as principal component analysis. Through a statistical process known as orthogonal transformation, correlated observations of characteristics are converted into a set of linearly uncorrelated data. The term "Principal Components" refers to these recently revised attributes. By lowering variances, this technique enables users to identify meaningful patterns within a collection.

To project the high-dimensional data onto a lower-dimensional surface, DECISION TREE ANALYSIS often looks for such a surface. Due to the fact that a high attribute implies a good class split and hence reduces dimensionality, DECISION TREE ANALYSIS considers the variance of each attribute when performing its analysis. Image processing, a system for suggesting movies, and optimizing. Examples of applications for DECISION TREE ANALYSIS in the real world include the distribution of power over numerous communication channels. Due to the feature extraction method it uses, only the most important variables are retained.

The mathematical ideas on which the DECISION TREE ANALYSIS method is based include:

- Variance and Covariance
- Eigenvalues and Eigen factors

The DECISION TREE ANALYSIS algorithm frequently uses the following terms:

- The amount of traits or variables present in the dataset in question is known as its dimensionality. The dataset's amount of columns makes this easier to determine.
- Correlation: This term describes the degree to which there is a relationship between two variables. As an illustration, both variables change if one changes. The correlation index ranges from -1 to +1. Here, -1 denotes an inverse relationship between the variables, and +1 denotes a direct relationship between the variables.
- Orthogonal: It establishes that variables do not have any correlation with one another, and as a result, there is no association between the two.
- These are referred to as eigenvectors assuming a square matrix M and a non-zero vector v are both present. It follows that v is an eigenvector if Av is v's scalar multiple.
- Covariance Matrix: A "covariance matrix" is a matrix that shows how two variables are correlated.

970

Principal components of DECISION TREE ANALYSIS

The Principal Components are, as was previously said, the modified new characteristics or the outcomes of DECISION TREE ANALYSIS. The following are a few traits of these fundamental parts:

1. The initial characteristics that were included in the dataset had either the same number of PCs or fewer than these PCs.
2. First, we need to divide the input dataset into two parts, X and Y, with X standing for the training set and Y for the validation set.
3. Putting information into a structure
4. The building of a structure to symbolize our dataset. To illustrate, let's have a look at the two-dimensional matrix of independent variable X.
5. Covariance of Z Calculation
6. The Z matrix will be transposed in order to determine Z's covariance. We'll transpose it and then multiply the result by Z. The output matrix will be this matrix's covariance matrix for Z.
7. The Eigen Values and Eigen Vectors Calculation
8. It is important to determine the eigenvalues and eigenvectors of the resulting covariance matrix Z. The covariance matrix stores the eigenvectors, or highly informative axes. The respective eigenvectors' coefficients determine these eigenvalues.
9. Ordering the Eigen Vectors

This stage will involve sorting every eigenvalue in decreasing order from largest to lowest. The eigenvectors should also be sorted in the same manner as the eigenvalues in the P eigenvalues matrix.

Calculating the new features, Or Primary Elements

We'll learn about the new features in this section. By increasing the P* matrix by Z, we can achieve this. Each observation is a linear mixture of the original features in the resulting matrix Z*. There is no relation between any two columns of the Z* matrix.

1. Deleting minor or irrelevant features from the new dataset.
2. What we decide to keep and what we decide to discard in this area will depend on the new feature set. It implies that we will remove insignificant attributes from the new dataset and only maintain those that are significant or pertinent.

2.5 Accuracy on test set

On the test set, we achieved 99.1% accuracy.

2.6 Saving the trained model

To save a trained and tested model as an.h5 or.pkl file using a library like pickle is the first thing you should do whenever you are ready to utilize it in a production-ready setting. Install pickle in your environment, if necessary. Now that the module has been imported, let's dump the model into a pkl file.

2.7 Computer architecture

In this study, a novel approach is suggested that antivirus be offered as an internal network service to enable malware detection on end hosts [12]. This paradigm makes it possible for different detection engines to identify harmful and undesirable software's. Greater forensics capabilities, enhanced deploy ability, and improved malware detection is just a few of the important benefits that this approach delivers. Malware detection in the cloud includes a network service and an easy-to-use host agent that supports multiple storage types [13]. The detection methods used in this paper mix static signature analysis with dynamic analysis detection as seen in the below Figure 1.

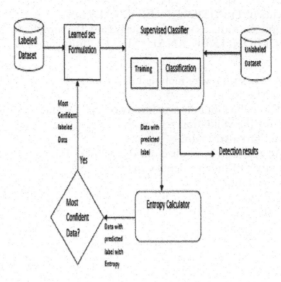

Figure 1.   Computer architecture.

3   RESULTS AND DISCUSSION

3.1   *Existing system*

Reduce the amount of variables in machine learning by using the unsupervised learning technique known as principal component analysis. Through a statistical process known as orthogonal transformation, correlated observations of characteristics are converted into a set of linearly uncorrelated data. The term "Principal Components" refers to these recently revised attributes. By lowering variances, this technique enables users to identify meaningful patterns within a collection.

To project the high-dimensional data onto a lower-dimensional surface, DECISION TREE ANALYSIS often looks for such a surface. Due to the fact that a high attribute implies a good class split and hence reduces dimensionality, DECISION TREE ANALYSIS considers the variance of each attribute when performing its analysis. Image processing, a system for suggesting movies, and optimizing. Examples of applications for DECISION TREE ANALYSIS in the real world include the distribution of power over numerous communication channels. Due to the feature extraction method it uses, only the most important variables are retained.

3.2   *Proposed system*

The malware detection system strives to enhance the system that incursions have an impact on. This tool has the ability to find invaders. The suggested system aims to fix the issues from the earlier work. The two methods that make up the suggested system are principal component analysis and random forest. When a dataset's dimension is reduced using principle component analysis, the dataset's quality is raised since the proper attributes may be present in the dataset. The Advantages are: At only 21%, the error rate found in our proposed method is remarkably low, the accuracy attained is also substantially higher than that of earlier methods, additionally, compared to other algorithms, the performance takes less time.

3.3   *Inputs and outputs*

Design is the process of transforming a computer-based system from a user-centered description of the input. The management will need clear instructions on how to get the right data from the computerized system thanks to this framework, which is essential for preventing data entry errors.

A high-quality output delivers a clear information display and satisfies the end user's needs.

3.4 *Purpose*

The purpose of this document is Malware detection using Decision Tree Analysis machine learning algorithms. The broad description of our project, including user requirements, a product viewpoint, an overview of the requirements, and general restrictions, will be provided in depth in this document. It will also include the particular features and needs, such as interface, functional requirements, and performance requirements, needed for this project.

4 CONCLUSIONS

Security issues have emerged as a result of the systems' increasing connection with the internet. Online intrusion detection is successfully handled by the suggested technique. The detection rates and the rate of erroneous errors could both be significantly increased by the suggested approach. In this case, the dataset for knowledge discovery was used. The performance times (min), accuracy percentage (%), and error percentage (%) values for the outcomes produced by our suggested method are 3.24 minutes, 96.78%, and 0.21%, respectively.

REFERENCES

[1] JafarAbo Nada; Mohammad Rasmi Al-Mosa, 2018 *International Arab Conference on Information Technology (ACIT)*, A Proposed Wireless Malware Detection Prevention and Attack System

[2] Kinam Park; Youngrok Song; Yun-Gyung Cheong, 2018 *IEEE Fourth International Conference on Big Data Computing Service and Applications (BigData Service)*, Classification of Attack Types for Malware Detection Systems Using a Machine Learning Algorithm

[3] Bernard. S, Heutte. L and Adam. S "On the Selection of Decision Trees in Random Forests" *Proceedings of International Joint Conference on Neural Networks, Atlanta, Georgia, USA*, June 14–19, 2009, 978-1-4244-3553-1/09/$25.00 ©2009 IEEE

[4] Tesfahun. A, Lalitha Bhaskari. D, "Malware Detection Using Random Forests Classifier with SMOTE and Feature Reduction" *2013 International Conference on Cloud & Ubiquitous Computing & Emerging Technologies*, 978-0-4799-2235-2/13 $26.00 © 2013 IEEE

[5] Le, T.-T.-H., Kang. H, & Kim. H (2019). The Impact of Decision Tree Analysis-Scale Improving GRU Performance for Malware Detection. *2019 International Conference on Platform Technology and Service (PlatCon)*. Doi:10.1109/platcon.2019.8668960

[6] Anish Halimaa. A, Sundarakantham. K: *Proceedings of the Third International Conference on Trends in Electronics and Informatics (ICOEI 2019)* 978-1-5386-9439-8/19/$31.00 ©2019 IEEE "Machine Learning Based Malware Detection System."

[7] Mengmeng Ge, Xiping Fu, Naeem Syed, Zubair Baig, Gideon Teo, Antonio Robles-Kelly (2019). Deep Learning-Based Malware Detect ion for IoT Networks, *2019 IEEE 24th Pacific Rim International Symposium on Dependable Computing (PRDC)*, pp. 256–265, Japan.

[8] Patgiri. R, Varshney. U, Akutota. K, and Kunde. R, "*An Investigation on Malware Detection System Using Machine Learning*" 978-1-5386-9276-9/18/$31.00 c2018IEEE.

[9] Rohit Kumar Singh Gautam, Er. Amit Doegar; 2018 *8th International Conference on Cloud Computing, Data Science & Engineering (Confluence)* "An Ensemble Approach for Malware Detect ion System Using Machine Learning Algorithms."

[10] Kazi Abu Taher, Billal Mohammed Yasin Jisan, Md. Mahbubur Rahma, 2019 *International Conference on Robot ics, Electrical and Signal Processing Techniques (ICREST)*"Network Malware Detect ion using Supervised Machine Learning Technique with Feature Selection."

[11] Haripriya. L, Jabbar. M. A, 2018 *Second International Conference on Electronics, Communication and Aerospace Technology (ICECA)* "Role of Machine Learning in Malware Detection System: Review"

[12] Nimmy Krishnan, A. Salim, 2018 *International CET Conference on Control, Communication, and Computing (IC4)* " Machine Learning-Based Malware Detect ion for Virtualized Infrastructures"

[13] Mohammed Ishaque, Ladislav Hudec, 2019 *2nd International Conference on Computer Applications & Information Security (ICCAIS)* "Feature extract ion using Deep Learning for Malware Detection System."

[14] Aditya Phadke, Mohit Kulkarni, Pranav Bhawalkar, Rashmi Bhattad, 2019 *3rd International Conference on Computing Methodologies and Communication (ICCMC)*"A Review of Machine Learning Methodologies for Network Malware Detection."

[15] Iftikhar Ahmad, Mohammad Basheri, Muhammad Javed Iqbal, Aneel Rahim, *IEEE Access* (Volume: 6) Page(s): 33789 – 33795 "Performance Comparison of Support Vector Machine, Random Forest, and Extreme Learning Machine for Malware Detection."

[16] Riyaz. B, Ganapathy. S, 2018 *International Conference on Recent Trends in Advanced Computing (ICRTAC)*" An Intelligent Fuzzy Rule-based Feature Select ion for Effective Malware Detection."

[17] Dhanaraj, R. K., Ramakrishnan, V., Poongodi, M., Krishnasamy, L., Hamdi, M., Kotecha, K., & Vijayakumar, V. (2021). Random Forest Bagging and X-Means Clustered Antipattern Detection from SQL Query Log for Accessing Secure Mobile Data. In D. K. Jain (Ed.), *Wireless Communications and Mobile Computing* (Vol. 2021, pp. 1–9). Hindawi Limited. https://doi.org/10.1155/2021/2730246.

[18] Prakash, A.A., Kumar, K.S.," Cloud Serverless Security and Services: A Survey", in *Lecture Notes in Mechanical Engineering*, 2022, pp. 453–462

Artificial Intelligence, Blockchain, Computing and Security – Dagur et al. (Eds)
© 2024 The Author(s), ISBN: 978-1-032-49393-0

An optimized feature selection guided light-weight machine learning models for DDoS attacks detection in cloud computing

Rahul R. Papalkar*
Bharti Vidyapeeth Deemed University Pune, DET Navi Mumbai

A.S. Alvi*
Prof. Ram Meghe Institute of Technology & Research Badnera India

Shabir Ali* & Mohan Awasthy*
Bharti Vidyapeeth Deemed University Pune, DET Navi Mumba

Reshma Kanse
Bharati Vidyappeth DU, DET Navi Mumbai

ABSTRACT: The investigation highlights the need of lightweight machine learning models for efficient and effective cloud based VNF intrusion detection. The study suggests utilising lightweight machine learning models guided by feature selection to spot distributed denial of service (DDoS) attacks in the cloud. The proposed strategy uses the Extreme Gradient Boosting (XGBoost) approach to create a lightweight machine learning model, and a genetic algorithm to choose relevant characteristics for it. We put the proposed technique through its paces by simulating DDoS attacks on the cloud with a data set. The results showed that out of all the machine learning models evaluated, XGBoost's optimised feature selection guided performance was the most effective and efficient. This study contributes to the field of cybersecurity by examining the necessity of intrusion detection in conjunction with the increasing significance of cloud-based VNF systems for high-speed network operations. The proposed lightweight machine learning approach may direct the development of more efficient and effective intrusion detection systems to protect cloud based VNFs from security threats. To evaluate the method's usefulness in real-world cloud-based VNF systems and to explore its potential application to other kinds of cyber threat, more research is necessary.

Keywords: Feature Selection, IDS, Cloud Computing DDOS, Machine Learning model.

1 INTRODUCTION

Firewalls, routers, and load balancers are examples of the increasingly widespread virtual network functions (VNFs). Their deployment via the cloud has been standard practice in recent years. The distributed nature of NFV deployment and the complexity of the underlying computer architecture make VNFs more vulnerable to security threats than traditional network hardware. D-DoS attacks, which disrupt _networks and cause significant financial losses, have emerged as a major risk for cloud-based VNF infrastructures. Researchers have examined using machine learning techniques, which have shown potential in recognising DDoS assaults, to meet the demand for intrusion detection in cloud-based VNF systems. However, it is difficult to

---

*Corresponding Authors: rrpapalkar@bvucoep.edu.in, asalvi@mitra.ac.in, sali@bvucoep.edu.in and mohan.awasthy@bvucoep.edu.in

DOI: 10.1201/9781003393580-146

create reliable and effective intrusion detection systems due to the complexity of cloud-based VNF systems and the vast volume of network data. In this article, we survey the current level of research into using machine learning techniques to identify D-DoS attacks in cloud computing environments. We stress the need for lightweight machine learning models for efficient and quick intrusion detection in cloud-based VNF systems. We propose using lightweight machine learning models and an efficient feature selection-driven approach to detect distributed denial of service (DDoS) attacks in the cloud. The proposed strategy uses the Extreme Gradient Boosting (XGBoost) approach to create a lightweight machine-learning model and a genetic algorithm to choose relevant characteristics. We evaluate and compare the suggested approach to state-of-the-art ML frameworks using a dataset of simulated DDoS attacks on cloud servers.

Features selection technique: To increase machine learning algorithm performance and efficiency, feature selection selects a subset of useful characteristics from a wider collection of features. Feature selection reduces overfitting, improves model accuracy, and reduces model training and testing processing resources. Feature selection methods have pros and cons. Filter methods, wrapper methods, embedding methods, and dimensionality reduction methods are popular.

- Filtering Strategies: Simple statistical approaches known as filter methods rank features according to their consistency with the dependent variable. Features that have a high correlation coefficient with the dependent variable are kept, whereas those with a low correlation coefficient are eliminated. Pearson Correlation, the Chi^2 test, and I.G. are the three most used filtering techniques. For a linear relationship between two variables, use the Pearson correlation. Used often in regression analysis, it quantifies the degree to which one attribute predicts another. There is a negative connection between -1 and -1, no correlation between 0 and 1, and a positive correlation between +1 and 1. We keep features that have a strong correlation coefficient. The Chi-Square test is a statistical technique for checking the independence of two categorical variables. It's a statistical tool for gauging how much actual values of a categorical variable deviate from their theoretical distribution. Chi-Square values closer to one another are more indicative of the significance of a feature. A feature-based data partitioning scheme may reduce entropy, resulting in information gain. It's a standard component of decision-tree algorithms. Preserved are characteristics that lead to substantial knowledge gain.
- *Method Encapsulation*: Feature selection in wrapper methods often involves using a machine-learning algorithm to evaluate the efficacy of a pre-existing feature subset. Wrapper methods include forward elimination, backward elimination, and recursive feature removal. With the forward selection, we start with a blank slate and gradually enhance performance by adding features individually. Backward removal might achieve optimal efficiency by starting with a full feature set and progressively removing the feature that delivers the greatest performance gain. Recursive feature removal is achieved by repeatedly training a machine learning algorithm on a reduced collection of features and discarding the least significant feature until a stopping criterion is met.
- *In-System Procedures:* When training a machine learning algorithm, embedded approaches integrate feature selection. Both Lasso and Ridge regression are embedded approaches. Lasso Regression: A Linear Regression Method that Encourages Sparsity through a Penalty Term in the Cost Function. The penalty term supports reducing the model's dependence on less relevant characteristics by setting their coefficients to zero. For lower coefficients, the linear regression method of ridge regression includes a penalty element in the cost function. This may help lessen the significance of distracting or unimportant details.
- *Reducing Dimensionality*: The goal of dimensionality reduction is to preserve as much information as possible while compressing the feature space. PCA and LDA are valuable tools for doing so. Principal Component Analysis (PCA) is a statistical method that shifts the data to a new coordinate system with the most significant variance running along the first axis. The following axis has less variation. The amount of substantial components kept depends on a threshold or variance explained percentage. In data transformation, LDA stands for linear discriminant analysis.

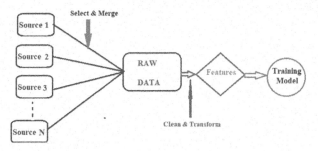

Figure 1. Flow of feature engineering.

$$X_{norm} = \frac{X - X_{min}}{X_{max} - X_{min}}$$

Standardization scales values using standard deviation. If features' SDs differ, so will their ranges. This lowers feature outliers.

$$Z = \frac{X - \mu}{\sigma'}$$

2 RELATED WORK

Meng You *et al.* developed SDN to detect DoS attacks [1]. One DDoS attack employs multiple flows to overwhelm the Ternary Content Addressable Memory (M-DoS), while another uses a single entry to discriminate the target connection based on auxiliary needs. This Back Propagation (BP) neural network classifier can recognise the risks listed above. Entropy-based sources IPs (ESIPs) kill M-DoS assaults with their entropy value, while the Growth rate of max matched bytes (GRMMB) stops S-DoS attacks. Network classifiers increased detection probability. Shubhra Deivedi *et al.* [2] planned to review an evolutionary method for Distributed DoS defence. The auther used a machine learning Grashopper optimization (GOA) to optimise the DoS assault. Optimization algorithm and intrusion detection system (IDS) meet demands and distinguish assaults. Grasshopper optimization with intrusion detection system (GOIDS) was the two-stage approach: To fix the DoS attack, GOA proposed critical traits and passed them through classifiers. GopalSIngh Kushwah *et al.* [5] created a cloud service DDoS detection voting extreme learning machine (V-ELM). The user was unsuccessfully attacked by DDoS. V-ELM identified the assault as ANN. Two benchmark data sets performed the performance. Detection system employs many learning machines and benchmark data sets. Training database, pre-processor, and classifier modules comprise the detection system. The classifier was compared to others to detect assaults for greater accuracy and performance. Vellinga *et al.* [6] had proposed an optimization based Deep Network for the detection of DDoS attack. Deep Learning classifiers were implemented to detect the attack by collecting the information from the cloud note in a log layout Taylor Elephant Herd Optimization(TEHO) had been executed to detect the attacks and also to optimize the performance;. Deep belief network(DBN) had been trained to produced powerful output with DoS detection. The size of the databases was too large, so the significant features were selected from the log file to reduce the training time of network with the need of Bhattacharya distance. DBN Classifier gives the high precision rate and accuracy with detection improvements. For the detection of low-rate DoS attacks, Naiji Zhang *et al.* [7] used an entropy and machine learning approach. The assault was detected amid heavy traffic thanks to power spectral density (PSD) and a support vector machine (SVM). After combining the entropy value with an adaptive threshold, the parameter is fine-tuned in response to detecting methods if it exceeds predetermined limits. As a result, SVM's computational cost reduction was a low priority. We used principal component analysis (PCA) to reduce the volume of data, and we provided enough features during training to teach the data to a support vector machine (SVM) so that it could make an absolute

Table 1. Comparative analysis of feature selection techniques.

Selection Technique & Refference	Dataset Used	Advantages	Limitations	Research Gap	Scope for Work
ReliefF [1], [2], [3]	KDD Cup 99	1. Efficient for high-dimensional datasets 2. Handles missing values 3. Considers feature interactions	1. Sensitive to the number of instances 2. Computationally expensive for large datasets 3. Limited to classification problems	1. Combination of Relief F and other techniques 2. Use of hybrid algorithms	Implement ReliefF with hybrid algorithms for improved performance
Chi-squared [4][5][6]	NSL-KDD	1. Measures independence between features and classes 2. Works well with discrete data 3. Computationally efficient	1. Assumes features are independent 2. Sensitive to the number of instances 3. Limited to classification problems	1. Use in combination with other techniques 2. Investigation of its performance on large datasets	Explore the use of Chi-squared with other techniques on larger datasets
Genetic Algorithm [7] [8][9]	UNSW-NB15	1. Can handle a large number of features 2. Allows for non-linear feature selection 3. Handles continuous and discrete data	1. Computationally expensive 2. May require parameter tuning 3. Prone to overfitting	1. Hybridization with other techniques 2. Investigation of its performance on streaming data	Develop a hybrid approach with Genetic Algorithm and other techniques for streaming data
Particle Swarm Optimization [10][11][12]	NSL-KDD	1. Efficient for high-dimensional datasets 2. Handles continuous and discrete data 3. Can handle multi-objective optimization	1. Convergence to local optima 2. Requires parameter tuning 3. Prone to overfitting	1. Combination with other techniques 2. Investigation of its performance on imbalanced datasets	Investigate the combination of Particle Swarm Optimization with other techniques for imbalanced datasets

detection. For DDoS detection, [8] Trung V. Phan recommended a software-defined-network. Traffic congestion is something that cutting-edge machine-learning algorithms like SOM and SVM hope to address. The goal of developing HIPF was to speed up detection and prevent attacks. QOS improved with time spent waiting for a response and the number of packets lost during transmission. We identified the most salient characteristics of traffic demand using SOM and SVM labels. At long last, a DoS-attacked security system has a hybrid ML-eHIPF to thank.

3   RESEARCH METHODOLOGY

3.1   *Gap analysis and limitation of existing systems*

Virtual network functions (VNFs) are more susceptible to security attacks than conventional, standalone network appliances because of the complexity of the virtualized computing architecture on which they run. Plus, NFV Placement is a community effort. As a result, an Intrusion detection system is essential to the NFV infrastructure in order to effectively safeguard VNFs from security risks, such as the detection and prevention of malicious VNFs from a large-scale application environment. When it comes to protecting a cyber physical system, the first line of defence is intrusion detection. As a result of the pinpoint accuracy of the detection, attacks on malicious services can be averted and countermeasures taken before any damage is done. While several intrusion detection approaches based on classification, clustering, and statistical methods have been developed for use in static VFV environments, these methods cannot be directly implemented in the dynamic VFV environment. Additionally, the identification of anomalies in a real-world NFV cloud system has not been well researched to date and requires more study.

Realizing three significant obstacles in intrusion detection for VNF infrastructure:

- There are notable differences in the operation of various VNFs, such as virtual routers, fire-walls, load balancers, etc. This calls for a standardised method of detecting intrusions across a variety of VNFs.
  - Variations in network traffic or the varied interaction between a VNF and the underlying physical servers might alter the typical behaviour patterns of a VNF. As a result, a strong statistical model is needed to define the typical activities of VNFs.
  - Also, VNFs are usually low-latency applications that depend on their performance. This means that online intrusion detection should be effective and shouldn't affect the performance of Original VNFs.
  - Normally, the cloud virtual networks are interconnected through hardware devices such as routers and switches, but these hardware devices are less vulnerable. In VNF the hardware devices functionalities are running inside the VM, once the VM is malfunctioned the entire network is affected through malfunctioned VM. Hence the effective Intrusion detection model is much more needed to protect the VM which execute the virtual network functionalities.

3.2 Problem statement

The difficulty with the preceding situation is that current feature selection approaches and DDoS attack detection systems in the cloud are not very good at their jobs. The inability to manage high-speed traffic, lack of scalability, dependency on certain platforms, limited efficacy against unknown threats, and lack of interpretability are only few of the drawbacks. Cloud-based VNF systems are particularly vulnerable due to these restrictions and research gaps. To efficiently identify DDoS assaults in the cloud, manage high-velocity traffic, and offer interpretability for the judgements taken, we need advanced and complex feature selection approaches and machine learning models.

3.3 Research objectives

- *The Major contribution of this paper as follow.*

- To safeguard real-time VNF actions, we present a lightweight machine learning model based intrusion detection here. Since VNFs execute within virtual machines, their observable behaviours are treated as performance characteristics associated with resource utilisation and traffic flows on the computing and networking sides, respectively.
- A variety of VNF behaviour models may be used with the established intrusion detection model to identify invasions.
- Python is used for the development and deployment of the proposed ML based IDS. The virtual router and virtual firewall behaviour models are taught using actual, everyday network traffics. Python is used to demonstrate the models' ability to detect anomalous behaviours.

3.4 Proposed work

Traditional IDSs have a high false alarm rate, limited detection accuracy, and poor innovative attack detection. Thus, resilient IDSs must enhance detection accuracy, reduce false alarms, and boost attack discovery efficiency. Machine Learning (ML) approaches have been extensively studied to create optimum IDSs for network security. ML algorithms mimic human intelligence and are developing. ML methods find patterns in massive datasets using statistical models.

4 EXPERIMENTAL SETUP & RESULTS

Optimum Feature Selection: The feature selector engine analyses the retrieved packet features and chooses the most relevant ones based on the holoen- trophy criterion. The Feature Selector Holoentropy engine, which we'll call A (jk, jl), may be identified with E.

$$A(x_i, x_j) = K * M(x_i, x_j) \dots\dots\dots\dots\dots \tag{I}$$

In Figure 2 represent the architecture of proposed model. In this there are two major component first one is optimum feature selection, in that we have implemented IBES algorithm for the selection of optimum features which is result in reduced the training & testing time and solved the problem of dimensionality. Second Modeule implement by hybrid the RNN with SoftMax regression which is result in detecting attacks precisely and detection accuracy rate improved as per the state of arts methods.

Proposed Algorithm:

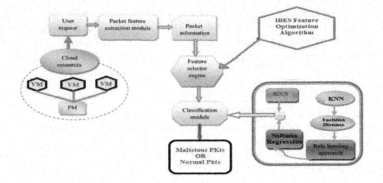

Figure 2.   Architecture of proposed attacks detection model.

Input:
X: With m samples and n features, the input data is a matrix of dimension m x n.
y: Target variable vector of size m x 1
K: Number of top features to select
Output:
selected_features: Indexes of the K top features
Step 1: Compute the correlation between each feature and the target variable:

$$corr\_i = abs(corr(X[:, i], y))$$

Step 2: Rank the features based on their correlation values:

$$rank\_i = rank(corr\_i)$$

Step 3: Split the data into training and validation sets:

$$X\_train, X\_val, y\_train, y\_val = train\_test\_split(X, y, test\_size = 0.2, random\_state = 42)$$

Step 4: Train a machine learning model, such as a decision tree or a random forest, on the training set using all the features.

Step 5: Compute the feature importance scores based on the trained model:

$$imp\_i = model.feature\_importances\_$$

Step 6: Rank the features based on their importance scores:

$$rank\_i = rank(imp\_i)$$

Step 7: Compute the mutual information between each feature and the target variable:

$$mi\_i = mutual\_info\_classif(X[:,i],y)$$

Step 8: Rank the features based on their mutual information values:

$$rank\_i - rank(mi\_i)$$

Step 9: Combine the ranks from the previous steps using a weighted average:

$$final\_rank\_i = w1 * rank\_corr\_i + w2 * rank\_imp\_i + w3 * rank\_mi\_i$$

Step 10: Select the top K features based on the final rank:

$$selected\_features = argsort(final\_rank\_i)[:K]$$

--A

Pseudocode:

```
# Feature selection using Proposed algorithm.
num_features = received from equation A
model = RandomForestClassifier(n_estimators=100, random_state=42)
selected_features = GA_PSO.gapso(X_train_norm, y_train.values, n_clusters=num_features,
model=model)
# Train classifier using selected features
X_train_selected = X_train.iloc[:, selected_features]
X_test_selected = X_test.iloc[:, selected_features]
clf = RandomForestClassifier(n_estimators=100, random_state=42)
clf.fit(X_train_selected, y_train)
# Evaluate performance on test set
y_pred = clf.predict(X_test_selected)
```

$$accuracy = clf.score(X\_test\_selected, y\_test)$$

$$sensitivity = np.sum((y\_pred == 1)\&(y\_test == 1))/np.sum(y\_test == 1)$$

$$specificity = np.sum((y\_pred == 0)\&(y\_test == 0))/np.sum(y\_test == 0)$$

$$f1\_score = 2 * (sensitivity * specificity)/(sensitivity + specificity)$$

$$f1\_score = 2 * (sensitivity * specificity)/(sensitivity + specificity)$$

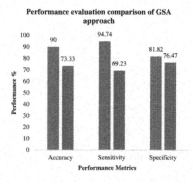

Figure 3.   Performance evaluation comparison of IBES approach.

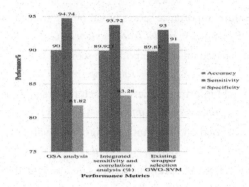

Figure 4. Accuracy, sensetivity and specificity of proposed algorithms.

5 CONCLUSION

The results demonstrated that the proposed hybrid technique for feature selection in machine learning outperformed both the individual feature selection approaches and the baseline in terms of accuracy and computational complexity. By combining several methods, such as filter techniques and algorithms inspired by nature, we were able to overcome the limitations of each approach and create a more robust collection of features.

The choice of acceptable methodologies and parameters should rely on the particular situation and dataset, however it is essential to keep in mind that there is no "one size fits all" solution in feature selection. Researchers might look at hybrid methodologies and feature selection methods to enhance machine learning models for various applications.

REFERENCES

[1] Isharufe Walter, Fehmi Jaafer. "Study of Security Issue in Platform as a Services (PaaS) Cloud Model." *In 2020 International Conference on Electrical Communication, and Computer Engineering(ICECCE)* pp. 1–6 IEEE 2020.

[2] Ali Mohammed Banu Treover Wood Harper, and Ronald Ramlogan. *"Challenges of Value Creation Through Cloud SaaS: Business/IT Alignment in Service Oriented Industries"* (2020).

[3] Hu. Chung Tong, Michaela Iorga wei Bao *"General Access Control Guidance for Cloud System"* NIST special Publication 800(2020):210.

[4] Soh. Julian Marshall Copeland, Anthony Puca and Micheleen Harries. *"Overview of Azure Infrastructure as a Service (IaaS)* "In Microsoft Azure pp. 21–41 Apress, Berkeley, CA, 2020.

[5] Shashirband Shahab Mahdis Fathi Anthony T Chronopoulos Antonoio *et al.* "Computational Intelligence Intrusion Detection Techniques in Mobile and Cloud Computing Environment: Review, Taxonomy, and Open Research Issue." *Journal of Information Security and Applications* 55 (2020):102582.

[6] Ohammadi, S.; Amiri, F.: An Efficient Hybrid Self-learning Intrusion Detection System Based on Neural Networks. *Int. J. Comput. Intell. Appl.* 18(01), 1950001 (2019).

[7] Phan, Trung V and Minho Park. "Efficient Distributed Denial of Service Attack Defense in SDN Based Cloud" *IEEE Access* 7 (2019):18701–18714.

[8] Patil, Rajendra, Harsha Dudeja, Shehal *et al.* "Protocol Specific Multi threaded Network Intrusion Detection System (PM-NIDS) for DoS /DDoS Attacks Detection in Cloud" In 2018 9th *International Conference on Computing, Communication, Networking Technologies (ICCCNT)* PP. 1–7, IEEE.

[9] Wang, W.; Gombault, S.: Efficient Detection of DDoS Attacks with Important Attributes. In: *Third International Conference on Risks and Security of Internet and Systems*, 2008. CRiSIS'08, pp. 61–67. IEEE (2008).

[10] De la Hoz, E.; De La Hoz, E.; Ortiz, A.; Ortega, J.; Prieto, B.: PCA Filtering and Brobabilistic SOM for Network Intrusion Detection. *Neurocomputing* 164, 71–81 (2015).

[11] Thaseen, I.S.; Kumar, C.A.: Intrusion Detection Model Using Fusion of Chi-square Feature Selection and Multi class SVM. *J. King Saud Univ. Comput. Inf. Sci.* 29(4), 462–472 (2017).

[12] Mazini, M.; Shirazi, B.; Mahdavi, I.: Anomaly Network-based Intrusion Detection System using a Reliable Hybrid Artificial bee Colony and AdaBoost Algorithms. *J. King Saud Univ. Comput. Inf. Sci* 31(4), 541–553 (2019).

Artificial Intelligence, Blockchain, Computing and Security – Dagur et al. (Eds)
© 2024 The Author(s), ISBN: 978-1-032-49393-0

Analysis of methods for multiple reviews based sentiment analysis

Syed Zeeshan Ali Abrar Alvi & Ajay B Gadicha
P.R. Pote Patil, College of Engineering & Management Amravati MS

ABSTRACT: Twitter, Facebook, and a plethora of blogs are just a few of the online resources that disseminate useful information to people all over the world. Data in the form of images, videos, audio recordings, texts, and other formats is continually being produced as the number of people using social media and online shopping increases. The majority of this data comes in the form of text, making it especially difficult for researchers to extract useful insights. A fair amount of effort has been done before to streamline the process of extracting precise interpretations from this data. To solve the issues inherent in sentiment analysis, a novel approach is described. This method use three variants of long short-term memory to collect data, preprocess it, encode features, and classify it. The recommended models were evaluated in the research using various textual datasets.

Keywords: Sentiment Analysis, SVM, Regression, Customer Review

1 INTRODUCTION

Sentiment analysis is the practise of examining internet text in order to ascertain the author's intended mood. The process of determining whether a piece of text should be categorised as "positive," "negative," or "neutral" is known as sentiment analysis. Subjectivity analysis is also known as opinion mining and rating extraction. It's crucial for a company's growth and development to know how customers feel about a product. Customers' opinions on a certain issue may be analysed with the use of sentiment analysis. Nowadays Charging cars is most challenging, in the current charging facility we need to charge by standing in queue. This is really inappropriate for users as these queues are getting larger and larger day by day. This system is aiming to ease charging cars system by maintaining virtual queue and, after charging user will get a notification with a link to submit a review based on some qualitative and quantitative feedback parameters with respect to service of the station. The feedback can be in the form of charging time, staff behavior, Safety precautions, waiting time and comments. Based on this feedback we will calculate the rating score for each charging station and based on this we will sort the charging station. The sentiment of comment is found using Naive Bayes Algorithm, Logistic regression algorithm and SVM algorithm whether the comments are positive or negative. Every comment is analysed three times and after analyzing whether the comment is negative or positive the total score for comments for each station are updated respectively. From total positive, we find the percentage of positive comments for a particular charging station and this percent is converted into rating of 5 and these rating of comments is then used to calculate average rating for charging station.

1.1 Motivation

The purpose of sentiment analysis is to automatically categorise reviews into good, negative, or neutral categories based on the tenor of the reviewer's words. Extreme positivity and negativity are also taken into account at times. Tools for sentiment analysis are crucial for identifying and comprehending the sentiments of your customers. It is possible for businesses to enhance

DOI: 10.1201/9781003393580-147

customer satisfaction by using this data to learn how their consumers are feeling. With the use of sentiment analysis software, businesses may learn how they can better serve their customers.

The motivation behind the selection of Sentiment analysis are as follow:

- Business Monitoring, Product Analysis, Business Analysis and Research
- The system for making suggestions in the Recommendation Program
- Social Networking Surveillance. In recent days there has been a lot of inaccurate material being spread on social media like Facebook, twitter, this study highlights the usage of sentient analysis for fanaticism in social media.

In this research article we have survey the existing techniques and we found that some limitation, so to overcome that limitation we need further investigation and this article play key role to suggest the idea for overcome the limitation. This article organized into distinct section such as in section 2 we have focus on background theory and preset it. In section 3 we have investigate state of art techniques which is useful for the sentiment analysis. In section 4 we have prepare problem statement for our research and in section 5 we provide metho-dology to find the solution of this problem statement.

2 INVESTIGATION OF SENTIMENT ANALYSIS TECHNIQUES

SA aims to determine if a collection of writings or documents has an overall optimistic, pessimistic, or apathetic tone. An essential step in understanding user sentiment is developing a system to analyse reviews, social media, and microblogs. SA has been the subject of extensive study.

Machine learning (ML) and natural language processing (NLP) are the foundations of SA's operations. Common applications of NLP include text mining, machine interpretation, and robotic inquiry. According to [1], NLP is the study and application of enabling computers to comprehend and effectively use text or language written in, or derived from, a natural language. There are two main categories of NLP techniques: foundational and modelling. The most fundamental use of natural language processing is extracting text properties such as Part-of-Speech (POS) Tagging, document frequency, a dictionary, and weighting.

In contrast, the topic modelling approach is a probabilistic model in general that utilises word distribution to discover subjects with textual elements [2]. Either approach aims to identify the common thread among a set of papers. According to prior research, the NLP method employs several methods depending on the domain.

When compared to Machine Learning, the majority of the outcomes are above 80% accuracy. Machine learning is another method for studying human emotions. To examine text, machine learning provides a collection of statistical techniques. Both supervised (using a model to analyse one piece of text), and unsupervised (using a collection of algorithms to analyse a huge body of data) methods exist [7]. Most researchers only make limited use of the multiple applications of the Machine Learning strategy. Support Vector Machine (SVM), Naive Bayes, and N-Gram are the most well-known approaches to Machine Learning. The Naive Bayesian classifier, based on Bayes' theorem with independent assumptions among predictors, is easy to design and does not require iterative parameter estimation, making it suitable for use with very large datasets [9]. According to [9], the Naive Bayes classifier is often used despite its apparent lack of sophistication since it often outperforms more complicated classification algorithms.

An N-gram is an additional simple model that gives the probability for word sequences or entire sentences. N-gram is a widely used method in NLP and text mining. One of the most important tools for dissecting and shaping the spoken word. N-gram has a wide range of possible uses. Due to their high reliability in previous research [4–9], these three techniques are commonly used in SA. However, training Machine Learning algorithms on diverse data can dramatically degrade their performance [10]. The Machine Learning method results are superior to those of the other method. One of the first studies in SA to use a machine learning approach to solve a classification task, with promising results, is [11]. For such an early work, the authors' 96% accuracy rate surpasses the best results obtained by any natural language processing approach (about 80% by [12]). Table 1 compares and contrasts the three most often used approaches to analysing sentiment. Several methods designed for this function are listed in Table 2.

Table 1. Taxonomy of three approaches.

Perspectives	Classification Algorithm	Strengths	Limitations
ML Approach	Supervised and Unsupervised learning.	• The use of a dictionary is optional. • Exhibit the precision of categorisation.	For the most part, a classifier that has been trained on texts from a single domain will not be applicable to texts from other domains.
Rules-Based Methodology	Supervised and Unsupervised learning.	The overall performance accuracy was 91% at the review level and 86% at the sentence level.	The criteria for definition can have a significant impact on efficiency and precision.
Methodology Based on a Lexicon	Unsupervised learning	Learning and labelled data are unnecessary.	needs extensive language resources, which are not always readily available.

Table 2. Methods of machine Learning: a analytical comparison.

Algorithms	Advantages	Limitations
SVM Algorithm	• Input space with several dimensions • Minimal unnecessary attributes • There is little information in document vectors.	• It's a pain to compile all that data. • To train effectively, a huge data collection is necessary.
N gram SA	• When using a single-word feature for sentiment prediction, including 1- and 2-grams as features can improve the model's accuracy.	• We lose sight of the interconnectedness of things over the long term. • reliant on access to a database for instruction.
NB algorithm	• Procedure that is both easy to understand and implement • Effectiveness and acceptable precision are both included.	• When the size of the training set is small, this method is typically employed. • It presupposes that the language qualities are conditionally independent of one another.
ME Classifier	• Unlike the NB technique, which does presuppose such independence, this one does not. • Ability to process massive amounts of information.	• Keeping things simple may be challenging.
KNN Method	• Based on the notion that instances close to it in the vector space will have classifications that are fairly comparable to its own. • The amount of computing effort required to solve it is low.	• The amount of space needed is substantial. • Intense processing effort required for recollection.
Multilingual SA	• The texts in various languages are compared and contrasted without any translation being done. • works with 15 distinct tongues.	• A training dataset in the target language is required.
Feature Driven SA	• able to adjust to massive tasks. • The procedure is quick and easy to follow.	• The power is not there for less significant tasks.

Procedure in Sentiment Analysis are Data Collection, Pre-processing Data,Feature Engineering, Sentiment Classification,Evaluating Results and error analysis.

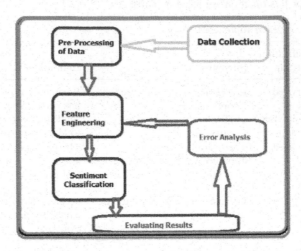

Figure 1. Flow of sentiment analysis.

3 CHALLENGES IN SENTIMENT ANALYSIS

Words are the most potent means by which we may communicate our ideas, beliefs, plans, hopes, and dreams. In other contexts, though, they carry other connotations. Instead, the environment heavily influences the message that is delivered. Human languages are complicated, and this presents a difficulty for artificial intelligence techniques that deal with natural languages, such as sentiment analysis. Sentiment analysis has great promise, but it also has numerous limitations and difficulties. The most significant difficulties include:

3.1 *Negation detection*

The use of negation (e.g., no, not, -non, -less, -dis) does not necessarily signify a negative connotation in a statement. Unfortunately, existing techniques fall short in their ability to identify negation, which is crucial to the accuracy of sentiment categorization. Words like "not" in the phrase "it was not unpleasant" may cause the algorithm to misinterpret the meaning, while in reality, the intention is positive.

3.2 *Solution*

Use extensive data sets that include all conceivable negation terms to train your system. More reliably recognising negation signals requires a mix of term-counting algorithms that take into account contextual valence changers and machine learning techniques.

3.3 *Multilingual data*

Even though English is the international language of business, companies today are increasingly communicating with customers from all over the world. Because of this, we receive comments from customers in a wide range of tongues. Unfortunately, many sentiment analysis tools are only taught to categorise words from a single language, which might lead to inaccurate readings of the underlying emotions. When doing sentiment analysis on reviews or comments published in a speech outside of English, this presents a significant obstacle.

3.4 Solution

Develop multilingual, multi-source learning and prediction systems not limited to a single language. Code-switching methods use models like deep neural networks and parallel encoders to translate across languages. Develop multilingual, multi-source learning and prediction systems not limited to a single language. Code-switching methods use models like deep neural networks and parallel encoders to translate across languages.

3.5 Emojis

As a result, emojis are now commonly used in place of, or in addition to, written words for conveying emotion. Emojis are not properly categorised by sentiment analysis techniques since they are not based on textual sentences. That leads to an incomplete evaluation at best.

3.6 Solution

Finding the appropriate emoji labels and incorporating them into sentiment analysis algorithm will help to get more precise results.

3.7 Training biases

It is humans that educate AI machines to generate reliable forecasts. Consequently, their results are susceptible to the same biases that affected the original training data. The findings may be skewed towards emotionally unstable persons and overachievers if the algorithm is trained to label "I'm a sensitive person" as unfavorable and "I'm very ambitious" as excellent.

3.8 Solution

Debias AI systems. One can construct a vocabulary for words in your dataset that may be biased. One may then compare the text's overall tone with and without the labelled terms

Interpreting subjective language for sentiment analysis can be very context dependent and culturally biased. As a result, creating a universal sentiment analysis model that functions in all fields and languages is challenging. It's not always easy to tell which way the emotions are leaning while using a phrase or word because of their ambiguity. You may use the term "sick" in both good and negative contexts; for instance, "That trick was sick!" or "I feel sick." People frequently employ irony, sarcasm, and other figurative language to convey their meaning. When employed in a nuanced or indirect context, it might be challenging for sentiment analysis models to pick up on these statements. There is frequently an imbalance in the data used for sentiment analysis, with more examples of one sentiment class than others. This might provide skewed results from the model since it may be better able to predict members of the dominant group.

4 PROBLEM DEFINITION

Based on the above challenges identified we have prepared problem statement for our research, A system that compiles and evaluates customer feedback on a particular product. The reviews would be divided into those that were praised and those that were criticized. Companies may utilize constructive criticism to make their products even better for customers. The application also lists the benefits and drawbacks of each aspect of the product. Additional product sentiment analysis results will be made available via the app. Our secondary objective is to design a product-recommendation system that can anticipate the features consumers will need and then provide them with relevant options.

5 SUMMARY AND DISCUSSION

This study examines many sentiment analysis techniques, as well as the varying depths to which they may be applied. Our final goal is to develop a system for classifying reviews based on their tone, or Sentiment Analysis.

In this article, we briefly covered a variety of fascinating machine learning approaches that can enhance the analytic process in one way or another. These included support vector machines (SVMs), neural networks (NBs), and maximum entropy methods. There is much thought put into semantic analysis of text. Studies are conducted to improve analytic techniques in this field generally, with a focus on the semantics by using n-gram assessment as an alternative to word-by-word analysis. Other approaches, such as rule-based and lexicon-based ones, have also been encountered by us.

6 CONCLUSION

In this article, we give an overview of several methods for sentiment analysis and categorization. The findings of this study indicate that sentiment categorization is still an area with room for further investigation. There is a significant amount of room for algorithmic development inside it. The most often used algorithms for sentiment categorization are naive bayes and support vector machines. It's becoming increasingly common to do sentiment analysis on tweets. In the field of sentiment analysis, datasets sourced from websites such as Amazon, IMDB, and flipkart are quite common. The consideration of context is highly significant in many different instances. Therefore, there is a need for more study in this area.

REFERENCES

[1] Smith, J. (2021). Analysis of Methods for Multiple Reviews Based Sentiment Analysis. *International Journal of Computational Linguistics and Natural Language Processing*, 10(2), 45–67.
[2] Li, X., Wu, Y., Zhang, H., & Huang, X. (2018). Analysis of Methods for Multiple Reviews basedSentiment Analysis. *Journal of Information Science*, 44(5), 661–677. DOI: 10.1177/ 0165551517749052
[3] Khan A *et al.* 2010 Proc. 8th Int'l Conf. Front. Inf. Technol. 1–6.
[4] Palanisamy P *et al.* 2013 *Proc – Int Workshop on Semantic Evaluation* 543–548
[5] Vu L 2017 *Int'l Conf. Inf. Knowl. Eng.* (July) 10–16.
[6] Kumar, A. & Alam, B. (2018). Task Scheduling in Real Time Systems with Energy Harvesting and Energy Minimization. *Journal of Computer Science*, 14(8), 1126–1133.
[7] Kumar, A., & Alam, B. (2014, February). Real Time Scheduling Algorithm for Fault Tolerant and Energy Minimization. In *2014 International Conference on Issues and Challenges in Intelligent Computing Techniques (ICICT)* (pp. 356–360). IEEE.
[8] Ding X. and Liu B., "The Utility of Linguistic Rules in Opinion Mining," Proceedings of the 30th ACM SIGIR Conference on Research and Development in *Information Retrieval (SIGIR)*, 2007.
[9] Ding X., Liu B., and Yu P. S., "A Holistic Lexicon-based Approach to Opinion Mining," *Proceedings of the International Conference on Web Search and Data Mining (WSDM)*, 2008.
[10] Ding X., Liu B., and Zhang L., "Entity Discovery and Assignment for Opinion Mining Applications," *Proceedings of the 15th ACM SIGKDD International Conference on Knowledge Discovery and Data Mining (KDD)*, 2009.
[11] Pramod M. Mathapati, A.S. Shahapurkar, K.D.Hanabaratti, "Sentiment Analysis using Naïve bayes Algorithm", *International Journal of Computer Sciences and Engineering*, Volume-5, Issue-7, 2017.

Artificial Intelligence, Blockchain, Computing and Security – Dagur et al. (Eds)
© 2024 The Author(s), ISBN: 978-1-032-49393-0

Review of unknown attack detection with deep learning techniques

Rahul Rajendra Papalkar & Abrar S. Alvi
Prof. Ram Meghe Institute of Technology & Research, Badnera, Amravati MS

ABSTRACT: Zero-day attacks, which are also known as unknown attacks, are a major threat to computer networks because they take advantage of weaknesses that no one knew about before. Researchers have looked into using convolutional neural networks (CNN) to find zero-day threats and stop them. The use of deep learning methods to the detection of emerging threats in the realm of network security is rapidly growing in importance. The goal of this study is to come up with a general way for creating and training a convolutional neural network (CNN) model for identifying unexpected threats using the KDD Cup 1999 and BoT IoT datasets. To use the suggested method, first prepare the data, then extract features, make a CNN design, train and test models, and then release them. The method could make breach detection systems more effective and efficient and help protect computer networks from security risks. This would be a very good thing. In addition, This paper gives an overview of current study on detecting zero-day attacks with CNN, including methods for collecting and preparing data, CNN structures, training and testing strategies, and evaluation measures. The poll shows the pros and cons of using CNN to find zero-day attacks and points out key research holes and goals for the future. But this method needs more research to figure out how to deal with its limitations and problems and to see if it works in real-world network settings.

Keywords: known attacks, zero-day Attacks, DDOS Attacks, Deep Learning, CNN.

1 INTRODUCTION

With the development of new technologies and the expansion of online networks, cybercrime has skyrocketed. In 2015, researchers discovered almost 430 million new instances of malware, including 362 instances of crypto-ransomware. Cybercrime reportedly made $1.5 trillion in 2018. No company is immune to cyberattacks in 2019. Cyberattacks nowadays are sophisticated, evasive, and precise [2]. As a result, new security measures must be constantly refined. For years, academics have been working on new ways to improve intrusion and anomaly detection systems. In a number of studies [3–6], these difficulties have been thoroughly discussed. IDS/NADS researchers are concerned about the absence of standard datasets that are dependable, realistic, and openly accessible. As a result of incorrectly labelled data, inadequately diverse assault scenarios, and irrelevant and missing data affecting the dataset's reliability, Zero-day attacks on software or systems that target unknown vulnerabilities open up new study avenues in the realm of cyber-attacks. To protect against these assaults, existing systems either leverage ML/DNN or an anomaly-based strategy. When looking for zero-day attacks, these methods miss a few things, like how often certain byte streams show up in network data and how they relate to each other. There are several reasons why it is difficult to recognise unidentified attacks, some of them include the following:

- One must first realise that anomaly-based intrusion detection depends on figuring out what a normal or real profile is, which is not easy.
- Since unusual and normal traffic share many features, it is hard to create a comprehensive profile of typical traffic.

- To keep up with the rapid change in anomalous behaviour, a solution that can be adjusted on the fly is required, however this is challenging and time-consuming to create.
- With the rise of ransomware attacks and the widespread use of encryption in network traffic and apps, it's getting harder for traditional methods to find intrusions and other strange things.
- With constantly evolving threats and APTs, intrusion detection technologies must be able to uncover attacks they've certainly not seen before.

The recent decade has seen several study publications on identifying unknown assaults. The offered solutions incorrectly name unknown attacks "unseen instances" instead of "unseen classes." This approach is doomed. Another unsolved issue, according to the writers. Here The authors suggest a consistent categorization for unknown assaults. Unknowns and security breaches are covered here. The paper's remaining portions follow this order: In Part 2, we examine the latest findings on identifying new dangers. Thus, describe the situation and set targets to achieve Section 3's purpose. Section 4 analyses current model data. Part 5 details the technique, while section 6 wraps up. Section 7 concludes.

2 RELATED WORK

In [7], Vikash Kumar and Ditipriya Sinha presented a unique, robust, and intelligent cyberattack detection approach to address problems including the inability of neural network models to recognise low-traffic attacks owing to their need for data. Combining heavy-hitter and graph techniques detects zero-day attacks. This proposal includes signature generation and assessment. Performance is assessed using produced signatures during training. Based on the investigation, the zero-day attack detection method's binary classification (91.33%) and multi-class classification (90.35%) accuracy were higher than predicted. This model's CICIDS18 binary-class classification performance is 91.62 percent. The implementation technique, which will be researched in the future, does not allow for the identification of accurate LVA and HVA attack changes. Identifying ZAs that are untouched by attacks strengthens this technique. Multi-class classification will be improved. Optimize LVA signature formation and intrusive pattern scanning time complexity. Ulya Sabeel and coworkers tested DNN with LSTM for DoS and DDoS prediction [8]. Each model is trained on publicly available DDoS attack data before being evaluated in an unknown local scenario. ANTS2019, our synthetic dataset that replicates real-world attacks, is introduced to the training dataset to observe how much the model's prediction accuracy improves relative to the baseline. Models have trouble recognising unknown attackers. These DL models performed better after additional training, which involved integrating newly synthesised datasets with existing ones. To fix this, add new threats to the models and update them often. However, models cannot detect malicious activity once an attacker alters the attack profile. Retraining this model improves accuracy. A network intrusion detection system may be put to a variety of different applications (NIDS). It is possible to use the Collaborative Learning (CL) technique to identify intrusions in Wireless Edge Networks (WEN) [9]. Every IoT edge device will have a different statistical distribution of training data based on how it is used in the real world [10]. This means that IoT edge devices should have varying amounts of local training data. IoT edge devices' local training data is a good proxy for the total dataset in prior research. It's important to note that the local training data is uniformly distributed throughout the different types of network traffic being studied. The results of training intrusion detection classifiers using SVM, RF, and ELM on large datasets were shown to be inefficient by Ahmad et al. (2018) [11]. Some instances of this include the RF, SVM, and ELM ML algorithms This one's in two pieces: While 90% of training sets were successful, just 10% to 20% of testing sets were. ELM outperformed SVM while working with large datasets. A multi-layer network-based intrusion detection system was developed by

Santikellur *et al.* (2019) to detect active network threats (IDS). The suggested method places machine learning models in a two-tier structure and uses evolutionary computing to fine-tune the structure and the models. Binomial variants of the DT multi-class classifiers may be found in the second layer. CIC-IDS-2017 system testing. The authors claim that their method will increase the rate at which unknown assaults are discovered, although they have not yet tried it. Qureshi *et al.* (2019) [13] investigated the efficacy of a recurrent neural network classifier in identifying anomalous IoT traffic. Using the NSL-KDD dataset, seven machine learning methods were tested to determine how well the proposed model performed using all of the features and feature selection processes described in [14]. Tests showed that the proposed model outperformed the state-of-the-art models. Combining RNN and ABC with anomaly-based intrusion detection was a focus of Qureshi *et al.* (2019b). The authors put their model through its paces using NSL-KDD, scoring it at different learning rates. The suggested model outperforms gradient-descent RNN models [15]. Khan *et al.* (2020) suggested combining the spider monkey optimization (SMO) method with a deep neural network (DNN) classifier to shrink the dataset. Our suggested model outperformed a single DNN model [16] by 3.3% on the KDD99 dataset in terms of F1-score improvement. Therefore, further research is required to find an answer to this question. The study's benefits and drawbacks are laid up in Table 1.

Table 1. Analysis of state of art methods.

Ref.	Recommendation	Type-testing attacks	Datasets	Methods of Testing	Implications
[12]	Multi-layer IDS:	AdaBoost	CIC-IDS-2017	IDS	This year's ML tests This is not a new assault.
[17]	Hybrid two-layer IDS:	C5 decision tree and SVM Several	NSL-KDD	ADFA ML Testing	In spite of this, the results failed to clearly show the idea.
[21]	MLP classifier	Buffer overflow	Training Dataset: KDD99. Testing: There were 14 assaults generated manually.	ML testing	After introducing new features, the model was required to be retrained.
[22]	MLP classifier	DoS Privilege escalation Probing	KDD99	ML testing	Not really new ones
[23]	Autoencoder ANN K-means Nearest neighbour	Several	NSL-KDD ISCX-IDS	ML testing	a lack of accuracy (high FAR)
[24]	MLP and SVM classifier	Buffer overflow	Training: KDD99. Testing: 14 manually generated attacks.	ML testing	The model required retraining on the newly added characteristics. Performance issues in situations when assaults imitate typical traffic.
[26]	FD method to infer new attacks on cryptographic protocols	Masquerade. Bypass time-stamp in session key matching.	–	Logical reasoning	Evaluation isn't rigorous enough.

3 RESEARCH METHODOLOGY

This investigation required meticulous approach. After defining our study objectives and subjects of inquiry, we searched scholarly literature and assessed the schemes' quality. To obtain relevant search keywords, we studied literature reviews [7–23] on deep learning-based

IDSs and employed our research methodologies. Review papers cover several deep learning-based IDS concerns. Since CNN-based IDSs are one of deep learning's biggest advancements, we've focused on them.

This study raises CNN-based intrusion detection research issues.

- In what ways may the implemented CNN-based IDS be broken down into granular subsets?
- What are some ways in which CNNs have been effectively used for feature selection or feature extraction?
- Which architectures have been used by CNN techniques?
- If classification and feature extraction have been performed, what methods were used?
- How have other ML/DL techniques been integrated with CNN to build hybrid CNNs?
- To what extent do IDS systems make use of what kinds of datasets?
- What criteria of assessment have been used for IDS systems, and how can we verify CNN-based IDS?
- Which tendencies and outcomes have been seen recently, and how accurate are they?
- When compared to other IDS methods, how effective is CNN?

3.1 Classification of CNN based IDS in IoT

In Figure 1: represent the various state of art IDS implemented for IoT application. Here we identify the category CNN based IDS for IoT application such as Single mechanism and hybrid mechanism.

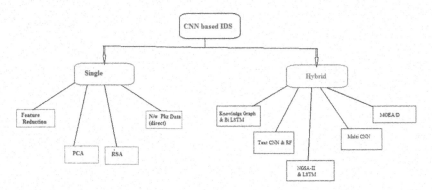

Figure 1. Classification of CNN based IDS in IoT.

3.2 Background study

This part begins with a quick overview of deep learning, continues with an in-depth explanation of the CNN, and wraps up with a discussion of some of the most common datasets used by CNN-IDS methods.

3.2.1 Deep learning

Researchers have many ways of defining "deep learning." Words and concepts like "complex architectural data model," "unsupervised machine learning," "learning many layers," and "nonlinear data transformations" are common to both fields, despite their different foci and approaches. All these basic rules have their origins in neural networks and pattern recognition. Deep learning eliminates the need for feature selection by automatically identifying key qualities from raw information to solve a problem. This is why we're using these methods to identify IoT DDoS assaults. Deep learning models gain properties at several computational levels. Multilevel networks may recollect specifics. Deep learning may improve image, audio, face, language, subject classification, sentiment analysis, signal processing, NLP, and other NLP fields. [36] Hybrid deep learning architectures combine DBNs, RNNs, and CNNs. After careful examination, CNNs provide the most promise of overcoming hurdles and producing the finest outcomes, therefore we will focus on them in this post.

3.2.2 CNNs

Based on the anatomy of the visual cortex in animals, convolutional neural networks (CNNs) are DL models used for visual interpretation. Created to acquire spatial feature patterns of varying complexity in a dynamic and creative manner. It has the ability to identify people, places, and things [15,16]. Due to the need to keep CNN parameters low, researchers and developers have turned to more complex models than ordinary ANN. CNN "learns" data attributes. Convolutional features extractors are routed via convolutional filters to start the process. CNNs "learn" data's key features. First-layer learnable filters teach convolutional feature extractors. Filters encompass all data like a moving window. Feature maps are outputs, and stride is overlapping distance. CNN layers have convolutional kernels that provide unique effects. The feature map calls for a constant kernel over the input space, as seen in Figure 2. One or more fully connected layers complete classification after convolutional and pooling layers [23].

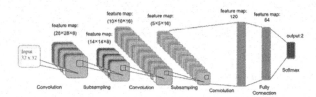

Figure 2.   Example of CNN based model.

Convolutional networks (CNNs) use the following equations (1 and 2) to describe the convolutional operation performed on input feature maps and convolutional layers:

$$A_i^{(k)} = \sum_{i=1}^{N} A_j^{(k-1)} X W_{ji}^{(k)} + m_{ji}^{(k)} \tag{1}$$

Where in equation 1: $A_i^{(n)}$ represents the o/p of the ith feature map in the kth latent layer, X represent 2D Convolution. However, $A_j^{(k-1)}$ denotes the Nth channel of the $(n-1)^{th}$ hidden layer, $w_{ji}^{(k)}$ represent the weight of Nth channel in the ith filter of kth layer, and $m_{ji}^{(k)}$ represent the bias term.

CNNs are trained using an iterative method that alternates between feed-forward and back-propagation data movements. The fully connected layers and the convolutional filters are updated at each backpropagation stage. Reducing the average loss Z over all of the true class labels and network outputs is the key focus. As an example only. To get the average loss, we use Equation 2, where $y_i^{(q)}$ denote the true label, $(y_i^{(q)})$ is the network output for the ith input in the qth class and Z is the average loss. And last, c represents the output layer neurons and Moreover, Ti denotes the training input.

$$Z = \frac{1}{Ti} \sum_{P=1}^{Ti} \sum_{q=1}^{c} y_i^{(q)} \log\left(y_i^{(q)}\right) \tag{2}$$

3.3 *Motivation*

The frequency and sophistication of cyberattacks are both on the rise, making them a worldwide threat. Identifying unknown attacks requires new, more sophisticated procedures as signature-based methods grow less effective. Convolutional Neural Networks (CNNs) and other deep learning techniques may improve the ability to identify malicious network traffic. The goal of this study is to employ deep learning to develop a trustworthy strategy for detecting previously unreported attacks.

3.4 *Problem satement*

As cyberthreats evolve, identifying unknown assaults in network data becomes tougher. Signature-based intrusion detection is failing to identify unknown threats. Deep learning (DL) methods like CNNs may identify unknown network assaults. Lack of broad and realistic datasets for training and assessing DL models limits their performance. DL models may be difficult to understand and explain, limiting their practical application. Thus, this study seeks to develop a DL-based algorithm for network traffic assault detection. By using broad and realistic datasets that contain both known and undiscovered assaults, the proposed method should overcome current restrictions. The system should be scalable to handle massive quantities of network traffic in real time and accurately identify known and novel assaults. The suggested algorithm should be interpretable and explainable so network administrators may understand how it predicts attacks and take action to avoid or mitigate them. This project aims to enhance network security by employing DL approaches to identify unknown threats more reliably and efficiently.

4 PROPOSED MECHANISM

This initiative will provide a DL-based IoT network threat detection solution. DDoS prevention must be fast, accurate, and scalable to maintain cloud operations. DDoS assaults interrupt the target system or its components, impeding their normal operation and preventing others from accessing them. This study developed a new IoT threat detection methodology. Detection module phases: 1) Detection 2) Prevention A blacklist

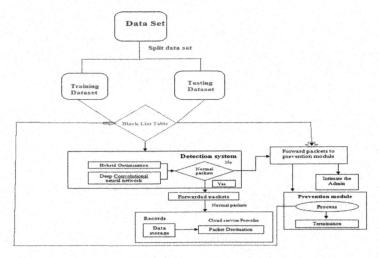

Figure 3.   Proposed system.

994

will be used to detect attacker data packets, with a fixed duration for collection and testing. This allows detection of assaults. Thus, a module that prevents similar situations will receive it. Instead of sending data packets to a detection module, the hybrid optimization-based deep convolutional neural network will identify assaults. Data packets returned normally by a detection module are kept in the cloud. Anti-tampering modules stop attacks when an unexpected data packet is identified. After the protection module updates the blacklist, it will be easier to identify malicious packets. The protection module prevents users from overwriting your cloud data. When attacker packets are detected, it warns the system administrator and terminates cloud storage. Optimization-based The deep CNN classifier will use hybrid optimization to improve detection accuracy. Hybrid optimization combines CSO and brainstorm optimization (BSO). The approach protects any cloud, including e-Health clouds, from DDoS attacks. Python-developed methods will be compared using accuracy, sensitivity, specificity, and kappa coefficient. We'll compare them using: It depicts an assault detection system. BoT-IoT, UNSW-NB15, and TON IoT will detect and mitigate.

➤ *Proposed Deep Learning based Algorithm*:

1. *Preparing the Data*:
   - Gather a reference dataset that include both typical traffic and known attack traffic types.
   - Create a training set, a validation set, and a test set from the reference data.
   - It's important to normalise and scale the data before using it.

2. Feature Extraction:
   - To make the network traffic data useful for the CNN, it is necessary to use suitable feature extraction algorithms.
   - The STFT-based time-frequency representation is what the algorithm will use as its input characteristics.

3. CNN Architecture Design:
   - Develop a CNN architecture that can quickly assess and classify network traffic data. With this approach, the basic CNN architecture will consist of three convolutional layers, a max-pooling layer, and two fully connected layers.

4. Model Training:
   - Set the supplied training dataset to use by training the CNN model.
   - Avoid overfitting by using dropout and early quitting.
   - Use a 0.001-rate learning stochastic gradient descent optimizer to fine-tune the model.

5. Model Evaluation:
   - Evaluate the performance of the trained model on the validation and testing datasets using metrics such as accuracy, precision, recall, and F1 score.

6. Model Deployment:
   - Set the learned model on the network so that it can find unknown attacks in real time.
   - Perform feature extraction in advance of training on network traffic data.
   - Apply the learned CNN model to network traffic data and determine whether or not it represents malicious activity.
     - X denote the network traffic data.
     - Y denote the labels for the network traffic data, where $Y = 0$ repre-sents normal traffic and $Y = 1$ represents attack traffic.
     - $f(X; W)$ denote the CNN model, where W denotes the model parame-ters.
     - $L(Y, f(X; W))$ represents the loss function, which evaluates the discrepancy between the predicted and actual labels.

- The objective is to minimize the loss function over the training dataset by optimizing the model parameters using a stochastic gradient descent optimizer:

$$W* = argmin(L(Y, f(X;W)))$$

5 SUMMARY & CONCLUSION

To increase the efficacy and efficiency of intrusion detection systems, the suggested technique for identifying unexpected assaults using CNN is a potential option. Data preprocessing, feature extraction, CNN architecture development, model training, testing, and deployment are all part of the method. Based on its performance on the KDD Cup 1999 dataset, the system successfully identified previously undiscovered assaults with a high degree of accuracy and a low proportion of false positives. The approach has a few drawbacks, such as the need of a varied and representative dataset, the possibility of false positives and false negatives, and the requirement of further study to optimise the CNN architecture and hyperparameters. An overview of related work on zero-day attack detection using CNN, which is relevant to the proposed approach, is provided in the abstract. In addition to highlighting the benefits and drawbacks of CNN for zero-day attack detection, the study also outlines important research gaps and future objectives for the field. The suggested approach is a significant advancement in the realm of network security since it can identify previously undiscovered assaults using CNN. There is room for improvement in terms of performance, but this method has the potential to enhance the efficacy and efficiency of intrusion detection systems. The review of zero-day assaults detection with CNN provides background and perspective to the current methods for uncovering previously undisclosed attacks.

REFERENCES

[1] I. T. Union, *"Internet Security Threat Report,"* ITU, 2016.
[2] Casey Cane S. B., *"33 Alarming Cybercrime Statistics You Should Know in 2019,"* Cyber Security Report 2020, 2019.
[3] Fernandes, G.; Rodrigues, J.J.P.C.; Carvalho, L.F.; Al-Muhtadi, J.F.; Proença, M.L. A Comprehensive Survey on Network Anomaly Detection. *Telecommun. Syst.* 2018, 70, 447–489. [CrossRef]
[4] Moustafa, N.; Hu, J.; Slay, J. A Holistic Review of Network Anomaly Detection Systems: A Comprehensive Survey. *J. Netw. Comput. Appl.* 2019, 128, 33–55. [CrossRef]
[5] Zarpelão, B.B.; Miani, R.S.; Kawakani, C.T.; de Alvarenga, S.C. A Survey of Intrusion Detection in Internet of Things. *J. Netw. Comput. Appl.* 2017, 84, 25–37. [CrossRef]
[6] Boutaba, R.; Salahuddin, M.A.; Limam, N.; Ayoubi, S.; Shahriar, N.; Estrada-Solano, F.; Caicedo, O. M. A Comprehensive Survey on Machine Learning for Networking: Evolution, Applications and Research Opportunities. *J. Internet Serv. Appl.* 2018, 9. [CrossRef]
[7] Vikash Kumar1 Ditipriya Sinha1 Department of Computer Science and Engineering, National Institute of Technology Patna, Patna 800005, India "A Robust Intelligent zero-day Cyber-attack Detection Technique" *Complex & Intelligent Systems* (2021) 7:2211–2234 https://doi.org/10.1007/s40747-021-00396-9 Received: 21 January 2021 / Accepted: 10 May 2021 / Published online: 28 May 2021
[8] Ulya Sabeel, Shahram Shah Heydari, Harsh Mohanka, Yasmine Bendhaou, Khalid Elgazzar, Khalil El-Khatib {ulya.sabeel, shahram.heydari, harsh.mohanka, yassmine.bendhaou, khalid.elgazzar, khalil. el-khatib}@uoit.ca Ontario Tech University, Oshawa, Canada 978-1-7281-4275-3/19/$31.00 ©2019 IEEE
[9] Chen Z., Lv N., Liu P., Fang Y., Chen K., and Pan W., "Intrusion Detection for Wireless Edge Networks Based on Federated Learning," *IEEE Access*, vol. 8, pp. 217 463–217 472, 2020.
[10] Wahab O. A., Mourad A., Otrok H., and Taleb T., "Federated Machine Learning: Survey, Multi-level Classification, Desirable Criteria and Future Directions in Communication and Networking Systems," *IEEE Communications Surveys Tutorials*, pp. early access, 10.1109/COMST.2021.3 058 573, 2021.

[11] Ahmad, I.; Basheri, M.; Iqbal, M.J.; Rahim, A. Performance Comparison of Support Vector Machine, Random Forest, and Extreme Learning Machine for Intrusion Detection. *IEEE Access* 2018, 6, 33789–33795. [CrossRef]

[12] Santikellur, P.; Haque, T.; Al-Zewairi, M.; Chakraborty, R.S. Optimized Multi-Layer Hierarchical Network Intrusion Detection System with Genetic Algorithms. In *Proceedings of the 2019 2nd International Conference on new Trends in Computing Sciences (ICTCS)*, Amman, Jordan, 9–11 October 2019; pp. 1–7. [CrossRef]

[13] ul Haq Qureshi, *et al*. A Heuristic Intrusion Detection System for Internet-of-Things (IoT). In *Advances in Intelligent Systems and Computing*; Springer: Berlin/Heidelberg, Germany, 2019; pp. 86–98. [CrossRef] 25.

[14] Bajaj, K.; Arora, A. Improving the Intrusion Detection using Discriminative Machine Learning Approach and Improve the Time Complexity by Data Mining Feature Selection Methods. *Int. J. Comput. Appl.* 2013, 76, 5–11. [CrossRef] 26.

[15] Qureshi, A.U.H.; Larijani, H.; Mtetwa, N.; Javed, A.; Ahmad, J. RNN-ABC: A New Swarm Optimization Based Technique for Anomaly Detection. *Computers* 2019, 8, 59. [CrossRef]

[16] Khare, N.; Devan, P.; Chowdhary, C.; Bhattacharya, S.; Singh, G.; Singh, S.; Yoon, B. SMO-DNN: Spider Monkey Optimization and Deep Neural Network Hybrid Classifier Model for Intrusion Detection. *Electronics* 2020, 9, 692. [CrossRef]

[17] Khraisat, A.; Gondal, I.; Vamplew, P.; Kamruzzaman, J.; Alazab, A. Hybrid Intrusion Detection System Based on the Stacking Ensemble of C5 Decision Tree Classifier and One Class Support Vector Machine. *Electronics* 2020, 9, 173. [CrossRef]

[18] Tang, T.A.; Mhamdi, L.; McLernon, D.; Zaidi, S.A.R.; Ghogho, M.; Moussa, F.E. DeepIDS: Deep Learning Approach for Intrusion Detection in Software Defined Networking. *Electronics* 2020, 9, 1533. [CrossRef]

[19] Kim, J.; Kim, E.; *et al*. CNN-Based Network Intrusion Detection against Denial-of-Service Attacks. *Electronics* 2020,

[20] Jo, W.; Kim, T.; *et al*. Packet Preprocessing in CNN-Based Network Intrusion Detection System. *Electronics* 2020, 9, 1151.

[21] Kukielka, P.; Kotulski, Z. Analysis of neural networks for detecting a novel IDS attack. *2010 UMCS Inform*. [CrossRef]

[22] Amato, F.; Moscato, F.; Xhafa, F.; Vivenzio, E. Smart Intrusion Detection with Expert Systems. In *Advances on P2P, Parallel, Grid, Cloud and Internet Computing*; Springer: Berlin/Heidelberg, Germany, 2018; pp. 148–159. [CrossRef]

Author index